THE WILEY
ENCYCLOPEDIA OF
PACKAGING TECHNOLOGY

THE WILEY ENCYCLOPEDIA OF PACKAGING TECHNOLOGY

MARILYN BAKKER, Editor-in-chief

DAVID ECKROTH, Managing editor

John Wiley & Sons

NEW YORK • CHICHESTER • BRISBANE • TORONTO • SINGAPORE

Library of Congress Cataloging in Publication Data:

The Wiley Encyclopedia of packaging technology.
   Includes index.
   1. Packaging—Dictionaries.   I. Bakker, Marilyn.
II. Eckroth, David.
TS195.A2W55   1986   688.8′03′21   86-4041
ISBN 0-471-80940-3

Printed in the United States of America

10 9 8 7 6 5 4 3 2

PREFACE

Packaging technology is a field that has grown out of many disciplines: engineering, metallurgy, chemistry, food science, and others. This Encyclopedia brings all of the disciplines together, for the first time, in one comprehensive reference work. The articles were contributed by 188 professionals who are applying their own disciplines to packaging, and each article was subjected to peer review. The realization of this project depended on the efforts of over 500 people. In its entirety, the book reflects the state of the packaging art today, from A to Z.

The alphabetical subject list, compiled with the valuable assistance of the Advisory Board, evolved from preliminary lists of categories and subcategories. The book covers raw materials and intermediate materials (eg, polymers and films, papers, foils); raw-materials conversion processes and products (eg, glass manufacture, plastics processing methods, tin-mill products); intermediate-materials conversion processes and products (eg, heat sealing, laminating, multilayer flexible packaging, can fabrication); packages and supplies (eg, corrugated boxes, sealing tapes); packaging machinery and equipment (eg, fillers, cappers, conveyors); design (eg, design of cushion systems, bottles); testing of materials, packages, and components; as well as other subjects such as standards and practices, laws and regulations, recycling and energy utilization.

The book is special because it contains at least something about virtually every aspect of packaging technology. It should satisfy the research and reference needs of a great variety of professional and student readers. The articles are rich in technical detail, far beyond what might be expected in a handy single volume. It is special in other ways as well. Many articles, particularly those that deal with the workings of packaging machinery, contain details that have never before been written, except in technical service manuals. In a small minority of those cases, because there is no background literature to cite, the Encyclopedia does not provide the customary references to original sources. On other subjects, the bibliography presents key references from the technical literature. On some subjects (eg, glass- and metal-can manufacturing), the literature is vast, and the significance of the Encyclopedia lies in the brevity and depth with which the subjects are treated.

The A–Z format that distinguishes an encyclopedia occasionally demands some guesswork and practice on the part of the reader, who might look for low density polyethylene under "L" instead of "P;" but an exhaustive index and profuse cross-references have been provided to help. Balance in coverage has been one of my principal goals. United States units are used throughout, but always with SI conversions. Most articles come from United States authors, but there have been many extraordinary contributions from abroad. The review process was designed to eliminate bias by soliciting the comments of competitors. Thanks go out to all of those anonymous reviewers for their valuable input.

This book is dedicated to all the present and future students, engineers, and designers who will benefit from this pioneering reference source, which sets a new standard for the packaging profession.

M. BAKKER
Editor-in-Chief

EDITORIAL STAFF

CONTRIBUTORS

Avalon L. Dungan, *L.J. Minor Corp.,* Trays, steam table

Dean D. Duxbury, *Pouch Technology, Inc.,* Retortable flexible and semi-rigid packaging

Jerry L. East, *Simplimatic Engineering Co.,* Conveying

David K. Eary, *Government Contract Services Co.,* Military packaging

Eugene E. Engel, *Permanent Label Corp.,* Decorating: silk screening

Mark B. Eubanks, *Sonoco Products Co.,* Cans, composite

William R. Evans, *Coca-Cola Bottling of New York,* Filling machinery, carbonated liquid

Frank B. Fairbanks, Jr., *Horix Manufacturing Co.,* Filling machinery, still liquid

M.C. Fairley, *Labels & Labelling Data and Consultancy Services Ltd,* Labels and Labeling

Harry M. Farnham, *Moor & Munger Marketing, Inc.,* Waxes

Faustel, Inc., Laminating machinery

Neil Ferguson, *I.C.C. Primex Plastics Corp.,* Corrugated Plastic

Ronald Foster, *Northern Petrochemical Technical Ctr.,* Ethylene–(Vinyl alcohol)

Theodore D. Frey, *Crown-Zellerbach,* Coextrusions for flexible packaging

Chester Gaynes, *Gaynes Testing Laboratories, Inc.,* Testing, packaging materials

James A. Gibbons, *Egan Machinery Co.,* Extrusion

Luigi Goglio, *Fres-Co System USA, Inc.,* Valves and vents

Gilbert E. Good, *KwikLok Corp.,* Closures, bread bag

Frank W. Green, *Point O'View, East,* Export packaging

James M. Gresher, *John D. Clarke Co.,* Carded packaging

Allan L. Griff, *Edison Technical Services,* Cans, plastic; Carbonated beverage packaging

K. R. Habermann, *Allied Fibers and Plastics,* Film, fluoropolymer

David L. Hambley, *Owens-Illinois, Inc.,* Glass container design

Randal J. Hasenauer, *Mobil Chemical Co.,* Film, oriented polypropylene

Jerome H. Heckman, *Keller and Heckman,* Food, drug, and cosmetic packaging regulations

Ilene R. Heller, *Keller and Heckman,* Food, drug, and cosmetic packaging regulations

Arthur A. Hirsch, *Arvey Corp.,* Health-care packaging

Joseph H. Hotchkiss, *Cornell University,* Canning, food

John S. Houston, *Amoco Chemicals Corp.,* Polystyrene

Norman T. Huff, *Owens-Illinois, Inc.,* Glass container manufacturing

Christopher Irwin, *Hoover Universal, Inc.,* Blow molding

John F. Jacobs, *Icore Corporation,* Checkweighing machinery

Daniel G. James, *Borden Chemical,* Film, flexible pvc

Nancy Janssen, *Food Processing Machinery & Supplies Assn.,* Can seamers

Richard H. John, *Windmoeller & Hoelscher Corp.,* Bag-making machinery

Montfort A. Johnsen, *Peterson-Puritan, Inc.,* Pressurized containers

Dee Lynn Johnson, *The 3M Center,* Indicating devices

Donald J. Jones, *Pharmagraphics, Inc.,* Inserts

F. W. Jowitt, *Metal Box, p.l.c.,* Cans, fabrication

Irving Kaye, *National Starch & Chemical Corporation,* Adhesives

William Kent, *Brown Machine Co.,* Thermoforming

B. Lee Kindberg, *American Hoechst,* Film, oriented polyester

Peter Kirkis, *Vulcan Industrial Packaging,* Pails, plastic

Ernest O. Kohn, *Metal Box, p.l.c.,* Cans, fabrication

Stephen F. Kravitz, *Sydney Schreiber Inc.,* Bands, shrink

Axel Kuhn, *Husky Injection-Molding Systems,* Injection molding

Pat Lancaster, *Lantech Inc.,* Wrapping machinery, stretch film

James M. Lavin, *Container Corporation of America,* Cans, composite, self-manufactured

Robert Le Caire, *Presto Products, Inc.,* Films, stretch

John F. Legat, *Tampoprint America Inc.,* Decorating: Pad printing

James Lentz, *Webcraft Technologies, Inc.,* Printing

Edmund A. Leonard, *Cornell University,* Economics of packaging; Specifications and quality assurance

John E. Lisi, *Owens-Illinois, Inc.,* Ampuls and vials

Robert E. Lisiecki, *Ex-Cello Corp.,* Cartons, gabletop

Hugh E. Lockhart, *Michigan State University,* Tamper-evident packaging

Ruskin Longworth, *E.I. duPont de Nemours & Co., Inc.,* Ionomers

Lee T. Luft, *Menasha Corp.,* Pallets, Plastic

Lawrence Lynch, *National Paperbox & Packaging Association,* Boxes, rigid paperboard

Don MacLaughlin, *Vercon Inc.,* Welding, spin

David Madison, *Sigma Systems,* Filling by count

Norma J. Maraschin, *Union Carbide Corp.,* Polyethylene, low density

Kenneth S. Marsh, *Kenneth S. Marsh & Associates,* Shelf life

E.L. Martin, *Printpack, Inc.,* Multilayer flexible packaging

Norman Martin, *Pneumatic Scale Corp.,* Bag-in-box, dry product

Robert E. Mastriani, *Arrow Converting Equipment, Inc.,* Roll handling

Ronald D. Mathis, *Phillips Petroleum Company,* Styrene–butadiene copolymers

Joseph P. McCaul, *SOHIO Chemical Co.,* Nitrile polymers

Robert W. McKellar, *The Gummed Industries Association Inc.,* Tape, gummed

Alfred H. McKinlay, *Consultant, Ind. Engr.,* Testing, shipping containers

Lynn McKinney, *Brown Machine Co.,* Thermoforming

Ronald G. McManus, *Noxell Corp.,* Testing, consumer packages

Ivan Miglaw, *Crown Advanced Films,* Film, non-oriented polypropylene

James M. Mihalich, *General Electric Plastics,* Polycarbonate

Jack Milgrom, *Walden Research, Inc.,* Recycling

Richard C. Miller, *Himont U.S.A., Inc.,* Polypropylene

Scot R. Mitchell, *Vercon Inc.,* Welding, spin

L.P. Mozer, *Phillips Chemical Co.,* Polyethylene, high density

George "Rocky" Moyer, *Rovema Packaging Machines,* Form/fill/seal, vertical

James F. Nairn, *Phoenix Closures, Inc.,* Closures

Edward M. Naureckas, *Coex Engineering, Inc.,* Multilayer plastic bottles

James K. Naureckas, *Coex Engineering, Inc.,* Multilayer plastic bottles

Eberhard H. Neumann, *Hoechst AG,* Polyesters, Thermoplastic

Norman F. Nieder, *Anheuser-Busch, Inc.,* Cans, aluminum

Frank R. Nissel, *Welex, Inc.,* Coextrusion machinery, flat

Thomas M. Norpell, *Phoenix Closures, Inc.,* Closures

Frank J. Nugent, *General Foods Corp.,* Vacuum coffee packaging

Richard H. Nurse, *American Hoechst,* Film, high density polyethylene

K.F. Oekel, *Bemis Co., Inc.,* Netting

Edward M. O'Mara, *Union Camp Corporation,* Slipsheets

Robert L. Oldaker, *American Hoechst,* Film, oriented polyester

Frank A. Paine, *Consultant in Packaging Technology and Management,* Distribution hazards; Laws and regulations, EEC

Robert J. Peache, *Wang Laboratories, Inc.,* Cushioning, design

Arthur C. Peck, *Permanent Label Corporation,* Decorating: Hot stamping

Alexander M. Perritt, *Perritt Laboratories, Inc.,* Child-resistant packages

Bert Peterson, *Vercon Inc.,* Welding, spin

William C. Pflaum, *W.C. Pflaum Co., Inc.,* Networks

Ed Pirog, *Crown Advanced Films,* Film, non-oriented polypropylene

Bruno Poetz, *Mauser Packaging Ltd.,* Drums, plastic

Philip E. Prince, *Inovpack,* Metallizing

F. Robert Quinn, *Union Camp Corporation,* Boxes, solid fiber

Robert F. Radek, *Selig Sealing Products, Inc.,* Closure liners

George Reingold, *Continental Packaging Co.,* Cans, steel

A. Bruce Robertson, *Allied Fibers and Plastics,* Film, fluoropolymer

Peter Rodgers, *Foam Fabricators,* Foam Cushioning

Richard Roe, *Brown Machine Co.,* Thermoforming

Sherman H. Rounsville, *American Hoechst,* Film, oriented polyester

Richard G. Ryder, *Klockner Pentaplast of America, Inc.,* Film, rigid pvc

Morris Salame, *Monsanto Company,* Barrier polymers

Don Satas, *Satas & Associates,* Coating equipment

Robert C. Schiek, *CIBA-GEIGY Corp.,* Colorants

John C. Sciaudone, *Tulox Plastics Corp.,* Boxes, rigid plastic

Charles H. Scholl, *Nordson Corporation,* Adhesive applicators

Herbert H. Schueneman, *Schueneman Design Associates,* Testing, cushion systems; Testing, product fragility

Arthur J. Schultz, *Steel Shipping Container Institute,* Drums and pails, steel

Raymond B. Seymour, *University of Southern Mississippi,* Additives, plastics

Richard L. Sheehan, *3M Corp.,* Tape, pressure-sensitive

David H. Shumaker, *Nordson Corporation,* Adhesive applicators

William R. Sibbach, *Jefferson Smurfit Corp.,* Lidding

J.P. Sibilia, *Allied Corp.,* Nylon

Jeffrey S. Siebenaller, *American Hoechst,* Film, high density polyethylene

H.J. Sievers, *The Kartridg Pak Co.,* Chub packs

Martin D. Sikora, *James River Corp.,* Paper

Donald T. Simmons, *The Marketing Link,* Instrumentation

Charles Simpson, *American Production Machine Co.,* Decorating: Offset printing

Michael A. Smith, *Phillips Chemical Co.,* Polyethylene, high density

J.A. Sneller, *The Proctor & Gamble Co.,* Decorating: In-mold labeling

Carol Spinner, *Packaging Corporation of America,* Contract packaging

Lawrence D. Starr, *Koch Supplies Inc.,* Thermoform/fill/seal

Ronald K. Steindorf, *Nicolet Paper Co.,* Glassine, greaseproof, parchment

Victor Suben, *Noxell Corp.,* Testing, consumer packages

Joseph T. Sullivan, *Rohm and Haas Company,* Acrylics

Leonard F. Swec, Boxes, corrugated

Frank J. Sweeney, *Sweeney Cooperage Ltd.,* Barrels

Colin Swinbank, *F. Inst. Pkg.,* Standards and practices

Thomas E. Szemplenski, *Marlen Research Corp.,* Aseptic packaging

Charles C. Taylor, *Charles C. Taylor & Associates,* Cellophane

James L. Throne, *Engineering Consultant,* Polymer properties

Michael F. Tubridy, *Allied Corp.,* Nylon

Iwan W. Turiansky, *Ethicon, Inc.,* Radiation, effects of

Harry J.G. van Beek, *Klockner Pentaplast of America, Inc.,* Film, rigid pvc

J. Robert Wagner, *Philadelphia College of Textiles and Science,* Nonwovens

Philip Wagner, *Dow Chemical Company,* Foam, extruded polystyrene

Edwin H. Waldman, *Keyes Fibre Co.,* Pulp, molded

Walter B. Wallin, *U.S. Department of Agriculture-Forest Service,* Pallets, wood

Walter E. Warren, *Salwasser Manufacturing Co., Inc.,* Case packing

Herman Weimer, *Signode Corp.,* Strapping

H.C. Welch, *Pharm Pack Co., Inc,* Pharmaceutical packaging

Richard S. Wheelock, *U.S.I. Film Products,* Bags, heavy-duty plastic

Douglas J. White, *Norden Packaging Corp.,* Tube filling

Al Work, *KLIKLOK Corp.,* Cartoning machinery, top-load

William D. Wright, *Western Polymer Technology, Inc.,* Coextrusion machinery, tubular

Ernst Wurzer, *Mauser Packaging Ltd.,* Drums, plastic

Roger W. Young, *John Dusenbery Company, Inc.,* Slitting and rewinding machinery

William E. Young, *W.E. Young Co., Inc.,* Sealing, heat

CONVERSION FACTORS, ABBREVIATIONS AND UNIT SYMBOLS

Selected SI Units (Adopted 1960)

Quantity	Unit	Symbol	Acceptable equivalent
BASE UNITS			
length	meter†	m	
mass‡	kilogram	kg	
time	second	s	
electric current	ampere	A	
thermodynamic temperature§	kelvin	K	
DERIVED UNITS AND OTHER ACCEPTABLE UNITS			
* absorbed dose	gray	Gy	J/kg
acceleration	meter per second squared	m/s^2	
* activity (of ionizing radiation source)	becquerel	Bq	l/s
area	square kilometer	km^2	
	square hectometer	hm^2	ha (hectare)
	square meter	m^2	
density, mass density	kilogram per cubic meter	kg/m^3	g/L; mg/cm^3
* electric potential, potential difference, electromotive force	volt	V	W/A
* electric resistance	ohm	Ω	V/A
* energy, work, quantity of heat	megajoule	MJ	
	kilojoule	kJ	
	joule	J	N·m
	electron volt^x	eV^x	
	kilowatt hour^x	kW·h^x	
* force	kilonewton	kN	
	newton	N	kg·m/s^2
* frequency	megahertz	MHz	
	hertz	Hz	l/s
heat capacity, entropy	joule per kelvin	J/K	
heat capacity (specific), specific entropy	joule per kilogram kelvin	J/(kg·K)	
heat transfer coefficient	watt per square meter kelvin	W/(m^2·K)	
linear density	kilogram per meter	kg/m	
magnetic field strength	ampere per meter	A/m	
moment of force, torque	newton meter	N·m	
momentum	kilogram meter per second	kg·m/s	

* The asterisk denotes those units having special names and symbols.
† The spellings "metre" and "litre" are preferred by ASTM; however "er-" is used in the Encyclopedia.
‡ "Weight" is the commonly used term for "mass."
§ Wide use is made of "Celsius temperature" (t) defined by

$$t = T - T_0$$

where T is the thermodynamic temperature, expressed in kelvins, and $T_0 = 273.15$ by definition. A temperature interval may be expressed in degrees Celsius as well as in kelvins.
^{x} This non-SI unit is recognized by the CIPM as having to be retained because of practical importance or use in specialized fields.

* power, heat flow rate, radiant flux	kilowatt watt	kW W	 J/s
power density, heat flux density, irradiance	watt per square meter	W/m²	
* pressure, stress	megapascal kilopascal pascal	MPa kPa Pa	
sound level	decibel	dB	
specific energy	joule per kilogram	J/kg	
specific volume	cubic meter per kilogram	m³/kg	
surface tension	newton per meter	N/m	
thermal conductivity	watt per meter kelvin	W/(m·K)	
velocity	meter per second kilometer per hour	m/s km/h	
viscosity, dynamic	pascal second millipascal second	Pa·s mPa·s	
volume	cubic meter cubic decimeter cubic centimeter	m³ dm³ cm³	 L (liter) mL

In addition, there are 16 prefixes used to indicate order of magnitude, as follows:

Multiplication factor	Prefix	Symbol	Note
10^{18}	exa	E	
10^{15}	peta	P	
10^{12}	tera	T	
10^{9}	giga	G	
10^{6}	mega	M	
10^{3}	kilo	k	
10^{2}	hecto	h[a]	
10	deka	da[a]	
10^{-1}	deci	d[a]	[a] Although hecto, deka, deci, and centi
10^{-2}	centi	c[a]	are SI prefixes, their use should be
10^{-3}	milli	m	avoided except for SI unit-multiples
10^{-6}	micro	μ	for area and volume and
10^{-9}	nano	n	nontechnical use of centimeter, as
10^{-12}	pico	p	for body and clothing measurement.
10^{-15}	femto	f	
10^{-18}	atto	a	

Conversion Factors to SI Units

To convert from	To	Multiply by
acre	square meter (m²)	4.047×10^{3}
angstrom	meter (m)	1.0×10^{-10}†
atmosphere	pascal (Pa)	1.013×10^{5}
bar	pascal (Pa)	1.0×10^{5}†
barn	square meter (m²)	1.0×10^{-28}†
barrel (42 U.S. liquid gallons)	cubic meter (m³)	0.1590
Btu (thermochemical)	joule (J)	1.054×10^{3}
bushel	cubic meter (m³)	3.524×10^{-2}
calorie (thermochemical)	joule (J)	4.184†
centipoise	pascal second (Pa·s)	1.0×10^{-3}†
cfm (cubic foot per minute)	cubic meter per second (m³/s)	4.72×10^{-4}
cubic inch	cubic meter (m³)	1.639×10^{-5}
cubic foot	cubic meter (m³)	2.832×10^{-2}
cubic yard	cubic meter (m)	0.7646
dram (apothecaries')	kilogram (kg)	3.888×10^{-3}
dram (avoirdupois)	kilogram (kg)	1.772×10^{-3}
dram (U.S. fluid)	cubic meter (m³)	3.697×10^{-6}
dyne	newton (N)	1.0×10^{-5}†
dyne/cm	newton per meter (N/m)	1.0×10^{-3}†
fluid ounce (U.S.)	cubic meter (m³)	2.957×10^{-5}
foot	meter (m)	0.3048†
gallon (U.S. dry)	cubic meter (m³)	4.405×10^{-3}

† Exact.

gallon (U.S. liquid)	cubic meter (m³)	3.785×10^{-3}
gallon per minute (gpm)	cubic meter per second (m³/s)	6.308×10^{-5}
	cubic meter per hour (m³/h)	0.2271
grain	kilogram (kg)	6.480×10^{-5}
horsepower (550 ft·lbf/s)	watt (W)	7.457×10^{2}
inch	meter (m)	2.54×10^{-2}†
inch of mercury (32°F)	pascal (Pa)	3.386×10^{3}
inch of water (39.2°F)	pascal (Pa)	2.491×10^{2}
kilogram-force	newton (N)	9.807
kilowatt hour	megajoule (MJ)	3.6†
liter (for fluids only)	cubic meter (m³)	1.0×10^{-3}†
micron	meter (m)	1.0×10^{-6}†
mil	meter (m)	2.54×10^{-5}†
mile (statute)	meter (m)	1.609×10^{3}
mile per hour	meter per second (m/s)	0.4470
millimeter of mercury (0°C)	pascal (Pa)	1.333×10^{2}†
ounce (avoirdupois)	kilogram (kg)	2.835×10^{-2}
ounce (troy)	kilogram (kg)	3.110×10^{-2}
ounce (U.S. fluid)	cubic meter (m³)	2.957×10^{-5}
ounce-force	newton (N)	0.2780
peck (U.S.)	cubic meter (m³)	8.810×10^{-3}
pennyweight	kilogram (kg)	1.555×10^{-3}
pint (U.S. dry)	cubic meter (m³)	5.506×10^{-4}
pint (U.S. liquid)	cubic meter (m³)	4.732×10^{-4}
poise (absolute viscosity)	pascal second (Pa·s)	0.10†
pound (avoirdupois)	kilogram (kg)	0.4536
pound (troy)	kilogram (kg)	0.3732
pound-force	newton (N)	4.448
pound-force per square inch (psi)	pascal (Pa)	6.895×10^{3}
quart (U.S. dry)	cubic meter (m³)	1.101×10^{-3}
quart (U.S. liquid)	cubic meter (m³)	9.464×10^{-4}
quintal	kilogram (kg)	1.0×10^{2}†
rad	gray (Gy)	1.0×10^{-2}†
square inch	square meter (m²)	6.452×10^{-4}
square foot	square meter (m²)	9.290×10^{-2}
square mile	square meter (m²)	2.590×10^{6}
square yard	square meter (m²)	0.8361
ton (long, 2240 pounds)	kilogram (kg)	1.016×10^{3}
ton (metric)	kilogram (kg)	1.0×10^{3}†
ton (short, 2000 pounds)	kilogram (kg)	9.072×10^{2}
torr	pascal (Pa)	1.333×10^{2}
yard	meter (m)	0.9144†

† Exact.

ABBREVIATIONS AND ACRONYMS

A	ampere
AAMI	Association for the Advancement of Medical Instrumentation
ABS	acrylonitrile–butadiene–styrene
ac	alternating current (*noun*)
a-c	alternating-current (*adjective*)
adh	adhesive
AF	aluminum foil
AFR	Air Force Regulation
AGV	automated guide vehicle
AM	aluminum metallization
AN	acrylonitrile
AN/MA	acrylonitrile–methacrylate copolymers
ANS	acrylonitrile–styrene copolymers
ANSI	American National Standards Institute
API	American Paper Institute
ASME	American Society of Mechanical Engineers
ASP	asphalt
ASQC	American Society for Quality Control
ASTM	American Society for Testing and Materials
avg	average
BATF	Bureau of Alcohol, Tobacco and Firearms
B&B	blow-and-blow
BBP	butyl benzyl phthalate
BCL	British Cellophane Limited

BHEB	*tert*-butylated hydroxyethylbenzene
BHT	*tert*-butylated hydroxytoluene
BIB	bag-in-box
BK	bleached kraft
BMC	bulk molding compound
BON	biaxially oriented nylon film
BOPP	biaxially oriented polypropylene film
bpm	bottles per minute; bags per minute
BSP	British standard pipe thread
Btu	British thermal unit
BUR	blow-up ratio
°C	degree Celsius
CA	controlled atmosphere
ca	approximately (*circa*)
CAD	computer-aided design
CAE	computer-aided engineering
cal	calorie (4.184 J)
Cal	food calorie (1000 cal)
CAP	cellulose acetate propionate; controlled atmosphere packaging
CELLO	cellophane
CFR	Code of Federal Regulations
CGMP	Current Good Manufacturing Practice
CGPM	Conference Generale des Poids et Mesures (General Conference on Weights and Measures)
CI	central impression

CIPM	Comité International des Poids et Mesures (International Committee on Weights and Measures)
cm	centimeter
CMA	Closure Manufacturers Association
CNC	computer numerical control
coex	coextruded
COF	coefficient of friction
COFC	Container on Flat Car
COREPER	Committee of Permanent Representatives (EEC)
CPE	chlorinated polyethylene
cpm	cans per minute
cPs	centipoise (10^{-3} Pa·s)
CPSC	Consumer Product Safety Commission
CPU	central processing unit
CR	child-resistant
CRC	child-resistant closure
CRT	cathode-ray tube
CT	continuous thread
d	diameter; day, 24 hours
D & I	drawing and ironing (cans)
dB	decibel
dc	direct current (*noun*)
d-c	direct-current (*adjective*)
DEHP	di(2-ethylhexyl) phthalate
DEG	diethylene glycol
DGT	diethylene glycol terephthalate
dia	diameter
DINA	diisononyl adipate
DLAM	Defense Logistics Agency Manual
DME	dimethyl ether
DMF	Drug Master File
DMT	dimethyl terephthalate
DOA	dioctyl adipate
DOD	U.S. Department of Defense
DOP	dioctyl phthalate
DOT	U.S. Department of Transportation
DOZ	dioctyl azelate
DPM	double package maker
DR	double-reduced
DRD	draw redraw (cans)
DSAM	Defense Supply Agency Manual
DWI	drawn and ironed (cans)
EAA	ethylene–acrylic acid
ECOSOC	Economic and Social Committee (EEC)
ECCS	electrolytic chromium-coated steel
EEA	ethylene–ethyl acrylate
EEC	European Economic Community
eg	for example (*est gratia*)
EG	ethylene glycol
EHMW	extra high molecular weight
EIA	Electronic Industries Association
ELC	end-loading construction
EMA	ethylene methacrylate
EMAA	ethylene–methacrylic acid
EMI/RFI	electromagnetic interference/radio frequency interference
EPA	Environmental Protection Agency
EPC	expanded polyethylene copolymer
EPE	expanded polyethylene
EPR	ethylene–propylene rubber
EPS	expanded polystyrene
ESC	environmental stress cracking
ESCR	environmental stress-crack resistance
ESD	electrostatic discharge
ESO	epoxidized soybean oil
est	estimated
ETO	ethylene oxide
ETP	electrolytic tinplate
EVA	ethylene–vinyl acetate
EVOH	ethylene–vinyl alcohol
F_{50}	value at which 50% of the specimens have failed

°F	degree Fahrenheit
FDA	U.S. Food and Drug Administration
FD&C	Food, Drug, and Cosmetic
FEP	fluorinated ethylene–polypropylene
ffs	form/fill/seal
FI	flow index
fl oz	fluid ounce (29.57 mL in the U.S.)
FPMR	Federal Property Management Regulations
FRH	full-removable head
FRP	fiberglass reinforced plastics
ft	foot
ft·lbf	foot-pound force (1.356 J)
G	specific deceleration:gravitational acceleration; specific acceleration:gravitational acceleration
g	gravitational acceleration (9.807 m/s²)
g	gram
gf	gram-force (0.0098 N)
ga	gauge
gal	gallon (3.785 L in the U.S.)
GMA	Grocery Manufacturers Association
GMP	Good Manufacturing Practice
GNP	Gross National Product
GPPS	general-purpose polystyrene
Gy	gray (10^{-2} rad)
h	hour; height
HBA	health and beauty aids
HCP	health-care packaging
HD	head diameter
HDPE	high density polyethylene
HDT	heat-deflection temperature
HF	high frequency
HFFS	form/fill/seal, horizontal
HIMA	Health Industry Manufacturers' Association
HIPS	high impact polystyrene
HLMI	high load melt index
HM	hot melt
HMW	high molecular weight
HNR	high nitrile resin
hp	horsepower (746 W)
HP	high pressure
HPP	homopolymer polypropylene
HRC	Rockwell hardness (C scale)
HRM	Rockwell hardness (M scale)
HRR	Rockwell hardness (R scale)
HTST	high temperature-short time
Hz	hertz (cycles per second)
IBC	intermediate bulk containers
ICAO	International Civil Aviation Organization
ie	that is (*id est*)
IM	intramuscular
IMDG	International Maritime Dangerous Goods
IML	in-mold labeling
IMO	International Maritime Organization
in.	inch (2.54 cm)
I/O	input/output
IPA	isophthalic acid
I.S.	individual section
ISO	International Standards Organization
ITRI	International Tin-Research Institute
IV	intrinsic viscosity; intravenous
J	joule (energy)
k	kilo (10^3)
K	Kelvin; the molecular weight of a resin
k.d.	knocked-down
kgf	kilogram-force (9.806 N)
kJ	kilojoule
km	kilometer
kPa	kilopascal (0.145 psi)

L	liter (volume)
lb	pound (mass) (453.6 g)
lbf	pound force (4.448 N)
LCB	long-chain branching
LCD	liquid crystal display
L:D	length:diameter
LDPE	low density polyethylene
LED	light-emitting diode
LLDPE	linear low density polyethylene
LMW	low molecular weight
LSI	large-scale integration
L:T	length:thickness
LVP	large-volume parenteral
M	mega (10^6)
M_N	number-average molecular weight
M_W	weight-average molecular weight
m	meter; milli (1/1000)
MA	modified atmosphere; methyl acrylate
MAN	methacrylonitrile
max	maximum
MBS	methacrylate–butadiene styrene
m/c	cylinder mold
MD	machine direction
MDPE	medium density polyethylene
MET	metallized
MF	machine-finished
MFR	melt:flow rate
mg	milligram
MG	mill-glazed; machine-glazed
MGBK	machine-glazed bleached Kraft
MGNN	machine grade natural Northern
mi	mile
MI	melt index
MIL	military
min	minute; minimum
MIR	multiple-individual-rewind
MMA	methyl methacrylate
MMW	medium molecular weight
mn	millinewton (2.25×10^{-4} lbf)
MN	meganewton (224,909 lbf)
mo	month
MOE	modulus of elasticity
MOR	modulus of rupture
MP	microprocessor; melting point
MPa	megapascal (145 psi)
ms	millisecond
MSW	municipal solid waste
MTB	Materials Transportation Bureau
MVTR	see WVTR
MW	molecular weight
MWD	molecular weight distribution
μ	micro (10^6)
μm	micrometer
N	newton (force)
na	not available
NASA	National Aeronautics and Space Administration; National Advertising Sales Association
NBS	National Bureau of Standards
NC	nitrocellulose
NCB	National Classification Board
NDA	New Drug Application
NEC	not elsewhere classified
NF	*National Formulary*
NK	natural kraft
nm	nanometer (10^{-9} meter)
NMFC	National Motor Freight Classification Committee
NODA	*n*-octyl *n*-decyl adipate
NPIRI	National Printing Ink Research Institute
NSTA	National Safe Transit Association
NWPCA	National Wooden Pallet and Container Association

Ω	ohm (resistance)
OD	optical density
OL	overlacquer
ON	oriented nylon
OPET	oriented polyester
OPP	oriented polypropylene
OPS	oriented polystyrene
OSHA	Occupational Safety and Health Administration
OTC	over-the-counter
OTR	oxygen transmission rate
Pa	pascal (pressure)
PA	Proprietary Association
PAN	polyacrylonitrile
P&B	press-and-blow
PBT	poly(butylene terephthalate)
PC	programmable calculator; polycarbonate
PCTFE	poly(chlorotrifluoroethylene)
PE	polyethylene
PEPS	Packaging of Electronic Products for Shipment
PET	poly(ethylene terephthalate); polyester
phr	parts per hundred of resin
PIB	polyisobutylene
PLI	pound force per lineal inch
PM	Packaging Materials
PMMA	poly(methyl methacrylate)
PMMI	Packaging Machinery Manufacturers Institute
PP	polypropylene
ppb	part per billion (10^9)
ppm	part per million (10^6)
PPP	poison-prevention packaging
PPPA	Poison-Prevention Packaging Act
PR	printing
proj	projected
PS	polystyrene
psi	pound (force) per square inch (6.893 kPa)
psig	psi gauge pressure
PSTA	Packaging Science and Technology Abstracts
PSTC	Pressure-sensitive Tape Council
PTFE	polytetrafluoroethylene
PVAc	poly(vinyl acetate)
PVC	poly(vinyl chloride)
PVDC	poly(vinylidene chloride)
PVF	poly(vinyl fluoride)
PVF_2	poly(vinylidene fluoride)
PVOH	poly(vinyl alcohol)
PX	post exchange
Q_{10}	change in reaction rate for a 10°C temperature increase
QA	quality assurance
QAI	quaternary ammonium inhibitor (nitrite)
qt	quart (946 mL in the U.S.)
R&D	research and development
RCF	regenerated cellulose film
RCPP	random-copolymer polypropylene
RD	root diameter
RDF	refuse-derived fuel
rf	radio frequency (*noun*)
r-f	radio-frequency (*adjective*)
rh	relative humidity
RIM	reaction injection molding
RM-HNR	rubber-modified high-nitrile resin
rpm	rotations per minute
RSC	regular slotted container
RVR	rim-vent release
s	second
SAN	styrene–acylonitrile
SB	styrene–butadiene
SBS	solid bleached sulfate
SCB	short-chain braching

SCF	Scientific Committee on Food (EEC)	UFC	Uniform Freight Classification Committee
SCR	semiconductor	UHMW	ultra high molecular weight
SIC	Standard Industrial Classification	UN	United Nations
SMC	sheet molding compound	UPS	United Parcel Service
SMMA	styrene–methyl methacrylate	USDA	United States Department of Agriculture
SP	special packaging	USP	*US Pharmacopeia*
sp gr	specific gravity	uv	ultraviolet
SPPS	solid-phase pressure forming		
SR	single-reduced	V	volt
SUS	solid unbleached sulfate	VA	vinyl alcohol
sq	square	VC	vinyl chloride
SVP	small-volume parenteral	VCI	volatile corrosion inhibitor
		VCM	vinyl chloride monomer
t	metric ton (1000 kg)	VDC	vinylidene chloride
TA	thread angle	VFFS	form/fill/seal, vertical
TAPPI	Technical Association of the Pulp and Paper Industry	vol	volume
		vs	versus
TD	thread-crest diameter		
TE	tamper-evident	W	watt (J/s)
tffs	thermoform/fill/seal	WD	wire diameter
TFS	tin-free steel	wk	week
T_g	glass-transition temperature	wt	weight
TH	tight head	WVTR	water vapor transmission rate
T_m	melting temperature		
TIS	Technical Information Service	XD	cross-direction
TOFC	trailer on flat car	XKL	extensible kraft linerboard
TPA	terephthalic acid		
TR	tamper-resistant	yr	year
UCB	Union Chimique Belge	ZCC	zero-crush concept
UCC	Uniform Classification Committee		

THE WILEY
ENCYCLOPEDIA OF
PACKAGING TECHNOLOGY

A

ACRYLIC MULTIPOLYMERS.   See Nitrile polymers.

ACRYLICS

Polymers based on acrylic monomers are useful in packaging as a basis for printing inks and adhesives and as modifiers for rigid PVC products.

Acrylic-Based Inks

Paste inks.   Acrylic solution resins are used in lithographic inks as dispersing or modifying letdown vehicles (see Inks; Printing). A typical resin (60% in oil) offers excellent dot formation, high color fidelity, exceptional print definition, non-skinning, and good press-open time. Set times are fast (ca 60–90 s), and a minimal level of starch spray (75% of normal) is effective. Coatings on cartons, fabrication stocks, and paper are glossy and exhibit good dry resistance.

Solvent inks.   Because of their resistance to heat and discoloration, good adhesion, toughness, and rub resistance, acrylics are widely used in flexographic inks on paper, paperboard, metals, and a variety of plastics (1). These inks also give block resistance, resistance to grease, alcohol, and water, and good heat-sealing performance (see Sealing, heat). With some grades, adding nitrocellulose improves heat sealability, heat resistance, and compatibility with laminating adhesives. This family of methacrylate polymers (methyl to isobutyl) has broad latitude in formulating and performance. Solid grades afford low odor, resist sintering, and dissolve rapidly in alcohol–ester mixtures or in esters alone (gravure inks). Solution grades (40–50% solids) are available, as well as nonaqueous dispersions (40% solids) in solvents such as VMP naphtha, which exhibit fast solvent-release and promote superior leveling and hiding. They are excellent vehicles for fluorescent inks.

Water-based inks.   The development of waterborne resins has been a major achievement. Their outstanding performance allows them to replace solvent systems in flexographic and gravure inks and overprint varnishes on corrugated and kraft stocks, cartons, and labels. The paramount advantage of aqueous systems is substantial decrease in environmental pollution by volatile organics. Aqueous acrylic colloidal dispersions (30% solids) and a new series of analogous ammonium salts (46–49% solids) are effective dispersants for carbon blacks, titanium dioxide, and organic pigments. Derived inks give crisp, glossy impressions at high pigment loading, good coverage and hiding, and water resistance. The relatively flat pH–viscosity relationship assures formulation stability on presses despite minor loss of volatiles. Adjusting the alcohol–water ratio controls drying rate, and quick-drying inks can be made for high speed printing. The resins are compatible with styrene–acrylic or maleic dispersants and acrylic or styrene–acrylic letdown vehicles. Blends of self-curing polymer emulsions are excellent overprint varnishes for labels and exhibit a good balance of gloss, holdout, slip, and wet-rub resistance.

Some aqueous acrylic solutions (37% solids) combine the functions of pigment dispersant and letdown resin and serve as ready-to-use vehicles for inks on porous substrates like kraft and corrugated stocks and cartons. They afford excellent color development, excellent heat-aging resistance in formulations, and fast drying. The flat pH–viscosity relationship gives the same benefit as the dispersants.

Acrylic Adhesives

Pressure-sensitive adhesives.   Solution copolymers of alkyl acrylates and minor amounts of acrylic acid, acrylonitrile, or acrylamide adhere well to paper, plastics, metals, and glass and have gained wide use in pressure-sensitive tapes (2). Environmental regulations, however, have raised objections to pollution by solvent vapors and are requiring costly recovery systems. This opportunity has encouraged the development of waterborne substitutes, such as emulsion polymers, which eliminate these difficulties and offer excellent adhesion, resistance to wet delamination, aging, and yellowing, and, like the solvent inks, need no tackifier. In packaging applications, the emulsion polymers provide high tack, a good balance of peel adhesion and shear resistance, excellent cling to hard-to-bond substrates, and clearance for food packaging applications under FDA Regulations 21 CFR 175.105, 21 CFR 176.170, and 21 CFR 176.180. Their low viscosity makes formulation easy, and the properties of the adhesives can be adjusted by adding surfactants, acrylic thickeners, and defoamers.

Resins are available that are designed specifically for use on polypropylene carton tapes (3). They are ready-to-use non-corrosive liquids applicable to the corona-treated side of oriented polypropylene film using knife-to-roll, Mayer rod, or reverse-roll coaters (see Coating equipment; Film, oriented polypropylene; Surface modification). A release coating is unnecessary because the adhesive does not stick to the untreated side and parts cleanly from the roll. The tapes are used to seal paperboard cartons with high speed taping machines or hand-held dispensers. Adhesion to the cartons is instantaneous and enduring. The colorless tape is well suited for label protection. The material adheres well to other plastics and metals.

Hot-melt adhesives.   These adhesives offer obvious advantages over solvent-borne or waterborne materials if equivalent performance is obtainable. Acrylic prototypes gave better color and oxidative stability than rubber-based products, but exhibit poor adhesion quality. New improved grades are providing an impressive array of adhesive properties and superior cohesive strength at elevated temperatures in addition to stability and low color. The action of the adhesives involves a thermally reversible cross-linking mechanism which gives ready flow at 350°F (177°C), rapid increase in viscosity on cooling, and a stiff cross-linked rubber at ambient temperature. The resins give durable peel adhesion, good shear resistance, resistance to cold flow, and excellent photostability in accelerated weathering. On commercial machinery these resins have displayed excellent coatability on polyester film at high line speeds (see Film, oriented polyester). A wide variety of possible applications, including packaging tapes, is envisaged for these materials.

PVC Modifiers

Acrylics have played a major role in the emergence of clear rigid PVC films and bottles (4,5). Acrylic processing aids provide smooth processing behavior in vinyl compounds when passed through calenders, extruders, blow-molding machinery, and thermoforming equipment (see Additives, plastics). One member of this group is a lubricant–processing aid that prevents sticking to hot metal surfaces and permits reduction in the level of other lubricants, thereby improving clarity. Other benefits of acrylics are low tendency to plateout and a homogenizing effect on melts to give sparkling clarity and improved mechanical properties. The usual level in vinyl compounds for packaging is about 1.5–2.5 phr.

In a second group are the impact modifiers, which are graft polymers of methyl methacrylate–styrene–butadiene used in the production of clear films and bottles. The principal function of impact modifiers is to increase toughness at ambient and low temperature. Levels of 10–15 phr, depending on modifier efficiency, are normal.

Many acrylics are cleared for use in food-contact products under FDA Regulation 21 CFR 178.3790. The regulation stipulates limits in the permissible level of modifiers relative to their composition. Processors should seek advice from suppliers on the makeup of formulations. Many modifiers are fine powders that may produce airborne dust if handled carelessly. Above 0.03 oz/ft^3 (0.03 mg/cm^3), dust is a potential explosion hazard and its accumulation on hot surfaces is a fire hazard. The recommended exposure limit to dust over an 8-h period is 2 mg/m^3. Eliminate ignition sources, ground equipment electrically, and provide local exhaust ventilation where dusting may occur (6). Workers may wear suitable MSHA-NIOSH respiratory devices as protection against dust.

BIBLIOGRAPHY

1. B. V. Burachinsky, H. Dunn, and J. K. Ely in M. Grayson and D. Eckroth, eds., *Encyclopedia of Chemical Technology*, Vol. 13, John Wiley & Sons, Inc., New York, 1981, pp. 374–398.

2. D. Satas, ed., *Handbook of Pressure-Sensitive Adhesives*, Van Nostrand Reinhold Co., 1982, pp. 298–330, 426–437.

3. W. J. Sparks, *Adhes. Age* **26**(2), 38 (1982).

4. J. T. Lutz, Jr., *Plast. Compd.* **4**(1), 34 (1981).

5. *Bulletin MR-112b*, Rohm and Haas Company, Philadelphia, PA, Jan. 1983.

6. American Conference of Governmental Hygienists, Cincinnatti, Ohio, *A Manual of Recommended Practice*, 1982; American National Standards Institute, New York, N.Y., *Fundamentals Governing the Design and Operation of Local Exhaust Systems*, ANSI Z-9.2, 1979.

J. T. SULLIVAN
Rohm and Haas Company

ADDITIVES, PLASTICS

Some polymeric materials, like fibers and film, contain few additives, but most plastic packaging materials contain some functional additive. There are a large number of additives and each must be selected in accordance with the product's end use. Most of these additives are more expensive than the polymers and are used in small amounts of about 1–2%. The additives which may be used in packaging are listed alphabetically in this article.

Antiblocking agents. Because they are electrically nonconductive, many packaging films tend to stick together. This cohesion or blocking tendency is reduced by the addition of organic amides, such as erucamide, and metallic soaps, such as zinc stearate. Packaging films may also be prevented from sticking to each other by the addition of flatting agents, such as silicas. The term denesting agents is also used to describe antiblocking agents. About 4000 tons (3630 t) of these additives is used annually in the United States (1).

Antifogging agents. Moisture tends to condense as droplets and obstruct the view of contents protected by packaging film. These droplets may be dispersed by the addition of nonionic ethoxylates or hydrophilic fatty acid esters, such as glyceryl stearate, which promote the deposition of continuous films of moisture.

Antimicrobials. The growth of microorganisms on the surface of cellulosics and plasticized poly(vinyl chloride) (PVC) films may be prevented by the addition of antimicrobials such as algicides, bactericides, and fungicides (see Film, flexible PVC). Copper-8-quinoleate, 2-n-octyl-4-isothiazolin-3-one, 10,1′-oxybisphenoxyarsine and N-(trichloromethylthiophthalimide) are used as antimicrobials (2).

Antioxidants. Many polymeric films, such as natural rubber and polypropylene, are degraded in the atmosphere (see Film, nonoriented polypropylene; Film, oriented polypropylene). This undesirable deterioration may be prevented by the addition of hindered phenols, such as butylated hydroxytoluene (BHT), which have been cleared by the FDA (3). Hindered phenols act as free-radical scavengers or polymer chain terminators. Other types of antioxidants, such as organophosphites (eg, tris (2,4-ditertbutylphenyl phosphite), act as hydroperoxide decomposers which when used with hindered phenols prevent the formation of chromophoric carbonyl groups. Thioesters, such as dilauryl thiodipropionate, have been used when odor is not a problem.

Different antioxidants are often used together for synergistic effects. Over 16,000 tons (14,500 t) of antioxidants is used annually by plastics industries in the United States.

Antistats. Polymeric films are nonconductors of electricity and tend to accumulate electrical charges which attract dust. This undesirable effect may be prevented by the addition of ethoxylated fatty amines and nonionic and quaternary ammonium compounds which migrate to the film surface (bloom) and cause the formation of a conducting layer of aqueous solution which permits the discharge of electrons. Over 2400 tons (2180 t) of antistats is used annually in the United States (4).

Blowing agents. Both chemical (CBA) and physical (PBA) blowing agents are used to produce packaging foams from plastics. The principal CBA is azobisformamide (ABFA), and one of the principal PBAs is methylene chloride (5). Over 4000 tons (3630 t) of ABFA derivatives is used annually in the United States for the production of polymeric foams. Polyurethane foams are produced by the use of PBAs such as fluorocarbons and *in situ* formation of carbon dioxide resulting from the reaction of water and organic isocyanates. The principal packaging foams are expanded polystyrene (EPS) and polyurethanes (PUR) (see Foam cushioning).

Catalysts (curing agents). The "catalysts" used for initiation of polymerization and curing of polyester prepolymers are actually initiators rather than catalysts. The principal initiator is benzoyl peroxide. These initiators are used to produce polymers, such as polyethylene, but their use in packaging films is extremely limited.

Amines, such as N-ethylmorpholine, and organotin catalysts, such as tin oleate, are both used as catalysts for the formation of polyurethane foam. The activity of the amine increases with the increased basicity of these catalysts, and the activity of the tin catalyst depends on the tin atom. The annual consumption of the amine and tin catalysts is 1400 and 2400 tons (1270 and 2180 t), respectively (6).

These additives catalyze the reaction between a polyol and an organic diisocyante (7). Potassium octoate is used as a catalyst in the formation of rigid isocyanate foams.

Colorants. Many packaging films are unpigmented, but some are colored by the addition of colorants (see Colorants). Special effects are obtained by the addition of metallic pigments, pearlescents, and fluorescents. Over 300,000 tons (272,160 t) of pigments is used annually by the U.S. plastics industry (8). The principal pigments are carbon black, white titanium dioxide, red iron oxide, yellow cadmium sulfide, molybdate orange, ultramarine blue, blue ferric ammonium ferrocyanide, chrome green, and blue and green copper phthalocyanines. The use of some of these colorants in food packaging has been questioned by the FDA. Polyethylene (HDPE) milk bottles are usually pigment-free, but pigments such as titanium dioxide may be added to screen out ultraviolet radiation (see Polyethylene, high density).

Coupling agents. Pigments and fillers do not bond well with polymers and may result in a reduction in mechanical properties. This problem may be solved by the addition of small amounts of coupling agents, such as organosilanes or organotitanates. Effective coupling agents actually form a bond with the filler which can be detected by infrared spectroscopy. Zircoaluminates, stearates, and castor-oil esters are also used as hydrophobic wetting agents or coupling agents (9).

Flame retardants. Most packaging materials are combustible. Their lack of flame resistance is disregarded in most packaging applications, but the flammability of bulky packaging materials, such as rigid foams, should be considered. Foamed polystyrene cups and plates are combustible, but one must balance the risk of fire versus the possible toxicity of flame-retardant additives (10).

One of the least toxic and most widely used flame retardants is alumina trihydrate (ATH) which acts as a heat sink by producing steam when heated. Over 805,000 tons (730,290 t) of ATH is used annually as a flame retardant by the U.S. plastics industry.

Poly(vinyl chloride) (PVC) is more flame-resistant than polystyrene or polyolefins, but the plasticizers present in PVC are combustible and flame retardants must be added to PVC if flammability is a problem. Antimony oxide and organophosphates and boron compounds may be added to PVC to reduce its combustibility. Zinc borate and molybdenum oxide act as smoke depressants.

Fragrance enhancers. Fragrance enhancers are added to films with undesirable odors and to films such as those used in trash bags which may absorb an undesirable odor from contact. These additives are also used in blow-molded plastic bottles to enhance the odor of the contents (11).

Heat stabilizers. Films and bottles made from PVC degrade when heated at moderately high temperatures or subjected to gamma ray sterilization or uv irradiation unless proper additives are present. These heat stabilizers retard the decomposition of PVC into hydrogen chloride and dark degraded polymers. Over 37,000 tons (33,570 t) of heat stabilizers is used annually by the U.S. plastics industry.

Barium–cadmium, organotin, and lead compounds account for 40, 27, and 26%, respectively, of the total heat stabilizers used by the U.S. plastics industry. Calcium–zinc and antimony compounds account for 5 and 1%, respectively (12).

The additives in PVC bottles used for cooking oil and other food products must have FDA clearance. Octyltin mercaptide, calcium–zinc compounds and methyltin have clearance as heat stabilizers. Epoxidized soybean and linseed oil, which are also cleared by the FDA, are used as secondary additives to supplement the effectiveness of metal-compound stabilizers. (Antimony compounds and methyltin are also used in PVC pipes for potable water.)

Substituted pyridine dicarbonate is being used as a heat stabilizer, and organotin compounds are being used as gamma shields. These stabilizers can be used for both rigid and flexible PVC (13).

The annual consumption of heat stabilizers in the United States is as follows: barium compounds, 15,000 tons (13,610 t); tin compounds, 10,000 tons (9,070 t); lead compounds, 10,000 tons (9,070 t); calcium–zinc compounds, 2,000 tons (1,814 t); and antimony compounds, 450 tons (408 t).

Impact modifiers. Poly(vinyl chloride) was synthesized in the nineteenth century, but because of its inherent brittleness, it was not used commercially until the late 1920s (see Poly(vinyl chloride).) Poly(vinyl chloride) was flexibilized by the addition of plasticizers, such as dioctyl phthalate, and impact modifiers, such as acrylonitrile elastomers (NBR) (see Nitrile polymers). Superior results are obtained by the addition of acrylic polymers (ACR) (see Acrylics). Over 8500 tons (7710 t) of ACR is added to PVC annually in the United States.

When 7.5 g of ACR is added to 100 g PVC, a product with good room-temperature impact resistance results, but the concentration of ACR must be raised to 12.5 phr to obtain a film or bottle with good impact resistance at 32°F (0°C) (14).

Poly(methyl methacrylate)-co-butadiene-co-styrene (MBS) and poly(methyl methacrylate)-co-acrylonitrile-co-butadiene-co-styrene (MABS) are also excellent impact modifiers for PVC. Over 20,000 tons (18,140 t) of MABS and MBS is used annually by the U.S. plastics industry. The U.S. plastics industry also consumes over 10,000 tons (9070 t) annually of ABS, chlorinated polyethylene (CPE), and polyethylene-co-vinyl acetate (EVA) as impact modifiers for PVC film. Hydroxyl terminated polyethers and styrene–butadiene block copolymers are also used as impact modifiers for polystyrene.

Low profile additives. Several different additives, such as polyethylene, cellulose acetate butyrate, poly(vinyl acetate), polyurethane, polycaprolactone, and poly(methyl methacrylate), are added to prepolymers of thermosetting polyesters as low profile additives (LPA) to assure a smooth surface. The use of thermoset polyesters in packaging is limited (see Thermosetting plastics).

Lubricants. The tendency for polymers, such as PVC and polyolefins, to stick to metal parts during processing is reduced by the addition of lubricants, such as fatty acid esters and amides and paraffin and polyethylene waxes. Metallic stearates such as zinc stearate are also used as lubricants and account for 40% of the 38,000 tons (34,470 t) of these additives used annually in the United States (15).

Mold-release agents. External mold-release agents or slip agents, such as calcium stearate, are applied to metal surfaces to prevent the sticking of plastic film or bottles during processing and fabrication. Slip agents may also serve as antifogging agents (16).

Plasticizers. Some thin films of polyolefins and bottles made from polyethylene terephthalate) (PET) are somewhat flexible and do not require plasticizers as additives (see Polyesters, thermoplastic). However, brittle polymers such as PVC are usually plasticized in order to obtain flexible films and containers. PVC was not commercialized until it was plasticized by the addition of tricresyl phosphate in the late 1920s. Because of toxicity, this plasticizer has been replaced by phthalic acid esters, such as dioctyl phthalate.

Phthalic acid esters account for 64% of the 632,000 tons (573,350 t) of plasticizers used annually by the U.S. plastics industry. Most of these plasticizers are used to flexibilize PVC (17).

According to the International Union of Pure and Applied Chemistry (IUPAC), a plasticizer is a substance incorporated into a plastic or an elastomer to increase flexibility, workability, and distensibility. A plasticizer may reduce the melt viscosity, lower the glass-transition temperature (T_g), or lower the elastic modulus of the plastic (see Polymer properties). This additive may be an external plasticizer, such as diethylhexyl phthalate (DOP), diisononyl phthalate (DINP), diisoheptyl phthalate (DIHP), or di(2-ethylhexyl) terephthalate (DOTP) (18), or it may be an internal plasticizer. Internal plasticization may be brought about by copolymerization of vinyl chloride with monomers, such as vinyl acetate, ethylene, or methyl acrylate.

Because the safety of DOP has been questioned, alternative materials are gaining favor, including citric acid esters. Some citric acid esters have been approved by FDA. Epoxidized oils, which are also used as heat stabilizers, and dioctyl adipate (DOA) are also being used in place of DOP. Low molecular weight polyesters are also being used as nonvolatile plasticizers (19).

Processing aids. Processing aids, such as ethoxylated fatty acids, are used as viscosity depressants in PVC plastisols, and flow promoters, such as polyisobutylene, are used to improve the processability of polyolefins. Acrylics which are incompatible with PVC at elevated temperatures actually improve the processability and clarity of PVC films (20).

Uv stabilizers. Many polymeric films, like human skin, are deteriorated in sunlight, and sunscreens must be used to prevent this photooxidation. The first sunscreen was phenyl salicylate which by way of a photo-Fries rearrangement forms a derivative of 2-hydroxybenzophenone. This sunscreen and derivatives of 2-hydroxybenzophenone have also been used as uv stabilizers in polymeric films.

Uv stabilizers may absorb high energy uv radiation and release it as lower energy radiation, or they may act as free-radical scavengers. Derivatives of 2-hydroxybenzophenone form a chelate through intramolecular hydrogen bonding when subjected to uv radiation. The chelate rearranges to a compound with a quinoid structure and the high energy radiation is transferred to low energy radiation by an energy release mechanism.

Over 2300 tons (2090 t) of uv stabilizers are used annually by the American Polymer Industry (21). Over 75% of these additives were used for the stabilization of polyolefins against sunlight. Polyarylates, which contain 2-hydroxybenzophenone groups, do not require the addition of uv stabilizers and can be used as additives to polycarbonates (see Polycarbonate) and poly(ethylene terephthalate) (PET) (22).

Zinc oxide has been used as a synergistic uv stabilizer (23).

Pigments, such as titanium dioxide in combination with zinc oxide, as well as carbon black, are also effective screens for ultraviolet radiation (24).

Nickel complexes and nickel salts act as quenchers or energy transfer agents and convert the excess energy from sunlight to harmless low heat energy. The efficiency of these quenchers, which are about 400% more effective than 2-hydroxybenzophenones, is independent of sample thickness.

Hindered amine light stabilizers (HALS), such as tetramethyl piperidine derivatives, have received FDA clearance. It is believed that HALS serve as free-radical scavengers. HALS are the most important class of uv stabilizers today.

BIBLIOGRAPHY

1. R. A. Lindar, *Plast. Compd.* **6**(4), 27 (1983).
2. C. R. Beiter, *Mod. Plast.* **60**(10A), 164 (1983).
3. J. W. Castagno, *Plast. Eng.* **40**(6), 29 (1984).
4. W. Olcott, *Mod. Plast.* **60**(9), 75 (1983).
5. R. B. Seymour, *Plast. Compd.* **3**(1), 42 (1980).
6. W. Olcott, *Mod. Plast.* **60**(9), 72 (1983).
7. A. von Hassell, *Plast. Technol.* **30**(8), 103 (1984).
8. A. S. Wood, *Mod. Plast.* **60**(9), 61 (1983).
9. R. B. Seymour, *Pop. Plast.* **28**(9), 14 (1983).
10. R. C. Nametz, *Plast. Compd.* **7**(4), 26 (1984).
11. V. N. Suran, *Plast. Technol.* **30**(6), 65 (1984).
12. R. R. McBride, *Mod. Plast.* **60**(9), 57 (1983).
13. P. Bogan, *Plast Technol.* **30**(8), 63 (1984).
14. A. von Hassell, *Plast. Technol.* **29**(8), 84 (1983).
15. R. Martino, *Mod. Plast.* **60**(8), 70 (1983).
16. T. R. Brelot and co-workers, *Polym. Modif. Add.* **1**(1), 11 (1984).
17. R. Juran, *Mod. Plast.* **60**(9), 68 (1983).
18. L. S. Krauskopf, *Plast. Compd.* **6**(1), 28 (1983).
19. P. Bogas, *Plast. Technol.* **30**(8), 70 (1984).
20. J. Frados, *Plast. Compd.* **6**(6), 42 (1983).
21. R. R. McBride, *Mod. Plast.* **60**(9), 60 (1983).
22. D. L. Love, *Mod. Plast.* **61**(3), 60 (1984).
23. R. J. Pierotti and R. D. Deanin in R. B. Seymour, ed., *Chapter 10 Additives for Plastics*, Vol. 2, Academic Press, New York, 1978.
24. W. S. Castor and J. A. Manasso in ref. 23, Chapt 11, Vol. 1.

RAYMOND B. SEYMOUR
University of Southern Mississippi

ADHESIVE APPLICATORS

Adhesive applicating equipment used in packaging applications is available in a vast array of configurations to provide a specific means of sealing containers. The type of adhesive equipment chosen is determined by a number of factors: the class of adhesive (cold waterborne or hot-melt), the adhesive applicating unit and pump style that is most compatible with the adhesive properties, and production line demands. The variables in the packaging operation are matched with the available adhesives and equipment to achieve the desired results.

Packaging Adhesives

Adhesives used in packaging applications today are primarily cold waterborne or hot-melt adhesives (see Adhesives).

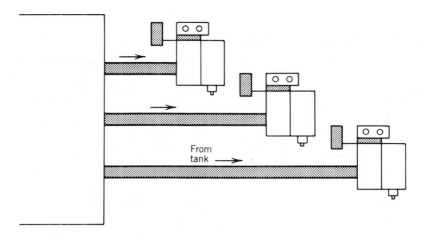

Figure 1. Noncirculating gun installation (parallel) system.

Cold waterborne adhesives can be broadly categorized into natural or synthetic. Natural adhesives are derived from protein (animal, casein) and vegetable (starch, flour) sources. Synthetic-based adhesives (primarily resin emulsions) have been gradually replacing natural adhesives in recent years. The liquid "white glue" is generally composed of protective poly(vinyl alcohol) or 2-hydroxyethyl cellulose colloids and compounded with plasticizers, fillers, solvents, or other additives (see Additives, plastic). Also, new copolymers have been developed and used to upgrade performance of cold emulsion adhesives in dispensing characteristics, set time, and stability.

Cold adhesives have good penetration into paper fiber and are energy-efficient, especially when no special speed of set is required. Hot-melt adhesives are thermoplastic polymer-based compounds that are solid at room temperature, liquefy when heated, and return to solid form upon cooling. They are blended from many synthetic materials to provide specific bonding characteristics. Most hot melts consist of a base polymer resin for strength, a viscosity control agent such as paraffin, tackifying resins for greater adhesion, and numerous plasticizers, stabilizers, antioxidants, fillers, dyes, and/or pigments.

Hot melts are 100% solid; they contain no water or solvent carrier. This offers several advantages: rapid bond formation and short set time because heat dissipates faster than water evaporates, shortened compression time, and convenient form for handling and long-term storage. Being a thermoplastic material, hot melts have limited heat resistance and can lose their cohesiveness at elevated temperatures.

Adhesive Applicating Equipment Classification

Both cold waterborne and hot-melt adhesive application systems are generally classified as noncirculating or circulating. Noncirculating systems are the most common (see Fig. 1). They are identified easily because each gun in the system is supplied by its own hose. The noncirculating system is often referred to as a dead end system because the hose dead ends at the gun. An offshoot of the noncirculating system is the internally circulating hot-melt system (see Fig. 2). There is circulation between the pump and manifold, but from the manifold to the gun it is the same as a dead end system.

Circulating systems are used to some extent in applications requiring a standby period, as in a random case sealing operation, when some setup time is needed. A circulating system is identified by the series installation of the hoses and guns (see Fig. 3). In the typical circulating installation, a number of automatic extrusion guns are connected in series with the hot melt hose. Molten material is siphoned out of the applicator tank and pumped into the outlet hose to the first gun in the series. The material then flows from the first gun to the second gun and continues on until it passes through a circulation valve and back into the applicator's tank. The circulation valve permits adjustment of the flow of material.

Cold-Glue Systems

Most cold-glue systems consist of applicator heads to apply adhesive either in bead, spray, or droplet patterns; fluid hoses to carry the adhesive to the applicator head from the tank; a pressure tank of lightweight stainless steel that can include a filter, quick-disconnect couplings for air and glue, a pressure

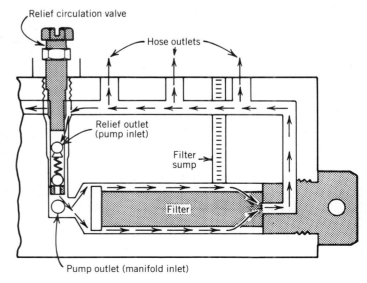

Figure 2. Internally circulating system.

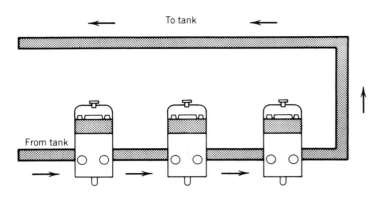

Figure 3. Circulating gun installation (series) system.

relief valve and an air pressure gauge; and a timing device to control the adhesive deposition.

The applicator heads are controlled by either an automatic pneumatic valve or a manually operated hand valve. The bead, ribbon, or spray patterns can be dispensed using multiple-gun configuration systems with resin or dextrin cold adhesives. Cold adhesive droplet guns dispense cold mastic and plastisols, and come in a wide variety of configurations for spacing requirements. In bead and ribbon cold glue extrusion, the tips either make contact or close contact with the substrate. Spray valves emit a mistlike pattern without touching the substrate's surface.

Hot-Melt Systems

Hot-melt application equipment performs three essential functions: melting the adhesive, pumping the fluid to the point of application, and dispensing the adhesive to the substrate in a desired pattern.

Melting devices. Tank melters are the most commonly used melting unit in packaging applications. Best described as a simple open heated pot with a lid for loading adhesive, a

significant feature of tank melters is their ability to accept almost any adhesive form. The tank melter is considered the most versatile device for accepting hot melts with varying physical properties of adhesion and cohesion.

In tank melters, tank size is determined by melt rate. Once melt rate objectives or specifications are determined, tank size is fixed. Larger tanks have greater melt rates than smaller tanks. Holding capacities range from 8 lb (3.9 kg) in the smaller units to more than several hundred pounds (>90 kg) in the larger premelting units (see Fig. 4).

The tanks are made of highly thermal conductive material such as aluminum, and are heated by either a cast-in heating element or a strip or cartridge-type heater. The side walls are usually tapered to provide good heat transfer and to reduce temperature drop.

Adhesive melts first along the wall of the tank as a thin film. Internal circulation currents from the pumping action assist in transferring heat throughout the adhesive held in the tank. Even when the hot melt is entirely liquid, there will be temperature differences within the adhesive. Under operating conditions, adhesive flows along tank surfaces and absorbs heat at a faster rate than it would if allowed to remain in a static condition. Even though adhesive in the center of the tank is cooler, it must flow toward the outer edges and pick up heat as it flows into the pumping mechanism.

Grid melters are designed with dimensional patterns resembling vertical cones, egg crates, honeycomb shapes, and slotted passages arranged in a series of rows. Such an arrangement creates a larger surface area for heat transfer. The grid melter is mounted above a heated reservoir and pump inlet (see Fig. 5).

The grid melting process is exactly the same as the tank melting process; however, the film of adhesive flows along with surfaces and flows through ports in the bottom of the grid. In this way, the solid adhesive will rest above the grid and force the molten liquid through the grid. The grid is designed for deliberate drainage of liquid adhesive to maintain a thin film adjacent to the heated surfaces. This thin film provides for a

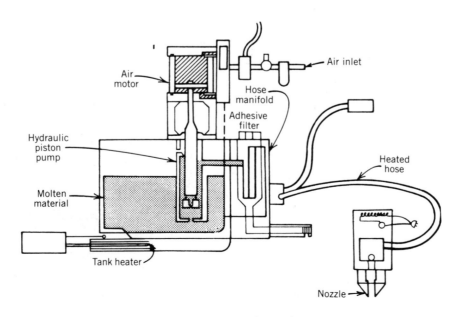

Figure 4. Tank-type hot melt unit.

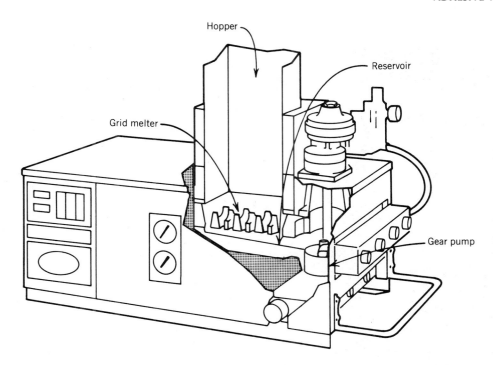

Figure 5. Grid melter hot melt unit.

greater temperature difference than normally found in a tank, and a much larger heat flow is attainable. The grid melter also achieves a much greater melt rate for a given size or area of melter. It is able to heat higher performance adhesives because it provides relatively uniform temperature within the melter itself, which also minimizes degradation. These features are achieved with some sacrifice of versatility since the adhesives must normally be furnished as pellets or other more restricted geometries.

Between each row of patterned shapes are passageways that open into a reservoir beneath the melter. The passageways ensure a constant and unobstructed flow of molten material to the reservoir below.

The reservoir has a cast-in heating system similar to that of most tank applicators. The temperature control can be separate from the grid melter. The floor of the reservoir is sloped so that molten material is gravity-fed toward the pump inlet.

Grid melters are available with optional hopper configurations. The major difference, other than capacity, is the ability to keep the material "cool" or "warm" before it reaches the grid. The cool hopper merely supplies adhesive and the material becomes molten at the grid. The molten time of the material is shortened before actual application. The cool hopper works well with hot melts of relatively high softening and melting points.

Warm hoppers are insulated and attached to the grid so that heat is radiated from the hopper through the hopper casting. Materials that are better formulated to melt in a zone-heating process, such as hot melts with medium to low melting points, and pressure-sensitive materials work well with the warm hopper design.

The tank capacities of both tank and grid melting devices can be extended with premelt tanks. They may be equipped with their own pumping devices or act on a demand signal

from a level sensor in the applicator tank; however, all perform like tank units to keep hot melt materials molten at controlled temperatures. One of the newer premelting devices is in the bulk melter unit (see Fig. 6).

Bulk melting systems are designed to dispense hot melt adhesives and other highly viscous thermoplastic materials in applications requiring a high volume or rapid delivery of material. Units can be used as direct applicators or as premelters as part of a central feed system.

The material is pumped directly from the drum or pail in which it is shipped. This provides ease of handling and lower material costs of bulk containers. In premelting applications, the bulk melter system preheats the material before it is pumped into heated reservoirs. The material is then pumped from the reservoirs to the application head on demand.

The electrically heated platen is supported by vertical pneumatic or hydraulic elevating posts. The platen melts the hot melt material on demand directly from the container and forces it into the pump inlet.

The platen can be a solid one-piece casting, or in the larger units, several grid or fin sections. It is important that the platen size match the ID (inside diameter) of the drum or pail. The platen is protected by one or more seals to help prevent leakage.

Pressurized melters, or screw extruders, are among the earliest designs used in hot melt applicators. Imitating injection molding machinery (see Injection molding), early screw-extruder and ram-extrusion handgun systems had limited success because they were designed only for continuous extrusion. The closed-system and screw-extrusion design allows for melting and pumping of high viscosity, highly degradable materials.

Present extruder equipment is now adapted to intermittent applications; it consists of a hopper feeder, a high torque dc-

Figure 6. Cutaway of drum showing bulk melter system.

drive system, a heated barrel enclosing a continuous flight screw, and a manifold area (see Extrusion; Extrusion coating). Heating and drive control systems can be independently controlled by a microprocessor. Temperatures and pressure are monitored by digital readouts. Adapted for high temperature, high viscosity, or degradation-sensitive adhesives, the new technological advances in extruder equipment give greater potential for adhesive applications such as drum-lid gasketing, automotive-interior parts, and self-adhering elastic to diapers (see Fig. 7).

Pumping devices and transfer methods. Once the hot melt material is molten it must be transferred from the tank or reservoir to the dispensing unit. Pumping mechanisms are either of piston or gear design.

Piston pumps are air driven to deliver a uniform pressure throughout the downstroke of the plunger. Double-acting piston pumps maintain a more consistent hydraulic pressure with their ability to siphon and feed simultaneously. Piston pumps do not provide complete pulsation-free output, but they are well suited for fixed-line-speed applications.

Gear pumps are available in several configurations: spur gear, gerotor, and two-stage gear pump.

Spur gear pumps have two counterrotating shafts that provide a constant suction and feeding by the meshing action of the gear teeth (see Fig. 8). They are becoming more common because of their versatility in handling a variety of high viscosity materials and their efficient performance in high speed packaging.

Gerotor gear pumps have a different arrangement of gears and larger cavities for the transfer of materials. The meshing action, which occurs upon rotation, creates a series of expanding and contracting chambers (see Fig. 9). This makes the gerotor pump an excellent pumping device for high viscosity hot melt materials and sealants.

The latest patented two-stage gear pump introduces an inert gas in a metered amount into the hot melt. When the adhesive is dispensed, exposing the fluid to atmospheric pressure, the gas comes out in solution which foams the adhesive much like a carbonated beverage (see Fig. 10).

All types of gear pumps provide constant pressure because of the continuous rotating elements. They can be driven by air motor, constant speed electric motors, variable speed drives, or by a direct power takeoff from the parent machine. PTO and SCR drives allow the pump to be keyed to the speed of the parent machine. As the line speed varies, the amount of adhesive extruded onto each segment of substrate remains constant.

Variations and modifications to the pumping devices incorporate improvements to their transfer efficiency and performance. Multiple-pump arrangements are also offered in hot melt systems to meet specific application requirements.

The transfer action of the pumping device moves the hot melt material into the manifold area. There, the adhesive is filtered and distributed to the hose or hoses. In the manifold there is a factory-set, unadjustable relief valve, which protects the system from overpressurization. The adhesive then passes

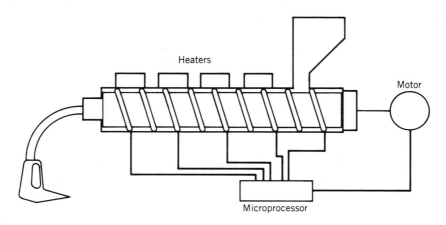

Figure 7. Screw extruder.

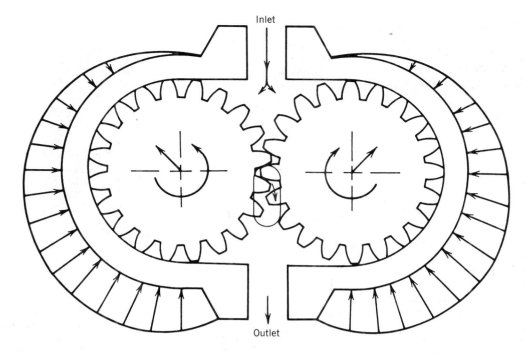

Figure 8. Spur gear pump.

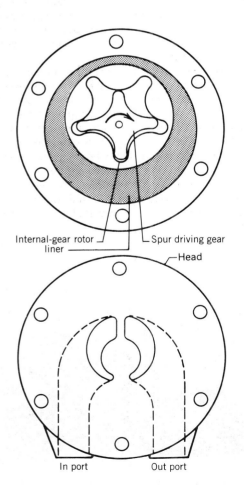

Figure 9. Gerotor gear pump.

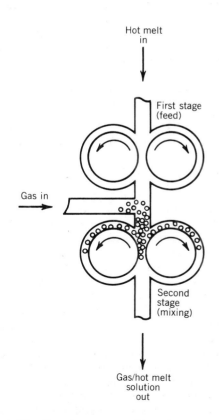

Figure 10. Foaming process of two-stage gear pump.

Figure 11. Cutaway of hot melt hose.

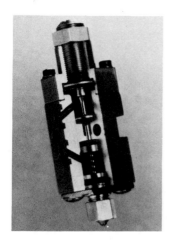

Figure 12. Cutaway of extrusion module.

through a filter to remove contaminants and is directed through the circulation valve. The circulation valve controls the hydraulic pressure in the system. The material circulates to the hose outlets and out the hose to the dispensing devices.

For hot melt systems requiring a fluid link between the melting–pump station and the point of application, hot melt hoses provide a pipeline for transferring the adhesive. Some methods of dispensing (eg, wheel-type applicators) do not use hoses and will be discussed later.

To be able to withstand operating hydraulic pressures up to 1600 psi (11.3 MPa), hot-melt hoses are constructed of air-craft-quality materials. They are flexible, electrically heated, and insulated, and come in various lengths to accommodate particular installation requirements.

Hot melt hoses are generally constructed of a Teflon inner-tube that is surrounded by a stainless steel wire braid for pressure resistance (see Fig. 11). Noncirculating hoses maintain temperature with a heating element spirally wrapped around the wire braid throughout the length of the hose. Circulating hoses sometimes use only the wire braid to maintain heat. Wrapped layers of materials such as polyester felt, fiberglass, and vinyl tape provide insulation. For abrasion resistance, the entire hose is covered with a nylon braid.

Hose temperature can be independently controlled and is monitored inside the hose by a sensing bulb, thermistor, or resistance temperature detector.

Dispensing devices. There are several methods of depositing adhesive onto a substrate once the material is in a molten state. The applicating devices can be categorized as follows:

1. Extrusion guns or heads (automatic and manual).
2. Web-extrusion guns.
3. Wheel and roll dispensers.

Extrusion guns are used on most packaging lines. This extrusion method entails applying beads of hot melt from gun and nozzle. The gun is usually fed from the melting–pumping unit through the hose or directly from the unit itself.

Most high speed applications use automatic guns which are triggered by timing devices or controllers on line with the parent machinery to place the adhesive on a moving substrate.

Automatic guns are actuated by pressure which forces a piston or plunger upward, lifting the attached ball or needle off the matched seat. Molten adhesive can then pass through the nozzle as long as the ball or needle is lifted off its seat by the applied pressure. The entire assembly can be enclosed in a cartridge insert or extrusion module (see Fig. 12). Either style, when fitted into or on the gun body, allows for multiple extrusion points from one gun head. Modular automatic guns with up to 48 extrusion modules are possible (see Fig. 13).

The pressure to actuate automatic guns is either electro-pneumatic by means of a solenoid or electromagnetic with a solenoid coil electrically signaled.

Gun temperatures can be thermostatically controlled with cartridge heaters to a maximum of 450°F (230°C). A maximum operating speed of 3500 cycles per minute (58.3 Hz) is possible.

Handgun extrusion is based on the same principles, but with manual rather than automatic triggering. A mechanical linkage operated by the gun trigger pulls the packing cartridge ball from its seated position to allow the adhesive to flow through the nozzle.

The extrusion nozzle used on the head or gun is the final control of the adhesive deposited and is used to regulate the bead size. It is designed for varying flow rates which are determined by the nozzle's orifice diameter and length. Classified as low pressure–large orifice or high pressure–small orifice, they provide different types of beads. Low pressure–large orifice nozzles are specified for continuous bead applications with the large orifice helping to limit nozzle clogging.

High pressure–small orifice nozzles are better adapted to applications requiring clean cutoff and rapid gun cycling. Drooling and spitting must be controlled for applications such as stitching.

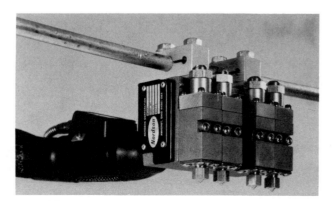

Figure 13. Automatic gun with four modules.

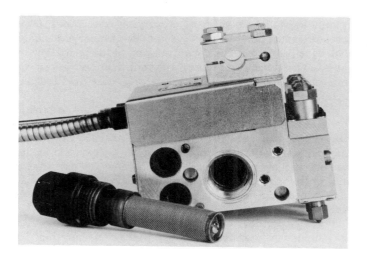

Figure 14. Heat Exchanger gun.

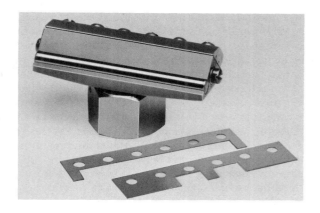

Figure 15. Two-inch slot nozzle with pattern blades.

Patterns can be varied further by selection of multiple-orifice designs, right-angle nozzles for differences in positioning, and spray nozzles for a coated coverage.

Heated in-line filters can be installed between the hose and gun to provide final filtering before adhesive deposition. Independent temperature control helps keep the temperature constant. However, further options are available to provide optimum control by combining a heat exchanger and filter integrally with the gun (see Fig. 14). These specialized guns can precisely elevate the temperature of the adhesive material and can hold adhesive temperature within $\pm 2°F$ ($\pm 1.1°C$) of the set point. This allows the rest of the system to be run at lower temperatures, thereby minimizing degradation of the material. The filter assembly incorporated into the service block catches contaminants not trapped by the hot melt system filter.

Recent development efforts to prevent nozzle clogging and drool have resulted in a zero cavity gun which replaces the traditional ball and seat assembly with a tapered needle and precision matched nozzle seat. In traditional guns, the ball and seat assembly interrupts adhesive flow some distance from the nozzle, allowing the adhesive left in the nozzle to drool from the tip and char to lodge in the nozzle orifice. With the zero cavity gun, no separate nozzle is needed. A microadjust feature adjusts the needle for precise flow control. When the needle closes into the nozzle seat, any char is dislodged. In addition, with no nozzle cavity area the cutoff is clean and precise.

Web-extrusion guns have adapted extrusion dispensing technology to deposit a film of hot melt on a moving substrate. Better known as slot nozzles or coating heads, they are well suited to continuous or intermittent applications. Mounted on an extrusion gun, a heated or nonheated slot nozzle extrudes an adhesive film of varying widths, patterns, and thickness. Pattern blades can be cut to desired patterns. Film thickness is adjustable by using different thickness blades, by stacking of blades in the slot nozzle, or by varying the adhesive supply pressure (see Fig. 15).

Web extrusion is well suited to coating applications such as labeling, tape/label, envelopes, business forms, and web lamination as in nonwovens.

For temperature-sensitive adhesives in continuous web extrusion, the slot nozzle is used with a heat-exchanger device to minimize temperature exposure of the material.

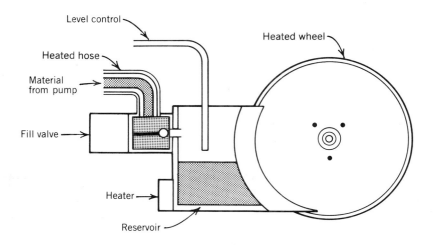

Figure 16. Wheel-type applicator.

Wheel and roll dispensers are the predecessors of present coating extrusion. Wheels or rolls are mounted in a reservoir of molten adhesive. The wheels or rolls are finely machined and may be etched, drilled, or engraved with desired patterns for specific pattern transfer. As the wheel rotates in the reservoir, it picks up the hot melt and transfers it to the moving substrate by direct contact (see Fig. 16). The reservoir may be the primary melting unit or fed by an outside melting device.

Roll coaters involve a series of unwinding and rewinding units for paper coating, converting, and laminating, plus wide web applications for tape and label applications.

Timing and controlling devices. Automatic applications require installation of one or more devices to control the placement of adhesives on the moving substrate. Such devices normally include a sensor or trigger to detect the presence of the substrate in the gluing station, and a timer or pattern control to measure the predetermined intervals between beads of adhesive or the substrate and to activate extrusion guns at the proper moments (see Fig. 17).

The sensor may be operated mechanically (as with a limit switch activated by the substrate or a cam on the packaging machine) or optically (as with photo eyes or proximity switches).

Timers or pattern controls may be used at constant line speeds to time delay extrusion intervals or produce stitched beads. When line speed varies, pattern controls equipped with line speed encoder or tachometer must be used to compensate for changes in line speed so that bead lengths remain the same. Such devices are highly reliable but are more complex and correspondingly more expensive than duration controls. They can usually control bead placement accurately at line speeds up to 1000 ft/min (5.1 m/s).

Some pattern controls can also be equipped with devices that electronically vary air pressure to the hot melt applicator to control adhesive output, thus maintaining constant bead volume as well as placement at varying line speeds. Other accessories can be obtained to count the number of packages that have been glued, check for missing beads, and allow the same device to control guns or different lines. Advanced controls are often modular in construction, user oriented, and include self-diagnostic features.

System selection. Choosing the correct system to produce the results desired on the packaging line is not difficult once the variables are identified. Some of the primary variables to consider in specifying equipment for a hot-melt system include rate of consumption, rate of deposition, adhesive registration, and control. Trained factory representatives for adhesive applicating equipment can identify and recommend the best system to fit those variables.

The melting device selected must be capable of handling the pounds-per-hour demand of the packaging operation. The unit must also have sufficient holding capacity to prevent the need for frequent refilling of the adhesive tank or hopper.

Adhesive consumption is affected by line speed, bead size, and pattern. The adhesive consumption rate and maximum instantaneous delivery rate of the pump must be matched to the application requirements (see below).

The pattern to be deposited will determine the dispensing device. Also, the pattern size and registration of the adhesive deposit must be matched with the cycling capabilities of that device.

The system must fit neatly into the entire operation with spacing considerations for mounting unit, gun, and hoses; location to point of application; and accessibility for maintenance.

Calculating Maximum Instantaneous Delivery Rate

The term MIDR (maximum instantaneous delivery rate) is used interchangeably with IPDR (instantaneous pump delivery rate). MIDR is the amount of adhesive that a pump would

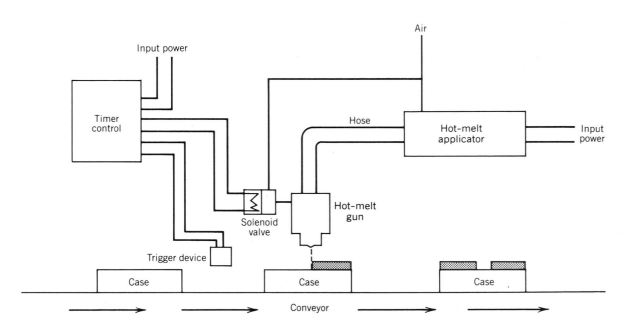

Figure 17. Timer application process. The timer control sequence is activated when the trigger device senses the leading edge of the case. The timing sequence controls preset adjustable delay and duration gun actuations.

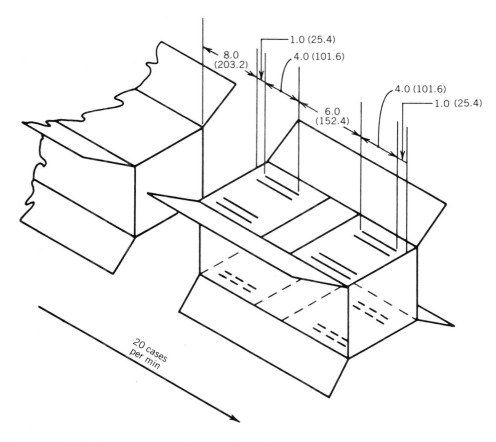

Figure 18. Calculating the MIDR for a specific application. Dimensions are given in inches (mm).

need to supply if its associated guns were fired continuously for a specified period of time. Useful units for measuring MIDR are pounds per hour (lb/h) and grams per minute (g/min).

To calculate the MIDR for a specific application, follow the steps in the example below. Refer to Figure 18 for a visual description of the physical parameters used in the calculation.

Given:

1. Four beads per flap, top and bottom sealing, for a total of 16 beads per case.
2. Case length of 16 in. (41 cm).
3. Production rate of 20 cases per min at 100 percent machine efficiency.
4. Bead length of 4 in. (10 cm), 1 in. (2.5 cm) from case end, with a 6-in. (15-cm) gap between beads.
5. Eight-in. (20-cm) gaps between cases.
6. Adhesive "mileage" of 700 lineal ft · lbf (949 J).

"Mileage" is a function of bead size and the specific gravity of the adhesive. The 700 lineal ft · lbf used above is based on a 3/32-in. (2.4 mm) half round bead (standard-size packaging bead) and melt density of 0.82 g/cm³ (melt density of standard packaging adhesive). Adhesives of this type yield approximately 30 in.³/lb (1 cm³/g).

Calculation:

1. Determine the total bead length in in./h (cm/h) and convert to ft/h (m/h).

$$(20 \text{ cases/min}) \times (4 \text{ in./bead}) \times (16 \text{ beads/case})$$
$$\times (60 \text{ min/h}) = 76{,}800 \text{ in./h } (195{,}000 \text{ cm/h})$$
$$(76{,}800 \text{ in./h}) \times (1 \text{ ft/12 in.}) = 6400 \text{ ft/h } (1950 \text{ m/h})$$

2. Determine the adhesive consumption rate at 100 percent machine efficiency using total bead length per hour and adhesive mileage:

$$(6400 \text{ ft/h}) \times (1 \text{ lb/700 ft}) \ (1 \text{ g/1.54 m})$$
$$= 9.14 \text{ lb/h } (4.15 \text{ kg/h})$$

3. Determine duty cycle using length of bead during machine cycle and machine cycle (length between case leading edges).

$$\frac{8\text{-in. bead length}}{24\text{-in. cycle length}} = 0.333$$

4. Determine the MIDR using the inverse of the duty cycle and the adhesive consumption rate from step 2.

$$\frac{1}{0.333} \times (9.14 \text{ lb/h}) = 27.42 \text{ lb/h } (207 \text{ g/min}) \text{ MIDR}$$

This example demonstrates the difference between the delivery rate and the adhesive consumption rate. When the guns are firing, the pump is delivering adhesive at a rate of 27.42 lb/h. This is the MIDR. Speed reducers are rated and selected according to the MIDR they are capable of providing. This figure should not be confused with the consumption rate, since the consumption rate is an *average* measure of consumption that includes the time when adhesive demand is zero.

Several equations may prove helpful in some applications: **Bead length, line speed, and duration.**

$$\frac{5 \times \text{bead length (in.) (cm)}}{\text{line speed (ft/min) (cm/s)}} = \text{duration (s)}$$

$$\frac{\text{line speed (ft/min) (cm/s)} \times \text{duration (s)}}{5} = \text{bead length (in.) (cm)}$$

$$\frac{\text{bead length (in.) (cm)} \times 5}{\text{duration (s)}} = \text{line speed (ft/min) (cm/s)}$$

BIBLIOGRAPHY

G. L. Schneberger, *Adhesives in Manufacturing,* Marcel Dekker, Inc., New York, 1983.

"Definitions of Terms Relating to Adhesives and Sealants." *Adhes. Age,* (May 31, 1983).

I. Kaye, "Adhesives, Cold, Water-Borne;" T. Quinn, "Adhesives, Hot Melt;" and C. Scholl, "Adhesives Applicating, Hot Melt" in *The Packaging Encyclopedia 1984,* Vol. 29, No. 4, Cahners Publishing, Boston, Mass.

1981 Hot Melt Adhesives and Coatings, Technical Association of the Pulp and Paper Industry, Norcross, Ga. (Short course notes presented at the TAPPI Conference, San Diego, Calif., July 1–3, 1981.)

"Calculating Maximum Instantaneous Delivery Rate," *Components Catalog, P&A Division,* Nordson Corporation, Norcross, Ga., 1984, p. A2.

D. H. SHUMAKER
C. H. SCHOLL
P&A Division, Nordson Corporation

ADHESIVES

Adhesive and sealant consumption for paper and packaging applications in the United States totaled 4,630,000 lb (2105 metric tons) in 1982 (1). Although a large component of that number is corn starch used in the manufacture of corrugated board, a significant portion was used in the forming, sealing, or labeling of almost every package in the marketplace. The principal uses include the forming and sealing of corrugated cases and folding cartons; the forming and sealing of bags; the winding of tubes for cores, cans, and drums; the labeling of bottles, jars, drums, and cases; the lamination of paper to paper board, and foil, and the lamination of plastic films for flexible packaging.

There are many types of packaging adhesives used, frequently for the same end-use constructions, with the choice dictated by cost, productivity factors, the particular substrates involved, or special end-use requirements. To help clarify this complex picture, it is useful to classify packaging adhesives into three physical forms: waterborne, hot-melt, and solvent-borne systems.

WATERBORNE SYSTEMS

This is the oldest, and still by far the largest volume class of adhesive used in packaging. These adhesives share the gen-

eral advantages of ease and safety of handling, energy efficiency, low cost, and high strength. Waterborne adhesives can be further divided into two categories, natural and synthetic.

Natural Waterborne Adhesives

The earliest packaging adhesives were based on naturally derived materials—indeed, almost all were until the 1940s—and they still constitute a large segment of the market. However, the last three decades have seen their gradual replacement by higher performance synthetics in many applications.

Starch. The largest class of natural adhesives is based on starch, and in the United States, this means corn starch. Adhesives are produced from raw flour or starch, but more frequently the starch molecule is broken down into smaller chain segments by acid hydrolysis. Depending on the conditions of that reaction, the resulting material can be a fluidity or thin-boiling starch or a dextrin. These can then be further compounded with alkaline tackifiers such as borate salts, sodium silicate, or sodium hydroxide, with added plasticizers or fillers.

The single largest use of starch adhesives is in the manufacture of corrugated board for shipping cases. The standard process involves suspending ungelatinized corn starch in a thin-carrier starch cook as a vehicle. When the bond line is subjected to heat and pressure, the corn starch gelatinizes almost instantly, forming a bond at very high rates of production. Additives are frequently used to improve adhesion, lower gel temperature, increase water resistance, and further increase set speed.

Other important uses of modified starches and dextrins are in the sealing of cases and cartons, winding of spiral or convolute tubes, seaming and forming of bags, and adhering the seams on can labels. Glass bottles are most frequently labeled with a special class of alkaline-treated starch adhesive called a jelly gum. These have the special tacky, cohesive consistency required by high speed labeling equipment. Specially modified starches based on genetically bred high amylose strains are also used as primary ingredients in the remoistening adhesive on gummed tape used for box sealing.

There are many strong points to recommend starch-based adhesives. They are regarded as very easy to handle, clean machining, easy to clean up, and they are inexpensive. Starch has excellent adhesion to paper, and being nonthermoplastic, it has outstanding heat resistance. The negatives are the relatively slow rate of bond formation, the limited adhesion to coatings and plastics, and poor water resistance.

Protein. Another class of natural adhesives is based on animal protein. Once very widely used, these materials now have specific narrow areas of use in packaging.

Animal glue. This is one of the earliest types of adhesives. It is derived from collagen extracted from animal skin and bone by alkaline hydrolysis. When used as a heated colloidal suspension in water, animal glues have an unusual level of hot tack and long, gummy tack range. However, because of fluctuating availability and cost, and the development of improved synthetics, there are only two significant uses of animal glue in packaging. One is as a preferred ingredient in the remoistening adhesive on gummed tape used for box sealing. The other is as the standard adhesive used in forming rigid set-up boxes.

Casein. This is produced by the acidification of skimmed cow's milk. The precipitated curds thus produced form the basis of casein adhesives. Casein is not manufactured in the United States because our milk price support program only applies to food products. The main sources are Australia, New Zealand, Argentina, and Poland.

There are two packaging applications where casein is used in large volume. One is in adhesives for labeling beer bottles. Here, casein provides the resistance to cold-water immersion required by brewers, together with removability in alkaline wash when the bottles are returned. The second use is as an ingredient in adhesives used to laminate aluminum foil to paper. Combined with synthetic elastomers such as neoprene or styrene–butadiene latices, casein provides a unique balance of adhesion and heat resistance (2).

Natural rubber latex. This is extracted from the rubber tree, *Hevea brasiliensis,* and is available in several forms of concentration and stabilization. The primary use in packaging is as a principal ingredient in adhesives for laminating polyethylene film to paper, as in the construction of multiwall bags. Natural rubber latex also finds application in a variety of self-seal applications, since it is the only adhesive system that will form bonds only to itself with pressure. This property is used in self-seal candy wraps (where it is called cold-seal), and press-to-seal cases, as well as on envelopes.

Synthetic Waterborne Adhesives

Synthetic waterborne adhesives are the most broadly used class of adhesives in general packaging. Most are resin emulsions, specifically poly(vinyl acetate) emulsions—stable suspensions of poly(vinyl acetate) particles in water. These systems usually contain water-soluble protective colloids such as poly(vinyl alcohol) or 2-hydroxyethyl cellulose ether, and may be further compounded by the addition of plasticizers, fillers, solvents, defoamers, and preservatives.

These emulsions are supplied in liquid form (the ubiquitous "white glue") in a range of consistencies ranging from milky fluids to thick pastes. They are used in a broad range of packaging applications, to form, seal, or label cases, cartons, tubes, bags, and bottles. In most of these uses they have replaced natural adhesives because of their great versatility. They can be compounded to have a broad range of adhesion not only to paper and glass, but also to most plastics and metals. They can be made very water insensitive for immersion resistance, or very water-sensitive for ease of clean up and good machining. They are the fastest-setting class of waterborne adhesives, facilitating increased production speeds. They are low in odor, taste, color, and toxicity, and have excellent aging properties. They are tough, with an excellent balance of heat and cold resistance. Finally, they are economical and reasonably stable in cost.

The utility of these emulsions systems has broadened in recent years with the greater use of copolymers of vinyl acetate. Copolymerizing vinyl acetate with ethylene or acrylic esters in particular has greatly improved the adhesion capabilities of these emulsions, particularly where adhesion to plastics is required. For example, very recently, cross-linking acrylic–vinyl acetate copolymer emulsions have replaced polyurethane solution systems for laminating plastic films for snack packages. The largest areas of use of vinyl emulsions adhesives, however, are in case and carton sealing, forming the manufacturer's joint on cases and cartons, and the spiral winding of composite cans.

The use of other synthetic waterborne systems is quite minor and specialized. Some synthetic rubber dispersions are used in film adhesives and, in conjunction with casein, for the lamination of aluminum foil to paper. There is increasing use of tackified rubber dispersions as pressure-sensitive masses on tapes and labels replacing solvent-borne rubber-resin systems.

Sodium silicate was once widely used in many paper packaging applications, ranging from corrugating to case sealing, but today the primary use of silicate adhesive is in tube winding, especially in convolute winding of large drums or cores where it produces a high degree of stiffness.

HOT MELT ADHESIVES

Hot melts have been the fastest growing important class of adhesives in packaging for the last two decades. Hot melts can be defined as 100% solids adhesives based on thermoplastic polymers, that are applied in the molten state and set to form a bond upon cooling. Their chief attraction is the extremely rapid rate of bond formation which can translate into high production rates on the packaging line.

The backbone of any hot melt is a thermoplastic polymer. Although almost any thermoplastic can be used, and most have been, the most widely used material is the copolymer of ethylene and vinyl acetate (EVA). These copolymers have an excellent balance of molten stability, adhesion, and toughness over a broad temperature range and compatibility with many modifiers. The EVA polymers are further compounded with waxes and tackifying resins to convert them into useful adhesives. The function of the wax is to lower viscosity and control set speed. Paraffin, microcrystalline, and synthetic waxes are used depending on the desired speed, flexibility, and heat resistance. The tackifying resins also function to control viscosity, as well as wetting and adhesion. These are usually low molecular weight polymers based on aliphatic or aromatic hydrocarbons, rosins, rosin esters, terpenes, styrene or phenol derivatives, or any of these in combination. The formulations always include stabilizers and antioxidants to prevent premature viscosity change and char or gel formation that could lead to equipment stoppage.

A second class of hot melts used in packaging is based on low molecular weight polyethylene, compounded with natural or synthetic polyterpene tackifiers. These lack the broader adhesion capabilities of the ethylene–vinyl acetate-based hot melts as well as their broader temperature resistance capabilities, but they are adequate for many paper bonding constructions and are widely used for case sealing and bag seaming and sealing.

A third type of hot melt is based on amorphous polypropylene, the by-product of the polymerization of isotactic polypropylene plastic. As a by-product, it is inexpensive, but these hot melts are extremely weak and are limited to applications such as lamination of paper to paper to produce water-resistant wrapping material or 2-ply reinforced shipping tape.

A more recent class of hot melt is based on thermoplastic elastomers: block copolymers of styrene and butadiene or isoprene. These find primary applications in hot-melt pressure-sensitive adhesives for labels and tapes. More recently, their broad adhesion and impact resistance are finding use in specialized applications such as the attachment of polyethylene base cups to polyester soft-drink bottles (3).

Even more specialized applications use hot melts based on polyamides or polyesters when specific chemical or heat-resistance requirements have to be met, but their high cost and relatively poor heat stability have precluded their widespread use to date.

All hot melts share the same basic advantages, based on their mechanism of bond formation by simple cooling. They are the fastest-setting class of adhesives—indeed, pre-set before both substrates can be wet is the most frequent cause of poor bonds with hot melts. Because of the wide range of polymers and modifiers used, they are capable of being formulated to adhere to almost any surface. With no solvent or vehicle to remove, they are generally safe and environmentally preferred. They are excellent at gap filling, since a relatively large mass of material can "freeze" in place, joining poorly mated surfaces.

However all hot melts also share the same weakness, which is a rapid fall off in strength at elevated temperatures. Therefore, properly formulated hot melts can be suitable for most packaging applications, but they are not appropriate for very hot fill or bake-and-serve applications.

SOLVENT-BORNE ADHESIVES

The smallest of the three classes of adhesives used in packaging, solvent-borne adhesives find use in specialized applications where waterborne or hot-melt systems are not suitable.

Rubber–resin solutions are still widely used as pressure-sensitive adhesives for labels and tapes. However, factors of cost, safety, compliance with clean air laws, and productivity have led to a strong movement toward waterborne or hot-melt alternatives. Such alternatives are available, and knowledgable observers predict a gradual disappearance of rubber–resin solvent-borne pressure-sensitives for labels and tapes over the next decade.

Polyurethane adhesives are widely used in flexible packaging for the lamination of plastic films. These multilayer film constructions find application in bags, pouches, wraps for snack foods, meat and cheese packs, and boil-in-bag food pouches. They have the ideal properties of adhesion, toughness, flexibility, clarity, and resistance to heat required in this area. However, here too, alternative systems are being introduced to eliminate the cost, hazards, and regulatory problems associated with solvent-borne systems. Cross-linking waterborne acrylic polymers have gained acceptance in the large snack-food laminating market for constructions such as potato chip bags. Polyurethane (100% solids) "warm melt" systems are being introduced for some of the more demanding food package applications.

Solvent-borne ethylene–vinyl acetate systems find application in some heat-seal constructions, such as the thermal strip to form the seam on form-fill-seal pouches, or on lidding stock for plastic food containers such as cream or jelly packs.

BIBLIOGRAPHY

1. A. Barker, "Adhesive Consumption May Rise 60% by Volume by 1985," *Adhes. Age,* **27,** 32 (Jan. 1984).
2. U.S. Pat. 2,754,240 (July 10, 1956), W. B. Kinney (to Borden Co.).
3. U.S. Pat. 4,212,910 (July 15, 1980), T. Taylor and P. Puletti (to National Starch & Chemical Corp.).

General References

C. V. Cagle, *The Handbook of Adhesive Bonding,* McGraw-Hill, New York, 1973.

I. Skeist, ed., *Handbook of Adhesives,* 2nd ed., Reinhold Publishing Co., New York, 1974.

R. G. Meese, ed., *Testing of Adhesives,* TAPPI Monograph Series No. 35, TAPPI Press, Atlanta, Ga., 1974.

E. W. Flick, *Handbook of Adhesive Raw Materials,* Noyes Publications, Park Ridge, N.J., 1982.

IRVING KAYE
National Starch & Chemical Corp.

AEROSOLS.   See Pressurized containers.

ALUMINUM CANS.   See Cans, aluminum.

ALUMINUM FOIL.   See Foil, aluminum.

AMPULS AND VIALS

Ampuls and aluminum-seal vials are glass containers primarily used for packaging medication intended for injection. Ampuls are essentially single-dosage containers that are filled and hermetically sealed by flame sealing the open end. Vials, which contain single or multiple doses, are hermetically sealed by means of a rubber closure held in place with a crimped aluminum ring.

An ampul is opened by breaking it at its smallest diameter, called the constriction. A controlled breaking characteristic is introduced by reproducibly scoring the glass in the constriction, or by placing a band of ceramic paint in the constriction. The ceramic paint has a thermal expansion that differs from the glass, thus forming stress in the glass surface after being fired. This stress allows the glass to break in a controlled fashion at the band location when force is applied. Medication is then withdrawn by means of a syringe.

Medication can be withdrawn from a vial by inserting the cannula of a syringe through the rubber closure. Since the rubber reseals after cannula withdrawal, multiple doses can be withdrawn from a vial.

Both ampuls and vials are fabricated from glass tubing produced under exacting conditions. The glass used for these containers must protect the contained product from contamination before use, and in the case of light-sensitive products, from degradation due to excessive exposure to light. In addition, the glass must not introduce contamination by interacting with the product.

Glasses

The most important property of a glass used to contain a parenteral (injectable) drug is chemical durability; that is, the glass must be essentially inert with respect to the product, contributing negligble amounts of its constituents to the product through long-term contact before use. The family of glasses that best meets chemical durability requirements is the borosilicates. These glasses also require higher temperatures for forming into shapes than other glass types.

Table 1. Compositions of Soda–Lime and Borosilicate Glasses, wt%

Constituent	Soda–lime	Borosilicate
SiO_2	68–72	70–80
B_2O_3	0–2	10–13
Al_2O_3	2–3	2–7
CaO	5	0–1
MgO	4	
Na_2O	15–16	4–6
K_2O	1	0–3
typical forming temperatures	1796–1895°F (980–1035°C)	2066–2264°F (1130–1240°C)

When glass–product interactions are far less critical, the soda–lime family of glasses can be used to fabricate vials. These glasses can be formed at lower temperatures than borosilicates, but do not nearly have their chemical durability. Typical compositions are shown in Table 1. Borosilicate and soda–lime glasses contain elements that facilitate refining, but borosilicates generally do not contain arsenic or antimony.

Both borosilicate and soda–lime glasses can be given a dark amber color by adding small amounts of coloring agents, which include iron, titanium, and manganese. The amber borosilicate and soda–lime glasses can then be used to package products that are light-sensitive.

The interior surface of containers formed from soda–lime glass is often subjected to a treatment that enhances chemical durability without affecting the desirable lower melting and forming temperatures typical of soda–lime glass. For very critical applications, borosilicate ampuls and vials can be treated to improve their already excellent chemical durability.

For pharmaceutical packaging applications (see Pharmaceutical packaging), the various types of glass have been codified into groups according to their chemical durabilities, as specified by the United States Pharmacopeia (USP) (1). The glasses are classified by the amount of titratable alkali extracted into water from a crushed and sized glass sample during steam autoclaving at 250°F (121°C). Thus borosilicate glasses are typical of a USP Type I glass, and most soda–lime glasses are typical of a USP Type III glass. There are some soda–lime glasses that are less chemically durable than Type III glass, and these are classified as USP Type NP.

USP Type III (soda–lime) containers that have had their interior surface treated to improve durability can be classified as USP Type II if they meet the test requirements. The test in these cases is performed on the treated container instead of a crushed sample, using a similar steam autoclave cycle.

The pharmacopeiae of other nations have also classified glass into groups according to their chemical durability. These classifications are generally similar to those specified by USP.

Forming Processes

Ampuls and vials are formed from glass tubing. The glass tubing is formed by processing in a glass furnace and by a tube-forming operation. The glass furnace operation consists of bulk batch preparation, continuous batch melting, and refining (see Glass-container manufacturing). The tube forming is done to exacting specifications in either a Danner process or

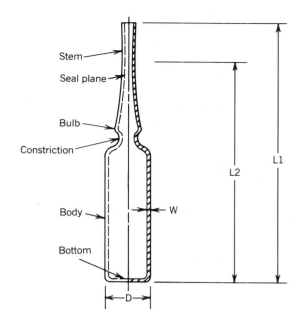

Figure 1. Standard long-stem ampul.

Capacity, mL	Diameter (D) mm	Width (W), mm	Length (L1), mm ± 0.50 mm	Length (L2), mm
1	10.40–10.70	0.56–0.64	67	51
2	11.62–12.00	0.56–0.64	75	59
5	16.10–16.70	0.61–0.69	88	73
10	18.75–19.40	0.66–0.74	107	91
20	22.25–22.95	0.75–0.85	135	120

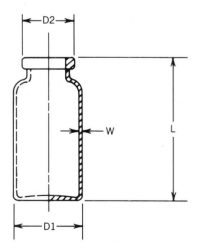

Figure 2. Standard tubular serum vial.

Capacity, mL	Diameter (D1), mm	Width (W), mm	Length (L), mm ± 0.50 mm	Diameter (D2), mm
1	13.50–14.00	0.94–1.06	27	12.95–13.35
2	14.50–15.00	0.94–1.06	32	12.95–13.35
3	16.50–17.00	1.04–1.16	37	12.95–13.35
5	20.50–21.00	1.04–1.16	38	12.95–13.35
10	23.50–24.00	1.13–1.27	50	19.70–20.20
15	26.25–27.00	1.13–1.27	57	19.70–20.20

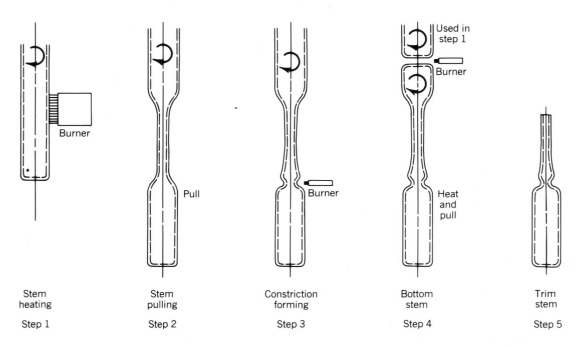

Figure 3. Ampul contour-forming sequence.

a downdraw process. The Danner process involves continuous streaming of molten glass onto an angled rotating sleeve that has an internal port for inflation air. The inflation air controls the tubing outside diameter (OD). The downdraw process is an extrusion process through an annular area. The inner core has an inflation air hole. The inflation air serves the same purpose as in the Danner process. In either process the tubing wall weight is controlled by adjusting the rate of glass withdrawal and supply. Typical ampul and vial tubing dimensions and tolerances are shown in the Figures 1 and 2.

The tubing is formed in a continuous line process. Various devices are used to support the tubing during pulling. A device, normally consisting of pulling wheels on belts and a cutting mechanism, is situated downstream to pull and cut the tubing. The tubing is cut to prescribed lengths and used in vertical- or horizontal-type machines for converting the tubing into vials or ampuls.

Many machines are rotary and either index or operate with a continuous action. The tubing is placed in the machines and is handled in a set of chucks. Heat is applied in the space between the chucks, and forming of the ampul or vial occurs throughout the machine rotation cycle.

Ampuls are formed on continuous-motion rotary machines. One sequence is shown in Figure 3. The process consists of sequentially heating and pulling (elongating the glass) to form the constriction, bulb, and stem contours of the ampul. The ampul contours are controlled primarily by proper temperature patterns in the tubing and by pulling rate of the tubing. Mechanical tooling of the glass can be used to assist in constriction contour forming. The forming process accurately controls the seal plane diameter which controls ampul closing after filling. After the basic ampul is formed on the machine, the ampul blank is separated from the tubing and is transferred to a horizontal afterforming machine. On the afterforming machine the ampul is trimmed to length, glazed, and treated if necessary. Also, the ampul constriction can be either color

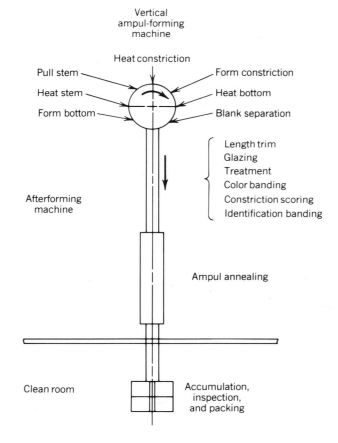

Figure 4. Tube converting for ampul manufacture.

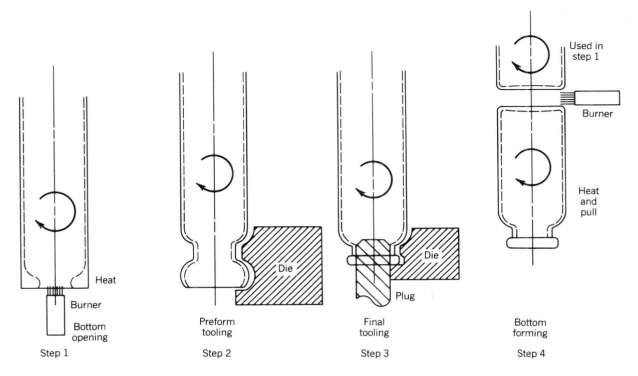

Figure 5. Vial contour-forming sequence.

banded with a ceramic-base paint or scored to control opening properties. The ceramic paint and scoring cause stress concentrations in the constriction which assist in obtaining desirable opening force and fracture characteristics. Identification bands are applied and the ampul is annealed to relieve the strains caused by the thermal forming of the ampul. The completed ampuls are then transferred into a packing area where the ampuls are accumulated, inspected manually or automatically, and packed into clean trays for distribution (see Fig. 4).

Vial forming is done on vertical machines that either index or have a continuous motion. A vertical forming sequence (see Fig. 5) consists of a parting (separation operation) wherein a narrow band of glass is heated to a soft condition and the vial blank and the tubing are pulled apart. After parting, the finish-forming operations occur. The finish forming consists of heating and mechanically tooling the glass in sequential steps. Normally, multiple heating and tooling operations are necessary to form the closely held tolerances of aluminum-seal finishes. The tooling is done with an inner plug to control the contour and diameter of the finish bore and with outer contoured round dies that control the contour and diameter of the finish outer surface. The vial bottom contours are formed in the lower chucks while finish forming occurs for another vial in the upper chucks. After tooling, the vial length is set by a mechanical positioner. The process then continually repeats itself until the whole tubing length is consumed. After fabrication, the vial blank is transferred to a horizontal afterforming machine. The operations that are normally performed on an afterformer are dimensional gauging, vial treatment, and annealing. The vials are then transferred to a packing area where they are accumulated, inspected manually or automatically for cosmetic conditions, and packed in clean containers (see Fig. 6).

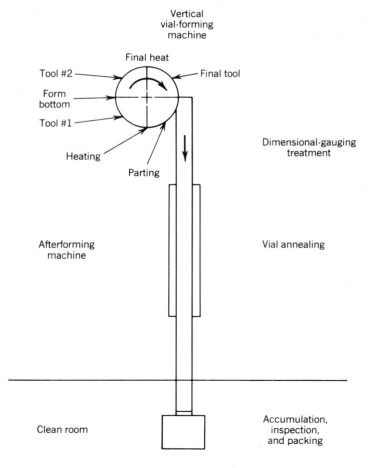

Figure 6. Tube converting for vial manufacture.

BIBLIOGRAPHY

1. *The United States Pharmacopeia XX*, The United States Pharmacopeial Convention, Inc., Rockville, Md., 1980.

R. P. ABENDROTH
J. E. LISI
Owens-Illinois, Inc.

ASEPTIC PACKAGING

Food processors are continually looking for better processing methods to improve the nutritional and organoleptic properties of food and reduce costs. Aseptic processing and packaging offers these advantages. This technique for processing food was developed in the 1940s, but its use in the United States was limited until recent years. Unlike conventional canning, aseptic processing, literally defined, is the continuous sterilization of the food independent from the container, the sterilization of the container and lid, and the filling and sealing in an absolutely sterile atmosphere. Unlike all other means of food packaging, when aseptic processing techniques are used, the processing of the food and the packaging are mutually dependent. A loss in sterility in any area of processing, packaging, filling, closing, or sealing adversely affects the sterility of the product. Aseptic processing and packaging of food products were originally developed to produce shelf-stable products such as dairy-based foods and other heat-sensitive foods, such as banana puree, that could not be conventionally retorted. It is now being applied to many different homogeneous food products, and even some food products containing discrete particulate matter.

Food Preservation

The main objective in the processing of foods is to prevent the food from becoming inedible from the time it is harvested until the time it is consumed.

Food can be preserved in many ways, such as extreme cold in freezing, heat as in pasteurization, extreme heat as in sterilization, deprivation of water or oxygen as in concentration or drying, and medium alteration such as excessive saltiness or acidity as in pickling.

In the 1800s Nicholas Appert developed canning (see Canning, food). It was a major breakthrough in the preservation of food products. Although there have been many technological advances since Appert's original invention, none has more significance than the development of aseptic packaging. Although conventional canning renders food products commercially sterile, the nutritional content and the organoleptic properties of the food generally suffer in the processing. All of the food product must reach a minimum safe temperature for a minimum safe time; and because the container is generally heated from the outside, the food next to the container wall reaches the maximum process temperature much sooner than the material in the center of the container. As a result, some of the product is held at an elevated temperature much longer than necessary to render it sterile, and the necessary time is used to penetrate the food matter in the container center. The problem is generally compounded in large containers because large containers require much longer processing times. Therefore, the quality of the food is poor in large containers as compared to that of the same food processed in small containers. In the

case of some foods, a truly acceptable product just cannot be made with any container processing. A few heat-sensitive foods, such as banana puree and dairy-based products, fall into this category and cannot be cooked in the can or retorted. With the advent of aseptic processing and packaging techniques, the problems with heat-sensitive and milk-base products have been alleviated. Many other products that have been traditionally retorted, and now are aseptically processed, are more nutritional and organoleptically more palatable.

Aseptic Processing

Processing and packaging in aseptic processing are mutually dependent. It is not enough to have a sterile container. The product must be aseptically processed, remain commercially sterile, and filled into a presterilized container in an atmosphere free of unwanted microorganisms. Prior to processing the food product, the equipment must be sterilized and maintained in a sterilized state. This is generally accomplished by subjecting the equipment to high temperature steam or, most often, continuous flow of high temperature water (300–320°F or 166–177°C) for a predetermined period of time. After sterilization of the processing equipment, the food product must immediately follow the sterilization media in order to remain sterile. Although the container itself or final food package can be sterilized in several different ways that include heat, hydrogen peroxide, radiation, or a combination thereof, the process system is always sterilized by high temperature steam, or high temperature water.

Unlike conventional canning where the food product is exposed to relatively high temperature for a long period of time, aseptically processed food is processed using high temperature–short time (HTST) techniques. In other words, the product is heated very quickly to a high temperature to sterilize the food, then cooled very quickly and filled, generally at ambient temperature. It benefits from even distribution of heat and the maintenance of more natural organoleptic properties of food.

The Dole system. The first aseptic packaging lines were developed over 40 years ago by the James Dole Corporation. There are over 60 commercial Dole systems installed in the United States, and approximately another 60 installations in other countries. The Dole system utilizes the metal can, which is presterilized using high temperature saturated steam to sterilize container and lid (see Fig. 1). Food products are packaged in cans that range in size from 4 oz to the large No. 10 size (see Cans, steel). Products such as pudding, cheese sauce, and soups are being aseptically processed and packaged on a commercial basis using the Dole container. For many years the Dole system was the only system commercially available in the world for products processed by aseptic processing techniques.

The desire to overcome increased container costs while improving upon the efficiency of aseptic processing lines has led to a multitude of new packaging alternatives within recent years. A number of systems were developed for new types of retail containers. Others were developed in order to reduce the overall transportation costs for reprocessed foods, such as the aseptic 55-gallon drum filler originally marketed by the Cherry Burrell Corporation in Cedar Rapids, Iowa, and subsequently by Fran Rica in Stockton, Calif. Tomato-based products and fruit purees are being commercially processed and contained in 55-gal (208-L) drums.

Figure 1. Heat-sensitive products have been packed in metal cans aseptically for over 40 years.

Aseptic bag-in-box. In the early 1970s, Scholle Corporation of Irvine, Calif., developed the first aseptic bag-in-box container for institutional and reprocessed acid (below pH 4.6) food products (see Bag-in-box, liquid product). Scholle currently has over 100 installations aseptically packaging products such as tomato paste, ketchup, diced tomatoes, stabilized fruits for yogurt and ice cream toppings, acidified diced peppers, and diced and sliced fruit (see Fig. 2). The bag is presterilized outside the sterile area by gamma radiation, and introduced into the sterile area at the time of filling. The bags consist of multiple layers of flexible materials sandwiched together to inhibit oxygen permeation and increase shelf life.

Various other companies have introduced similar bag-in-box structures. Fran Rica in Stockton, Calif., and Liqui-Box in Worthington, Ohio, have similar machines using similar principles for bag material and sterilization. Federal regulations restrict all of these machines to acid foods only.

Form/fill/seal. In the United States, growth in aseptic processing was somewhat stagnant for many years, until the FDA cleared the use of hydrogen peroxide as a sterilant for packaging material. Since then, more packaging alternatives have become available, and aseptically processed foods are already starting to demand more shelf space in the grocery store. This trend will continue as still more packages become available and the learning curve in aseptic processing methods improves. The first retail package to enter the United States market after FDA clearance was the Tetra Brik (Tetra Pak, Stamford, Conn.). The Tetra Pak machine is a form/fill/seal machine which sterilizes the packaging material with hydrogen peroxide and heat. Rolls are continually fed into a vertical machine which sterilizes, forms, fills, and seals the package. The brick-shaped container consists of five to seven plies. The typical construction is polyethylene/printing/duplex paper/polyethylene/polyethylene/aluminum foil/polyethylene (see Laminating). By 1980, close to 2000 Tetra machines had been installed worldwide, but none were in commercial operation in the United States, where the process had to await FDA clearance. The clearance for the use of hydrogen peroxide as a sterilant for polyethylene came in January 1981. A subsequent rule issued in 1984 cleared all olefin polymers as food-contact services in aseptic packaging systems using hydrogen peroxide as a sterilant. Since 1981, developments in aseptic processing have taken place at a rapid rate. Many aseptic packaging lines have been installed for juice drinks that use the Brik-Pak system as well as other systems based on paperboard/plastic laminates. Another similar package is the Combibloc. The Combibloc system uses a container that is preformed outside the aseptic zone of the machine from seamed and prescored sleeves rather than rolled stock. It is sterilized, like the Brik-Pak, by a combination of hydrogen peroxide and heat. A number of these machines have been installed in the United States for homogeneous food products such as milk and juice drinks.

Two other form/fill/seal machines currently in use and available to processors in the United States are the Bosch, from the FRG, and the Benco, from Italy. Both machines use paperboard/plastic laminates and hydrogen peroxide and heat as a primary means of sterilization of the packaging materials.

Figure 2. Aseptically filled bags are shipped in 55-gal (208-L) drums or 60-gal (227-L) corrugated boxes. Courtesy of Scholle Corporation.

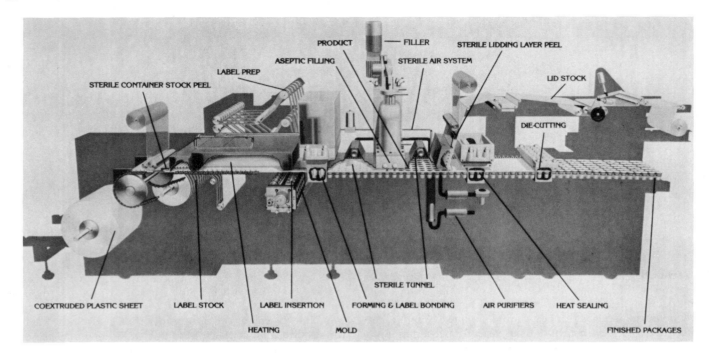

Figure 3. An aseptic thermoform/fill/seal system. Courtesy of Continental Packaging Company, Inc.

Low acid foods, such as puddings, are currently being supplied to the retail market in the Bosch package.

In 1982, Continental Can Company introduced the Conoffast package. Instead of using hydrogen peroxide as a sterilization media, this system relies on the sterile product-contact surface created by the temperatures reached by thermoplastic resins during the coextrusion process (see Coextrusion, flat)

used to produce the multilayer packaging material. During container production, the multilayer package material is fed into the machine where it is delaminated under sterile conditions (see Fig. 3). This removes the outer layer of material and exposes a sterile product-contact surface. The material is then thermoformed into cups (see Thermoform/fill/seal). The lid material, which is also delaminated, is sealed onto the cup

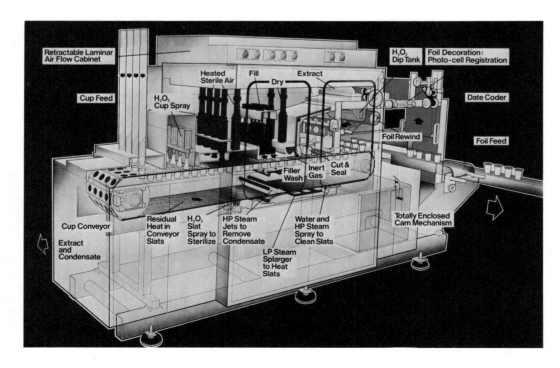

Figure 4. An aseptic system that uses preformed cups. Courtesy of Metal Box.

after filling. The sterility of the forming, filling, and sealing areas is maintained by sterile air under positive pressures.

Preformed containers. In cooperation with Boise Cascade, the James Dole Corporation has now introduced a machine for aseptically filling composite cans (see Cans, composite). Using the same basic principle as the original Dole line, where steam was used as the sterilization media for the container, the composite can utilizes hot air.

Two other packaging systems use preformed containers. These are the Metal Box system, which utilizes hydrogen peroxide as the sterilization media (see Fig. 4), and the Rampart Packaging–Mead "Crosscheck" machine which utilizes hot citric acid as a sterilization media. As of this writing, both of these systems are awaiting regulatory clearance for use with low acid foods in the United States. Metal Box has several installations outside of the United States aseptically packaging low acid products, some with particulate matter.

Summary

There are many reasons for the interest in aseptic processing and packaging. Some of the more paramount include relative energy costs, more intense competition, higher consumer awareness in terms of more nutritional food products, generally with less preservatives, and most importantly, the recent multitude of new packaging alternatives available. For whatever the reasons, aseptic processing and packaging is certainly emerging as the processing and packaging method of the 1980s. One must remember, however, that aseptic packaging of food products is dynamic. New methods of processing and packaging are being introduced every day, and before this is published other alternatives will most assuredly be available. One must also remember that to produce a commercially sterile product, an aseptic system must meet three basic requirements. The product must be sterile; the package or container in which the product will be placed must be sterile; and the environment in which the product and package will be brought together must be sterile. This is the only area of food packaging where the food processing and packaging are mutually dependent.

BIBLIOGRAPHY

General References

C. E. Bayless and W. M. Waites, *J. Food Technol.* **17,** 467 (1982).

D. Bernard, "Microbiological Considerations of Testing Aseptic Processing and Packaging Systems," proceeding of *Capitalizing on Aseptics* conference sponsored by National Food Processors Association, October 11–12, 1983, Washington, D.C.

D. F. Blake and C. R. Stumbo, *J. Food Sci.* **35,** 26 (1970).

V. R. Carlson, *Aseptic Processing CB-201,* a publication of Cherry Burrell Corporation, Cedar Rapids, Iowa, 1977.

C. P. Collier and C. P. Townsend, *Food Technol.* **10,** 477 (1956).

P. W. de Ruyter and R. Brunet, *Food Technol.* **27,** 44 (1973).

K. Ito and K. Stevenson, *Food Technol.* **38,** 60 (Mar. 1984).

J. E. Manson and J. F. Cullen, *J. Food Sci.* **39,** 1084 (1974).

NFPA Criteria for Aseptic Systems for Aseptic Systems for Packaging Low Acid Foods, NFPA, Washington, D.C., 1985.

K. Stevenson, "Establishing Critical Factors for Filling and Packaging Systems," proceeding of *Capitalizing on Aseptics* conference sponsored by National Food Processors Association, October 11–12, 1983, Washington, D.C.

T. E. Szemplenski, "Equipment and Systems to Aseptically Process Food Products Containing Discrete Particulate Matter," proceedings of *Capitalizing on Aseptics* sponsored by National Food Processors Association, October 11–12, 1983, Washington, D.C.

B. Swaminathan, "General Microbiological Considerations in Aseptic Processing," proceedings of *Aseptic Processing and the Bulk Storage and Distribution of Food* sponsored by Food Science Institute, Purdue University, March 15–16, 1978.

A. A. Teixeira, and J. Manson, *Food Technol.* **37,** 128 (Apr. 1983).

R. T. Toledo, "Container and Equipment Sterilization in Aseptic Packaging," proceedings of *Aseptic Processing and Bulk Storage on Distribution of Food"* sponsored by Food Sciences Institute, Purdue University, March 15–16, 1978.

T. E. Szemplenski
Marlen Research Corporation

ASSOCIATIONS.   See Networks.

B

BAG CLOSURES. See Closures, bag.

BAG-IN-BOX, DRY PRODUCT

When the bag-in-box concept is applied to dry products, it generally involves a bag inside a folding carton (see Cartons, folding). In order to appreciate the impact of the bag-in-box (BIB) concept for dry products, one must understand the history of the folding carton. The turn of the century marked the first use of the folding carton as a package when National Biscuit Company introduced the "Uneeda Biscuit" (soda cracker). Instead of opting for the conventional bulk method of selling crackers, Nabisco decided to prepackage in smaller boxes, using a system that would prolong freshness. The paperboard carton shell with creased score line flaps had recently been developed, along with a method for bottom and top gluing on automatic machinery (see Cartoning machinery). Waxed paper (see Paper; Waxes) was to be added manually to the inside of the carton. So was born the "lined carton."

The evolutionary process eventually culminated in two basic methods of producing lined cartons. The first was a machine to automatically open a magazine-fed side-seamed carton and elevate it around vertically indexed mandrels where glue is applied to the bottom flaps and folded up against the end of the mandrel with great pressure. The result is a squarely formed open carton with a very flat bottom surface capable of being conveyed upright to a lining machine which plunges a precut waxed-paper sheet into it by a system of reciprocating vertical mandrels. The lining is overlapped at the edges and sealed together to form an inner barrier to outside environmental factors. It protrudes above the carton top score line by a sufficient amount of paper to be later folded and sealed at a top-closure machine. Straight-line multiple-head filling machines are used to fill premeasured product into the carton. Initial fill levels are often above the carton-top score line and contained within the upper portion of the lining which eventually settles with vibration before top sealing takes place. This fact has relevance with respect to the bag-in-box concept.

The second method involves the use of a double package maker (DPM) which combines the carton forming/gluing operation with a lining feed mechanism which wraps the lining paper around a solid mandrel prior to the carton feed station. In this instance the carton blank is flat and is side-seamed on the DPM. The lined carton is then discharged upright onto a conveyor leading to the filler and top closing machine.

These packaging lines are typically run at up to 80 packages per minute (in some special cases 120/min). They are considered to be very complex machines requiring skilled operating personnel and are usually restricted to a single size.

Although the time reference is rather vague, it would appear that the bag-in-box concept began with the refinement of vertical form/fill/seal (VFFS) machinery in the 1950s (see Form/fill/seal, vertical). Packaging machinery manufacturers and users saw an alternative to the DPM in the horizontal cartoner coupled with VFFS equipment. The idea of automatically end-loading a sealed bag of product into a carton offers the following important advantages compared to lined cartons: simplicity (ie, fewer and less-complicated motions); flexibility (ie, size changes more easily and quickly accomplished); higher speed (ie, up to 200 packages per minute is theoretically possible with multiple VFFS machines in combination with a continuous-motion cartoner); lower package cost (ie, higher speeds and lower priced machinery); improved package integrity (ie, bags are hermetically sealed using heat-seal jaws); reduced personnel (ie, possible for one operator to run line); requires less floorspace (ie, more compact integrated design); and wider choice of packaging materials (ie, unsupported as well as supported films can be handled).

Although bags are sometimes inserted manually, the high-speed methodology (see Fig. 1) employs multiple VFFS machines stationed at right angles to the cartoner infeed, dropping filled and sealed bags of product onto an inclined conveyor which carries them to a sweep-arm transfer device for placement into a continuously moving bucket conveyor.

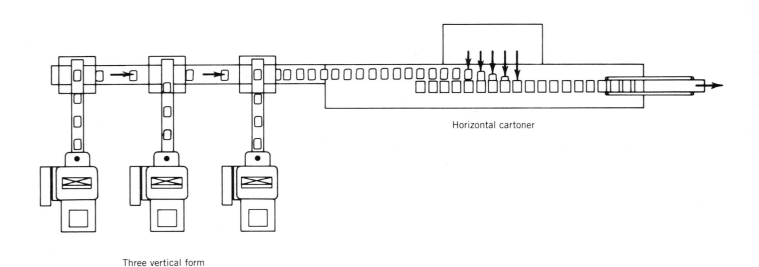

Horizontal cartoner

Three vertical form fill and seal machines

Figure 1. Bag-in-box packaging system with horizontal cartoner.

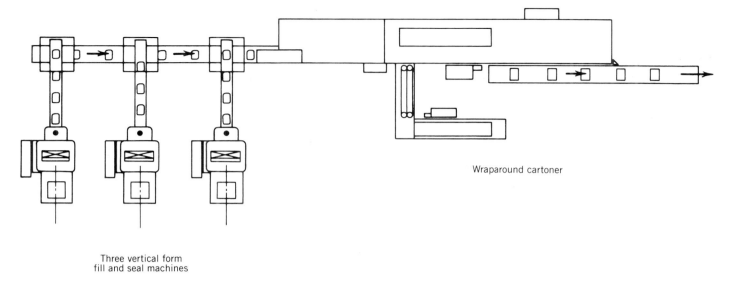

Wraparound cartoner

Three vertical form
fill and seal machines

Figure 2. Bag-in-box packaging system with wraparound cartoner.

The VFFS machines are electrically synchronized with the cartoner and the transfer device is mechanically driven by the bucket conveyor. When the bag reaches the carton-loading area it is gradually pushed out of the bucket by cammed push rods through guides and into the open mouth of the box which is contained within chain flights traveling at the same speed adjacent to the bucket. A guide plate drops into the bucket from overhead to confine the bag during insertion into the box. Once the bag is in the box the end flaps are glued and rail closed before entering compression belts for discharge to the case packer.

When difficulties occur on this system it is usually at the insertion station, caused by misshaped or rounded bags with a girth larger than the carton opening. An attempt is made to condition the bag on the inclined conveyor by redistributing the product within the bag more evenly and, once in the bucket, by vibrating tampers to flatten it. However, if there is too much air entrapment in the bag, or a high product fill level, or an improperly shaped bag, these devices become futile. Increasing the carton size would be a simple solution, but the packager is often not free to do this. Marketing departments are generally reluctant to change the size of a carton that has been running satisfactorily on DPM equipment. The main problem is that the BIB manufacturer must allow more clearance of bag to box than is required on a close-fitting lining. The usual BIB bag-sizing rule of thumb is a gusseted bag having a width 3/8 in. (9.53 mm) less than the carton face panel and 1/4 in. (6.4 mm) less than carton thickness. This can vary somewhat, but the bag must be small enough to transfer positively into the bucket and subsequently into the box without interference.

One manufacturer has attempted to deal with the problem by wrapping the carton blank around the bucket after the bag has been top loaded into it (see Fig. 2). In wraparound cartoning, flat blanks are used and glue is applied to the manufacturer's joint and side seamed against the mandrel by rotating-compression bars. The mandrel/bucket is withdrawn leaving the bag in the box ready for flap gluing and closing. This approach is more forgiving than end-loading, but is still sus-

ceptible to bag sizing problems because additional clearance must be allowed to compensate for the gauge of the three-sided bucket walls. Speeds up to 140 per minute are possible.

A more recent innovation has gone a long way toward overcoming bag-insertion problems. The "vertical load" concept (see Fig. 3) relies on simple gravity and special bag-shaping techniques to drop the bag directly into the box from a film transport-belt-driven type of VFFS machine. The carton shell is formed from a flat blank and bottom glued on a rotary four-station mandrel carton former. The upright carton is conveyed to a starwheel-timed flighted chain indexing device to position it squarely under the VFFS rectangular forming tube. Two VFFS machines with electronically synchronized motor drives operate independently of the carton former. A prime line of empty cartons initiates the operation of each VFFS machine.

The bag-forming parts consist of a rectangular forming shoulder and tube which has been manufactured to produce a bag with cross-sectional dimensions 1/4 in. (6.4 mm) less than carton face panel and 1/8 in. (3.2 mm) less than the side panel. This very close fit is made possible by a combination of several mechanisms. First a gusseting device creates a true flat-bottom bag using fingers that fold in the film from the sides as contoured cross-seal jaws close on the bag. The bottom of the rectangular forming tube is within 1/4 in. (6.4 mm) of the seal jaws and acts as a mandrel around which the bag bottom is formed. The bag is thereby given a sharply defined rectangular shape which is maintained as it is filled with product and lowered through a shape-retaining chamber. This chamber is vibrating so as to present a moving surface to the bag to reduce frictional contact and, more important, to settle the product before the top seal is made. Since the bag is confined and not allowed to round out, head space between product and top seal is kept to a minimum, thereby reducing air entrapment to manageable levels. Flap spreaders ensure that there is unobstructed access into the box.

The shaped filled bag slips freely into the box allowing the four corners of the bag to settle snugly into the bottom, making maximum use of available volume. The bag cut off is deter-

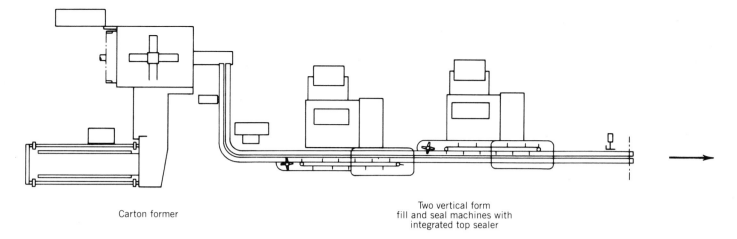

Figure 3. Bag-in-box packaging system with mandrel carton former.

mined by product fill level and, if necessary, can be made so that top seal protrudes over the score line by several in. (cm) when fully seated in the box which is neatly pressed down at the next station. While the carton is contained in the flighted chain it is indexed through a top sealer where hot-melt glue is applied to the flaps which are railed over before passing under compression rollers. The top sealer is integrated mechanically and electrically with each VFFS machine, and because the operation is performed immediately after bag insertion, the carton never has a chance to bulge, which is eventually important for efficient case packing.

This packaging system is rated at up to 100 cartons per minute and two lines can be mirror-imaged for higher speeds. A recent adaptation of the vertical-load system integrates a carton erector for side-seamed blanks with one VFFS and a top-and-bottom gluer into a single compact module rated at up to 50 boxes per minute.

Another BIB system close couples a carton former (preglued or flat blanks) with a pocket conveyor which indexes the box to a series of bag insertion, filling, sealing, and carton top/bottom gluing stations. The bag is formed using vertical form/fill techniques utilizing a rectangular forming tube. A unique gripping mechanism engages the bottom seal of the bag and pulls it into the box from underneath. The bag top is left open to be filled with product at succeeding stations. This system offers the advantages of multiple-stage filling for optimum accuracy, checkweighing, and settling before the bag top is sealed. This is a very effective way of dealing with the problem of high fill levels. Outputs of 65 to 80 packages per minute are claimed for this type of machinery.

Related concepts. A growing field of opportunity is institutional bulk packaging of large quantities of product in 10–20 lb (4.5–9.1 kg) sizes which usually require corrugated-box materials (see Boxes, corrugated). The VFFS unit is an expanded machine capable of producing large deeply gusseted bags and is usually set up for vertical bag loading. Speeds are in the range of 15–30 cartons per minute. In a related process, not technically "bag-in-box," horizontal pouch (three-side-seal) machines are close coupled to a cartoner to automatically insert one or multiple pouches into a carton at high speeds.

BIBLIOGRAPHY

General Reference

S. Sacharow, *A Guide to Packaging Machinery*, Magazines for Industry, Inc., 1980.

Norman H. Martin, Jr.
Pneumatic Scale Corporation

BAG-IN-BOX, LIQUID PRODUCT

Bag-in-box is a form of commercial packaging for food and nonfood, liquid and semiliquid products consisting of three main components: (*1*) a flexible, collapsible, fully sealed bag made from one or more plies of synthetic films; (*2*) a closure and a tubular spout through which contents are filled and dispensed; and (*3*) a rigid outer box or container, usually holding one, but sometimes more than one, bag (see Fig. 1).

The bag-in-box concept appeared in the United States in the late 1950s. As early as 1957 (*1*), the package was intro-

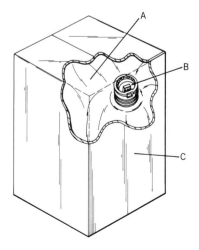

Figure 1. A bag-in-box package for liquid products consists of a bag (A), a closure and tubular spout (B), and a rigid outer container (C).

duced into the dairy industry in the form of a disposable, single-ply bag for bulk milk. By 1962, it had gained acceptance as a replacement for the returnable 5-gallon (19-L) can used in institutional bulk-milk dispensers (2). One of the first nonfood items to be offered in a bag-in-box package was corrosive sulfuric acid used to activate dry-charge batteries.

During this early period, bags were manufactured from tubular stock film by labor-intensive methods. Initially, the physical properties of monolayer films (chiefly low-density polyethylene homopolymers) limited applications, and filling equipment was slow and often imprecise. This situation changed significantly with the introduction in the 1960s of ethylene–vinyl acetate (EVA) copolymer films that provided added sealability and resistance to stress and flex cracking. By 1965 faster, dual-head fillers became available featuring semi-automatic capping capabilities. Developments in automated box forming and closing kept pace. Also by 1965, proprietary bag-manufacturing equipment capable of making bags from single-wound sheeting sealed on all four sides had been developed (see Sealing, heat). In the mid-1970s, filling machines that automatically loaded filled bags into boxes came on-line. This was followed by totally automatic filling equipment that accepted a continuous feed of bags in strips (3), separated, filled, and capped them, then placed them in outer boxes. Meanwhile, barrier films (4) (see Barrier polymers) with improved handling and storage characteristics were being developed. Multilayer films of polyethylene coextruded with poly(vinylidene chloride) (PVDC) to provide an oxygen barrier had become commercially available as Saranex (Dow Chemical Co.) in the early 1970s (see Coextrusions, flexible; Vinylidene chloride copolymers). The application of this barrier film permitted packaging of oxygen-sensitive wines and highly acidic foods such as pineapple and tomato products. Beginning in 1979, multi-ply laminates (5) (see Laminating; Multilayer flexible packaging) combining foil or metallized polyester substrates were introduced and in wide use by 1982 (see Metallizing; Film, polyester). Such laminations are thermally or adhesively bonded and in some instances by hot melt extruding the adhesive layer (see Adhesives; Extrusion coating). The most commonly used barrier film in the United States today is a three-ply laminate consisting of 2 mil (51 μm) EVA/48 gauge (325 μm) metallized polyester/2 mil (51 μm) EVA. The barrier properties of metallized polyester are directly proportional to the optical density of the metal deposit. Multilayer coextruded films combining the barrier properties of ethylene–vinyl alcohol copolymer (EVOH), the strength of nylon, and sealability of linear low density polyethylene are being successfully used for some bag-in-box applications. It is the sensitivity to moisture by EVOH that is limiting the films' wider use (see Ethylene–vinyl alcohol; Nylon; Polyethylene, low density).

Currently, fully automatic filling machines with as many as six heads (6) handling up to 40 two-liter (~2 qt) bags per minute are in operation. Films have been refined and specialized to meet tight packaging specifications for such procedures as hot fill at 200°F (93.3°C) temperatures. Bags are also presterilized by irradiation for filling with a growing number of aseptically processed products for ambient storage without preservatives (see Aseptic packaging). Outer boxes have become not only stronger, but more attractive and appealing as consumer sales units.

Manufacturing Process

In general, large producers of bag-in-box packaging design, develop, and manufacture packages to specifications meeting customer requirements. Containers vary in capacity from small, consumer and institutional sizes (or 2–20 L), to large, process and transportation packs of 55–300 gal (or 200–1000 L).

Bag. The principal considerations in choosing a film or laminate for the bag construction are strength and flexibility, with permeability and heat resistance added critical factors in an increasing number of applications. In the case of laminates, the bond between the layers of dissimilar materials must be maintained at a high level. Minimum requirements of over 1.1 lbf (500 gf) per inch (2.54 cm) (ie, ~193 N/m) is not uncommon. Films and laminates must be abuse resistant and must also, of course, be compatible with the product being packaged.

Once the appropriate film compositions have been determined for a specific application, the typical manufacturing procedure is as follows: Referring to Figure 2, two or three pairs of rolls of single-wound sheeting (A) are unwound on a machine where the webs advance intermittently, holes are punched (B) and spouts are sealed (C) into one of the duplex or triplex film sheets at predetermined points depending on finished size, and the bags are formed by sealing the two films together along the sides (D) then at the ends (E). Therefore, the bags for liquids are of a "flat" nature, not gusseted. Because precise time and temperature must be maintained to generate the seal between the thin films, the uneven thickness resulting from wrinkles or folds must be avoided because the resulting "darts" will not be fully sealed. A removable closure is applied to the spout (F), and after passing by the draw rolls (G) the bags are either cut apart as the final operation, or are perforated (H) for subsequent machine separation at time of filling.

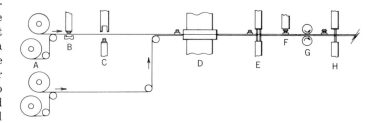

Figure 2. Typical bag-in-box manufacturing procedure.

Bag size. The box size must be measured first, and then the exact sizing of the inner bag can be determined. The bag must occupy virtually all of the interior space of the box without unfilled corners or potentially damaging excess head space resulting from an oversize box.

Spout and closure. The spout is the filling port of the bag. Together with the closure, they are designed to mate with filling heads and must be able to withstand the mechanical shock of the closure being removed and replaced during the filling operation without damage to the spout or closure. Spouts are generally molded of polyethylene with a thin flexible flange to which the bag film is sealed. The spout has han-

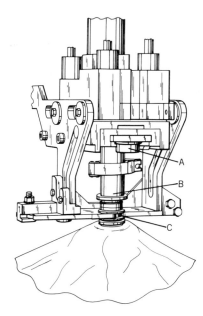

Figure 3. Filling head in fill position. The closure (A) has been removed to permit the filling nozzle (B) to come down and enter the spout (C).

dling rings for holding the spout during the filling sequence, and the closure likewise has rings for the same purpose. Figure 3 shows the filling head in the fill position. The closure (A) has been removed and lifted up and away at an angle permitting the filling nozzle (B) to come downward and enter the spout (C) which is being held firmly in position.

Because the design of bag-in-box packages provides for the contents always to be in contact with the spout, the spout-closure fit must be leakproof. When high oxygen-barrier properties are essential, the spout and closure appear as the weak area in the bag because the materials used are highly permeable. This can be minimized by certain design considerations and by the application of barrier coatings.

There are many designs for spouts and closures, depending on function. In packages destined for consumer use, the closure typically is a simple, one-piece, flexible valve which opens and closes as a lever is activated. Such a combination is shown in Figure 4. Two layers (A,B) of film are sealed to the spout sealing flange (C). Around the tubular wall (D) are one or more spout handling rings (E). The closure also has a handling ring

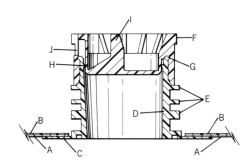

Figure 4. A flexible-valve closure. See text.

(F) which facilitates its removal by the filler capping head. The closure is retained on the spout by a mating groove and head (G), and the liquid seal is achieved by a pluglike fit (H) between the two components. The one-piece closure is molded from a resilient material and becomes a dispensing valve by flexing the toggle (I), creating an opening to an orifice (J), and providing a path for the contents to exit.

For food-service use, such as in restaurants, the closure may have a dispensing tube attached or be compatible with a quick connect–disconnect coupler leading to a pump. For other uses, just a cap may need to be removed prior to emptying of contents.

Box. In some dairy applications, the bags are transported in returnable plastic crates (see Crates, plastic). Most typically a wide variety of materials in many forms may be used for the nonreturnable box of bag-in-box packages. For smaller sizes [1–6 gal (4–23 L)], the outer box is usually made of corrugated board in a conventional cubic configuration. For larger sizes [30–55 gal (114–208 L)], rigid plastic and metal containers, or even cylindrical drums, may be employed. The required strength must be designed into the box as dictated by the specific application.

Outer boxes may be manufactured with built-in handholds or locked-in-place handles, and some have special wax or plastic coatings for moisture protection. Most boxes are built with punch-out openings for easy access to the spout and closure.

Filling

Filling bag-in-box packages may be a manual or semi- to fully automatic operation and is adaptable to a wide range of standard industrial processing procedures, including cold, ambient, high temperature, and aseptic filling. The basic design of a typical filling machine incorporates a flow meter, filling head or heads, an uncap–draw vacuum–fill–recap sequence, and filled bag discharge (see Filling machinery, still liquid). Bags may be manually loaded into the film head or automatically fed in strip form into more sophisticated models. Advanced filling equipment can also be provided with such devices as a cooling tunnel where hot-filled bags are agitated and cooled by jets of chilled water; a specialized valve to allow passage of liquids with large particulates; steam sterilizing and sterile air chambers for aseptic filling; or other modifications as determined by application.

Five-gallon (19 L) bags can be filled at speeds that range from four per minute (1200 gal/h or 76 L/min) to 20 per minute (6000 gal/h or 303 L/min). Low speed filling is done on single-head, worker-attended fillers; high speed filling on multihead equipment, comparing favorably with line speeds of conventional rigid-container operations. Complete systems including box former, conveyors, automatic bag loading, and top sealers are available to support the automated large-capacity fillers.

Shipping and Storage

Bag-in-box packaging offers significant weight- and space-saving economies. Before filling, components are shipped flat; after filling, the basic cubic shape of most bag-in-box outer boxes occupies less space and tare weight than cylindrical metal containers of comparable volume. Limitations include restrictions on palletizing and stacking height, where content weight may exceed outer box ratings, especially those constructed of corrugated board; vulnerability of uncoated boxes to humidity and moisture; possibility of flex cracking of the bag from the effects of long, transcontinental shipments; and damage potential to surrounding packages from a leaking unit.

Dispensing

Dispensing may be accomplished in one of three basic ways: uncapping and discharging contents; attaching one or more packages to a pumping system; or activating a small volume, user-demand closure often referred to as a dispensing valve.

In single-bag packages, the spout–closure is contained within the outer box for protection and withdrawn prior to use through a perforated keyhole opening in the box. During dispensing, the bag collapses from atmospheric pressure as contents are expelled without the need for air to be admitted. When completely empty, bag-in-box package components, except those outer boxes or containers specifically designed for reuse, are fully disposable. Corrugated board and polyethylene are easily incinerated, and metallized and foil inner bags compact readily in landfills.

Applications

With advancements constantly being made in bag film capabilities, along with filling and dispensing techniques, practically every commercial product is either being considered for or is now available in bag-in-box packages. Major users include the dairy industry, especially in the packaging of fluid milk for restaurant/institutional use, and of soft ice cream mixes; the wine producers, who offer 4-L (3.8-qt) bag-in-box packages to consumers and 12- and 18-L (11.4- and 17-qt) sizes to restaurant/institutional outlets; and the fast-food markets, where condiments are quickly dispensed onto menu items from bag-in-box packages and soft drinks are prepared from fountain syrup pumped from a bag-in-box arrangement (7) which eliminates the need for recycling and accounting for metal transfer containers.

BIBLIOGRAPHY

1. U.S. Pat. 2,831,610 (Apr. 28, 1958), H. E. Dennis (to Chase Bag Company).
2. U.S. Pat. 3,090,526 (May 21, 1963), Robert S. Hamilton and co-workers (to The Corrugated Container Company).
3. U.S. Pat. 4,120,134 (Oct. 17, 1978), W. R. Scholle (To Scholle Corporation).
4. "Films, Properties Chart" in *Packaging Encyclopedia 1984,* Vol. 29, No. 4, Cahners Publication, pp. 90–93.
5. John P. Butler, "Laminations and Coextrusions" in *Packaging Encyclopedia 1984,* Vol. 29, No. 4, Cahners Publication, pp. 96–101.
6. U.S. Pat. 4,283,901 (Aug. 18, 1981), W. J. Schieser (to Liqui-Box Corporation).
7. U.S. Pat. 4,286,636 (Sept. 1, 1981), W. S. Credle (to Coca-Cola Company).

General References

Glossary of Packaging Terms (Standard Definition of Trade Terms Commonly used in Packaging). Compiled and published by The Packaging Institute, U.S.A., New York.

C. J. Bond
Liqui-Box Corporation

BAG-MAKING MACHINERY

Heavy-duty bags, ie, shipping sacks, of multiwall paper or single-wall mono- or coextruded plastic are used to package such dry and free-flowing products as cement, plastic resin, chemicals, fertilizer, garden and lawn-care products, and pet foods. These bags typically range in capacity from 25 to 100 lb (11.3–45.4 kg), although large plastic bulk shipping bags may hold as much as a metric ton (see Bags, paper; Bags, heavy-duty plastic; Intermediate bulk containers).

Although there are dozens of variations in heavy-duty bag constructions, there are only two basic styles: the open-mouth bag and the valve bag. The former is open at one end and requires a field-closing operation after filling. Valve bags are made with both ends closed, and filling is accomplished through an opening called a valve. After filling, the valve is held shut by the pressure of the bag's contents.

Multiwall-bag Machinery

Traditionally, multiwall bags are manufactured in two operations on separate equipment lines. Formation of tubes takes place on the *tuber.* Closing of one or both ends of the tubes to make the bags is done on the *bottomer.* Multiwall bags have two to six plies of paper. Typical constructions are three and four plies. Polyethylene (PE) film is often used as an in-between or innermost ply to provide a moisture barrier.

Tube forming. The tuber (Fig. 1) starts with multiple giant rolls of kraft paper of a width that will finish into the specific bag width. At the cross-pasting station, spots of adhesive are applied between the plies to hold them together. The material is then formed into a tube that is pasted together along the seam. The tube may be formed with or without gussets. During seam-pasting, the edges of the various plies form a shingle pattern. When they are brought together to form a seam, these edges interweave so that each ply glues to itself. This provides optimal seam strength.

Flush-cut tubes are cut to the appropriate sections by a rotating upper and lower knife assembly. With stepped-end tubes, perforating knives are used to cut stepping patterns on both ends of the tube. The tube sections are then snapped apart along perforations that were made prior to cross-pasting. This snapping action is accomplished by sending the tubes through two sets of rollers, with the second set moving slightly faster than the first. Once the tubes have been flush-cut or separated, they proceed to the delivery section of the line.

About one-third of the multiwall-bag market is accounted for by bags with an inner, or intermediate, ply of plastic film. Flat film, used as an inner or intermediate layer, is formed into a tube along with the paper plies and pasted or, if necessary, hot-melt laminated in place. Another possibility is the insertion on the tuber of open-mouth film liners, the open end of which can project beyond the mouth of the paper sack. Stepped-end and flush-cut tubes are usually made on differently equipped tubers, but there is also a universal model which can be adapted to produce either type.

Flush-cut vs stepped end. Flush cutting is the most inexpensive tubing method in terms of both original equipment investment and tubing productivity, but these gains are lost in the subsequent bag-making operations. The bottom of a flush-cut tube is normally sewn, and sewing is also the traditional method of field closure for many products such as seed and animal feeds. There was some use of flush-cut tubes as valve

Figure 1. A universal tubing machine. Figure insert shows components: 1, flexoprinter; 2, unwind stations with reel-change arrangements; 3, automatic web brake; 4, web-guider path rollers; 5, web guider; 6, vertical auxiliary draw; 7, perforation; 8, cross pasting; 9, longitudinal register rollers; 10, seam pasting and auxiliary draw; 11, tube forming; 12, cut/register regulator; 13, cutting and tear-off unit; 14, variable-length drive unit, 15, packet delivery unit; 16, take-off table.

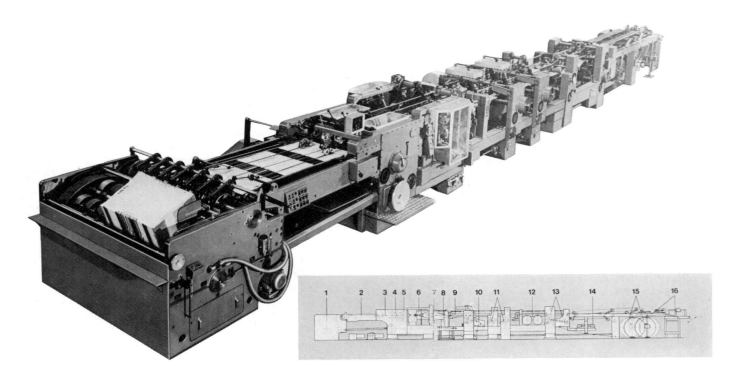

Figure 2. A valve bottomer. Figure insert indicates components: 1, rotary or double feeder; 2, tube aligner; 3, diverter for removing incorrectly fed tubes; 4, diagonal creasing and needle vent hole arrangements; 5, bottom center creasing stations with slitting arrangement for bottom flaps; 6, bottom opening station; 7, bottom creasing station; 8, unwind for valve patch; 9, bottom turning station; 10, valve unit; 11, bottom pasting; 12, bottom closing station; 13, bottom capping unit or intermediate pressing station; 14, flexo printers for bottom caps; 15, unwinds for bottom caps; 16, delivery, optionally with incorporated counting and packeting station.

bags, particularly in Europe, but they are not widely used today because a pasted flush-cut bottom is structurally weak. The gluing of the bottom takes place only on one ply. To compensate for this weakness, a patch would normally be added to the bottom of the bag. Sewing is a widely used bottoming method in the United States because there is a great deal of flush-cut tubing and sewing equipment in place, and replacing it in many instances would result in only a marginal return on investment. Unfortunately, sewing has many drawbacks. It is labor-intensive and, because the equipment has a large number of delicate moving parts, maintenance and repair costs are high. Also, the needle holes created by sewing weaken the bag, allow sifting, and make the bag more accessible to rodents and other pests.

The stepped-end tube makes the strongest bag. The ends of the bag have shinglelike stepping patterns which intermesh at the gluing points. In the bottoming process, ply one is glued to ply one, ply two to ply two, etc. Generally speaking, all bag manufacturers have their own stepping-pattern designs.

Bottoming equipment.   The finished tube sections are converted into bags by closing one or both of the tube ends in any of the following three ways:

(*1*) One end of the tube is shut, forming a sewn open-mouth (SOM) bag. After filling, the top of the bag is closed by means of a portable field sewing unit.

(*2*) A satchel bottom is formed on each end of the tube, with one of the bottoms provided with an opening or valve through which the bag is filled by insertion of the spout of an automatic filling machine or packer. The valve is closed by the pressure of the bags contents. Additional means are available to make the bag more sift-proof. In a valve bottomer (Fig. 2), tubes advance from a feeder to a tube aligner and a diverting unit for removing incorrectly fed tubes. The tubes pass through a series of creasing stations, and needle holes may be added under the valve for proper venting of the bag during filling. At the opening section, the tube is opened and triangular pockets are formed. Valves are inserted at a valving station.

Valves are automatically formed by a special machine unit and then inserted into the bottom. They may be inserted and folded simultaneously along with the bottom or preformed and automatically inserted. Preformed valves permit the use of a smaller valve size in proportion to the bottom of the bag. In Europe, reinforcing patches are customarily applied to both ends of the bag for added strength. The bags are discharged to a press section where they are conveyed in a continuous shingled stream. Powerful contact pressure of belts (top and bottom) ensures efficient adhesion. In most instances, the final station is an automatic counting and packeting unit.

(*3*) Stepped-end tubes with gussets and a special step pattern can be converted into pinch-bottom bags on which beads of hot-melt or cold adhesives (see Adhesives) are applied to the steps in the bottom (see Fig. 3a). These, in turn, are folded over and pressed closed to make an absolutely sift-proof bottom (Fig. 3b). Beads of hot melt applied to the steps at the top of the bag are allowed to cool and solidify. After the bag is filled, a field-closure unit reactivates the hot-melt adhesive, folds over the top of the bag, and presses it closed.

In most instances, bags are collected in packets or bundles palletized for shipment to the end user. However, it is also possible to collect the bags on reels for efficient loading of automatic bag-feeding equipment in the field. The reeled bags form a shingled pattern held in place by the pressure of two

Figure 3.   (**a**) adhesives being applied; (**b**) pressing station (bags are folded over and pressed closed).

plastic bands that are wound continuously around the reel along with the bags.

Recent developments.   With conventional equipment, it usually takes two bottomers to keep pace with one tuber. This fact has generally discouraged the development of in-line multiwall bag-making systems in the United States. For example, tubers for cement bags typically operate at speeds from 270 to 320 tubes/min, whereas old-style bottomers run at 120–150 bags/min. Recently introduced bottoming equipment can achieve speeds up to 250 bags/min, enabling one-to-one operation of tuber and bottomer on an in-line system. The tuber operates at less than maximum output, but the in-line system still produces more finished bags because of the increased efficiency resulting from the bottomer being continually fed with

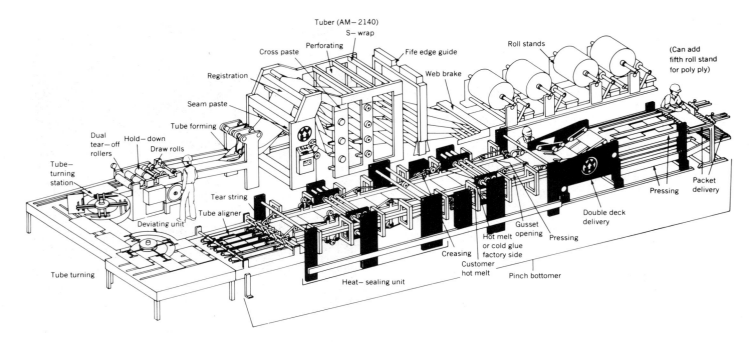

Figure 4.   In-line tube forming and pinch bottoming.

fresh tubes. Tubes where the paste has dried become stiff and difficult to handle. As paper is unwound from a roll, it quickly loses its moisture content and becomes less workable. These types of problems are alleviated with in-line bottoming.

In-line tube-forming and -bottoming also lend themselves to significant improvements in manpower utilization. The U-shaped in-line pinch-bottoming system shown in Figure 4 is capable of reducing the manpower requirements from nine to four. The key to the system is a unique turning station which rotates the axis of the tube by 90° for proper alignment with the bottomer. The "factory end" of the bottomer may be heat-sealed in-line (see Sealing, heat). On the "customer" end of the bag, hot melt or cold glue can be applied, or this end of the bag can be flush-cut for sewing in the field. A sewn top with a pinch bottom offers strength and sift-proofness in this bottom style while allowing the customer to retain existing closing equipment. For a consumer product such as pet food, the pinch bottom allows the bag to be stacked horizontally on the shelf, still presenting a large graphics display area for the shopper.

Another recent development is an out-of-line double feeder-equipped pinch bottomer for the manufacturers just starting to produce pinch-bottom bags and those not anticipating having the volume to fully utilize the more productive in-line system. The trend for bag users to reduce inventories and place more small orders is expected to continue indefinitely. For the converter, this has meant decreased productivity because of a disproportionate amount of time being spent in changeovers. New CNC (computer numerical control) bottoming equipment promises to reduce changeover time from an average of about 3 h to about 30 min. All gross adjustments of machinery for a particular set-up are stored in the microprocessor and made on the machine by way of stepping motors. Although minor fine-tuning is still required, the starting adjustment point of each operator is the same, and settings are optimized according to a logical sequence designed into the control (see Instrumenta-

tion). In addition to faster set-ups, standardization of tuning procedures should result in more consistent and improved product quality.

Plastic Bag Machinery

The procedure for making all-plastic heavy-duty bags is similar to the procedure for multiwall bags, ie, various bottoming techniques are used to transform a tube into a finished bag, generally either an open-mouth or valve bag, with or without gussets. The three basic differences are described below:

(1) Plastic bag making almost always uses a single ply of material, either mono-extruded film, coextruded film, or woven fiber instead of the multiple plies used in paper shipping sacks.

(2) All bag-making operations are performed on a single converting line. If the bag is made from flat sheet, the tubing and bottoming operations are integrated into a single bag-making line. Bags are often made from tubes of blown film or circular woven fibers, and no tubing step is necessary.

(3) The plastic-bag-making line may incorporate in-line printing, although the outer ply of kraft paper used in a multiwall bag is typically preprinted off-line.

Woven-bag machinery.   Economy of raw materials and toughness are two features that make the woven plastic bag an attractive packaging medium for goods mainly intended for export.

A typical line for converting woven high density polyethylene (HDPE) or polypropylene (PP) material into heavy-duty shipping sacks includes the following: unwind units for sheet or tubular webs, jumbo or normal size; a flexographic printing machine (see Printing) designed for in-line operation; a wax-application unit (see Waxes) to apply a hot-melt strip across uncoated material at the region of subsequent cutting to prevent fraying; and a flat and gusseted tube-forming unit. The flat sheet of coated or uncoated material is longitudinally folded into tubular form. Some machines have the ability to do this without traditional tube-forming parts. A longitudinal seam is sealed by an extruded bead of plastic. Output of the extruder is matched to the web speed by a tacho-generator.

A PE liner unit can be arranged above the tube-forming section to apply a PE liner to the flat web automatically. The principal element of this unit is a welding drum with rotating welding segments that provide the reel-fed PE with a bottom weld at the correct intervals. A Z-folding device enables a fold to be made in the crosswise direction for the provision of a liner which is longer than the sack. In addition, there is a crosscutting unit that cuts the outer web and the PE insert, usually by means of heated rotating knives. In the bottoming unit, cut lengths are transferred to the bottoming equipment by conveyor. Bottoming is accomplished either by sewing or the application of a tape strip. Instead of folding the tape over the open end of the sack, the sack end can be folded once or twice and the tape applied in flat form over the folds. The delivery unit collects finished sacks into piles for manual or automatic unloading.

Plastic valve sack machinery. Plastic valve bags operate by the same principle as multiwall valve bags. Upon filling, the pressure of the product closes a valve that has been inserted in either the bottom or the side of the bag. If the material is granular (not pelletized), channels along the bottom of the valve sack would allow some of the product to sift out. These channels can be made sift-proof by closing them off with two beads of hot wax during the bottoming operation. Only 5–10% of the plastic valve sacks made in the United States require this feature. Therefore, most plastic valve sacks are produced on high speed lines that produce sacks at about twice the speed of the sift-proof machinery.

A typical system for the production of pasted PE bags from either flat film or blown tubes (Fig. 5) consists of the following equipment:

1. An unwind unit for flat film or tubing incorporating automatic tension and edge-guide controls.
2. A tube former in which folding plates form flat film into a tube. A longitudinal seam is bonded by an extruded PE bead.
3. A rotary cross cutter in which the formed tube is sepa-

rated into individual lengths by the perforated knife of the rotary cross cutter. Fraying of woven materials can be eliminated with a heated knife which bonds the tapes together.

4. A turning unit in which, after the cross cutter, the tubes are turned 90° to bring the cut ends into position for the following processes.
5. A tube aligner and ejector gate in which exact alignment of tubes in longitudinal and cross direction is achieved by means of stops affixed to circulating chains and obliquely arranged accelerating conveyor bands. Photocells monitor the position of the tube lengths. In the event of misalignment, leading to malformed bottoms and, therefore, unusable sacks, the photocell triggers an electropneumatic gate which, in turn, ejects the tube length from the line.
6. Pasting stations in which each tube end is simultaneously pasted by a pair of paste units using a special adhesive.
7. An enclosed drying system evaporates and draws off solvent from the adhesive.

8–10. Creasing, bottom-opening, and fixing of opened bottom in which a rotating pair of bars hold the tube length ends by suction and the rotary movement pulls the tube ends open sufficiently to enable rotating spreaders to enter and complete the bottom-opening process. The diagonal folds of the pockets are fixed by press rolls to avoid subsequent opening of the pockets.

11. A valve-patch unit forms the valve from rolls of flat film and places it in the leading or trailing pocket, as required.
12. The bottom-closing station, where, after the valve is positioned, the pasted bottom flaps are folded over, one to the other, and the sack bottom is firmly closed.

13–17. The bottom-patch unit, bottom-turning station, flexoprinting units for bottom patches, pasting stations with drying, and unwind for bottom-patch film, in which patches are formed from two separate rolls of film, flexo-printed (if required), and pasted to both sack bottoms. The bottom geometry is checked by photocells and faulty sacks are ejected through a gate. Just before they reach the delivery section, the bot-

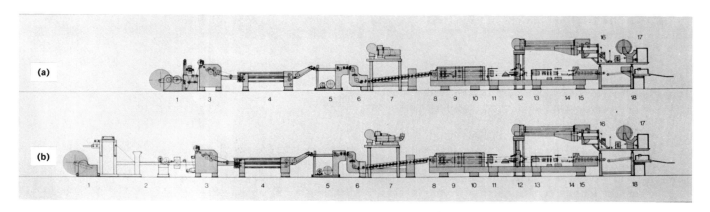

Figure 5. Systems for making all-plastic heavy-duty bags from (**a**) tubular film or from (**b**) tubular or flat film. Components are described by number in the text.

toms are turned from a vertical to horizontal processing plane.

18. Delivery with counter and packeting station, in which good adhesion of the cover patch to the sack bottom is assured by applying pressure to the shingled sacks with staggered spring-loaded disks. Having reached a predetermined count, the conveyor accelerates the shingled sacks to the packing station, where the counted sacks are collected into packets and discharged. To accommodate a user's automated filling line, equipment is also available to wind the plastic valve sacks onto reels.

Continuous bag-forming and -filling. Plastic valve bags have been used extensively, especially in Europe, for products such as plastic resin. However, continuous systems for forming, filling, and closing flat and gusseted plastic bags are becoming increasingly popular in the resin market. Such systems typically use prefabricated tubing for high strength. The tubular material is usually preprinted with random printing. Since resin weight varies from day-to-day, depending on ambient conditions and other factors, random printing allows the bag length to be adjusted according to the prevailing resin volume–weight relationship. In this manner, a tight and graphically appealing package is formed.

An integrated system for forming, filling, and closing of shipping bags would contain the following stations: unwind unit; compensator roller; hot-emboss marking unit; sealing station for bottom seam; bag shingling; separation of bags; introduction of bag-holding tongs; bag filling; supply of the filling product; sealing station for closing seam; bag outfeed conveyor; and control panel. Such a system can produce up to 1200 filled sacks/h.

One-way flexible containers. One-way bulk shipping containers are becoming very popular in Europe. These oversized bags are designed for handling by forklift trucks equipped with one of several specially designed transport devices. Called intermediate bulk containers, they range in capacity from 1100 lb to about a metric ton (0.5–1 t), and are constructed of woven PP or HDPE. Tubes are generally woven on a circular loom because elimination of the longitudinal seam gives the bag exceptional strength. The advantage of this bag is that it represents an exceptionally economical and efficient method of handling bulk quantities. Acceptance of the concept has been relatively slow in the United States because it requires bag producers, product manufacturers, and product customers all to invest in special equipment for bag making, product filling, or handling.

Electronic Controls

Today's bag-making equipment is following the overall industrial trend toward the use of programmable microprocessor control systems of increasing complexity. Ancillary equipment such as printing presses and extruders already have a high level of control, and other units on the bag-making line are quickly being adapted to the computer. The first objective in conversion to programmable control is replacement of cumbersome mechanical logic. The next is storage of set-up and processing parameters for subsequent reuse. Microprocessors are being used for controlling temperatures, web tension, surface-tension treatment, adhesive application, ink, and registration. The most recent stage of automation has been provision of multiple outputs so that lines may be monitored or controlled by hierarchal computers.

RICHARD H. JOHN
Windmoeller & Hoelscher Corporation

BAGS, HEAVY-DUTY, PLASTIC

All-plastic, heavy-duty bags account for less than 10% of total heavy-duty bag usage in the United States today, but they are essential for some products that require the properties that only an all-plastic bag can provide. In the paper bag business, the heavy-duty category begins with 25-pound (11.3-kg) capacity. In the all-plastic segment of the industry, 20-pound (9.1-kg) capacity is considered the start of the heavy-duty range (1). The words bag and sack are often used interchangeably, but although bag is used for the entire size range, sack generally applies only to the heavy-duty category. The word bag is used in this article to reflect modern usage in the United States. It refers here only to all-plastic, heavy-duty bags (see also Bags, paper; Bags, plastic). Woven polyolefin shipping sacks belong to another family.

Uses

All-plastic bags are generally used for free-flowing products, although some are used for compressed products such as peat moss or insulation batts. They are particularly useful for packaging products that need a very effective moisture barrier, either because the product is hygroscopic or stored outside. Hygroscopic products include rock salt, fertilizers, and some food products. Their suitability for short-term outside storage has made all-plastic bags the package of choice for consumer lawn and garden products and for insulation stored outside at construction sites. Plastic bags are also preferred for acid-producing products, eg, composted steer manure.

All-plastic bags for shipping plastic resins eliminate the risk of fiber contamination, which can present problems in precision molding or extruding operations (2). In many cases, the bag can save labor by going into a digester such as a banbury along with the product, eg, carbon black, titanium dioxide, zinc oxide, silicas, and other products for compounding (see Additives, plastics).

All-plastic bags should not be used for products that must breathe. Industrial sulfur, for example, must breathe to prevent spontaneous combustion. Some dry and semimoist pet foods are packaged in all-plastic bags, but some pet foods, depending on grease content, mold in a bag that does not breathe. One must also make sure that the product does not penetrate the film wall and react adversely with the printing inks.

Films

All-plastic bags are produced from tubular or flat film. The film is generally single-ply, although some bags, particularly those used for lawn and garden fertilizers, are made as duplex or two-ply bags, either monofilm or coextruded (see Coextrusion machinery, flat; Coextrusion machinery, tubular; Coextrusions for flexible packaging). About 60% of the heavy-duty bags used in the United States today are monofilm. Monofilm bags are almost always low density polyethylene: conventional (LDPE), linear (LLDPE), or blends of the two (see Polyethylene, low density). Coextrusions are generally LDPE–LLDPE, but high density polyethylene (HDPE) or modified HDPE is sometimes included in the structure for special properties (see Polyethylene, high density). Before the advent of LLDPE, minimum gauge thickness was 4 mil(102 μm). Now it is 2.8 mil(71 μm) in single-wall bags because linear LDPE's greater strength and toughness compared to conventional LDPE have permitted downgauging, which results in lower bag costs.

The property improvements afforded by LLDPE can be attained by blending or coextrusion, and both methods are used. There are indications that coextrusions are stronger, but percentages of each product in the film product plays a greater role. Coextrusion is also used to combine the properties of dissimilar resins. The physical properties of unmodified HDPE monofilm (see Film, high density polyethylene) are not adequate for heavy-duty bags, but combined with a low density polyethylene structural layer, it imparts heat resistance for products that are hot-filled and improves tensile strength.

For maximum physical properties, all-plastic bags may be made of nonwoven polyolefins (see Nonwovens) or of cross-laminated HDPE film (Valeron film, the Van Leer organization). Some of the earliest heavy-duty bags in Europe were made of PVC film (see Poly(vinyl chloride) (3).

Constructions

Most heavy-duty bags are preformed, but the form–fill–seal concept widely used in small packages is now applied to heavy-duty bags as well (see Bag making machinery).

Preformed bags have two styles: open-mouth and valve bags. Open-mouth bags are shipped to the user with one end completely open. The product is gravity-filled through a spout, and the open end is sewn or heat-sealed. Relatively few bags are sewn, because sewing requires a stiff film of about 6 mil(152 μm), and the majority of open-mouth bags are ≤ 5.0 mil (127 μm). Heat sealing is the most common practice, with continuous-band or hot-air sealers (see Sealing, heat). Most open-mouth bags have a pillow configuration, but side- or bottom-gusseted bags are also available. Filled gusseted bags are more stable on pallets, and the flat sides and ends can be printed for product identification.

Valve bags are closed on both ends, but there is a valve in one corner of one end. The product is filled through a spout by forced air or by augers. Plastic bags can be filled easily by augers, but where forced air is used, trapped air can cause the bag to balloon. This difficulty can be overcome by putting a vacuum line on the spout or by modifying the bag. Satchel-style valve bags do not require sealing or sewing because the

Table 1. Typical Constructions of All-Plastic, Heavy-Duty Bags

| Style | Dimensions, in. (mm) | | | Gauge, mil (μm) | Typical applications |
	Width	Length	Ends		
handle bags, max 40-lb (18.1 kg) capacity	13 (330)	22.5 (570)		6 (152)	25-lb (11.4 kg) salt
	15 (380)	25.5 (650)		6 (152)	40-lb (18.1 kg) salt
open mouth	15 (380)	25 (635)		5 (127)	50-lb (22.7 kg) decorative stone
	13.5 (345)	22.5 (570)		5 (127)	20-lb (9.1 kg) fertilizer
	16 (410)	26.5 (675)		5 (127)	40-lb (18.1 kg) fertilizer
	15.5 (395)	25 (635)		3 (76)	20-lb (9.1 kg) peat, potting soil, manure
	20 (510)	28 (710)		5 (127)	40-lb (18.1 kg) peat, potting soil, manure
	18 (455)	29.5 (750)		3 (76)	1-ft^3 (0.03 m^3) soil conditioner
	21 (535)	35 (890)		3 (76)	2-ft^3 (0.06 m^3) bark/mulch
	24.5 (620)	41.5 (1055)		3.5 (89)	3-ft^3 (0.08 m^3) bark/mulch
valve bags	21 (535)	23 (585)	5.5 (140)	7 (178)	50-lb (22.7 kg) plastic resin
	23 (585)	27.5 (700)	5.5 (140)	5 (127)	50-lb (22.7 kg) carbon black
bale bags	18.5 (470)	52 (1320)	14.5 (370)	5.5 (140)	6-ft^3 (0.17 m^3) peat moss
	12.5 (320)	37 (940)	9.5 (240)	5 (127)	2-ft^3 (0.06 m^3) peat moss
	16 (410)	66 (1675)	14.25 (360)	5 (127)	insulation batt
	23 (585)	69 (1755)	15.5 (395)	5 (127)	insulation batt

valve is closed by internal pressure from the product. They can be made more sift-proof with an extended sleeve that is folded or sealed closed.

Valve bags have an important advantage compared to open-mouth bags in that they square when filled. They are also called square-bottom bags. This contributes to stability on pallets, and helps guard against shifting in shipment. Valve bags have traditionally been less popular than open-mouth bags because of slow filling speeds, but modern automated lines have improved their economics considerably. Filling speeds have been improved, and multispout fillers require less labor than open-mouth filling lines. The popularity of valve bags is increasing, but they are still in the minority. In the United States today, roughly 85% of plastic, heavy-duty bags are open-mouth; about 15% are of the valve variety. For a sampling of typical heavy-duty bag constructions and applications, see Table 1.

Most all-plastic bags are printed by flexography, which is simpler and less expensive than rotogravure, which is also used to some extent in the United States and abroad (see Printing).

Bale Bags

A bale bag is a special type of open-mouth bag that is used to package compressed products. They are filled by a baling machine that rams the product into the bag. This bag style has long been used for peat moss, and it has more recently opened the important new market for fiberglass insulation batts, blowing wool, and cellulose insulation. Compared to paper sacks, the all-plastic bags offer better protection against mechanical damage, and they permit outdoor short-term storage at construction sites.

Economics

All-plastic, heavy-duty bags are generally more expensive than all-paper bags. Their cost-competitive position versus multiwall bags depends on the multiwall construction required for adequate properties. Generally speaking, an all-plastic sack is competitive if the paper plies in the competing multiwall bag add up to a basis weight total of at least 200 lb (90.7 kg) and a plastic barrier layer is required in the form of a film or a coating.

Plastic bags sometimes compete by offering indirect cost savings. The use of plastic bags for lawn and garden products is an example. Producers, wholesalers, and retailers save valuable indoor storage space by storing the products outside in plastic. They can sometimes offer savings to packers in warehouse storage because they are lighter than multiwalls and take less space. Compared to paper, plastic bags offer better moisture and chemical resistance and better impact resistance. They resist propogation of snags and can be easily patched.

BIBLIOGRAPHY

1. *Specification B-10,* Flexible Packaging Association, Washington, D.C., 1971.
2. "Shipping plastics in Plastics," *Packag. Dig.,* 66 (May 1981).
3. D. J. Flatman, "Sacks made from Plastics Film" in F. A. Paine, ed., *The Packaging Media,* John Wiley & Sons, Inc., New York, 1977, p. 3.74.

R. S. WHEELOCK
U.S.I. Film Products

BAGS, MULTIWALL.   See Bags, paper.

BAGS, NET.   See Netting.

BAGS, PAPER

Billions (10^9) of paper bags are produced annually for packaging industrial and consumer products. Ongoing improvements in barrier plies, coatings, closures, printing technologies, and packaging machinery contribute to paper packaging's continuing prominence and growth potential in the marketplace. A paper bag, or sack, is generally defined as a nonrigid paper container made by forming a tube with one or both ends closed, with an opening to introduce the product to be packaged. They are usually classified into two types: small paper bags, with capacities of less than 25 lb (11.3 kg), generally for consumer products; and multiwall bags, generally for industrial products, with capacities of 25–110 lb (11.3–49.9 kg). Supermarket carry-out bags are not included here.

Types of Bags

There are three basic types of paper bags, classified by the number of layers of paper involved in their construction: single-ply, duplex (two-ply), and multiwall (three or more plies). Two-ply bags are often called multiwalls as well. The strength and vapor-barrier performances of multiwall bags are improved by the inclusion of plastic plies. Of the three types, multiwalls are the most common.

Advances in multiwall packaging since the 1950s have developed from the combination of paper, foil, and plastic or other plies for improved performance. Products requiring moisture-vapor, gas, flavor, and odor barriers as well as strength and protection against infestation all benefit from these developments. Other relatively recent improvements include nonskid-coating systems, load-locking adhesives, and palletizing procedures, which were developed for safer handling and stacking, reducing shipping damage and product loss. Among the materials currently used as performance and barrier plies are LDPE, HDPE, and PP films. A multiwall bag can be custom-tailored for its application. The ply material can be selected to provide specific characteristics. Cost is also factor in selecting the type of ply material used. The range of available materials gives the packager a choice of several options, which can be utilized on existing packaging-machinery lines. Additional significant advances in recent years include new coatings for the outer sheets which allow for improved graphics for consumer appeal. Although most multiwall bags today contain only kraft paper, approximately one-fourth of the multiwall bags produced in the United States contain a plastic ply or component.

Styles

The most common bag styles in current use are sewn open-mouth, sewn valve, pasted open-mouth, pasted valve, and pinch-bottom open-mouth (see Fig. 1) (see Bag-making machinery):

1. Sewn open-mouth (SOM)—The mouth of the empty bag is fully open to allow for easy, rapid filling, with the bottom sewn. When filled, the mouth of the bag can be closed by sewing. Taping, tying, pasting, or stapling are alternative methods, although sewing is the preferred method. This style is generally used for granular products or large particulates.

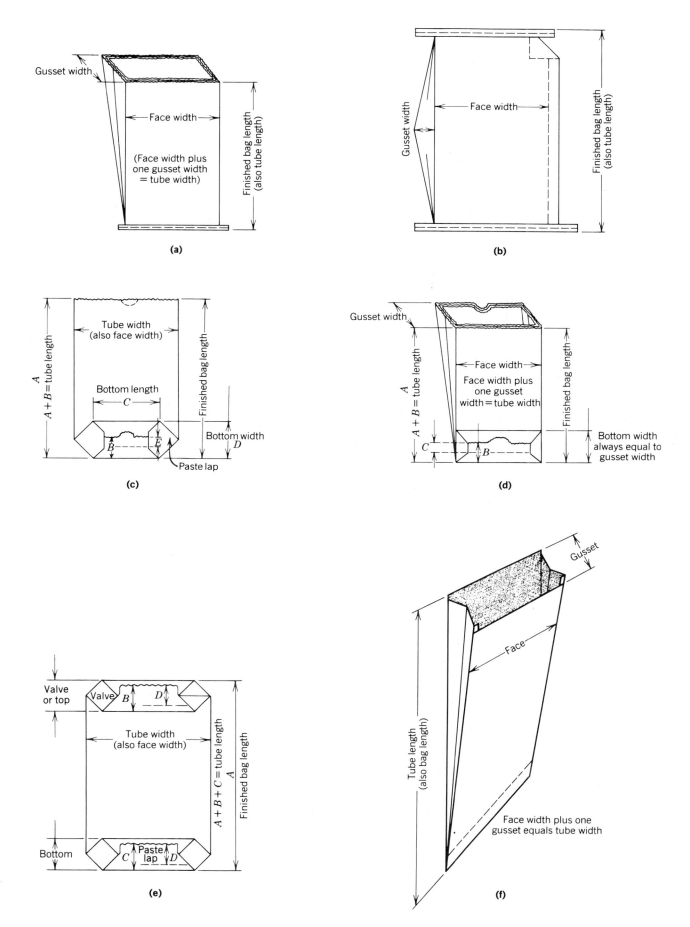

Figure 1. Common bag styles in current use: (**a**) sewn-open mouth; (**b**) sewn valve; (**c**) and (**d**) pasted open-mouth; (**e**) pasted-valve stepped-end; and (**f**) pinch-bottom open-mouth.

2. Sewn valve (SV)—The bottom and top are presewn. Filling is accomplished through an integral valve, usually located on a corner of the bag. Valve-style bags are generally preferred for smaller particulates.

3. Pasted open-mouth (POM)—The bag is filled like the sewn open-mouth (SOM), folded, and sealed by pasting.

4. Pasted-valve stepped-end (PVSE)—The bottom and top are pasted closed, with filling accomplished through an integral valve sleeve. Unlike the sewn valve, the bag, when filled, takes a particularly "squared" configuration, facilitating printing, space-efficient and safe palletizing, storage, and shipment.

5. Pinch-bottom open-mouth (PBOM)—This bag is made by stepping each ply, folding, and gluing one end. After filling, the other full-open-mouth end is folded and glued with a preapplied hot-melt adhesive to provide a total closure. This style is preferable where complete product protection or siftproofness is desired.

In addition to these five important styles, there is a subcategory, the paper baler or master container for small bags, generally a variation of the pasted-open-mouth style bag.

Of the important bag styles, pasted-valve bags currently dominate the market, followed by sewn open-mouth, pinch bottom, pasted open-mouth, and sewn valve. The dominance of pasted-valve bags is due to the preponderance of smaller particulate products packaged, eg, cements, and the generally lower labor costs required to fill and close. The stepped-end feature on both pasted-valve and pinch-bottom bags creates a stronger bag bottom. (Stepped-end means that the plies are not flush-cut, but are offset, appearing like steps in profile.) In effect, the step pattern permits each ply of a panel to be adhered to its counterpart of the opposing panel.

The fastest-growing bag style in recent years is pinch-bottom open-mouth bags. Through the use of a hot-melt adhesive reactivation closing unit, instead of the traditional sewing method, the pinch-bottom bag offers a more secure seal that is less prone to leakage and infestation.

Bag Filling

Today a wide range of bag-filling equipment is available which can be as simple as manual equipment for a small packaging operation to high speed, automated lines for large-volume packers. Basically, there are two types of filling equipment: one for open-mouth bags, and one for valve bags.

Open-mouth bags, which have an open top and preformed bottom, are generally gravity-filled through a spout connected to a preweighing or volumetric measuring device, ie, open-mouth packer. Once filled, the bags are closed by one of several types of equipment (see below).

Valve bags are closed when the pressure of the contents of the bag closes the paper valve constriction. Valve bags are filled through a tube connected to a preweighing or measuring device, ie, valve packer. For valve bags, there are five types of filling technology:

1. Impeller—A valve packer in which an impeller wheel forces the product through the valve spout into the bag.

2. Belt—The product flows into a space formed by a belt and groove in a pulley. The movement of the belt and rotation of the pulley forces material through the valve spout into the bag.

3. Auger—The product is forced into the bag by a horizontal screw, ie, auger, turning in the valve-spout tube.

4. Gravity—Gravity is used with very free-flowing materials.

5. Air flow—Proper dispersion of air is used to activate or fluidize the product and, by positive pressure, causes it to move freely through the valve spout into the bag.

Closures

There are four basic methods employed in closing open-mouth style multiwall bags:

1. Sewing—A common the technique for closure, the open end of the bag is stitched with cotton or plastic thread or a blend of both.

2. Tape under sewing—In this technique, the bag is topped with a width of paper tape to increase strength, with the sewing accomplished through the tape and the paper walls.

3. Tape over sewing—After sewing, this closure is reinforced with a width of tape covering the stitching, usually sealed by glue or heat-sealing. This reduces the problems of sifting and moisture contamination, particularly for fine powder products.

4. Heat-sealing—By the use of a series of spring-loaded heating elements, the inner polyethylene liner of a multiwall bag is sealed at the end to reduce product loss due to leakage, puncturing, and moisture absorption.

Printing

The potential for improved graphics was an added bonus when paper bags replaced textile bags. Utilizing improved papers, coatings, and printing technologies, the paper bag offers superior opportunities for appealing, quality graphics. Graphics have become more important owing to the increasingly competitive marketing of many industrial, agricultural, and chemical products, as well as consumer products which are packaged in smaller bags. Consumer bags demand increasingly intricate process printing, and industrial shipping bags, which may have simpler brand identification, also require improved printing.

Flexographic printing (see Printing) has become the industry norm and is in constant evolution, but new technologies in printing have also been adapted to the bag industry. Flexography was initially utilized for simple identification printing, using highly fluid inks (see Inks). Equipment has been improved and modified to the point where flexographic printing quality is approaching that of the rotogravure method, allowing photographic reproductions with use of coated papers. When highest impact is needed to market the product, rotogravure printing on high finish coated paper is used, resulting in excellent photographic reproductions.

Materials

Paper. The basic material used for multiwall bags is kraft paper (see Paper). Kraft, which in German means strength, is the common term used to designate the paper made from wood pulp produced by a modified sulfate pulping process. The natural unbleached color of kraft paper is brown, but it can be produced in lighter shades by use of semibleached or bleached-sulfate pulps. Kraft paper is usually made in 25–60 lb (11.3–27.2 kg) basis weights, ie, the weight of 500 sheets of paper, size 24 × 36 in. (61 × 91.4 cm).

A chief use of kraft paper is in the multiwall-bag industry. A special type of natural kraft is shipping-sack kraft, commonly called multiwall-bag kraft. It generally has specifications higher than other kraft paper grades, such as that used for "grocery kraft." For many years, extensible paper, achieved by a process in which natural multiwall-bag kraft paper is processed to increase its stretch in the machine direction, has been used in bag manufacture. Recently, improved sheet has been produced by the "free dry" process. It demonstrates excellent strength characteristics due to vastly improved MD and XD stretch characteristics.

There are a variety of finishes available on kraft paper. The three important categories are: *machine finish,* which is neither high nor low finish and is designed for good balance between nonslip qualities and good printing surface; *rough or uncalendered,* which is achieved without the use of calendering and is used in bags to provide a rough surface to prevent slippage; and *high finish,* which is obtained by the use of more nips on a calender or ironing stack, providing a better printing surface.

Plastics. The explosion in the development of plastics since World War II has had a tremendous impact on the paper-bag industry. Although plastic bags have taken over some areas previously served by paper (see Bags, heavy-duty plastic), plastic developments have greatly enhanced the multiwall paper bag, providing opportunities to "custom design" packages cost-effectively. In addition to a variety of ply materials (films and foil), developments in coatings provide additional opportunities for the packager. Polyethylene was the first coating used extensively with kraft, achieving rapid popularity because of the excellent water-vapor and moisture resistance it provided. A number of other coatings are currently employed, including polypropylene, PVC, and PVDC.

Markets

There are five broad markets for paper bags, listed here in order of size:

1. Agriculture and food—This includes a broad range of livestock and poultry feeds, many bakery products such as flour, dough improver, and bakery mixes, dry-milk products, seeds, and pet food. In sizes under 25 lb (11.3 kg), countless consumer items are packaged in small paper bags, including the traditional commodity-type products such as sugar, rice, flour, and grits, as well as a growing number of specialty items such as coffee, cookies, and pet food.

2. Building materials—Cements and mortar mixes, insulation, and numerous other building materials fall into this category.

3. Chemicals—These include fertilizer, plastic resins, salt, and water softeners.

4. Minerals—These include clays, limestone, and other mineral products.

5. Absorbents and miscellaneous products—Cat litter and other absorbents as well as various products not fitting into the other categories fall into this class.

History of the Multiwall Industry

Until the middle of the 19th century, most packaging of goods for transport was done in barrels or crates. By 1850, cotton and burlap textile bags had begun to replace barrels, and textile bags dominated packaging into the 20th century. The scarcity of cotton during the Civil War spurred the development of single-wall open-mouth bags made of manila rope fiber. It is not conclusively known when the first machine-made paper bags were introduced into the United States, but the first patent for a machine to manufacture single-wall paper bags was granted in 1852. Around the turn of the century, a cotton textile bag with a valve in one corner was developed for packaging salt; it was later used for sugar, rice and other grain products, and cement. Multiwall paper sacks were in use in Norway before 1920. Made of a tube of four or five walls of kraft paper, the bags were tied at the bottom, then filled and tied at the top. World War I shortened the supply of jute and cotton, and thus kraft paper began to make headway as a substitute. Another scarcity after World War I was manila rope. It became so scarce that manufacturers of manila rope began mixing kraft pulp with the manila fiber. For flexibility, two plies of the manila–kraft combination paper were used as a substitute for the old single-wall package.

After 1925, the sewn multiwall bag was introduced all over the world. The stepped-end bag was originally developed in 1928 in Europe, but its use in the United States did not grow until after World War II. Multiwall technology began to advance significantly in the late 1950s, and today the multiwall-bag industry in the United States produces ca 3.5 billion (10^9) multiwall bags per year.

Although the market for paper bags has leveled off, paper bags retain a very strong position as a favored packaging container. As innovations in paper packaging continue, the paper bag may make inroads in markets now dominated by other types of packaging.

Bemis Company, Inc.

BAGS, PLASTIC

Plastic bags, available in virtually all shapes, sizes, colors, and configurations, have replaced paper in most light-duty packaging applications. Paper has been more difficult to replace in heavy-duty applications (see Bags, paper; Bags, heavy-duty plastic). Light-duty plastic bags are generally described in one of two ways: by the sealing method or by application. This article explains the methods used to produce plastic bags and defines the various types of plastic bags in terms of their intended use. A plastic bag is defined here as a bag manufactured from extensible film (see Films, plastic) by heat-sealing one or more edges and produced in quantity for use in some type of packaging application.

Methods of Manufacture

By definition, all plastic bags are produced by sealing one or more edges of the extensible film together. The procedure by which this heat-sealing occurs (see Sealing, heat) is typically used to identify and categorize types of plastic bags. There are three basic sealing methods in use today: sideweld, bottom-seal, and twin-seal.

Sideweld seal. A sideweld seal is made with a heated round-edged sealing knife or blade which cuts, severs, and seals two layers of film when the knife is depressed through the film material and into a soft rubber back-up roller. The

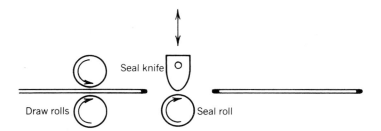

Figure 1. Sideweld-seal mechanism.

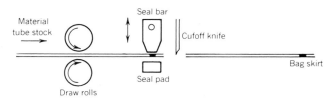

Figure 3. Bottom-seal mechanism (top heat only).

materials are fused by a combination of pressure and heat (see Fig. 1).

The sideweld seal is the most common bag-sealing method. Typical high usage bags, eg, bread bags and sandwich bags, are produced using the sideweld technique. The term sideweld is derived from the fact that many of the bags produced in this fashion pass through the bag machine with the length (or depth) of the finished bag perpendicular to the machine floor. The film fed into the machine is either prefolded, ie, J stock, or folded during the in-feed process (see Fig. 2).

Bottom seal.   The bottom-seal technique seals the bag at the bottom only. Tube stock is fed into a bag machine, the single seal is produced at the bottom of the bag, and the bag is cut off with a knife action that is separate from the sealing action. A bottom seal is generally made by a flat, heated sealing bar which presses the layers of film to be sealed against a Teflon (Du Pont)-covered rubber pad, ie, seal pad (see Fig. 3), or another hot-seal bar (see Fig. 4). A separate cutoff knife is used to separate the bag from the feedstock while the seal is made or immediately thereafter. Both bottom-seal mechanism designs produce a bag with only one seal, unless the tube has been manufactured by slit-sealing (see below). The small amount of unusable, wasted film between the edge of the seal and the cutoff point, called the skirt of the bag, is an important factor in the total cost of the bag. Any disadvantages caused by the presence of the unwanted skirt are usually offset by greater control of the sealing process. The sideweld method actually melts through the plastic film, and overheating of the film resins can change the physical structure of the plastic molecules when the seal cools. In contrast, the bottom-seal method controls the amount of heat and the dwell time, ie, the time that the heat is applied to the film, to produce a seal that does not destroy the film or change its physical properties. In addition, the total amount of film sealed together is usually

larger since the seal bar has a fixed width and none of the film material is melted or burned away.

The bottom-seal method is commonly used to produce HDPE merchandise bags and LDPE industrial liners, trash bags, vegetable and fruit bags, and many other types of bags supplied on a roll. In contrast to the sideweld method, designed primarily for high speed production of bags made from relatively light-gauge films ie, 0.5–2.0 mil (13–51 μm), bottom-seal methods are often used to produce bags from film from 0.5 to ≥6 mil (13–≥152 μm) at slower production speeds. Bags manufactured by bottom-seal methods are delivered through and out of the bag machine with the length (or depth) of the bag parallel to the machine direction. Since all of the bags are produced from tube sock, multiple-lane production of bags is limited only by the widths of the machine and the bags and the film-handling capability of the bag machine (see Fig. 5).

Twin seal.   The twin-seal method employs a dual bottom-seal mechanism with a heated or unheated cutoff knife located between the two seal heads (see Fig. 6). The unique feature of the twin-seal mechanism is that it supplies heat to both the top and bottom of the film material and makes two completely separate and independent seals each time the seal head cycles. Like the bottom-seal method, the twin-seal technique can supply a large amount of controlled heat for a given duration. This makes the twin-seal useful in sealing heavier-gauge films as well as coextrusions and laminates. Since two seals are made with each machine cycle, the twin-seal method can be used to make bags with the seals on the sides of the finished bag, ie, like sideweld bags, or on the bottom of the bag in some special applications, such as retail bags with handles. Many special applications call for the use of a twin-seal-type sealing method, but it is most often used in production of the plastic "T-shirt" grocery sack.

Slit seal.   Another type of sealing method, the slit seal, involves sealing two or more layers (usually only two) of film together in the machine direction through the use of a heated

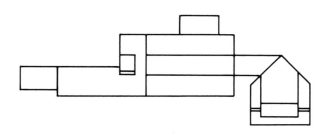

Figure 2. Sideweld process (top view).

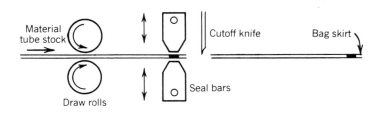

Figure 4. Bottom-seal mechanism (top and bottom heat).

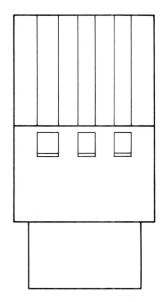

Figure 5. Bottom-deal bag machine, 3-level.

Figure 7. Slit sealer (hot-knife type). (**a**) side view; (**b**) top view.

variety of special bags that have been created by bag producers to meet commercial and consumer needs. These bags are usually described as specialty bags, but they are all variations of the standard bags described above. Specialty bags include rigid-handle shopping bags, sine-wave handle bags, pull-string bags, patch-handle bags, double-rolled bags, square-bottom bags, round-bottom bags, deli bags, etc. Examples of both common and specialty bags are shown in Figure 8.

knife, hot air, laser beam, or a combination of methods. The slit-seal technique is usually used to convert a single large tube of film into smaller tubes. In the production of grocery sacks, for example, a single extruded 60-in. (152-cm) lay-flat tube of film (see Extrusion) can be run through two slit sealers in line with the bag machine. This results in three tubes of 20-in. (51-cm) lay-flat material being fed to the bag-making system (see Fig. 7).

Applications

Most plastic bags are characterized in terms of their intended use, eg, sandwich bags, primal-meat bags, grocery sacks, handle bags, bread bags, etc. To provide some order to this user-based classification and definition system, it is convenient to separate the bags into commercial bags and consumer bags.

Commercial bags. A commercial plastic bag is used as a packaging medium for another product, eg, bread. Typical commercial bags and seal methods are listed in Table 1.

Consumer bags. Consumer bags are purchased and used by the consumer, eg, sandwich bags, trash bags. The plastic bag is the product. Typical consumer bags and seal methods are listed in Table 2.

In addition to the conventional bags, there is an enormous

Table 1. Commercial Plastic Bags

Sealing method	Types of bag
sideweld	bread bags, shirt and millinery bags, ice bags, potato and apple bags, hardware bags
bottom seal	vegetable bags on a roll, dry-cleaning bags, coleslaw bags, merchandise bags
twin seal	primal-meat bags, grocery sacks[a]

[a] With dual bottom seal.

Table 2. Consumer Plastic Bags

Typical sealing method	Types of bag
sideweld	sandwich bags, storage bags
sideweld or bottom seal	trash bags, freezer bags, can liners
bottom seal	industrial liners

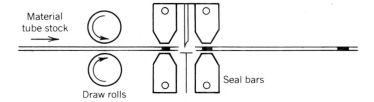

Figure 6. Twin-seal mechanism.

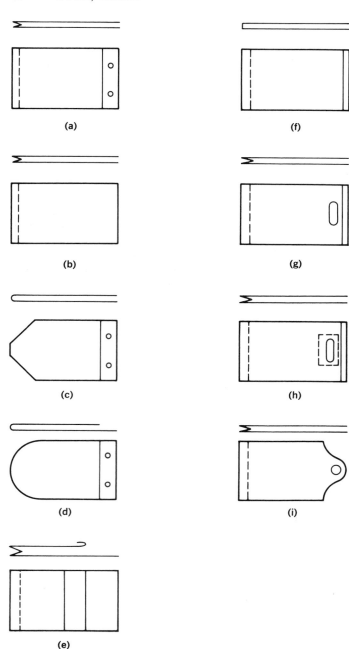

Figure 8. Bags, plastic, (**a**) conventional wicket bag; (**b**) conventional bag; (**c**) square-bottom bag; (**d**) round-bottom bag; (**e**) sandwich bag; (**f**) trash bag; (**g**) handle bag; (**h**) patch-handle bag; and (**i**) sine-wave bag.

BIBLIOGRAPHY

General References

"Bag Making: Inline or Off-Line," *Plast. Technol.* **30**(2), 55 (Feb. 1984).

L. R. Whittington, *Whittington's Directory of Plastics,* 2nd ed., Technomic Publishing Co., Inc., Lancaster, Pa., 1978.

Polyethylene Film Extrusion, U.S. Industrial Chemicals Co., New York, 1960.

L. L. Claton
FMC Corporation

BANDS, SHRINK

Most shrink bands are made of oriented plastic films that shrink around a container when heat is applied. They are used as labels, tamper-evident neckbands, combination label/neckbands, and devices for promotional packaging (see Fig. 1). These are called "dry" bands, in contrast to cellulose bands that are applied wet and shrink to a tight fit as they dry. Dry bands are much more versatile than wet bands, and they have become the predominant form by a wide margin.

Films

Most dry bands are made of PVC or vinyl copolymers (see Film, PVC; Poly(vinyl chloride)). They can be made of flat film that is stretched (oriented) and seamed, or from tubular film that is stretched in the blowing process (see Extrusion). The degree to which the band shrinks is determined when the film is stretched. To obtain a band that will eventually shrink 60%, the film must be stretched 60% (eg, a film that is 100 mm wide is stretched to 160 mm). In effect, stretching the film programs memory into the material (see Films, shrinkable). That memory is recalled by applying heat after the band is placed on the container, and the band shrinks to its original dimensions. The degree of shrink required is determined by the shape of the container and the amount of container coverage required; for example, to label a tapered container, with a wider radius at the bottom than at the top, bands with relatively high shrink capability accommodate the drastic difference in radius. Depending on the grade of film used, the material can yield a controlled accurate shrink of 65% or more with current technology.

The most common film thicknesses for most packaging applications are 1.5, 2, and 3 mil (38, 51, and 76 μm). Uniaxially oriented (preferential) film shrinks in height or width; biax-

Figure 1. An assortment of containers with shrink bands.

Table 1. Examples of Shrink-Band Applications

Product	Capacity oz (mL)	Line speed, bottles/min	Film gauge, mil (μm)	Shrink, %	Film size	
					Start	Finish
mouthwash, neck band only	12 (355)	150	3 (76)	50	2.25 in. (57 mm) flat width (fold to fold)	1.38 in. (35-mm) dia
ketchup, combination neck band and label	16 (473)	250	1.5 (38)	60	4.65 in. (118 mm) flat width (fold to fold)	neck: 1.57-in. (40-mm) dia, label: 2.76-in. (70-mm) dia

ially oriented film shrinks in both directions. Uniaxial orientation is preferred for printed bands because they do not wrinkle and graphics are not distorted.

Application

Shrink bands are sized just large enough to allow them to be placed over the length of the container (labels) or over the cap and neck (neck seals), either manually, or by automated high speed applicating machinery. Once applied, the container and band are exposed to a heat source (usually 285–320°F, or 140–160°C) long enough (usually 2–3 s) to activate the band's memory and cause it to shrink tightly and smoothly around the container or cap. Different time/temperature cycles are required for different types of film. If the container has relatively low heat resistance (eg, LDPE or PVC), a film is selected that shrinks at a relatively low temperature. Films with low temperature shrink properties are also used in packaging alcohol, paints and lacquers, and heat-sensitive food products. Automatic band-applicating machines are available in two basic categories: constant motion (containers and applicating devices are in continuous motion), and intermittent motion (containers are stopped while bands are applied). Some machines can apply up to 400 bands per minute depending on the size of the container and band.

Examples. Two examples of shrink-band applications are shown in Table 1.

Labels

Designed as a label, dry shrink bands offer many advantages. They can be used on plastic, glass, metal, or paperboard containers. A single shrink band offers 360° total surface coverage, including the shoulder, if desired. They are suitable for full-color printing (see Printing), and the natural shiny appearance of the film adds a high gloss look to printed graphics. Transparent flat film is generally printed on the inside surface before forming into bands. On the container the printing is protected from abrasion. Unidirectional orientation prevents distortion of the graphics. Printing may be done by flexography or rotogravure, depending on the number of colors and quantities.

The fact that shrink labels can be removed completely is an advantage for some products (eg, air fresheners, decorative containers). They leave no residue because they cling, rather than stick, to the container. Because shrink bands conform smoothly to most container shapes, they eliminate the need to redesign an existing container in order to introduce a new

product to the line or give a new look to an old product. If a multiproduct line uses a generic container, shrink bands eliminate the need to inventory multiple preprinted containers for each product. Because dry heat-shrinkable bands withstand temperature extremes ranging from about −20 to 200°F (−29 to 93°C), they can retain their appearance and remain on the container after chilling or heating.

Neck Bands

Neck bands provide tamper-evident (TE) seals on pharmaceutical, health-care, and food products (see Tamper-evident packaging). The inherent properties of properly made, properly applied shrink bands (see Fig. 2) offer a number of advantages.

Unlike some other TE devices (eg, inner seals), the band is readily visible to the inspector on the packaging line and to the consumer. If an attempt is made to tamper with a container that has a perforated band, the band splits apart along the perforation. If the band is solid, without perforations or easy-open tabs, the band must be cut or torn to be removed. For TE applications, 3-mil (76-μm) film is normally used, which is less extensible (more brittle) than the more common 1.5-mil (38-μm) film. Most important, however, is that once broken, a dry shrink seal cannot be reexpanded or properly reapplied to a container. Additional heat will shrink the band further. Wet bands can be reexpanded with water.

Although shrink bands lend themselves to most container shapes, not all container/cap arrangements are well suited for optimum TE effectiveness. Containers with a neck ring (bead) make the most effective use of shrink bands as TE devices. The neck bead provides the tension needed to break the band when the cap is twisted or opened. TE shrink bands can be made with a variety of tear-strip styles to facilitate their removal from a container. They provide reliable tamper evidence without making the package difficult to open. For containers where

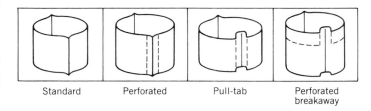

Standard   Perforated   Pull-tab   Perforated breakaway

Figure 2. Some types of tamper-evident shrink bands.

Figure 3.   Shrink band on a promotional package.

the positioning of a TE shrink band may be difficult, preformed bands are available. Preforms are formed around a mold and have an overlapping piece at the top. When the band is applied over the container cap, the overlap catches at the top of the cap, preventing the band from being placed any lower on the container. This assures accurate band placement. Shrink bands are also used as a combination label and TE seal. They are perforated at the line where the cap joins the container so that when the cap is removed, the perforation breaks and the label remains on the container.

Promotional Packaging

With shrink bands, a manufacturer can group two or more individual packages together to form a single unit, as in a "two-for-one" or multipack offer. A premium or sample can be attached to the package as a purchase incentive (see Fig. 3). One of the major advantages of transparent shrink bands for these and other promotional packages is that the identity of the product and its brand is not obscured.

BIBLIOGRAPHY

General Reference

Packaging **28**(6), 87 (1983). Special tamper-evident packaging supplement.

STEPHEN F. KRAVITZ
Sydney Schreiber Inc.

BAR CODE

A bar code is a series of black-and-white bars and spaces that represent a series of characters or symbols. Its purpose is to code information in a form that is easy to read by machine. Bar codes are read by sweeping a small spot of light across the printed bar-code symbol. Because a bar code cannot be read if the sweep wanders outside the symbol area, bar heights are chosen to make it easy to keep the sweep within the bar-code area.

The three advantages of bar coding over manual data collection are (*1*) *speed:* data are entered into the computer more rapidly; (*2*) *accuracy:* bar-code systems are almost error free. They do not rely on people to type correctly; and (*3*) *reliability:* bar-code formats are designed with various forms of error checking built into the code.

In manual data collection, errors can occur when the information is written or when the data are entered. A key-entry operator commonly makes one error for every 300 characters typed. Bar-coded information can be sent to a computer as the data are being read. Because the data are read quickly, fewer people are required to do the data collection, and the timeliness of the data often allows managers to make money-saving decisions. The investment in a bar-code system is often paid back in less than a year.

The Structure of Bar Code

Many different bar-code formats have been developed, each designed to meet specific requirements. Bar-code formats can be either continuous or discrete. Discrete bar-code formats store data as individual sets of bars and spaces that correspond to individual characters. Continuous codes store data as one set of bars. Some of the most popular formats are described below. For information about other codes, see references 1 and 2.

Code 39.   Code 39, developed in 1975 by David Allais and Ray Stevens of Interface Mechanisms (now Intermec), is the code used most often for industrial applications (1). It has been selected as the official Department of Defense bar code format and is now the standard format of the United States government (3). The Department of Defense now requires its contractors to mark all items with Code 39 bar code, which identifies the product by National Stock Number and government contract number. Code 39 is also used by the automotive and health-care industries. The name "Code 39" relates to the fact that three of the nine bars are always wide. The complete character set includes 43 data characters: 10 digits, 26 letters, space, and the 6 symbols −, ., $, /, +, and %. Code 39 is a discrete code and can be read in either direction.

Universal Product Code (UPC).   In 1970, a grocery industry *ad hoc* committee was formed to select a standard code for product identification. The UPC symbol was adopted on April 3, 1973. Several additions and enhancements were added to the format (4). The European Article Numbering (EAN) code format, similar to UPC, was adopted in 1976 (5).

UPC symbols have 12 characters; EAN symbols have 13. The first character of a UPC symbol is a number related to the type of product (0 for groceries, 3 for drugs, etc); in an EAN symbol, the first two characters designate the country of origin of the product. UPC is a subset of the more general EAN code. Scanners equipped to read EAN symbols can read UPC sym-

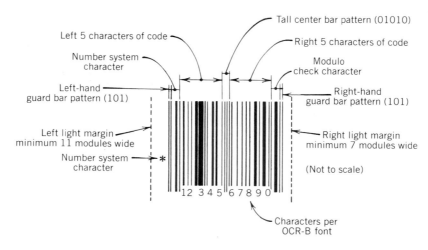

Figure 1. UPC Version A symbol (2). Courtesy of North American Technology, Inc.

bols as well, but UPC scanners do not always read EAN symbols.

The UPC format was developed for the automatic identification of grocery products, but UPC-formatted bar codes are also used on record jackets, liquor bottles, and many other non food items. UPC can be printed on packages using a variety of commercial printing processes. The format allows the symbol to be scanned with any package orientation. If a standard-size UPC symbol is too large for the product, the problem can be solved by truncating or reducing the overall height of the symbol; but this reduces the symbol's ability to be scanned in any orientation (see Figs. 1 and 2).

Two of Five Code. The Two of Five Code was developed by Gerry Woolf of Identicon Corporation in 1968. It is used primarily for warehouse inventory handling, photofinishing, airline ticketing, baggage handling, and cargo handling. The Two of Five Code format is a discrete code since the white spaces between the characters are not part of the code. Because the white spaces carry no information, their dimensions are not critical (see Fig. 3).

In 1972, David Allais of Intermec proposed a modification of the Two of Five format to increase the density of the data stored. The resulting format, named the Interleaved Two of Five Code, has been widely accepted in warehouses and heavy industries, particularly the automotobile industry (1). It is used for numeric labeling of corrugated shipping containers and to mark shipping boxes in the grocery industry. In the Interleaved Two of Five Code, both the bars and spaces carry information. The odd-positioned digits are coded in the bars and the even-positioned digits are coded in the spaces. The format requires the coded symbols to have an even number of

digits. A leading zero is often added to those with an odd number of digits. It does not contain alpha or symbol characters.

Codabar. The Codabar format (1) was developed by Monarch Marking Systems, Division of Pitney Bowes, for use in retail price labeling (see Fig. 4). After UPC was chosen for grocery marking and OCR-A format (6) was chosen by the National Retail Merchants Association for department store product labeling, Codabar was proposed as the format for a wide variety of nonretail applications. It is widely used by libraries and in the health-care industry (eg, for blood bags). It has also been used to identify photofinishing envelopes. Codabar is a variable-length, discrete, and self-checking code. The character set consists of all the numbers and the symbols −, $, :, /, ., and +.

Plessey Code. The Plessey Code (variants are called MSI Code, Telxon Code, and Anker Code) was developed in 1975 by Plessey Company Limited of Dorset, United Kingdom. It is widely used for shelf markings in grocery stores.

Bar-Code Printing

The print quality of a bar-code label is critical to the success of a bar-code system, but it is a factor that is often neglected. Poor print quality results in bar codes that are difficult to read, have reduced data accuracy, and have unhappy users. There are as many ways to print bar code as there are for printing text. The processes for bar-code printing, however, fall into two categories: commercial printing processes, and on-site printing processes.

Commercial printing processes. Commercial printing (off-site printing) has the advantage of low cost, high volume label production. Commercial printing can replicate the same bar-code symbol over and over again, but its flexibility for changing the information carried by the bar code symbol is limited.

Most commercial bar-code printing processes rely on photographic methods to produce an accurate imprint of the bar-code symbol on a printing plate. Making a printing plate for bar-code printing requires a very accurate film master of the bar-code symbol. The film-master manufacturer uses special photocomposing equipment to produce this accurate master. The printed bar-code symbol varies significantly in dimension from the bar-code symbol of the film master. Most of this varia-

Figure 2. Truncated UPC symbol (2). Courtesy of North American Technology, Inc.

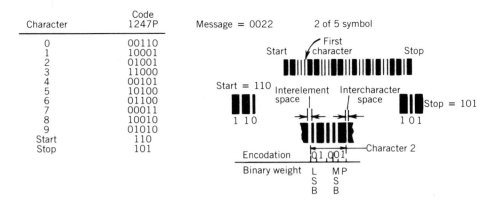

Character	Code 1247P
0	00110
1	10001
2	01001
3	11000
4	00101
5	10100
6	01100
7	00011
8	10010
9	01010
Start	110
Stop	101

Figure 3. 2 of 5 coding conventions (2). Courtesy of North American Technology, Inc.

tion is a result of errors introduced by the printing process used to print the bar-code symbol. The film master manufacturer compensates for this expected variation by reducing or increasing the bar width in the symbol master.

Any conventional printing process can be used to print bar-code symbols directly on product packaging along with the graphics and text. Printing processes that can be used for bar-code printing include flexography, rotogravure, letterpress, letterset and screen printing (see Printing). Direct off-site printing has the added advantage of eliminating the cost of labeling equipment, maintaining a label material inventory, and staff time spent applying the labels. During a press run, several labels can be printed at different locations on the package. Some experienced printers have developed expertise in producing a label with little printing error.

Off-site printing of discrete bar-code labels is another alternative. Off-site label vendors can supply bar-code labels with a variety of information on a multitude of label stocks; for example, rolls of sequentially numbered labels printed on vinyl-label stock (see Labels and labeling). The use of off-site label printing eliminates the need for on-site label printing equipment, and it provides bar-code printing expertise and accountability for print quality. Commercially printed bar-code labels offer distinct advantages over label printing on site. These advantages include direct bar-code printing, high volume production, printing expertise, intermixed graphics, responsibility, quality, and flexibility of symbol size.

Commercial printers can print bar codes directly on containers and cans thus eliminating the cost of printing and applying individual labels. When bar-code symbols are printed as part of a press run, the symbol can be accurately positioned on the container. Individually printed labels cannot be so positioned.

On-site printing processes. Commercial printing techniques are satisfactory if the information to be coded is known in advance. When a food manufacturer packages a product, the identification number for the product is known. Since many thousands of the same product are manufactured, a printing process that uses a film master, printing plates, and high speed printing presses is cost effective. However, it is often impossible to predefine the codes that will be needed. Even if the codes can be predefined, it may be too costly to maintain an inventory of preprinted labels. On-site label printing is the system of choice when the nature of the symbol's data prevents

ordering in advance. Many applications require coding random information such as lot number, weight, shift, or operator identification. Such information cannot usually be predicted and requires bar-code labels that can be printed on site on demand. The major on-site label printing methods are dot-matrix printing, electrostatic and xerographic printing, formed-character printing, ink-jet printing, laser printing, rotary-encoder printing, and thermal printing (see Code marking and imprinting).

Bar-Code Scanners

Bar-code labels are read by bar-code scanners. The familiar scanners used in many grocery stores use a moving light beam from a helium-neon laser to read UPC symbols on packages (7). The UPC symbol is read as the package is drawn across a glass plate in the checkout counter. The value decoded from the bar code symbol, which is the item's product number, is transmitted to a computer in the store. The product number is compared to a list of product numbers stored in the computer. When a match is found, a description of the product, its price, and taxability is transmitted back to the checkout terminal for processing.

The four basic scanner types are hand-held contact scanners, stationary fixed-beam scanners, stationary moving-beam scanners, and video-image scanners. Scanners work by illuminating the bar-code symbol with a light source. The source can be a lamp, LED, laser, or any other light source. The reflected light from the symbol is received by a light detector, which

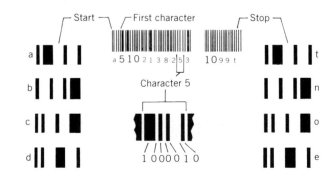

Figure 4. Sample Codabar symbol (2). Courtesy of North American Technology, Inc.

converts the light to an electrical signal. As the light moves across the bar-code symbol, the relative intensity of reflected light changes and this causes the electrical signal to change. The analogue signal from the light detector is decoded by measuring the time duration of each oscillation of the signal. The time measurements are processed by a programmed microprocessor, which extracts the information coded in the bar-code symbol. All four scanner types follow the same basic steps in decoding bar-code symbols, but they differ in the way they optically scan the bar-code symbol.

Hand-held contact scanners. The most widely used bar-code reading device is the hand-held contact scanner. It is the most popular because of its portability and low cost. A hand-held contact scanner requires an operator to find the bar-code symbol and to move the scanner smoothly over the symbol. The operator must sweep the scanner tip over the entire symbol at a relatively constant speed. Without motors or moving parts, hand-held scanners are inexpensive, small, and rugged. A hand-held scanner consists of a pen-shaped housing, a light source, and a photodetector. The light is projected out from the scanner's tip. The reflected light from the bars and spaces reenters the tip and is received by the photodetector. The detector converts the reflected light in to an analogue signal, which is decoded and passed to a computer. Hand-held scanners require a smooth scanning motion, but the speed can be 3–50 in./s (8–127 cm/s).

Stationary fixed-beam scanners. Stationary fixed-beam scanners are closely related to contact scanners, but they use a stationary beam of light to read a moving bar-code symbol. Fixed-beam scanners have no moving parts; they simply illuminate a small area with a spot of light. When a bar-code symbol passes through the beam, the scanner detects the reflected light and reads the bar code. As with a contact scanner, the light is converted to an analogue signal, digitized, decoded, and transmitted to the computer. Fixed-beam scanners use broad-beam illumination. A fairly large area of the bar-code symbol is illuminated by the reader's light source, but only a small portion of the symbol is focused onto the photodetector. The principal advantage of stationary fixed-beam scanners is that they are less expensive than other systems using moving beams. The principal disadvantage is the closeness required between the scanner and the symbol. This distance must not vary greatly from symbol to symbol. Fixed-beam scanners are used for in-line verification of printed bar codes, envelope sorting, badge reading, test-tube tracking, and package identification. They can be used in any application where the orientation and location of the bar-code symbol can be closely controlled.

Moving-beam fixed scanners. Moving-beam fixed scanners use a moving beam of laser light to read the bar-code symbol. Industrial versions of moving-beam fixed scanners move the light back and forth in a straight line; supermarket versions move the light in a cross-hatch or star-burst pattern. The cross-hatch or star-burst pattern allows the UPC label to be read in any orientation as long as the label faces the scanner. Compared to other bar code scanners, moving-beam fixed scanners are expensive. Most are designed only to read UPC-formatted bar code, since they are used primarily for retail checkout in grocery and drug stores. Some newer moving-beam fixed scanners provide omnidirectional scanning, and they can also scan symbols that do not directly face the scanner. These systems use holography to produce a three-dimensional scanning pattern that permits the scanner to "see" a label on the side of a package. The holography scanner is quite expensive.

Video-image scanners. Most bar-code scanning systems use a light source and a single photodetector. The operator, symbol, or light source supplies the scanning movement. Video-image scanners, on the other hand, operate like television cameras. The symbol is illuminated by either a photoflash or photoflood lamp and the reflected symbol image is focused onto a linear photodiode array. These arrays are made up of many tiny photodetectors. Each photodetector is periodically sampled by a microprocessor to produce a video signal of the symbol image. The video signal is conditioned and then decoded. Video-image scanners are available for either fixed or hand-held applications. Because of their limited field-of-view, they have been used in applications with fixed-length bar-code formats (eg, UPC). These image scanners are often used in small retail stores as a less costly alternative to moving-beam laser scanners.

BIBLIOGRAPHY

1. Uniform Symbol Descriptions for Code 39, Interleaved 2 of 5, Codabar, and other codes (Code 128, Code 93, and Code 11) available from Automatic Identification Manufacturers, 1326 Freeport Rd. Pittsburgh, Pa. 15238.

2. R. Adams and C. K. Harmon, *Reading Between the Lines: An Introduction to Bar Code Technology,* North American Technology, Inc., Peterborough, N.H., 1984.

3. MIL-STD-1189, available from Naval Publications and Forms Center, 5801 Tabor Avenue, Philadelphia, Pa. 19120.

4. UPC-symbol specifications, guidelines manual, symbol location guidelines and shipping container symbol-specification manual, available from Uniform Product Code Council, Inc., 7051 Corporate Way, Suite 201, Dayton, Ohio 45459-4294.

5. EAN (European Article Numbering) information, available from International Article Numbering Association E.A.N., Rue Des Colonies, 54, Kolonienstraat, Brussels, 1000 Belgium.

6. OCR-A—an abbreviation commonly applied to the character set contained in *ANSI standard X3.49-1975,* available from American National Standards Institute, Inc., 1430 Broadway, New York, N.Y. 10018.

7. "Automated Food Store Checkout: A History of its Early Development," *Bar Code News,* (Sept./Oct. 1983).

Russ Adams
Bar Code News

BARRELS

A barrel, or cask, is a cylindrical vessel of wood that is flat at the bottom and top, with a slightly bulging middle. The three primary parts of a barrel are heads (bottom and top), staves (sides), and hoops (rings that bind the heads and staves together) (see Fig. 1). Specifications are contained in the Department of Transportation regulations (1).

In architecture and physics, the arch is probably the strongest possible structure. The more pressure or weight exerted on the top (keystone) of the arch, the stronger the arch becomes.

The wooden barrel is designed according to the double-arch principle of strength. Like an egg shell, it is doubly arched, both in length and girth. The bend in the stave's length is the first arch and the bilge circumference of the stave's width is the second arch. These arches impart great strength.

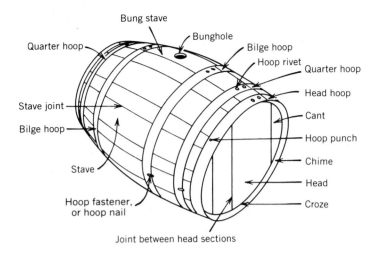

Figure 1. The wooden barrel.

Table 1. Wooden Casks for International Shipments of Alcoholic Beverages

Beverage	Name	Gallons[a]	Liters[a]
Wine			
sherry	butt	137.5/140	500
	hogshead	67.5/69	250
port	pipe	145/147.5	53
	hogshead	72.5/74	265
vermouth	hogshead	67.5/70	250
burgundy	hogshead	57.5/60	215
Spirits			
rum	puncheon	137.5/144	520
	hogshead	67.5/72.5	255
	barrel	50	180
brandy	hogshead	77.5/79	280
	quarter case	39/40	140
beer	butt	135	500
	hogshead	67.5	245
	barrel	45	165
	kilderkin	22.5	82
	firkin	11	41
	pin	6	20

[a] Gal (U.S.) and L sizes are not equivalent.

There are three basic cooperage operations: logging the timber; milling the logs; and assembling the barrel staves and heading material. Saws reduce the logs to length, and produce edge-grained pieces of cylindrically shaped and jointed wood for the staves, and flat pieces of wood for the heading. In recent times, staves have been quarter sawn as opposed to earlier cylindrical sawn staves. The quarter-sawn straight staves are planed interiorly and exteriorly throughout their thickness to achieve a stave of cylindrical width. After the wooden material has been air- and/or kiln-dried to approximately 12% moisture content, the staves and heads are assembled into steel-hoop-bound barrels. Assembly operations include: setting up staves; steaming and winching staves to achieve the belly, bilge, or circumference arch; heating to make wood pliable and give one last drying after being steam bent; tapping out for uniform thickness; trussing to tighten stave joints; crozing interior grooves in each end of the staves where the heads will be inserted; heading up by inserting heads in the croze at each end of the staves; hooping up by driving riveted-steel hoops onto exterior of staves; boring for testing, lining, and future filling; bunging up the bored hole; and rolling the barrel out to the marketplace.

Dozens of species of timber from all over the world, have been used to make tight (for liquid) and slack (for nonliquid) cooperage. Hardwood barrels for spirits and wine include oak timber from Limousin and Nevers in France; Alastian and Italian oak; and fork-leafed American white oak, found principally in the slow growing forest regions of the United States of Missouri, Indiana, Tennessee, Kentucky, and Arkansas. Virtually all of the wooden barrels made in the United States today (1–2 million (10^6)/yr) are 50-gal (189-L) capacity barrels used by the bourbon whisky trade. Barrels for bourbon are charred interiorly about 1/16 in. (1.6 mm) of their 1 in. (25.4 mm) thickness to bring out the tannin in the wood. Tannin aids in the coloring and flavoring of spirits and wine.

Wooden barrels have had numerous names, depending on their size and use. A small sampling of these names include: pickled-pigs-feet kit; fish pail; one-quarter; one-half, and full beer ponies; hogshead; salmon tierce; tallow cask; rum puncheon; and port wine pipe. A list of common international cask sizes is presented in Table 1 (2).

Just as the wooden barrel replaced the crude basketry used centuries ago, many other types of containers have replaced the wooden barrel: steel and fiber drums, plastic pails, aluminum and steel cans, fiber glass and cement tanks, etc; aluminum and stainless steel replaced wood for beer barrels (3).

To date, no industrial engineer has come up with a blueprint to replace the strength of a wooden barrel. In tests involving high stacking, they can perform better than steel drums.

BIBLIOGRAPHY

1. *Code of Federal Regulations,* Title 49, Sect. 178.155–178.161.
2. F. A. Paine, *The Packaging Media,* Blackie & Son Ltd., Glasgow and London, John Wiley & Sons, Inc., New York, 1977.
3. H. M. Broderick, *Beer Packaging,* Master Brewers Association of the Americas, Madison, Wisc., 1982.

Frank J. Sweeney
Sweeney Cooperage Ltd.

BARRIER POLYMERS

The packaging of foods and beverages, specifically of "shelf-stable" products requiring extended storage at ambient conditions, is probably the most severe challenge to polymeric materials. Traditionally, glass and metal, with essentially no permeation, have been the only materials available to food processors for this class of packaging. But high barrier polymeric materials, with distinct advantages over glass and metal, are now serious contenders as alternative packaging materials. Some of the advantages over glass or metal, or both, are cost, light weight, shatter resistance, transparency, microwave reheat, convenience and consumer preference, design options, and (versus metal) direct food contact. Generally speaking, the chief disadvantages of polymers have been in the area of barrier properties and heat resistance at reasonable cost. This article deals with the barrier properties of polymers (permeation and absorption), and specifically with the class of polymers known as "barrier polymers." It also gives some guidelines as to the degree of barrier from the point of view of molecular structure (see also Polymer properties).

Transport Properties

As has been reviewed in the past (1, 2), there are many factors to be taken into account when assessing the barrier properties of a particular polymeric structure. Although direct permeability (ie, the actual mass transport of a gas or liquid through the polymer) is of utmost importance, there are at least two other important barrier-related mechanisms of importance. These are (1) the degree to which a polymeric film can "absorb" or "scalp" specific food or beverage flavors and (2) the amount of small molecule "residuals" that might be in the plastic which could migrate into the product via diffusion and cause a taste or odor change. This article discusses permeability and absorption in detail, since migration is a function of the type and level of residual, assuming that only pure polymers are used (or that residuals cause no taste or other problems). Together, these are the transport-related properties of interest in food and beverage packaging as well as in packaging non food products, such as solvents.

What, specifically, is defined as a barrier polymer? Obviously, any packaging material that is used successfully could be called a barrier of sorts (polyethylene used for detergents, etc.), but it has been generally accepted that the term barrier polymer be limited to those polymers which exhibit a high degree of *oxygen* barrier. Simply put, a barrier polymer is any polymer that protects the product over its intended shelf life (see Shelf life) when used alone (or in combination with non-barrier layers) and has an O_2 permeability of <10 cm$^3 \cdot$ mil/ (100 in.$^2 \cdot$ d $\cdot$ atm) [38.9 cm$^3 \cdot \mu$m/(m$^2 \cdot$ d $\cdot$ kPa)] at 23°C.

In most cases, a polymer showing an O_2 permeation of <10 will also be a good barrier to CO_2, as well as many organic vapors and odors. In addition, the absorption (scalping) mentioned above is low, and migration is negligible (provided the polymer is relatively pure). Water permeation, however, cannot be correlated to O_2 permeation directly.

Factors Influencing Transport Properties

Previous publications (3–6) have indicated some of the physical and chemical parameters which control the three transport mechanisms that make up a barrier. Some of these are

1. Effect of the actual chemical substituent on the polymer backbone (ie, the repeat units making up the polymer chain).
2. The degree of packing, crystallinity, and orientation of the molecular chains. Close packing results in low free volume and better barrier, as does high crystallinity. Orientation of the chains and crystallites results in better barrier by creating a more tortuous diffusion path.
3. The susceptibility to moisture and other possible interactants with the chains. Moisture can act as a "plasticizer" and thus reduce barrier by loosening the chains.
4. The nature of the polymer surface contacting the product (ie, hydrophilic vs hydrophobic, etc).

5. Additives (see Additives, plastics) used in manufacturing or modifying the polymer. These often reduce barrier by loosening the chains.

Chemical nature of polymers. This is probably the most important parameter determining ultimate barrier. Table 1 gives an example of various polymers, as well as their O_2 permeability, which is a direct result of the chemical structure. Since polymer chains must "move aside" or "open up" to allow permeation, the weaker the forces holding the chains together, the more rapidly permeation will occur. These chain-to-chain forces are determined in great part by the chemical nature of the backbone.

Degree of Packing, Crystallinity, and Orientation. As the polymeric chains cool from the molten state or coalesce from solution, they begin to pack, and the degree of packing depends upon the backbone structure. If they pack perfectly, they can form crystallites which are completely impervious to permeability. Thus, the greater the crystallinity, the better the barrier. Polymers showing little or no crystallinity are called amorphous. Polymers that are or can be crystalline are actually semicrystalline, combining both amorphous and crystalline areas. Table 2 gives some examples of what crystallinity can do for barrier. Note that not all highly crystalline polymers are good barriers, since their residual amorphous regions may be so "floppy" that gas still goes through, although with a more tortuous path. Also, not all amorphous polymers are poor barriers, since amorphous chains can be held together almost as strongly as crystallites. Orientation is helpful usually only in the context of crystallinity. If the chains are "stretched" during forming and cooling, one can actually induce a greater level of crystallinity, and, since the crystallites are "lined up" and more ordered, they cause a higher degree of tortuosity. This results in lower permeation.

Susceptibility to Moisture. Some chemical species are adversely affected by moisture absorption, and thus barrier is reduced. This is caused by water entering the polymer and interacting with the bonds that hold the chains together. Thus, the polymer swells and the chains become floppy and allow gas permeation to increase. Since water is highly polar it will only react in this way with polar polymer groups that depend upon hydrogen bonds (H bonds) for barrier. The most common polar groups are hydroxyl (—OH) and amide

$$(-\overset{\overset{\textstyle O}{\|}}{C}NH-).$$ Other polar groups, although they might absorb water, are not affected, since their barrier properties do not depend upon H bonds. Examples of unaffected groups are

ester $(-\overset{\overset{\textstyle O}{\|}}{C}O-)$ and cyano (—C≡N). Table 3 illustrates the effect of moisture and how it can turn a barrier polymer into a nonbarrier polymer. Also illustrated is the advantage of making a polymer less polar, and thus less affected by water. This is shown with the copolymer (ie, a polymer composed of two different starting monomers) of ethylene and vinyl alcohol (2a and 2b) versus pure poly(vinyl alcohol) (7) (see Ethylene–vinyl alcohol; Water-soluble polymers.). Water-sensitive polymers, when used as a barrier, are normally sandwiched between good water barriers (such as polyethylene) in order to protect them from the water in the product or in the air.

Table 1. The Influence of Chemical Nature on Permeability

Polymer	Structure	O_2 permeation[a]	Comments
polyethylene ($d = 0.92$)	$+CH_2CH_2\rightarrow_n$	480[1865]	low polarity; very little cohesion between chains, and they open easily to allow the gas through
polypropylene	$+CH_2CH\rightarrow_n$ | CH_3	150[583]	chains are a little stiffer, but still very little cohesion or attraction
poly(methyl methacrylate)	CH_3 | $+CH_2C\rightarrow_n$ | $C{=}O$ | OCH_3	17[66]	chains are much stiffer due to steric hindrances; polarity (ester group) results in chain-to-chain attraction superior to hydrocarbon polymers
poly(vinyl chloride)	$+CH_2CH\rightarrow_n$ | Cl	8.0[31]	chain-to-chain attraction very high due to chlorine electrons (fluorine also good); movement of chains restricted
poly(vinyl alcohol)	$+CH_2CH\rightarrow_n$ | OH	<0.01[<0.04]	the "ultimate" polymeric barrier due to actual H bonds between chains

[a] $cm^3 \cdot mil/(100\ in^2 \cdot d \cdot atm)$ [3.886 $cm^3 \cdot \mu m/(m^2 \cdot d \cdot kPa)$], 23°C, 0% rh.

Polymer Surface. In some cases, the nature of the surface determines permeability as much as the polymer bulk itself. An example is polyethylene. This polymer is very nonpolar, and thus is a very poor barrier to nonpolar liquids such as hydrocarbon liquids. By treating the surface with fluorine gas, however, the fluorine adds on to the chains at or near the surface, and by virtue of its bulk and polarity prevents hydrocarbon liquids from entering the polymer (see Surface modification). Thus the surface-treated polyethylene is converted into a barrier polymer towards hydrocarbon liquid (8). Oxygen, however, being much smaller, is not affected.

Table 2. Chain Order, Crystallinity, and Orientation

Polymer	Morphology	O_2 permeation[a]	Comments
1a. polyethylene, low density ($d = 0.92$)	50% crystalline	480[1865]	
1b. polyethylene, high density ($d = 0.96$)	75% crystalline	110[427]	higher crystallinity than 1a reduces the available sites for permeation
2a. poly(ethylene terephthalate)	10% crystalline	10[39]	
2b. poly(ethylene terephthalate)	50% crystalline	5.0[19]	higher crystallinity than 2a reduces the available sites for permeation;
2c. poly(ethylene terephthalate)	50% crystallized and oriented	3.0[12]	crystallinity and orientation produces more tortuous path for permeation
3a. polypropylene, atactic	low crystallinity	250[971]	CH_3 group (see Table 1) is in many different positions; packing is thus poor
3b. polypropylene, isotactic	higher crystallinity	150[583]	CH_3 groups are all in same position; packing is good, as is crystallinity
3c. polypropylene, isotactic	oriented	100[389]	orientation "lines up" crystallites; better barrier; more tortuous path

[a] $cm^3 \cdot mil/(100\ in^2 \cdot d \cdot atm)$ [3.886 $cm^3 \cdot \mu m/(m^2 \cdot d \cdot kPa)$], 23°C, 50% rh.

Table 3. Effect of Moisture

Polymer	Conditions	O_2 permeation[a]
1a. poly(vinyl alcohol)	0% rh	0.01[0.039]
1b. poly(vinyl alcohol)	95% rh	>25.0[>97]
2a. ethylene–vinyl alcohol, 70% VOH	0% rh	0.017[0.066]
2b. ethylene–vinyl alcohol, 70% VOH	95% rh	3.0[12]
3a. polycaprolactam (nylon-6)	0% rh	1.5[5.8]
3b. polycaprolactam (nylon-6)	95% rh	5.0[19]
4a. cellophane	0% rh	0.17[0.66]
4b. cellophane	50% rh	>5.0[>19]

[a] $cm^3 \cdot mil/(100 \ in^2 \cdot d \cdot atm)$ [3.886 $cm^3 \cdot \mu m/(m^2 \cdot d \cdot kPa)$], 23°C at specified rh.

Additives and Modifiers. Depending upon the chemical nature, as well as the manner in which the additive is blended, one can either enhance or reduce barrier. An example of the latter is pure (rigid) PVC (see Poly (vinyl chloride)), with an O_2 permeation of 8(SI = 31.1), versus plasticized (soft) PVC with a permeability of 150 (SI = 582.8). The latter contains a liquid plasticizing agent which softens the polymer, and thus allows the chains to be floppy and permit permeation at a high rate. On the other hand, a blend of nylon (a good gas barrier) (see Nylon) with polyethylene (a poor gas barrier) (see Polyethylene) results in a composition with gas barrier superior to polyethylene, provided the blend is made correctly so that the nylon forms discrete barrier sheets within the polyethylene (9) (see Surface modification).

Barrier Polymers

The number of polymers that are true barriers is quite limited. These are presented in Table 4, along with examples. Most barrier polymers contain at least one of the following structures:

Hydroxyl	(—OH)
Cyano	(—C≡N)
Halogen	(—Cl or —F)
Ester	(—$\overset{\overset{\displaystyle O}{\|}}{C}$O—)
Amide	(—$\overset{\overset{\displaystyle O}{\|}}{C}$NH—)

All of these structures create strong chain-to-chain forces thus restricting the movement of the chains and preventing permeation. Note that although all are good gas barriers, some show poor water barrier, which is due to polarity. The best water

Table 4. Barrier Polymers

Polymer	O_2 permeation[a]	Water barrier
1. poly(vinyl alcohol)[b]	<0.01[<0.04](0% rh)	poor
2. poly(acrylonitrile)[b]	0.04[0.16]	good
3. ethylene–vinyl alcohol, 70% VOH	0.017[0.066](0% rh)	poor
4. PVDC homopolymer[b]	0.10	excellent
5. cellophane[b]	0.17[0.66](0% rh)	poor
6. ethylene–vinyl alcohol, 60% VOH	0.17[0.66](0% rh)	poor
7. PVDC copolymer (90% VDC)	0.25[0.97]	excellent
8. PVDC copolymer (80% VDC)	0.50[1.9]	excellent
9. poly(acrylonitrile) copolymer[c,d], 70% ACN	1.0[3.9]	fair
10. poly(acrylonitrile) copolymer[e], 70% ACN	1.1[4.3]	fair
11. PVDC, plasticized	1.3[5.1]	excellent
12. polyamide (nylon-6)	1.5[5.8](0% rh)	poor
13. polyamide (nylon-6, 6)	2.5[9.7](0% rh)	poor
14. epoxy, thermoset (Bis A/Amine)[b]	3.0[12]	poor
15. poly(ethylene terephthalate) film	3.0[12]	good
16. poly(chlorotrifluoroethylene)	3.0[12]	excellent
17. polyamide copolymer	4.5[17]	poor
18. poly(vinylidene fluoride)	4.5[17]	excellent
19. poly(ethylene terephthalate) bottle	5.0[19]	good
20. polyamide (nylon-6, 10)	6.0[23]	fair
21. poly(vinyl chloride)[d]	8.0[31]	good
22. polybisphenol-epichlorohydrin	7.0[27]	fair
23. poly(ethylene terephthalate), amorphous	10.0[39]	fair
24. polyacetal	10.0[39](0% rh)	poor
25. poly(vinyl chloride) bottle[e]	12.0[47]	good

[a] $cm^3 \cdot mil/100 \ in^2 \cdot d \cdot atm$ [3.886 $cm^3 \cdot \mu m/(m^2 \cdot d \cdot kPa)$] at 23°C, 50% rh unless noted.
[b] Cannot be melt-extruded and/or must be applied from solution or emulsion.
[c] Polymer used to manufacture beverage containers (Monsanto). Not commercially available.
[d] Pure polymer, no modifiers.
[e] Contains an impact modifier (rubber).

Table 5. Permeability of Lower-Barrier Polymers

Polymer	O_2 permeation[a]	Water barrier
poly(vinyl fluoride)	15.0[58]	excellent
poly(methyl methacrylate)	17.0[66]	poor
poly(methyl methacrylate)[b]	25.0[971]	poor
polyamide (nylon-11)	26[100]	fair
poly(ethylene terephthalate) copolymer	26[100]	poor
polystyrene copolymer (25% acrylonitrile)	65[250]	poor
cellulose nitrate	100[389](0% rh)	poor
styrene/AN/rubber	100[3890]	poor
polyethylene ($d = 0.96$)	110[4270]	excellent
polyurethane, elastomer	135[5250](0% rh)	poor
polypropylene	150[583]	excellent
polysulfone	200[777]	poor
polycarbonate	225[874]	poor
polybutene	330[1280]	excellent
polyethylene/vinyl acetate	350[1360]	fair
polystyrene	420[1630]	poor
polyethylene ($d = 0.92$)	480[1870]	excellent
polytetrafluoroethylene	500[1940]	excellent
polyethylene ionomer	550[2140]	good
SBR rubber	1,500[5830]	good
polybutadiene	2,500[9710]	poor
polymethyl pentene	4,000[15,500]	poor
silicone elastomer	>90,000[>350,000]	poor

[a] $cm^3 \cdot mil/(100\ in^2 \cdot d \cdot atm)$ [$3.886\ cm^3 \cdot \mu m/(m^2 \cdot d \cdot kPa)$] 23°C, 50% rh unless noted.
[b] Rubber modified.

barriers are the polyolefins (polyethylene, polypropylene) and the halogens, such as PVDC (see Vinylidene chloride copolymers) and fluoropolymers (see Film, fluoropolymer).

Table 5 gives the permeability of some common nonbarrier polymers.

Factors Other than Direct Permeability

Table 4 shows the oxygen and water barrier of the barrier polymers. Thus, one can estimate how much O_2 will enter the package, how much CO_2 will be lost (CO_2 is generally 3–6 times higher than O_2), or whether moisture from the air is a problem. The other extremely important factor is the phenomenon known as dilute solution absorption. This is the ability of a polymer surface to pick up chemical entities from the solution (food, beverage, etc) that are there in dilute solution (flavors, top notes, additives, etc) and thus possibly affect the taste, odor, or efficacy of the product. Although most barrier polymers (by nature of their tight chains) are good dilute solution absorption barriers as well, absorption must be taken into account. In a very broad sense (the like-dissolves-like generality), a polymer with polarity close to that of the ingredient in solution could pick it up quite rapidly. As an example, a small amount of hexane (nonpolar) dissolved in alcohol is rapidly absorbed into polyethylene (nonpolar), whereas, a small amount of water is not. Likewise, a small amount of ketone dissolved in water is absorbed into PVC, whereas alcohol (highly polar) or hexane (nonpolar) are not absorbed.

The test used to evaluate this property is simply to dissolve the organic of interest at a low level (1000 ppm) in either water or a mixture of water and alcohol. The solution is then placed in contact with the polymer to be studied. This is conveniently done by placing the solution into a blow-molded container of the polymer, or a coextruded package with the polymer of interest contacting the liquid. If a container cannot be made directly of the plastic, strips can be immersed into the liquid. The package is then sealed and either stored for 30 d at 120°F (49°C) which, in general, is equivalent to a year at 73°F (23°C), or can be stored at any other convenient time and temperature (see Testing, permeation and leakage). The package is then opened and an analysis is made of the liquid for the organic in solution. The loss is then attributed to absorption into the polymer (provided a hermetic seal is assured). Table 6 gives the results (6) of such a test run of 15 different model compounds, ranging from very low polarity to high polarity, in blow-molded containers of polyethylene, poly(vinyl chloride) (PVC), poly(ethylene terephthalate) (PET) (see Polyesters, thermoplastic) and poly(acrylonitrile copolymer (see Nitrile polymers). The last three are barrier polymers (nos. 25, 19, and 10 of Table 4). As might be expected, the better the barrier, the lower the absorption of the organic. Had this test been run with a water-sensitive barrier, however, such as nylon or ethylene–vinyl alcohol (EVOH), the losses would be high owing to the swelling effect of water. None of the four polymers tested shows water sensitivity.

Barrier Coatings, Laminations and Coextrusions

Two or more layers (combining nonbarrier and barrier) have been used for film and sheet for many years. Typically, the barrier layer is a coating from emulsion or solvent (eg, PVDC latex) or applied during melt extrusion of the sheet or film (see Coextrusion, flat; Coextrusion, tubular). The most recent advances have been made in producing bottles and thermoformed containers with up to seven layers. Typically, the barrier layer is either PVDC resin or EVOH, which can be applied directly during melt extrusion of the package. Very often (depending upon the nonbarrier substrate) adhesives are needed to bind the layers (see Multilayer flexible packaging).

The barrier properties of a structure of two or more layers can be calculated using the series equation:

$$\frac{L_1}{P_1} + \frac{L_2}{P_2} + \frac{L_3}{P_3} + \ldots = \frac{L_T}{P_T}$$

where L_1, L_2, and L_3 are the thicknesses of the layers, and P_1, P_2, and P_3 are their permeation rates. L_T and P_T are the values for the whole package. For example, assume a 5-layer structure of PE/adhesive/EVOH/adhesive/PE (P/A/E/A/P). Since there are only three actual "layers" as far as permeation is concerned (assuming the same adhesive on both sides), this reduces to:

$$\frac{L_P}{P_P} + \frac{L_E}{P_E} + \frac{L_A}{P_A} = \frac{L_T}{P_T}$$

Assume that each PE layer is 15 mil (381 μM), each adhesive layer is 1 mil (25.4 μM), and the EVOH is 3 mil (76.2 μM). From Tables 4 and 5, $P_P = 110$ [427], $P_E = 0.017$ [0.066] (dry), and since most coextrudable adhesives are modified polyolefins, it can be assumed $P_A = 100$ [3890]. An error in the P of the nonbarrier layers has little influence on P_T:

$$\frac{30}{110[427]} + \frac{2}{100[389]} + \frac{3}{0.017[0.066]} = \frac{35}{P_T}$$

Table 6. Dilute Solution Absorption Testing, 30 days at 120°F (49°C)[a]

Test compound	Percent loss after conditioning			
	LDPE[b]	PVC[c]	PET[d]	ACN copolymer[e]
hexane	100	5	2	<1
octane	95	10	2	2
dipentene (limonene)	98	15	3	<1
citral	83	30	12	4
menthone	95	60	3	<1
benzene	85	65	10	3
chloroform	99	20	9	5
carvone	80	40	4	<1
anethole	100	75	25	<1
menthol	30	25	<1	<1
acetone	35	30	10	2
octanol	35	25	5	<1
methyl salicylate	69	20	7	<1
2-propanol	10	5	2	1
ethanol	5	2	2	2

[a] Initial solutions: 1000 ppm of test compound in water or water–alcohol mixture.
[b] Table 5.
[c] Table 4, no. 25
[d] Table 4, no. 19
[e] Table 4, no. 9

Solving, $P_T = 0.20$. The actual gas-transmission rate of the PE alone would be (P_P/L_P) or 3.7 cm³/(100 in.² · d · atm) [0.566 cm³/(m² · d · kPa)] whereas the composite package is P_T/L_T or 0.0057 (cm³/(100 in.² · d · atm) [0.00087 cm³/(m² · d · kPa)] or a reduction of 99%. Note that had the EVOH been used alone, one would get the same gas transmission for the package (0.017/3). This shows how highly controlling the barrier layer is. The rest of the system is only the carrier. It was assumed, of course, that the EVOH remained at 0% rh which, in practice, would not be the case. In 90 d at 73°F (23°C), 50% rh on one side and 100% on the other, the PE would transmit about 0.3 g water in the EVOH, which could be 5% on a weight basis of the EVOH. At 5% H_2O gain, the P of EVOH to O_2 is 0.07, and inserting that into our model structure, one gets a P_T that is four times the rate of the structure with dry EVOH. Had the barrier layer been a nonwater-sensitive material (PVDC, ACN copolymer, or PET), one could neglect the water, and P_T would be unchanged although higher owing to the inherent higher permeation level of these other materials when dry or wet.

Typical commercial examples of coatings and multilaminations are (1) oriented PET (OPET) containers coated with PVDC emulsion, and (2) coextruded polyolefin/EVOH containers. The PVDC-emulsion-coated OPET containers are currently being tested for wine and some other products which cannot be packaged in uncoated PET due to O_2 permeation. The O_2 permeation rate of a typical 2-L OPET bottle is about 30 cm³/yr in air at 25°C. This translates to about 20 ppm/yr based on the product. By applying a coating of an emulsion-based PVDC copolymer, this can be reduced to about 15 cm³, or 10 ppm/yr. A greater reduction is very difficult to achieve because the coating, from emulsion, is very thin (about 0.5 mil or 12.7 μm). The 50% reduction is, however, sufficient for some products. On larger containers, the O_2 permeation is lower due to a superior area-to-volume ratio. In a 4-L size, the permeation of an uncoated bottle is only 16 ppm/yr, and this is reduced to 8 ppm/yr with the coating.

Coextruded bottles now being used for ketchup and other oxygen-sensitive products have outer layers of polypropylene and an inner layer of EVOH. Two adhesive layers are used, since adhesion between EVOH and polypropylene is poor. The resulting O_2 permeation of the container is very low, but it does depend somewhat on relative humidity conditions. With a water-based product inside (such as ketchup) and 50% rh air outside, the permeability of the EVOH is 0.07 cm³ · mil/(100 in² · d · atm) [0.27 cm³ · μm/(m² · d · kPa)]. Since almost 3 mil (76 μm) of EVOH is used, the resulting O_2 permeation of a 26-oz (780-mL) package is only 2–3 ppm/yr, or about 1.0–1.7 cm³/yr. To compare this to the coated OPET, one could consider a hypothetical 2-L coextruded bottle. The O_2 permeation would be only 1.3–2.0 ppm/yr, compared to the 10 ppm/yr for the coated OPET bottle.

Although the number of commercial barrier polymers is limited today, it appears to be an emerging technology and there will be new materials. Molecular engineering, the ability to custom-design molecular structures to act as specific barriers, should enable polymers to compete to an even greater degree with traditional packaging materials.

BIBLIOGRAPHY

1. W.A. Combellick, "Barrier Polymers," in J.I. Kroschwitz, ed., *Encyclopedia of Polymer Science and Engineering,* Vol. 2, John Wiley & Sons, Inc., New York, 1985, pp. 176–192.

2. S.P. Nemphos, M. Salame, and S. Steingiser, "Barrier Polymers," in M. Grayson and D. Eckroth, eds., *Kirk-Othmer Encyclopedia of Chemical Technology,* 3rd ed., Vol. 3, John Wiley & Sons, Inc., New York, 1978, pp. 480–506.

3. M. Salame, *SPE Trans.* 1 (4), 153 (1961).

4. M. Salame, *Polym. Prepr. Am. Chem. Soc. Div. Poly. Chem.* 8, 137 (1967).

5. H.G. Yasuda, H. Clark, and V. Stannett, "Permeability," in N. M. Bikales, ed., *Encyclopedia of Polymer Science and Technology,* Vol. 9, John Wiley & Sons, Inc., New York, 1968, pp. 794–807.

6. M. Salame and E. Temple, *Adv. Chem. Seri.* **61**(135), 61 (1974).
7. Kuraray, Inc., Resin Literature Brochure, EVAL ® Resins.
8. Air Products Co., Airopak ® System, Company Brochure, 1981.
9. DuPont Selar ® Nylon Barrier Resin

MORRIS SALAME
Monsanto Company

BEVERAGE CARRIERS.   See Carriers, beverage.

BLISTER PACKAGES.   See Carded packaging.

BLOW MOLDING

The first attempt to blow mold hollow plastic objects, over a century ago, was with two sheets of cellulose nitrate clamped between two mold halves. Steam was injected between the sheets which softened the material, sealed the edges, and expanded it against the mold cavity (1). The highly flammable nature of cellulose nitrate limited the usefulness of the technique.

In the early 1930s more suitable materials, cellulose acetate and polystyrene, were developed which led to the development of some automated equipment by the Plax Corp. and by Owens-Illinois, based on glass-blowing techniques (1, 2). Unfortunately, the high cost and poor performance of these materials discouraged rapid development. They offered no advantage over glass bottles. The development of low density polyethylene in the mid-1940s provided the needed advantage. The "squeezability" of this material gave the plastic bottle a feature glass could not match.

The real beginning of blow molding came in the late 1950s with the development of high density polyethylene and the availability of commercial blow-molding equipment (1). High density polyethylene solved many of the problems of low density polyethylene. Most importantly, it provided greater stiffness. Commercial equipment gave more firms the opportunity to start blow molding. Until that time, all blow molding was done by a select few, using proprietary technology.

The acceptance of high density polyethylene for packaging bleach, detergent, household chemicals, and milk in the 1960s and 1970s, and the acceptance of poly(ethylene terephthalate) for packaging carbonated beverages in the 1980s has expanded the blow-molding process to unprecedented levels. It is one of the high growth technologies (see Fig. 1.).

The process is used primarily for bottle production, but blow molding is being discovered and used more and more by designers for industrial-part applications. Examples include automotive rear deck air spoilers, seat backs, toy tricycles and wheels, typewriter cases, surfboards, flexible bellows, and fuel tanks.

This new activity has led to more and more resins being considered for blow molding. Although most thermoplastic resins can be blow molded, the most commonly used are high density polyethylene (HDPE), poly(vinyl chloride) (PVC), poly(ethylene terephthalate) (PET), and polypropylene (PP) (see Polyethylene, high density; Poly(vinyl chloride); Polyesters, thermoplastic; Polypropylene). For industrial-part applications other resins often considered are acetal, poly(acryonitrile-*co*-butadiene-*co*-styrene), polyamide (see Nylon), polycarbonate (see Polycarbonate), polyester elastomers, mod-

Figure 1.   An array of extrusion and injection blow-molded bottles.

ified poly(phenylene oxide), polyurethane, and high and ultrahigh molecular weight polyethylene.

BLOW-MOLDING PROCESSES

Blow molding is a process to produce hollow objects (2). Air, or occasionally nitrogen, is used to expand a hot preform or parison against a female mold cavity. Today three fundamental blow mold process methods are in use: injection blow molding; extrusion blow molding; and stretch blow molding.

Injection blow molding uses an injection-molded "test tube" shaped preform or parison (see Injection molding). This process is generally used for small bottles, usually less than 16-fl oz (473-mL) capacity. The process is scrap-free with extremely accurate part-weight control and neck-finish detail. On the other hand, part proportions are somewhat limited and handled containers are not practical. Tooling costs are also relatively high.

Extrusion blow molding uses an extruded-tube parison. This is the most common process, used most often for bottles larger than 8-fl oz (237-mL) capacity. Tanks as large as 1300 gal (4920 L) weighing 260 lb (118 kg) have been blow-molded. Compared to injection blow molding, tooling costs less, and part proportions are not severely limited. Containers with handles and offset necks are commonplace. On the other hand, flash or scrap resin must be trimmed from each part and recycled. Operator skill is more crucial to the control of part weight and quality (see Table 1).

Stretch blow molding uses either an injection-molded, an extruded-tube, or an extrusion blow-molded preform. It uses resins with molecular structures that can be biaxially oriented. Stretch blow molding is generally used for bottles between 16 fl oz (473 mL) and ½ gal (1.89 L) in size. It can be

Table 1. Injection versus Extrusion Blow Molding

Injection blow molding	Extrusion blow molding
used for small heavier parts, typically ≤16 fl oz (473 mL)	used for larger parts, typically ≥8 fl oz (237 mL)
best process for general purpose polystyrene; most resins can be and are used	best process for poly(vinyl chloride); many resins can be used provided adequate melt strength is available
scrap-free—no flash to recycle, no pinch-off scars, no post-mold trimming	much fewer limitations on part proportions permitting extreme dimensional ratios; long and narrow, flat and wide, double walled, offset necks, molded in handles, odd shapes
injection-molded neck provides more accurate neck-finish dimensions and permits special shapes for complicated safety and tamper-evident closures	low cost tooling often made of aluminum; ideal for short run or long run production
accurate and repeatable part weight control	adjustable weight control ideal for prototyping
excellent surface finish or texture	

used for smaller bottles, however, and some bottles as large as 6.5 gal (24.6 L) have been molded. The process manipulates the molecular structure of certain resins by orienting the molecules biaxially. This orientation enhances the stiffness, impact, and barrier performance of the bottle permitting a reduction in weight or a lower cost grade of material (3, 4). The technique is limited to extremely simple bottle shapes.

Injection Blow Molding

In the injection blow-molding process, melted plastic resin is injected into a parison cavity around a core rod. The resulting injection-molded "test tube" shaped parison, while still hot, is transferred on the core rod to the bottle blow mold cavity. Air is then passed through the core rod expanding the parison against the cavity, which in turn cools the part (5, 6).

Early injection blow-molding techniques, all two-position methods, were adaptations of standard injection molding machines fitted with special tooling. The primary difficulty with all of the two-position methods was that the injection-mold and blow-mold stations had to stand idle during part removal. This led to the 1961 invention in Italy by Gussoni of the three-position method, which uses a third station for part removal. By the late 1960s, the concept had been developed by commercial machinery builders, and it forms the basis for virtually all injection blow molding today (see Fig. 2). One or more additional stations can be added for various purposes: detection of a bottle not stripped or temperature conditioning of the core rod, decoration of the bottle between bottle blow-mold and stripping stations, or temperature conditioning or preblowing of the parison between parison-injection and bottle blow mold stations.

Another process somewhat related to injection blow molding is displacement blow molding, used to some extent for small containers. The process provides some of the advantages of injection blow molding with less molded-in stress. A premeasured amount of melted resin is deposited into a cupel the shape of a parison preform. A core rod is inserted into the cupel, displacing the resin, packing it into the neck-finish area.

Injection blow molding is used to a great extent for pharmaceutical and cosmetics bottles because the bottles are frequently small, precise neck finishes are frequently important, and the process is more efficient than the extrusion blow molding alternative. The resins most commonly used are high density polyethylene (HDPE), polypropylene (PP), and polystyrene (PS). In the case of PS, the process provides a degree of orientation which enhances impact resistance. All of these resins have been injection blow molded for over 20 yr. PVC, another common blow molding resin, was not injection blow molded until the late 1970s. The use of PVC had to await the development of resins that could take the heat of the process without degradation. Development of more streamlined machinery also was a factor. PET is now being injection blow molded on a limited scale, but it will soon play a more important role and replace some PVC applications.

Extrusion Blow Molding

In the extrusion blow molding process, melted plastic resin is extruded as a tube into free air. This tube, also called a parison, is captured by the two halves of the bottle blow mold. A blow pin is inserted through which air enters and expands and then cools the parison against the cold-mold cavity. Unlike injection blow molding, flash is a by-product of the process which must be trimmed and reclaimed. The excess is formed when the parison is pinched together and sealed by the two halves of the mold.

Extrusion blow molding is divided into two categories: continuous extrusion and intermittent extrusion. These in turn are divided into other subcategories.

Continuous Extrusion. The parison in this process is continuously formed at a rate equal to the rate of part molding, cooling, and removal. To avoid interference with parison formation, the mold-clamping mechanism must move quickly to capture the parison and return to the blowing station where the blow pin enters. The process has three subcategories:

The first is the **rising-mold method** (see Fig. 3). The parison is continuously extruded directly above the mold cavity. When it reaches the proper length, the mold rises quickly to capture the parison and returns downward to the blow station. After the bottle is blown, the mold opens, the part is removed, and the process repeats.

The second is the **rotary-wheel method.** Up to 20 clamping stations are mounted to a vertical or horizontal wheel (see Figs. 4 and 5). A parison is captured, bottles are molded and cooled, and a cooled bottle is removed, all simultaneously, as the wheel rotates past the extruder. The method can provide high production yields, but a disadvantage is the complexity and setup of the multiple mold clamps. It is usually not suited for short production runs.

The third is the **shuttle method** (see Fig. 6). With this method, a blowing station is located on one or both sides of the extruder. As the parison reaches the proper length, the blow mold and clamp quickly shuttle to a point under the extrusion head, capture and cut the parison, and return to the blowing station. With dual-sided machines, the clamps shuttle on an alternating basis. For increased production output, multiple-extrusion heads are used.

The continuous extrusion process, although well suited for most resins, is the best process for PVC resins. PVC can degrade rapidly if overheated slightly. The relatively slow uninterrupted flow of material in this process reduces the tendency for "hot spots" to occur which would damage the material.

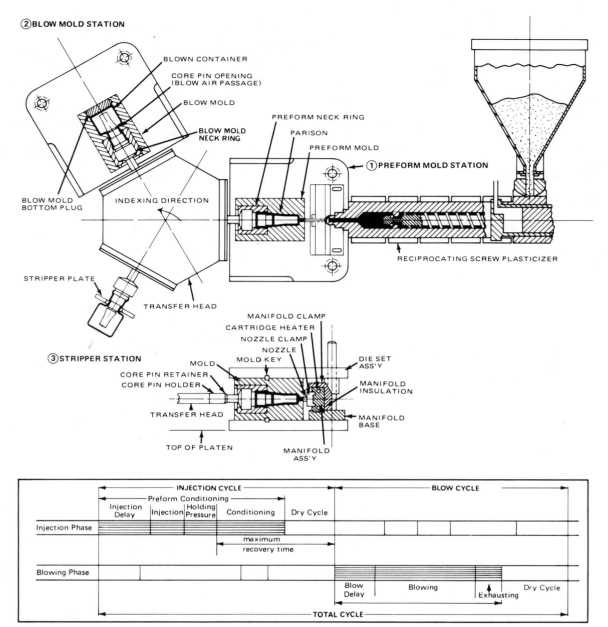

Figure 2. Process description of a typical three-station injection blow molding machine. Courtesy of Rainville Operation Hoover Universal.

Generally, the process is used for bottles of 1-gal (3.8-L) or less capacity.

Intermittent extrusion. In this process the parison is quickly extruded after the bottle is removed from the mold. The mold-clamping mechanism does not have to transfer to a blowing station. Blow molding, cooling, and part removal all take place under the extrusion head (see Fig. 7). This also allows the clamping system to be more simple and rugged. The stop/start aspect of the extrusion method makes this process suitable for polyolefins but not for heat-sensitive materials. The process also has three subcategories.

The first is the **reciprocating screw method.** Normally this method is used for bottles between 8-fl-oz (237-mL) and 2.5-gal (9.5-L) capacity. After the parison is extruded, the screw moves backward accumulating melt in front of its tip.

When the molded bottle has cooled, the mold opens, the bottle is removed, and the screw quickly moves forward pushing plastic melt through the extrusion head forming the next parison. Up to twelve parisons have been extruded simultaneously.

The second subcategory is the **ram accumulator method.** No longer in widespread use, the method is intended for heavy parts weighing ≥5 lb (2.27 kg). It is not used for packaging applications.

The third subcategory is the **accumulator-head method** (see Fig. 8). This method has replaced the ram accumulator for the production of heavy industrial parts and large drums (see Drums, plastic). The reservoir, tubular in shape, is a part of the extrusion head itself. Plastic melt that enters the head first is also first to leave. A tubular plunger quickly extrudes

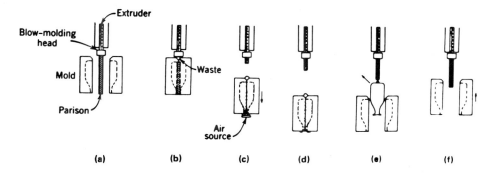

Figure 3. Continuous extrusion with rising molds. (**a**) Complete extrusion of parison; (**b**) close mold; (**c**) blow; (**d**) cool; (**e**) eject; (**f**) raise mold.

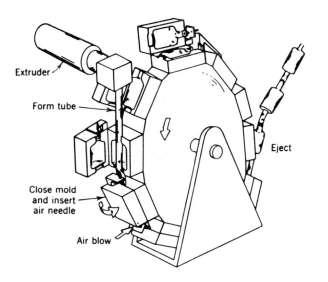

Figure 4. Multicavity continuous-tube wheel-type blow molding machine. Individual bottles are usually separated from the continuous parison tube at the time of mold opening.

the melt from the head annulus with a low uniform pressure which helps to avoid the stresses found in other systems.

Another process somewhat related to extrusion blow molding is the **extrusion–molded-neck process.** Still used today by Owens-Illinois, this is a proprietary process that traces its roots to glass-blowing technology. It is a unique approach where the bottleneck finish is injection molded and the bottle body is extrusion blow-molded.

Extrusion blow molding is used to produce bottles for milk, bleach, shampoo, antifreeze, and a multitude of other products. The process is the only practical way to mold a bottle with a handle. Intermittent extrusion, using the reciprocating screw, is used for virtually all milk bottles produced in the United States, over 5.3 billion (10^9) per year. Many parisons can be extruded at one time, which is an important factor in achieving the production rates required by dairies. The equipment is simple and rugged, and can be run easily by dairy personnel. The resin used is HDPE, the most common of all blow molding resins.

Continuous extrusion is often used to produce small- and medium-sized bottles for shampoo in HDPE, table syrup in PP, and edible oil in PVC. Continuous extrusion is the only suit-

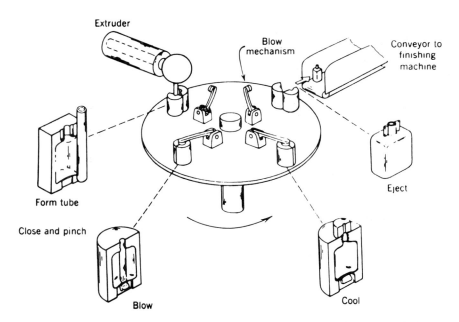

Figure 5. Horizontal rotary blow molding machine. A four-station machine is illustrated; the turntable indexes each mold intermittently.

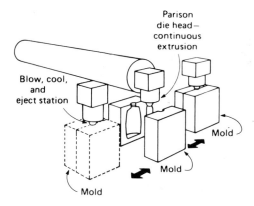

Figure 6. Shuttle continuous-extrusion blow molding machine. Molds on this dual-sided system alternatingly move to capture a parison.

able extrusion-blow method for heat-sensitive resins such as PVC and acryonitrile copolymers (AN). A special thermoplastic copolyester, PETG (Eastman Chemical Co.), has also been tailored to the process. The common grades of PET cannot be extrusion blown.

Stretch Blow Molding

The major stretch-blown resin is PET, and the major application is carbonated beverage bottles (3, 6) (see Polyesters, thermoplastic). Three other resins are stretch blown: PVC, PP, and AN. In the stretch blow-molding process, which is based on the crystallization behavior of the resin, a parison or preform is temperature conditioned and then rapidly stretched and cooled. For best results the resin molecules must be conditioned, stretched, and oriented at just above the glass-transition temperature (T_g) where the resin can be moved without the risk of crystallization (see Figs. 9 and 10) (7).

The advantage of the process is the improvement in resin properties, such as bottle impact strength, cold strength, transparency, surface gloss, stiffness, and gas barrier (3, 6). This in turn permits lighter weight, lower cost bottles, lower cost resins with less impact modification, and sometimes the use of resins that would normally not be suitable for particular products.

In the **one-step approach,** the stages of parison production, stretching, and blowing take place in the same machine. The machines are called "one-stage" or "in-line" machines. In

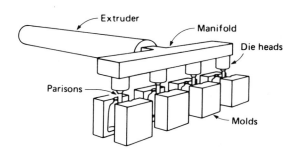

Figure 7. Intermittent-extrusion blow molding machine. Parisons are extruded quickly; molding, cooling, and part removal all take place under the extrusion die head.

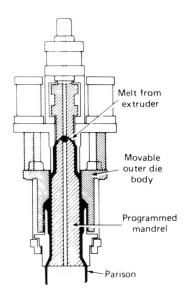

Figure 8. Typical accumulator head.

the two-step approach, parison production is done separately from parison stretching and blowing. The machines are called "two-stage," or "reheat-blow" machines. The main advantage of the one-step approach is energy savings. The parison is rapidly cooled to the stretch temperature. With the **two-step approach,** the parison is completely cooled to room temperature and reheated to the stretch temperature (see Fig. 10). On the other hand, the two-step approach can be productively

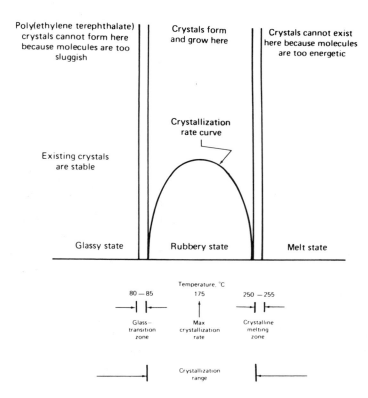

Figure 9. Molders diagram of crystallization behavior. Courtesy of Society of Plastics Engineers.

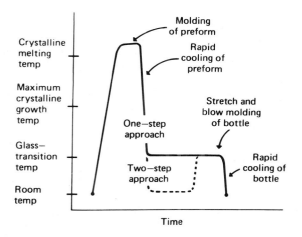

Figure 10. Basic stretch blow process. Courtesy of J.S. Schaul.

more efficient; a minor breakdown in one of the steps does not halt the other, and preforms and bottles can be made in separate plants. The optimum balance of product and preform design versus production output is also easier to achieve with the two-step approach. Limits on parison production, for example, will not force a compromise in parison design to achieve higher bottle production. Each and every bottle design, for optimum performance, has a unique parison design and temperature conditioning requirement which may or may not fit, for optimum productivity, the assumptions used in the design of the one-step equipment. The process uses either injection-molded, extruded, or extrusion blow-molded parisons with a one- or two-step approach.

Injection molded parison. The parisons are virtually the same as those used in injection blow molding. Both one-step and two-step process approaches are used. With the two-step process, the parison is injection-molded in a separate machine, sorted, and later placed in an oven for temperature conditioning and blow molding. A rod is most often used inside the parison, in combination with high air pressure, to complete the stretch (see Fig. 11). Injection stretch blow molding is most often used for PET resin.

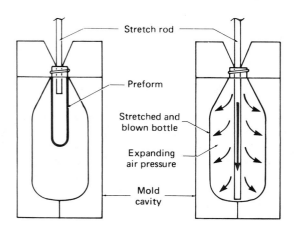

Figure 11. A temperature-conditioned preform is inserted into the blow-mold cavity, then is rapidly stretched. Often a rod is used to stretch the preform in the axial direction with air pressure to stretch the preform in the radial direction.

Extruded parison. Both one-step and two-step approaches can be used. In the one-step approach, a tube is extruded and fed directly into an oven for conditioning. After conditioning, the tube is cut into parison lengths. Mechanical fingers grab both ends stretching the parison. Two mold halves close where air pressure expands the stretched parison against the mold cavity. The two-step approach differs in that the extruded tube is cooled and cut to length. Later the cut tubes are placed in an oven for conditioning. PP is most often used with this stretch blow-molding method, although some PVC is used as well.

Extrusion blow-molded parison. The parison is shaped and temperature conditioned in a preform cavity in the same way a bottle is extrusion blow-molded. From this preform cavity, the parison is transferred to the bottle cavity where a rod and air pressure combine to stretch and expand the resin. Although the one-step approach is the most common, a two-step approach is feasible. PVC is most often stretch blow-molded using this method.

Multilayer Blow Molding

All materials, whether metal, glass, or paper have certain strengths and weaknesses, advantages and disadvantages. Many times two or more materials can be layered and combined to economically overcome weaknesses. Examples are chromium-plated steel, laminated-windshields for automobiles and wax- or polyethylene-covered paperboard. Multilayer blow molding is a process where the strengths of two or more resins are combined to package a product far better than any of the resins could individually.

A few important characteristics or requirements of many bottles are cost, strength, clarity, product compatibility, and gas barrier. Polyethylene and polypropylene, for example, are relatively low cost resins, cleared by FDA for food contact, and excellent water vapor barriers. They are also poor oxygen barriers. As such, these materials are not suited for packaging many oxygen-sensitive foods requiring long shelf life. Poly-(ethylene–vinyl alcohol) (EVOH) (see Ethylene–vinyl alcohol), on the other hand, is a relatively high cost material that provides an excellent barrier, but it is sensitive to water which can deteriorate its properties. A thin layer sandwiched between two layers of polyethylene or polypropylene can solve the problem.

All of the basic blow-molding process methods have been used with multilayer blow molding. In each case, additional extruders are used for each resin. The continuous coextrusion process has been used for bottles up to 1.5 gal (5.7 L) in size. The accumulator head process has been used for drums and tanks up to 130 gal (492 L) in size. The coinjection process has been used for small bottles, 16 fl oz (473 mL) in size.

Related to the coinjection process is a thermoform-insert process (8). With this method, a coextruded sheet is thermoformed into an insert which is placed on the core rod of an injection blow molder just prior to parison/preform molding. The hot resin from the injection step softens the material in the insert permitting it to be blow molded in the next or blow station.

A common problem of multilayer materials is that the different layers do not stick to each other. An adhesive layer is often required to create the bond. As a result, three or more extruders are required. With the HDPE/EVOH example above, five layers are actually required: HDPE/adhesive/EVOH/adhesive/HDPE (see also Coextrusion, flat; Coextrusion, tubular).

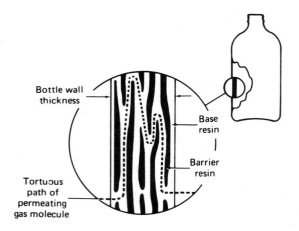

Figure 12. DuPont's blended resin method.

In Europe and Japan the multilayer blow-molding process has been a commercial success, but until recently the problem and cost of scrap prevented broad application in North America. The reclaiming of scrap material, particularly with the extrusion process, often requires the layers to be separated into their basic components.

DuPont has recently developed a novel approach to gain some of the advantages of a coextruded blow-molded bottle without the cost and complexity of multiple extruders and expensive scrap reclaim systems. Only minor modification of the basic equipment is required. A barrier resin is tailored for physical blending with a base resin much the same way that colorant is blended with a resin; however, instead of producing a homogenized mix, the barrier resin is permitted to laminate in the base resin (see Fig. 12). Several random multiple layers of barrier resin of various lengths and sizes are created. Although no single layer completely covers the entire shape of the bottle surface, barrier is created by the tortuous path the permeating gas must follow. A modified nylon material, DuPont Selar, is the first to be used for this approach; when blended with HDPE, it imparts resistance to solvents and solvent-containing chemicals (see Surface modification).

Aseptic Blow Molding

Aseptic blow molding is a process where the bottle is extrusion blow-molded in a commercially sterile environment with highly modified equipment. In many cases the product filler is combined with the blow molder.

Aseptic blow molders are generally divided into two subgroups. The first is a *blow and hold* method. The bottle is molded and sealed in the blow molder. Moving plates in the mold close off and seal the blow pin opening after the bottle is molded. Next the bottle is stored for a period of time, which could be from a few minutes to several days. While in storage, the bottle is not usually in a sterile environment. At the filler, the outside of the bottle is resterilized and the top seal is cut off. Then the bottle is filled and resealed. Although commercially acceptable, this approach has a minor process flaw. The seal created by the blow molder, but later cut off by the filler, cannot be tested. If it is faulty, the bottle can theoretically be contaminated while in storage. On the other hand, productivity is relatively high because the output of the filler is not directly tied to the output of the blow molder.

The second subgroup is a *blow, fill, and seal* method which

can be accomplished two ways. In one approach the bottle is blown and filled in the mold cavity. In 1–2 seconds after the bottle is blown with air, a measured amount of product is forced in, filling the bottle. A few seconds later another mold cavity closes onto the parison just above the main bottle cavity, sealing the filled bottle inside. Bottles made by this process can have a twist-off cap that can be opened without a knife or scissors. In another approach, the bottle is molded by conventional means and then immediately transferred to a fill station. The molding, filling, and sealing of the bottle take place in the same environment.

In general, modifications to the equipment include the use of special stainless steel and plated materials throughout. The molding/filling area of the machine is enclosed in a cabinet. Sterilized air with positive pressure and laminar flow characteristics is maintained inside. All internal surfaces, passageways, hoses, blow pins, valves, and so forth are sterilized with special "clean-in-place" fixturing. Once the process begins, nothing can be touched with the human hand.

Although bottles as large as 2.5 gal (9.5 L) have been aseptically molded, the process is generally used for bottles and vials ≥32 oz (946 mL).

MACHINERY AND TOOLING

Extruder and Screw Design

In all blow-molding processes, plastic resin must be melted. The quality of the melt and productivity of the blow-molding equipment depend heavily on extruder screw design. The designs are the same as those used with profile extrusion and extrusion-blown films technologies (9) (see Extrusion). Three basic approaches are commonly used: the single metering screw or compression screw, some fitted with high shear mixing tips; the barrier screw (10); and twin-conical counterrotating, intermeshing screws. A few continuous-extrusion systems are also fitted with a melt screenpack or breaker plate. Some of the large industrial blow molders are fitted with a grooved barrel feed zone. The tapered grooves improve the processing of powdered ultrahigh molecular weight polyethylene (11).

Injection Blow Mold Tooling Design

The injection blow-molding process requires two molds; one for molding the preform (parison), the other for molding the bottle. The preform mold consists of four major components: preform cavity, injection nozzle, neck-ring insert, and core-rod assembly. The blow mold typically consists of three major components: bottle cavity, neck-ring insert, and bottom-plug insert (see Fig. 13 (**a–e**)) (12).

A variety of materials are used to construct the parison cavity and core rods. For nonrigid polyolefin resins, the parison cavity is made of prehardened P-20 tool steel. The parison cavity for rigid resins and the parison neck-ring inserts for most resins are made of A-2 tool steel. The core rod, for greater strength, is made of L-6 tool steel. All cavity surfaces are highly polished and chromium plated. The only exception is the neck-ring insert for polyolefin resins, which is occasionally sand blasted with a 120 grit.

Aluminum, steel, or beryllium copper is used for the bottle cavity and neck ring. For polyolefin resins, #7075 aluminum is used. Surface finish is usually #120 grit sandblast which helps improve the venting of trapped air. For rigid resins, A-2 tool steel is used. Surface finish is highly polished with chro-

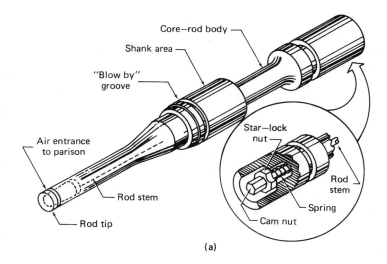

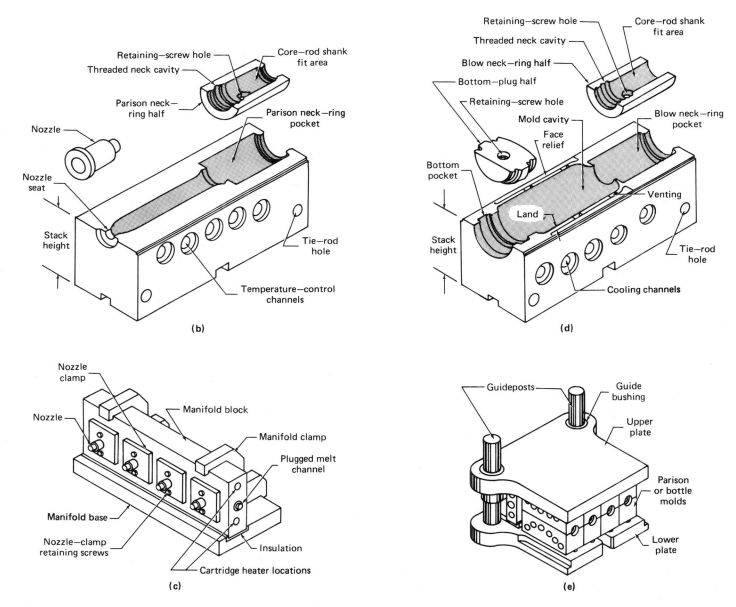

Figure 13. (**a**) Typical bottom-blow core rod showing its principal elements. The core-rod tip mechanism which closes the air passage during the parison-injection cycle is shown in the enlarged view at left. (**b**) Exploded view of one half of a parison-mold cavity, with nozzle and neck-ring details. (**c**) Injection manifold for injection molding of parison. Individual nozzles are clamped to the manifold block which houses a hot runner for the melt. (**d**) Exploded view of one half of an injection blow mold, with details of bottom plug and neck ring. (**e**) Die set for maintaining position and alignment of injection blow mold cavities. Courtesy of *Plastics Engineering*.

mium plating. Cast beryllium copper is often used for either application when minute detail is required. As with the parison cavity, water lines are usually drilled as close as possible to each other, perpendicular to the cavity axis.

Both the parison and bottle molds are mounted onto a die set which in turn is mounted to the platens of the injection blow molder. Keyways in two directions, on each of the upper and lower platens, are used to precisely position the cavities. Guide posts and bushings are then used to maintain precise alignment between the plates. To speed blow molder setup the entire die set and mold assembly is exchanged during a job change. It is considered false economy to reuse the die set with another mold set.

Injection blow mold tooling must be held to very precise tolerances. Dimensions often must be held to a range of ±0.0005 in. (±0.013 mm). If tools are poorly made, processing suffers and bottle quality is inconsistent and poor. Consider for a moment how closely the several core rods must be located fore and aft, and left and right of centerline with the parison and bottle cavities. If too tight, the mold could be damaged or the assembly bind. If too loose, resin could flash around the shank area of the core rod or the core rod could shift sideways causing uneven parison wall distribution. Consider further the many parts and sections of the mold setup that must fit together interchangeably. Several core rods must fit the pocket of the parison or bottle cavities which are stacked side by side next to each other on a die set. This need for precision is the most important factor in the high cost of injection blow mold tooling. Once the tools are properly made and set up, however, the process can be extremely productive and trouble-free.

Extrusion Blow Mold Tooling Design

Extrusion blow molding requires only one mold; but since the parison is formed from an extruded tube and flash is a by-product from the molding process, several tooling elements combine to produce the finished bottle. These elements include the extrusion head, head tooling, blow-pin tooling, mold-cavity tooling, and trim tooling.

Extrusion head. There are three basic styles: the "spider" or axial flow head; the side feed or radial flow head; and the accumulator head. A spider head (13), in its simplest form, has a central torpedo positioned in the melt-flow path supported generally by two spider legs. The smooth and direct path with little or no place for material to hang up and degrade makes this head ideal for PVC resin. Cross-sectional areas of the melt-flow path from the extruder, over the torpedo, and over the spider legs are carefully balanced to ensure even and consistent pressure and flow. Occasionally the spider-head style is used for polyethylene to facilitate very rapid color changes, but offset spider legs are required to mask polyethylene's elastic memory of the legs. This requirement raises the manufacturing cost of the head considerably.

The side feed head is used most often for polyethylene. Melt enters the head from the side, divides around the mandrel, and rewelds. As the melt moves downward and enters the pressure-ring area, tremendous back pressure is created, which helps to ensure the resin reweld. The head is simple and rugged, but color changes are slow. During a color change a fine trail of the previous color, which comes from reweld areas of the head, is often observed on the parts for several hours of production. The problem has been reduced in recent designs by careful attention to streamlining and polishing of the flow

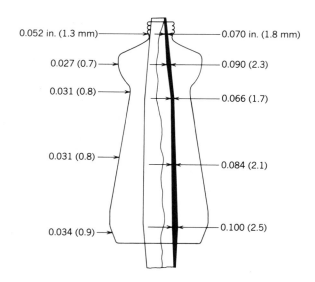

Figure 14. Parison programming distributes resin efficiently by controlling parison thickness. Courtesy of McGraw-Hill, Inc.

path and, in some cases, the addition of a bleed screw to "pull off" the trail of bad color.

The accumulator head (see Fig. 8) is considered a subcategory of intermittent extrusion blow molding. It is the combination of an extrusion head with a first-in/first-out tubular-ram melt accumulator.

All heads can be fitted with parison programming. A parison programmer is a device that can change the gap or relationship between the head tooling die and mandrel while the parison is extruded. The wall of the parison is ringed with sections of thinner and thicker material. These rings are located to correspond to specific sections of the bottle (see Fig. 14) (14). For many blow-molded parts, parison programming can reduce part weight and cost and improve performance and strength.

Head tooling. There are two basic styles: convergent and divergent. As the name implies, convergent tooling has a land angle that converges toward a point. With divergent tooling the land angle flairs outward. Figure 15 illustrates basic nomenclature.

Special shaping of the head tooling can be used to redistribute material in the parison. As the parison is extruded, shaped tooling produces axial stripes of thinner or thicker material that correspond to specific sections of the blow-molded part. Usually the die is shaped or "ovalized" while the mandrel is kept round.

Blow pin tooling. The air which expands the parison against the mold cavity must usually enter through blow pin tooling. There are four basic styles: simple tube, needle, ram down or calibrated prefinish, and pull-up prefinish.

With the tube style, the mold is closed around the parison which in turn is closed and sealed around the tube. The tube usually enters the cavity through the center of the extrusion mandrel.

A hypodermic needle is pierced through the parison from some remote position in the mold after the mold has closed. The location of the hole on the part is usually inconspicuous or hidden from view.

The tube and needle approaches are used primarily for industrial parts. When they are used for bottles, a post-machin-

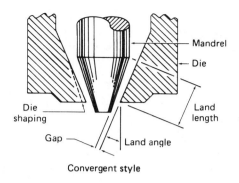

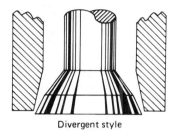

Figure 15. Head-tooling nomenclature.

ing or facing of the neck finish is required. Plastic chips falling into the bottle led to the development of bottle-prefinishing systems, in which the mold neck ring and blow pin work together to form and size the neck finish to specification before the bottle is removed from the mold cavity.

With the ram down or calibrated prefinishing system the critical top-sealing surface of the neck finish is held flat. Immediately after mold closing the blow pin moves into the mold cavity, gathering a small amount of additional material to fill in and pack the neck-finish area. The blow pin moves until the cutting or shear sleeve cuts through the parison flash contacting the striker plate or shear steel mounted in the neck ring of the mold (see Fig. 16). The flash above the neck is easily separated, leaving the finish clean and flat. The blow pin has a tip which forms the inside diameter, but a precise dimension can not always be held for the entire length. Because the neck-finish portion of the bottle is usually the heaviest, the blow pin is often cooled with circulating water. This helps ensure faster production cycles with accurate finish dimensions.

With the pull-up prefinishing system, patented by Hoover Universal, Inc., a very precise inside diameter can be formed. As in the simple-tube system, the mold closes on the parison which in turn closes and seals around the pull-up blow pin. At the end of the molding cycle just before mold opening, the blow pin "pulls up", shearing the inside diameter. This system is used almost exclusively for plastic milk bottles.

Mold-cavity tooling. Flash is a by-product of extrusion blow molding. In the design of the mold-cavity tooling, provision for this flash must be made. Where the flash will exit from the cavity, clearance or a flash pocket is created. The depth of this flash pocket is critical, particularly with large diameter thin-walled parisons. If it is too deep, the flash will not be adequately cooled which may cause trimming problems later. If it is too shallow, the flash will hold the mold open, creating a thick pinch-off which may cause trimming problems. The pinch-off divides the cavity from the flash pocket. The pinch-off has a dual role: It must seal the open ends of the parison and at the same time cut the flash from the bottle. Compression pinch-off is often used where some plastic material must be squeezed into the cavity to ensure a good weld (see Fig. 17).

The cavity defines the shape of the part. Its dimensions are enlarged slightly to compensate for resin shrinkage. Polyolefin shrinkage is slightly higher in extrusion blow molding than in injection blow molding, particularly in the neck finish. Vents in an extrusion blow mold cavity are often deeper as well, for some polyolefin molds as deep as 0.003 in. (0.08 mm). Cavity surfaces for polyolefins are sandblasted. Only PVC molds have a high polish cavity surface. Extrusion blow mold cavities differ from injection blow mold cavities for two major reasons. The heavier neck-finish area is often not cooled as well, which promotes the additional shrinkage. Also, because lower parison-expand air pressures are used, typically 90 psi (620 kPa), cavity-surface imperfections are not as noticeable. With injection blow molding, pressures as high as 150 psi (1034 kPa) are sometimes used.

Unlike typical injection blow mold cavities, extrusion blow mold cavities are cooled in parallel from a manifold system. When possible, the neck-ring insert, bottom insert, and mold body are cooled with an individual circuit which permits each

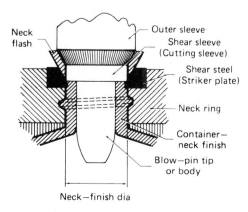

Figure 16. Calibrated neck finish.

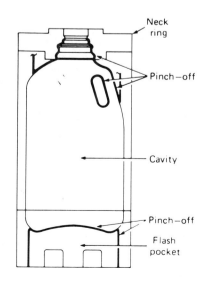

Figure 17. Typical extrusion blow mold cavity.

section to be set differently. This promotes even and uniform cooling of the part. Water lines are straight drilled holes parallel to the cavity axis with several right-angle turns for greater turbulent flow. The lines are ⅜ in. (9.5 mm) to 9/16 in. (14.3 mm) in diameter located approximately 1.5 diameters below the cavity surface. In bottle molds, spacing between the lines is between 3 and 5 diameters.

Trim tooling. Trim tooling removes flash from the blow-molded part after it is molded. Automatic flash trimming is used for high speed bottle production. There are two basic approaches: in-mold trim and downstream trim. In-mold trim is most often used to remove the trail or bottom flash from small nonhandled bottles. A sliding or pivoting plate is fitted into the flash pocket of the mold. A cylinder moving the plate pulls the tail flash from the bottle prior to mold opening and part removal.

The flash on handled bottles is most often removed in a downstream trim press. In this press the bottle is held in a contoured nest where a punch, also contoured, knocks off the tail, handle, and top flash. The flash cut from the bottle by the mold pinch-off is held in place by a very thin membrane.

Wide-mouth bottles are also trimmed downstream. A common approach is to use a "fly cutter" to remove a dome and trim the finish inside dimension; another is to mold a special pulley-shaped dome above the part. Using a spin-off trimmer the special dome engages a belt drive which rotates the bottle against a stationary knife blade. The knife blade, which is fitted into a sharp groove directly above the finish, progressively cuts the dome from the bottle as it rotates.

With the exception of large industrial blow molds cast from aluminum, typically #A356, most extrusion blow molds today are cut from #7075 or #6061 aluminum or from #165 or #25 beryllium copper. Beryllium copper is preferred for PVC because of its corrosion resistance and considerable hardness. Compared to aluminum, however, it weighs more, costs more, and takes more time to machine. Thermal conductivity of the beryllium copper alloy is slightly poorer as well (see Table 2). For polyolefin blow molding, some mold makers have combined the materials by inserting beryllium copper into the pinch-off area of an aluminum cavity, creating a lightweight,

easy-to-manufacture mold with excellent thermal conductivity and hard pinch-off areas.

Unlike injection blow molds which are mounted onto a die set, all extrusion blow molds are fitted with hardened steel guide pins and bushings. The guide pins ensure perfect matching of the two mold halves.

The remaining hardware in the process (dies, mandrels, blow-pin cutting sleeves, and neck-ring striker plates) are all made from tool steel.

Process Variations

Many related operations have been used to improve blow-molding production. In-mold labeling, fluorination surface treatment, and internal cooling systems are significant.

In-mold labeling. This process was developed by Procter & Gamble in cooperation with several major custom blow molders. A label with a heat activated adhesive is automatically placed into the mold cavity and held by a vacuum. The expanding hot parison activates the adhesive which creates a 100% bond (see Decorating, mold labeling).

Fluorination surface treatment. This process improves the barrier of polyethylene to nonpolar solvents (15) (see Surface modification). The barrier is created by the chemical reaction of the fluorine and the polyethylene which forms a thin 200–400-Å (20–40-nm) fluorocarbon layer on the bottle surface. Two systems are available for creating the layer. The in-process system uses fluorine as a part of the parison expand gas in the blowing operation. With it, a barrier layer is created only on the inside. The post-treatment system requires bottles to be placed in an enclosed chamber filled with fluorine gas. The method forms a barrier layer on both the inside and outside surfaces.

Internal cooling systems. Normally a blow molded part is cooled externally by the mold cavity, which means that heat must travel through the entire wall thickness. Because plastic resins have poor thermal conductivity, molding cycle times of heavy parts can be considerable. Internal cooling systems are designed to speed the mold-cooling time, and thus reduce costs, by removing some of the heat from the inside. Several systems have been developed, but they amount to three basic approaches: liquefied gas, supercold air with water vapor, and air-exchange methods.

With the liquefied-gas system, immediately after the parison has been expanded, liquid carbon dioxide or nitrogen is atomized through a nozzle in the blow pin into the interior of the bottle (16). The liquid quickly vaporizes, removing heat, and it exhausts at the end of the cycle. In practice, this method has improved production rates by 25–35%. A disadvantage is the cost of the liquefied gas, which can eradicate the cycle-time cost savings if consumption is not precisely controlled.

The supercold air system with water vapor works much the same way. Here, very dry subzero blowing air expands the parison. The expand air is allowed to circulate through the bottle and exhaust. Immediately after the parison has expanded, a fine mist of water is injected into the cold air stream. As it flows, the water mist turns into snow. As the snow circulates through the container, it melts and then vaporizes. At the end of the molding cycle, the water mist is stopped, permitting the circulating air to purge and dry the interior before the mold is opened and the part removed. Production rates can be improved as much as 50% (17).

The air-exchange system is a far simpler method. Here

Table 2. Blow-Mold Tool Materials

	Hardness, Rockwell C (HRC) Brinell (HB)	Tensile strength, psi (MPa)	Thermal conductivity, W/(m · K)
Aluminum			
A356	HB-80	37,000 (255)	149.8
6061	HB-95	40,000 (276)	169.9
7075	HB-150	67,000 (462)	129.7
Beryllium copper			
25			
and	HRC-30	135,000 (931)	105.2
165	(HB-285)		
Steel			
0-1			
and	HRC-52-60	290,000 (1999)	34.6
A-2	(HB-530-650)		
P-20	HRC-32 (HB-298)	145,000 (999)	36.7

plant air, after the parison has been expanded, is allowed to circulate through the bottle and exhaust (18). Differential pressure inside the bottle is maintained at 80 psi (552 kPa) to ensure the parison retains contact with the mold cavity. Production rate improvements are a modest 10–15%.

The better internal cooling systems are often not justifiable with today's equipment because most blow-molding machines do not have the additional extruder plasticizing capacity to support the faster production rate. This is particularly true for the heavier bottles which would benefit most. As a result, only the low cost air-exchange systems have reached any degree of popularity.

Product-Design Guidelines

Good product design begins with a clear understanding of process and tooling limitations. With injection blow molding, for example, parison core rod strength generally limits its length to a maximum of ten times its diameter. This means the bottle can be no more than ten times the finish diameter in height. Limits in parison wall thickness and temperature conditioning restrict the blow-up ratio to three times diameter. That is, the diameter or largest dimension must be three times or less than the finish diameter. Bottle ovality is also restricted. Problems in maintaining a consistent wall distribution in an oval bottle limit the major diameter to no more than two times the minor diameter.

Extrusion blow molding is not quite so limited. The blow-up ratio nevertheless is held to four times parison diameter or less. This rule applies not only to overall size shape, but also isolated sections as well. For example, bottle-handle designs that are deeper than they are wide across the mold parting face are difficult to mold and often thin and weak.

In general, most blow-molded articles perform better if the designer remembers to radius, slant, and taper all surfaces. Square, flat surfaces with sharp corners do not work very well. Corners become thin and weak; heavy side panels become thick and distorted. Wall thickness can very considerably from side panels to corners. Flat panels are never flat, and flat shoulders offer little strength. Likewise, highlight accent lines should be "dull," with a radius of ≥0.060 in. (1.5 mm). If they are any sharper, the parison does not penetrate, and trapped air marks along the edge become obvious.

Ribs do not always stiffen. Often, blow-molded ribs create more surface area to be covered, which in turn thins the wall. A bellows or accordion effect is created, which flexes easier. If flexing is to occur, then where might the "hinge" points be? The design can be altered to interrupt the hinge action.

The Society of Plastics Industry has identified 21 recommended standard procedures for testing bottle performance. The most important are vertical compression or top load strength, drop impact resistance, product compatability and permeability, closure torque, and top load stress crack resistance.

Blow molding process conditions can influence not only bottle dimensions but also bottle volume. Seven conditions have been identified that produce significant changes in bottle volume. HDPE bottles shrink over time. About 80–90% of the shrinkage takes place in the first 24 h. Lighter weight bottles are not only bigger inside from less plastic, they also bulge more. A 5-g weight reduction in a 1-gal (3.8-L) bottle results in a volume increase of about 12 mL; 5 mL for the plastic and 7 mL for the bulge. Faster cycle times and lower parison-expand

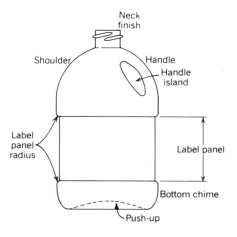

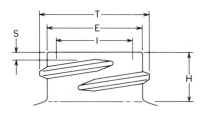

Figure 18. Bottle and finish nomenclature. T, diameter of thread; E, diameter of root of thread; I, diameter of inside; S, distance of thread start to top of finish; and H, height of finish.

air pressure, melt temperature, and mold temperature reduce bottle volume. Storage temperature is very important. After 10 d, significant volume changes can occur in bottles stored at 140°F (60°C). Figure 18 illustrates basic bottle and finish nomenclature.

BIBLIOGRAPHY

1. G. P. Kovach, *Forming of Hollow Articles* in E. C. Bernhardt, ed., *Processing of Thermoplastic Materials*, Robert E. Krieger Publishing Co., 1959 (reprinted 1974), pp. 511–522.

2. R. Holzmann, *Kunststoffe* **69** (10), 704 (1979).

3. K. Stoeckhert, *Ind. Prod. Eng.* 4 62 (1980).

4. K. J. Presswood, *Oriented PVC Bottles: Process Description and Influence of Biaxial Orientation on Selected Properties*, 39th Annual Technical Conference, Society Of Plastics Engineers, Inc., Brookfield Center, Conn., May 1981, pp. 718–721.

5. R. J. Abramo, *Fundamentals of Injection Blow Molding*, 37th Annual Technical Conference, Society Of Plastics Engineers, Inc., Brookfield Center, Conn., May 1979, pp. 264–267.

6. H. G. Fritz, *Kunststoffe* **71** (10), 687 (1981).

7. J. S. Schaul, *Drying and Injection Molding PET for Beverage Bottle Preforms*, 38th Annual Technical Conference, Society of Plastics Engineers, Inc., May 1980, p. 536.

8. S. Date, *Co-Pak Multilayer Plastic Containers*, 5th Annual International Conference On Oriented Plastic Containers, Ryder Associates, Inc., Mar. 1981, pp. 37–48.

9. S. Collins, *Plast. Mach. Equip.* 15 (May 1983).

10. R. A. Barr, *Screw Design for Blow Molding*, 39th Annual Technical Conference, Society of Plastics Engineers, Inc., Brookfield Center, Conn., May 1981, pp. 734–735.

11. J. Sneller, *Mod. Plast. Int.*, 48 (Mar. 1982).

12. J. R. Dreps, *Plast. Eng.*, 34 (Jan. 1975).

13. D. Boes, *Kunststoffe* **72** (1), 7 (1982).

14. B. T. Morgan and Co-workers, "Blowmolding" in *Modern Plastics Encyclopedia* 1969–1970, McGraw-Hill, Inc., New York, p. 525.

15. "Surface Treatment Improves Polyethylene Barrier Properties." *Package Eng.*, 64. (Nov. 1981).

16. E. Jummrich, *Ind. Prod. Eng.* **2** 180 (1981).

17. L. B. Ryder, *Plast. Eng.*, 22 (Jan. 1980).

18. L. B. Ryder, *Plast. Eng.*, 32 (May 1975).

General References

Glossary of Plastic Bottle Terminology, Plastic Bottle Institute of The Society of the Plastics Industry, New York, 1980.

Operator's Guide—Controlling Shrinkage of HDPE Bottles, Dow Chemical Co., Midland, Mich., 1979.

W. W., Bainbridge and B. Heise, *Design and Construction of Extrusion Blow Molds*, SPE, Palisades Section, National Symposium—Plastic Molds/Dies, Design Retec, Brookfield Center, Conn., Oct. 1977, pp. 21-1 to 21-7.

C. C. Davis, Jr., *Materials for Plastics Molds and Dies*, SPE, Palisades Section, National Symposium—Plastic Molds/Dies, Design Retec, Brookfield Center, Conn., Oct. 1977, pp. 1-1 to 1-17.

R. D. DeLong, *Injection Blow Mold, Design And Construction*, SPE, Palisades Section, National Symposium—Plastic Molds/Dies, Design Retec, Brookfield Center, Conn., Oct. 1977, pp. 22-1 to 22-8.

J. R. Dreps, *Plast. Eng.*, 32 (Feb. 1975).

M. Hoffman, *Plast. Technol.* 67 (Apr. 1982).

C. Irwin, *Plast. Mach. Equip.*, 57 (Sept. 1980).

W. Kuelling and L. Monaco, *Plast. Technol.*, 40 (June 1975).

C. Lodge, *Plast. World*, 45 (Aug. 1984).

P. D. Marsh, *Barrier Properties Through Fluorination*, SPE Chicago Section, High Performance Container Technology Technical Conference, Nov. 1983, pp. 49–54.

B. Miller, *Plast. World*, 30 (July 1983).

D. L. Peters, *Plast. Eng.* 21 (Oct. 1982).

J. Szajna, *Food Drug Packag.* 14 (May 1983).

C. Irwin, "Blow Molding" in J. Kroschwitz, ed., *Encyclopedia of Polymer Science and Engineering*, 2nd ed., Vol. 2, John Wiley & Sons, Inc., New York, 1985, p. 447.

Christopher Irwin
Hoover Universal Inc.

BOTTLE DESIGN, PLASTIC

The designer of a plastic bottle (see Blow molding) requires detailed information about the product and how it will be marketed, packed, and stored. About the product, in addition to shelf-life requirements, the designer must know if it requires special containment or exclusion barriers (see Barrier polymers; Multilayer plastic bottles), and whether it is subject to regulations. Input from the marketing department should include information about competition and shelf display; color, surface, and clarity required; convenience features; sizes and annual volume; unit cost; and distribution.

Input from the production department should include details about filling, (see Filling, carbonated liquid; Filling, still liquid), capping (see Capping machinery; Closures), labeling (see Labels and labeling), and handling; bulk-delivery or reshippers; desired line speed; filling temperature; and storage and distribution. All of these factors affect the design of the bottle: the fill level; the shape and color; labeling and decora-

tion (see Decoration); the need for a handle; the type of closure required; and whether heat resistance is required.

The performance of the bottle is determined in large part through proper resin selection, but there are many ways to influence performance through processing and design modifications. For a step-by-step description of the design process and examples of process and design modifications, see ref 1.

BIBLIOGRAPHY

1. J. Szajna, Designing Effective Plastic Bottles," *Food and Drug Packaging* **47**(5), 14–18 (May 1983).

BOTTLES, GLASS.   See Glass.

BOTTLES, PLASTIC.   See Blow Molding.

BOXES, CORRUGATED

The manufacture of corrugated paperboard and boxes began near the end of the 19th century and grew rapidly early in the 20th century. Approval by the railroads to replace wood boxes for shipping many commodities was the key to the growth of this industry. Corrugated boxes are lightweight and inexpensive, can be mass produced in many sizes and weights, and take up little storage space before use. It is estimated that 90% of all manufactured goods in the United States are shipped in corrugated boxes. Annual corrugated board consumption is approaching 300 billion (10^9) ft^2 (2.78×10^{10} m^2), almost triple the usage in 1960.

Board Construction

A corrugated box is made from two or more sheets of linerboard and one or more fluted sheets of corrugating medium.

Linerboard grades.   Most of the liner for corrugated board made in the United States today is unbleached kraft. Kraft paper is made on a fourdrinier machine from fibers consisting primarily of softwood sulfate pulp. Some mills add lesser portions of hardwood sulfate pulp and recycled kraft paperboard pulp. The fourdrinier paper machine has a moving wire-mesh belt on which the pulp is deposited and through which the pulp water drains.

Bleached kraft linerboard is more expensive than unbleached, but is occasionally used when the container is displayed and high quality printing is required. A compromise between fully bleached and unbleached linerboard is "mottled white" liner. In this case, the top layer of an unbleached liner contains bleached pulp, giving a white, though somewhat mottled, appearance. Many products sold in the container in retail stores use a printed mottled white outside liner.

Linerboard comes in a variety of weights, which range in the United states from 26 to 110 lb/1000 ft^2 (127 to 537 g/m^2). The most commonly used linerboard is 42 lb/1000 ft^2 (205 g/m^2) unbleached kraft. Both U.S. and metric standard weights are used abroad, but the metric weights are not direct conversions of U.S. weights. The range of standard weights is shown in Table 1.

The choice of liner weight for many applications is the result of rail and motor freight packaging classification rules. The important characteristics of linerboard are its stiffness,

Table 1. Standard Linerboard Weights

U.S. standard weight, lb/1000 ft^2(g/m^2)	Metric standard weight, g/m^2 (lb/1000 ft^2)	% of U.S. production
26 (127)	125 (26)	5
33 (161)	150 (31)	11
38 (186)	175 (36)	7
42 (205)	200 (41)	50
69 (337)	300 (61)	21
90 (439)	400 (82)	4
other		2

Table 2. Standard U.S. Corrugated Flutes

	Flutes per linear foot	Flutes per linear meter	Flute thickness, in. (mm)	
A-flute	33 ± 3	108 ± 10	3/16 (4.8)	
B-flute	47 ± 3	154 ± 10	3/32 (2.4)	
C-flute	39 ± 3	128 ± 10	9/64 (3.6)	
E-flute	90 ± 4	295 ± 13	3/64 (1.2)	

bursting strength, uniform moisture content, and surface finish.

Medium grades. The most widely used grade of semichemical corrugating medium is 26 lb/1000 ft^2 (127 g/m^2). It gets its name from the way the pulp is prepared: virgin hardwood, usually sawmill chips, is treated with sulfite and soda ash, cooked briefly, then mechanically ground into fibers, giving a high pulp yield. It is made on fourdrinier machines, and is available in heavier weights for special applications. Corrugated box stacking strength (compression strength) is more sensitive to medium weight than to liner weight, and 33–40 lb/1000 ft^2 (161–195 g/m^2) grades are frequently used for that reason.

About 20% of the medium used is made on cylinder machines from recycled paper and board. The source of the recycled material may be box plant clippings or in some cases good quality post-consumer waste such as used boxes.

The important characteristics of corrugating medium are its "runnability" on a corrugator and its resistance to flat crush. Runnability is the ability to conform to the fluted rolls at high speed without breaking. The tips of the flutes must also be capable of forming a good bond to the linerboard facings with a minimum amount of aqueous adhesive.

Corrugated flutes. The corrugations, or flutes, impressed in the medium give corrugated board its strength and cushioning qualities. In corrugated containers, flutes are usually vertical to give maximum stacking strength. They come in four standard sizes (see Table 2).

C-flute is the most common size, but each type has its own special quality. A-flute is used when stacking strength or cushioning is the primary concern. B-flute gives better crush resistance, folds more easily, and is stronger at the score lines than A- or C-flute. C-flute has qualities that fall between A and B, making it a good compromise and explaining its wide usage. E-flute is a specialty grade that is easiest to fold, can be printed very well, and is used for some folding cartons.

Corrugated boards. Corrugating medium faced with linerboard is called corrugated fiberboard. If it is lined on one side only, it is single-face board. If lined on both sides, it is single-wall or double-faced board. Additional mediums and liners yield double-wall and triple-wall board. Figure 1 illustrates these combinations.

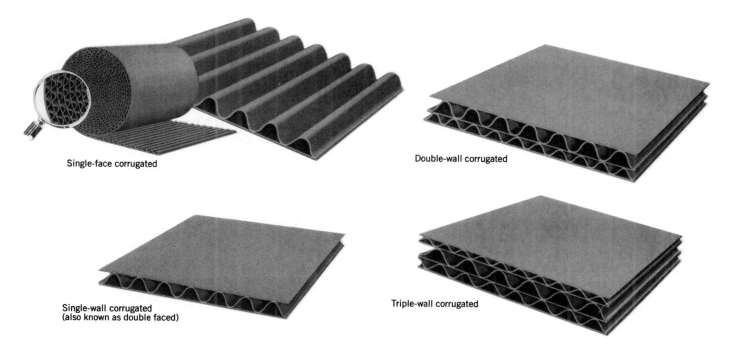

Figure 1. Corrugated boards (1).

Table 3. Freight Regulations[a]

| | | Corrugated fiberboard | | | | |
| | | Single-wall | | Double-wall | | Triple-wall |
Max weight of box and contents, lb (kg)	Max inside dimensions (length, width, and depth added), in. (cm)	Min combined weight of facings, lb/1000 ft² (g/m²)	Min bursting test of combined board, psi (MPa)	Min combined weight of facings including center facing, lb/1000 ft² (g/m²)	Min bursting test of combined board, psi (MPa)	Min combined weight of facings including center facings, lb/1000 ft² (g/m²)
Part A.	*For all fiberboard boxes other than double thickness score-line boxes*					
20 (9)	40 (102)	52 (254)	125 (0.862)			
30 (14)	50 (127)	66 (322)	150 (1.034)			
40 (18)	60 (152)	75 (366)	175 (1.206)			
65 (29)	75 (191)	84 (410)	200 (1.379)	92 (449)	200 (1.379)	
90 (41)	90 (229)	138 (674)	275 (1.896)	110 (537)	275 (1.896)	
120 (54)	100 (254)	180 (879)	350 (2.412)	126 (615)	350 (2.412)	
140 (64)	110 (279)			222 (1084)	500 (3.446)	
160 (73)	120 (305)			270 (1318)	600 (4.136)	
275 (125)	120 (305)					264 (1289)
Part B.	*For double thickness score-line boxes*					
225 (102)	60 (152)	138 (674)	275 (1.896)	110 (537)	275 (1.896)	
300 (136)	60 (152)	180 (879)	350 (2.412)	126 (615)	350 (2.412)	

[a] Refs. 1, 2, and 3.

Single-face board is used primarily as a protective wrap and cushioning material. It is especially useful as an interior packing for fragile products such as glass. Single-face board represents less than 1% of the volume of the corrugated industry.

Single-wall board is the backbone of the corrugated container industry. About 90% of all corrugated boxes are made of it. The most common grade consists of two 42-lb/1000 ft² (205-g/m²) kraft liners and 26-lb/1000 ft² (127-g/m²) semichemical medium in C-flute. Single-wall board is also used to make pads, tubes, partitions, and other forms of inner packing for boxes.

Double-wall board, which has an additional layer of liner and medium, represents almost 9% of corrugated volume in square footage. It is used when greater strength and rigidity are required.

Triple-wall board adds yet another layer of liner and medium, for a total of seven layers of paperboard. It is used for packing large and heavy items, but still represents less than 1% of industry volume.

Corrugated boards are usually designated by the minimum Mullen bursting test of the board (TAPPI test method T 810). The common 42-26-42 single-wall board is "200 test board." This came about because rail and truck freight regulations specify the Mullen test as well as minimum combined weight of facings (liners) and maximum inside box dimensions for various weights of box and contents (see Table 3). These Mullen test designations are taken from Rule 41 of the rail regulations (3) and from Item 222 of the truck regulations (4) (see Laws and regulations, U.S.).

All corrugated boxes made to conform to the rail and truck regulations must carry a printed certificate identifying the boxmaker and the appropriate board grade as shown in Figure 2.

Since only minimum total weight of facings is specified, the linerboard weights and medium weights can be varied to some

extent for each "test board," depending on the end use. Balanced board with equal liner weights on each side is more common than unbalanced board.

Board Treatments

Wet strength. Linerboard containing a small percent of a thermosetting resin added to the pulp has considerably improved wet strength. This can be measured by tensile and bursting strength tests on sheets soaked in water. A corru-

Figure 2. Boxmaker certificate. 200 psi = 1.38 MPa; 84 lb/1000 ft² = 410 g/m²; 75 in. = 1.91 m; 65 lb = 29.5 kg.

gated box containing a heavy load which gets wet in use will break open much less easily when wet strength liners are used. Major applications for wet strength liners are government overseas shipping containers for various supplies and iced poultry, fish, and produce boxes. The latter are also wax treated as discussed below.

Paper machine surface treatments. Linerboard can be surface treated on the paper machine with several types of chemicals to give special properties. One such treatment which has had considerable use is colloidal silica for skid resistance. Skid-resistant boxes are needed for better running on certain automatic packing lines and to improve the stability of some pallet loads.

Corrugator treatments. A series of aqueous emulsion surface treatments have been developed by chemical suppliers for application on corrugators for special properties. Usually the amount applied is very small (in the range of 1 lb/1000 ft² or about 5 g/m²) to minimize cost and not add too much moisture to the board.

Water repellency, skid resistance, scuff resistance, oil and grease resistance, and improved sealability and release properties can be obtained by these treatments. For example, scuff resistance may be important in packaging expensive products with highly polished surfaces such as appliances or office furniture. Some food products might impart oil or grease stains which can be prevented by a treatment on the inside of the box. Combination with a skin packaging film can be enhanced by an appropriate surface treatment. Release properties are important in packaging products that tend to stick to a plain board, such as synthetic rubber and some foods.

Another type of corrugator treatment that formerly was done in a separate operation is impregnation with paraffin wax. The large market for fresh iced poultry boxes requires three board treatments. Wet strength linerboard from the mill has already been mentioned. Wax impregnation with about 5–10 lb/1000 ft² (24–49 g/m²) of paraffin inside each component or in selected components is now accomplished with hot roll coaters on the corrugator. The wax impregnation helps to delay and minimize the absorption of water by the corrugated board during the relatively brief shipment cycle of iced poultry. The same treatment is needed for several other fresh foods such as fish, produce, and fruit. The third board treatment needed for iced poultry is curtain coating, discussed below.

Separate board treatment operations. The most extensively used type of coating for corrugated board is curtain coating. Water-resistant board is obtained in this manner by coating the finished printed blank as the final step before the box is set up. Proprietary wax blends are usually used for curtain coating. They may contain paraffin wax, microcrystalline wax, ethylene–vinyl acetate copolymer, and a petroleum resin. Usually about 5–10 lb/1000 ft² (24–49 g/m²) of coating is applied. This is sufficient to give a continuous waterproof layer which has enough flexibility to withstand folding at the scores.

A wet strength, wax-impregnated, two-side curtain-coated box is widely used for iced poultry and fish. The curtain coating is the primary barrier to liquid water in these applications. However, since these boxes are die-cut and have some exposed edges, the wet strength and wax impregnation are also needed to resist exposure to melting ice. Curtain-coated boxes are also used for some types of wet produce and for fresh meat.

Another separate operation used to make water-resistant corrugated board is wax dipping or cascading. In this process a finished box with a glued joint is completely saturated with paraffin wax, either by dipping or by cascading the molten wax through and around the board flutes. Considerably more wax (about 40–50 wt % of the board) is picked up in these processes compared with wax impregnation on the corrugator and curtain coating both sides. A dipped or cascaded box is very stiff because most of the crystalline wax is impregnated into the paperboard. Such a box will withstand water spraying and excessive exposure to water for short periods of time. It is most often used for fresh vegetables that are hydrocooled, such as celery.

Polyethylene-coated linerboard is also a good moisture barrier for corrugated boxes. However, this extrusion-coated board is expensive and not used extensively. In a few cases where a good water-vapor barrier is needed or where the excellent release characteristics of polyethylene are needed on the inside box surface, this coated board is effective.

Many other special treatments of corrugated board are possible and some have been used in limited quantities. Generally, the performance improvements of the board treatment must be weighed against the increased cost in order to justify a commercial use.

Box Designs

Slotted boxes. The regular slotted container (RSC) is the most common configuration among corrugated containers. It is economical to produce and makes efficient use of board. The design has four flaps of equal length at the top and bottom, thus being made from a rectangular blank. The outer flaps meet; the inner flaps do not. As a result, parts of the top and bottom of the box have only a single thickness of corrugated board (see Fig. 3).

The RSC is used for hundreds of products, especially in the food industry, where economy is important and normal handling is involved. For uses where more protection or special features are needed, there are several variations on this basic design, all of which require more board for the same size box.

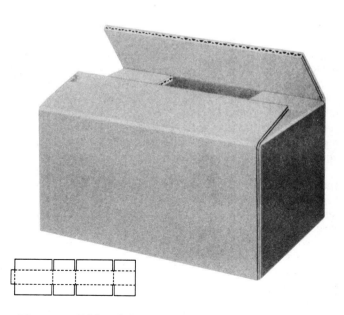

Figure 3. RSC-style box. International box code 0201 (5).

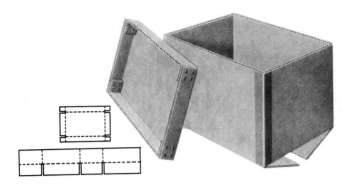

Figure 4. Half slotted box with cover (HSC). International box code 0312 (5).

The variations include inner flaps that meet and outer flaps that overlap from a small amount to a full overlap.

Half-slotted boxes. The half-slotted box (HSC) is an RSC with only one set of flaps plus a cover (Fig. 4). The cover may be a telescoping half-slotted style or a design-style tray as illustrated. Either way, this style is used when the cover is to be removed and replaced. A typical application is a combination shipper and shelf package.

A half-slotted inverted cover may come down only partway to allow for settling of the contents, such as textile or paper products. It may also be a full telescope cover to give extra wall strength. Fresh fruits and vegetables are often shipped in this style box.

Design-style trays and telescoping two-piece boxes. Slotted and scored flat blanks can be formed into design-style trays or boxes by the user with gluing or stitching machines. When used as two-piece boxes with telescoping covers, they are similar to the half-slotted boxes and covers in the previous section. They differ from the half-slotted boxes and covers in that the tops and bottoms have only a single thickness of board and the side walls are at least partially reinforced with the overlapping flanges.

This is a more efficient use of board for good stacking strength. Paper products are frequently shipped in this style of box as well as poultry, fruit, and many other products. Sometimes die-cut self-locking trays are used for produce packing or for box covers when it is necessary to set them up by hand.

The design-style tray itself is often used for shipping canned products, such as beverages and other foods. The depth of the tray is kept low enough so that the can labels can be read on a supermarket shelf. The use of trays instead of boxes for canned foods was increased when plastic shrink film and equipment became available to unitize the group of cans.

Bulk boxes. The three-piece double-cover box is basically a tube with telescoping covers on each end. It can be used to ship especially heavy or tall items, often in conjunction with a pallet. It is also used to ship textile and food products in bulk. For large appliances and machinery, special covers which interlock with flanges at the ends of the tube are used. Double-cover boxes with six or eight sides are also used to minimize bulge and take advantage of the stacking strength obtained from the extra corners.

A major use for bulk boxes exists in the chemical industry, especially for plastic resins in the range of 1000–1500 lb (454–680 kg) of product. Originally, large double-wall HSCs or dou-

ble-cover boxes were lined with an extra tube of double-wall board around the inside perimeter of the box to give the necessary stacking strength. It was discovered that if the double-wall liner is glue laminated to the double-wall HSC, the box is stronger in compression and bulge resistance as well as easier to set up (see Fig. 5). Most bulk boxes used for resins, other chemicals, and some food products are now laminated in this fashion. The half-slotted outer body usually has overlapping bottom flaps and short flanges at the top. A separate tray-style telescoping cover completes the package. The lamination of the liner to the body is done with flat blanks having correspondingly placed vertical body scores. Some heavy-duty bulk boxes even require two liners separately laminated to the body.

Certain products, such as synthetic rubber bales, exert such outward force on the side walls of a bulk box that two-cell or multicell laminated bulk boxes have been designed for adequate bulge resistance and stacking strength.

Bliss and recessed end boxes. Two designs for special applications are the Bliss box and the recessed-end box. The Bliss box comes in two styles, plus a third that combines the first two (see Fig. 6). The Bliss box has good stacking strength because all four corners are reinforced. It also uses less board than most slotted boxes because the bottom is of single thickness and there is less flap area. However, it is a three-piece die-cut box which requires special set-up equipment at the point of use. Machines using hot melt adhesive are available to set up Bliss boxes. These boxes are used for bulk packs of meat, explosives, fresh fruits and vegetables, and articles of concentrated weight.

A recessed-end box is made by stitching a scored body sheet to two flanged endpieces. It is used to ship long, fragile items (fluorescent bulbs) or items of the same width or girth but varying lengths. The body sheets can easily be cut to fit.

Folders. One-, two-, and three-piece folders are used to ship books, apparel, and other products via express or parcel post. All three are scored sheets with tucks of specified lengths. All are easy to store and set up by taping or with staples.

A tuck folder is a special one-piece folder which provides easy reclosure for toys and hardware, for example. A five-panel folder and wraparound blank are closely related to the above folder designs. When set up, the five-panel folder has

Figure 5. Bulk box compression test (1).

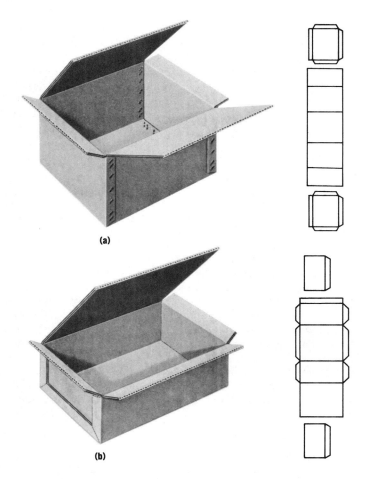

Figure 6. Bliss boxes; (**a**) No. 2 Bliss box, Int. box code 0605, (**b**) No. 4 Bliss box, International box code 0601 (5).

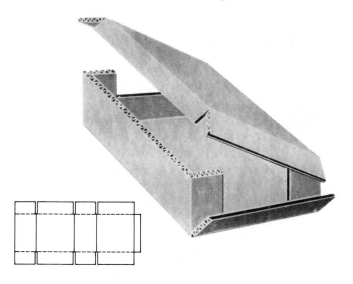

Figure 7. Wraparound blank (1).

extra thicknesses of board at its ends, giving additional strength for packaging long, narrow, dense items, such as canes, rods, umbrellas, and light fixtures.

Wraparound blanks. Wraparound blanks are prescored and slotted for use in automatic gluing equipment. They provide an extra-tight pack for canned goods or jars, preventing product damage while using a minimum amount of board. As the name implies, the blank is automatically wrapped around the collated product which has been placed on the bottom panel. The machine effects both the joint and flap closures (see Fig. 7.)

Slide boxes. Slide boxes are small-sized multiple piece boxes used to ship books or similar items via express or parcel post, or as inner packs. A double-slide box (DS) provides at least one thickness of board on all sides, and two thicknesses on four sides. A triple-slide box (TS) has double board thickness on all six sides. It is used for freight shipment of fragile items.

Displays. Corrugated board is uniquely suited to the construction of temporary displays for consumer products in retail stores. The three general types are counter displays, floor displays, and display/shippers. Display/shippers are cases of product including the materials for the display. Designer ingenuity, die-cut pieces which enable manual setup, and multicolor graphics all combine to give eye-catching displays for many products. Usually these are one-time orders in conjunction with a promotional marketing campaign.

Inserts and interior packing pieces. Inserts and interior packing pieces are used to keep items in place inside a container. They may also cushion items from shock or simply separate them.

Some basic types of interior packing are tubes, partitions, and pads. Tubes may provide additional stacking strength and may be taped, glued, or simply scored and bent. Partitions are often just slotted sheets, interlocked at right angles to form cells. Beverage bottle cases are the most common application. Other types of dividers can be designed from scored sheets and scored and slotted sheets. Separate sheets to cover the bottom or go between layers of product are called pads.

Trays and other folded sheets can be formed inside a box in many ways to give clearance for the product or to suspend the product away from the box panels. These include corner protectors which are often used for furniture and large appliances.

Die cuts allow simple adaptation of irregular sizes to regular containers. These are especially useful for facilitating assembly-line packing methods for odd-shaped items. The ingenuity of the box plant designer in protecting a difficult-to-pack product with the least amount of material and number of pieces is often the key to success for corrugated packaging.

Manufacturer's Joint

Most boxes, such as slotted containers, are folded at two score lines so that the ends of the blank are brought together to form the manufacturer's joint. In this way a single joint is formed where one end panel meets one side panel. As the name implies, most manufacturer's joints are made in the box plant. However, some large box users have found it advantageous to buy flat blanks and form their own boxes with joints, as with wraparound or Bliss box machines.

Manufacturer's joints are glued, stitched, or taped (Fig. 8). According to the carrier regulations, all the types of joints may be used for single- or double-wall boxes, but triple-wall box joints must be made with stitches or staples for greater strength. Both glued or stitched joints are made using the

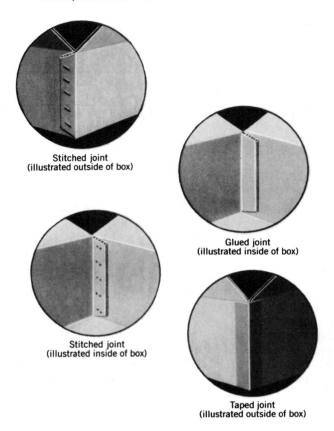

Figure 8. Manufacturer's joints (1).

manufacturer's joint tab, which is an extension of the end panel or the side panel at one end of the blank. The tab overlaps the joint by at least 1.25 in. (3.2 cm). If glued, the glue should be water resistant. Hot melt adhesive is generally used only for light-duty boxes that are not over 65-lb (29.5-kg) gross weight and must cover at least 25% of the area of contact between tab and panel. Carrier regulations require hot melts to consist of 100% solids and to be able to withstand temperatures in a range of −20 to 165°F (−29 to 74°C). Hot melt adhesives are usually used on box-forming machines in customer plants starting with flat blanks.

Glued joints on slotted boxes that use a cold aqueous resin adhesive and run on high speed folder–gluers are considered to be the least expensive manufacturer's joint and are the most common.

In making a stitched joint, stitches from a continuous coil of metal wire join the tab to the adjacent panel of the box at 2.5-in. (6.3-cm) intervals. For triple-wall boxes, stitches or preformed metal staples are 1 in. (2.5 cm) apart and the tab overlaps by 2 in. (5.1 cm). Stitched or stapled joints are normally used for large boxes with dense loads because there is great resistance to breaking of the joint. The cost of stitched or stapled joints is greater than that of glued joints.

A glued or stitched tab is usually placed inside the box so that the outside panels are smooth for printing and ease of material handling. If an inside tab causes problems with the contents, the tab can be placed on the outside of the box.

Taped joints are usually made with laminated, reinforced tapes. For light-duty boxes up to 65-lb (29.5-kg) gross weight,

it is permissible to use 2-in. (5.1-cm) wide kraft paper tape having a basis weight and bursting strength of 60 lbf (267 N). For boxes of more than 65 lb (29.5 kg) gross weight, 3 in. (7.6 cm) wide reinforced tape is usually used, with various weight requirements, depending on the reinforcement. Glass thread reinforcement of laminated kraft paper is commonly used. Taped joints save on board because no overlapping tab is needed. However, it is still more expensive than a glued joint in most cases and may interfere with printing design. Its main advantage is to give both a smooth inside and outside surface as well as good strength.

Corrugated Manufacture

Modern box plants contain a variety of equipment for corrugating, printing, scoring, slotting, die-cutting, slitting, folding, gluing, taping, stitching, coating, impregnating, and performing special operations on corrugated board. More than half the plants in the United States are supplied with corrugated sheets made in another location, and are known as sheet plants. The larger plants with corrugators usually have the same types of finishing equipment as sheet plants, so almost all produce finished boxes. There are a few corrugating operations that simply specialize in producing sheet for sheet plants.

Corrugators. A modern corrugator is the largest and most expensive piece of equipment in a box plant. Really a series of machines, it makes fluted medium, glues it to a liner, adds another liner, makes double- or triple-wall board if needed, and then scores and cuts the corrugated sheet to produce the raw material of the corrugated container (see Fig. 9). Starch adhesive is used to glue the flute tips to the liners.

Corrugators vary considerably in width and speed, but a typical width is 87 in. (221 cm). Normal practice is to run several orders or multiple widths of the same order to make maximum use of the width of the machine.

Printer–slotters. The finished sheet produced by a corrugator is cut to size, and the scores defining the flaps and depth of the box are usually made. The next step includes scoring for the length and width panels, the cutting of slots to separate the flaps, the cutting of the manufacturer's joint tab, and the printing of necessary information or decoration.

Printer–slotters perform all these functions in a single operation. Although they are gradually being replaced by the more advanced flexo folder–gluers (see below), these machines are still in wide use. A simple version is the two-color unit, made in a variety of sizes.

In addition to scoring, slotting, and printing, these machines usually have an automatic feeder, an ink washup device, and an automatic stacker. Resilient letterpress-type printing plates mounted on rotary cylinders are used for both oil-based inks and flexographic water-based inks. The flexo inks have largely replaced oil-based inks because they dry much faster.

Folder–gluers, tapers, and stitchers. These machines fold box blanks and make the manufacturer's joint. The fastest are folder–gluers which are automatically fed, use belts and guides to complete the two folds, apply a water-resistant adhesive to a 1.25-in. (3.2-cm) wide tab, square the folded box while the adhesive is still wet, and apply pressure to set the joint. Glued joints are now more prominent than taped or stitched joints because of lower cost and the uniform strength of the overall adhesive application on the tab.

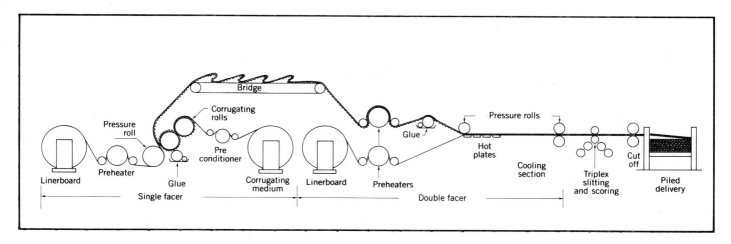

Figure 9. Schematic of a corrugator (simplified and condensed).

Tapers can handle a wide variety of box sizes and are economical for short runs. Folding on an automatic folder–taper is done in a stationary position with guides and mechanical wings. After folding, the box is squared on a roller assembly and taped. The cutting, moistening, and positioning of the tape on the box joint is automatically accomplished.

Stitchers can be hand fed with manually folded blanks as well as being combined with automatic folders for a completely automatic folding and stitching operation. Box blanks are folded and squared, then fed through the stitcher, which cuts staples from a wire roll and inserts them one at a time into the blanks. Motion is not continuous because the blank must be stopped for each stitch to prevent the board from tearing. Most stitchers insert staples at a 45-degree angle, providing a strong joint whether the box corrugations run vertically or horizontally.

Flexo folder–gluers. The introduction of flexographic printing, using fast-drying easy-to-clean inks, made it possible to bring the printing operation in line with the folder–gluer. The flexo folder–gluer takes scored blanks from a corrugator and can perform all of the following to produce finished boxes: feeding, multicolor printing, scoring, slotting, glue tab cutting, additional die cutting, folding, gluing, and stacking piles of finished boxes.

The only two machines now needed to produce long runs of slotted boxes with one glued joint are the corrugator and flexo folder–gluer. The increasing efficiency of both these basic machines is revolutionizing the fiber box industry, making much of the older equipment obsolete.

Die cutters. Die cutting of corrugated blanks has traditionally been done on flatbed presses, which can give great accuracy even for the most complicated die-cut design. In recent years rotary die cutters with cylindrical dies have become more popular because of greater productivity.

Modern flatbed machines have automatic blank feeders, flexographic printing capability, steel rule wooden dies for cutting and scoring, automatic waste stripping capability, and automatic stacking units. These machines can produce several thousand die-cut sheets per hour.

Rotary die cutters can process large sheets with outputs exceeding those of flatbed machines. They are used in many new installations in conjunction with flexo folder–gluers. Sep-arate rotary die cutters with flexographic printing units and complete automation are now quite common. When used with a flexo folder–gluer, it becomes possible to produce uniform, accurately made boxes in high volume, which run well on automatic filling and sealing equipment.

Curtain coaters and wax impregnators. Advances in curtain coating and wax impregnating techniques opened the fresh meat and produce markets to corrugated containers. Curtain coaters feed box blanks horizontally through a vertical "curtain" of liquid coating, usually a molten hot melt. The heated liquid flows downward through a narrow straight orifice across the width of the machine. Coating that falls beyond the edges of the blanks and between the blanks is recirculated back to the reservoir above. Corrugated blanks are automatically placed on a fast-moving horizontal belt which carries them under the curtain and are coated on the top side. Adjusting the speed of the belt and the thickness of the curtain determines the weight of the coating.

Separate wax impregnating machines are used to produce heavily treated boxes (40–50% wax based on the weight of the board) which must withstand very severe water exposure for short periods of time. These specialty machines vary considerably in their design. It is usually desirable to treat a finished box with the joint already made. One type of machine is known as a cascader. It pours molten wax down through the knocked-down boxes held vertically on their edges while they pass through a waxing and cooling tunnel. Another type of machine passes the knocked-down boxes through a tank of molten wax in a process called dipping.

Partition makers. Partition slotting and assembly machines are found in many box plants that supply the glass bottle industry. The special slotting machines for partitions provide the accuracy and flexibility for this type of interior packing that is made in many sizes. They are readily adjustable so that slots of correct width are made to match flute size. For high volume runs, automatic partition assemblers are used to bring the slotted pieces together and collapse them.

Auxiliary equipment. Several types of auxiliary equipment are needed in every box plant and some specialized machines may also be used, such as automatic sheet stackers and counters, strappers for unitized loads, automatic conveyors, fork-lift trucks, slitters, laminators, starch glue preparation equipment, waste-handling equipment, die-making equipment, automatic feeders, and label-gluing machines.

The endless challenge to increase productivity and reduce manufacturing cost continually spawns the development of new and better boxmaking and auxiliary equipment.

Industry Trends

User automation.   For many years most users of corrugated boxes handled the boxes manually. This included the setup, filling, and closing. The development of mechanical case sealers has a long history, but automatic case openers, palletizers, and machinery associated with filling boxes are of more recent origin. Today, many box users with substantial volume have automated packaging lines from beginning to end. Most users have at least some automation, such as a case sealer.

Automatic machinery for handling corrugated boxes is not as forgiving of box defects as manual handling. Poor quality in the form of warping, dimensional variations, improper joints, and weak bonds is not tolerated by many of the box-handling machines. Boxmakers have had to make significant improvements in quality control and product uniformity. Voluntary standards have been developed by the Packaging Machinery Manufacturers Institute and the Fibre Box Association on dimensional tolerances for corrugated boxes to be run on automatic packaging equipment.

Further automation is expected in the distribution of many corrugated shippers to retail outlets. A Universal Product Code case symbol has been developed for imprinting boxes (6) (see Barcoding). This bar symbol is similar to the code now used on retail packages for pricing and inventory control. The manufacturer and the product will be identified by the case symbol. This is an important step in the development of automatic warehousing and inventory control, especially in distribution centers for large retail chains.

Box plant productivity. Productivity keeps improving through continued investment in converting and printing machinery.

The biggest boost to productivity has come from the gradual development of a practical, continuous corrugator. This equipment is moving out of the prototype stage and will ultimately bring still further improvement in productivity levels. It can include automatic splicers, fingerless single-facers, bridge speed control, direct-drive knives, automatic slitters and stackers, and computer control of the entire operation. The objective is to change orders without stopping the machine and maintain higher average speed. Waste is also minimized by continuous operation. Dramatic improvements in square footage per machine hour and per man-hour have been made on these modernized corrugators.

Rotary die cutting combined with flexographic printing is another example of maximizing productivity of intricate box designs. When compared with the older method of separate printing and flatbed die-cutting operations, the difference in output per man-hour is very impressive. A fairly recent development is the growing use of preprinted linerboard (7).

The large volume being handled in modern plants has led to conveyorization of floor areas between the major production machines. Large stacks of sheets or boxes can be moved more quickly through the plant on rollers which may or may not be mechanized.

Competitive Threats

Plastic uses in packaging pose the most significant threat to corrugated market growth. Plastic film wrapped tray packs are a good example. A corrugated tray containing canned products overwrapped with plastic film has replaced a full corrugated case for some canned foods and beverages. The loss of corrugated volume has been estimated at less than 1%, but it is a proven alternative to corrugated boxes for certain applications.

Other plastic developments, such as containers for produce (see Corrugated plastic), foam inner packing (see Foam, expanded plastic), and film overwrap (see Films, shrink; Films, stretch) of pallets and textile bales, have also displaced corrugated for a few specific applications.

The use of bulk bins for shipping a variety of foods and chemicals has no doubt had a negative effect on total corrugated usage. In many cases the bulk bins are made of corrugated, but the corrugated usage per unit weight of product is less than if smaller, corrugated boxes were used. In other cases, large bulk bins made of other materials have eliminated corrugated entirely.

Retail Packages

The use of corrugated boxes as retail packages for some consumer products is growing. On a store shelf the graphics become more important to attract customers. Preprinted litho labels and preprinted linerboard are offered as higher quality printing on boxes by a number of box plants. The litho labels are glue laminated to one or several box panels on special laminating machines. Preprinted linerboard is made on large flexographic or gravure presses with accurate registration, which must coincide with the box dimensions on the corrugator.

Corrugated Uses

The Fibre Box Association collects statistics on total shipments, in square feet, of corrugated boxes and sheets from a high percentage of the corrugated manufacturers throughout the United States. In addition, the total shipments are characterized by end use category, using Standard Industrial Classifications. A summary of the major end-use industries and the percent of their usage is given in Table 4.

Food products.   Food products represent by far the largest end-use market for corrugated boxes. This market has grown not only with population but also in the penetration of corrugated into packaging a greater number of food products. One example is the increased use of boxes for shipping fresh meat and poultry. This has replaced the cutting of meat carcasses at the retail store and the use of wood boxes for fresh poultry.

Canned, bottled, and frozen foods such as fruits, vegetables, and seafoods are a longstanding traditional market for corrugated, in which a collated group of cans, bottles, or cartons are placed into a box in high speed automated packaging operations. Cans and cartons usually require only lightweight, low-cost RSCs which can be automatically set up, loaded, sealed, and palletized. Glass and plastic bottles usually require corrugated partitions or dividers for protection and necessary stacking strength in the palletized loads.

Fresh fruits, vegetables, and seafoods, like fresh meat and poultry, has been another relatively recent fast-growth mar-

Table 4. Major Corrugated End Use Industries[a]

SIC	Industry	% of corrugated usage
20	food products	37.2
21	tobacco products	0.6
22	textile mill products	1.8
23	apparel and fabrics	1.7
26	paper products	13.3
27	printing and publishing	1.6
28	chemicals	6.6
29	petroleum products	0.8
30	rubber and plastic products	6.9
31	leather products	0.4
	Subtotal, nondurable goods	*70.9*
24	wood products	0.8
25	furniture	3.0
32	stone, clay, and glass products	8.4
33	primary metal products	0.5
34	fabricated metal products	3.2
35	nonelectrical machinery	1.7
36	electrical machinery	4.7
37	transportation equipment	1.5
38	instruments, clocks	0.6
39	miscellaneous manufacturing	4.1
	Subtotal, durable goods	*28.5*
42	motor freight transportation	0.5
90	government	0.1
	Total	*100.0*

[a] Percentages based on total corrugated shipments in 1984 of 266,565,000,000 ft² (24,765,000,000 m²).

ket for corrugated boxes. Some of the largest volume items in this category are oranges, grapefruit, melons, lettuce, tomatoes, apples, peaches, cabbage, celery, and fresh fish. Usually heavier boards, occasional special designs, and moisture-resistant treatments are needed to withstand the rough treatment and protect the product (Fig. 10).

Another important food end use is the beverage industry, including juices, soft drinks, and alcoholic beverages. The bev-

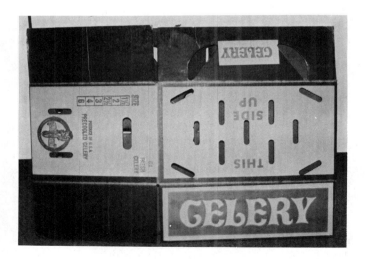

Figure 10. Waxed celery box (1).

erage market has outpaced growth in many other markets, which has been beneficial for corrugated volume. Because the entire food industry is relatively recession-proof, it contributes stability to the corrugated industry as a result of its dominant use of corrugated.

An unusual new market for corrugated is bulk liquids in a plastic bag in a box (see Bag-in-box, liquid product). Institutional deliveries of milk and wine are being made in such containers. The box must be designed to allow access to a dispensing valve built into the plastic bag. It must have good strength and quality to protect the rather fragile package of liquid inside.

Textile products. Textile mill and fabric products is a long-established corrugated use which has not grown as rapidly as other use industries. In some cases large bales of synthetic fiber have been wrapped in plastic film, replacing corrugated bale wrap. Considering the volume of apparel and other fabric products which do not need as much protection in shipment as food products, for example, the textile industry is not as intensive in corrugated usage as many other industries.

Paper. The paper and publishing industries is the second largest corrugated market after food. The major products in this category are sanitary paper products, printing paper in sheets, folding cartons (knocked down), and stationery. As the per capita consumption of all paper products has increased over many years, this has also been a major growth industry for corrugated shippers. For the most part, the boxes are low cost RSCs.

Chemicals and allied products. Chemicals, including drugs, cleaning compounds, toiletries, paints, rubber, plastics, and petroleum products represent another important end-use market. The largest product line in these categories is plastic products of all kinds. The growth of molded, thermoformed, and extruded plastic products for many uses has been a rapid growth market for corrugated boxes. Another large market is household chemical products, such as soaps, detergents, polishes, toiletries, and paints. Petroleum products such as lubricating oil and other miscellaneous chemicals also comprise a sizeable market. Finally, synthetic resins and fibers, particularly in bulk boxes, have created a fast-growing specialty corrugated market.

Durable goods. Furniture and wood products represent an important corrugated end use. Some portions of the household furniture market pose unusual problems in corrugated container design. Many special containers have been developed for approval in truck and rail shipments by common carriers. Items such as expensive tables and chairs must be protected from rough handling because of their vulnerability to damage in shipment. This has led to many special die-cut inserts and box designs, literally custom-made for each item.

Glass and ceramics. Glass and ceramic products have long been an important market for corrugated. This is in addition to food and other household products packaged in glass bottles. Virtually all glassware and pottery is shipped from the manufacturer to market in corrugated boxes, usually with corrugated inserts or special designs to protect the breakable product. Expensive sheet glass also requires highly protective containers, sometimes involving combinations of corrugated and wood.

Machinery. Machinery of all types, transportation equipment, and metal products combine to represent a major corrugated market. The most significant of these are electric appliances, automotive parts, tools, and other household products. A trend in the shipment of some machinery and automotive parts is to attach a very rigid triple-wall corrugated box to a wood skid for containing the heavy product. The triple-wall gives resistance to impact and bending. This type of container, as well as large appliance containers, may also be fitted with inside corner posts for rigidity and stacking strength. The corner posts are made of wood or glued corrugated combinations.

Miscellaneous. A miscellaneous category of markets includes toys and sporting goods. A recent development in these markets as well as some housewares is the use of lithographed printed labels glued to the outside of the corrugated box. The box then becomes a more attractive retail package on store shelves.

Corrugated displays in retail stores is a growing market because of the economy in a limited quantity for a one-time use, compared with more expensive materials. Sometimes corrugated is used in combination with wood, plastic, or metal for displays. High quality multicolored printing to advertise the product is also important. Complex die-cut designs have been developed to enable the same corrugated unit to act as both a shipping container and store display for the product.

Another new corrugated market is a U.S. Postal Service mail tray to replace fabric bags. These consist of nestable trays with open-ended sleeves to protect letters during shipment. They are shipped back and forth between major distribution points and local post offices.

BIBLIOGRAPHY

1. *Fibre Box Handbook,* Fibre Box Association, Chicago, 1984.
2. Technical Association of the Pulp and Paper Industry (TAPPI), Atlanta, Ga.
3. *Uniform Freight Classification,* Uniform Freight Classification Committee, Chicago.
4. *National Motor Freight Classification,* American Trucking Associations, Washington, D.C.
5. Federation Europeene des Fabricants de Carton Ondule, Paris.
6. "A U.S. Case History: Barcoding Corrugated," *Paperboard Packag.* **69**(11), 22–30 and (12), 18–23 (1984).
7. "Equipment Problems Overcome, Pre-Printed Liner Demand Grows in U.S." *Boxboard Containers,* **92**(3), 21–25 (October 1984).

General References

"State of the Industry Report on Corrugated Containers," *Paperboard Packag.* **69**(10), 33–48 (1984).

J. F. Hanlon, *Handbook of Packaging Engineering,* 2nd ed., McGraw-Hill Book Company, New York, 1984.

The Packaging Encyclopedia, Cahners Publishing, Des Plaines, Ill., annual.

LEONARD F. SWEC
Consultant

BOXES, RIGID-PAPERBOARD

Rigid-paperboard boxes are also called "setup" boxes. Unlike folding cartons (see Cartons, folding) they are delivered to the packager setup and ready to use. The rigid box was originally used by the Chinese, who were among the earliest to discover a process for making strong and flexible paper from rice fiber. The first known use for a paper box was tea. The word *box* generally means a receptacle with stiff sides as distinguished from a basket, and so-called because it was first made from a tree called Box or Boxwood.

Boxes for gift-giving became popular over 2000 years ago when the Roman priests encouraged people to send presents during the seasons of rejoicing. The paperboard box of current use originated in the 16th century, with the invention of pasteboard. In Europe, one of the earliest types of paper boxes was commonly known as a band box. It was a box highly decorated by hand and was used to carry bands and ruffles worn by the Cavaliers and Ladies of the Court. It was not until 1844 that setup boxes were manufactured in the United States. Starting with a machine that cut the corners of the box, Colonel Andrew Dennison soon found that manufacturing boxes by hand was tedious work and developed the Dennison Machine which led to the creation of the Dennison Manufacturing Company. The Colonel's invention was revolutionary, but until the Civil War, most consumer products were packaged in paper bags or wrapped in paper. There were only about 40 boxmakers in the country and most boxes were made by hand. For these 40 craftsmen the box business was merely an adjunct to other lines of business, which varied from printing to the manufacture of the consumer items they would eventually pack.

In 1875, John T. Robison, who had worked with Colonel Dennison and others, developed the first modern scoring machine, corner cutter, and shears. These three machines still form the machinery basis for most box shops, but it was not until the end of World War II that significant progress was made to improve the production of machinery for the industry. Today's machinery takes a scored piece of blank boxboard through to the finished covered box.

Manufacturing Process

The process starts when sheets of paperboard are sent through a machine known as a scorer. The scorer has circular knives that either cut through or partially cut through the paperboard and form the box blanks from the full sheet. The scorers must be set twice, once for the box and once for the lid, since there is usually a variation of 1/8 to 3/8 in. (3.2–9.5 mm) between the box and the lid. After the individual box blanks are broken from the full sheet they are stacked and prepared for corner cutting. Once the corners are cut the basic box blank is ready. Diecutting, usually performed on a platen press and an alternative method of cutting blanks, is economically justifiable for large orders.

The blanks are now ready for staying. For small quantities the box blanks can be sent through a single stayer where an operator must first bend all four sides of the blank to prebreak the scores. The single-staying machine will then glue a strip of 7/8-in. (22-mm) width kraft paper of the required length to each corner of the box. For greater quantities the boxmaker uses a quad stayer. The quad stayer feeds the box blanks automatically under a plunger which has the same block size as the box or cover to be formed. With each stroke of the machine the box blank sides are turned up and stay paper is applied to all four corners at one time.

After staying, the box is in acceptable form, requiring only an attractive outer wrap. In most operations, the paper box wrap is placed onto the conveyor gluer which applies hot glue to the back of the wrap and then places it on a traveling belt. The wrap is held in place by suction under the belt. As the

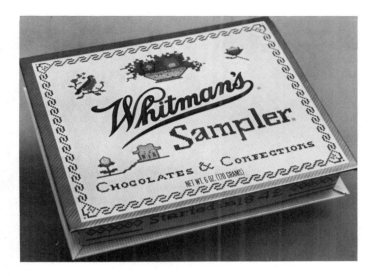

Figure 1. A new version of a traditional candy box.

wrap travels on the belt it is either removed from the belt by a machine operator, who manually spots the paperboard box on the wrap, or it is automatically spotted by machine. After spotting, the box and wrap move on to another plunger mechanism where the wrap is forced around the box. Simultaneously, nylon brushes smooth the paper to the four sides of the box. Just before the plunger reaches the bottom of the stroke, the wooden block splits allowing metal "fingers" to push the paper in the box. The wooden block then closes together and completes its downstroke where felt-lined blocks press the sides and assure the gluing of the paper to the inside as well as the outside of the box.

The manufacturing process described above is for the simplest box, but the setup box can accomodate unusual requirements with regard to windows, domes, embossing, platforms, hinges, lids, compartments, and other variations. Standard variations of the rigid box include the telescope box, the ended box, the padded-cover box, special shapes (eg, oval, heart-shaped), slide tray, neck or shoulder style, hinged cover, slanted side, full telescope, box-in-box, specialty box, interior partition, extension bottom, three-piece, slotted partition, and interior platform. This versatility is extremely valuable in meeting the merchandiser's demands for quality, quantity and convenience. Figures 1 and 2 show a familiar candy box and an unusual configuration for cosmetics.

Materials

Four primary materials are needed for manufacturing the rigid box: chipboard (for the rough box), stay paper (to hold the sides of the box together), glue (to hold the outer wrap to the box), and outer wraps (for the decorative appearance).

There are four common types of chipboard used in boxmaking: plain, vat-lined book-lined, and "solid news." Plain chipboard, is made entirely of waste paper, with the minimum of selection or de-inking of waste. Vat-lined chipboard provides a cleaner appearance. It is primarily chipboard with a liner (made of low grade-white waste paper) that is applied on the board machine at the mill. Book-lined chipboard is chipboard with a liner of book or litho paper pasted to one or two sides as a separate operation. "Solid news" differs from plain chipboard

in that the waste used in this sheet is a little more selected and is made mostly from newspaper waste which has been de-inked. Other boards are available, including glassine-lined, foil-lined, and folding grades of boxboard, but the four listed above are those that are primarily used.

To properly serve the customer, a boxmaker must stock various sizes and weights of board. Setup boxboard is measured in basis weight: a 50-lb (22.7-kg) bundle of 50 basis weight boxboard contains 50 standard 26 × 40-in. (66–102-cm) sheets; a 50 lb (22.7 kg) bundle of 40 basis weight boxboard contains 40 sheets. The larger the basis weight, the thinner the sheet.

Stay paper is almost always ⅞-in. (22-cm) wide kraft or white paper. Glues are either animal-or starch-based, formulated to dry fast or slow. They are available in dry form for mix-it-yourself, or flexible form to melt as used. Most glue is hot melted to give a faster drying and lay-flat quality to the paper with minimal warping (see Adhesives).

The final ingredient in box manufacturing is the outer wrap. There are thousands of stock papers, most of them available in 26-in. (66-cm) or 36-in. (91.4-cm) rolls. The wraps are generally paper, but foil and cloth wraps are also used. Through the use of artwork, photography, and good printing, a boxmaker can also produce a distinctive custom-made wrap.

Applications

The rigid (setup) box has stood the test of time and competes well within its selected markets. It protects, it builds image, it displays, and it sells. The set-up box possesses unique qualities that satisfy the specific needs of all four segments of the marketing chain: the consumer, the retailer, the marketer, and the product packager. The boxes are delivered set-up and ready to load. Small "market-test" or emergency quantities can be prepared quickly at low cost, and individual custom designs are available without expensive investment in special tools, jigs or dies. The rigid "feel" creates consumer confidence and the manufacturing process permits utilization

Figure 2. The high-fashion look of a lacquered wooden box is copied with style in the glossy finish and fine detail work of this rigid box designed for Lancôme cosmetics. Especially noteworthy is the registration of the gold-stamped border on the wrap of each separate drawer. Brass drawer-pulls complete the illusion. Shelves are sturdy and wrapped, allowing the drawer to slide in and out smoothly.

Table 1. Use Survey of Industry Sales, Percent

Industry	1975	1977	1979	1981	1983
textiles apparel and hosiery	7.5	4.8	4.5	5.5	3.4
department stores and specialty shops	9.9	10.6	8.3	7.1	7.2
cosmetics and soaps	1.8	5.1	4.3	3.9	1.4
confections	13.3	8.1	15.7	18.0	17.2
stationery and office supplies	9.7	13.2	15.0	9.3	19.1
jewelry and silverware	5.2	14.0	7.6	9.7	6.3
photographic products and supplies	2.2	2.7	3.5	4.3	3.6
shoes and leather	4.8	1.1	0.4	0.2	0.4
drugs, chemicals, and pharmaceuticals	6.9	7.3	13.1	7.8	6.7
toys and games	2.3	2.6	1.4	1.8	2.2
hardware and household supplies	7.0	6.2	3.4	4.0	6.7
food and beverages	2.9	0.3	5.4	3.9	1.5
sporting goods	1.9	1.0	1.7	1.2	0.4
other major customers					
electronics	3.4	2.6	4.3	3.5	3.4
educational	3.9	6.5	3.8	4.4	2.7
other	12.2	7.1	1.1	7.0	7.9
Miscellaneous customers	5.1	6.8	6.5	8.4	9.9
Total	*100*	*100*	*100*	*100*	*100*

of varied overwraps to reflect product quality. The boxes are easy to open, reclose, and reuse without destroying the package. Rigid boxes are recyclable and made from recycled fibers. Production runs can be small, medium, or large, and volume can fluctuate without excessive economic penalities. In addition, these boxes provide lasting reuse features for repeat advertising for the seller. They are customized to provide product identity. Rugged strength protects the product from plant to consumer. Reinforced corners and dual sides provide superior product protection and minimize damage losses.

The major markets for rigid boxes are listed in Table 1, along with percentage figures that tend to fluctuate from year to year. They compete well in some of these markets but with changing technologies, industry forecasters predict a decline in rigid box use for department stores, textiles, personal accessories, and hardware and household items. To offset that decline, there appears to be an increase in use by manufacturers of computer software, confections, educational material, and electronic supplies.

Larry Lynch
Julia Anderson
National Paperbox and Packaging Association

BOXES, RIGID-PLASTIC

The common types of plastic boxes are similar in design and construction to paperboard set-up boxes (see Boxes, rigid-paperboard) and folding cartons (see Cartons, folding). The most widely used plastic boxes are injection-molded from crystal general-purpose polystyrene (GPPS) (see Injection molding; Polystyrene). Other box manufacturing processes involve extrusion, sometimes followed by fabrication or thermoforming (see Thermoforming).

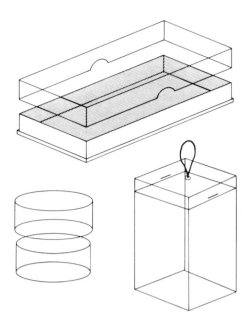

Figure 1. Nonhinged rigid-plastic boxes.

Injection-molded boxes. Rigid injection-molded boxes are available in a wide range of sizes, styles, and shapes, usually manufactured from tooling owned by the box manufacturer. Box styles include hinged and telescoping types. Stock shapes include rectangulars, rounds, squares, and special novelty designs. As custom molded containers, rigid-plastic boxes can be made in virtually any geometric shape subject only to the limitations of the injection-molding process (see Fig. 1).

Transparent polystyrene boxes are used as consumer packages for hardware items, cosmetic kits, fishing tackles, writing instruments, and games and toys, and in various industrial markets as packages and as in-plant storage containers. Crystal polystyrene is often chosen for retail products because its transparency provides point-of-purchase impact. For retail and industrial applications, transparency eliminates the need for double labeling since the product and its identification are visible through the walls of the container. This feature provides a cost benefit to the packager, which helps compensate for a box cost that is sometimes higher than the cost of alternative packaging. Reusability is another benefit for retail and industrial products, since the products are often packaged in multiple quantities.

Rigid injection-molded boxes are available in plastics other than crystal polystyrene, including but not limited to polypropylene (see Polypropylene), cellulosics, PVC (see Poly(vinyl chloride) and, if transparency is not required, impact polystyrene. The alternative materials are chosen for their special functional benefits such as superior durability. In the case of polypropylene, a one piece living-hinge construction reduces manufacturing costs while providing an excellent reusable hinge (see Fig. 2).

Hot stamping and silk screening (see Decorating) are the most popular methods used to decorate rigid injection-molded-plastic boxes. In high volume custom applications, mold engraving is often used to provide legal or sales decoration as the box is molded, thereby eliminating the added expense of secondary decorating. Although engraving is usually less attractive than the other methods because it lacks color contrast, it

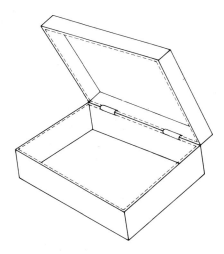

Figure 2.   Rigid-plastic box with living hinge.

is extremely cost effective once the initial tooling cost has been incurred.

Rigid injection-molded hinged boxes are made in cover and base sections, and then the two sections are mechanically attached at the hinged area. The most widely used hinged construction is a ball-and-socket style (see Fig. 3). For larger boxes, separate metal or plastic hinges are used if greater hinge strength is required. Substitute raw materials for molded boxes such as K-resin (Phillips Chemical Co.) (see Styrene–butadiene copolymers) and cellulose propionate are also used to effect more durable boxes and stronger hinges.

The rigid-plastic box market in the United States is dominated by four manufacturers. Annual sales total is approximately \$12 million (10^6).

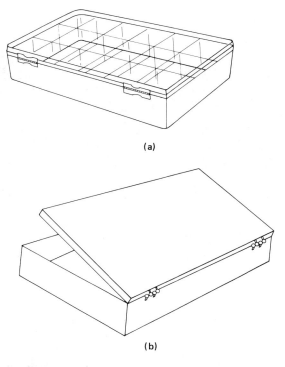

(a)

(b)

Figure 3.   **Hinged rigid-**plastic boxes. (**a**) Metal-hinged box. (**b**) Ball-and-socket hinge.

Extruded boxes.   Plastic boxes are also made from extruded-plastic tubing in virtually any geometric shape, although the most popular cross section is round. In such cases, the semirigid extrusion is cut the desired length, sealed with a permanent bottom closure, and supplied with a removable top closure, which is usually friction fit to the tube. Standard containers of this type are used for cosmetics, hardware, soft goods, toothbrushes, and scores of other applications. The extruded container's closure can be made to hang for display and storage purposes or it can be made to dispense multiples of products one at a time. The extruded-plastic box offers the benefits of transparency (cellulosic materials), reusability and little, if any, tooling cost.

Extruded containers can be decorated in a variety of ways including offset printing, silk screening, flexographic printing, and hot stamping.

There are two major extruded-box manufacturers in the United States with total annual sales of approximately \$9 million (10^6).

Thermoformed boxes.   Plastic boxes made by this process are usually made in one piece from plastic sheet with an integral hinge. These cost-effective constructions are often made in tray form and combined with a paperboard base or cover. The paperboard, in addition to its function of positioning the packaged product, provides inexpensive decoration. Most thermoforming processes are capable of manufacturing thermoformed boxes. However, a handful of major thermoformers provide design services that the smaller companies cannot offer.

Fabricated boxes.   Fabricated plastic boxes are made from clear sheet, either GPPS, PVC, acetate, or polyester (see Polyesters, thermoplastic). The sheet is cut, scored, folded, and glued much like the fabrication of a folding carton or a set-up box. With such products, decoration is relatively inexpensive since the sheet can be printed before the fabrication is performed. Applications for transparent boxes include soft goods, gift items, cosmetics, and toiletries.

Two fabricated box styles are commonly used. The fabricated plastic carton is identical in construction to a paperboard folding carton with integral end-flaps as closures. There are also fabricated open-end seamed-sleeves, usually made in rectangular or square shapes. They are made to slide over set-up paperboard boxes or thermoformed-plastic bases. In some instances, the ends are closed with injection-molded caps and plugs. In addition to the benefits of transparency and excellent decorating techniques, fabricated cartons and sleeves can be shipped in a collapsed form, which minimizes shipping costs. Although these products are not new, their use has increased substantially since the early 1970s, when European product and machinery design ideas were introduced in the United States.

BIBLIOGRAPHY

General References

Packaging Encyclopedia, Cahners Publishing, Boston, Mass.
J. F. Hanlon, *Handbook of Packaging Engineering,* 2nd ed., McGraw-Hill, New York, 1984.
Modern Plastics Encyclopedia, McGraw-Hill annual, McGraw-Hill, New York.

J. C. Sciaudone
Tulox Plastics Corporation

BOXES, SOLID-FIBRE

Solid-fibre containers are used almost exclusively for applications in which container return and reuse are desirable and where return can be controlled by the distributor. Without such control, the impetus to use the multi-trip shipping containers, which are more costly than corrugated boxes, would not exist.

As a rule of thumb, the solid-fibre box costs two to three times as much as a comparable-size general-purpose corrugated shipper. However, the solid-fibre container can be used an average of 10 to 15 times before retirement. The economics are obvious, but only in a "closed-loop" distribution system.

In the United States, annual shipments of solid fibre total about 900 million (10^6) ft^2 (83.6 km^2). That is less than 1% of those recorded by corrugated board for the same general use category (1). Nevertheless, solid fibreboard possesses intrinsic performance values that have assured a continuity of acceptance dating back to the late nineteenth century, when solid fibre and corrugated board began to supplant wooden cases in world commerce.

Solid fibreboard differs from corrugated board in several significant respects (see Boxes, corrugated). As its name implies, the former is a solid (nonfluted) structure consisting of two or more plies of containerboard. Four plies generally are used to manufacture board from which solid-fibre boxes are to be produced.

Solid-fibre sheets are constructed by gluing roll-fed containerboard plies together on a machine called a laminator. The plies are bonded under controlled pressure to form a sheet that comes off the line as a continuous strip that is subsequently cut to predetermined lengths.

Caliper of the finished board is the result of the number and thickness of individual plies. It varies according to the needs of the customer market. For most applications, finished sheet thickness ranges between 0.035 and 0.135 in. (0.089 and 0.343 cm). (In the industry lexicon, 0.035 in. is called 35 points.) For special heavy-duty applications, solid fibreboard of 250-point thickness can be produced. This, however, is the exception. For the largest market (shipping containers), board thickness averages 70 to 80 points (0.18–0.20 cm).

When the continuous web of solid fibreboard exits the laminator, it is cut to length. Converting equipment prints, die-cuts, and, if needed, coats the material with polyethylene or other protective finish. The converted board is then set up or assembled into containers, usually at the customer's plant.

Brewery use of solid-fibre shipping containers encompasses a variety of constructions, both closed-top and open-top. A common feature is a die-cut handhold at each end to facilitate carrying.

Breweries are the largest users of solid-fibre shipping containers, largely because of that industry's relatively manageable distribution controls. Overall, shipping containers take about 40% of the total annual volume of solid fibreboard, and 80% of all solid-fibre shippers are used to carry bottled beer. The rest of the solid-fibre box business is shared mainly among producers of meat, poultry, fish, and fresh produce (for so-called wet shipments). The U.S. government also uses solid fibre containers for "C" rations.

Soft-drink bottlers once were heavy users of returnable solid-fibre shippers. This market, however, has diminished for two reasons. One is competition from molded plastic crates (see Crates, plastic). The other is the decline of returnable soft-drink bottles.

Most industry authorities agree that local or state deposit bills that mandate return of empty bottles or cans will have little long-term effect on solid-fibreboard use trends. Consensus is that only a nationally enforced litter law will bring about a meaningful increase in sales to the beverage industry.

A principal difference in construction between solid-fibre and corrugated boxes is that rail and highway shipping-authority rules limit solid-fibre boxes to only two styles of manufacturer's joints: stitched and extended-glued. Standard methods for testing solid-fibre boxes have been developed and published by the American Society for Testing and Materials (ASTM) and the Technical Association of the Pulp and Paper Industry (TAPPI). Common-carrier requirements for solid-fibre box performance (burst strength, size, and weight limits) have been established and are fully detailed in Rule 41 of the Uniform Freight Classification (rail shipment) and Item 22 of

Figure 1. Solid-fibre beverage cases and tote boxes. (**a**) 1. One-piece bin box with locking feature. 2. Three-piece 24/12-oz (355-mL) beverage case. 3. One-piece vegetable box. 4. Attaché-style tote box. (**b**) 5. One-piece 4/1-gal (3.785-L) beverage case. 6. Attaché-style tote box. 7. One-piece lidded tote box. 8. One-piece self-locking tote box. 9. Three-piece 24/12-oz (355-mL) beverage case.

the National Motor Freight Classification (truck shipment) (see Laws and regulations).

After containers, the largest use of solid fibreboard is for slip sheets (see Slip sheets). They account for about 45% of annual solid-fibre sheet production. Slip sheets are gaining wide acceptance in materials-handling applications, chiefly as replacements for bulkier and more expensive wood pallets. Designed for forklift handling and requiring minimal warehouse space, solid-fibre slip sheets are easy to use and store.

The remaining 15% of annual solid-fibreboard volume goes into a variety of applications. Point-of-purchase displays display is one. Furniture is another (sheets are affixed to the wooden framework to provide a firm backing for upholstery). Mirror backing is another. Automative applications include nonloadbearing interior bulkheads. Drums, railcar dunnage, multi-tier partitions for glass-container shipments, and wire-and-cordage reels are other miscellaneous uses in which solid fibreboard's strength, damage resistance, and machinability are cost-effective.

A major source of information on solid fibreboard is the Fibre Box Association, with headquarters in Chicago, Ill. Available information includes annual statistical reports on shipments, inventory, etc, and the *Fibre Box Handbook* (1). This comprehensive reference includes industry definitions, details on box constructions (see Fig. 1) and testing procedures, and a full discussion of all regulations affecting solid-fibre boxes.

BIBLIOGRAPHY

1. *Fibre Box Handbook,* Fibre Box Association, Chicago, Ill.

General References

A. R. Lott, "Solid and Corrugated Fibreboard Cases" in F. A. Paine, ed., *The Packaging Media,* John Wiley & Sons, Inc., New York, 1977.

ROBERT QUINN
Union Camp Corporation

BOXES, WIREBOUND

Wirebound boxes and crates are high strength-to-weight ratio, resilient containers designed to support heavy stacking loads even under adverse humidity and moisture conditions. They are available worldwide and are used for a variety of agricultural and industrial products. Wirebounds are fabricated and delivered in flat form to conserve shipping and warehouse storage space and may be easily assembled, as required, by the user with simple hand tools or automated equipment when volume warrants.

The basic material used in wirebound construction is wood integrally combined with steel binding wires fastened to the wood elements by staples. In a relatively small number of design applications corrugated fiberboard is substituted for some wood components and plastic strapping replaces steel binding wire. The many design options available by combining the various components of the composite container present an opportunity to custom design wirebounds for specific products, weight, sizes, and distribution hazards. The high speed automated machinery used to fabricate wirebounds is custom built for specific size ranges and styles of wirebounds.

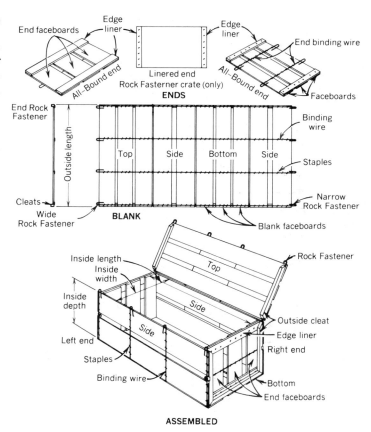

Figure 1. Standard reference for All-Bound wirebound crates.

Wirebound Styles

The most frequently used style of wirebound is the All-Bound (Fig. 1) used in crate form for shipment and storage of fresh fruits and vegetables, and in box form (Fig. 2) for industrial and military applications. The openings between faceboards in the Figure 1 style All-Bound provide needed ventilation for such produce items as sweet corn, celery, beans, cabbage, citrus, etc (1). Outside the United States this style wirebound is often called the Bruce box. When the All-Bound is intended for industrial or military products it is normally specified with heavier face material, extra binding wires, and/or interior cleats (compare Figs. 1 and 2). The All-Bound is characterized by binding wires on all six faces of the made-up container, and a cleat framework in the vertical plane. Both assembly and closing is accomplished by engaging the wire loops at the ends of each binding wire. These loops are called Rock Fasteners. Two simple hand tools are required to engage the Rock Fasteners: the Bon Ender for assembly of the All-Bound and the Sallee Closer for final closure. If high volumes of All-Bounds are to be assembled or closed at one location, automated equipment is available. A variation of the All-Bound, the Rock Fastener box, is less frequently used. The Rock Fastener box ends do not have binding wires and may be of solid plywood, linered or battened construction.

A completely different style of wirebound, the Wirebound pallet box (Fig. 3) is used for bulk storage and/or shipment of industrial products including auto and plane parts, chemicals, castings and forgings, etc (2). The bottom horizontal cleats are

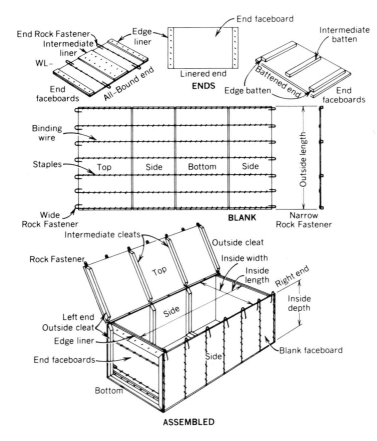

Figure 2. Standard reference for All-Bound wirebound boxes. When linered end and battened end are used, the wirebound style is Rock Fastener box.

either fastened to or lock onto a two- or four-way pallet base for fork truck handling and stacking. Cleats may be located inside the pallet container as shown or outside if the nature of the load requires a smooth interior.

Although All-Bounds and Wirebound pallet boxes are often manufactured, inventoried, and sold as stock containers, many wirebounds are custom designed and manufactured on order for specific applications. These applications include cast-iron bathtubs, insulators, garden tractors, machine tools, etc (3). Net weights carried range between 100 and 6000 lb (45 and

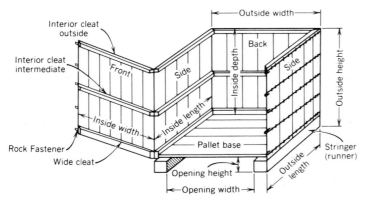

Figure 3. Wirebound pallet box.

2722 kg). There is no size limit since blank sections may be joined together if required. Wirebounds for military use are generally covered under Federal specifications PPP-B-585-C, *Boxes, Wood, Wirebound* and Military specification MIL-B-0046506-C, *Boxes, Ammunition Packing, Wood, Wirebound.*

Wirebound Material

Any of the deciduous wood species in Group IV or III (4) may be used for wirebound containers. Selected Group II and I woods, from higher density coniferous trees, are permissible in specific design situations. The faceboard members may be generated directly from the log on high speed veneer lathes equipped with a backroll and spur knives, or from resawing sawmill-supplied rough cut lumber using band saws. Thicknesses from the first method generally range from 1/9 in. to 3/16 in. (2.8 to 4.8 mm) and the resawn material 1/4 in. to 3/8 in. (6.4 to 9.5 mm). In general, the thinner veneer faceboards are used in wirebounds for produce and the heavier resawn faceboards for industrial and military applications. Both forms of faceboards are dried to 15–20% moisture content before fabrication to prevent subsequent mold growth. Wirebounds for long-term military storage or export are also dipped in a wood preservative (5). The cleat structure is sawn from air dried planed lumber on specialized saw equipment.

The steel wire used for both binding the wirebound and to form staples has unique characteristics that permit machine runability and functional performance (6). Depending on its use in the wirebound, the low carbon steel wire has a tensile strength between 45,000 and 125,000 psi (310 and 861 MPa) and an elongation of 0.5 to 10%. A simple testing device, the Rockaway Wire Tester, is used by wire suppliers and wirebound manufacturers to ensure wire performance.

BIBLIOGRAPHY

1. *Wirebound Boxes and Crates for Fresh Fruits and Vegetables,* Package Research Laboratory, Rockaway, N.J. 1980.
2. *Engineered Wirebound Pallet Containers,* Package Research Laboratory, Rockaway, N.J., 1972.
3. *Versatile Wirebounds,* Package Research Laboratory, Rockaway, N.J., 1983.
4. *Wood Handbook,* U.S. Department of Agriculture, Forest Products Laboratory, Madison, Wisc., revised 1974.
5. Federal Specification, *Wood Preservative, Water Repellent TT-W-572B,* May 28, 1969.
6. Wire for Wirebound Boxes and Crates, Package Research Laboratory, Rockaway, N.J., 1979.

H. H. Dinsmore
Stapling Machines Company
Division of Rockaway Corporation

BOXES, WOOD

The use of wooden boxes and crates dates back to the Industrial Revolution, when the building of roads and railways led to their development as the first "modern" shipping containers (1). (See also Boxes, wirebound; Pallets, wood.) They are still used today for products that require the strength and protection that only wood can provide. The difference between a box (or case) and a crate is that a box is a rigid container with closed faces that completely enclose the contents. A crate is a

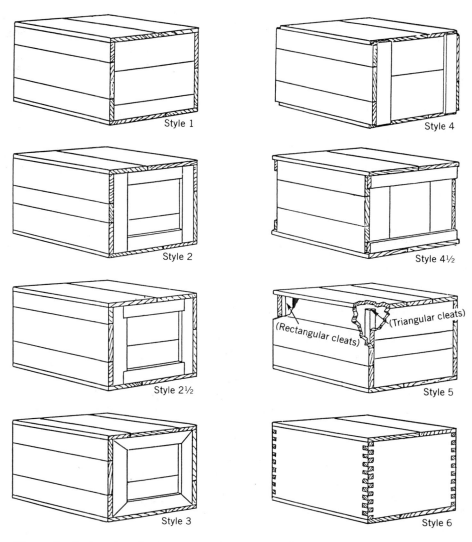

Figure 1. Styles of wooden boxes (3, 5). Style 1: uncleated ends. Style 2: full-cleated ends, butt joints. Style 2½: full-cleated ends, notched cleats. Style 3: full-cleated ends, mitered joints. Style 4: two exterior end cleats. Style 4½: two exterior end cleats. Style 5: two interior end cleats. Style 6: lock corner.

Table 1. Commercial Box Woods[a]

Group 1		Group 2	Group 3	Group 4
Alpine fir	Magnolia	Douglas fir	Black ash	Beech
Aspen	Noble fir	Hemlock	Black gum	Birch
Balsam fir	Norway pine	Larch	Maple (soft	Hackberry
Basswood	Redwood	North Carolina	or silver)	Hickory
Buckeye	Spruce	pine	Pumpkin ash	Maple (hard)
Butternut	Sugar pine	Southern	Red gum	Oak
Cedar	Western	yellow pine	Sap gum	Rock elm
Chestnut	yellow pine	Tamarack	Sycamore	White ash
Cottonwood	White fir		Tupelo	
Cucumber	White pine		White elm	
Cypress	Willow			
Jack pine	Yellow			
Lodgepole	poplar			
pine				

[a] Ref 4.

rigid container of framed construction. The framework may or may not be enclosed (sheathed) (2).

Boxes

Box styles.   Wooden boxes are of either "nailed construction" or "lock-corner construction." Of the eight basic box styles shown in Figure 1 (3), Style 6 is the only lock-corner construction. The others are nailed (cleated).

Wood.   Specifications for wooden boxes refer to the categories developed by the United States Forest Products Laboratory, which relate to strength and nail-holding power (see Table 1). Groups 1 and 2 are relatively soft; Groups 3 and 4 are relatively hard. For a given box of a given style, the thickness of the wood and cleats depends on the type of wood.

Fastenings.   The strength and rigidity of crates and boxes are highly dependent on the fastenings: nails, staples, lag screws, and bolts. Nails are the most common fastenings in the construction of boxes. The size and spacing of the nails depends on the type of wood (5).

Loads.   The type of load is determined by the weight and size of the contents and its fragility, shape, and capacity for support of, or damage to, the box. Loads types are classified as Type 1 (easy); Type 2 (average); or Type 3 (difficult). Descriptions and examples of each load type are contained in Ref. 5.

Crates

A wood crate is a structural framework of members fastened together to form a rigid enclosure which will protect the

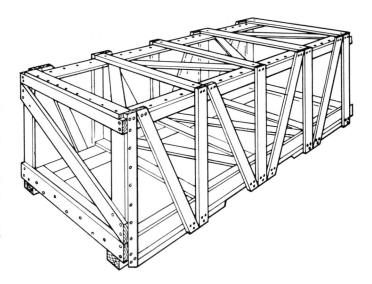

Figure 2.   Simple wooden crate.

contents during shipping and storage. The enclosure is usually rectangular, and may or may not be sheathed (4). A crate differs from a nailed wood box in that the framework of members in sides and ends must provide the basic strength. A box relies for its strength on the boards of the sides, ends, top, and bottom. A crate generally contains just a single item, and its

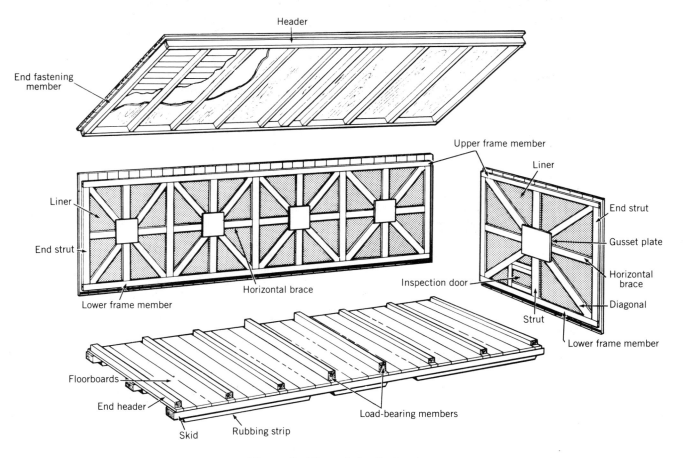

Figure 3.   Plywood-sheathed crate.

dimensions are not subject to standardization. The function of a crate is to protect a product during handling and shipping at the lowest possible cost. A simple enclosing framework is shown in Figure 2 (4).

Some products and shipping conditions require greater protection. The value of the contents or the likelihood of top loading may dictate the use of a sheathed crate (6). The sheathing can be lumber or plywood. A plywood-sheathed crate is shown in Figure 3 (4).

BIBLIOGRAPHY

1. F. A. Paine, *The Packaging Media,* John Wiley & Sons, Inc., New York, 1977.

2. *ASTM D-996: Standard Definitions of Terms Relating to Packaging and Distribution Environments,* ASTM, Philadelphia, Pa.

3. National Wooden Box Association, *Specifications for Nailed Wooden and Lock Corner Boxes for Industrial Use,* Washington, D.C., 1958.

4. *Wood Crate Design manual, Agriculture Handbook No. 252,* U.S. Department of Agriculture, Forest Service, Feb. 1964.

5. *Federal Specification PPP-B-621C,* Oct. 5, 1973.

6. American Plywood Association, *Plywood Design Manual: Crating,* Tacoma, Wash., 1969.

C

CANNING, FOOD

Canning may be defined as the packaging of perishable foods in hermetically sealed containers that are to be stored at ambient temperatures for extended times (months or years). The objective is to produce a "commercially sterile" food product. Commercially sterile does not mean that the food is free of microorganisms, but rather that the food does not contain viable organisms which might be a public health risk or might multiply under normal storage conditions and lead to spoilage. Canning processes do not necessarily kill all microorganisms present in a food, and it may be possible to isolate viable organisms from canned foods. The food product may be made commercially sterile either prior to or after filling and sealing. Three conditions must be met for canning safe and wholesome food:

1. Sufficient heat must be applied to the food to render it commercially sterile.
2. The container must prevent recontamination of the product.
3. The filled and sealed container must be handled in a manner which prevents loss of integrity.

Canning was invented as a means of food preservation in 1810 in response to a prize offered by Napoleon. The original containers were corked glass; handmade tinplate "canisters" (shortened to "cans") were introduced shortly afterward. Cans used prior to 1900 were manually produced from a cylindrical body, an end unit or disk, and a top ring. All seams were formed by dipping in hot solder. The food was filled through the hole in the top ring and a plate containing a small hole was soldered over the opening. Cans were heated to exhaust the headspace so that a partial vacuum would be created after sealing and cooling. A drop of solder was used to seal the small hole and the can was then thermally processed.

Around the turn of the century, the process for manufacturing the three-piece open-top can became widely available. This container used the same double-seamed ends that are in use today. The second end was not put on until the can was filled which meant that food no longer had to be forced through the hole in the ring.

Today the tin-plated steel double-seamed can is still the predominant food canning package (see Can seamers; Cans, steel). Glass is also used for some products (see Glass container design). Very recently, flexible pouches (see Retortable flexible and semi rigid packages), rigid plastics (see Cans, plastic), and thin aluminum (see Cans, aluminum) cans have been used to can foods. Processes have also been developed in which the food and container are commercially sterilized separately, often by different methods, and the container filled and sealed without recontamination (see Aseptic packaging).

Food canning accounts for just over 30% of U.S. metal can shipments (1) and just under 30% of glass container shipments. Approximately 1700 canning plants process about 36 billion (10^9) pounds (16.33×10^6 metric tons) of food per year (2). The importance of canning in marketing food products varies widely. Virtually all tuna is canned, as is 90% of the tomato crop (2). Other foods rely less on canning.

Process Description

The processing of canned food must produce a commercially sterile product and minimize degradation of the food. The container must also withstand the process and prevent recontamination of the product after processing and up to the time of use, often months or years after processing. The most common sequence of events in canning is that the food product is prepared for canning, the container is filled and hermetically sealed, and the sealed container is thermally processed to achieve commercial sterility. The thermal process necessary to commercially sterilize a canned product depends on the acidity of the food.

Role of pH. High acid foods such as fruits and fruit juices, pickled products, and products to which acid is added in sufficient amounts to give a pH of 4.6 or lower require considerably less heat treatment than low acid foods (pH >4.6). Low acid foods include most vegetables, meats, fish, poultry, dairy, and egg products. High acid foods may be processed at boiling water temperature (212°F or 100°C) after sealing. High acid liquid foods, such as fruit juices, may also be sufficiently processed by "hot-filling" the container with product near the boiling point and allowing slow cooling after sealing. Low acid foods (pH >4.6) must be processed at temperatures above the boiling point of water. Most often this is accomplished in a pressurized vessel called a retort or autoclave, which contains water or steam at 250°F at 15 psi (121°C at 103 kPa).

Time and temperature requirements. The process or scheduled process refers to the specific combination of temperature and time used to render the food commercially sterile. Several factors affect this process including the nature of the product, shape and dimensions of the container, temperature of the retort, the heat-transfer coefficient of the heating medium, the number and type of microorganisms present, and the thermal-death resistance of these microorganisms.

The relationship between the heat destruction of specific organisms, heating time, and temperature has been intensely studied since the 1920s and equations have been derived for several organisms, the most important of which is *Clostridium botulinum* (3). This spore-forming organism is found in soil, is ubiquitous, and grows in anaerobic environments such as canned foods and produces a deadly toxin. Its spores are also highly heat resistant. When commercially sterilizing a canned product, it is the temperature profile of the coldest spot in the container which must be known before the correct process can be calculated. This is accomplished by placing a thermocouple inside the can, usually at the geometric center for products heated by conduction or in the lower portion of the can for products heated by convection. The temperature is recorded during heat processing and used to calculate the proper process time under the given conditions of product, container size and geometry, and retort temperature. This information coupled with the thermal death characteristics of *Cl. botulinum* or a more heat-resistant organism is used to determine the correct process time. Often these processes are described in terms of F values.

F value. An F value is the time in minutes to heat-inactivate a given number of a certain microorganism at a fixed temperature (4). If the temperature is 250°F (121°C) and the organism is *Cl. botulinum*, the F value is called F_o. This value is the number of minutes required to kill a given population of *Cl. botulinum* spores at 250°F (121°C). Combinations of times and temperatures other than 250°F and F_o minutes can inactivate the same number of spores; temperatures lower than 250°F for longer time periods or temperatures greater than

250°F for shorter times have an equal ability to inactivate the spores. An $F_o = 2.45$ min reduces the population of *Cl. botulinum* spores by a factor of 10^{12} (5). In practice, F_o values of greater than 3 are used as a safety measure. In order to prevent overprocessing (overcooking) of the food, the spore inactivation (called lethality) is summed up during the time the coldest spot in the can is coming up to the retort temperature. Some viscous conduction-heated foods may never completely reach the common retort temperature of 250°F (121°C) yet still receive the proper F_o treatment. The Food and Drug Administration regulations require that these tests and calculations be carried out only by recognized authorities (6).

Recent years have brought renewed interest in thermal processes which commercially sterilize fluid foods in continuous-flow heat-exchange systems before packaging (7). The thermal death calculations described above still apply to these processes, and proper F values must be achieved (8). The continuous-flow commercial sterilization procedures have the advantage that products can be heated and cooled more rapidly for shorter times with equal lethality. This can give a higher quality product. These products must be filled into presterilized containers (see Aseptic packaging).

Canning Operations

The canning process requires several unit operations that normally take place in a set sequence (9).

Product preparation. As soon as the raw agricultural product is received at the canning plant, it is washed, inspected, sorted to remove defective product, and graded. Often, the edible portion is separated from nonedible as in the case of peas or corn. Fruits and vegetables are subjected to a blanching operation by exposing them to either live steam or hot water at 190–210°F (88–99°C). Blanching serves to inactivate enzymes which would otherwise cause discoloration or deterioration in the product. It also softens, cleans, and degases the product. Peeling, coring, dicing, and/or mixing operations may be carried out next. These operations prepare the product for filling into the can.

Container preparation. Containers must be thoroughly washed immediately prior to filling. Cans are washed inverted so that any foreign objects and the excess water can drain out. The container is now ready for filling. Accurate and precise filling is necessary to meet minimum labeled fill requirements yet leave sufficient headspace for development of the proper vacuum after closure. Too large a headspace results in an underweight container, while overfilling can result in bulging or domed ends after processing. Excessive headspace may also suggest that large amounts of oxygen remain in the can which accelerates product deterioration and can corrosion. Liquid or semiliquid products including small pieces are filled by automated equipment. Larger, more fragile products such as asparagus, are packed by hand or by semiautomated equipment. In most products, brine, broth, or oil is added along with the product. This liquid excludes much of the air between the particles and provides for more efficient heat transfer during thermal processing.

Vacuum. Proper application of the closure after filling is one of the most critical steps in the canning operation. The two-step seaming operation must not only produce a sound, well-formed double seam at speeds of several hundred cans per minute but must also produce an interior vacuum of 10–20 in. Hg (34–68 kPa) (10). This vacuum reduces the oxygen content

and retards corrosion and spoilage, leaves the can end in a concave shape during storage, and prevents permanent distortion during retorting. Proper internal vacuum can be achieved by several methods. Containers that are sealed while the food is at or near the boiling point develop a vacuum when the product cools. This preheating or hot fill also serves to sterilize the container when high acid foods are packaged. Products which are cool when filled can be heated in the container prior to sealing with the same result as the hot fill. This is often termed "thermal exhaust." Internal vacuum may also be achieved by mechanical means. The filled, unsealed container is fed into a vacuum chamber by means of an air lock and the closure sealed while under vacuum. This system has the disadvantage that flashing of the liquid may occur if air is entrapped in the food or high levels of dissolved air are found in the liquid. The most common method of producing internal vacuum is by displacing the air in the headspace with live steam prior to and during double seaming the cover. The steam in the headspace condenses and forms a vacuum as the container cools.

Retorting. In conventional canning operations of low acid foods, the sealed containers are next thermally processed at 250°F (121°C) in retorts. Recent regulatory agency rule

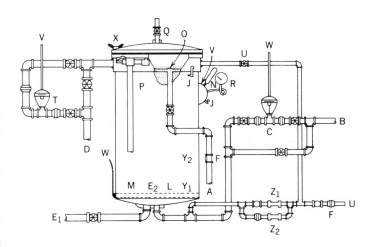

Figure 1. Vertical still retort (21 CFR, Part 113).

A—Water line.
B—Steam line.
C—Temperature control.
D—Overflow line.
E_1—Drain line.
E_2—Screens.
F—Check valves.
G—Line from hot water storage.
H—Suction line and manifold.
I—Circulating pump.
J—Petcocks.
K—Recirculating line.
L—Steam distributor.
M—Temperature-controller bulb.
N—Thermometer.
O—Water spreader.
P—Safety valve.

Q—Vent valve for steam processing.
R—Pressure gauge.
S—Inlet air control.
T—Pressure control.
U—Air line.
V—To pressure control instrument.
W—To temperature control instrument.
X—Wing nuts.
Y_1—Crate support.
Y_2—Crate guides.
Z—Constant flow orifice valve.
Z_1—Constant flow orifice valve used during come-up.
Z_2—Constant flow orifice valve used during cook.

changes allow specific flexible containers to be processed at 275°F (134°C).

There are several distinct types of commercially manufactured retorts for thermally processing canned food (11). Although all, by necessity, operate at pressures above 15 psi (103 kPa), the design characteristics of each type are considerably different. At least six design variables exist: (1) discontinuous (batch) types versus continuous container processing; (2) the heating medium used to transfer heat to the container; (3) the agitation or nonagitation of containers during processing; (4) the layout of the pressure vessel (vertical vs horizontal); (5) the method used to load and unload the containers from the retort; and (6) the cooling procedures used after thermal processing.

The simplest retorts are batch (discontinuous) retorts that use pure steam as the heating medium and do not have provisions for mixing (agitation) of the container contents during processing. These retorts are termed still retorts (Fig. 1). Temperature inside still retorts is maintained by automatic control of the steam pressure.

Loading and unloading the containers from discontinuous still retorts are accomplished by preloading containers into crates, baskets, cars, or trays. "Crateless" systems randomly drop containers into the retort vessel which is filled with water to act as a cushion and prevent container drainage (Fig. 2). The water is drained prior to processing. The orientation of the retort depends on the type of container handling system. Systems using crates or baskets and the crateless systems, by necessity, use vertical vessels whereas car handling necessitates a horizontal orientation.

Glass, semirigid, and flexible containers must be processed in still retorts which have been designed to accommodate the fragility of these containers at retort temperatures and pressures. These retorts operate at pressures greater than the 15 psi (103 kPa) of steam required to reach 250°F (121°C) in order to counterbalance the internal pressure developed in the container. This is termed "processing with overpressure." The pressure buildup inside individual containers during processing would result in the loss of seal integrity in heat-sealed containers (see Sealing, heat) and could loosen the covers of glass containers or permanently distort semirigid plastic containers. Four design changes in still retorts must be made to process with overpressure:

1. Either steam or air overpressure must be automatically controlled. Pressures of 25–35 psi (172–241 kPa) are typical.

2. Control of the retort temperature must be independent of retort pressure.

3. Mixed heating media of either steam–air, water–air, or water–steam are used in place of pure steam. Heat transfer is less efficient in these mixed heating media. For this reason some means of circulating or mixing the heating medium is necessary. For steam–air mixtures, fans may be provided. Water–air and water–steam systems use circulating pumps.

4. Provisions to prevent stress on the containers due to motion during processing are made.

These designs may be incorporated into either vertical or horizontal retorts depending on how the containers are handled. Glass containers are typically loaded into crates or baskets for processing in vertical water–air or water–steam retorts. Flexible retortable pouches are often loaded into trays and cars (which also serve to maintain the proper shape of the pouch) and moved into horizontal retorts for water–steam or steam–air processing (see Fig. 3).

Still retorts, whether designed for metal cans or other containers, are batch (discontinuous) systems. The hydrostatic retort is technically a still retort (product is not agitated) that continuously processes containers. The retort operates at a constant temperature (and pressure) as the containers are carried through the retort by a continuously moving chain (Fig. 4). The required 15 psi (103 kPa) of steam pressure inside the retort (or steam dome) is maintained by two columns of water which also serve as pressure locks for incoming and outgoing containers. These columns of water (called feed and discharge legs) must be greater than 37 ft (11.3 m) high in order to maintain at least a minimum 15-psi (103-kPa) steam pressure. Hydrostatic retorts have the highest throughputs, are efficient in their use of floor space, steam, and water, can process a variety of container sizes and types (including flexible), and

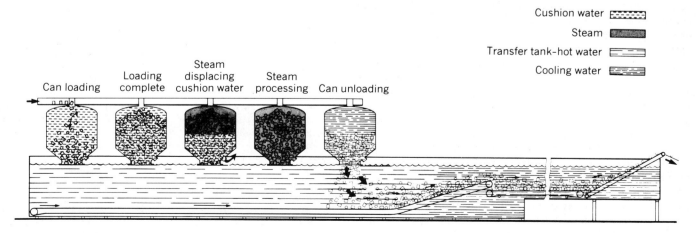

Figure 2. FMC crateless retort system. Courtesy of FMC Corporation, Food Processing Machinery Division.

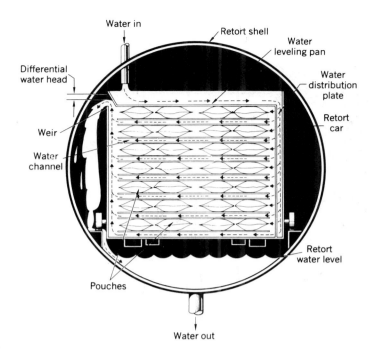

Figure 3. FMC convenience foods sterilizer, showing water flow and pouch restraints. Courtesy of FMC Corporation, Food Processing Machinery Division.

are highly automated. They have the disadvantage of high capital costs and are therefore applicable only to high volume operations.

The heating time necessary to ensure that the coldest spot in the container receives the proper lethality depends some-

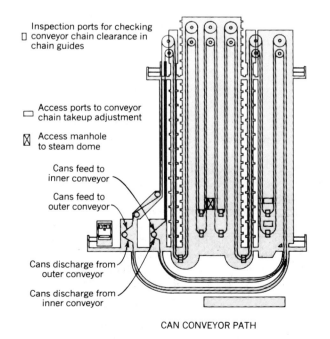

CAN CONVEYOR PATH

Figure 4. Flow diagram of a hydrostatic sterilizer for canned foods. Courtesy of FMC Corporation, Food Processing Machinery Division.

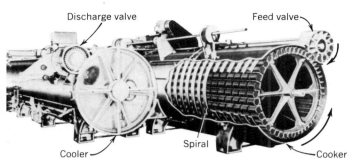

Figure 5. Cutaway view of a continuous rotary cooker–cooler's turning wheel and interlock. Courtesy of FMC Corporation, Food Processing Machinery Division.

what on the consistency of the product. For viscous products such as canned pumpkin and baked beans, the primary heat-transfer mechanism is conduction. Products which have a thin consistency or are packed with brine (canned peas) are heated by convection. The transfer of heat to the center of the container can be greatly facilitated if internal mixing occurs in the can during retorting. This results in a shorter processing time and higher quality product.

Retorts which are designed to increase convection heating by container motion during processing are termed agitating retorts. Both end-over-end and axial rotation are used but the latter predominates. Glass, semirigid, and flexible containers are not agitated because of fragility. Agitating retorts may be either batch or continuous types. Continuous retorts predominate because they have greater efficiency and throughput but are less easily adapted to changes in container size.

The continuous rotary cooker–cooler (see Fig. 5) has become widely used for large volume operations in which convection-heated products such as vegetables in brine are packed in metal cans. This system feeds individual cans into and out of the pressurized vessel by means of rotary pressure lock valves. Cans are rotated around the inside of the vessel's shell by means of an inner rotating reel and a series of spiral channel guides attached to the shell (Fig. 5). This system provides for intermittent agitation of the cans by providing rotation about the can axis during a portion of the reel's rotation inside the vessel. This system has the disadvantage that container size cannot be easily changed.

Regardless of retort design, consideration must be given to cooling containers after processing. For glass, flexible, and semirigid containers, cooling with overpressure is necessary. These containers would fail due to the internal pressure developed during heating should the external pressure drop. Even metal cans may buckle and panel if brought to atmospheric pressure while the contents are at 250°F (121°C). In batch-type still retorts, overriding air pressure with water cooling is used. Hydrostatic retorts (continuous still retorts) cool containers by removing heat from the water in the discharge leg. If further cooling is necessary, an additional cooling section is added and cool water is cascaded over the containers. Continuous agitating (rotary) systems cool under pressure by transferring containers to pressurized cooling vessels by means of rotating transfer valves (Fig. 6). These second-stage vessels are maintained at elevated air pressures while the containers are

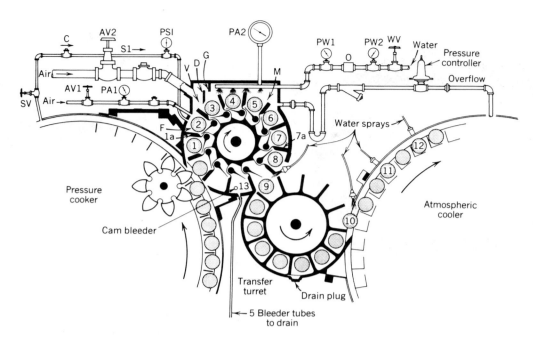

Figure 6. Transfer valve between cooking and cooling vessels of a continuous cooker–cooler.

R	Rotor containing can pockets and ejector paddles.	PW1	Downstream water pressure gauge.
		O	Orifice for flow control.
SV	Forespace steam valve.	Overflow	Release line for excess air or water.
M	Micro-cooling Chamber.	S1	Steam line.
F	Forespace for pressure equilization.	C	Check valve.
D	Water dam.	PSI	Pressure vacuum gauge.
G	Splash guard.	1.	Hot can.
Air	Pressure regulated air supply.	1a.	Valve-leading edge.
V	Vestibule.	2.	The can receive some water splash.
AV1	Forespace regulated air inlet valve.	3.–7.	The cans are fully exposed to water flow.
AV2	Micro-cooling space air inlet valve.		
PAI	Forespace air pressure gauge.	7a.	Valve-trailing edge.
PA2	Micro-cooling space air pressure gauge.	8.	Can is about to leave Micro-Cooler valve
Water	Water inlet and perforated distributor.		
WV	Water inlet regulator valve.	9.–12.	Cans are subject to sprays of water.
PW2	Upstream water pressure gauge.	13.	Drain pocket.

Courtesy of FMC Corporation, Food Processing Machinery Division.

cooled with water. A third-stage atmospheric cooler may also be incorporated.

In addition to the pressurized heating and cooling vessels, all retort systems require a set of precise instruments and controls. Regulations require a direct reading mercury-in-glass thermometer as well as temperature recording devices. A continuous temperature controller must be installed. For retorts using pure steam, this may be a pressure controller; processing with overpressure requires a direct temperature controller. Retorts require reliable sources of steam, air, and water. A pressure reading device is required as well as an accurate recording timing device so that the scheduled process can be insured and the proper records maintained. All instruments must undergo periodic calibration.

Regulation

The canning of foods is carefully regulated by the FDA, or in the case of canned meats and poultry, the USDA. These agencies recognize the serious public health implications of improperly processed foods. The FDA has developed a complete set of regulations commonly referred to as the Good Manufacturing Practices (GMPs) for canning foods. These regulations govern the type of equipment used to can foods and the procedures, the frequency of inspection of containers and equipment, and the records which must be kept, and they provide for the filing of individual processes prior to production. The regulations pertaining to food canning are contained in Title 21 of the Code of Federal Regulations (CFR) under the following sections: (1) 21 CFR Part 108, "Emergency Permit Control"; (2) 21 CFR Part 113, "Thermally Processed Low-acid Foods Packaged in Hermetically Sealed Containers"; and (3) 21 CFR Part 114, "Acidified Foods."

Part 108 stipulates that food canning plants must register their establishments and specific processes with the FDA. This section also contains provisions for issuing emergency permits to firms that the FDA believes do not fully meet the regulations. Part 113 is the most extensive section and details the equipment, procedures, process controls, establishment of correct process, critical factors, and necessary records for canning low acid foods. This section also details the procedures to be used in evaluating the integrity of the double seams. Part 114 describes the GMP requirements for packaging high acid foods (pH ≤4.6). This section includes general provisions as well as specific requirements for production and process control.

The USDA's Food Safety and Inspection Service (FSIS) has regulatory authority over canning poultry and meat products and has promulgated a series of regulations under Title 9 of the CFR. Current FSIS regulations are considerably more general than FDA regulations and have not kept pace with changes in canning technology. Meat and poultry canning operations are subject to continuous inspection in a manner similar to other FSIS regulated plants. These regulations are contained in sections 318.11 and 381.149, which deal with the cleaning of empty containers, inspection of filled containers, coding, use of heat-sensitive indicators (see Indicating devices), and incubation of processed products. In 1984, FSIS proposed a more detailed set of regulations similar to those promulgated by the FDA for canning low acid foods. The sections of the CFR which will deal, in part, with meat and poultry canning, and related requirements will be 9 CFR 308, 318, 320, and 381.

Recent and Future Trends

During the last 5–10 yr, the food canning industry has undergone substantial changes, most notably in the area of containers (12). A major trend is to move away from tinned and soldered metal cans into tin-free steel with welded side seams or two-piece drawn–redrawn cans (13) (see Cans, steel). A majority of the cans produced in the United States today are lead-free because solder is not used. These trends are due to the unfavorable cost and availability of tin and the desire to make cans without lead. Cans made today are also significantly lighter in weight than cans made just a few years ago.

In the future, metal as well as glass cans will have increasing competition from plastics and composite materials (see Retortable flexible and semirigid containers) (14). Although the retort pouch, in its present form, has not been a large success, second- and third-generation pouches may increase the use of these multilaminate flexible, retortable containers. The same thermal-processing technology used for pouches is being used to process large one-half and one-fourth steam-table trays (15) (see Trays, steam table). These large flat containers hold foods used in institutional feeding that would be difficult to place in conventional cans. The containers are reheated before opening by submerging in boiling water and are kept warm on steam tables.

The most dramatic trend will be the further development of thermal processes in which the food is commercially sterilized before packaging. This allows foods to be thermally processed in continuous-flow heat-exchange systems which can result in higher quality products and allow the use of less expensive containers based on paperboard or thin plastics. This technology is currently used for juices, drinks, and milk. However, a great deal of active research is underway to apply the same principles to other products such as soups, stews, and vegetables (8).

BIBLIOGRAPHY

1. S. R. Friedman in W. C. Simms, ed., *The Packaging Encyclopedia—1984,* Cahners Publishing, Boston, **29**(4), 334 (1985).
2. A. Lopez, *A Complete Course in Canning,* 11th ed., Book 1, The Canning Trade, Baltimore, 1981, p. 9.
3. I. J. Pflug and W. B. Esselen in J. M. Jackson and B. M. Shinn, eds., *Fundamentals of Food Canning Technology,* AVI Publishing Co., Westport, Conn., 1979, pp. 10–94.
4. N. N. Potter, *Food Science,* 3rd ed., AVI Publishing Co., Westport, Conn., 1978, pp. 177–193.
5. Ref. 2, p. 330.
6. *Establishing Scheduled Processes,* Code of Federal Regulations, Title 21, Part 113.83, U.S. Printing Office, 1983, Washington, D.C., p. 112.
7. D. Wernimont, *Food Eng.* **55**(7), 87 (July 1983).
8. A. A. Teixeira and J. E. Manson, *Food Technol.* **37**(4), 128 (Apr. 1983).
9. Ref. 4, pp. 550–557
10. Ref. 2, pp. 217–219
11. *Canned Foods: Principles of Thermal Process Control, Acidification and Container Closure Evaluation,* 4th ed. The Food Processors Institute, Washington, D.C., 1982, p. 162.
12. B. J. McKernan, *Food Technol.* **37**(4), 134 (Apr. 1983).
13. "Welded Can Expected to Capture 3-Piece Can Market," *Food Prod. Manage.* **104**(12), 12 (June 1982).
14. J. Haggin, *Chem. Eng. News,* **62**(9), 20 (Feb. 27, 1984).
15. "The Optimum Container for Mass Feeding," *Food Eng.* **54**(8), 59 (Aug. 1982).

General References

A. Lopez, *A Complete Course in Canning, 11th ed., Books 1 and 2,* The Canning Trade, Inc., Baltimore, 1981, pp. 556.

J. M. Jackson and B. M. Shin, *Fundamentals of Food Canning Technology,* AVI Publishing Co., Westport, Conn., 1979, pp. 406.

Canned Foods: Principles of Thermal Process Control, Acidification and Container Closure Evaluation, Fourth Edition, The Food Processors Institute, Washington, D.C., 1982, pp. 246.

J. H. Hotchkiss
Cornell University

CANNISTERS, COMPOSITE.   See Cans, composite.

CAN SEAMERS

Seamers are used to make double seams on cans. In can manufacturing, seamers are called double seamers; in a canning plant, they are often called closing machines. A double seam consists of a "first-operation" seam to curl the cover around the can flange and a "second-operation" seam to form and iron out a tight hermetic seal between cover and can. A hermetically sealed container is defined as a container designed and intended to be secure against the entry of microorganisms and to maintain the commercial sterility of its contents after processing (1) (see Canning, food; Cans, fabrication; Cans, steel).

Classification

Seamers evolved to meet production requirements and are now classified as hand seamers, semiautomatic seamers, and automatic seamers.

Hand seamers. Hand seamers are single-station machines, either pedal or motor driven; the can is placed on the seaming station by hand and the seaming cycle is initiated by a hand-operated lever. These seamers are used primarily in gift shops for packaging nonfood items.

Semiautomatic seamers. Semiautomatic seamers are motor-driven, single-station machines. The can is placed on the seaming station, and the seaming cycle is automatic after manual actuation of the starting button or lever. These seamers are used where production speeds are very low or for test packs in the laboratory.

Automatic seamers. These can be single-station or multiple-station machines, where cans are transferred on a conveyor to reach the seaming station. Cans are spaced on the infeed conveyor by means of a can transfer chain with evenly spaced lugs or feed chain fingers. The transfer chain is timed to the seaming heads in the seamer.

Seamers are further classified into head-spin and can-spin seamers. On a head-spin seamer the can stands still while the first- and second-operation seaming rolls are mounted on a rotating seaming head. On a can-spin seamer the can rotates about its center, while lever-mounted seaming rolls engage the rotating can to form the seam. Some models of can-spin seamers are equipped with seaming rails instead of seaming rolls. Here the seam is formed by running the rotating can against a rail that is tooled with a complementing shape of the required seam. Such seamers are known in the industry as rail seamers.

Functions

Seamers are designated to describe their use.

1. Atmospheric—for closing filled cans.
2. Steam vacuum—for filling headspace of cans with steam at time of can and cover makeup, prior to seaming.
3. Under cover gassing—for filling headspace of cans with carbon dioxide or nitrogen gas at time of can and cover makeup, prior to seaming.
4. Vacuum—seamer enclosed where seaming is done in vacuum; seamer has no cover feed, so cover must be placed on can prior to being conveyed into vacuum seamer.
5. Clincher—seamer that forms a loose first-operation seam. Used with vacuum seamer without cover feed.
6. Aerosol—seamer that attaches dome-shaped covers on aerosol cans.
7. Can shop—seamer that attaches the first end on a three-piece can; name derived because of use in canmaking only.

Seamers are designed for a particular can-diameter range, can-height range, and speed range. Speed is expressed in either cans per minute (CPM) or cans per hour (CPH). The seamer manufacturer determines seamer speed by the number of seaming heads, rotational speed of seaming heads, can and cover diameter, and type of can and cover. These factors determine the speed of seam formation in linear velocity given in inches (cm) per minute. This value in inches (cm) per minute changes over the years as technology in the manufacturing process of cans and covers changes. The seamer user determines the seamer speed by production requirements.

Cannery seamers are generally used in conjunction with can-filling machines that must run in timed relationship with can-closing machines. For that reason, can fillers are driven through a drive from the seamer. The can spacing of the filler must be matched closely to that of the seamer's can transfer chain to avoid excessive spill at transfer of can from filling station to transfer chain. In all filler–seamer connections, the design and manufacture of filler connecting parts are done by the seamer manufacturer.

Once a seamer is in service it can be converted to run any can and cover within the designed diameter and height range by fitting new can and cover contacting parts. These parts are called seamer change parts.

Optional Attachments

Most machines may be fitted with optional attachments listed below:

Filler drive disconnect. This is an overload clutch fitted in drive line between seamer and filler to disconnect filler drive from seamer in case of a severe can jam or mechanical failure in the filler, which could damage the seamer. Clutch is special for one-position engagement to retain filler–seamer timing when reengaged.

In-motion timer. An in-motion timer is fitted in drive line between seamer and filler to time transfer of cans from filler stations to seamer transfer-chain fingers while machines are running. Timing change is accomplished by changing the rotational position of output shaft relative to input shaft.

Marker. Also called code-embossing devices, markers are used to mark covers with canner's identification codes. Markers are designed as rotating markers and reciprocating-impact markers. Two types of marking dies are available. One is called "debossing," where the characters are indented into the covers. These are primarily used on beverage cans. The second is called "embossing," where the characters are raised on the covers. These are used primarily on sanitary food cans. Mechanical markers are recommended for seamer speeds up to 1200 cans per minute. For higher speeds, ink-jet markers are available. These are not built by the seamer manufacturer.

Worm can infeed. Worm can infeed was built as part of the seamer-infeed conveyor table. Worm is driven by seamer to accelerate cans from random infeed to seamer-conveyor feed-chain finger spacing. On some seamer models the worm feeds cans directly into a can transfer turret on the seamer.

Accelerator star. This serves the same purpose as the worm can infeed.

Topper. Also called a packing device, a topper is a unit mounted above the seamer infeed conveyor. The topper has cups attached to a drive chain or belt. Cups reach into the headspace of cans to compact the product. The topper is driven in timed relationship with the seamer-conveyor can transfer chain.

Belt packer. Used primarily in vegetable canning, the belt packer is a unit mounted above the seamer infeed conveyor to compact product in cans and cut off any product hanging over the side of cans. This unit requires no timing with can spacing of transfer chain, but the belt must be run at the same linear speed as the can transfer chain.

Liquid-nitrogen injector. A liquid-nitrogen injector is mounted above the seamer infeed conveyor to inject a drop of liquid nitrogen into the headspace of each can. As cans enter the seamer, the nitrogen displaces the air in the headspace of the cans and builds internal pressure in the can during seaming. This unit is used primarily where product is canned in thin-walled cans so that they can be stacked on pallets after seaming without collapsing.

BIBLIOGRAPHY

1. *Canned Foods: Principles of Thermal Process Control, Acidification, and Container Closure Evaluation,* The Food Processors Institute, Washington, D.C., 1982.

NANCY JANSSEN
Food Processing Machinery
and Supplies Association

CANS, ALUMINUM

Over 98% of all aluminum cans are drawn-and-ironed (D & I) cans used for beer and soft drinks (see Cans, fabrication; Carbonated-beverage packaging). The other 1–2% is accounted for by small shallow-draw food cans. In the United States, production of aluminum beverage cans has reached about 60×10^9/yr. Virtually all beer cans are aluminum, and over 85% of all soft-drink cans. Until 1965, the three-piece soldered can was the only can used for beer and beverages (see Cans, fabrication; Cans, steel). It was generally made of 75-lb per base box (16.8 kg/m^2) tinplate (see Tin-mill products). Aluminum does not have the strength of steel per unit weight and cannot be soldered. Therefore, early in the development of the aluminum can, attention was focused on improving the properties of aluminum and perfecting two-piece D & I technology. More recently, the emphasis has been on saving metal, because the cost of metal is the single largest component of final product cost. One of the chief reasons for the success of aluminum cans has been their scrap value for recycling (see Energy utilization; Recycling). This article pertains to the technical developments that have led to today's aluminum beverage can.

The significant developments took place in the late 1960s. Earlier, in 1958, Kaiser Aluminum attempted to make a 7-oz (207-mL) aluminum can using a 3003 soft-temper aluminum of the type used for aluminum-foil production, but the effort was unsuccessful. In the universe of aluminum alloys, the 3-series alloys contain a small percentage of manganese as the principal alloying element. The success of the aluminum can depended on the development of the 3004 alloy for the can body, which contains manganese along with a slightly lower amount of magnesium. (The softer 5182 alloy for can ends contains a higher amount of magnesium.) Reynolds Metals Co. began making production quantities of 12-oz (355-mL) seamless D & I cans in 1966 using a 3004 alloy. The walls of these first cans were straight, with a top diameter of "211": 2 11/16 in. (68.26 mm). The starting gauge was 0.0195 in. (0.495 mm); can weight was 41.5 lb/1000 (18.8 kg/1000).

Reynolds introduced the first necked-in cans (see Fig. 1) at about the same time, which reduced the top diameter from 211 to 209 ie. 2 9/16 in. (65.09 mm). This represented a breakthrough in technology and container performance, particularly as it related to cracked flanges. Cracked flanges were a serious problem in both double-reduced tinplate and straight-

Figure 1. A necked-in can.

walled aluminum cans. The introduction of carbide knives at the slitter essentially eliminated cracked flanges on tinplate cans. Eliminating them on aluminum cans required a change from die flanging to spin flanging, and necking-in before flanging, which does not stretch the metal beyond its elastic limits.

In 1968, a new harder-temper (H19) 3004 alloy was introduced for aluminum cans. Although the 3004–H19 combination had been available since the 1950s for other purposes, it was not until 1968, when Alcoa and Reynolds were in commercial production with full-hard-temper can sheet, that it could be used to effect significant weight reductions. Since then, weight/1000 cans has decreased from 41.5 lb (18.8 kg) in the mid-1960s to 34 lb (15.4 kg) in the mid-1970s and to less than 30 lb (13.6 kg) in the mid-1980s.

Gauge reductions have increased the point where design techniques have become critical to sustaining the can's ability to hold the product. With few exceptions, U.S. brewers pasteurize beer in the can. This generates high internal pressures, and most cans used for beer must be designed so that they have a minimum bottom buckle strength of 85–90 psi (586–620 kPa), depending on the carbonation level. Brewers are also asking for cans with minimum column strength, ie, vertical crush, of 300 lbf (1330 N).

A revolution has taken place in bottom profiles (see Fig. 2). The original D & I bottom had a rather generous bottom-heel radius. In order to meet the 90-psi (620 kPa) minimum bottom buckle-strength requirement, a 211 can with this configura-

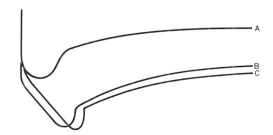

Figure 2. Can-bottom profiles: A, original bottom profile; B, Alcoa B-53, V-bottom profile; C, Alcoa B-80. Courtesy of Alcoa.

tion would have to be made with a starting gauge of 0.016–0.0165 in. (406–419 μm). The next profile development, basically the Alcoa B-53 design, is widely used throughout the beer industry today because it allows the starting gauge to go as low as 0.013 in. (330 μm), in most cases without sacrificing the 90-psi (620-kPa) minimum bottom buckle strength. The newest entry is the Alcoa B-80 profile, which allows starting-gauge reduction to ca 0.0126 in. (320 μm).

An "expandable bottom" design, still on the drawing boards at the aluminum and can companies might permit use of a 0.010-in. (254 μm) starting gauge for pasteurized product. In contrast to the current dome profiles, an expandable bottom is essentially a flat bottom with small pods located near the perimeter to provide stability to the pressurized can, reduce drag, and increase mobility. A totally flat bottom would drag too much on the filling line. The "expandable" bottom is designed to flex outward during pasteurization, relieving some of the generated pressure.

Sidewalls are also being redesigned (see Fig. 3). In the 1970s, the so-called nominal thinwall, ie, the area of the can that has been thinned most, generally ranged from 0.0052 to 0.0053 in. (132–135 μm). More recent versions of the D & I can have reduced the nominal sidewall to 0.0045 in. (114 μm). A reduction of this magnitude represents substantial cost reduction. It also means a corresponding reduction in the overall column strength (vertical crush) of the can, not below the minimum 300 lbf (1330 N), but in terms of overall operating average. Column strength is very critical to the brewers, who ship long distances by truck and rail. Extensive testing by the can companies and the beer and beverage industry in general has shown that the 300-lbf (1330-N) minimum is satisfactory. Because of the reduction in body-wall thickness, dents that were acceptable before have now become critical owing to their influence on reducing the can's column strength. Can makers and brewers are monitoring their handling systems for empty and full cans to minimize denting wherever possible.

In the 1980s, can suppliers have reduced costs further by double "necking-in" (see Fig. 4). These configurations reduce costs primarily because of the diameter reduction of the lid. In 1984, further activity with respect to necking-in began to occur. Cans with three or four die necks are now being run commercially (see Fig. 4). In Japan, a can with eight necks is being tested. Metal Box (UK) has introduced a spin-neck can (see Fig. 4) which essentially produces the same 206 top diameter as the triple- or quadruple-neck can, ie, 2 6/16 in. (60.33 mm).

Figure 4.   Double-, triple-, quadruple-, and spin-neck cans.

Another advantage of aluminum is the "split gauge." The industry used to sell coils in 0.0005-in. (12.7-μm) increments, eg, 0.0130 in. (330 μm), 0.0135 in. (343 μm), etc. A new pricing structure introduced in 1983 allows can-stock buyers to order gauge stock in 0.0001-in (2.54-μm) increments. A can manufacturer can reduce costs by taking advantage of these slight gauge reductions. In addition, there has been a change in the gauge tolerance as rolled by the aluminum mills. In the 1970s, the order gauge was subject to a ±0.0005-in. (12.7 μm) tolerance; today, it has been reduced to 0.0002 in. (5.1 μm). This permits further gauge reduction because it allows the can manufacturer to reduce the order gauge without changing the minimum bottom buckle strength.

BIBLIOGRAPHY

General References

J. F. Hanlon, *Handbook of Package Engineering,* 2nd ed., McGraw-Hill, Inc., New York, 1984.

Drawn & Ironed Aluminum Cans, Aluminum Co. of America, Pittsburgh, Pa., 1975.

Tin Mill Products Manual, U.S. Steel Corp., Pittsburgh, Pa., 1970.

Brewing Industry Recommended Can Specifications Manual, United States Brewers Association, Inc., Washington, D.C., 1981.

N. F. NIEDER
Anheuser-Busch, Inc.

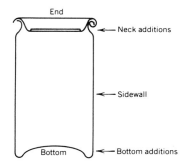

End
← Neck additions
← Sidewall
Bottom
← Bottom additions

Figure 3.   Design elements.

CANS, COMPOSITE

The "composite can" is broadly defined as a can or container with body and ends made of dissimilar materials. In commercial practice the composite can has several more focused descriptions: cylindrical or rectangular shape; rigid paperboard (or plastic) body construction; steel, aluminum, or plastic end closures; generally employing inside liners and outside labels; and generally delivered with one end attached and one end shipped separately to be attached by the user. Today the most common form is the cylindrical paperboard can with a liner, a label, and two metal ends. There are other packaging forms similar to traditional composite cans that are sometimes

Figure 1. Examples of composite cans in commercial use.

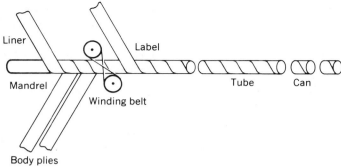

Figure 2. Spiral-wound composite can fabrication.

called composites, but which are more closely related to folding cartons. These are single-wrap fiber cans made from blanks and mainly designed for the users' in-plant production (see Cans, composite, self-manufactured).

The composite can is not a new package. Early applications included refrigerated dough and cleansers. As technology improved, motor oil and frozen juice concentrates were converted to composites. In the last decade, snack foods, tobacco, edible oils, shortening, powdered beverages, pet foods, and many other items have been added to the list. Figure 1 illustrates some of the products currently available in composite cans.

Since their introduction, composite cans have generally been marketed and used as a lower-cost packaging form relative to metal, plastic, and glass. This emphasis has overshadowed other positive attributes such as the frequent use of recycled materials, weight advantages, noise reduction, improved graphics, and design flexibility. Because of its early applications in cleansers and oatmeal, the composite can once suffered from a low performance image associated with the term "cardboard can." This term cannot begin to describe the current and potential properties of the paperboard and other materials that go into today's composite cans.

Composite cans are normally available in diameters of 1–7 in. (3–18 cm) and heights of 1–13 in. (3–33 cm). Dimensional nomenclature for composites has been adapted from metal cans, and nominal dimensions are expressed in inches and sixteenths of an inch (see Cans, steel). Hence, a 404 diameter can has a nominal diameter of 4$\frac{4}{16}$ in. (10.8 cm). Likewise, a height of 6$\frac{10}{16}$ in. (16.8 cm) is expressed as 610, etc.

Manufacturing Methods

Composite-can bodies are produced by two basic methods: spiral winding and convolute winding. Figure 2 shows a schematic drawing of the spiral process. Multiple webs including a liner, body plies, and label are treated with adhesive and wound continuously on a reciprocating mandrel. The resulting tube is trimmed and the can bodies are passed on to flanging and seaming stations. Figure 3 depicts the convolute process, wherein a pattern is coated with adhesive and entered onto a turning mandrel in a discontinuous process. Trimming and finishing operations for the convolute and spiral systems are virtually the same. Most composite-can manufacturers favor

the spiral process in situations where long production runs and few line changeovers are involved.

Body Construction

Paperboard. The primary strength of the composite can is derived from its body construction, which is usually paperboard (see Paperboard). Body strength in composite cans is an attribute that has improved over the years and can be varied to meet many application demands.

In the early years of composite can development, it was common for can manufacturers to start with a readily available body stock such as kraft linerboard or tube-grade chip. These boards are adequate for most applications, but new boards with special qualities have also been developed for more demanding end-use requirements.

Research and development efforts in the combination paperboard field encompass a number of areas of expertise. Examples are engineering mechanics concepts used to develop structural criteria and to develop tests to assure that the paperboards possess the necessary resistance to bending, buckling, and creep; surface chemistry used to predict the resistance of the board to penetration of adhesives, coatings, and inks and, similarly, protection from environmental conditions such as rain, high humidity, and freezing temperatures; and process engineering used to optimize paperboard manufacturing and converting and to assist in quality assurance programs (see Specifications and quality assurance).

The scientific and engineering efforts have, in many instances, supplemented the artisanlike judgment of yesterday's papermaker. However, in many cases the new technological approaches have been blended with the papermaker's art to achieve the best of both worlds. As a result, the following advances have taken place: super-high-strength board that can be converted into composite cans with reasonable wall

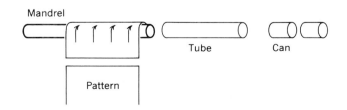

Figure 3. Convolute composite can fabrication.

Table 1. Adhesives Used in Composite Can Manufacture

Adhesives	Properties
poly(vinyl alcohol–acetate) blends	good initial tack, good runnability, moderate to good water resistance
dextrin	fast tack, poor water resistance
animal glue	good tack, vulnerable to insect attack
polyethylene	requires heat, good dry bond, moderate water resistance
hot melts	require heat, difficult to handle, good water resistance, good water bond

thicknesses that can resist implosion when subjected to near-perfect vacuum; resin-treated paperboards that retain their structural integrity when thoroughly wet; chemically treated paper that resists penetration by water over long time intervals; and paperboards that can be distorted, rolled, and formed in high-speed converting equipment without "creeping" back to their original shape.

Adhesives. The adhesives (see Adhesives) and coatings used in the manufacture of composite cans have also been improved to provide better heat and water resistance plus increased operating efficiency. Most product applications require precision gluing equipment to control the amount and position of the adhesive on each web. The most commonly used adhesives in today's composite can production are listed in Table 1.

Liners. Like all successful packages, the composite can must contain and protect the products. For that reason, continuous improvement is sought in liner materials. By combining materials such as LDPE (see Polyethylene, low density), HDPE (see Film, high density polyethylene; Polyethylene, high density), PP (see Film, nonoriented polypropylene; Film, oriented polypropylene; Polypropylene), ionomer (see Ionomers), PVDC (see Vinylidene chloride copolymers), and PET (see Film, oriented polyester; Polyesters, thermoplastic) with aluminum foil (see Foil, aluminum) or kraft (see Paper; Paperboard), the barrier properties of composites can be matched with a broad range of product requirements (see Laminating; Multilayer flexible packaging). The polymers may be included as film or coatings, or both. Tables 2 and 3 illustrate the physical properties of some of the more common liner films and complete liner constructions.

The laminates shown in Table 3 vary significantly in cost. Depending on can size, the difference in cost per thousand units can be substantial. The foil-based laminates, which provide virtually 100% water and gas barrier, are becoming quite expensive. If less than total impermeability is acceptable, it pays to investigate coated- or plain-film alternatives. For example, packers of frozen juice concentrates have gradually moved away from foil liners to laminations of PE or ionomer/PE and kraft.

End closures. A critical process in the manufacture of a composite can is double seaming the metal end (see Can seamers). Since a composite can body is typically thicker than a metal can for any given package size, metal-can specifications for finished seam dimensions cannot be followed. Figure 4 shows composite double-seam profiles that are correctly and incorrectly made. Careful attention should be given to compound placement, selection of first- and second-stage seamer rolls, seamer setup, chuck fit, and base plate pressure if a satisfactory double seam is to be achieved.

In the future, end-seam configurations will probably be substantially different. Present double seams tend to creep under a continuous pressure, and minimum abuse will cause most composites to leak at pressures exceeding 15 psi (103.4 kPa). The present trend in end development is to combine mechanical and chemical systems to improve double-seam integrity.

A variety of steel and aluminum ends with solid panel,

Table 2. Physical Characteristics of Commonly Used Liner Films[a]

Physical characteristics	Polypropylene			Polyester	LDPE	Ionomer
	Oriented	Oriented PVDC coat	Nonoriented			
Properties						
tear strength, gf (N)	3–10[b] (0.03–0.10)	3–10 (0.03–0.10)	50–MD[b] (0.5) 300–XD[b] (3)	12–27[b] (0.12–0.27)	50–150[b] (0.5–1.5)	50–150[b] (0.5–1.5)
burst strength, psi (kPa)				55–80 (379–551)	10–12 (69–83)	10–12 (69–83)
WVTR, g·mil/(100 in.²·24 h) [g·mm/(m²·d)]	0.75 [29.5]	0.3–0.5 [11.8–19.7]	1.5 [59]	1.5 [59]	2.0–3.0 [78.7–118.1]	2.0–3.0 [78.7–118.1]
O₂ rate, cm³·mil/(100 in.²·24 h) [cm·m/m²·d)]	160 [630]	1–3 [3.9–11.8]	160 [630]	3.0–4.0 [11.8–15.7]	500 [1970]	250–300 [985–1180]
elongation, %	35–475	35–475	550–1000	60–165	100–700	400–800
Product resistance[c,d]						
strong acids	G	G	G	G	G	G
strong alkalies	G	G	G	G	G	G
grease and oil	G	G	G	G	G	E

[a] Source: Sonoco Products Company, Hartsville, S.C.
[b] gf/mil (0.386 N/mm) thickness.
[c] G = good.
[d] E = excellent.

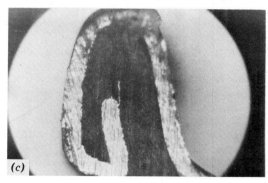

Figure 4. Composite-can double seams. (**a**) Loose; (**b**) correct; and (**c**) tight.

Table 3. Physical Characteristics of Composite Can Liners

| Composite[a] | WVTR[b] at 100°F (37.8°C) and 90% rh | | O$_2$ permeability[c], cm^3/(m^2·d) |
	Flat	Creased	
100 ga PP/adh/100 ga AF/LDPE/25# MGNN kraft	< 0.001	<0.001	<0.001
100 ga PP/adh/35 ga AF/LDPE/25# MGNN kraft	0.06		<0.02
12# ionomer/35 ga AF/LDPE/30# XKL kraft	0.01		<0.02
1# PET slipcoat/35 ga AF/casein/25# MGNN kraft	0.09	0.93	<0.001
1# PET slipcoat/35 ga AF/LDPE/25# MGNN kraft	0.06	0.09	<0.001
12# HDPE/20# MGNN kraft	15.35		153
14.4# HDPE/20# MGNN kraft	12.09		126

[a] key: # = lb/ream = 454 g/ream.
  100 ga = 0.001 in. = mil = 25.4 μm.
  MGNN = Machine-grade natural Northern.
  XKL = extensible kraft linerboard.
[b] ASTM Test Method E 96-80.
[c] ASTM Test Method D 3985.

Recent Developments

The composite can for shortening, introduced in 1979, illustrates the improved containment capacity of fiber cans. Traditionally in metal, shortening can now be packaged in composite cans with a PP/foil/kraft liner. The can body consists of two plies of paperboard with a foil/kraft outer ply. The label can be supplied convolutely after filling, or it can be wound as the outer ply. It has been shown that powdered milk and other oxygen-sensitive food products can remain stable for at least one year. It is expected that more powdered and granular foods will be packed in composites in the future through the use of nitrogen-purge systems or similar processes.

Certain products, such as tobacco and snacks, are already removable tape, or other easy-opening features are available, as well as plastic-end closures with easy-opening and sifter tops. In composites, the most expensive components are usually the metal ends. With this in mind, gastight, puncture-resistant membrane closures for composites have been test-marketed and evaluated by several companies (see Fig. 5). They are considerably less expensive than aluminum full-panel removable ends, and they eliminate the cut-finger hazard posed by both the center panel and score residual on rigid ends. In addition, the membrane end eliminates the metal fines that can be produced by can openers.

Labels. The outer label on a composite can supplies additional package protection and, more importantly, enhances the can's aesthetic appeal and provides required consumer educational-instructional information. Composite labels include coated papers, foil/kraft laminates, and film constructions based on polyethylene or polypropylene. Flexographic, rotogravure, or offset printing (see Printing) are used, depending on cost and quality requirements.

Figure 5. Composite with peelable membrane closure.

packed with nitrogen, and the concept is being considered for coffee as well. It would eliminate the vacuum (see Vacuum coffee packaging) that has held back the use of composite cans thus far. Coffee and other vacuum-packed products, eg, powdered milk and nuts, are now being tested in composite cans. These cans must be capable of holding a 29.5-in. Hg (100-kPa) vacuum for periods of up to one year and performing not only under static conditions but under actual abuse conditions of packing and distribution.

The packaging of single-strength juice and juice drinks represents an exciting new opportunity for composite cans in both hot-fill and aseptic processes (see Aseptic packaging). Extensive technical work begun in the 1970s has resulted in can constructions and filling techniques that have proven successful in commercial canning and distribution systems.

M. B. EUBANKS
Sonoco Products Company

CANS, COMPOSITE, SELF-MANUFACTURED

There are two basic types of self-manufactured composite cans produced in-house by packagers. One is the traditional spiral-wound or convolute composite can, the familiar form used for motor oil, orange juice, and cocoa-based products (see Cans, composite). A more recent development is the utilization of folding-carton material for the production of self-manufactured paperboard cans using composite materials in the folding-carton base stock. An even newer innovation is a paperboard can that is hermetically sealed and capable of holding a gas environment.

Composite cans. A typical example of self-manufacturing of a traditional composite can is a system known as Sirpack. This system, from Sireix (France), installed in the packager's manufacturing facility, produces composite cans that can be round, square, rectangular, or oval. The cans are produced from a continuous form-and-seal process from four reels of material in a horizontal fashion. The materials are shaped around a forming mandrel and sealed, generally by a hot-melt adhesive (see Adhesives). The inside liner is heat-sealed for excellent moisture and liquid tightness (see Sealing, heat). The reels of material are slightly staggered to each other so that the sealing lines on the four materials do not superimpose. The inside materials that provide moisture and liquid tightness are generally made from a plastic film or aluminum foil (or both), according to the protection needed for the product. The outside laminate can be printed by web rotogravure, or web offset (see Printing).

The bottom is generally a metal end, but it can also be made of a composite material matching that of the sidewall. It can also be one of the many plastic closures available. Generally, the top is a heavy injection-molded plastic that is heat-sealed to the upper edges of the container. In most cases the cover incorporates an easy-open device allowing a separation or opening of the lid and can be reclosed after removing a portion of the product.

The main advantages of the in-plant Sirpack system are that it permits the option of various sizes (material options) and the ability to save the conversion cost normally paid to the converter of the composite cans. Added to this are other savings, such as the savings in floor space and warehousing of empty composite cans prior to utilization in the filling and closing process. The machinery is relatively compact and easy to maintain. It has been designed for in-plant production with the average mechanic in mind.

Paperboard cans. Several companies provide systems for in-plant manufacture of paperboard cans. They include folding-carton companies such as Westvaco (Printkan), International Paper (CanShield), and Sealright (Ultrakan). All of their packages are low-to-moderate barrier paperboard cans that can be manufactured in the packager's facilities. Another paperboard can, Cekacan, offered in the United States by Container Corporation of America, provides a high barrier, gastight supplement to the paperboard-can list. All the paperboard cans are formed from a flat blank and assembled into either a straight-wall can or tapered-wall can. In some cases, eg, CanShield, the use of paper-cup technology has been employed to develop the vertical-wall paperboard can.

A paperboard can is described as a semirigid container with the body sidewall fabricated from a single sheet of folding-carton-based material, wound once and sealed to itself, with either or both ends closed by a rigid or semirigid closure. The typical paperboard can has three pieces: a single-sheet single-wound body, a single sheet of base material fixed to one end, and a closure. The system is almost always made in line with the packager's filling and closing operations, but it is also possible to manufacture to storage. The can body can be cylindrical, rectangular, or combinations thereof, but most are cylindrical. Commercial or prototype paperboard cans range in size from 2 to 10 in. (0.8–3.9 cm) high, and up to 6 in. (2.4 cm) in dia, but mostly are confined to a maximum 5.25-in. (2-cm) dia and 3–10-in. (1.2–3.9-cm) heights.

In the CanShield construction, the paperboard sidewall is rolled into a cylindrical shape and the two edges are overlapped and sealed. Continuous thermofusion along the seal is effected by bringing the coating on one or two edges of the blank to a molten state, ie, by direct contact with a heated plate or more recently by blowing hot air onto the edge, bringing the two edges together, and applying pressure. Usually a polyolefin, eg, polyethylene, is used on the surface of the board to provide the adhesive factor. A disk of paperboard with a diameter approximating that of the cylinder is crimpfolded around its perimeter to form an inverted shallow cup. This base piece is positioned in one end of the hollow cylinder so that the bottom edge is about ¼ in. (0.64 cm) below the edge of the cylinder. The edge of the inner periphery is heated and crimpfolded over to lock and seal the disk in place. In this manner, the outer peripheral of the base is sealed to the inner perimeter of the cylinder wall. The segment of the outer wall extended beyond the inner disk is then heated and folded over to come in contact with the inner side of the disk. A spinning mandrel applies pressure to the base of the cylinder to effect the final seal.

The result is a primary seal between the bottom disk and the sidewall and a secondary seal, wherein the sidewall is folded over, capturing the disk with an additional seal. For the rectangular version, the base piece is forced against the body wall under pressure using an expanding mandrel to seal the base to the body. Here too, the material for sealing is usually polyethylene, and hot air is used to bring the material to a molten state to act as a sealant. On the round containers, the top is usually rolled out and the closure, eg, a foil membrane, is adhered to the top rolled edge after the product has been placed into the paperboard can. On the rectangular version, the top rolled edge is generally closed by a rim closure that clamps onto the peripheral of the opening and is sealed into place by induction or by glue. The rim then acts as a holder for a full panel closure.

The Ultrakan concept is similar to the CanShield in that the body wall is wrapped around a mandrel and the two edges are overlapped and heat-sealed to each other. The bottom disk is inserted in the container, and the body wall and bottom disk are heated and crimped or rolled together. Thermoplastic hot-melt adhesive may be used for added security and seal strength. The interior edge of the sideseam can be skived to enhance WVTR or greaseproofness of the container, or both. The top of the Ultrakan container can be finished in a variety of ways: rolled outward (to accept a membrane seal); flared (for a variety of seamed metal ends); rolled inward (for special thermoformed or injection-molded plastic closures); or gently flared (for insert rotor/dispenser style closures). The Ultrakan system also provides the option of customizing by special bottom techniques which offer dispensing features for granulated products, powders and paste, or semiliquid sauces or condiments.

More recently, a high barrier paperboard can has been introduced from Sweden (Cekacan). By incorporating the use of polyolefin laminates (see Laminating) along with foil and a special means of sealing the package, a hermetic seal has been demonstrated, making the package virtually impermeable to gas, liquid, fat, etc (see Table 1). The Cekacan system involves both a can-forming operation and a can-closing operation. In the forming operation, the sidewall is wrapped around a mandrel and butt-seamed (not overlapped). Just prior to the wrapping operation, a foil-laminated tape is induction-sealed to one edge of the blank. With the seam butted, and induction sealer affixes the tape to the interior of the can in such a way as to provide a continuous hermetic seal along the longitudinal seam.

The package is transfered to an end-closing device wherein a top or bottom closure is affixed. In this case, the closure is inserted into the can with the closure sidewalls flanged to the vertical position. Through the use of pressure and induction sealing, the disk is hermetically sealed into place. The package is then discharged for filling and brought back to the second piece of equipment, which inserts and hermetically seals the final closure. Closures are available that provide easy opening without compromising the gas-tight integrity of the package. Through the use of a butt seam, held together with sealable tape, there are no discontinuous joints to bridge. Ends are inserted into and fused by induction sealing to the smooth interior wall. During the induction-sealing process, the fluid flow of the internal coating, usually polyolefin, fills any short gaps that may occur. The equipment is simple to operate and does not require special expertise.

A special attribute of all the in-plant paperboard-can packaging systems is the reduction of materials storage and handling. Since the body walls are shipped flat along with the bottom disk and top closures, a minimal amount of storage space is necessary. The average space needed to contain the paperboard can in its flat form represents approximately 97% savings over a similar number of composite or metal cans or glass jars. Additional savings are realized by the reduced cost of shipping the container components to the plant and also in the weight of the final product.

Materials. Basic paperboard-can-body structures are made of laminations of paperboard, aluminum foil, and polyolefins. End structures are analogues which might omit the paperboard for some applications. The generalized structure is paperboard/bonding agents/aluminum foil/polyolefin (outside-to-inside). Engineering the components to each other and the structure to the package has been a significant advance. The material components must be functional, economic, structurally sound, and compatible with the contained product.

The paperboard component of the lamination is not critical to the hermetic function of the Cekacan, but it is essential to the commercial value of the system. The exterior surface must be smooth and printable and the interior surface must be sufficiently tied to ensure adhesion to the adjacent layer. Since the paperboard-can body is composed of a single-ply material (as opposed to the multiple plies in composite cans), ranging from 0.016- to 0.032-in. (406–813-μm) finished caliper, the appropriate finished caliper must be chosen to meet the physical stress. At the same time, the economics of the additional caliper board must be weighed against the cost.

Structures can be engineered for each product's specific requirements. Interior polyolefins may be polyethylenes or polypropylenes. They can be applied by extrusion, coextrusion, ex-

Table 1. Comparative Water-vapor Transmission[a]

Product	Conventional package	Rate of moisture pickup at 75°F (24°C) and 100% rh, wt % per wk	
		Conventional package	Cekacan
dehydrated sweetened beverage powder	26.5-oz (751-g) composite can	0.03	0.0075
powdered soft-drink mix	34.0-oz (964-g) composite can	0.02	−0.0025
sweetened cereal product	12.0-oz (340-g) bag-in-box	0.2	0.04
snack	7.5-oz (213-g) composite can	0.01	−0.01

[a] Courtesy of the Center for Packaging Engineering, Rutgers, The State University of New Jersey.

trusion–lamination, or adhesion–lamination (see Extrusion coating; Laminating). In the case of a gas-tight container, a crucial variable is the bonding of the interior polyolefin to the aluminum foil or paper substrates. This adhesion must be maintained above preestablished minimums in converting, body erecting, sealing, and operation. This is a very demanding requirement.

J. M. LAVIN
Container Corporation of America

CANS, FABRICATION

The metal can is one of the oldest forms of packaging preserved food for long periods. The traditional method of manufacture is to start with a flat rectangular sheet of tinplate, roll the blank into a cylinder, and solder the resultant longitudinal joint line to form the side seam. Circular ends are joined mechanically to flanges formed at each end of the cylinder by a rolled seam, known as a double seam (see Can seamers). One, the maker's end, is fitted by the can manufacturer to form the body. The other, fitted by the filler after putting the contents into the can, is known as the packer's end. Since the container is made from three separate items, it is known as a three-piece can. Its construction has remained basically unchanged for over 150 years, although advances have been made in automating and speeding up the original hand process, in gradually reducing the metal content through more sophisticated design geometry, and more recently, in changing the method of making the side seam from soldering to welding (see Fig. 1).

Since the early 1970s a different concept of can making has gained acceptance in commercial production. In this the body and one end is formed in one entity from a flat circular blank by press-forming techniques (1). Sealed with the usual packer's end, it is known as a two-piece can. Two methods of forming are identified: drawing and ironing (D&I in the United States, DWI in Europe) and draw redraw (DRD). The techniques themselves are not new (2). D&I, for instance, was used in World War I for making shell cases. What distinguishes

Figure 1. Three-piece can side seams. (a) Soldered; (b) cemented; (c) welded. Courtesy of *Proceedings, 3rd International Tinplate Conference*, 1985.

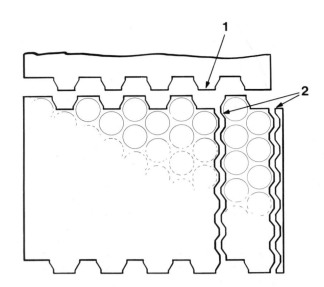

Figure 2. Scrolled sheet showing layout of blanks and scrolled edges for material utilization. 1, Primary scrolling used when cutting coil into sheets for coating. 2, Secondary scrolling used for cutting sheets into strips for feeding into the shell press.

them in can making is the use of ultrathin metal in very high speed production to yield outputs counted in hundreds of millions (10^8) per year.

Can Types

All processes convert flat sheet into finished cans, supplied with a loose end for the packer, according to this basic scheme: prepare plate; make body; form features; and apply finishes. The order may vary, depending on the process used. The manufacture of three-piece and two-piece DRD cans starts with the finishing step. Coil is usually cut into sheets, to be coated on one or both sides and decorated if appropriate. The coatings are called enamels in the United States, and lacquers in the United Kingdom. If the starting point is a circular blank, as for DRD or ends, the edges of the sheet are scrolled for economy in material usage (see Fig. 2). Alternatively, precoated coil may be fed directly into the first blanking operation. In the manufacture of two-piece D&I cans, plain coil, as supplied from the mill, is the starting point.

Three-piece manufacture is readily adaptable to making cans of any diameter and height. The production equipment is amenable to changes in size and is capable of production speeds of up to 500 cans per minute. Where the use of lead solder is no longer acceptable, a change to welding can be made at minimal cost on an existing line, since only the equipment which makes the cylinder needs to be changed. Three-piece manufacture is the choice of the smaller- to medium-sized operation requiring flexibility, for producing can sizes required in relatively modest quantities, or to suit a variety of fill products requiring changes in coating specification.

Two-piece manufacture is basically suited to a single can size, requiring outputs of at least 150 million (1.5×10^8) per year. DRD, using precoated sheet or coil, is used for food cans, and predominantly in the shallower sizes (h/d < 1). The process is being used, however, for food cans with h/d of 1.5 in the popular 3-in. (7.6-cm) diameter sizes (eg, 300 × 408¾ in Eu-

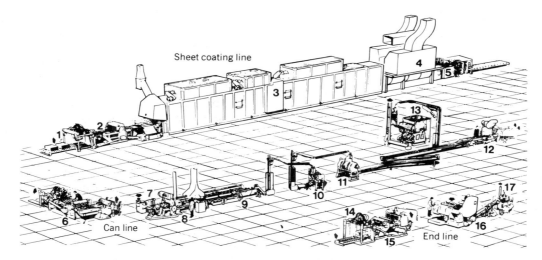

Figure 3. Three-piece can manufacture. Sheet-coating line: 1, sheet feeder; 2, roll coater; 3, oven; 4, cooler; and 5, unloader and sheet stacker. Can line: 6, slitter; 7, blank feed; 8, bodymaker; 9, side seamer; 10, flanger; 11, beader; 12, end seamer; and 13, tester. End line: 14, sheet feeder; 15, secondary scroll shears; 16, shell press; and 17, compound liner. Courtesy of Metal Box Ltd.

rope, 300 × 306 in the United States) (for explanation of can dimensions, see Cans, steel). The application of enamels to both sides permits the use of electrolytic chromium-coated steel (ECCS), which has a surface that is too abrasive to be used uncoated (plain). Because economics demand the use of the thinnest possible plate, the high strength needed dictates the use of double-reduced (DR) grades (see Tin mill products). Generally utilizing two presses with a minimum of peripheral equipment, DRD provides a compact installation of relatively low capital cost for the achievable outputs. For that reason, it appears to be a preferred method for self-manufacture of relatively shallow containers by packers (eg, cans for tuna fish). However, material property requirements are high, and the standard of enameling is critical to maintain integrity as the metal is formed.

D&I converts plain coil into a fully finished can in a totally integrated, fully automatic process. The high capital cost of the full range of equipment needed predicates a high annual output. This, together with the desirability of keeping the line running once harmonious operation has been achieved, usually results in a 24-hour per day, 7-day per week operation with annual production measured in hundreds of millions (10^8). Basically a single-can-size process, it is best suited for the production of beverage cans, which are made worldwide in a few standard sizes for use on high speed filling lines.

Either tinplate or aluminum may be used, usually on a dedicated line. In recent years, due to the uncertain economics relating to metal costs, new lines are built as "swing lines," basically capable of handling either metal. They nevertheless require some downtime to make the change and normally operate continuously on one metal.

Three-Piece Can Manufacture

The manufacture of three-piece cans proceeds as follows: cut up coil into rectangular sheets; coat and decorate sheets; slit into rectangular blanks; form cylinder and side seam; repair coating at seam; separate cylinders (if appropriate); neck

or bead (if appropriate); form flanges at both ends; fit maker's end; and test and palletize (see Fig. 3).

Slitting. Large rectangular sheets of tinplate, precoated if required on one or both sides by roller coating techniques, are slit by pairs of shear rolls (3), first longitudinally into strips, and then transversely into rectangular blanks of the appropriate size.

Because welded can bodies have considerably less overlap at the side seam than soldered cans, slitting demands greater precision than is needed for soldered can blanks. Slitting rolls can be positioned very accurately by applying internal pressure at the bore to expand them so as to move freely on the drive shaft (4). After precise location using electronic measuring methods, release of the oil pressure binds the roll securely on the shaft and avoids the application of forces which might distort the setting.

Bodymaking, soldered cans. The solder commonly used consists of 98% lead and 2% tin. As more and more countries impose lower limits of lead content in food, the soldered can is being phased out in Western countries. Nevertheless, many soldering lines around the world will continue in operation for some time. The machine consists of a bodymaker which forms the flat blank into a cylinder, coupled to the side seamer where the joint is soldered. Blanks are still commonly transferred manually in stacks from the slitter, although equipment exists for automatic transfer. In the bodymaker, the edges are prepared and hooks are formed, which mechanically couple the two sides when the cylinder is formed. For high speed operation, the body cylinder is usually formed by a pair of wings, and the side hooks are flattened once they have engaged. The body then passes to the side seamer, where flux is applied and the seam area pre-heated by gas jets before passing over a longitudinal solder roll where the solder is applied. More gas jets ensure that the solder flows well into the joint, after which a rotating wiper mop removes excess solder, mainly in the form of blobs. The can is then cooled by air jets before discharge into the downstream finishing operations. As lead con-

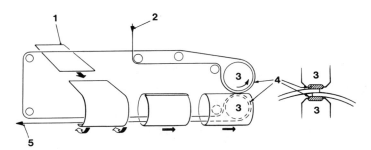

Figure 4. Manufacturing stages in forming a welded cylinder. 1, Blank feed; 2, wire feed; 3, welding rolls (electrodes); 4, copper wire (welding contact); and 5, used wire (to scrap chopper).

tamination has become of increasing concern, the wiper-mop unit has been the subject of considerable improvement to contain the lead dust generated by its operation. By its very nature, the process requires the use of tinplate. If the tin-coating weight is less than 25 lb/base box (2.8 g/m^2), soldering difficulties arise. It is nevertheless possible to use lighter coating weights, down to 18 lb/base box (2 g/m^2), if steel makers and can makers collaborate to ensure that there is an adequate level of free (unalloyed) tin available (see Tin mill products).

Bodymaking, welded cans. The seam is made by a resistance-welding process using the lost-wire-electrode principle (5). Body blanks are fed from the bottom of the hopper through double blank detectors into the forming rolls (see Fig. 4).

The two laps of the curled cylinder are held in a Z bar, and the cylinder is pushed along the bar by one or more driven chains into the welding rolls. Overlap accuracy, which is dependent on gauge and temper of the plate, has to be controlled to avoid bodymaker wrecks. The two overlapping edges of the cylinder are bonded by a-c resistance-seam welding, using approximately 4000 A at 6–8 V. The two overlapping edges must be free of contamination on both sides, to eliminate resistance variations which would lead to welding faults. A significant amount of energy is lost in heating other parts of the welder, which therefore need water cooling. In high humidity locations this can lead to problems with condensation. Each resistance-welding spot, called a nugget, is achieved by one half of the a-c waveform. Welding speed at the electrical supply frequency is limited, to ensure some overlap between the nuggets. To achieve higher welding speeds (up to 200 ft/min (61 m/min)), higher frequencies (up to 520 Hz) have been employed via frequency transformers.

Earlier systems used a large overlap and raised the steel temperature to the melting point with light roll pressure to join the metal. The latest welders use a small overlap (0.3–0.5 mm) with metal temperatures just below the melting point and increased roll pressure to forge the two laps together. To ensure reproducible welding conditions for every part of the seam, the welding contact is made by copper wire wrapping around both welding rolls and moving with the tinplate. Any contamination of the welding electrodes by tin is thus continuously removed from the contact area. After use the wire is either cropped or rewound for recycling.

Having dealt effectively with the problem of tin contamination, the system paradoxically requires a minimum of tin coating, around 10 lb/base box (1.0 g/m^2), although there are developments for reduced amounts. ECCS or blackplate can be

processed only if the oxide films are removed from the edges of the blanks. This is called edge cleaning. An older technique using welding rolls without an intermediate copper-wire electrode is still in use, but is now of less importance in high speed can manufacture, mainly because of the frequent need for changing the electrode rolls.

The integrity and quality of the seam weld is usually tested by mechanical means (eg, Ball test). For more detailed examination, weld cross sections are metallurgically inspected for any sign of separation between laps or radiographically examined for welding faults. Bodymakers have also been fitted with weld monitors to continually monitor welded seam quality. Usually these rely on measurements of voltage or current fluctuations between the welding rolls, and cans made outside preset limits are rejected. Recently, measurements of weld deformation and weld temperature have been explored to enhance the performance of these monitoring systems.

Renewed effort is going into welding the side seam by means of a laser beam (6). The principle of can welding by laser was demonstrated in the late 1970s, but welding speeds appeared to be too low to justify commercial exploitation. Developments in the United States, Europe, and Japan are directed at increasing welding speeds and making the laser welding process commercially viable. Apart from the elimination of costly copper wire, the method offers pure butt welding, with advantages to seaming and necking as well as versatile decoration.

The welded seam interrupts an otherwise smooth inside-enameled (or plain) surface. It contains exposed iron and iron oxide, as well as tin, on both sides of the weld. To protect the product from contamination and/or the weld from attack by the product, the side seam needs to be coated in most cases. Further information is provided below in the section on coating.

Bodymaking, cemented cans. Cans used only for dry products can be made with cemented side seams. Before the almost universal use of two-piece cans for high output beverage can production, one method of making them was to cover the longitudinal edges of the blank with nylon strip, which was fused after forming the cylinder. The process has a number of proprietary names (eg, Miraseam, A-Seam). An advantage was complete protection of the raw edges of the blank. It could be used only with TFS, since the melting point of tin is close to the fusion temperature of the plastic. The bodymaker used is an adaption of the soldering configuration.

Completing the body. The plain cylinder must be furnished with a flange at each end for attachment of the closure. For food processed in the can, where the can may be subjected to external pressure during processing or remain under internal vacuum during storage, the surface of the body may be ribbed for strength. This process is known as beading. Cylinders for shallow containers may be made in a length suitable for two or three cans to obtain maximum efficiency from the forming machine. In this case, the first operation would be parting the cylinders, traditionally at partial cuts (scores) cut in the flat blank before forming. More recently, however, as can-trimming methods have been developed for two-piece cans, there has been a move towards adapting these to part unscored cylinders. Because all this equipment is the same however the cylinder is joined, it is relatively easy to eliminate lead by substituting welding machines for soldering in existing lines.

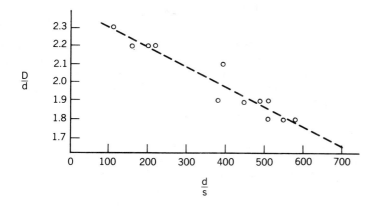

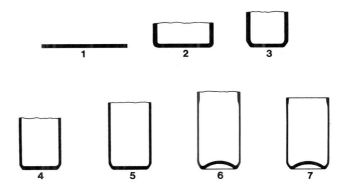

Figure 5. Limiting draw ratio. The maximum blank diameter which can be drawn into a cup without metal failure. The relationship illustrates the significance of sheet thickness (gauge). D = blank diameter; d = punch diameter; s = sheet gauge. Courtesy of International Tinplate Conference, 1976.

Figure 7. Stages in forming a drawn and ironed (D&I) body. 1, Circular blank; 2, cup; 3, redraw; 4, first ironing stage; 5, second ironing stage; 6, third ironing and dome forming (note thickening at top where flange is to be formed); and 7, trimmed body.

Two-Piece Can Manufacture

Metal-forming methods. Both methods of making two-piece cans use forming methods that depend on the property of metal to "flow" by rearrangement of the crystal structure under the influence of compound stresses, without rupturing the material.

Drawing. In drawing, as applied to can manufacture, a flat sheet is formed into a cylinder by the action of a punch drawing it through a circular die. The tendency of the material to wrinkle as the outer diameter is reduced must be supressed by the application of pressure through a draw pad. Some thickening is inevitable, but the process is essentially one of diameter reduction at constant metal thickness so that the surface area of the cylinder is equal to the surface area of the blank from which it was formed. This factor forms the basis of design, in particular the calculation of metal utilization. The amount of diameter reduction achievable is governed by the properties of the material and the surface friction interactions between the tooling and the material influenced by lubrication. It has been the subject of much research. One of the classical relationships is shown in Figure 5.

The diameter of the cup produced in the initial draw may be further reduced by a similar redraw operation, with a draw sleeve fitting between punch and the inside diameter of the cup, instead of the draw pad. The rule of constant area determines that the reduction in diameter is accompanied by a corresponding increase in height (see Fig. 6). The redraw oper-

ation can be repeated once more, provided the progressive reductions fall between definite limits, to avoid metal failure.

Wall ironing. In pure ironing, as used in the D&I process, a cylindrically drawn cup is fitted onto a punch and forced axially through a die, whose diameter creates a gap with the punch, which is smaller than the wall thickness of the cup. The process, similar to extrusion, thus results in a reduction in wall thickness at constant diameter, and the governing regime is one of constant metal volume. In other words, the volume of metal in the ironed cylinder is equal to that in the ingoing cup, and thus to the volume of metal in the original blank. In can making, the process is repeated two or three times in line, the punch passing sequentially through the dies in a single stroke. The most convenient way of fitting the drawn cup onto the punch is by a redraw operation ahead of the first ironing die (see Fig. 7).

The amount of reduction at each stage is determined by the material properties, governed by the need to avoid metal failure. The very high friction, under extreme surface pressure, makes special demands upon lubrication, which is combined with copious flood cooling to maintain the critical punch-to-die gap. A similar effect can be obtained in drawing if the gap between the draw die and its punch is less than that of the metal being drawn. It is common practice to make this gap equal to the nominal thickness of the plate, to control thickening due to diameter reduction. This process is called sizing. The gap can be further reduced to produce a definite thinning of the wall, relative to the base material (a combination of drawing and ironing).

D&I can manufacturing. The procedure for making D&I cans is roughly as follows: unwind plain coil; lubricate; blank disks and form cups; redraw; iron walls; form base; trim body to correct height; wash; and treat (if appropriate). Then, for *beverage cans:* coat outside (if appropriate); decorate; coat inside; neck and flange open end. (If tinplate is used, the order of the last two may be reversed.) For *food cans:* coat outside during washing; flange open end; form body beads; coat inside. (In some lines, the order of the last two is reversed.)

The closures, or ends, are made from precoated sheet on multiple-die presses, with beads impressed to withstand the internal and external forces. They are finished by curling the edge to assist in double-seaming, and placing a ring of lining compound to act as a gasket in the finished seam (see Fig. 8).

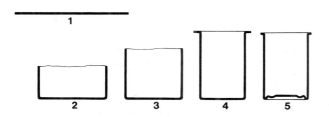

Figure 6. Stages in forming a draw redraw (DRD) body. 1, Circular blank; 2, cup; 3, first redraw (through die); 4, second redraw (operation stopped to leave flange); and 5, form base and trim flange.

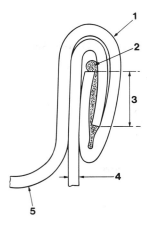

Figure 8. Double-seam closure. Showing fixing of end to cylinder, and function of lining compound: 1, end curl, folding round flange on cylinder; 2, lining compound; 3, seam overlap; 4, cylinder; and 5, end.

The components of a D&I line are shown in Figure 9. Coils are shipped with the axis vertical, for safety and to avoid damage to the laps. A down-ender is used to bring the axis horizontal and transfer the coil to a coil car, for distribution to one of a number of dereelers (unwind stands), each feeding one cupping press. Dereelers may have two arms, one aligned towards the coil car and the other feeding the press, for rapid coil change through rotation of the arms into the alternative positions. This is not only for renewing the spent coil, but also to remove from service coils found to be defective.

The coil is passed through a lubricator, where drawing lubricant is applied to both sides by dipping in a tank, and removing the excess in a pair of rubber-covered rolls. In older lines the lubricant in the tank is replenished as required from a container of prepared lubricant, mixed to the correct consistency. In newer lines, there has been a tendency to use an external circulating system, to permit temperature control, continuous filtration, and constant lubricant quality.

Steel coil may additionally be passed through an inspection unit for continuous monitoring of gauge thickness and detection of pinholes, interlocked to stop the press when out of specification material is detected. The cupping press (7), or cupper, is a purpose-built double-acting press, stamping out a number of circular blanks and drawing them into shallow cups in one operation. Depending upon can size and available coil width, as many as 12 cups may be produced in one working stroke. Presses may have the drive at the top (Minster) or underneath (Standun), and operate at up to 200 strokes per minute.

The process demands great precision due to (1) the extremely small gap between blank punch and die (1/10 metal thickness on material of ≤0.012 in. (≤0.30 mm)); (2) the small amount of metal (shred) between adjacent blanks (about 0.04 in. or 1 mm); and (3) maintaining the draw pad parallel to the die face, to avoid wrinkling the very thin and relatively hard material.

The wall ironing machine (bodymaker in the United States) (8,9) converts the cups into a cylinder with the correct thickness distribution along the wall and forms the base into a shape designed to resist internal pressure in service. The cup is redrawn onto a punch, passed through three consecutive ironing rings to thin the walls, and into a subpress at the end of the stroke, the doming station, to form the base. At the start of the return stroke, spring-loaded stripper fingers, assisted by compressed air fed through the ram, remove the can body from the punch into a conveyor which takes it sideways out of the machine. At this stage, the body height depends on manufacturing tolerances of plate and tooling, and the top displays the eared structure characteristic of metal-forming operations.

The machine runs at about the same speed as the cupper, so that one wall ironer is needed for every cup produced at one cupping stroke. This is only a rough guide, as in practice ironers tend to run a little faster than the cupper. Twin-ram machines are also available to halve the number of units required

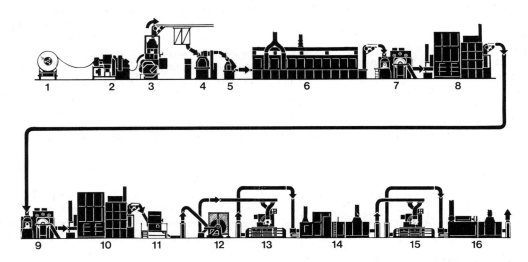

Figure 9. D&I can manufacture. Diagram showing all the equipment that may be required. 1, Uncoiler (dereeler); 2, lubricator, and optional coil inspection; 3, cupping press; 4, wall ironer (bodymaker); 5, trimmer; 6, can washer; 7, external coater; 8, pin oven; 9, decorator; 10, pin oven; 11, necker–flanger; 12, tester; 13, internal spray machines; 14, curing oven (IBO); 15, optional second spray; and 16, IBO. Courtesy of Metal Box Ltd.

but with the attendant risk of doubling the loss of output at every stoppage. The tooling uses tungsten carbide inserts in the ironing dies, together with a tungsten carbide punch when used for tinplate. A tool-steel punch is adequate for aluminum.

Associated with every wall ironer is a trimming machine to which the eared bodies are transferred, ideally held by the base either magnetically or by vacuum. Two types of trimmer are used: rail and roll. In the rail trimmer, the end of the can is trimmed between a mandrel mounted on a rotating turret and a stationary rail. A number of mandrels are fitted, and the turret is in continuous motion. In the roll trimmer, cans are indexed to a position where the end is inserted between two rolls and rotated about its axis, and the rolls are closed to perform a trimming action similar to a can opener. In one widely used machine, the cutting rolls are mounted in an easily removable cartridge for refurbishing and precision-setting in a toolroom environment.

After the cans have been trimmed to a consistent height, they can be handled in bulk, to be conveyed to a washer. Cans are conveyed on a flatbed open mesh conveyor, open end down through a series of chambers containing spray nozzles above and below. Top conveyors are used in each spray chamber to restrain the cans from falling over. After washing off the drawing and ironing lubricants with detergent and rinsing with plain water, aluminum cans commonly receive an etching treatment to make the surface receptive to the organic coatings applied in the finishing operations. A final rinse in deionized water ensures stain-free drying. Alternatively, the final rinse may contain an organic coating, called a wash-coat. This is mainly used to protect the outside surface of a tinplate food can, underneath the paper label applied by the filler. The cleaned and dried can is now ready to receive its internal and external finishes.

One metal-forming operation, however, remains to be done at some stage during the finishing operations: necking and flanging in the case of a beverage can, or beading and flanging for the food can. The latter operations are similar to the corresponding operations on the three-piece can, although the machinery may differ in detail due to the presence of the integral bottom and the absence of any thickening around the circumference due to a body seam. Spin flanging is invariably used to avoid splits in the axial metal grain structure caused by ironing.

The top of the beverage can is necked in to yield metal-cost saving through the use of smaller ends. The diameter reduction may be achieved by die-necking, followed by spin flanging. This method was commonly used for the original "single-step" reduction, from 211 to 209 for the most common beverage can. It was also applied for a double-step reduction from 211 to 207½. However, as cans become thinner and the neck diameter is reduced further to accept a 206 end, multiple-die necking, although practiced, creates problems. A method of simultaneously creating the neck and the flange, known as spin neck and flange, is finding increasing application (10). It creates an aesthetically pleasing neck contour, with minimal strain on the split-prone flange and very low axial force on the can. For the smaller necks, one or two die preneck stages are still necessary (11). Development of an alternative single-stage preneck operation by spinning is at an advanced stage. The final forming operations are carried out either before or after internal protection depending on local circumstances. In general, aluminum requires an enameled surface to prevent

pick-up (galling) on the tooling. With tinplate either sequence is used.

DRD can manufacturing. The steps in DRD can manufacturing are as follows: cut up coil into scrolled sheets; coat and decorate sheets (alternatively, coat coil); blank disks and form cups; redraw (once or twice, depending on can size); form base; trim flange to correct width; and test and palletize.

As in D&I, a multitool cupper (7) is used to cut blanks from wide coil or sheet and form them into shallow cups. The cups are reformed by one or two further draws to progressively reduce the diameter and increase the height. By stopping the final draw at an appropriate point, a flange is left in the can, but as with all metal-forming operations the irregular edge has to be subsequently trimmed. The base is impressed with a shape contoured to withstand the processing requirements.

All these operations are usually carried out on a multilane, multistage, progression press of a capacity to match that of the cupper. However, due to the critical nature of the die-trimming operation, separate machines employing different principles are available. When simultaneously drawing and ironing, it is convenient to draw the body right through the die in order to achieve thickness control in the flange area. In this case, the body is trimmed and flanged in conventional D&I equipment. For the taller cans, the walls may then be beaded in the usual machine.

Variants in metal-forming equipment are the Metal Box rotary press line (12) (Fig. 10) and the Standun opposed-action press. The latter features a split draw pad with separate pressure controls, which is claimed to permit greater draw reductions and provide control of wall thickness through stretch rather than ironing, which is said to be less severe on the coating. Equipment is being introduced for smaller output and easier change in can size (eg, by Alfons Haar in the FRG).

Coating

With the exception of three-piece cans for certain products, organic coatings are applied to both inside and outside surfaces of the can. Internally they provide a protective barrier between the product and the metal; externally there may be a plain coating to provide protection against the environment, or decoration to give product identity, as well as protection.

The coating is applied "wet"; that is, the resin is suspended in a carrier for ease of application and allowed to coalesce or flow out. The coating is then baked ("stoved", in the UK), first to remove the carrier, which may be an organic solvent or a mixture, predominantly water, and then to polymerize (cure) the resin. Two methods of application are in general commercial use: roller coating, if physical contact is possible (flat sheets or the outside of two-piece cans); and spraying, if physical contact is impossible or difficult. Spraying is predominantly for the inside of two-piece D&I cans, but DRD cans are sometimes sprayed as well. Drying and curing are normally carried out by forced convection using hot air. Alternative methods of application and baking are available where special conditions warrant their use.

Ultraviolet curing permits application of virtually solvent-free "wet" coating by conventional means. It requires the use of special materials amenable to this form of polymerization, which demand care in storage and handling, and complete polymerization in the interests of health and safety. Its attraction lies in the absence of solvent in difficult environmental situations, and extremely short curing times coupled with rel-

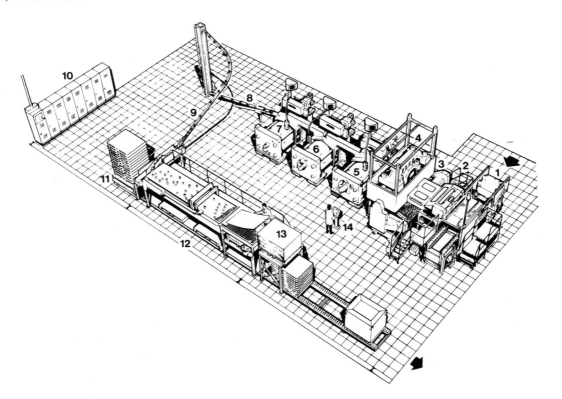

Figure 10. Rotary Press DRD line. 1, Infeed conveyor (for stack of coated sheets); 2, lubricator; 3, sheet feeder; 4, cupping press; 5, first redraw press; 6, second redraw press; 7, trimming and base forming press; 8, can sampling chute; 9, can conveyor (a tester is usually inserted here); 10, electrical controls, includes press synchronization; 11, pallet feed; 12, palletizer; 13, layer pads; and 14, master control station. Courtesy of Metal Box Ltd.

atively low energy demand. It is thus used for very high speed operations or where space is limited, as for instance between two print color applications to the same side ("Inter-deck" drying).

Powder coating, where the resin is applied "dry" in the form a fine powder, is directed in many cases by creating an electrostatic field. It is used mainly where unusually heavy coating is required, notably in the protection of welded side seams. The absence of solvents avoids the otherwise excessive cost of drying and eliminates the blistering, which would be caused by trapped solvent trying to escape through a thick film. Curing is usually by infrared radiation or high frequency induction heating, as hot air could disturb the uncured coating.

A new development, electrophoretic deposition, provides a means for electrically depositing a resin film on a metallic substrate from an aqueous suspension (13). Originally used at low speed for protecting automobile bodies, it has now been developed to be applied at can production speeds with practicable voltage and current demands. Two-piece cans may be coated both inside and out in a purpose-designed machine, using a system of process cells (14,15). Compared to spray application it gives far more even distribution of coating over the can wall and base, providing a saving of the quantity used without sacrificing the required maximum coating thickness. Its throwing power enables it to coat regions inaccessible to spray, useful for the severe profiles which have to be used to obtain adequate container strength with thin plate.

Thermal transfer of the complete design from printed paper

to a plain coating on the can provides a method of obtaining very high quality decoration in two-piece cans if the extra cost is justified (16). It is used in the UK under the trade name Reprotherm (Metal Box Ltd.) for promotional designs on single-service beer cans.

Coating Equipment

Offset coating. Offset coating employs the principle of offset, whereby a metered quantity of coating is applied to a rubber-covered roller, to be transferred to the metal substrate (see Fig. 11) (see Coating equipment; Printing). In roll coating, metering is accomplished by a series of steel or rubber rollers which pick up the coating from a trough and ensure even distribution and weight control through a combination of relative surface speed and the pressure or gaps between them. Another method, gravure coating, employs a pattern of cavities etched into a steel roll, which are filled as the roll dips into a trough. The excess is removed, and the precise amount filling the cavities is then given up to the transfer roll. A third method, rotary-screen printing, is a new development which utilizes the old established principles of silk-screen printing. The screen is in the form of a roll, fed internally with the coating material, which is forced through the screen by an internal doctor blade onto the transfer roll. Patterns governing shape and weight of the applied coating can be formed in the screen by photographic methods.

Roll coating is the most common method for coating in the flat, used for both coil and sheets. In sheet coating, it is possi-

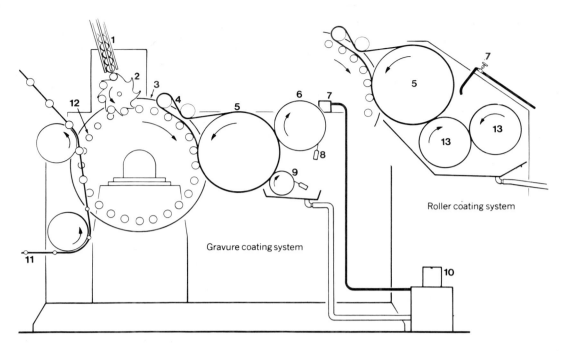

Figure 11. Offset coating, showing application to two-piece can bodies. 1, Body infeed; 2, starwheel; 3, mandrel wheel; 4, pre-spin; 5, offset cylinder; 6, gravure roll; 7, coating material feed; 8, doctor blade; 9, stabilizer roll; 10, coating material tank, with pump; 11, oven pin chain; 12, mandrel; and 13, metering rolls.

ble to cut a pattern into the transfer roll and coat only the area to be used for blanks. In a large-scale operation this could yield significant savings in material, but requires the sheet to be accurately registered with the transfer roll. Rotary-screen coating permits a better definition of the pattern in the applicator itself without cutting the transfer roll, thus leaving a sharper, cleaner edge between coated and uncoated areas. Other advantages claimed are the ability to apply double the coating weight of roller coating, possibly saving one coating operation, and the ability to deposit different coating weights on adjacent areas of the sheet. Roller coating is also used for the outside of two-piece cans. For this application, however, the more precise method of gravure coating is gaining favor.

Decorators operate in a similar fashion (17) but are considerably more complicated, since they apply four or sometimes five colors with extreme accuracy of registration to build up the total picture. Thus a can decorator uses the same can-handling systems as the coater, replacing the coating head with printing equipment. Since the whole design has to be applied in one rotation of the can, the picture is built up on the offset roll, now divided into discrete "blankets" of a length to correspond to the circumference of one can.

Spraying. The method normally employed is the airless spray system, where atomization of the enamel into droplets of appropriate size is achieved by the use of high liquid pressure, about 650–850 psi (4.48–5.86 MPa). A spray gun consists of a spray nozzle designed to give the desired spray pattern, a needle valve to cut off the flow without dripping, and the means to activate the valve, usually a solenoid. Since viscosity of the enamel has a decisive influence on the spray pattern and discharge quantity, it is heated and continuously circulated through the gun. The quantity deposited in the can is deter-

mined by precise timing of the valve; a typical spray time is approximately 100 milliseconds. In modern practice, the gun is at a fixed position at the mouth of the can, although machinery exists for moving it axially inside the can during spraying. This is called a lancing gun. The can is indexed to a position in front of the gun and rotated at about 2000 rpm, to ensure an adequate number of rotations while spraying is in progress. It is beneficial to keep rotating the can on its way from the spray machine to the oven to assist flow-out and even distribution.

The pattern of coverage and number of operations depends on the type of product to be packed (beer or carbonated soft drinks), the can material (steel or aluminum), and the can size (h : d ratio). Coverage can be total, where one spray covers the total area, or zonal, where one spray is directed toward the base and lower sidewall of the can, and a second covers the upper sidewall up to the open end. Two coats may be applied in consecutive indexing stations on the same machine (wet on wet) or in separate machines with intermediate drying (wet on dry).

Internal side-seam protection for welded cans. The bare metal that exists in the weld area must be protected. Roller coating, spray, and electrostatic powder coating are used. The applicator is mounted on an extension of the welding arm, which carries the supply of coating material from a connection fitted before the can cylinder is formed around it. Roller coating permits a low pressure fluid supply, but lack of space makes the applicator components extremely small, and conformity with the area to be coated is difficult to achieve. Spray application requires a high pressure fluid supply. Although good initial coverage can be achieved, liquid enamel tends to retract from sharp edges, so that high application weights are needed for adequate cover of the cut edge of the blank. The overspray, which escapes from the gaps between the cans, poses collection problems, especially in avoiding external contamination through drips.

Powder is fluidized with air for conveyance through the welding arm and electrostatically charged to achieve deposition in the welded area. This method probably provides the best protection, but is expensive in material and difficult to run in production.

BIBLIOGRAPHY

1. Brit. Pat. 621,629 (June 16, 1949), J. Keller.

2. U.S. Pat. 760,921 (May 24, 1904), J. J. Rigby (to E. W. Bliss Company).

3. U.S. Pat. 2,355,079 (Aug. 8, 1944), L. L. Jones (to American Can Company).

4. Brit. Pat. 1,574,421 (Sept. 10, 1980), J. T. Franek and E. W. Morgan (to Metal Box Ltd.).

5. Brit. Pat. 910,206 (Nov. 14, 1962), Soudronic AG.

6. Fr. Pat. 2,338,766 (Aug. 19, 1977), E. E. V. V. Saurin and E. V. Gariglio.

7. Brit. Pat. 1,256,044 (Dec. 12, 1971), E. Paramonoff and H. Dunkin (to Standun Inc.).

8. U.S. Pat. 3,270,544 (Sept. 6, 1966), E. G. Maeder and G. Kraus (to Reynolds Metals Company).

9. U.S. Pat. 3,704,619 (Dec. 5, 1972), E. Paramonoff (to Standun Inc.).

10. Brit. Pat. 1,534,716 (Dec. 6, 1978), J. T. Franek and P. H. Doncaster (to Metal Box Ltd.).

11. Brit. Pat. 2,083,382B (Mar. 24, 1982), J. B. Abbott and E. O. Kohn (to Metal Box Ltd.).

12. Brit. Pat. 1,509,905 (May 4, 1978), J. T. Franek and P. Porucznik (to Metal Box Ltd.).

13. Brit. Pat. 455, 810 (Oct. 28, 1936), C. G. Sumner, W. Clayton, G. F. Morse, and R. I. Johnson (to Crosse & Blackwell Ltd.).

14. U.S. Pat. 3,922,213 (Nov. 25, 1975), D. A. Smith, S. C. Smith, and J. J. Davidson (to Aluminum Company of America).

15. Brit. Pat. 1,604,035 (Dec. 2, 1981), T. P. Murphy, G. Bell, and F. Fidler (to Metal Box Ltd.).

16. Brit. Pat. 2,101,530A (Jan. 19, 1983), L. A. Jenkins and T. A. Turner (to Metal Box Ltd.).

17. Brit. Pat. 1,468,904 (Mar. 30, 1977), (to Van Vlaanderen Container Machinery Inc.).

General References

A. L. Stuchbery, "Engineering and Canmaking," *Proceedings of the Institution of Mechanical Engineers,* **180,** 1, 1167–1193 (1965–1966).

J. T. Winship, *Am. Machinist,* Special Rep. No. 721, 155 (Apr. 1980). Explains how metal containers are made.

The Metal Can, Open University, Milton Keynes, UK 1979, 52 pp.

C. Langewis, Technical Paper MF80-908, Society of Manufacturing Engineers, Detroit, Mich., 1980.

E. Morgan, *Tinplate and Modern Canmaking Technology,* Pergamon Press, 1985. Contains a broad summary of printing processes.

W. A. H. Collier in *Proceedings of Cold Processing of Steel,* The Iron and Steel Institute and the Staffordshire Iron and Steel Institute, Bilston, U.K., October 1971 and Mar. 1972, pp. 70–77. Describes drawing, forming, and joining of steel containers.

J. D. Mastrovich, *Lubrication* **61,** 17 (Apr./June 1975). Describes aluminum can manufacture.

Proceedings of 1st International Tinplate Conference, International Tin Research Council, London, Oct. 1976. Full set of papers and discussions issued in book form which provide the best source of reference for can-making technology. Papers include G. F. Norman, "Welding of Tinplate Containers—An Alternative to Soldering," Paper 20, pp. 239–248; J. Siewert and M. Sodeik, "Seamless Food Cans Made of Tinplate," Paper 13, pp. 154–164; and W. Panknin, "Principles of Drawing and Wall Ironing for the Manufacture of Two-Piece Tinplate Cans," Paper 17, pp. 200–214.

Proceedings of 2nd International Tinplate Conference, International Tin Research Council, London, Oct. 1980. Papers include G. Schaerer, "Food and Beverage Can Manufacture," Paper 17, pp. 176–186.

Proceedings of 3rd International Tinplate Conference, International Tin Research Council, London, Oct. 1984. Papers include W. Panknin, "New Developments in Welding Can Bodies;" G. Schaerer, "Soudronic Welding Technigues—A Promoter for the Tinplate Container;" and R. Pearson, "Side Seam Protection of Welded Cans."

Developments in the Drawing of Metals, Conference organized by The Metals Society, London, May 1983. Section on deep drawing and stretch forming, pp. 76–125. Papers include P. D. C. Roges and G. Rothwell, "DWI Canmaking: The Effect of Tinplate and Aluminum Properties;" E. O. Kohn, "The Use of Spinning for Re-forming Ultra-thin Walled Tubular Containers;" and three papers on aspects of deep drawing.

A. M. Coles and C. J. Evans, *Tin and Its Uses,* International Tin Research Institute, No. 139, 1984, pp. 1–5.

F. L. Church, *Mod. Metals* **37,** 18 (July 1981).

N. T. Williams, D. E. Thomas, and K. Wood, *Metal Constr.* **9,** 157 (Apr. 1977); **9,** 202 (May 1977).

A. G. Maeder, *Aerosol Age* **18,** 12 (Nov. 1973). Still the best first introduction to the subject.

D. Campion, *Sheet Metal Ind.* **57,** 111 (Feb. 1980); **57,** 330 (Apr. 1980); **57,** 563 (June 1980); **57,** 830 (Sept. 1980).

S. Karpel, *Tin Int.* **54,** 208 (June 1981).

Metal Decorating and Coating, Annual Convention of National Metal Decorators Association (NMDA), October, annual. Reviewed in *Modern Metals.*

E. A. Gamble, *J. Oil Colour Chem.* **10,** 283 (1983).

E. O. Kohn
F. W. Jowitt
Metal Box p.l.c.

CANS, PLASTIC

Cans are defined here as open-mouthed cylindrical containers, usually made of aluminum (see Cans, aluminum) or tin-plated steel (see Cans, steel). This form can also be made of plastics, by injection molding (see Injection molding), blow molding (see Blow Molding) or forming from sheet (see Thermoforming). Such cans have been made for many years, and have taken small shares of some markets, but are not dominant in any. Plastics are much less rigid than metals, thicker walls are needed for equivalent performance, and container weight and cost can become excessive. For carbonated beverages (see Carbonated beverage packaging), cans are rigidified by internal pressure, but other properties such as burst strength, creep, and gas barrier become more critical.

Tensile strength is needed, especially in beverage cans, and can be achieved through selection of appropriate plastics plus orientation (stretching). Orientation in the circumferential (hoop) direction is hardest to do, and this explains why blow molding from a narrow parison may give better performance than thermoforming or direct injection to size. Heat resistance may be needed: either around 122°F (50°C) to resist extreme warehousing conditions; or 140°F (60°C) for 30 min (pasteurization); or 185–212°F (85–100°C) for a few seconds (hot filling); or 257°F (125°C) for 20–40 minutes (retort sterilization cycle). A can for processed foods must withstand the internal pressure generated when a closed can is heated. In a sterilizing retort, compensating overpressure may be needed outside the can to prevent failure. Not all retorts are capable of such overpressure (see Canning, food). Gas and/or moisture barrier and chemical resistance are needed for most packages. The can shape is advantageous for barrier, as it offers a low surface-to-content ratio, but it also introduces the possibility of failure from chemical attack in some places, particularly the stressed flange area and the bottom edge.

Product-design centers on these two features: the flange and the base. A precise flange is needed to guarantee a perfect seal, which is absolutely critical for sterilized cans, and certainly desirable for contained liquids, especially under pressure. If a plastic end is used, heat-sealing is possible, but the plastic end will be less rigid and may require a heavier flange. Common metal ends may be used, but flanges must be very flat. Also, no matter what end is used, stresses and later stress cracking in that region must be anticipated and avoided.

Vertical sidewalls are desired for easy transport in filling lines. Tapered walls do allow nesting, which has storage advantages, but for mass applications the filling speed is more important. Another design concern is the necked-in end now customary for beverage cans, which allows tighter six-packing and cheaper ends. This can be done with plastics, but mold design is more complicated to permit the undercut needed.

Potential applications. The beer and soft drink markets beckon like gold to the alchemists, but mass use is still well in the future. There are thermoformed 250-mL PET cans used in the UK (Plastona), but they are nonvertical (they nest), and they are not coated to enhance barrier. In the United States, Coca-Cola has been working with "Petainers," which are cans drawn in solid phase from molded PET cups developed in Europe by Metal Box and PLM A.B. These can be coated with PVDC (see Vinylidene chloride copolymers) for improved barrier, but such use would require special attention to recycling, both in-plant and at consumer level. In Italy, some beverage cans are made by thermoforming, and some by blowing bottles and cutting/flanging the tops. The latter should give the best properties for given weight because of more orientation.

For heat-processed foods, polypropylene is the preferred material because it has the highest heat resistance of the commodity thermoplastics. Polycarbonate (see Polycarbonate) and other engineering plastics have been suggested, but are much more costly. Both PP and PC are poor oxygen barriers, however, and would need a barrier layer for many foods. American Can's "Omni" is a coinjection-blown PP/EVOH/PP can, first used by Hormel for single-service heat-and-eat meals (see Multilayer plastic bottles) and many companies have produced and offered such cans thermoformed from multilayer sheet (see Coextrusion, flat; Coextrusion for semi-rigid packaging). Metal Box has also worked with extruded tubes with top and bottom flanged ends, and resin suppliers have made cans on

pilot lines, but there is still no large-scale commercial use. One of the problems is that other forms of plastic (eg, trays, bottles, pouches and bowl shapes) are competing for the same markets.

In motor oil, the long-neck HDPE bottle is taking over and the can shape is somewhat old-fashioned now. Wide-mouth blow-molded HDPE oil cans have been in the market for many years, but with very limited success. All-plastic paint cans appear from time to time, and the consumers are said to welcome an easy closure, but market penetration is still low (around 1% in 1984), due to intense competition in metal, plus considerable captive metal production. They are generally injection-molded, but they can also be blow-molded. For frozen orange juice, composite cans have most of the market (see Cans, composite) although Tropicana injection molds its own polystyrene cans, and some injection-molded HDPE cans are in the market as well. Injection-molded HDPE cans are used for cake frostings, and dry beverages are sometimes seen in plastic cans, notably the heavy injected PP cans used by Star in Italy for coffee.

BIBLIOGRAPHY

General Reference

"Serving up a Better Package for Foods," *Chemical Week*, 100–104 (Oct. 16, 1985).
Allan L. Griff
Edison Technical Services

CANS, STEEL

In 1795, France was not only involved in a revolution but also at war with several hostile European nations. The French people as well as the armed forces were suffering from hunger and dietary diseases. Consequently, a prize of 12,000 francs was offered by the government to any person that developed a new means for the successful preservation of foods. Napoleon awarded this prize to Nicholas Appert in 1809. Mr. Appert's discovery was particularly noteworthy because the true cause of food spoilage was not discovered until some 50 years later by Louis Pasteur. Appert had nevertheless recognized the need for utter cleanliness and sanitation in his operations. He also knew or learned the part that heat played in preserving the food, and finally, he understood the need for sealed containers to prevent the food from spoiling. The containers he used were wide-mouthed glass bottles that were carefully cooked in boiling water (1,2).

A year after the recognition of Appert's "canning" process, in 1810, Peter Durand, an Englishman, conceived and patented the idea of using "vessels of glass, pottery, tin (tinplate), or other metals as fit materials." Thus the forerunners of modern food packages were created. The original steel cans had a hole in the top end through which the food was packed. A disk was then soldered onto the top end. This disk had a small hole in it to act as a vent while the can was cooked. The vent hole was soldered and closed immediately after cooking (2).

Durand's tinplate containers were put together and sealed by soldering all the seams. The techniques were crude but nevertheless, with good workmanship, afforded a hermetic seal. A hermetically sealed container is defined as a container that is designed to be secure against the entry of microorganisms and to maintain the commercial sterility of its contents after processing. Commercial sterility is the inactivation of all pathogenic organisms and those spoilage organisms that grow in normal ambient distribution and home-storage temperatures. No technological advance has exerted greater influence on the food habits of the civilized world than the development of heat treatment and the use of hermetically sealed containers for the preservation of foods (1,3).

Foods canned commercially by modern methods retain nearly all of the nutrients characteristic of the original raw foods. Several investigations showed that good canning practices and proper storage and consumer preparation improve the retention of the nutritive value of canned foods (4–6).

The steel container provided a reliable lightweight package that could sustain the levels of abuse that were common in packing, distribution, and sale of products. It has also been necessary to improve and develop new concepts to keep up with the advances in packing procedures, materials handling, and economic pressures.

Evolution of the Can

A three-track evolutionary road developed. One uses solder to seal all the can seams. This evolved from the hole-in-cap container to one that holds evaporated or condensed milk. These "snap-end cans" retain the vent hole in the top that permits filling the can with a liquid. They are then sealed with a drop of solder or solder tipping. This style of milk container is still on market shelves. The second track combines the soldering of the side seam with the mechanical roll crimping of the ends onto the body. The attachment procedure, known as double seaming, was patented in 1904 by the Sanitary Can Company. This invention significantly improved opportunities to increase the speeds of can manufacture and packing operations. Today double seamers or closing machines seal cans in excess of 1500 cans per minute with filling equipment capable of matching this task. The third track utilizes a press operation that stamps out a cuplike structure with an integrated body and bottom, with the lid or top attached by either soldering or double-seaming. These began as shallow containers like sardine cans. These drawn or two-piece cans have evolved into taller cylindrical cans that are popular for many food products and carbonated beverages. The primitive shallow cans are fabricated on simple presses. The taller two-piece cans go through multiple press operations like the draw–redraw technique or through a drawn press-and-wall-iron operation (1).

The growth of the steel can caused the manufacturing function to change. At first, the packers manufactured their own cans. This was understandable, since the same craftsmen and equipment were necessary to seal the can as well as make them. When the double-seamed can, or "sanitary can," was accepted, the can manufacturing function coalesced into large manufacturing organizations. This came about because the double-seamed can lent itself to mechanized production whereas the all-soldered can remained a hand operation. Hence, it was economically attractive to invest large sums of money for improving the sanitary can. A few large packers could afford to be self-manufacturers, and they invested large sums to keep up with the developing technology. Recent economic conditions and technology with aluminum beverage cans have engendered the building of sophisticated can manufacturing plants by the major breweries. Similar considerations were pursued by the food industry through the forma-

tion of manufacturing cooperatives. In the nonfood industry, some companies, particularly paint manufacturers, made their own cans. Some continue to do so.

Shapes and Sizes

A wide variety of shapes and sizes have grown out of the tremendous usage of steel containers (see Tables 1 and 2).

Can Performance

Food cans are expected to have tightly drawn-in ends. A can with swelled ends or with metal feeling springy or loose is not merchantable. Consumers associate such appearances with spoiled contents. Gases produced by microbial action can indeed cause such appearances. It is also important to remove air from food cans to retard adverse internal chemical reactions. Therefore, end units must be properly engineered to cope with their many environments. They must act as diaphragms that expand during thermal processing and return to their tight drawn-in appearance when vacuum develops upon cooling.

Ends used for carbonated beverages and aerosol products need little or no vacuum accommodation but rather resistance to high internal pressure. An end unit will permanently distort or buckle when the internal pressure exceeds its capability. This also renders the can unmerchantable because the distortion can affect the double-seam integrity and permit leakage. The necessary strength is built into the end units by the use of adequate basis weight and temper of materials and design of end profile (see Fig. 1).

Cans with inadequate body strength will panel because of internal vacuum or collapse under axial load conditions. Paneled cans have a drawn-in body wall and therefore distortion of the cylinder wall. The condition may appear as a single segment, many flat segments, or panels that develop around the

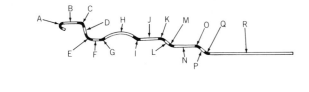

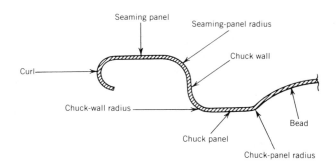

Figure 1. End profiles: A, curl; B, seaming panel; C, steaming panel radius; D, chuck wall; E, chuck-wall radius; F, chuck panel; G, chuck-panel radius; H, bead; I, bead-edge radius; J, first expansion panel; K, first expansion-panel radius; L, first expansion-panel step; M, first expansion-panel step radius; N, second expansion panel; O, second expansion-panel radius; P, second expansion-panel step; Q, second expansion-panel step radius; and R, center expansion panel.

circumference of the cylindrical can body. The axial load capacity of a can is greatest when the cylinder wall is in no way deformed. A casual dent or designed structure that breaks the integrity of the straight cylinder will, in most cases, reduce the axial load capability of the container. Panel resistance and axial load capability are direct functions of basis weight and temper of the steel. Panel resistance can also be enhanced by the fabrication of beads in the body wall. This in effect produces shorter can segments that are more resistant to paneling. However, such beads reduce axial load resistance by acting as failure rings. The deeper the bead is, the greater the paneling resistance but the greater the reduction in axial load capability. Many shallow beads can provide additional paneling resistance with less reduction in axial load capability, but labeling problems are often associated with cans having such bead configurations. Consequently, there are many bead designs and arrangements, all of which are attempts to meet certain performance criteria (7).

The steel container with proper material and structural specifications possesses good abuse resistance, within limits. Excess abuse causes obvious damage. Severely dented cans are unacceptable in the marketing of canned products. There are also insidious events that run parallel with excessive abuse. The double seams may flex momentarily, permitting an equally short-term interruption of the hermetic seal. This lapse can permit the entrance of microorganisms that cause spoilage. This leakage can also admit air that accelerates adverse chemical reactions within the container. Many damaged cans are in the food-distribution chain, and the safety of using such products is frequently questioned. To prohibit the sale of all containers that have insignificant amounts of damage would be a waste of large amounts of very acceptable food (8).

Can Corrosion

The steel container is not chemically inert and therefore can react with its environment and its contents. Steel's major ingredient, iron, is a chemically active metal that readily takes part in reactions involving water, oxygen, acids, and a host of other elements and compounds that can partake in oxidation–reduction reactions.

The application of tin to the surface of sheet steel significantly increases the corrosion resistance of steel. Nevertheless, the potential for corrosion persists. The mechanism of corrosion is an electrochemical reaction analogous to the action of a galvanic cell, where dissimilar metals in a solution of an electrolyte, when connected, set up an electrical current. Specialists in the electrochemistry of tinplate recognize several different types of internal can corrosion. Different electrochemical processes are responsible for these different types of corrosion, that is, the dissimilar metals' couple reacts to the

Table 1. Steel Can Styles and Sizes

Style	Dimensions^a, in. (cm)	Capacity	Some uses	Convenience features
aerosol cans (1)	202 × 214–300 × 709 (5.4 × 7.3–7.6 × 19.2)	3–24 oz (85–680 g)	foods, nonfoods	designed for fit of standard valve cap
beer-beverage cans (2)	209/211 × 413 (6.5/6.8 × 12.2) 209/211 × 604 (6.5/6.8 × 15.9) 207.5/209 × 504 (6.3/6.5 × 13.3)	12 or 16 fl oz (355 or 473 cm^3)	soft drinks, beer	easy-open tab top, unit of use capacities
crown cap, cone top can	200 × 214–309 × 605 (5 × 7.3–9 × 16)	4–32 oz (113–907 g)	chemical additives	tamperproof closure, easy pouring
easy-open oblong can	405 × 301 (11 × 7.8) #¼ oblong	4 oz (113 g)	sardines	full-paneled easy-open top
flat, hinged-lid tins (3)	112 × 104.5 × 004–212 × 205 × 003.75 (4.4 × 104.5 × 0.64–7 × 5.9 × 0.6)	12–30 tablets	aspirin	easy opening and reclosure
flat, round cans	213 × 013 (7.1 × 2.1)	1½ oz (43 g)	shoe polish	friction closure
flat-top cylinders (4)	401 × 509–610 × 908 (10.3 × 14.1–16.8 × 24)	1–5 qt (946–4730 cm^3)	oil, antifreeze	unit of use capacities; tamperproof since it cannot be reclosed
	211 × 306 (6.8 × 8.6)	8 fl oz (237 cm^3)	malt liquor	
	211 × 300 (6.8 × 7.6)	8 oz (227 g)	cat food	
	300 × 407 (7.6 × 11.3)	15 oz (425 g)	dog food	
hinged-lid, pocket-type can			tobacco, strip bandages	firm reclosure

Table 1. (*Continued*)

Style	Dimensions[a], in. (cm)	Capacity	Some uses	Convenience features
key-opening, nonreclosure can (5)			sardines, large hams, poultry, processed meats	contents can be removed without marring product
key-opening, reclosure cans (6)	307 × 302–502 × 608 (8.7 × 7.9–13 × 16.5)	$\frac{1}{2}$–2 lb (0.2–0.9 kg)	nuts, candy, coffee	lugged cover reclosure
	401 × 307.5–603 × 712.5 (10.3 × 8.8–15.7 × 19.8)	1–6 lb (0.45–2.7 kg)	shortening	lid is hinged
	211 × 301–603 × 812 (6.8 × 7.8–15.7 × 22.2)	$\frac{1}{4}$–5 lb (0.11–2.2 kg)	dried milk	good reclosure
oblong F-style cans (7)	214 × 107 × 406–610 × 402 × 907 (7.3 × 3.7 × 11.1–16.8 × 10.5 × 24)	$\frac{1}{16}$–1 gal (237–3785 cm³)	varnish, waxes, insecticides, antifreeze	pour spout and screw cap closure
oblong key-opening can (8)	314 × 202 × 201–610 × 402 × 2400 (9.8 × 5.4 × 5.2–16.8 × 10.5 × 61)	7 oz–23.5 lb (0.2–10.7 kg)	hams, luncheon meat	wide range of sizes, meat-release coating available
oval and oblong with long spout (9)	203 × 014 × 112–203 × 014 × 503 (5.6 × 2.2 × 4.4–5.6 × 2.2 × 13.2)	1–4 fl oz (30–118 cm³)	household oil, lighter fluid	small opening for easy flow control
pear-shaped key-opening can (10)	512 × 400 × 115–1011 × 709 × 604 (14.6 × 10.2 × 4.9–27.1 × 19.2 × 15.9)	1–13 lb (0.45–5.9 kg)	hams	easy access through key opening feature, meat-release coating available
round truncated			waxes	screw cap simple reclosure
round, multiple-friction cans (11)	208 × 203–610 × 711 (6.4 × 5.6–16.8 × 19.5)	$\frac{1}{32}$–1 gal (118–3785 cm³)	paint and related products	large opening, firm reclosure, ears and bails for easy carrying of large sizes
round, single-friction cans	213 × 300–702 × 814 (7.1 × 7.6–18.1 × 22.5)	to 10 lb (4.54 kg)	paste wax, powders, grease	good reclosure
sanitary or open-top can (three-piece) (12)	202 × 214–603 × 812 (5.4 × 7.3–15.7 × 22.2)	4 fl oz–1 gal (118–3785 cm³)	fruits, vegetables, meat products, coffee, shortening	tamperproof, ease of handling, large opening
slip-cover cans			lard, frozen fruit, eggs	simple reclosure
spice can, oblong (13)	wide range	1–16 oz (28.4–454 g)	seasonings	dredge top, various dispenser openings
square, oval, and round-breasted containers (14)			powders	perforations for dispensing, reclosure feature
two-piece drawn redrawn sanitary can	208 × 207/108 (6.4 × 6.2/3.8)	3–5$\frac{1}{2}$ oz (85–156 g)	food	improved can integrity, stackability
	307 × 111 (8.7 × 4.3)	6$\frac{3}{4}$ oz (191 g)		
	211 × 214 (6.8 × 7.3)	7$\frac{1}{2}$ oz (213 g)		
	404 × 307 (10.8 × 8.7)	1$\frac{1}{2}$ lb (680 g)		

[a] 202 × 214 means the can is 2$\frac{2}{16}$ in. (5.4 cm) dia and 2$\frac{14}{16}$ in. (7.3 cm) high. For rectangular cans, the first two sets of numbers are base dimensions, and the third set is height.

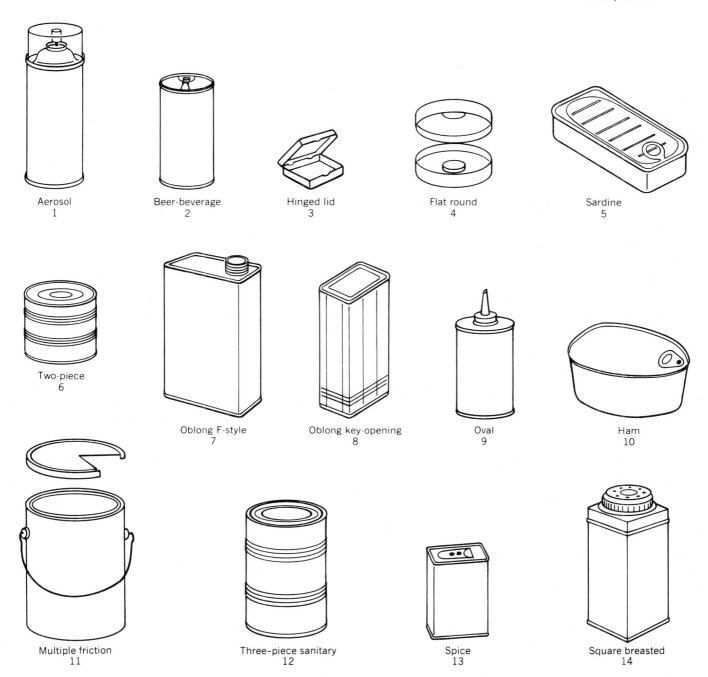

Aerosol
1

Beer-beverage
2

Hinged lid
3

Flat round
4

Sardine
5

Two-piece
6

Oblong F-style
7

Oblong key-opening
8

Oval
9

Ham
10

Multiple friction
11

Three-piece sanitary
12

Spice
13

Square breasted
14

Table 2. Popular Can Sizes

Dimensions	Popular name	Inches	(mm)	Capacity[a], (oz)	(cm³)
202 × 214		2.13 × 2.88	(54 × 73)	4.60	(137)
211 × 413	12 oz	2.69 × 4.81	(68 × 122)	12.85	(380)
300 × 407	#300	3.00 × 4.44	(76 × 113)	14.60	(432)
303 × 407	#307	3.19 × 4.38	(81 × 111)	16.20	(479)
307 × 409	#2	3.44 × 4.56	(87 × 116)	19.70	(583)
401 × 411	#2½	4.06 × 4.69	(103 × 119)	28.60	(846)
404 × 700	40 oz	4.25 × 7.00	(108 × 178)	49.55	(1465)
603 × 700	#10	6.19 × 7.00	(157 × 178)	105.10	(3108)

[a] Completely filled.

chemistry of the container contents. The wide variety of products packed in tinplate cans gives rise to a bewildering number of variables that influence corrosion, and many comprehensive studies have been published (9–11). Under normal conditions tin forms the anode of the couple, going into solution at an extremely slow rate and thus providing protection to the canned food for two or more years. Under some conditions, iron forms the anode with resultant failure due to perforations or the development of hydrogen and subsequent swelled cans (hydrogen springers). Under still other conditions, as when depolarizing or oxidizing agents (ie, nitrates) are present, the removal of tin will be greatly accelerated with a consequent significant reduction in shelf life. Hydrogen is formed by two distinct processes: (1) at exposed steel areas which are protected cathodically by the tin–steel couple current and (2) as the steel corrodes, either because it is not completely protected by the tin or after the tin has been consumed. When perforations occur, they are usually the result of the same process in which hydrogen is developed except the steel is consumed in a localized area (9–11).

Product Compatibility

Products can be loosely categorized in terms of their susceptibility to chemical reaction with the can. Oils and fatty products do not generally react with the metal surface of can interiors. However, when small quantities of moisture are present, either by design or accident, adverse reactions can develop. Highly alkaline products, usually nonfoods, will rapidly strip off tin or organic coatings but will not corrode the base steel. Acid products are corrosive, and highly colored acid foods and beverages are also susceptible to color reduction or bleaching due to the reaction of the product with tin. Foods with pH above 4.6 often have sulfur-bearing constituents (eg, protein) and react with both tin and iron. This has been recognized as a problem for as long as such products have been canned, and the cause has been sought by many investigators. These low acid foods form a dark staining of the tin surface and react with the iron to form a black product that adheres to both the can interior and the food product or can cause a general graying of the food product and liquor. This is often referred to as sulfide black. It can be very unsightly but harmless. Although sometimes exclusively a can headspace phenomenon, any interior container surface can be affected. In addition to container and product appearances, some food products and beverages are highly sensitive to off-flavors caused by exposure to can metals (12).

The sulfide black condition generally occurs during or immediately following heat processing, but it occasionally develops during storage. It is probably an interaction of the volatile constituents from the food product and/or oxidation–reduction agents in the food product with an oxidized form of iron from the tinplate. The staining of the tinplate is not part of this reaction (13,14).

Can Steel

Tinplate has been discussed as a substrate for steel containers with respect to its role as a structural material as well as its chemical activity. These performance characteristics influence the choice of gauge, tin coating weight, temper, and steel chemistry. They interplay in fulfilling the needed performance of a steel container. For structural integrity, basis weight or gauge is a prime consideration, but any increment of basis weight is an increment in container cost. Very often temper or design of can components can provide added strength without need to add basis weight. The chemistry and general metallurgy of steel plate has considerable influence on the performance of steel containers. Before the advent of continuously annealed and cold-rolled doubly reduced steel plate, which influence plate stiffness or temper, it was necessary to add ingredients to the steel to produce high tempers. Rephosphorized steel was necessary to fabricate beer can ends with sufficient buckling resistance. When the market opened for canned carbonated soft drinks, rephosphorized steel was not acceptable because of the corrosive nature of soft drinks and the very high susceptibility of this plate to corrosion. Test-

pack experience indicated that some formulations could perforate the end plate in four to six weeks. The availability of continuously annealed plate without the corrosion sensitivity properties of rephosphorized steel permitted the canning of soft drinks. It was still necessary to use expensive heavier basis weight material for the high carbonation drinks. The added stiffness afforded by cold-rolled doubly reduced plate permitted the canning of all soft drinks with carbonations ranging from 1.0 to 4.5 volumes in cans having the same basis weight ends. These same economies have been applied to other products where container fabrication could be adapted to the degree of stiffness characteristic of this plate (15).

Can Fabrication

The side seam of three-piece cans must be sealed without flaw to afford a nonleaking and, when necessary, a hermetic seal (see Cans, fabrication). Solders used in canmaking are generally composed of lead and tin. They are categorized as soft solders and have relatively low melting points, usually below 700°F (371°C). There are special needs that require the use of 100% tin, and other uses that require such additives as silver and antimony. Tinplate cans are easily soldered because the tin solder alloy readily fuses with the tin on the surface of the steel. In addition to providing a hermetic seal, solder also contributes to the mechanical strength of the seam by forming a metallurgical bond with the tinplate. Solder also has a certain amount of ductility and can be plastically deformed within stress limits that depend upon the ratio of lead and tin in the alloy. This characteristic permits the soldered laps of the body to be flanged and then incorporated into the double-seaming process. The ratio of tin and lead, the speed of the can manufacturing line, the temperature of the molten solder, and the length of cooling all play an important role in good soldering operations. Solders with tin contents that range between 2 and 30% are not compatible with high speed manufacturing operations because of their low melting points and long solidification time. These are often referred to as eutectic points and mush periods (see Figs. 2 and 3).

An important part of soldering is the preparation of the surface that will be bonded with the solder. Metals have a natural oxide surface and the oxide will not accept solder. Proper surface cleansing is accomplished with the use of a flux. A flux also lowers the surface tension and permits the solder to spread easily and adhere to the plate surface. These functions are accomplished without any corrosive action on either the container or the manufacturing equipment. Fluxes are selected for specific soldering operations and vary with the type of plate, the method of soldering (eg, molten pot and roll, jet soldering, wire solder), the seam or joint construction, and the method of flux application. Some examples of active ingredients are hydrogenated rosin, ethanolamine hydrochloride, rosin–oleic acid, and chlorinated paraffin.

The side-seam structure plays an important role in container performance. A soldered lap seam, similar to the structure that Durand used, has been demonstrated to have excellent pressure resistance and burst strength. However, this structure also presents difficulties in can manufacturing and container performance. It is difficult to clamp the two edges of the body when soldering the side seam. Also, the raw edges that are exposed in this type of side-seam construction are subject to corrosion on both the exterior and interior of the can. A lock-and-lap seam is one in which the edges of the body

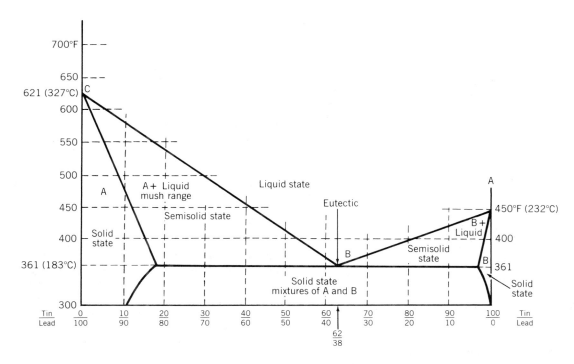

Figure 2. Tin–lead phase diagram.

blank are folded back so the two edges, when engaged, act like a clamp and hold the two edges together during soldering. The hooks at both ends of the seam are cut away to reduce the number of folds of metal that would be included in the double seam. These terminal areas are called laps. The raw edges of the side seam are enveloped within the interior of the seam except in the lap area. For special uses like beer and beverage cans, a full solder fillet is required. This is accomplished by the use of vent slits in the side-seam construction (see Fig. 4). A full solder fillet cannot be attained without venting the expanded vapors and gases that develop pressure within the seam cavity. By venting, the molten solder can be infused throughout the seam cavity. Such a structure completely embeds the raw edges in the solder and also provides a bridge of solder upon which organic protective coatings can be applied.

Still another side-seam configuration can be fabricated to give additional strength to cans that must hold high pressures (eg, aerosols). This variant is known as an interrupted lock-and-lap side seam. This design interrupts the locking hooks of both body edges to provide tabs that perform like the lapping surfaces of the lap seam. Since the advent of welded seams, this seam construction is in limited use.

Welding has been considered for canmaking for many

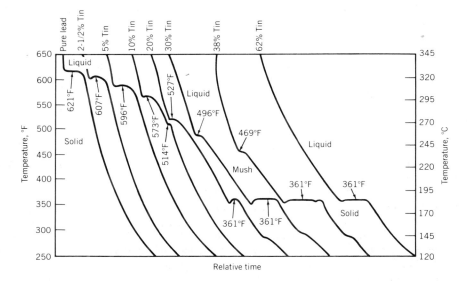

Figure 3. Typical cooling curves for tin–lead alloys.

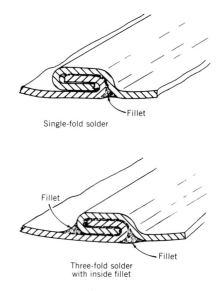

Single-fold solder

Three-fold solder
with inside fillet

Figure 4. Filleted and nonfilleted side seams.

years. It gained reasonably good acceptance in the manufacture of steel pails. Because of inconsistencies in weld integrity, however, the process was considered unacceptable for the manufacture of high quality leak-free cans. This was particularly true of food cans and attendant concerns of spoilage and public health. The available equipment also did not lend itself to high speed production. Developments in electrical energy sources and high speed welding equipment gave rise to at least two reliable systems: the Conoweld roll-welding system developed by Continental Can Company and the wire weld system developed by Soudronic A.G. Both are resistance-welding systems. The materials to be welded together are heated by an electric current passing through the materials. The resistance of the overlapped side seam produces the required welding heat. The side-seam bond is a solid-state weld, which means that no melting of metal occurs. To obtain a solid-state weld, the heated material must be pressed together or forged. This forging capability is provided by the wheel in the Conoweld process. Because welded seams are basically lap seams, the raw-edge problem exists. Therefore, the welding process must be designed to correct this problem (15,16).

Organic sealing materials can be used as side-seam sealants. They are used in situations where solders are not compatible with the product, when no high temperature thermal process is needed, when special wraparound lithography is desired, and when tin-free steel is used. For the most part, polyamides and organisols are used as the basic resin of organic sealing materials. They perform satisfactorily for many dry and nonfood products in a wide variety of can sizes, shapes, and styles. These sealing materials are thermoplastic to permit application and they are by nature brittle. Hence, they have poor fabricating qualities and are subject to cracking on shock or impact with resulting bond failure. Economic pressures to use tin-free steel gave rise to other organic sealing systems. American Can Company developed the Mira Seam, a specially formulated material that provides the necessary sealing capabilities as well as bonding properties to the chemically treated steel. The Japanese developed a system that is also applicable to tin-free steel. (Continental Can Company

developed the Conoweld welding process to accomplish the same thing.)

The popularity of steel cans has been due in large measure to the double seam's ease of fabrication and robust resistance to physical abuse. This seam structure is fabricated by two rolling operations that engage parts of both the can and the end to form an integrated seam with the qualities to effect a hermetic seal. Since metal-to-metal contact does not produce good sealing capability, a rubberlike material, known as sealing compound, is applied to the end unit in an area that is incorporated in the double seam. This material acts like a sealing gasket. Compound placement and compatibility with the product are important in producing a reliable seal.

Coatings

The shelf life and general quality-keeping attributes of steel cans have been dramatically improved through the use of protective coatings. For the most part these are organic materials, and they are applied to the flat sheet by roller coaters. Inorganic protective coatings that often contain phosphates, chromates, or both are usually applied to tin-free steel at the tin mill to prevent rust and provide a good surface for organic materials. If tin or iron is detrimental to a product, a suitable protective coating can be applied. These coatings can prevent bleaching discoloration by tin to dark pigments of foods such as strawberries, blueberries, blackberries, and cherries. Also, protective coatings can reduce the metal exposure for metal-sensitive products such as beer and certain soft drinks. Other products would react with the unprotected can interior to cause corrosion or cause discoloration of both container and product. Finally, a protective coating can mask unsightly discoloration of the can interior.

Organic protective coatings were first applied to can interiors to protect the red fruit colors. Heavy tin applications onto steel plate compensated for the corrosiveness of the product. However, staining of the tinplate and sulfide black discoloration, though known since the start of the canning industry, was not accommodated until 1922 when G. S. Bohart of the National Canners Association incorporated a small amount of zinc oxide into the can-coating formula. This type of can lining is known as a "C" enamel. There was sufficient zinc added to react with the sulfur-bearing components to preempt the same reaction from occurring with the tin or iron of the can. Zinc produces a white product with no deterioration of product appearance. Severe unexplained sulfide black discoloration, notwithstanding the use of "C" enamels, has occasionally caused substantial economic loss either through widespread occurrence throughout a pack or the involvement of a very expensive raw product such as shrimp. As cited earlier, several studies have been conducted to determine and explain the cause of these severe incidents. Cured meat products have also given rise to severe problems. This became more acute with the advent of phosphate curing methods. Ham and other cured canned products become more corrosive and liberate more tin and iron into the canned medium. As a result, severe sulfide black discoloration develops. The use of aluminum sacrificial anodes controls this problem. An aluminum disk or wafer is clinched onto the bottom or top of the can, and the attachment must be such that there is an electric coupling between the metal of the can and the aluminum anode. The aluminum reacts preferentially with the tin and/or iron, and the product is innocuous. Zinc oxide enamels are not able to control this problem. Interestingly, powdered metallic aluminum or its compounds do not perform when incorporated in an organic coating because no electric coupling is established (17–19).

Very complex organic protective-coating systems are required for the successful canning of beer and carbonated soft drinks. Although beer is not corrosive, it is highly sensitive to iron. Off-flavors can be detected with iron contamination levels at 0.5 ppm or less. Carbonated soft drinks comprise a wide range of products that vary markedly in their degree of corrosiveness. Low air content in the canned product is essential, and control over depolarizers, such as azo-type food dyes, limits corrosiveness. However, low metal exposure, particularly in the side-seam and double-seam areas are essential to reduce corrosion currents. Hence, perforation and shelf-life problems are eliminated or controlled. The steel beverage can is fabricated with roller-applied base coats, usually an epoxy or polybutadiene material. During can manufacture, the side-seam stripe is spray-applied and cured by the heat of the soldering operation. Finally, a spray top coat is applied and subsequently baked. Vinyl materials are the usual resins used in spray-applied lacquers, a term usually used when referring to spray-applied materials. Recent development work in protective coatings has been directed to means that will significantly reduce the load of solvents that are vented into the atmosphere during the can manufacturing process. This work has led to the electrodeposition of powder side-seam stripes and the use of water based and high solids coatings that greatly reduce the solvent levels spewed into the environment. Ends used for these very metal-sensitive products usually have two roller-applied coatings. To be sure that metal exposure of cans are kept within specified limits, cans are tested with the use of Enamel Raters that quantify the level of metal exposure. These devices are used by quality-assurance groups of both manufacturers and users of these cans (7,20,21).

It has been apparent since the early days of canning green vegetables that the bright green darkens as a result of the required thermal process, that is, the bright green chlorophyll undergoes chemical change to an olive green pheophytin. There have been several studies conducted to inhibit this chemical change. They all demonstrate an inhibition capability by using additives containing magnesium ions and careful control of pH by the addition of hydroxyl ions. All of these procedures cause some toughening of the product which probably discourage any commercial pursuits. In 1984, Continental Can patented a process, "Veri-green", which incorporates additives to the product as well as the protective coating. The materials added are specific for particular products, and pretesting is necessary (22–26).

Decoration

Decorative lithographic designs afford an external protective coating. The process of application is known as offset lithography. Lithography is a printing process based on the fact that oil and water do not mix. The decorative design to be printed on the tinplate is etched onto a plate, known as a master plate, in such a manner that the image area to be printed is ink-receptive whereas the portion to be blank is water receptive. The master plate is attached to a large drum and as it revolves, it picks up the ink and acidified water. Since the ink and water are immiscible, only the image area accepts the ink which is then deposited on the container metal plate and a decorative design is achieved. A lithographic design is a system or a series of coatings and inks printed on plate in a particular sequence. The order of laydown is determined by the purpose of the coat, the kind of coating or ink, and the baking schedule for each coat or print. Any and all of these coatings and inks used in metal decorating must cure and dry independently of the surface to which they are applied. For example, paper and cloth are fibrous and absorbent so that any applied ink or coating would be absorbed to some degree into the material. Since metal is nonabsorbent, the coatings and prints cure by internal chemical reactions involving oxidation, polymerization, or both. The usual sequence of laydown is as follows: (a) size coat (roller coat) if necessary, (b) base coat (roller-coated or printed), (c) ink print or prints, and (d) varnish coat. The print process is usually carried out with sheet stock prior to slitting into can body blanks or scroll shearing into end stock. The equipment used to apply the lithography includes coaters, single-color presses, tandem single-color presses, and two-color presses. The sequence and type of equipment used depends upon the design and the ultimate use of the can. With the advent of two-piece cans, less elaborate designs are used on beer and soft-drink cans because it was necessary to use presses that print completely fabricated cans. The finishing varnishes are usually the external protective coatings. They not only protect against corrosion, but must also be rugged enough to resist scuffing and abrasion.

Technological Developments

The steel container has grown despite competitive pressures. During war years, tin and iron become controlled materials and other more available packaging materials usually make inroads into the steel-can market. These same constricting conditions often give rise to the development of methods that reduce the consumption of the controlled materials. During World War II, electrolytic deposition of tinplate was commercialized, which resulted in substantial reductions in tin consumption. Competition from aluminum in the beverage-can business gave impetus to pursue the use of tin-free steel to afford more favorable economics. Successful cost reduction with no reduction in performance, however, has been the cause for the continued growth of the steel can. The advent of cold-rolled doubly reduced plate permitted significant reductions in basis weight. To accommodate the fabricating characteristics of this material, new approaches in end manufacture and double seaming had to be developed. Another direction to cost reduction was to reduce materials usage. This can be accomplished by using smaller diameter ends and necking-in the can bodies to accommodate the reduced-diameter ends. This saving can be directed to one or both ends. Many soft-drink, beer, and aerosol cans exhibit these structural innovations. Now, flexible packaging materials are becoming very competitive with steel containers. New composite structures permit the packaging of many fruit drinks by aseptic methods (see Aseptic packaging). The retortable pouch has also become an important package as the standard for individual military rations in lieu of the steel can (see Retortable packages).

The use of convenience features has been an important stimulus to the growth of steel cans. Key opening lids and rip strips have been part of fish and meat cans for generations. This style was also standard for coffee and shortening cans, but economic pressures caused a change to less expensive open-top food cans. The more recent development of the integral rivet and scored end solved many can opening problems, particularly in the beverage and snack-food businesses. The rivet is formed from a bubblelike structure that is fabricated in the end unit by press operations. Other approaches to tab attachment were never successful in meeting severe carbonation and product-loss standards. The huge success of the integral rivet concept in the beverage industries throughout the world led to very objectionable litter problems. Many laws and ordinances have been legislated prohibiting the use of these easy-opening cans unless the tabs are nondetachable. Newly designed units meet these requirements. Integral rivet ends have usually been aluminum because of obviously greater ease in fabrication. The use of steel cans with aluminum ends, however, presents some serious performance problems. Food products that contain salt and need thermal processing will cause perforations through the aluminum during high temperature exposure. Corrosion of carbonated soft drinks in such cans is accelerated because of the bimetallic container. This same container has a positive influence on the shelf life of beer because of the sacrificial nature of aluminum and the more tolerable aluminum–beer reaction products than the iron counterparts. The severe constrictions in the use of these end units for food products have led to the production of steel easy-opening units. These are much more difficult to fabricate than aluminum and have not gained the expected interest (27).

Early in the 1970s health authorities and the FDA became concerned about the increment, if any, of lead and other heavy metals that are picked up by foods packed in soldered cans. The manufacturers of baby foods and baby-formula foods or ingredients, such as evaporated milk, were the first to be asked to reduce the level of lead in their canned products. Better care during soldering operations and ventilation resulted in major reductions but did not eliminate all lead in the respective foods. All food canners now seek cans with reduced lead-contamination potential. Many have sought lead-free cans. These pressures caused the adoption of welded and two-piece cans for food.

No containers can yet outperform the steel can on its own terms despite its limitations and the incursions of competitive packaging materials (28).

BIBLIOGRAPHY

1. *Canned Foods, Principles of Thermal Process Control, Acidification and Container Closure Evaluation,* 3rd ed., Food Processors Institute, Washington, D.C., 1980, pp. 7, 141.
2. *Canned Food Reference Manual,* 3rd ed., American Can Company, McGraw-Hill, New York, 1947, pp. 25–29.
3. S. C. Prescott and B. E. Proctor, *Food Technol.* 4, 387 (1937).
4. E. J. Cameron and J. R. Esty, *Canned Foods in Human Nutrition,* National Canners Association (NFPA), Washington, D.C., 1950.
5. G. A. Hadaby, R. W. Lewis, and C. R. Ray. *J. Food Sci.* 47, 263–266 (1982).
6. B. K. Watt and A. L. Merrill, "Composition of Foods—Raw, Processed, Prepared" in *U.S. Department of Agriculture Handbook,* 8th ed., Washington, D.C., 1963.
7. Continental Can Company, New York, Unpublished information.
8. *Safety of Damaged Canned Food Containers,* Bulletin 38-L, National Canners Association (NFPA), Washington, D.C., 1975.
9. R. R. Hartwell, *Adv. Food Res.* 3, 328 (1951).
10. R. P. Farrow, J. E. Charboneau, and N. T. Loe, *Research Program on Internal Can Corrosion,* National Canners Association (NFPA), Washington, D.C., 1969.
11. N. H. Strodtz and R. E. Henry, *Food Technol.* 8, 93 (1954).
12. J. S. Blair and W. N. Jensen, *Mechanism of the Formation of "Sulfide Black" in Non Acid Canned Products,* presented on June 12, 1962 at the 22nd Annual Meeting of Food Technologists, Miami Beach, Fla.
13. J. E. Chabonneau, *The Cause and Prevention of "Sulfide Black" in Canned Foods,* National Food Processors Research Foundation, Washington, D.C., 1978.
14. "Tin Mill Products" in *Steel Products Manual,* American Iron and Steel Institute, New York, 1968.
15. U.S. Pat. 3,834,010 (Sept. 10, 1974), R. W. Wolfe and R. E. Carlson (to Continental Can Co.).
16. Soudronic AG, Ch-8962, Bergdietikon, Switzerland.
17. R. G. Landgraf, *Food Technol.* 10, 607 (1956).
18. O. C. Johnson and L. G. Frost, *Natl. Provis.* 14 (1951).
19. O. C. Johnson, *Some Practical Hints on Canning Problems,* Continental Can Company, New York, 1964.
20. E. L. Koehler, J. J. Daly, H. T. Francis, and H. T. Johnson, *Corrosion* 15(9), 477t (1959).
21. Enamel Rater, Wilkins Anderson, Milwaukee, Wisc.
22. F. A. Lee, *Basic Food Chemistry,* The Avi Publishing Co., Westport, Conn. 1975, pp. 163–165.
23. U.S. Pat. 2,189,774 (Feb. 13, 1940), J. S. Blair (to American Can Co.).
24. U.S. Pat. 2,305,643 (Jan. 5, 1942), A. E. Stevenson and K. Y. Swartz (to Continental Can Co.).
25. U.S. Pat. 2,875,071 (May 18, 1955), Malecki and co-workers (to Patent Protection Corp.).
26. U.S. Pat. 4,473,591 (Sept. 25, 1984), W. P. Segner and co-workers (to Continental Can Co.).
27. R. L. Hart and G. C. English, *Mater. Perform.* 16, 23 (1977).
28. M. Bakker, "The Competitive Position of the Steel Can" in *Technology Forecast,* Westport, Conn., 1984.

George Reingold
Continental Can Company

CANS, TIN.   See Cans, steel.

CAPPING MACHINERY

In categorizing capping machinery that applies closures to bottles and jars, the best place to start is with the closure itself (see Closures). The different types of machinery for applying these closures have features in common (eg, straight-line vs rotary). This article provides a basic description of machinery used for continuous-thread (CT) closures, vacuum closures, roll-on closures, and press-on closures.

Cappers for CT Closures

There are four basic types of cappers for CT closures: hand cappers and cap tighteners; single-spindle (single-head) intermittent-motion cappers; straight-line continuous-motion cappers; and rotary continuous-motion cappers.

Torque control.   All of the automatic cappers apply the closures mechanically, but they differ in their approaches to torque control. In general, torque control is achieved with chucks or spinning wheels (rollers). Straight-line continuous-motion cappers generally control torque mechanically, but all of the other types use either pneumatic or mechanical means, or combinations thereof.

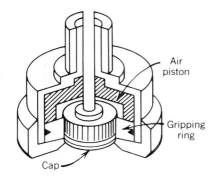

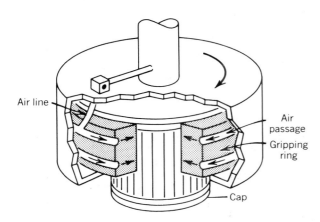

Figure 1. Pneumatic chuck (1).

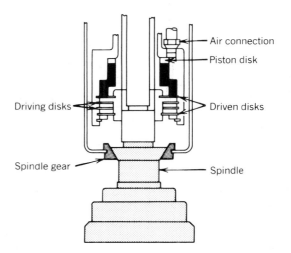

Figure 2. Pneumatic clutch (1).

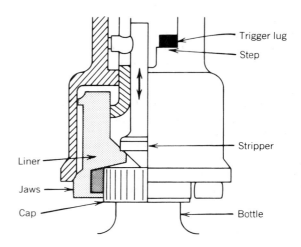

Figure 3. Mechanical chuck with jaws (1).

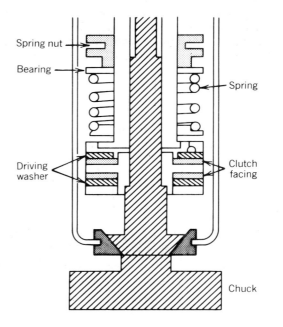

Figure 4. Spring-loaded clutch (1).

A pneumatic chuck contains a round flexible ring with a hole in the center (1). The cap enters the hole when the ring is in its relaxed state. When air pressure is applied, the ring compresses and grips and holds the cap while it is moved to the bottles and screwed on. The air pressure may be applied by the downward movement of an air piston, or it may be directed into the space between the ring and the wall of the chuck (see Fig. 1).

The amount of torque is controlled by a pneumatic clutch operated by pressure from low level pneumatic lines (see Fig. 2). It contains two or more sets of disks that are pressed against each other when air is applied, connecting the chuck and the drive shaft. When the cap is screwed on to the point that the torque being applied equals the force being applied to the disks by air pressure, the disks start to slip and the drive shaft is disconnected.

A typical mechanical chuck has jaws that close around the skirt of the cap to maintain a grip until the closure application is complete (see Fig. 3). A mechanical chuck can be controlled by a pneumatic clutch, but it can also be controlled by a spring-loaded clutch, or a barrel cam. The pneumatic clutch is similar to the clutch used for a pneumatic chuck. A spring-loaded clutch (see Fig. 4) uses a spring to disconnect the chuck from the power source when the pre-set amount of torsion has been applied to the cap (1). Torque can be increased by compressing the spring by screwing down the collar on top of the spring; it can be decreased by moving the collar to loosen the spring. The chuck opens to release the bottle when the torsion on the cap matches the torsion on the spring.

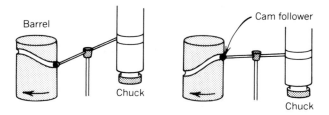

Figure 5. Barrel cam arrangement (1).

On some capping machines, the heads are raised and lowered by a follower riding on a barrel cam (see Fig. 5). Manufacturers of capping machinery have different approaches to torque control based on the principles described above, or combinations thereof.

Hand cappers/cap tighteners. With operating speeds of up to 20 caps/min, these are used for very low production volumes or unusual caps. Hand cappers work with some kind of hand-held chuck. Cap tighteners, a step up in automation, are useful for pump and trigger-cap applications, retorqueing after induction sealing, and short production runs in general.

Single-spindle intermittent-motion cappers. Using either pneumatic or mechanical torque control, these cappers can theoretically apply up to 60 caps/min, and they are useful for relatively short production runs. Their versatility is limited by the intermittent motion that would tend to cause spillage from widemouth containers.

Straight-line continuous-motion cappers. On these machines, torque-control clutches are built into multiple spindles that turn the rollers (disks) that apply the caps (see Fig. 6). In contrast to the single-spindle cappers, the bottles never stop and the speed of capping is limited only by the speed of the conveyor. The capability of the machines has traditionally been quoted as 60–300 caps/min (in contrast to rotary machines which can be more than twice as fast), but by using multiple spindles speeds can be increased greatly; for example, with eight spindles at four stations, production can be as high as 600/min or more, depending on the size of the cap and the container. The major advantage of straight-line (vs rotary) cappers is that they generally do not require change parts, and cap and container sizes can be changed with very little downtime.

Rotary continuous-motion cappers. Rotary cappers use many heads in combination in a rotary arrangement that permits very high speeds: 40–700/min (see Fig. 7). The speed of application of one cap by one head is limited, but rotary cappers can achieve 700/min through the use of 24 heads. (Production speeds are always related to the size of container and cap. Rotary machines are available with more than 24 heads, but 700/min is a rough upper limit).

Extra features: Most of the automatic machines can be supplied with optional equipment for flushing with inert gas. There was a time when application torque could be measured only by testing removal torque, but today's capping machinery is equipped with sensors that provide continuous measurement and constant readouts.

Cappers for Vacuum Closures

There are three basic types of vacuum closures: pry-off side-seal; lug; and press-on twist-off. The pry-off side-seal closure, the earliest type of vacuum closure, has a rubber gasket to maintain the vacuum in the container. This type of closure has been displaced in the United States by the other two types. The vacuum is generally achieved by a steam flush (2), which also softens the plastisol to facilitate sealing.

Lug closures. Lug closures have 4–6 lugs that grip onto special threads in the bottle finish as well as a flowed-in plastisol liner (gasket). The lugs mate with the threads with a half turn. This is almost always done by straight-line machinery that incorporates two belts moving at different speeds. Maximum production speed with this type of closure and machinery is about 700/min.

Press-on twist-off closures. These closures are used primarily on baby food jars, but they are being used now on other products as well. One of the reasons for their increasing popularity is the very high production speeds attainable. Like lug closures, they have a plastisol liner, but the liner is molded for high precision. Unlike lug closures, the skirt of the closure is straight. The seal is achieved when the plastisol softens and conforms to the bottle threads. Unlike threaded closures, which require multiple turns, and lug closures, which require a half turn, press-on twist-off closure require no turn at all. They are applied by a single belt that presses the closure onto the bottle at very high speed (eg, over 1000/min on baby food jars). These closures are held on by vacuum, which is achieved by sweeping the headspace with steam. The use of these clo-

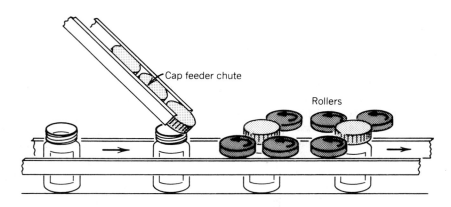

Figure 6. Roller screw capping (1).

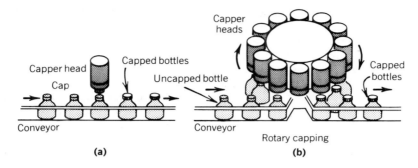

Figure 7. (**a**) Straight-line and (**b**) rotary capping (1).

sures was limited at one time by a requirement of reduced pressure of at least 22 in. of mercury (74.5 kPa) to hold them on. Not all products benefit from that degree of vacuum. A recent development is the ability to use these closures with a much lesser degree of vacuum.

Cappers for Roll-On Closures

A roll-on closure has no threads before it is applied. An unthreaded (smooth) cap shell (sleeve) is placed over the top of

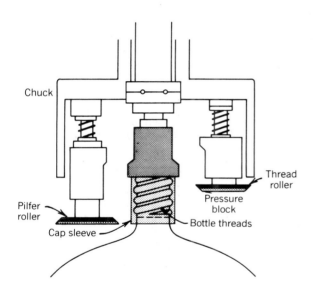

Figure 8. Applications of roll-on closures (1).

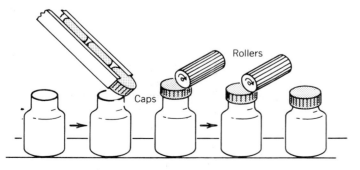

Figure 9. Roller application of press-on closures (1).

the bottle, and rollers in a chuck (see Fig. 8) form the threads to conform with the threads on the bottle finish. The chucks are raised and lowered by capper heads. These closures are available with and without a pilfer-proof band that is a perforated extension of the skirt. If there is such a band, it is rolled on by a special roller in the chuck. Operating speeds of the machines are in roughly the same range as rotary cappers for CT closures: up to 700/min.

Cappers for Press-On Closures

The familiar crown closure for carbonated beverage bottles is one type of press-on closure. They are applied by rotary machines that can operate with production rates of over 1000/min (see Fig. 9). Other types of press-on closures are snap-fit caps, with or without a tamper-evident band; dispensing caps, which are often secondary closures; and overcaps.

Like screw caps, press-on caps can be applied by chucks or rollers; but the chucks and rollers do not need twisting action, nor clutches for torque control. There are several approaches to controlling the amount of pressure applied by chuck-type cappers. The capper head can be spring operated, with tension controlled by a collar adjustment; or a pneumatic clutch can control it with air pressure; or a barrel cam can be used. On a roller-type press-on capper, the cap passes under one (shallow cap) or more (deeper cap) rollers that press the cap on tight.

BIBLIOGRAPHY

1. C. Glenn Davis, "Bottle Closing" (Packaging Machinery Operations course book) Packaging Machinery Manufacturers Institute, Washington, D.C., 1981.

2. *Canned Foods: Principles of Thermal Process Control, Acidification, and Container Closure Evaluation,* The Food Processors Institute, Washington, D.C.

CAPS. See Closures.

CARBONATED BEVERAGE PACKAGING

Carbonated beverages comprise the biggest single packaging market, with over 100 billion (10^9) beer and soft-drink containers filled each year in the United States. The packages cost substantially more than the contents, and are actually the beverage producer's largest expense item. Performance requirements are quite severe, as the contents are under pressure, they must hold their carbonation, and must withstand summer storage and (for beer) pasteurization temperatures that may reach 55–60°C. Furthermore, the highly competitive nature of this business has brought container design and manufacturing to a highly refined level in order to keep cost to a minimum. Even $0.001 per container becomes big business when multiplied by the billions (10^9) of containers made and used.

Package Types

There are four major categories of carbonated beverage containers: plastic bottles, nonreturnable glass, refillable glass, and metal cans.

Plastic bottles. These are mainly 2-L soft-drink bottles, which first appeared in the mid-1970s and in just a few years took over around one-third of the soft-drink sales volume. The bottles are stretch-blowmolded (see Blow molding) from PET polyester (see Polyesters, thermoplastic). The stretching (orientation) is needed to get maximum tensile strength and gas barrier, which in turn enables bottle weight to be low enough to be economical. Because material cost is about two-thirds of total manufacturing cost, such weight savings are essential. Typical 2-L bottle weights are around 60 g PET, with some as low as 50 g, plus a high density polyethylene base cup (18–24 g), a label (3 g), and a closure. One design uses no base cup; instead, the bottle itself forms five petal-like feet at the base. It uses slightly more PET (65–70 g) but total weight is less because of the absence of the base cup.

Recently, a 3-L PET bottle has been introduced, with weights around 80–85 g PET plus 22–28 g for the base cup, or around 90 g without the base cup. It is too early to predict whether the 3-L size will be as successful as the 2-L, but there is much optimism, despite fears that its inability to fit in refrigerator doors or on most shelves will discourage its purchase.

Half-liter PET has been offered since 1979 but has never caught on because it does not provide enough economic incentive to change, and in small sizes, the advantages of safety and light weight are not as significant. Instead of 0.5-L, 16-oz (473-mL) PET is now offered, to compete more directly with 16-oz (473-mL) glass, and to save a gram or two of bottle weight because of the smaller volume (6% less than the half-liter). Typical bottle weights for these small PET containers are 28–30 g plus a 5-g base cup. Some cupless designs have been developed but have only minor commercial use thus far.

In the deposit-law states, small plastic bottles are more successful because some consumers do not like the handling and return of glass and are unwilling to pay more for metal cans. In these nine states, however, which comprise over 20% of the population, the 2-L PET bottle has been the real winner, as many people prefer to pay one deposit instead of six (see Recycling).

It is believed that an improved plastic barrier to CO_2 loss (see Barrier polymers) would allow lighter (hence cheaper) bottles and thus spark the large-scale use of single-service sizes for soft drinks, and also open up the beer market to plastic. Such containers are the object of much development, both in can and bottle forms, but there has been no commercial activity beyond a few small test markets (see Cans, plastic).

Plastic bottles of all sizes usually carry full wraparound labels, either a shrunk polyolefin sleeve or a glued polypropylene/paper laminate (see Labels and labeling). Some plastic bottles have foamed plastic labels, which have a desirable non-slip feel and offer a little thermal insulation. These used to be very large, extending down over the base cup for maximum advertising effect (Owens-Illinois Label-Lite) but lately have been in conventional wraparound form and size (Owens-Illinois Plasti-Grip).

Nonreturnable glass. These are mainly 10- or 16-oz (296- or 473-mL) size, with a Plastishield (Owens-Illinois and licensees) foam-plastic protective label, or a paper/polyolefin roll-fed or an all-plastic shrink sleeve (see Bands, shrink). Typical bottle weights are around 130–140 g and 180–185 g, plus 3–4 g for the label. The glass industry has strongly promoted these containers in competition with metal cans, both in retail markets and in vending machines. Glass has an apparent cost advantage which carries through to retail level and has managed to hold on to its market share in this size. However, the advantage in container cost is eroded by labels, closures, secondary packaging, filling speeds, and occupied space, plus the possibility of breakage and costs of clean-up. In all of these aspects, cans have an edge.

Prelabeling has become standard. The labels are preprinted and come to the bottler with no need for a further printing operation. Patch labels are cheaper and better for short runs, but are disliked by retailers as the bottles can be turned to obscure the labels.

The old crown closure that had to be pried off has all but disappeared for nonreturnables and is seen only on refillable glass to allow use of bottles with corresponding threadless necks (see Closures). The nonreturnables and even some refillables use roll-on aluminum screw caps on threaded necks, with a tamper-evident ring (see Tamper-evident packaging). The threaded screw-off crowns that are crimped over a fine-threaded neck remain, but in diminishing numbers despite low cost. They are harder to open and are thus used mainly for beer (stronger hands), and their lack of tamper-protection has discouraged their use.

The use of plastic closures for both glass and plastic bottles increased in 1984 after a decade of trial use and design refinement. The two most popular designs use a separate liner, but there are linerless designs as well. These closures weigh around 3 g in standard 28-mm size and are slightly more expensive than competitive aluminum roll-ons. All unscrew and fit over the same threads as the roll-ons and have some visible indication of tampering. Initial successes have been with soft drinks; a few have been tried for beer, but they have not been accepted due to their inability to withstand pasteurization.

Refillable glass. The old standby, the refillable glass bottle, still accounts for about a third of the soft-drink business, and about a fifth of the beer market. The main sizes are 16-oz (473-mL) for soft drinks, and 12-oz (355-mL) for beer, with some soft drinks also in sizes around a quart. It is usually the cheapest way to buy soft drinks on a unit price basis (except where 2-L bottles are being discounted), but most consumers prefer to pay a little more for convenience. In deposit states, refillables do better but remain a minor factor, after 2-L PET and metal cans. Typical bottle weight is around 300 g for the 16-oz (473-mL) soft-drink bottle, and 250 g for a 12-oz (355-mL) beer bottle that is used in bars and hotels where economy is foremost and delivery/return is not a problem.

Labels on major-brand soft drink refillables may be permanently silk-screened. Paper/plastic patch labels are used on most refillable beer bottles. Full wraparound labels are seldom seen because the containers are usually sold in six-packs or cases which cover most of the label area.

The economics of refillables are very different from other containers, as the bottler must support a "float" of containers and cases (which may cost as much as the containers themselves), as well as a larger fleet of delivery trucks (less compact, must stop to pick up as well as deliver), and the appropriate cleaning and washing machinery. For this reason, soft-drink franchises have been a classic home for local investor/entrepreneurs and this spirit persists in the soft-drink industry today despite the predominance of nonreturnable packaging.

Refillables are often promoted for their environmental benefits. The concept of refill/reuse may be laudable as it supports a resource conservation ethic, but the actual use of refillables brings its own environmental problems: water pollution from washing, air pollution from less efficient truck usage, and sanitation problems in both shops and homes.

Proponents of nonreturnables also point to the success of aluminum recycling. Opinions and emotions are very strong in these areas, but an impartial examination of the facts leads to no firm conclusion. In fact, much depends on locality; that is, the nature of the water supply, the extent of recycling possible, the degree of urbanization, and the number of trips made by each refillable. There is no easy answer to this question and neither refillables nor nonreturnables offer a clear advantage on environmental grounds.

Metal cans. The 12-oz (355-mL) aluminum can with easy-open end has become the primary small carbonated-beverage package, despite apparently higher container costs and retail price compared to either glass or plastic (see Cans, aluminum). Container weight and design have been the subject of much development work, with the weight of a modern can now down to around 18 g including the end. There is at least one necked-in ridge at the seam to allow wall-to-wall contact, which helps on the filling lines, makes firm six-packs, and uses smaller, cheaper ends. Double- and triple-necked cans are sold to get even smaller ends, and a new spin-neck design has a conical top section which achieves the same effect.

Nondetachable ends are now quite common, with a ringlike tab to pull, but without the ring coming off to create litter and a safety hazard. They are mandatory in certain states and are often used in other areas to avoid manufacture and stocking of both types, and to present an environmentally-supportive image. A reclosable end for a flat-topped can has been announced but is not yet on the open market. Another reclosable metal can had a special polypropylene bottle-sized closure. It was introduced in 1983, but later withdrawn.

Steel cans were much more common than aluminum as recently as 10 years ago, but their use has declined, especially for beer, and the old three-piece can with soldered side seam and separate top and bottom ends is seldom seen any more in this market (see Cans, steel). There are some 3-piece cans with welded side seams and even more two-piece steel cans, drawn from steel much as aluminum cans are made (see Cans, fabrication). These may weigh as little as 37 g (with 5½ g aluminum end). Although the steel can is heavier, the material is cheaper, and the steel industry had expected to keep more of the beverage market than it did with the two-piece cans. But fabrication is more expensive, coatings are more critical (to prevent rust), and aluminum did a fine job of selling its recyclability on a consumer level. (Steel can be recycled, too, of course, and is easy to separate from solid waste streams by magnets. However the economics are not favorable, and it does not always pay to get the cans back into the new-material stream, which is what really counts.) Formerly, the need for short runs of preprinted cans for house brands and other low-volume products kept three-piece steel in the running, but the growth of cooperative large-volume canning, improved machinery to coat and print finished cans, and the disappearance of many smaller brands have all contributed to steel's decline.

If steel is doing so poorly in beverages, why is it still the main can material for foods, juices, coffee and many other products? The answer is that the internal pressure of carbonated beverages stiffens the filled container, and makes steel's great advantage in rigidity of little importance. For the other products, this advantage counts, and makes aluminum more costly on an equal-performance basis.

Beer vs Soft Drinks

The beer market is approximately the same size as the soft-drink market and uses similar containers, but there are also some important differences. The market for large-size containers such as the 2-L PET bottle is very small for beer, and will remain so because beer quickly goes stale once the container is opened and exposed to the oxygen in the air. Thus, it must be consumed quickly and cannot be reclosed and finished later, and the large-size market is limited to parties where rapid consumption can be expected. But even at these occasions, there is some preference for cans and small bottles (what one does not drink now, one can drink later), and on the other end, some competition from kegs and even a plastic sphere which is set in a waterproof carton surrounded with ice. These bulk packages are also used in bars, of course, and hold a fairly steady 13% of the annual United States beer market of around 58 billion (10^9) 12-oz (355-mL) equivalents (estimated 1984 sales). The remainder is divided among cans (mostly aluminum, 34.5×10^9), and nonreturnable and refillable glass (12.4 and 3.6×10^9, respectively).

Similar figures for soft drinks are as follows:

	12-oz equivalents, 10^9
Unpackaged soft drinks (fountain sales)	26.0
PET bottles	19.5
Metal cans	31.3
Nonreturnable glass	12.5
Refillable glass	18.7
Total	*108.0*

Almost all soft-drink cans are actually 12 ounces (355 mL) in capacity, as are most beer cans; but some beer is sold in 8-, 10-, and 16-oz (237-, 296, and 473-mL) sizes, and even a 32-oz (946-mL) size has been introduced. The nonreturnable glass includes 12-oz (355 mL) and 32-oz (946-mL) for beer, and 16-oz (473-mL) and 28–32-oz (829–946-mL) for soft drinks. Refillables are 12-oz (355-mL) for beer, and 16- and 32-oz (473- and 946-mL) for soft drinks.

The latest growth idea in both beer and soft-drink marketing is the 12-pack of cans (and bottles), boxed in carry-home secondary packaging, and offering an economical compromise between the six-pack and the 24-unit case (see Carriers, beverage). Secondary packaging is an important aspect of beverage packaging, with shrink-film, plastic can-holding rings, paperboard carriers, molded-plastic bottle carriers, and corrugated and molded-plastic cases all vying for their share, and making their contribution to the total system price of the primary package.

Plastic containers have not yet entered the huge single-service beer market because no container has been offered that is cheap enough and still able to withstand pasteurization conditions, with a good-enough oxygen barrier to assure desired shelf life under the most unfavorable storage conditions. The shelf-life problem is made worse by the presence of some oxygen in the beer as brewed and in the headspace of the container. In effect, any oxygen permeability at all puts pressure on the brewing operation to tighten their oxygen-excluding procedures even more.

Despite these problems, many companies are working to develop a plastic beer container, as the sheer size of this market gives these efforts a huge potential. Plastic beer containers do exist in the UK, where PVDC-coated (see Vinylidene chloride copolymers) 2-L PET bottles for products that are not pasteurized in the bottle are sold. Plastic beer cans have also been tested in the UK. Elsewhere in Europe, PVC bottles have been used at soccer games and institutions where breakage and cleanup are serious problems. In Japan, 2-L and 3-L elaborate beer bottles are made from PET, but they are virtually gift items and far too expensive for any mass market. The Japanese also have small (11½-oz or 340-mL) PET beer bottles on the market on a limited scale. In Sweden, the "Rigello" container was used for beer for 15 years, but has recently been discontinued. This was a unique can/bottle made by welding two thermoformed halves together, surrounding with a paperboard cylinder, and topping with an injection-molded polyethylene closure.

Beer and soft drinks differ in carbonation content, which affects pressure requirements and rate of CO_2 loss. Pressure is typically expressed in volumes or in g/L of carbon dioxide. One volume equals approximately 2 g/L. At room temperature, each volume produces about one atmosphere (0.1 MPa) of internal pressure, but this changes with temperature, so that a 4-volume beverage such as a cola rises to 7 atmospheres (0.7 MPa) pressure at 100°F (38°C) and to 10 atmospheres (1 MPa) at maximum storage/pasteurization temperatures. The carbonation levels of some common beverages are

	Volumes of CO_2	g/L
Club soda and ginger ale	5	10
Common cola drinks	4	8
Beer	3	6
Citrus and fruit soft drinks	1½	3

The beer industry differs from the soft-drink industry in still another, very important way: there is no franchise system. In the franchise system, a parent company licenses a large number of local bottlers, who run independent businesses under the supervision of the parent, and buy their flavor concentrate from that parent. In the beer industry, on the other hand, there is great concentration, with 90% of the beer made by the top 10 companies. Most of these have some cap-

tive container capacity, so that introducing a new container may mean idling of existing capacity, and not just a simple cost comparison. The distribution systems also differ greatly. Soft drinks are much more local and even the big franchises and cooperative canning plants are still regional. Brewers, however, distribute over wider areas, and some ship to more than half the country from a single location. Such distances make container compactness important, discourage breakable glass and less-than-perfect closures, and make refilling (but not recycling) less economical.

Deposit Laws

No discussion of carbonated beverage packaging would be complete without comment on the deposit laws which are in effect in nine states at the present time. They are Connecticut, Iowa, Maine, Massachusetts, Michigan, New York, Oregon, Delaware, and Vermont. In these states, all beverage containers carry a deposit, typically 5 or 10¢. Despite many complaints, the industry and the public have learned to live with such laws, and argue over the relative merits and troubles that they bring. It is fairly well agreed, however, that the cash value for discarded containers does keep most of them off the highways and gets more of them into the recycle streams. The aluminum-can industry, which has a widely-publicized recycle system in operation in both deposit and nondeposit states, now claims that more than 53% of all aluminum beverage cans are recovered in this way! There were some attempts to repeal the deposit laws in a few states, all unsuccessful, and attention then turned to the financial side: compensation for handling of the containers, who is obliged to refund money to whom, the scale of deposits, exempt containers, the status of refillables, and the like.

On a national scale, there has been a deposit law proposed now for over 10 years, with no short-term likelihood of passage, but always a possibility. In the next few years, pro-deposit supporters plan campaigns to change the laws in other states, notably California, where such a law was narrowly defeated a few years ago. If a few more large states do indeed change their laws, the pressure for a nationwide law to avoid fragmented and often conflicting regulations among states will certainly increase.

BIBLIOGRAPHY

General References

Beer Packaging, Master Brewers' Association of the Americas, Madison, Wisc., 1982
Plastic Containers for Beer, Soft Drinks, and Liquor, Edison Technical Services, Bethesda, Md., 1980.

A. L. GRIFF
Edison Technical Services

CARDED PACKAGING

Marrying plastic materials with paperboard to produce visual, self-vending packages is one of the most important and fastest-growing methods of merchandising products today. The rapid growth of self-service retailing created a demand for innovative packaging that protects the product and also provides sales appeal in terms of product visibility and instructions for use. The styles and forms of visual carded packaging are too numerous to discuss in detail, but they can be grouped in two categories: blister packaging and skin packaging.

Blister Packaging, Components and Assembly

Components. Three of the many styles of blister packages are shown in Figure 1. The basic components of a blister pack are the preformed plastic blister, heat-seal coating, printing ink, and paperboard card.

Preformed plastic blister. An important key to package success is the selection of the right plastic film for the blister: property type, grade, and thickness. Consideration must be given to the height and weight of the product, sharp or pointed edges, impact resistance, aging, migration, and cost. The plastic must also be compatible with the product. Heat-sealing properties, ease of cutting and trimming the formed blisters, and other factors influencing production and speed of assembly must be taken into account.

Heat-seal coatings. Heat-seal coatings (see Sealing, heat) provide a bond between the plastic blister and the printed paperboard card. These solvent- or water-based coatings can be applied to rolls or sheets of printed paperboard using roll coaters, gravure or flexographic methods, knives, silkscreening or gang sprays (see Coating equipment). Whatever the system, it is essential that the proper coating weight be applied to the paperboard card for optimum heat-sealing results. In addition, the heat-seal coating must be compatible with the paperboard card and the plastic blister. For example, the type of coating used with an acetate blister is not the same as one used with polystyrene (PS). With proper coatings, strong fiber-tearing bonds can be obtained. The heat-seal coating also protects the printed areas of the card and provides a glossy finish.

Printing inks. Printing inks provide graphics and aesthetic appeal, and must also provide a bond between the paperboard card and the heat-seal coating. Inks may be applied to the paperboard by letterpress-, gravure-, offset-, flexography-, or silk-screen printing processes (see Inks; Printing). The inks must be compatible with the heat-seal coating and paperboard card, must resist high heat-sealing temperatures, abrasion, bending, and fading, and must be safe for use with the intended product. Inks should not contain excessive amounts of hydrocarbon lubricants, greases, oils, or release agents. Qualification tests should always precede production runs.

Paperboard. Paperboard (see Paperboard) for blister packaging must be selected according to the size, shape, and weight of the product as well as the style of package to be produced. It provides the base or main structural component of the package. Paperboard for blister packaging ranges in caliper, ie, thickness, from 0.014 to 0.030 in. (0.36–0.76 mm), but 0.018 to 0.024 in. (0.46–0.61 mm) is the most popular range. The surface must be suitable for printing by the required process and inks and also compatible with the heat-sealing-coating process. Paperboard must be able to meet the stresses imposed by the printing processes, ie, primarily delamination of offset-printing presses, and still provide good fiber-tearing bonds when heat-sealed to a plastic blister. Clay coatings are added to paperboard to enhance printing results and heat-seal-coating holdout. Heat-sealing and printability are both important considerations in blister packaging, and the paperboard must offer the best workable compromise.

Assembly. The normal sequence of assembly involves loading the blister with product, placing a paperboard card over the blister, and heat-sealing the package. This can be a simple manual operation, semiautomatic, or fully automated. All heat-sealing methods mate the blister and card under constant pressure for a specified time during which heat is applied. The mating surfaces fuse and bond, setting almost instantaneously when heat input stops. Heat-sealing machines and methods should provide the necessary production rate and should also be able to be adjusted to maintain constant conditions of dwell time, temperature, and pressure. Heat-sealing machines usually employ heated dies, produced when an electric current is passed through a resistance-type heating element in the die (see Sealing, heat). Some heat-sealing machines utilize impulse, electronic, or high frequency heating. Both direct (heat applied through the blister) and indirect (heat applied through the paperboard card) heating techniques are successful. Direct sealing is faster than the indirect method and minimizes scorching or warping of the paperboard card.

Skin Packaging, Components and Assembly

Skin packaging is a special form of visual carded packaging. It differs from blister packaging in that the product itself

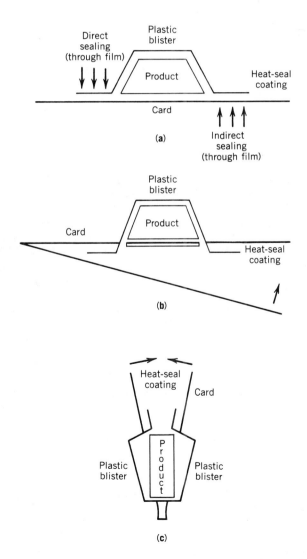

Figure 1. Blister packaging: (**a**) conventional surface seal; (**b**) foldover-card variation; (**c**) hinged blister/foldover-card variation.

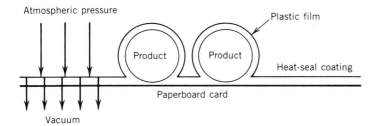

Figure 2. Skin packaging.

becomes the mold over which the heated plastic film or "skin" is drawn by vacuum and heat-sealed to a paperboard card (see Fig. 2). As in blister packaging, there are four principal components in a skin package: the plastic film, the heat-seal coating, the printing ink, and the paperboard card.

Plastic film. There are three types of flexible plastic film used in skin packaging: LDPE (see Polyethylene, low density), PVC (see Film, flexible PVC), and ionomer (see Ionomers). Which film is used depends on the application requirements. The film must be compatible with the heat-seal coating on the paperboard card to ensure a fiber-tearing bond or an otherwise acceptable heat seal. Normally, skin-packaging films are heated, draped and formed, and bonded to the paperboard card in one operation.

Heat-seal coatings. Heat-seal coatings for skin packaging must be compatible with the inks, paperboard, and plastic film used in the package. They may be solvent- or water-based. Methods of application are the same as those for blister packaging.

Printing inks. Inks for skin packaging must provide a bond between the paperboard surface and the heat-seal coating. Inks should not contain any compounds that may inhibit heat-sealing, such as waxes, oils, release agents, etc, and must be resistant to heat and fading and compatible with the product.

Paperboard. Paperboard for skin packaging must be selected according to the caliper, stiffness, and other strength characteristics necessary to support the product. High porosity paperboard should be used to allow proper drawdown of the film and good contact with the card. Paperboard for skin packaging is generally not clay-coated. Clay coatings provide good printing surfaces and holdout properties for heat-seal coatings, but they also decrease porosity and interfere with vacuum drawdown during heat-sealing on skin-packaging equipment. Certain inks and heat-seal coatings may also decrease porosity, and in that case, the paperboard must be perforated to increase air flow through the sheet during vacuum drawdown and heat-sealing.

Assembly. After proper selection and processing of the skin-packaging components, assembly of the package can take place. In effect, the product to be packaged is laminated between the paperboard card and the plastic film. The product is usually positioned on the printed heat-seal-coated paperboard card, which then moves onto the base or platen of the skin-packaging machine which contains air passages connected to a vacuum system. Plastic film is held in a frame above the product/paperboard card. The film is heated to a softening temperature, and at the proper time in the heating cycle the frame carrying the film is dropped, allowing the film to drape over the product and paperboard card. Vacuum is then applied

through the platen and card, bringing the film in contact with the card. The residual heat in the plastic film activates the heat-seal coating, which fuses and forms a fiber-tearing bond. Skin-packaging machinery is available in many designs ranging from manual to fully automated. Control of preheat, timing cycle, and postheat are all critical variables which must be controlled.

Plastic Films

Blister films. There are three principal types of rigid plastic films used in blister packaging: cellulosics, styrenics, and vinyls. A copolyester is being used now as well (see Polyesters, thermoplastic). The most popular cellulosic films are acetate, butyrate, and propionate. All three films have excellent clarity and thermoforming characteristics and heat-seal well to properly coated cards. Sealing temperatures are generally higher for the cellulosics than they are for the other films. Cellulosics do not have exceptional cold strength, but they are reasonably shock and craze resistant.

Oriented polystyrene (OPS) has low resistance to impact and shatters easily. Low temperature performance is also poor. Impact polystyrene has good impact resistance and cold-temperature properties. The styrenics seal well to coated paperboard under the proper conditions. Clarity of OPS is excellent (see also Styrene–butadiene copolymers).

Both plasticized and unplasticized vinyl films (PVC) are made (see Film, flexible PVC; Film, rigid PVC). The amount of plasticizer affects cold-temperature resistance and impact strength. PVC heat-seals well to properly coated paperboard. Vinyls vary in clarity from excellent to good, and in color from slightly yellow to slightly blue. Thermoforming characteristics are excellent to good.

Skin-packaging films. Flexible films that conform to product shape are used in skin packaging. The flexible films are LDPE, PVC, or Surlyn (DuPont) ionomer. LDPE is the least expensive of the group. It is generally not as clear as the others, requires more heat, and because it shrinks more upon cooling, board "curl" can result. LDPE is strong in both impact and tensile properties, and it adheres well to heat-seal-coated paperboard cards. PVC film was used in the past for skin packaging more than it is today, having been replaced to a great extent by ionomer. PVC clarity is excellent, and the slight yellow or blue tint is not objectionable. It heat-seals well to coated paperboard and conforms well to intricately shaped products. Compared with polyethylene (PE), it requires less heat and shrinks less. Ionomer, a relatively recent entry in the skin-packaging field, offers excellent clarity and color, fast heating, and good adhesion to a properly coated paperboard. Preheating time for ionomer is the same as for PVC and much faster than for PE, and it has exceptional strength. It is more expensive, but because of its strength, relatively thin ionomer films can replace heavier-gauge alternatives and be cost-effective on an "applied" basis.

Heat-seal Coatings

Heat-sealable coatings for blister- and skin-packaging cards are perhaps the most critical component in the entire system. The appearance and physical integrity of the package depends upon the quality of the heat-seal coating.

Blister-card coatings. A successful blister-board coating must have good gloss, clarity, abrasion resistance, and hot tack and must seal to the various blister films. Hot tack is particularly important because the product is usually loaded into the blister and the board heat-sealed in place face down onto the blister. The package is ejected from the heat-seal jig and the entire weight of the package must be supported by the still-warm bond line. A relatively low heat-seal temperature is desirable for rapid sealing and to prevent heat distortion of the blister film. Blister-board coatings are still predominantly solvent-based vinyls because of their superior gloss. Some inroads are being made by water-based products, but these must be carefully evaluated for hot tack, gloss retention, adhesion to specific inks, and sealability to selected blister films. The rheology and flow properties of the coating must be appropriate for roll-coater application and holdout on clay-coated solid-bleached-sulfate (SBS) board. The final test of a blister-board coating is a destruct bond to the printed board.

Skin-board coatings. Heat-seal coatings for skin board must have special properties unique to this application. In skin packaging, the board is placed face up on a vacuum plate and the product positioned on the board. The film is heated to the proper temperature and dropped down on the product, and a vacuum is pulled through the board. The hot film conforms to the product and adheres to the heat-seal coating on the board. Here again, the skin film must form a fiber-tearing, destruct bond to the printed board. Unlike the clay-coated SBS board used for blister cards, skin board is either uncoated SBS or combination, ie, recycled board. Because these substrates are porous, they have poor holdout for coatings. Heat-seal coatings for skin board must have good holdout nevertheless, and must be heat-sealable to the films at low temperatures. Gloss and clarity of the coating must be sufficient to permit accurate identification of graphics after printing. The appearance of the final package is more a function of the skin film than of the coating, but the coating must not detract from the clarity of the film. Conventional solvent-based coatings tend to soak into the board and interfere with porosity. Mechanical perforation is usually necessary when this type of coating is used. Water-based coatings are now available with excellent holdout and porosity, thus eliminating the need for mechanical perforation.

Another package that presents a coating problem is the foldover blister card. The heat-seal coating is applied on the back of the board, which is die-cut and folded around the blister. The coated surfaces of the board are heat-sealed to themselves. This type of card is usually combination board which is extremely porous and has poor holdout for coatings. When conventional solvent-based products are used, several passes through the roller coater are often necessary before adequate coating buildup is obtained. Fortunately, there are water-based coatings available now which have good holdout on this type of porous board. These products offer economies through lower material costs and faster production rates.

Several different types of heat-sealable coatings are used for carded packaging. Nitrocellulose (NC) lacquers have long been noted for their gloss and abrasion resistance. Their use is limited to the cellulosic blisters, however. NC lacquers have good coatability and leveling by most application methods, including roller coating.

Gel lacquers are used by some board converters for both skin and blister board. These coatings, based on ethylene–vinyl acetate (EVA) copolymers, have good gloss and adhesion to most films used in carded packaging. Skin-board coated with a gel lacquer must be mechanically perforated for adequate porosity. The chief difficulties in using gel lacquers are that they must be heated for application and that most require the use of aromatic solvents for thinning and cleanup.

Some of the most widely used coatings for blister cards are solvent-based vinyls. These lacquers have excellent gloss and abrasion resistance and are especially favored for use with PVC blisters. They are also used for skin packaging with PVC film, but the board must be perforated after coating. Adhesion to olefinic skin films, eg, LDPE and ionomer, is marginal at best. Most vinyls lend themselves to roller coating.

Among the newer coatings for blister board are water-based acrylics (see Acrylics). These approach the solvent-based vinyls in gloss and abrasion resistance, but they require more care in coating to achieve optimum appearance. Water-based EVA dispersions have become quite popular for coating skin board. They have good sealability to most skin-packaging films and their high porosity eliminates the need to perforate the coated board. Most of these products are solvent-free, thus eliminating the odor and fire hazards associated with solvent-based products.

Paperboard

Solid bleached sulfate (SBS). The basic raw material for SBS paperboard is bleached wood pulp (see Paperboard). On a fourdrinier machine, a single layer of pulp is laid down on the wire and becomes the base for building the sheet. Internal sizing is added at the wet end of the machine. Surface sizing is added at the size press (both sides) and wet calenders (one side or both sides). This may be followed by clay coatings on one side or on both sides. SBS paperboard is produced in this manner for various packaging end uses, including folding cartons, and as a base stock for laminations.

Recycled paperboard. Recycled paperboard is generally made with multiple layers. This may be accomplished on machines such as multicylinders and vats, multifourdrinier and headboxes, multicylinders and headboxes, fourdriniers with multiple headboxes, etc. Fiber furnish is normally recycled fiber from waste collection. The multi-ply concept allows numerous variations in sheet construction, eg, a sheet may contain a recycled center ply or plies with a white liner on both sides. Clay coatings can also be added for printability and holdout. Recycled paperboard is suitable for setup boxes (see Boxes, rigid paperboard), some folding cartons, visual carded packaging, and other packaging applications.

Paperboard variations. Visual carded packaging places many demands on the paperboards, which is always the base upon which the final package is built. The finished package is comprised of a variety of materials (see Fig. 3).

First, there is a wood-pulp base sheet with internal and surface sizing on both sides. The next layer (or layers) is clay coating, followed by printing ink, heat-seal coating, and plastic film or blister. The main objective is to produce a package in which the product is visible, securely contained, with the base sheet as the weakest link.

Depending on the package style, the selection of paperboard can be crucial. For a blister pack (see Fig. 4a) with intricate process printing on one side and line printing on the other (see Fig. 4b), a double-clay-coated one-side sheet would be selected. This sheet should accept multicolor printing, usually by the offset process, and withstand heat-seal coating, die-cutting, and heat-sealing to a plastic blister, adhering with such force

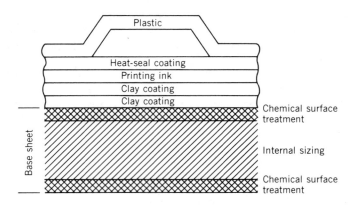

Figure 3. Cross section of a carded package.

as to tear fiber from the paperboard base sheet when the blister is removed.

Another variation is a blister sealed to a die-cut card (see Fig. 4c). Heat sealing of the blister flange to the back of the card may require clay coating or special heat-seal coating for the uncoated side of the paperboard. For a hinged or foldover card, which is die-cut and folded over with the blister protruding through the die-cut opening (see Fig. 4d) which is to be heat-sealed, a coated two-sided sheet would probably be selected. This would permit printing on one side and application of a heat-seal coating on the other side which would be held out by the clay coating on the paperboard. Fiber tear is not so important. If the foldover card is not heat-sealed, it can be glued or stapled. The sandwich or double card (see Fig. 4e) is a variation of Figure 4d without the hinge.

Machinery

Blister-packaging machinery. Blisters are formed of 5–15-mil (127–381-μm) film in epoxy or water-cooled aluminum molds. After vacuum-forming (see Thermoforming), the blisters are die-cut and placed in a heat-sealing tool. After the product is loaded into the blister, a solvent- or water-based coated card is placed face down on the tool. Locating pins ensure accurate register of card to the blister. The heat-seal tool is then moved into the heat-seal area where pressure and temperatures of 250–400°F (121–204° C) are applied for 1.5–3 s and the film is bonded to the card.

When heat is applied through the card (indirect method), it is advisable to use a milled upper sealing die to apply pressure

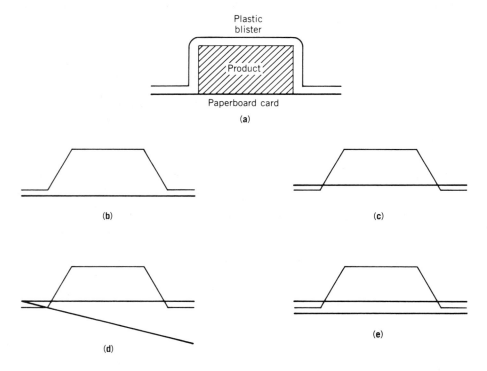

Figure 4. Blister packs. (**a**) Basic configuration. (**b**) Blister adhered to single card. Paperboard requirements: clay-coated on one side for multicolor printing, line printing on uncoated side; rigidity; flatness; heat seal on face down through board to fiber-tear end point. If multicolor printing is required on both sides, clay-coated two side may be used. (**c**) Blister sealed to die-cut card. Paperboard requirements: same as (**b**). Heat seal on backside of card may require clay coating or special heat-seal coating for uncoated side of paperboard. (**d**) Foldover card. Paperboard requirements: clay-coated on two sides for multicolor offset printing on one side and heat-seal-coating holdout on opposite side; rigidity; flatness; good heat seal. Foldover style may be hot-melt- or resin-glued or stapled, in which case clay-coated one side board may be used. (**e**) Sandwich or double card. Paperboard requirements: clay-coated on two sides for multicolor offset printing on one side and heat-seal coating on opposite side; rigidity; flatness; good heat seal. For resin glue, hot-melt, or staples, clay-coated on one side may be used.

and heat only in the blister flange area on the back of the card. This speeds sealing time and reduces card warping. Manual and semiautomatic sealers are available as multistation rotary machines, shuttle-type machines, and indexing belt-type sealers. The indirect sealing tools are constructed of wood or aluminum with cork or rubber mounted to them in the configuration of the bottom blister flange.

Direct (through the flange) sealing tools use Teflon (DuPont) to prevent the film from sticking to the heated tool. Fully automatic machines are multistation rotary or belt type. They can be equipped with in-line forming/cutting, blister feeding, and card feeding. Speeds vary from 10 packages/min on smaller sealers to 150 packages/min on large automatic sealers. Blisters are formed by small users on small-bed, multipurpose vacuum-forming machines, and on large-bed or in-line vacuum-forming machines by large users. Multipurpose vacuum-forming machines can be used for both blister forming and skin packaging. Depending on speeds and blister configurations, options on machines may include blister blowoff (air used to lift blister off molds), water-cooled molds (higher speed operations), film assists (on higher blisters to eliminate film webbing), and plug assists (for deep-draw applications).

Roller-type and ram-type die-cutters are used to die-cut sheets of blisters apart after forming. For accurate cutting, care must be taken in rule-die construction to allow for variations in film gauge and shrinkage rates of different films after removal from the vacuum former. In the blister-packaging process, forming molds, blister-cutting dies, and heat-sealing tooling are required. Platform blister molds are sometimes used to reduce the number of cutting dies and blister sealing tools required for a multiple-item product line. Platform molds are designed to have a common blister flange and reduce setup times.

Skin-packaging machinery. Skin packaging is a vacuum-forming process where the product being packaged acts as a forming mold. Sheets of multiple products are processed and die-cut or slit apart into individual packages. Skin packaging is used for visual retail display and industrial product protection. The principal objective of industrial product protection is to immobilize the product for handling and shipping. Skin packaging costs much less than dunnage materials used in industrial packaging and it requires much less storage space. Since industrial products are generally relatively large and heavy, 7.5–15-mil (191–381-μm) PE and treated corrugated are commonly used. Display skin packaging typically uses ionomer film and coated printed boxboard, which is somewhat denser than corrugated board.

In the skin-packaging process, products are positioned on a boxboard or corrugated sheet and automatically conveyed or placed manually on the machine's vacuum platen. A heater bank is energized, and when the film is softened to a pliable, thermoformable state, the film frame descends, stretching film over the product. The vacuum system initiates the exhaust of air below the film and between the products. Vacuum draws the film in tight conformity to the product, encapsulating each in a tough, transparent plastic skin. Film is simultaneously heat-sealed to the surface of the board. When the film frame opens, the packaged products are discharged from the machine, which automatically rethreads the film for the next cycle. When the film frame is closed, the film is cut, and the frame elevates to its raised position, permitting entry of the preloaded sheet for the next cycle.

For industrial applications using heavy-gauge PE film, turbine pumps are used. They are capable of removing large volumes of air with a softer vacuum, which is generally required on odd-shaped or undercut items. Display packaging using ionomer film generally uses positive-displacement pumping systems, which consist of a pump and a storage tank which, when activated, can evacuate air between the film and substrate rapidly with a much stronger force. If certain items in a product line have deep recesses or undercuts, a combination system may be needed. The positive-displacement system is also used in blister forming.

Heating ovens are constructed with either metal-sheathed elements or nichrome ribbon. A film frame is required to hold the film during the heating cycle. These two-piece clamping frames or vacuum frames can be operated manually or automatically with automatic film cutoff for higher volume applications.

Since skin packaging is a sheet-processing system, the board converter and the machinery manufacturer should be involved at the early stages of a new program. The board converter and the machinery manufacturer can often lay out sheets that maximize printing-press and skin-packaging-machine efficiency and substantially reduce material costs.

Skin-packaging equipment is available in a wide variety of sizes and configurations, ie, from 18 × 24 in. (45.7 × 61 cm) to 30 × 96 in. (76.2 × 244 cm). Manual and semiautomatic systems are capable of running 2–3 sheets/min on smaller sizes and 1½–2 sheets/min on larger machines. Fully automatic machines can run 3–8 cycles/min. These speeds are for 3–5-mil (76–127-μm) ionomer display packages. Speeds of 1–2/min are generally achieved with 10-mil (254-μm) PE for industrial-packaging applications. Selection of size, infeed systems, and options is dictated by application and volume. Roller-type die cutters are used to die-cut sheets apart and die-cut hanger holes.

Blister Packaging vs Skin Packaging

The choice between blister and skin packaging is not always clear-cut. When the size of the product and blister is small, eg, a ballpoint pen, in relation to the card size, total blister-pack cost is generally less. If the product and blister are large in relation to the card, total skin-packaging cost is generally less. For this reason, large items such as antennae, hand tools, kitchen implements, and plumbing accessories are generally skin-packaged. Totally automatic blister-sealing lines have been available for 20 years and are used for high volume items such as razor blades, batteries, and pens.

Until recently, round items such as batteries, screwdrivers, and fasteners were blister-packaged since they moved on the card during the skin-packaging process. These now can be skin-packed through use of magnets and recessed moving platens. Fully automatic skin-packaging systems are a relatively recent development.

J. M. Gresher
John D. Clarke Company

CARRIERS, BEVERAGE

Beverage carriers are designed to do more than just provide carrying convenience. In broad terms, they perform three basic functions: unitization, product protection, and marketing vehicle. Unitizing bottles or cans into multiples of 6, 8, 12, or 24 minimizes distribution costs because 6 or more individual primary packages can be moved as a single unit. A properly designed carrier protects the product by forming a tight group of bottles or cans. This reduces vibration and increases vertical compressive strength. Cushioning material such as paperboard may also be placed between bottles to improve impact resistance. Printed beverage carriers act as marketing aids by providing the consumer with information about the product. The convenience of a multipack also encourages the consumer to purchase more product in a single transaction.

Can Multipacks

More than 50% of all beverage cans are sold as six-packs. The most widely used unitizing device is the ring carrier. The Hi-Cone (Illinois Tool Works) plastic ring carrier slips beneath the rim of the cans and grips them tightly throughout the distribution cycle. It provides the most cost-effective and efficient package for 6 cans (see Fig. 1). The Hi-Cone system consists of two elements: the plastic carrier cut from extruded polyethylene sheet and an applicating machine. The speed of the line matches that of the filler, often as high as 1800 cans (300 six-packs) per minute. Rolls of interconnected Hi-Cone rings are wound onto the applicating machine, a high speed continuous motion machine that consists of a can collator and a unit that joins the cans and rings. After the cans have been collated, the plastic rings are stretched over the can necks and cut into the desired multiples.

Paperboard twelve-packs of cans have become popular since the late 1970s, and they now account for over 30% of all can packaging in the United States. The majority of these use a preglued paperboard sleeve. Machinery for the twelve-pack sleeve system removes the preglued sleeve from a magazine and snaps it open, forming a tube. The collated cans moving in a parallel and continuous motion are pushed from both sides into the open ends of the tube. The ends of the tube are folded into place and secured either by glue or mechanical lock. A twelve-pack system operates at speeds up to 1800 cans (150 twelve-packs) per minute.

Hi-Cone plastic neck rings are also available for twelve-packs. A plastic band around the perimeter of the cans secures them and provides a handle. This system is less expensive than paperboard, but it requires a separate Universal Product Code number on the cans (see Barcoding) and does not have the graphic impact of a printed paperboard carrier.

Shrink-film systems are starting to be used for six- and twelve-packs of cans on a large scale (see Fig. 2). These systems utilize a thin polyolefin shrink film, either clear or printed, to tightly contain the cans in a 2×3 or 4×3 configu-

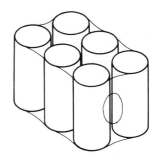

Figure 2. Shrink-wrapped multipack.

ration (see Films, shrink). Most shrink-wrapping machines in the past were of a reciprocating-motion shutter type that limited the speed to approximately 35 cycles per minute. Kisters (FRG) developed a continuous-motion machine that has cycle speeds of up to 100 per minute. This equipment can register printed film in a two-lane configuration that permits an output of 200 six-packs per minute. Although this increase in output makes the package more economical to produce, the equipment is quite large and is best suited for high volume-dedicated operations.

Returnable Bottle Multipacks

Materials and designs for glass-bottle carriers are determined by the primary container. Returnable glass is packaged in carriers that can make a round trip from the bottler to the consumer and be returned to the bottler for refilling. Carriers for nonreturnable glass are discarded by the consumer, generally after one use.

The most common multipack for returnable bottles is the preformed basket carrier fabricated from virgin or recycled paperboard (see Fig. 3). Those used by the brewery industry are generally in compliance with Rule 41 of the Uniform Freight Classification for rail shipment, which requires bottle separation by paperboard to prevent breakage. Soft drinks are usually shipped by truck and therefore do not require compliance carriers (see Carbonated beverage packaging). Prefabricated basket carriers are preferred for returnable glass because they are convenient for the consumer and they are easy to handle in the bottling operation. Basket carriers arrive at the bottling plant either fully erected and loaded with empty bottles from the glass manufacturer or in a knocked-down state from the carrier manufacturer. If the carriers are previ-

Figure 1. A plastic ring carrier.

Figure 3. Basket carrier for returnable bottles.

ously erected and loaded, the bottler removes the bottles for cleaning and filling, and returns them to the carriers with a drop packer (see Case loading). Knocked-down paperboard basket carriers frequently delivered to the bottling plant are erected by mechanical snap-open devices such as those manufactured by R. A. Pearson and Mead Packaging. The carriers are then placed into corrugated shipping cases, and the bottles are drop-loaded into the baskets.

In several parts of the United States, rigid injection-molded, high density polyethylene basket carriers are being used for returnable bottles (see Injection molding; Polyethylene, high density). These carriers generally have solid-wall construction, which permits the application of graphics. Plastic carriers have an advantage over paperboard carriers in high humidity areas such as the U.S. Gulf-state regions where paperboard can lose significant strength. Although the initial cost of plastic basket carriers is significantly higher than paperboard carriers, they may be used for as many as 50 trips, which reduces the cost per trip and justifies the investment. However, plastic carriers do not permit high quality graphics or frequent logo-type design changes.

Nonreturnable Bottle Multipacks

Nonreturnable bottles use a wider variety of carriers because fewer restrictions are imposed by shipping regulations, the markets are diverse, and less product protection is required. The major carriers used are paperboard wraparound, paperboard sleeves, plastic clips, shrink film, and basket carriers.

Paperboard carriers. Paperboard basket carriers, as described above, are generally used for products that command a high quality image, eg, premium, superpremium, and imported beers.

Wraparound paperboard carriers are probably the most widely used system of paperboard packaging for beverage bottles in the world. This is because of their relatively low cost, large graphic presentation, product protection, and efficient application (see Fig. 4). Wraparound paperboard carriers begin with a die-cut blank made of virgin wet-strength kraft paperboard (see Paperboard). These are shipped to the bottling operation, where they must be folded and locked or glued on a machine in the bottle line.

The elements of the wraparound packaging machine include blank feed, product infeed, folding, and locking or gluing. The bottles are collated into a grouping such as 1×2, 2×3, 2×4, or 2×5 on a continuous-motion machine. The flat

Figure 4. Wraparound paperboard carrier.

paperboard blank is removed from a stack magazine by vacuum or by mechanical lugs and positioned above the collated bottles. The blank is then folded along the score lines and wrapped tightly around the bottles. The bottom of the wrap is either locked or adhesively secured. The most common wraparound machines are supplied by Mead Packaging, Manville, Kliklok, and Federal Paperboard with machine speeds of 100–200 carriers per minute. Paperboard wraps may be designed to comply with Rule 41 by separating the bottles with paperboard partitions or using die-cut tabs at the base of the carrier. End panels may also be added to beer carriers for light protection.

Most breweries utilize the full enclosure of a paperboard preglued sleeve when they wish to protect their product from light degradation. This type of carrier has a gable-shaped top to conform to the shape of the bottle and uses a combination of divider tabs from the bottom of the carrier and die-cut holes at the top to separate the bottle necks, thus complying with Rule 41. The packaging machine removes the sleeve from the magazine and snaps it open to form a tube. Collated bottles are conveyed in parallel with the moving carrier and are pushed into the sleeve as fingers from beneath push the separation tabs up from the sleeve base. Glue is applied to the end tabs, which are then folded down and secured to form a six-sided light-resistant carrier.

Plastic clips. The introduction of the Owens-Illinois foamed-polystyrene-clad bottles (see Foam extruded polystyrene) and plastic bottles has spawned the growth of plastic clip-on style carriers in the United States. Since these bottles do not require any added separation for protection, they only need to be contained in a grouping by neck-retaining carriers. Contour-Pack (Owens-Illinois), the most common clip-on carrier, is thermoformed (see Thermoforming) from HDPE sheet. The sheet is die-cut with notched openings to retain the bottles at their necks and formed with a skirt to conform to the perimeter of the bottle shoulders to form a solid unit. Injection-molded carriers are also available. Applicating machinery for plastic clip-on style carriers removes the formed nested carriers from a hopper and places them over the bottles being conveyed under the machine. Pressure bars complete the process by snapping the locks on the necks of the bottles. Hartness International also manufacturers clip-on applicating equipment.

Shrink films. Shrink-film systems using thermoplastic films that shrink with the application of heat were commercially introduced in the United States during the late 1970s. Carriers produced with this material may use either printed or clear film, depending upon the intended market and cost guidelines. Kisters equipment, mentioned above, is typical of the shrink-wrap machinery available for bottles as well as cans.

BIBLIOGRAPHY

General References

G. C. Geminn and J. Hirschy in H. M. Broderick, ed., "Product Packaging" in *Beer Packaging,* Master Brewers Association of the Americas, Madison Wisc., 1982, pp. 385–399.

S. Sacharow, *Handbook of Packaging Materials,* Avi Publishing Co., Inc., Westport, Conn., 1976.

S. Sacharow, *Principles of Package Development,* Avi Publishing Co., Inc., Westport, Conn., 1972.

Secondary Glass Packaging, Voluntary Specifications Guideline For Paperboard & Corrugated Board Packaging Systems, National Soft Drink Association, Washington, D.C., 1979.

A. L. Brody and J. V. Bousum in W. C. Simms, ed., "Multi-Packing" in *The Packaging Encyclopedia 1982,* Cahners Publishing Co., Chicago, Ill., 1982, pp. 275–281.

The Folding Carton, Paperboard Packaging Council, Washington, D.C., 1975.

R. W. Heitzman, ed., *Packaging Digest Machinery/Materials Guide 1984,* Delta Communications, Inc., Chicago, Ill., 1984.

"A Broader Packaging Universe," *Packag. Dig.,* 21(1), 60–68 (Jan. 1984).

J. V. Bousum
Container Corporation of America

CARTONING MACHINERY, END-LOAD

The cartoning machine, located near the end of the packaging line, produces a package that in most cases brings the product to the consumer. For this reason the package must be attractive and adequately protect the product. Cartoning machines handle an almost infinite variety of products; solid objects like bottles, jars, and tubes, and also breakfast cereals, rice, pasta products, and similar free-flowing items which may be filled into the carton by scales or volumetric means. This article is devoted primarily to horizontal and vertical cartoners handling solid products.

There are two basic types of cartoning machines: semiautomatic and fully automatic. By definition, a semiautomatic cartoner is one with which the operator manually places the product into the carton; a fully automatic cartoner is a machine that automatically loads the product into the carton even though an operator may place the product in a bucket or flight of the intake conveyor (1). Both types are available in horizontal or vertical modes, that is, the carton is carried through the machine lying horizontally or standing upright in the conveyor carrying the carton.

Semiautomatic Vertical Cartoners

In most cases the semiautomatic vertical cartoner is arranged as shown in Figure 1. The tubular carton is fed from a horizontal magazine, expanded and transferred into the carton conveyor. The bottom of the carton is closed by tucking or gluing and then conveyed past one or more operators who manually place the product in the carton. The machine then closes the top of the carton by tucking or gluing.

The semiautomatic vertical cartoner is usually used for products with low production volume where changeover to another size is frequent. It is also well suited for packages containing a variety of different items. A semiautomatic vertical cartoner usually has a wider size range than a fully automatic machine. Common operating speed is up to 120 packages per minute, although many machines of this type operate at much slower rates because of the need to place the product manually into the moving carton.

The semiautomatic cartoner can be equipped with a number of attachments such as a leaflet feed mechanism, code impressor, or printing mechanism for lot numbers, expiration dates, or prices. On-machine printing is usually located on the end panel or tuck flap of the carton rather than on any of the body panels because adequate backup pressure is required in order to obtain a good, legible impression (see Printing of packages).

Many packages require a leaflet or coupon to be placed inside the carton with the product. This may be accomplished by an operator stationed along the carton conveyor who can place a prefolded leaflet into the carton next to the product. For marketing reasons this is not always acceptable because the product can be removed by the consumer without remov-

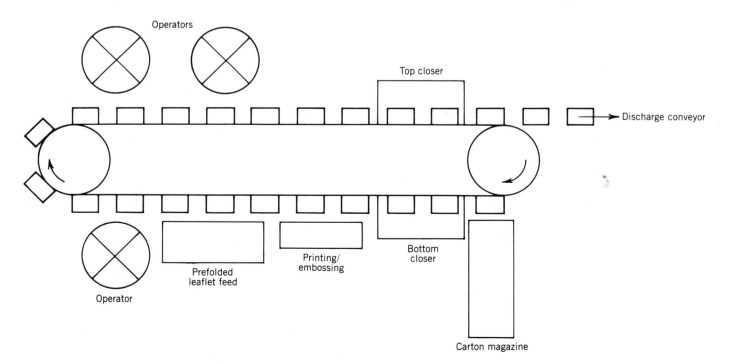

Figure 1. Semiautomatic vertical cartoner.

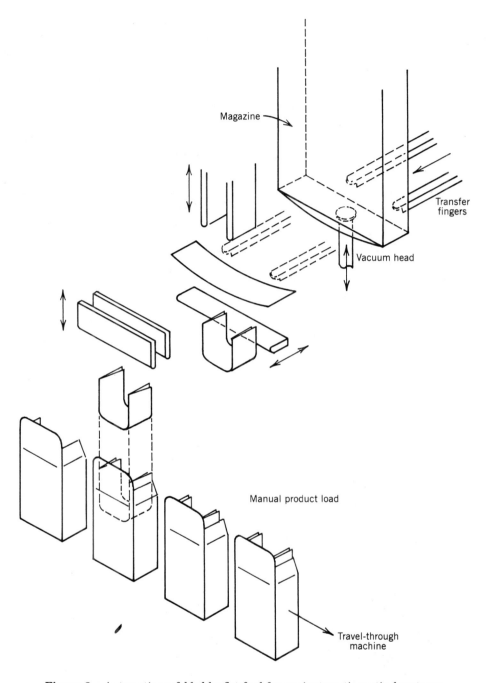

Figure 2. Automatic prefolded leaflet feed for semiautomatic vertical cartoner.

ing the leaflet. To overcome this, the prefolded leaflet is often folded around the product in a U-shape. If the leaflet is to be readily visible when the carton is opened, the ends of the leaflet must be inserted first. It is much easier to wrap the leaflet around the product and insert the assembly into the carton with the ends of the leaflet trailing; consequently, many cartons are run through semiautomatic vertical cartoners upside down so the cap of the bottle (or top of the bottle) with the leaflet is at the bottom of the carton as it runs through the machine.

An automatic mechanism can be installed to feed a prefolded leaflet from a magazine and partially insert it into the carton. When the product is loaded by the operator, the leaflet is pushed down to the bottom of the carton (see Figure 2). A pick-and-place unit may be used to transfer a prefolded leaflet or coupon from a magazine into the carton, but generally such a unit does not orient the coupon in the carton; it merely drops it into the moving carton, usually before the product is inserted.

Semiautomatic Horizontal Cartoners

The semiautomatic horizontal cartoner is similar to the vertical cartoner except that the carton is carried through the machine lying on its back panel. Because the product is inserted into the carton manually, it is classified as semiautomatic. Some models have the carton conveyor arranged in such a way that the carton is inclined downward away from the operator, enabling the product to be loaded without lifting, that is, just slid into the carton by gravity. Some of the slant-type horizontals can be adjusted by a crank so the carton is carried anywhere from the horizontal to 20° to suit the type of product being handled. The semiautomatic horizontal machine, like the vertical, can be equipped with many different attachments, code impressors, printers, and hot melt adhesive systems.

Fully Automatic Horizontal Cartoners

In general, this class of packaging machine is similar to Figure 3. It consists of a product infeed conveyor, carton feed, carton conveyor, loading mechanism, and closing system. Fully automatic machines can operate at speeds from 50 packages per minute to well over 600 packages per minute for certain items, although most are designed for operation in the range of 150–300. This type of cartoner usually has less flexibility in size range than the semiautomatic vertical machines but has the advantage that all operations are automatic; that is, no operators are required to place the product into the carton. Each mechanism is critical to the efficient operation of the machine, but probably the most important is the carton feed.

Carton feeds.   There are many types of carton feeds used on cartoning machines, the selection dependent upon the speed of operation required, the size of the carton to be handled, and the style of carton, (ie, preglued tubular style or flat blanks). The tubular carton, preglued by the carton manufacturer, is the most common although flat blanks also have advantages.

The basic carton feed for tubular cartons consists of a magazine to hold the supply of unexpanded cartons, a vacuum head, and a transfer system to place the open carton into a conveyor. The magazine can be vertical, horizontal, or inclined. In most cases the cartons are retained in the magazine by ledges, which are small projections from the side of the magazine, usually located in the cut-out areas of the carton flaps so the movement of the carton by the vacuum head releases the carton from the magazine. The carton must be pulled free of the stack before expansion can start (2). A simple carton feed is illustrated in Figure 4. The suction head, moving downward, pulls the carton from the ledges, and expansion begins when the carton contacts the beveled expander block. As the vacuum head reaches the bottom of its stroke, it straddles or goes between the carton conveyor chains, and as the vacuum is released, the carton is transferred into the conveyor lugs which are moving around the chain sprocket. Expansion is completed

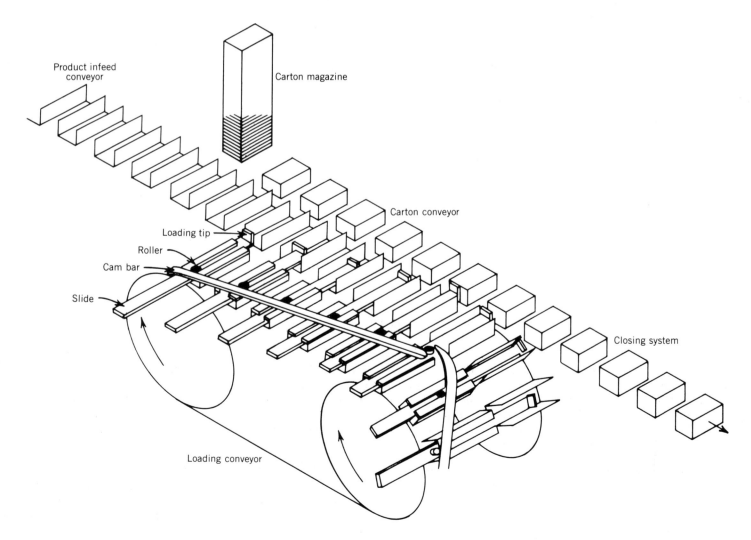

Figure 3.   Fully automatic horizontal cartoner.

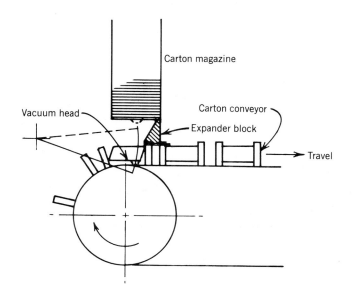

Figure 4. Basic carton feed for tubular cartons.

This blade supports the weight of the stack to enable the bottom carton to be pushed by reciprocating fingers into the expansion position. Expander blades are inserted inside the carton, frequently from both ends of the carton, to start the expansion. An operated finger partially expands the carton by pressing down against the side panel while the carton is still held by the blades. Forming of the cartons is completed by the lugs of the carton conveyor as they come around the sprocket.

When a machine is to handle large, flat cartons such as those required for pizza pies, the carton may be transferred from the magazine in a manner similar to that described above, but expansion is performed by top and bottom vacuum heads. The bottom panel may be held in position by short-stroke vacuum heads moving up and down. When firmly held by vacuum, a second pivoted vacuum head may contact the top panel and expand the carton by moving in an arc corresponding to the height of the carton as shown in Figure 6. In some cases the top vacuum head may move almost 180° in an arc to completely overexpand the carton, which is advantageous when handling wide cartons because it breaks the score lines and produces a neat, square carton. This type of feed requires more time to expand the carton so the operating speed is somewhat lower than that for a smaller size.

Rotary motion carton feeds are used extensively for high speed operation. A typical feed is illustrated in Figure 7. The only reciprocating motions are a short-stroke vacuum head and stripper fingers. The carton is bowed by a vacuum cup and pushed downward a very short distance so the flat carton can be nipped by a set of feed rolls which drive the carton down in time with a lugged pair of wheels. These wheels mesh the

as the lugs level out. The flaps on the loading side of the carton are usually guided outward so they mesh with the mouth of the bucket carrying the product to create a funnel for loading the carton.

Another type of carton feed (Fig. 5) bows the bottom carton from under the stack in the magazine to provide space for a blade to enter between the bottom carton and the one above.

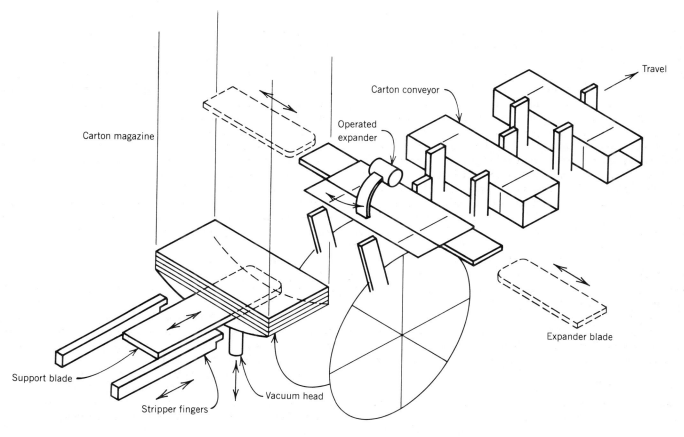

Figure 5. Carton feed with expander blades for tubular cartons.

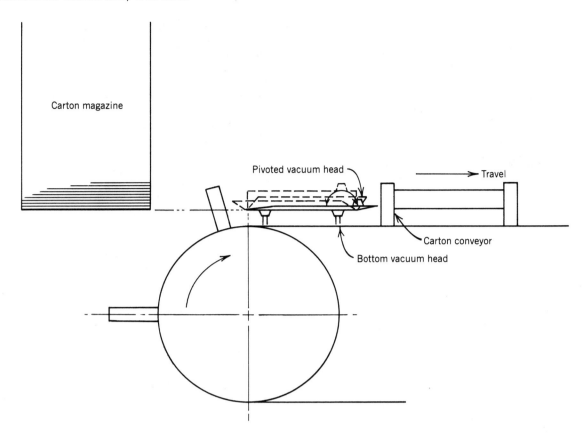

Figure 6. Carton feed for large flat cartons.

carton with a second wheel carrying a number of vacuum heads, one for each pitch of the wheel. While the carton is firmly gripped by the vacuum cups an expander plate, controlled by a fixed cam, gradually expands the carton over several positions. This system allows the carton to be overexpanded to take the "fight" out of the score lines. While still on the vacuum wheel, the leading side flap is operated forward, the trailing side flap plowed backward, and the end panels guided open; as the carton meshes with the carton conveyor the flaps spread around the end of the product bucket, creating a natural funnel for entry of the product into the carton.

Air expansion of cartons is an excellent system, particularly for large cartons, although it also has been used successfully on very small cartons. The unexpanded tubular cartons are usually placed in a horizontal or slightly inclined magazine. In some feeds a metering wheel assisted by air flow from the top brings the flat carton to the horizontal position, where it is picked up by feed chain lugs which transport it to the opening station. The end flaps are guided slightly apart, and the carton is carried past air manifolds which pass a high volume of air at low pressure into the carton. The volume of air expands the carton completely while still in the transport chain conveyor. This style of carton feed has no reciprocating motions, vacuum pump, vacuum lines, or cups to wear or clog.

Side Seam Gluing

Most cartoning machines in operation handle tubular cartons with the side seam or manufacturer's joint preglued on high speed in-line gluers at the carton manufacturer's plant.

The carton must be carefully packed on edge in the corrugated shipper by the manufacturer with each tier in the shipper separated by a piece of carton board. If cartons are packed too tightly, they tend to lose their prebreak or become warped. Preglued cartons tend to lose their prebreak in direct proportion to the time they are stored in the shipping container and develop a "set", making the carton more difficult to open on the cartoning machine.

Many cartoners can be equipped to handle flat, unglued carton blanks with the cartons shipped in stacks on a pallet. The elimination of pregluing by the carton manufacturer and the cost of the corrugated shipper results in direct savings. Indirect savings to the packager result from savings in storage space, usually no additional labor required on the cartoner, and improved cartoner efficiency due to better score breaking (3).

There are two basic systems for on-machine side seam gluing. A separate gluing machine located at the cartoner can feed flat blanks from a magazine, break the scores, apply hot melt adhesive to the side seam, fold and compress the carton joint, and discharge the carton into the magazine of a conventional cartoning machine. This system can handle a substantial size range and has the advantage that it can be added to a cartoning machine at a later date.

The second major system is the wrap around style which feeds a flat carton blank from a magazine, forms it, glues the seam around a three-sided mandrel carrying the product, and transfers the filled and formed carton from the mandrel into a second conveyor where the ends of the carton are closed. Be-

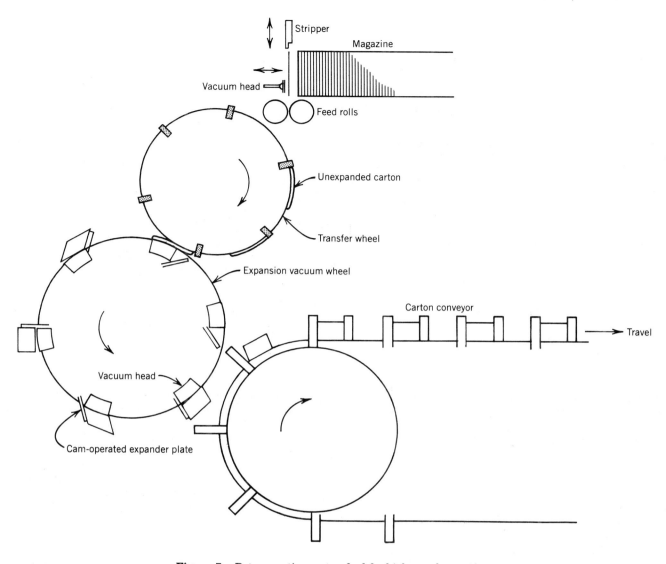

Figure 7. Rotary motion carton feed for high speed operation.

cause the carton is formed and glued around the mandrel, the package is extremely square. The product is already inside the carton when the carton is formed, so there is no transfer of the product into the carton as there is on conventional cartoners. This system requires less clearance between the product and the carton, resulting in the greatest box capacity utilization of any cartoning system. The carton blanks require fewer carton converter operations, larger quantities can be shipped on a pallet requiring less storage space in the packager's plant, a slightly smaller carton can be utilized, and quite often a lighter caliper of board can be handled than on conventional cartoners running tubular cartons. Savings of as much as 25% can be obtained in some cases.

Carton Loading

The conventional method of loading a carton on a fully automatic horizontal cartoner is illustrated in Figure 3. The loader consists of a pair of parallel chains running at the same speed as the carton conveyor. A series of slide bearings is fastened to the chains on the same pitch as the carton. On each

bearing is a slide with a suitable loading head to gradually insert the product into the carton as a roller on the slide rides against a fixed cam bar. The angle of the cam bar usually does not exceed 38–40°, which provides a smooth transfer of the product from the product bucket into the carton. Loading is accomplished over several positions and therefore at a fraction of the machine operating speed.

Product Infeeds

There are a host of different automatic infeeds available, all designed for a specific type of product, such as bottles, jars, tubes, pouches, stacks of tissues, spaghetti products, pencils, and shotgun shells. Because of space limitations only a few of the most common will be described.

Bottle infeeds. Bottles are usually brought into the cartoner on a platform chain, timed by a star wheel or screw, and transferred into a pocketed wheel and then into the product bucket with the bottle still standing on end as shown in Figure 8. The trailing wall of the product bucket pushes the bottle which usually rides up on a wedge-shaped guide tipping the

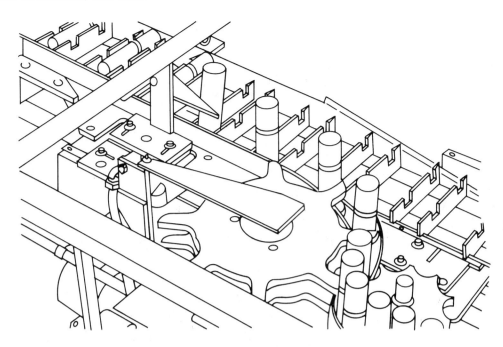

Figure 8. Standard bottle infeed. Courtesy of R. A. Jones Co., Inc.

bottle against guides which gently lay it horizontally in the bucket. Bottle transfers are usually completely adjustable for the entire range of bottles to be run, although it may be necessary to use a different timing wheel or screw for some sizes.

Tube transfers. A tube filling machine may be driven by the cartoner to obtain the necessary timing between the two machines, although it is now common to drive the machines separately and time the tubes on infeed belts by means of clutches driving the belts. Tubes are frequently filled in tandem so the cartoner infeed is also in tandem, accurately delivering each tube into its product bucket despite the small difference in timing between the two tubes. Usually the transfer into the bucket is by means of a shuttle sweeping the tubes off a dead plate.

Bag-in-box. Bags are usually made and filled on a vertical form/fill/seal machine, frequently at the rate of two bags per cycle (see Form/fill/seal, horizontal and vertical). As with tube fillers, the form/fill/seal machines may be driven by the cartoner but normally are synchronized electrically with the cartoner. The bags drop onto an inclined conveyor and may be conditioned while on the conveyor to distribute the product within the bag. The conveyor elevates them to the level of the carton conveyor buckets where they will rest on a dead plate. An overhead conveyor, driven in time with the buckets, sweeps each bag into its bucket. At the loading position of the cartoner the bags may be confined within the carton dimensions on the leading side and the top by an overhead conveyor. To obtain high speed it is not uncommon to have two or even more form/fill/seal machines feeding one cartoner.

Multipackers

Another form of cartoner is the multipacker, a machine which accumulates a number of identical products and loads them into a carton which frequently is used as a display as illustrated in Figure 9. Cartons used for multipacks can be set-up trays, trunk-style top-loaded cartons, or end-loaded tu-

bular cartons as shown in Figure 9. The latter has the advantage of excellent control of the assembled unit packages during loading, it results in a strong package to protect the contents, and by folding the end flaps with a special folding sequence, it results in a trunk-style carton for display purposes. Figure 9 shows the sequence of folding and the application of glue. The inner end panel is folded first, then the side flaps, and finally glue is applied to the outer end panel so the panel adheres to the side flaps. For display purposes the "ears" can be removed by tearing along the perforated score line or can remain to provide a wider display flap. The side seam (manufacturer's joint) of the multipack carton is spot glued by the carton supplier, providing enough strength to hold the carton together during expansion. The joint can be broken on the multipack machine by a slitting mechanism or, if only a few spots of adhesive are required to provide the necessary strength, by the machine which exerts pressure along the top panel. The length of the glue flap is less than the carton depth which provides room for the flap to be depressed enough to break the glue spots.

The multipacker operates at relatively low speeds except for the accumulator which may take the output of more than one machine. The motion of the cartoner may be intermittent or even cyclic, operating only when the assembly of packages has been collated. Because the multipacker takes the output of one or more machines it must be very efficient; when the multipacker is down the entire line must stop. If the multipacker is designed to take the output of more than one machine, it must be designed to operate at reduced speed if one of the machines feeding it goes down. A separate collating mechanism may be built to take the discharge of each machine feeding the multipacker with each collator feeding the complete assembly into a sequenced pocket of the cartoner infeed. If one of the collators goes down, the pocket for the inoperative collator will be empty but the cartoner will merely not feed or skip a carton for that bucket and will continue to run. On the

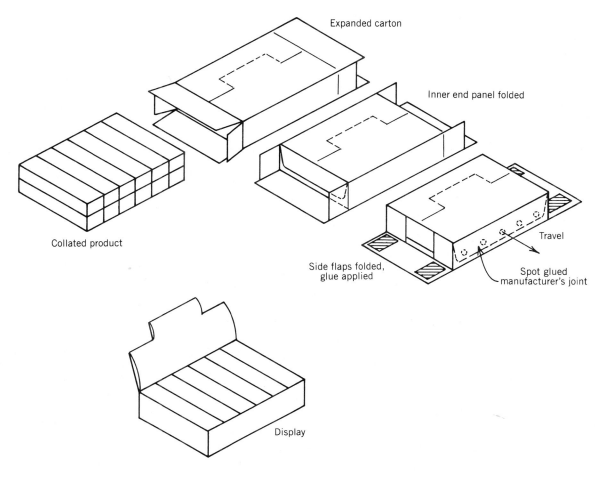

Figure 9. Multipacker.

other hand, one collator may take the output of two or more machines. As an example, let us assume four machines are feeding an accumulator and four products are required to be placed in the multipack carton, one from each machine. If one of the machines goes down, only three products would be available for the multipacker. To overcome this it is necessary to combine the total output into a single line, but this could result in very high line speed on the incoming line to the collator. For this reason a multipacker may be equipped with more than one accumulator.

Leaflet Feeds

One of the most important attachments on a cartoning machine is a mechanism to place a leaflet or coupon into the carton with the product. The leaflets may be for advertising purposes or may be essential to provide the technical information regarding the product (eg, pharmaceuticals). Feeds are available to handle prefolded leaflets which are frequently used with semiautomatic vertical cartoners (see Fig. 2), die cut sheets which the mechanism folds prior to insertion, or leaflets from roll stock, which are cut to length, folded, and then inserted into the carton with the product. Leaflets from roll stock have some decided advantages over die cut sheets: higher speeds can be obtained since the web is in continuous motion, and since the leaflets are on rolls, the chance of an incorrect leaflet on a given run is eliminated.

A typical mechanism to feed and fold die cut sheets is illustrated in Figure 10. The sheets are stacked in a magazine with the weight of the stack usually supported by freely turning wheels and the rear of the stack resting on a pair of very sharp needles which penetrate several leaflets. A set of start rolls with rubber segments pulls the bottom leaflet from under the stack, the sheet above being held back by the needles. The grain of the paper is parallel to the direction of feed so the tears from the needles are very small. Generally the start rolls are driven with a variable driving mechanism which enables the leaflet to be picked up at a slower speed. As the leading edge of the leaflet reaches the outside of the magazine, it is usually nipped by a set of feed rolls which are running faster than the start rolls, pulling it ahead. The folding of the leaflet is accomplished by operated blades which usually move just enough to drive the folded leaflet into another set of feed rolls.

The folding of the leaflet may be done by buckle folding, eliminating the need for folder blades. The leaflet enters a slot until it contacts an adjustable stop. As the feed rolls are still driving, the leaflet is forced ahead, which causes the paper to buckle and enter another slot leading to a set of feed rolls at the desired location. Multiple folds can be made easily with this type of unit. If a certain fold is not required, the stop in that slot is located forward to block off the slot to prevent entry of the leaflet.

Roll-fed leaflet mechanisms are equipped with either a

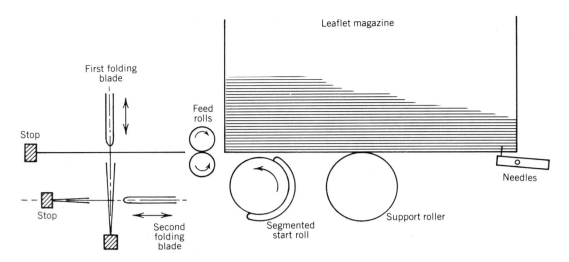

Figure 10. A typical mechanism to feed and fold die cut sheets.

blade folding unit or a buckle folder. Cross folds can be made if required, usually by folding the web parallel to the direction of flow by passing it over a plow before cutting the sheet to length. Cross folds can also be made on a blade folder mechanism.

After the leaflet is folded it must be transferred into the carton with the product. The most common system used with the horizontal cartoner is to insert the folded leaflet into the bucket, carrying the product either in a slot underneath the bucket, clamped to the bottom of the bucket with a chain synchronized with the bucket, or in some cases, held in position by a clamp which is an integral part of the bucket. While held under the bucket the projecting end of the leaflet is folded up and back over the product and held in this position by guides until the loading conveyor transfers the assembly of product and leaflet into the carton. In most cases the cap of the bottle leads as it enters so the folded leaflet is readily visible when the package is opened.

When handling leaflets, particularly with pharmaceutical products, it is essential that the leaflet is carried into the carton with the product. As the loading tip contacts the base of the product, the clamp or chain holding the leaflet to the product bucket releases and the product pushes the leaflet into the carton. However, if the sides of the leaflet are of unequal length the leaflet could slip alongside the product and not enter the carton properly. To overcome this, some horizontal cartoners can be equipped with a clamp conveyor which is similar to the loading conveyor described earlier, but mounted on the rear of the machine. The slide enters the open carton from the rear and passes completely through the carton until the spring-loaded tip contacts the leaflet and clamps it against the leading surface of the product as shown in Figure 11. The slide is withdrawn at the same rate the loading slide pushes the product into the carton, keeping the leaflet securely clamped to the product. Because the clamp conveyor slide must pass through the carton, it enters the carton several positions ahead of the loading position. This type of mechanism is usually designed for the full range of carton length, providing the same length of stroke whether the carton being handled is short or long. A similar mechanism is sometimes used for certain products that may tend to topple when loaded, such as a group of wrapped chocolate bars stacked on edge.

Carton Closing

Tucking. The tucking operation is accomplished over a number of positions on the machine: side (dust flap) folding, prebreak of the tuck flap score line, possible slitting of the dust flaps to assure clearance if lock slits are used, alignment with the carton, first and usually second tucking, followed by a final contact to seat the lock slits. Usually these operations are done with the tucking parts on a bar carried by parallel cranks with each tucking position on a separate pitch of the machine. Tucking by means of belts is also done successfully, eliminating the need for cranks, although the trailing side flaps must still be folded over by a separate mechanism prior to tucking. The final tuck for locking the lock slit is accomplished by a separate rotary mechanism.

Gluing. A glue end carton may be closed by single or double gluing as shown in Figures 12 and 13. When single gluing, the two outer end flaps are glued together; when double gluing, the inner end flap is glued to the side flaps and then the outer end flap is glued to the inner. Single gluing results in a slight crack between the glued end flaps and the folded side flap which may be acceptable for a bottle or other solid object. Cartons for food products or facial tissue are usually double glued to protect the contents.

There are two general types of adhesive used, cold glue or hot melt. Cold glue may be either dextrin or resin base, both obtaining their set by evaporation of the water in the adhesive. In general, dextrin glues require approximately 30 seconds to set and resin 20 seconds, but in either case a compression conveyor is required to keep the carton under pressure until the glue has set. High packaging speeds result in long compression conveyors, particularly when running thick cartons such as facial tissue boxes.

The use of hot melt adhesives has greatly reduced floorspace requirements because the glue sets by temperature reduction instead of water evaporation. A compression system is still required, but usually the carton can remain in the carton conveyor since only a very few seconds are required for the glue to set, just enough time to square up the flaps.

Hot melt adhesive is applied by guns or intaglio rollers (see Adhesive applicators). When using guns there are three definite heated components, that is, the glue reservoir, the hose,

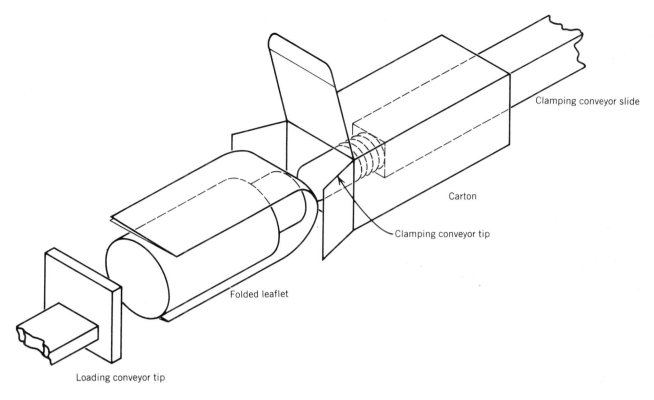

Figure 11. Leaflet feed with clamp conveyor.

and the gun itself, each controlled by individual thermostats to maintain precise temperatures. The glue gun can apply a strip of glue, an interrupted strip, or dots. For single gluing (ie, gluing the outer end flaps together), a single strip of glue is usually applied, but if a thick carton is being run a multiple orifice nozzle may be used to apply two or even more lines simultaneously. Obviously a separate gun is required for each end of the carton although only one reservoir is needed. When double gluing a carton, two guns are required at each end of the carton, one to fire an interrupted line to apply glue to the inside of the inner end panel to glue to the side flaps and the other to apply a continuous line to glue the outer end panels to the inner. Intaglio rollers can apply the dots of adhesive in the

required pattern for double gluing. In this case the glue tank and the intaglio roller are heated.

Detectors

Cartoning machines, particularly the fully automatic horizontal cartoner, are usually equipped with a variety of detection systems. Some of the many detectors available are listed below:

Skip carton.   The carton feed does not feed the carton for an empty product bucket; instead, it "skips" a carton.

No carton.   If a carton fails to feed or expand for a properly filled article bucket the machine will stop. In some cases, usually when the carton conveyor consists of buckets instead of

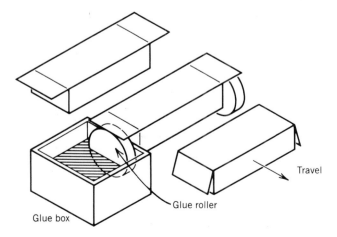

Figure 12.   Single gluing; that is, the two outer flaps are glued together.

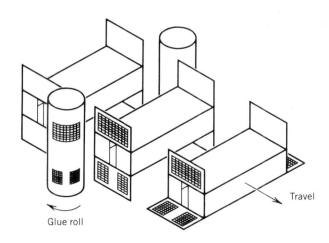

Figure 13.   Double gluing; that is, the inner second flap is glued to the side flaps, and the outer end flap is glued to the inner flap.

lugs, the product can be transferred into the carton bucket and discharged at the end of the carton conveyor, which enables the cartoner to continue to run.

Skip load.   If a carton is not present in the carton conveyor the loading slide is not cammed forward and the contents of the product bucket remain in the bucket and are usually discharged from the bucket as it goes around the loader sprocket. The machine can continue to run.

Missing leaflet.   If a leaflet is not placed properly in the product bucket the machine will stop. In some cases the presence of the leaflet can be detected in the carton. This is a more advantageous system because a leaflet may be lost during the transfer into the carton even though it was in the product bucket. However, it is much more difficult to sense the leaflet after it has been placed in the carton.

Low product level.   If the incoming products reach a low point the cartoner can be arranged to stop. When handling glue end cartons, particularly with hot melt adhesives, the machine will come to a timed stop, a position where the carton just glued is held under compression.

Overloads

To protect the machine most cartoners are equipped with several overload devices, frequently merely a detent that contacts a microswitch if an overload is sensed. Most bottle feeds have an overload device on the timing wheel and sometimes also on the transfer wheel. Another common overload is a detent in the side-flap-folding drive mechanism. The operated folder or star wheel may contact the product if it did not enter the carton completely or, as sometimes occurs when handling a leaflet with many folds, is pushed out of the carton by the springiness of the leaflet. The overload will stop the machine before the product or machine parts are damaged. Some cartoners have an overload mechanism on the main drive to protect the machine in the event of an overload in any portion of the machine not already protected. This type of safety is an excellent feature although it must be set in such a way that it does not throw out because of the starting torque of the machine yet will sense an overload during normal running.

Many cartoners equipped with detectors and overloads have indicator lights on the control panel to show the operator which detector stopped the machine. Usually the light operates through a holding relay to keep it on as the machine may coast past the microswitch as it comes to a stop.

Microprocessors

Centralized microprocessors are frequently used to control and monitor the operation of cartoning machines. Encoders, working with programmable limit switches, have greatly simplified operating, set-up, and maintenance procedures. They have replaced electromechanical sequencers and shaft driven cams which have been common on cartoners. Programmable devices can be adjusted remotely, can be set up and changed very quickly, provide better resolution, and better repeat accuracy, and generally can operate at higher speeds.

BIBLIOGRAPHY

1. S. Knapp, *The Packaging Encyclopedia*, Cahners Publishing Co., Boston, Mass., 1984, pp. 218–222.
2. C. Glenn Davis, *Packaging Machinery Operations—Cartoning*, Packaging Machinery Manufacturers Institute, Washington, D.C., 1980, pp. 2/2–2/7.
3. P. A. Toensmeier, *Impact and Future: The Packager Side Seam Gluing of Folding Paper Cartons*, A Packaging Technology and Marketing Study/Report, Patrick A. Toensmeier, Hamden, Conn., 1979, pp. 11–15.

W. F. DENT
Ansell, Incorporated

CARTONING MACHINERY, TOP-LOAD

Top-load cartoning employs flat paperboard blanks die-cut by the carton manufacturer (see Fig. 1) to produce a specific size and shape of finished package. Since no pregluing is required and flat blanks can be shipped in stacks on a pallet, top-load cartons are economical, and because they allow product placement through the largest opening of the carton, they can greatly simplify the loading operation (see Cartoning machinery, end-load; Cartons, folding).

Carton Forming

The heart of any top-load packaging operation is the carton-forming machine (see Fig. 2). Generally, these incorporate a slightly inclined horizontal magazine from which individual die-cut blanks are fed using vacuum cups. Carton blanks are retained by tabs or small projections that extend from the sides at the front of the magazine, so that movement of a blank, using a vacuum cup, releases it from the magazine. The vacuum cups, mounted on a reciprocating feed bar, move individual carton blanks in an arc from the magazine and deposit them in a registered position on top of a forming cavity. The carton blank is rotated about 90° through this feed cycle with the vacuum cups in contact with its inside surface. As the feed bar moves upward to feed the next blank, a plunger, or mandrel, moves downward and forces the blank through the forming cavity (see Fig. 3). The plunger is designed so that the carton body conforms to its shape, being guided and manipulated by a series of metal fingers and plows installed within the forming cavity. The formed carton is stripped or removed from the plunger as it begins moving upward using mechanical traps.

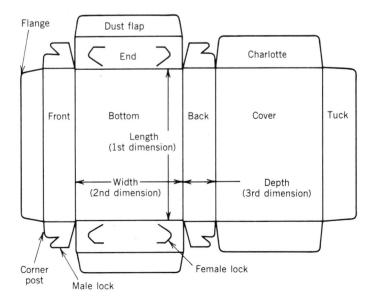

Figure 1.   Carton diagram.

Figure 3. Lock-style carton-forming head.

Figure 2. Top-load carton-forming machines; (**a**) back, and (**b**) front.

Special tooling is designed and manufactured for each carton size and style. This tooling, commonly referred to as a forming head, consists of various hopper components, the forming cavity, and a plunger or mandrel. By interchanging this tooling, any given forming machine can erect different carton styles, shapes, and sizes within a specified size range. For standard designs such as simple trays or hinge-cover cartons, this tooling can generally be changed in 10–15 min. The carton body can be formed utilizing locks, adhesive, or heat-sealing.

Lock forming. Dozens of different lock designs are available for forming trays and cartons to meet various packaging requirements. Most often, the lock design consists of a vertical and/or horizontal slit in the upright panels or walls of the carton body through which a specially shaped corner post or tab is inserted during the forming operation (see Fig. 4). The opening of the die-cut slits in the vertical walls and folding and insertion of the locking tabs are accomplished using specially designed and fabricated metal fingers or guides which are an integral part of the forming cavity. Normally, a mechanical actuation is also incorporated in the plunger to pull in the locking tab and ensure its positive engagement.

Glue forming. Glue forming of flat die-cut paperboard blanks can be accomplished using either hot-melt or cold-vinyl adhesives (see Adhesives). Hot-melt adhesive is generally applied in one of two ways. The simplest method employs open reservoirs containing hot-melt adhesive which are mounted in the machine directly below the forming cavity. After a blank is fed, applicator blades mounted on a shaft running across the top of each reservoir are mechanically actuated upward with the blades rising from within the adhesive. These apply a series of dots or lines of adhesive to the underside of a flap located in each corner of the carton blank. As the carton is plunged through the forming cavity, these flaps are folded inside the vertical walls of the carton body and spring-loaded rollers provide compression using the plunger as backup. This compresses the adhesive and dissipates heat to allow quick setup for reasonably high forming speeds.

The other method commonly used involves feeding individ-

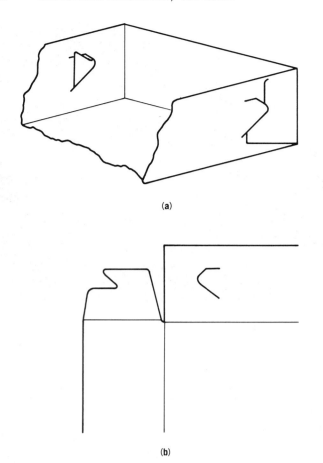

Figure 4. Common stripper-lock design (a) assembled; (b) unfolded.

ual carton blanks onto a short section of horizontal conveyor, where hot-melt adhesive is applied using guns or nozzles which spray the adhesive on the inside of the vertical body panels. After adhesive application, the blanks are shuttled into position on top of a forming cavity and then plunged vertically through the cavity and compressed in the manner described above.

Cold-vinyl carton forming, which was virtually eliminated with the advent of hot-melt adhesives, is used only infrequently today for packaging applications; however, there is renewed interest in connection with "ovenable" cartons. This method is also used to produce paperboard scoops, boats, and clamshells for the fast-food and food-service industries. A carton blank is fed from a hopper and deposited on top of a forming cavity. Adhesive-application nozzles mounted on the forming cavity are then mechanically actuated and deposit a small dot or dots of cold-vinyl adhesive as required inside the carton blank. The adhesive application is effected using specially designed nozzles with adhesive supplied under pressure from a single remote tank. A plunger forces the blank through a forming cavity, but in this case it deposits the formed trays into a nest or stack in order to keep them under pressure until the adhesive has set. This method allows forming at speeds up to about 80 strokes/min in order to satisfy the substantial production volumes required by the fast-food industry.

Heat-seal forming. Heat-seal forming utilizes special coatings on the paperboard as a bonding medium in the carton-

forming operation. Overall board coatings that can be heat-seal-formed include single-side polyethylene, double-side polyethylene, polyester, and polypropylene. Special pattern-applied hot-melt adhesive coatings placed only in the area to be bonded are also used. A special system designed to force air over an electrically heated quartz element is incorporated in the forming cavity. Air at temperatures ranging from 400 to 800°F (200–425°C), depending upon the carton coating, is forced through specially designed nozzles and directed over specific areas of the carton blank, where it melts the coating on the boardstock. The carton blank is then plunged through the forming cavity, and the board coating serves to bond the appropriate carton flaps. Very high speeds, ie, up to a max of about 90 strokes/min, can be achieved, depending upon variables such as coating thickness, carton size, style, etc. For heat-seal-forming operations, some degree of water cooling is needed to prevent heat buildup in various parts of the forming head. The extent of water cooling generally depends upon the bonding medium used, but becomes most extensive when double-polyethylene coatings are involved. Heat-seal forming is employed primarily in the frozen-food industry, where polyethylene coatings are required for moisture-barrier protection or when some degree of leak resistance is required for very wet products.

Top-load-forming Capabilities

Various carton-forming machine models are produced, each designed for a given size range and maximum speed rating. The same basic model can usually be equipped for lock forming or modified with special attachments for glue or heat-seal forming. Machines can be equipped for single-head forming, using a single-forming cavity and plunger, or double-head forming. The latter uses a twin-cavity design to feed and form two carton blanks simultaneously. Under certain conditions, it can form as many as four carton blanks at once. Forming speeds generally range from 20 to 120 strokes/min. Using multiple forming heads, a single machine can produce more than 300 cartons/min. A wide range of carton sizes can be accommodated in styles ranging from a simple rectangular corner-lock tray to a myriad of shapes including triangles and complex shadow-box structures. The key to top-load carton forming lies in the design of the forming head, which can be a relatively simple device or a complicated unit requiring complex cams for flap folding and manipulation of the paperboard panels.

Carton Conveying

After forming, the top-load carton is typically carried on a conveyor for loading either manually or automatically. For slow-to-moderate-speed hand packing, simple flat-belt or polyacetal tabletop chain conveyors are frequently employed and offer the most economical approach. For this type of operation, one end of the conveyor is generally placed below the forming cavity with no electrical or mechanical connection between the conveyor and the forming machine. After forming, trays or cartons drop onto this conveyor and are carried downstream for product loading. The alternative to this method is a lugged or flighted conveyor generally attached to and mechanically driven by the forming machine (see Fig. 5). This offers the advantage of pacing the operators and is required to achieve adequate carton control in any high speed operation. The flighted-conveyor system also permits control

Figure 5. Flighted carton-packing conveyor.

of the carton cover during the packing operation either in a vertical position or folded back almost 180°. Packing conveyors should be designed so that the bottom of the carton is approximately 30–34 in. (76–86 cm) from the floor; this helps optimize the efficiency of operators who are placing product by hand.

Product Loading

Manual loading. For hand-pack operations, product can be presented to the operator in many different ways. These range from tote bins to product conveyors running up to the carton conveyor. The most efficient method involves bringing the product in on a flat belt or tabletop chain conveyor running parallel and adjacent to carton flow, with the bottom of the product elevated just slightly above the top of the carton. This allows for simple sweep loading of product into the largest opening of the carton, as well as the most efficient and reliable hand-packing operation.

Automatic loading. Automatic loading of products into top-load cartons can be accomplished using a number of different standard and highly customized systems. Free-flowing products, such as individually quick-frozen vegetables, are often filled automatically using a volumetric system integrated mechanically with a carton-conveying system. Where net-weigh filling is desirable, a variety of different systems can be employed. These are normally interlocked electrically with the carton-forming and conveying system in order to sense the presence of a formed carton and to signal the scales to dump. For net-weighing or automatic loading of some products into top-load cartons, the carton conveyor must sometimes operate on an intermittent-motion basis. This allows the carton to be stopped, or dwell momentarily, below a filling device to provide sufficient time to completely load the product charge.

Nonfree-flowing products, ranging from typewriter ribbons to bare frozen hamburger patties and including many other products, such as wrapped candy, spark plugs, overwrapped baked goods, frozen fish sticks, etc, have been automatically loaded into top-load cartons. The loading technique used de-

pends on the product and can vary from a simple mechanical shuttle to advanced units which automatically align, accumulate, group, and transfer product into the carton. The loading or transfer of such products is most frequently accomplished using a vacuum pickup system employing high volume, low pressure air flow.

Carton Closing

The final important consideration in any top-load cartoning application employing a hinge-cover design is carton closure. For slow-to-moderately high speed operations, the closing machine can be an independent unit with its own drive motor. Equipped with a special infeed assembly, these accept cartons at random from an upstream packing conveyor and automatically time them into the flights of the closing machine. For high speed operation in excess of 200 cartons/min, it is desirable to eliminate this infeed section and drive the entire packaging line from the carton-forming machine. This method requires the use of flighted packing conveyor and ensures positive carton control from the forming machine discharge through final carton closure.

Dust-flap style closure. For dust-flap style cartons which require closure of only a single tuck panel, a static plow folds down the leading dust flap while a rotating paddle assembly operating in time with the carton-conveying chain serves to kick in the trailing dust flap. After this has taken place, the cover is plowed down using guide rods and rollers, the front tuck score is prebroken, and final closure occurs. Dust-flap style cartons can be closed either by locking the front tuck using a series of static plows and guides or by applying hot-melt adhesive to the inner surface of the tuck which is then folded against the front panel of the carton, proceeds through compression rollers, and is bonded.

Tri-seal style closure. The hinge-cover carton design most frequently used is the tri-seal style, also called the three-flap or charlotte style. For this carton style, there are three primary closer designs available. For slow-to-moderate-speed operations, where space is a limiting factor, a very compact machine design is available which indexes the carton and then transfers it at right angles into a single hot-melt-adhesive application station where adhesive is applied to all three carton flaps simultaneously. This type of closer generally uses an open-reservoir-style hot-melt-adhesive application system with blades rising from within the adhesive to apply solid lines or dots of adhesive as required. It can also be equipped with gun- or nozzle-style applicators which spray adhesive onto the carton flaps. After adhesive application, the entire carton is elevated vertically through a compression unit and discharged at the top of the machine.

For most applications, tri-seal cartons are closed using either a right-angle or straight-line configuration. In the right-angle operation, after indexing into the flights of the closer, the front tuck is closed, the carton proceeds through a 90° direction change and the charlottes, ie, end flaps, are closed. With the straight-line closer design (see Fig. 6), after closure of the front tuck, the carton body is turned 90°; as it moves parallel to the direction it enters the closer and proceeds broadside leading through the charlotte-closing mechanism. To achieve proper cover-to-carton body registration, it is best to fold down and trap the trailing charlotte before front-tuck closure. This is generally executed with an overhead rotating tucker paddle positioned adjacent to a transfer area where a

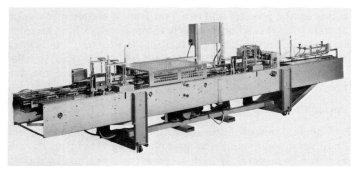

Figure 6. Independent straight line tri-seal carton closer.

new set of lugs then traps the trailing charlotte against the rear end of the carton and holds it in this position through the front-tuck closure section.

Lock closure. Closure of tri-seal cartons can be accomplished using locks, adhesive, or heat-sealing. Lock closure is effected by utilizing specially designed plows, guides, and tucking fingers and is employed most often in the frozen-vegetable industry for plain-paperboard shells which are later overwrapped with printed paper.

Adhesive closure. Adhesive closure can be accomplished with either hot-melt or cold-vinyl adhesive. Hot-melt is most frequently employed, with the adhesive commonly applied using an open-pot system. Depending upon the pattern required, a thin wheel or wheels is used to apply adhesive on the carton flaps. Where desirable, special patterns can be applied utilizing enclosed wheel-applicator systems or nozzle-type systems. Cold-vinyl adhesive was virtually displaced by hot melts, but its use has recently been prompted again by the requirement for ovenable packaging. Hot melts are not suitable for this application because they tend to soften and release during cooking. Systems have been developed for polyester- and polypropylene-coated boardstocks that utilize an atomized-spray application of cold-vinyl adhesive followed by the application of electrically generated hot air through a nozzle. The hot air accelerates the water-evaporation process and allows use of compression sections only slightly longer than those employed with hot-melt systems at reasonably high speeds. This system has been used with ovenable coated paperboards in applications where the product must be reconstituted in either microwave or conventional ovens.

Heat-seal closure. Heat-seal closure can be used for cartons made from boardstock coated inside and outside with polyethylene or heat-sealable waxes (see Waxes). Here again, the board coating is activated or melted using electrically generated hot air at very high temperatures and then proceeds through compression to bond the flaps as the coating solidifies. For wax-coated cartons, the compression section must consist of refrigerated bars to prevent smearing of the wax and allow quick setup of the bond within a relatively short compression section.

Top-load cartoning systems can be equipped with a wide variety of options, including many types of coding devices, leaflet feeders, labelers, and sensors to facilitate the effective control of the entire packaging system.

A. R. WORK
Kliklok Corporation

CARTONS, FOLDING

Folding cartons are containers made from sheets of paperboard (see Paperboard) which have been cut and scored for bending into desired shapes. In common usage, the term is not used in reference to shipping containers, either from corrugated or solid stock (see Boxes, corrugated; Boxes, solid fibre), or setup boxes (see Boxes, rigid paperboard). Although there is evidence that paper was used by Egyptian merchants to wrap goods for their customers as early as 1035 A.D., the folding-carton industry did not begin until Colonel Andrew Dennison began producing commercial folding cartons in the 1840s. By the mid-1890s, automatic machines were in widespread use for the production of cartons (1). From these beginnings in the 19th century, the folding-carton industry has grown into an economic segment with 1984 sales of \$3.2 billion (10^9) that utilized 2.7 million (10^6) short tons (2.45 million (10^6) metric tons) of paperboard, excluding milk packaging (2).

Merchandise displays at supermarkets or drug, hardware, automotive, and department stores demonstrate the extent to which folding cartons are used today. Cereal, crackers, diapers, facial tissue, detergent, dry mixes, frozen food, ice cream, butter, bacon, bar soap, candy, cosmetics, sandpaper, spark plugs, toys, cigarettes, canned beverages, carryout foods, and pharmaceuticals represent the broad range of products for which folding cartons are commonly utilized. The use of folding cartons is widespread because of the ability of this packaging format to satisfy the functions of protection, utility, and motivation. Protection from crushing, bending, contamination, sifting, grease, moisture, and tampering can all be built into folding cartons. For the packer, utility is achieved through high speed automatic packing (see Cartoning machinery). For the end user or consumer, utility is provided by opening, reclosing, and dispensing features. In some cases, the carton even serves as the cooking utensil. High quality graphic reproduction, excellent billboard presentation of the graphics design, and the ability to take on unique and varied shapes provide the carton user with the means to motivate the consumer to purchase products packaged in folding cartons.

Paperboard Selection

Successfully meeting the needs of a folding-carton user begins with choosing the paperboard best suited for the job. In general, this means selecting the grade with the lowest cost per unit area that is capable of satisfying the performance requirements of the specific application. Economics and performance dictate careful selection of paperboard grades for each use.

Selection criteria. A variety of criteria are commonly used in the selection of paperboard grades. The Technical Association of the Pulp and Paper Industry (TAPPI) has published standardized test methods for many of these criteria (3) (see Testing, packaging materials). TAPPI Standard Methods are widely used and accepted by the industry. The most important and widely used criteria are shown below.

FDA/USDA compliance. This is a nondiscretionary criterion for food products and is dependent upon the type of food and the type of contact anticipated between the food and the paperboard or coatings on the paperboard.

Color. Color is typically chosen for marketing reasons. The side of the paperboard that becomes the outside of the carton is generally white, but the degree of whiteness varies among grades. Board color on the inside of cartons varies from white to gray or brown. Depending on the materials-selection and processing strategies of suppliers, board shade can be blue-white or cream-white. These shades are noticeably different and can limit substitution of grades.

Strength. It is possible to establish minimum strength levels for each carton application that allow the package to satisfactorily withstand the rigors of packaging machinery, shipping, distribution, and use by the consumer. Strength properties commonly used to predict suitability of board for a given use include stiffness, tear strength, compressive strength, plybond strength, burst strength, tensile strength, elongation, and tensile energy absorption. Strength criteria normally define the basis weight and thickness of paperboard that is used to produce a carton.

Printing characteristics. Following the selection of a specific graphic design and printing method for the carton, a paperboard is selected based on these criteria: smoothness; coating strength; ink and varnish gloss; mottle resistance; and ink receptivity. Not all criteria are important for every printing technique.

Barrier. The most common barrier requirements are for cartons to provide protection against moisture and grease. The choice of a barrier material and application method influences board choice. For example, if wax is to be applied to the carton, a board with a treatment that holds the wax on the board surface can have economic and processing advantages over an untreated board (see Waxes). Materials and application methods are described below.

Paperboard types. In the United States, the three most widely used types of paperboard are identified as follows:

Coated solid bleached sulfate (SBS). 100% virgin, bleached, chemical furnish, clay coated for printability.

Coated solid unbleached sulfate (SUS). 100% virgin, unbleached, chemical furnish, clay coated for printability.

Coated recycled. Multiple layers of recycled fibers from a variety of sources, clay coated for printability.

The estimates of U.S. industry indicate that coated SBS and coated recycled boards are the most popular, with 35% and 39% market shares, respectively. Only in the last five years has coated recycled eclipsed coated SBS. Coated SUS has a 13% share with the remainder split among other recycled grades (2). Boards incorporating mechanical pulps have been used elsewhere in the world for folding cartons, but have had no significant penetration in markets in the United States.

Overall treatments or coatings are applied to webs of paperboard to provide specific functions. Clay-based coatings to provide high quality printing surfaces are the most common treatment applied on the paperboard machine. Grease-resistant fluorochemicals are applied on board machines as well, either as furnish additives, surface treatments, additives to clay coatings, or in combination. Mold-inhibiting chemicals are also applied to boards designed for bar-soap packaging, to prevent moisture in the product from initiating mold growth.

Surface treatments applied on other-than-board production equipment are discussed below under Carton Manufacturing Processes.

Paperboard is the overwhelming choice as the substrate for folding cartons. In recent years, however, a very small segment has developed that utilizes plastic sheet as a substrate. These cartons are normally produced from clear, impact grades of PVC sheet (see Film, rigid PVC) using specialized heated-scoring techniques to achieve acceptable folding characteristics. Unique product visibility is the primary reason for the use of this more costly substrate for specialty folding-carton applications, such as cosmetics and soft goods (see Boxes, rigid plastic).

Carton Styles

As the demand for cartons grew, so did demands for additional features. These demands catalyzed the development of

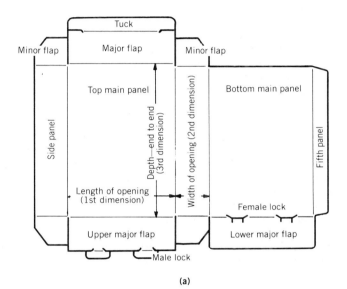

(a)

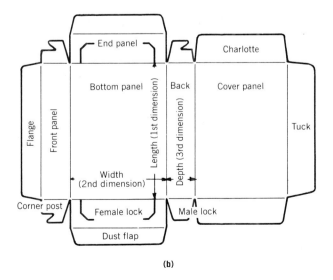

(b)

Figure 1. Carton terminology. (**a**) Tube-style carton. (**b**) Tray-style carton.

new and unique ways to cut and fold sheets of paperboard to produce cartons. The records of the U.S. Patent and Trademark Office contain many thousands of patents granted to protect folding-carton structures.

Three broad classifications are commonly used to categorize folding-carton styles: tube (end load); tray (top load); and special construction. Figure 1 describes accepted terminology for the various parts of tube (**a**) and tray (**b**) cartons (4), as well as the order in which dimensions are listed in carton specifications. Compliance with this convention prevents confusion.

Tube style. Tube (shell) constructions are the most common style in use today. Figure 2 shows a typical sealed-end carton in various stages of production and filling. These cartons are characterized by a fifth panel glue seam in the depth direction, yielding a side-seamed shell that folds flat for transportation. The cross section of the carton opening is normally rectangular, and product may be loaded either horizontally or vertically (see Cartoning machinery). Tube-style cartons are well suited for very high speed automated filling lines, but they are also used for manual filling applications.

Figure 3 shows the treatment of end flaps on a tuck-end carton. Other end treatments are in common use, including zippers and similar opening features for sealed-end cartons. Internal shelves and panels are often included to position the product. This is particularly done when the product is irregular in shape or the carton is much larger than the product for improved graphic presentation or shoplifting deterence. When heavy granular products are packaged, bulging of main panels can be a problem. The use of bridges connecting the two main panels reduces the board strength required to resist bulging. These bridges can be made from paperboard or paper, attached during the gluing operation.

Tube-style cartons are commonly used for granular or pourable solid products such as detergent, cereal, and dry mixes. Dispensing is often accomplished for these products by paperboard or metal pour spouts. Full-end opening is preferred when inner bags are employed (see Bag-in-box, dry product). Large products packed one to a carton, such as pizzas, TV

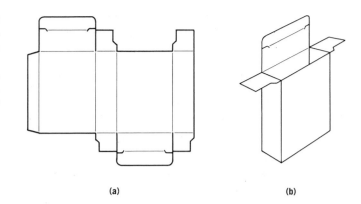

Figure 3.   Tuck-end carton: **a,** Blank; **b,** Setup carton.

dinners, pot pies, cosmetics, and pharmaceuticals are packed in end-opening tube-style cartons. End-loaded cartons are also designed for opening and product removal through the main panel; a cream cheese carton is a good example of this approach. Gabletop milk cartons also fall into the tube or shell category (see Cartons, gabletop). They incorporate liquid-tight sealing and a reclosable pour spout. From ice cream to spark plugs, tube-style cartons satisfy many diverse packaging needs.

Tray style. Tray or top-load cartons are characterized by a solid bottom panel opposite the product-loading opening. As shown in Figure 1(**b**), panels are connected to each edge of that bottom panel. Tray cartons are especially useful for manual or automatic loading of multiple products. Figure 4 contains schematic drawings of a tray carton blank, the carton set up for loading, and the completely closed and sealed carton. In this example, the front and back panels are connected to the end panels using mechanical locks. Panels are also commonly connected using adhesives or heat sealing (see Adhesives; Sealing, heat).

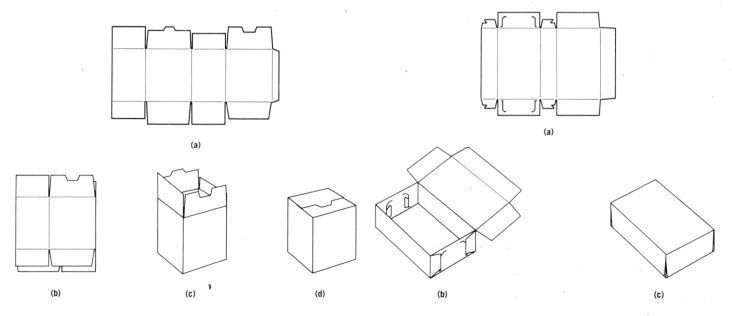

Figure 2.   Sealed-end carton: **a,** Blank; **b,** Side-seamed shell; **c,** Carton erected for loading; **d,** Filled and sealed carton.

Figure 4.   Locked-corner hinge-cover carton: **a,** Blank; **b,** Carton erected for loading; **c,** Filled and sealed carton.

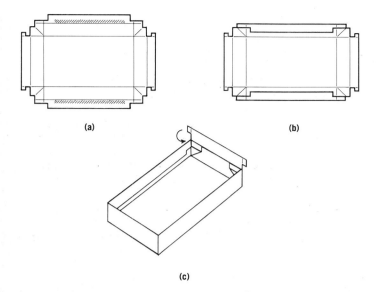

Figure 5. Web-corner tray: **a,** Blank; **b,** Sidewalls glued; **c,** Final panel folding.

Where additional resistance to leakage is desired, web corners are employed. Figure 5 shows a web-corner tray with folded double sidewalls that provide finished sidewall edges. A similar, slightly larger tray could be used to cover the tray following product loading, yielding an extremely crush-resistant package. Figure 6 shows a six-corner carton that can easily be set up by hand. Diagonal scores permit the tray to be glued and delivered in a collapsed form. Web corners could be incorporated with additional modifications.

Tray cartons that require no gluing at the point of use (eg, the six-corner carton of Fig. 6) are used extensively for manual loading. The cake, pie, or pastry carton employed by the local bakery is the best example of the utility of these designs. Garments and other dry goods are often packed, especially when purchased as gifts, using two-piece cartons that comprise two collapsible glued trays. Tray cartons are also widely used for products that can be automatically packaged. Cookies, doughnuts, and fish sticks are examples of products commonly packaged in tray-style cartons with attached covers. Cartons similar to that shown in Figure 4 are used for products of this type. Most display cartons for smaller candy packages also fall into the tray category.

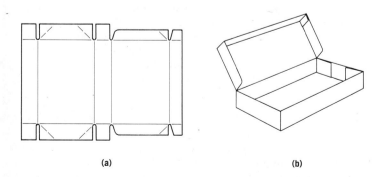

Figure 6. Six-corner carton: **a,** Blank; **b,** Carton erected for loading.

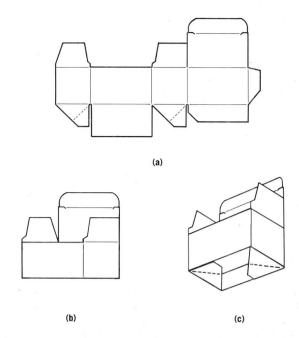

Figure 7. Automatic-bottom carton; **a,** Blank; **b,** Carton glued for shipping; **c,** Bottom view of erected carton.

Special construction. Special construction is a classification employed for cartons that do not fit tray or tube descriptions, or that represent sufficient departures from normal tray or tube practice. A blister package that employs a combination of heat-seal-coated paperboard and a clear thermoformed plastic blister is a good example of a special construction (see Carded packaging). The automatic-bottom carton shown in Figure 7 combines elements of a side-seamed tube carton with those of a top-load carton and requires no manipulation to

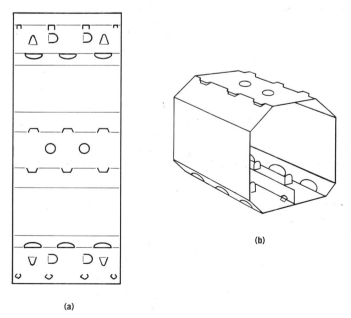

Figure 8. Bottle-carrier carton: **a,** Blank; **b,** Erected carton (bottles omitted for clarity).

form the carton bottom. When the collapsed shell is opened, the bottom panels lock into place. This carton is used extensively for fast-food carryout packaging. It is also popular for hardware items.

The bottle-carrier carton in Figure 8 is an example of the many wraparound carrier cartons used for multipacks of bottles, cans, or plastic tubs (see Carriers, beverage). These cartons are either locked or glued after being wrapped around the primary packages.

The tube, tray, and special carton styles depicted and described here are broadly representative of the great variety in shapes and sizes produced by the folding-carton industry. Customization of design for function or appearance is a significant advantage of folding cartons.

Carton Manufacturing Processes

After a paperboard grade has been selected for a specific carton style and use, a variety of manufacturing options are available for converting that board into cartons. Although it is a highly unusual carton that requires each one of the steps or stages described below, all are commonly employed to produce folding cartons in today's market.

Extrusion coating. This technique involves the coating of one or both sides of the paperboard web with a relatively thin (generally less than 0.001-in. (25.4-μm)) layer of a thermoplastic polymer (see Extrusion coating). Low density polyethylene (LDPE) is the most commonly used extrusion coating for folding cartons and provides a cost-effective means of obtaining excellent protection against liquid water strike-through as well as good water-vapor barrier. LDPE is also used as a heat sealant (see Sealing, heat) particularly when two-side coatings are employed. When the use temperature of the package exceeds 150°F (65°C), HDPE or PP can be used to raise the acceptable use temperature to 250°F (121°C). These two polymers also provide improved grease resistance. Coating board with PET can raise the use temperature to over 400°F (204°C), suitable for most "dual-ovenable" applications. Coextrusion coating is not yet extensively used in the folding-carton industry.

Laminating. The earliest means of significantly enhancing the properties of paperboard was the combination with other materials through lamination (see Laminating). The most commonly used laminating adhesives are wax (see Waxes), water-based glues (see Adhesives), or thermoplastic polymers. Materials laminated to paperboard include high quality printing paper for enhanced graphics capabilities (see Paper), grease- or water-resistant paper for improved barrier, aluminum foil for barrier or aesthetics (see Foil, aluminum), and film (sometimes metallized) for barrier or aesthetics (see Metallizing).

Printing. Prior to the printing operation, paperboard is handled in web form. A decision must be made to continue in web form or convert the web to sheets before printing and die cutting. This choice is primarily dictated by the printing technique chosen (see Printing). Sheeting is most often done at the carton-producer's facility. A small segment of the industry purchases board sheeted at the paperboard mill.

Sheet-fed offset *lithography* is the most popular printing technique used to produce folding cartons. Its primary advantage is its ability to accommodate wide variations in the size of the sheet to be printed and thus, the sizes and shapes of the cartons on those sheets. The process is economical for producing short- to medium-length runs of high quality process-printed cartons of varying sizes. Printing plates can easily be remade for every order, permitting graphics changes to be accomplished at relatively low cost. This is especially useful for printing sheets of mixed graphic designs with common carton shape and size. A different distribution of graphic designs can be printed for each order.

Web-fed *rotogravure* is the second-most-popular carton printing technique and is well suited for large orders and repeat orders not requiring many copy changes. Excellent print quality and color consistency may be obtained, but a greater sensitivity to board smoothness exists than with lithography. The web-fed nature of the process yields higher output than a sheet-fed process. Cylinders cost more than offset plates, but can often be used for several million (10^6) impressions before reengraving is required. *Letterpress* and sheet-fed rotogravure represent technologies more popular in the past; they are being phased out. Web-fed offset lithography is in early stages of use for folding cartons in the United States, with limited application for fixed-size cartons requiring frequent copy changes. *Flexography* has limited usage for folding cartons. Fidelity on paperboard substrates limits this technique to line work at this time. The high quality flexography seen in flexible packaging is not yet available for folding cartons. *Varnishes* or overcoats providing varying levels of smoothness, gloss, rub resistance, coefficient of friction, and water resistance are used in rotogravure, offset, and flexo operations for folding cartons.

Cutting and Creasing. Following the printing operation, individual cartons are cut from webs or large sheets and creased or scored along desired folding lines. Reciprocating flat-bed or platen cutting is almost invariably used to cut and crease sheets printed by offset lithography. In this technique, an accurately positioned array of steel cutting knives and scoring rules (see Fig. 9) is pressed against a printed sheet of paperboard. The knives penetrate through the paperboard to cut out the pattern of the carton. Rules force the board to deform into channels in the counter plate producing controlled lines of weakness (scores) along which the board will later predictably bend or fold. Alternatively, scores can be produced by cutting partially through the paperboard or by alternating uncut segments with completely cut-through segments.

Until recent years, knives and rules were separated and held in place by hand-cut blocks of dimensionally stable hardwood plywood. Hundreds of individual blocks were required for dies incorporating 10 or more carton positions. Greater accuracy and consistency as well as substantially reduced die preparation time is achieved through the growing use of computer-controlled laser die cutters. The laser beam is used to cut slots in large sheets of the same special plywood. Knives and rules are cut and bent automatically or manually and placed by hand into the slots. Crease or score quality has been improved through the use of computer-controlled counter-plate machining and accurate pin-registration systems.

Transport of printed sheets in a sheet-fed platen cutter is accomplished with mechanical grippers that hold the leading edge of the sheet. Small nicks in the cutting knives result in uncut areas that permit transport of full sheets from the cutting and creasing station to the stacking station. Large stacks are then removed from the machine and unwanted pieces of board, called broke, are stripped from the cut edges of the cartons, yielding stacks of printed carton blanks. Newer sheet-fed platen cutters incorporate automatic stripping of broke

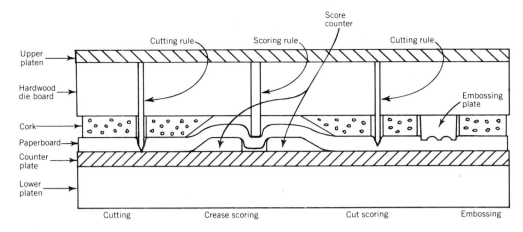

Figure 9. Flat-bed cutting-die schematic.

between the cutting and stacking stations. Platen cutters are also employed to cut and crease paperboard printed in web form. In the past, these cutters were often placed out of line from normally faster-running presses. Speed increases permit economical in-line placement, which is common today.

Rotary cutting and creasing offers the advantage of higher speeds than reciprocating platen cutting, but at greater cutting die cost. Matched machined cylinders used for fixed carton sizes are most often placed in-line with printing operations. Both electrical discharge and mechanical machining are used to produce the knives, rules, and score channels in these matched cylinders.

Rotary cutting dies have also been developed for sheet and web cutting and creasing that are produced by pattern-chemical-etching of thin metal plates. Cutting and creasing patterns are coated with chemical-resistant materials and chemicals are used to reduce the thickness of the plate in the unprotected areas, resulting in raised rules. Creases are formed by pushing the paperboard with the rule of one plate into a channel formed between two rules of the second plate, a configuration quite analogous to that of platen creasing. Cutting, however, is quite different; cutting rules on opposing plates are offset slightly from each other. As these rules rotate, approaching each other closely but not touching, they compress the board. Compressive forces within the board cause it to rupture, yielding cut edges. For sheet-fed cutting, these etched flexible dies are mounted on large cylinders which, like sheet-fed offset plate cylinders, leave a gap between leading and trailing edges. This gap accommodates sheet feed-up and variable repeats. For web cutting, however, leading and trailing edges must be butted to correspond to the continuously printed webs. Die mounting techniques as well as carton layout on the web are the keys to the successful operation of this approach.

Two additional specialty converting steps are accomplished on cutting and creasing equipment: foil stamping and embossing. Foil stamping involves the use of heat and pressure to transfer a thin metallic or pigmented coating from a carrier film to the carton surface to obtain patterned decorative effects. When this is done in combination with embossing, reflectance and gloss are combined with raised image effects for enhanced graphic presentation. Embossing alone can generally be accomplished on standard die-cutting equipment. Foil stamping and detailed, deep embossing requires the ability to heat the stamping and embossing plates. This is most commonly accomplished on small sheet-fed devices, often on cartons that have already been cut and creased.

Waxing, hot melt application.   Molten wax blends applied to individual cartons can provide gloss, grease- and moisture-resistance, and heat-sealing capability. These blends are mainly composed of paraffin and microcrystalline waxes with smaller amounts of polyethylene and EVA copolymers (see Waxes). Earliest overall application methods included dipping for two-side coverage and curtain coating for one-side coverage (see Coating equipment). High quality waxed finishes today are produced by "printing" molten wax using a flexography-like application method (see Printing). Patterned applications are used to reduce wax consumption as well as facilitate gluing. Additional gloss can be obtained by reheating the coated wax film and quenching with water in-line with wax application. The most effective wax coatings for cartons are those that form continuous wax films on the surface of the paperboard. As noted earlier, paperboard selection can impact the ease and cost with which this goal is attained.

The use of wax as a carton coating has decreased in recent years as LDPE coatings have become more popular. Wax availability, relative cost, increased extrusion-coating capacity, and increased flexibility of substrate choice with extrusion coating have all contributed to this trend. From a technical standpoint, LDPE offers a slight improvement in moisture protection, but it cannot yet duplicate the gloss and depth of a high-quality wax finish. In the high growth area of premium-priced frozen entrees and meals, the products are almost exclusively packaged in waxed cartons, primarily for aesthetic reasons.

Hot melts (see Adhesives) can be preapplied at this or a later production point using knurled wheels or timed guns. The hot melt is later heat-activated on the packaging machinery to effect sealing. Although hot melt application on the packaging machine is common, some carton users find it advantageous for the carton manufacturer to preapply the hot melt adhesive.

Windowing or couponing.   When product visibility is desired, a hole is cut out of the carton blank. To protect the product or prevent it from spilling out of the carton, pieces of transparent film are glued to the inside of the carton blank covering the open area. Special windowing equipment is used to apply an adhesive pattern around the edge of the opening, cut a rectangular piece from a roll of film, and press it in place. Registered application of printed films or printed coupons in roll form to interior or exterior surfaces adds value and function to the carton. Devices are also available that adhere coupons supplied in sheet form.

Gluing. Although more and more packaging machinery is designed to accept flat carton blanks (see Cartoning Machinery), gluing still represents a major and important converting operation. The simplest operation converts a flat blank into a side-seamed tube or glued shell (see Fig. 2(**b**)). Carton blanks are removed one at a time from a stack and carried by sets of endless belts. Stationary curved plows move one or more panels of the blank out of the original plane to either prebreak scores or form the glue seam. Prebreaking of scores assists packaging-machine operation, since the force required to bend a previously bent score is greatly reduced. Sealing is accomplished with cold glues, hot melts, or heat sealing of waxes or polymers. Side-seamed cartons are discharged into a shingled delivery that provides compression and time to set the bond; case or bulk packing for shipment follows.

Gluers in which the cartons move in a continuous straight line, transported by belts, are known as straight-line gluers. Although straight-line gluers are most commonly used to produce glued shell-type cartons, attachments provide the ability to produce automatic-bottom as well as certain collapsible-tray styles. Paper or paperboard bridges can be attached to main panels during straight-line gluing. For simple styles, the feeding of carton blanks into the gluer does not need to be timed into specific folding actions. Complicated folding devices may dictate that blank feeding be timed, which generally reduces speeds. Compound folds in both directions on the blank cannot be handled by straight-line machines.

For more complicated carton and collapsible-tray styles, right-angle gluers are employed. As the name implies, midway through the machine the travel direction of the blank is changed by 90 degrees. All parallel folds can be made in the direction of blank travel, resulting in simplified machine setup and more positive and accurate folding. Generally speed is limited by the transfer section, which changes blank travel direction. Right-angle gluers combine flexibility and precision in the manufacture of complex folding cartons. Straight-line gluers with special attachments and right-angle gluers have largely replaced in-fold gluers. In-fold gluers are characterized by timed feeding, transport by chains rather than belts, and folding accomplished by timed rotating and reciprocating folding fingers. Although such gluers are quite versatile, their use is limited by slow speeds and complicated setup.

Setup and nested tapered trays are also produced in folding-carton manufacturing plants for shipment to customers. These trays are produced on plunger-type gluing equipment which is designed to accept either blank or roll feeds. Blank-fed machines first apply adhesive, then form the tray as a moving plunger forces the previously printed and creased blank through a stationary folding and forming device. Roll-fed machines incorporate printing as well as cutting and creasing units in line prior to gluing. Nested trays are not as space efficient as unglued blanks; they do, however, have application in uses for which it would be uneconomical or impractical to operate a forming device at the location of use. Paperboard french-fry scoops and sandwich containers used by fast-food outlets are good examples of these trays.

BIBLIOGRAPHY

1. D. Hunter, *Papermaking—The History and Technique of an Ancient Craft,* Dover Publications, Inc., New York, 1978, pp. 471, 552, and 577.
2. F. Scharring, *Paperboard Packag.* **69** (8), 42 (Sept. 1984).
3. *1984–1985 Catalog—TAPPI Official Test Methods, Provisional Test Methods, Historical Test Methods,* TAPPI Press, Atlanta, Ga., 1984.
4. *Kliklok Packaging Manual,* Kliklok Corporation, Greenwich, Conn., 1983 revision, p. I.

General References

Paperboard Packaging, Magazines for Industry, Inc., Cleveland, Ohio. A monthly trade magazine for solid and corrugated paperboard containers.

Boxboard Containers, Maclean Hunter Publishing Corp., Chicago, Ill. A monthly trade magazine for solid and corrugated paperboard containers.

Package Printing, North American Publishing Co., Philadelphia, Pa. A monthly trade magazine including printing and die-cutting articles.

The Folding Carton, Paperboard Packaging Council, Washington, D.C., 1982. Short booklet.

J. Byrne and J. Weiner, "Paperboard. II. Boxboard," *IPC Bibliographic Series Number 236,* The Institute of Paper Chemistry, Appleton, Wisc., 1967. The most complete historical bibliography of paperboard and its use in folding cartons.

J. Weiner and V. Pollock, "Paperboard. II. Boxboard," *IPC Bibliographic Series Number 236 Supplement I,* The Institute of Paper Chemistry, Appleton, Wisc., 1973. Update of previous bibliography.

T. H. Bohrer
James River Corporation

CARTONS, GABLETOP

The gabletop folding format is one of the oldest and most basic end closures possible for a paperboard package. The first patent dates back to 1915 (1), but 20 years passed before the first commercial installation began to operate at a Borden Company plant, after the patent was acquired by Ex-Cell-O Corp. Today a number of manufacturers supply machinery to make gabletop cartons for milk and other still liquids.

Early gabletop milk packages were precision-cut folding boxes with an adhesively sealed side-seam and bottom closure and a stapled top closure (see Fig. 1). Semiformed cartons were dipped in hot paraffin for sanitization and moistureproofing prior to filling. Tops were stapled. The first packages had no convenient opening device. Subsequent designs had convenience openings based on secondary patch seals adhesively secured to either the inside or outside of a side panel. The secondary patch was eventually eliminated in favor of an integral pouring spout.

The modern gabletop carton retains a simple basic geometry but includes design refinements acquired over 50 years of development and commercial use. The transition from the wax-coated carton to precoated paperboard came in 1961, necessitating several new developments in package and materials technology (see Extrusion coating; Paperboard). The use of precoated board eliminated paraffin, wire, and adhesives from a filling plant's inventory. Also eliminated were the asso-

Figure 1. Early gabletop container. It is paraffin coated with a patch-type opening device.

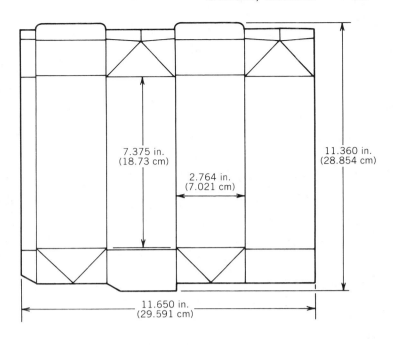

Gabletop carton

Figure 3. Typical profile for the quart (946-mL) series.

ciated mechanical systems, including carton-coating chiller units, wax melters, adhesive applicators, and related instrumentation. Precoated blanks simplified the form/fill/seal process and permitted the design of faster, more-efficient equipment with filling rates up to 300 cartons per minute. To retain the essential pouring-spout feature, an antisealant or *Abhesive* was developed. This allowed the carton top seal to be tightly sealed, yet easily opened (see Fig. 2).

Figure 2. Current gabletop container. It is polyethylene coated with a pitcher-spout opening device.

A typical blank for a quart (946.25 mL) gabletop container is shown in Figure 3. With a panel width of 2.764 in. (7.02 cm) and a body height of 7.375 in. (18.73 cm), the apparent contained volume of 56.34 in.³ (923.25 cm³) falls short of the quart volume of 57.75 in.³ (946.35 cm³). The needed extra volume is found in the bulge of the side panels after filling, which leaves the filled product line below the top horizontal score, providing "headspace" necessary to compensate for foam generated during filling and to allow for a certain amount of splash as the filled container is conveyed through the top heat-sealing machine function.

The standard square cross section of the quart (946.25 mL) carton is used for a full range of containers from 6 fl. oz (177.4 mL) through the Imperial quart (1182.8 mL). Other cross sections in the same carton format have panel widths of 2.240 in. (5.69 cm), 3.3764 in. (8.576 cm), and 5.531 in. (14.05 cm), with container volumes from 4 fl. oz (118.3 mL) through one gallon (3.785 L). For quart-series containers, the typical paperboard structure consists of 195–210 lb (88.5–95.3 kg) per ream paperboard with a coating of 0.0005 in. (12.7 μm) polyethylene on the outside surface and 0.001 in. (25.4 μm) polyethylene on the inside surface. Other structures that include aluminum foil, ionomer (see Ionomers), and other barrier materials are also possible (Fig. 4).

Coated-paper containers for liquids provide relatively short shelf life. It is possible, however, to tailor the container to its contained product. Liquids with high solids content, such as milk or fountain syrups, are relatively easy to contain, since there is little product penetration of cut edges or random flaws in the coating. Other products require near hermetic seals, dictating a continuous high barrier such as aluminum foil, and the elimination of cut edges in the finished containers.

This can be accomplished by a number of mechanical techniques including skiving and hemming of the cut edge of the

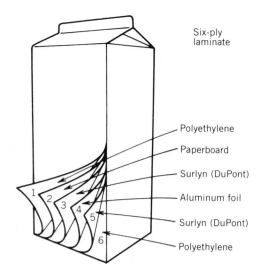

Figure 4.   A typical six-ply laminate.

side seam, and refining the folding of the bottom closure to protect the cut edge from liquid contact. These techniques permit the successful packaging of oils and alcohol-bearing liquids with little problem. Special treatment of the paperboard may also be necessary to assure package stability. Today's form/fill/seal equipment (see Fig. 5) contributes to shelf life by enclosing the processing line. Some machines feature air filtration and carton sanitization systems.

At its inception, the gabletop packaging system bordered on the brink of failure because the boxmaker's inability to deliver precision-cut blanks necessitated the development of high speed rotary converting equipment. Now an industry standard, this type of equipment delivers thousands of blanks per minute, cut and creased to machine-tool accuracy. Rigid quality control is essential to the system. Blanks are routinely checked for profile, cut, and crease alignment, and other critical factors affecting container durability and machinability.

In the near term, manual procedures will be augmented by

an automatic inspection system. Currently available is a new device, controlled by a small computer, with the capability of making several hundred evaluations of cut and crease accuracy, symmetry, and formation. Data are displayed instantly, recorded and printed out on demand.

At its peak, the gabletop carton was the premier package for milk and other fluid products in the United States. Process refinements permitted its use for other likely products, such as fruit juice and fountain syrups, and a few unlikely products, such as candy and epsom salts. In the United States, the plastic blow-molded bottle (see Blow molding) has become the container of choice for gallons (128 fl oz or 3.785 L) of milk but the paperboard gabletop carton is still favored for smaller sizes. Worldwide, the carton is the popular package for still-liquid products, with over 30 billion (10^9) sold each year.

BIBLIOGRAPHY

1.   U.S. Pat. 1,157,462 (Oct. 19, 1915) J. Van Wormer.

R. E. LISIECKI
Ex-Cell-O Corporation

CASE LOADING

The corrugated case is still the most universally accepted method of packaging and shipping products from one destination to another (see Boxes, corrugated). It offers excellent product protection in both storage and shipping with the additional benefit of full-panel graphic identification or advertisement exposure. The terms *case packing, case loading,* and *casing* all refer to the method of placing product into corrugated shipping containers. This can be accomplished by the very fundamental method of hand loading or by semiautomatic or automatic case-loading machinery. This article is structured to demonstrate a normal sequence of conversions from hand loading to fully automatic loading. Cases, whether manually or automatically loaded, are classified as top-, side- or end-load. The top-load case has flaps in the largest panel and is the most expensive because of the large flap area. An end-load case, with flaps on the smallest panel, is the least expensive. Proper machinery selection depends on many variables such as type of package, style of case, production rates, automatic versus semiautomatic machinery, floor space, rate of investment return, etc. The following basic case-loading methods are discussed below: hand loading; horizontal semiautomatic case loading; horizontal fully automatic case loading; vertical drop-load and gripper-style case loading; wraparound case loading; and tray former/loader.

Hand Loading

This is the simplest version of case loading. It requires limited machinery involvement, but it is highly labor intensive. As packages are delivered to the packing area, personnel manually open, load, and seal the corrugated cases utilizing a variety of closing methods including cold glue, hot-melt glue (see Adhesives), tape (see Tape, pressure sensitive), or metal staples (see Staples). Cold and hot-melt adhesive applicators are most common (see Adhesive applicators). Replacing the case-closing operation with an automatic top-and-bottom case-sealing machine is the first step in automation. This unit can always be utilized later as the packaging line converts to fully

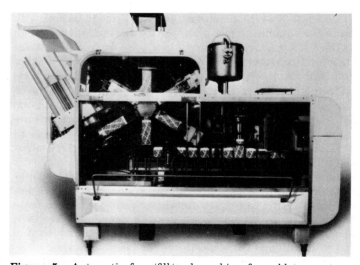

Figure 5.   Automatic form/fill/seal machine for gabletop cartons. Courtesy of Ex-Cell-O Corp.

automatic. To decrease labor and increase production, the manual loading operation is replaced with a semiautomatic case-loading machine integrated with the existing top-and-bottom case sealer.

Horizontal Semiautomatic Case Loader

With this type of equipment, product is loaded from the side (see Fig. 1). This method offers considerable flexibility in obtaining the desired case pattern, which is defined as the position or orientation of cartons grouped inside the case. It is ideally suited for handling cartons, cannisters, cans, or any product that takes a rigid or semirigid shape. The most common product handled is consumer-type cartons, which are loaded in a variety of configurations and counts. The semiautomatic case loader requires an operator to manually open a flat premade corrugated case, fold the bottom flaps, and place it on the machine-loading tube or funnel ready for package insertion. All other machine functions are performed automatically. Typically, an operator can open approximately 10–12 cases/min. Higher case rates would require fully automatic case-loading machinery, described below. Package speeds in excess of 500/min are obtainable, but the number of cartons per case and the dexterity of the operator dictate if it can be done semiautomatically.

The basic unit includes an infeed conveyor to receive packages and deliver to the machine accumulator section, where the product is grouped or stacked to the prescribed case pattern and loaded into the already opened and formed corrugated case. Single and multitier case patterns are easily accommodated. Multitier applications require the cartons to be stacked prior to loading. Functionally, the infeed conveyor delivers cartons to a lifter plate, where the prescribed number is accumulated. Through a pneumatically operated cylinder, the cartons are lifted and deposited on stacker bars. The lifter returns and continues this cycle until the correct number of cartons have been grouped in front of the loading tube. A pneumatically operated side-ram cylinder pushes the final load into the case. Many machine configurations and accessories are available including multilane units for higher package production rates. Standard upstream filling and packing machinery usually discharge product in a single lane; the use of multilane casers would require some type of package lane dividing systems. Many packages must be repositioned to coincide with the case pattern and this is accomplished through rail twisters, upenders, turn-pegs, etc. Converging equipment is also available if the output from several upstream packaging machines must be converged into the caser single-lane infeed. After final loading, the filled case is lowered onto a short discharge conveyor ready for final case sealing. The existing top-and-bottom case sealer can be utilized, or any other type of sealer. This approach to case loading is ideal for lower-case rate applications. It requires minimal capital investment, but is a major step in automation. At this point, the horizontal semiautomatic case loader has eliminated the hand-load operation and reduced the personnel to one operator. The next sequence in automation replaces the operator with an automatic corrugated-case erector, forming a fully automatic case-loading system.

Steps Toward Fully Automatic Case Loading

There are normally two approaches to automatic case loading: a fully automatic integrated system including a case erector, loader, and sealer or a case erector/loader utilizing an existing case sealer. Most case-loading machinery is manufactured in modular design to allow the proper equipment selection for the application. Consideration should always be given to how existing equipment can be used in conjunction with new equipment. A decision to automate, and how to do it, is based on many considerations: case rates in excess of 10–12/min; high package-production rates; packages more easily loaded automatically; large-size cases more easily handled automatically; labor reduction; floor space reduction; and increased line production and efficiency.

Case erector/loader. This machine is equipped with a flat-corrugated storage magazine which will, on demand, extract a case from the magazine, open it, fold in bottom flaps, and automatically place it on the caser loading tube for final package insertion (see Fig. 2). After loading, the filled case is lowered onto a discharge conveyor and transferred into a new or existing case-sealing machine. Case extraction and opening are the most critical functions of an automatic erector. Corrugated cases have a built in memory (resistance) and proper blank scoring will increase opening efficiency. Although more expensive, experience has proven that equipment offering mechanisms to prebreak or restrict the case back panel during opening are well worth the investment. Generally, vacuum/pneumatic mechanisms appear to function well up to approximately 20 cases/min. Higher case rates normally require a mechanical/vacuum/pneumatic combination with several stations for case extracting, opening, and loading. Both horizontal and vertical case magazines are available. The automatic case erector can be added to an existing semiautomatic case loader with its separate sealer, or it can be offered as part of a new integrated system interfaced with the existing case sealer. The case erector eliminates the operator who would otherwise open cases manually; therefore, the erector flat-case storage magazine must have sufficient capacity for at least 30–60 minutes supply. If floor space permits, additional magazine capacity is encouraged. Vertical style magazines are offered with bulk-storage feed systems where several stacks of cases are loaded on a floor-level conveyor and on demand, feed

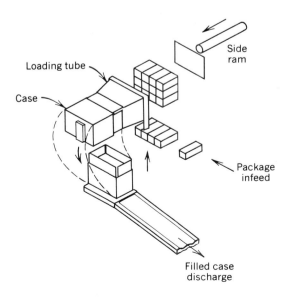

Figure 1. Horizontal semiautomatic case loader.

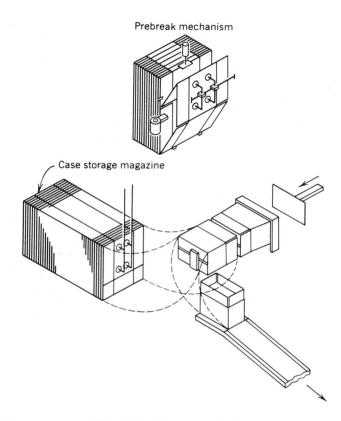

Prebreak mechanism

Case storage magazine

Figure 2.   Case erector/loader.

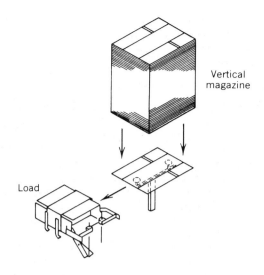

Vertical magazine

Load

Figure 3.   Vertical magazine.

automatically to the magazine (see Fig. 3). A variety of case erectors are offered for various case rates up to 30 cases/min.

Horizontal Automatic Caser Erector/Loader/Sealer

The final approach in automation incorporates a completely integrated system. A new automatic case extractor/sealer can be integrated with an existing semiautomatic caser to form a fully automatic line, but the case erector/loader/sealer is nor-

mally purchased new as a part of the complete packaging line. Case extracting/loading functions are the same as those discussed above, but after loading, the filled case is transferred horizontally through the glue-application section and into the compression unit using heavy-duty continuous-motion cleated chains. Minor case flaps are folded closed and major flaps are opened ready for adhesive application. After gluing, stationary plow rods fold in major flaps as the case is deposited into the intermittently driven side-sealing compression unit (see Fig. 4). A secondary set of top chains may be employed to ensure that the case is presented squarely to the compression unit. Vertical compression units for use in overhead filled-case conveying systems reduce initial floor-space requirements. The compression-section length is a function of the type of adhesive used and its corresponding drying time. Both hot-melt and cold-glue adhesives are commonly used (see Adhesives). Hot-melt adhesive has a faster set-up time and requires a relatively short compression section, usually 4–5 ft (1.2-1.5

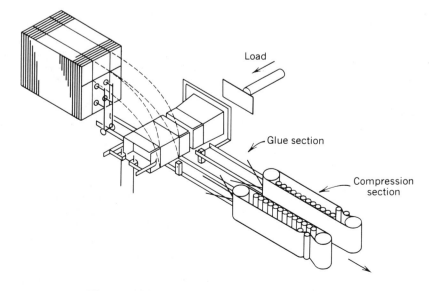

Load

Glue section

Compression section

Figure 4.   Horizontal fully automatic case loader.

m). Cold-glue adhesive takes longer to set and requires more compression-section length. The hot-melt adhesive unit takes less space, but it is somewhat more expensive than the cold-glue system. PVC sealing tapes are becoming an attractive sealing method for various reasons, and most automatic machines can be equipped with tape heads in place of glue heads (see Tape, pressure sensitive). Compression-section length can usually be reduced, because no drying time is required. As with all automatic machines, the flat-corrugated case magazine storage capacity should be large enough to ensure that an operator is not constantly replenishing the supply hopper. A complete automatic system offers many advantages including higher case rates, increased line efficiencies, labor reduction, and the relatively new technology of programmable logic controllers (PLC). The machine functions are now computerized and programmed accordingly. This information can be coordinated into the main control center, providing valuable information to the production department. This new electronic technology offers many specialized options, such as canned-message-display troubleshooting units that read out in alphanumeric language. In summary, the horizontal fully automatic case opener, loader, and sealer is capable of receiving product from upstream packaging equipment and delivering that product to the shipping department in a sealed corrugated case. This is all accomplished in a relatively small area at speeds of ≥30 cases/min.

Vertical Case Loaders

This method of case loading is used primarily in the beverage, glass, can, and plastic container industries, where fragile or irregular-shaped containers require some special packing considerations. As with the horizontal case packers, the product is delivered to the machine infeed conveyor from upstream filling equipment to the accumulator section. Tabletop chain is commonly used in delivering the product to reduce backpressure during the load cycle and for infeed washdown applications. Cylindrical-type products are divided automatically into several lanes using oscillating or vibrating dividers to form the accumulated load pattern. Irregular-shaped containers such as blow-molded plastic bottles must be divided by special equipment. When all lanes are filled in the accumulator area, a formed corrugated case is positioned underneath the loading mechanism ready for depositing. At that point, retractable shifter plates in the accumulator area move out allowing the containers to drop vertically through fingers into the cells of the case (see Fig. 5). Special fingers guide and reduce side shock to the containers during the load cycle. The use of cells or corrugated partitions inside the case to eliminate container contact is based on product-protection requirements. Usually glass containers have partitions, and plastic containers do not. Case rates of up to 25/min are achieved. The vertical case loader can be interfaced with many different kinds of corrugated-case erecting equipment. Manual case erecting and placement under the load area tied into a case sealer is one alternative. Another, used by the glass and plastic bottle industries, is to ship empty bottles in cases to the filling plant where they are emptied, filled, and loaded back into the re-shipper cases utilizing a top case sealer. A third method is to incorporate an automatic case erector, vertical loader, and sealer. The machinery selection is based on floor space, capital investment, type product to be handled, and most important, case-rate requirement.

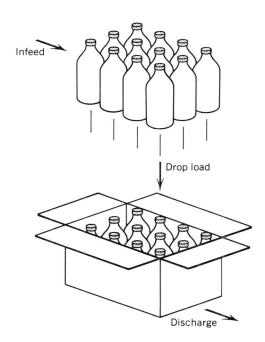

Figure 5. Vertical drop-load case packer.

Another vertical load method for fragile or heavy containers uses equipment that utilizes vacuum or mechanical grippers to lower the containers into the case (see Fig. 6). This approach is ideal for containers that can be gripped at the top, such as glass or plastic bottles. The containers are delivered and accumulated in the same manner as with the vertical drop loader, but a special plate or load head incorporating vacuum or mechanical grippers moves down and picks up approximately 12 containers at a time and places them in the opened corrugated case. Generally, both the drop-load and vacuum or mechanical gripper-style machines are offered in multi-load station modules to obtain speeds of 40–50 cases/min with automatic case erectors.

Equipment is also available that loads cartons, cans, tapered cannisters, etc, vertically up through the bottom of a

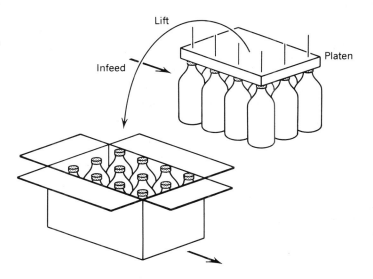

Figure 6. Vertical gripper-style case packer.

case. This method is usually limited to top-load cases, but it does offer the advantage of eliminating possible package repositioning. A case is extracted and opened in the conventional method over the accumulated product. On demand, the product is lifted up into the case, which is then transferred horizontally into the sealing section. For the most part, both horizontal and vertical case-loading equipment are of the intermittent-motion design, which is somewhat speed limited. Requirements for higher speed have led to the development of faster filling machinery and continuous-motion casing equipment that runs in excess of 2400 cpm and 1200 bpm. This special machinery is an integrated system handling multiple cases. Continuous-motion horizontal case-loading equipment using both premade RSC and wraparound blank cases (see Boxes, corrugated) have exceeded the 50/min range. These rates apply to some special tray forming/loading applications as well. Because upstream filling equipment for cartons and flexible packages has not achieved the high rates of the can and bottling fillers, the requirement for continuous-motion carton-type casers has been limited.

Automatic Wraparound Case Loading

An entirely different approach to case loading utilizes a five-panel corrugated blank instead of a flat premade corrugated case with the manufacturer's joint already glued. Vertical or horizontal corrugated blank-storage magazines are employed that extract the blank and position it between chain lugs by either vacuum or mechanical mechanisms. During this motion, both side panels are folded into a vertical position forming half the case. The blank is then positioned in front of the loading machine where the product is either pushed onto the blank or dropped vertically (see Fig. 7). After loading, the top panel is folded down over the product and final flap folding and gluing is completed. Depending on the type equipment, the manufacturer's joint is then glued and folded down to one of the vertical panels for final sealing. Wraparound casers are usually larger and more complex than loaders for premade cases owing to the additional functions that must be performed, but depending on the size of the case, there may be some economical board-cost advantages using a five-panel blank. Pneumatically operated machines can achieve speeds up to 20 cases/min; higher rates of approximately 30 cases/min would require a more mechanical/pneumatic design. Continuous-motion wraparounds are available for special applications with speed requirements in the 40–50/min range. Both horizontal and vertical blank storage magazines are available as well as vertical compression sections suited to overhead case-conveyor systems (see Conveyors). Selection of a wraparound versus a premade case machine requires close scrutiny of the particular application requirements including type of

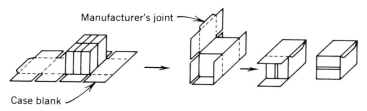

Figure 7. Wraparound case loader.

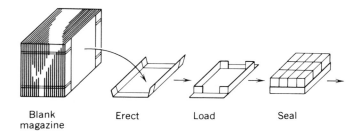

Figure 8. Tray former/loader.

case, potential board savings, machine cost, floor space, case rate, size case, etc.

Tray Former/Loader

The concept of replacing a corrugated case with a shrink-wrapped tray was developed as part of an ongoing effort to reduce packaging material costs. The tray former/loader (see Fig. 8) with a shrink wrap is ideally suited for those products eg, cans and bottles, that do not require the complete product protection provided by a full case. Corrugated trays that are 1–2 in. (2.5–5.1 cm) high are preferred, but there are a host of different tray designs suited to the product, distribution, display, etc. Many machine alternatives are available including equipment to make a tray and present it to a vertical or drop-load caser. There are also integrated systems that extract a blank from the magazine, fold up two or three sides of the blank, index to the load area (where the product is deposited horizontally or vertically), and transfer, folding remaining panels and gluing to form the final tray. At this point the filled and sealed tray is indexed into a shrink-wrapping machine (see Wrapping, shrink) where film (see Films, shrink) is completely wrapped and shrunk around the complete load. The film unitizes and holds the product in the tray and provides protection from the external environment. The cost for the corrugated tray and shrink film is considerably less than a full corrugated case, but consideration must be given to product protection and warehouse stacking strengths because now the product must bear the full vertical load, warehouse identification, etc. Tray rates in excess of 30/min are obtainable, but at these higher rates multiple shrink machines may be required.

Other Concepts

The case-loading field is extremely broad, and many other different types of machines are available for special markets and applications too numerous to describe in detail. For example, there is automatic case-opening/bottom-gluing machinery that presents an opened case with the bottom flaps glued for the manual loading of large or irregular-shaped items. This same equipment can also present an opened/glued case to a vertical or drop-load caser, increasing the operating speeds of hand-packed lines. Automatic case openers are available that bring a case down vertically over the accumulated product and then fold and seal the flaps or that lift the accumulated product vertically up through the bottom of a corrugated case, folding and sealing flaps accordingly. There are case loaders that roll the product into the case and manual machines that are used for handling irregular-shaped products in a wraparound blank. All these variations and concepts are necessary in providing the unique machinery to meet the changing markets and products of today.

BIBLIOGRAPHY

General References

The PMMI 1984–1985 Packaging Machinery Directory, Packaging Machinery Manufacturers Institute, Washington, D.C.

S. Sacharow, *A Guide to Packaging Machinery,* Harcourt-Brace Jovanovich, New York, 1980.

WALTER WARREN
Salwasser Manufacturing Company, Inc.

CASKS.   See Barrels.

CELLOPHANE

Cellophane is a family of transparent films that is manufactured by chemically regenerating a dissolved cellulosic compound into a continuous thin sheet. This compound, called viscose, was discovered and developed in England in 1892, but 20 years passed before a Swiss chemist, J. E. Brandenberger, known today as the father of cellophane, succeeded in developing a workable viscose-control technique and film-casting method. Late that year, viscose-based cellulose film was commercialized under the trade name Cellophane and was first produced in France by La Cellophane SA (1).

From that beginning, an international cellophane industry grew. At its peak, there were over 50 producers with close to 1.5 billion (10^9) lb (6.8×10^5 metric tons) of annual capacity. With the introduction of low cost petroleum-based thermoplastic films in the early 1960s, cellophane's near-monopoly in transparent packaging markets began to erode (see Films, plastic). However, even with this formidable competition, cellophane remains an exceedingly viable flexible packaging material. Today, 41 producers have an estimated worldwide annual capacity of 900 million (10^6) lb (4.1×10^4 t). These producers represent some of the largest chemical companies in the free world, including E. I. du Pont de Nemours & Co., Inc. in the United States and its subsidiary Ducilo SA in Argentina; Olin Corp. in the United States; British Cellophane, Ltd. (BCL), a part of Courtaulds, in the UK and Canada; Sidac, a unit of Belgium's UCB, in the UK, Spain, and Belgium; Wolff, a part of Bayer AG in the FRG; and Celanese Corp. in Mexico and Colombia. It is expected that cellophane demand should be in good balance with supply during the mid-1980s.

Manufacturing Process

The cellophane manufacturing process is highly complex, but this very complexity allows cellophanes to be tailor-made to meet many diverse market requirements (2). Cellophane's primary building block is a specialized dissolving-grade sulfite wood pulp that has a high alpha-cellulose content. This type of pulp is most commonly made from soft woods, but also from eucalyptus and hard woods. The pulp, in sheet or roll form, is steeped in caustic soda and then shredded. Next, the alkali cellulose is aged to begin the molecular depolymerization pro-cess, after which carbon disulfide is added to the swollen, shredded alkali cellulose, creating sodium cellulose xanthate.

Viscose is created by adding dilute caustic to the xanthate. In so doing, the cellulose is dissolved into solution. After filtering, the mixed viscose stream is moved to ripening tanks. The length of aging time at this point depends on the pulp's specific characteristics. The degree of depolymerization and viscosity are also closely monitored. When the viscose has the proper viscosity and ripeness, it is moved through another filter and into a feed tank. At this point, the viscose is a complex dispersion of cellulose xanthate and sodium hydroxide. It resembles a light orange heavy sugar syrup and is ready to be made into film. Each state of molecular depolymerization is very important and must be exactly timed. Also, the various stages in the process are conducted under carefully controlled temperatures. If such care is not taken, an overripe viscose produces a weak film and the high probability of wet-end breaks on the casting machine. Conversely, an under-aged or "green" viscose poses significant filtration and pumping problems and can result in hazy film.

Film-ready viscose is pumped to the front, ie, wet, end of the casting machine, where it enters the casting hopper. A hopper is a horizontal die through which the viscose is pushed by pump action, forming a film. This piece of equipment has adjustable, thin, highly machined lips which, along with film takeoff speed, control film thickness. The opening of the hopper is submerged in a sulfuric acid bath. The thin sheet of viscose is extruded into the acid, changing into cellulose gel and a by-product, sodium sulfate. Following the primary acid bath, the gel sheet passes through up to ten tanks in a festoon fashion. Each tank has a particular function, starting with the removal of excess sodium sulfate and ending with the addition of plasticizers (softeners), eg, glycerol, propylene glycol, polyethylene glycol, triethylene glycol, or ethylene glycol.

After the softeners are added to the film, it moves into the dry end of the casting line. Here, the film is dried to the proper moisture level on a series of steam-heated rollers. During the drying process, a certain amount of latitudinal shrinkage occurs. If this shrinkage is inhibited mechanically in an effort to maximize film width and if the primary plasticizer is propylene glycol, the film will have a greater propensity to husband coating solvents than if it is dried in a relaxed, uninhibited manner. Once dried to 7–8% moisture, the cellophane is wound on steel cores and is ready for coating, slitting, or both.

The coating of cellophane is a relatively simple process compared with base film manufacture. There are still some long horizontal machines, but they have been largely replaced by vertical tower coaters. With this method, the raw cellophane passes through a dip tank holding the coating formulation, a mixture of resins and other ingredients dispersed in solvents. The solvents are driven off at elevated temperatures as the film moves up one side of the tower. Coming down the back side of the machine, the film is rehumidified to the desired moisture level of 7–8%. Just prior to winding, polymer (PVDC) coatings (see Vinylidene chloride copolymers) are often top-coated on one or both sides to ensure good packaging machine sealing-jaw release. This is done on a gravure station using sodium lauryl sulfate as the agent and 2-propanol as the vehicle. Some producers use an internal ingredient that is part of the basic polymer formulation to obtain good sealing-jaw release. Throughout viscose preparation, film casting, and

Table 1. Comparative Cellophane Types

Coating	Two-side coating	One-side coating	BCL	DuPont	Olin	Sidac	Applications
nitrocellulose heat-sealing, moisture-proof	X		MS	MSD	MST	MS	used for low cost bags and tray and carton overwraps for shelf-stable products; moisture barrier is damaged by solvents, greases, and oils, and when sharply creased or flexed
nonheat-sealing, moisture-proof	X	X	MF	MD	MT MBO	MF DM	two-side coated for confectionery twist wraps; one-side coated for extrusion laminations and coatings when nonblocking outer surface is required
heat-sealing, less moisture-proof	X	X	QMS, G	LSD	LST MSBO	LMS, QMS DMS	two-side coated wraps and bags for products that must breathe, eg, fresh produce and glazed baked goods
PVDC heat-sealing, solvent-based	X	X	MXXT/S, Z	K	V	MXXT, XS DXS	used for bags, overwraps, and laminations; excellent moisture and oxygen barriers; can be run on high-speed form/fill/seal equipment when one side is treated for sealing-jaw release; one-side coated for extrusion laminations and coatings
heat-sealing, water-based	X X	X	MXXT/A MXXT/W MXDT/W				extra-high moisture barrier with no residual solvent content; MXXT/W has superior heat sealability and jaw release; one-side coated for extrusion laminations and coatings
heat-sealing, colored, water-based	X		MXXC/W				color pigments contained in coatings; cleared for food contact by the FDA.

coating, extensive recovery systems are used to minimize chemical waste.

Softeners. Softener selection has a great influence on cellophane's performance characteristics. Glycerol, the original plasticizer, is used when optimum dimensional stability is desired, as when the film is to be used in unsupported form as a box overwrap or to withstand the high temperatures of thermoplastic extrusion coating at a later converting stage (see Extrusion coating). However, when used alone, glycerol does not give cellophane its best durability potential for bag and pouch applications (see Multilayer flexible packaging). Propylene and ethylene glycols are better in that respect, but ethylene glycol is not cleared for food use in the United States by the FDA, and propylene glycol has the propensity for holding onto solvents if they are not carefully removed during coating. Polyethylene glycol has less tendency to hold solvents, but it provides lower durability than the other glycols. Triethylene glycol is used for quick-setting heat seals in nitrocellulose-coated cellophanes, but its drawback is the possibility of film blockage in high temperature storage. Since none of these soft-

Table 2. Cellophane Gauge Designation

	SI units		English units		
Gauge, 10 g/m²	Thickness, μm	Coverage, m²/kg	Gauge in.²/lb/100 (or in.²/100 lb)	Thickness, mil	Coverage, in.²/lb
280	19.8	35.7	250	0.78	25,000
306	21.6	32.7	230	0.85	23,000
320	22.9	31.3	220	0.90	22,000
335	23.6	29.9	210	0.93	21,000
340	23.9	29.4	207	0.94	20,700
350	24.9	28.6	200	0.98	20,000
360	25.4	27.8	195	1.00	19,500
391	27.4	25.6	180	1.08	18,000
440	31.0	22.7	160	1.22	16,000
445	31.2	22.5	150	1.23	15,000
460	32.3	21.7	153	1.27	15,300
500	35.3	20.0	140	1.39	14,000
600	42.7	16.7	116	1.68	11,600

eners used alone provide cellophane with an optimal balance of desired characteristics, they are often used in combination. When the need for one particular characteristic far overshadows all others, a single plasticizer system is employed. All contribute to maintaining the proper moisture concentration in the sheet. This is very important because cellophane's physical strength depends on moisture. The average cellophane has 7–8% moisture content, and the plasticizer concentration can be 11–18%.

Coatings. Coatings give cellophane its broad packaging functionality. Uncoated cellophane has only a few packaging uses, mainly as decorative wraps. It does have some industrial applications, such as release sheets, pressure-sensitive tape base, and as a substrate for roll-leaf coatings. Without moisture-proof coatings, moisture loss creates film brittleness and shrinkage. Excessive moisture pickup causes cellophane to lose its excellent inherent gas and aroma barrier. The need for heat sealability in coatings is very important if cellophane is to be used on high speed automatic packaging equipment (see Sealing, heat). The first moisture-proof coating, a nitrocellulose lacquer, originated at DuPont in 1927. It is universally designated by the letter M. A heat-sealable version, introduced later, is designated MS throughout the world. Other grades are designated by different letter combinations in different countries (see Table 1). Nitrocellulose coatings consist of pyroxylin, plasticizer, resins, and waxes. The combined ingredients are applied to the film using a vehicle of toluene, ethyle acetate, and ethyl alcohol. Although the coatings provide a good moisture barrier in flat sheet form, this performance feature is compromised by sharp creasing or contact

Table 3. Cellophane Properties[a,b]

Property	ASTM test	Nitrocellulose, nonheat-sealed	PVDC-solvent coated, heat-sealed	Uncoated, nonheat-sealed	Nitrocellulose, heat-sealed
specific gravity	D 1505	1.5	1.5	1.5	1.5
yield, in.^{2}/lb[c]		19,500	19,500	19,500	19,500
haze, %	D 1003	3.0	3.5	3.0	3.0
tensile strength, psi[d]	D 882				
MD		18,000	18,000	18,000	18,000
XD		8,000	8,000	8,000	8,000
elongation, %	D 882				
MD		20	22	16	16
XD		50	60	60	60
tensile modulus, 1% secant, psi[d]					
MD		750,000	750,000	800,000	800,000
XD		400,000	420,000	400,000	400,000
bursting strength, psi[d]	D 774	40	40	40	40
service-temperature range, °F (°C)		0–350 (−18–177)	0–350 (−18–177)	0–350 (−18–177)	0–350 (−18–177)
heat-seal range temperature, °F (°C)			165 (74)		170 (77)
oxygen permeability, cm^3·mil/(100 in.·d·atm)[e]	D 1434				
at 50% rh		2–3	0.5	3–5	2–3
at 95% rh		10–25	0.5	50–80	10–25
water-vapor transmission rate WVTR at 100°F (38°C) and 90% rh, g·mil/(100 in.^{2}·d)[f,g]		0.40	0.45	110	0.50
COF, face-to-face, back-to-back	D 1894	2–3	2–3	4.5	2–3

[a] 1-mil (25.4-μm) thickness.
[b] Test conditions: 73°F (23°C) and 50% rh unless otherwise stated.
[c] To covert in.^{2}/lb to cm^2/kg, multiply by 1.422 × 10^{-3}.
[d] To convert psi to MPa, divide by 145.
[e] To convert cm^3·mil/(100 in.^{2}·d·atm) to cm^3·μm/(m^2·d·kPa), multiply by 3.886.
[f] To convert g·mil/(100 in.^{2}·d) to g·mm/(m^2·d), multiply by 0.3937.
[g] Nitrocellulose-coated cellophane's WVTR is increased when film is sharply creased or attacked by solvents, greases, or oils.

with grease and printing solvents. These types of coatings are still widely used today, partly because of their relative economy. Nitrocellulose-coated cellophanes are used to package dry, reasonably shelf-stable products such as confectionery items, tobacco products, and selected baked foods, and are used in laminated form for condiments, drugs, and prophylactic devices.

Most cellophane carries a polymer coating of PVDC (see Vinylidene chloride copolymers). It was introduced in 1949 by DuPont and designated as K film. Throughout the world, most polymer-coated cellophanes are made with a solvent-based system, ie, toluene and tetrahydrofuran, and producers take great care to ensure that retained solvents are within regulatory limits. An aqueous-based PVDC coating system, used by BCL, eliminates the possibility of retained solvents and provides better water-vapor barrier than the solvent coatings. The water-vapor barrier of nitrocellulose-coated and polymer-coated cellophanes is independent of film gauge. Most two-side coated films carry a standard coating weight of just under 3.0 g/m^2. PVDC-coated cellophanes are used to package most classifications of food products and a wide variety of nonfood items. Primary markets include snack foods, candy, cookies, crackers, nuts, cheese, pasta products, and some processed meats (see Multilayer flexible packaging).

Many cellophane producers make a number of specialty products, eg, metallized, white-opaque, and colored films. These are normally coated. Mill-made reinforced cellophane laminations (see Laminating machinery) are made by Olin, using cellophane in combination with BOPP (see Film, oriented polypropylene) or aluminum foil (see Foil, aluminum). The Japanese produce a PVC-copolymer-coated cellophane having characteristics that fall between those of nitrocellulose and PVDC-coated films.

Cellophane is made in over 20 thicknesses, ranging from 0.77 mil (19.6 μm) to 1.70 mil (43.2 μm). In the United States, the gauge of a film is expressed in in.^{2}/lb/100 (or 1 mil = 195 ga = 19,500 in.^{2}/lb) (see Table 2). In Europe, the same gauge would be expressed as g/m^2 × 10 (or 1 mil = 25.4 μm = 360 ga = 36 g/m^2) (see Table 2).

Performance

In judging a flexible packaging material's application performance, four principal categories of analysis should be considered: package-forming efficiency, product protection, efficiency in converter processes, and merchandising appeal. Properties of selected grades of cellophane are shown in Table 3.

Package-forming efficiency. Cellophane offers excellent contact heat stability. It does not distort when in contact with high temperature sealing mechanisms. Its quick heat conductivity rapidly passes sealing energy from the sealing mechanism to the surfaces to be heat-sealed. These two characteristics plus coating formulations give the films very wide sealing ranges. Cellophane, because of its moisture content, has an exceptionally low electrostatic level, preventing hang-ups on metal packaging machine parts. Its inherent stiffness allows it to move readily, without support, over areas of packaging machines such as elevator shafts on overwrap equipment (see Wrapping machinery). All of these characteristics make cellophane the most efficient package-forming film available.

Product protection. Cellophane offers an excellent gas and odor barrier, and polymer-coated films provide good moisture

barrier as well. Although cellophane's durability is moisture-dependent, its availability in a wide number of gauges makes it possible to match most product needs. However, some particularly arduous distribution systems require cellophane laminations with thermoplastic films, which are well able to meet the durability requirement.

Efficiency in converter operations. Cellophane's characteristics have been used as bench marks for almost all other films. Cellophane coatings readily accept a wide variety of printing inks (see Inks; Printing) and adhesives (see Adhesives) designed for specific packaging needs. Flat-sheet gauge uniformity has always been one of cellophane's strong points. It has exceptional memory when stretched during converting processes which allows tight printing registration control. Cellophane's predictable shrinkage in printing press and laminator driers minimizes converter waste.

Merchandising appeal. Cellophane combines clarity and sparkle with a feeling of richness. Its stiffness allows attractive upright pouch and bag display at retail. During normal distribution cycles, cellophane's atmospheric dimensional stability is more than adequate for carton overwraps. It is only when overwrapped cartons are subjected to extremes of low and high humidity that cellophane's performance becomes somewhat marginal. Because cellophane is impervious to grease penetration, package graphics remain fresh and unstained.

The versatility of cellophane is illustrated by the small sampling of typical structures shown in Table 4.

In summary, cellophane's highly complex manufacturing process converts wood pulp, a renewable resource, into the most versatile family of films available to the packagers of the world. It is also one of the few transparent films that is biodegradable in landfill operations. Although cellophane has been

Table 4. Typical Cellophane Laminations[a]

Application	Lamination
candy	K cello/K cello
	BOPP/adh/K cello
	K cello/adh/coex BOPP
	K cello/wax/K cello
chewing gum	M cello/wax/paper/wax/foil/wax
cheese, natural	MS cello/wax/foil/wax
	K cello/wax/foil/wax
	BOPP/LDPE/K cello/MDPE
cheese, processed	wax/K cello/wax
	V cello/adh/BOPP/adh/V cello
	K cello/K cello
	K cello/adh/coex BOPP
condiments, single-service	MSBO cello/LDPE
	MBO cello/LDPE/foil/LDPE
drugs and pharmaceuticals	MSBO cello/LDPE
freeze-pop confections	MSBO cello/LDPE
pasta products	K cello/K cello
	K cello/adh/coex BOPP
snack foods	K cello/K cello
	Acrylic BOPP/K cello
	BOPP/adh/K cello
	BOPP/LDPE/K cello
	K cello/LDPE/PVDC BOPP
	K cello/adh/coex BOPP

[a] See Table 1 for cellophane-type description.

in existence for many years, it will continue to be used in significant volumes for many years to come.

BIBLIOGRAPHY

1. C. H. Ward-Jackson, *The 'Cellophane' Story: Origins of a British Industrial Group*, British Cellophane Ltd., Bridgwater, Somerset, UK, 1977. Contains an excellent bibliography.
2. G. C. Inskeep, ed., and P. Van Horn, "Cellophane," *Ind. Eng. Chem.* **44**(11), 2512-2521 (Nov. 1952). This report, a primary source for the technical description of the cellophane manufacturing process, contains a good bibliography.

CHARLES C. TAYLOR
Charles C. Taylor & Associates

CELLULOSE, REGENERATED.   See Cellophane.

CHECKWEIGHERS

Automatic, in-line checkweighers have long been used to perform basic weight-inspection functions. These units, as distinguished from static, off-line check scales, perform 100% weight inspection of products in a production process or packaging line. Furthermore, these machines perform their functions without interrupting product flow and normally require no operator attention during production.

These charactertistics allow the machines to be used in several different ways. The applications can be broadly categorized as weight-regulation compliance, process control, and production reporting. This article describes a number of checkweigher applications in these categories. The emphasis is on problem solutions and benefits, although some description of the machine hardware and features employed is necessary.

Weight-Regulation Compliance

The earliest use of in-line checkweighers was to help producers guard against shipment of underweight products, and today various regulatory agencies set standards for weight compliance of packaged goods. Most of these standards are adopted from recommendations made by the National Bureau of Standards (NBS) in their various handbooks. In general, these standards recommend that the average product weight shipped by the producer be equal to or greater than the declared weight and that no unreasonable underweights be shipped. The in-line checkweigher provides assurance to the producer that these general requirements are met. With proper adjustment of checkweigher and process, an additional benefit of reduced product giveaway can be enjoyed.

The equipment required to carry out the basic weight-compliance requirement consists of a weighing element (weighcell), associated setup controls and indicators, a product transport mechanism for continuous- or intermittent-motion product movement, and a reject device for diverting off-weight items out of the production stream. A typical machine comprising these elements is shown in Figure 1.

The central portion of this machine shows a weighcell consisting of a strain-gauge loadcell that is connected to the integrally mounted electronics enclosure. Associated with the weighcell is a product transport, in this case consisting of a pair of stainless-steel roller chains. Shown on the left (up-

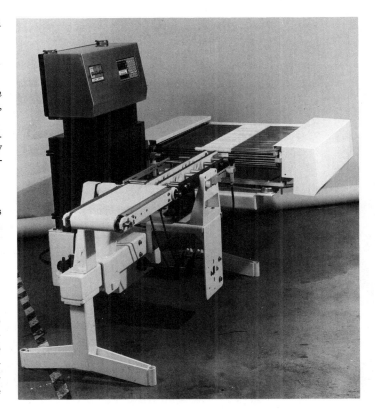

Figure 1.   A typical checkweigher.

stream) side of the unit is an infeed belt conveyor, which serves as a product-spacing (speed-up) device and assures that only one product at a time passes over the weighcell. On the right (downstream) side of the unit is a channelizer product-reject device. This unit receives signals from the checkweigher and provides a gentle lateral displacement of rejected product by carrying it to the side on sliding carrier plaques. Other commonly used reject devices include air blasts, air pushers, swing gates, and drop-through mechanisms.

Rejection of underweight products (and in some cases overweight products such as critical pharmaceutical packages or expensive products) provides the required consumer protection. The added benefit to the producer of reduced product giveaway frequently provides additional incentive to install the checkweighing machine. The information from the typical modern checkweigher allows the producer to control the target weight of the process and assures minimum average product weight, and it minimizes rejected underweight products. The following application example illustrates these points.

Net-contents weight.   Company A prepares a variety of expensive frozen-food products. These products are clearly labeled showing net-contents weight and come under close scrutiny from regulatory agencies. Some ingredients, such as chunky pieces, are of somewhat nonuniform piece weight. Until an accurate in-line checkweigher was installed, the producer was forced to overfill most packages to prevent underweight shipments. The company installed a number of checkweighers that provide the required controls. Now, by means of digital displays and easily adjustable reject cutpoints on the checkweigher, the processor is able to closely monitor

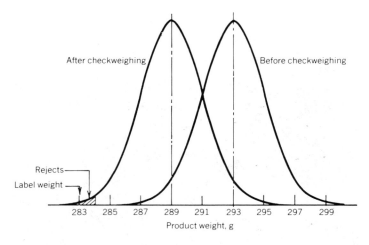

Figure 2. Weight distributions before and after checkweighing.

the filling process and establish the best compromise between product giveaway and excessive rejects.

Product-weight distribution. Figure 2 illustrates a typical product-weight distribution curve before and after a suitable checkweigher is installed.

Before checkweighing, the mean product weight was maintained at approximately 293 g to insure against underweight shipments. After checkweigher installation, the mean product weight is reduced to approximately 289 g. This results in less than 1% underweight rejects but saves an average of 4 g per package. The economics of this checkweigher installation is illustrated below:

Package labeled weight	283 g
Average overweight per package	10 g
Possible reduction in average overweight	4 g
Annual production	900,000 kg
Value of product	$0.75/kg
Cost of checkweigher (including freight and installation)	$10,000

Results

Product savings	1.4%
Product saved per year	12,720 kg
Savings per year	$9540
Payback period, approx	1 yr

Process Control

Many products that are sold on the basis of a declared (labeled) weight are packaged by a volumetric filler or count or other nongravimetric processes. These processes frequently yield products of varying weight because of product density changes. Density changes typically occur with hygroscopic products under conditions of changing humidity. Also, mate-

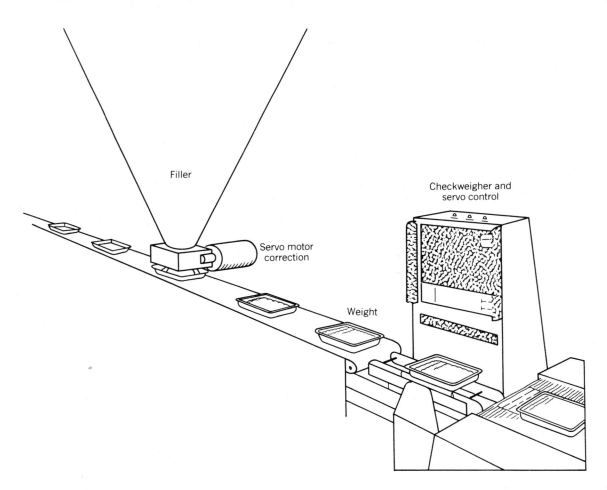

Figure 3. Checkweigher feedback-control to filler.

rial-handling methods can cause bulk-density changes in other products. Frequently, the density changes occur in the form of a drift or periodic change. These changes lend themselves well to automatic-feedback control from a weight sensor (checkweigher) back upstream to the process controller. Figure 3 diagrams a checkweigher/filler feedback-control system.

This feedback loop continually adjusts the process setpoint and maintains delivered product weight within acceptable limits, even under conditions of density variations. This in turn reduces the long-term weight standard deviation of the packaged products. This allows the process setpoint to be set at a lower average value, while minimizing the number of underweight (rejected) packages. The result is product savings. Two applications are described to illustrate the use of checkweighers with feedback for process control.

Feedback control. Company B packages a low density, hygroscopic product using a volumetric filler. Spot checks revealed that moisture variations from batch to batch and humidity changes throughout the day resulted in substantial density changes. These changes often went unnoticed until a large number of off-weight packages were produced, causing either excessive product giveaway or an excessive number of underweights being rejected by the old checkweigher. In the former case, the giveaway represented considerable revenue loss, and in the latter case resulted in product waste due to the impracticability of recycling and repackaging the rejected product.

The company purchased a replacement checkweigher equipped with feedback-control features for signaling their existing filler. The checkweigher is located as close as possible to the filler to achieve maximum responsiveness of the feedback loop. In this case, the feedback signals are electronically adjusting the setpoint of the filler. However, other control arrangements are possible including signaling a servo motor attached directly to the control shaft of a filler. This company was able to justify purchase of the checkweighing system solely on the basis of product savings. Figure 4 illustrates how long-term standard deviation reduction allows lowering of av-

erage setpoint with resultant product savings. Short-term standard deviation resulting from package-to-package filler errors is inherently beyond the control capability of the checkweigher.

Company C packages sliced luncheon meats. The "stick" or "loaf" of luncheon meat is fed through a slicer that counts a preset number of slices per draft to be packaged. Because of density changes of the meat, the correct slice count did not always yield the declared weight of the draft within allowable limits. With no on-line way to detect these density changes, the company had to supply an extra number of slices to assure labeled weight or take a chance on packaging underweights with correct slice count. It installed an in-line checkweighing machine that very accurately weighs each draft of sliced meat just after it leaves the slicer. Error signals, representing departures of delivered weight from target weight, are fed back to the slicer controls. The controls in turn cause slice thickness to be adjusted so that the preset slice count results in correct draft weight within very narrow limits. The result was control over the slicing process that economically justified the checkweigher quickly.

Production Reporting

Automatic production reporting by a checkweigher can provide running analysis of packaging-line performance. Checkweigher-generated production data can be transmitted to printers, computers, terminals, and other data receivers. Whatever the reporting medium, the result is virtually instantaneous indications of production-line performance or automatically kept records of production data. Productivity gains and material savings result from such uses. Here are two applications of the checkweigher for production recording:

Company D requires that a statistical sampling procedure be carried out on each of its production lines. Historically, quality assurance personnel periodically removed a number of product samples from the production line, weighed them on manual scales, recorded the weights, and returned the samples to the production line. Using the raw data they had collected, quality-assurance personnel manually calculated statistical information about the samples. From this information, they prepared a product sampling report. Because of the frequency of the sampling procedure and the number of lines involved, a full-time quality-assurance (QA) person was employed to perform this task. Furthermore, the manual nature of the weighing and recording resulted in reports of questionable accuracy.

Microprocessor-controlled checkweigher. The company purchased a modern microprocessor-controlled checkweigher, that is capable of accumulating production-weight data in real time and performing the required sampling calculations for transmission to a printer in the QA supervisor's office. Data reported to the supervisor include date and time of report, product code, operator identification number, average weight, standard deviation, and total weight for both accepted and rejected product over the selected sample size. Additionally, verification of checkweigher setup parameters such as target weight and reject cutpoint settings are recorded. Installation of the checkweighing system results in cost savings in personnel for acquiring the sample data and virtually instantaneous feedback of production data to QA and production personnel.

Company E ships large quantities of product to customers for further processing. To corroborate material-amount re-

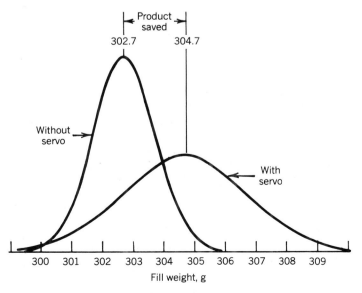

Figure 4. Filler setpoint reduction with servo system.

ports between supplier and customer, accurate total production records were required. In addition to actual weight of shipped product, an accounting of underweight (rejected) packages would yield important production-efficiency data.

The company installed an automatic checkweighing system and appropriate data-collection capabilities along with a printer for location in the accounting office. On command, accumulated data are transmitted from the checkweigher to the printer, which provide subtotals and totals of average weight, total weight of product shipped and rejected, package counts, and a figure for "yield" (shipped product divided by total production) on the basis of both weight and package count. The result of this installation is a true and accurate summary report for the accounting department of product shipped to each of their customers, thereby minimizing protests. The additional benefit is a more accurate report of actual production efficiency.

The application examples presented here describe a few of the many kinds of packaging lines that can and should benefit from the use of in-line automatic checkweighers. A few additional applications are suggested below:

Piece-count verification in packages and cases.

Product-safety assurance by detecting undesirable underfills or overfills.

Multizone classification of natural products such as chicken parts, fish, etc.

Detection of missing components in "recipe" food packages.

A survey of product weights in any processing/packaging line may reveal opportunities for substantial cost savings through use of an automatic checkweigher.

BIBLIOGRAPHY

General References

C. S. Brickenkamp, S. Hasko, and M. G. Natrilla, *Checking the Net Contents of Packaged Goods*, National Bureau of Standards, Washington, D.C., 1979.

C. Andres, "Microprocessor-Equipped Checkweighers," *Food Process.* **40**, 152 (Jan. 1979).

C. Andres. "Special Report: Expanding Capabilities of Checkweighers," *Food Process.* **42**, 58 (Mar. 1981).

N. W. Rhea, "Are You Ready For Checkweighing?," *Mater. Handl. Eng.* **5**, 74 (May 1983).

Automatic Checkweigher Test Procedure, Scale Manufacturers Association, Inc., Washington, D.C., Jan. 1977.

"Checkweighing Scales," *Mod. Mater. Handl.* **37**(3), 66 (Feb. 1982).

U.S. Department of Commerce/National Bureau of Standards, *NBS Handbook 130, 1979 (Draft)*, U.S. Government Printing Office, Washington D.C., 1979.

U.S. Department of Commerce/National Bureau of Standards, *NBS Handbook 133, Checking the Net Contents of Packaged Goods*, U.S. Government Printing Office, Washington D.C., 1981.

U.S. Department of Commerce/National Bureau of Standards, *NBS Handbook 44, Specifications, Tolerances, and Other Technical Requirements for Commercial Weighing and Measuring Devices, (Fourth Edition)*, U.S. Government Printing Office, Washington D.C., Dec. 1976.

J. F. Jacobs
Ramsey Engineering Company
Icore Products

CHILD-RESISTANT PACKAGING

Child-resistant (CR) packaging, a term synonymous with poison-prevention packaging (PPP) and special packaging (SP), can be defined as a container that precludes entry by children under the age of five years but not adults to hazardous substances such as drugs, household cleaning agents, and pesticides. Packaging of this nature may take a number of forms, including bottles, drums, pouches, and blister packs. Many children have been saved from bodily harm and even death as a result of CR packaging.

Historical Aspects

Designs for CR packaging can be traced back to 1880 when the first U.S. patent was issued for a CR package. The U.S. Congress began to take direct interest in 1966, in response to public concern about the large number of children gaining access to harmful substances in the home. As a result of Congressional hearings that year, the commissioner of the FDA appointed Dr. Edward Press to be chairman of a committee to review the "state of the art of safety packaging." The Press committee, as it came to be known, was comprised of members from U.S. industry and government. The committee reviewed the 63 patents on CR packaging that had been awarded between 1880 and 1966 and decided that the most realistic and practical approach to this problem was to establish a performance standard, using children to test the units. A series of closure studies was conducted, involving more than 1000 panelists: adults from 18 to over 65 years of age and children between the ages of 18 and 52 months. From the data obtained in these studies a Protocol to evaluate child-resistant packaging was derived. This protocol, submitted to the FDA in 1970, is reflected in the Protocol cited in the CFR (1), with the exception that the Press committee protocol contained two three-minute test periods for adults, one before a demonstration and one after a demonstration. The current Protocol, as it appears in the CFR, cites just one five-minute test period for adults without a demonstration.

The U.S. Poison-Prevention Packaging Act (PPPA), signed into law December 30, 1970, was under the jurisdiction of the FDA. The protocol for the evaluation of poison-prevention packaging appeared in the Federal Register in July 1971 (2). FDA standards began to appear in the Federal Register in 1972, first for aspirin, and then for controlled drugs, methyl salicylate, and furniture polish. In May 1973, jurisdiction was transferred from the FDA to the newly formed U.S. Consumer Product Safety Commission (CPSC), and additional products came under regulation. The CPSC was given the responsibility for medicines and household substances, and in 1979 the EPA was delegated to administer the compliance of pesticides.

Household pesticides of a hazardous nature were regulated in March 1981 (3). The initial EPA regulation was more stringent than the CPSC regulation, but much of the initial draft regulation was deleted to a point where the EPA regulation (4) approximates the CPSC regulation regarding testing requirements and standards. The major difference between the two regulations, at this point, is that the CPSC regulation makes provisions for a noncompliance package that can be used by older adults and/or households without children, and the EPA regulation does not.

Although there is worldwide awareness of child-poisoning problems in the home, only three other countries have followed the lead of the United States in regulating the need for certain hazardous substances to require CR packaging. These countries are Canada (5), the FRG (6), and the United Kingdom (7). Regulations for these countries are not as extensive as those for the United States, in that the U.S. regulations cover drugs, household-type cleaning agents, pesticides, and other products, whereas the regulations of the other countries are aimed primarily at pharmaceutical products. Their CR packaging standards and test procedures are comparable, however.

Members of the International Standards Organization (ISO) have been working on a proposed child-resistant packaging standard since 1981, and it is anticipated that the standard should be enacted by 1988. A standard of this nature could have quite an impact on the types of packaging used for items imported by member countries. There are some differences in the ISO proposed standard and the U.S. Protocol, which are discussed below.

Effect of Regulation

Fifteen types of substances regulated by the CPSC are listed in Table 1. Regulations were placed upon these substances because of their harmful nature to small children and the large number of ingestions recorded. The table was assembled by members of the CPSC staff from FDA data, which were obtained from approximately 400 U.S. poison-control centers. It shows that the Poison-Prevention Packaging Act has saved many small children from harm. This is obviously due to the use of child-resistant packaging for these regulated substances. Similar data for household pesticides have not yet been processed by the EPA.

Testing Procedures

The test procedures for the evaluation of child-resistant packaging (protocols) are similar in the four countries with regulations in force (1, 4, 5, 6, 7). They all involve the evaluating of test packaging with human subjects. The panelists consist of 200 healthy children with no obvious handicaps, between the ages of 42 to 51 months old, equally divided between age in months and sex with an allowable 10% preponderance, and 100 normal adults, 18–45 yr of age, 70 females and 30 males.

The children are tested in familar surroundings, such as a nursery school, day-care center or day camp. They are tested in pairs under the observation of a tester. First, the children are each given a unit and asked to open it, and allowed 5 min to achieve this task. If they are unsuccessful in opening the unit in the first 5 min, they are given a nonverbal demonstration of how to open the unit, and if in the first 5 min they have not used their teeth, the tester tells them they can use their teeth if they wish. They are then allowed a second 5 min to open the unit.

Opening patterns are recorded, and a passing unit is one which has not been opened by at least 85% of the child panelists in the first 5 min of the evaluation and at least 80% of the panelists in the course of the full 10-min test period. Adults are tested individually in the adult phase of the evaluation and are allowed 5 min to open and resecure the test unit (if appropriate) merely utilizing the directions that appear on the package. A passing unit is one which at least 90% of the adults can successfully open and, if necessary, close.

In the case of unit packages, such as blister packs (strip packs), an individual unit failure for the child test is one which has more than eight blisters or packets opened or the number of individual units which contain a toxic dose of product. A toxic dose, in this case, is that quantity of a substance that would cause bodily harm to a 25-lb (11.3-kg) child. To pass the adult phase of the test, a unit package must have at least one unit accessed by at least 90% of the adult panelists.

The child standards of 85% before demonstration and a total of 80% unsuccessful panelists after demonstration, as stated, are cited in the regulations, but packagers should use a more stringent in-house standard to be assured that a package can satisfy the regulations after variations in the molds, handling on the packaging line, and distribution. Companies that have aimed for an unsuccessful-child pattern of less than 87% after the full 10-min test have encountered problems with the U.S. federal compliance agencies.

The Fifth Draft of the proposed ISO standard for CR packaging differs in two major aspects from the procedures cited above. First, the ISO draft standard incorporates a sequential test pattern that appears in the FRG standard (7). This se-

Table 1. A Comparison of the Number of Ingestions of Regulated Substances by Children Under 5 Years Old

Regulated substance	Effective year	Ingestions during the effective year	Ingestions during 1982	Decrease since effective year
aspirin	1972	8146	1753	78%
controlled drugs	1972	1810	541	70%
methyl salicylate	1972	161	49	70%
furniture polishes	1972	697	229	67%
illumunating and kindling preps.	1973	1736	452	74%
turpentine	1973	777	110	86%
lye preparations	1973	508	69	86%
sulfuric acid[a]	1973	7	1	86%
methanol[a]	1973	37	21	44%
prescription drugs (oral)	1974	4180	2251	46%
ethylene glycol	1974	138	59	57%
iron preparations	1977	359	163	55%
paint solvents	1977	641	180	72%
acetaminophen	1980	1511	1118	26%

[a] Since these figures are based on a few cases, the percent decrease may not be reliable.

quential testing pattern was devised because of the difficulty of obtaining large numbers of children of the appropriate test age in Europe. The sequential testing is basically a statistical approach to conducting the package evaluation in order that, if desired, fewer than 200 children may be employed to conduct the packaging evaluation. The FRG version of this method takes into consideration the variability between each group of children, but does not consider the variability of age of the children, the sex of the children, or the number of sites from which the children are obtained. On the basis of tests with hundreds of thousands of panelists in the United States, it has been observed that ability to gain access to the contents of child-resistant packaging is dependent on more than the population of a group of children.

As a result, strict adherence to the following procedures has been recommended for the ISO standard, if a sequential test pattern is to be utilized:

Age distribution should be rigidly adhered to since trends demonstrate an increasing ability to open units according to the age of the child panelists.

Sex distribution should be rigidly adhered to since male versus female success patterns differ according to the type of package under test.

More than five test sites should be utilized for each test. Variations in the abilities of children from different test sites have been observed as a reflection of teaching methods, discipline patterns, and socioeconomic distribution of the children.

A number of testers should be employed. Variability was found in the results obtained from tester to tester in a study conducted by the CPSC.

Because the CPSC is aware of the above test parameters, package evaluation conducted under contract to the CPSC requires that testing be conducted at no less than five sites, employing no less than four testers, and the age and sex distribution must be very explicit. Package evaluations, utilizing the CPSC format and employing only 50 child panelists, have resulted in tests that require as many or more sites than is necessary to conduct a typical study with 200 child panelists. This is due to the fact that the 50-child test does not allow for a 10% variation in distribution of each sex and age group. Thus, there does not appear to be any real time or personnel savings utilizing the 50-child CPSC test procedure over the 200-child CPSC test procedure. This may also apply to any sequential test.

The second major difference in the draft ISO standard versus the U.S. protocol is that the ISO standard incorporates an additional test group consisting of 100 elderly adults of ages 60–75 yr. Twenty of these adults are to be more than 65 yr old, and the ratio of females to males shall be 7 : 3. The rationale for the inclusion of an elderly adult test in this standard is based on reported difficulties both in Europe and the United States that this group encounters when trying to open some of the CR packaging types that are being marketed.

The CPSC is aware of the older-adult problem and is currently reviewing the merit of the addition of an older-adult test as part of PPP protocol. A lower-pecentage success standard (70–80%) would be employed than for the adults of 18–45 yr, at least for the ISO standard.

In contrast to U.S opinion, some Europeans consider unit packaging in the form of typical blister packs or strip packs to be inherently child-resistant. The theory is that a child would only access one tablet or capsule at a time from such a package, and that by the time a parent found the child gaining access to the package contents, the child would not have gained access to a toxic dose or the child would become bored and stop opening the units. In the United States this premise has been found to be totally incorrect. It has been observed in the course of some unit-packaging evaluations that over 80% of the children can gain access to a toxic dose if the packaging is not designed for child resistance. Children have been observed to remove as many as 34 tablets from unit packaging in a 10-min test period. Further, it has been observed that a number of children, once they realized how to open the package, did not want to leave the test site until all the tablets or capsules were removed from the test package, even after the 10-min test period was over. Based upon these observations, in the United States only CR unit-dose packaging is utilized for regulated products. Marketing groups in many U.S. companies consider the blister-pack-type unit to be an excellent vehicle for marketing tablets and capsule products, because dose levels can be compartmentalized, the container is light and easy to carry, and it displays the product well. Problems have been observed with this type of packaging, however, with products that have a high toxic level per tablet or capsule. In cases where a toxic dose can be found in one, two, or three individual blisters, it is difficult to find a package that passes the child test and is still acceptable for the adult population. This problem has led many companies to turn to alternative packages.

Mechanical Testing

In 1972, the American Society for Testing and Materials (ASTM) initiated efforts in developing standards for PPP by establishing a subcommittee to deal with this subject. Since its formation, the ASTM D 10.31 subcommittee has published a number of mechanical tests for child-resistant closures. These standards have been employed by many closure manufacturers and manufacturers of end-products to be assured that packaging satisfies quality-control specifications (see Specifications and quality assurance). They are contained in the ASTM volume of standards on packaging (8).

Along these lines, members of the CPSC staff, with the assistance of the National Bureau of Standards, have developed a mechanical device that measures the forces necessary to open and close a variety of different types of closures and plots them electronically. This device is used to assist CPSC personnel in evaluating CR packaging for compliance purposes.

Mechanical devices have been found to be useful in measuring the characteristics of established closures for quality control and compliance indication once the units have been demonstrated to be effective in the child testing. However, a mechanical device that has the ingenuity and the tenacity of small children has not yet been invented. Therefore, it is necessary to employ children and adults to evaluate new and modified packaging designs.

Enforcement

CPSC and EPA are the two federal agencies in the United States that are responsible for enforcing the CR regulations: the CPSC for household products and medications, and the EPA for pesticides utilized in and around the home. Both agencies have their own legal staffs, which can act with the Justice Department against companies whose products are not in compliance with packaging regulations. The U.S. Federal Government relies on product samplings from retail stores and warehouses, complaints from the public, and the number of child ingestions to determine which products should be evaluated for compliance purposes. Producers of products found not to be in compliance are approached by the agency, informed of their problem, and advised to improve their package. If the package is still found not to be in compliance, the Federal Agency involved can go to the extent of halting further production of the package form and ordering product recall. This, of course, can result in loss of revenues and undesirable publicity for package and product manufacturers. Voluntary recalls of products by manufacturers have occurred in the past because of packages not satisfying the regulation.

The Canadian, British, and FRG regulations are based on registration and certification of CR packaging used for specific products. Once this is established, it is up to the companies manufacturing the products to make certain the packaged product is CR. Compliance efforts to the degree utilized in the United States are not employed.

The United States has recognized that CR characteristics of packaging can change because of dimensional problems encountered in the manufacture of components of the packaging to render the packaging non-CR. These usually have been the result of a poor quality-control program by the package manufacturer and/or the package user. Rigid compliance sampling and testing has resulted from problems observed with faulty CR packaging.

Features Affecting CR Effectiveness

In the course of evaluating many CR packages, it has become apparent that certain features can affect the CR characteristics of a package.

Closures generally perform better on plastic or metal containers than on glass containers.

Smaller sizes (18–24 mm) of closures are generally easier for children to open than larger sizes (≥33 mm).

Different-size bottles may perform differently with exactly the same closure size.

Different closure liners (see Closure liners) perform differently in the same closure.

In recognition of these factors, as well as other bottle/closure arrangement characteristics, the EPA regulations published in 1981 (3) include a testing scheme to be carried out if modifications or changes were made after original testing in the following respects: packaging shape; packaging material; volume of package; closure material; and/or cap liners. These requirements, though realistic, were later deleted (9). Nevertheless, these features should be considered in the design and modification of CR packaging.

Classification

Some of the designs of CR packaging in the late 1960s and early 1970s were based upon the need for strength to open them. Because these units prevented child entry, but adult entry as well, they were not of much value. Other designs requiring two dissimilar simultaneous motions in order to activate or open a unit, such as push-and-turn, squeeze-and-

turn, pull-and-turn, and turn-and-push, were developed. These designs have been quite successful since small children appear to lack the ability to readily perform the two motions at the same time.

In addition, there are units that require implements, such as screwdrivers or scissors, to open. This concept has been effective since the protocol disallows giving child panelists implements not included with the unit in the course of the test pattern. These and other types of units appear in the ASTM classification standard.

The ASTM D 10.31 subcommittee has established what can be considered the official classification system for child-resistant packaging. This ASTM standard, the D 3475-76 Classification of Child-Resistant Packages, is based on the forces required to open the packages (10). Nine major types of packages are included in this classification system, each with a number of subgroups within the types. Examples of the different types of packages along with their producers are presented in the standard. A total of 54 different kinds of packaging appear in this document with as many or more manufacturers of packaging. The types of packaging presented in the classification system are Continuous Thread Closure, Lug Finish Closure, Snap Closure, Unit Packaging—Flexible, Unit Packaging—Rigid, Unit Reclosable Packaging, Aerosol Packaging, Nonreclosable Packaging—Semirigid (Blister), and Mechanical Dispensers.

BIBLIOGRAPHY

1. *Code of Federal Regulations, Title 16, part 1700*, CPSC Regulation, U.S. Consumer Product Safety Commission, Washington, D.C.
2. *Fed. Regist.*, (July 20, 1971).
3. *Fed. Regist.* **46**(41), 15106 (Mar. 3, 1981).
4. *Code of Federal Regulations, Title 40, Part 162*, EPA Regulation, U.S. Environmental Protection Agency, Washington, D.C.
5. *CSA Standard Z76-1979, Child-Resistant Packaging*, Canadian Standards Association, 1979.
6. *DIN 55 559, Child-Proof Packaging/Requirements Tests*, FRG Standard, 1980.
7. *BS 5321, Reclosable Pharmaceutical Containers Resistant to Opening by Children*, British Standard, 1975.
8. "Closures, Child-Resistant 15.09," in *Annual Book of ASTM Standards*, American Society for Testing and Materials, Philadelphia, Pa., 1983.
9. *Fed. Regist.* **47**(179), 40659 (Sept. 15, 1982).
10. *Child-Resistant Packaging*, ASTM Technical Publication 609, American Society for Testing and Materials, Philadelphia, Pa., 1976.

General References

R. L. Gross and H. E. White, *Identification of Selected Child-Resistant Closures*, U.S. Consumer Product Safety Commsion, 1978, U.S. Government Printing Office, Stock No. 052-011-00194-5.

Safety Packaging in the 70s, proceedings of a conference sponsored by the Scientific Development Committee, The Proprietary Association, New York, 1970.

P. Van Gieson, "ASTM History in Child-Resistant Packaging," *ASTM Stand. News*, 26 (Apr. 1983).

A. M. PERRITT
Perritt Laboratories, Inc.

CHUB PACKAGING

The term "chub" as related to packaging originated in the processed meat industry as a term used to describe large "chubby" sausages similar to bratwurst or kielbasa. The term "chub package" is used to describe sausage-shaped packages used for a wide range of semiviscous products (see Fig. 1).

The chub package is made by forming a web of flexible packaging film around a mandrel, sealing the overlapped edges to form a tube. A clip or closure is applied to the bottom of the tube as product is introduced through the interior of the mandrel. Finally, a clip or closure is applied to the top of the tube, forming a closed finished package (see Fig. 2). Alternative methods of chub package formation involve the use of preformed tubes of flexible packaging film which are filled and closed at the ends but retain the cylindrical sausage shape of the classic chub package.

Automatic form/fill/seal equipment such as the Kartridg Pak Chub Machine (see Fig. 3) provide a means to automatically and continuously produce chub packages. Package size ranges from about 0.6 in. (15 mm) to 6 in. (150 mm) in diameter and up to about 48 in. (1220 mm) in length. Depending on package size, production rates of up to 100 packages per minute are common. Semiautomatic means of producing chub packages are also available.

Chub-packaged products have found wide acceptance in the consumer and industrial marketplace. Almost any pumpable viscous product can be chub packaged. A listing of chub-packaged products appears in Table 1.

Many types of flexible packaging film are used to produce chub packages. Films are generally selected on performance and cost considerations. Aesthetic considerations such as finished package appearance and printability may also influence packaging film selection.

Chub-packaged products such as fresh pork sausage, ground beef, cold pack cheese, or slurry explosives are packaged in essentially the same form in which they are delivered to the user. Some chub-packaged products such as soups, thermoplastic poultry rolls, and imitation cheese are packaged at elevated temperatures, usually not higher than 194°F (90°C), in a viscous state and set or become solid when chilled. Products such as liver sausage, thermoset poultry rolls, precooked pork sausage, and cooked sausage are heat processed after

Figure 1.   A chub package.

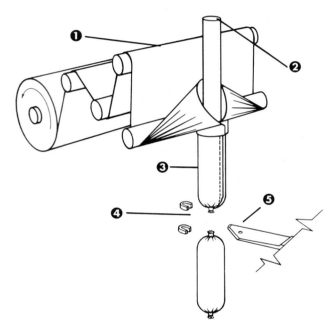

Figure 2.   Formation of chub package. **1**, Film is formed around mandrel; **2**, the overlapped edges are sealed **3** to form a tube and the product is introduced through the mandrel interior; **4** closures or clips are applied to the ends of the package; and **5** the finished package is discharged.

Table 1.   Chub-Packaged Products

barbecue beef	epoxy resin (2-part)	pizza topping
barbecue sauce	explosive slurries	pork roll
beef roll	frosting	pork sausage
bologna	frozen bread dough	poultry rolls
butter	frozen juice	precooked pork sausage
caulking compound	fruit preserver	processed cheese
chili	ground beef	refried beans
chocolate fudge	ham roll	sandwich spread
citrus pulp	ham salad	sauce bases
cooked salami	ice cream	scrapple
cookie dough	lard	snuff
cornmeal mush	mashed potatoes	soup
cream cheese	mushroom gravy	tamales
dental impression material	pet food	vegetable shortening
drywall compound	pie dough	
elongated hard-boiled eggs	pizza sauce	

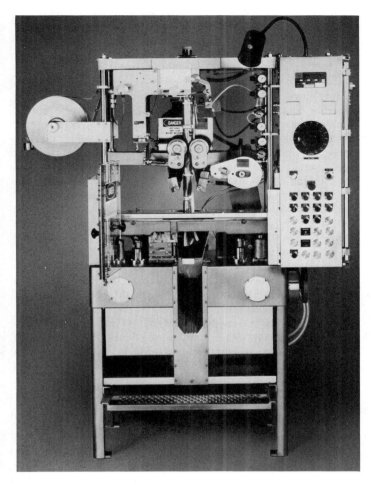

Figure 3. Automatic chub-packaging machine. Courtesy of The Kartridg Pak Co.

packaging. Chub packages have not been aseptically processed commercially.

General References

The Wonderful World of Chub Packaging, sales brochure, The Kartridg Pak Co., Davenport, Iowa, 1983.

H. J. SIEVERS
The Kartridg Pak Co.

CLOSURE LINERS

The proper choice of a closure liner (see Closures) can often make the difference between the success and failure of a product. Most closure liners contain two basic components: a backing and a facing. The backing provides compressibility, resiliency, and resealability, and the facing provides barrier protection. In some liners, one component serves both functions, and in some two-component liners, each component contains more than one material.

Selection of the proper lining material generally involves one or more of the following considerations: adequate chemical resistance to the product; sealing ability against volatile loss; sealing ability against moisture loss or gain; sealing ability against air or oxygen; compliance with FDA regulations; free-

dom from odor and taste contamination; ability to withstand sterilization by various methods; avoidance of "over packaging" by selection of the most economical material for the required performance; sealing ability for products prone to leakage; sealing for products which develop pressure; sealing for hazardous and highly corrosive products; sealing for products which are vacuum packed; compatability with the container and the closure; and filled product stability. Selection is typically based upon field experience with a similar product, and knowledge of the properties of specific materials and their reaction to various products. Performance requirements are often appraised differently by individual packagers. The differences are based on marketing approach, merchandising practice, and the rate of product turnover and expected shelf-life. Responsibility for the selection always rests with the packager.

Liner Materials

The state of the art in liner technology can best be described as "crowded." Available choices of liners currently number in the hundreds and new structures continue to appear. Although there are more than 40 basic structures, they generally contain one or more of the following: polyethylene, PVC, PVDC, polyester, aluminum foil. Along with pulp, chipboard, or newsprint backings, these five substrates, in combination with various waxes and adhesives, comprise the basic "standard" materials list. The many options for facings fall into about six major categories: coated paper; paper laminations; unsupported foil; coated foil; film/foil; and plastic film.

Each of these categories contains subgroups: eg, paper laminations include paper/foil, paper/OPET, paper/foil/PVC, paper/PVDC, paper/foil/PE/, paper/foil/PVDC, etc. The coated-foil category includes foil-PE, foil-PVC, foil-PVDC, foil-PP, foil-ionomer, and foil-EAA, etc. Manufacturers of lining materials generally use their own trade names and grade designations to describe structures: for example, PET-coated aluminum foil is available in the United States under at least four separate brand designations.

Extruded lining materials have gained wide acceptance in the past ten years. These include solid monolayer (eg, LDPE, EVA) and coextruded plastics; monolayer foam plastics; and coextruded solid/foam structures. The polymers are generally either polyolefins or PVC, which provide outstanding chemical resistance and relatively low water-vapor transmission rates (WVTR). They can be tailored to density, thickness, width, color, and surface texture. The solid/foam coextrusions typically consist of solid polymeric skins on the top and bottom, with a foamed core in the middle. Most are combinations of homopolymer and copolymer polyolefins. All of the synthetic materials, when used as backings, can be combined with all of the common facings used with pulp backings.

Innerseals

Innerseals had been used for many years for barrier and leakage protection before they attracted attention as tamper-evident packaging materials (see Tamper-evident packaging). There are three types of innerseals: induction, pressure-sensitive, and glued-glassine. Heat-induction seals are particularly useful as tamper-evident devices. Potential users can select from over 50 structures. Typically, heat-induction foils are specified in one of three ways: wax-adhered (temporarily) to a separating base material; permanently adhered to chipboard;

or coated unsupported foil for use in dispensing closures. The heat-induction (see Sealing, heat) foils vary with respect to foil gauge, thickness of coating/film, type and thickness of backing material, foil alloy, level of adhesion to various types of container materials, melting and sealing temperature, strength, appearance, destruction properties, and cost. Because seals applied by heat-induction cannot be removed without breaking the seal, they are particularly useful as tamper-evident features.

Pressure-sensitive innerseals provide functional sealing and a measure of tamper-evidence to glass and plastic containers without the need for heat induction or a secondary adhesive application. Typically, these structures consist of a foamed polystyrene (see Foam, extruded polystyrene) with a surface adhesive that seals under torque-activated pressure. Although these structures are not recommended for oils, hydrocarbons, and solvents, they do provide an inexpensive contribution to seal integrity. Glassine (see Glassine, greaseproof, parchment) the oldest of the three innerseal types, is generally wax-bonded (see Waxes) to clay-coated pulp board and adhered to the container with adhesive. Unlike heat-induction structures, the bond between glassine and pulp is broken when the consumer removes the closure.

Before the regulations for tamper-resistant packaging went into effect, glassine was commonly used in the pharmaceutical industry. With the general trend toward heat-induction foils, the largest remaining market for glassine is in instant coffee and a variety of powdered drink mixes and nondairy coffee creamers.

The performance characteristics of all closure lining materials are based on the following criteria: WVTR, gas transmission rate and blocking properties; chemical resistance; stability and compatability with the container, the closure, and the product being filled; and containment of the product.

When evaluating available structures for a specific application, the selection process does not necessarily result in an obvious solution. The primary objective should be to achieve an equitable balance between meeting the performance criteria and remaining cost-effective.

BIBLIOGRAPHY

General References

Technical data books with structural breakdown, FDA compliance, and suggested uses and limitations of closure liners can be obtained from suppliers such as: Cap Seal Division, 3M Co., St Paul, Minnesota; Selig Sealing Products, Oakbrook Terrace, Illinois; Tekni-Plex Inc., Brooklyn, New York; and Sancap Inc, Alliance, Ohio.

Information about extruded and coextruded closure liners may be obtained from suppliers such as: J. P. Plastics, Naperville, Illinois and Tri-Seal International, Blauvelt, New York.

R. F. RADEK
Selig Sealing Products, Inc.

CLOSURES, BOTTLE AND JAR

The cork stopper and the continuous-thread cap represent two epochs in closure evolution. The cork stopper began its slow ascendency 25 centuries ago, attaining its broadest use by the middle of the 19th century. With the arrival of the standardized continuous-thread cap and the introduction of plastic closures, both in the 1920s, the modern closure era was underway.

Cork provided an incomparable friction-hold seal. A material of high cellular density, cork is compressible, elastic, highly impervious to air and water penetration, and low in thermal conductivity; a natural panacea for the elementary problems of closure. Historical antecedents of the cork stopper are found within the Roman Empire. The art of glass-blowing matured commercially there, resulting in a vast commerce of vases, jars, bottles, and vials. Bottles and jars were more common during the Roman Empire than at any period before the 19th century (1). The use of cork floats and buoys by the Romans, with subsequent applications as bungs (large stoppers) for casks, suggests the likelihood of fabricated cork bottlestoppers. Yet with the fall of the Roman Empire, glass-blowing and the use of the cork stopper declined until after the Renaissance. Other sealing methods used at the same time as the rise of the cork stopper in the 16th century include a Near Eastern method of covering the container with interlaced strands of grass, or strips of linen, and applying a secondary seal of pitch. Western Europeans used glass stopples and various lids of glass and clay before the common use of cork. Wax was a very common closure, inserted into the neck and covered with leather or parchment. Raw cotton or wool, sometimes dipped in wax or rosin, was also employed, frequently covered with parchment or sized cloth, which was then bound to the neck (2).

The aftermath of the Industrial Revolution was characterized by a heightened quest for technical sophistication. Steam and, later, electric power provided quicker realization of more complex goods. The ethic of the economy of scale, "the more you produce, the less costly it is to produce it," became the momentum for the sudden explosion in manufacturing technology. During this time the world was colonized on vast scales, and population doubled in 150 years. These forces of urbanization and industrialization created an unprecedented demand for bottled goods by the 19th century. Closure evolution of this period reflects the search for a practicable seal through a variety of mechanical devices. By mid-19th century cork was the predominant closure, providing a friction seal for foods, beverages, and patent medicines.

Many attempts were made then to attune the concept of threaded closure to the demands of a new industry. A major contributor was John Mason. His 1858 patent of the Mason Jar redesigned glass threads to accomplish a tighter and more dependable seal. Developing closure and container industries were to remain a chaos of varying pitches, lengths, and thicknesses throughout the century as manufacturers tried to perfect some particular feature which would require their exclusive manufacture. Intermediate to the development of the standardized continuous thread cap was the Phoenix band cap (see Fig. 1), a popular closure among food packagers because it could provide a hermetic seal in a world of imperfect finishes. Invented in France by Achille Weissenthanner in 1892, the closure provided adjustability to finishes by means of a slit-and-tongue neckband, an improvement over the original neckband closure patent of 1879 by Charles Maré of France (3). This cap also introduced the custom which later became standard in the industry: measuring closures in millimeters (mm).

By the end of the century some form of external mechanical fastener, such as Henry W. Putnam's bailed clamping device

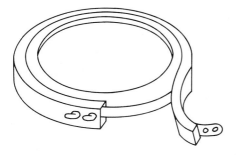

Figure 1. The Phoenix band closure.

known as the "Lightning Fastener" was the leading closure for beer and ale. Internal stoppers provided the seal for most carbonated soft drinks, most notably Hiram Codd's glass-ball stopper and Charles Hutchinson's spring stopper (4).

In the early 1900s Michael J. Owens successfully automated the production of jars and bottles, which in turn created latent market demand for inexpensive, easy to use, standardized closures. The crown cap, devised by William Painter around 1890, provided a solution for the beverage industry. Shortly after WWI glass and cap manufacturers, through the Glass Container Association, designed and standardized the shallow, continuous-thread cap. Subsequent forces shaping today's closure were the emphasis on package styling created in the 1930s and the development of thermoplastic-molding technologies. During the 1970s, plastic closures showed a 60% market increase (5). In the 1980s the forces of consumer demand for convenience, society's need for access control, and industry's need for cost-efficient innovation continue to redefine the closure.

Closure Functions

A closure is an access-and-seal device which attaches to glass, plastic, and metal containers. These include tubes, vials, bottles, cans, jars, tumblers, jugs, pails, and drums. About 80 billion (10^9) closures are produced annually in the United States (6). The closure works in conjunction with the container to fulfill three primary functions: to provide protective containment through a positive seal; to provide access and resealability according to varying requirements of convenience and control; and to provide a vehicle for visual, audible, and tactile communications.

Protective containment. Protective containment and seal are achieved when closure and container are integrated to form a unified protection system for the product during its cycle of use. Protective containment has a two-fold meaning: containing the product so that neither the contents nor its essential ingredients escape; and providing a barrier against the intrusion of gases, moisture vapor, and other contaminates.

A positive seal. A packaged product is vulnerable to many forms of natural deterioration, including migration of water or water vapor, contamination by oxygen or carbon dioxide, and assaults by microbiological life. The packaged product is further challenged by extremes of heat and cold, dryness and humidity, and by physical stresses imposed upon it during the distribution cycle (7).

A positive seal is attained when the contact points of the closure and the top of the container (its "land" surface) are pressed together to form a seal. Frequently a resilient lining material, compressed between the closure and the container, provides a tighter, more secure seal. A liner may be made of paper, plastic, or metal foil, and is often a composite of many materials (8). A seal may also be formed by caps containing flow-in compounds where a gasket is devised by pouring a liquid sealing compound into the closure. A variety of linerless thermoplastic closures utilize molded-in sealing devices. These embossed or debossed features press against the land surface and provide a seal when the closure is applied and tightened. Sealing specifications may range from mere containment to the preservation of highly sensitive food, pharmaceutical, household, and industrial products. Three common types of seal applications are sterilized, vacuum, and pressurized.

Two closure methods provide containment and seal: friction-fitting closures, including snap-ons, stoppers, crowns, and press-ons; and thread-engagement closures, including continuous-thread and lug caps. A positive seal depends upon such factors as the type of product, closure, container, and seal desired, the resiliency of the liner, the flatness of the sealing surface, and the tightness or torque with which the closure is applied (9).

Access. Contemporary closure design is shaped by the demands of a pluralistic marketplace where strong consumer preferences for convenient access exist alongside legal mandates for access control. Many packages today are ergonomically designed systems capable of easy opening and dispensing, and also affording critical access control. Closure technology has always sought to provide "a tight seal with easy access", but today's simultaneous demands for easy access and access control are the most polarized in the industry's history. Access to a product can be said to exist on a continuum of convenience. This may range from the knurls (vertical ribbing) on the side of a continuous-thread cap, designed to provide assistance in cap removal, to what can be called convenience closures, which have a variety of spouts, flip-tops, pumps, and sprayers to facilitate easy removal or dispensing of the product.

Control. Concurrent to greater demand for convenient, often one-handed, access to a product, legal mandates and consumer preferences press for more access controls. These access controls are of two major types: tamper-evident (see Tamper-evident packaging) and child-resistant (see Child-resistant packaging). Regulated tamper-evident (TE) closures may be breakable caps of metal, plastic, or metal/plastic composites. In one variety the closure itself is removable but a TE band remains with the neck of the bottle. In another, the TE band is torn off and discarded. Another system not specifically addressed in the regulations incorporates a vacuum-detection button on the closure. Other TE systems include paper, metal foil, or plastic innerseals affixed to the mouth of the container. The FDA has stated 11 options for making a package tamper-evident, two of which apply to closures (10). Child-resistant closures (CRCs) are designed to inhibit access by children under the age of five. This is frequently accomplished through access mechanics involving a combination of coordinated steps which are beyond a child's level of conceptual or motor skills development. Of these closures, 95% are made of plastic; the remaining 5% combine metal with plastic (11).

Verbal and visual communications. The closure is a focal point of the container. As such, it provides a highly visible position for communications, an integral aspect of today's packaging. Three communication forms include styling aesthetics, typography, and graphic symbols. Since the closure is handled and seen by the consumer every time the product is used, the audible, visual, and tactile message (often subconscious) becomes very important to the packager.

Styling Aesthetics. Aesthetics are an important consideration because package design has the same basic goals as advertising: to promote brand awareness leading to brand preference. The closure and the container provide a visual symbol of the product, creating imagery through aspects of styling. Three important aspects are form, surface texture, and color. The form of a closure can be utilitarian to suggest value, or it can assume elaborate and elegant forms to suggest luxury. The surfaces of glass, metal, and plastic can provide a variety of surface textures unique to the materials used. Metal caps and decorative overshells of steel, aluminum, copper, or brass, can be burnished, painted, screened, or embossed. Plastics can be molded in vivid colors, anodized to assume metallic sheens, or printed, hot-stamped, screened, or embossed. Glass can provide the kind of design statement exclusive to the glass arts, creating imagery of luxury or elegance. Many closures today are styled simply, with brand-identification or functional embossments (eg stacking rings) appearing on the closure top.

Color is the most pervasive form of closure decoration. A closure may be purely functional in form yet, with color, it can take on dramatic significance. With the advent of color-matching systems in industry, such as the Pantone Matching System (Pantone, Inc.), the closure, container, label, and point-of-purchase display can be coordinated to produce a strong emotional reaction. The emotion may be one of action and excitement, as the hot primary colors used in soap and detergent packaging, or cool and subdued colors that characterize many cosmetics and fragrances. In addition to its decorative aspects, color can provide functional assistance. Color contrast directs the eye to areas of emphasis, and this direction can be important in teaching the mechanics of container access and use. In a crowded environment of dispensing options found at point-of-purchase, color can help identify a closure as one that the consumer already knows how to use. A great many dispensing closures today, for example, differentiate the spout from its surrounding fitment by strong color contrast. Closure color can also be used to identify the flavors of a food product or beverage and help to differentiate these flavors quickly within a product line.

Typography. Common forms of written communications found on a closure may include brand identification, a listing of ingredients, nutritional information, access instructions, or consumer advisories. These can be printed, screened, hot stamped, or molded onto a closure. For purposes of impact at point of purchase, a brand name frequently appears on the closure top.

Graphic symbols. A graphic symbol frequently found on the closure is a company or product logo. Another common graphic is the arrow, a symbol which has gained importance with the advent of safety, convenience, and control mechanisms of modern closures. Arrows direct the consumer to proper disengagement of the closure, indicate engagement points where access is possible, or signify the direction of dispension (eg in the control tips of spray-type mechanisms). The scannable bar code is a more recent functional graphic to appear on closures (see Barcoding) (12).

Methods of Closure

Removable closures attach to containers by two principal methods: thread engagement and friction engagement. Threaded closures include continuous-thread caps, lug caps, and metal roll-on caps. Friction-fit closures include crowns, snap-fit, and press-on types. Thread engagement is the most widely used method of attaching a closure to a container (13).

Thread-engagement types. Three closure types provide a seal through thread engagement: continuous thread (CT), lug, and roll-on caps. CT designs attain a seal through the attachment of a continuously threaded closure to a compatibly threaded container neck. The lug cap uses an abbreviated thread design, with access and reclosure accomplished in one-quarter turn. The roll-on is supplied as a blank unthreaded metal shell which then becomes a closure on the capping line when it is compressed to conform to the finish of a bottle. The press-twist closure also has its threads formed after it is applied to the container.

The CT Closure. Threaded closures were standardized in the 1920s and continue to prosper due to the basic soundness of their principle, which offers a mechanically simple means of generating enough force for effective sealing, access, and resealing (14). Today the CT design is manufactured in plastic, tin-free steel, tin plate, and aluminum. Some CT control-closures combine metal and plastic by using a regular CT metal cap and a plastic overshell. The CT closure provides a seal for the container by engagement of its threads with the corresponding threads of the container (see Fig. 2). As the thread structure, (or "finish"), is designed on an inclined plane, the engagement and application torque cause the threads to act like the jaws of vise, forcing closure and container into contact to form a positive seal. Critical sealing applications typically include a liner placed between the closure and the container. When tightened, the material is compressed between the sealing surfaces to form a seal. The thermoplastic "linerless closure" employs a variety of molded devices which can also provide a positive seal. All CT closures are designated by diameter as measured in mm followed by the finish series number. A closure with the designation number "22-400" refers to a closure with an inside diameter of 22 mm designed in the 400 finish of a shallow continuous thread.

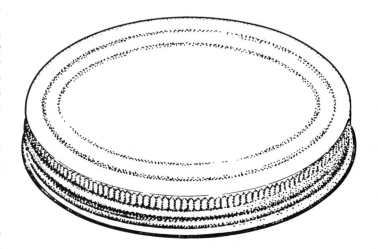

Figure 2. A metal CT closure.

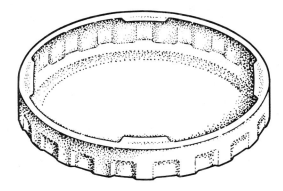

Figure 3. Lug closure.

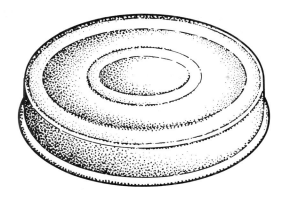

Figure 5. Press-on/twist-off vacuum closure.

Lug Cap. The lug cap operates on the principle of thread engagement as does the CT cap. The thread structure of the lug is not continuous, but consists of a series of threads which may be inclined or horizontal depending upon finish variations. The lugs of the cap are positioned under corresponding threads on the container finish (see Fig. 3). When tightened, the engagement of two, four, or six lugs pulls the closure and lining material onto the container. The lug cap is the most popular steel vacuum-closure today (15). Frequently a flow-in plastisol inner gasket is used as the liner. The lug cap is used extensively for vacuum packs in the food industry, and is suitable for use on many products packaged in glass containers. The lug cap's design allows application and removal with a one-quarter turn. This not only means consumer convenience, but also quick capping. The common finish designations for lug caps are 120, 140, and 160 referring to 2, 4, and 6 lug finishes respectively (16).

Roll-On. The aluminum roll-on cap was an innovative method of closure that won immediate acceptance within the packaging industry in the 1920s (17). Although frequently categorized separately from threaded closures, the roll-on nonetheless utilizes thread engagement to accomplish seal and reseal. What makes this closure unique is the capping process. A lined, unthreaded shell (or "blank") is furnished to the packer. During capping the blank is placed on the neck of a container and the capping head exerts downward pressure, which creates a positive seal as the liner is pressed against the container finish. Next, rollers in the capping head shape the malleable aluminum shell to conform to the contour of the container thread (see Fig. 4).

The roll-on closure is used in the food, carbonated beverage, and pharmaceutical markets where pressure sealing is required. It is considered one of the most versatile sealing devices for normal, high and low pressure seals. A widely used version today is the tamper-evident roll-on. A tear band perforated along the bottom of the closure skirt is tucked under a locking ring during capping by a special roller. When opened by the consumer, the band separates from the closure to provide visible evidence of tampering. The roll-on cap can be applied at high speeds, approaching 1200 bottles per minute. In the standard bottle finishes, the 1600 series designates roll-on.

Press-Twist. The "press-twist" is another closure which attains its threads on the capping line (see Fig. 5). Primarily used for baby foods it also provides closure for sauces, gravies, and juices. Applied in a steam atmosphere, the plastisol side gasket in the heated cap forms thread impressions when pressed against the glass finish. As it cools, permanent impressions are formed in the compound so the cap can be twisted to open and reseal similar to a CT cap.

Friction-fit closures. Many bottles are sealed with simple metal or plastic closures that are pressed onto the top and held in place by friction. The four basic types of friction-fit closures are crowns, snap-fit caps, press-on caps, and stoppers.

Crowns. The crown beverage cap was a major innovation in friction-fitting closure. It has been widely used since the turn of the century for carbonated beverages and beer. Crowns are made of tin-free steel and tin plate. Matte-finish tin plate is used for soft drinks and a brighter finish is used for beer. The crown has a short skirt with 21 flutes which are crimped into locking position on the bottle head. The flutes are angled at 15° in order to maintain an efficient seal (see Fig. 6) (18).

The crown contains a compressible lining material, which over the years has included solid cork, composition cork or

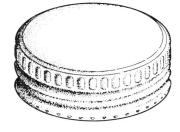

Figure 4. Roll-on closure.

Figure 6. Crown closure.

plastic liners, and foil and vinyl spots. Though simple in concept, the crown provides a friction-fit sufficient to seal pressurized beverages. The flared cap skirt in conjunction with the smoothness of the bottle neck provides easy access through the prying motion of the bottle opener. As convenient access came to be demanded in the marketplace, easy-open crowns were developed. The twist-off beverage crown first appeared on beer bottles in 1966 and gained in popularity (19). Designed for convenient access, it could be twisted off by hand or removed with a bottle opener. Bottlers switched to this cap because no special capping equipment was required for the new closure. In 1982 many companies moved away from this cap because of potential tampering. The crown is currently capable of application speeds exceeding 1000 units a minute.

Snap-Fit Caps. The snap-fit caps are simple lids that can be pressed onto the tops of bottles. They may be held in place by the friction of a tight fit or by supplementary flanges, ridges, or grooves that grasp the bottle finish (20). The skirts of some metal snap-fit caps are rolled under to form a spring action as the cap is pushed against the lip of the bottle. Some types of snap-fit caps have ridges on the inside of the caps that match the grooves in the ridges on the bottle. The simplest form is designed to fit so snugly that the friction between the surfaces of the cap and bottle is sufficient to hold the cap in place. Snap-fit caps may be made of metal or plastic. They are used for such food products as jellies and for over-the-counter medicines such as headache remedies. An important variation is the press-on TE closure, frequently used on milk and juice products. When these caps slide over a ridge near the bottom edge, they fit so tightly that they cannot be pulled off. The bottle is opened by pulling a tear tab located above the ridge. This separates the top of the cap from the bottom portion that was locking it in place.

Press-on Vacuum Caps. Sealing for the press-on vacuum closure is obtained from atmospheric pressure when air is withdrawn from the headspace of the container by steam or mechanical sealing methods. Sometimes further security for this type of seal is provided when the edge of the cap is forced under a projection of the glass finish or held under a similar projection by snap lugs. In many vacuum caps, it is the pressure of a gasket against the top of the container finish which provides a seal. In others, atmospheric pressure alone is adequate. A recent press-on variation is the composite cap, a gasketed metal disk and plastic collar used in sealing dry-roasted nuts and seafood products. Its plastic collar is used as a TE device.

Types of Closure

It is impossible to place all closures into clean-cut categories where there is no overlap of functions. Yet despite these limitations a classification can provide focus for understanding contemporary closure trends. As defined by their utility, the four classes of contemporary closures are: containment, convenience, control, and special purpose.

Containment closure. Though all closures provide containment, a containment closure is here defined as a one-piece cap whose primary function is to provide containment and access on vast production scales. CT caps (for general-purpose sealing), crowns and roll-ons (for sealing of pressurized beverages), lug and press-on caps (for vacuum sealing of foods) are within this class of containment closure.

Convenience closure. Closure development in recent years has been in response to consumer preferences for convenient access to the product. Convenience closures provide ready access to liquids, powders, flakes, and granules for products that are poured, squeezed, sprinkled, sprayed, or pumped from their containers. There are five types of convenience closures: spout, plug-orifice, applicator, dispensing-fitment, and spray and pump types.

Fixed-Spout Closures. A spout is a tubular projection used to dispense liquid and solid materials. It may be fixed or movable, and capable of dispensing a product in a wide ribbon or a fine bead depending on size and configuration of the orifice. Fixed-spout caps incorporate a cylindrical or conical projection into the center of a threaded or friction-fitting closure. Spouts on reusable containers are often sealed by a small sealer tip on the end of the spout. On some sealed spouts, dispensing control can be attained by cutting the spout at various heights, thereby providing different orifice sizes. A more contemporary form of fixed-spout closure is molded with a smaller sealed spout on the top of the cap. Called "snip-tops", they are one of the most inexpensive forms of dispensing closure (see Fig. 7). Contemporary "dripless pour spouts" have recently been introduced to provide "No-Mess" dispensing of viscous, sticky products packaged in large containers, such as liquid detergents and fabric softeners. Large pour spouts are protected by screw-on overcaps which double as measuring caps. Upon reclosure, the measuring overcap is designed to drain residual product directly into the container through the spout.

Movable-Spout Closures. Also referred to as turret, swivel, or toggle types, the movable-spout concept features a hinged spout which can be flipped into operating position and reclosed with the thumb alone to provide one-handed access and reseal (see Fig. 8). Most movable spouts are two-piece constructions, though a one-piece swivel spout design requiring one manufacturing operation has recently been introduced. Newer refinements of this type include the incorporation of a tear band across the spout to provide tamper evidence. Valve-spout closures, such as the "push-pull" closure, are opened and closed in a straight-line, vertical fashion (see Fig. 9). "Twist spouts"

Figure 7. Snip-top closure.

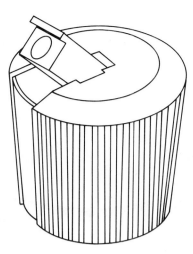

Figure 8. Flip-spout closure.

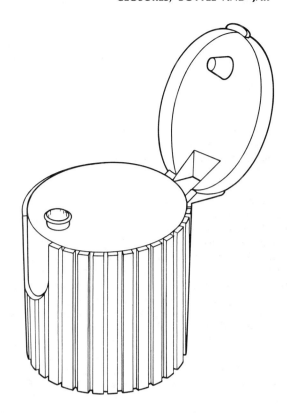

Figure 10. Hinged plug-orifice closure.

employ a tapered flange design and open and close by a twist-ing motion.

Plug-Orifice Closures. These closures first aided in the dis-pensing of personal-care and cosmetic products, and are now used in conjunction with multilayer high-barrier plastic bot-tles for convenient dispensing of food products (see Multilayer plastic bottles). To some, the hinged-top designs represent the wave of the future in food packaging (21). The closure consists of a dispensing orifice incorporated into a screw-on base clo-sure, and a plug, or "spud", hinged within the top of the closure or molded into a flip-up hinged cap (see Fig. 10). In the poly-propylene plastic versions, the plug and orifice provide a fric-tion-fitting seal which produces an audible "snap" when en-gaged, an instance where a closure can communicate its sealed state by sound. The top of a "snap-top hinged closure" swings open on two or three external hinges. The "disk closure" is another plug-orifice type, a two-piece design consisting of an orifice closure base and a plug fitment hinged to a round disk which is set into the closure top. By pressing upon the access point, the disk fitment swings up upon its hinges, deactivating

the plug seal from the orifice for one-handed dispension. Some of these designs also produce audible "snap" upon engagement and disengagement.

Applicators. There are many different kinds of convenience applicators, many specialized for particular product applica-tions. Four major types are brushes, daubers, rods, and drop-pers. Brush caps range from small cosmetic brushes to large applicators used for applying adhesives. Sponge, cotton, felt, or wool pads affixed to applicator rods are used to apply a wide variety of household and cosmetic products and are known as dauber caps. Glass and plastic rods are used in the drug and cosmetic industries, such as the balled-end rod used to apply medicines. Glass and plastic droppers, with straight, bent, and calibrated points, are frequently used to provide precise dos-ages for medicinal products. The three components of the drop-per are an elastomeric bulb, the cap, and the pipette.

Fitment Closures. Fitments and fitment closures are de-signed to regulate the flow of liquids, powders, flakes, and granules. Fitments are inserted into the neck of the container or are permanently attached. Fitment closures incorporate regulating devices into screw-on or press-on caps. Those which plug inside the neck finish include dropper and flow-regulat-ing fitments. "Dropper tips", used with squeeze-type bottles, dispense liquid in increments of one drop and are usually cov-ered with a protective overcap. "Pour-out fitments" control the splashing of liquids by retarding their flow, a frequent prob-lem when precise, small-volume pouring is required from cum-bersome containers. Those fitments or closures which regulate solid materials include sifter and shaker designs. Most contain a number of sifter holes in which powdered material can be

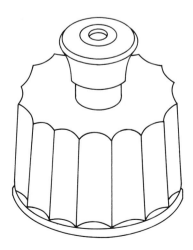

Figure 9. Push-pull closure.

dispensed evenly. Others incorporate options which permit the material to be shaken-out, poured-out, or spooned. "Shaker caps" and "powder sifter caps" dispense powdered or granular products, sometimes incorporating revolving fitments which provide containment. A wide variety of sliding panels or hinged covers provide the consumer with sprinkling, pouring, or spooning options. Many variant designs combine spouts, sifters, dial disks and sliding covers.

Spray and Pump Dispensers. "Regular sprayers" operate on a basic pump-principle and function with a piston, accumulator, or cylinder. Dispensing a heavier particle size, these sprayers are often used for household and personal-care products. Generally, the smaller the orifice size, the finer the spray pattern. "Fine mist sprayers" dispense in finer particle size as required by some personal-care products. "Trigger sprayers" are larger and more complex in design, offering convenient dispensing for large volumes of liquids. The bulb-and-piston-drive trigger-sprayer units emit spray patterns ranging from a fine mist to a stream. The amount of product delivered by piston-driven "pump dispensers" depends upon its viscosity; the more viscous, the more strokes to prime and the lower the output per stroke (22). Regular dispensing pumps dispense in volumes from less than 0.5 cm^3 to slightly more than 1 cm^3 per stroke in water. Large-volume dispensers range in capacity from 1/8 oz (3.7 mm) to 1 oz (29.6 mm) per stroke.

Control closure. The first "clerkless" food store appears to have been opened in 1916 by Clarence Saunders. By 1930 there were 3000 of them, soon to be known as "supermarkets" (23). These stores raised new problems of hygiene as more and more products were available in unit packs that had to be capable of withstanding repeated handling, attempts at sampling, and occasional malicious intrusion. The need for consumer product safety grew along with this new concept in food retailing, a need which would become a matter of increasing concern for the U.S. Congress as the number and variety of products increased. Among the legal mandates developed to protect the public against harmful substances are those which specify access controls for containers.

Tamper-Evident Closures. Tamper-evident caps have been in use for years, though earlier they were referred to as "pilferproof caps". Today these metal and plastic caps provide visible evidence of seal disruption and are used for over-the-counter (OTC) drugs, beverages, and food products. The two kinds of TE closures are "breakaway" or "tear band" closures used for pressurized and general sealing applications, and TE vacuum designs for vacuum-sealing applications. The closure user can also fulfill tamper-evident requirements through the use of innerseals which cover the container mouth (see Tamper-evident packaging).

Breakaway caps. In 1982 the FDA established requirements for tamper-evident packaging for over-the-counter (OTC) drug products. The agency defined such packaging as "having an indicator or barrier to entry which, if breached or missing, can reasonably be expected to provide visible evidence to consumers that tampering has occurred" (24). The FDA did not issue rigid standards of compliance but instead listed 11 suggested and approved methods from which a packer may choose. Regulation #6 describes paper or foil bottle seals covering the container mouth under the cap as a means to provide tamper-evidence. Regulation #8 describes breakable caps as another option. The FDA defined these caps as being plastic or metal that either break away completely

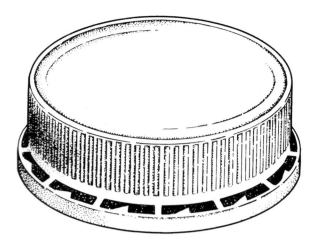

Figure 11.   Mechanical breakaway closure.

when removed from the container or leave part of the cap on the container. The breakaway cap is the most common form of tamper-evident closure. With other options, such as shrink bands and strip stamps, a packager needs additional operations and equipment to achieve tamper evidence. Two forms of tamper-evident caps are mechanical breakaway and tear bands.

Mechanical breakaway. These are threaded caps with perforations along the lower part of the skirt which form a "break line" in the closure (see Fig. 11). When the closure is twisted for removal, the band, which is locked to the finish by crimping or rachets, separates from the closure along the break line. The cap is removed and the lower part of the skirt remains on the container neck. The breakaway cap can be efficiently applied, is highly visible, familiar to consumers, and is durable enough to maintain its integrity throughout distribution. Metal closures of this type frequently crimp the band to the container neck for a friction hold. Variations in this type of TE closure include different band designs and methods of off-torque resistance; for example, ratchets on the band that lock to mated protrusions on the finish. TE bands on some plastic closures are shrunk by heat to form a tight fit around the container neck.

Tear bands. These types, frequently called tear tabs, employ a locked band to prevent cap removal (see Fig. 12). Access is accomplished by completely removing the band from the

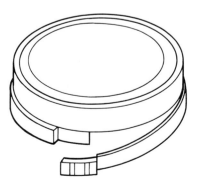

Figure 12.   Tear band.

container. Frequently a protruding tab is evident for the consumer to grasp and commence tearing. Many non threaded TE closures utilize this type, such as the press-on friction fit closures found on milk containers. The closure is removed by tearing off the lower skirt, which overrides a bead on the container finish. Most of the removable-band types are made of plastic, usually polyethylene.

TE vacuum caps. Marketing leverage, rather than legal mandate, accounts for the expansion of TE into food packaging (25). These measures are not referred to as "tamper-evident" in label or closure communications, but are placed in a more positive light, such as "Freshness Sealed" or "Safety Sealed". The two major types of TE vacuum closures are vacuum button and vacuum tear-band caps.

A popular TE option for food products packaged in glass containers under vacuum is the "button-top closure". These include lug versions used for jellies, sauces, and juices, and the threaded-seal version popular with the baby-food industry. A safety button, or coin-sized embossment on the top of the cap, pops-up as the jar is opened and its vacuum is lost. Accompanying this is the "pop" which serves as audible evidence of an undisrupted seal. When capped the embossed button is held down by vacuum pressure, providing the consumer with visual evidence that the container has not been opened. Another type, the "vacuum tear-band closure", is a two-component closure used for the packaging of nuts and condiments. It consists of a metal vacuum lid inserted into a plastic tear-band closure skirt. Protrusions molded into the plastic collar provide friction-fitting resealability for the container.

Child-resistant closures. Alarmed by the increasing number of children being harmed through accidental poisoning, Congress acted in 1970 to pass the Poison Prevention Packaging Act. This act, Public Law 91-601, established mandatory child-resistant closures for rigid, semirigid, and flexible containers of such compounds as the Consumer Product Safety Commission (CPSC) deemed dangerous to children (see Child-resistant packaging). Studies revealed that the accidental-poisonings curve in children peaked at 18–24 mo, but the manual dexterity curve increased with age, peaking in 4–5 yr olds. Therefore, protocol testing requires testing of children aged 42–51 mo. Four major premises were considered in the development of child-resistant closures (26): children 42–51 mo of age could not perform two deliberate and different motions at the same time; children of that age could not read, nor could they determine alignments, but they can learn quickly by watching; children are not as strong as adults, but through ingenious use of teeth, table edges, or other tools around them, their persistence would give them leverage to make up for their strength; and although their hands and fingers are smaller than adults, childrens teeth and fingernails are thin and sharp and can slide under and into gaps. Childhood deaths involving all household chemicals have declined 75% since the first regulation under the act was passed. Packages designed to protect children, however, have come under fire for restricting access by the handicapped and the elderly. Most OTC drug packagers agree that CR devices are generally more burdensome than TE devices. Efforts are underway to develop new CR closures.

Child-resistant measures are defined by the CPSC as packaging that is designed to be significantly difficult for children under five years of age to open within a reasonable time, yet not difficult for adults without overt physical handicaps to use

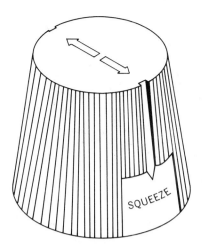

Figure 13. Child-resistant "squeeze-and-turn" closure.

properly. Under current regulations, a package fails to be child-resistant if more than 20% of a test panel of 200 children are able to gain access, or if more than 10% of a test panel of 100 adults are unable to open and properly resecure the test package (27). The three most frequently used child-resistant closure types are press-turn, squeeze-turn, and combination-lock. The "press-turn" cap is removed by applying downward force while simultaneously turning the cap. "Squeeze-and-turn" caps employ a free-rotating soft-plastic overcap which engages an inner threaded cap or disengages a locking mechanism when sidewall pressure is applied (see Fig. 13).

The "combination-lock" caps use interrelated components formed into the cap which must be oriented before the cap can be removed. A common low cost variety of this closure is the one-piece "line-up, snap-off cap" (see Fig. 14). A slight interruption of the thread on the container serves as an engagement point for these caps. A protrusion on the cap fits under the single thread. When the cap is turned it becomes "locked" onto the container. As with TE closures, many packagers add CR closures to packages not required to have them, even though the cost may generally add 1–5% to the cost of the package.

Special-purpose closure. Special-purpose closures are those which are of specialized application or premium design. These include aesthetic closures, special-function closures, stoppers, and overcaps.

Aesthetic Closure. The aesthetic closure is an important sales-promotion aspect of the package. It is designed to communicate clearly and powerfully by imagery. Original private-mold glass stoppers used in fragrance bottles, some with lavish sculptural representations, are an example of this type. These are frequently the most expensive forms of closure.

Special-Function Closures. These closures serve a specialized function in the marketplace. There are, for example, closures that vent containers which sustain a pressure build-up. Such pressures can cause the container to rupture, or violently expel the product when the cap is removed. Since venting closures can leak when the container is not in an upright position, they must be used in controlled circumstances. Many manufacturers require a "hold-harmless" agreement as a condition of sale for such closures. Another special-function clo-

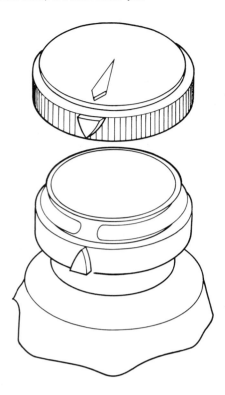

Figure 14. Child-resistant "snap-off" closure.

sure is the twist-off closure for injectables, a two-piece aluminum cap used on parenteral vials.

Stoppers. The wine and champagne industry is the largest user of stoppers. Cork stoppers are standardized by size and grades, the latter according to the degrees of product vintage (28). Stoppers of natural rubber, synthetic silicone rubbers, and thermoplastic materials provide closure in some chemical and biological applications. Rubber plug closures are crimped onto ampules with metal bands and allow for the insertion of a hypodermic needle in medical uses.

Overcaps. The overcap is a secondary cap designed to protect the primary closure, dispenser, or fitment of a container. Metal or plastic overcap designs attach to the container by friction-fit or thread engagement, and are used to protect aerosol and dispensing fitments. Overcaps frequently double as measuring caps for mouthwash, liquid detergents, and fabric softeners.

Sealing Systems

Though often the smallest aspect of a package, the seal is responsible for keeping the entire concept intact. If the seal is not maintained by the closure, liner, and container working together, the success of the product is at stake.

Liners. Today's lining material is either a single substance (usually paperboard or thermoplastic) or a composite material. Synthetic thermoplastic liners include foamed and solid plastics of varying densities. A composite lining material consists of a backing and a facing. The backing, usually made of cellulose or thermoplastic, is designed to provide the proper compressibility to affect the seal and proper resiliency for resealing. Facing materials, representing the side of a composite liner that comes into direct contact with the product, are numerous, as are the variables of product chemistry with which they must contend. Generally, facing materials are thermoplastic-resin-coated papers, laminated papers of foil or film, or multilayer types devised for special applications (see Closure liners).

Innerseals. The innerseal affords TE protection by sealing the mouth of the container. Three common types are inserted by the closure manufacturer into the cap (29). A waxed-pulp backing and glassine innerseal is common within the food industry. After the filling operation the container runs under a roller system which applies an adhesive to the lip of the container, and then the cap is applied. Upon removal, the glassine adheres to the container while the pulp backing remains in the closure. Pressure-sensitive innerseals, generally a foamed polystyrene, adhere to the lip upon application and require several hours to set. Heat-induction innerseals are plastic-coated aluminum foils, often adhered to a waxed pulp base liner. After the cap is applied, the container passes under an electromagnetic field which causes the aluminum to generate heat. The plastic facing on the aluminum subsequently melts and adheres to the container.

Linerless closures. Plastic linerless closures provide a positive seal in certain circumstances, foregoing the need for intermediary materials and secondary liner-insertion operations. To many packagers, the cost savings provided by the linerless closure can be considerable. The seal of a linerless closure is achieved by molded embossments forming diaphragms, plugs, beads, valve seats, deflecting seal membranes, or rings which press upon, grasp, or buttress the sealing surfaces of the container. Over a dozen types of linerless closures are in common use, each designed to provide a seal at one or more critical sealing surfaces of the container, which include the land surface, the inside edge of the land surface, or the outside edge of the land surface. Some form of land seal in conjunction with a valve or flange represents one type of effective linerless closure design. The land is typically the most consistent sealing surface. A land-seal ring can bite into plastic container finishes or deflect on glass finishes. An inner buttress can correct ovality problems in plastic containers by forcing such off-round finishes back into proper shape.

Closure Materials

Closures are made of plastics, metal, or glass.

Plastic closures. Molded plastic closures are divided into two groups: thermoplastics (eg, polyethylene, and polypropylene, and polystyrene), and thermosets (eg, phenolic resins and urea components) (see Polymer properties). Thermoplastic materials can be softened or recycled by heat; thermoset materials cannot be recycled once they are molded.

Thermoplastics. In general, thermoplastic closures offer the packager light weight, versatility of design, good chemical resistance to a wide range of products, and economical resins and manufacturing processes. Their relative flexibility is essential to contemporary closure design with its emphasis on convenience and control devices. Thermoplastics provide good application and removal torque. They maintain a good seal and tend to resist back-off. Unlike thermosets, thermoplastics can be pigmented in the full-color spectrum in strong, fade-resistant intensities.

Most thermoplastic closures are produced by injection

molding (see Injection molding), although some are made by thermoforming (see Thermoforming). Polypropylene and polyethylene account for about 90% of all thermoplastic closures.

Polypropylene. Polypropylene (see Polypropylene) has unusual resistance to stress-cracking, an essential characteristic of hinged closures. In thin hinged sections, it has the quite remarkable property of strengthening with use. The homopolymer has limited impact resistance, but it can be modified for better performance. It has excellent resistance to acids, alkalies, oils and greases, and most solvents at normal temperatures. It has the best heat resistance of the polyolefins, with a high melting point suitable for sterilized products, but it becomes embrittled at low temperatures. Polypropylene has better printability than polyethylene but both are inferior to polystyrene or thermosetting plastics in that respect. As a relatively rigid molded material it has outstanding emboss and deboss potential for closure communications.

Low Density Polyethylene. LDPE (see Polyethylene, low density) is resilient and flexible. It is relatively tasteless and odorless, although some organoleptic problems are more prevalent with LDPE than with polypropylene. It provides outstanding moisture protection, but it's not a good gas barrier. LDPE's economy as a closure material is provided by low-cost resins and relatively short injection-molding cycle times. Though it is considered to have good resistance to stress cracking, problems may occur in the presence of certain chemicals such as detergents. Communications embossments or debossments are good but limited by the softness of the material.

High Density Polyethylene. Compared to LDPE, HDPE is stiffer, harder, and more impermeable (see Polyethylene, high density). It is tasteless, odorless, and impact-resistant, but it will stress-crack in the presence of some products such as detergents unless it is specially formulated. Its heat resistance and barrier properties are superior to LDPE. HDPE resin is more expensive than LDPE, but it is still considered a relatively low-cost material. A particular drawback to HDPE closures is a potential for warpage and loss of torque.

Polystyrene. Polystyrene (see Polystyrene) is used for about 10% of the closures produced today. Polystyrene homopolymer is attacked by many chemicals, is very brittle, has relatively low heat resistance, and does not provide a good barrier against moisture or gases. Many of the disadvantages of polystyrene are overcome by rubber modification and/or copolymerization.

Thermosets. Phenolic and urea compounds have a wide range of chemical compatibility and temperature tolerances. Some thermosets can sustain sub-zero temperature without embrittlement, and survive at temperatures higher than 300°F (149°C) (30). The density and rigidity of thermosetting plastics give the material its heavy weight and guard against slippage over threads, a problem with softer thermoplastics such as LDPE. Thermosets cannot provide the color range or intensity of thermoplastics, but they accept vacuum metallizing decoration in silver and gold with superior adhesion qualities. During the molding process, thermosets undergo a permanent chemical change and cannot be reprocessed as thermoplastics can. Thermoset closures are manufactured by compression molding (see Compression molding). Cycle time for thermosets is generally longer than thermoplastics, (30–120 s), depending upon thickness of the product and additives.

Phenolics. Phenol–formaldehyde closures are hard and dense. They are the stiffest of all plastics, but are relatively brittle and low in impact strength. The properties of phenolics depend to a large extent upon the filler material used. Wood flour improves impact resistance and reduces shrinkage. Cotton and rag fiber additives increase the impact strength; asbestos and clay additives improve chemical resistance. Phenolics are resistant to some dilute acids and alkalies and attacked by others, especially oxidizing acids. Strong alkalies will decompose phenolics, but they have excellent solvent resistance. Their heat resistance is outstanding. Phenolics cost less than ureas, and are easier to fabricate, but they are limited in color to black and brown unless decorated.

Urea. Urea–formaldehyde is one of the oldest plastic packaging materials, first used in the early 1900s. The resin produces extremely hard, rigid closures with excellent dimensional stability. It has the highest mar resistance of plastics discussed, but is the most brittle. Urea compounds are odorless and tasteless, with good chemical resistance. They are not affected by organic solvents but are affected by alkalies and strong acids. They show good resistance to all types of oils and greases. They will withstand high temperatures without softening. Urea compounds are available in white and a wide range of colors, but with muted intensities compared to thermoplastics. Urea compounds, like phenolic resins, do not build up static electricity which leaves them free of dust. They are the most expensive of the plastic closure materials.

Metal. Metal caps, the strongest of closures, are used today for general, vacuum, and pressurized applications. Tin-plate and tin-free steel (see Tin mill products) are used in the production of continuous thread, and vacuum press-on closures, lugs, overcaps, and crown caps. The largest market for steel closures is vacuum packaging. Aluminum closures are primarily continuous thread caps and roll-on designs.

Steel Closures. Steel closures are of two materials: tinplate and tin-free steel. Tinplate closures are plated steel with a thin coating of tin on both sides that helps protect the base steel from rust and corrosion. There are limitations to tin's protective abilities, however, for tinplate is suceptible to rust when exposed to high humidity. Additional coating operations offer increased protection. Tinplate is graded according to temper. Temper T-1 is soft, and T-6 is quite hard. Closures can be fabricated in any temper, but are predominantely T-2 to T-5. The more common base weights used include single-reduced 80- and 90-lb (36.3- and 40.8-kg) with a cost-reducing trend toward double-reduced 55- and 65-lb (25- and 29.5-kg) plate (15).

Tin-free steel shows promise of becoming the dominant steel closure material. Crown closures are now made primarily of tin-free steel produced as single-reduced stock in a 90-lb (40.8-kg) plate weight for conventional crowns and a lighter, 80-lb (36.3-kg) plate for twist-off crowns.

Aluminum Closures. Light weight, malleability, and resistance to atmospheric corrosion characterize aluminum closures. Some products more corrosive to aluminum than tinplate require special coatings for optimum protection (31). The composition of aluminum alloys varies according to intended use with up to 5% magnesium and lesser elements such as manganese, iron, silicon, zinc, chromium, copper and titanium (32).

Metal-Overshell Closures. Steel, aluminum, copper, or brass shells slip over plastic closures to form composite "overshell" caps. Freed from finish contours, the smooth, often-tall sidewalls of polished and burnished metals provide aesthetically-pleasing characteristics much in demand by the cosmetic and fragrance industries. There is a greater willingness to pay a premium for appearance in these industries because the closure assumes a greater role in sales promotion.

Glass. Glass stoppers are used in commercial glassware and in cosmetic and fragrance packaging. Frequently a polyethylene base cap assists in friction-fitting the stopper into the bottle. Stoppers for the premium fragrance industries represent superlative designs in molded glass.

Closure Selection

Selection. Who selects and specifies the closure depends upon the size, nature, and organization of a company. It may be a president or general manager, a brand manager, package engineer, purchasing agent, or a packaging committee. General guidelines for closure selection are provided below by "The 5 Cs of Closure" (33).

Containment. The essential requirements of containment are product compatibility and the ability to provide functional protection. This objective is reached by evaluated choices in closure method, type, material, and sealing system. Determining the sealing system, for example, may involve decisions as to whether lining materials or a linerless closure will resist permeation of the product to standards. Other important variables arise in the interaction of closure and container and how they affect the efficacy of engagement and seal. Torque considerations include seal pressure (the amount of pressure exerted on sealing surfaces), and strip torque (the torque at which a closure slips over the container threads). As torque is affected by different coefficients of friction between the liner surface and the container, as well as by materials used in closure manufacture, each closure system should be individually evaluated to ensure it meets applicable performance criteria.

Convenience. Opportunities for convenient dispensing may begin with a containment closure that provides a reduced number of turns, or broader "knurls" on the wall of the closure skirt to provide surer opening and closing. Convenience closures provide many options including simple spouts, plug-orifice snap caps, and, at the mechanically complex end of the continuum, variable-dispersion sprayers and pumps. The method of closure engagement, the requirements of containment, the type of sealing system required, and the premium placed upon convenience will determine options in dispensing closures.

Control. A variety of substances are mandated by law to be packaged as tamper-evident or child-resistant. Cost and sealing needs will determine options in control closures, as well as whether secondary sealing systems are required. More consumer complaints result from inadequate opening and closing of product containers than any other package function. Careful review and testing of control-closure and lining system can prevent potential access problems with elderly or handicapped consumers.

Communications. The shelf appearance of a closure is perceived as a reflection of product quality. The closure communicates this by style and brand signature. In addition it often gives a detailed list of ingredients, and sometimes instructs on disengagement. The larger the closure the more it may aug-

Table 1. Common Finishes and Descriptions

Finish designation	Description
120	2-lug Amerseal quarter-turn finish
140	4-lug Amerseal quarter-turn finish
160	6-lug Amerseal quarter-turn finish
326	pour-out snap cap CT combination
327	snap cap CT combination
400	shallow CT finish
401	wide sealing surface on 400 finish
405	depressed threads of 400 finish at mold seam
410	medium CT concealed-bead finish
415	tall CT concealed-bead finish
425	8 to 15 mm shallow CT
430	pour-out CT
445	deep S CT finish
450	deep CT Mason finish
460	home-canning jar finishes
600	beverage crown finish
870	vacuum side seal pry-off
1240	vacuum lug-style finish
1337	roll-on pilferproof finish
1600	roll-on finish
1620	roll-on pilferproof finish
1751	twist-off vacuum seal

ment the label in communicating ingredients or nutritional information. Today's closure not only communicates visually, but audibly as well. Steel vacuum button closures "pop" to confirm the freshness of a product, and polypropylene plug-orifice types "snap" when the seal is engaged. Determining the kinds of communications required, and selecting the graphic options which maximize readibility and impact, specify the final appearance, or point-of-purchase impact, of the closure.

Cost. Some cost considerations depend on production requirements. Thermoplastic-mold costs, for example, are generally more expensive than those for thermosets, but faster cycles and resin economy may prove more economical in larger volumes. A cost savings can be realized through the use of linerless closures if they can maintain seal integrity. Lightweighting the closure by selection of an appropriate material has helped to reduce transportation costs for packagers. Another cost consideration is whether a "stock" or a "privately-tooled" closure is required. A privately tooled closure is far more expensive to produce, but, again, the packaging concept or production volumes may "economize" it. There is a trend toward specifying stock closures with market-tested designs. Many manufacturers and distributors offer extensive stock-closure lines to the packer.

Closure Specification

A closure is designated by a series of numbers and/or letters. An example is the designation 48–400. The first number refers to the inside diameter of the closure as measured in millimeters. Common closure diameters range from 22–120 mm. The second set of numbers, the 400, is the finish designation.

The "finish" of a closure is its thread design, and the size, pitch, profile, length, and thickness of the engagement threads on plastic and metal closures and containers. Today there are over 100 glass-finish designations for a great variety of glass containers (see Table 1). Series designations for the most popular CT closures are 400 and 425 for shallow continuous thread designs, 410 for medium CTs, and 415 for tall CTs. For all glass finishes, tolerances have been established by the Glass Packaging Institute, whose closure committee became the Closure Manufacturers Association (CMA) in 1980. Voluntary standards for closures have been issues by CMA which include closures for both glass and plastic container finishes (see Table 2) (34).

Closures developed for glass containers were used for plastic containers when the latter were introduced, but it was soon realized that the contour of a glass bottle thread is not an optimum profile for the plastic bottle. It does not provide accurate closure centering on the finish nor does it permit higher capping torques required to provide a positive seal on plastic containers (35). The Dimensional Subcommittee of the Society of the Plastics Industry developed specific finish dimensions, tolerance, and thread contours for blown plastic bottles. The two basic contours are the M-style and the L-style. Where a typical glass thread is rounded in contour, the M-style thread engaging surfaces are angled at 10° and the L-style is angled at 30°. Both contours increase sealing abilities for closures on plastic bottles.

Four critical closure dimensions are represented by the four letters, T, E, H, and S (see Fig. 15). T is the dimension of the root of the thread inside the closure. E is the inside dimension of the thread in the closure. H is the measurement from the inside top of the closure to the bottom of the closure skirt. S is the vertical dimension from the inside top of the closure to the starting point of the thread. These critical closure dimensions and tolerances for metal and plastic closures designed for glass and plastic containers are represented in the voluntary standards for closures as issued by CMA.

Closure Trends

Closure concepts seem to have changed more in the last 40 yr than in the last 4000 yr. Yet changes in state-of-the-art concepts, materials and manufacturing processes do not represent the real driving forces behind today's closure developments. The industry is consumer-driven. More and more, a premium is readily paid for convenience. The industry has also been awakened, sometimes with great shock and alarm, to the powers of human foible, which demands a redefinition of access control.

Functional trends. Today's consumer has been characterized as oriented toward health, diet, appearance, longevity, and convenience (36). Households with two working spouses, increased single households, and retiree households all com-

Table 2. CT Closure Finishes and Pitch

Finish	Description	Sizes, mm	Threads per in., pitch, (per cm)
400	shallow continuous thread	18, 20, 22, 24	8 (3.2)
		28, 30, 33, 35, 38, 40	6 (2.4)
		43, 45, 48, 51, 53, 58	6
		60, 63, 66, 70, 75, 77	6
		83, 89, 100, 110, 120	5 (2.0)
410	medium continuous thread	18, 20, 22, 24	8
		28	6
415	tall continuous thread	13, 15	12 (4.7)
		18, 20, 22, 24	8
		28	6
425	shallow continuous thread	8, 10	14 (5.5)
		13, 15	12
430	pour-out continuous thread	18, 20, 22, 24	8
		28, 30, 33, 38	6
445	deep "S" continuous thread	45, 56, 58, 63, 73, 75	6
		77, 83	5
450	deep CT Mason finish	70, 86, 96, 132	4 (1.6)
455	CT for 455 glass finish	28, 33, 38	8
460	Home canning jars	70, 86	4
470	CT for GPI 470 glass finish	70, 86	4
480	CT for GPI 480 glass finish	24, 28, 33, 38	6
485	deep "S" fitment cap	28, 33, 35, 38, 40	6
		43, 48, 53, 63	
490	deep "S" larger "H" fitment cap	18, 20, 22, 24, 28, 30	8
		33, 35, 38, 43, 48, 63	
495	CT for GPI 495 glass finish	28, 33, 38	8
SP 100	CT for plastic SP-100 finish	22, 24, 26, 28, 38	8
SP 103	CT for plastic SP-103 finish	26	8
SP 200	CT for plastic SP 200 finish	24, 28	6
SP 444	CT for plastic SP 444 finish	24, 28, 33, 38, 43, 45, 48	6
		53, 56, 58, 63, 73, 75	6
		70	4
		83	5

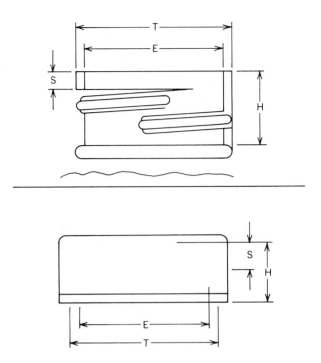

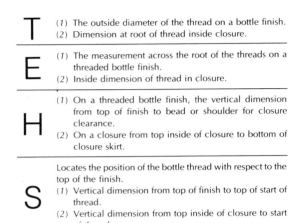

T
(1) The outside diameter of the thread on a bottle finish.
(2) Dimension at root of thread inside closure.

E
(1) The measurement across the root of the threads on a threaded bottle finish.
(2) Inside dimension of thread in closure.

H
(1) On a threaded bottle finish, the vertical dimension from top of finish to bead or shoulder for closure clearance.
(2) On a closure from top inside of closure to bottom of closure skirt.

S
Locates the position of the bottle thread with respect to the top of the finish.
(1) Vertical dimension from top of finish to top of start of thread.
(2) Vertical dimension from top inside of closure to start of thread.

Figure 15. T, E, H, and S dimensions

mand a market for convenience packaging (37). The dispensing closure is no longer a functional appendage, but is seen as an integral part of the total package (38). Today's convenience closure is time and labor-saving. It prevents spills, leaks, and drips. It provides measured-dose dispensing, and can visually signal tampering (39). As plastic containers continue to penetrate the food market, the closure will broaden squeeze-dispensing. Other functional trends include the expansion of TE food packaging, larger closure sizes, increased use of stock caps to avoid private-mold costs, and new concepts in linerless closure design. Since innovative packaging can increase market share, special emphasis is being placed on improved tamper-evidence, child-resistance, and convenience designs. These functions will no doubt become more and more integrated into one cap. Closures are now being marketed which provide for both TE and CR.

Material trends. United States shipments of metal and plastic closures in 1983 came to 76.8 billion (10^9) units. Metal caps accounted for 52.9 billion units, and shipments of plastic closures amounted to 23.8 billion units (6). Shipments of plastic child-resistant closures numbered 2.8 billion (10^9) units in 1983, a rise of 18% over the year before (6). These figures reveal that metal caps comprise two-thirds of the total closure units manufactured, a figure due largely to the high volumes in crown-cap production. Crown shipments, however, showed a decrease of 25%, and plastic beverage closures showed an increase of 178% during the same time period (11). One major beverage-company executive believes the entire industry will convert to plastic closures within five years because plastics provide an opportunity to lightweight containers, simplify capping, and improve color matching (40).

The ductility of plastics accounts for the fast progress of plastic closures, which will undoubtedly take a still-larger share of the market in years ahead (41). Polypropylene represents the largest volume and highest growth plastic material

with 200 million (10^6) lb (90,700 t) used in the production of plastic closures in 1984, an increase of 33% within four years (42). Polystyrene has shown slight growth in recent years, with 72 million lb (32,700 t) used in closure production in 1984. Closures accounted for 66 million lb (29,900 t) of HDPE consumption, an amount that has remained relatively stable over recent years. LDPE (37 × 10^6 lb or 16,800 t) and PVC (35 × 10^6 lb or 15,900 t) have both remained stable in recent years. As for thermosets, 15 × 10^6 lb (6800 t) of phenolic resins and 11 × 10^6 lb (5000 t) of urea compounds were used in closure production in 1984 reflecting little growth in recent years.

Metal thread-engagement closures will continue to assume a position in the food and pharmaceutical industries due to new fabricating, plating, and light-weighting technologies which will keep steel and aluminum closures competitive.

BIBLIOGRAPHY

1. H. McKearin and K. M. Wilson, *American Bottles and Flasks and their Ancestry,* Crown Publishers, New York, 1978, p. 17.

2. *Ibid.,* pp. 210–212.

3. H. Higdon, ed. *The Phoenix Flame,* (a house magazine of the Phoenix Cap Company, Chicago, now Phoenix Closures, Naperville, Ill., **XV,** 32, (Feb. 1940).

4. A. Lief, *A Close-Up of Closures,* Glass Container Manufacturers Institute, New York, 1965, p. 16.

5. Closure Manufacturers Association, "Closures for Bottles, Cans, Jars," *Packaging Encyclopedia 1984,* **29,** 153 (March 1984), Cahners Publ. Co., Boston, Mass.

6. "Closures for Containers," *Current Industrial Reports, M34H (85)-1,* U.S. Department of Commerce, Washington, D.C., April, 1985, p. 1.

7. R. J. Kelsey, *Packaging in Today's Society,* St. Regis Paper Co., Ridgewood, New Jersey, 1978, pp. 20–38.

8. T. Tang, "Closures, Liners and Seals," *Packaging Encyclopedia 1984,* **29,** 158–160 (March 1984).

9. J. F. Hanlon, *Handbook of Package Engineering,* McGraw-Hill, New York, 1971. Sect 9, pp. 5–6.

10. "Tamper-Evident Packaging Requirements," *Fed. Regist.* **47,** (215), 50442 (Nov. 5, 1982).

11. J. B. Carroll, *Memorandum from the Closure Manufacturers Association,* McLean, Va. April 19, 1985, p. 1.

12. B. Knill, *Food and Drug Packaging,* **48,** 5 (Dec., 1984).

13. "Quantity and Value of Metal and Plastic Closures," *Current Industrial Reports, M34H(83)-13* U.S. Department of Commerce, Washington, D. C., Oct. 1984, p. 1.

14. Ref. 5, p. 152.

15. *Steel Cans,* Report by the Committee of Tin Mill Products Producers, American Iron and Steel Institute, Washington, D.C., 1984, p. 8.

16. Ref. 9, Sect 6, p. 20.

17. Ref. 4, p. 29.

18. P. Zwirn, "The Crown is Still King," *Canadian Packaging,* 37(6), 24 (June, 1984).

19. "The Best Ideas in Packaging," *Food and Drug Packaging,* **39,** (Nov., 1984).

20. C. G. Davis, *Packaging Technology,* **27,** 27 (Oct./Nov. 1982).

21. R. Heuer, *Packaging,* **29,** 34 (April, 1984).

22. R. L. Harris, "Closures, Dispensing Systems," *Packaging Encyclopedia 1984,* Cahners Publishing Co., Boston, Mass, **29** 157 (March 1984).

23. T. T. Williams, *A History of Technology,* Clarendon Press, Oxford, 1978, p. 1411.

24. "Tamper-Resistant Packaging Requirements," *Fed. Regist.* **47,** (215), 50444 (Nov. 5, 1982).

25. H. Forcinio, *Food and Drug Packaging* **29,** 35 (Sept. 1984).

26. *Child Resistant Packaging,* American Society for Testing and Materials, Philadelphia, Pa., pp. 20–21.

27. "CPSC Replies to Queries on CR Rules Compliance," *Food and Drug Packaging* **40,** 20 (March 1985).

28. Ref. 9, Sect. 9, p. 21.

29. *Options for Successful Medical/Pharmaceutical Packaging, Bulletin PB-484,* Phoenix Closures, Inc., Naperville, Ill., April 1984, p. 2.

30. L. Roth, *An Introduction to the Art of Packaging,* Prentice-Hall Inc., Englewood Cliffs, New Jersey, 1981, p. 153.

31. Ref. 9, Sect. 7, p. 16.

32. Ref. 9, Sect. 7, p. 31.

33. "The 5 C's of Closure," *Marketing Communications Bulletin,* Phoenix Closures, Inc., Naperville, Ill., Dec. 1, 1984.

34. *CMA Voluntary Standards,* Closure Manufacturers Assoc., McLean, Va. 1984.

35. J. Szajna, *Food and Drug Packaging* **29,** 12 (May 1984).

36. A. J. F. O'Reilly, *Food and Drug Packaging* **29,** 72 (Sept. 1984).

37. H. K. Foster, *Food and Drug Packaging* **29,** 46 (Nov. 1984).

38. Ref. 21, p. 36.

39. B. Miyares, *Food and Drug Packaging* **29,** 72 (Nov. 1984).

40. R. Graham, *Food and Drug Packaging* **29,** 42 (Sept. 1984).

41. H. Peter Aleff, "Comparison of Thermoplastic with Thermoset Closures, *Packaging Technology* 12(2), 18 (April 1982).

42. "Materials 1985," *Mod. Plast.,* **63**(1), 69 (Jan. 1985).

General References

H. L. Allison, "High Barrier Packaging: What are the Options?," *Packaging,* **30,** 25 (March, 1985).

J. Agranoff, ed., *Modern Plastics Encyclopedia,* **60,** McGraw-Hill, Inc., New York, 1984. Note: The Engineering Data Bank section contains extensive data on properties of plastics, design data, chemicals and additives information, and data on molding machinery.

L. Barail, *Packaging Engineering,* Reinhold Publishing Corporation, New York, 1954.

The Closure Industry Report, Closure Manufacturers Association, McLean, Va., Fall, 1983 and Summer, 1984.

E. F. Dorsch, "Closures and Systems," *Packaging Encyclopedia,* **30,** 126–128 (1985) Cahners Publishing Company, Boston, Mass., 1985.

R. C. Griffin, Jr., and S. Sacharow, *Principles of Packaging Development,* The Avi Publishing Co., Westport, Conn., 1972.

K. Hannigan, "Baby Food Closure Grows Up," Chilton's *Food Engineering,* Chilton Company, Radnor, Pa., Jan., 1984.

G. Jones, *Packaging: A Guide to Information Sources,* Gale Research Company, Detroit, Mich., 1967.

Kline's Guide To The Packaging Industry, Charles H. Kline & Co., Inc., Fairfield, N.J., 1979.

E. A. Leonard, *Packaging Economics,* Books for Industry, New York, 1980.

"Mack-Wayne Introduces New One-Piece Dispensing Closure and New Breakaway Closure," *Packaging Technology,* 14(1), 34 (July/Aug., 1984).

"Packaging: Big Changes Ahead," *Prepared Foods,* **153,** 60–70 (June, 1984).

"Plastics Hit the Market with Tamper Evident Closures," Chilton's *Food Engineering,* **56,** 60 (April, 1984).

N. C. Robson, "The State of the Anti-Tampering Art," *Packaging Technology,* **13,** 26 (June, 1983).

S. Sacharow, *A Guide to Packaging Machinery,* Books for Industry, New York, 1980.

S. Sacharow, *A Packaging Primer,* Books for Industry, New York, 1978.

B. Simms, ed., "Closures, Dispensers and Applicators," *Packaging Encyclopedia,* Vol. 29, Cahners Publishing Company, Boston, Mass. 1984.

J. F. Nairn
T. M. Norpell
Phoenix Closures, Inc.

Drawings for figures 1–14 courtesy of S. Kiefer and B. Zemlo, Phoenix Closures, Inc.

CLOSURES, BREAD BAG

Packaging bread in polyethylene bags became an almost universal method of bread packaging in 1963. Customer preference for easy opening and reclosing the package influenced the development of bag closures at the very start of bread bagging. Closing by tape, which had been used in connection with early bagging methods, was quickly discarded when the bulk of the bakery production lines changed to the new polyethylene bagging and closing equipment. Tape was neither convenient for the consumer nor fast and dependable for the bakers. Two types of closures and automatic equipment systems have emerged as the standards of the baking industry. Developed simultaneously, these are the wire tie and the plastic-clip closure.

Wire ties. The predominant automatic wire-tie equipment now being used by the United States baking industry is from

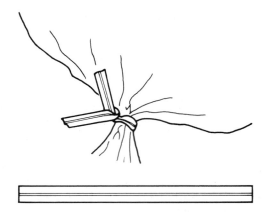

Figure 1. A wire tie for a bread bag.

the Burford Corporation, Maysville, Oklahoma. Different models are required to attach to the different types of installed bread and bun baggers. This type of automatic bag-closing equipment is limited to about 60 packages per minute. The wire ties used in automatic application are 4-in. (10.2-cm) long. The wire comes on reels of 6,000 ft (1,829 m), 18,000 closures per reel, and is cut to length as it is applied to the package. Colors are available for color-coding purposes (see Fig. 1).

The wire tie has a left- or right-hand twist, depending on the production equipment. Wire tires are available with all-plastic, laminated plastic/paper, and all-paper covers over the wire core. The quality of the wire tie must be carefully maintained or separations of the laminations occur leaving the bare wire exposed. Source of the wire ties is Bedford Industries, Worthington, Minn.

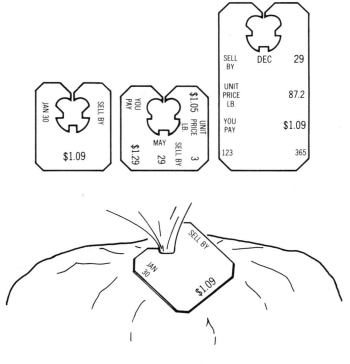

Figure 2. A plastic-clip closure for a bread bag.

Plastic-clip closure. Plastic-clip closures are produced from medium-impact polystyrene (see Polystyrene). The material has a resiliency that allows it to return to its original shape after many reuses by the consumer (see Fig. 2). They are furnished with various aperture sizes to accomodate the different polyethylene bag widths and thicknesses. The proper selection of aperture size can result in an almost air-tight package. Colors are available for color-coding purposes. The plastic-clip closures are provided in reels of 4000 or 5000 closures per reel.

The Striplok closures and closure-applying equipment are manufactured by the Kwik Lok Corporation, Yakima, Wash. The equipment is simple and dependable, with production speeds up to 120 bakery bags per minute. The Kwik Lok Corporation manufactures two basic machines that attach to any of the common bagging machines from Formost Packaging Machines, Inc., Woodinville, Wash.; United Bakery Equipment, Compton, Calif. and AMF, Inc., Union Machinery Division, Richmond, Va. By mounting an imprinter on the closing machine, imprinting can be done on the plastic clip. Many bakeries are required to print on the package the price per pound, the unit price of the package, and the "sell by" date (see Code marking and imprinting). Printing on the polyethylene bag lacks legibility and is often obscured by the package graphics. Three sizes of Striplok closures are available to facilitate compliance with regulatory coding and dating requirements.

G. E. Good
Kwik Lok Corporation

COATING EQUIPMENT

Coating equipment is used to apply a surface coating or adhesive for lamination or for saturation of a fibrous web. Such equipment consists of three main components: (1) a coating head; (2) a dryer or other coating solidification station; and (3) web-handling equipment (drives, winders, edge guides, etc).

Coatings when applied must be sufficiently fluid to be spread into a uniformly thin layer across the web. Therefore, coatings are applied as solutions in organic solvents, as aqueous solutions or emulsions, or as molten or softened solids. Solutions and emulsions require drying to obtain a solid coating, whereas hot melts are solidified by cooling. Some coatings may be applied as reactive liquids which are polymerized by irradiation or by application of heat. Different coating equipment is required for various coating types.

This article does not pertain to extrusion coating (see Extrusion coating). Extrusion coaters are designed for higher viscosity materials. A typical extrudable coating might have a viscosity of 2000 P (200 Pa · s) at 250°F (121°C). Hot melt coaters usually handle lower viscosity coatings, but at somewhat higher temperatures. A typical hot melt viscosity might be 100 P (10 Pa · s) at 320°F (160°C). Coating equipment discussed in this article is suitable for coating various materials in roll or sheet form: paper and paperboard, films, foils, metal coil and sheets, textiles, and nonwoven fabrics.

Coating may be applied directly to the substrate, or it may be cast to another surface and later transferred to the substrate of choice. Cast coating is utilized when the substrate is sensitive to the vehicle or when it may be damaged by exposure to oven temperatures. Thus, pressure-sensitive label

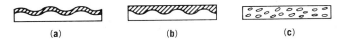

Figure 1. (a) Coating of uniform thickness. (b) Coating to a uniform total thickness. (c) Saturation.

stock is usually manufactured by cast coating; that is, the adhesive is first applied to the silicone release coated paper, dried, and then laminated to the paper stock. A glossy paper surface finish may be achieved by casting and drying a clay coating against a chrome-plated cylinder.

Coating Heads

The coating head accomplishes two functions: applies the coating to the substrate and distributes a metered amount uniformly over the surface. Metering may be combined with the coating function or it may be carried out separately following the coating deposition. The coating head may be designed to deposit a coating of a constant thickness regardless of web irregularities (Fig. 1a); it may coat to a constant total thickness filling the web irregularities (Fig. 1b); or the coating may be used to impregnate a fibrous web (Fig. 1c), which is called saturation. Coating equipment has been reviewed in several books (1–3). Most coating equipment falls into one of the three major categories: roll coaters, knife, blade or bar coaters; and extrusion or slot orifice coaters.

Roll coaters. Roll coaters are most widely used and they may be subdivided according to several construction types: direct, reverse, gravure, and calender coaters.

Direct roll coaters. The web carrying roll and the coating applicator roll rotate in the same direction and at the same speed in the direct roll coating process (Fig. 2). The coating is split about evenly between the roll and the substrate, and it undergoes several stages in this process: laminar flow at the beginning of the coating nip, cavitation, cavity expansion, and filamentation, as shown in Figure 3 (4). There are many different coating heads that utilize the direct roll coating principle. A squeeze-type roll coater is shown in Figure 2. Direct roll coaters can be also used for coating sheeted materials (Fig. 4). The coating is metered by a gap between accurately machined metering and applicator rolls. Sheets are fed through a nip formed by applicator and carrier rolls.

Kiss roll coaters apply the coating to the web from a pan (Fig. 5). This type of coating is not accurate and therefore kiss roll coaters usually employ a metering device, such as a wire wound rod, to remove the excess coating. Kiss roll coaters can

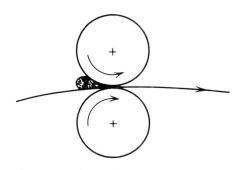

Figure 2. Squeeze coater.

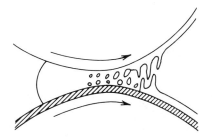

Figure 3. Coating splitting.

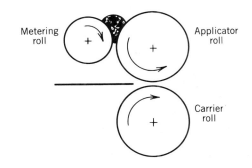

Figure 4. Direct roll sheet coater.

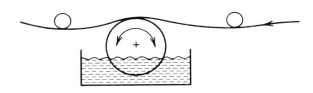

Figure 5. Kiss roll coater.

be also run in the opposite direction of the web travel. In such cases the web wipes the roll clean.

Transfer roll coaters have been used in the past for clay coating of paper, but most have been replaced by air knife and blade coaters. A second generation of transfer coaters, of much lighter construction than the original paper coating machines, have been designed and are offered for application of lightweight coatings including 100% solids radiation-curable coatings (Fig. 6). The rubber-covered applicator roll is run 5–25 times faster than the transfer roll, and this decreases the amount of coating transferred to the substrate by that factor. Direct roll coaters are used widely for such applications as paper sizing, paper color coating, overcoating of blister packaging board, and heat-seal coating.

Reverse roll coaters can handle a wide range of coatings and viscosities. Solvent solutions, aqueous coatings, and hot melts are coated on these machines. Reverse roll coaters are expensive because the rolls must be accurately machined. The rolls are usually made from chilled iron and are chrome-plated for corrosion resistance. There are several designs of reverse roll coaters: feed location may be varied from nip to pan, three or four rolls may be used, and the geometrical arrangement of rolls may vary. The coating is delivered to the nip between metering and applicator rolls, and dams are used to contain the coating. A gap set between these two accurately machined

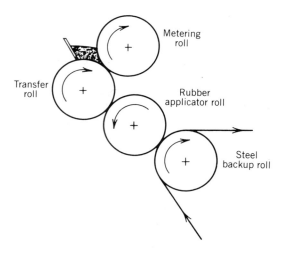

Figure 6. Transfer coater.

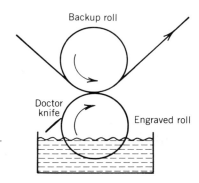

Figure 8. Direct gravure coater.

rolls determines the amount of coating carried by the applicator roll. The main feature of the reverse roll coaters is that the applicator roll rotates in the opposite direction to the web travel and transfers the coating by wiping (Fig. 7). The web usually runs faster than the applicator roll, and the amount of coating deposited depends on the ratio between the speed of these two rolls (wipe ratio) as well as on the gap between metering and applicator rolls.

Gravure coaters. The most important part of a gravure coater is the engraved roll. The roll is wetted with the coating, excess is removed by a doctor knife and the coating remaining in the engraved cells below the roll surface is transferred to the web at the gravure roll–backup roll nip (Fig. 7). Three different engraving patterns are used for gravure coating rolls: pyramidal, quadrangular (truncated pyramid), and trihelical. The amount of coating depends on the engraving depth and the cell density, referred to as screen or ruling (ie, the number of cells per unit length). The coarser the screen, the larger are the diameter and depth of cells and the higher the deposit is. Wet coating weight may vary from 0.5 lb/1000 ft² (2.4 g/m²) for a 200 pyramidal screen (200 cells per inch or ca 8

cells/mm) to 13 lb/1000 ft² (63 g/m²) for a 24 trihelical roll with various combinations between these two extremes. The fluid viscosity must be sufficiently low to allow the transfer of the coating from the cells to the web at nip pressures.

In the direct-gravure arrangement (Fig. 8) the engraved roll contacts the web directly. In offset gravure the coating is first transferred to a rubber-covered roll or an apron and then transferred to the web. Offset gravure allows the use of higher nip pressures and is therefore more suitable for coating rough-surface substrates. It also allows the coating to level on the roll surface before its transfer to the web.

Gravure coating is widely used for various decorative and functional lightweight coatings on plastic films, paper, and other substrates used for packaging. Most of the gravure-applied coatings, adhesives, and inks are solutions in solvents, but gravure coaters are increasingly used for waterborne coatings and they have also been adapted for hot melt applications and for 100% solids reactive coatings. Gravure coaters are relatively inexpensive and reliable machines.

Calender coating. The calendering process involves squeezing a polymeric coating between steel rolls into a thin sheet and then laminating it to the substrate. Rubber and PVC are most often used in calender coating. The calender coating process is essentially the same as the calendering process used in the production of PVC film (see Film, semirigid PVC), except that in the coating process another substrate is coated during the process. Although calendering is a very important process for heavyweight coatings, such as used for artificial leather, it is rarely used for manufacturing of packaging materials.

Knife and bar coaters. Knife and bar coaters are metering devices that remove excess coating and allow only a predetermined amount to pass through.

Knife-over-roll coaters. A knife-over-roll coater consists of a straight-edged knife placed against a roll. The coating weight is adjusted by setting a gap between the roll and the knife. An excess of coating is delivered to the bank before the knife, and the desired amount is metered by the gap. An accurately machined steel roll and knife are used. In another version of a knife-over-roll coater, a rubber-covered roll is used and the coating weight is adjusted by varying the pressure of the knife against the rubber surface. The knife-over-roll coater is a widely used machine most suitable for thicker coatings with a higher viscosity range. Easily leveling solvent coatings are best. Waterborne coatings have a tendency to form streaks.

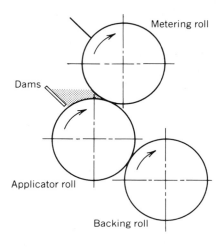

Figure 7. Nip fed reverse roll coater.

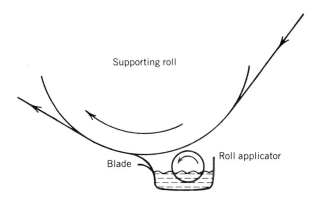

Figure 9. Blade coater.

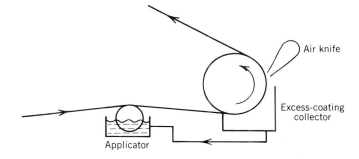

Figure 10. Air-knife coater.

Other knife coaters. Knife coaters in various other configurations are used: floating knife, knife-over-blanket, knife-over-channel, and inverted knife. These methods are generally used in textile-coating operations.

Blade coating. Flexible-blade coating is the dominant process along with air-knife coating for applying pigment coatings over paper and paperboard. Pigment coatings give a smooth and printable surface. Blade-coating processes are suitable for high speed applications such as required in paper converting. The coating head consists of a coating applicator, such as roll or fountain, which applies an excess of coating to the paper. The excess coating is removed by a blade (Fig. 9). Several modifications of blade coaters are available employing different blade designs, different methods of applying and regulating blade pressure, and different ways of applying the coating to the paper. Machines capable of applying coating to both sides of the web are available.

Air-knife coaters are being replaced by more recently developed blade coaters for paper-coating applications. Their main disadvantages are lower speeds and the use of lower solids content coatings. Air-knife coaters do, however, apply a uniform thickness coating, regardless of the surface roughness. Thus, for coating paperboard which has a rough surface and which is coated at speeds well below 1500 ft/min (457 m/min), air-knife coating remains the preferred method.

In air-knife coating an excess of material is applied to the web surface, usually by a kiss-roll applicator, and the excess is removed by the air knife. The air knife consists of a head provided with a narrow slot. Pressurized air is forced into the head and is accelerated at the slotted nozzle. The escaping air stream impinges upon the coated web and removes excess coating. It removes the coating evenly from high and low spots. A schematic diagram of an air-knife coater is shown in Figure 10.

Wire-wound-rod coater. A wire-wound-rod coater (Meyer rod) is a simple metering device widely used in applying lightweight coatings over film and paper packaging materials. The coating is applied by an applicator, usually a kiss roll, and the excess of coating is removed by a rod wound spirally with a stainless-steel wire. The rod wipes the surface clean, except what escapes through the spaces between the wires. The larger the wire diameter, the heavier is the coating. The rod is rotated to help to dislodge any particles that might get trapped between the wire and the web and to assure uniform wear of the wire.

Slot-orifice and curtain coaters. These methods are variations of the extrusion coating process, which is widely used to apply coatings of thermoplastic polymers, especially polyethylene, on paper and other substrates (see Extrusion coating).

Slot-orifice coater. The equipment is similar to an extruder: It is equipped with an orifice of an adjustable width through which a coating is extruded. A smoothing roll might follow the orifice on some equipment. Slot-orifice coaters are most often used for hot melts, but they may also be used for application of aqueous coatings. These machines are of lighter construction than extruders, which can handle polymers of much higher viscosities. A slot-orifice coater is widely used for manufacturing hot melt pressure-sensitive adhesive packaging tapes (5) and for applying waterproof coatings over paper.

Curtain coater. A curtain coater consists of a slotted head through which a liquid curtain is allowed to fall down onto the moving web or sheets. The excess coating is collected in a pan underneath and returned to the head (Fig. 11). Curtain coaters are especially suitable to coat sheets, such as corrugated paperboard, and may be used for various coating types including hot melts.

The slot is adjustable usually between 40–80 mil (1–2 mm) in width. The flow rate is regulated by the slot width and by the liquid level (in weir-type coaters), or by pressure and pump speed in enclosed-head coaters. Coating thickness and coating speed also depends on the distance between the slot and the conveyor. The liquid curtain accelerates as it falls down, be-

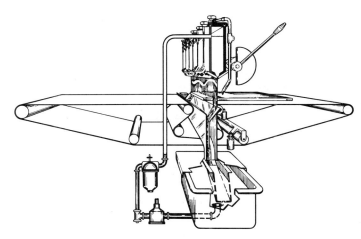

Figure 11. A schematic diagram of a curtain coater. (Courtesy of Ashdee Div., George Koch Sons, Inc.)

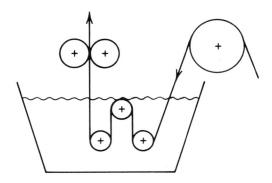

Figure 12. Sawtooth saturator.

Table 1. Drying Equipment

Heat transfer	Web handling
convection dryers	
parallel air flow	idler-supported dryers
impingement air	conveyor dryers
air foil	catenary dryers
through dryers	U-type dryers
infrared radiation dryers	arch dryers[a]
near infrared (electric)	tenter frame dryers
far infrared (electric or gas)	floater dryers
conduction dryers	
hot roll dryers	

[a] An arch dryer including winding, coating, and laminating stations is shown in Figure 13.

coming thinner. The conveyor rate must be equal to or slightly faster than the curtain falling rate. Curtain coaters are simple and inexpensive devices suitable for deposition of heavy coatings. The fluid must form a stable curtain. Fluids up to 150 P (15 Pa · s) viscosity may be used.

Saturators

The saturation or impregnation process is used to treat porous fibrous webs with a polymeric binder to improve the web's strength, its water or grease resistance, or other properties. The process consists of immersion of the web into a coating bath, or applying an excess of coating on both sides and then squeezing or scraping the web to remove the excess. The coating does not remain on the surface but penetrates the web. Packaging papers and paperboard may be wax impregnated to achieve the required level of moisture barrier. Some rubber-saturated papers may be used for packaging or labeling applications that require increased strength and tear resistance.

Saturating machines consist of a web-immersion section and a metering section. Several saturating arrangements are used. Figure 12 shows a conventional sawtooth saturator followed by squeeze-roll metering. Other types of metering arrangements are inflatable bars, bar scrapers, doctor blades, and similar devices.

Drying

Coatings applied as solutions or emulsions must be dried in order to remove the liquid vehicle. Heat and mass transfer

take place simultaneously during the drying process, and the heat is transferred by convection in air dryers, by radiation in infrared radiation dryers, or by conduction in contact drum dryers. The drying equipment also has a means of vapor removal and recirculation and heat-exchange equipment to conserve energy. Figure 13 schematically shows the air flow in a convection dryer. If a coating from a solution in an organic solvent is used, the solvent vapor must be removed from the exhaust in order to satisfy the environmental laws and to decrease solvent costs. Solvent adsorption on activated carbon or incineration is used. Drying equipment may be subdivided according to the heat-transfer mechanism or according to the method of web handling as listed in Table 1.

Extruded and hot melt coatings do not require drying and are solidified by chilling. Such coating machines require considerably less space than the coating lines with drying ovens. Such coaters are often used to manufacture pressure-sensitive adhesive label stock (5).

Some coatings are applied as reactive monomers or polymer–monomer blends and may be cured by either ultraviolet or electron-beam irradiation. Such irradiation units are incorporated into the coating lines (see Curing).

Web Handling

Coating machines usually apply the coating or the adhesive to the packaging material supplied as a continuous web on a

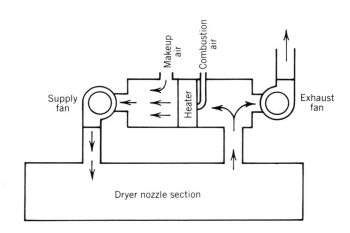

Figure 13. Air flow in a convection dryer.

Figure 14. Coating-lamination line with an arch dryer. (Courtesy Faustel, Inc.)

Figure 15. Carton blank coater. (Courtesy of International Paper Box Machine Co.)

roll and the finished product is rewound after completion of the operation (see Roll handling). Unwind and rewind stands, web-carrying equipment consisting of driven rolls, automatic tension-control devices, edge-guiding equipment to keep the web properly tracking on the machine, and other accessories complete the coating machine.

Some packaging materials are coated as sheet, requiring sheet-feeding and sheet-handling devices. A carton-blank coater is shown in Figure 15.

BIBLIOGRAPHY

1. D. Satas, ed., *Web Processing and Converting Technology and Equipment,* Van Nostrand Reinhold Co., Inc., New York, 1984.
2. H. L. Weiss, *Coating and Laminating Machines,* Converting Technology Co., Milwaukee, Wisc., 1977.
3. G. L. Booth, *Coating Equipment and Processes,* Lockwood Publishing Co., Inc., New York, 1970.
4. R. C. Meyers, *J. Polym. Sci. Part C* **35,** 3 (1971).
5. D. Satas, ed., *Handbook of Pressure-Sensitive Adhesive Technology,* Van Nostrand Reinhold Co., Inc., New York, 1982.

D. SATAS
Satas & Associates

CODE MARKING AND IMPRINTING

The challenge of package imprinting and coding has intensified as state and federal regulations impose more formal restrictions on product packaging. The increased threat of regulations mandating the clear imprinting of sell-by statements, freshness dating, ingredients, prices, batch or date-of-manufacture coding, and other variable information on packages has brought about a marked change in the traditional perspective on overprinting and on identification systems and the technology of product marking in general. Due in part to consumer pressure, new legislative proposals are resulting in new regulations that require the printing of place of manufacture, recall control and batch codes, and other information. The United States is not alone in its current preoccupation with information printing. The EEC has enacted similar laws, and continues to revise its standards for package information (see Laws and regulations, EEC; Standards and Practices). Code printing can offer benefits in generating tighter inventory and production controls, which is especially important in view of the ever-present threat of product recall. Accurate coding enables the manufacturer to identify where, when, and by whom the product was packaged, as well as to pinpoint area of distribution. In addition, and beyond the legal issue, coding offers manufacturers a method of communicating an intrinsic commitment to reliable product quality at all levels, from ingredients to freshness standards. Voluntary legible coding reassures the consumer that the product is fresh (1).

Nevertheless, the day-to-day reality in the face of coding mandates that manufacturers are expected to "deliver the goods" by whatever means at their disposal, whether those means are economically reasonable. High productivity can be a hindrance to efficient printing. Faster production lines, new packaging technologies, Universal Product Codes, (see Bar coding), and packaging advances in general demand modern and compatible printing systems. Mandatory coding generates enormous and complex manufacturing changes that are being addressed by manufacturers of printing systems and printing equipment supplies through the development of responsive technology. This article identifies and details printing systems and coding supplies developed specifically to address needs intensified by package-coding mandates. Evidence suggests that continued change in package printing is probable, and in that light, it is cost expedient to prescribe printing requirements for applications in anticipation of projected demands. A manufacturer, for example, who must apply variable information to satisfy regulations for some states would do well to apply the information to all packages in anticipation of future regulations. This discussion highlights solutions to the coding dilemma that are reasonable within cost-efficient, reliable, uncomplicated parameters, and convey a futuristic as well as routine perspective. This dualistic approach can be successful only within the context of the specific coding application, and in relation to particular variables, including production rate and flow, print quality and legibility, as well as existing equipment. Ideally, a reasonable coding/imprinting system is designed and engineered for maximum versatility in response to these variables.

Outside Versus In-plant Printing

There are two basic kinds of information to be imprinted: constant and variable. Constant preprinted information generally includes the company and product name. Variable information typically includes freshness dating, weights, and price. Ingredients are also considered variable, although such information generally changes less frequently. Although the types of information are fairly straightforward, the package materials themselves may generate greater deviations from a norm. The most common surfaces (substrates) include cellophane (see Cellophane), plastic films (see Films, plastic), and metallic (see Metallizing) and waxed (see Waxes) surfaces.

There are three primary ways to apply both constant and/or variable information. One is to have a commercial printer preprint all packages. There are two advantages to this: print quality is high, and production costs for long runs are low. The two disadvantages, however, are high cost for short runs and lack of versatility. Once an inventory of preprinted packages is printed and maintained, the investment has been made. Should information change, preprinted packaging material becomes obsolete. The broader the product line, the greater the increase in short and costly press runs, and the more common the inventory control irregularities and delivery problems. Preprinting of packages is generally impractical for variable information.

The second imprinting option is to have constant information printed outside by a commercial printer, and all variable information added in-house as part of the packaging operation. Many manufacturers employ some mix of in-house and commercial printing: commercial for quality or decoration, or to take advantage of bulk printing discounts; and in-house printing for versatility and inventory reduction. However, the additional management of information and control of printing compatibility factors are, in themselves, an additional cost.

The third option is complete in-house printing, which enables a manufacturer to purchase blank packaging film and print all packaging information in-house. This is particularly suitable for manufacturers of many short-run products, and it generates special benefits for manufacturers printing variable information or those with packaging lines designed for high production but that have fixed, dedicated cut-off lengths. In-house printing incurs initial capital investment in equipment, but unlike the other options, it puts the entire packaging process in the control of the product manufacturer. Current technology permits virtually effortless integration of overprinting systems in existing packaging lines. A baker of rolls and buns for fast-food chains, for example, can print product ingredients on all packages in-house and change the freshness code regularly. By simply changing printing plates, the baker can imprint the ingredients of the kind of roll being baked, which satisfies the legal obligation. For convenience and additional savings, the bakery simply snaps in a new plate when ingredient formulations change. The only inventory requirement is the stocking of one standard type of bag for all products.

There are a number of considerations to address when selecting a coding system. These include the size and location of the legend on the package; whether the imprint will be an integral part of the package design; type styles and colors; and the type of machinery to be used. Production demands must also be considered: the versatility and flexibility needed for quick information change; the ability to handle various package sizes; or the need for variable speeds. Multitrack feed requirements and frequency of information change as well as future production expectations may also affect coding system selection. If an existing packaging machine is being equipped with an imprinter, installation of chain drives from the imprinter to the packaging machine can be expensive and cumbersome. An electronically controlled, independently driven imprinter is easier to install and requires minimal conversion for most common packaging lines.

Maintaining print quality takes consistent supplies performance. Coding supplies to achieve the best results with minimal downtime for setup and maximum daily production time are critical. In summary, specifications for package overprinting are unique to the manufacturer and product needs.

Imprinting Methods

Different package printing systems offer diverse capabilities for speed, colors, handling, and other in-line coding requirements. Whether the system is standard or custom designed, the processes for imparting ink (see Inks) from the printing plate to the package surface are essentially the same. There are two primary methods of printing: contact and noncontact. The most common is contact printing, a technique in which the printing elements actually touch the surface being printed. The second method, noncontact printing, is a relatively new technology. Each has benefits and disadvantages (see Table 1).

Table 1. Package Coding Technologies—Comparative Features

Method	Capital equipment costs	Cost/print	Cycle rate
Contact printing			
wet ink	low	low	low
hot stamp	low–medium	medium	low–medium
dry ink	low–high	low	low–high
Noncontact printing			
laser	very high	medium	high
ink jet	high	medium	high

Contact printing.   There are three forms of contact printing: wet-ink, hot stamping (see Decorating), and instant-dry inks.

Wet Ink

Wet-ink direct application is the most traditional printing method, but unless properly applied and carefully maintained, wet inks can smudge. Wet inks can also dry rapidly in the ink reservoir, necessitating regular thinning of the ink and cleaning procedures that are potentially labor intensive. Wet inks also require drying time, a potential hindrance to production flow. Disposable ink capsules have been developed that overcome some of the shortcomings of wet ink including ink spillage, regular cleanup, and complexity of use.

Hot Stamping

A second alternative is hot stamping, also known as foil printing. This process uses dry inks bonded to a thin layer of carrier film. When the foil is sandwiched between heated metal type and the packaging material, the ink is transferred from the carrier to the packaging material. With this method the packager obtains crisp, highly legible results that cannot smudge or rub off and require no drying. One operator can oversee product feed, check registration, and replace empty foil rolls.

Dry printing with foils offers many advantages over other methods, including clean, dry prints, minimal cleanup, and automatic production. Although initial equipment investments may be higher than for some other systems, the printed results are impressive. And, when the printed information serves as package design or presents important consumer information, the superior appearance of foil printing may justify the slightly higher purchase and operating costs. In addition, dry printing considerably reduces the downtime required to change foil.

Both wet-ink and dry-stamping printing systems are available in off- or in-line models. Off-line systems typically are more labor-intensive, but printer breakdown does not necessarily require production line shutdown. With in-line imprinting, unit capabilities must be adequately matched to current line speeds and, ideally, anticipated line speeds. Quality of product and service availability must be major considerations when imprinters are installed in-line because of their potential to upset production if the imprinting operation breaks down or requires maintenance.

An interesting example of hot stamping's versatility is its use as a replacement for pressure-sensitive labels on flexible cellophane packages (see Cellophane) of baked goods for vending machines. The labels require a large inventory and significant downtime to change the type of label. Hot stamping imprints product information and the company's logo directly onto the cellophane package as a high quality print, comparable to commercial printing. Yet, with all the benefits that can be derived, there are disadvantages to hot stamping. These include downtime to change foil rolls, higher equipment-investment costs and relatively high print costs. For example, a $\frac{1}{8} \times 1$ in. (3×25 mm) date code would cost approximately \$0.05–0.06/1000 for foil versus less than \$0.02/1000 for instant dry ink, the alternative form of contact printing.

Dry Ink

Although the technology came as recently as 1980, dry ink is being used with great success by a number of major food manufacturers (2,3). The ink is contained in a totally dry impregnated heat-activated roll, which eliminates mess and cleanup. Aside from the benefit of being clean, and virtually instantly dry, dry inks increase speed and print on all common packaging materials. With these inks, systems can operate continuously at speeds up to 1000 cycles per minute, without splattering or spraying. Ink-roll changes are less frequent, as traditional rolls can print over 250,000 ($\frac{1}{8} \times 1$ in.) impressions before requiring replacement, and replacement can be scheduled to coincide with predetermined downtime. Dry-ink printers can also be installed at any position with no danger of ink spillage. Because the ink from a dry-ink roll is gradually released, and ink rolls can be changed at convenient times, dry-ink technology contributes to increased productivity. Even though the dry-ink systems are new, they have been developed for easy tie-in with popular packaging machines. With the faster production speeds now common to the food industries, these new inks represent a significant advance, virtually eliminating smear and related problems discussed earlier (4).

Noncontact printing. In contrast to the three methods of contact printing, noncontact printing offers two major types of imprinting: laser and ink jet.

Laser Printing

Imprinting equipment manufacturers and industry in general are beginning to realize the potential of laser technology. The principle of laser printing involves reflecting an intense light beam off mirrors, through a stencil, and onto the material to be marked. The laser beam prints in one of three ways: it either changes the color of the material; affects the pigment; or most often, removes a precoating. Laser printing is reliable, applies a predictably permanent mark, and prints at extremely fast rates. In the electronics industry, for example, laser printing can process components at speeds up to 72,000 parts per hour with consistent high quality. Even though lasers offer these benefits, it is doubtful that they will be used in the food industry because lasers change the color of the packaging film or burn off preprinted ink. In addition, the size of the laser-print area, which is usually confined to $\frac{1}{4} \times \frac{1}{2}$ in. (6×13 mm) is generally too small to be useful to the food industry.

Ink Jet

In ink-jet printing, wet ink is formed into droplets. These droplets are then electrostatically deflected, giving the up-and-down composition in marking the character. The degree of deflection depends on the amount of charge on each droplet. The ink droplets are applied by a high pressure nozzle. By coordinating the deposit of the droplets to the printing surface with the motion of the substrate (eg, on the conveyor line) the left-to-right spray of droplets that form the dot-matrix printing characters is achieved. Given current economic restrictions, this method requires high throughput to be cost-effective. Ink-jet printing is mainly used in the beverage industry.

In conclusion, as packaging machinery becomes more productive, imprinting systems must respond not only with faster printing capabilities but also with integration equipment and supplies technology that minimize the complexities of setup, maintenance, or routine service. That involves the engineering of equipment with fewer mechanical parts, less pneumatics for certain environments, and versatile power source options.

BIBLIOGRAPHY

1. "Pepperidge Farm Adds New Ingredient to Production Recipe: Markem 904 Package Coders," *Packaging Digest*, 53 (Sept., 1983).
2. B. Dickson, *Canadian Packaging*, 24 (Oct., 1983).
3. "Code Printer Graduates to Big Time Candy Line", *Food and Drug Packaging*, 1 (Aug. 20, 1981).
4. J. Gardetto, *Baking Industry*, 120 (Sept., 1981).

KEVIN CAFFREY
Markem Corporation

COEXTRUDED BOTTLES.   See Multilayer plastic bottles.

COEXTRUSION MACHINERY, FLAT

Multilayer coextruded flat film and sheet are produced on single slot T dies. The overall process is similar to that used for single-layer products of the same dimensions (see Extrusion). The specialized design considerations for coextrusion are discussed below.

Machinery

Extruders. Each product component requires a separate extruder. Several layers may be produced by the same extruder using suitable feed block or die connections. Systems range from two extruders for a simple BA or ABA structure to five or six extruders for high barrier sheet (see Coextrusions for semirigid packaging).

Since all of these extruders feed one die, the area behind the die can become a crowded place. Extruders are therefore built as narrow as possible. Vertical gearboxes permit tuck-under motors which reduce space requirements and provide good service access to the motor and other components.

Smaller extruders can be mounted overhead at various angles. Larger machines present an access and height problem when located overhead. The most effective and accessible arrangement is usually a fan layout of larger extruders with some small machines overhead (Fig. 1).

Thermal expansion requires that all but one and sometimes all machines be mounted on wheels with expansion capability both axially and laterally. Height adjustments must also be

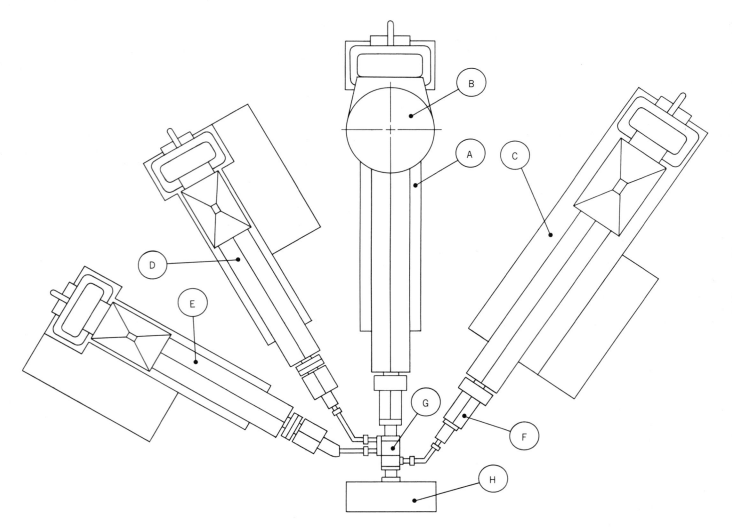

Figure 1. A typical layout for coextrusion (1). A, Recycled layer extruder; B, crammer feeder; C, virgin layer extruder; D, glue layer extruder; E, barrier layer extruder; F, static mixer; G, feed block; and H, sheet die.

provided to permit accurate alignment to the interconnecting piping.

PVDC requires special corrosion-resistant extruder construction (see Vinylidene chloride copolymers). High nickel cylinder lining and Z nickel screws are essential to avoid corrosion and polymer degradation. The optimum ratio is 24 L/D for this heat-sensitive material. Everything associated with the PVDC extruder is critically streamlined, and all flow surfaces normally contacting or possibly contacting PVDC must be nickel. No screens or breaker plates are used. A long conical tip with matching adapter assures streamlined flow. A PVDC extruder can run other barrier materials equally well with different suitable screw designs.

All other extruders can be of conventional materials of construction. A ratio of 30 L/D is desirable for best performance in most cases. These can also be vented for devolatilizing when necessary to remove entrapped air or moisture. Venting cannot be used with high back pressures or low screw speeds. Screen changers are used on most extruders to avoid laborious disassembly when screens are plugged. Good screening is essential in barrier sheet extrusion to avoid plugging critical flow passages and pinholes in some layers. Remote-control hopper shutoffs help quick start-up and shutdown.

Ethylene–vinyl alcohol (EVOH) requires predrying. Otherwise, it degrades during extrusion with a reduction of melt viscosity. An increase in melt index will disturb layer distribution (see Ethylene–vinyl alcohol).

Scrap or recycled material is often used as a 100% constituent of one layer. This may require special feed handling such as a grooved feed section or the addition of a crammer feeder. If the scrap contains PVDC, the extruder must also include the appropriate materials of construction and streamlined design.

Good mixing with special screws is necessary to homogenize the components.

Melt quality and uniformity are absolutely critical for good multilayer coextrusion. Small variations which are invisible in single-layer sheet can cause severe disturbances in coextrusion resulting from layer interactions. It therefore cannot be assumed that an extruder which works well in single-layer service is suitable for critical coextrusion work.

Cylinder cooling system. Melt viscosity is the major factor controlling layer distribution. The ability to control melt temperature level upward and downward to some extent is essential to permit layer distribution control. This requires conservative speed extruder operation. Complex sheet extruders should therefore always be larger for a given capacity than those used for single-layer extrusion. Closed-loop liquid cooling rather than air cooling is desirable on larger extruders to achieve desired melt temperatures.

PVDC extruders of any size should be liquid-cooled for fast cooling in case of problems. Automatic fail-safe liquid cooling is often used to cool the extruder in case of a power failure.

Layer uniformity and stability can be improved by two devices which are relatively new to extrusion.

Gear pump. Gear pumps provide positive output delivery systems for extruders. They permit accurate control of the content of each layer and ensure that all layers are present in the preselected proportions. The pump is run at an accurately controlled speed. The extruder speed is automatically regulated to maintain a constant feed pressure into the gear pump. All variations in extruder output are therefore automatically compensated. The regrind extruder, which is subject to the largest variability, should be fitted with a gear pump.

Static mixer. Static mixers play an important part in stabilizing melt uniformity. This is particularly important with viscous polymers such as PP and HDPE which have long stress-relaxation times.

The mixer also provides an extended residence time at low shear rate for stress relaxation after the high shear in the extruder and gear pump. The static mixer is best installed as the last element prior to the feed block.

Piping. The extruder output is conveyed to the feed block or die through a feed pipe. There are a number of important considerations regarding this technical plumbing (Fig. 2):

It should be as short as possible with a minimum of bends.

All bends should be smooth to avoid material hang up.

The internal diameter should be large enough to avoid large pressure drops, which cause a rise in melt temperature, but not so large as to create stagnation.

The wall thickness should not only take operating pressure into consideration but also act as a good heat sink and distributor. Polymer-filled pipes can be subject to enormous thermal expansion pressures during heating.

Heating must be very uniform to avoid hot and cold spots.

Low voltage density heaters with almost complete pipe coverage are desirable. Heater tapes are dangerous owing to the possibility of poor uniformity of heat distribution.

Control thermocouples must be carefully located to sense the actual pipe temperature.

The construction material must suit the polymer to be conveyed.

Pipes should be easily and quickly disconnected for cleaning and access. Longer pipes should be sectionally assembled to help with this. C clamps are ideal for coupling feed pipes. These also permit rotational motion and compensation for minor misalignment without leakage.

Methods

Two different methods are used to coextrude flat film and sheet: multimanifold die and feed-block coextrusion.

Multimanifold die. The molten polymer streams are fed to separate full-width manifolds in a T die. They are merged prior to exiting from a common slot. These dies are complex and expensive but provide for accurate adjustment of individual layer profiles. The number of layers is limited by the die design, and five appears to be the practical upper limit. The layer capability can be increased by using a feed block on one of the manifolds. Layer adjustment is tedious owing to the great number of adjustment points, and these dies are usually limited in use to single-purpose applications.

Feed-block coextrusion. The product is coextruded on a conventional single manifold T die preceded by a feed block in which the layers are formed. This is the most frequently used process for complex structures. It has been the object of many patents and much litigation.

Feed blocks combine the polymer layers in the structure arrangement desired for the finished sheet, in a narrow width and a relatively thick cross section. This makes the layer assembly fairly easy. Thereafter, laminar and nonturbulent flow in the die is necessary to maintain the desired structure.

Viscosity matching of components is essential, that is, the viscosities of the separate components must be alike. Higher viscosity material displaces lower viscosity material at the edges of the die. Even materials having apparently identical viscosity may not flow evenly because of interfacial slip or die surface drag. In spite of this, viscosity matching works very well for many complex structures. Viscosity differences can often be compensated by temperature adjustments. Feed blocks also incorporate mechanical compensating devices. Consequently, layer distribution uniformity in the 1% range is attainable across the sheet.

There are three major feed-block systems in use commercially. Each has its advantages and disadvantages. All use the principle of nonturbulent laminar flow through the die to achieve good results. The difference between the systems is in how the layers are assembled before the die.

The *Dow system (Dow Chemical Co.)* uses a square die entrance with the height of the die manifold. The layers are assembled in one plane through a series of streamlined flow channels. Details are covered by secrecy agreements and therefore cannot be disclosed. This coextrusion technology has been developed around Dow's Saran PVDC and excels in this field. The Dow feed block incorporates a number of flow adjustments. These feed blocks are available from Dow machinery licensees who include most sheet extrusion system builders. (1)

Figure 2. Feed block and piping.

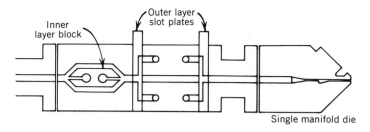

Figure 3. Welex feed block (2).

The *Welex system* uses a circular flow passage which is the usual die entrance configuration for single-layer extrusion (see Fig. 3). The layers are assembled sequentially in and around the cylindrical flow. The system is modular so that further layers can be added to a given feed block at any time. Inner and outer layer feed components are fixed but are removable for correction or adjustment (2).

The *Cloeren system* uses a rectangular die entrance with the height of the die manifold and a width of about 4 in. (100 mm) (see Fig. 4). The block is essentially a miniature multi-layer sheet die with movable separating vanes. These permit adjustment of the relative flow gaps during operation. The number of layers is predetermined by the number of flow channels. The feed to each channel can be selected by interchangeable inlet plugs. They are available from Cloeren Company or through builders of sheet extrusion systems. Cloeren also builds multimanifold dies similar in structure to the feed block. These can be fed by a feed block into the central manifold. Although this may seem complex and expensive, it offers a solution to structures with widely differing melt viscosities (3).

Encapsulation and Lateral Adjustment

It is generally desirable to limit the width of the barrier layer to less than the full sheet width. This is essential with PVDC to avoid contact with the die surfaces and much of the feed block. This not only permits the use of normal materials of construction but totally eliminates degradation of PVDC by stagnation on a metal surface. This problem is solved by encapsulating or totally surrounding the barrier layer with other polymers so that it is floated through the die.

Accurately adjustable width of the barrier layer permits major savings in barrier-material cost and reduces recycling

problems. Preferably, this layer should extend only to the tooling width in the thermoformer. Edge trim from the extrusion operation and from thermoforming should be single layer.

All three feed-block systems incorporate adjustments for this purpose. This is usually achieved by a flow mechanism in the feed block which adjusts the flow width of the barrier layer. Other layers such as the glue layers can be similarly controlled.

Other Equipment

Controls. The most important aspect of multilayer sheet and film extrusion is control. A system includes a tremendous number of variables and adjustments which affect the layer thickness and distribution. Since 100% inspection is impossible, reliance is placed on the consistency and stability of operation.

A microprocessor control system is ideal to help maintain good control and to alarm process deviations. It can also be easily programmed for rapid product changes and for critical start-up and shutdown procedures.

Gauging. Single-layer gauging and control have reached a high state of perfection. Tolerances better than 1% are readily achieved. Measurement of individual layers is possible in certain cases. In thin transparent structures, selective infrared absorption bands permit the separate measurement of widely differing polymers. This method does not work on thick and opaque products, and it does not give the layer location within a structure. Continuous nondestructive-layer measurement remains under intensive development.

Downstream equipment. Multilayer sheet and film use conventional sheet and film takeoff equipment. Nickel-plated rolls are preferably used when processing product containing PVDC to avoid damage to chrome plating in the event of a breakdown. Multiple-edge trimming is sometimes used to separate single-layer from multiple-layer material to reduce scrap recycling problems.

BIBLIOGRAPHY

1. U.S. Pat. 3,557,265 (Jan. 19, 1971), D. F. Chisholm et al. U.S. Pat. 3,479,425, (Nov. 18, 1969), L. E. Lefevre et al. (to Dow Chemical U.S.A.).
2. U.S. Pat. 3,833,704 (Sept. 3, 1974), U.S. Pat. 3,918,865 (Nov. 11, 1975), and U.S. Pat. 3,959,431, May 25, 1976, F. R. Nissel (to Welex Incorporated).
3. U.S. Pat. 4,152,387 (May 1, 1979), and U.S. Pat. 4,197,069, (Apr. 8, 1980), P. Cloeren.

General References

L. B. Ryder, "SPPF Multilayer High Barrier Containers," *Proceedings of the Eighth International Conference on Oriented Plastic Containers,* Cherry Hill, N.J. 1984, pp. 247–281.

F. Nissel, "High Tech Extrusion Equipment for High Barrier Sheeting," *Proceedings of the Second International Ryder Conference on Packaging Innovations,* Atlanta, Ga., Dec. 3–5, 1984, pp. 249–274.

J. A. Wachtel, B. C. Tsai, and C. J. Farrell, "Retorted EVOH Multilayer Cans with Excellent Barrier Properties," *Proceedings of the Second International Ryder Conference on Packaging Innovations,* Atlanta, Ga., Dec. 3–5, 1984, pp. 5–33.

F. R. NISSEL
Welex, Incorporated

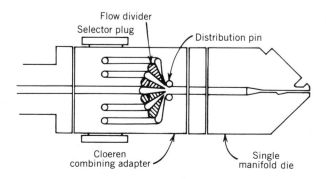

Figure 4. Cloeren feed block (3).

COEXTRUSION MACHINERY, TUBULAR

Tubular coextrusion for packaging applications is generally referred to as blown-film coextrusion, distinguishing it from other similar tubular processes that produce products such as pipe and heavy-wall tubing. Blown-film coextrusion therefore refers to the process of forcing more than one molten polymer stream through a multimanifold annular die to yield a film consisting of two or more concentric plastic layers (see Extrusion; Films, plastic). The laminar characteristic of polymer flow permits the maintenance of discrete layer integrity such that each polymer in the film structure can fulfill a specific and individual purpose (see Coextrusions for flexible packaging).

Coextruded film structures are designed to incorporate one or more of the following objectives: heat sealability; barrier against gas or moisture transmission; high strength (ie, tensile, impact, and tear); color differential; surface frictional properties; adhesive between layers; stiffness (modulus of elasticity); optical quality (clarity and gloss); and reclaim carrier. Combinations of these properties can be achieved by the arrangement of polymer layers in which each polymer exhibits the specific desired property.

Process Equipment

A tubular coextrusion process fundamentally consists of the extruders, die, air ring, collapsing mechanism, haul-off, and winder. These elements are similar to those of single-layer film extrusion except for the die, which must contain more than one flow manifold, ie, layer channel, for extrusion (see Fig. 1).

The added complexity of multilayer die components, coupled with the inherently superior quality requirements for coextruded films, make the die the highest design priority of the extrusion system. The most critical die-design considerations for multilayer applications are (1) structural integrity, ie, the hardware's ability to withstand typical internal pressures of 3000–6000 psi (21–41 MPa); (2) dimensional integrity, the interlocking of and machining precision related to

mating parts defining flow-stream concentricity; (3) polymer-flow distributive quality, in order to utilize a range of diverse materials; and (4) reduction of design-flow restriction, permitting extrusion of high viscosity polymers.

Closely related to the die issue is the frequent need to rotate (or, preferably, oscillate) the die assembly for the purpose of randomizing film-thickness variations across the entire windup width. In the seal section, where there is an interface between fixed and oscillating members, polymer pressure is large, ie, typically 5000 psi (34.5 MPa). Because the seal must act against this force, the seal design must be well qualified, and thrust-bearing and seal-maintenance costs are likely to be high.

Alternative methods sometimes used for thickness randomization are oscillating haul-off assemblies, rotating winders, or rotating extrusion systems. Each method poses some significant technical difficulty worthy of extensive selection and design consideration.

Specialized Process Design

Because coextruded films often employ polymers uncommon to those used in single-layer extrusion, some unusual process-design criteria, discussed below, are added for multilayer systems.

Degradable polymers. Most of the gas-barrier resins are vulnerable to temperature degradation. This imposes a need for specialized die streamlining and extruder-feedscrew design. The feedscrew configuration is critical in minimizing melt temperature and ensuring uniformity of temperature and viscosity across the melt-flow stream.

High modulus of elasticity (film). Many barrier and high strength polymers exhibit modulus, ie, stiffness, characteristics that cause unique web-handling and winding difficulties. The elimination of web wrinkles and flatness distortions becomes a critical design objective related especially to collapsing geometry, idler and nip-roll size, and line-drive quality. The handling of stiff webs usually entails realtively high equipment costs because higher tension levels are required, along with more precise tension control. Use of highly accurate regenerative d-c drive equipment is usually advisable for coextruded films. The high modulus webs also dictate greater hardware rigidity and tighter roll-alignment tolerances.

Four-side treatment capability. Because of layer-thickness structure considerations in coextrusion, the surface to be printed may be extruded as the inner layer of the bubble. This shifts corona-treatment requirements (for subsequent ink adhesion) from the outer to the inner layer. Two treater stations are sometimes installed on coextrusion systems; one is for the treatment of the inner layer downstream of the web separation.

High film-surface COF. For many high speed sealing applications, as well as such products as stretch film, multilayer film surfaces are abnormally tacky. One of the principal advantages of the coextrusion process is its ability to create such properties with relatively low additive content concentrated in individual layers. However, very high resultant COF values can cause unusual web-handling difficulty, especially in relation to the bubble-collapsing function. This aggravated geometric problem, ie, flattening a cylinder (bubble) into a single plane, can be alleviated with the use of very low friction collapsing means. In contrast to the more conventional wood-slat configuration, low friction systems employ rollers on ball bear-

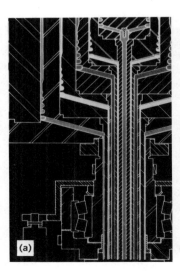

Figure 1. Typical rotating tubular 5-layer configuration: (**a**) die structure; (**b**) extruder inlet arrangement.

ings or air-cushion surfaces to minimize film-surfacing drag forces during collapsing.

Quality-Control Requirements

Multilayer films, because of their enhanced physical properties, frequently command premium selling prices; however, these films also necessitate several added cost factors and engineering complications related to process design. The cost differential is due in part to the film's added value, exemplified by more stringent thickness-uniformity and winding-quality standards. These are described below.

Temperature control. In addition to plant-space problems associated with multiple extruders, tighter film-quality specifications dictate improved temperature control in a smaller control console. Most recent coextrusion-system plans employ microprocessors to save space and take advantage of digital-control logic. Achieving process-temperature stability is often essential for coextruded films, in contrast to the fluctuations and errors normally tolerated in single-layer processes.

In-line blending. Because of the many types of raw material used in coextrusion, the purchase and storage of specialized-resin blends is impractical. There is an advantage to in-line blending of additives, and investment plans for a complex coextrusion system generally include a high priority for blending equipment. This is also logical in view of the relatively high per-pound (kg) cost of the special resins and additives involved, and the particularly high quality demanded of multilayer products.

Layer-thickness control. Individual layer thickness must be carefully monitored either by tedious off-line measurement or in-line by gravimetric (weigh-feeding) extruder loading. A difficult technical objective unique to coextrusion, layer-thickness control is a key process-control priority that provides the opportunity to achieve cost savings or the liability to waste raw material and produce defective film. Although layer-thickness measurement can be achieved with spectrophotometers, weigh feeding seems the most practical and reliable means of controlling layer percentages.

Roll-winding quality. Roll-conformation requirements associated with multilayer films are usually severe, representing a more costly and complicated winder configuration. In-line slitting is a common cost-saving requirement, encouraging the use of advanced web-handling technology in the categories of alignment precision, web spreading concepts, and tension and speed control (see Slitting and rewinding machinery). Typical high multilayer line speeds, often 200–500 ft/min (61–152 m/min), dictate the incorporation of automatic cut and transfer mechanisms. Manual roll transfers are not practical at these speeds and with multiple slits. Additionally, the broad range of film elasticity, stiffness, thickness, and surface tack encountered in coextruded applications demands extraordinary winder versatility and performance quality.

Economic Factors

Most blown-film coextrusion systems operate in an output range of 200–1000 lb/h (91–454 kg/h). A typical average rate is 300 lb/h (136 kg/h). Although some 2-, 4-, and 5-layer systems exist, a common installation utilizers three extruders, even when producing 2-layer products. Usual extruder combinations include 2.5-in (6.4-cm) dia and 3.5-in. (8.9-cm) sizes,

although many 4.5-in. (11.4-cm) extruders are also used. Some lines operate at 2000 lb/h (907 kg/h) with 6-in. (15.2-cm) extruders.

In the United States, the coextrusion industry consists of a large population of in-line multilayer bag operations in addition to those requiring film winding. Investment levels and process-quality requirements are usually not as high for the bag operations. A three-layer 300-lb/h (136-kg/h) in-line bag extrusion system, for example, costs approximately $300,000; a film-winding version with the same output specification would probably cost at least $400,000.

Operating costs for coextrusion systems are similar to those of single-layer extrusion except for the higher initial investment, ie, typically 50% higher for coextrusion of the same output category. Energy costs are equivalent ($0.03–0.05/lb or $0.07–0.11/kg) to those of single-layer extrusion, and manpower requirements vary only slightly. Labor costs per unit weight are often higher for coextrusion, not as a function of manpower requirements but because of more elaborate processes require greater skills.

A frequent important economic incentive for the manufacture of coextruded films is that premium film pricing reduces the cost percentage of raw materials, eg, resin. Therefore, multilayer products are generally reputed to offer higher profit margins than their single-layer counterparts.

Scrap reclaim can often be an economic disadvantage with coextrusion. It may be limited or prohibited by incompatibilities between the polymers of corresponding layers, a complication particularly prevalent among specialty food-packaging films which contain gas-barrier resins. In these cases, reclaimed scrap may only be eligible for insertion into a thin adhesive layer, thus severely limiting reclaim percentages. Conversely, some coextruded films are designed specifically to exploit high scrap-input potential. In these cases, high loadings of scrap or reprocessed resin are sandwiched between skin layers of virgin polymer.

BIBLIOGRAPHY

General References

W. J. Shrenk and R. C. Finch, "Coextrusion for Barrier Packaging"; R. C. Finch, "Coextrusion Economics"; *Papers presented at the SPE Regional Technical Conference (RETEC)*, Chicago, Ill., June 1981, The Society of Plastics Engineers, Inc., Brookfield Center, Conn., pp. 205–224 and pp. 103–128.

R. Hessenbruch, "Recent Developments in Coextruded Blown and Cast Film Manufacture," *Paper presented at COEX '83*, Düsseldorf, FRG, Schotland Business Research, Princeton, N.J., 1982, pp. 255–273.

N. S. Rao, *Designing Machines and Dies for Polymer Processing with Computer Programs*, Macmillan, Inc., New York, 1981.

R. L. Crandell, "CXA—Coextrudable Adhesive Resins for Coextruded Film"; G. Burk, "On Line Measurement of Coextruded Coated Products by Infrared Absorption"; *Papers presented at TAPPI Coextrusion Seminar*, May 1983, TAPPI, Atlanta, Ga., pp. 89–90.

Properties of Coextruded Films, TSL # 71-3, E. I. du Pont de Nemours & Co., Inc., Wilmington, Del.

C. D. Han and R. Shetty, "Studies of Multi-layer Film Coextrusion," *Polym. Eng. Sci.* **16**(10), pp. 697–705 (Oct. 1976).

D. Dumbleton, "Market Potential for Coextrudable Adhesives," *Paper presented at COEX '82*, Düsseldorf, FRG, 1982, Schotland Business Research, Princeton, N.J., 1982, pp. 55–74.

G. Howes, "Improvements in the Control of Plastics Extruders Facilitated by the Use of Microprocessors, *Paper presented at the TAPPI Paper Synthetics Conference '81*," TAPPI, Atlanta, Ga., pp. 21–31.

"Coextrusion Coating and Film Fabrication," TAPPI Press Report 112, Atlanta, Ga., 1983.

W. D. WRIGHT
Western Polymer Technology, Inc.

COEXTRUSIONS FOR FLEXIBLE PACKAGING

Coextrusion technology has been one of the leading growth areas for new packaging materials in the 1970s and 1980s. Coextrusion has replaced many of the maturing technologies in packaging, such as solvent coating and lamination. Coextruded films have replaced many coated paper products such as wax-coated glassine (see Glassine, greaseproof, and parchment). They are expected to replace laminations or components of laminations in the future. The use of coextrusion coatings and laminations is expected to expand rapidly in the 1980s.

A coextruded film is best defined as a multilayer film in which each distinct layer is formed by a simultaneous extrusion process through a single die (see Multilayer flexible packaging). The principal reason for development and use of the technology for packaging has been the quest for cost improvements. A coextrusion allows one to prepare a multifunctional packaging material in one manufacturing step as opposed to the traditional multistep processes of coating and lamination. For example, a single-step coextrusion of high density polyethylene and ethylene–vinyl acetate can replace an adhesive lamination of oriented polypropylene film to low density polyethylene film in some packaging applications. Such a lamination requires four manufacturing processes: polypropylene film manufacture, polypropylene film orientation, polyethylene film manufacture, and adhesive lamination. The potential cost savings are apparent. Coextrusion technology has been driven by the potential to reduce the costs of packaging materials. An optimum packaging system can be developed with minimum costs in materials.

Advantages of Coextrusions Over Blends

The layers of a coextruded film are generally composed of different plastic resins, blends of resins, or plastic additives. The difference between a coextruded film and a resin blend lies in the existence of distinct layers in the coextruded film as opposed to the blend. Figure 1 illustrates the differences. The

---

HDPE–LDPE

blend

HDPE

LDPE

coextruded

Figure 1. Cross section of film composed of resin blends and coextrusion.

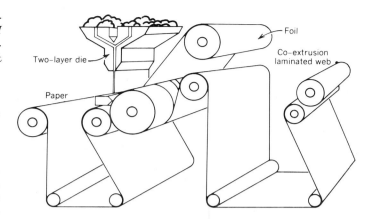

Figure 2. A typical coextrusion lamination process.

functional properties related to packaging can be vastly different. The temperature differential of 54°F (30°C) between layers of the coextruded film (HDPE–LDPE) allows for heat sealing on most packaging machines. A blend of high density polyethylene and low density polyethylene would cause jaw sticking and distortion on many machines.

Principal Manufacturing Processes

There are five principal manufacturing processes utilizing coextrusion technology. They are: cast-film coextrusion, blown-film coextrusion (see Coextrusion, tubular), coextrusion coating (see Extrusion coating), coextrusion lamination (see Lamination), and cast-sheet coextrusion (see Coextrusion, flat).

The coextrusion processes for cast film, coating, and lamination are similar. Essentially the same basic techniques are used with the exception that the cast-film process extrudes the molten film directly onto a chill roll. In coextrusion coating the film is cast directly onto a substrate such as paper or foil. A coextrusion lamination simply casts the film between two substrates such as foil and paper. A typical example of a coextrusion lamination process is illustrated in Figure 2. Coextrusion coating and lamination processes are generally characterized by larger die gaps and higher melt temperatures to promote adhesion. Resins with higher melt indexes are generally utilized.

Cast-film die options. The main focal points of technology in the cast processes are the designs of the dies and the melt-flow properties of the resins.

There are two types of designs used. They are a multimanifold die and a single-manifold die with an external combining adapter. A schematic of each die design is shown in Figure 3.

In the single-manifold die, separate resin melt streams are brought together in a common manifold. The resin streams are combined in a combining adapter (or feedblock) prior to the die where distinct layers are maintained. In a multimanifold die, resin streams flow in separate channels and are combined inside the die after attaining full width. The important advantages of the single manifold are capital costs, flexibility of operation, thinner layers, and a larger number of possible layers. Coextrusions of several hundred layers have been reported (1). The design of feedblocks and control of laminar flow

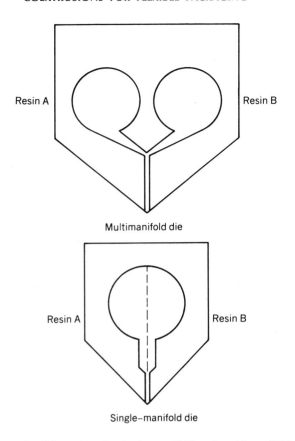

Figure 3. Schematic of a single-manifold and multimanifold die.

of the various components are critical to successful operations of this process.

A potential disadvantage of the single-manifold die is the need to select carefully materials with relatively similar melt-flow properties. The multimanifold system allows for easier processing of dissimilar materials. In addition to a broader range of melt properties, a greater differential of temperatures between layers is possible. In practice, almost all coextrusion is done with the combining adapter and the single-manifold die.

Blown-film coextrusions.   The blown-film coextrusion process is illustrated in Figure 4. In this process, separate resins are extruded into a circular, rotating die. The molten-resin streams are blown into a bubble, cooled by air rings, and collapsed in the primary nip. The tubular film is generally slit for specific packaging applications. The die design for blown film, in addition to being circular, is different than the cast process in that separate melt streams are combined near the die exit or external to the die.

Compared to cast films, blown films generally have more balanced physical strength properties, higher moisture barrier, and greater stiffness. Optical properties such as clarity and gloss, however, are generally inferior to those of the cast process because of the slower rate of crystallization in the blown process. (Blown films are more crystalline than the corresponding cast films.)

Principal Raw Materials

Polyolefins (polyethylenes and polypropylene) are the key polymers used in coextruded packaging films. This class of

material is preeminent because of low cost, versatility, and easy processability. LDPE–LLDPE resins (see Polyethylene, low density) are used extensively in coextruded structures for their toughness and sealability. HDPE resins are selected for their moisture barrier and machinability characteristics (see Polyethylene, high density). Polypropylene is chosen for its ability through orientation to provide machinable films with high impact and stiffness properties.

Although the polyolefins are the workhorse grades for coextruded packaging, they are almost always combined with other resins to achieve multilayer functionality. Copolymers of ethylene–vinyl acetate (EVA), ethylene–acrylic acid (EAA), and ethylene–methacrylic acid (EMA) are regularly used as skin layers for their low-temperature sealing characteristics. These resins are also often used as "tie layers" to adhere together two layers of dissimilar polymers with low interlaminar bond strength. Other polymers such as nylon, poly(vinylidene chloride) (PVDC), and ethylene–vinyl alcohol (EVOH) are selected for their ability to protect the products from deterioration or loss of flavor (see Nylon; Poly(vinylidene chloride); Ethylene–vinyl alcohol]. Other polymers such as polycarbonate (see Polycarbonate) or polyester (see Polyesters, thermoplastic) may be used as skin layers to provide unusual thermal integrity for packaging machine performance.

Structures

Coextruded flexible packaging applications include coextruded films, laminations, and coatings. In general, coextruded films are preferred to coextruded laminations and coatings because of their cost-effectiveness in use. Because lamination and coating require an extra value-added stage, they tend to cost more.

Coextruded multilayer films can be divided into three cate-

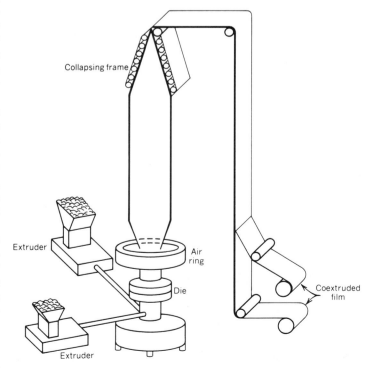

Figure 4. Blown-film coextrusion process.

gories: single-resin, unbalanced, and balanced films. Many films that are based on the performance properties of a single resin are coextruded for performance or cost reasons. Unbalanced structures typically combine a functional layer with a heat-seal resin. Balanced structures generally have the same heat-sealable resin on both sides of the film.

Single-resin structures. Single-resin films are coextruded for a variety of reasons. Many commodity film applications may not appear to be multilayer films, yet they actually have three or more distinct layers. Bakery, produce, and trash-bag films, for example, are often three-layer structures. The core material may contain pigment or recycled material, while virgin skin layers control surface quality and machinability. Single-resin coextrusions can also provide a differential coefficient of friction on the two surfaces.

Unbalanced structures. Typical of the unbalanced structures are films designed for vertical form/fill applications with a fin seal. A base resin such as high density polyethylene is augmented by an ethylene–vinyl acetate skin layer for sealability. For horizontal wrappers a polypropylene skin layer is sometimes selected for its higher thermal resistance. In another important unbalanced application, cast polypropylene, which has a limited sealing range, is combined with more sealable polyethylene for single-slice cheese wrappers (see Film, cast polypropylene).

There are multilayer films using only one polymer (A/A/A), unbalanced coextruded films with two or more polymers (A/B/C), and balanced multilayer structures with two or more polymers (A/B/C/B/A).

Balanced structures. Balanced coextruded structures typically have a core resin selected for its functionality plus two skin layers which are heat sealable. Oriented polypropylene films, for example, are increasingly coextruded instead of coated to attain machinable surfaces (see Film, oriented polypropylene). Frozen-food films are typically constructed with an EVA skin layer for enhanced sealability. Heavy-wall bags are regularly coextruded with LLDPE cores for impact strength and LDPE skins to limit the film's elongation under load. Primal meats are packaged in PVDC shrink film with EVA skins for seal integrity.

Two main applications which appear to be shifting from monolayer films to coextrusions are overwrap and stretch wrap (see Wrapping machinery, stretch film). Horizontal overwrap machines typically use an MDPE film or an LDPE–HDPE blend. Coextrusions can provide comparable overwrap machinability at lower gauge. Stretch wrap is difficult to produce as a single-layer structure without blocking. By splitting stretch wrap into a multilayer structure, its LLDPE core can be provided with controlled tackiness on the surface layer.

BIBLIOGRAPHY

1. T. Alfrey, E. F. Gurnee, and W. J. Schrenk, *Polym. Eng. Sci.* **9**, 400 (1969).

General References

Coextrusion Patents

U.S. Pat. 3,222,721 (Dec. 14, 1965), M. Reynolds, Jr. (to Anaconda).

U.S. Pat. 3,223,761 (Dec. 14, 1965), G. E. Raley (to Union Carbide Corp.).

U.S. Pat. 3,321,803 (May 30, 1967), H. O. Corbett (to USI).

U.S. Pat. 3,308,508 Mar. 14, 1967), W. J. Schrenk (to The Dow Chemical Co.).

U.S. Pat. 3,398,431 (Aug. 27, 1968), H. O. Corbett (to USI).

U.S. Pat. 3,320,636 (May 23, 1967), H. O. Corbett (to USI).

U.S. Pat. 3,400,190 (Dec. 3, 1968), H. J. Donald (to The Dow Chemical Co.).

U.S. Pat. 3,477,099 (Nov. 11, 1969), R. E. Lee and H. J. Donald (to The Dow Chemical Co.).

U.S. Pat. 3,479,425 (Dec. 18, 1969), L. E. Lefevre and P. Dreidt (to The Dow Chemical Co.).

U.S. Pat. 3,476,627 (Nov. 4, 1969), P. H. Squires (to E. I. du Pont de Nemours & Co., Inc.).

U.S. Pat. 3,448,183 (June 3, 1969), D. S. Chisholm (to the Dow Chemical Co.).

U.S. Pat. 3,440,686 (Apr. 29, 1969), H. O. Corbett (to USI).

U.S. Pat. 3,524,795 (Aug. 17, 1970), N. R. Peterson (to The Dow Chemical Co.).

U.S. Pat. 3,557,265 (Jan. 19, 1971), D. S. Chisholm and W. J. Shrenk (to The Dow Chemical Co.).

U.S. Pat. 3,583,032 (June 8, 1971), L. O. Stafford (to Beloit).

U.S. Pat. 3,365,750 (Jan. 30, 1968), H. J. Donald (to The Dow Chemical Co.).

U.S. Pat. 3,611,492 (Oct. 12, 1971). R. Scheibling (to Siamp-Cedap).

Die Design

R. J. Brown and J. W. Summers, *Plast. Eng.* **37** (9), 25 (Sept. 1981).

Melt Rheology

J. H. Southern and R. L. Bullman, *J. Appl. Polym. Sci.* **20**, 175 (1973).

J. D. Han, *J. Appl. Polym. Sci.* **17**, 1289 (1973).

A. E. Everage, *Trans. Soc. Rheol.* **17**, 629 (1973).

General Extrusion Processes

G. R. Moore and D. E. Kline, *Projection and Processing of Polymers for Engineering,* Prentice-Hall, New York, 1984

THEODORE D. FREY
JAMES A. ALBRIGHT
Crown-Zellerbach Corporation

Table 1. Typical Coextruded-Film Structures

Outside layer	Core layer	Inside layer	Remarks
LDPE	white LDPE + recycle	LDPE	virgin skin layers control surface quality
HDPE	HDPE + recycle	EVA	EVA provides rapid fin seal machinability
EVA	LLDPE + recycle	EVA	EVA increases lap seal cycle time
LDPE	LLDPE + recycle	LDPE	LDPE limits film's elongation under load
EMA	OPP	EMA	oriented polypropylene sealability poor without coextruded or coated skin layers

COEXTRUSIONS FOR SEMIRIGID PACKAGING

This article pertains to flat semirigid coextruded sheet which is a minimum of 0.010-in. (0.25-mm) thick (see Coextru-

Table 1. Barrier Materials

Resin	O$_2$ Transmission rate[a]	Water-vapor[b] transmission rate	Mid-1985 price, $/lb ($/kg)
EVOH (Eval F, Kuraray)	0.035 [0.136]	3.8 [1.50]	2.41 [5.31]
PVDC (Saran 5253, Dow Chemical)	0.15 [0.583]	0.10 [0.04]	1.02 [2.25]

[a] cm^3 · mil/(100 in.^{2} · d · atm) [cm^3 · μm/(m^2 · d · kPa)] at 73°F (23°C), 75% rh.

[b] g · mil/(100 in.^{2} · d) [g · mm/(m^2 · d)] at 100°F (38°C), 90% rh.

sion machinery, flat). These coextruded sheet structures are thermoformed to produce high barrier plastic packages (see Barrier polymers; Thermoforming). A similar concept is used to produce high barrier plastic bottles except that the bottles are formed from coextruded multilayer tubes instead of flat sheet (see Blow molding).

The production of coextrusions for semirigid packaging was made possible by technology developed in the late 1960s and early 1970s (1, 2). Utilization of this technology was initially limited to "simple" structures such as two-layer systems (a general purpose polystyrene cap layer on a high impact polystyrene base layer) for drink cups. Commercialization of high-barrier coextrusions occurred in the 1970s in Europe and Japan. Large-scale commercial barrier coextrusion applications did not surface in the United States until the 1980s. For purposes of this discussion, barrier materials are defined as those that exhibit an oxygen transmission rate of less than 0.2 cm^3 · mil/(100 in^2 · day · atm) [0.777 cm^3 · μm/(m^2 · d · kPa)] (see Barrier polymers). Other techniques that can be used to produce multilayer barrier structures are coating and lamination (see Coating equipment; Laminating). Some advantages coextrusion offers versus these other two methods are thicker barrier layer capability, single-pass production, barrier layer sandwiched between cap layers, and generally lower cost. The potential markets for packages formed from these high-barrier coextrusions include both low- and high-acid food products sterilized by aseptic, hot-fill, or retort methods. These markets obviously represent a significant opportunity for barrier coextrusions.

Barrier Materials

Based on the barrier definition above, only two commercially available thermoplastic resins can be considered as barrier resin candidates for these extrusions. These are ethylene–vinyl alcohol (EVOH) (see Ethylene–vinyl alcohol) and poly(vinylidene chloride) (PVDC) (see Vinylidene chloride copolymers). The barrier properties of specific grades of these

two materials are listed in Table 1. The resins identified in the table are currently the highest barrier commercially available coextrudable resins of their respective polymer classes. Other formulations of both resin types are available offering certain property and processing improvements at the sacrifice of barrier properties.

The most significant technical issue concerning the use of EVOH as a barrier material is its moisture sensitivity. The material is hygroscopic, and its barrier properties are reduced as it absorbs moisture. The importance of this property to the food packager is dependent upon the sterilization process, food type packaged, and the package storage conditions. The most severe conditions are encountered during retort processing (see Canning, food). Special consideration to coextrusion structure design and post-retorting conditions may be required to achieve the desired oxygen barrier for packages produced from EVOH coextrusions (3).

PVDC is not moisture sensitive and does not exhibit the deterioration of barrier properties shown by EVOH. The challenges associated with using heat-sensitive PVDC are faced by the coextruded sheet producer. Equipment and process design are critical to the production of coextrusions containing PVDC. Concern relating to the reuse of scrap generated in the production of coextrusions based on PVDC is a real economic issue. Development of new material forms and recycle-containing structures is underway with commercialization targeted for 1985 (4). In the meantime, resin manufacturers are working on the development of other types of barrier materials for coextrusion applications (5).

Structural Materials

The materials generally used to support the barrier resins in coextrusions are listed in Table 2. The maximum process temperature listed is the highest sterilization temperature that packages based on these resins should experience. Polystyrene, polypropylene, and the polyethylenes are the predominant structural materials used in coextrusions for semirigid

Table 2. Structural Materials

Resin	Maximum process temperature, °F (°C)	Mid-1985 price $/lb ($/kg)
polystyrene	195 (90.6)	0.49–0.51 (1.08–1.12)
polypropylene	260 (127)	0.43–0.47 (0.95–1.04)
high density polyethylene	230 (110)	0.44–0.50 (0.97–1.10)
low density polyethylene	170 (77)	0.40–0.44 (0.88–0.97
polyester, thermoplastic (heat-set)	>260 (>127)	0.63–0.67 (1.39–1.48)
polycarbonate	>260 (>127)	1.69–1.81 (3.73–3.99)

packaging applications. Structural resin selection is dependent upon use requirements, coextrusion processability, and container-forming considerations.

Polystyrene (see Polystyrene) exhibits excellent coextrudability and thermoformability. It can be used in applications requiring low temperature processing and in some hot-fill applications. Polypropylene (see Polypropylene) is also excellent from a coextrusion-processing standpoint, but it requires special forming considerations. Deep-draw containers from polypropylene-based sheet are most commonly formed using solid-phase forming techniques. Polypropylene can be retorted; but some grades exhibit poor low temperature impact characteristics which limit their use in applications requiring resistance to refrigerated or freezing temperatures.

High density polyethylene (see Polyethylene, high density) offers a significant improvement in low temperature properties compared to polypropylene, but its suitability in applications requiring retort processing is marginal. Low density polyethylene would be incorporated in coextrusions requiring good heat sealability (see Sealing, heat) for applications involving low-temperature-fill conditions.

Although coextrusions based on crystallizable polyester (see Polyesters, thermoplastic) and polycarbonate (see Polycarbonate) are not commercially available at this time, these materials are included as structural materials because of their future potential in retort applications. The success of these relatively expensive materials will be dependent on the cost and performance achieved. Considerable developments of coextrusion and forming techniques need to be completed prior to commercialization of coextrusions based on polyester and/or polycarbonate.

Applications

Three representative commercially coextruded structures are shown in Table 3. The transition layers in these structures are materials used to ensure the integrity of the coextrusion. The technology of transition layers is complex and maintained as proprietary by coextrusion manufacturers. The first structure, which uses polystyrene as both cap layers, finds use in form/fill/seal applications because of the particularly good thermoformability of polystyrene (6) (see Thermoform/fill/seal). The second structure has one polystyrene cap layer to maintain thermoformability and one polyolefin cap layer. The polyolefin layer in this case would be the food-contact layer. This structure would comply with the current FDA regulations for aseptic H_2O_2 package sterilization (see Aseptic packaging). The resins that comply with current FDA regulations for H_2O_2 sterilization are polyethylenes, polypropylenes, polyesters, ionomers (see Ionomers), and ethylene vinyl acetates (EVA). Petitions have been submitted for FDA clearance of polystyrene and ethyl methyl acrylate (EMA) as food-contact layers as well. Containers formed from this structure, with polypropylene as the food-contact surface, can also be hot filled (7).

The last structure shown in Table 3 has the most potential of those listed because it can be used in applications including retort processing. The primary market target for coextrusions with polypropylene as the cap layers is processed foods currently in metal cans (8, 9).

In addition to the food-packaging markets, barrier coextrusions can be utilized in the medical (see Health care packaging), pharmaceutical (see Pharmaceutical packaging), and in-

Table 3. Commercial Coextrusions

Structure	Application
polystyrene	form/fill/seal
transition	preformed containers
barrier	hot fill
transition	
polystyrene	
polystyrene	form/fill/seal
transition	preformed containers
barrier	H_2O_2 aseptic
transition	hot fill
polyolefin	
polypropylene	preformed containers
transition	H_2O_2 aseptic
barrier	hot fill
transition	retort
polypropylene	

dustrial packaging markets where barriers to oxygen, moisture, and hydrocarbons are required.

Economics

Simply utilizing resin prices to calculate a material cost for a coextruded sheet structure can be unreliable in determining the economics of barrier plastic packages. Using material prices only to compare the economics of several coextruded sheet structures based on different resins can result in erroneous conclusions. Items such as required equipment costs, coextrusion output rates, package-forming method and rates, amount of scrap generated, amount of scrap reutilized, container design, and container performance are some of the cost considerations that can be dissimilar for different coextruded sheet structures. Economic comparison of various coextruded barrier packages with alternative packaging materials should be based on a total packaging systems analysis. The current commercial applications and market tests underway show that packages from coextruded sheet offer economic and/or performance advantages versus other packaging materials.

BIBLIOGRAPHY

1. U. S. Pat. 3,479,425 (Nov. 18, 1969), L. E. Lefeure and P. Braidt (to Dow Chemical Company).

2. U. S. Pat. 3,557,265 (Jan. 19, 1971), D. Chisholm and W. J. Schrenk (to Dow Chemical Company).

3. K. Ikari, "Oxygen Barrier Properties and Applications of Kuraray EVAL Resins," presented at Coex 1982, sponsored by Schotland Business Research, Inc., Princeton, N. J.

4. W. J. Schrenk and S. A. Marcus, "New Developments in Coextruded High Barrier Plastic Food Packaging," presented at SPE–RETEC, Cleveland, Ohio, April 4–5, 1984.

5. R. McFall, "New High Barrier Polyester Resins for Coextrusion Applications," presented at Coex 1984, sponsored by Schotland Business Research, Inc., Princeton, N. J., Sept. 19–21, 1984.

6. "Cheese Invades Europe," *Packag. Dig.*, 45 (Feb. 1981).

7. "Industry/Newsfocus," *Plast. Technol.*, p 114 (Sept. 1984).

8. "Campbell's Plans for Plastics: 'mm, mm, good'," *Plast. World*, 6 (June 1984).

9 S. A. Marcus, *Food Drug Packag.*, 22 (Aug. 1982).

General References

S. E. Farnham, *A Guide to Thermoformed Plastic Packaging*, Cahners Publishing Company, Boston, Mass., 1972.

S. Sacharow and R. C. Griffin, *Basic Guide to Plastics in Packaging*, Cahners Publishing Company, Boston, Mass., 1973.

J. A. Cairns, C. R. Oswin, and F. A. Paine, *Packaging for Climatic Protection*, Newnes–Butterworths, London, 1974.

Proceedings of Coex '81, '82, '83, and '84, Schotland Business Research, Inc., Princeton, N.J.

R. J. Kelsey, *Packaging in Today's Society*, St. Regis Paper Company, New York, 1978.

Proceedings of the Seventh International Conference on Oriented Plastic Containers, Ryder Associates, Inc., Whippany, N.J., 1983.

L. B. Ryder, *Plast. Eng.*, May 1984.

S. Hirata and N. Hisazumi, *Packag. Japan*, 25 (Jan. 1984).

A. Brockschmidt, *Plast. Technol.*, 67 (Sept. 1984).

R. J. Dembowski
Ball Corporation

COLLAPSIBLE TUBES. See Tubes, collapsible.

COLORANTS

Colorants for packaging materials fall into two broad categories: pigments and dyes. Used for both decorative and utilitarian purposes, their diversity is at least as broad as the diversity of packaging materials. This article focuses on their use in inks, plastics, and paperboard. The emphasis is on pigments, which are far more prevalent than dyes in packaging applications.

Pigments are black, white, colored, metallic, or fluorescent organic or inorganic solids which are insoluble and remain essentially unaffected by the medium into which they are dispersed or incorporated. They are small in particle size, generally in the range of 0.01–1.0 μm diameter. Pigments produce color by selective absorption of light, but because they are solids, they also scatter light. Light scattering is undesirable in a transparent material, but desirable if opacity is the goal. Some organic pigments that are extremely small in particle size scatter very little light and therefore act like dyes; for example, Benzimidazolone Carmine HF3C, with a particle size of 0.05–0.07 μm. Some colorless pigments, relatively large in particle size (up to 100 μm) are used as fillers or extenders.

Organic pigments are characterized by high color strength, brightness, low density, high oil absorption, transparent and translucent properties, bleeding in some solvents, and heat and light sensitivity. In the world at large, the major user of organic pigments is the printing-ink industry. In packaging, they are useful for numerous applications such as printing on cartons, labels, and flexible bags. Naphthol reds, for example,

are used for soap- and detergent-carton printing because of bleed resistance. Barium lithol is a most important red for packaging flexo and gravure inks, and alkali blue is used in glycol-type inks (see Inks; Printing).

Compared to organic pigments, inorganic pigments are more opaque, less bright, and weaker in tint; but they are more resistant to heat, light, chemical attack, bleed, migration, and weathering. They have higher density, lower cost, and less antioxidant effect. The major use of inorganic pigments is the paint industry. In packaging, they are useful for printing on cartons, bags, and glass bottles. Examples are molybdate orange for gift wrap and vinyl film, titanium dioxide for glass beverage bottles, and cadmium reds for plastics. Metallic pigments such as gold, platinum, and silver help vivify colorants for glass bottles.

Dyes are intensely colored solubilized organic substances that are retained by the medium which they color by chemical bonding, absorption, or mechanical retention. Dyes produce color by absorption of light, without affecting transparency and high optical purity. The major user of dyes is the textile industry. In packaging, dyes are used to some extent in inks for special effects, for coloring paperboard, and to produce tinted transparent plastic containers or films.

Pigments in Packaging

Properties of pigments are a function of the chemical composition as well as other physical and chemical parameters such as particle size, particle shape, particle-size distribution, and the nature of the pigment's surface. Particle size affects a

Table 1. Some Factors Involved in Selection of a Pigment

For printing ink	For plastics	For paper and paperboard
color	color	whiteness
(a) masstone	nature of resin	brightness
(b) tintone	end use	opacity
(c) printone	toxicity	rheological
density	heat resistance	properties
rheological behavior	resistance to	bulk
opacity	migration	specific gravity
oil absorption	(a) bleeding	transparency
texture	(b) crocking	refractive index
chemical resistance	(c) bronzing	use cost
(a) acid	(d) plate-out	color
(b) alkali	lightfastness	color migration
solvent resistance	weatherability	flocculation
heat resistance	dispersibility	gloss
oil, fat, grease,	electrical	mechanical
soap resistance	properties	properties
lightfastness	allowance for	(a) ink
resistance to	additives such	absorbency
sterilizing	as antioxidants,	(b) sheet
bake stability	uv absorbers	strength
pearlescence	morphological	abrasion
iridescence	properties	
viscosity	filtration	
bulk	characteristics	
transparency	effect on mechanical	
use cost	properties	
particle size	use cost	
	tensile strength	

number of pigment properties. Lightfastness improves with increasing particle size, and oil absorption and strength decrease. Hue is also affected by particle size; for example, an orange pigment usually appears yellower as the size decreases. Narrower particle-size distribution leads to cleaner hue, higher gloss, and lower oil absorption and viscosity. A pigment's light absorption, light-scattering power, and particle size contribute to determining the hiding power of the pigment. Opacity is also affected by refractive index differences between the pigment and the dispersing medium. Selection of a pigment for a specific application depends on a great many physical properties and characteristics. Some of the factors involved in pigment selection are listed in Table 1. Table 2 lists most of the pigments used in packaging materials.

Table 2. Listing of Pigments for Utilization in Packaging Materials

Common name	Colour Index name number	CAS Registry Number	Application[a]	Color permanency,[b] indoor fadeometer, max h,		Plastic applicability,[c]		Some other data
				Masstone	Tint	Wide use	Limited use	
White pigments								
zinc oxide[d]	White 4 77947	[1314-13-2]	2			A,B,C,D, E,F,G, H,I,J, K,L, M,N,O	P	refractive index 2.01; embrittles oleoresinous film
lithopone	White 5 77115	[1345-05-7]	1			A,B,C,E, G,H,I, J,K,L, M,N, O,P	D	refractive index 1.84
titanium dioxide, (anatase, rutile)[d]	White 6 77891	[13463-67-7]	1,2,3	250	250	A,B,C,D, E,G,H, I,J,K, L,M, N,O, P,Q	F	refractive index 2.76 (rutile) refractive index 2.55 (anatase)
zinc sulfide	White 7 77975	[1314-98-3]	1,2			A,B,C,D, G,H,I, J,K, L,M	N,O,P	refractive index 2.37
calcium carbonate[d]	White 18 77220	[471-34-1] [1317-65-3]	1,3					refractive index 1.48–1.65; brightness 85–95% (nat.) 92–98% (syn.)
kaolin clay, bentonite[d]	White 19 77004	[1332-58-7] [8047-76-5]	1,3					refractive index 1.56; low brightness
blanc fixe, process white[d]	White 21 77120	[7727-43-7]	1					refractive index 1.64
aluminum hydrate[d]	White 24 77002	[1332-73-6]	1					refractive index 1.57; rheology modifier for inks
talc, French chalk[d]	White 26 77718	[8005-37-6] [14807-96-6]	1					refractive index 1.54–1.59
silica[d]	White 27 77811	[7631-86-9] [14808-60-7] [61790-53-2] [63231-67-4]	1					refractive index 1.45–1.55; brightness 91–96% (syn.), < 90% (nat.)
Black pigments								
aniline black	Black 1 50440	[13007-86-8]	1	120	20–30			gives deep matt black or velvety finish
lamp and vegetable black	Black 6 77266	[1333-86-4]	1					
carbon black[d] furnace black channel black	Black 7 77266	[1333-86-4]	1,2	>240	>240	A,B,C,D, E,F,G, H,I,K, L,M, N,O, P,Q	J	excellent stability to light, chemicals and heat; good uv absorption

Table 2. (*Continued*)

Common name	Colour Index name number	CAS Registry Number	Application[a]	Color permanency,[b] indoor fadeometer, max h,		Plastic applicability,[c]		Some other data
				Masstone	Tint	Wide use	Limited use	
iron titanate brown spinel	Black 12 77543	[68187-02-0]	2			A,B,C,D, F,G,H, J,K,L, M,N,O	E,I,P,Q,	excellent heat, light, and chemical resistance
iron copper chromite black spinel	Black 23 77429		2			A,B,C,D, F,G,H, J,K,L, M,N,O	E,I,P,Q	certain plastics embrittled by iron
manganese ferrite black spinel	Black 26 77494	[68186-94-7]	2			A,B,C,D, F,G,H, J,K,L, M,N,O	E,I,P,Q	certain plastics embrittled by Mn and Fe
copper chromite black spinel	Black 28 77428	[68186-91-4]	2			A,B,C,D, F,G,H, J,K,L, M,N,O	E,I,P,Q	excellent chemical and heat resistance
Red pigments								
Naphthol Red FRR	Red 2 12310	[6041-94-7]	1,3	20–80	15–20			printing inks for packaging; excellent chemical resistance
Toluidine Red	Red 3 12120	[2425-85-6]	3	40–140[D]	5–20[F]		D,H,J	
Chlorinated Para Red[d]	Red 4 12085	[2814-77-9]	1,3	40–120[DL]	5–30[F]	N,O	D,H,J	bleeds in organic solvents and overstripes
Naphthol Carmine FB	Red 5 12490	[6410-41-9]	1	60–120	20–40			excellent chemical resistance; gravure inks for packaging
Naphthol Red F4RH	Red 7 12420	[6471-51-8]	1,3	80–160[F]	60[F]		C,E,J,K, L,M, N,O,Q	packaging printing inks; excellent chemical resistance
Naphthol Red FRLL	Red 9 12460	[6410-38-4]	1,3	80–120	40–60			excellent chemical resistance
Naphthol Red FRL	Red 10 12440	[6410-35-1]	1,3	60–80[F]	25–35			
Naphthol Red (medium shade)	Red 17 12390	[6655-84-1]	1,3	30–80[F]	10–30[FL]		A,D,E,H, J,K,N, O	excellent chemical resistance; packaging printing ink
Naphthol Red (light yellow shade)	Red 22 12315	[6448-95-9]	1,3	40–80[DFL]	15–30[FL]		A,D,E,H, J,K,N, O	printing ink for packaging; superior chemical resistance
Naphthol Red (dark blue shade)	Red 23 12355	[6471-49-4]	1,3	60[DFL]	30[FL]		A,D,E,H, J,K,N, O	printing inks for packaging
Pyrazolone Red	Red 38 21120	[6358-87-8]	1	50–75[DF]	15–50[F]	A,P	C,D,E,H, J,K,N, O	metal decorating and packaging printing inks
Dianisidine Red	Red 41 21200	[6505-29-9]	1				K,P,Q	packaging printing inks
Permanent Red 2B (barium)	Red 48:1 15865:1	[7585-41-3]	1,2,3	10–30[DFL]	10–20[FL]	K,L,M, N,O,P	A,D,E,J, Q	carton and label printing inks; excellent brightness
Permanent Red 2B (calcium)	Red 48:2 15865:2	[7023-61-2]	1,2	20–100[DFL]	10–50[FL]	E,K,P,Q	D,J,L,M, N	printing inks for labels and cartons; bright and good tint strength
Permanent Red 2B (strontium)	Red 48:3 15865:3	[15782-05-5]	1,2	10–30[D]	5–30[F]			excellent brightness; solvent-based printing inks

Table 2. (*Continued*)

Common name	Colour Index name number	CAS Registry Number	Ap-plica-tion[a]	Color permanency,[b] indoor fadeometer, max h,		Plastic applicability,[c]		Some other data
				Masstone	Tint	Wide use	Limited use	
Permanent Red 2B (manganese)	Red 48:4 15865:4	[5280-66-0]	1	80–120[DFL]	30–40[DFL]			poor alkali and soap resistance
Lithol Red (sodium)	Red 49 15630	[1248-18-6]	1	5–10	2–5			excellent tint strength
Lithol Red (barium)	Red 49:1 15630:1	[1103-38-4]	1	5–40[DL]	2–20[FL]		P	resination increases transparency; poor alkali and soap resistance
Lithol Red (calcium)	Red 49:2 15630:2	[1103-39-5]	1	2–5[DL]	2–5[FL]			excellent brightness and tint strength
Red 2G (calcium)	Red 52:1 15860:1	[17852-99-2]	1,2	10–15[D]	5–10[FL]	N	B,C,D,E, H,J,K, O,P,Q	process magenta for printing inks
Red Lake C[d] (barium)	Red 53:1 15585:1	[5160-02-1]	1,2	5–50[DFY]	1–25[F]	A	C,D,E,H, J,K,L, N,O	standard warm red; foil coatings
Lithol Rubine[d] (calcium)	Red 57:1 15850:1	[5281-04-9]	1,2	15–50[DFL]	5–25[FL]	E,K	C,D,H,J, L,M, N,O,P, Q	standard process magenta; foil coatings
Pigment Scarlet (barium)	Red 60:1 16105:1	[15782-06-6]	1,2	25–50[D]	20–30[B]	A,C,E,K, L,N,O, P	D,H,J,M, Q	printing ink for gloss labels, waxed papers, metal decorating; foil coatings
Anthosine Red 3B (Ba, Na)	Red 66 18000:1	[68929-13-5]	1					metal decorating printing inks
Anthosine Red 5B (Ba, Na)	Red 67 18025:1	[68929-14-6]	1					metal decorating printing inks; transparency
Rhodamine Y (PTMA)	Red 81:1 45160:1	[12224-98-5]	1	15–30[D]	5–10[F]			brilliant, color purity, good tint strength
Rhodamine Y (SMA)	Red 81:3 45160:3	(63022-06-0)	1	15–30[D]	5–10[F]			process magenta printing inks
Rhodamine Y (PMA)	Red 81:x 45160:x	[63022-07-1]	1	15–30[D]	5–10[F]			poor alkali and soap resistance
Alizarine Red B	Red 83 58000:1	[72-48-0]	1	120	30		A,D,E,H, J,N,O, P,Q	metal decorating inks, butter and soap packages
Thioindigold Red	Red 88 73312	[14295-43-3]	2	120–160	80–120	P,Q,	D,H,J,K, L,M, N,O	clean color with excellent fastness
Phloxine Red (Lead)	Red 90 45380:1	[1326-05-2]	1	<20[DF]	<20[F]			poor chemical, light, solvent and heat resistance
Synthetic Red[d] iron oxide	Red 101 77491	[1309-37-1]	1,2			A,E,H,J, K,L,N, P,Q	C,F,G,M	foil coatings, Fe embrittles certain plastics
Molybdate Orange	Red 104 77605	[12656-85-8]	1,2	20–160[D]	20–160	D,H,J,P, Q	A,B,K,L, M,N,O	poor alkali and acid resistance
Cadmium Sulfoselenide Red	Red 108 77202	[58339-34-7]	2	500	200[GF]	A,B,C,D, E,G,H, I,J,K, L,M, N,O,P, Q	F	bright, clean, intense colors
cadmium sulfoselenide Lithopone Red	Red 108:1 77202:1	[58339-34-7] and [7727-43-7]	2	500	150[GF]	A,B,C,D, E,G,H, I,J,K, L,M, N,O,P, Q	F	sensitive to mineral acids

Table 2. (*Continued*)

Common name	Colour Index name number	CAS Registry Number	Application[a]	Color permanency,[b] indoor fadeometer, max h,		Plastic applicability,[c]		Some other data
				Masstone	Tint	Wide use	Limited use	
Naphthol Red FGR	Red 112 12370	[6535-46-2]	1,2,3	60–160	40–60			excellent brightness, very good lightfastness; paper coatings
Mercadium Red	Red 113 77201	[1345-09-1]	2			A,B,C,D, E,F,G, H,J,K, L,M, N,O,P, Q		poor lightfastness when light and moisture present
cadmium mercury Lithopone Red	Red 113:1 77201:1	[1345-09-1] and [7727-43-7]	2			A,B,C,D, E,F,G, H,J,K, L,M, N,O,P, Q		low tint strength
Quinacridone Magenta Y	Red 122 73915	[980-26-7]	1,2	140–160[D]	80–120[F]	K,L,M,P, Q	A,B,C,D, H,J,O	soluble in nylon and certain plastics
Perylene Vermilion	Red 123 71145	[24108-89-2]	1,2			A,B,C,E, J,K,P, Q	L,M,N	transparent; good fastness properties
Disazo Red	Red 144	[5280-78-4]	1,2	160	100–140	C,E,H,J, K,L, M,N, O,P,Q	D	high performance pigment
Naphthol Carmine FBB	Red 146 12485	[5280-68-2]	1,2,3	60[D]	30[D]	C,E,H,J, K,L, M,N, O,P,Q	D	packaging and metal decorating inks, paper coatings
Perylene Red BL	Red 149 71137	[4948-15-6]	1,2	40–80[D]	20–80	A,B,C,E, H,I,J, K,L, M,N, O,P,Q	G	metal decorating
Disazo Scarlet	Red 166	[12225-04-6]	1,2			C,E,H,J, K,L, M,N, O,P,Q	D	high performance pigment
Brominated Anthanthrone Red	Red 168 59300	[4378-61-4]	1			J,K,L,M, P,Q		metal decorating printing ink
Rhodamine 6G	Red 169 45160:2	[12224-98-5]	1	30[D]	10[F]			gravure printing inks, excellent tint strength
Naphthol Red F5RK	Red 170 12475	[2786-76-7]	1	80–120[D]	60[D]	C,L,M, N,O,Q	A,E,F,I, J,K,P	brilliant, excellent chemical resistance
Benzimidazolone Maroon HFM	Red 171 12512	[6985-95-1]	1	120	120	C,K,L,N, P,Q	A,B,E,G, J	foil coatings; very transparent
Benzimidazolone Red HFT	Red 175 12513	[6985-92-8]	1,2	160[F]	120[F]	C,K,L,N, P,Q	A,B,E,G, J	highly transparent; inks for packaging and metal decorating
Benzimidazolone Carmine HF3C	Red 176 12515	[12225-06-8]	1,2	80[D]	40[F]	C,K,L,N, O,P,Q	A,E,F,I,J, M	transparent, bright, chemical resistant; packaging and metal decorating inks
Anthraquinoid Red	Red 177 65300	[4051-63-2]	2	120	70–100	C,D,G,H, J,K,L, M,P,Q	B,E,I	transparent

Table 2. (*Continued*)

Common name	Colour Index name number	CAS Registry Number	Application[a]	Color permanency,[b] indoor fadeometer, max h,		Plastic applicability,[c]		Some other data
				Masstone	Tint	Wide use	Limited use	
Perylene Maroon	Red 179 71130	[5521-31-3]	2			A,B,C,E, J,K,P, Q	L,M	excellent fastness properties
Naphthol Rubine F6B	Red 184		1	60[F]	15[F]			printing ink for packaging
Benzimidazolone Carmine HF4C	Red 185 12516	[61951-98-2]	1,2	60[D]	30[F]	C,K,L,N, O,P,Q	A,E,F,I, J,M	process magenta for metal decorating
Naphthol Red HF4B	Red 187 12486	[59487-23-9]	1,2	80[F]	40[F]	K,L,N,P, Q	C,E,I,J, M	bright, transparent; packaging and metal decorating inks
Naphthol Red HF3S	Red 188 12467	[61847-48-1]	1,2	80[D]	60[D]	K,L,N,P, Q	C,E,I,J,M	superior chemical resistance; packaging inks
Perylene Scarlet	Red 190 71140	[6424-77-7]	2					transparent, dull tints
Rubine Red (calcium)	Red 200 15867	[58067-05-3]	1,2	20–70[FL]	10–15[FL]			oil-based printing inks, poor soap, solvent and alkali resistance
Quinacridone Scarlet	Red 207	[1047-16-1] and [3089-16-5]	1,2	120–320[D]	80–120[F]			high performance pigment
Benzimidazolone Red HF2B	Red 208 12514	[31778-10-6]	1,2	80[F]	40[F]			bright medium red; packaging and metal decorating inks
Quinacridone Red Y	Red 209 73905	[3089-17-6]	1	120[D]	120[F]			soluble in nylon
Naphthol Red F6RK	Red 210		1	60[D]	30[F]	C,J,K,L, M,N, O,Q	A,E,F,I, J,K,P	packaging printing inks
Perylene Red Y	Red 224 71127	[128-69-8]	2	>500	>500			transparent; very good strength and fastness properties
Orange pigments								
Dinitraniline Orange[d]	Orange 5 12075	[3468-63-1]	1,3	40–80[DFL]	5–10[DFL]			good chemical resistance
Pyrazolone Orange	Orange 13 21110	[3520-72-7]	1	10–60[DFL]	5–10[FL]	K,L,N,O, P,Q	A,C,D,H, J,M	
Dianisidine Orange	Orange 16 21160	[6505-28-8]	1,2	25–75[DF]	5–50[F]	J,P,Q	K,L,M	
Persian Orange Lake (A1)	Orange 17:1 15510:2	[15876-51-4]	1					inks for waxed bread wrappers
Pure Cadmium Orange; Cadmium Sulfoselenide Orange	Orange 20 77202	[12556-57-2]	2	500	100[GF]	A,B,C,D, E,F,G, H,J,K, L,M, N,O,P, Q		lightfastness needs protection from moisture
Cadmium Sulfoselenide Orange Lithopone	Orange 20:1 77202:1	[12556-57-4] and [7727-43-7]	2	500	100[GF]	A,B,C,D, E,F,G, H,J,K, L,M, N,O,P, Q		

Table 2. (*Continued*)

Common name	Colour Index name number	CAS Registry Number	Application[a]	Color permanency,[b] indoor fadeometer, max h,		Plastic applicability,[c]		Some other data
				Masstone	Tint	Wide use	Limited use	
Mercadium Orange	Orange 23 77201	[1345-09-1]	2			A,B,C,D, E,F,G, H,J,K, L,M, N,O,P, Q		fades in presence of light and moisture
Mercadium Lithopone Orange	Orange 23:1 77201:1	[1345-09-1] and [7727-43-7]	2			A,B,C,D, E,F,G, H,J,K, L,M, N,O,P, Q		
Disazo Orange	Orange 31	[5280-74-0]	2					
Diarylide Orange	Orange 34 21115	[15793-73-4]	1,2,3	20–60^D	15–30^F	K,L,N,O, P,Q	A,C,D,H, J,M	bleeds in some overstripes
Benzimidazolone Orange HL	Orange 36 11780	[12236-62-3]	1	120^D	80^F	A,C,E,J, K,L, M,N, O,P,Q	B,G,I	
Naphthol Orange	Orange 38 12367	[12236-64-5]	1,2	80–120^D	10–40	K,L,N,O, P,Q	C,E,I,J, M	printing inks for metal decorating and packaging
Perionone Orange	Orange 43 71105	[4424-06-0]	1	160^D	120^D	J,K,L,M, P,Q		
Ethyl Red Lake C (barium)	Orange 46 15602	[67801-01-8]	1,2	5–30^{FL}	2–20^{FL}			metal decorating inks
Quinacridone Gold	Orange 48	[1047-16-1] and [1503-48-6]	2	120–320^D	80–120^F			
Quinacridone Deep Gold	Orange 49	[1047-16-1] and [1503-48-6]	2	120–320^D	80–120^F			
Benzimidazolone Orange HGL	Orange 60		1,2			E,J,K,P, Q	A,B,C,F, H,L, M,N,O	transparent
Tetrachloro-isoindolinone Orange	Orange 61		2	200–500	200–500			high performance pigment
Benzimidazolone Orange H5G	Orange 62		1	>160	60–160			oil-based printing inks
Orange GP	Orange 64		2					
Yellow pigments								
Arylide Yellow G	Yellow 1 11680	[2512-29-0]	1,3	60–120^{DFG}	20–40^F			printing inks requiring alkali resistance, aqueous dispersions for paper
Arylide Yellow 10G	Yellow 3 11710	[6486-23-3]	1,3	120–200^{DF}	20–60^F			printing inks requiring alkali resistance, aqueous dispersions for paper
Diarylide Yellow AAA	Yellow 12 21090	[6358-85-6]	1	10–60^{DFL}	2–30^{FL}		A,C,D,E, H,J,K, M,N, O,P	
Diarylide Yellow AAMX	Yellow 13 21100	[5102-8-30]	1,2	20–60^F	10–40^F	A,P	D,E,G,H, I,J,K, L,N,O	

Table 2. (*Continued*)

Common name	Colour Index name number	CAS Registry Number	Application[a]	Color permanency,[b] indoor fadeometer, max h, Masstone	Tint	Plastic applicability,[c] Wide use	Limited use	Some other data
Diarylide Yellow AAOT	Yellow 14 21095	[5468-75-7]	1,2	10–60[DFL]	5–40[FL]	G,K	A,C,D,E, H,J,L, M,N, O,P,Q	
Permanent Yellow NCG	Yellow 16 20040	[5979-28-2]	1,2	80–120[DF]	30[F]	C,E,K,L, M,N, O,Q	A,I,P	printing inks for packaging
Diarylide Yellow AAOA	Yellow 17 21105	[4531-49-1]	1,2	20–80[DF]	10–40[F]	G,K,L, M,P,Q	A,C,D,H, I,J,N, O	
Chrome Yellow (primrose, light, medium)	Yellow 34 77600 77603	[1344-37-2]	1	10–160[D]	10–160[FG]	D,H,J,P, Q	A,B,I,K, L,M, N,O	
Cadmium Zinc Yellow (primrose, lemon, golden)	Yellow 35 77205	[12442-27-2]	2	400	100[GF]			
Cadmium Zinc Yellow Lithopone	Yellow 35:1 77205:1	[12442-27-2] and [7727-43-7]	2	400	80[GF]			
Cadmium Yellow	Yellow 37 77199	[1306-23-6]	2	400	80–100[GF]	A,B,C,D, E,F,G, H,I,J, K,L, M,N, O,P,Q		
Cadmium Lithopone yellow	Yellow 37:1 77199:1	[1306-23-6] and [7727-43-7]	2	400	80[GF]	A,B,C,D, E,F,G, H,I,J, K,L, M,N, O,P,Q		
Synthetic Yellow[d] iron oxide	Yellow 42 77492	[12259-21-1] [51274-00-1]	2					
nickel antimony Titanium Yellow rutile	Yellow 53 77788	[8007-18-9] [71077-18-4]	2	1000[F]	1000[F]	A,D,E,F, G,H,J, K,L, M,N, O,P,Q	B,C,I	
Diarylide Yellow AAPT	Yellow 55 21096	[6358-37-8]	1	35–60	25–40			printing inks for waxed food wrappers
Arylide Yellow 4R	Yellow 60 12705	[6407-74-5]	1	60–100	35–50[F]			alkali resistance and lightfast printing inks
Arylide Yellow RN	Yellow 65 11740	[6528-34-3]	1,3	80–150[DF]	60–100			lightfast and alkali resistance printing inks; aqueous dispersions for paper
Arylide Yellow GX	Yellow 73 11738	[13515-40-7]	1,3	70–120[DF]	30–40[F]			lightfast and alkali resistance printing inks; aqueous dispersions for paper
Arylide Yellow GY	Yellow 74 11741	[6358-31-2]	1,3	70–120[DF]	20–60[F]			lightfast and alkali resistance printing inks; aqueous dispersions for paper

Table 2. (*Continued*)

Common name	Colour Index name number	CAS Registry Number	Application[a]	Color permanency,[b] indoor fadeometer, max h,		Plastic applicability,[c]		Some other data
				Masstone	Tint	Wide use	Limited use	
Diarylide Yellow H10G	Yellow 81 21127	[22094-93-5]	1	60–160	30–60[F]	C,E,K,L, N,O,P, Q	A,M	printing inks for packaging
Diarylide Yellow HR	Yellow 83 21108	[5567-15-7]	1,2	70–240[DF]	20–60[DF]	P,Q	C,D,E,H, J,K,L, M,N	
Disazo Yellow G	Yellow 93	[5580-57-4]	1,2	120	80	A,K,L, M,N, O,P		
Disazo Yellow R	Yellow 95	[5280-80-8]	1,2	120	80	A,K,L, M,N, O,P		metal-free printing inks
Permanent Yellow FGL	Yellow 97 11767	[12225-18-2]	1	120[D]	120[F]	A,C,E,J, K,L, M,N, O,Q	B,G,N,I	heat, lightfast, and alkali resistant printing inks
Arylide Yellow 10GX	Yellow 98 11727	[12225-19-3]	1	80–120[F]	40–60[F]			
FD&C Yellow No. 5[d] aluminum lake	Yellow 100 19140:1	[12225-21-7]	1			C,F,G,K, L,O,P	E	colorant for food, drugs, cosmetics and food contact surfaces; metal decorating
Fluorescent Yellow	Yellow 101 48052	[2387-03-3]	2					
Diarylide Yellow GGR	Yellow 106	[12225-23-9]	1	30–50[F]	20–30[F]			printing inks for packaging
Tetrachloro-isoindolinone Yellow G	Yellow 109		2	200	100–200	H,J,K,L, M,P,Q	A,D,G, N,O	
Tetrachloro-isoindolinone Yellow R	Yellow 110		2	200–500	200–500	E,H,J,K, L,M,P, Q	A,D,G, N,O	
Diarylide Yellow H10GL	Yellow 113 21126	[14359-20-7]	1	60[F]	30[F]	C,E,K,L, M,N, O,P,Q	A,G,I	printing inks for packaging
Diarylide Yellow G3R	Yellow 114		1	40[F]	15[F]			oil-based ink for packaging
Azomethine Yellow	Yellow 117	[21405-81-2]	1					food packaging
zinc ferrite brown spinel	Yellow 119 77496	[12063-19-3] [68187-51-9] [61815-08-5]	2					
Diarylide Yellow DGR	Yellow 126		1	25–35[F]	10–20[F]			
Diarylide Yellow GRL	Yellow 127		1	30–50[F]	20–30[F]	A,K,L,N, O,Q	C,D,E,H, J,M,P	inks for packaging and metal decorating
Disazo Yellow GG	Yellow 128		2					
Quinophthalone Yellow	Yellow 138		2					
Isoindoline Yellow	Yellow 139		2			C,D,E,J, K,L, M,N, P,Q		
Nickel Yellow 4G	Yellow 150	[68511-62-6]	2					

Table 2. (*Continued*)

Common name	Colour Index name number	CAS Registry Number	Application[a]	Color permanency,[b] indoor fadeometer, max h,		Plastic applicability,[c]		Some other data
				Masstone	Tint	Wide use	Limited use	
Benzimidazolone Yellow H4G	Yellow 151		1,2	160[F]	120[F]	A,C,E,J, K,L, M,N, O,P,Q	F,G,H,I	
Diarylide Yellow YR	Yellow 152	[20139-66-6]	1	20–40[F]	10–20[F]			lead chromate replacement
Benzimidazolone Yellow H3G	Yellow 154		1,2	160[F]	120[F]	C,E,J,K, P,Q	A,F,H,L, M,N,O	inks for metal decorating
nickel niobium Titanium Yellow Rutile	Yellow 161 77895	[68611-43-8]	2					
chrome niobium titanium buff rutile	Yellow 162 77896	[68611-42-7]	2					
manganese antimony titanium buff rutile	Yellow 165 77899	[68412-38-4]	2					
Green pigments								
Brilliant Green (PTMA)	Green 1 42040:1	[1325-75-3]	1,3	10–15[D]	5–10[F]			poor alkali and soap resistance
Brilliant Green (PMA)	Green 1:x 42040:x	[68814-00-6]	1,3	10–15[D]	2–10[F]			excellent brilliance, color purity
Permanent Green (PTMA)	Green 2 42040:1 and 49005:1	[1328-75-3] and [1326-11-0]	1,3	10–15[D]	5–10[F]			poor alkali and soap resistance
Permanent Green (PMA)	Green 2:x 42040:x 49005:1	[68814-00-6] and [1326-11-0]	1,3	10–15[D]	5–10[F]			for lustrous appearance in printing
Malachite Green (PTMA)	Green 4 42000:2	[61725-50-6]	1,3					poor alkali and soap resistance
Phthalocyanine Green	Green 7 74260	[1328-53-6]	1,2	120–320[DFL]	120–160[FL]	A,B,D,E, G,H,I, J,K,L, M,N, O,P,Q	C	standard green for printing inks
Nickel Azo Yellow (green gold)	Green 10	[51931-46-5]	2	>70	>70	E,N,O,P	C,D,F,H, I,J,K, L,M,Q	very lightfast pigment
Chrome Green	Green 15 77510 77603	[1344-37-2] and [25869-00-5]	2,3	60–80[DG]	20–40	J	C,D,H,I, N,O, P,Q	
cobalt chromite green spinel	Green 26 77344	[68187-49-5]	2	>160	>160			outstanding chemical and light stability
Phthalocyanine Green (Cl, Br)	Green 36 74265	[14302-13-7]	1,2	160–320[FL]	120–160[FL]	A,B,D,E, H,I,J, K,L, M,N, O,P,Q	C,G	high performance pigment
cobalt titanate green spinel	Green 50 77377	[68186-85-6]	2					outstanding chemical, light and heat stability
Blue pigments								
Victoria Blue (PTMA)	Blue 1 42595:2	[1325-87-7]	1,3	20–40[D]	5–10[F]			outstanding brilliance
Victoria Blue (SMA)	Blue 1:2 42595:3	[68413-81-0]	1,3	30–40[D]	5–10[F]			SMA salt is stronger than PMS/PTMA but not as clean
Victoria Blue (PMA)	Blue 1:x 42595:x	[68409-66-5]	1,3	15–40	5–10[F]			printing inks with lustrous appearance

Table 2. (*Continued*)

Common name	*Colour Index* name number	CAS Registry Number	Application[a]	Color permanency,[b] indoor fadeometer, max h, Masstone	Tint	Plastic applicability,[c] Wide use	Limited use	Some other data
Permanent Blue (PMA) Peacock	Blue 9:x 42025:x	[68814-07-3]	1,3					excellent color purity and strength
Phthalocyanine Blue Alpha (red crystallizing)[d]	Blue 15 74160	[147-14-8]	1,2	120–320[FLZ]	120–160[FL]	A,B,C,D, E,G,H, I,J,K, L,M, N,O,P, Q	F	excellent transparency, chemical resistance, and lightfastness
Phthalocyanine Blue Alpha (R, NC)	Blue 15:1 74160 74250	[147-14-8] [12239-87-1]	1,2	120–320[FLZ]	120–160 [FL]	A,B,C,D, E,G,H, I,J,K, L,M, N,O,P, Q		
Phthalocyanine Blue Alpha (R, NCNF)	Blue 15:2 74160 74250	[147-14-8] [12239-87-1]	1,2,3	120–320[FLZ]	80–160[FL]	A,B,C,D, E,G,H, I,J,K, L,M, N,O,P, Q		
Phthalocyanine Blue Beta (G, NC)	Blue 15:3 74160	[147-14-8]	1,2	120–320[FLZ]	120–160[FL]	A,B,C,D, E,G,H, I,J,K, L,M, N,O,P, Q		standard process cyan.
Phthalocyanine Blue Beta (G, NCNF)	Blue 15:4 74160	[147-14-8]	1,2	120–320[FLZ]	80–160[FL]	A,B,C,D, E,G,H, I,J,K, L,M, N,O,P, Q		
Phthalocyanine Blue (metal free)	Blue 16 74100	[574-93-6]				A,G,K,L, M,N, O,P		
Fugitive Peacock Blue (Ba)	Blue 24 42090:1	[6548-12-5]	1					
Iron Blue, Milori Blue[d]	Blue 27 77510	[25869-00-5]	1	>160[G]	20–80[G]	K	B,C,D,H, J,L,N, O,P,Q	
Cobalt Blue[d]	Blue 28 77346	[1345-16-0] [68186-86-7]	2	>100[F]	>100[G]	B,C,D,E, F,G,H, J,K,L, M,N, O,P,Q	I	outstanding lightfastness, chemical resistance
Ultramarine Blue[d]	Blue 29 77007	[57455-37-5]	1,2	>240	>240	A,C,E,G, H,K,L, M,N, O,P,Q	F,I,J	
cobalt chromite blue green spinel	Blue 36 77343	[68187-11-1]	2	>100[F]	>100[G]	B,C,D,E, F,G,H, I,J,K, L,M, N,O,P, Q		
Alkali Blue G	Blue 56 42800	[6417-46-5]	1					for printing inks needing alkali and soap resistance
Indanthrone Blue	Blue 60 69800	[81-77-6]	2	120–160[F]	120–160[F]	O,P,Q	B,C,D,E, H,J,K, L,M,N	
Alkali Blue, Reflex Blue	Blue 61 42765:1	[1324-76-1]	1	2–10[FL]	2–10[FL]			poor alkali and soap resistance
Victoria Blue (CFA)	Blue 62 42595:x		1	20[D]	10[F]			

Table 2. (*Continued*)

Common name	Colour Index name number	CAS Registry Number	Application[a]	Color permanency,[b] indoor fadeometer, max h,		Plastic applicability,[c]		Some other data
				Masstone	Tint	Wide use	Limited use	
Violet pigments								
Rhodamine B (PMA)	Violet 1:x 45170:x	[63022-09-3]	1	<20[D]	<20[F]			brilliant, color purity and tint strength
Methyl Violet (PTMA)	Violet 3 42535:2	[1325-82-2]	1,3	15–30[DFL]	5–10[FL]		A,D,H,J, N	printing inks with lustrous appearance
Fugitive Methyl Violet	Violet 3:3 42535:5	[68308-41-8]	1	2–5[FL]	2–5[FL]			
Methyl Violet (PMA)	Violet 3:x 42535:x	[67989-22-4]	1,3	15–30[DG]	5–10[F]			standard purple for printing inks
cobalt violet phosphate	Violet 14 77360	[13455-36-2]	2					decolorizer for clear and white plastics
Ultramarine Violet and Pink[d]	Violet 15 77007	[12769-96-9]	2			A,D,E,G, H,I,J, K,L, M,N,O	B,C,P,Q	counteracts yellowing in plastics on heating
Manganese Violet[d]	Violet 16 77742	[10101-66-3]	2					
Quinacridone Violet	Violet 19 73900	[1047-16-1]	1,2	120–320[D]	80–120[F]	I,J,K,L, M,N, O,P,Q	A,B,C,D, E,H	high performance pigment
Crystal Violet (CFA)	Violet 27 42535:3	[12237-62-6]	1	20[D]	10[F]			
Benzimidazolone Bordeaux HF3R	Violet 32 12517	[12225-08-0]	1,2	60	40	K,L,N,O, P,Q	C,E,I,J, M	inks for packaging and metal decorating
cobalt lithium violet phosphate	Violet 47 77363	[68610-13-9]	2					decolorizer for clear and white plastics
cobalt magnesium red-blue borate	Violet 48 77352	[68608-93-5]						decolorizer for clear and white plastics
Brown pigments								
Monoazo Brown (copper)	Brown 5 15800:2	[16521-34-9]	1,2	>160	>100			amber effects in plastics
magnesium ferrite	Brown 11 77495	[12068-86-9]	2					
Diazo Brown	Brown 23		1,2			K,L,P,Q	A,M	high performance pigment.
chrome antimony titanium buff rutile	Brown 24 77310	[68186-90-3]	2	>1000	>1000	A,C,D,E, F,H,J, K,L,N, O,P,Q	G,I,M	outstanding light, heat and chemical stability
Benzimidazolone Brown HFR	Brown 25 12510	[6992-11-6]	1,2	80[F]	80[F]	K,L,N,O, P,Q	C,M	printing inks for metal decorating; highly transparent
zinc iron chromite brown spinel	Brown 33 77503	[68186-88-9]	2					excellent heat and light resistance
iron chromite brown spinel	Brown 35 77501	[68187-09-7]	2					excellent heat and light resistance
Metal pigments								
Aluminum Flake[d]	Metal 1 77000	[7429-90-5]	1,3			B,C,D,E, G,J,K, L,M, N,O,P, Q		nonleafing for decorative inks, paper coatings
copper powder, bronze powder[d]	Metal 2 77400	[7440-50-6] [7440-50-8] [7440-66-6]	1,2			B,C,D,G, J,K,L, N,O, P,Q	E,I,M	decorative printing inks for packaging, labels

[a] Key to application: 1 = printing inks; 2 = plastics; 3 = paper and paper coatings.
[b] Key to permanency failures: F = fades; D = darkens; L = loses gloss; B = turns bluer; G = turns gray or greener; Y = turns yellower; Z = bronzes.
[c] Key to plastic: A = ABS; B = acetal; C = acrylic; D = amino resins; E = cellulosics; F = fluoroplastics; G = nylons; H = phenol–formaldehyde; I = polycarbonate; J = polyester or alkyd; K = polyethylene, low density; L = polyethylene, high density; M = polypropylene; N = polystyrene, general purpose; O = polystyrene, impact-resistant; P = flexible vinyl; Q = rigid vinyl.
[d] Pigments having an FDA status.

Special-effect pigments. Nacreous pigments, like basic lead carbonate and titanium-coated mica are used for luster effects for cosmetic containers. Fluorescent pigments and dyes are used in gravure inks for carton printing and special effects in gift-wrap printing. They are also used as colorants for blow-molded bottles, closures, tubs, cartons, and pails. Other special-effect pigments include luminescent and phosphorescent pigments, as well as metallics (eg, aluminum flake and various copper bronzes which vary in shade depending on their chemical composition).

Dyes in Packaging

When dyes are used for coloration, as in plastics, they must be checked for migration, heat stability, lightfastness, and sublimation.

For plastic materials, dyes that are widely used are azo dyes, anthraquinone dyes, xanthene dyes, and azine dyes which include induline and nigrosines. Azo dyes such as Solvent Red 1, 24, and 26, Solvent Yellow 14 and 72, and Solvent Orange 7 are colorants for polystyrene, phenolics, and rigid PVC. Better heat stability and better weatherability are obtained from anthraquinone dyes such as Solvent Red 111, Blue 56, Green 3, and Disperse Violet 1 in the coloration of acrylics, polystyrene, and cellulosics. Basic Violet 10 is a xanthene dye used in phenolics. Solvent Green 4, Acid Red 52, Basic Red 1, and Solvent Orange 63 are used for acrylics, polystyrene, and rigid PVC. Azine dyes produce exceptionally jet blacks and are used in ABS, polypropylene, and phenolics. A perinone dye, Solvent Orange 60, has good light and heat stability for ABS, cellulosics, polystyrene, and rigid PVC. ABS, polycarbonate, polystyrene, nylon, and acrylics may be colored with quinoline dye. Methyl Violet and Victoria Blue B, two basic triphenyl-amine dyes find limited use in phenolics.

For printing inks, five dye families are of particular interest: azo, triphenylmethane, anthraquinone, vat, and phthalocyanine. *Certified food colorants* are used in packaging applications where the printed surface is in direct contact with food. Dyes for paper include acid, basic, (including resorcine and alizarine), and direct dyes. Basic directs are used to color containers, boxboards, wrapping paper, and multiwall bags.

Colorants for Printing Inks

Printing inks are used for a broad range of packaging items such as cartons, bags, labels, metal cans, rigid containers, and decorative foils. Table 2 highlights the pigments used for printing (see Inks). Important qualities of pigments for printing inks are contribution to printing properties (ie, rheology and bleeding), print appearance (ie, the color has sales appeal), and useful service life (ie, resistance to fading and chemicals). In general, most ink pigments can be utilized in most types of inks; but there are minor differences in a pigment's performance that must occasionally be considered. For example, pigments for lithographic inks must not bleed in water or very mild inorganic acidic (phosphoric acid) solutions. Glossy finishes require small-particle-size pigments. Most printing-ink pigments are organic.

Extender pigments are essentially transparent or translucent and contribute neither to hiding power nor color. They are formulated into inks to extend the covering power of strong pigments and enhance the working properties (eg, to increase viscosity without affecting color).

Colorants for Plastics

In selecting a colorant for plastics, in addition to the colorant characteristics listed in Table 1, the resins, its properties and compatibility, and the method by which it is processed are critical factors. Because PVC and its copolymers liberate acid at high temperature, acid-sensitive colorants like cadmium reds and ultramarine blue must be used with care. Hansa yellow tends to crock (smudge) in polyethylene. In polyethylene, impurities in pigments such as cobalt and manganese should be avoided. Polyacetals and thermoplastic polyesters are sensitive to moisture. Molten nylon acts as a strong acidic reducing agent and can decolorize certain dyes and pigments.

Colorants must have enough heat stability to withstand processing temperatures, which are sometimes very high, eg, ca 600°F (316°C) for injection-molded polycarbonate. Some organic pigments processed at 350°F (177°C) for 15 min begin to show signs of darkening, but some cadmium pigments can withstand 1500°F (816°C) without noticeable color change. When processing temperatures are particularly high, the choices for colorants are few. Colorants may also interact with other additives used in the plastic such as heat and light stabilizers; for example, sulfur-containing cadmium pigments react with the nickel-bearing stabilizers used in some film-grade polypropylenes.

Dispersion. Dispersing the pigment to develop color strength and maximum optical properties is an extremely important issue that is continually being addressed. Dispersion of dyes is not as great a problem because of their solubilizing nature. Pigments need to be properly wetted by vehicles or the plastic medium. Poor dispersion can lead to processing difficulties for thin films. Agglomeration and aggregation in pigments are due to many reasons, such as different particle sizes and shapes, presence of soluble salts or impurities, and improper grinding. Improvements in pigment dispersion have been obtained through surface treatments of pigments and better grinding techniques.

Pigments for Paper and Paperboard

Two reasons for the use of pigments, fillers, and extender pigments for paper goods are to load or fill the sheet during manufacture to increase bulk and improve such properties as opacity, printability, and brightness; and to coat the paper to provide opacity, black out defects, or color and provide a receptive surface for printing. Table 3 lists some pigments and extender pigments used in the manufacture of paper goods for packaging.

Supply Options

Colorants are supplied in a number of different forms.

Dry powder. Pigments and dyes are sold in a dry powder form. Pigments are ground to suitable working particle sizes for ink manufacture or for dispersion into a plastic. The maximum working particle size of most dry pigments is about 44 μm. These pigments should pass through a 325-mesh (44-μm) screen with less than 1% retention.

Presscake form. This is an undried form in which the presscake may typically contain 25–50% solids. Presscake is used for the preparation of water-based inks.

Flushed colors. Pigments are dispersed in a varnish or mineral oil forming a paste which has a pigment content of ≥30%. The flushing operation involves exchanging water in a

Table 3. Some Pigments for Use in Paperboard and Paper

Filler	Composition	Refractive index	Bright-ness	Use
titanium dioxide (rutile, anatase)	TiO$_2$	2.7 for rutile 2.55 for anatase	98–99 98–99	board, waxing stock, board liner, paper, specialties
clay (kaolin)	aluminum silicate	1.87–1.98	80–85	board, papers, specialties
calcium carbonate (natural and precipitated)	CaCO$_3$	1.56	95–97	printing, cigarette papers
talc	magnesium silicate	1.57	70–90	board, printing papers
gypsum	CaSO$_4$	1.57–1.61	80–90	boards, specialties
diatomite (natural)	diatomaceous earth	1.40–1.46	65–75	improves bulk and drainage of board stocks

presscake for the organic vehicle by a kneading action in a Sigma blade mixer.

Chip dispersion. Pigments dispersed in resin with little or no solvent content. Usually prepared by milling on a two-roll mill.

Resin bonded pigment. This is a dry flush in which the vehicle phase is a resin.

Easy-dispersing pigments and stir-in pigments. Pigments treated with surfactants or polymeric materials to make them readily dispersible in various ink vehicles, particularly gravure-ink types.

Color concentrates. Colorant dispersed in resin which is let down with virgin resin to make final product. Color concentrate is supplied in chip or pellet form. Liquid- and paste-color concentrates are used for vinyls.

Slurry. Titanium dioxide and other white pigments or extenders are supplied to the paper industry in this manner.

Regulatory Requirements

In the United States , the FDA and USDA administer the laws of interest to colorists. The Federal Food, Drug and Cosmetic Act, the Federal Hazardous Substances Act, and the Poison Prevention Packaging Act of 1970 are relevant FDA statutes. The Meat Inspection Act and the Poultry Inspection Act of the USDA relate to colored plastics where food contact is a concern. The Food, Drug, and Cosmetic Act (1938) was modified by the Food Additives Amendment (1958) and the Color Additives Amendment (1960). Pigments that have an FDA status are noted in Table 2 (see Food, drug, and cosmetics regulations).

The regulations do not deal with colorants in printing inks. Four groups of colorants are presently permitted in the coloring of plastics for food, drugs, and cosmetics: (1) Certified colorants are those in the list of FD&C certified dyes and alumina lakes; (2) Purified nonaniline colors include iron oxides, carbon black, and titanium dioxide for use in and on plastics; (3) Use of a noncertified colorant may be petitioned. Responsibilities for compliance with regulations rests with user. (4) A colorant is not subject to the color-additive amendment if there is an impermeable barrier between colorant and food, drug, or cosmetics, and no chance of contact.

Lead pigments and other toxic pigments have been eliminated from inks for food packaging. For nonfood packaging, the three major lead pigments for inks are chrome yellow, molybdate orange, and phloxine red. No other heavy metal pigments are used in significant quantities.

BIBLIOGRAPHY

General References

T. C. Patton, ed., *Pigment Handbook,* Wiley-Interscience, New York, 1973

T. B. Webber, ed., *Coloring of Plastics,* Wiley-Interscience, New York, 1979.

M. Ahmed, *Coloring of Plastics, Theory and Practice,* Von Nostrand Reinhold Co., New York, 1979.

Modern Plastics Encyclopedia, Vol. 56, McGraw-Hill, New York, Oct. 1980.

R. P. Long, *Package Printing,* Graphic Magazines, Inc., Garden City, N.Y., 1964.

Pigments, Vol. 4 of *Raw Materials Data Handbook,* NPIRI, 1983.

R. C. SCHIEK
CIBA-GEIGY Corporation

COMPOSITE CANS. See Cans, composite.

COMPRESSION MOLDING

Compression molding is used to produce parts from thermosetting plastics that cannot be processed by thermoplastic processing methods (see Blow molding; Extrusion; Injection molding; Thermoforming). In compression molding, polymerization takes place in a closed mold. Under heat and pressure, the material fills the mold cavity, and with continued heat and pressure, the part hardens as cross-linking takes place (see Polymer properties). The irreversibility of the process rules out thermoplastic-processing methods, but it also imparts excellent heat resistance and dimensional stability to the part. These properties are essential for many nonpackaging applications (eg, electrical components, automotive and aircraft body panels). The use of compression molding in packaging is very limited. Thermosets rarely provide the best balance of properties and compression molding is not economical for high volume production.

Compression molding has some advantages compared to injection molding, including lower tooling costs and fewer stresses in the part; but design flexibility and production rates are limited (1). For many years, the only significant packaging application for thermosets was closures (see Closures). At one time, phenolic and urea–formaldehyde molding compounds were the only plastics available for the purpose. Thermoplastics have replaced them to a great extent, but in 1984, United States consumption of these compounds for closures still totaled about 24×10^6 lb (10,900 metric tons) (2).

There has been a resurgence of interest in thermosets and compression molding in recent years because of the heat required for "ovenable" packages. Compression-molded glass-filled thermoset polyesters, which had been used for some time for heat-and-serve airline trays, moved into the consumer market with the advent of frozen dinners that could go from the freezer to the convection or microwave oven, and then to the table (3).

BIBLIOGRAPHY

1. G. A. Tanner, "Compression Molding," *Modern Plastics Encyclopedia,* Vol. 61, no. 10a, 1984–1985.
2. *Mod. Plast.* **62**(1), 69 (Jan. 1985).
3. "Dish Sales Sizzle for Ovenable Frozen Dinners," *Plast. World* **42**(9), 59 (1984).

CONTRACT PACKAGING

Contract manufacturing, which includes contract packaging, is defined as the availability of labor, production machinery, and/or technical expertise for hire on a fixed-cost basis, outside the confines of the manufacturing function of any given company. Contract packaging is used by many industries to expand in-house packaging capacity. Contract packagers specialize in many instances, eg, foods, pharmaceuticals, chemicals, etc.

Historically, contract packagers have been primarily labor contractors, providing the work force for assembly of one or more components of a finished product, eg, shrink wrapping, skin and blister packaging, kit packaging, bagging, display setup, adding premiums, etc. The industry has evolved from labor contracting only to the provision of a full range of packaging services, from product and/or process design and product formulation, to a complete packaging service which begins with product specification and ends with delivery of the product to the point of final customer distribution. Today, full-service contract packagers can offer a high level of packaging and quality control expertise to replicate or exceed available internal production capacities.

As the scope of contract (service) packagers has increased, so have the reasons for use of service packaging increased. Today service packagers are utilized to : (*1*) Expand geographical coverage (ie, order fulfillment capabilities), maximizing transportation and service parameters by being located in the market of distribution. (*2*) Provide a full-scale packaging operation for a user that has elected to concentrate on research and development and marketing (by assigning the production of the product on a long-term basis to a service manufacturer, capital resources are released to product development and marketing). (*3*) Create surge capacity for seasonal and promotional production (ie, inventory control). (*4*) Expand internal packaging capacity to match sales needs. For test-market production, contract packagers provide capacity and expertise for short runs of capital- and time-intensive products without interruption of ongoing production processes. For more mature products, contract/service packagers provide the opportunity to increase production capacity without large investment in capital or personnel that might outweigh the benefit of increased sales. (*5*) Eliminate the time lag between new product idea conception and actual production. Lead time in the machine industry today worldwide is 18–36 weeks. A contract/service manufacturer will often have existing capacity which, with minor modification, can be used to produce a product immediately, allowing for rapid entry into the market place. (*6*) Promote flexibility. (Service packagers are dedicated to the ability to change rapidly to meet client needs. Machinery and services are designed for easy modification, to allow for rapid change.) (*7*) Provide resources where expandable capacity is needed to support efficient production equipment, by combining the products of many customers. (The manufacture of aerosols and semiconductor chips are two examples of high speed, high volume, high overhead production processes which are routinely contracted.) (*8*) Maintain reserve capacity to be utilized on short notice, in case of equipment failures, transportation disturbances, floods, fire, or equipment breakdowns. (Many United States firms with offshore production facilities, eg, in Puerto Rico, maintain reserve capacity in the domestic United States.) (*9*) Provide technical expertise and counsel to client companies. (In each of their areas of expertise, service packagers are experts who have seen a broad spectrum of problems over time. Working in cooperation with in-house technical personnel, contract/service packagers can be valuable resources for problem solving.)

In summary, service packagers are utilized for their flexibility and orientation to service. They are valuable partners in the production supply chain, as on-call branch plants.

The use of service packagers also facilitates financial planning. Service charges are fixed expenses, which vary only for specification changes. The use of a service packager allows a company to set its manufacturing budget and not be subject to the variances which occur in in-house manufacture. The use of a service packager also allows for the direct assignment of costs to the specific brand or product line, increasing financial control. The use of a service packager can also free up valuable corporate assets for other uses, particularly product sampling and promotion.

BIBLIOGRAPHY

General Reference

Directory of Contract Packagers, Packaging Institute, U.S.A., Stamford, Conn.

C. B. SPINNER
PCA east, inc.

CONTROLLED ATMOSPHERE PACKAGING

Modification and control of gaseous environments surrounding food products has been employed to limit their biological activity since early in this century. In commercial practice, altering the environment surrounding food products complements refrigeration to retard rates of biological and biochemical deterioration significantly. For most of its history, modified/controlled atmospheric environment has been used solely in bulk (ie, warehouse, ship hold, train, or truck) to extend commodity food shelf life and reduce spoilage and waste. Since the early 1950s, oxygen reduction within packaging has been applied to aid refrigerated preservation of processed meat, cheese, and similar food products. Sporadic ventures in the United States and more widespread empirical applications of modified/controlled atmosphere food packaging in Western Europe and Japan have taken place since the early 1970s.

Vacuum packaging of foods such as processed meats (eg, frankfurters and bacon), cured cheese and nuts represents a binary thesis: packaging in air accelerates oxidation and microbial growth; packaging under vacuum retards oxidation and aerobic microbial growth. The vacuum may be displaced by inert gases such as nitrogen in situations in which the package might collapse or the product might be damaged from differential pressure of air on the outside, for example, in packaging coffee, dry milk, hydrogenated vegetable shortening, etc.

Altering the composition of the gaseous environment which a food product contacts reduces the rate of microbiological growth and enzymatic action, reduces the rate of oxidation, and also retards the loss of product water content through respiration or differential water-vapor pressure.

Modified atmosphere (MA) implies introducing a gas that is different from normal air (eg, evacuation, nitrogen flushing, etc) as a one-time change. Controlled atmosphere packaging (CAP) means actually controlling the *total* gaseous environment the product experiences. In commercial practice, it usually means lowering the oxygen and raising the CO_2 and ensuring that O_2 does not diminish to extinction while the respiratory gases are swept out. Controlled atmosphere packaging implies constant control over all gases: O_2, CO_2, H_2O, and ethylene and other trace gases.

Reducing the O_2 content lowers the respiratory rate of living foods, such as meats, fruits, and vegetables, especially at refrigerated temperatures. Increasing the CO_2 level drives aerobic respiration reactions in reverse and thus reduces respiratory rates of fresh foods. Elevating the H_2O level also reverses aerobic respiration and suppresses respiratory rates. Simultaneously, maintaining the H_2O level high decreases losses of H_2O from the product and thus helps to retain initial food product quality. An alternative is to dry the surfaces to reduce surface microbiological activity and thus adversely influence product quality.

Suppression of aerobic respiration reduces the rate of propogation of microorganisms and also reduces the rate of aerobic enzymatic reaction. Concurrent with control is the potential problem of oxygen extinction or anaerobiosis. In respiring products, absence of oxygen drives metabolic pathways that result in off flavors and other undesirable effects. This is not uncommon in vegetables and fruit. Anaerobic conditions can also permit the growth of anaerobic toxin-producing organisms in products with a water activity level above 0.85 (eg, meats). In addition, reduction of oxygen reduces the rate of biochemical oxidation reactions such as rancidity, browning, flavor changes, etc.

Thus, controlled atmosphere is not merely the removal of oxygen, but rather the careful control of the gas(vapor) environment experienced by the food product in a conscious effort to significantly alter its shelf life. Controlled environment would represent a dynamic action in which the atmosphere is constantly altered to meet the food shelf-life demands. Although practical in bulk storage, such systems have not yet been developed in practice in packaging.

Maintenance of controlled levels of O_2, CO_2, H_2O, and other minor constituents such as carbon monoxide, ethanol, ethylene, and nitrogen oxide leads to significant reductions (2–10 times) in food deterioration rates under optimum temperature conditions. Foods such as apples, pears, oranges, tomatoes, red meats and fish are bulk-stored under controlled atmosphere conditions and conventionally distributed after prolonged storage. Lettuce, strawberries, and cherries are distributed under modified atmosphere conditions to reduce spoilage waste and water loss. Cut lettuce is packaged under modified atmosphere to achieve 10 days refrigerated storage. The package is a bag made of high EVA copolymer polyethylene film.

Primal-cut red meats are vacuum packaged in irradiated coextrusions of polyolefins and high oxygen-barrier resins, and distributed for back-room retail-store cutting and air packaging to reduce water loss and to extend microbiological shelf life. In vacuum packages, the meat color is purple myoglobin pigment not desirable at retail level. Processed meats are packaged in vacuum or inert gas flush conditions after manufacture to reduce spoilage and moisture loss. Tomatoes, bananas, and citrus fruits are commercially held under CA, including controlled ethylene, to control the ripening processes and specify the shelf life. The product may then be packaged with a residual from the CA providing shelf life benefits.

In Western Europe, the new CA/MA packaging techniques have been developed for retention of refrigerated quality of fresh food products, pizzas, some pastries, red meats, fish, fruits, and ready-to-heat vegetables. Further, extensive application of CA for packaging soft baked goods is underway in Western Europe.

The techniques are based on three parameters: high gas and water-vapor barrier packaging materials; precise mixture within the package of key gases according to the product; and maintenance of controlled temperatures.

Prepackaging retail cuts of meat has permitted the introduction of centralized meat packaging in several regions of Western Europe and a handful of chains in the United States. Generally distributed in supermarkets under refrigeration, MA-packaged retail red meat cuts have extended shelf life to up to 14 days. PVC, PVDC, and EVOH are used as oxygen barriers to make multilayered flexible or semirigid materials (see Barrier polymers). Allied with high barrier properties is the achievement of effective, reliable seals during packaging processes for both pouches and thermoformed trays (see Sealing, heat).

Refrigerated Products

MA packaging systems for food products distributed at retail level under refrigerated temperatures include the foods described below.

Red meats. High oxygen may be used to induce the desired cherry-red color. This MA packaging system is used extensively in the UK, the FRG, and Denmark. The package is usually a thermoformed base of semirigid plastic plus oxygen barrier plus heat sealant, closed with a flat flexible web also containing a high oxygen barrier plus heat sealant. Vacuum packaging, analogous to that used for primal cuts, has been introduced for red meats in both the United States and Western Europe. The product is placed on a bottom sheet or tray and a top flexible oxygen barrier is heated. The top sheet is draped over the product and assumes the product contour due to vacuum and is simultaneously sealed to the base plastic. Few shelf-life differences are reported between vacuum and high-CO_2 package environments. In the UK, Denmark, and the FRG, Multivac thermoform/fill/vacuum-seal equipment (see Thermoform/fill/seal) is used to package retail cuts of fresh red meat under high oxygen/CO_2 conditions. True centralized fresh red meat packaging of retail cuts is commercial. Combinations of 40–45% oxygen to maintain the red oxymyoglobin color of the surface and 20% carbon dioxide to suppress microbiological growth surround the products. Shelf lives of up to seven days for ground beef and significantly longer for steaks, roasts, and chops, are achieved under refrigeration. A major British retail store chain has all its retail meat centrally packaged in modified atmospheres in this manner, in semirigid thermoformed plastic bases and flexible laminations containing an antifoggant.

"White" meats. CA-packaged fresh meats such as pork, veal, and poultry are kept for seven days in atmospheres such as 50% N_2 and 50% CO_2 in sealed thermoformed plastic trays.

Organ meats. Highly perishable offal meats such as liver, kidneys, etc, may be kept under refrigeration for 6–10 days packaged with about 40% CO_2 and 50–60% O_2 for color retention. The alternative of refrigeration in air permits less than three days shelf life, and freezing is, of course, expensive.

Processed meats. Delicatessen products such as pâtés can be kept in good condition up to 21 days in MA packaging of 20–50% CO_2 and 50–80% N_2. The alternative is thermal processing, which totally alters the character of the product.

Fish. Fresh fish is especially susceptible to microbiological and enzymatic deterioration. For 4–8 days refrigerated shelf life, CA packaging with 80% CO_2 and 20% O_2 at 37°F(3°C) is used with commercial success. This system is a packaging extension of bulk distribution systems. Oxygen is considered necessary to avoid problems of growth of toxic anaerobic microorganisms indigenous to the waters.

Fruits and vegetables. Fruits, vegetables, and flowers can be kept for several days by MA packaging in 3–5% CO_2, 2–5% O_2 and 90% N_2. MA packaging of fruits and vegetables is a direct extension of bulk storage and distribution of apples, pears, and lettuce, used widely for these products in the United States and Western Europe. In bulk, the atmospheres are controlled by natural respiration coupled with scrubbing excess CO_2 and controlled leakage of O_2 or by injection of scrubbed combustion gases into the storage areas. The packaging applications are much less sophisticated. The principal method is filling into sealed polyethylene bags, which permits CO_2 and H_2O elevation and O_2 depression, with control over distribution as the oxygen-extinction-limiting factor.

Pizzas. In France, pizzas are packaged in sealed thermoformed high-oxygen-barrier plastic trays. Mixtures of 50% CO_2 and 50% N_2 (ie, no O_2) extend refrigerated shelf life up to 21 days. In the United States, pizza dough shells and premade pie crust dough are being packaged under modified atmosphere to provide extended shelf life at both room temperature and under refrigeration.

Precooked Entrees. All-plastic thermoforms with heat-seal closures are used in France for controlled atmosphere packaging of main-course foods such as beef stew. The atmospheres introduced are elevated CO_2 and no oxygen. Because the products have been cooked, relatively little change occurs over time in the internal gaseous environment. Parfried potatoes for French frying are also packaged and distributed in this manner in the Netherlands.

Coextruded high-barrier plastic laminates in film form are employed in France as inside liners of paperboard trays to permit CA within packages of entrees and side dishes. A preprinted paperboard blank is erected by machine to form the tray. A coextruded sheet liner is heated and drawn into the paperboard shell. This "skin liner" provides the barrier to moisture and gases and also helps maintain package shape. After filling, a closure fabricated from the same plastic combination is heat sealed while a CA mixture is introduced into the package. This system has been used for entrees, and with corrugated fiberboard, for large quantities of fresh red meat.

Because of the use of high-barrier materials, CA packaging is more expensive than conventional methods such as PVC-film (see Film, flexible PVC) overwrapping of foamed polystyrene trays (see Foam, extruded polystyrene). On the other hand, the changed distribution process permits retailers to buy prepackaged fresh products rather than having to prepare and package them in the store with resulting waste and labor costs.

Most CA packaging of fresh and precooked foods is being performed on thermoform/fill/seal (TFFS) systems, typically using PS/EVOH/PE for the base sheet and PVC/PVDC/LDPE or PET/PVDC/LDPE for the heat-seal closures. The combinations are generally high O_2/water-vapor barriers. TFFS systems used in Western Europe include those from Multivac, Waldner, Kraemer and Grebe (Tiromat), Tourpac, and Dixie Union. The approximate price of a machine producing 12,000 packages/hour is about $100,000, significantly higher than that of a simple manual store backroom overwrapping machine.

Nonrefrigerated Products

Soft baked goods. Considerable application of CA to soft baked goods has taken place in Western Europe. Soft baked goods include bread and sweet goods, which can further be classified into yeast- and chemically leavened, simple and iced, etc. Soft baked goods are flour-based products with highly aerated structures derived from grain gluten interspersed with hydrated starch. Sweet goods contain dissolved sugar and, usually, lipid structural laminants. Quality is expressed by masticatory properties, which in turn, are based on maintaining the moisture content and slowing an inexorable conversion of amorphous starch into the crystalline form which generates staling.

Retention of water content is crucial in starch retrogradation as well as in other elements of baked product deterioration. Water also stimulates microbiological propagation on the product surfaces. Masticatory properties depend on elevated moisture content which fosters surface mold growth. The speed of staling (one to three days from baking to perceptible onset) is a limiting factor. Elevated rh extends the time, but simultaneously generates environments suitable for mold growth. The application of chemical additives such as sorbic acid or benzoates to retard mold propogation is limited by regulation, flavor, and consumer resistance. Elevated CO_2 or markedly reduced O_2 permits high rh within packages with little or no mold growth.

The British Flour Milling and Baking Research Association has reported the scientific basis for CA packaging of soft baked goods: CO_2 above 20% exerts a highly beneficial effect on mold suppression; below 20% concentration, the effect, although present, is not as pronounced; lowering the O_2 concentration *per se* has almost no effect until the level is reduced to below 1%; neither CO_2 elevation nor O_2 reduction *per se* has a measurable direct effect on staling; ethanol at very low levels (about 0.1%) suppresses mold growth and retards staling; and the ability to employ an elevated water vapor without mold growth imparts a perception of retardation of starch retrogradation and consequent staling. The result is an extended textural shelf life.

In France, some baked bread is wrapped in polyamide (nylon) film (see Nylon) and exposed to infrared radiation to destroy microorganisms on the surface and thus permit longer shelf life. To retain crust texture, minute holes may be die cut in the packaging film to release surface moisture.

For conventional CA packaging of baked goods, two systems are used: one, by the UK's Forgrove Rose, involves a horizontal form/fill/seal (see Form/fill/seal, horizontal) flexible film wrapper with counter-current CO_2 gas flow; and the other, a Multivac twin-web thermoform/fill/seal system for semirigid materials. The systems use CO_2 blanketing; vacuum evacuation of the package followed by CO_2 back flush; ethanol injection; or partial baking to expel the air coupled with surface infrared radiation. Products benefited include white, French, and rye breads; hamburger buns; iced chocolate cakes; and partially baked "brown-and-serve" products.

Distributing bread and cake in atmospheres containing up to 60% by volume carbon dioxide has an antimold effect which increases with the concentration in the atmosphere and with reduction in storage temperature. CA differs from inert atmospheres such as nitrogen, which increase mold-free shelf life at very high concentrations only by reducing the amount of oxygen present to a very low level. In systems in which the initial concentration of carbon dioxide gas is above 50% and the gas is lost slowly during storage, the increase in mold-free shelf life is approximately equal to the time taken for the carbon dioxide concentration to fall below 20%.

Deoxidizing agents

"Ageless" is one trade name in Japan for porous pouches of oxidizing agents such as iron filings engineered to absorb oxygen in sealed packages to extend mold-free shelf life of packaged products. Deoxidizers have found a variety of uses in Japan for protection of packaged confections, rice cakes, biscuits and crackers, seafoods, oily products, cake mixes, etc., and in the U.S. for coffee. Deoxidizers compete with vacuum packaging and CA systems, because they eliminate oxygen in the package through chemical and biochemical reactions. Deoxidizers are influenced by types of food packed and by environmental conditions. The absorptive speed of deoxidizers is influenced by temperature and relative humidity within the package.

BIBLIOGRAPHY

General References

"Controlled Atmosphere Preservation in Flexible Packages," *Proceedings,* University of Delaware, Department of Food Science and Human Nutrition, Newark, Delaware, 1984.

"Controlled Atmosphere Packaging," *Proceedings,* Institute of Packaging, UK, 1984.

"Vacuum Packaging of Meat and Meat Products," *Citations from the Food Science and Technology Abstracts Data Base;* U.S. Department of Commerce National Technical Information Service, 1983.

"Controlled Atmosphere," *Food Eng.,* (June 1984).

A. L. Brody, *Proceedings of International Conference on Controlled Atmosphere Packaging,* Schotland Business Research, Princeton, N.J., 1984.

S. P. Burg and E. A. Burg, "Fruit Storage at Subatmospheric Pressures," *Science* **153,** 312 (1966).

R. B. Duckworth, *Fruit and Vegetables,* Pergamon, Oxford, UK, 1966.

G. Finne, "Modified and Controlled Atmosphere Storage of Muscle Foods," *Food Technology,* (Feb. 1982).

N. F. Haard and D. K. Salunkhe, *Post Harvest Biology and Handling of Fruits and Vegetables,* Avi, Westport, Conn., 1975.

A. Hirsch, "A New Approach to Fresh Meat Packaging is Inevitable," *Meat Processing,* (June 1975).

R. A. Lawrie, *Meat Science,* Pergamon Press, Oxford, UK, 1966.

R. Martin, *Proceedings of the First National Conference on Modified and Controlled Atmosphere Packaging of Seafood Products,* National Fisheries, Institute, Washington, D.C., 1981.

F. A. Paine and H. Y. Paine, *A Handbook of Food Packaging,* Leonard Hill, London, 1983.

S. J. Palling, *Developments in Food Packaging,* Applied Science, London, 1980.

E. B. Pantastico, *Post Harvest Physiology, Handling and Utilization of Tropical Fruits and Vegetables,* Avi, Westport, Conn., 1975.

N. Pintauro, *Food Packaging,* Noyes, New York, 1978.

D. K. Salunkhe, *Storage, Processing and Nutritional Quality of Fruits and Vegetables,* CRC, Boca Raton, Fla., 1974.

D. K. Salunkhe, and B. B. Desai, *Postharvest Biotechnology of Fruits,* CRC, Boca Raton, Fla., 1984.

D. A. L. Seiler, "Preservation of Bakery Products," *British Flour Milling and Baking Research Association Bulletin,* No. 4, 1983.

A. A. Taylor, "Extending Shelf Life in Retail Packaged Meats," *Meat Processing,* (March 1984).

F. M. Terlizzi, R. R. Perdue, and L. L. Young, "Processing and Distributing Cooked Meats in Flexible Films," *Foods Technology* **38,** (March 1984).

A. L. BRODY
Schotland Business Research, Inc.

CONVEYING

The conveyor plays an integral and irreplaceable role in modern manufacturing and packaging processes. Fundamentally, in spite of its many variations, the conveyor's function is to move objects from one point to another, whether they are to be processed in some manner or simply transported. The primary objective in the design or selection of the conveyor is that it function as efficiently and with as little maintenance as possible. Factors to consider are materials, load capacity, type of drive, speed range, and compatibility of the conveyor, not only with the objects being transported, but also with the other equipment in the production line.

The conveyor can take almost any form the particular task demands. Bucket conveyors individually contain objects and move them along an established path. Each bucket is spaced on a moving chain so that its loading and unloading is synchronized with a particular production process. Other conveyors have extending arms that are spaced on a chain and are loaded to carry larger parts or assemblies to different stages in a manufacturing process, or through a packaging cycle that makes them ready for shipment to a warehouse or market. In all cases, the conveyor must be compatible with the handling requirements of the specific product.

Depending upon the load to be carried, conveyors that handle shipping cartons can be a light-duty reinforced rubber belt; steel, rubber, plastic-coated or all-plastic rollers; or a roller chain, driven by a constant-speed a-c motor. For heavier palletized loads, a heavy-duty carbon-steel roller conveyor with a carbon-steel frame is typically used. A standard pallet load measures about 44-in. (1.12-m) wide and 56-in. (1.42-m) long and can weigh more than a ton (2000 lb). A third major conveyor type primarily employs a constant speed, a-c motor-driven heavy-duty reinforced rubber belt to transport products in loose form. Belt conveyors suitable for direct food contact are designed with FDA-accepted PVC belting. Driven by hydraulic or constant-speed a-c motors, they feature stainless steel construction to facilitate washdowns or other means of sanitizing the system.

Many industries have developed specialized conveyors to suit particular needs. Often they travel at very slow speeds. Overhead trolley conveyors, for example, are used in heavy industrial applications, such as in automotive plants and in other heavy manufacturing facilities. Drag-type chain conveyors connect under an automobile's chassis and intermittently and very slowly pull the vehicle along its manufacturing and assembly cycle.

Flat-Topped Conveyors

One of the most significant areas of conveying technology, from the standpoints of extremely wide usage and technological development, is the TableTop (Rexnord Inc.) chain (see Fig. 1). These conveyors are composed of individually connected flat slats that form a smooth surface. They function at any level, including substantial elevations, although the most common height is usually at waist level.

Flat-topped conveyors can run at very high speeds and are widely used, for example, in the filling and packaging of bottles and cans, where production requirements constantly press at the existing conveyor speed boundaries. These chains are

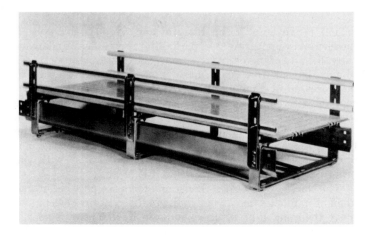

Figure 1. Flat-topped conveyor.

also used for many industrial conveying requirements, such as the manufacture and packaging of transmission components, engine pistons, and miscellaneous small manufactured parts. They can even interface with robotic manufacturing and packaging.

Flat-topped conveyors use plastic or metal chain. Plastic chain, usually an acetal resin for low friction, or polypropylene for low- or high-temperature applications, is the choice for conveying metal cans or plastic containers and other nonabrasive products. Glass containers generally are conveyed on carbon- or stainless-steel flat-topped chains. Sometimes plastic chains can be used mainly as an economy measure.

Flat chains can be straight running or can side flex, because of their individual linkage design, to go around corners. They are driven primarily by hydraulic motors or a-c variable-frequency or d-c variable-speed drives. A mechanical belt drive technique uses a pulley-and-belt method. In operation, two halves of the pulley move into and away from the belt to adjust running speed. In another approach, a variable friction drive varies the centerline between two disks to change the output speeds.

The widths of flat-topped conveyors can range from 3¼ to > 8 in. (8.3 to > 20.3 cm). Multiple strands running on a common drive establish the width for a given process. Determined by the amount of storage time required, these multiple-strand conveyors measure up to 4 ft (1.22 m) wide. Multiple sprockets on the single-drive shaft provide the input power. Speeds of the flat-topped designs range up to 500 ft/min (152 m/min).

On a high speed container packaging line (eg, 1500 containers per minute), the chains running in a single-file mode can move at rates up to 400 ft/min (122 m/min). Some high speed food packaging lines operate at 650 units per minute; for example, filling and packaging ketchup bottles. Baby-food jars on flat-topped conveyors can be packaged at a rate of 1200 containers per minute. Operating speeds are determined largely by the capacity of the filler.

Packaging technology now includes methods of stabilizing lightweight, sometimes oddly shaped containers by drawing the product down on the flat-topped chain through use of a vacuum plenum. This permits conveying the containers at much higher speeds. The same concept is applied to the operation of elevators and lowerators.

In one of the latest conveying innovations, containers are moved in mass or single file on a thin film of air. Pressurized air is blown into a plenum and small holes are drilled at an angle into a stationary plate. The containers then are supported by air pressure, with their motion over the plate induced by air coming out of the holes' raked angle. A major advantage of this low-friction system is the reduction in the number of moving parts.

Cable Conveyors

Another basic conveyor design employs a cable as the transporting vehicle. The cable conveyor is used as an inexpensive method of moving lightweight containers, primarily empty cans and plastic bottles, in a single-file mode only, at speeds up to 800–1000 containers per minute. The method is also used in the food industry for handling full containers, generally at speeds under 500 containers per minute. In operation, a ⅜-in. (9.5-mm) diameter, nylon-coated steel cable driven by a constant-speed a-c electric motor is supported on 3-in. (7.6-cm) diameter sheaves along its length. The containers are seated on top of the cable and held in position by adjustable side guide rails made of stainless steel, aluminum, or plastic.

The method provides relatively low-cost conveying, notably for conditions where lengthy distances must be economically traversed in a plant environment. A capacity for long pulls can minimize transfer points. Although about one-third the cost of flat-topped conveyors, the system cannot be used for accumulating containers.

High Speed Conveyor Technology

The beer and beverage industry is a prime example of reliance on advanced conveying and packaging technology to meet steadily increasing demands for greater speeds, production efficiencies, and marketing innovations. The developments in current high speed conveyor systems are a response to this need. Today's state-of-the-art high production conveyor systems can achieve smooth, damage-free travel of containers from delivery to final packaging. The design challenge is to sustain high production rates by integration of the various conveyor stages, in spite of any temporary interruptions in container flow that may occur at intermediate points.

High speed can- and bottle-conveyor technology is a combination of up-to-date mechanical, electromechanical, and electrical/electronic techniques, resulting in production capabilities of up to 2000 cpm and plastic or glass bottle rates of 200–1000 bpm. The high production derives from container transport and accumulation techniques combined with microcircuit-controlled, start-stop and speed modulation that compensates for intermittent container-flow variations.

The fulcrum of a high speed container packaging line is the filler station, which dictates the flow parameters of the remainder of the system (see Fig. 2). For smooth continuous production, all functions upstream and downstream of the filler must be designed to assure an uninterrupted supply of empty containers into the filler, and filled containers out of it. The conveyor system then must isolate discontinuities in container flow so that the filler will neither be short of container input from the upstream side, nor slowed or shut down because of containers that are backed up downstream. Continuous movement of the containers into and from the filler is the best indication that the conveyor system is functioning at its designed capacity.

The high-production conveying system handles primary and secondary packages. The empty primary packages must be loaded efficiently into the system and transported through the various processing functions that normally include rinsing, filling, capping, and labeling. When these steps are completed, the containers are packaged in groups, either in cases or as multipacks in trays, at the conveyor's exit end. These secondary packages then are routed to loading points for transport to the warehouse or for final shipment to the marketplace.

With current conveyor technology, a well-designed, high speed line can operate routinely at efficiencies of ≥95%. Modulation of chain speeds, and provision for accumulation at selected intervals in the line of container travel, help prevent disruptions in flow or stoppages at the filler station (see Fig. 3). Proper accumulation at stages in the system also requires consideration of the type of container being processed.

The provision of optimum conveyor "pull lengths," to limit back pressures on the containers and loads on the chains, also is a major consideration.

Back pressure is a multiple of the product weight times the coefficient of friction between the chain and product. Factors affecting chain pulls include the type of chain, container

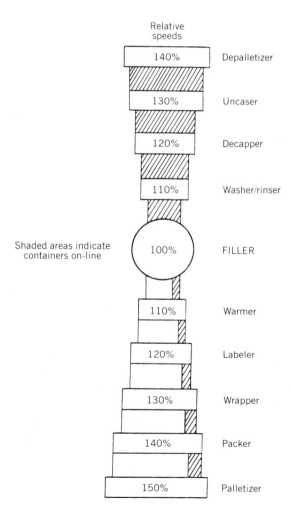

Relative speeds

140%	Depalletizer
130%	Uncaser
120%	Decapper
110%	Washer/rinser
100%	FILLER
110%	Warmer
120%	Labeler
130%	Wrapper
140%	Packer
150%	Palletizer

Shaded areas indicate containers on-line

Figure 2. Conveying sequence.

Figure 3. In-line accumulation tables help provide an uninterrupted supply of empty containers to the fillers.

weight and whether the process is dry or wet, depending on the product being handled. Smooth side transfers to and from the accumulation areas and coordination of conveyor speeds with number of lanes also are critical to the maintenance of the high production rates.

Handling Lightweight Containers

Conveyors for lightweight aluminum cans and plastic bottles must handle the containers gently, with as little container-to-container contact and pressure as possible. This is accomplished by combining the accumulation techniques and container traffic patterns with sensing and interlock devices that prevent jamming of units and assure smooth, shock-free flow, especially under push-through conditions, as in side transfers and accumulation areas.

Conveyor surfaces that reduce friction, detector techniques that maintain container-separation gaps, dimensional precision in fabrication of conveyor components, and interlocked motion controls also respond to these requirements.

The container material, type, and shape are equally important factors in the planning of the conveyor. Flow paths and accumulation areas must be sensitive to the increased jamming potential of nonround shapes. Lightweight plastic bottles, particularly without base cups, are more likely to tip over. Both of these container types might function best with vacuum conveyors that stabilize the packages by pulling them into contact with the conveyor surfaces.

Stages in Conveying

Infeed. At a high production filling and packaging line, a continuous supply of containers may be provided by multiple infeed lines. The latest designs use a programmable controller, which contains the electronic logic for operation of the solenoids, motors, clutches, brakes, and other control components to monitor the supply lines and determine the routing of container-loaded pallets to the conveyor system. Automatic pallet-handling systems continuously feed full pallets to the first station to maintain the infeed system.

The containers are brought to the production line either in reshipper cases, or more probably in a high speed line, as bulk pallets of layer-on-layer of containers. If the containers are received at the beginning of the conveying cycle in bulk loads, they normally are raked from the top down, tier by tier, onto the conveyor by a bulk depalletizer (see Palletizing). If the containers arrive at the input point in reshipper cases, they are automatically decased and usually placed on flat-topped conveyors. Depalletizing is often done at a height of ≥12 ft (3.7 m), where the containers are deposited on an elevated mass conveyor. The containers move from the mass conveyor to an accumulation bed in which they move into a single-file combiner using two fixed converging guides or a motor-driven guide. For automatic infeed of palletized loads, photoelectric sensors determine whether the containers are in position to be transferred onto the conveyor by a continuous sweep bar which could effect a straight-line discharge or a right-angle transfer.

To prevent impact shocks downstream, notably at the high-speed filler, a comparatively slow, usually double- or triple-7½-in. (19-cm) wide mass conveyor is used to spread the large volume of containers over a wider surface area. The high production rate then can be maintained and the conveyor velocity is limited to up to 50 ft/min (15.2 m/min). Experience has shown this rate to be the maximum manageable at this stage of the production line. If the containers do not continue to move forward normally as they are unloaded, the blocking of a photoelectric detector stops the depalletizer sweep mechanism until the detector circuit clears again. A lowerator positive-feed device then conveys the containers from the high discharge level to a surge table at waist level which acts as a cushion to prevent downstream shocks. The lowerator is stopped by a limit switch in the event of a jamming condition at the surge-table station.

Single filing. The maximum 50 ft/min (15.2/min) speed must be significantly increased at this point to reach the higher speeds required by the filler. This is done in progressive steps with multiple-speed step-up chains and guide configurations that incrementally increase the container speeds, as they move out of the accumulation area, to about 200 ft/min (61 m/min) on the single-file discharge end of the station.

Conventional methods of single filing containers have been limited to round units, with the volume of containers from multiple lanes being directed into a single lane with the aid of converging guide rails (see Fig. 4).

With a new computer-controlled, low pressure technique, however, nonround shapes now also can be conveyed at high speed by photoelectric monitoring of the gaps between the containers at the output end of the station (see Fig. 5). Chain speeds then are adjusted automatically to maintain the desired gap between the containers and the optimum single-file exit rate. In addition to being much more tolerant of nonround container shapes, this method is significantly quieter and assures gentler handling of less-sturdy containers.

Rinsing cycle. Before moving through the filling stage, the containers are transported to rinsers. If the containers are cans, they remain at the elevated level and are transported, single file, from the bulk depalletizer to the filler area where they go through a 180° inversion. They are then rinsed with high pressure air and/or water, conveyed to the filler, and dropped by gravity for the filling operation. If they are bottles, the rinsers may be either a push-through or captive-type inline design.

In the push-through rinser, the bottles are held between two power-driven belts and are pushed through a fitting that allows them to be rotated 180° and cleaned with high-pressure

Figure 4. Single filing with guard rails.

Figure 5. Single filing with low pressure combiner.

air and water. In situations where glass containers are hot-filled (eg, with ketchup) hot water tempers the glass prior to filling. The containers are turned back to their normal upright position, deposited back onto a single-file chain, and side transferred to a mass conveyor. The bottles are then again combined into single file and moved into the filler station.

In the alternative method, the bottles are held captive during the process. The bottles are independently gripped, inverted, washed, blown dry, and deposited in single file. The transfer is positive throughout, with no force applied to push the bottles ahead. In addition, a space is maintained between them to help ensure smooth, gentle handling.

Where plant space is at a premium, or stability of bottles is a problem, the rinser and filler can be synchronized to act in tandem. A sensing device feeds the filler's accelerations and decelerations to the rinser station's conveyor motor. The rinser's discharge then functions directly as the filler's infeed, and eliminates an accumulation requirement.

As the containers leave the rinser, they are accumulated again on a widened conveyor and again are brought back to single-file flow. In many systems, sensing devices scan the containers before they enter the filling stage to detect foreign objects or debris.

Synchronizing container flow. The containers' entrance to the filler station must be accurately timed to accommodate a difference between the container diameter and the valve spacing on the filler's rotating turret. A worm-type infeed to the filler acts as an indexing device that runs at a slightly slower speed than the conveyor that transports the containers. The worm makes contact with the containers, which are bearing against each other at the single-file outlet point, and gradually spaces them by holding back their entrance into the filler disk as the chain pulls them forward. The bearing of the containers on the roots of the rotating worm stabilizes the transition of the fast-moving containers into the filler, and as a holdback device, provides the necessary gap between the containers that synchronizes them with the valve centerlines for proper filling. Running the conveyor slightly faster than the worm provides the needed precision holdback function.

Maintaining filler input. Conveyors play a critical role at the filler. To ensure a steady supply of containers, the primary equipment upstream and downstream must be capable of operating at higher production rates than the filler itself (see Fig. 2). On the upstream side, for example, an accumulation area could become empty as its container storage feeds the filler, owing to an interruption of container delivery to the station. On reactivation of the upstream container feed, if the station were running at the same speed as the filler, its entire output would be taken by the filler, and the depleted accumulation area would not be replenished. To maintain a constant flow and efficiency at the filler, therefore, the supply of containers through the upstream stations must be increased sufficiently to refill the accumulator. This helps assure continuous filler operation in spite of a temporary stoppage.

The percentage increase in speed depends on the amount of compensation that is designed into the accumulation system. Accumulator storage of three minutes, for example, is considered adequate to compensate for temporary stoppages in a 1000 container-per-minute line. If more accumulation is needed at any point to handle the production, an off-line bidirectional accumulation table can be utilized. Photoelectric and mechanical sensors activate the table in the proper direction to maintain product flow.

Clearing filler output. A similar condition prevails, but in reverse, downstream of the filler station. The objective there is to keep the accumulation areas sufficiently empty in the event of a temporary interruption of container flow. Without compensation, in such a circumstance, filling up of the accumulation area would cause the containers to back up into the filler station. This condition is prevented by running the downstream conveyors at higher speed relative to the filler, to sufficiently empty the container storage areas and assure continued flow from the filler.

Post-filling operations. As the containers come off the filler, usually closures are applied to the containers, which move through a full-bottle detector which checks their liquid level. Any rejected units are shuttled to a side conveyor. Those containers that are filled with refrigerated liquid at about 36°F (2°C) such as carbonated beverage, and which will be labeled and/or packed in corrugated boxes must be warmed to ambient temperature to avoid excessive "sweating" or condensation. A warming cycle may not be required, however, if the containers are packed in wooden or plastic shipping crates, which can withstand moisture. Also, a "warm fill," as on some state-of-the-art beverage lines, eliminates the warming requirement. The single-file line of bottles moves back onto a mass conveyor. If a label is to be added, the bottles are again combined down to single file and enter a turret-type labeling unit, moving out again onto a mass conveyor, which can be split to feed two packing stations.

Secondary packaging. In the next stage of the conveying cycle, containers are grouped in secondary packages.

For loading on a palletizer at the exit of the conveyor system, the boxes can be mechanically manipulated in numerous ways, depending on their size, shape, and the particular equipment that is used. Sensing devices, such as limit switches or photoelectrics, however, generally are used to control the flow of cases to the palletizer to avoid jamming at the palletizer stage and loss of accumulation area upstream.

During the transport to the case-packing stage (see Case loading), a condition of either low or zero pressure, depending on the handling requirements of the product, can be designed into the system. Screw adjustment of the pressure on the conveyor rollers is a common method. Where zero pressure is desired, this can be accomplished with pneumatic or photoelectric sensors that modulate the roller drive to maintain a separation gap between the cases. Conveyor systems generally elevate the cases for the palletizing function, mainly to clear the plant floor areas by routing the conveyors overhead. Also, the relatively heavy pallet loads ultimately can be handled more readily in a gravity, rather than a lifting, mode.

The high-speed container conveyor and packaging systems are on the threshold of even greater productivity, through application of increasingly complex speed modulation, accumulation and interlocked controls.

System design. The following are factors involved in the design of a conveyor system: know the container material, shape, weight and center of gravity; establish the conveyor system's speed of operation; know the space available and the production configuration; determine the best locations for the major pieces of equipment and their interrelationships; know the location points of supply for the system; establish the routing of finished products to warehouse and/or marketplace; determine the production volume needed by the marketing department; establish the cost justification for the system; calculate the probable return on the investment; locate points of labor requirements, and most efficient use of the labor supply; establish access routes of materials needed for system operations, such as primary and secondary packaging, labeling and maintenance; know the power, air, and/or hydraulic services that are available for systems operation; as system evolves, determine the logic requirements of the electrical power and control system; and determine the system's basic pacing factors—whether the design is based on maximum process speed or on the projected sales volume.

BIBLIOGRAPHY

General References

Conveyor Equipment Manufacturers Association, *Belt Conveyors for Bulk Materials: A Guide to Design and Application Engineering Practices,* Cahners Publishing Co., Boston, Mass., 1966.

C. Hardie, *Materials Handling in the Machine Shop,* Machinery Publishing Co., London, UK, 1970.

W. G. Hudson, *Conveyors and Related Equipment,* John Wiley & Sons, Inc., New York, 1954.

L. Jones, *Mechanical Handling with Precision Conveyor Chain,* Hutchinson & Co., London, UK, 1971.

H. C. Keller, *Unit-Load and Package Conveyors; Application and Design,* Ronald Press Co., New York, 1967.

Materials Handling in Industry, rev. ed., British Electrical Development Association, Inc., London, UK, 1965.

D. K. Smith, *Package Conveyors: Design and Estimating,* Griffin, London, UK, 1972.

J. L. EAST
Simplimatic Engineering Company

CORONA TREATMENT.   See Surface modification.

CORRUGATED BOXES.   See Boxes, corrugated.

CORRUGATED PLASTIC

Traditional corrugated packaging board is paper-based and in many instances, treatment of these materials with waterproofing or chemical-resistant compounds is necessary to suit specific applications (see Boxes, corrugated). Methods have been devised to form a corrugated plastic profile, which suit the markets where traditional materials were inadequate. Theoretically, almost any plastic material can be formed into a corrugated profile, but costs can be prohibitive. In the United States market, polypropylene copolymer (see Propropylene) and high-density polyethylene (see Polyethylene, high density) are the materials in common use, although there is a small quantity of polycarbonate (see Polycarbonate) material imported for specialized outdoor applications. The board is marketed under the following trademarks: Cor-X (I.C.C. Primex Plastics Corp., Garfield, N.J.), Coroplast (Coroplast, Inc., Granley, Quebec), Corrulite (Southbay Growers, Inc., Southbay, Fla.).

Cor-X and Coroplast are extruded profiles. Corrulite is laminated from three separate sheets and has the characteristic S-shaped flute of standard fiber board. In all cases, the formability and mechanical properties are very similar, though the printability of the extruded sheet is superior. The plastic corrugated board has advantages over standard fiber board, but certainly cannot be used as a substitute for all applications. The packaging designer should consider the following opposing criteria:

Advantages	Disadvantages
Long life	Cost
Chemical resistance	Formability
Insulation	Temperature resistance
Multiple color choice	Ultraviolet degradation
Strength : weight ratio	
Waterproof	

Test data. Plastic corrugated board utilizes test data from both the fiber and plastics industries, and this can lead to confusion or, at least, lack of pertinent data. One example of this can be illustrated by extruded board, which is categorized by thickness, and not by test or flute specifications, yet laminated board is identified by $lb/1000 ft^2$ (g/m^2). It is also possible to vary the weight : thickness ratios on the plastic corrugated board. Mechanical properties of extruded PP copolymer board are shown in Table 1.

Table 2 compares the performance of plastic and paper boxes.

Printing techniques. Plastic corrugated board is supplied in various base colors by blending pigment into the plastic resin. It can be printed by screen printing or flexography (see Decorating; Printing) if the extruded board is flat enough. The polyolefins are nonabsorbent and have poor adhesion surfaces unless they are corona treated or flame treated prior to printing (see Surface modification). Minimum film thickness of ink is essential to expedite ink drying (see Inks). Therefore, ink viscosity should be approximately 10% higher than normal. Squeegee pressure is normal with the squeegee medium sharp to sharp. Halftones and transparencies are possible using direct emulsion screens or indirect photo-films on a fine monofilament fabric (245–305 mesh). When force-drying, care should be taken to keep the oven temperature below 110°F (43°C), and to prevent sharp variations in air temperature. Corrugated plastic tends to be relatively rigid and, therefore, is best printed on flat bed types of automatic and semiautomatic equipment.

Table 1. Mechanical Properties Extruded Profile of Polypropylene Copolymer[a]

Property		Value		
thickness	in. (mm)	0.157 (4.0)	0.157 (4.0)	0.196 (5.0)
weight	lb/ft^2 (g/m^2)	0.143 (700)	0.159 (775)	0.205 (1000)
impact strength[b]				
73.4°F (23°C)	lbf/in. (N/cm)	90.7 (159)	97.9 (171)	126.5 (222)
32°F (0°C)	lbf/in. (N/cm)	88.5 (155.0)	100.3 (175.7)	129.4 (226.6)
−4°F (−20°C)	lbf/in. (N/cm)	62.7 (109.8)	69.4 (121.5)	89.6 (156.9)
tensile strength[c]				
load	lbf (N)	62	68.6	88.6
yield point	lbf/in. (N/cm)	661.5	732.4	945.9
point of failure	lbf/in.^{2} (N/cm^2)	3417	3783.2	4886.3
elongation	%	166.3	165.8	165.3
compression strength[d]				
flat				
load	lbf (N)	36 (160.1)	39.5 (175.7)	51.4 (228.6)
compression	lbf/in.^{2} (N/cm^2)	9.2 (6.3)	10.2 (7.0)	13.2 (9.1)
strain	%	1.04	1.06	1.08
vertical flute				
load	lbf (N)	87.7 (390.1)	97.1 (431.9)	125.4 (557.8)
compression	lbf/in.^{2} (N/cm^2)	280.3 (193.3)	310.2 (213.9)	401.1 (276.5)
strain	%	2.37	2.52	2.71
horizontal flute				
load	lbf (N)	6.4 (28.5)	7.1 (31.6)	9 (40.0)
compression	lbf/in.^{2} (N/cm^2)	21.3 (14.7)	23.6 (16.3)	30.5 (21.0)
strain	%	1.7	1.7	1.4

[a] Tests conducted by Tokan Kogyo Co., Ltd., Japan, on extruded profile.
[b] DuPont Impact Tester to ASTM D 781-59T. Test specimen 1.9685 × 1.9685 in. (50 mm × 50 mm).
[c] Instron Material Tester to ASTM D 828-60. Test specimen 9.8425 × 0.5905 in. (250 mm × 15 mm).
[d] Tensilon Material Tester to ASTM D 695-69. Test specimen 1.9685 × 1.9685 in. (50 mm × 50 mm).

Table 2. Comparative Tests Between Plastic and Paper[a]

Test #1 Box size	in. (mm)	12.4 × 9.4 × 11.8		
		(315 × 240 × 300)		
material		*Polypropylene*		*Paper*
board thickness	in. (mm)	0.157 (4)	0.197 (5)	0.205 (5.2)
board weight	lb/ft^2	0.150 (730)	0.191 (930)	A Flute[b]
compression load	lbf (N)	485 (2157)	1455 (6472)	661 (2940)
Test #2 Box size	in. (mm)	23.6 × 19.7 × 16.1		
		(600 × 500 × 410)		
material		*Polypropylene*		*Paper*
board thickness	in. (mm)	0.157 (4)	0.197 (5)	0.299 (7.6)
board weight	lb/ft^2 (g/m^2)	0.150 (730)	0.191 (930)	double wall[c]
compression load	lbf (N)	717 (3189)	1482 (6592)	1753 (7798)
distortion	in. (mm)	0.630 (16)	0.787 (20)	0.512 (13)
Test #3 Box size	in. (mm)	15.7 × 9.4 × 8.8		
		(400 × 240 × 225)		
material		*Polyethylene*		*Paper[b]*
board thickness	in. (mm)	0.150 (3.8)		0.191 (4.85)
compression load	lbf (N)	794 (3532)		708 (3149)
distortion	in. (mm)	0.472 (12)		0.630 (16)

[a] Testron #2000, compression speed 0.472 in./min (12 mm/min). Ten samples cases at 68°F (20°C).
[b] A Flute (B-240)(B-240)(SCP-135)(B-240).
[c] Double-wall corrugated (B-240)(SCP-135)(SCP-135)(SCP-135)(B-240).

Forming methods. Standard box-making techniques can be used to fabricate corrugated plastic board. Generally, flat-bed presses using cam action or single stroke are used to die cut, score, crease or fold the material. Three-point or four-point, single-side bevel-edge rule is used for cutting. Six-point creasing is used for creasing parallel with the flutes, three-point for creasing across the flutes, to obtain a 90° bend. The packaging designer must bear in mind that the polyolefins have a "memory" and, unlike paperboard, will generally attempt to return to their previous shape. This characteristic calls for modified bending and creasing techniques, but difficulties can be overcome.

High frequency welding has been the most successful method of joining the material. Due to the nature of the polymer, glues are not generally successful; but lap joints have been accomplished using corona-treated board with silicone-type or hot-melt adhesives (see Adhesives). Metal stitching can be used, but this creates a weak spot immediately surrounding the staple (see Staples).

Conductive containers. Changes in the electronic industry over the past decade have resulted in a requirement for different packaging materials. Plastic resins have been formulated to prevent, or dissipate, a static-electricity charge which would normally build up in the material. In the past, a carbon-loaded film was printed onto the surface of corrugated paperboard, but this had a tendency to slough off easily. The sloughing rate can be reduced by dipping the entire material, but this is another area in which plastics have an advantage. If the carbon is introduced into a polymer before extrusion, the wear factor is sharply reduced and the board can be used in near "clean room" conditions. The electronic industry has requirements for conductive containers for dip tubes, kitting trays, stackable tote boxes, multitrip shipping containers, dividers, covers and lids.

NEIL FERGUSON
I.C.C. Primex Plastics Corporation

CRATES, PLASTIC

Plastic crates have been replacing wooden and wire crates for the delivery of milk and soft drinks. The concept was accepted quickly in the United States dairy industry, where wire had already replaced wood to a great extent. Direct replacement of wood for soft-drink transport has been a more gradual process. Plastic crates are made by injection molding (see Injection molding). In the United States, the resin is high density polyethylene (see Polyethylene, high-density), but in the United Kingdom and some other parts of the world the standard resin is polypropylene (see Polypropylene).

CRATES, WOOD. See Boxes, wood.

CUSHIONING, DESIGN

Some products do not require cushioning to protect them from environmental shock and vibration hazards during distribution. This is best determined through fragility assessment (see Testing, product fragility). This article deals with those items that require some degree of cushioning to be able to survive the handling and transportation environment (see

Distribution hazards). Many design decisions are based on the results of testing the cushioning materials themselves (see Testing, cushion systems). A cushion in a package mitigates the force of impact by providing space through which a body in motion is brought to rest and dissipates some of the energy of impact. Proof of the cushion design's ability to protect the product must be through laboratory testing that subjects the packaged product to anticipated "worst-case" handling and transportation hazards, such as drops and impacts. In addition to providing protection from mechanical shocks the cushion must be capable of attenuating damaging vibration inputs. This is most commonly accomplished by a cushion system with a resonant frequency different from that of critical components of the product (see Testing, product fragility). The effects of vibration must also be verified by testing, because many variables determine the actual resonant frequency of a cushion system within a shipping container. Additional testing may also be needed to prove the ability of the package to survive static and dynamic compression, heat, humidity, and other hazards (see Testing, shipping containers).

Cushion Characteristics

Although many factors are used to assess the success of cushion design, certain characteristics of the cushion are particularly important at the materials-selection stage; the material's ability to mitigate shock pulses from shock levels that will cause product damage to lower levels that will not, attenuation of damaging vibration inputs from the transportation environment, resistance to compressive creep during storage, resistance to buckling under dynamic loading, and resistance to deformation under the load of the product. Design data are available from multiple sources. The most commonly used are U.S. Department of Defense MIL-HDBK-304B (1) and test data made available by manufacturers of cushioning materials (2–4). The designer must be able to apply the test data available to the problem at hand.

Shock transmission. The ability of a material or device to function as a cushion is usually presented as a dynamic cushion curve. These curves are developed through such test methods as ASTM D 1596 (5). They provide the designer with the level of shock transmitted at various static loadings as a function of cushion thickness for a particular drop height (Figure 1). Typically the designer chooses a set of curves for various materials at a specific drop height, based on the anticipated environmental hazards. Combining the shock transmission values with the product's fragility, and the product's available bearing surface area, the designer is able to decide which cushioning materials are able to function effectively for the design contemplated.

Vibration attenuation. This may be done by choosing a cushion system having a different natural frequency than the critical components of the product, or by shifting the natural frequency of the cushion by various means, such as changing static loading or the shape of the cushions. Vibration data are developed through test methods outlined in MIL-HDBK-304B (6). The data are typically presented as transmissibility-versus-frequency plots (Figure 2). They give the designer information regarding the range of frequencies through which possibly damaging amplification of the input vibration may be expected. Due to the manner in which these data are generated a considerable shift in resonant frequency may be experienced in an actual shipping package. Because of this shift, the

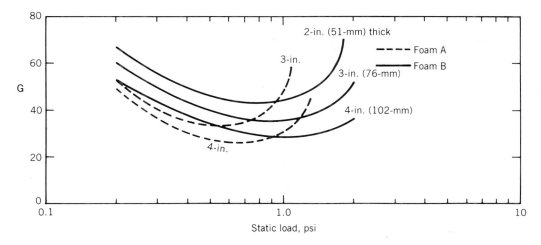

Figure 1. Dynamic cushion curve, 30-in. (76-cm) drop. G is a dimensionless ratio of specific deceleration : acceleration due to gravity. To convert psi to kPa, multiply by 6.893.

completed package should always be subjected to laboratory tests if vibration damage due to a critical component resonance is a concern.

Creep resistance. Excessive creep would increase shock transmission by reducing the thickness of the cushion. It also leads to looseness in the package, which can alter the overall ability to resist damage through loss of overall compressive strength of the package (through reduction in support of the shipping container), altered vibration response, and secondary impacts during drops. Creep data are developed to allow the designer to anticipate and compensate for changes in cushion thickness due to long-term static loading. The data generally follow a format of percent loss in thickness versus time. They may also be presented as strain: in./in. (cm/cm) versus time. In either case the designer can quickly determine if the loss in

thickness at a given static load over a period of time presents a problem. In most cases a 5–10% loss may be expected at normal loadings. In cases where the originally selected cushion thickness is minimal, this loss in thickness may be unacceptable. In these cases the designer can opt to increase the original cushion thickness by the anticipated loss.

Buckling resistance. A ratio of cushion thickness to bearing area determines the ability of the cushion to resist buckling. When buckling occurs, the product may not return to its original position in the package prior to the next impact. When using cushions with small bearing areas the designer should be aware of the possibility of buckling. In this mode the cushions have a tendency to buckle rather than compress throughout their thickness giving unpredictable results. A widely used formula for determining if there is adequate area to avoid buckling is Minimum bearing area = $(1.33 \times$ Design thickness of cushion$)^2$ (7). When the bearing area exceeds this minimum the possibility of buckling is generally eliminated.

Resistance to deformation under load. This is assessed by compressive stress-strain information, which may also be combined with shipping-container compressive-strength data to help determine stacking limits for the finished packages (see Testing, shipping containers). Although compressive stress-strain curves are available from the same sources mentioned above, their use has been limited by the availability of dynamic data. In cases where the cushion is to contribute to the overall compressive strength of the package the compressive strength of the cushion must be determined as a part of the completed package rather than relying on test data for the material itself.

Design Constraints

The design of the cushion system is determined by many factors. Some of these are described below.

Product size. The overall linear measurements of the product, such as length, width, and height, or diameter and height, affect the design of the cushion system and may help define the shipping environment.

Product weight. The design must take into account the gross weight of the product in lb (kg) and the location of the center of gravity of the product. If much of the product weight

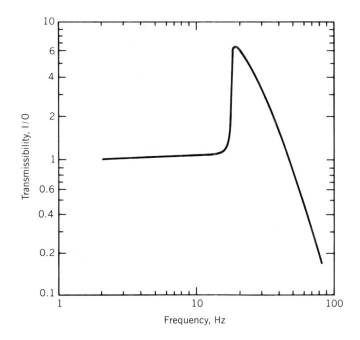

Figure 2. Vibration transmissibility plot of Foam B.

Table 1. Approximate Fragility of Typical Packaged Articles, G

Extremely fragile	
Missile-guidance systems, precision-test instruments	15–25
Very delicate	
Mechanically shock-mounted instruments and some electronic equipment	25–40
Delicate	
Aircaft accessories, electric typewriters, cash registers, and office equipment	40–60
Moderately delicate	
Television receivers, aircraft accessories	60–85
Moderately rugged	
Major appliances	85–115
Rugged	
Machinery	>115

is substantially off center, the cushioning medium will have to be distributed accordingly.

Product fragility. The designer must know what level of mechanical shock causes damage. Damage may be cosmetic, functional, or both. In any case, it represents an unacceptable change in the product. Mechanical shock fragility is sometimes expressed as the G-factor of a product, where G is the dimensionless ratio between a specific deceleration and the acceleration due to gravity. Table 1 contains values that are often used as indicators. They should not replace actual product data as determined through testing (8) (see Testing, product fragility).

Distribution environment hazards. To use the fragility information the designer must know what types of hazards are present in the shipping environment. Several sources of this information are available. Among the most useful are MIL-HDBK-304B (9) and Forest Products Laboratories Reports (10). The data presented in these sources can help the designer determine the realistic levels of shock and vibration to be expected. These data are often presented in formats such as those shown in Figures 3 and 4.

In some cases the data provided for environmental hazards may be superseded by the need to pass specific test criteria. For example, the designer may find that shipments are made via a parcel delivery service that requires test procedures set forth by the National Safe Transit Association. Other test criteria determining design drop are performance-oriented test specifications such as ASTM D-4169 "Standard Practice for

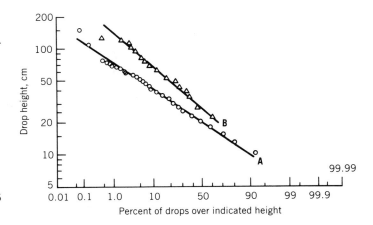

Figure 4. Drop height vs percent over indicated height. A, 43 lb (20 kg) plywood box 19 × 19 × 19 in. (48 × 48 × 48 cm), 862 drops above 3 in. (7.6 cm) (8). B, 25 lb (11.3 kg) fiberboard box 16 × 12 × 20 in. (41 × 30 × 51 cm), 312 drops above 6 in. (15.2 cm) (6).

Performance Testing of Shipping Containers and Systems" (11). This recently issued performance test provides good correlation between the type of shipping mode, degree of hazard presented, and tests needed to determine the ability to successfully ship through that environment.

Design Procedure

Cushion design proceeds according to the following general procedure.

1. Obtain product information: size; gross weight, including center of gravity; and product fragility, based on both mechanical shock and vibration.
2. Obtain information about environmental hazards/test criteria needed for "proof testing" the design.
3. Obtain other data that influence the final design, but are secondary in importance to the cushion design. Typically these might include cost constraints, storage/warehousing/handling constraints, and marketing/sales constraints.
4. From available cushion-curve data, determine which materials satisfy the shock-mitigation needs of the product.
5. From vibration-transmissibility curves, determine which of the materials selected in step 4 satisfy the vibration-attenuation needs of the product.
6. From the selections made in steps 4 and 5 calculate the size (bearing area) and thickness (amount of cushioning) dictated by the product.
7. Determine if the chosen static stress causes excessive compressive strain or undesirable compressive creep.
8. Determine the approximate cost of the cushion system. Compare this with guidelines if applicable.
9. Determine the effect on storage, handling, and transportation. See if minor modifications may be made to improve the design in these areas without affecting the performance of the cushion system.
10. Prepare a sample package for testing.
11. Subject the packaged product to performance tests.
12. Redesign or approve, and document as applicable.

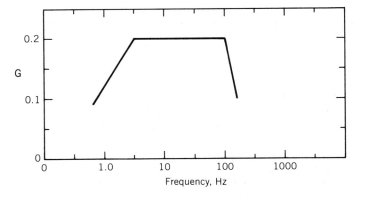

Figure 3. Typical environmental vibration envelope.

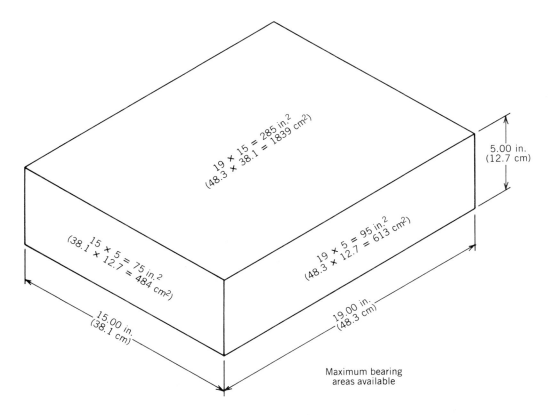

Figure 5. "Product."

Example problem. The foregoing procedure is illustrated by the following example. The product is a 19 × 5 × 15 in. (48 × 13 × 38 cm) rectangular prism with no protrusions (see Figure 5). It has a gross weight of 30 lb (14 kg).

Fragility assessment has revealed that the product is subject to damage when incurring mechanical shocks higher than 40 G in magnitude (see Figure 6).

Critical resonant frequency is 90 Hz. The company plans to produce 1000 of these per quarter. The shipping environment is primarily common carrier. One of the carriers requires that a package of this weight must withstand a series of ten drops (six faces, one corner, three edges) from a height of 30 in. (76 cm) if freight claims are to be honored.

Design Data

Product size: 9 × 5 × 15 in. (48 × 13 × 38 cm)

Product weight: 30 lb (14 kg)

Design drop height: 30 in. (76 cm)

Mechanical shock fragility: 40 G

Critical resonant frequency: 90 Hz

The next step is choice of the right foam (see Foam cushioning). Consulting cushion-curve data (see Fig. 1), the designer sees that a 3-in. (7.6-cm) thickness of Foam B transmits the required ≤40 G at static load values of 0.5–1.5 psi (3.4–10.3 kPa). With respect to Foam B, the curves show that a 2-in. (5.1-cm) thickness does not transmit a value low enough to be considered further; a 4-in. (10.2-cm) thickness transmits the required 40 G or less over a wider range, but its higher cost is not warranted. The most cost-effective cushion of Foam B is obtained by choosing the point at which the 40 G line intercepts the 3-in. (7.6-cm) cushion curve at the highest static loading: in this case, 1.5 psi (10.3 kPa). Similar information is available for Foam A. In practice, many materials may be examined using curves supplied by vendors or sources such as MIL-HDBK-304B.

The next step is the calculation of the required foam-bearing area. This is the area of the foam that distributes the static load of the product to the shipping container. The bearing area on any face is calculated by dividing the weight of the product by the static load selected from the cushion curve: in this case, 1.5 psi (10.3 kPa).

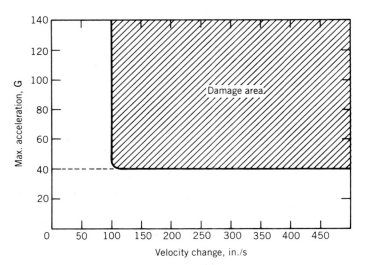

Figure 6. Damage boundary plot. To convert in./s to cm/s, multiply by 2.54.

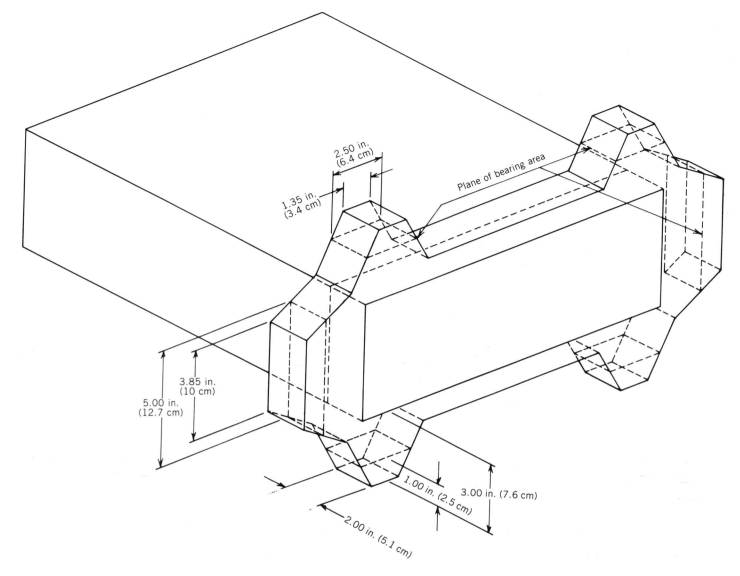

Figure 7. Die-cut cushion set [end pads not shown].

Required bearing area = 30 lb/1.5 psi = 20 in.² (129 cm²) (In this calculation allow lb to be the same as lbf).

As noted above, the static-load value could have been anywhere between 0.5 and 1.5 psi; but the use of values lower than 1.5 psi would increase the cost of the cushioning unnecessarily. For example, a choice of 0.5 psi would indicate a need for 60 in.² (387 cm²). The 20 in.² bearing area derived from the calculation above must be large enough to prevent buckling. Its adequacy is shown by the following calculation (see discussion above):

Minimum bearing area = (1.33 × 3 in.)² = 3.99² = 15.92 in.² (102.7 cm²). Figure 2 shows that the resonant frequency of the foam and its attenuation range complement the critical resonance of the product. In other words, the foam will effectively attenuate in the range of frequencies that may cause product damage.

It is good practice to anticipate how the unit will be packed and unpacked in placing the cushion pads. This gives the de-

signer insight into how much room may be needed to reach into the box and remove the cushioned product. Because of this and the fact that the product is symmetric, with its mass centered, the cushion pads are placed toward the outer corners of the product. To provide even support for the unit two cushion pads are used for each of the vertical faces of the product (front, back, left and right sides) and four pads for the top and bottom. Four acceptable solutions are described below.

Solution 1. Using die-cut slab material made from Foam B, each of the end cushions must have 10 in.² (64.5 cm²) of bearing area per pad for the front and back, left and right sides, top and bottom. The shape of the trapezoidal cushion itself is determined by noting the available material slab thicknesses of 2, 3, and 4 in. The most economical thickness is 2 in. (5.1 cm). A 2 × 5 in. or two 2 × 2.5 in. pads result. In order to place this near the outside of the product when viewed from the end or sides, the corners may be cut at 60° angles. This angle is not critical, but does allow the cushions to compress more easily at the beginning of impact, and compress more slowly as the duration of the pulse increases. At least 1 in. of material is used to form the framework holding the cushion pads together. The effective bearing area should be within the outline of the product in each axis. To make the best use of material possible and hold cost to a minimum, an attempt should be made to obtain the left and right end cushions that are heat sealed to the basic frame from the material that would otherwise be scrap. End cushions are generally available in the area cut out for the end profile of the product (see Figure 7).

Solution 2. Many rectangular products may be cushioned using simple corner pads manufactured from many materials. These may include molded EPS, die-cut foams etc. In the case of the above choice of materials we can use two types of corner pads: hinged one-piece or heat-sealed filled-corner type. The concept of bearing area applies in that each corner pad should have 5–15 in.^{2} (32–97 cm^2) available for each axis (see Figure 8).

Solution 3. One of the major drawbacks to the use of corner pads is the difficulty in packing the product without disturbing the placement of the corner pads. One of the drawbacks to die cut cushions is their indulgent use of materials. Both of these problems may be overcome by forming a die cut sheet of corrugated fiberboard into a folder around the product and bonding minimum sized cushions to the surfaces of the corrugated (see Figure 9).

Solution 4. When volume warrants the investment in tooling (sometimes as low as 50 units per week for high value products), many items can be cushioned using molded foams. These may be of several types; expandable polystyrene (EPS), polyurethanes (both self-skinned and supported by a polyethylene outer layer), expandable polyethylene, and a few new types of proprietary moldable copolymers. The last category represents the response of plastics suppliers to produce foams with the appearance qualities of EPS with a higher degree of resilience. One of the major advantages of moldable materials is their ability to reduce the amount of material needed as a framework holding the cushion pads together to a minimum. It also allows such options as full encapsulation at reasonable cost. The same rules for designing these cushions pads apply (see Figure 10).

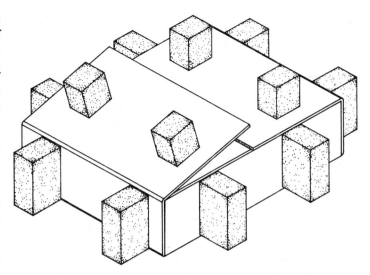

Figure 9. Foam pads bonded to corrugated fiberboard folder.

Trends

With the recent availability of moldable resilient foams, some of the high volume applications of resilient slab stock and bun foams will be assumed by these materials, in spite of their higher cost. EPS has been underused as a cushioning material, in part due to the need to meet test criteria that call for repeated exposure to high drop levels. As more knowledge of the shipping environment is developed, newer specifications such as ASTM D- 4129 require that the package be tested in a manner that more closely replicates the environment; that is, more drops at lower heights combined with specifically oriented design height drops.

Microcomputer programs that simplify the repetitive mathematics of the design process are available. Some of these programs also have mathematical models of cushion curves and recommend material selection and cost. This type of modeling can also be programmed into some of the commercially avail-

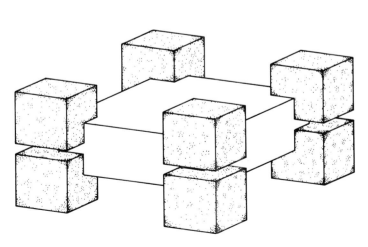
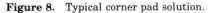

Figure 8. Typical corner pad solution.

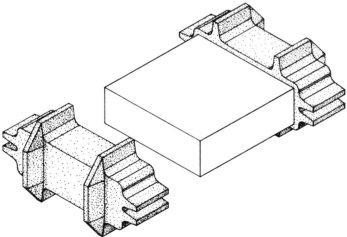

Figure 10. Molded end caps.

able formula-handling software for many microcomputers. Computer-aided design (CAD) systems with software developed for three-dimensional modeling will allow the cushion designer more freedom to experiment with cushion designs before they must be committed to expensive handmade prototypes. When this is combined with finite element modeling (FEM) software, the designer will be able to perform a portion of the vibration and shock testing with the computer model, thus discarding unworkable designs early in the process.

BIBLIOGRAPHY

1. U.S. Department of Defense, *Military Standardization Handbook, Package Cushioning Design,* MIL-HDBK-304B, 1978. Contains an exhaustive bibliography and works very well as a guidebook.

2. Arco Chemical Company, *Expanded Polystyrene Package Design,* Philadelphia, Pa., 1984.

3. BASF Wyandotte Corporation, *Styropor Protective Packaging, Properties and Design Fundamentals,* Technical Center, Jamesburg, N.J., June, 1985.

4. Dow Chemical U.S.A., *Product and Design Data for ETHAFORAM Brand Polyethylene Foam,* Functional Systems and Products, Midland, Mich., Feb., 1986.

5. *ASTM D 1596, Standard Test Method for Shock Absorbing Characteristics of Package Cushion Materials,* American Society for Testing and Materials, Philadelphia, Pa., 1984

6. Ref. 1, chap. 4.

7. Ref. 1, p. 45.

8. *ASTM D 3332, Standard Test Methods for Mechanical-Shock Fragility of Products, Using Shock Machines,* American Society for Testing and Materials, Philadelphia, Pa., 1984.

9. Ref. 2, Chap. 2.

10. F. E. Ostrem and W. D. Godshall, *An Assessment of the Common Carrier Environment,* General Technical Report U.S. Forest Products Laboratory, FPL 22, Madison, Wis., 1979. The complete series of Forest Products Laboratories Reports cover virtually all area of cushioning testing and as well as a historical perspective.

11. *ASTM D 4169, Standard Practice for Performance Testing Shipping of Containers and Systems,* American Society for Testing and Materials, Philadelphia, Pa., 1984.

12. C. M. Harris and C. E. Crede, *Shock and Vibration Handbook,* McGraw-Hill, Inc., New York, 1976. An excellent reference for the mathematics of cushion design.

ROBERT PEACHE
Wang Laboratories, Inc.

D

DATABASES. See Networks.

DATING EQUIPMENT. See Code marking and imprinting.

DATING REGULATIONS. See Shelf life.

DECALS. See Decorating.

DECORATING

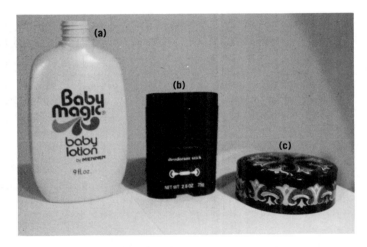

Figure 1. (**a**) Therimage heat transfer, (**b**) and (**c**) Nissha heat transfer, (**d**) and (**e**) hot stamping, (**f**) hot stamping and silk screening, (**g**) silk screening, and (**h**) pad printing.

The use of plastic containers for all types of consumer products is increasing dramatically from year to year. One of the reasons for their popularity is design flexibility. A number of resins can be processed by a variety of techniques to produce unusual shapes. Pigments (see Colorants) can provide a broad range of visual effects. In addition, plastic containers provide the package designer with many decorating options. Package graphics convey information, but they can do much more. They catch the eye of the consumer and project product quality. Metals are lithographed (see Cans, aluminum, Cans, fabrication; Cans, steel; Printing), glass bottles and jars are screen printed (see Glass container design; Glass container manufacturing), and separate labels can be used on all types of containers (see Labels and labeling). Plastic containers are decorated by heat transfer labels or labeling, hot stamping, in-mold labeling, offset printing, pad printing, and screen printing. The method depends on factors that include package shape and productivity requirements (see Fig. 1).

HEAT-TRANSFER LABELING

Heat-transfer labeling is a decorating process that permits the single-pass application of single- and multi-colored copy and designs which have been preprinted on a paper or polyester (see Film, oriented polyester) carrier by a combination of heat, dwell time, and pressure. Preprinting is accomplished by gravure printing, silk screening, flexography, and more recently, selective metallization (see Metallizing; Printing).

The more significant advantages of this process, when compared to other decorating processes, include tighter color-to-color registration, fewer holding fixtures, fewer decorating machines, lack of drying or curing ovens, fewer operators, and lower scrap rates. These advantages are primarily a result of the fact that this is a dry and indirect process of application as opposed to direct methods such as silk screening. As a result it is not subject to ink smearing and part spillage resulting from multiple handling of the individually applied passes. The tighter registration is accomplished because the side to be decorated must be fixtured only once, and dimensional variation from fixture to fixture is not a factor in color-to-color registration. Disadvantages include limited flexibility with copy or design changes owing to label lead times, and costs involved with maintaining label inventories (1) (see Labels and labeling).

Two types of labels in common use today are wax-release and sizing-adhesive. Wax-release labels are applied at linear speeds as great as 750 in./min (19 m/min). Application is accomplished through a combination of heat and pressure, with the greater emphasis on heat. This type of label system depends upon the inks and/or lacquer to form the bond to the part to be decorated. Sizing-adhesive labels are applied at linear speeds of 50–150 in./min (1.3–3.8 m/min) with application more a function of dwell time and pressure. Sizing-adhesive labels depend upon the sizing, which is heat reactive, to form the bond to the part to be decorated. The bond with sizing-adhesive labels is usually much stronger than the bond that can be achieved with wax-release labels.

All labels consist of a series of coating layers which are deposited or printed on a paper or polyester carrier web. Paper carriers are generally used with wax-release labels and polyester carriers are normally used with sizing-adhesive labels. Paper costs less than polyester, but polyester offers a significantly finer print surface as well as resistance to tearing during application. The first coating layer is commonly called the release coat, which allows the subsequent layers to break cleanly from the carrier. The second layer may include a protective lacquer coating as well as the actual decoration. The final layer normally includes a bonding or adhesive coating that is formulated specifically for the substrate to be decorated (2). Since wax-release systems depend on the inks and/or lacquer to form the bond, a separate adhesive coat is not required (3).

As an adjunct to developing a bond to polyethylene and polypropylene substrates, most label systems require an oxidized decorating surface. This can be accomplished by flame treatment or the corona-discharge treatment methods (see Surface modification).

The selection of appropriate coatings is done by the label manufacturer. It is important, therefore, to provide the manufacturer with as much information as possible, such as durability testing requirements and samples of the exact substrate to be decorated.

Transfer is achieved when heat and pressure are applied to the unprinted side of the carrier, causing the inks and/or adhesive coatings to become fluid and mechanically bond to the part. As the pad or roller is separated from the label and item being decorated, the carrier cools and the inks and/or adhesive coatings bond to the substrate.

The choice of the equipment used for application depends on the release coating that is used. Hot stamping machinery is widely used as the means of applying sizing-adhesive labels. Equipment exists for roll-on as well as direct-hit application.

Specialized machines and handling equipment have been developed specifically for use with wax-release label systems. These specialized machines feature modular tooling systems. When combined with the relatively high decorating rates, this becomes the most flexible and widespread system in use today.

Registration or positioning of the copy on the carrier in relationship to the part to be decorated is accomplished by use of a metering-hole system which is prepunched in paper carriers during printing or by a reference mark that is normally printed outside the copy area for detection by a photoelectric label advance system during decoration. Registration can vary from ± 1/64 to ± 1/16 in. (±0.4–±1.6 mm) depending on the registration system used, the accuracy of the tooling and machinery, and the dimensional consistency of the parts to be decorated.

Recent developments. Several new developments in the area of heat-transfer technology offer a much wider range of options to packaging designers. One is a sizing-adhesive-type label manufactured by Nissha Printing Co. Ltd. of Japan. This label, which offers a high-gloss, scuff-resistant finish, not only includes the effects achieved through gravure printing, but also the highly desired mirror-metallic effect of hot stamping. The mirror-metallic effect is made possible by selectively metallizing the carrier with the desired copy or design. Other features are also available such as tortoise-shell, woodgrain, or marbleized finishes. An additional advantage of this process is that surface oxidation of polyethylene and polypropylene substrates is not a requirement for developing satisfactory label adhesion. The combination of these special effects along with the exceptionally tight color-to-color registration that this label offers can change the entire image of a product (see Fig. 1).

Another development is offered by Dennison Mfg. Co., a pioneer and leader in the field of heat-transfer technology. This innovation combines the pad-printing process with gravure-printed transfers. This process, which uses a flexible heated silicone pad, can be used to decorate sculptured or recessed surfaces. The preprinted carrier is routed past a preheater to a heated platen. The silicone pad presses against the carrier allowing the heat-softened label to adhere to the pad. It is then pressed against the part to be decorated where it bonds upon removal from the pad (4).

A third breakthrough in transfer technology is a universal decorating machine offered by Permanent Label Machinery Corp. for printing items of noncircular cross section.

The Acrobot as modified for heat-transfer application is a patented, computerized machine that allows 360° decoration on noncylindrical parts. Shape tables which are stored on cas-

Figure 1.   Nissha heat-transfer labels for decorated plastic tubes.

sette, disk, or ROM are used to control the various axes of motion during application. This machine has opened up a new dimension to package designers by removing the traditional copy design and layout restrictions intrinsic to noncylindrical shapes. By accurately controlling label application velocity and maintaining tangential contact of the surface to be labeled to the print roller or pad, multifaceted or elliptical shapes can be continuously labeled in a single pass as if they were cylindrical. Prior to the advent of this machine, these shapes could only be considered by the use of exotic and expensive tooling and machinery that was designed specifically for the shape to be labeled (5).

BIBLIOGRAPHY

1. W. S. Anderson, Jr., *The 1983 Plastics Design and Processing Manual,* Lake Publishing Corp., Libertyville, Ill., 1982.

2. S. Glazer, *Modern Plastics Encyclopedia,* McGraw Hill Inc., New York, 1983.

3. W. La Voncher, *Decorating Plastics RETEC,* Society of Plastic Engineers, Cherry Hill, N.J., 1983.

4. "Toiletries Packagers Warm up to Heat Transfer," *Packaging,* 60–61 (Feb. 1984).

5. U.S. Pat. 4,469,022 (Sept. 4, 1984), N. E. Meador (to Permanent Label Corporation).

DOUGLAS H. CONTRERAS
Permanent Label Machinery Corp.

HOT STAMPING

The term hot stamping, as applied to decorating, is a very broad category. One definition of hot stamping is the application of characters and/or designs to a surface through the use of heat and pressure to press a web (foil) onto the surface for a finite period of time. The web is removed leaving the desired characters and/or designs permanently attached to the surface.

The process known as heat transfer would appear to fall under the above definition, but there is a difference. In hot stamping, the decoration is on the hot die; in heat transfer, the web is preprinted by various processes and then registered to the surface to be decorated. Heat transfer and hot stamping have many common characteristics and there is some overlapping of techniques.

Materials. The web (foil) has several layers of material which are built up by successive application of ink and/or coatings in one or more passes (see Fig. 1a) (1). Each layer has a special function (2). The manufacture of hot-stamp foil is a highly technical and difficult process, and successful hot stamping depends on selecting the proper foil for the particular job. The material to be printed (substrate) and the kind of artwork (fine or broad line) are the primary factors in selecting the correct foil.

Polyester is the most common carrier material in thicknesses of 0.0005–0.001 in. (13–25 μm) (see Film, oriented polyester). The thickness, which affects the printing characteristics, is chosen based on the end process. Other carrier materials, such as cellophane, are also used. The release coat, which can be clear or a translucent color, also controls the speed and printing characteristics of the foil. The terms *tight* and *loose* refer to how the print medium transfers from the carrier to the item to be decorated: a tight foil, used for fine line copy, breaks cleanly at the edges of the die; a loose foil, used for broad areas, tends to bridge over small surface imperfections in the die or the part. Foils are available over the complete range from loose to tight. The protective lacquer is the most important factor in both product resistance and abrasion resistance since it is the outermost part of the decoration when applied to the part. For a mirror-metallic effect, a tint can be added to the protective lacquer to get the exact metallic color desired.

The print medium can be either a continuous color or a reflective metallic layer. If a continuous color is desired, it is generally applied in one or more gravure-type-coating steps. The thickness of all the various coated layers is critical to proper printing performance. In vacuum metallizing for a reflective metallic layer, a thin coating of aluminum is deposited on the transparent (or translucent) protective lacquer (see Metallizing). This results in a mirrorlike surface which is still somewhat flexible. The sizing, a heat activated adhesive, is the layer which sticks to the item to be decorated. It is very important to consider the substrate to be printed before a foil sizing is selected. The foil manufacturer is the best source for suggesting what foil will adhere to a particular substrate. Technicians generally acknowledge that the only sure way to determine if a particular foil works on a particular substrate is to try it.

Methods. There are two general methods of hot stamping: direct stamping and roll-on stamping (see Fig. 1b and 1c).

Direct stamping is the process of making contact with the entire area to be printed at one time. The time of contact, called dwell time, is generally controlled by either a timer or a pressure switch. The die contacts the entire part and after proper dwell, leaves the entire part as one motion. With this method, the die is contoured to match the part to be printed (including flat).

Roll-on stamping is the process of making line contact between the part and the die by rolling the part across the die

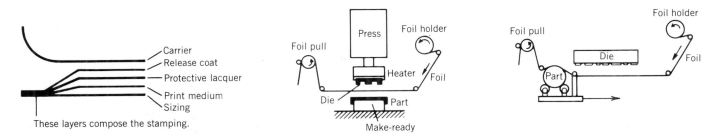

Figure 1. Hot stamping. (**a**) Hot-stamp foil contains many layers. (**b**) Direct stamping. (**c**) Roll-on stamping.

surface. The dwell time is controlled by the deflection of the die (or part) and the speed of traverse from the beginning to the end of the print. The sketch shows a flat die; it can also be curved or even a continuous round. In most cases, foils that are made for direct hit stamping are not good for roll-on stamping and vice versa.

Before making the selection of a foil, it is necessary to know what material is to be used for the die. The two most common die materials are metal and silicone rubber. Any material can be used if it can be molded, etched or engraved, and then heated repeatedly without distorting. Many factors must be considered including the length of run, detail of the printed copy, hardness and smoothness of the substrate, and the skill of the technician who sets the machine. Different printed effects can be obtained by choosing a harder or softer die. The surface finish (degree of polish) and heat conductivity of the die are also significant factors.

The hot-stamp machine is the most important factor in the hot-stamp process. The important factors in selecting a machine are rigid construction and accurate controls. There are many good hot-stamp machine manufacturers, and most make a variety of sizes. The maximum clamping force needed and the size of part that can be handled govern size selection. As the size increases, the power required to run the machine increases, and the maximum speed of operation decreases. The clamping force required is directly proportional to the surface area being printed. Metal dies require more force per area transferred than rubber dies. Roll-on stamping can be done with low clamping forces because the area being transferred at any particular instant is small.

The hot-stamp process is becoming increasingly popular for the decoration of plastic parts because of the advantages which are unique to this process. Hot stamping is a dry process; that is, no wet ink or post-treatment (drying) is required. The printed part is complete as it leaves the print station. Hot stamping can provide a mirror-metallic print; and the addition of simulated gold or silver adds marketing appeal. Recent developments in the foils, dies, and the machines to do hot stamping are making it possible to hot stamp items which could not be decorated this way just a few years ago. Hot stamping is a developing technology which can be used to advantage on high-speed production lines. A recently patented, computer driven hot-stamp machine (3) now permits 360° decoration on noncircular parts without the need for costly gears or cams.

BIBLIOGRAPHY

1. L. Kurz, *Application Techniques, Hot Stamping Process*, D-8510, Furth, FRG, Aug. 1981.
2. A. Panzullo, "Hot Stamp Foils Update," *paper presented at the Regional Technical Conference of the Society of Plastics Engineers*, Cherry Hill, N.J., Oct 5, 1983, pp. 39–44.
3. U.S. Pat. 4,469,022 (Sept 4, 1984), N. A. Meador (to Permanent Label Corp.).

ARTHUR C. PECK
Permanent Label Machinery Corp.

IN-MOLD LABELING

In-mold labeling (IML) is a relatively recent addition to the category of predecoration that is available for blow-molded

Figure 1. A fabric-softener bottle decorated by in-mold labeling.

plastic bottles. The first commercial shipments of in-mold labeled HDPE bottles in the United States were made by Procter and Gamble (P&G) in February 1979.

The driving force that caused P&G to seek in-mold labeling as a commercial process was economic. None of the other predecoration techniques provided the color and printing fidelity of die-cut, rotogravure printed, paper labels at comparable unit cost (see Labels and labeling). The basic in-mold label system utilizes a paper label that has a heat-activated coating on the back side. The label is placed in the mold at the beginning of the blow-molding cycle before the mold closes around the parison (see Blow molding). The label is held in position by vacuum ports. When the parison of plastic is extruded and the molds close and air is injected to inflate the parison, the heat from the parison activates the label adhesive coating. During the mold-close cycle, the combination of the high inject air pressure and the cold-mold surface causes the adhesive to set and the label to adhere to the bottle surface in a heat-seal fashion. Selection of a heat-activated label adhesive system is a critical step in the IML process. Ideally, the adhesive should be compatible with an in-line printing operation so that the label graphics and adhesive system can be accomplished in one pass through the printing press. The activation temperature of the adhesive is also critical to the success of IML in that it must activate at a temperature low enough to have initial tack when activated by the hot parison. The reactivation (release) temperature is also important: it must be lower than the softening point of the bottle; allow label removal for plastic bottle scrap recovery systems; and high enough so that it is not reactivated by hot-fill product systems or elevated temperatures encountered in transport or warehousing.

Most of the initial work conducted suggested that the bottle shape and the size/shape of the label would have a bearing on the initial success of the system. On bottle and label combinations, where less than satisfactory results were obtained initially, adjustments in bottle material distribution and/or a combination of plastic distribution and label configuration have usually resulted in acceptable labeled bottles.

Some disagreement exists within the industry as to whether or not IML permits bottle weight reduction without a compromise in impact and top load resistance. The IML bottles evaluated at P&G meet the non-IML specifications at lower gram weights. There is also some disagreement about the effect of the IML process on cycle time. The P&G bottles produced via the IML process are run as fast as, or faster than, non-IML bottles. To date, both reciprocating and continuous motion (wheel type) blow molding machines have been successfully fitted with IML labeling mechanisms for use with HDPE bottles.

Die-cut paper labels were used in the initial IML development. It is reasonable to assume that labels cut from a continuous roll or labels fed from a transfer web could also be utilized if proper labeler transfer mechanisms are developed that would be compatible with the blow-molding machines.

To date, although almost no published data exist, it is reasonable to assume that laminate label structures and nonpaper label materials can also be utilized in the IML/blow-molding process. The effects that these label materials and structures may have on the finished blown plastic item is unknown.

The concept has already been broadened to include small PET liquor bottles. It is assumed that PVC bottles could also be decorated via IML by using an adhesive system that would adhere to PVC and activate at the molding temperatures of the bottle resin.

IML label systems will be developed that will dramatically change the design of plastic bottles. The IML process already utilizes label sizes and shapes that are not compatible with conventional paper label/waterborne adhesive labeler systems. In addition, it is reasonable to assume that IML label systems will be developed that will be compatible with food processing systems. This will improve the potential for plastic packages as replacements for traditional glass bottles and metal cans.

BIBLIOGRAPHY

General References

D. C. Beckmann, "Pre-decorating of Blown Plastic Containers in Mold Labeling," *TAPPI 1983 Synthetic Conference*, Sept. 26–28, 1983, TAPPI, Atlanta, Ga., 1983, pp. 487–491.

M. J. Roman, "Prelabeling Trends in Plastic Bottles," *Soap Cosmet. Chem. Spec.* 55 (Sept. 1984).

C. Lodge, "In-mold Labeling: Next Value Added for Blow Molders," *Plast. World,* 45 (Aug. 1984).

A. Brockschmidt, "In-mold Labeling and Decorating Technology Gains in Sophistication," *Plast. Technol.* 29(3) 23, 25, 27 (March 1983).

A. Brockschmidt, "In-mold Labeling of Bottles: Is it the Way to Go?" *Plast. Technol.* 30(13), 63 (Dec. 1984).

JACK SNELLER
The Procter & Gamble Company

OFFSET CONTAINER PRINTING

The dry offset process is the most satisfactory method for the high speed, large-volume printing of multicolored line copy, halftones, and full process art on preformed containers (see Table 1). Dry offset is used primarily to print products such as tapered cups, tubs, and buckets as well as tubes, jars, aerosol and beverage cans, bottles, and closures (lids, can ends, caps).

Dry offset printing is similar to offset lithography in that a rubber blanket is used to carry the image from the printing plate to the container surface (see Printing). As in letterpress, the image area is raised above the surface of the plate. Ink is distributed through a series of rollers onto the raised surface of the plate (see Inks). The plate transfers the image to the blanket, which then prints the entire multicolor copy taken from one to as many as six plate cylinders on the container in one operation. The "dry" denotation of this offset system serves to differentiate it from the offset system which uses the incompatibility of water and inks to dampen the surface of the plate or substrate to prevent ink transfer.

Every dry offset-printing system contains two distinct sections: the print section, where the high quality image is reproduced on the printing blanket; and the material handling section, where the container to be decorated is prepared and then positioned for contact with this blanket for the act of printing.

Print section. One to six colors can be printed in a single pass over the container, with all colors applied simultaneously by the same blanket (see Fig. 1). The inking station on each color head contains the fountain and the individual set of rollers (A–D). The number of rollers and their arrangement guarantees the finest, most uniform distribution of ink to the individual plate cylinder (E). Ink laydown on the container is so minimal that, on average, one pound of ink gives coverage on approximately 150,000 in². (or 213 m²/kg ink).

The offset blanket is inked in turn by each plate cylinder, each plate cylinder having all of the copy for one color. The plate cylinder holds the printing plate, which is secured to the cylinder by clamps or through magnetism. Each plate cylinder has very fine micrometer-adjustment: 0.001 in. (25.4 μm) in any direction. The inked printing plates (E) deposit their image in sequence and in registration on a common printing blanket (F). The printing plate is prepared by a photochemical process, providing extremely fine reproduction of delicate artwork. The paste type ink used in this process, ultraviolet or conventional, allows wide latitude in choice of colors that are resistant to scuffing and moisture.

The blanket platen holds the rubber printing blanket, which is secured through the use of either "sticky back" material or ratchet clamps. The blanket transfers all of the images and copy to the container in one pass. Various blanket materials and thicknesses are available for varying printing requirements on different containers.

Material-handling section. The container shape and its tolerances are important, not only for the mechanical nature of its handling, but also for the quality of print that can be transferred to it.

The material-handling section of the decorating line is customized for the individual style container. The indexing or constant motion turret holds the specific container tooling, commonly called mandrels (or spindles), during all operations including printing. The material-handling section or the individual container tool can be swiveled for container taper, and

Table 1. Selected Vocabulary Used in Connection with Dry Offset Printing.

Halftone	The printing plate gives the illusion of tones of one color by means of a dot pattern or screen.	*Photo-engraving*	Printing plate produced by photochemistry for relief printing.
Makeready	Adjustment of printing pressure of plate to blanket to accomplish better contact, or more controlled, and therefore higher quality printing.	*Process color*	Duplication of full-color original copy by means of optically mixing two or more primary colors to create another. Black is used for highlighting.
Mandrel	Workholding tooling for material handling of container (called *spindle* in tube decorating).	*Reverse plate*	Printing plate which gives the effect of white letters, which are really the white of the container, with the plate "printing" the background color. This is the way to compensate for dry offset's inability to print white ink on a colored surface.
Offset	Printing process where image is lithographed or letterpressed onto rubber printing blanket, from which it is transferred to paper or container surface.		

positioned for printing pressure. Because of the pressure required for printing, containers must be supported by a mandrel or inflated with air pressure.

Containers can be cleaned by rotating them in front of a vacuum destaticizer which uses ionized air, or with surface treatment, which is accomplished by a flame burner head system or an electric corona discharge (see Surface modification). When required, surface treatment promotes the adhesion of ink to the substrate. The use of flame treatment has the added advantage of producing a warm container surface, beneficial in both transfer of ink from the blanket to the container and in

cure reaction time in ultraviolet drying. When required, stations for orientation are provided which use a container lug, side seam, or bottom notch to register the container prior to the print station. After printing, the parts pass through the print curing unit (for "on mandrel" drying) or are transferred to a conveyor for transport through an appropriate dryer oven (ultraviolet or infrared) (see Curing).

Printer production speeds are high: tapered plastic containers such as yogurt cups are handled by indexing equipment up to 400 parts per minute; constant motion two-piece beverage decorating equipment operates at up to 1200 cans per minute.

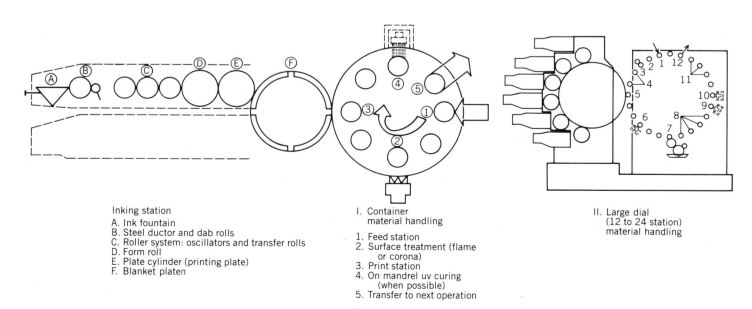

Inking station
A. Ink fountain
B. Steel ductor and dab rolls
C. Roller system: oscillators and transfer rolls
D. Form roll
E. Plate cylinder (printing plate)
F. Blanket platen

I. Container material handling
1. Feed station
2. Surface treatment (flame or corona)
3. Print station
4. On mandrel uv curing (when possible)
5. Transfer to next operation

II. Large dial (12 to 24 station) material handling

Figure 1. Dry offset container printing.

Auxiliary equipment for automatic operation such as feeders, take off devices, and dryer ovens are all important adjuncts to an offset-printing system for containers. No system is universal. Most are designed for size changeover among product lines or families of containers, but few permit the modification required to handle various types of containers (eg, jars, tubes, and bottles).

The state of the art. Ultraviolet-curing technology has been a leading factor in the progress made by the dry offset process over the last 15 years. The first ultraviolet dryers were conveyor type, giving extra curing time with existing material handling systems. The improvement of the uv-drying systems led to the realization of the true potential of ultraviolet: the "on mandrel" curing of the decoration while the container is still on its respective mandrel, prior to transfer. Due to the reduction in container handling and the elimination of many of the transfers required by conveyor systems, production rates and overall efficiency rose dramatically. For example, an 8 oz (237 mL) yogurt container that was printed and dried at 300 containers per minute using a uv-conveyor system can now be "on mandrel" cured and handled at speeds up to 400 per minute.

Another important improvement is in the quality and detail of the copy. Most of it is multicolor line copy, but the use of intricate halftones and full four-color process work is increasing. That customers are demanding this quality of work is certainly the impetus; that it is being accomplished is the result of improvements in printing presses, color separation and platework, ink technology, and printing-room personnel. Printing presses use advanced technologies, not only in their electronics (eg, programmable controllers) but also in printing adjustment capabilities and operator controls. Platemaking companies have improved the color-separation techniques used to accommodate the dry offset process, and they have added photographic makeready systems to help print more intricate artwork. Most importantly, for without it all of the above improvements amount to nothing, the pressmen on the job floor have learned to implement this technology into the everyday workplace.

For the dry offset process, the near future seems to hold a continuation of forces already in progress. Increased electronics, used in controlling and monitoring the mechanical printing process, involve larger, more powerful programmable controllers. Production speeds have continued to increase as a result of this redesign of the printing presses.

Although new container designs and their continuing weight reduction sometimes slow the progress to higher production speeds, at some point the limitation of mechanical indexing equipment will be met. Whether plastic containers, such as the metal beverage cans, will embrace the continuous-motion material-handling systems that have propelled can decorating to speeds up to 1200 containers per minute remains to be seen.

BIBLIOGRAPHY

General References

Decorating Polyolefin Molded Items, U.S.I. Chemicals Corporation, Division of National Distillers and Chemical Corporation, New York, 1980, pp. 12, 13, 20, 21, 27, and 28.

R. J. Kelsey, "Choosing the Right Decoration Process," *Food and Drug Packaging,* 10 (Sept. 1983).

"Processing Handbook," *Plast. Technol.* 247 (mid-Oct, 1968).

Wilfag/Polytype Information Review, Wifag Research Department, Berne, Switzerland, May 1974, pp. 34, 40.

E. C. Arnold, *Ink on Paper 2,* Harper and Row, New York, 1972 and G. A. Stevenson, *Graphic Arts Encyclopedia,* McGraw Hill, New York, 1968. These are two fine books on printing in general. No specific work on offset container decorating exists.

Technical papers on finishing process, Association for Finishing Process of SME (Society of Manufacturing Engineers), One SME Drive, PO Box 930, Dearborn, Mich., 48128.

CHARLES SIMPSON
American Production Machine Co.

PAD PRINTING

Pad printing, the unique process for decorating three dimensional objects, was invented hundreds of years ago in Europe to print watch-face plates.

The printing plates were made of steel and engraved by hand or by the use of a pantograph.

The transfer pads were made of gelatin. They were very fragile and sensitive to humidity, and therefore limited in their durability. A good operator was able to produce 25–30 good parts per hour, and from a production stand point, that was a ridiculously small amount. All these factors limited the pad-printing process and restricted its further commercial and industrial use.

The first step toward large-scale use of the process was made when Tampoprint/Germany developed the silicone rubber pad as it is used today. Then related items were eventually developed: printing plates, inks, and special machines.

The process. Pad printing is a deceptively simple process, especially if one discounts the highly developed parts handling system that is often part of the equipment. The basic machine consists of a parts holding device, a soft, oddly shaped silicone pad which picks up the ink image from an engraved, flat steel plate called cliche. The cliche is engraved with the design to a depth of approximately 0.001 in. (25.4 μm). All the processing steps (except the placement of the substrate) are automatic and generally operator independent. The pad-printing printers usually operate with a reciprocal motion of the pad, which alternately picks up and delivers the image to the substrate, and an ink spatula keeps the cliche filled with ink.

Rotary-pad printers, used for decorating round objects, which can turn around a given axis, operate with a gravure-like small cylinder, and the design is transferred by a rotating pad.

The events that take place during pad printing in their sequence of operation are as follows:

In pad printing, which is often called an indirect gravure process, ink is deposited on an etched metal plate (cliche), where it fills the etched portions of the plate with ink and is cleared from the non-etched portions by a "doctor blade". The etched portions represent the imprint.

The ink is picked up in total by a silicone pad, which positions itself over the item being printed, descends and deposits ink on the substrate. The pad is so soft and flexible that it conforms to almost any shape and is so resistant to ink that it deposits every bit on the substrate, leaving it clean and ready for the next printing cycle.

Multicolor work is done in the usual sequence, one color at a time, but with the great potential advantage of printing wet on wet.

Tools of the Trade.   Pad printing utilizes the following components:

Cliché. The most conventional element in pad printing is the cliché, ie, the engraved printing plate. Like any gravure cylinder, it is produced by photographic etching. If the plate material is steel, as it should be for long runs and extremely fine details, it is coated with a photosensitive emulsion, exposed and etched with fairly conventional equipment.

There are also photosensitive polymers (plastic cliché, express cliché) available that can be easily processed by the users of pad-printing equipment. These "plastic materials," similar to the ones used in flexography and letter-press printing, provide nearly the same quality as steel, with the exception of reduced production life cycle.

Inks. The inks used in pad printing are closely related to the inks used in screen printing. Pad printing also borrowed a wealth of inks from screen printing which have been reformulated for the various substrates, with slight modifications in pigmentation and tack. Higher pigmentation is necessary to compensate for the limited thickness of the ink deposit common in pad printing. The added tack assures the unique transferring capabilities of the process.

Pad. The silicone pad, equivalent to a printing plate, is made in several durometer sizes and geometrical shapes. The durometer of the pad is directly proportional to its life and inversely proportional to its wrap-around capabilities on complicated three dimensional forms. In addition the durometer also determines the amount of contact force necessary to transfer the design. Higher durometers require higher pressures [ranges are available from 10 ozf. to 500 lbf./in.^{2} (4.3–3400 kPa)]. The size of the pad is determined by the size of the image to be transferred and its shape is dictated by the shape of the substrate.

There are more than 400 shapes and sizes available to accomplish the necessary roll-off effect to release the image.

Physics.   The softness of the pad may explain the contour-following capability of the process, but it does not reveal the reasons for its flawless transferring capability. Anyone who has tried to use a rubber stamp to transfer a design manually would attest to the fact that the transfer is usually severely distorted, with extremely poor ink distribution. In addition, there is a curious paradox in the ink transfer: the ink is tacky enough to adhere to the pad in the first place, although it separates completely from the pad and adheres to the substrate.

As the cliché is wiped with the doctor blade during pad printing, the ink surface within the engraved areas is exposed to the atmosphere and changes rapidly due to the loss of solvents. The surface becomes more viscous and tacky than the rest of the ink layer. The smooth surface of the silicone pad has a unique property: a high critical surface tension relative to most of the inks used for decorating. Because of this, initial wetting of the pad by the ink is easily accomplished when the pad is pressed against the cliché. Although the ink has very weak lateral adhesion to the pad (it can easily be wiped off at any time) it has sufficient vertical strength to stay with the pad as it lifts off the cliché.

Depending on the cohesiveness of the ink, only a portion (approximately 75%) of the ink leaves the cliché. The cliché retains the ink due to adhesion and the vacuum created by the upward motion of the pad. In pad printing just as in gravure printing, the ink deposit is much more dependent on the cohe-

siveness of the ink than on the depth of the engraved design. The ink on the surface of the pad rapidly changes its properties, and it loses solvents from the exposed surface becoming tacky and viscous; but it rewets the interface between the pad and the ink, reducing the adhesion between them. When the pad is moved over to and pressed down on the substrate to be decorated, the adhesion is greater between the substrate and the ink than between the pad and the ink.

In pad-printing inks, there is a delicate balance among the solvents, the drying, and the process.

The most frequently occurring problems in pad printing are either that the ink is not picked up by the pad or not released from it. To correct these problems, air is sometimes blown across the surface of the cliché (increasing the tackiness) to promote the ink's adhesion to the pad, and sometimes air is blown on the pad to promote ink adhesion to the product. In either of these cases, too much air causes the ink to dry so much that it does not transfer to either the pad or to the product. The fast drying characteristics of pad-printing inks make it possible to print wet on wet.

The thin deposits of ink, coupled with fast drying, allow for four-color process printing, usually without the need for any drying system between various colors.

Even with the appropriate "switching" in adhesion, the ink transfer occurs because of the unique properties and geometrical design of the pad. As in any printing process, the key to successful ink transfer is the point-by-point or line-contact separation between the substrate and the printing plate. In screen printing, for example, it is accomplished by off contact; in gravure, through the tangential contact of the substrate and the printing cylinder. In pad printing, the shape of the pad is designed so that it rolls away from the surface by means of a geometrical deformation. Ideally, no portion of the pad presents a flat surface to the substrate that could separate as an area rather than a line. In other words, the contact angle between the pad and the substrate is never zero within the image area.

The zero angle is excluded because the pad is a solid surface; any entrapped air between it and the substrate will prevent ink transfer. Since pad printing is used almost exclusively for three-dimensional decoration, changes in the angle of the printed surface could result in air entrapment, unless careful consideration is given to the design of the pad's shape. There are hundreds of differently shaped pads available, there is no "standard" pad in the industry. Beyond using a pad already tried for a given shape, there are no guarantees inherent in any design. To date, the manufacturers of pad-printing equipment must also design the pads since they are the most crucial element for the successful use of pad-printing equipment.

Summary.   Compared to other methods of product decoration, pad printing is clearly the most suitable for three dimensional objects. Its capability to wrap around and print close to edges and corners makes it in some cases the only logical alternative to manual decoration. Its basic limitations are that of image size (most useful in the range of 0.01–50 in.^{2} or 0.06–323 cm^2) and ink deposit thicknesses (0.0005–0.001 in. or 13–25 μm). Large, solid print areas can only be printed with the help of a screened cliché. Owing to the few elements required by pad printing (cliché and pad) nearly all the processing parameters can be built into the equipment, making its operation extremely easy and efficient.

Pad-printing machines lend themselves well to automation not only for handling the products, but for incorporating other necessary auxiliary functions such as corona treatment, drying, sorting, and packaging.

BIBLIOGRAPHY

General References

P. Wasserman, "Looking at Pad Transfer Printing," *Screen Printing,* 122 (Oct. 1979).

"Rotary Printing for Mennen Speed Sticks," *Packaging Digest,* 83 (June 1983).

J. F. Legat, "Pad Printing—The Magic is Gone," talk presented to the Society of Plastics Engineers, Decorating Division, RETEC, Itasca, Ill., Oct. 17, 1984.

J. F. Legat, "Pad Printing," talk presented to the Society of Plastics Engineers, Decorating Division, Elmhurst, Ill., Sept. 10, 1984.

JOHN F. LEGAT
Tampoprint America, Inc.

SCREEN PRINTING

The ancient Chinese and Egyptians employed open screens and inked brushes for applying ornamental decorations to fabrics, wallpaper, and walls, although no one place of origin is credited with the discovery of screen printing. Their screens were like stencils made of impervious materials cut to form an open design. The early Japanese screens consisted of oil-treated heavy paper sheets through which openings were cut in rather intricate detail. To keep the isolated parts of the cut paper from dropping out, the Japanese painstakingly glued a spiderlike network of human hair across the openings, keeping the integral parts of the screen intact.

The modern screen-printing process, no matter how rudimentary or automated, consists of a screen stretched over a wood or metal frame, a very viscous ink which is flooded upon the screen and pressed through the screen mesh to the ware with a rubber squeegee (see Inks). A fixture is normally used to hold the ware to be printed and a method for curing the ink once printed is required.

Screens are made of a woven fabric of silk, polyester, nylon, or wire cloth stretched tightly over a metal or wood frame. The mesh count of the screen fabric can range from 80–520 per linear inch (30–200/cm). The screen is then completely coated on both sides with a light-sensitive emulsion. A positive image of the graphics to be screen-printed is placed upon the outside of the emulsion-coated screen. The unit is then exposed to intense light, curing (hardening), the emulsion not shielded from the light by the positive. The positive is then removed and the noncured emulsion is washed away. The nonblocked mesh area of the screen allows ink to pass through the fine mesh of the screen while the cured area remains closed to the passage of ink. A fairly viscous ink is used to flood the entire surface of the screen and is contained by the screen frame. Only when pressure from a rubber blade or squeegee is applied and passed over the screen surface is the ink forced through the unblocked print area. Just as an artist selects a brush for size, shape, and firmness to create varying effects, the screen printer has a number of variables to consider in the selection of the ideal screen and squeegee for a particular job.

The selection of the appropriate ink is just as important as screen and squeegee construction. The screen printer can select from a wide array of epoxy, vinyl, acrylic, enamel, and ultraviolet inks. The decision must take into consideration the chemical formulation of the substrate to be printed since various ink formulations are required to obtain a good bond between the substrate and the ink. In addition, it is necessary to pretreat polyolefin containers by flaming or electrostatic oxidation to obtain a good resin-to-ink bond (see Surface modification). The effect of the product contained within the package when it comes in contact with the screening is a further consideration in ink selection. No other printing process offers the depth, richness, and spectrum of color obtainable with screen printing.

The proper design of fixtures and the selection of equipment enable the screen printer to print effectively a wide variety of oval, round, flat, and tapered parts. Due to the relatively low cost of fixtures and screens and the speed with which they can be made, screen printing is the overwhelming choice for small runs; yet screen printing can remain highly competitive with alternative printing methods on larger production runs which use fully automated equipment capable of producing at speeds in excess of 100 pieces per minute.

The curing systems used for silk-screen inks range from air drying, electric and gas-fired ovens, to ultraviolet chambers (see Curing). The ware to be cured can be placed on a metallic or fabric belt, into specially designed baskets or on pins. Generally speaking, the conventional inks are cured at temperatures of 135–220°F (57–104°C) for 10–20 min. Ceramic inks used to print glass need to be cured in high temperature ovens at 1100°F (593°C). Organic inks for glass are cured at 250°F (121°C) for two-part inks with a catalyst and 500°F (260°C) for one-part inks.

Recent developments that are providing screen printers with improved process capabilities attest to the longevity of the screen-printing process. They include the introduction and perfection of ultraviolet inks with improved product-resistant capabilities and superior scuff-resistance. The ultraviolet inks are more energy efficient and save valuable floor space needed to dry conventional inks. In addition, the compact ultraviolet-curing ovens can be placed between stations on automatic equipment enabling screen printing to be more competitive on multicolor work. Automated screen printing equipment designed to print more than 100 parts per minute on rounds and ovals is being marketed by several suppliers. In addition, a patented computer controlled screen printer designed to print nonround cross sections as if they were round, is being marketed. This machine concept further expands the shape printing capabilities of screen-printing technology.

BIBLIOGRAPHY

General References

J. I. Biegeleisen, *The Complete Book of Screen Printing Production,* Dover Publications Inc., New York, 1963.

F. Carr, *Guide to Screen Process Printing,* Pitman Publishing Corp., New York, 1962.

R. Mastropolo, personal interview, Sept. 28, 1984.

J. Agranoff, ed., *Modern Plastics Encyclopedia,* McGraw-Hill Publications Company, New York, 1983–1984.

EUGENE E. ENGEL
Permanent Label Machinery Corp.

DISPOSABLES, MEDICAL. See Healthcare packaging.

DISTRIBUTION HAZARDS

Damage in transit is one of the oldest problems in packaging. The complete eradication of damage is not a realistic goal, because every hazard, including accidents, that packages meet cannot be anticipated. Protection is normally required against the average hazards encountered, and not against the most severe that might occur on any particular journey. In practice, the absence of damage over a long period of time to any specific product usually indicates excessive packing. This article deals only with hazards encountered in the distribution chain (see Cushioning, design; Testing, cushion systems; Testing, product fragility).

Mechanical Hazards

Typical mechanical hazards are listed in Table 1 (1). The circumstances listed in the table can be grouped as follows: hazards of loading and unloading, hazards of movement, and hazards of warehousing.

Hazards of loading and unloading. The two main hazards are drops and impacts of one container against another. In general, the drop hazard is the most damaging for goods up to about 50 kg (110 lb) in weight. A preferred weight range for packages is 10–25 kg (22–55 lb) where the packages are neither too heavy to handle, nor light enough to be thrown. The frequency and height of drops depends on the handling conditions. Both are relatively low where loading platforms are at the same level as the truck or rail floor, and hand trucks or mechanical handling equipment are available. Export hazards are normally somewhat greater than hazards encountered in domestic shipments because of the additional transfer points where drops may occur and because the handling facilities at ports worldwide are not always well equipped (see Export packaging).

Hazards of movement. These are the hazards encountered during transport. *Rail transport.* There are three main hazards: shunting shocks, which occur when trains are assembled in sidings and marshaling yards; snatching, which occurs when starting and stopping loose-coupled cars; and vibration under stacking loads limited by the height of the cars. The damage from shunting shocks is dependent not only on the nature of the packaged goods, but also on the method of loading and bracing the packages within the car. Snatching shocks are eliminated where loose-coupled cars are replaced by close-coupled vehicles fitted with vacuum brakes. Vibration hazards are particularly important in shipping domestic appliances and light equipment. The nature of the vibration depends on the speed of travel and the condition of the track, as well on the type of railcar. Vibration can cause loosening of screws and nuts and allow movement of parts that should remain in position. This can be avoided by proper design of the equipment, and packaging should be considered before, not after, the product has been designed.

Road transport. The principal hazards are vibration and bouncing of the load. Apart from the condition of the road, the frequency of any vibration depends on the loading and spring characteristics of the vehicle concerned. The amplitude depends both on the road surface and the speed of travel.

Water transport. The hazards of transport are strongly dependent on stowage conditions. The goods are often stacked 6–

Table 1. Mechanical Hazards of Distribution[a]

Basic hazard		Typical circumstances
impact		
(a) vertical	(i)	package dropped to floor during loading and unloading on to or off nets, pallets, vehicles, landing boards etc
	(ii)	package rolled over or tipped over to impact a face
	(iii)	fall from chutes or conveyors
	(iv)	result of throwing
(b) horizontal	(i)	rail or road vehicle stopping and starting
	(ii)	swinging crane impacts wall etc
	(iii)	arrest by stop or other packs on chute or conveyor
	(iv)	arrest when cylindrical package stops rolling
	(v)	result of throwing
(c) stationary package impacted by another		all above where circumstances cause the falling pack to impact another
vibration	(i)	from handling equipment (in factory, depot and at transhipment points)
	(ii)	engine and transmission vibration from road vehicles
	(iii)	running gear—suspension vibration on rail
	(iv)	machinery vibration on ships
	(v)	engine and aerodynamic vibration on aircraft
compression	(i)	static stacks in factory, warehouse and store
	(ii)	transient loads during transport in vehicles
	(iii)	compression due to method of handling, eg, crane grabs, slings, nets, squeeze clamps etc
	(iv)	compression due to restraint
racking or deformation	(i)	uneven support due to poor floors, storage etc
	(ii)	uneven lifting due to bad slinging, localized suspension etc
piercing, puncturing, tearing, snagging		hooks, projections, misuse of handling equipment, or wrong method of handling

[a] Ref. 1.

10 m (20–33 ft) high and subjected to low frequency vibration from the engine and propellers. In addition, pitching and rolling of the vessel can result in appreciable stresses, particularly in the lower levels of cargo.

Air transport. The capacity of a freight plane is limited by volume and weight, and operators want package weight to be reduced to a bare minimum. The major hazards are relatively high frequency vibration from the aircraft engines, and the low pressures and temperatures associated with flying at relatively high altitudes in unheated and unpressurized cabins. Leakage of fluids in bottles or sprayers can occur. If the package must also be carried by secondary transport, such as road or rail, it must be designed to withstand greater rigors than would be encountered in air transport alone.

Hazards of warehousing. These hazards are generally not serious in countries where modern warehouses are available and stacking is limited to a safe height. The possibility of rudimentary warehouses and severe stacking hazards must be considered in export shipments, however.

Climatic Hazards

Typical climatic hazards are listed in Table 2 (1). These can be grouped as follows: exposure to liquid water, exposure to humidities that can cause product deterioration, and temperature changes. Exposure to liquid water may be encountered from rain, sea spray, or condensation. Any humidity different from that in which the product is at equilibrium is potentially

Table 2. Climatic Hazards of Distribution^a

Basic hazard		Typical circumstances
high temperature	(i)	direct exposure to sunshine
	(ii)	proximity to boilers, heating systems, etc
	(iii)	indirect exposure to sun in sheds, vehicles etc, with poor insulation
	(iv)	high ambient air temperature
low temperature	(i)	unheated storage in cold climates
	(ii)	transport in unheated aircraft holds
	(iii)	cold storage
low pressure		change in altitude, particularly in unpressurized aircraft holds—aircraft pressurization failure
light	(i)	direct sunshine
	(ii)	uv exposure
	(iii)	artificial lighting
liquid water (a) fresh	(i)	rain during transit, loading and unloading, warehousing and storage
	(ii)	puddles and flooding
	(iii)	condensation and ship sweat, etc
(b) polluted	(i)	salt sea spray—deck cargo, lightering surf boats, etc
	(ii)	salt water puddles on docks etc
	(iii)	bilge water and seawater in holds
	(iv)	industrially polluted puddles and spray, eg, at chemical works
dust storms, etc		exposure to wind driven particles of sand, dust, grit etc
water vapor		humidity of the atmosphere, both natural and artificial

^a Ref. 1.

damaging, whether high or low. Temperature changes affect materials in many ways. A fall in temperature protects most foodstuffs, but it can have an adverse effect on some emulsion-type products. Plastics become more brittle at lower temperatures, ie, more prone to breakage. Sudden temperature changes can also cause condensation of moisture on packages and products. Relative humidity, moisture content, and temperature are interrelated.

Effects of moisture. The changes brought about by the influence of moisture take at least five forms:

Physical changes such as hardening of leather from drying out, or loss of crispness in crackers from moisture gain.

Physicochemical changes such as lump formation in salt and sugar, or hydrate formation.

Microbiological changes such as the growth of molds or bacteria if moisture content rises above a critical level or water droplets form in a package.

Chemical changes such as rusting of steel, corrosion or tarnishing of other metals, autooxidation, and browning reactions in foodstuffs.

Enzymatic changes, particularly in unprocessed foodstuffs.

All of these changes depend not just on moisture content, but on temperature as well (see Insulation).

Corrosion. Corrosion is the result of a reaction between a metal and the environment in which it is placed. It is an electrochemical reaction brought about by water and some contaminating substance that acts as an electrolyte on the bare metal surfaces on which the moisture has been deposited. Atmospheric water vapor can start the reaction, particularly at high temperatures. All metals do not behave and corrode alike, or for the same reasons. The most frequent causes of corrosion of metal goods in packaging are (a) contamination by corrosive influences during processing and before applying temporary protection, (b) the use of packaging materials containing corrosive substances, (c) inadequate protection against the mechanical hazards of transport, and (d) inadequate protection against climatic hazards (1). It is very important, therefore, that metals be properly cleaned before packaging, that packaging materials be specified adequately, and that adequate protection be provided against mechanical and climatic hazards.

Food deterioration. Many products deteriorate if their moisture content moves outside certain limits. The factors that cause food deterioration can be divided into two classes: (a) factors inherent in the product that cannot be controlled by packaging alone, and (b) factors that are dependent on the environment, that can be controlled by proper packaging. Class (a) includes changes due to temperature, such as the softening of chocolate or the breaking of emulsified fluids. Biochemical changes such as browning of meat cannot be controlled by packaging alone. Class (b) includes spoilage due to mechanical damage, changes due to moisture content of the food, absorption of and interaction with oxygen and other gases, and flavor loss or gain. (For discussions of the requirements of food packages, see Food packaging; Shelf life; Testing, permeability and leakage.)

Biological Hazards

Biodeterioration is damage caused by living organisms, most commonly bacteria, molds, insects, and rodents. Typical biological hazards are listed in Table 3 (1).

Rodent control can generally be achieved by clean storage conditions. Since moisture is the most fundamental need of

Table 3. Biological Hazards of Distribution[a]

	Typical circumstances
(a) microorganisms fungi molds bacteria	Are ubiquitous and adapt themselves to varied conditions. Require moisture and generally will not grow at relative humidities of less than 70%. Will grow over a wide range of temperatures.
(b) insects beetles moths flies ants termites	In general high temperatures are more favorable for development than low ones and, below 15°C, development is unlikely. A relative humidity of 70% is very favorable for most insects but some will develop at below 50% rh. Infestation usually starts from eggs laid on packaging materials, penetration then being made by the small newly hatched insects. Migration from adjacent packs or from natural habitat (particularly in tropical localities) may occur.
(c) mites	As for insects, but they are less tolerant of dry conditions (few survive and develop slowly at about 60% rh) and they develop over a lower temperature range.
(d) rodents	May be present in warehouses, transit sheds, storage areas, holds etc. Will attack most material to keep in condition, and softer materials for making nests (or for food).

[a] Ref. 1.

both bacteria and molds, exclusion of moisture is the most fundamental method of control. This can be done by packing in a sealed moisture-resistant barrier, if the contents are dry enough to preclude condensation. Desiccants can be used to prevent moisture lost by the product from condensing in the

Table 4. Rate of Penetration of Typical Packaging Materials by Insects[a,b]

Packaging material	Thickness (mil) mm	Average number of weeks before penetration[c]
cellulose film	0.9 (0.023) 1.4 (0.036) 1.6 (0.041)	3 3 3½
polyethylene film	1.5 (0.038) 2.0 (0.050) 3.9 (0.100)	3 3 3
PVC/PVDC copolymer film	1.5 (0.038) 2.0 (0.050)	3 4
polyethylene teraphthalate	1.0 (0.025)	6

[a] Ref. 1.
[b] 11 types of boring insect including *Rhizopertha dominica*.
[c] Envelope tests . . . food inside the envelope.

Table 5. Contamination Hazards[a]

	Contamination by other goods
(a) by materials of adjacent packs	Obliteration of marking, printing etc, by rusty metalwork—strapping, wire bands. Effects of damp packing materials, especially hessian, on nonwater resistant materials, adhesives and metal parts.
(b) by leaking contents of adjacent packs	Damage to containers of liquids, powders and granulated substances may result in leakage of the contents. The effect of the resultant contamination on adjacent packs can range from the spoiling of external appearance to complete disintegration of a pack and its contents, depending on the nature of the contaminant, the packing materials and the contents of the pack contaminated.
(c) radioactivity	

[a] Ref. 1.

package if the temperature drops. Moisture barriers must not be used when a damp product is packed, because the moisture must be allowed to escape as quickly as possible. Stacks should contain air spaces. Some molds thrive on some packaging materials (eg, starch and glue), and the use of nonnutritive synthetic materials can prevent their growth, as can the exclusion of oxygen from a package. Sterilization is, of course, another effective way to control bacteria and molds. Some insect larvae can bore into lead when seeking a pupation site, but in general, the stronger and thicker a packing material is mechanically, the more resistant it will be to insects. Smooth surfaces are preferable to rough ones, and unnecessary folds should be avoided. Table 4 shows rates of insect penetration through a variety of packaging materials.

Other Hazards

In addition to mechanical, climatic, and biological hazards, we must also mention the hazards of contamination by other goods (Table 5), as well as the hazards of pilferage.

BIBLIOGRAPHY

1. F. A. Paine, ed., *Fundamentals of Packaging,* The Institute of Packaging, Melton Mowbray, Leicestershire, UK, 1981.

General References

J. A. Cairns, C. R. Oswin and F. A. Paine. *Packaging for Climatic Protection,* Newnes-Butterworths, London, 1974.

F. E. Ostrem and W. D. Godshall, "An Assessment of the Common Carrier Shipping Environment," General Report FPL 22, Forest Products Laboratory, USDA, Madison, Wisconsin, 1979.

F. E. Ostrem and B. Lebovicz, "A Survey of Environmental Conditions Incident to the Transportation of Materials," Report PB-204442, General American Transportation Corp., Niles, Ill.: 1971.

J. P. Phillips, "Package Design Considerations for the Distribution Environment," *Proceedings of the 1979 International Packaging Week Assembly,* Packaging Institute, Stamford, Conn., Oct. 1979.

J. R. Winne, "What Really Happens to Your Package in Trucks and Trailers," *Proceedings of the Western Regional Forum of the Packaging Institute*, Packaging Institute, Stamford, Conn., 1977.

Frank A. Paine
Consultant

DRUG PACKAGING.   See Pharmaceutical packaging.

DRUMS, PLASTIC

Chemicals and other industrial products are shipped mainly in pails and drums. In most parts of the world, 1–6-gal (4–23-L) open-top plastic containers, generally injection-molded (see Injection molding), are called pails (see Pails, plastic). In North America, 1–6-gal (4–23-L) closed-head, ie, bung-type, blow-molded containers (see Blow molding) are called pails as well, or jerrycans. The term jerrycan is often used in other countries to describe bung-type containers in the 1–16-gal (4–61-L) size range. In general, however, the word drum applies to open- and closed-head containers larger than 6 gal (23 L). In North America, the standard drum sizes are 15, 20, 30, 35, 55, and 57 gal (57, 76, 114, 132, 208, 216 L). In Western Europe and most other parts of the world, standard drum sizes are 30, 60, 120, 210, and 216 L (roughly 8, 16, 32, 55, and 57 gal).

Polyethylene liners for steel and fiber drums were blow-molded and rotationally molded in the United States and Western Europe in the 1950s, but there were no self-supporting plastic alternatives until the 1960s. In 1963, U.S. production of all-plastic 5-gal (19-L) pails and 16-gal (61-L) drums began. In Western Europe, production of 16-gal open-top and bung-type drums started at about the same time, and a 32-gal (121-L) open-top drum soon followed. The introduction of all-plastic 55-gal (208-L) drums did not come until the early 1970s. The development of large all-plastic drums took many years because they required special resins and processing equipment. Market acceptance has also required special designs.

Plastic containers have excellent performance characteristics. They are strong, lightweight, durable, corrosion-free, and weather-resistant, and, in accordance with international transport regulations, are authorized to carry a great number of hazardous materials. Most plastic drums are used for chemicals, but they are also used in the food-processing industry for the shipment and storage of products that include concentrated fruit juice, vegetable pulps, condiments, etc.

Resins

Self-supporting plastic drums are made of extra high molecular weight (EHMW) high density polyethylene (see Polyethylene, high density). The molecular weight of these resins is so high that their flow rates cannot be expressed in terms of melt index (MI) as measured according to ASTM 1238, Condition E (44 psi or 303 kPa). The MI of relatively low molecular weight HDPE injection-molding resins ranges from 1 to 20 g/10 min; the MI of higher molecular weight blow-molding bottle resins is less than 1 g/10 min. Measuring the flow of EHMW resins requires higher pressure (Condition F, 440 psi or 3 MPa), and the values obtained are expressed in terms of high load melt index (HLMI). The HLMI of resins used for self-supporting drums range from 1.5 to 12 g/10 min. Design trends in plastic drums are related to the availability of EHMW resins. In the United States, 10 HLMI became the standard resin for drums with separate plastic- or metal-handling rings. The development of drums with integral handling rings required higher molecular weight resins (HLMI 1.5–3) that were available in Western Europe before they were produced in the United States.

All properties of polyethylene depend on three important factors: molecular weight (length of molecule chains), density (degree of crystallinity), and molecular weight distribution (distribution of longer and shorter chains). As molecular weight increases (and MI or HLMI decreases), toughness and resistance to stress cracking increases. The trade-off for these improved properties is difficulty in processing. HDPE has crystallinity of 60–80% at density of 0.942–0.965 g/cm³. (In contrast, LDPE has crystallinity of 40–50% at density of 0.918–0.930 g/cm³.) As density increases, toughness and stress crack resistance decrease, but stiffness, hardness, and resistance to oils and chemicals improve. Molecular weight distribution is related primarily to processing.

Chemical resistance. Chemical resistance is particularly important in drum design. All 55-gal (208-L) drums use EHMW high density resins, but there are variations within that category. Drums used for chemicals that are compatible with HDPE generally use relatively high molecular weight, eg, 1–3 HMLI, and relatively high density, ie, >0.95 g/cm³, resins. Where stress cracking is a potential problem, relatively low molecular weight, eg, 6–10 HLMI, and relatively low density, ie, <0.95 g/cm³, resins give better performance. Although no chemical dissolves polyethylene, particularly high molecular weight polyethylene, the effects of certain chemicals include strong swelling action by penetration into the container wall, stress cracking, oxidation, degradation by destruction of the macromolecules, or permeation through the container wall. Resistance tests should be made based on laboratory samples and on containers in use.

The resistance to stress cracking must be examined before using self-supporting plastic drums. The ESCR test prescribed in ASTM D 1693 can be used for material selection, but tests must be performed on finished containers. This can be done by storing a drum filled with 5% surfactant for more than 3 mo at temperatures higher than 40°C. Stress cracking may occur if there are stresses in the wall. Tensile and compressive stresses can be avoided by using an optimum wall-thickness distribution in production or in the design of the blow-molding tool. They may also occur when the products contain surface-active substances such as wetting agents at concentrations up to 20%. HDPE's resistance to stress cracking depends on the density and the molecular weight of the raw materials. As density increases, ESCR decreases. As molecular weight increases, ESCR increases. High molecular weight blow-molding resins with densities of 0.947–0.954 g/cm³ have a high degree of stress-crack resistance and are ideal for drums (1).

HDPE drums can safely package a wide variety of chemicals. Uses within the chemical industry include dairy, agricultural, electronic, specialty, photographic, and oil-well applications, as well as those for organic and inorganic chemicals and natural flavorings.

Permeability. HDPE is susceptible to permeation by certain chemicals. Special attention should be paid to inorganic chlorinated hydrocarbons, (eg, per- and trichloroethylene), and aromatic hydrocarbons, (eg, benzene, toluene, and xylene). Normally, inorganic chlorinated or aromatic hydrocarbons cannot be packed in HDPE drums owing to permeation, but the permeation rate can be reduced by several methods of surface modification (see Surface modification). Within the HDPE family, high molecular weight grades have relatively high resistance to permeation. The risk is also reduced by using relatively thick walls.

Uv resistance. The service life of HDPE containers cannot be accurately predicted because it largely depends on climatic conditions. Different colorings (see Colorants), especially black (with carbon black), blue, green, white, and gray, increase resistance to weathering and protect the product from light. Depending on climatic conditions, additional uv stabilizers must be added.

Design

Closed-head drums. Closed-head drums, also called tight-head drums, are available in different designs, with 2-in. (51-mm) and ¾-in. (19-mm) bungs, in 15-, 30-, and 55-gal (57-, 114-, and 208-L) sizes (see Fig. 1). They are used where handling equipment is available to accommodate them. This is important because large self-supporting drums are designed to replace composite steel drums (steel with plastic liners or coat-

Figure 2. Handling closed-head L-ring drums with parrot-beak equipment. Courtesy of Mauser Werke GmbH.

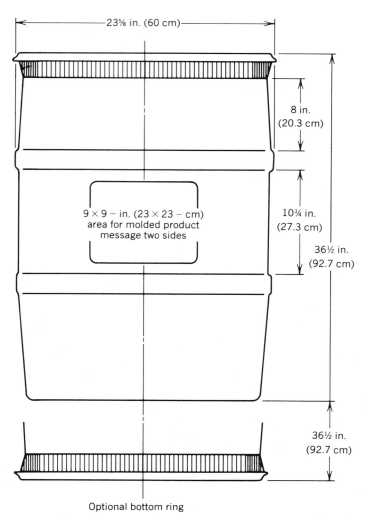

9 × 9 – in. (23 × 23 – cm) area for molded product message two sides

23⅝ in. (60 cm)

8 in. (20.3 cm)

10¾ in. (27.3 cm)

36½ in. (92.7 cm)

36½ in. (92.7 cm)

Optional bottom ring

Figure 1. Dimensions of a 55-gal (208-L) drum. Courtesy of Container Corporation of America.

ings), for which mechanical handling equipment is available worldwide. In the United States, plastic drums were initially designed with metal handling rings so that the drums could package heavy liquids, eg, up to 1.8 g/cm³ or 825 lb/55 gal, and still be handled by traditional steel-drum handling equipment, particularly the "parrot's beak."

A different approach was taken in Europe and other parts of the world, where the L-ring drum in 30- and 55-gal (114- and 208-L) capacities has become the standard. L-ring drums are produced by blow-molding the drum and integral rings in one step. The use of this configuration required the development of a wider parrot's beak, and at first, it was approved only for liquids with densities less than 1.2 g/cm³ or 550 lb/55 gal. The L-ring is now approved for heavier liquids (see Fig. 2).

Open-top drums. For general applications, open-top drums are provided in 8-, 15-, 30-, and 55-gal (30-, 51-, 114-, and 208-L) sizes (see Fig. 3). The advantages of the standard open-top drum are easy handling, absolute tightness, good stacking properties, and high radial rigidity. They are used most often for water-based products such as glues, softeners, and liquid soaps, as well as for foodstuffs. The standard open-top drum can be cleaned easily and reused.

Shipment of Hazardous Materials

International regulations for transport of dangerous goods (hazardous materials in the United States) is one of the central issues regarding the use of plastic drums because of heightened worldwide concern for environmental protection and the increasing number and volume of dangerous goods being shipped. HDPE is a safe and durable material for shipping dangerous goods.

United Nations regulations. International regulations concerning packaging tests for hazardous materials are contained in the *United Nations Transport of Dangerous Goods* (2) and in the *International Maritime Dangerous Goods Code* (IMDG Code), which regulates substances shipped overseas. The recommendations do not specify how a package is to be made. They stipulate package-performance tests for each dangerous

Figure 3. An open-top 55-gal (208-L) HDPE drum. Courtesy of Mauser Werke GmBH.

substance and modification of the test procedures based on the degree of hazard and some physical properties of the substance. They require a certification mark to show that the package has passed the tests. Self-supporting plastic drums can be used for a wide variety of dangerous substances. A typical drum marking under these regulations would be

1H1/Y.1.8/150/84/D/Mauser 824

where UN = United Nations; 1H1 = plastic drum, small opening; Y.1.8 = Group II products up to 1.8 g/cm³; 150 = test pressure, kPa; 84 = year of production; D = FRG; Mauser = producer; and 824 = registration no.

U.S. regulations. The Materials Transportation Bureau (MTB) of the Department of Transportation (DOT) has expanded Specification 34 (CFR, Title 49, Part 178.19) to include 55-gal (208-L) plastic drums (3). Previously, only drums of up to 30 gal were included, and 55-gal (208-L) drums needed special exemptions. The revision eliminates the need for those exemptions. They are now marked DOT-34-55. Some familiar commodities have been written into the regulations with no restrictions, but many of the chemical listings in Part 173.24(d) prescribe chemical compatibility and permeation tests. Reference to specific types of polyethylene has been deleted from the most recent revision. Minimum wall thickness is specified, but this does not create significant differences between drum construction in the United States and other countries.

European regulations. European regulations for rail and road transport of hazardous materials in plastic containers up to 250 L (66 gal) include testing procedures based on test data obtained with model liquids such as acetic acid, nitric acid, water, white spirit, surface-active agent, and *n*-butyl acetate. On the basis of the dangerous properties of these materials, all other dangerous substances can be approved and assimilated as long as their individual requirements are taken into account in testing. Plastic containers are included in the packaging regulations as follows: up to 15 gal (57 L) for Class-1 dangerous substances (very dangerous); up to 55 gal (208-L) for Classes 2 (dangerous) and 3 (less dangerous). Apart from the requirements concerning approval of the containers, quality assurance in production plays an important role in the safe transportation of dangerous goods.

Plastic drums will continue to gain importance, particularly if worldwide standardization occurs.

BIBLIOGRAPHY

1. D. L. Peters, P. E. Campbell, and B. T. Morgan, "High Molecular Weight High Density Polyethylene Powder for Extrusion Blow Molding of Drums and Other Large Parts," *Proceedings, 42nd SPE Annual Technical Conference and Exhibition,* 1984, Society of Plastics Engineers, Inc., Brookfield Center, Conn., 1984, p. 939.

2. "General Recommendations on Packing," *United Nations Transport of Dangerous Goods—Recommendations of the Committee of Experts on the Transport of Dangerous Goods,* 3rd rev. ed., United Nations, New York, 1984, Chapt. 9.

3. *Fed. Regist.* **49**(116), 24684 (June 14, 1984) and **49**(199), 40033 (Oct. 12, 1984).

BRUNO POETZ
ERNST WURZER
Mauser Werke GmbH

DRUMS/PAILS, STEEL

Despite centuries of innovation, no one has devised a more useful and adaptable medium-sized container for liquids and semisolids than a cylindrical container. Its shape, based on the circle, provides maximum strength; when fully laden, it can be tipped over and rolled. Early ocean shippers employed heavy, easily breakable clay and ceramics. They knew smelting and metalworking, but evidently could not produce metal containers much larger than pots and pans. From the Middle Ages until recent times, the standard container material was wood, usually oak, formed into metal-bound, stave-constructed barrels and kegs (see Barrels, wood). The wooden barrel had no weight advantage but was far stronger and could be manufactured anywhere of materials widely available at low cost. Design differed little from the early jars, featuring the same bilged sides for maximum strength. In almost every trading nation, the wooden barrel reigned as the universal shipping container for liquids and semisolids until the late 19th century, when the first steel barrels appeared in Europe.

Impetus for the invention and development of the modern steel drum had begun years earlier with the Great Oil Rush of 1859. Oil-drilling technology improved so rapidly that by 1869

U.S. wells were producing 4,800,000 barrels (7.6×10^5 m³) a year. That introduced a much tougher problem: how to store and ship the oil. The only container available was the wooden barrel. Demand soon outstripped the production capacity of the cooperage firms, and oak became scarce and very expensive. The immediate solution was the development of pipelines, railroad tank cars, and the tank truck. But for smaller packages, oil-industry shippers still had to struggle along for another 40 yr with the time-honored wooden barrel. Kerosene was the most important barreled product. The wooden barrel's chief problem was leakage. Designed to hold 50 gal (0.189 m³), it commonly lost enough per trip to arrive at its destination containing only about 42 gal (0.159 m³), a figure that has remained the standard "barrel" measurement of the oil industry.

A few years after their development in Europe, the first steel barrels were produced commercially in the United States by Standard Oil at Bayonne, N.J., in 1902. Constructed from 12–14-gauge terne steel, they were heavy, clumsy, bilged affairs with riveted or soldered side seams, and were anything but leakproof. Extremely rugged, many of them lasted 20–30 yr. They were also expensive compared with wooden barrels. Despite the need, steel barrels were slow to catch on; yet technical developments came rapidly. In about 1907, the welded side seam was introduced, which curtailed the leakage problems of riveted barrels and reduced costs. Shortly afterwards, the first true 55-gal (208 L) drum was introduced, its characteristic straight sides contrasting markedly with the bilged barrel. Rolling hoops were introduced soon thereafter, both expanded and attached, the latter utilizing an I-bar section. The mechanical flange was invented after 1910. These improvements were far-reaching and gave the new container added advantages. Among them was lighter weight, which reduced shipping costs and reduced the amount of steel required. The new drum design permitted use of 16- and 18-gauge steels instead of the far heavier gauges used previously. These drum developments were followed in 1914 by the first true steel 5-gal (18.9-L) pail featuring the first lug cover.

The use of steel containers grew slowly before 1914, despite their cost, weight, and safety advantages over wooden containers. The advent of the First World War marked the beginning of the end for the wooden barrel, and the eventual dominance of the steel drum and pail. Wartime demand also spurred many improvements in manufacturing techniques and equipment. After the war, many innovations appeared, including pouring pails, agitator drums for paints, and new, colorful decorating techniques. Manufacturers began to use steel containers for products other than petroleum products and chemicals. Toward the end of the 1930s, the steel-container industry started gearing up for a second wartime effort. This time, however, the demands were far more stringent, requiring fuel containers for a highly mechanized war on more than one front. Innovation took a back seat to production considerations as war machines on the ground and in the air consumed vast amounts of fuel and chemicals. The 55-gal (208-L) 18- and 16-gauge drums were indispensable to the fuel supply of island bases and assaults in the Pacific, frontline mechanized operations in Europe, and to air and ground operations in East Asia. Apart from its ruggedness, the fact that a cylindrical drum could be rolled by one man was an important feature.

Although a downturn in steel drum and pail production occurred at the end of World War II, it was of short duration.

Resumption of business created a demand from industry, agriculture, and consumers. The acceleration in chemical and pharmaceutical product development and output provided new markets for steel drums and pails. Demand for paints, lacquers and varnishes, adhesives, inks, foodstuffs, and other products made the steel shipping-container industry the second-largest user of sheet steel, outranked only by the automotive industry. Despite the introduction of competitive containers made of other materials, U.S. production increased from 2.4 million (10^6) drums of all types per year in 1922 to 40 million (10^6) 55-gal (208-L) drums alone in 1984.

Drums range in size from 13 to 110 gal (49–416 L), but the 55-gal drum accounts for about 80% of annual production. Over 75% of all new drums are used for liquids, and the rest for viscous and dry products. About 80% of the market is accounted for by five broad product categories: chemicals; petroleum products; paints, coatings, and solvents; food and pharmaceutical products; janitorial supplies, cleaning compounds, and soaps.

Drum and Pail Construction

Steel drums (13–110-gal (49.2–416-L) capacity) and pails (1–12-gal (3.8–45.4-L) capacity) are generally fabricated from cold-rolled sheet steel in a broad range of gauges, ie, thicknesses. They consist of a cylindrical body with a welded side seam and top and bottom heads. The lighter gauges, ie, 30–22 are generally used for pails and smaller drums. Thicker steels, ie, 20–16 gauges, are used for larger, reusable drums. Steel gauge numbers are listed in Table 1 (1,2).

Most drums are made of commercial-grade cold-rolled sheet steel, but stainless steel, nickel, and other alloys are used for special applications. Only about 35% of all new drums are lined with interior protective coatings, but the percentage is much higher, ie, 80%, for drums used for chemicals. Over the years, the cost and weight of steel drums have been reduced owing to technological advances. This was made possible by improvements in cold-rolled steel and greater uniformity in steel-thickness control. Changes have been made that have resulted in improved chime construction. Until the early 1960s, most tight-head 55-gal (208-L) drums were made of 18-gauge steel. There has since been a shift to the 20/18-gauge

Table 1. Sheet Steel Thickness[a] vs Gauge No.

Gauge no.	Minimum thickness	
	in.	mm
12	0.0946	2.40
16	0.0533	1.35
18	0.0428	1.09
19	0.0378	0.960
20	0.0324	0.823
22	0.0269	0.683
24	0.0209	0.531
26	0.0159	0.404
28	0.0129	0.328
29	0.0115	0.292

[a] Sheet steel thickness is measured at any point no less than ⅜ in. (9.53 mm) from the edge.

Table 2. Common Steel Drum Sizes, Dimensions[a], and Gauges

Capacity, gal (L)	Inside diameter, in. (cm)	Overall height in. (cm)	Common gauges	Applications
55 (208.2)	22½ (57.2)	33¾ (85.7)	all 18	
			20/18	for certain hazardous materials
			all 20	for nonhazardous materials
30 (113.6)	18¼ (46.4)	28⅞ (73.3)	all 20	for liquid products
16 (60.6)	13¹⁵⁄₁₆ (35.4)	26⅞ (68.3)	all 20	for hazardous materials
			all 20–all 26	for various materials

[a] Dimensions are approximately the same for both tight-head (TH) and (full-removable) open-head (FRH).

steel drum: 20 gauge for the sidewalls and 18 gauge for the ends. Now, 55-gal (208-L) drums of all-20-gauge construction are used for packaging nonhazardous materials (see Table 2).

Styles

There are two basic styles of drums: closed-head (or tight-head), with permanently attached top and bottom heads, and full-removable head (or open head), in which the removable top head or cover is secured by using a separate closing ring with either a bolted or lever-locking closure (see Fig. 1). Expanded rolling hoops, ie, swedges, in the drum body stiffen the cylinder and provide a low friction surface for rolling filled containers. Steel drums are not designed to be used as pressure vessels. Tight-head drums (and pails) have their top and bottom heads mechanically rolled (seamed) in multiple layers to the body using a nonhardening seaming compound to form a joint (chime). Two openings, one 2 in. (51 mm) and the other ¾ in. (19 mm), for filling and venting are usually provided in the top head, although side openings and other opening combinations and sizes are sometimes used. The openings are fitted with mechanically inserted threaded flanges conforming with American National Pipe thread standards. Threaded plugs for insertion in the flanges are made of steel or plastic and have resilient gaskets where appropriate. On full-removable-head drums, the top of the body sidewall is rolled outward to form a hollow curl (false wire) to which the top head or cover is attached using a gasket of resilient material and a separate closing ring.

Steel pails are generally of the same configuration and style as the large capacity steel drums, but are usually of thinner metal and may have only one expanded body hoop. A bail handle or carrying grip is often provided for handling purposes. A common closure for open-head pails is a lug cover which is crimped in place around the top curl and is removed by lifting the lugs (see Fig. 1(**c**)).

Protection and Linings

Most steel pails and drums are fabricated from steel treated to resist rusting owing to moisture in the air. Steel is a nonpermeable, biodegradable material that is compatible with most chemicals and petroleum-based products.

Interior. Linings are used for protection against acids, alkalies, and some organic chemicals. Phenolics provide protection against certain acids, and epoxies offer protection against alkalies. Linings consisting of varying percentages of epoxy and phenolic materials are most commonly used today. In some instances, the needed protection is supplied by a flexible or semirigid polyethylene liner insert.

Exterior. New steel containers can be painted, lithographed, or silk-screened to provide an attractively decorated and durable finish. Enamels are sprayed or roller-coated, baked, and oven-cured to give a scuff-resistant exterior coating. Black is generally the standard color, but other colors are available as well. Product and manufacturer information for merchandising or to satisfy transportation needs is applied by lithography, silk-screening or stenciling.

Standardization

National standards for steel pails and drums have been developed in the United States within the American National Standards Institute (ANSI) by a Committee on Steel Pails and Drums sponsored by the Steel Shipping Container Institute. These dimensional standards have received international acceptance and have provided many advantages in the areas of filling, handling, storage, and shipping. A partial list of the standards included in *ANSI MH2-1985* (2) are given below for a 55-gal (208-L) steel drum:

DOT-17E. UFC-Rule 40. NMFC-Item 260 Tight Head (Double Seam Chime) Drums

UFC-Rule 40. NMFC-Item 260 Full-Removable-Head (Double Seam Chime) Drums

DOT-5B. Tight Head (Double Seam Chime) Drum

DOT-17C. Tight Head (Double Seam Chime) Drum

DOT-17H. UFC Rule 40. NMFC-Item 260 Full-Removable-Head (Double Seam Chime) Drums

DOT-17E. UFC-Rule 40. NMFC-Item 260 Tight Head (Round Seam Chime) Drums

UFC-Rule 40. NMFC-Item 260 Full-Removable-Head (Round Seam Chime) Drums

DOT-17H. UFC-Rule 40. NMFC-Item 260 Full-Removable-Head (Round Seam Chime) Drums

Regulations

All steel pails and drums used in the United States usually comply with the minimum requirements of the specifications of one or more of the three main regulatory groups: the Department of Transportation (DOT) (1), the railroads' Uniform Classification Committee (UCC) (3), and the highway carriers' National Classification Board (NCB) (4). Products classified as hazardous materials, such as acids, flammables, explosives, and poisons, etc, as well as products designated as hazardous wastes or hazardous polluting substances by the EPA, are regulated by the DOT and the majority must be packaged in a

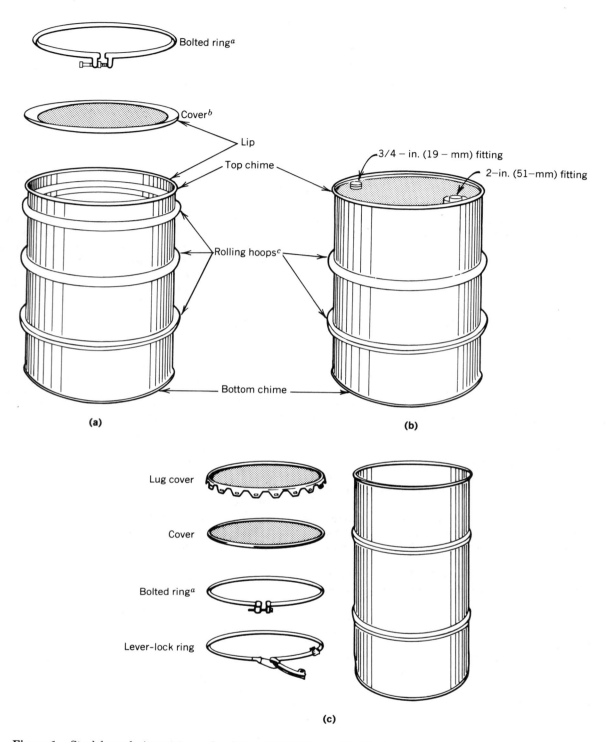

Figure 1. Steel drum designs: (**a**) open-head 55-gal (208-L) drum; (**b**) tight-head 55-gal (208-L) drum; (**c**) open-head 16-gal (606-L) drum. [a]Lever-lock ring may be used; [b]Lip of cover is turned down to fit over lip of drum; [c]Two or three rolling hoops may be used depending upon size and material packaged. Hoops are equally spaced horizontally.

DOT specification container. (see Fig. 2). DOT regulations also prescribe whether a container may be used more than once. Some specification containers for hazardous products may be used only once. Full details of regulations and specifications covering shipping containers for all hazardous products appear in the *Code of Federal Regulations* (CFR) (Title 49, Parts 100–199; Title 14, Parts 100–199). These regulations govern shipments by land, sea, and air. Container specifications for products not requiring a DOT specification container are covered by the *Uniform Freight Classification* (UFC) (3) and the *National Motor Freight Classification* (NMFC) (4) tariffs. Recommended international regulations governing the transport

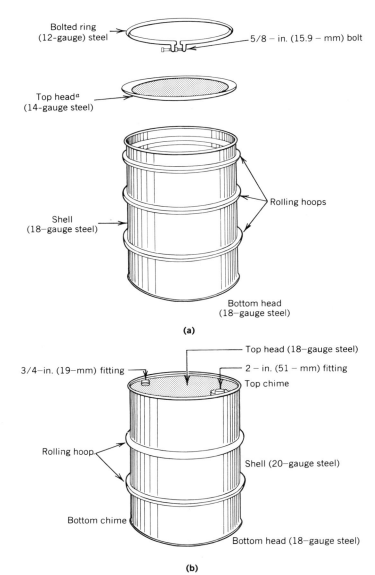

Figure 2. U.S. DOT specification drums; (**a**) DOT-17H, full-removable-head 55-gal (208-L) drum; (**b**) DOT-17E, tight-head 20/18-gauge 55-gal (208-L) drum. ^aLip of cover is turned down to fit over lip of drum.

of hazardous materials have been developed by The Committee of Experts on the Transport of Dangerous Goods, acting under the direction of the United Nations Economic and Social Council (5) (see Standards and practices). The *International Maritime Dangerous Goods Code* (IMDG Code) developed by The International Maritime Organization (IMO) for water movement parallels the UN Recommendations. Similar regulations for movements by air have been adopted by International Civil Aviation Organization (ICAO). The UN Recommendations are not mandatory, but they have been adopted by these international bodies and a number of countries and they facilitate international trade. European road and rail regulations, ADR/RID, are in general conformance with the UN Recommendations as of May 1, 1985. In the UN publication covering these requirements (5), Chapter 9 contains "General Recommendations on Packing," which details packing re-

quirements types of packagings, and marking and testing requirements.

The DOT Regulations permit containers produced in the United States to be marked in accordance with the IMDG Code (if they are in conformance with the provisions of that code) in addition to any required DOT specification markings. The UN regulations require commodity-related hydrostatic and drop tests as well as stacking tests which in some cases differ from DOT requirements. Shippers must ensure they are using containers tested and marked in accordance with the minimum requirements for the material to be transported.

Steel Pails

About 80 million (10^6) new steel pails are currently produced in the United States each year; sizes range from 1 to 12 gal (3.8–45.4 L). About 80% is accounted for by the popular 5-gal (18.9-L) pail. They are made in four basic configurations: full open-head, straight side, lug cover; full open-head, nesting, lug cover; tight-head, straight side; and tight-head dome top (see Fig. 3). They are constructed of 29-gauge or heavier steel, and have a carrying-handle feature. Pail heights vary by volume, but diameters are 6 and 8 in. (15.2 and 20.3 cm, respectively) for 1–2½-gal (3.8–9.5-L) sizes, 11.4 in. (28.6 cm) for 3–7 gal (11.4–26.5 L), and 13 $^{15/16}$ in. (35.4 cm) for larger capacities. The two open-head designs account for about 75% of the total pail production.

Table 3 provides information on the 5-gal (18.9-L) pails now commonly produced in the United States.

Both types of open-head pails have a liquid-tight, welded side seam on the pail body and a bottom affixed by double seaming (see Can seamers). Two side "ears" are welded or riveted to the body, and a galvanized wire bail handle is attached. Handles are furnished with or without a grip, which can be wood or contoured plastic. The straight-sided pail normally has one strengthening body bead, ie, swedge, to add

Table 3. Five-gal (18.9-L) Steel Pail Specifications, Gauges and Types^a

Type of Pail	CFR Title 49	Specification	Gauge, min
commercial pails			
tight-head	178.116	DOT-17E	24
tight-head			26
tight-head	178.132	DOT-37B60	28
lug cover	178.131	DOT-37A80	24
lug cover	178.131	DOT-37A60	26
lug cover		STC	28
lug cover		STC	29
nesting, lug cover	178.131	DOT-37A80	24
nesting, lug cover	178.131	DOT-37A60	26
nesting, lug cover		STC	28
nesting, lug cover		STC	29
lug cover	178.135	DOT-37C80	28/26
lug cover, nesting	178.135	DOT-37C80	28/26
special pails			
tight-head and open-head	178.80	DOT-5	22
tight-head and open-head	178.82	DOT-5B	24
tight-head and open-head	178.99	DOT-6C	22
tight-head and open-head	178.115	DOT-17C	24
open-head	178.118	DOT-17H	24/24/20
open-head with polyethylene liner	178.133	DOT-37P	26/26/24
open-head overpack	178.134	DOT-37M	26

^a Ref. 6. Courtesy of SSCI.

Figure 3. Designs for 5-gal (18.9-L) steel pails: (**a**) open-head, straight side, lug cover; (**b**) open-head, nesting, lug cover; (**c**) tight-head, straight side; (**d**) tight-head dome top. Courtesy of SSCI.

rigidity to the top of the cylinder. The tapered type usually has a second bead that, in nesting, rests on the top curl of the pail below and limits nesting depth.

The tight-head pail, accounting for some 25% of sales, is often used for the shipment of low viscosity or free-flowing liquids. It embodies a welded side seam, double-seamed top and bottom ends, and a carry handle, usually a D-ring, of galvanized wire spot-welded to the head. This container can be fitted with a variety of pouring and venting apparatus. An offshoot of the tight-head pail is a domed-top or utility pail, especially popular in 2½- and 5-gal (9.5- and 18.9-L, respectively) sizes for petroleum products.

Pails are used for liquids, viscous products, powders, and solids. Pail markets include paint and printing inks; chemicals; adhesives, cements, and roofing materials; petroleum products; janitorial supplies, eg, cleaners, waxes, etc; abrasives; cosmetics; fasteners and stampings; foods; insecticides; marine supplies; pharmaceuticals; powdered metals; and scores of other products and materials.

In addition to varying capacities, steel thicknesses, and container construction a host of options, fittings, and accessories are available to design a pail to a buyer's exact requirements.

Open-head pails can take two types of covers. The lug cover, usually incorporating 16 wide lugs around its circumference, can be applied at production-line speeds by automatic crimping equipment, although hand-operated and semiautomatic crimping tools are also available. It is opened with standard hand tools. The second cover, a ring seal, is best for resealing purposes. It consists of a formed disk that sits on the top curl of the container and is clamped to it either by a separate ring band or with rings that lock by lever action or bolt-tightening.

Another option is a combination of steel thicknesses. Cover and ends can be made of steel of different thickness than the body for different requirements of strength and economy. Lids and bottom ends can be strengthened by using embossed circumferential beads to provide increased rigidity.

On both tight-head and open-head pails, a wide range of opening sizes, pouring spouts, and cap closures is available. To cut costs, covers can have a simple, threaded pouring nozzle topped by a screw cap. Even simpler are pails furnished with just a dust cover over the pour opening to keep the interior clean, with the user clinching on or pressing in a pouring fitting after filling. Various metal and plastic pouring devices are offered, mostly of the pull-up style; these are covered by a cap during shipment. Some pour fittings incorporate vent openings that eliminate the need for a separate vent opening on the cover. Tamperproof seals, consisting of a steel cap clinched directly onto the pail fitting, are often used.

BIBLIOGRAPHY

1. U.S. DOT Regulations, *Code of Federal Regulations*, Title 49, Part 178, 1985.

2. *ANSI MH2-1985,* American National Standards Institute, New York.

3. *Uniform Freight Classification,* Uniform Classification Committee Chicago, Ill.

4. *National Motor Freight Classification,* National Motor Freight Classification Board, Alexandria, Va.

5. *United Nations Transport of Dangerous Goods—Recommendations of the Committee of Experts on the Transport of Dangerous Goods,* 3rd rev. ed., The United Nations, New York, 1984. This is available from the United Nations Bookseller Office, United Nations Building, New York, as well as from certain other governmental offices and commercial publishers.

6. *SSCI Guide for Export Shippers of Hazardous Materials,* rev. ed., Steel Shipping Container Institute, Union, N.J., 1985.

ARTHUR J. SCHULTZ, JR.
Steel Shipping Container Institute

E

ECONOMICS OF PACKAGING

Packaging economics is an imprecise term which has three usages:

1. *Packaging in macroeconomics.* This pertains to packaging as a part of gross national product (GNP) and total employment. This aspect of packaging as a component of the national economy and international trade balance is a matter of interest to economists in government and industry; to businesses that produce and sell packaging materials, machinery, and services, and to their trade associations; and to companies that buy packaging materials, especially the large and multinational corporations, and the professional societies therein.

2. *Packaging-industry economics.* This pertains to the cost structures and profitability of industry categories and companies that supply packaging materials and machinery; design, printing, and decoration for packages; and contract-packing services. This aspect is mainly of interest to buyers and sellers of packaging components and services; to the investment community, including banking; to labor; and to professional societies and trade associations related to packaging and its functions.

3. *Packaging-development economics.* This pertains to the technical process of analyzing and controlling in detail the cost structure for packaging any specific product, including the origination of package design and engineering, adaptation to its distribution and in-use environment, and product-package cost relationship. This aspect is the every-day concern of professional practitioners who, in the role of packaging specialists, are engaged in the creation, production or distribution of particular packages for particular products; and of educational institutions where appropriate skills are taught. In this area, economics is one of several sets of criteria for success, along with the technicalities of product protection, marketability, manufacturing feasibility, and quality control. Aside from the finished product/package combination the packaging development *process* has its own economics. Its steps are common to most projects.

Packaging is an activity diffused through nearly all economic entities of society, in both goods industries and service industries. On the supply side, manufacturing industries produce packaging materials and packages for use by other manufacturing industries. Service industries provide design and technical skills for package manufacturers and users, and packing services for manufacturers, wholesalers, and retailers. On the demand side, nearly every grower and manufacturer requires packaging for containment, protection and preservation of products through distribution to the consumer. Packaging materials and opertions are manufacturing functions, whereas distribution (warehousing and transportation) is a service function, as are wholesaling and retailing.

The kind and level of protection needed and provided are technical matters specific to the nature of each product and its distribution pattern. Cost components for containment, pro-

tection, and preservation therefore vary widely from rugged to delicate products, and from inert to perishable. The combination of technical protective needs and packaging costs suffice to define the economics of packaging industrial products.

Packaging for consumer products performs one or more marketing functions in addition to the technical functions, such as multicolor decoration and convenience features to communicate a favorable brand image, which usually represents a cost increment. A further increment of cost accrues to food and drug packaging, because they are heavily regulated by the Federal and state governments, and in some cases by local governments. The net effect of regulation is to require relatively expensive packaging materials for reasons of purity, sanitation in packing operations, and considerable mandatory label information. Some areas of regulation are nonuniform geographically, which limits economies of broad-scale distribution for regional and national brands.

Packaging in Macroeconomics

The framework for analyzing the United States' national economy was developed by the Executive Branch of the Federal government. The ongoing process of monitoring the state of the economy is a responsibility of a few Executive Departments, principally Agriculture, Commerce, and Labor. A focus on business and industry is maintained by the Department of Commerce, through its Bureau of the Census, specifically in the Office of Industrial Economics.

In broadest terms, the National Income (GNP) is conceived as the aggregate of moneys put into three sectors of activity over any one-year period: Goods Industries, Service Industries, and Government. Table 1 shows this distribution for 1982 (1).

As a basis for gathering, publishing, and analyzing business statistics, every industry within the Goods and Service sectors is identified by a Standard Industrial Classification (SIC) number. Twice each decade, in years ending in "2" and "7", the Department of Commerce takes a census of manufactures, and publishes the findings in the form of reports by SIC numbers over two to three years following. Since packaging materials originate from a number of basic industries, are converted by another group of processing industries, and are assisted in many cases by service industries, the list of SICs which contributes to the total U.S. expenditures for packaging

Table 1. National Income by Industry, 1982[a]

Item	$ 10^9	%
Goods industries		
agriculture	763.8	30.6
mining	68.4	2.0
construction	39.8	1.6
manufacturing	548.9	22.0
Service industries	1317.8	52.0
transportation, communications, utilities	199.9	8.0
wholesale trade	152.4	6.1
retail trade	209.7	8.4
finance, insurance, real estate	369.8	14.8
services	386.0	15.5
Government	363.5	14.8
Rest of world	47.3	1.9
Total	2492.4	100.0

[a] *Without capital consumption adjustment.* Note: Numbers may not add up to the total due to rounding. Courtesy of the Bureau of Economic Analysis.

is long and not always unequivocal. Complicating the situation are such facts as these:

Some companies in basic materials industries also convert their primary products into finished packaging forms. Examples are petroleum refiners who produce plastic resins and convert some of the resin into molded containers; and paperboard manufacturers who convert part of their board into coated and printed carton stock, and folding cartons.

Some industries make both packaging and nonpackaging materials and the two are not separately reported. SIC 3353 reports aluminum metal sheet, plate, and foil. Of these, foil is commonly used for packaging, whereas the others are not. On the other hand, SIC 3479 reports metal foils, including aluminum, copper, and others, of which only aluminum is used for packaging. Packaging machinery is reported as part of two industries: SIC 3551 Food Products Machinery, and SIC 3559 Special Industrial Machinery, NEC (Not Elsewhere Classified).

Some important packaging materials are not yet uniquely identified by SIC number, because they are "young" parts of an SIC industry. Plastic bottles, tubes, and sheet for thermoforming are all parts of SIC 3079 Miscellaneous Plastics Products, which also includes injection molded containers and foamed plastics.

Data are not published in any detail on industries which are represented primarily by one company.

These points are noted to indicate that the SIC system is continually developing, in tune with the industries it classifies, but since the system must follow industry developments, there is and will continue to be a lag in precise reporting in some circumstances, especially small and emerging sub segments of a field so diverse as packaging.

Thus, to obtain an approximation of the economic dimensions of the packaging "industry," one must assemble a picture from many pieces, as a jigsaw puzzle. Many of the pieces are quite accurately defined by unequivocal SIC statistics. This is the case if the entire industry output goes to packaging uses, such as metal cans, glass containers, and corrugated/solid board boxes. In many other industries, such as paper and paperboard, plastic films, aluminum foil and sheet, and molded plastics, packaging is only one of many uses. In such cases, SIC data are of limited value. There are many trade associations and trade publications, however, which speak of and for packaging industrial segments, and much statistical and economic information is available from them (see Networks).

By combining data from government publications and trade sources, it is possible to estimate the total expenditures for packaging materials as shown in Table 2.

Some indeterminate part of an additional $4 billion ($10^9$) in materials is consumed in packaging end uses. These are such items as tags, inks, tapes, adhesives, cushioning forms of paper and plastic, cotton (for plugs), strapping (metal and plastic), and waxes. Those which are used in the manufacture and delivery of the items in Table 2 appear in the dollar values of those items. Additional quantities are used by buyers of packaging materials, who use them for packing agricultural, industrial, and consumer products, for example, adhesives and waxes for sealing cartons and corrugated boxes, string for sewing multiwall bags, and inks for imprinting pack dates and plant codes on packages of finished goods.

Table 2. Major Packaging Materials Industry Shipments, 1983[a]

Product	SIC	Units shipped	$\$ \times 10^9$
metal cans	3411	units, 93×10^9	10.7
glass containers	3221	units, 45.3×10^9	5.5
plastic bottles	3079[b]	units, 16×10^9	1.6
folding cartons	2651	236,000 t	3.0
Containers and food	2654		2.8
corrugated boxes	2653	23.5×10^9 m^2	12.7
setup boxes	2652	363,000 t	0.5
closures	3466, 3079[b]	units, 60×10^9	3.4
bags	2643		2.0
fiber cans, tubes	2655[b]		0.5
drums	2655[b], 3412		1.0
wood packaging	2441, 2249		0.7
flexibles, paper	2621[b]		3.0
films, laminations	2641, 3079[b]		2.5
aluminum foil	3353[b], 3497[b]	163,300 t	0.5
		Total	$50.4

[a] Refs. 2–5.
[b] Part only of indicated SIC.

Most tags are used for marking in wholesaling and retailing. Strapping is usually applied to hold a group of packages together on a pallet for handling and shipping, or to hold unpackaged heavy goods and machinery on skids; a contribution to damage resistance, but not strictly a packaging medium. Tapes made of paper and glue (gummed tapes) are used for sealing corrugated boxes in small-quantity packing operations. Automated glue-sealing of box flaps is much less expensive in large-scale packing operations. The paper stock used to make the tape is included in SIC 2641 Paper Coating. Film-base tapes are always coated with a pressure-sensitive adhesive, and are used in small amounts for packaging end uses. These tape films and plastic strapping are included in SIC 3079 Miscellaneous Plastic Products.

There is redundancy in the reporting of flexibles, films, and foil, in that many laminations are made of paper, films, and foil in combination. The foil category includes gauges up to 0.006 in. (152.4 μm), which includes grades used to make formed trays commonly applied in packing frozen foods to be oven-heated. The truly flexible-packaging gauges of foil for laminating are generally not thicker than 0.00035 in. (9 μm). The converting industry that makes flexible packaging also uses a significant amount of adhesives in combining layers for paper, film, and foil. The amounts and dollar values of such adhesives are built into the output of SICs 2641 and 3079, from which flexible materials cannot be accurately identified. The amount of paper used in packaging conversion during 1983 is reported to have been 5.9×10^6 tons (5.35×10^6 t) (6). This would include overwraps, liners, labels, and laminated flexible materials. A comparison of packaging materials industry sales against gross national product (Table 3) indicates that packaging accounts for 1.5–2% of GNP, declining slightly.

When expenditures for packaging materials are converted from current dollars to constant (1972) dollars to correct for inflation, and compared with population growth over the en-

Table 3. Packaging Materials as a Percent of GNP

	1960	1970	1975	1980	1982	1983
PM[a]	10.5	18.9	29.5	46.1	48.4	50.4
GNP[a]	506	982	1529	2626	3069	3305
$\frac{PM}{GNP}\%$	2.08	1.92	1.93	1.76	1.58	1.52

[a] In billions (10^9), current dollars.

tire period, it is shown that real value of packaging has grown about 80% more than population, but only half as much as GNP similarly corrected to real GNP. (see Table 4).

Table 4. Population and Packaging Growth

	1960	1983	Increase	Increase, %
population, 10^6	179.3	233.4	54.1	30.2
GNP[a]	737.	1534.7	797.7	108.2
PM[a]	15.2	23.7	8.5	56.1

[a] In 1972 dollars, billions (10^9).

In 1972 dollars, per capita costs for packaging in 1960 were $84.77; in 1983, $101.54. This represents an increase of 19.8%. For the same period, real per capita income rose 60% from $4110 to $6575.

Several interpretations can be made from Tables 3 and 4:

In terms of current dollars, gross national expenditures multiplied about 6.5 times 1960 to 1983, whereas outlays for packaging materials multiplied 5 times.

In constant (1972) dollars, GNP doubled, and expenditures for packaging increased about 1.5 times.

Population grew by $\frac{1}{3}$ in the same time span. The use of packaging grew per capita, but by much less than per capita income. Manufacturers have concentrated on controlling the cost of packaging their products to keep the rate of increase below the level of real growth.

The interpretation in the last statement is a very real one. Product manufacturers normally spend 15–25% of their technical packaging expense budgets seeking means of reducing packaging costs. The basic reason is with few exceptions, packaging does not add value to the products it contains. Its primary mission is to preserve product value through the rigors of distribution and sale, after which it is discarded. The few exceptions where packaging adds value are

1. The package is reusable by intention. This is most commonly a promotional package designed with the marketing objective of promoting short-term sales. An example is packing tea bags in a decorated metal tea canister; too expensive for routine packaging, but affordable for a one-time "special" to gain market share and build brand loyalty.

2. The package adds a function, such as applying the product. Examples are roll-on deodorants, spray cans of paint, and prepacked disposable hypodermic syringes of injectable medicines.

3. The package conveys perceived value by style and design, as in fragrance cosmetics.

Even in these instances, the package is engineered for minimum cost consistent with meeting its objectives. The objectives of the reusable package and the cosmetic package are market-oriented. The medical injectable is safety-oriented, and in some states regulations require that all in-hospital medication must be packed in single doses. The roll-on and spray packages are of course market-oriented. Competition is the force which impels manufacturers to minimize their packaging costs, even in "luxury" consumer products. This subject is examined in the last section of this article.

Packaging Industry Economics

In the first section, packaging is viewed as a piece of the picture of the national economy, wherein a group of industries supplies nearly all other enterprises (agricultural, manufacturing, and service) with packaging materials, and in turn the buyers package their products for distribution throughout the economy.

In this section, the focus is applied to the economics of packaging materials themselves, as products of selected manufacturing industries. The major packaging vehicles, called "media," are

glass bottles and jars, bottles with a neck finish smaller than 38-mm are "narrow-mouth", jars with finish larger than 38-mm are "wide-mouth;"

metal cans, made of steel or aluminum;

cartons, made of paperboard;

cups, tubs and trays, made of paper, paperboard, plastic, or aluminum;

bags, pouches, and envelopes, made of paper, plastic film, aluminum foil, or combinations thereof (including multiwall);

corrugated boxes, made of three or more layers of kraft paperboard;

drums and barrels, made of steel, paperboard or plastic; and

bulk shippers, made of corrugated board, wood, textiles or plastic sheet, or metal.

Of the above groups, glass and metal packages are considered "rigid;" cartons, cups, tubs, and trays, "semirigid;" bags, pouches and envelopes, "flexible" packaging. Multiwall bags, corrugated boxes, drums and barrels, and bulk shippers are called "shipping containers".

With the exception of drums, barrels, bulk shippers, and multiwall bags, not one of the above is used alone. Combinations of two are necessary.

1. Bottles, cans, cups, tubs, and trays require additional components to complete their function as packages. Bottles need closures and labels, cans need a closing end and labels, the others need lids or flexible lidding material, or covers, outer bags, or overwraps. All of these are called "primary packages", because they directly contain and protect units of product, either by weight, volume, or count.

2. Primary packages are packed into shipping containers for distribution. All shipping containers are called "secondary packaging". The most common is a corrugated box, but the term includes other concepts such as paperboard trays with plastic film overwrap, kraft-paper bundle wrap, and molded returnable carriers (as for soda bottles).

Table 5. Employment in Packaging Materials Industries, 1983[a]

SIC	Industry	1983 Shipments, $\times 10^9$	Units shipped	Total employment	Packaging end-uses, %	Employment for packaging products
3221	glass	$ 5.5	43×10^9	51,000	100	51,000
3411	cans	10.7	93×10^9	43,400	100	43,400
2611	pulp	3.2	5.44×10^6 t	15,000	38	6,000
2621,31,61	paper/paperboard	33.4	59.9×10^6 t	188,000	38	71,400
2651	folding cartons	4.0	2.36×10^6 t	43,000	100	43,000
2652	setup boxes	0.5	na	13,000	100	13,000
2654	sanitary containers	2.8	na	25,900	100	25,900
2655	fiber cans, etc	1.6	na	15,400	100	15,400
2653	corrugated boxes	12.7	22.95×10^9 m²	94,000	100	94,000
2643	bags	5.8	na	54,200	100	54,200
2821	plastics	17.3	19,000 t	53,600	27	14,500
3079[b]	plastic products	2.2	na	29,000	100	29,000
	plastic bottles	1.6	16.2×10^9	15,400	100	15,400
Total		*101.3*				*476,200*
Total manufacturing						19,000,000
% of Manufacturing employment for packaging materials						2.51
Total U.S. employment						106,000,000
% of total employment for packaging materials						0.45

[a] Ref. 7.
[b] Estimated, injection-molded, extruded, and thermoformed packages and components.

The significance of distinguishing the primary and secondary packaging is that they are most often independently variable components of total packaging material costs. Thus, when one wishes to compare the costs of packaging a given product in several packaging media, the materials calculation must be carried at least through the secondary packaging. Pretzels, for instance, can be packed in cartons, plastic-coated paper bags, or in plastic-film bags. Looking at the primary packaging materials only, one finds that the film bag is least expensive, the plastic-coated paper bag a little more expensive, and the carton considerably higher. The obvious question is, "Why would anyone use a primary package other than a film bag?". The answers lie in adding the cost of the secondary package, the corrugated shipper, which must be stronger and heavier to protect bags than to protect cartons. It likely must also be larger (and more costly) for bags of a given net weight than for cartons, since the product is less compact in bags. The net result is that the sum of primary and secondary packaging for pretzels is nearly the same for all three primary packages, and choices are made on other criteria such as printability, retail shelving and display, and promotional uses of the package. Specific-product packaging procedures and costing are discussed in greater detail below. The intent of this section is to explain the economics of a few important branches of the packaging materials industries, and the implications thereof to package users.

Table 5 is an expansion of Table 2, introducing an insight to the effect of labor costs on packaging materials costs. One can compare glass containers and metal-cans industries, for example. It appears that in 1983 the glass industry produced 43×10^9 units (and packaged and sold them) with 51,000 employees (2), and the metal-can industry produced and sold 93×10^9 cans with 43,400 people, or 15% less than the glass-industry employment. The comparison of cost per employee is shown Table 6.

At first glance, it appears that productivity in the metal-can industry is more than twice that of the glass industry, but a correction must be made for the number/cost of personnel in glass plants who make and pack corrugated boxes for delivering the ware, which is not done to any extent in the can industry. The numbers are not separately reported in the industry statistics, but can be estimated by referring to the statistics for corrugated boxes (SIC 2653) which indicate that 94,000 persons are employed to make and convert 247×10^9 ft² (22.95×10^9 m²) of corrugated board.

The Department of Commerce (8) estimates that for each person in the U.S. population, 100 corrugated boxes were made and used in 1983. Dividing the 247×10^9 ft² (22.95×10^9 m²) by 235×10^6 population yields an average of 1050 ft² (97.55 m²) of board per capita; or an average box usage of 10.5 ft² (0.976 m²) of corrugated board; very close to the 10.72 ft² (0.996 m²) required to case 12 1-qt (0.946-L) jars. Then, noting that 24.7×10^9 average boxes employed 94,000 in their manufacture, the glass industry is estimated to have employed 13,600 of their 51,000 total for making 3.58×10^9 cases to hold an average of 12 glass containers each. The actual number is probably somewhat less, because glass bottles to be used for carbonated beverages, malt beverages, and baby food are usually not delivered in corrugated cases. An estimate of 40,000 glass-industry employees for actual glass-making (excluding 11,000 for box-making) brings the productivity closer to, but not matching, that of the metal can industry.

Two other factors to be noted are the weight differences and energy-to-make differences between glass and can packages. If

Table 6. Labor Component for Glass and Metal Packages

	1983 shipments		Total employment	Shipments per employee	
	$ \times 10^9$	Units $\times 10^9$		$	Units
Glass	5.5	43	51,000	108,000	843,000
Cans	10.7	93	43,400	246,500	2,140,000

the glass jar weight for a given quantity is 428 g; the can weight is 98 g. The energy input to make and form glass and steel are about the same per cm^3 (9). Glass has a specific gravity of 2.5, and steel a sp gr of 7.5, thus, the volume of glass in the jar is 428/25 = 171.2 cm^3 and the volume of steel in the can is 98/7.5 = 13.1 cm^3, a ratio of 13:1 in favor of steel.

The energy-consumption ratio is not necessarily identical to the energy-cost ratio, because glass- and can-making use different energy sources. Along with coopering, glass-container manufacturing is the oldest packaging-materials industry in what is now the United States; a glass plant was built at Jamestown, Virginia in 1608 (8), using local wood for fuel. The fact is that glass bottle-blowing was largely a craft throughout the 17th and 18th centuries, and most of the 19th. The first bottle-blowing machine was built in 1903. During the craft years, wood was the common fuel, but with the birth of the petroleum industry at the end of the 19th century, with its by-product of clean natural gas, glass plants were built in the early 20th century at locations convenient to supplies of high-purity sand and natural gas. Since the jump in energy costs which started in 1973, the glass industry has halved its energy costs by reducing plant-heat losses, briquetting the sand/soda/lime/cullet mix, preheating the briquettes by recirculating stack gases, and judiciously balancing gas, oil, and electricity as heat sources for glass furnace operations.

Because the glass industry is vertically integrated from basic raw material to finished packaging material (10,11) plants can easily determine their energy usage and costs per thousand units produced. Calculating the energy used in making cans is more difficult, because the steel industry supplies can sheet along with a multitude of other steel products from rolling mills which may or may not perform the basic reduction of the metal to prepare the starting ingots. In any case, the principal fuel for steel-making is coke, which is made from coal at considerable energy cost. The final stage of forming cans from steel sheet represents about 10% of the total energy input, but trim waste can increase the net energy used per 1000 cans by 15%, depending on can size. (see Energy utilization).

Aluminum-can manufacturing consumes twice the energy (9) per cm^3 as glass and steel, most of it in basic metal reduction. Two factors make aluminum cans affordable:

Aluminum sp gr is 2.5, the same as glass, and ⅓ that of steel. It draws well to thin-wall sections, so can be formed into lightweight containers, suitable for pressurized products such as carbonated drinks and malt beverages, which support the thin wall against denting.

Aluminum cans are recycled, so the metal can be reformed without the penalty of repeating basic reduction.

Many interesting comparisons can be made. Using the data in Table 5, for instance, dollar shipments per employee can't be calculated (see Table 7). Table 7 shows that the manufacture of setup boxes (such as shoe boxes, candy boxes, and department-store gift boxes) is outstandingly low in productivity. The industry has in fact reduced from 800 plants to 300 over the last 20 yr (12), and gift boxes are no longer freely distributed by retail establishments. Many retailers have converted to folding-carton structures.

Another useful comparison is that of glass and plastic bottles (see Table 8).

Table 7. Selected Industries, Shipments per Employee

SIC	Industry	Shipments : Total employment, $
2651	folding cartons	93,000
2652	setup boxes	38,500
2654	sanitary food counters	108,000
2655	fiber cans, tubes	104,000
	plastic bottles	104,000

Dollar shipments per employee (see Table 8) are very similar, but unit shipments appear much more favorable in the plastic-bottle industry. Dividing shipments by units shows that the average price (or value) of a glass unit is higher than plastic by 29%, whereas the productivity per employee is lower by 25%. This observation, although factual, is not the whole story. The shipments per employee are almost the same on a dollar basis, because the plastic-bottle industry ships more units at a lower price per unit. The price advantage has been the key to their success. If, however, the employees who make and pack corrugated boxes in the glass industry, say 11,000, are not included in the total employment, then the 43 × 10^9 glass containers are made by 40,000 employees, and their productivity is 1,075,000 per employee—almost identical to that in the plastic-bottle industry.

Table 8. Glass and Plastic Bottles

Item	Glass	Plastic	Glass ex-Casing
shipments	$ 5.5 × 10^9	$ 1.6 × 10^9	$ 4.3 × 10^9
units	43 × 10^9	16.2 × 10^9	43 × 10^9
employment	51,000	15,400	40,000
shipments per employee	$108,000	$ 104,000	$ 107,500
units per employee	840,000	1,052,000	1,075,000
average price per bottle	$ 0.128	$ 0.099	$ 0.100

To see how this comparison looks in dollar shipments per employee in the glass industry without the casing operation, it is necessary to subtract $28 per 1000 bottles, or $ 1.2 × 10^9. The results are shown in the last column of Table 8.

Package Development Economics

This section moves from the broader pictures of national and industrial economics to the "microscopic" scene in which choices are made and moneys are expended for the concrete purpose of developing specific packages for specific products. Sellers and buyers of packaging materials, machinery, and services are equal partners in the process; the sellers provide the means, the buyers acquire the performance.

Two aspects of packaging development economics must be examined for the successful accomplishment of a development project. The first is an overview of the economic options, and the second is a definition of the steps through which the project must advance to yield a satisfactory packaging specification. The product of every packaging development is a specification for packaging which can be procured within an economic target range; will protect the product contained, within a defined distribution and selling environment; and will perform required marketing functions in competition.

Economic options. The process for comparing package costs is illuminated below by example. Since food and beverages use 50% of all packaging, the example is a familiar beverage product.

Table 9.   Product X, Basis for Cost Calculations

Product: a dry, moisture-sensitive granular substance
Density: 0.4798 g/cm^3
Net package weight: 1 lb (454 g)
Package size: 454/0.4798 = 46.2 cm^3 (1 qt)
Primary package options: glass jar   composite can
                          plastic jar   paperboard carton
                          metal can   flexible-foil bag
Type of shipping container: corrugated box
Primary packages/shipping container: 12
Shipping containers required for 1000 packages: 83.3
Other requirements: reclosability of primary package

Let it be assumed that the product is a dry, moisture-sensitive granular substance, such as a powdered beverage mix. It is a solid of sp gr = 0.4798, such that 1 lb (454 g) requires a package of 1-qt (946.3-cm^3) capacity. The choice of packages includes a glass jar, metal or composite can, moisture-proof carton, plastic jar, or barrier flexible package, such as a reclosable bag, all based on the further assumption that the product is used a little at a time. One further condition must be stated: the number of primary packages per shipping container (a very typical number is 12). Thus, the problem is to rank the costs of several packaging media for 1 lb (454 g) of product X in reclosable moisture-barrier packages, 12 packages per corrugated box. These data are summarized in Table 9.

Comparative costs, based on 1983 dollars, are shown in Table 10. The table shows that glass packaging is the most expensive, followed by plastic-jar and metal-can packs in that order. Compared to metal, composite cans are about 10% less expensive. Cartons with barrier liners and foil-laminated bags are both considerably less expensive than the other options.

Product X in glass jars. The design is a round 1-qt (0.946-L) jar made of clear, uncolored, flint soda-lime glass (see Glass container design; Glass container manufacturing). The 70-mm plastic screw closure (see Closures) contains a diaphragm which seals across the finish of the jar to provide a tamper-evident feature before first opening (see Closure liners; Tamper-evident packaging). Jar diameter: 9.5 cm. Jar height: 20 cm with closure. Space for label: height (8 cm) × length (31 cm) = 248 cm^2, complete wraparound with 1-cm overlap (see Labels and labeling).

The corrugated shipping container for 12 jars is a "regular slotted case" (RSC) with "nest" of corrugated partitions (dividers) to prevent the jars from coming in contact with each other (see Boxes, corrugated). Each cell formed by the partitions is 9.5 cm × 9.5 cm square and 20 cm full height, making the filled and capped jars fit snugly, but not squeezed, in their separate cells. The overall dimensions of the case are 40 × 30 × 20 cm (inside measurements), which allow for the thickness of the partitions. The total area of corrugated board required for this case is 10.72 ft^2 (0.996 m^2). The cost of a case is based on an industry-wide price of $350/ton ($385/t) of board, which represents 55% of the case cost with one- or two-color printing. Glass makers do not price the glass and the case separately. This is a practice of long standing, based on the fact that most glass vendors have their own box-making facilities. They also deliver to buyers within about 500 miles (805 km) at no charge. Both factors are used as selling points to compensate in part for the higher cost of glass relative to the other media.

Product X in plastic jars. The shape, closure, and case construction is the same as for glass jars. Jar cost is based on a polyolefin resin, with translucency, but not total transparency. The jar weighs about one-eighth as much as glass (plastic sp gr 0.90–0.96 vs glass sp gr 2.5). Jar height is the same as glass (20 cm), but because the walls are about one-third as thick, jar diameter is 9 cm. This reduces label length to 28.5 cm.

Plastic jars are priced without the shipping container, so the container cost is entered separately. Because the jar diameter is smaller, the RSC is smaller: 38 × 28.5 × 20 cm^3, board area 9.94 ft^2 (0.923 m^2). Partitions are used in plastic bottle and jar cases not to prevent contact (plastic is more resilient than glass), but to provide vertical compression strength. It is common practice in distribution to stack products up to 18-ft (5.5-m) high in warehouses. Under those conditions, the compression strength of glass containers is more than adequate; but plastic containers may distort, depending on the specific design, wall thickness, and weight of product. Partitions are included to add "column strength."

As for the cost difference between the glass and plastic jars themselves, the glass jars cost roughly $220/1000 ($250 minus about $30 for the cases) and the plastic jars roughly $175/1000. The cost of raw materials is much lower for glass than for plastic. The 428-g glass jar is made of material (sand, soda, lime, and broken glass) that costs about $0.05/lb ($0.11/kg), which amounts to $47/1000 jars. The 53-g polyolefin jar is made of resin that costs about $0.50/lb ($1.10/kg), which

Table 10.   Product X, Comparative Package Costs, $ × 1000.

	Primary package	Closure/ reclosure	Label	Corrugated box	Total
glass jars	250.00	35.00	9.50	^{a}	294.50
plastic jars	175.00	35.00	8.70	28.00	246.70
metal cans	175.00	25.00	9.65	16.80	226.45
composite cans	157.00	25.00	^{b}	16.80	198.80
paperboard cartons	53.10	12.75	^{c}	14.60	80.45
flexible-foil bags	70.00	^{d}	^{e}	11.70	81.70

^{a} Glass prices normally include the corrugated box unless specifically excluded by the buyer.
^{b} Labels are part of spiral-wound composite cans as delivered.
^{c} Paperboard cartons are preprinted.
^{d} Foil bags do not require a separate closure. They are folded at the top after filling and heat-sealed or glue-sealed.
^{e} Foil bags are preprinted.
Note: Costs are estimates based on requirements of 10–20 million (10^6) packages. Actual costs depend on volumes ordered, location of delivery within the continental United States, and specification details.

amounts to $58/1000 jars. The difference of $11/1000 jars is far outweighed by relative energy costs: 1000–1500°C to melt glass ingredients vs about 300°C to melt the resin (see Blow molding; Energy utilization). Further, a glass plant recycles at least 15% of all containers made to maintain a minimum of broken glass in the melt, which hastens solution of the other ingredients.

Product X in metal cans. Current specifications for cans of about 1-qt capacity call for a grade of steel sheet known to the trade as "75TU-CCO" or "0.05TP". The 75TU refers to weight and temper; the CCO or 0.05TP to either of two rust-resistant coatings (see Cans, steel; Tin mill products). This grade of sheet steel costs about $0.37/lb ($0.816/kg). A 1-qt can with a 4.0625-in. (10.3-cm) diameter is 5.25-in. (13.3-cm) high. It weighs 98 g. Allowing for trim waste in cutting the round ends, one pound (454 g) of steel yields 4.3 cans (see Cans, fabrication). This amounts to steel cost of $86/1000 cans, which represents about 50% of the cost of the finished can with a loose end.

The metal top end of the can, once removed, cannot serve as a reclosure. Therefore, a plastic snap cap, as provided with coffee cans, is specified. Can labels cover the entire cylindrical body, which in this instance has a circumference of 13.5 in. (34.3 cm) and a height of 5.25 in. (13.3 cm). The can label is slightly larger than the jar label and slightly more expensive.

The corrugated case is more compact than a case for jars, because cans occupy space more efficiently than the tapered jar shapes. In addition, can cases are made with an end-loading construction (ELC), which has less overlap area in the closing flaps (see Case loading), and they do not require partitions. The 12 cans require only 6 ft^2 (0.557 m^2) of corrugated board, which is 40% less than the jar requirement.

Product X in composite cans. To the casual observer, a composite can looks the same as a metal can because the shape is cylindrical, the ends are metal, and the body is fully labeled (see Cans, composite). Actually, the spiral-wound body is made of paperboard plus plastic film and/or aluminum foil, and a printed outer layer which provides the labeling. Compared to steel, there is a cost advantage of about 10% in the body, which also eliminates the need for a separate label. Corrugated box dimensions are assumed to be the same as for metal cans because steel and composite cans have the same dimensions within a tolerance of 1/16 in. (1.6 mm).

Product X in paperboard cartons. A construction analogous to a cereal box is employed. The nonbarrier carton contains a laminated paper/plastic bag liner for moisture barrier (see Bag-in-box, dry product) that is reclosable by folding. The carton is made from clay-coated recycled chipboard in a widely used thickness of 0.018 in. (0.46 mm) (see Cartons, folding). Separate labeling is not required because the clay coating provides a smooth white printing surface.

Since the bag liner takes space in the carton, the carton must be larger than an unlined carton for comparable product volume. For that reason, this carton is dimensioned to a volume of 1200 cm^3, about 25% over the required minimum for the product. The carton dimensions are 20 × 12 × 5 cm, and the bag is a 25 × 15 cm "pillow pouch" (see Form/fill/seal, horizontal; Form/fill/seal, vertical). The carton is formed from 140 in.^{2} (903 cm^2) of paperboard, very close to 1 ft^2 (929 cm^2).

The bag is formed from 750 cm^2 of a flexible laminate of paper/polyolefin film (see Laminating; Multilayer flexible packaging), a specification which costs $0.11/1000 in.^{2} ($0.171/ m^2), which amounts to $12.83/1000 bags. The corrugated case is relatively small, because cartons leave no wasted space in the case. Dimensions are 30 × 20 × 24 cm^3, formed from 5.2 ft^2 (0.483 m^2) of corrugated board.

Product X in flexible foil bags. A barrier bag is made of a heat-sealed construction (see Sealing, heat) with a square bottom so that it can stand up. The material is a stiff paper laminated to aluminum foil with an interior layer of plastic film for reinforcement of the foil barrier and for heat sealing the top after filling and folding. The bag has most of the features of the bag-in-carton, but built into a single structure. Reclosure is achieved by folding down the top of the bag. The difference between this and the carton pack is that the total package shrinks as the contents are used, and the void space in the carton increases as the bag contents are reduced.

As barrier packages, the lined carton and flexible bag are about equal in cost, and they represent an almost irreducible minimum for a printed reclosable stand-up system. The corrugated box for 12 bags is even less expensive than the carton case, because the waste space required in the carton for the bag liner is eliminated along with the carton. The bags are folded down before sealing to the volume of the product. The net result is that the finished sealed bags fold to a height of 16 cm in contrast to the 20-cm cartons. This is reflected in the lower cost of the bag case.

Product/package cost ratio. All packers are dedicated to minimizing packaging costs as a matter of economic principle and competitive necessity, but alleged "overpackaging" of foods and beverages, the essential commodities of life, is a recurrent issue (13–15). Cases are cited in which the package costs more than the value of the product contained, such as sugar-free soft drinks, malt beverages, some ready-to-serve soups, and some cereals.

At any given time, the cost of a certain package type, say a can, jar, carton, or bag of a given capacity, is nearly an absolute, within a range for costs of structural features to adapt to a specific product. For example, Table 10 shows the cost of a 1-qt (946 ml) glass jar pack, with all components, as $0.2945. If one fills it with water, which may cost 1¢, the product/package cost ratio is about 1 : 30, which is to say that the package costs 30 times as much as the contents. The ratio could be improved to 1 : 25 if a plastic jar pack were used instead (Table 10), or to 1 : 20 if a composite can were chosen.

For the same packages holding milk, which has a value of about $0.50 per quart, the ratio would be quite different: 5 : 3 for the glass pack, 2 : 1 for the plastic jar pack, and 5 : 2 for the composite-can pack. With the cost of a given pack essentially fixed, the value of the product contained is the determining factor in the ratio. The more expensive the product, the more favorable the package ratio: a $700 refrigerator is packaged in a corrugated box and bolted to a light wooden platform. The $5 package is 0.71% of the retail price.

The least-expensive package possible is a paper bag or a polyethylene film bag. The latter provides a barrier to water and humidity, and is used as a bread package for that reason. It is also used for loose frozen food and a few high-bulk cereals, such as puffed rice. Unless the product has shape, as a bread loaf, the bags do not stack well on retail shelves, nor do they carry fatty products, such as cheeses and snacks, or marketing-based communications, such as separable coupons, complex reclosures, or hard articles which could puncture. For these reasons, the simplest, cheapest package will not contain and protect and market all products.

Packaging development economics. There is a difference between package development and packaging development. The term package development pertains to the structure and performance of a primary container and its shipping container. The product of this process is a detailed specification for materials and structure of the components. Packaging development involves much more than that. The term includes the package specifications, but also matters related to manufacturing and packing facilities, marketing and distribution (see Specifications and quality assurance).

Both package development and packaging development can be initiated by either a vendor or a packer. A vendor that perceives a market opportunity can initiate the process on speculation, or under contract to a packer. A packer initiates the process to improve an existing package or develop a package for a new product. In either case, the development steps are essentially the same (16):

Step 1. Establish Criteria

Activity. Define the product or product line; define protective requirements, business objectives; cost constraints; use of established production facilities; time.

Cost or cost basis. There are little or no outlays. Personnel in the packer's organization is assigned to study the product/ product line, assess its fragility (see Testing, product fragility) or climatic sensitivity (see Distribution hazards), package options, and in-house processes. Product samples are laboratory tested.

Step 2. Identify Concepts

Activity. Select packaging options that meet the criteria of Step 1. In a consumer products company, Marketing leads in Steps 2–5. In an industrial products company, Operations and Sales share leadership. Step 2 provides direction for a designer or vendor to execute Steps 3 and 4.

Cost or cost basis. In-house work hours, concepts solicited from all functions concerned.

Step 3. Design, Phase I

Activity. Develop sketches illustrating a broad range of possible embodiments of the options selected in Step 2; usually 20–30 ideas.

Cost or cost basis. Design services are provided by several sources: vendors, commercial design organizations, some advertising agencies, freelance design specialists, and in-house design departments of large companies. Vendors who are established suppliers generally provide design at no charge up to some limiting percent of sales volume. The same is true for advertising agencies. Design houses and freelance designers work on a contract basis. Charges for Phase I are based on the need for creativity and innovation. A typical contract, given good direction and criteria, would cover a two-month assignment for $15,000–35,000. If design is done in-house, cost basis is in-house work hours.

Step 4. Design, Phase II

Activity. Select two or three embodiments based on criteria established in Step 1; make refined illustrations (renderings), and actual-size models. Packer is responsible for screening

process. Marketing leader obtains guidance from other functions in determining practicality, cost estimates, other pros and cons of the 20–30 options.

Cost or cost basis. Phase I designer proceeds to Phase II, at $100–200/hour for designer time, $30–70/hour for technicians' time. Depending on number and complexity of drawings and models, Phase II might take one or two months. Cost might amount to $150/drawing, $500/model, plus $5,000–10,000 for designer supervision.

Step 5. Final Selection

Activity. Select one embodiment based on criteria of Step 1 plus results of Step 4.

Cost or cost-basis. No direct costs other than in-house personnel. Packer's team has responsibility for decisions, which represent a commitment to larger expenditures in approaching production readiness.

Step 6. Design, Phase III

Activity. Prepare specifications for making test samples of the selected embodiment, in as many sizes as called for by the business plan.

Cost or cost-basis. This step is called the "mechanical phase" of package development. The designer or vendor prepares blueprints to size for all components while the client's packaging function prepares specification texts. Designer's fees follow those of Phase II. Vendors do not charge if an order is forthcoming.

Step 7. Sample Evaluation

Activity. Run samples through manufacturing and distribution, or simulations thereof. If necessary, repeat Steps 6 and 7 to a point of satisfaction.

Cost or cost-basis. The vendor charges for sample tooling on a cost-plus-overhead basis, which provides a few hundred samples to the client. The costs differ widely, depending on the packaging medium. A glass bottle mold or an injection mold might cost $5,000–8,000; a thermoforming mold about half as much; a carton die as little as $500–1000. Costs multiply, of course, with the number and kind of components required to make finished packages.

For the first time in the project, packing-line machinery becomes involved, and modifying attachments or parts may be needed. The machinery manufacturer usually makes these to specification. Costs are a function of the number and complexity of the parts needed, and based on material, labor, engineering overhead, and a margin. If the supplier installs the parts, there is a charge for a field engineer's time and expenses.

Step 8. Issue Specifications

Activity. All parties who supply package components, machinery or parts, and graphics are given information which enables them to execute their assignments.

Cost or cost-basis. Manufacturing and packing specifications are issued by the client's packaging function, at the cost of work-power, with negligible outlays.

Step 9. Tool for Manufacturing

Activity. Prepare graphic design and color/decoration, printing plates, component molds, dies or tools, and packing-machine parts.

Cost or cost-basis. This is the highest-cost step. Every outlay is expensed except the possible cost for packing-line machinery and installation, which is capitalized. An all-new line for bottling or canning costs as little as $100,000 or as much as $2.5 million depending on the desired rate of production. Cartoning lines can range from $10,000 to $500,000. Printing plates and cylinders run $200–2000 per color. There is no average in this area.

Step 10. Start Production

Activity. Check first deliveries of materials through manufacturing, and adjust packing process as necessary. Train quality-assurance personnel in any new procedures.

Cost or cost basis. At the start of a new package and packaging facility, the component vendor(s), machinery supplier(s), and client are represented to oversee a quality audit of components and performance of both package and process. The client incurs cost for the machinery supplier's field engineer, as in Step 7.

Costs and timetables. When an existing in-production specification is to be applied to a new product, the effort is a package development, and the process is shorter. Steps 2–5 are eliminated. Step 9 is limited to graphic design, color/decoration, and printing plate preparation, because the criteria in Step 1 include a decision that established facilities will be used for packing.

The costs and time span for each of the 10 steps vary widely. Every package and packaging development involves a degree of innovation, even if only in graphics and decoration for product identity; but there are norms which can be anticipated, based on experience. If, however, a true invention is required, a hitherto-unknown solution to a problem, the timetable is indeterminate. It is therefore important for packaging developers to recognize the difference between need for an innovation and need for an invention. The management of every packer wants innovation; but generally does not want to get involved with the expenses and delays of invention. That is usually left to vendors. The costs described above assume innovation only.

Steps 5–10 have a widely-variable timetable. One cycle through Steps 5–7 could be as short as six weeks or as long as six months, depending principally on the time for decision-making in Step 5 and sample-making in Step 7. A repeat of Steps 6 and 7, if necessary, would cost 1–4 weeks for a repeat of Step 6, plus the time for the first cycle of Step 7. Step 8 is short, 1–2 months. Step 9 is the longest. Tooling to modify existing machinery and mold-making for package components is ordinarily a 20–24 week process. New machinery fabrication takes 6–12 months, depending on complexity. Printing preparations are faster, due to the growing use of computerization in graphics development. Most of the total cycle (3–4 months) is taken by the first creative efforts and decision making.

Step 10 usually requires a 1–4 weeks. Production of a new package is done first at the facilities of the component supplier(s), with the client's packaging specialist(s) on site to check quality. Each component is cleared for shipment to the packing plant as it becomes available; then the assembly, filling, and sealing are started at slow speed in the packing plant, with increasing speed to a predetermined target based on output requirements.

Packaging-Function Economics. The packaging-development process just described is a project outline which is repli-

cated for as many projects as a packer undertakes, for any of several objectives:

to launch a new product or product line;

to improve the costs of an established product line/package;

to promote sales of an established product with an incentive-oriented package, such as a reusable container, a coupon pack, a sample pack, a multipack, a bonus pack (more product at regular price), a holiday pack, or a new pack size added to an established product;

to respond to opportunities in new technology, such as new packaging materials which can add shelf-life, or higher-speed production, or patentability; and

to meet changing regulatory requirements, such as those related to coding, dating, packaging materials, tamper-evidence.

In every business enterprise where packaging is involved, these activities gravitate into a *packaging program,* which is centered in an individual if the business is small, or a departmental function, if the business is large enough to keep two or more people occupied fully with packaging developments. The work is essentially technical in nature, and can report into any of a number of business functions, such as Research & Development, Operations, Purchasing, or Marketing; the choice is determined by the emphasis of the business. For example, where protection and efficient delivery at minimum cost is the basic objective, as in an auto-parts business, the packaging activity is most likely to be found in Operations. In a drug-manufacturing business, on the other hand, where cost is secondary to quality assurance, maximum sanitation, and regulatory compliance, the packaging function is most likely in R & D. In a highly-competitive consumer-products line, as cosmetics and HBA (health and beauty aids), Marketing has responsibility for packaging development.

In any case, the packaging development function is budgeted on an annual basis for workpower, project expenses, overhead, and capital. Skilled technical service is currently budgeted at about $50,000/yr to cover a salary and benefits, per capita, plus overhead for staff services and office supplies, for a total of about $100,000 per person-year. Projects are budgeted at expected costs for materials, travel, special parts and dedicated testing equipment, plant charges for trial runs, and laboratory testing services. The capital part of the department budget is applied to laboratory and pilot plant equipment for experimental packing, storage, and handling facilities. Division of budget by category of objectives is likely to be in the neighborhood of

	%
Cost-reduction objectives	20
New-products package development	50
New-technology evaluation	15
Established-package improvement	15

BIBLIOGRAPHY

1. D. H. Dalton and J. H. Boyd, "Growth Trends in the Service Sector," *U.S. Industrial Outlook 1984,* U.S. Dept. of Commerce, Bureau of the Census, March 1984, p. 22

2. Ref. 1, pp. 6-5, 6-7, 6-9, 5-10, 5-11.

3. *1984 Carton Capsules,* Paperboard Packaging Council, Washington, D.C.

4. "Building & Forest Products, Including Paper," *Standard & Poor's Industrial Surveys, Basic Analysis,* Sec. 2, Sept. 27, 1984.

5. *1982 Census of Manufactures,* U.S. Dept. of Commerce, Bureau of the Census, SIC 3353, Preliminary Report, April 1984.

6. *Pulp, Paper and Board (M26A),* Bureau of the Census, USDC Subscriber Services, Washington, D.C., 1984.

7. Ref. 1, pp. 6-3, 5-3, 5-4, 5-12, 11-1, 6-10.

8. Ref. 1, p. 5-8.

9. *Energy Requirements for Various Materials,* Composite Container Corp., Medford, Mass., 1980.

10. I. Boustead and G. Hancock, *Energy and Packaging,* John Wiley & Sons, Inc., New York, 1981.

11. *Billions of Bottles,* Glass Container Manufacturers Institute now Glass Packaging Institute, McLean, Va., 1959.

12. "Set-up Boxes," in ref. 1, p. 5-11.

13. L. Densford, "USDA Report Compares Cost of Food, Packs," *Food & Drug Packaging Magazine,* **45,** 1 Sept. 3, 1981.

14. "Food's Demand for Packaging," *Food Engineering Magazine,* 64 (Oct. 1981).

15. D. Udell, "Food Packaging: Looking Out for the Bum Wrap *Nutrition Action Magazine,* CPSI Subscriptions, Washington, D.C., (Sept. 1979).

16. E. A. Leonard, "The Economics of Design," *Packaging Economics,* pp. 91–106, Libr. of Congress No. 80-67989; Magazines for Industry, div. of Harcourt Brace Jovanovich, Duluth, Minn, 1980.

General References

R. J. Kelsey, *Packaging in Today's Society,* Libr. of Congress No. 78-055347, St. Regis Paper Co., 1978.

H. J. Raphael and D. L. Olsson, "Management of the Packaging Function," American Management Assn., from *package Production Management,* AVI Publ. Co., Westport, Conn., ISBN 0-8144-2205-5, 1976.

E. A. Leonard, *Packaging Economics,* Libr. of Congress No. 80-67989, Harcourt Brace Jovanovich, Duluth, Minn., 1980.

B. E. Moody, *Packaging in Glass,* Hutchinson & Co., Ltd., London, 1963.

C. A. Lewis, "Packaging in Today's Economy," in *New Directions in Packaging,* American Management Assn., ISBN 0-8144-2142-3 1970, Chapt. 1.

Carton Capsules, an annual statistical summary of the folding carton industry, issued in March of each year by the Paperboard Packaging Council, 1101 Vermont Ave., N.W., Washington, D.C. 20005

E. A. Leonard, *How to Improve Packaging Costs,* ISBN 0-8144-2252-7, AMACOM, New York, 1981.

The Packaging Encyclopedia, Cahners Publ. Co., Cahners Magazine Div., Boston, Mass., annual publication.

Current Industrial Reports, U.S. Dept. of Commerce, Bureau of the Census, Industrial Div., Washington, D.C. 20233

J. C. Harrington, "A Tryal of Glasse," *The Story of Glassmaking at Jamestown,* The Dietz Press, Inc., Richmond, Va., 1972.

The Rauch Guide to the U.S. Packaging Industry (data for 1984, 1985, and projections to 1990), Rauch Associates, Inc., Bridgewater, NJ, 1985.

E. A. Leonard
Cornell University

ELECTROSTATIC DISCHARGE PROTECTIVE PACKAGING

Electrostatic discharge (ESD) has become a recognized problem in the electronics industry. Many devices are susceptible to damage from this hazard. Additionally, there are components that are susceptible to the electrostatic fields associated with static charges as well as components that can be damaged by electromagnetic interference (EMI) and radio frequency interference (RFI). "ESD protective packaging" is perhaps a misnomer because protecting these different components and devices could entail consideration of factors in addition to ESD. The various materials and packaging forms available that address these problems are quite numerous. There are films, foams, bubbles, boxes, bins and tubes with more introduced almost daily. Unless a design is mandated by government or military specifications, selecting appropriate materials, controls, and forms is by no means an easy or straightforward task. By examining common materials' classifications, the scope of the problem can be more thoroughly understood. These classifications are usually based on certain test procedures that may reflect on in-use performance requirements.

Antistatic, Static-Dissipative, and Conductive Materials

Most plastics are electrically nonconductive. This insulative property means that they are easily charged by rubbing or separating them from themselves or other materials. This type of charge generation is termed triboelectric charging. Charges generated on insulators are a significant problem because they can be quite large in magnitude and cannot be dissipated by grounding. To make plastic materials less susceptible to charging and to provide a means of dissipating those charges that might be created, they are altered in one of several ways to make them more electrically conductive. The measurement of this conductive ability is resistivity (either surface or volume). Resistivity is inversely proportional to conductivity (ie, the lower the resistivity, the better the conductivity and vice versa). For materials with surface-conductive properties, the resistivity measurement is ohms per square. For volume-conductive materials it is ohms-centimeter per centimeter (see definitions in refs 1, 8, and 9).

The tests most often used to determine surface or volume resistivity are ASTM D 257 and D 991. For surface resistivity measurements, the test procedure consists of measuring the resisitivity of a defined area. This is done by using a fixture that applies a known voltage across a confined surface area. Several commercial fixtures are available to accomplish this objective. Most antistatic packaging materials have a component added to them that attracts and retains a thin layer of moisture on the surface. This moisture layer is the conductive agent.

The ability of these materials to dissipate a charge is dependent on the surface-moisture layer. Surface resistivity tests should always be performed on materials that have been preconditioned at a specific temperature and relative humidity so that results of tests performed at different times will correlate. Most specifications call for testing materials preconditioned at 15% rh and 73°F (23°C).

For classification as antistatic, a material should have a surface resistivity of 10^9–10^{13} Ω/sq. Depending on the specification, this range can vary somewhat, but 10^9–10^{13} is a commonly referenced range. Values of 10^5–10^9 Ω/sq are normally considered static dissipative.

Volume-conductive materials (those with a conductive agent included throughout the material structure) are classified in a similar manner. To be considered conductive, these materials must have a volume conductivity less than 10^4 Ω-cm.

Another method used to classify materials is through the measurement of static decay. Method 4046 of Federal Test Methods Standard 101 describes the required apparatus and procedures for performing this test. The test consists of measuring how long it takes a material charged with 5000 V to dissipate that charge to zero or some percentage of the original charge specification. Usually, to pass the test, the charge must decay to zero or 10% in less than 2 s. This test also should be conducted at 15% rh and 73°F (23°C) on samples preconditioned at those levels.

The Packaging of Electronics Products for Shipment (PEPS) committee of the Electronic Industries Association (EIA) has been working on a packaging materials standard. Interim Standard Number 5 has been issued and a revised version is expected. In the revision, the definition of antistatic has been changed to describe materials that do not generate triboelectric charges in use and handling. This means that the electrical properties of the materials are no longer a determining factor.

Antistatic materials are normally used when the primary objective is the reduction or elimination of triboelectric charges in handling and use. Their protective ability (ie their ability to prevent damage from external ESD events) varies considerably. For devices that are highly ESD sensitive, additional protection should be provided. Antistatic materials do not provide shielding from electrostatic or electromagnetic fields. When this protection is called for, additional protection must also be provided. Antistatic films, foams, bubbles, etc. are often used along with other types of ESD-protective materials because of their ability to reduce triboelectric charge generation.

Shielding

Materials that provide shielding from electrostatic and/or electromagnetic fields can be separated into two categories: static shielding and electromagnetic interference/radio frequency interference (EMI/RFI) shielding. The difference between the two classifications is that a material can provide static shielding and not provide EMI/RFI shielding. The most commonly referenced procedure for determining EMI/RFI shielding capabilities is described in MIL B-81705. The test measures the ability of a material to attenuate rf signals of 1–10 Ghz. In order to qualify, the material must provide 25 dB attenuation (approximately 99%) over this frequency range.

Static shielding test procedures are not as well defined. One method of classifying static shielding is the requirement that the materials have surface resistivities below 10^5 (per EIA IS5, 1983) or volume resistivities less than 10 Ω-cm per mil (25.4 μm) thickness when tested per ASTM D 991. A problem with this means of classification is that many static shielding materials are composite structures with the shielding layers buried in the structures. Accessing these layers for resistivity measurement can be a major problem. EIA IS5 provides an alternative measurement technique for materials with inaccessable shielding layers. This procedure uses a capacitive probe inside a bag. The bag is "zapped" with a specified discharge pulse and the differential voltage across the internal probe is measured. The results are then expressed in terms of internal voltage or attenuation of the original pulse.

Although intuitively attractive because of its apparent simulation of a possible real-life event, this procedure has some shortcomings. The bags are not normally sealed, the leads from the probe can act as antennas to the high frequency

pulse, changing the pulse parameters can significantly alter results, r-f noise on the input pulse will effect results, and readout instrumentation must be capable of accurately capturing and displaying these high frequency, large-magnitude (typically 1,5 and 10 kV) pulses. Even with these shortcomings, the procedure can yield information for comparing materials. It should not be assumed that the absolute values obtained from this procedure reflect anticipated performance in actual use.

Conclusions

ESD and other associated hazards do represent a significant problem and challenge to the protective-package designer. In addition to preventing damage from physical shock and vibration, and other traditional distribution hazards, designers must now consider ESD and other electrical hazards. Component sensitivities to these ESD-related hazards are not always available. Definition of the magnitude, frequency, and likelihood of these events occurring is lacking. Dependable, accurate performance information on ESD protective-packaging materials is not readily available. Procedures for verifying design performance have not yet been standardized. However, industry organizations such as ASTM, EIA, and the Electrical Overstress/Electrostatic Discharge Association (EOS/ESD) have committees and task groups working on such problems. Perhaps these efforts will soon provide the means for resolving the issues of ESD-protective packaging.

BIBLIOGRAPHY

1. A. Q. Testone, *Static Electricity In The Electronics Industry,* Testone Enterprises Inc, Kulpsville, Pa., 1980.

2. *DOD Handbook 263 Electrostatic Discharge Control Handbook for Protection of Electrical and Electronic Parts, Assemblies and Equipment (Excluding Electrically Initiated Explosive Devices),* 1980.

3. *DOD Standard 1686 Electrostatic Discharge Control Program for Protection of Electrical and Electronic Parts, Assemblies and Equipment (Excluding Electrically Initiated Explosive Devices),* 1980.

4. *EIA Interim Standard Number 5 Packaging Material Standards for Protection of Electrostatic Discharge Sensitive Devices,* 1985.

5. *MIL B-81705 Military Specification-Barrier Materials, Flexible, Electrostatic-Free, Heat Sealable,* 1974.

6. *ASTM D 257 D-C Resistance or Conductance of Insulating Materials,* 1978.

7. *ASTM D 991 Volume Resistivity of Electrically Conductive and Antistatic Elastomers,* 1983.

8. W. Armstrong, "ESD Protective Packaging" *Evaluation Engineering Magazine,* 80–81 (April, 1983).

9. W. Armstrong "ESD Materials Head-Off Damage," *Packaging Magazine,* 44-47 (February, 1984).

W. R. ARMSTRONG
Sealed Air Corporation

EMBOSSING.   See Code marking and imprinting.

ENERGY UTILIZATION

Although engineers have been concerned with the energy consumption of plant and machines for many years, recent interest has concentrated on energy use by large industrial systems covering the operations of many factories. This interest arises primarily from the increasingly close scrutiny of industrial operations by governments and environmental pressure groups. Nowhere has this probing been so persistent as in some sectors of the packaging industry.

In principle, the calculation of energy requirements for such systems should be little different from that for individual machines: Decide what operations should be included in the system, and then total the energy consumed by each component operation. In practice, however, the situation is complicated by two main factors. First, it is often difficult to decide which operations should be included within the system. For example, when calculating the energy used to produce a beverage can, certain operations, such as metal extraction, sheet rolling, and converting sheet to bodies, must obviously be included. But should the energy required to construct the machines which perform these operations be included? Should the energy required to produce inks, lacquers, and varnishes be included? Indeed, does their inclusion or exclusion significantly affect the results? Because of the large number of choices that can be made, it is hardly surprising that few published analyses of apparently identical systems choose precisely the same system.

The second complication is that few industries closely monitor the energy consumption of individual machines. Thus, even if a satisfactory system can be defined, it is seldom possible to obtain a complete set of data for the performance of all component operations. Consequently, a variety of mathematical techniques have been devised to allow total energy consumption by a factory to be partitioned between the individual internal operations. Such partitioning exercises usually require some starting assumptions, and the results of the calculations depend on them.

Because of these problems, it is seldom possible to compare the results of two different analyses directly. The differences between the systems must first be identified and allowed for. As a result, calculation of energy requirements and interpretation of results have become specialized activities. Nevertheless, the nonspecialist can quickly learn how to identify the main differences between different analyses and correct for them so that useful preliminary comparisons can be made. To do this it is necessary to know how systems can be defined.

Choosing a System

The methodology of energy analysis is discussed in detail elsewhere (1), but the essential first step is to specify the system to be examined. Recent interest has centered on systems that trace all raw materials and fuels back to their extraction from the earth. The energy consumption of such a system is referred to as a gross energy requirement and represents the total energy resource that must be extracted from the earth to support the system. Gross energy requirement is the sum of the contributions from nine main components within the system; outlined below.

1. The main production sequence. This is directly responsible for converting input materials to output products and is a continuous, unbroken chain of operations. Of all the components in a production system, these are the simplest to identify since they represent the chronological sequence of operations through which the materials pass and form the backbone of the system. It is usual to start by identifying the main groups of operations involved. For example, the production of an aluminum can body would follow the sequence: mining, Bayer process, smelting, casting, hot rolling, cold rolling, and conversion to can body. Each of these groups of operations would then be further subdivided until the level of detail matches that for which numerical performance data are available.

2. Subsidiary supply industries. These feed materials to the main processing operations. In can production, for example, they provide the inputs of ink, lacquers, varnishes, and solvents to the conversion process. The production sequences that manufacture these inputs from raw materials in the ground represent the subsidiary supply industries. Note that each subsidiary production sequence is similar in character to the main production sequence, except that the final operation feeds into the main production chain.

3. Ancillary operations. These do not directly contribute to the processing of materials but are nevertheless regarded as essential to the operation of any commercial enterprise. Such operations include heating, lighting, offices, salesmen's cars, and so on. In some processes this ancillary energy is negligible when compared with the energy consumption of the main processing operations (eg, in the primary metal-producing industries). In other processes where the process energy is relatively low, such as plastics conversion, the ancillary energy can be significant indeed, amounting to as much as 50% of the total plant energy (2). Frequently, however, the energy of ancillary operations will be included automatically in the performance data for a factory since many operators are unable to distinguish the energy consumption of such operations from the process energy.

4. Transport. Transport operations occur in all extended systems where materials and products must be moved between factories. They are especially important in these packaging systems which include the distribution of packaged products. Such energies depend on the type of transport, the size of the vehicles employed, and the distances covered (3).

5. Capital plant. This refers to the provision of buildings and machinery needed to carry out the operations within the system. The energy required to construct such facilities is high, but because of the high throughput of products during their useful lives, the energy contribution per unit of throughput is small and frequently no significant error is introduced by its omission (4). One important exception to this is road transport, where the energy required to construct the vehicles is a significant contributor to the energy per ton-mile (5). The contribution of this capital energy is frequently ignored, however, because of the difficulty in calculating a sufficiently accurate value.

6. Maintenance. Maintenance refers to the supply of materials and facilities for maintaining plant and machinery. Often the energy associated with maintenance is subsumed under the overall energy and materials requirements of the operation and tends to be a small contributor to the total energy requirement. Again, road transport is an exception where maintenance energy can represent up to 10% of the total energy to operate a vehicle (6).

7. Labor. Although it significantly contributes to production cost, labor is insignificant in energy terms. The energy needed to keep a human alive is only about 9490 Btu (10 MJ) per day, and this is trivial compared with the energy of most machines (7). As a result, labor energy is usually omitted from energy calculations. Because of the low energy requirement of labor, it follows that labor-intensive practices will, in general, be more energy efficient than the corresponding automated practices even though they may be less cost effective.

8. Fuel-producing industries. Energy is derived from raw materials in the ground, and these must be extracted, processed, and delivered to the ultimate consumer. Energy is used in this processing. Since inputs and outputs to the fuel-producing industries can be described in energy terms, it is usual to refer to the efficiency of fuel production as the ratio of energy out to energy in. Thus coal, which involves little processing other than mining, has a production efficiency on the order of 92%. This means that 92% of the energy extracted from the earth is delivered to the consumer. In contrast, the thermal generation of electricity is relatively inefficient at about 30%; that is, for every 100 units of energy extracted as primary fuel from the ground, only 30 units are delivered as electricity to the consumer. The precise efficiencies of fuel production vary from one country to another, so great care must be taken when comparing data from different countries. In particular, great care must be taken when hydroelectricity is involved. There is much disagreement on how to calculate the production efficiency for such electricity; literature values range from 65 to 98%. As will be seen later, this can significantly affect the results of energy analysis.

9. Feedstock. Many materials used in packaging are themselves fuels. Plastics, for example, are derived from crude oil, and paper and board are based on wood. When such materials enter a production system, their energy content, or feedstock energy, must be added to the total energy requirement of the system even though this energy is not consumed in the same way as other fuels. The consumption of 2.2 lb (1 kg) of oil, for example, is equivalent to an energy consumption of some 41,000 Btu (43.2 MJ). If the system includes some form of materials reclamation, such as the production of refuse-derived fuel, a part of this feedstock energy can be reclaimed and credited to the system. It is important to remember that the whole of the feedstock energy cannot be reclaimed because of materials losses and chemical changes during processing.

Calculating Energy Requirement

The energy requirement of any defined system is calculated by converting the fuel inputs to each component operation to units of energy, using the gross calorific (high heat) value of the fuels since this represents the maximum energy available from the fuel. Occasionally, calculations use net calorific (low heat) values, and these tend to be about 5% lower. Table 1 gives typical gross calorific values and production efficiencies for various fuels. Note that these values are intended for guidance only; they must not be regarded as definitive values with universal applicability.

Presenting Results

The results of energy calculations should retain some distinction between the different operations within the system and between the different types of fuels employed. Table 2 gives an example of an analysis of aluminum hot metal production from raw materials in the ground (8). The production system has been broken down into 16 production operations. Complete disaggregation of energy by fuel type would require tables listing all the fuels in Table 1 and would be cumbersome. However, for most practical purposes, the list can be simplified by arranging the fuels into three groups: electricity, oil fuels, and other fuels. Electricity is separately identi-

fied because, with few exceptions, it is consumed by all operations and it is also the fuel with the lowest production efficiency. Oil fuels are grouped together because they are all derived from the same source, crude oil, and they all possess production efficiencies that lie in a narrow band at about 83%. The category "other fuels" includes natural gas, manufactured gas, coal, and coke. Unlike oil fuels, they possess markedly different production efficiencies. Their energies are aggregated in the table, but fuel consumptions were treated separately before aggregation and could be recovered from the detail of the calculations.

The advantage of a detailed energy breakdown, such as that in Table 2, is that it allows results to be adjusted to different assumptions. For example, if the aluminum smelter were operated with hydroelectricity (efficiency 80%, see Table 1), the electricity production energy would decrease from 121.19 MJ/kg (52,155 Btu/lb) to 15.67 MJ/kg (6744 Btu/lb), reducing the overall total from 294.71 MJ/kg (126,830 Btu/lb) to 189.19 MJ/kg (81,419 Btu/lb), a decrease of 36%. Clearly, such modifications can only be made if the results are initially presented in sufficient detail, and it is advisable to be wary of results that quote only a single number (eg, 126,830 Btu/lb (294.71 MJ/kg) in Table 2) for the gross energy requirement.

Sources of Energy Data

Unfortunately, the results of energy analyses are scattered throughout the literature, sometimes in the most unlikely places, or are contained in commercial reports which receive only limited circulation. Reference 1 gives an extensive summary of the literature up to 1978, and some government reports (9,10) provide useful additional data. Reports are continually appearing, especially in the beverage-container sector. Of these, refs. 11–13 provide useful sources of more recent data. These reports provide a wealth of useful data for the

Table 1. Typical Gross Calorific Values and Production Efficiencies of Fuels [a]

Fuel	Quantity	Delivered energy, MJ [b]	Production efficiency, %
coal	1 kg	28.01	95.0
coke	1 kg	25.42	86.6
natural gas	100,000 Btu (1 therm)	105.44	87.5
manufactured gas	100,000 Btu (1 therm)	105.44	71.9
heavy fuel oil	1 L	40.98	82.7
medium fuel oil	1 L	40.92	82.8
light fuel oil	1 L	40.18	82.9
gas oil	1 L	37.84	83.6
kerosene	1 L	36.53	84.0
diesel	1 L	37.71	83.5
gasoline	1 L	35.97	84.0
LPG [c]	1 kg	50.00	84.9
electricity (thermal)	1 kW·h	3.60	27.5
electricity (hydro)	1 kW·h	3.60	80.0

[a] Production efficiencies vary from one country to another, especially for electricity, so the values shown here should be considered as a guide only.
[b] To convert MJ to Btu, multiply by 948.8.
[c] LPG = liquefied petroleum gas.

Table 2. Energy Required to Produce 1 kg of Aluminum Hot Metal in the UK in 1977 from Raw Materials in the Ground [a]

Production operation	Electricity, MJ		Oil fuels, MJ			Other fuels, MJ			Total energy, MJ
	Production and delivery energy	Direct energy use	Production and delivery energy	Direct energy use	Feedstock energy	Production and delivery energy	Direct energy use	Feedstock energy	
bauxite mining	0.4	0.13	0.04	0.31	0	0.18	2.07	0	3.13
Bayer process	1.46	0.49	5.51	27.02	0	0.76	9.33	0	44.57
transport	0	0	0.17	0.82	0	0	0	0	0.99
smelter	121.19	62.68	0.77	4.29	0	0.14	1	0	190.07
fluorspar mining	0.13	0.04	0.02	0.08	0	0	0	0	0.27
hydrogen fluoride production	0.31	0.1	0.07	0.33	0	0.04	0.47	0	1.32
aluminum fluoride production	0.26	0.09	0.06	0.27	0	0.03	0.39	0	1.1
brine extraction	0.21	0.07	0	0	0	0	0	0	0.28
Solvay process	0	0	0.2	1.01	0	0.03	0.51	0	1.75
limestone mining and calcination	0.09	0.03	0.02	0.1	0	0.05	0.58	0	0.87
brine electrolysis	0.24	0.08	0.01	0.04	0	0	0	0	0.37
cryolite synthesis	0.19	0.06	0.04	0.21	0	0.02	0.29	0	0.81
petroleum coke production	0.42	0.14	2.8	13.66	25	0	0	0	42.02
pitch production	0	0	0.18	0.86	6.12	0	0	0	7.16
Totals	124.9	63.91	9.89	49	31.12	1.25	14.64	0	294.71

[a] From ref. 8. To convert from MJ/kg to Btu/lb, multiply by 430.4.

many different materials used in beverage packaging. Table 3 illustrates typical energy data for a number of materials commonly used in packaging. Although this is a very restricted list, it does give some indication of the range of values.

Using Energy Data

Frequently, data such as those given in Table 3 are used to compare the energy requirements of different materials. It is doubtful, however, whether such comparisons are meaningful, for three main reasons.

First, the data of Table 3 are presented as energy per unit mass, but in many applications energy per unit volume is more appropriate. On a mass basis, tinplate with a gross energy requirement of 49.84 MJ/kg appears more "energy efficient" than polypropylene film with an energy requirement of 74,349 Btu/lb (172.76 MJ/kg). Their relative positions are re-

Table 3. Typical Gross Energy Required to Produce 1 kg of Various Materials Used in Packaging from Raw Materials in the Ground [a]

Packaging material	Electricity, MJ		Oil fuels, MJ			Other fuels, MJ			Total energy, MJ
	Production and delivery energy	Direct energy use	Production and delivery energy	Direct energy use	Feedstock energy	Production and delivery energy	Direct energy use	Feedstock energy	
low density polyethylene resin	7.28	2.76	7.08	36.82	49.95	0	0	0	104.35
polyethylene terephthalate (PET) resin	20.01	7.59	16.09	78.57	60.74	0	0	0	183
aluminum hot metal	124.9	63.91	9.89	49	31.12	1.25	14.64	0	294.71
tinplate	8.99	3.41	1.23	6.01	0	4.1	26.1	0	49.84
kraft paper	16.9	6.41	4.03	19.67	0	0.05	16.63	17.89	81.58
can-sealing compound	7.49	2.84	12.35	60.3	96.7	0	0	0	179.68
paperboard	24.65	9.35	5.25	25.63	0	0.05	16.63	17.89	99.45
glass containers	3.45	1.31	1.77	8.63	0	0.61	5.93	0	21.7
lacquer	16.21	6.15	20.6	100.58	151.54	6.23	73.38	0	374.69
cellulose film	31.4	11.91	18.77	91.65	0	0.89	17.93	19.29	191.84
polypropylene film	37.38	14.18	11.21	54.71	55.28	0	0	0	172.76
shrinkwrap film	43.68	16.19	12.77	62.36	52.45	0	0	0	187.45

[a] To convert from MJ/kg to Btu/lb, multiply 430.4.

versed, when calculated on a volume basis; tinplate requires 6095 Btu/in.^{3} (392 kJ/cm^3), and polypropylene film requires 2410 Btu/in.^{3} (155 kJ/cm^3). Comparisons must justify the choice of normalizing parameter.

Second, the data of Table 3 refer to the production of primary materials, that is, materials produced from virgin raw materials in the earth. Increasingly, however, materials recycling is being practiced, especially in the packaging industries, and the energy associated with processing recycled materials is invariably lower than the energy required to produce the material from virgin raw materials. For example, the energy required to produce primary aluminum hot metal using the system described in Table 2 is almost 121,100 Btu/lb (300 MJ/kg), but the energy to produce aluminum hot metal from recycled post-consumer scrap is only about one tenth of this value. It is impossible to achieve a 100% scrap recovery rate, so the practical value for the production of aluminum lies between these two extreme values. The precise value depends on the recovery rate and the melt losses occurring during treatment. The energy savings occurring when other materials are recycled are less dramatic than those for aluminum recycling but are, nevertheless, worthwhile.

The third reason why simple comparisons are often unhelpful is that energy analysis is ultimately concerned with the production and use of commodities rather than just with the production of materials (ie, systems rather than products). The energy associated with a system depends on such factors as package design, method of handling at all stages of its life, and ease of recycling, some of which are independent of the nature of the material. Furthermore, many practical systems are not simple linear sequences of operations but are networks with a complex interchange of materials between the component operations. In such cases it is often possible to calculate the energy required to operate the system, but it is impossible to partition this energy satisfactorily from the output of the different component operations.

BIBLIOGRAPHY

1. I. Boustead and G. F. Hancock, *Handbook of Industrial Energy Analysis,* Halstead Press, a division of John Wiley & Sons, Inc., New York, 1979.
2. Rubber and Plastics Research Association (RAPRA), *Energy Use in the Plastics Industry* (MLH276), Industrial Energy Thrift Scheme Report No. 17, Department of Industry, H.M.S.O., London, 1980.
3. Ref. 1, pp. 192–219.
4. Ref. 1, pp. 178–180.
5. A. B. Makhijani and A. J. Lichtenberg, *Environment* **14**(5), 10 (June 1972).
6. Ref. 1, pp. 203–205.
7. Ref. 1, pp. 180–181.
8. I. Boustead and G. F. Hancock, *Resource Recovery Conserv.* **5**, 303 (1981).
9. Research Triangle Institute, *Energy and Economic Impacts of Mandatory Deposits,* NTIS Report No. PB-258 638, Research Triangle Park, NC., Sept. 1976.
10. *Resource and Environmental Profile Analysis of Nine Beverage Container Alternatives,* Final Report, U.S. Environmental Protection Agency, Washington, D.C., 1974.
11. I. Boustead and G. F. Hancock, *Energy and Packaging,* Halstead Press, a division of John Wiley & Sons, Inc., New York, 1981. (Report prepared for the UK Department of Industry.)
12. Franklin Associates, *Total Energy Impacts of the Use of Plastics Products in the US,* Prairie Village, Kans., Jan. 1981.
13. Franklin Associates, *Comparative Energy Impacts Associated with the Delivery of Beverages in Eight Single-Serving Containers 1981–1990,* Report for Alcoa, Prairie Village, Kan., Jan. 1982.

I. Boustead
The Open University,
Milton Keynes, UK

ETHYLENE–VINYL ALCOHOL COPOLYMERS (EVOH)

Ethylene–vinyl alcohol copolymers are hydrolyzed copolymers of vinyl alcohol and ethylene. The polymer poly(vinyl alcohol) has exceptionally high gas-barrier properties, but it is water-soluble and difficult to process. By copolymerizing vinyl alcohol with ethylene, the high gas-barrier properties are retained and improvements are achieved in moisture resistance and processability.

The resulting EVOH copolymer has the following molecular structure:

$$---(CH_2CH_2)_m---(CH_2CH)_n---$$
$$| $$
$$OH$$

ethylene unit   vinyl alcohol unit

EVOH copolymers are currently produced by Kuraray Company, Ltd. (EVAL resins) and Nippon Gohsei (Soarnol), both Japanese firms. EVAL Company of America (EVALCA), a joint venture between Kuraray and Norchem, Omaha, Nebraska, has announced its intention to build an EVOH manufacturing facility in the United States with initial production planned for 1986. Solvay, a European resin producer, has announced its intention to construct an EVOH production facility in Italy. Table 1 lists the range of EVOH resins available.

Table 1. Range of EVOH Resins

Property	Range
melt index, g/10 min	0.7–20
density, g/cm^3	1.13–1.21
ethylene content, mol %	29–48
melting point, °C	158–189

EVOH copolymers are highly crystalline in nature and their properties are highly dependent upon the relative concentration of the comonomers. Figure 1 shows the relationship between comonomer concentration and copolymer moisture and gas-transmission properties. Exceptional barrier properties permit the use of extremely thin layers of these materials in multilayer food packaging structures.

EVOH Copolymer Properties

The following are general characteristics of EVOH copolymers.

Gas barrier properties. The most outstanding characteristic of EVOH resins is their ability to provide a barrier to gases such as oxygen, carbon dioxide, or nitrogen. The use of EVOH copolymers in a packaging structure enhances flavor and quality retention by preventing oxygen from penetrating the pack-

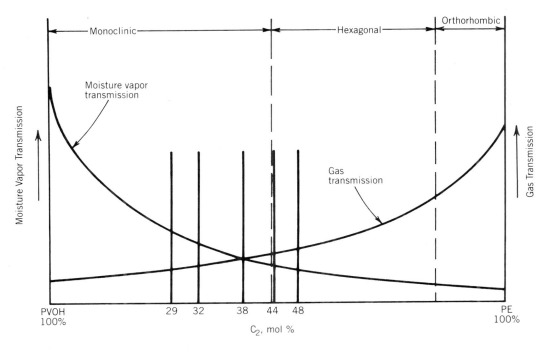

Figure 1. Comonomer concentration vs barrier properties.

age. In those applications where gas-fill packaging techniques are used, EVOH resins effectively retain the carbon dioxide or nitrogen used to blanket the product (see Controlled atmosphere packaging.) Table 2 compares the gas-transmission properties of EVAL copolymers with other commonly used packaging materials (see Barrier polymers).

EVOH resins, as indicated by the presence of hydroxy groups in their molecular structures, are hydrophilic and absorb moisture. As moisture is absorbed, the oxygen-barrier properties of these resins are adversely affected as shown in Figure 2. However, by proper package design and the use of high moisture-barrier resins, such as polyolefins, to encapsulate the EVOH resin, the relative humidity of the barrier resin can be controlled. Figure 3 shows how the relative humidity of the barrier layer can be determined. As the example in Figure 3 indicates, even when wet packaging is considered, the EVOH barrier-layer relative humidity can be controlled at a level of 70% to 78%. At these humidity levels, EVOH resins still offer the highest barrier properties.

Mechanical and optical properties. EVOH resins have high mechanical strength, elasticity, surface hardness, excellent abrasion resistance, and excellent weatherability. These films are highly antistatic in nature, preventing the build-up of dust and/or static charges. EVOH films have a very high gloss and low haze resulting in outstanding clarity characteristics. Table 3 compares the mechanical and other properties of EVOH resins.

Oil and organic solvent resistance. EVOH resins have a very high resistance to oils and organic solvents. The weight percent increase in EVOH resins after being immersed for 1 year at 20°C in various solvents or oils is 0% for solvents such as cyclohexane, xylene, petroleum ether, benzene, and acetone, 2.3% for ethanol, 12.2% for methanol, and 0.1% for salad oil. This property makes EVOH resin an excellent choice for the packaging of oily foods, edible and mineral oils, agricultural pesticides, and organic solvents.

Fragrance and odor protection. Packaging structures containing EVOH resins as a barrier layer are highly effective in retaining fragrances and preserving the aroma of the package contents. Undesirable odors are prevented from entering the container or if odorous materials are being packaged, the odor can be kept within the container. Table 4 compares packaging structures containing EVOH resins and those without a barrier layer as to retention of aromas.

Processability. EVOH resins are thermoplastic polymers easily processed on conventional fabrication equipment with-

Table 2. Gas-Transmission Rate

	$(cm^3 \cdot mil)/(100\ in.^2 \cdot 24\ h \cdot atm)^a$							
	O_2	N_2	CO_2	He	O_2	N_2	CO_2	He
EVAL EP-F grade	0.010	0.001	0.046	9.3	0.039	0.004	0.179	36.135
EVAL EP-E grade	0.091	0.008	0.41	23.8	0.354	0.031	1.593	92.475
high barrier PVDC	0.15	0.012	0.75	27.3	0.583	0.047	2.914	106.075
oriented nylon 6	1.78	0.70	11.9	116	6.916	2.720	46.238	450.721
oriented PET	2.3	0.46	6.4	180	8.937	1.787	24.867	699.394

^{a} 77°F(25°C), 0% rh, to convert to $cm^3 \cdot \mu m/(m^2 \cdot d \cdot kPa)$ multiply by 3.886.

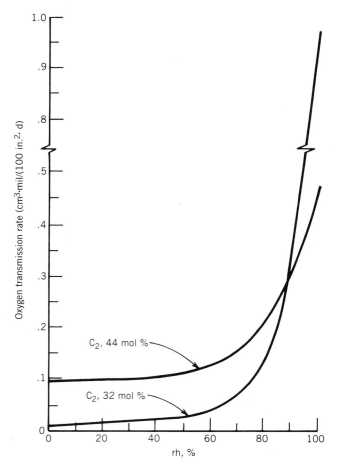

Figure 2. Oxygen-transmission rate 77°F or 25°C. To convert $cm^3 \cdot mil/(100 \ in.^2 \cdot d)$ to $(cm^3 \cdot \mu m)/(m^2 \cdot d)$, multiply by 393.7.

For a structure of N layers, the relative humidity of the barrier layer in position i is determined by:

$$1/2(Pi - 1 + Pi) = Po - \left[\left(\sum_{J=1}^{i-1} \frac{1}{A_J} + \frac{1}{2Ai} \right) \times \left(\frac{Po - Pn}{\sum_{J=1}^{n} \frac{1}{A_J}} \right) \right]$$

Where:

$1/2 (Pi - 1 + Pi)$ = rh of barrier layer
Po = rh of outside environment
Pn = rh of inside environment
i = Position of barrier layer (number of layers from outside)
N = Number of layers
$A = \dfrac{MVTR}{thickness}$

LDPE/paper/primer/tie/EVOH/tie/PP/ionomer

Layer	1	2	3	4
Thickness	0.65 mil (16.5 μm)	0.3	1.5	0.5
MVTR	1.2 G/100 in² (18.6 G/m²)	3.8	0.69	2.4

Example #1

$Po = 65\%$
$Pn = 100\%$ Barrier rh = 71.8%

Example #2

$Po = 70\%$
$Pn = 100\%$ Barrier rh = 76.8%

Figure 3. Relative humidity of barrier layer

out special modifications. This stability allows the reuse of process scrap. Regrind layers containing up to 20% EVOH resin are being used in many rigid containers.

Regulatory approvals. EVOH resins are cleared by the FDA for direct food contact as specified in 21 C.F.R., Section 177.1360, of the Federal Food, Drug, and Cosmetic Act. This clearance applies to all conditions except for high temperature retort applications. Clearance for retort applications is pend-

Table 3. Properties of Ethylene–Vinyl Alcohol Copolymer Resins

Property	Ethylene, mol % (applications)					
	32 (sheet, bottle)	32 (film, sheet, bottle)	32 (cast film)	38 (film, sheet, bottle)	44 (cast film, sheet)	48 (biaxially oriented film)
melt index, g/10 min	0.6	1.3	4.4	1.5	5.5	15.0
density, g/cm³	1.19	1.19	1.19	1.17	1.14	1.12
melting point, °C	181	181	181	175	164	156
crystallization temp, °C	161	161	161	151	142	134
glass-transition point (T_g), °C	69	69	69	62	55	48
tensile strength, psi (MPa)	11,790 (81.3)	11,220 (77.4)	10,650 (73.4)	9,870 (68.1)	8,520 (58.8)	
tensile yield strength, psi (MPa)	12,370 (85.3)	10,365 (71.5)	8,380 (57.8)	8,875 (61.2)	7,385 (50.9)	
elongation, ultimate, %	220	230	270	255	280	
elongation, yield, %	10	10	8	12	15	
Young's modulus, psi (MPa)	385,000 (2655)	385,000 (2655)	385,000 (2655)	340,000 (2345)	298,000 (2055)	
Izod impact strength notched bar, ft·lbf/in. (J/m)	2.4 (128)	1.7 (91)	1.0 (53)	1.5 (80)	1.0 (53)	
Rockwell hardness, HRM (ASTM E 18-74) 100	100	100	97	95	88	

Table 4. Aroma Retention

Construction	Thickness mil[a]	Days to aroma leakage			
		Vanillin (vanilla)	Menthol (peppermint)	Piperonal (heliotropin)	Camphor
PET/EVOH/PE	12/15/50	15	25	27	<30
OPP/EVOH/PE	18/15/50	30	<30	27	<30
PET/EVOH	12/15	<30	<30	30	<30
PET/PE	12/50	2	16	5	<30
OPP/PE	17/50	6	2	1	13

[a] To convert mil to μm, multiply by 25.4.

ing. The use of EVOH film in direct contact with meat and poultry is also cleared by the USDA.

Packaging Structures. In the United States today, EVOH resins are used primarily for food packaging. EVOH-containing multilayer structures can provide a cost performance advantage over metal and glass containers.

Rigid containers and flexible films are commercially available. Both of these types of structures use the multilayer concept to protect the barrier layer (EVOH) from the effects of moisture. This concept is also used to produce an economical structure by using relatively inexpensive materials, such as polyolefins, as the bulk of the structure. Most multilayer structures used today have five or six layers. Using feedblock coextrusion technology, seven and nine layer structures have been produced for special applications.

When using EVOH in multilayer structures, it is necessary to use an adhesive layer to gain adequate bonding strength to the other polymers. Commercially available adhesive resins such as Admer (Mitsui Petrochemical Industries Ltd.), Plexar Norchem Modic (Mitsubishi Petrochemical Co. Ltd.), or Bynel (DuPont Company) are suitable for use with EVOH resins.

Rigid containers. Two of the most common methods of producing rigid containers, such as tubs, bottles, cans, etc, are thermoforming and blow molding. Most thermoformed containers are produced in a two-step process. First, a coextruded multilayer sheet is produced. This sheet, containing a barrier layer (EVOH), is then either melt-phase formed or solid-phase pressure formed into the final container. Figure 4 shows typical multilayer structures used to produce thermoformed containers. There are two developmental methods for producing rigid containers that do not use the intermediate sheet concept. These are the rotary thermoforming process and the coextruded profile process.

Flexible films. Flexible films being used in the U.S. today are produced by several methods. These include emulsion coating, laminating, and coextruding either tubular or flat (cast) films. Converting operations such as vacuum forming and heat sealing are used to fabricate packaging structures from these films. Figure 4 also shows typical multilayer flexible film constructions.

Although monolayer EVOH films are not widely used in the U.S., they are used in Japan and Europe for laminating and other applications.

There are two other manufacturing processes under development for producing multilayer films. One of these is a biaxially oriented multilayer film, the other is coextrusion barrier coating.

Table 5 lists film properties of EVOH copolymers.

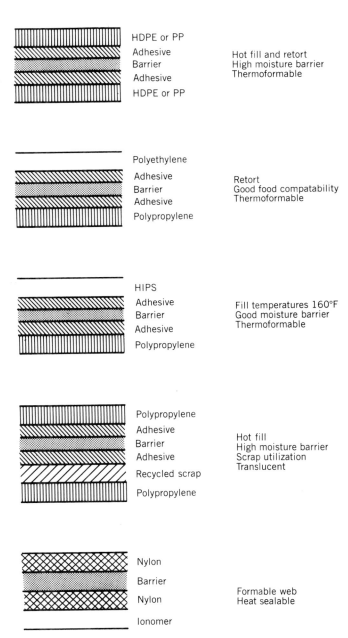

Figure 4. Barrier constructions.

Table 5. Film Properties of Ethylene–Vinyl Alcohol Copolymers (1-mil or 25.4-μm film), Based on EVAL Grades

Property	ASTM test		Ethylene, mol %		
			32	38	44
specific gravity	D 1505		1.19	1.17	1.14
yield, in.^{2}/(lb·mil) [m^2/(kg·mm)]			23,263 [1303]	23,460 [1314]	24,283 [1360]
haze, %	D 1003		1.5	1.5	1.7
tensile strength, psi (MPa)	D 882				
		MD	12,800 (88.3)	22,755 (156.9)	9,960 (68.7)
		XD	5,690 (39.2)	5,690 (39.2)	5,690 (39.2)
elongation, %	D 882				
		MD	300	170	325
		XD	235	290	235
tear strength, gf/mil (N/mm) initial	D 1004		400 (154)	400 (154)	500 (193)
bursting strength 1 mil (25.4 μm), psi (kPa)	D 774		31 (214)	33 (228)	23 (159)
water absorption, 24 h, %	D 570		8.6	7.6	6.7
change in linear dimensions at	D 1204	MD	−2.7	−6.9	−1.6
140°C for 1 h, %		XD	−0.9	+1.4	−1.2
service temperature °F, range (°C, range)			355–460 (179–238)	320–480 (160–249)	355–460 (177–238)
heat-seal temperature °F, range (°C, range)			360–440 (182–227)	330–470 (166–243)	355–450 (179–232)
oxygen permeability cm^3·mil/(100 in^2.d·atm) [cm^3·μm/m^2.d·kPa)]	D 1434				
23°C, 0% rh			0.010 [0.039]	0.091 [0.354]	0.017 [0.066]
23°C, 50% rh			0.019 [0.074]	0.121 [0.470]	0.030 [0.117]
23°C, 95% rh			0.077 [0.299]	0.170 [0.661]	0.058 [0.225]
water-vapor transmission rate, g.mil/(100 in^2.d) [g·mm/(m^2.d)] 100°F(38°C), 90% rh			3.8 [1.5]	1.4 [0.55]	2.1 [0.27]

Table 6. EVOH Applications

Fabrication process	Application	Structure
cast coextrusion	processed meats, natural cheese, snacks	PET/EVOH/EVA nylon/EVOH/nylon/ionomer OPP/EVA/EVOH/EVA
blown coextrusion	bag-in-box, red meat	LLDPE/EVOH/LLDPE
lamination	tea, condiments, snacks	OPP/EVOH/LDPE
coextrusion coating	aseptic packaging	LDPE/paper/EVOH/PP/ionomer
thermoforming	yogurt, juices, vegetables (retort) fruits (retort)	PP/EVOH/PP PS/EVOH/PE
coextrusion blow molding	ketchup, sauces, salad dressings, agrichemicals	PP/EVOH/PP HDPE/EVOH
coextruded profile	cosmetics pharmaceuticals	LDPE/EVOH/LDPE EVOH/LDPE

Applications

Since the introduction of EVOH resins in Japan in the 1970s, and the United States in the early 1980s, the use of these polymers has continued to grow. Table 6 is a listing of typical commercial applications utilizing EVOH resins as a barrier layer. When using EVOH resins with polymers except nylon, the use of a tie layer between the EVOH and the polymer is desirable.

Economics

To determine the most economical barrier polymer, the price of obtaining the barrier needed for a given application at a given set of use conditions *and* a given packaging structure must be compared. Factors such as price/pound of the polymer, polymer yield, oxygen-transmission rate, and the use of recycled scrap must be considered.

When the low barrier cost of EVOH resins is combined with their recyclability, and the new developments in coextrusion technology, the packaging specialist has available a means to design a packaging structure that not only will compete with glass and metal containers on a performance basis, but will also provide a definite economic advantage.

BIBLIOGRAPHY

General References

M. Salame, "The Use of Low Permeation Thermoplastics in Food & Beverage Packaging," *Coatings and Plastics Preprints of the 167th Meeting of the American Chemical Society, Div. of Organic Coatings and Plastics Chemistry*, **34**(1), 516 (April 1974).

R. Foster, "Plastics Barrier Packaging—The Future is Bright," *Proceedings of the Society of Plastics Engineers, Inc.*, Regional Technical Conference, April 1984.

C. R. Finch, "Coextrusion Economics," *Coextrusion Comes of Age*, Society of Plastics Engineers, Extrusion Div., Regional Technical Conference, Chicago Section, June 1981.

K. Ikari, "Oxygen Barrier Properties and Applications of Kuraray EVAL Resins," *Proceedings of the Second Annual International Conference on Coextrusion Markets and Technology*, Schotland Business Research, Inc, Princeton, N.J., Nov. 1982.

H. Lanquetot, "New Barrier Structures for Rigid Containers," *Proceedings of the Third Annual International Conference on Coextrusion Markets and Technology*, Schotland Business Research, Inc, Princeton, N.J., Oct. 1983.

RONALD FOSTER
Northern Petrochemical Technical Center

EXHIBITIONS

Suppliers and users of packages and packaging equipment come together at exhibitions. Some of the major exhibitions are devoted solely to packaging; for example, PACK EXPO and INTERPACK. Packaging is also in important component of many shows held for user industries, for example, INTER BEV and the INTERNATIONAL EXPOSITION FOR FOOD PROCESSORS. Some of the major events are listed below.

United States Exhibitions

PACK EXPO (International Packaging Week) (The largest trade show devoted solely to packaging.)
Emphasis: Packaging
Season: Fall/winter
Frequency: Biennial (. . . 1984, 1986 . . .)
Contact: PMMI, 1343 L Street NW, Washington, DC 30005

WEST PACK (Western Packaging Exhibition)
Emphasis: Packaging
Season: Fall/winter
Frequency: Biennial (. . . 1985, 1987 . . .)
Contact: Clapp & Poliak Associates, 708 Third Avenue, New York, NY 10017

NEW ENGLAND PACKAGING SHOW
Emphasis: Packaging
Season: Spring/summer
Frequency: Annual
Contact: Bayside Exposition Center, 200 Mt. Vernon St., Boston, MA 02125

INTER BEV
Emphasis: Soft drink industry
Season: Fall/winter
Frequency: Biennial
Contact: National Soft Drink Association, 1101 16th Street, N.W., Washington, DC 20036

INTERNATIONAL EXPOSITION FOR FOOD PROCESSORS
Emphasis: Food processing
Season: Fall/winter
Frequency: Annual
Contact: Food Processing Machinery & Supplies Association, 1828 L Street, N.W., Washington, DC 20036

FOOD EXPO
Emphasis: Food technology
Season: Spring/summer
Frequency: Annual
Contact: Institute of Food Technologists, 221 North LaSalle Street, Chicago, IL 60601

FOOD AND DAIRY EXPO
Emphasis: Food and dairy industries
Season: Fall/winter
Frequency: Biennial
Contact: Dairy and Food Industries Supply Association, 6245 Executive Blvd., Rockville, MD 20852

AMERICAN MEAT INSTITUTE CONVENTION AND EXPOSITION
Emphasis: Meat industry
Season: Fall/winter
Frequency: Annual
Contact: American Meat Institute, P.O. Box 3556, Washington, DC 20007

PHARM TECH CONFERENCE AND LITERATURE DISPLAY
Emphasis: Pharmaceutical industry
Season: Fall/winter
Frequency: Annual
Contact: Pharmaceutical Technology, 320 North A Street, Springfield, OR 97477

INTERPHEX
Emphasis: Pharmaceutical and cosmetics industries
Season: Fall/winter
Frequency: Annual
Contact: Cahners Exposition Group, 999 Summer St., P.O. Box 3833, Stamford, CT 06905

PRIME PACK
Emphasis: Pharmaceuticals and cosmetics packaging
Season: Spring/summer
Frequency: Annual
Contact: Cahners Exposition Group, 999 Summer St., P.O. Box 3833, Stamford, CT, 06905

MD&DI MANUFACTURING CONFERENCE (East)
Emphasis: Medical devices and *in vitro* diagnostics
Season: Fall/winter
Frequency: Annual
Contact: Expocon Management Associates, Inc., 163 Main St., Westport, CT 06880

MD&DI (Regional)
Emphasis: Medical devices and *in vitro* diagnostics
Season: Spring/summer
Frequency: Alternates West Coast, Midwest
Contact: Expocon Management Associates, Inc., 163 Main St., Westport, CT 06880

CONVERTING MACHINERY/MATERIALS CONFERENCE AND EXPO
Emphasis: Converting
Season: Fall/winter
Frequency: Annual
Contact: National Expositions Co., Inc. 14 West 40th St., New York, NY 10018

INTERNATIONAL PULP AND PAPER INDUSTRY EXHIBIT
Emphasis: Pulp and paper industry
Season: Spring/summer
Frequency: Annual
Contact: Technical Association of the Pulp and Paper Industry (TAPPI), One Dunwoody Park, Atlanta, GA 30338

NPE (NATIONAL PLASTICS EXHIBITION)
Emphasis: Plastics
Season: Spring/summer
Frequency: Triennial (. . . 1985, 1988 . . .)
Contact: Leslie & Leslie, Inc., 175 W. 93rd Street, New York, NY 10025

MATERIAL HANDLING INSTITUTE (Exposition, Show and Forum)
Emphasis: Materials handling
Season: Spring/summer
Frequency: Even years (exposition and forum); odd years (show and forum)
Contact: Material Handling Institute, Inc., 1326 Freeport Road, Pittsburgh, PA 15238

International Exhibitions

INTERPACK (FRG)
Emphasis: Packaging
Season: Spring/summer
Frequency: Triennial (. . . 1984, 1987 . . .)
Contact: Düsseldorf Trade Shows, 500 Fifth Avenue, New York, NY 10110

PAKEX (UK)
Emphasis: Packaging
Season: Spring/summer
Frequency: Triennial (. . . 1986, 1989 . . .)
Contact: Industrial and Trade Fairs Ltd., Radcliffe House, Blenheim Court, West Midlands, 391 2BG, UK

MACROPAK (Netherlands)
Emphasis: Packaging
Season: Spring/summer
Frequency: Triennial (. . . 1985, 1988, . . .)
Contact: Royal Netherlands Industries Fair, P.O. Box 8500, NL-3503 RM, Utrecht, Holland.

EMBALLAGE (France) (International Packaging Show)
Emphasis: Packaging
Season: Fall/winter
Frequency: Biennial (. . . 1986, 1988 . . .)
Contact: International Trade Exhibitions in France, Inc., 8 West 40th St., New York, NY 10018

IPACK—IMA (Italy)
Emphasis: Packaging
Season: Spring/summer
Frequency: Annual
Contact: Ipack-Ima, 62, via C. Ravizza, I-20149, Milano, Italy

SCANPACK (Sweden)
Emphasis: Packaging
Season: Fall/winter
Frequency: Triennial (. . . 1985, 1988 . . .)
Contact: Svenska Messan, Box 5222, S-40224 Goteborg, Sweden.

JAPAN PACK (Japan)
Emphasis: Packaging
Season: Fall/winter
Frequency: Biennial (. . . 1985, 1987 . . .)
Contact: Japan Packaging Machinery Manufacturers Association (JPMMA), No. 2 Nanoh Bldg. 20-1, Nishi-Shimbashi 2-Chome Minato-Ku, Tokyo 105, Japan

J. M. Corso
Omni Package Engineering Corp.

EXPORT PACKAGING

Export shipments have been a major business for thousands of years. Even today, export cargo is carried by surface ships in the holds and on decks. Since the worldwide acceptance of containerization, a large percentage of exports is moved in cargo containers. A smaller, but growing, quantity of high value and priority items is shipped by air as break bulk or in air containers.

There are no standards, rules or regulations, codes, references, or guidelines for the export packaging of specific products. In a sense, every item, destination, and form of shipment dictates the requirements. When packaging for export, every detail must be carefully considered. The products and packages will be completely out of the control of the shipper, and they may be "somewhere out there" for weeks. Export shipping is the test. The package fails at its weakest point. It is almost impossible to overpack. Export shipments must arrive in good order. They may have taken months to arrange and to produce. They cannot be easily replaced or repaired at destination.

Hazards of Export Shipments

The packaging engineer must understand and meet the requirements and hazards of various forms of shipments (see Distribution hazards).

1. Break bulk cargo in the holds and on the decks of oceangoing surface vessels of many types, sizes, flags, and ages.
2. Unitized loads in the holds of surface ships.
3. Containerized loads on container ships, on the decks and in the holds of cargo vessels.
4. Transport by lighters and barges; on open boats through surf to sandy beaches.
5. Roll on, roll off (Ro-Ro).
6. Containerized air shipments plus ground handling and transportation.
7. Individual packages and unitized loads shipped by air freighters and in the holds of passenger planes.
8. Miscellaneous combinations of modes of transportation, handling, and intermittent storage in foreign places and unknown conditions.

Export packaging encounters at least two domestic movements which involve the usual hazards of handling, loading, dropping, compression, and moisture, plus the hazards of transportation. One major exposure is during the movement to the seaport or the gateway airport. The second involves the customs procedures and storage of the country of debarkation, multiple handling, unforeseen conditions of transshipment and eventual delivery. Products must be packed for the toughest part of the journey and survive all of the exposures.

Only about 30% of the cargo losses can be classed as fortuitous losses resulting from sinking, stranding, fire, collision, seawater, or heavy weather. One ship is lost, somewhere in the world, almost every day. Approximately 70% of the losses may be preventable:

10%	water damage—fresh water, sweat, condensation, and salt water
20%	theft, pilferage, and nondelivery
40%	handling and stowage—container damage, breakage, leakage, crushing, contact with oil and other cargo, contamination, and failure of refrigeration or other equipment

During transit, cargo must withstand the conditions of rough seas, turbulent air, and substandard roads.

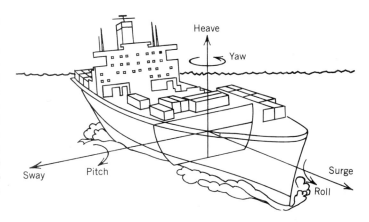

Figure 1. The six directions of motion of a ship at sea.

At sea, a ship may move in six different directions at the same time. A top-loaded container may travel 70 ft (~20 m) in each direction with each 40° roll as often as eight or nine times per minute. Figures 1 and 2 diagram the movements involved. The center point is least subject to these movements, but the shipper does not control the location and stowage of the cargo or of individual containers.

Specific Conditions

With the above as a brief introduction to the hazards of international cargo movements, the packaging engineer should further study this subject and the conditions that may exist at the ports of embarkation and debarkation.

Brief descriptions of port conditions and equipment are provided in *Ports of the World* (1,2) and similar publications. Information can be obtained from the cargo carriers regarding their equipment, methods of handling, routes, stops, transshipment, and other details. They can also provide some information about local conditions at the port facilities. Some marine underwriters may have constructive suggestions. The consular staffs and the commercial development departments of each country are well informed. Foreign sales agents can provide details about the customers and their facilities, needs, and preferences. Too often the shipper is uninformed and re-

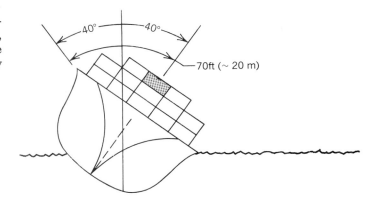

Figure 2. The distance (~20 m) traversed by a top-loaded container in a 40° roll.

lies on gossip rather than acquiring a realistic understanding of this complex subject.

Product Analysis

Every facet of the products to be exported must be analyzed, and the potential problems, even marginal ones, must be identified. Export handling and shipping expose the weaknesses if they are overlooked.

Corrosion and mildew. All metals must be protected against water and moisture for surface transportation and even for air shipment from or to humid and tropical areas. All natural materials (eg, leather, cotton, wool, and paper) must be protected against moisture, fungus, and mildew.

Problems can begin because the products are not thoroughly cleaned for fingerprints, cutting or cooling oils, dirt, and foreign matter, or because of unnoticed rust, mildew, or local moisture in the packages. In transit and in storage, the moisture and mildew, condensation, or sweat may find the weak point in the packaging and cause irreversible damage. If the product surface is critical or if the item is very expensive or choice, the damage may be total.

When the problems are identified, corrosion inhibitors (special oils and petrolatums) or properly used vapor-phase inhibitors and waterproof wrapping are applied. For electronic equipment and very critical metal surfaces or hidden areas, desiccants with sealed moisture-vapor barriers are used. If possible, the critical sections or elements are removed from the larger units and packaged separately (inside the larger box or a sheathed crate). A small vapor-phase inhibitor or moisture-vapor barrier package provides better protection more economically than a thorough package for an entire unit. Containerization does not assure protection against water damage or moisture. Air shipment does not offer full protection because of entrapped atmosphere, and the cargo is on the ground more than in the air.

Pilferage and nondelivery. It is essential to recognize that the exported products may be days or weeks in conditions where pilferage is easy. Therefore, the packages should not identify the contents as worth pilfering for private use or resale. Identification of the product on the outside of the package by brand name, manufacturer's name, or the shape and size of the box is an open invitation to steal. A damaged package exposes the contents for pilferage.

Cigarettes, liquor, cameras, stereos, jewelry, furs, small appliances, and many other items are ideal targets for pilferage. Many consignees use code identification because their names on the packages would encourage pilferage. Some even have the items delivered to a "front" to avoid identification.

With the potential problem of pilferage identified, steps should be taken to assure special handling and accountability. Packages should be stored in separate locked areas and on board in lockers or safes.

Nondelivery occurs when the packages are stolen, destroyed, lost, unloaded and left at the wrong destination, or misdelivered. Adequate, permanent, and prominent markings help to avoid some of these problems. Unitizing or using master overpacks is also helpful.

Breakage. The internal weaknesses of electronic products; the fragile construction, design, or materials of many products; any part which extends; and any item that is not in balance are all subject to breakage or physical damage. It may

be necessary to protect the package from the product because of a dense weight load or severe imbalance.

Any product that is hard to handle, tends to fall over, or is especially awkward for any reason must be given special consideration. Often, production equipment requires special handling, and provision must be made in the package. Skids and other devices may help to keep the equipment on its base and handled properly.

Cushioning, blocking, and bracing are all essential (see Cushioning design). Built-in lifting eyes or other devices, special skids and pallets, and guide marks for the stevedores are helpful. On large items the balance and the lifting points should be marked. Occasionally, an experienced exporter includes special instructions and photographs with the documentation and on the outside of the packages or crates.

Contamination. Some products can be spoiled or damaged because they can absorb the odors or fumes from the other items that may be in the same container or nearby in the holds or warehouses. Often the people who work with these items ignore or forget these problems.

Precautions can be taken to specifically request the carriers not to expose the items to adverse conditions. Sealed moisture-vapor barriers may give further assurance. Packages should be marked to indicate that they should not be stowed with potentially harmful cargo.

If photographic supplies and other sensitive items are exposed to excess heat, light, or moisture, damage can result. This type of problem must be identified, the product shielded in the packages, and full instructions provided to the carriers.

Hazardous materials. The shipper must identify any hazardous materials. These should be separated and given special documentation and packing under strict rules and specifications. It is essential that each hazardous material is classified and properly identified on the packages, with the correct and legal labels.

It cannot be assumed that compliance with domestic regulations will assure foreign acceptance of the shipments. This requires special study. Improperly labeled cargo may not be shipped and can cause great confusion (see Standards and practices).

Marks and Symbols

It is essential that proper markings and symbols are used. Even a poor package, adequately closed and marked, has a good chance of being delivered in reasonable order. An excellent package that is not adequately marked may never reach its destination.

Only those markings which are essential and appropriate should be used. Any other markings, or too many, can be confusing and serve no purpose. New, clean packages with no advertising or other printing should be used. The selected markings can be printed by the manufacturer of the packages or stenciled permanently. Crayon, chalk, marking pens, tags, and cards should not be depended upon.

Code marks for the name of the consignee and of the shipper (if used) are best for products that might be pilfered (see Code marking and imprinting). These and port marks should be large, clear, and applied on the side, end, and top of each package (Fig. 3). Required weight and dimension information should be clear.

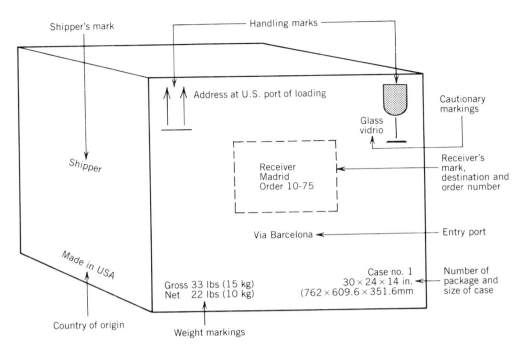

Figure 3. Assignment of codemarks for shipping.

Handling precautionary instructions should be printed or stenciled on the outside in the language of the destination country (Fig. 4). Many times the cargo handlers cannot read the language of the country in which they are working. Pictorial precautionary markings may be the most helpful (Fig. 5). In general use, they are recognized and have replaced a wide variety of symbols used in the past. Regardless of the markings, it is essential to package adequately and protect the products being shipped because the precautionary marks are frequently ignored.

When a large number of units are being moved in a single shipment, it helps to corner-code mark them with a distinctive symbol or to color code the opposite top corners with triangles or stripes. If a series of shipments is being sent to the same destination to be collected and staged for further inland shipment, the same method of coding should be used. This helps to reduce the number of stray packages.

When a number of packages are being shipped together, they should be numbered on the packages with the same numbers as the documentation. If there are eighteen packages in a shipment, they should be numbered on the packages as $1/18$, $2/18$, $3/18$, . . . $18/18$. Thus, if a package is missing, it can be checked on the documents for size, weight, and contents.

Packaging for Export

Some exporters use their domestic packages for export. This is not good practice, but it may be economical. Preparation of domestic packages for export includes the following:

1. Waterproofing by liners, overwrapping, or overbagging.
2. Master packing small to medium sized packages.
3. Unitizing a quantity of packages with suitable strapping (qv), adequate pallets (qv), and shrink or stretch films (qv).

Unless conditions are thoroughly understood, this applies even for air shipments because of ground handling, storage, and transshipment.

Packaging for break bulk. This is the traditional form of shipment for small and large items stowed in holds or on the decks of surface ships. The individual packages are handled by the ship's gear individually or on stevedore pallets. They are pushed or moved manually, or sometimes by roller conveyor or lift truck, out of the square of the hatch to wings. They are stacked with other cargo and used to help brace the loads against movement as the vessel rolls and pitches. There is a loading plan which specifies that the lightweight items should go on top of the heavier and more rugged cargo and that special classes of products should be given special treatment. The plan must also consider late cargo and the sequence of discharge at ports of call. Packages must withstand a static load of similar material 20 ft (6.1 m) high without distortion or rupture throughout the intended voyage. This applies in the lateral as well as the vertical direction. Problems of shock and vibration may be 50 times those normally experienced in domestic transit. Moisture is usually encountered during a sea voyage at deckside, in customs, and in lighters. Wooden boxes, sheathed crates, and cleated "export" plywood are appropriate for break bulk shipments when properly constructed, packed, and secured.

Unitized loads. In addition to the details already covered, it is necessary to note that the outside boxes in unit loads may be chaffed by other cargo during the ship's movement. The 20-ft (6.1-m) static load rule applies. It may be necessary to provide a wood or heavy-duty cover to distribute the superimposed weight and to protect the individual packages.

Containerized loads. Cargo containers provide physical protection so the 20-ft (6.1-m) guideline does not apply. The

English	French	German	Italian	Spanish	Portuguese	Swedish	Japanese	Chinese	Arabic
Handle With Care	Attention	Vorsicht	Manne-giare con Cura	Manéjese Con Cuidado	Tratar Com Cuidado	Varsamt	取扱注意	小心處理	بانتباه
Glass	Verre	Glas	Vetro	Vidrio	Vidro	Glas	ガラス	玻璃製品	زجاج
Use No Hooks	Manier Sans Crampons	Ohne Haken handhaben	Non Usare Ganci	No Se Usen Ganchos	Nao Empregue Ganchos	Begagna inga kroka	手鈎無用	勿用鈎子	عدم استعمال خاطيف
This Side Up	Cette Face En Haut	Diese Seite oben	Alto	Este Lado Arriba	Este Lado Para Encima	Denna sida upp	天地無用	此面向上	هذه الجهه فوق
Fragile	Fragile	Zerbrech-lich	Fragile	Frágil	Fragil	Omtaligt	破物注意	易碎貨物	قابل للكسر
Keep in Cool Place	Garder En Lieu Frais	Kuehl auf-bewahren	Conservare in luogo fresco	Mantén-gase En Lugar Fresco	Deve Ser Guardado Em Lugar Fresco	Forvaras kallt	冷暗所蔵	保持冷凍	احفظ بمكان بارده
Keep Dry	Proteger Contre Humidite	Vor Naesse schuetzen	Preservare dall umidita	Mantén-gase Seco	Nao Deve Ser Molhado	Forvaras torrt	水気厳禁	保持乾燥	احفظ بمكان جاف
Open Here	Ouvrir Ici	Hier offnen	Lato da Aprire	Abrase Aqui	Abra Aqui	Oppnas har	取出口	由此開啟	افتح هنا

Figure 4. Precautionary handling instructions.

moisture problem can sometimes be intensified because of minimum air circulation. Containers should be secured and loaded evenly from end to end. Containers should be thoroughly cleaned and inspected before use. There are special containers for unusual loads and conditions. The packages may be removed at portside and then face another domestic shipment to the consignees. Customs inspectors in some foreign ports do not restuff containers the way they were shipped. They sometimes repack with the larger and heavier items on top of the smaller items, for example.

Lighters, barges, and open boats. Some modern barges may be similar to large containers. Lighters and open boats may provide the most difficult tests of the trips. The handling is doubled and may be crude or rough, partly because of greater exposure to sea water, fog, mist, and spray. Sound packaging is required.

Roll on, roll off (Ro-Ro). In this method of shipment, highway trailers and other units on wheels can be loaded and secured in the vessel for relatively short voyages. These modern ships offer good conditions, and the handling is minimized; however, the loading, unloading, customs, and storage are portside.

Containerized air shipments. If the containers are properly loaded and precautions are taken against moisture in the packages and on the ground, this may be the safest method of shipment. Domestic packaging can often be used if the delivery is known to be normal.

Individual units by air. Unit loads by air freighters and individual packages unitized by the airlines for shipment on air freighters may be moved automatically and have little or no handling or moisture problems. Individual packages in the holds of passenger planes, however, may experience many rough handlings. In the cargo space, the handler usually cannot stand up straight and often resorts to crude methods of moving packages. The small items may be rolled or thrown and will not be thoroughly secured.

Miscellaneous modes. Packages shipped to inland points may experience a good deal of rough handling, storage under unexpected conditions, and exposure to pilferage and corrosion.

Guidelines

If shipments can be fully controlled by the shipper and the consignee by containerization, documentation, representation at both ends, and complete understanding by all parties, there can still be serious trouble unless adequate packaging is used and properly implemented. It is almost impossible to do packaging that is too good for export shipment.

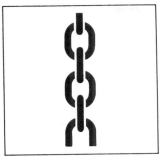

Sling here

Fragile. Handle with care.

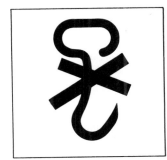

Use no hooks

This way up

Keep away from heat.

Keep dry

Center of gravity

U.S. STANDARDS

Do not roll

Hand truck here

Keep away from cold

Figure 5.   Pictorial representations of precautionary markings.

BIBLIOGRAPHY

1. *Ports of the World,* Benn Publications, Ltd., London, annual.
2. *Ports of the World, A Guide to Cargo Loss Control,* Insurance Company of North America, Philadelphia, Pa, 13th ed 1984.

General References

Preservation, Packaging and Packing of Military Supplies and Equipment, 2 Volumes, Defense Supply Agency, Department of the Army, the Navy, the Air Force, Washington, D.C.

FRANK W. GREEN
Point O'View

EXTRUDED POLYSTYRENE FOAM.   See Foam, extruded polystyrene.

EXTRUSION

Molding and extrusion are the basic techniques of forming polymers into useful shapes. The molding process, which is normally intermittent, can fix three dimensions (height, width, length) of an object. The continuous extrusion process through a die can fix only two (height, width). These two processes are usually complementary rather than competitive and produce a wide variety of products as diverse as pigmented pellets, threaded closures, and refrigerator liners (1).

The extrusion process in which an Archimedean screw rotates within a cylindrical barrel is probably the most important polymer processing technique used today. It is used to manufacture continuous profiles such as fibers, tubing, hose, and pipe; to apply insulation to wire; and to coat or laminate paper or other webs (see Extrusion coating). This article, however, deals primarily with single-screw extruders for compounding polymers and producing pellets, for producing rigid or foam sheet, and for making blown or cast films.

The extrusion principle was first employed about 1795 for the continuous production of lead pipe. The first patents on an Archimedean screw extrusion machine were granted to Gray in England (2) and Royle in the United States (3). During the nineteenth century the machinery became refined for manufacturing rubber, gutta-percha, cellulose nitrate, and casein products. Modern extrusion technology as applied to synthetic thermoplastic polymers began in about 1925 with work on PVC (see Poly(vinyl chloride)). The first screw extruder designed specifically for thermoplastic materials appears to have been made by Paul Troester in Germany in 1935 (4).

Single-Screw Extruders

Modern single-screw extruders designed to process thermoplastic resins normally are < 1 in.–12 in. dia., although larger extruders have been built. The most common diameters for production-sized machines are 2–8 in. (51–203 mm). Figure 1 shows an 8-in. vented extruder with a barrel length of 32 dia., and a 600-hp (447.4 kW) drive. The main features of an extruder are shown cross-sectionally in Figure 2.

The solid polymer fed to the extruder may be in the shape of powder, beads, flakes, pellets, or combinations of these forms. The extruder conveys, melts, mixes, and pumps the polymer at high temperature and pressure through a specially shaped die. The die's configuration, and the solidifying or cooling process, determine the shape of the product.

All extruders consist of a barrel, a screw, a drive mechanism, and controls. The heart of the extrusion process is the screw. It is fashioned with a helical thread or threads, and varying channel depth. The function of the screw is to convey material and generate pressure in order to produce pellets or other shapes. In the case of a solids-fed screw the function is expanded to include solids conveying, compression, and melting. Rotation of the screw accomplishes all these functions.

Successful operation of an extruder depends on the design of the screw. The depth and length of each zone of the screw is determined by the product to be run. Barrier flights and/or mixing sections are sometimes built into the screw to improve its efficiency in melting and delivering a homogeneous polymer to the die at the proper temperature and pressure. The profile of the melting process in a mixing screw is shown in Figure 3.

An extruder interior schematic is shown in Figure 4. Here the solid resin is introduced into the hopper, and through the action of the rotating screw is conveyed into the heated barrel.

Figure 1.   An 8-in. (203-mm) 32:1 L/D vented extruder.

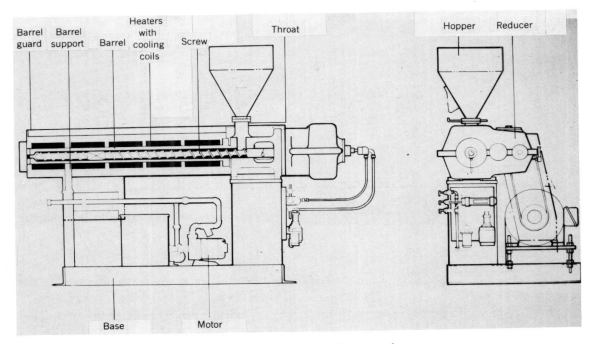

Figure 2. A cross section of an extruder.

The screw in this section is feeding or conveying the solids and hence is quite deep. The geometry of the screw is such that the depth decreases in the transition zone, and the solids start to melt. Melting results from the shearing action of the screw as motor horsepower is converted into frictional heat. Barrel heaters are used for start-up and to supplement the melting process. The melt continues to be pumped toward the discharge, or die end, of the extruder through the metering section of the screw. (Metering is pumping at a given rate in a uniform manner, within close temperature and pressure tolerances.) The die then forms the polymer in the desired shape. Downstream cooling equipment solidifies and maintains that shape.

Screws are cut from alloy steel. The tops of the conveying flights are hardened or surfaced with special alloys for extended life. The clearance between screw and barrel is close. In operation, the screw floats in the barrel on a layer of melted polymer.

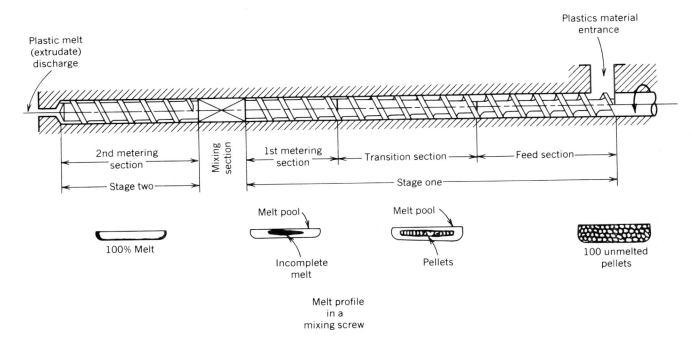

Figure 3. Melt profile in a mixing screw.

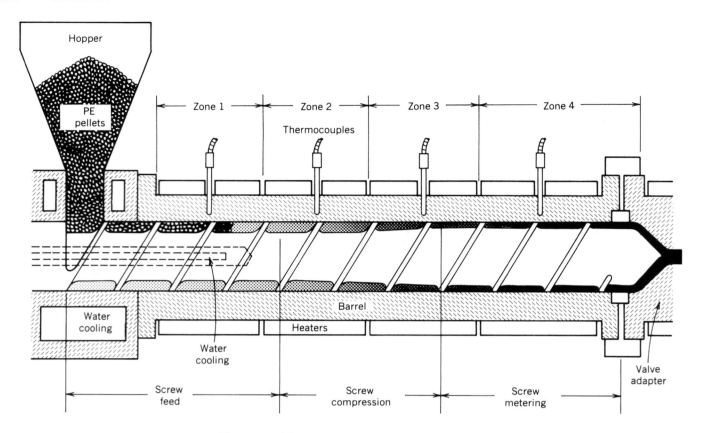

Figure 4. Schematic of the interior of an extruder.

The barrel, or hollow cylinder in which the screw rotates, is manufactured from machined steel and built to withstand pressures of 7500–10,000 psi (51.7–68.9 MPa). Barrels are also lined with special alloys or hard-surfaced to extend life. The length of a barrel is defined as a multiple of its diameter (ie, L : D ratio of 32 : 1 = a barrel : screw 32 dia. long). The length of an extruder barrel is determined by the polymer and process involved.

Extruders must be heated and cooled. Electrical heating or fluid heating can be used on the barrel. In electrical heating, resistance heating elements in various forms surround the barrel. Tubes for cooling fluid, cast in aluminum, also contain heating elements. Some barrels are built with jackets through which heating and cooling fluid can be circulated. Extruder barrels are usually divided into zones of specific lengths, each of which can be set at a desired temperature. The zones are controlled by instruments or by microprocessors. Thermocouples are normally used to sense temperature, and to signal the action of the controller (see Instrumentation).

The drive mechanism consists of a motor and a gear reducer. The motor is usually a variable-speed d-c drive system with the ability to run a speed range of slow to fast in a ratio of approximately 1 : 20. The gear reducer is used to lower the speed of the motor output shaft (eg, 1750–2000 rpm top speed) to the desired screw rotation speed. Most extruders operate in a variable screw speed range of up to 200 rpm, but speeds considerably lower are used with certain polymers, especially on very large machines. The screw rotation speed depends on the diameter of the extruder, the polymer to be extruded, and the production rate desired. The limiting factor in a given-size

extruder is quality of the product, which is dependent on the melt quality. As screw speed increases and more rate is achieved, the melt quality deteriorates because of nonuniform mixing or excessive temperatures, or because of degradation of the polymer from excessive heat generated by high shear.

There are many variations of the basic extruder design, employed to perform specific operations. The extruder and the process are in a constant state of evolution to suit requirements of new polymers being developed for new products.

Compounding

Extrusion compounding as it relates to packaging consists of preparing polymers for use in specific product applications. Mixtures of polymers, filled polymers, pigmented polymers, and a host of other polymer additives constitute a huge market. Compounding with a single-screw extruder consists of mixing and dispersing one or more minor constituents (eg, pigments, stabilizers) into a major constituent, a polymer. The product of a compounding line is pellets. These are used by the converting industry. Converting is the process of melting the pellets and producing extruded sheet, film, injection-molded parts, etc.

Compounding can be separated by functions: resin-plant extruding, blending, reclaim, and devolatization.

Resin plant extruders take the products from a polymerization operation and make pellets. The feed to these machines can be powder, granules, other irregular shapes, or even a melt. A melt extruder has a molten feedstock and only generates pressure, whereas a plasticating extruder has the job of turning solids into a molten mass and then generating pres-

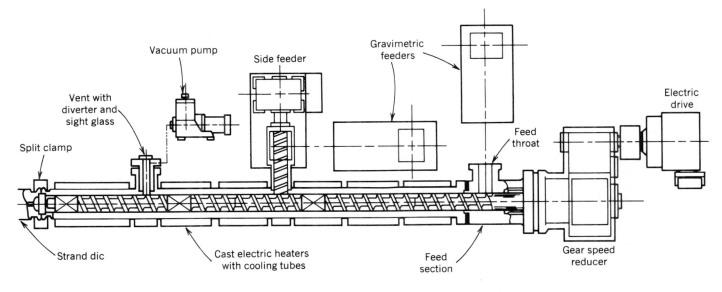

Figure 5. Cross section of a side-fed extrusion system.

sure. During pelletizing, stabilizers and processing aids are combined with the polymer (see. Additives, plastic).

A blending extruder is used to mix feedstocks of compatible polymers, or different viscosities of the same polymer. The blending operation tailors the physical properties to meet a specific end use.

Reclamation of polymers is a rapidly growing segment of the plastics industry. The problems are numerous because the feedstock for the extruder comes in many sizes and shapes. Some examples are polypropylene battery cases, polyester x-ray film, soft-drink bottles, bread wrappers, unusable foam products, fibers, filaments, and off-grade film of all kinds. Specially designed extruders must be used for processing these materials.

Some polymers require extraction of moisture or gases from the melt before a satisfactory product can be produced from the die. To accomplish this, a vent hole with vacuum pump is introduced along the extruder barrel, and the screw is specially designed. Devolatizing extruders are used where residual monomer, water, or other unwanted materials must be removed from the extrudate.

The machinery required for the aforementioned processes must be specifically designed for the application. A wide variety of design features are utilized (5).

Specially designed extruders are used for the addition of short fibers to thermoplastic polymers (6). This is done to improve the physical, mechanical, and structural properties of the virgin plastic. Cost-per-unit volume can also be improved. The extruder uses a three-stage screw. The polymer is melted in the first stage. The short fiber, usually chopped fiberglass, is screw conveyed into the side of the extruder barrel. The glass and polymer are mized in the second stage. The third stage allows venting and pressure generation for the die at the exit. A cross-sectional view of a side-fed extrusion system is shown in Figure 5. Figure 6 shows a 3.5 in. (89 mm) extruder with side feeder.

Another unique design is used for reclaiming. Because the bulk density of most of the scrap-plastic items is low, use of a dual-diameter extruder has become prevalent. The feed end of the screw has a larger diameter so there is a greater volume for the entering light fluffy feedstock. This facilitates high production rates.

Vertical extruders meet special needs such as limited floor space, or other plant-layout requirements.

Blown and Cast Film

Film is a relatively thin (usually ≤10 mil (≤254 μm) flexible web made from one or more polymers, either blended or coextruded (see Coextrusion, flat; Coextrusion, tubular) but not to be confused with a fabricated extrusion-coated or laminated web (see Laminating; Multilayer flexible packaging).

Films are used in flexible packaging (for overwraps, bags) as industrial wraps (stretch and shrink films), in medical and health care products (disposable diapers, backings, hospital bed liners), in agriculture (mulch films), for sacks (drum liners, garbage bags), and as laminates (aseptic container stock). (Some packaging overwrap films are produced by casting a solvent solution of PVC resin on a stainless steel belt and evaporating the solvent as the belt travels through a heated chamber. This process normally uses a pump to distribute the solution through the die. Solvent casting is beyond the scope of this discussion.)

The same basic extrusion processes are used for producing both blown and cast film. The first two steps, melting and metering, are part of the extrusion process described above.

Forming. In the forming process, the polymer is squeezed through a die as it leaves the extruder, to form a thin uniform web. The cast-film process produces a flat web. In the blown-film process the die shapes the polymer into a tube. This latter process is more versatile because it can produce not only tubular products (bags), but flat film as well, simply by slitting open the tube. Key to the success of both processes is the die, which must distribute the polymer uniformly.

Orientation. In the orienting stage of the blown-film process, the tube is blown up into a bubble that thins out or "draws down" the relatively thick tube to the required product gauge (thickness). In certain blown-film processes, the polymer is blown downward to produce films with special proper-

Figure 6. A 3.5 in. (89-mm) 40:1 L/D vented extruder with a 3-in. (76-mm) side feeder.

ties. The ratio of the diameter of the blown bubble to the diameter of the die is called the blow-up ratio. Most LDPE (see Polyethylene, low density) blown films used in packaging are made using blow-up ratios between 2.0 and 2.5 : 1. The blow-up ratio is changed depending on the characteristics of the resin being extruded and the properties desired in the film. In cast film, the molten polymer is also drawn down to the desired finished gauge. Drawdown ratios between 20 : 1 and 40 : 1 are typical (If a polymer exits a die at 40-mil (1-mm) thick and finishes up 1-mil (25 μm) thick, the overall drawdown ratio is 40 : 1.) In the orientation process the long molecules of the polymer line up in the stretching direction, which improves the film's strength in that direction. A key difference between the two filmmaking methods lies in the manner of orientation.

Because both edges are free in cast film, it is drawn down only in the direction the material exits the die (machine direction). Because cast film is drawn in one direction only, it usually exhibits excellent physical properties in the machine direction, and poor properties in the cross-machine direction. The cast-film process is shown in Figure 7. In the blown-film process, the extruded tube is stretched in two directions: as it is blown into a bubble and as it is drawn from the die in the machine direction by the adjustable speed drive system. This results in strength properties that are more uniform and can be balanced depending on the blow-up ratio and the takeoff speed. Figure 8 illustrates the blown-film process.

Certain films can be biaxially oriented to enhance properties for specific packaging uses such as shrink films (see Films, shrink) or overwrap. The blown process produces a thick tube that is then reheated and blown out while increasing the takeaway speed to maximize orientation in both directions. The cast process extrudes a thick, flat sheet which is chilled, then reheated and stretched in the machine direction and then reheated and stretched across its width by means of a tenter frame. The most common plastic materials to be biaxially oriented are polystyrene (see Polystyrene), homopolymers and copolymers of polypropylene, usually coextruded (see Film, oriented polypropylene); and polyester (see Film, oriented polyester). These films are then coated to enhance heat seal or barrier properties.

Quenching. After the polymer has been extruded, it must be solidified into finished film. In blown film the quenching (or cooling) process is achieved by convection; by blowing air on the outside and sometimes on the inside of the bubble. Air rings at the die exit direct and distribute air uniformly to the bubble. In cast film the web leaves the die and is deposited on

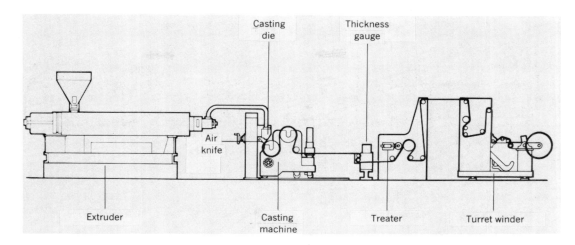

Figure 7.   Cast-film line

the surface of a driven cooled roll. There are usually several rolls in series (normally called chill rolls), arranged to cool the polymer by conduction, or direct contact.

Conduction cooling is quicker than convection cooling and this has an effect on the clarity of the film. Because convection cooling (quenching of blown films) is relatively slow, more and larger crystals form in the film, as compared to those formed in the casting method. Because interfaces between crystals scatter light, blown film tends to be more hazy than cast film. This "haze factor" normally rules out use of blown film where clarity is very important, such as food overwrap applications. Because conduction cooling is more uniform and rapid than convection cooling, cast film has less gauge variation than blown film. This superior flatness means the film can be handled

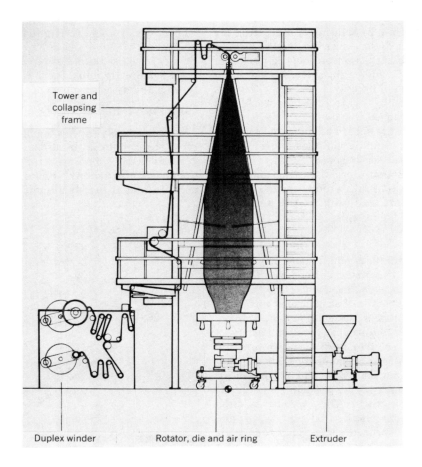

Figure 8.   Blown-film line.

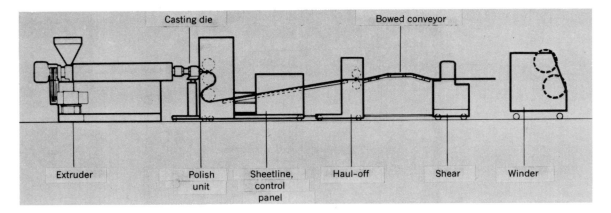

Figure 9. Sheet line.

better in subsequent converting operations such as multicolor printing and laminating. These operations are performed at high speeds and cast film is preferred to minimize scrap.

Gauge randomization. In practice, perfectly flat film cannot be made, due to die geometry and machine tool constraints. Blown-film thickness variations of ±7% and cast-film thickness variations of ±3% are typical. Variations in thickness are frequently evidenced by gauge bands. If relatively small variations become significant at the film winder or at a later converting process (printing, laminating) adjustments must be made to distribute them. In the blown method, variations in the film are usually randomized by rotating or oscillating the die to distribute the gauge variations over the finished web. In cast film, because a flat web with two free edges is produced, the downstream winder with edge trim slitters is normally oscillated across the film, winding only a portion of the cast web (see Slitting and rewinding). This generates waste film that must be recycled or scrapped.

It is common now to measure sheet and flat-film thickness automatically after the die, compute and average thickness, and use the signal to control screw speed and thus control thickness in the linear direction. Special casting and sheet dies which operate in conjunction with a computer and a thickness-measuring gauge automatically control the film thickness across the width of the die as well.

Rigid Sheet Extrusion

Film thicker than 0.010 in. (0.25 mm) is normally defined as sheet (see Films, plastic). It is thermoformed (see Thermoforming) into objects that hold their shape, a property that film does not possess. Extruded sheet is thermoformed into cups, lids, containers, packaging blisters, automotive panels, signs, and windows. The machinery required for the manufacture of sheet usually extrudes the polymer horizontally into a nip formed by two hardened cooling rolls that define the final product thickness and surface. Additional rolls and a conveyor for more cooling and pull rolls and a winder or shear complete the sheet extrusion line (see Figure 9). The extruder is often vented to remove low levels of moisture from polystyrene and ABS. Sheet is traditionally extruded horizontally from a die similar to a flat-film die, but with specially designed interior flow surfaces to suit the particular polymer. Restrictor bars are usually used for added gauge uniformity (see Figure 10). Sheet dies are often more massive, to minimize distortion.

The takeoff unit for extrusion of sheet usually consists of a cooling and polishing unit (C & P unit) having three driven, highly polished, chrome-plated rolls; a roller conveyor; and a pair of driven rubber-covered pull-off rolls. The C & P unit serves three functions: cooling, polishing, and gauge control. In some cases one or two of the chrome-plated rolls are embossed to yield a sheet with specific surface qualities. Roll diameter is contingent on the output of the extruder, the linear speed of the equipment, and the level of heat transfer required. High-capacity multiple-zone temperature control units are often built into the C & P unit, which must be of rugged construction to eliminate vibration. Sheets up to approximately 0.050-in. (1.3-mm) thick can be wound onto rolls; thicker sheet is cut to desired lengths. In some cases the sheet is pulled directly into a thermoforming machine, providing an in-line, pellet-to-part operation.

Foam Sheet Extrusion

Extruded polystyrene foam (see Foam, extruded polystyrene) sheet material used for making egg cartons, meat and vegetables trays for fast-food packaging, and similar applications continues to find new uses, ranging from decorated picnic ware to coated and laminated sheets. Polyolefin foams are

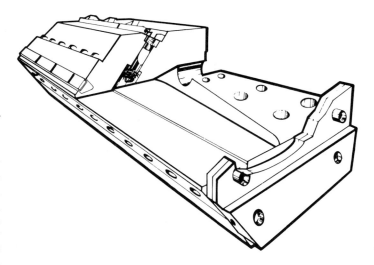

Figure 10. Cutaway of sheet die with restrictor.

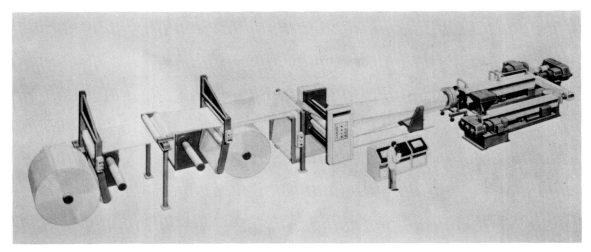

Figure 11. Tandem-extruder foam-extrusion line.

used for packaging materials (see Foam cushioning), insulation, and wire coverings. Most of the world's extruded foam is produced on tandem-extrusion equipment. Although the extrusion process is relatively straightforward, special equipment and controls are needed (see Figure 11).

The first extruder has a long barrel and is used to mix a nucleating agent uniformly throughout the melt of a base polymer. The nucleator, typically a selected filler, controls foam cell quality. In effect, it creates imperfections in the polymer melt, forming nucleation centers for cells to originate.

About two-thirds down the primary extruder barrel, the gas blowing agent is introduced. At this point the melt is homogeneous and at a pressure of 3500–4000 psi (24.1–27.6 MPa). Fluorocarbons are the usual agents, often blended with hydrocarbons to reduce costs. Carbon dioxide blended up to 35% with either fluorocarbons or hydrocarbons reduces material costs still further.

The product mix is fed through a screen changer for filtering. Then the mix, still under pressure, is fed into a larger extruder that cools and discharges the product under conditions to allow extrusion through the annular die. This is achieved through use of a low-speed screw rotating in a barrel cooled by high flow rates of water. Foaming occurs only outside the die lips.

The foamed tube is expanded to 3.5–4 times its diameter and extruded horizontally over an internal cooling and sizing mandrel that cools and orients the foam and supports the tube as it leaves the die. After the tube passes along the mandrel and a slitting unit, the two webs are pulled through nip rolls. From the nip rolls the webs are wound on either dual-spindle turret type or cantilevered winders. Large-diameter reels are required to handle foam sheet.

Sheet weight per inch (or centimeter) is governed by the amount of blowing agent incorporated in the mix. Sheet thickness is determined by adjustment of the die lips and the take-off speed of the nip rolls. Sheet orientation is controlled by a combination of die gap, blow-up ratio, and line speed.

Accurate metering equipment is needed for a good finished product. The blowing agent system calls for sophisticated controls to safely handle high-pressure gas products on the production line. The difficulties of operating a two-extruder system have been reduced through the use of microprocessors

that automatically monitor and control a multitude of functions on the extrusion line.

BIBLIOGRAPHY

1. J. M. McKelvey, *Polymer Processing,* John Wiley and Sons, New York, 1962.
2. Brit. Pat. 5056 (1879), M. Gray.
3. U.S. Pat. 325,360 (1885), V. Royle and J. Royle, Jr.
4. Z. Tadmor and I. Klein, *Engineering Principles of Plasticating Extrusion,* Reinhold, New York, 1970.
5. E. C. Bernhardt, *Processing of Thermoplastics Materials,* Reinhold, New York, 1959.
6. U.S. Pat. 4,006,209 (Feb. 1, 1977), J. J. Chiselko and W. H. Hulbert (to Egan Machinery Co.).

General References

J. H. Du Bois and R. W. John, *Plastics, Fifth Edition,* Van Nostrand Reinhold Co., New York, 1974.

R. Barr, "Solid Bed Melting Mechanism: The First Principle of Screw Design," *Plastics Engineering* Vol. 37, No. 1, p. 35ff (1981).

C. I. Chung, "A Guide to Better Extruder Screw Design," *Plastics Engineering* Vol. 33, No. 2, p. 28ff (1977).

C. Rauwendaal, "Optimal Screw Design for LLDPE Extrusion," *Plastics Technology* Vol. 30, No. 9, p. 61–63 (Aug. 1983).

J. A. GIBBONS
Egan Machinery Company

EXTRUSION COATING

Extrusion coating is a process in which an extruder forces melted thermoplastic through a horizontal slot die onto a moving web of material. The rate of application controls the thickness of the continuous film deposited on the paper, board, film, or foil. The melt stream, extruded in one or several layers, can be used as a coating or as an adhesive to sandwich two webs together.

Equipment for extrusion coating and laminating lines is normally associated with product groups, with some overlapping between groups. Substrates or web handling characteris-

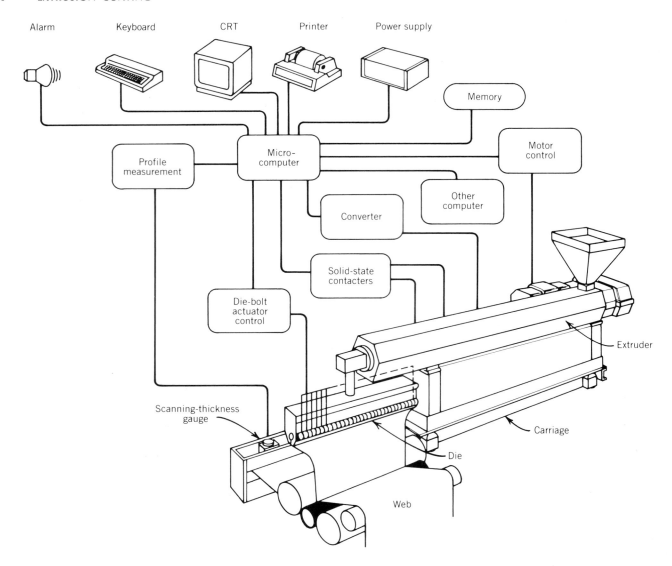

Figure 1. Simple extrusion-coating line.

tics distinguish the difference among plastic films, paper, and paperboard combinations.

Three types of lines for extrusion coating and laminating are: thin-film or low tension applications at operating web tension levels of 8–80 lbf (35.6–356 N); paper and its combinations in the middle range of 20–200 lbf (89–887 N); and high tension for paperboard applications at 150–1500 lbf (667–6672 N).

In extrusion laminating, a film of molten polymer is deposited between two moving webs in a nip created by a rubber pressure roll and a chrome-plated steel chill roll. In this continuous operation, rolls of material are unwound, new rolls are automatically spliced on the fly, and the surface of the substrate is prepared by chemical priming or other surface treatment to make it receptive to the extrusion coating, and to help develop adhesion between the two materials (see Fig. 1).

Pressure and temperature on the web and extrudate combine to produce adhesion. The substrate normally provides the mechanical strength to the resultant structure, and the polymer provides a gas, moisture, or grease barrier.

As materials, especially for food packaging, become more complex with ever increasing performance standards, coating lines become more complicated. The requirements for new extrusion coating lines are high productivity, extreme flexibility, and labor-saving computerized and robotized equipment. Modern extrusion coating lines must be able to process the new specialty resins that offer greater adhesion, allowing line speeds to be increased.

Applications

Products from extrusion coating/laminating lines have six main market classifications: liquid packaging; flexible packaging; board packaging; industrial wraps; industrial products; and sacks.

Liquid packaging.   Liquid packaging utilizes a single web-coated lightweight board, or a combination of board, plastic, and aluminum foil, for semirigid containers for milk, juices, water, oils, processed foods, sauces, cheese products, and aseptic packaging of liquids. The polyethylene-coated milk carton was largely responsible for the emergence of the extrusion coating industry in the 1960s, and as more commercial uses were found for polyethylene-coated materials, the industry grew rapidly (see Polyethylene; Cartons, gabletop).

In the 1980s, aseptic packaging is making strong inroads in replacing traditional metal and glass containers (see Aseptic packaging). The sterile flexible "paper bottle" which extends the shelf life of dairy products for months without refrigeration has emerged as a major alternative form of packaging (1,2) (see Shelf life).

Although aseptic packaging systems differ, most of them use paperboard–foil–plastic composite material which is formed to shape, sterilized, and filled with a sterile liquid or semiliquid product under sterile conditions. Customized extrusion-coating lines, complete with in-line laminating stations, are used to produce an almost unlimited variety of shapes, sizes, and printing options for aseptic and other packaging materials.

Flexible packaging. The flexible packaging classification covers the combination of plain, printed, or metallized films, papers, polymers, and foil, used for protection, unitizing, dispensing or holding of commodities. These include medicine and pharmaceutical supplies, foods, chemicals, hardware, liquids, notions, sterile products, and primal meats. Flexible packages also include wrappers for fast food, the bag for "bag-in-box" containers, and the multilaminated web for Glaminate tube packaging (3) (see Bag-in-box, dry; Bag-in-box, liquid; Tubes, collapsible; Multilayer flexible packaging).

Flexible packaging lines are processing progressively thinner substrates of polyester, oriented polypropylene, and metalized materials (see Film, polyester; Film, oriented polypropylene; Metallizing). The light-gauge preprinted substrates used for snack foods require minimum tension to assure that preprinted webs are not distorted. Machines to create these new structures are becoming increasingly complex. Thinner coating layers are more difficult to extrude on the coating machine, and often the coating head itself must be engineered to handle a variety of coating materials (see Fig. 2).

Flotation drying, using air on both sides to float the web, is widely used in flexible package manufacturing because it handles light films well. Improved drying efficiency, compared with roll-support dryers, allows higher line speeds.

Board packaging. In board packaging, heavyweight boards are coated, laminated, and then formed into boxes (folding cartons) for packaging detergents, tobacco, liquor, frozen foods, and bakery products (see Cartons, folding). "Ovenable" trays and ice cream cartons are also in this category. The plastic-coated containers protect against grease, moisture, and gas. Release characteristics can also be provided.

Industrial wraps. Industrial wraps cover the range of heavy or reinforced papers, films, or boards used for products in which the extrusion-coated material may be added to other media used for products such as composite cans, drum liners, soap wrappers, and sheet overwraps for a variety of baled materials (see Cans, composite). The coated product is not necessarily used as a unit container, but as a wrapper or part of a protective structure.

Industrial products. The industrial products classification takes in extrusion coated or laminated material serving various industry requirements. Products include photographic-base papers, substitutes for bitumen coatings, base papers for silicone coatings, insulation backing, automative carpet coating, and metallized film balloons. Also in this category are functional laminates like credit cards and printed circuit boards, decorative laminates such as wallpaper, and disposables like tablecloths and various hospital and surgical supplies. As with industrial wraps, these products may or may not be associated with packaging.

Sacks. Sacks covers materials for multiwall paper bags and plastic-coated raffia. Intermediate tension lines produce coated scrim that is woven from tapes of oriented polyethylene or polypropylene. These materials are used for heavy-duty sacks and tarpaulins, and have applications in building, recreational, and agricultural areas (see Bags, paper; Bags, heavy-duty plastic).

A miscellaneous classification would include coated foam used in the fast-food industry.

Machinery

Obviously, no single coating line can produce all the foregoing products. Today, various types of machinery are manufactured to produce coated and laminated products using substrates ranging from 0.4 mil (10 μm) film to 246 lb/3000 ft^2 (400 g/m^2) board, with coating weights 4.3–49 lb/3000 ft^2 (7–80 g/m^2), at widths 76.2–787 mil (300–3100 μm), and at speeds 1.1–33 ft/s (0.33–10 m/s).

The most important consideration in web processing or web

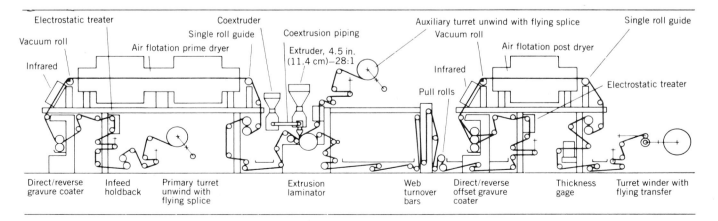

Figure 2. High speed extrusion-coating line used to produce flexible packaging-type materials.

handling equipment is the determination of the practical range for the system—the maximum and minimum unwind and winder roll diameters, maximum and minimum web-tension forces, splicing speeds, core diameters, and other process needs.

The unwind basically takes material in roll form and processes it continuously over a series of idler or driven rolls with a suitable amount of tension in order to minimize wrinkling yet not produce deformation. A dancer or transducer roll can be used for tension control; dc regenerative drives and electric or pneumatic brakes are used where applicable. Similar considerations apply to the infeed holdback when levels of tension required differ from those of the in-line operations. These sections tend to isolate tension transients from the unwinding roll.

Electrostatic treatment and flame treatment is available for enhancing surface tension or wettability conditions of the inert substrates prior to applying aqueous solutions. The direct/reverse gravure coater can be used either for priming, coating, or printing. Chemical priming is used mostly in flexible packaging lines to promote adhesion between the extrudate and substrates such as cellophane, polyester, ionomer, nylon, or polyolefin films (see Film articles). Infrared preheating and vacuum rolls provide the means to dry the PVDC-coated web and effect proper web handling.

Air-flotation, driven-roll, idler-roll, or drum-support dryers are selected depending on the strength, support, or tension required for the substrates. Recirculation of heated air in the dryers is a common energy conservation practice in all these dryers. The single-roll web guide at the dryer exit and chill or pull rolls are needed for special web processing requirements.

The extrusion laminator along with the extruder and die system is the heart of the process. The backup chill roll, rubber roll, and large-diameter chill roll form a three-roll system. Two-roll laminators can be used for heavy substrates or paperboard applications. As the moving web enters the nip section it is coated, laminated, or both. Ozone in close proximity to the entering web is used for oxidation of the molten polymer for improved adhesion in high bond level applications. Most of the heat is removed from the coating or laminate by the chrome-plated chill rolls. Chill rolls normally are steel and are constructed with a double-shell arrangement and spirally baffled. Outer shells of aluminum have been used for high coating weights. High velocity chilled water is circulated to maintain a temperature rise between inlet and outlet of 2–4°F (1–2°C).

The coated or laminated structure is normally edge trimmed at the laminator by razor, score, or shear cutting. Trim removal systems are installed just after the laminator. Slitting can also be done just prior to winding at a turret or single-drum winder.

Auxiliary unwinds can be located on, near, or over the extrusion laminator to provide a secondary substrate for laminating at the nip where the extrudate acts as an adhesive. These unwinds can consist of single-position or turret assemblies with flying splices for aluminum foil, oriented polypropylene or polyester film, paper, or paperboard substrates.

Web turnover systems, pull rolls, coaters, infrared heating, dryers, and web processing steps after the extrusion laminator are designed according to product needs. Thickness measuring devices include infrared and scanning of clear webs.

There are two basic winding techniques. The turret winder or center wind system is used for most flexible packaging materials. Tension is controlled by a dancer or transducer roll. The same design criteria for unwinds also applies to winders. The type of web, operating speed, tension range, and roll buildup must be properly controlled to wind up a satisfactory roll. Paper and paperboard products can be wound by surface methods on a single-drum winder.

All-plastic constructions require more advanced web controls. Many converters utilizing traditional wood cellulose substrates are specifying that their new coating lines must be able to handle all-plastic films. Wider tension ranges and air flotation dryers are two principal requirements of these convertible systems.

Other features being incorporated into various lines include dc-regenerative unwinds and infeed holdback drives for prices and low level tension, direct/reverse gravure coaters for aqueous PVDC coating, infrared preheating, and vacuum rolls for web controls.

In the production of photographic-base papers, exacting specifications and special criteria for pigmented polymers are needed to produce coated materials that constantly provide high quality photographs. The concept of tandem operations or coating two sides of a substrate in one pass can be applied to many flexible packaging lines that produce combinations of paper, extrusion lamination to aluminum foil, and extrusion coating a polymer for heat sealing. Higher operating line tensions can be used in producing structures with paper for granulated or powdered mixes and freezer-wrap or sugar-pouch materials. Polyethylene is not the only resin used for lamination or coating. Polypropylene, ionomer, nylon, ethylene–acrylic acid (EAA), ethylene–methacrylic acid (EMA), and ethylene–vinyl acetate (EVA) can also be part of a converter's inventory of resins.

Single-unit pilot coating lines feature an entire coating system preassembled and prewired at the factory and mounted on a structural steel base. These lines can be completely enclosed and have applications for the development of products such as the retort pouch, aseptic packaging, vacuum packaging, and other extended shelf-life products used to replace conventional glass and metal-can packages; they can also be used in the development of many types of medical-grade extrusion coatings. (3)

Stainless steel is used when extreme cleanliness is required. The "clean room" machines are designed so that any metallic particles generated by machine friction are either contained or swept away by laminar air flow. Stainless steel is also used when lines are frequently washed with solvents that could remove conventional paint.

A typical pilot coating line consists of an unwind, coating heat, air-flotation dryer, dryer exit tension control, cooling station, extrusion coater, and rewinder, all aligned on a one-piece steel frame. Pilot coating lines are designed to handle narrow web widths and can be built so that components are cantilevered instead of being supported by traditional sideframes. The spindles, idler rolls, force transducers, and air-flotation bars are all mounted on a vertical backplate.

Extrusion-coating lines are experiencing increased automation. Raw material and roll stock can now be selected from a controlled inventory, delivered to the line, and handled through robotics (see Robots). The entire operation can be monitored and controlled by computer. (see Instrumentation/controls).

Drives are also coming under computer process control, and

there have been advances in digital drives and in energy-efficient ac inverters. While a number of different drive systems have been installed and operated, the multimotor dc system is predominantly used for extrusion coating equipment. These drives can consist of as many as ten motors in one line with a single control to bring the web up to operating speed. The tandem follower is another drive or computer feature whereby the extruder will increase or decrease in rate with line speed in order to maintain a fixed coating weight as the line speed is changed.

BIBLIOGRAPHY

1. W. Schoch, "Aseptic Packaging", *Tappi,* 56 (Sept. 1984).
2. L. J. Bonis, *Correlation of Coextruded Barrier Sheet Properties with Packaged Food Quality,* Composite Container Corporation, Medford, Mass. (presented as Proceedings at the 1984 Polymer, Laminations and Coatings Conference, sponsored by TAPPI, Atlanta, Ga., Sept. 24–26, 1978, pp. 319–328).
3. M. Schlack, "Extrusion Coaters Gear for New Packaging Action," *Plast. World,* 42 (July 1984).

General References

H. L. Weiss, *Coating and Laminating Machines,* Converting Technology, Inc., Milwaukee, Wisc.

H. L. Weiss, *Control Systems for WEB-FED Machines,* Converting Technology, Inc., Milwaukee, Wisc.

MICHAEL G. ALSDORF
Extrusion Group, Egan Machinery

F

FIBERBOARD. See Boxes, solid fiber.

FILLERS. See Additives, plastics.

FILLING MACHINERY, BY COUNT

Accurate measuring is imperative in packaging to avoid costly overage and shortages that are now prohibited by law. As a basis for measurement, most packaging lines use either the weight of the package (see Checkweighing machinery) or the number of pieces in the package. This article pertains to methods used to produce packaging containing a specified number of pieces. Modern automatic counting systems are based on concepts that evolved in ancient times. They are all comprised of three basic functions: parts representation, parts detection, and product handling.

Parts Representation

Counting is used to determine the amount of a specified batch, and the method used to achieve this begins by selecting a basic system of representation. Systems of representation are used in all forms of counting. Units can be represented by fingers and toes, or knots in a rope, or, as in the packaging industry, they can be pulses of electrical current generated from specially designed detection units. Over the years, humans have engineered quick and accurate ways of counting, but none is as accurate as a single-file count of an individual unit of product.

Parts Detection

The next step in counting is detecting the unit of product to be counted. A person detects the product either by sight or by touch. Machines are designed to operate on the same principles, and use either optical systems (sight) or nonoptical systems (touch).

Optical systems. An optical system operates much like a human eye. A photosensitive receiving device is established and the unit of product to be counted is passed within detection distance of it. There are many models of optical systems used in packaging, but most are based on either a simple digital photocell system or a very intricate electronic-analogue-detection unit.

In the photocell system, the breaking of a light beam fed to the photocell receiver indicates that a unit of product has passed through the detection zone. The break is then recorded as the counting of one unit of product. This method of detection is perfect for most product applications that meet specifications for light-blocking systems, but it is limited by the fact that the light source fed to the receiver must be completely blocked out before detection is recognized. This method is not very efficient with transparent, overlapping, or bicircular objects such as clear plastic, two pieces of material riding on top of each other, or objects with holes, which may trigger the photocell more than once.

Another approach to optical detection is the analogue photo-optic system. This system is fast, flexible, and accurate

since certain parameters must be established and met before detection is recorded. The Photo Optic Shadow Detector (Sigma Systems, Inc.), for example, detects the dimensions of a specified shadow made by the unit of product when fed through the detection zone. Each shadow of a detected unit must meet certain parameters before being recorded. The parameters can be entered into the computer portion of the counting system to compensate for objects such as flat washers and O rings, which would trigger a photocell system twice, or clear plastic, which would not block out a light source but would cast a shadow that could be detected.

Nonoptical systems. The nonoptical methods of parts detection are similar to human touch. The touch methods generally involve escapement devices, electrical contact, or magnetic-field contact. The escapement device is usually fully mechanical and is similar to an analogue machine. Each unit of product must be of a specific shape and size. The product is fed into a receiving device that fits those exact parameters and then is discharged and recorded as a specific unit of product. This is equivalent to placing dominoes in a box made just for dominoes. If only ten dominoes fit, the unit of product would be recorded as ten units. This system is very accurate, but it offers little flexibility. It is commonly used for high speed counting of uniform products such as pills and tablets.

Another nonoptical method is the electrical-contact method, in which an electrical switch is triggered each time a part comes in contact with the switch. The part is then recorded as one unit of product. The response time and the ability of the product to actuate the mechanical switch-triggering device greatly affects the accuracy and flexibility of this type of system. The *magnetic system* is another touch-system method. Each unit of product to be counted must come in contact with or disturb the magnetic flow being transmitted from a magnetic source. As each unit of product is fed through the magnetic field, it is recorded as a counted unit.

Coupled with the detection system is a process known as *discrimination* (eg, a farmer counting cows knows how to exclude sheep). This process has hampered the automated counting system greatly, for once this vital function leaves the dependability of the human senses, accuracy often suffers. Only two of the detection systems mentioned above are able to discriminate: the photo-optic-analogue system and the escapement-device system. In both, certain parameters must be met before a unit of product is recorded; the other systems record any item that is detected. Engineers have been working for years to improve the efficiency of the detection system to ensure an accurate count.

Product Handling

In almost every counting system, the process of getting the product to the detection zone and then moving it away from the zone must be achieved, whether the detection zone goes to the product or the product comes to the zone. This system of product movement is known as product handling. The product is usually brought to the detection zone, and in packaging, all of the methods bring the product to the zone in single file. There are various ways to do this. One is vibratory feeding, which is designed to vibrate a track or bowl filled with product which in turn causes the product to vibrate along a designated path or ramp. The tracks or ramps narrow as the product nears the detection zone in order to create a single file. These are the most flexible of feeding systems since most objects lend themselves to vibration.

Another method is the *belt* or *V-belt system*. The product is placed in a master container and discharged onto a belt system. The width of the belt track is narrowed to allow only one part to pass at a time in order to achieve single-file feeding. Product handling plays a very important role in the counting process, for even the most refined counting systems will be inaccurate if the product is not presented to the detection zone in a manner acceptable to the detection device.

There are many approaches to product feeding, but most counting systems allow free entry and exit of product through the detection zone. In packaging, however, it is sometimes necessary to retain all or part of the amount counted for a specific function. The retention is known as accumulation or partial accumulation. The accumulation functions are usually determined or predetermined by a manual function in which an accumulation parameter, ie, the amount of product desired, is assigned to the counting unit. The counting unit usually must meet the assigned parameters before permitting the accumulation functions to discharge the retained product. This function is essential when a manufacturer wants to place a predetermined amount of product into a specific-size container, eg, accumulating 20 tablets, then discharging them into a package or bottle. It also plays an important role in the packaging process since most packaging machines are intermittently cycled by the signal received from the counting system when the accumulation function discharges product. Most containers are aligned under the accumulation chute to receive the allotted amount of product.

As electronic technology advances, the ability to count and discriminate parts will improve. However, the attainable operating speeds will depend on the speed with which the unit of product can be fed into the detection mechanism and the speed of the product-handling system. The performance of any counting system must be objectively evaluated on its ability to count accurately and its adaptation to the intended use.

DAVID MADISON
Sigma Systems, Inc.

FILLING MACHINERY, DRY-PRODUCT

The objective of dry-product filling machinery is to fill a container with a product to a specified weight or volume, meeting legal or formula requirements. This should be accomplished at an economic balance of fill accuracy, product cost, package cost, production rate, and overall equipment and operating cost.

The accuracy of fill is measured by the standard deviation known as sigma. This is a calculated statistical number and indicates the shape of a normal curve (see Fig. 1). The majority of weighing plots approximate a normal curve so sigma is used to indicate the shape of this plot. A commonly used index of performance is plus or minus three sigma, which indicates that 99.73% of all weights measured fall within this range. As the value of sigma is improved or decreased it indicates that the average weight can be targeted closer to the label or formula weight with less risk of over or under weights.

Sigma can be determined for the overall system over a period of time or as a machine capability parameter. In statistical weight procedures, target weights are calculated based on a system-performance sigma. Determination of machine performance is made by comparing the machine-capability sigma with the system-performance sigma.

Overfill, or "giveaway," is determined by target weight and is a function of filling equipment accuracy.

The basic elements of a dry-product filling system are product feed, product measurement, and package- or container-forming or handling system.

Product-Feed Systems

Product-feed systems commonly utilized in dry-product filling equipment are

1. *Augers.* For free-flowing and nonfree-flowing products.
2. *Gravity flow.* For free-flowing products.
3. *Vibratory feeders.* For free-flowing, nonfree-flowing and friable products.
4. *Belts.* For free-flowing, nonfree-flowing, friable, and sticky products.
5. *Screw-type units.* For free-flowing and nonfree-flowing products.
6. *Vibrating-bin outlet.* For free-flowing and nonfree-flowing products.
7. *Cascade-filling systems.* For free-flowing, nonfree-flowing, and sticky products.
8. *Vacuum-filling systems.* For free-flowing and nonfree-flowing powders.

Flowability is basically measured by the angle of repose of a product on a flat plate after a gravity fall from a fixed orifice from a given height. There are standard tests to give a quantitative value to flowability.

Product measurement can be determined by either volumetric or gravimetric methods. If density and flow characteristics are consistent, volumetric fillers give excellent results. Normally, volumetric fillers are less expensive, easier to maintain, and operate at higher production rates than weighing units.

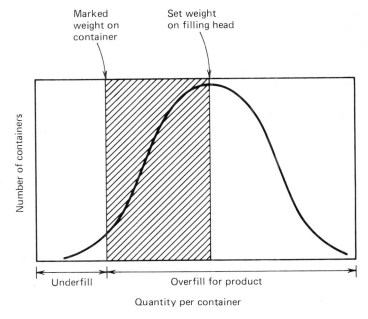

Figure 1. Tolerance weight (shaded area) added to nominal fill weight stated on container.

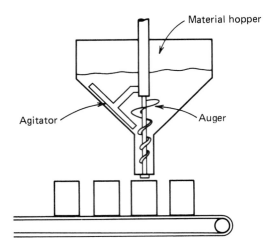

Figure 2. An auger filler.

Weighing units give improved fill accuracy when density and flow characteristics vary. Problems develop with weighing units if density varies too widely. High densities cause slack container fills, which can exceed legal or quality limits. Low densities can cause container overflow. Variable container vibration is frequently used to package a wider range of densities.

Examples of volumetric measuring methods include varying auger rotation by time or angular measurement; varying time or flow rate of gravity flow systems, vibrators, screw feeders, and belts; utilizing container as volume measurement and varying fill density by vibration or vacuum; and varying intermediate flask volumes.

Augers. Augers provide a method for uniformly feeding both free-flowing and nonfree-flowing powders, flakes, granules, and other types of products (see Fig. 2). Augers are normally mounted in the vertical position but have also been designed to operate on an angle or in a horizontal position. Augers may be designed with rotating agitators in the feed hopper to give positive product into the auger screw area and prevent "rat holeing." Variable screw pitches are utilized to provide product compaction and minimize product density variations. Augers for free-flowing products are equipped with spinner plates at the discharge of the auger to prevent product runout. The discharge of the auger can be provided with a compartmented divider head to split the main flow into multiple streams. The auger drive can be intermittent or continuous, depending on the application. A timed cycle or controlled rotation through a preset number of revolutions are normally used to control volumetric fills.

Continuous-feed augers are used on continuous-motion packaging machinery where pockets pass, at a constant rate, under the auger discharge. These pockets are a part of a rotary table or affixed to a chain drive. They are designed to overlap so there is no product spillage. Fill volume is regulated by changing the speed of the auger or the rate of the pockets passing through the auger flow.

Integrated auger systems are used on multiple lane machines such as horizontal and vertical pouch machines, thermoform machines, and form/fill/seal units (see Form/fill/seal, horizontal; Form/fill/seal, vertical; Thermoform/fill/seal).

Each auger is usually controlled by its own clutch brake system, and fill volume is individually set.

Augers can be used as feeders for net-weigh systems. The auger rate can be varied to provide a bulk and dribble feed, or dual augers are utilized to provide this feature. Cutoff points are set by the weight-sensing unit.

Gravity-flow systems. Gravity-flow designs are the simplest of feeding methods and are used on free-flowing products such as coffee, tea, nuts, rice, sugar, and salt. Flow rate is controlled by the angle of flow, product characteristics, pressure or "head" of the feed system, and orifice sizing. Normally a bulk and dribble rate is used to achieve improved weight accuracy. Gravity-flow systems can feed other volumetric units such as pockets, and if these are continuously moving under the flow, the fill is determined by the flow rate and pocket velocity. Gates can be used as product-shutoff devices, and these can operate on a timed or product-level-sensing control system.

Vibratory feeders. Vibrating-feed systems are used on free-flowing and relatively nonfree-flowing products to achieve controlled feed rates (see Fig. 3). These units can handle a wide range of products, such as powders, friable materials like snack foods, large discrete particles such as those in the vegetable industry, and abrasive materials, to name a few.

Feed rates are controlled by varying vibration frequency or amplitude. Vibrators can be driven electrically, mechanically, or hydraulically. In many installations, vibrators are mounted in series to obtain more consistent flow rates. Fill volumes can be varied by changing the flow rate to the vibrator, time of feed, frequency, amplitude of vibration, or pitch angle of the vibrator tray. Vibrators are sometimes equipped with shutoff gates at the discharge of the pan to prevent product from sliding off after the fill weight has been reached.

Vibrator shapes can be a flat pan, tube, serrated, or V-shaped, depending on the product and the rate to be fed. In many instances the pans are coated with a material such as Teflon (DuPont Co.) to prevent product buildup. Some pans are

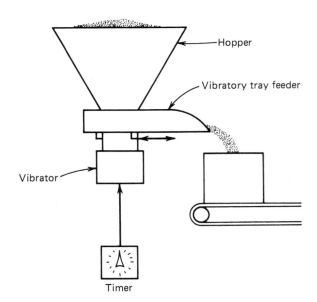

Figure 3. A timed vibrator filler.

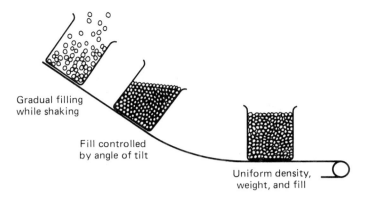

Gradual filling
while shaking

Fill controlled
by angle of tilt

Uniform density,
weight, and fill

Figure 4. A cascade-filling system.

constructed entirely of plastic to minimize and facilitate cleaning. In some designs, where buildup is a factor, the pans are designed to clip apart for easy replacement and immersion in cleaning solutions. Vibrators are commonly used to feed scale systems because of their versatility and ease of control.

Belt feeders. Belts are used to feed free-flowing, nonfree-flowing, friable, and sticky products. They can be provided with scrapers and washing systems. There are a wide range of belt materials and sizes to handle an extremely broad range of products. Belt materials range from tough multiple construction for use on items such as minerals and abrasive chemicals to sanitary belting for food products. Lightweight belts of polyester film (see film, polyester) and fiberglass-coated belts are used on sensitive products similar to those in the pharmaceutical industry. Belts can be supplied with molded pockets and side walls and can operate horizontally or on an incline. They can function on an intermittent or continuous basis and controls are basically similar to those outlined for augers or vibrators.

Weigh-belt systems use weighing devices to measure the mass flow across the belt. A wide range of weigh-belt designs is commercially available. The most common design supports the belt on a load cell, and either totalizes the weight across the belt or varies the rate to maintain a constant mass flow.

Screw feeders. There are two basic types of screw feeders: the horizontal type that delivers a consistent flow rate and flexible feeders that are used to move product from a floor hopper to the filling system. Some horizontal units are designed to vibrate to improve the consistency of the product feed. There are designs where the side walls of the feed hopper are made of an elastomeric material and flexed to improve the flow of difficult materials. The screws can be solid with constant or varying pitch similar to an auger. Some screws are made of a coil of round or flat wire, depending on product requirements. Feed rates are varied by changing the rotational speed of the screw.

Vibrating-bin discharge. Vibrating-bin-discharge systems are primarily used for filling large containers such as drums or bags. The design normally consists of a relatively shallow angle-dished bottom that is flexibly connected to a bin. A vibrating system is mounted to this section, and when activated, provides a uniform stream of product from the bottom orifice. A gate can be provided to obtain a positive shutoff. Container fill is volumetric or by weight control systems.

Cascade-filling systems. In this system the product is fed in waterfall fashion to the open-mouth containers (see Fig. 4). The containers fill as they move along a conveyor in the path of the falling product. The containers are tilted at a variable angle and vibrated to settle the product. The final volume is determined by the container speed, the angle of container tilt, and the vibration amplitude and frequency. The product overfill is recycled back to the original product stream. This system provides high filling rates for free-flowing and nonfree-flowing products.

Vacuum-filling systems. There are two basic types of vacuum-filling systems. In one type the container is evacuated and the product flows through an orifice into the container (see Fig. 5). The vacuum is pulled through a fine mesh screen inside the gasket that seals the container. The vacuum is pulsed to compact the product within the container. The amount of vacuum and the number of pulses determine the final weight. The base weight is established by the amount the vacuum screen protrudes into the container. This may be varied by changing the gasket thickness. The containers must have a relatively constant volume and be rigid enough to withstand the vacuum without flexing to obtain reasonable weights.

In another form of vacuum-filling system, the product is filled into a rotating pocket that has been evacuated (see Fig. 6). The pocket is mounted on a wheel that moves in a vertical plane. The vacuum holds the product in the pocket and discharges it to a container or funnel that moves below the wheel. The volume is adjusted by varying the telescoping pocket depth.

Weighing Systems

A weighing system basically senses an imbalance due to the addition of product in a weight receptacle. The feed to the weighing system is critical to its performance and could consist of many of the product-feed systems outlined above. The weighing or sensing elements vary considerably in detail. In its simplest form the imbalance reaches a set point that un-

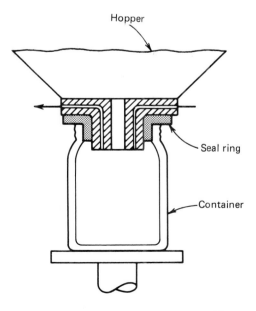

Hopper

Seal ring

Container

Figure 5. A vacuum-filled container filler.

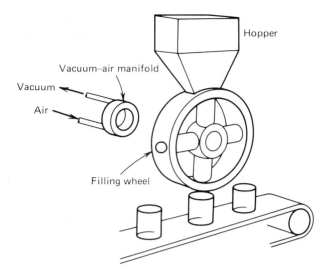

Figure 6. A vacuum-filled pocket filler.

latches a mechanical stop to discharge the product. Sensing elements used in advanced weighing systems are

1. Linear-voltage-differential transformers commonly known as LVDTs.
2. Low capacity strain gauges.
3. Load cells.
4. Pneumatic-pressure-differential switches.
5. Proximity switches.
6. Photo electric sensors.
7. Limit switches.
8. Reed switches.
9. Mercury switches.

Most weighing systems dampen the action of the weight-sensing element with an oil-filled mechanical dashpot. Electronic systems require constant voltage systems, which are usually designed into the circuitry. Most units also have tem-

perature compensation circuits to maintain accuracy under fluctuating environmental conditions.

Almost all of the net-weigh systems operate on a bulk- and dribble-feed principle (see Fig. 7). The majority of the product is fed at a high rate until the bulk set point is reached. The bulk feed is halted, and the dribble rate is continued until the final-weight set point is reached and the feed is stopped. With this system there is always some material in suspension when the final set point is reached. For improved accuracy this amount should be kept to a minimum.

Weighing can take place in an intermediate hopper (net weighing) or in the final container itself (gross weighing). Gross weighing is commonly used where the product buildup on a weigh bucket would severely impair the accuracy of the weighing system. In order to obtain accurate gross weights, the container weights must be consistent or the empty containers must be weighed and the final weight compensated for the weight of the empty container (see Fig. 8).

Computer-combination net weighing is a relatively new weighing concept that was introduced commercially in 1971. The system utilizes multiple weigh buckets that are filled to a portion of the total net weight. A microprocessor analyzes individual bucket weights and selects the combination of weigh buckets that yields the weight closest to the target weight. Since the product is completely weighed when the selection is made, there is no variation owing to product in suspension or change in product characteristics. This system provides extremely accurate weights, but is particularly effective where discrete pieces of varying size and weight are being packaged. They have shown remarkable weight improvement in the snack, vegetable, and shellfish food-packing areas.

These units are available for automatic or manual product feed. Rates of up to 120 weighings per minute are available with automatic feed, depending on product characteristics. Manually fed systems are currently designed to operate at rates of 25 weighings per minute.

In the 120-per-minute systems the weigh buckets are

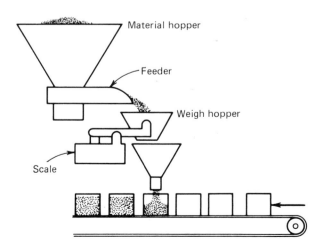

Figure 7. A net-weighing filler.

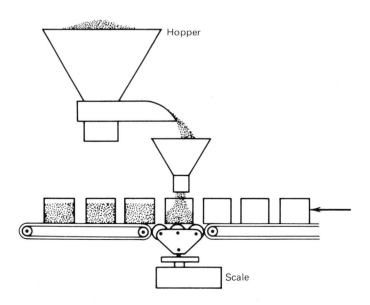

Figure 8. A gross-weighing filler.

mounted in a circular position around a central discharge point. Automatically fed straight-line systems are available now as well. The manually fed systems are in-line configurations that discharge to a belt conveyor.

The majority of the sensing elements used in modern electronic weighing systems provide electrical outputs that are directly proportional to the weight applied. Multiple cutoff points can be used to control product feed and improve accuracy. The use of electronic circuitry in conjunction with microprocessor controls and computer interface can provide auxiliary functions such as:

Automatic checkweighing. The final weight can be sensed, recorded or displayed, and refilled to a proper weight if below an underweight reject point (see Checkweighing).

Tare weighing. The empty weigh container can be checked for tare between weighings, and the fill weight compensated for a change in tare.

Feed timing. Overall cycle timing can be monitored and feed rates adjusted for optimum performance. Bulk and dribble times can be electronically compared to preset values.

Informational readouts. Weight performance can be fed to a computer interface to generate statistical and operating data. This data can be indicated utilizing a CRT, LED, LCD system, or hard copy. Actual weights from each bucket can be displayed along with performance data for a predetermined period of time. Based on actual performance, target weights can be automatically recomputed and the machine sensing unit reset to meet these new targets. When tare-weight changes exceed a given limit, a signal can be given that the buckets need to be cleaned. Total average weight, product giveaway, and machine efficiency can be displayed and recorded. Normal curve distributions can be visually displayed on a CRT with each weighing indicated. These electronic systems can monitor other machine functions not associated with the weighing equipment.

Bulk-Product Feed

The primary feed system that delivers product to the packaging equipment is critical to system performance. Both for volumetric and gravimetric filling equipment, changes in density, flowability, and particle size have a detrimental effect on fill performance.

A major cause of changes in product characteristics comes from variation in production processes or natural variations in agricultural products. These variations may be minimized by the use of blending systems. Blending can be done in batch units, belts, or blending screws. Even with these blending systems, segregation can occur in bins feeding the packaging equipment. Static-blending storage systems and desegregation cones are used to minimize product variations. Breakage in mechanical feeders causes changes in product characteristics and the product handling system should be designed to take this into account. Another factor that affects fill accuracy is the level of product in the feed hopper. Adequate level control is required to maintain consistent feed.

Packaging Equipment

The variety of packaging equipment utilizing the systems outlined is extensive. A partial listing of major types is given below:

Vertical form/fill/seal machines.
Horizontal form/fill/seal machines.
Rotary form/fill/seal machines.
Rotary volumetric fillers.
Rotary net-weighing fillers.
In-line volumetric net-weigh machines.
Thermoforming equipment.

In many cases there may be multiple feeding systems used on one machine. For instance, there are indexing pouch machines and vertical form/fill/seal equipment that have up to five separate feeding systems to provide a blended product in the final package.

Gravimetric equipment utilizes the volumetric feed systems outlined earlier and controls the feed cutoff by a weight-sensing unit. There is equipment available where a combination of volumetric and gravimetric units are utilized both in series and in parallel. One type of machine fills the container below its label weight with an auger volumetric unit. The container is indexed to a weighing platform and the fill is completed to the target weight. To improve the production rate the container can be preweighed after the bulk fill and this weight stored in a memory circuit to control the final weight addition. The container tare weight has to be consistent to obtain accurate fills.

An example of a parallel system is a rotary filling unit that fills adjustable volumetric flasks. One flask discharges product to a weigh bucket prior to filling the container. This weight is used to control all flask volumes to meet the required weight. This system combines the simplicity of a volumetric system with accuracies approaching that of a complete scale system.

Within a given category of packaging machines there are significant design differences. For instance, there are rotary scale systems where the scale elements are fixed and the containers move under the scale discharge. In another design the scales rotate with the container. There are advantages and disadvantages to each design. Product differences would dictate the proper selection.

The physical installation of certain types of filling equipment is critical to fill performance and efficiency. Equipment and building vibration has a major effect on accuracy. Location within the packaging layout affects efficiency and also fill variation. Start-stop operation of a filling machine gives poor weight performance.

Selection of filling equipment for a particular package and product is a major study in itself and the subject of many articles. Factors to be considered, among others, are production rate, product cost, package type and cost, machine efficiency, maintenance cost, equipment cost, operating cost, and layout limitations.

Careful analysis of all variables is necessary to determine the optimum system for any given situation.

BIBLIOGRAPHY

General References

E. R. Ott, *Process Quality Control*, McGraw-Hill, Inc., New York, 1975.

J. M. Juran in J. M. Juran, F. M. Gryna, Jr., and R. S. Bingham, Jr., eds., *Quality Control Handbook*, McGraw-Hill, Inc., New York, 1974.

C. S. Brickenkamp, S. Hasko, and M. G. Natella, *Checking the Net Contents of Packaged Goods*, U.S. Department of Commerce, National Bureau of Standards, U.S. Government Printing Office, Washington, D.C., 1981.

D. L. Winegar, *Package Engineering Encyclopedia*, Cahners Publishing Co., 1984, pp. 247–250.

J. Blackwell, "Machinery, Filling, Dry Products" in *Package Engineering Encyclopedia*, Cahners Publishing Co., 1981.

F. C. Lewis, *Package Engineering Encyclopedia*, Cahners Publishing Co., 1984, pp. 247–250.

LVDT Transducers for Weight, Dimension, Pressure in *Catalog A-100 June 1980*, Automatic Timing and Controls, Co., King of Prussia, Penn., 1980.

"Weighing and Proportioning," in M. Grayson and D. Eckroth, eds. *Kirk-Othmer Encyclopedia of Chemical Technology*, 3rd ed., Vol. 24, Wiley–Interscience, New York, 1984, pp. 482–501.

ROBERT F. BARDSLEY
Bard Associates

FILLING MACHINERY, LIQUID, CARBONATED

The method for filling carbonated liquids, primarily beer and soft drinks, differs from other filling techniques (see Filling machinery, still liquid) in that it is accomplished under pressure and uses the container as part of the control of net contents. Carbonated beverages, which tend to foam, require filling techniques that ensure the retention of the required carbonation levels in different sizes of cans and bottles. It is imperative that the carbonated liquid be processed in a way that prevents excess foaming, which would result in uncontrolled filling levels as well as impaired closing of the vessel.

The filling machine (filler) consists of: a rotating bowl with a valve and CO_2 pressurization control; filling valves which attach to the perimeter of the bowl; a table top which contains the controlling technique for feeding in and taking away the container; a closing section which applies either a bottle closure or a can end; and a drive system that keeps all components in proper synchronization with each other.

The bowl must control the pressure upon the liquid within it, yet be at a pressure lower than the system that feeds it. In this way, a continuous supply of product is assured, with the least amount of turbulence. The level within the bowl is maintained by a simple float valve or similar device.

The filling valves embody the applied technology involved in filling carbonated products. Almost all of the machinery manufacturers of note employ the same principles, with some variation in the mechanical interpretation of the concepts. To understand the valve's function, it is necessary to follow it through the various filling stages described in the following paragraphs.

The vessel is presented to the filling valve to assure complete intimacy between both surfaces. This is controlled by a pneumatic pressure system that holds the vessel firmly against the valve, yet tenderly enough to avoid crushing thin metal cans or light plastic bottles.

The container rotates with the filler valve, and a mechanical trip actuates the valve to permit CO_2 from the upper part of the bowl (above the liquid level) to pressurize the container. The pressure in the container is now equal to or somewhat lower than the pressure in the bowl.

The valve is then actuated into the next stage, permitting the product to flow into the prepressurized container gravimetrically. During the filling phase (as fluid enters the vessel) it is essential that the CO_2 in the container be displaced. This is accomplished through a vent tube, which is normally in the center of the valve and protruding downward into the vessel. As the product fills the vessel, it ultimately rises to seal off the vent tube or ball check. This stops the filling process, since pressures in the bowl above and in the container below have reached an equilibrium.

While the container is still in an intimate seal with the valve, another external latch actuates an internal chamber in the valve which closes the connection to the bowl (both the liquid portion and the CO_2 charge above it) and simultaneously vents the container to atmosphere. This step maintains control of the product in the container during the depressurizing step and assures that the product will be virtually foam-free when the container is removed from contact with the valve.

The table top, which employs an exit star wheel (not generally used on can fillers), sweeps away the lowered package from the rotating bowl, and transfers it into the closing section at table top height. If the container is a can, it is closed in the closing machine by an "end" which is rolled and seamed in place (see Can seamers), integrally with the can body, with forces great enough to withstand high internal pressures. Bottles can be closed with a pry-off or twist-off crown; a rolled-on aluminum closure, which is threaded in place using the threaded portion of the bottle as a mandrel; or a prefabricated threaded closure applied by standard technique (see Capping machinery; Closures).

Carbon dioxide dissolves more readily in cold water than in hot water; thus, to keep foaming at a minimum and filling speeds at a maximum, it has been common to run beer and soft drinks as close to 32°F (0°C) as is practicable. In recent years, to save the energy consumed by refrigeration, fillers have been introduced that operate at ambient temperatures. Controlling the CO_2 in the carbonating and filling stages means these operations must be accomplished at above-normal pressures, which complicates the internal parts of the filling valve and reduces rates of speed. The production rates of modern fillers have increased in recent years to approximately 2000 12-oz (355-mL) cpm and reportedly 1500 16-oz (473-mL) glass bpm. These increased rates of output are made possible by the use of programmable computers and new advanced instrumentation. Precise fill heights in the containers are controlled by the length of a vent tube in a bottle filler and ball check technique in a can filler. However, due to minor variations in the dimensions of glass and plastic bottles, there is more variability in volumetric content than in the exact volumetric measurement technique sometimes employed in "still" liquid filling.

W. R. EVANS
Coca-Cola Bottling of N.Y.

FILLING MACHINERY, LIQUID, STILL

This article deals with the filling of noncarbonated liquids intended for packaging and distribution into rigid and semi-rigid preformed containers, eg, glass bottles, sanitary cans, plastic bottles, and preformed paper cartons. Form/fill/seal packaging is not discussed (see Form/fill/seal, horizontal; Form/fill/seal, vertical; Thermoform/fill/seal), and the information relates only indirectly to such applications as the filling of paper cups in vending machines.

Liquid-filling machines are classified here in terms of two fundamental characteristics: filling principle employed (see Table 1) and container-positioning method utilized (see Table 2). Except for a few specific restrictions (discussed below), the two characteristics are independent, but certain combinations are not commercially available. Packagers can select from a wide range of fillers, however. Among the members of the U.S. Packaging Machinery Manufacturers Institute (PMMI) alone, 45 companies offer liquid fillers (1), and this figure does not include machines made in other countries.

METHODS OF FILLING

In Table 1, methods of fill are divided into two primary categories: the sealed-container system in which the filling device seals positively against the container, and the un-sealed-container system in which the container is left open to the atmosphere during the fill process.

Sealed-container Filling Systems

In sealed-container filling, there are seven distinctly identifiable types of fillers. All sealed-container fillers fill to a controlled level in the container.

Balanced-pressure fillers. The first three of the seven mentioned above are balanced-pressure fillers, in which product flows through a valve into the container from a tank of liquid located above the container, and air from the container is vented back to the headspace in the tank through the same valve. The typical embodiment of such a filling system, *gravity filling*, one of the simplest and most reliable, is illustrated in

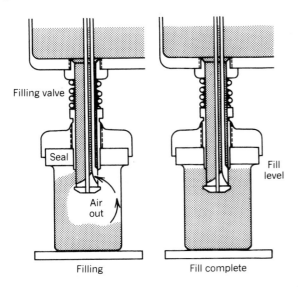

Figure 1. Pure gravity filling. Courtesy of Horix.

Table 1. Methods of Container Filling

Type of container	Filling methods
sealed container	balanced pressure
	gravity
	gravity–vacuum
	counterpressure
	unbalanced pressure
	vacuum
	prevacuumizing
	gravity
	pressure
unsealed container	level sensing
	piston volumetric
	cylinder vertical, closed ends
	cylinder vertical, open-end inlet
	cylinder horizontal, single-ended
	cylinder horizontal, double-ended
	rolling-diaphragm volumetric
	displacement-ram volumetric
	volume cup
	turbine-meter volumetric
	positive-displacement-pump volumetric
	weight
	gross weight
	net weight
	time
	controlled pressure head
	constant-volume flow
	overflow

Figure 1. As the filling takes place through the sleeve-type valve illustrated, liquid flows from the tank through the liquid port into the container, and air within the container flows up the vent tube to the top of the tank. The container fills product to an exact level determined by the position of the vent port relative to the bottle. Any liquid in the vent from a previous filling cycle is returned to the product in the tank by the air flowing up the vent tube. If necessary, air or vacuum may be used to clean the vent before the start of fill.

A modification of pure gravity filling is the *gravity–vacuum filler*. In such a system, a low vacuum is maintained in the headspace in a sealed tank. When the container is brought into sealing contact with the filling valve and the valve is opened, the pressure of air in the container helps force any product in the vent back into the tank, accelerating the start of the filling process. The use of gravity–vacuum fillers also prevents the loss of product that would occur if a chipped or slightly broken container were filled. This savings is of particular advantage when filling more expensive products.

To fill thin-wall plastic containers, such as 1-gal (3.8-L) milk bottles, a pulsating vacuum in the tank is sometimes

Table 2. Container Positioning Methods and Configurations

Positioning method	Configuration
manual	
automatic, in-line	single or dual lane
automatic, rotary	single or dual lane

used to cause the container walls to flex in and out, assisting the foam in moving up the vents. The pulsations must be timed so that a container is not flexed inward at the position at which it is ready to break away from the filling valve, because this could cause underfill.

Thousands of different styles of filling nozzles, utilizing single ports, multiple ports, screens, sliding tubes, or check valves, are offered by various manufacturers. All of these styles are designed to achieve the maximum production rate with the fewest number of filling valves and provide greater accuracy of fill. The selection of nozzle type is probably best left to the judgment of the machinery manufacturer, based on experience and product testing.

Counterpressure fillers for carbonated beverages are discussed elsewhere (see Filling machinery, carbonated liquid).

Unbalanced-pressure fillers. Unbalanced-pressure fillers utilize a difference in pressure between that on the liquid to be filled and on the vent that permits air in the container to escape during the filling process. The usual combinations are listed in Table 1. The use of unequal pressure permits higher rates of product flow than possible with the balanced-pressure fillers. Unequal pressure is particularly advantageous when filling containers with small openings, viscous products, or large containers. Unbalanced-pressure filling has the disadvantage of requiring an overflow-collection and product-recirculation system, in contrast to the relative simplicity of balanced-pressure fillers. Higher liquid-flow rates do not necessarily result in faster filling because the additional foam generated by rapid entry of product into the container must be drawn off through the overflow system to obtain accurate filling-level control.

A schematic diagram of a typical *vacuum filler* is shown in Figure 2. The supply tank may be located either above or below the container to be filled. After the filling valve seals against the container and the valve opens, the vacuum on the vent draws the liquid into the container up to the filling level. Usually, a substantial quantity of liquid is drawn into the vent, which leads to an overflow tank. Product is recovered in the overflow tank and then recycled.

The *prevacuumizing filler* is a special form of vacuum filling. On such a filler, vacuum is first drawn in the container, evacuating the air. The valve then permits liquid to enter the container. Because such a system is complex and expensive, it is normally only used when liquid is being added to solids already in the container. Certain solids, such as peach halves, trap air. Such air entrapment may be eliminated by use of a prevacuumizing filler.

In an unbalanced-pressure *gravity filler*, the product-supply tank and the overflow tanks are open to the atmosphere, but the product tank is located above the container and the overflow tank is located below the container, permitting the differential pressure achieved by the difference in elevation to cause product flow. Such a filler is necessarily rather restricted in its ability to adapt to varying products and containers since the pressure difference is established solely by the product-tank and overflow-tank locations. Fillers of this type are not very common.

A *pressure filler* is very similar to a vacuum filler except that pressure is applied to the product. This may be achieved either by pressurizing the headspace over a tank or by direct pumping of the product to the filling valve. In the most common form of pressure filling, the product is pressurized and the overflow tank open to the atmosphere. Such a system allows unbalanced fill without vacuum. This is desirable when vacuum cannot be drawn on the product. For example, drawing a high vacuum on alcoholic beverages can reduce the alcoholic content of the beverage. Applying a vacuum to a hot product, such as juice at 200°F (93°C) causes the liquid to flash. If desired, both the product and the vent can be maintained above atmospheric pressure, but with a higher pressure on the product. Such a filler is often used for filling lightly carbonated products, such as certain wines, using the pressure to retain the low carbonation in the product.

Unsealed-container Filling Systems

Level-sensing fillers. Level-sensing fillers fill containers to a controlled level without sealing the container, as shown in Figure 3. Such a filling technique eliminates product recirculation and allows filling to a level in plastic containers which

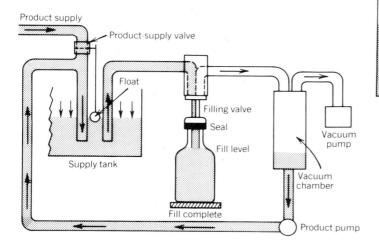

Figure 2. Pure vacuum filling. Courtesy of PMMI.

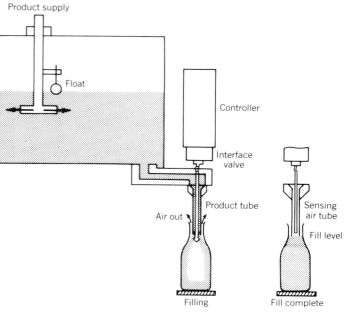

Figure 3. Level-sensing filling. Courtesy of PMMI.

would bulge out or flex inward if pressure or vacuum were applied to the sealed container. A level-sensing filler uses some type of sensing means, typically a flow of very low pressure air. Rising liquid level in the container blocks air flow, triggering a control system that shuts off product flow to the container. Such control mechanisms, which are required at each filling nozzle, are expensive, but high rates of fill may be achieved because there is no product overflow and no foam to be removed. Electronic sensing devices are also used in filling open-top containers with hard-to-clean products such as paint.

Piston volumetric fillers. At present, unsealed-container fillers are the most common, and most unsealed containers are filled volumetrically. In view of this popularity, many different kinds of volumetric fillers are available. For volumetric filling, piston fillers are most widely used. Table 1 indicates four subclasses of piston fillers, depending on the orientation of the pistons, ie, vertical or horizontal, and inlet arrangement.

Cylinder vertical, closed ends. One station of a typical vertical-cylinder rotary filling machine is illustrated in Figure 4. The valve(s) controlling the product flow between the supply tank, measuring chamber, and dispensing nozzle may have either a rotary or a reciprocating motion. The rotary style, which is more common, is illustrated. The product is drawn into the cylinder from the liquid-supply tank when the piston moves upward. The valve then rotates to permit the premeasured volume in the cylindrical chamber to flow into the container. Usually, either a direct mechanical drive from a cam track or an air cylinder is used to stroke the piston. If an air cylinder is used to drive the piston, controls are usually such that the piston does not cycle if there is no container in place. This eliminates moving the liquid back and forth between the measuring chamber and the supply tank, a situation which is usually undesirable and may cause product breakdown with foods such as mayonnaise. It is not easy to uncouple a mechanically driven piston.

Cylinder vertical, open-end inlet. In an alternative design, product enters vertical volume chambers through cylinders open at their upper ends, with a cam drive located below. A nonrotating plate with an orifice allows product to enter the open-ended cylinders at the appropriate position during the rotation of the filler bowl. Such a configuration is mechani-

cally complex and generally considered difficult to clean. Fillers of this type have the advantage, however, of being able to handle products containing sizable solids in suspension.

Cylinder horizontal. Measuring cylinders may be mounted horizontally. Usually single-ended, they are very similar to vertical cylinders in operating principle. They are frequently found on in-line, large-volume fillers, in order to avoid excessive height for such machines. A volume cylinder may also be double-ended, with inlets and outlets at both ends. Product under pressure flows into one end, causing a floating piston to move and expel the liquid in the opposite end of the cylinder into the container. In some fillers, the double-acting cylinder may be used for a single fill, and the first half of the fill comes from one end. The flow pattern is then reversed to discharge product from the other end of the cylinder for the second half of the fill while the first end is being filled for the next cycle.

Rolling-diaphragm volumetric fillers. The volumetric fillers described above normally have some type of a seal, such as V- or O-rings, between the pistons and cylinder walls. An alternative method for measuring volume is to use a rolling diaphragm; a typical arrangement can be seen in Figure 5. The diaphragm provides an absolute seal and also eliminates the friction contact of a seal with a cylinder wall. Such sliding causes abrasion and particle generation, which is usually not important, but minimizing particulate generation is very important in the packaging of intravenous solutions and injectable drugs. Rolling-diaphragm volumetric fillers often employ flexible diaphragms or pinch valves to control product flow to and from the measuring chambers.

Displacement-ram volumetric fillers. Another method for dispensing a specific volume of liquid is to use a displacement ram. The ram enters one end of a cylinder through a seal of some type, but the displacement ram does not touch the inside of the cylinder wall. An external air cylinder drives the ram, which displaces a controlled volume from the cylinder into the container. Such fillers are easy to clean.

Volume-cup fillers. Cup-type volumetric fillers operate by first transferring the product from an open tank into measuring cups of precise volume. Depending upon the design, each cup may be filled to a level matching that of the tank or the cup may fill to overflow and then rise above the level of liquid in the tank. Each valve then opens at the bottom, permitting liquid to flow from the cup into a container. Although such fillers are appropriate only for low viscosity liquids which do not cling to the sidewalls of the cups, for suitable applications the volume-measuring method is accurate, inexpensive, and reliable. Various volume-adjustment systems are provided. These fillers are usually rotary and are often designed to be easily changed, by means of a change of filling valves, to gravity or gravity–vacuum filling.

Turbine-meter volumetric fillers. The amount of liquid dispensed from a nozzle can be measured by placing a turbine flowmeter in the line ahead of the nozzle. Such meters, which include an electronic counting and control system to start and stop flow, are accurate but expensive, and generally used only for filling large containers, ie, ≥5 gal (≥19 L).

Positive-displacement volumetric fillers. Viscous liquids may be moved directly from a supply system through a positive-displacement pump into a container. Volumetric measurement is achieved by accurately controlling the number of revolutions made by the pump. An auger may be used as the pump mechanism.

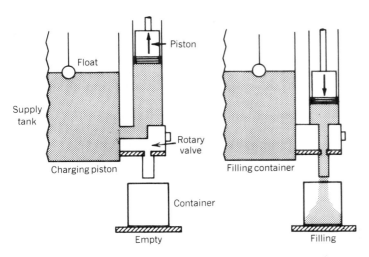

Figure 4. Piston volumetric filling. Courtesy of PMMI.

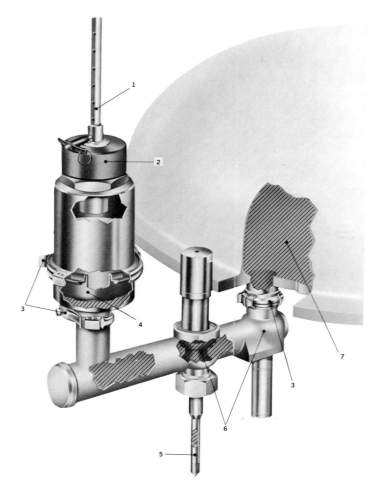

Figure 5. Rolling-diaphragm volumetric filling. Courtesy of Horix. **1** Volume adjustment rod; **2** precision volume adjustment; **3** quick-disconnect design; **4** Volume-chamber diaphragm; **5** product tube; **6** valve seats; and **7** product supply manifold.

Weight fillers. On *gross-weight fillers,* each fill station is fitted with a weighing device, typically a beam scale, which acts to shut off product flow when a predetermined weight has been reached. Each scale is set for the maximum weight of container and contents. Adjustment for other containers and content weight may be made by adding special weights to each filling platform. By using a load cell as the weighing device, the weight is electronically measured continuously, and liquid flow is stopped at a predetermined weight. With microprocessor controls, the change from one weight to another is quite simple. *Net-weight fillers* employ more advanced load-cell weighing devices. The tare weight of each container is measured, and then product fill proceeds until the proper net weight is in the container. Most weight fillers are manufactured outside the United States.

Time-fill fillers. Time-fill filling consists of delivering liquid under pressure to an orifice that is open for a controlled length of time. Such a filler may be either of the controlled-pressure-head type or the constant-volume-flow type. On multistation *controlled-pressure-head* fillers, which are most frequently in-line, all orifices are opened for approximately the same length of time. Minor adjustments may be made at each station to compensate for individual orifice characteristics. Until quite recently, such time-fill fillers depended for their accuracy on maintaining a precise pressure head on the liquid at the orifice. This was achieved with pressurizing tanks or a liquid level established by flowing the product over a dam. Small pressure fluctuations are now acceptable, because a microprocessor controls the time each orifice is open, based on measured pressure variations. Such fillers are designed for easy cleaning and changeover from one product to another. To date, such fillers have not generally been used for high speed production runs. They are probably best suited to filling pint (473 mL) or smaller containers with free-flowing liquids.

The *constant-volume-flow* type of time filler is almost always adapted to a rotary filler. In a typical embodiment (see Fig. 6), product is delivered continuously to the filler at a constant flow rate, using constant-displacement pumps or their equivalent. The amount of product entering each container is proportional to the length of time a nozzle is under the liquid ports. The time is determined by the rotational speed of the filler. Constant-volume-flow time fillers are relatively simple and inexpensive. They are capable of reasonable accuracy, particularly with products of medium to high viscos-

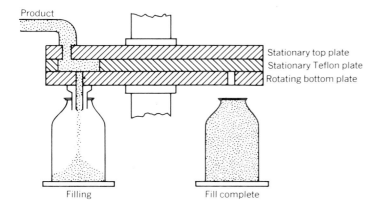

Figure 6. Constant-volume-flow time filling. Courtesy of PMMI.

ity. Leakage between plates is hard to control with low viscosity liquids. The product flow must be simultaneously altered any time the filler is stopped or started. It is difficult to do this and maintain consistent filling accuracy. Since a no-container/no-fill mechanism is usually not provided, any missing containers cause product overflow, which must be collected and either returned for reuse or discarded.

Overflow fillers. Some products can be filled by filling open containers, usually sanitary cans or widemouth glass bottles, to overflow. The liquid may flow from a pipe or over a barrier. In more advanced overflow fillers, the liquid flow into open containers is directed by moving buckets synchronized with the movement of containers. The headspace in the container, typically very small, may be established in various ways. When brine is added to pickles, for example, the headspace is usually created by displacement pads which enter the container and establish the desired headspace. If solids are present, the headspace pads also ensure that the solids are properly down into the container. Another method of establishing headspace is to tilt the containers slightly, permitting liquid to pour out. With some overflow fillers, an upwardly directed curtain of air prevents the overflowing liquid from contacting the outside of the containers. With careful adjustment of liquid flow rate and container speeds, the amount of fluid that is overflowed and recirculated may be limited to a very small proportion. Tilted-container overflow fillers are relatively inexpensive. They are frequently used for filling juice, but they normally cannot be used to fill narrow-neck containers.

CONTAINER POSITIONING

Filling machines may be characterized by the way they deliver containers to the liquid-dispensing mechanisms and remove them after filling (see Table 2).

Manually Loaded Fillers

The oldest and simplest method of container delivery and removal is by manual means. Filling occurs after one or more containers are in place. The containers may be raised to the filling valves or the valves lowered to the containers, or no relative motion may be required. Because of the amount of labor involved, manual filling is usually limited to small production runs.

In-line Fillers

The simplest automatic fillers are single-lane in-line machines. In a typical machine of this type, containers, standing on a conveyor, are delivered to the filler. One or more containers back up behind a stop or gate. The barrier then opens, and a controlled number of containers move under the filling heads, where they are positioned by another barrier of some type. Conveyor motion is usually intermittent, with the conveyor stopped during the fill cycle to prevent tipping the containers. After the filling is completed, the positioning barrier opens, and the filled containers leave while unfilled packages enter.

The size and shape of a container are factors that determine how many containers may be filled simultaneously on an in-line intermittent-motion filler. Increasing the number of filling stations increases the total output of the machine, but this approach runs into space limitations, and, at a certain point, an additional valve is not cost-effective. Sixteen stations appear to be the maximum commercially feasible number of valves. If the containers are not straight-sided, there usually is difficulty in backing up any significant number behind a positioning stop; even with straight-sided containers, the process is limited by container dimensional tolerances. The size of the containers determines the position of the last container in a row relative to the first container. If the lead container is properly positioned under a nozzle, the accumulation of container dimensional tolerances may cause the trailing container not to be positioned under a nozzle.

An in-line filler may be used for in-case filling. A multiple array of filling valves is used to fill all of the containers in a case at the same time. As with individual containers, two or three cases may be backed up for simultaneous filling. Such filling is possible only if the cases have reasonably consistent dimensional control.

A variation of in-line intermittent-motion filling is occasionally used to raise the containers relative to the filling nozzles. This is done (for example, in filling mayonnaise or peanut butter) to change the position of the container relative to the nozzle during fill. On such a filler, the containers are moved by means of a pusher mechanism from the infeed conveyor onto a platform under the filling heads, and the platform is then raised. After the containers have been filled and the platform lowered, successive unfilled containers may be used to push the filled packages onto a discharge conveyor running behind and parallel to the infeed conveyor, or individual mechanisms may move the containers between the infeed and discharge conveyors. Some manufacturers offer a rising-platform device with a straight-through conveyor arrangement.

Dual-lane straight-line fillers permit more efficient utilization of filling stations. Containers move on two parallel conveyors. The filling nozzles, in a row, fill in one lane while container movement occurs in the other lane. The use of dual-lane fillers is generally considered only if the limit of the number of valves in a single lane has been reached. They are typically used for small containers because fill time is short relative to the time required for container movement.

Valve utilization is the percentage ratio of actual filling time, ie, the time the valve is fully open, to the total cycle time for that valve from the beginning of filling one container to the start of filling the next container. Valve utilization is low on in-line fillers, ie, 25–50% of cycle time, dependent on conveyor speed, container diameter, and actual unit-filling time required for the fill.

Rotary Fillers

The most common system for filling containers are moderate to high speeds is rotary filling. Containers arrive continuously on a conveyor and are spaced by some means into a rotating infeed star which delivers them, properly separated, to filling stations on the main rotary assembly. The timing mechanism normally employed to take containers coming at random to the filler is a feed screw; for large containers and low-speed operation, timing fingers, escapement wheels, or like devices may be used. The discharge from the main rotary is usually by means of a second star wheel with the same diameter as the first. However, if the liquid level is high in the container, either a large-diameter star or a tangential conveyor should be used to prevent spill.

On rotary fillers, valve utilization is almost independent of container size but is a function of filler diameter. For example, a 22-in.- (56-cm-)-pitch-diameter gravity filler for 32-oz (946-mL) containers has a valve utilization of 49%, whereas a similar gravity filler having the valves on a 60-in. (152-cm) pitch diameter has a valve utilization of 73%. The time that the valve is closed is necessary for the transfer of containers into and out of the rotary section and for the relative movement between container and filling valve needed to open and close the valve.

The concept of dual-lane rotary fillers, involving two lanes (inner and outer) on the main rotary, was proposed and patented many years ago. Except, perhaps, for large-diameter carbonated-beverage fillers, the system is not practical. Two infeed stars and two discharge stars (or a tangential system) would be needed, and such an arrangement greatly reduces valve utilization. The fact that valve spacing is limited in the inner row is a disadvantage as well, unless unequal production speeds are desired. In one configuration study, the valve utilization on an 80-valve, dual-lane, 60-in.- (152-cm-) outer-lane-pitch-diameter machine is only 66%, compared with the normal 85% utilization of a 48-valve, 60-in.- (152-cm-) pitch-diameter filler. Adding 67% in number of valves adds only 29% to the production capability.

Relative Motion

The relative motion between the container and the liquid-dispensing device can significantly affect filler performance. The various relative motions possible are none; raise container; lower valve; and raise container and lower valve.

In sealed-container filling (see Table 1), relative motion between the container and the valve is required to bring the valve into contact with the container. With manual container placement, the container can be raised against the valve, or the valve(s) can be lowered to one or more containers on platforms.

The valve utilization of rotary sealed-container fillers can be increased by lowering the valve (see Fig. 7). With a descending valve, the valve can begin to enter the container at position 1, and it must be clear of the container at position 1'. If the container is to be raised up to the valve, the rise may not begin until position 2 because space must be allowed for the bottle rest to clear the infeed star. Likewise, the container rest and container must be in the full-down position by position 2'. As an example, with 40 valves on a 60-in. (152-mm) pitch

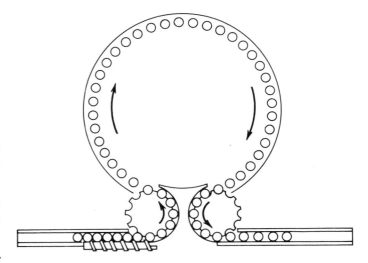

Figure 7.   Rotary-filler arrangement.

diameter, the arc 1–1' is 327°, whereas the arc 2–2' is 306°. This difference of 6% may be advantageous, but movement of each valve imposes other difficulties on filler design, often including the need for a flexible liquid connection to each valve. Such flexible connections often make cleaning more difficult. The combination of raising the container and lowering the filling valve is very rarely used, as it requires mechanisms for moving both elements, with all of the disadvantages of each system.

In-line, sealed-container fillers are rather slow and not very common, except for filling containers in a case. Most of those available operate by lowering the valves to the container, which permits the containers to remain on the conveyor. As with sealed-container fillers, level-sensing fillers require relative motion between the filling valve and the container. Rotary level-sensing fillers typically raise the container; in-line fillers bring the nozzle down to the package.

Volumetric, weight, and time-fill fillers do not require relative motion between the dispensing nozzle and the package. Thus, for simplicity and economy, most of these fillers operate with the nozzle located above the container opening. There are, however, two principal exceptions. First, in filling containers with relatively small openings, eg, long-neck plastic bottles, greatly increased rates of flow are possible if the product flow emerges from the side of the tube and is directed against the container wall (see Fig. 1). Filling straight down into that style of container from above may cause air entrapment with consequent possible splashout or overflow at the end of the filling cycle. The lost time, if any, required to raise the container or lower the filling valve may be more than offset by the faster flow rates possible. Second, viscous products such as mayonnaise and peanut butter are best filled volumetrically using a flow tube open at the bottom. To prevent air entrapment, the flow tube is just above the surface of the product in the container as it rises during filling.

CONSIDERATIONS IN THE DESIGN AND SELECTION OF FILLERS

A great many factors must be considered in selecting filling machinery. These include such obvious factors as operating and maintenance costs, efficiency, reliability, size, speed, and materials of construction. Several considerations, discussed below, are very specific to liquid-filling machinery and very important to the satisfactory operation and use of such equipment.

Accuracy. Accuracy of fill is important for two reasons. First, it is necessary in order to comply with state or other regulations as to labeled content. The regulatory agencies generally refer to the filling tolerances recommended in the *National Bureau of Standards Handbook 44* (2). In addition to meeting legal requirements, however, packagers of most products desire to keep overfill to a minimum, thus minimizing product giveaway. Even with relatively inexpensive products, the amount given away with inaccurate overfilling on high speed production lines can usually justify the higher cost of purchasing more accurate fillers.

Changeover and cleaning. As a part of operating costs, the time and labor required for cleaning and preparing the filler for daily operation, changeover from one product to another or one container to another, and cleanup at the end of each day's production should be carefully evaluated. User needs may range from fillers that operate 24 h per day with no need for product or container-size change to 30-min production runs, after which the filler must be completely sterilized and adjusted for a different container size. Comments are made above concerning the influence of filling techniques. The manufacturers' descriptions of container-positioning methods contain comments on cleanability, container-size flexibility, and ease of adjustment. Specific aspects of each filler design must be carefully considered in choosing the proper liquid filler for a particular application.

No-container/no-fill system. In general, it is desirable to have a filler equipped with some type of no-container/no-fill system. In most sealed-container fillers, the filling-valve mechanism is designed to automatically provide a no-container/no-fill system in conjunction with the action of the container and the filling valve coming together. Likewise, with most of the unsealed-container filling techniques, a relatively simple container-detection device can be used to prevent product flow if no container is present. The constant-volume time filler does not permit start or stop of product flow, but means such as drains in the container platform to catch the product flow may be provided. Alternatively, particularly on in-line fillers, a control system may be provided to ensure that a full row of containers is present.

Drive location. The location of the main drive for the filler should be considered in filler selection. Usually, the main motor drive, or, alternatively, a lineshaft synchronization system, is located near floor level. Thus, the heavy drive components are conveniently supported and reasonably accessible for routine maintenance. Because splash of corrosive products may significantly harm drive components even if they are protected against splash, some filler manufacturers locate the main drive above the product. This may not be desirable for filling food or drug products because of the potential for product contamination from drive lubricants.

Equipment standards. Various industry-consensus standards are applicable in the United States:

ANSI/ASME 2.1-1982	Food, Drug, and Beverage Equipment, Food, Drug and Beverage Equipment Committee, American Society of Mechanical Engineers
ANSI B155.1-1986	Safety Requirements for the Construction, Care, and Use of Packaging and Packaging-related Converting Machinery, American National Standards Institute
3-A Sanitary Standard 17-06	3-A Sanitary Standards for Fillers and Sealers of Single Service Containers for Milk and Fluid Milk Products, International Association of Milk, Food, and Environmental Sanitarians; U.S. Public Health Service; The Dairy Industry Committee

The first two standards are general but applicable to fillers. The 3-A standard is specifically for milk fillers.

The packaging of certain products is regulated by government agencies. The packaging of fluid-milk products in the United States is usually regulated by local, ie, county or state, authorities. They normally adopt the 3-A Standards without modification, but a few jurisdictions impose their own requirements, usually stricter. If fresh milk is packaged in flexible plastic containers, the container is usually the legally specified measuring device. In such cases, volumetric tolerances normally follow the recommendation of the National Bureau of Standards (2).

The packaging of liquids in meat and poultry establishments subject to inspection by the USDA Food Safety and Inspection Service is governed by guidelines issued by the department (3,4). Equipment for packaging food and drug products is regulated in the United States by the FDA (CFR Title 21, Chapt. I), and the packaging of distilled spirits by the Bureau of Alcohol, Tobacco, and Firearms (BATF).

BIBLIOGRAPHY

1. *The PMMI 1984–1985 Packaging Machinery Directory*, Packaging Machinery Manufacturers Institute, Washington, D.C., 1984.

2. *Specifications, Tolerances, and Other Technical Requirements for Commercial Weighing and Measuring Devices, National Bureau of Standards Handbook 44*, NBS, U.S. Dept. of Commerce, Washington, D.C., 1985.

3. *Accepted Meat and Poultry Equipment*, Meat and Poultry Inspection Technical Services, Food Safety and Inspection Service, U.S. Dept. of Agriculture, Washington, D.C., May 1984, pp. 1–8.

4. *Guidelines for Aseptic Processing and Packaging Systems in Meat and Poultry Plants*, Meat and Poultry Inspection Technical Services, Food Safety and Inspection Service, U.S. Dept. of Agriculture, Washington, D.C., June 1984.

General References

C. G. Davis, *Product Filling, Vol. 1 of Packaging Machinery Operations*, Packaging Machinery Manufacturers Institute, Washington, D.C., 1978.

K. A. Crandall, *Significance of Timing, Flow Rate and Environmental Factors in Liquid Filling by Weight,* Crandall Filling Machinery, Inc., Buffalo, N.Y., undated.

F. B. Fairbanks, Jr.
Horix Manufacturing Company

FILM, COEXTRUDED.   See Coextrusions for flexible packaging.

FILM, FLEXIBLE PVC

Poly(vinyl chloride) (PVC) is an outstanding raw material for packaging. It is used either alone or in combination with other materials for bags, bottles, blisters, cans, and closures. Not a new material, plasticized PVC was used as a rubber sheet substitute during the World War II shortages (1). When blended with certain additives it can be processed into thin transparent film. PVC's ability to accept and respond to additives (see Additives, plastic) makes it such a unique and popular flexible packaging material (see Poly (vinyl chloride).

ASTM sets the upper limit on "film" thickness at ≤10 mil (≤254 μm) (2), and it characterizes as nonrigid (flexible) plastics those that have an elastic modulus of ≤10,000 psi (≤69 MPa) at 73°F (23°C) and 50% rh. The PVC packaging films discussed in this article fit these definitions (see also Film, rigid PVC).

Methods of Manufacture

Essentially all flexible PVC packaging films are manufactured from externally plasticized resin that is extruded through a film-forming die. Two basic methods are used: blown-film extrusion and film casting. (Heavy-gauge flexible films are also made by calendering, but not for packaging applications. They are used for wall coverings, shower curtains, etc.) The most popular method in the United States and Europe is blown film extrusion (see Extrusion). PVC resin is blended with selected additives using either a high intensity mixer (preferable) or a ribbon blender. The mixing time varies with the blender type, the type and amount of additives, and the adsorption characteristics of the PVC resin. Mixing is complete when the compound (the resin–additive blend) becomes free-flowing and the dry additives are completely dispersed. Heat of mixing generated by a high intensity mixer must be removed to prevent degradation if the compound is to be stored. Mixer manufacturers offer a variety of heat-transfer-equipment designs for this purpose. The compound in this state is normally called dry blend (3).

Dry blend may be pelletized for later use, or fed directly to the blown-film extruder. Processing conditions and screw and die design are critical with heat-sensitive PVC. Films used for food contact applications must utilize regulated stabilizer systems that offer somewhat limited effectiveness. Successful extrusion of PVC film requires control of shear in the extruder as well as streamlined melt flow in the die. The melt exits the annular die, the tube is inflated, cooled, and collapsed. It is then slit and the film is wound into rolls. Both surface and center winders are used. Secondary slitting is sometimes performed when film-width precision is needed (see Slitting/rewinding).

Flexible PVC films are also made by film casting, by one of two methods. One method, solvent casting, utilizes a solution of PVC and additives in an appropriate solvent. The solution is pumped through a slot die and cast on a smooth continuous belt. The solvent evaporates leaving a thin, flexible film that can be stripped from the belt, slit, and wound into rolls. Solvent recovery is a costly element of this method. The other method, popular in Japan, utilizes compound feedstock and an extruder similar to that employed for the blown-film process. The melt is pumped through a slot die and cast on a smooth, chilled rotating drum. Winding is as previously described. Both methods feature good output, but lack the width flexibility of the blown-film process.

Although flexible PVC is not considered to be a crystalline material, it can be preferentially or biaxially oriented to provide shrink characteristics. Machine-direction orientation can be achieved by stretching heated film in a narrow gap between two rolls, the second of which is revolving at a higher speed than the first. Preferential or biaxial orientation can be obtained by bubble inflation under closely controlled film temperatures and strain rates.

Composition

The ingredients used in manufacturing flexible PVC films are dictated by the applications they serve (food or nonfood contact, packaging equipment used, type of packaged-product protection required), and the manufacturing method employed. The single largest application of flexible PVC film requires direct food contact. The U.S. Food and Drug Administration (FDA), complying with the provisions of the Food Additives Amendment of 1958, lists ingredients that may be used with or without limitations for manufacturing food-packaging films. Responsibility for adherence to these regulations rests with the packaging-film manufacturer. Restriction on allowable raw materials often creates processing difficulties for the film producer. This is particularly true in the case of stabilizers.

Resin.   The major ingredient in flexible PVC films is PVC resin. Typically, the resin selected is a medium molecular weight homopolymer (K value of about 66–68) that is highly adsorptive and low in gels (4) and contamination, and offers very good heat stability compared to general-purpose resins. Resins used are nearly always classified as film-grade and command a premium price.

Plasticizers.   Plasticizers are the major additives, and their prime purpose is to impart flexibility. The type(s) used are dictated by the film application and cost. For food-packaging applications, only extensively tested, FDA-sanctioned plasticizers may be used (5). Many that are sanctioned have regulated limitations. For example, di(2-ethylhexyl)phthalate (DOP or DEHP), a prior sanctioned plasticizer, may be used only in films designed for packaging high water-content foods, such as fresh produce. Many of the adipates, which are popular choices for fresh-meat wrap, have limitations that pertain to film thickness, meat-fat content, and food-storage temperature. For nonfood applications, the plasticizer choice seems unlimited, but the overwhelming majority that are used are nontoxic and often FDA-sanctioned. Most flexible PVC-film formulations use an oil epoxide, typically epoxidized soybean oil (ESO). This additive, frequently called a secondary plasticizer, imparts some flexibility; but more importantly, it provides considerable processing heat stability. Its effectiveness depends on its oxirane oxygen content, degree of saturation, and the compound stabilizer system used. It is the latter choice which determines whether ESO-stabilization synergism will develop (6).

Heat stabilizers. Heat stabilizer selection for flexible PVC-film production is a highly complex matter. It involves not only the end-use application, but the interaction with other additives and processing requirements. One of the more popular is an organometallic salt based on barium and cadmium. Although not sanctioned for food contact, it does offer excellent heat stability in transparent flexible PVC film production (7). Combined with a phosphite chelator and ESO, it is one of the industry workhorses. Liquid barium/cadmium/zinc phosphite systems are also effective (8). For food-contact applications stearates of calcium, magnesium, and zinc are most often used. They are relatively inefficient and usually must be boosted with additives that include oil epoxides and phosphites.

Lubricants. Lubricants are included in PVC compounds primarily to improve processing characteristics. In flexible PVC compounds, the primary purpose of a lubricant is to reduce sticking of the melt (or semimelt) to metal surfaces. A secondary effect is to reduce melt viscosity. Montan ester waxes, paraffins, and low molecular weight polyethylenes are popular lubricants for flexible PVC. Fatty acids, esters, and metallic soaps are frequently used as internal lubricants for flexible film production (9,10).

Most other components of a flexible PVC film are included in relatively low concentrations to impart a specific physical property. Esters of multifunctional alcohols that offer the proper hydrophilic/hydrophobic balance provide antifog and antistatic properties. A number of inorganic additives (eg, clays and talc) and organic compounds (particularly fatty acid amides) are offered for slip and antiblock characteristics. Both FD&C colors and general purpose pigments (see Colorants) are used for tinting and coloring.

Markets

Food packaging. Most flexible PVC film is used for packaging food products, particularly fresh red meat (see Table 1). Properly formulated flexible 0.5–1.0 mil (13–25 μm) PVC film is perfectly suited for fresh red meat. It features the high oxygen permeability necessary to preserve fresh-cut meat's bright red color, which is the color of oxymyoglobin, oxygenated fresh meat surface pigment. Inadequate permeability or excess surface microorganisms result in the production of the undesirable brown metmyoglobin (11). In addition, it offers an unsur-

passed combination of toughness and resiliency, which minimizes in-store rewraps resulting from harsh package handling. Its merchandizing appeal remains unchallenged. These flexible films adapt well to simple in-store hand-wrapping stations, complex automatic shrink wrappers, or the increasingly popular automatic stretch wrappers. Stretch wrappers can wrap more than 30 packages per minute and automatically handle different-size packages, and they use less energy than the older shrink equipment.

The packaging of fresh fruits and vegetables in flexible PVC is performed primarily in-store. Films are specifically formulated to provide high carbon dioxide and oxygen permeability, and control the ripening process of the "living" produce (see Controlled atmosphere packaging). They provide excellent product protection, even for the very sensitive mushroom. Universities and growers are searching for additional applications for these interesting films.

Virtually all poultry processors in the United States use PVC stretch film for chilled, tray-packed poultry parts. In printed or unprinted form, the 0.75–1.0-mil (19–5-μm) film fills the stringent demands of centralized packing. PVC is tough enough to take abuse, and extensible enough to prevent the crushing or collapse of the packaged parts. Good heat sealability in a wet environment is another advantage offered by flexible PVC (see Sealing, heat).

Institutional packaging is the second-largest food wrap application for PVC films. The majority of these films are 0.40–0.60-mil (10–15-μm) thick and are wound on rolls and placed in dispenser boxes fitted with cutter blades. They are used in institutional kitchens and restaurants, and by caterers for overwrapping food plates, trays, sandwiches, glassware, and utensils. The films are sometimes perforated in sheet lengths prior to winding, which eliminates the need for a cutter blade. A small amount is oriented for shrink packaging. Outstanding cling, stretch, and clarity are the major features.

Properly formulated 0.75–1.25 mil (19–32-μm) PVC films are effective for frozen-food storage. Food locker services use them for frozen-meat packaging, and some printed films are used by centralized frozen-food processors. The film is usually unoriented and offers good transparency, sealability, and satisfactory low-temperature flexibility. Very thick films (6.0–9.0 mil or 152–229 μm) are used in boxed-beef operations to protect barrier bags from bone puncture. Good icing release is a feature of specially formulated PVC films that make them useful in packaging baked goods.

Other applications. Pharmaceutical packaging (see Food, drug and cosmetic regulations; Pharmaceutical packaging) also requires the use of FDA-sanctioned raw materials. Flexible PVC films are used in individual capsule and tablet skin packaging taking advantage of PVC's good formability, clarity, and sealability. Shrink versions find application in tamper-evident fittings and overwraps.

Flexible PVC films have greater competition from other materials in the nonfood packaging area. Oriented PVC films compete with polyolefins in the overwrap shrink-film markets (see Films, shrink). These films, which typically range in thickness of 0.5–1.0 mil (13–25 μm), are put up either as flat or centerfolded, depending on the packaging equipment used. PVC offers a controlled shrink-force range coupled with good sealability, clarity, and gloss.

Flexible PVC was one of the first stretch films used for pallet unitization and it continues to hold a niche in this mar-

Table 1. Food Packaging Applications of Flexible PVC Film

Application	Estimated U.S. consumption, 1983, 10^6 lb (metric ton)
fresh red meat	110.5 (50,122)
poultry packaging	7.5 (3,402)
fruits and vegetables	17.5 (7,938)
fish (fresh)	6.5 (2,948)
frozen foods	2.5 (1,134)
institutional packaging	45.0 (20,412)
baked goods	2.0 (907)

ket (see Films, stretch). It features good stretch and outstanding tack and optics and is offered in thicknesses of 0.65–2.0 mil (16.5–51 μm). As textile and rug-roll overwraps, they provide both shipment protection and merchandise visibility. When formulated with weather-resistant plasticizers and uv screens, they provide very good outdoor weatherability, and pigmented versions conceal tempting merchandise.

One of the newest applications for oriented flexible PVC is in the bundling of multiple aseptic packages (see Aseptic packaging). Ranging in thickness 0.75–1.50 mil (19–38 μm), this film provides a relatively high shrink at low temperatures, resulting in a firm, unitized pack. PVC seals easily, offers good economics, and merchandising appeal. Heavy-gauge (2.0–7.5 mil or 51–191 μm) flexible PVC has been used as a component of windowed cards for about 20 years. Toughness, formability, good dielectric sealability, and merchandising appeal are important qualities. Hardware and electrical items, flashlights, and automobile parts are typically packaged this way.

An interesting nonpackaging application is the use of high-clarity, glossy flexible PVC films as laminating films for paperback book covers. The combination of low cost, flexibility (resists cold crack) and adaptability to the Trio-Bond lamina-

tor (uses water-based adhesives, no solvents or heat) have made PVC competitive with acetates and lacquer coatings.

Physical Properties

Not surprisingly, of all the PVC additives, the type and level of plasticizer used in a flexible PVC film exercises the most influence over its physical properties. Its most obvious effect is on modulus (stiffness). PVC film can be stiff, like cellophane; but by simply adding plasticizer, one can produce a film that is nearly elastomeric. Some plasticizers are more efficient than others, which means that one can use less of a more efficient plasticizer to achieve a specific modulus reduction. A discussion of the selection process for some specific flexible PVC-film applications and the resultant film properties is germane to packaging applications.

Flexible PVC film to be used as a wrap for fresh red meat must offer good stretch characteristics (low modulus), high oxygen permeability, moderately low water vapor transmission rate (WVTR), good low temperature properties, heat sealability, and resilience. It must also be very transparent, glossy, and reasonably priced, and it must be manufactured from FDA-sanctioned raw materials. The first plasticizer to

Table 2. Properties of 1-mil (25.4 μm) Flexible PVC Film

Property	ASTM test		Flexible PVC meat package stretch type	Flexible PVC dispenser film	Flexible PVC shrink bundle film
specific gravity	D 1505		1.23	1.27	1.3
yield, in.^{2}/(lb·mil) [m^2/kg·mm)]			22,400 [1,254]	21,600 [1210]	21,400 [1198]
haze, %	D 1003		1.2	1.0	2.5
light transmission, %			>90	>90	>90
tensile strength,	D 882				
psi (MPa)		MD	5,000 (34.5)	5,500 (37.9)	18,000 (124)
		XD	4,500 (31)	5,500 (37.9)	5,500 (37.9)
elongation, %	D 882				
		MD	275	300	90
		XD	375	325	275
tear strength,					
gf/mil (N/mm)		MD	300 (116)	325 (125)	335 (129)
initial	D 1922	XD	450 (174)	500 (193)	575 (222)
propagating					
water absorption, 24 h, %	D 570		0	0	0
change in linear	D 1204				45
dimensions at 212°F		MD	na	na	10
(100°C) for 30 min, %		XD			
service temperature			−20–150	0–150	10–150
°F (°C), range			(−29–66)	(−18–66)	(−12–66)
heat-seal temperature			290–320	290–340	280–330
°F (°C), range			(143–160)	(143–171)	(138–166)
oxygen permeability cm^3·mil/(100 in.^{2}·d·atm)[cm^3·μm/ (m^2·d·kPa)], 73°F (23°C), 50%rh	D 1434				
23C, 50% rh			860 [3342]	340 [1321]	na
water vapor transmission rate: g.mil/(100 in.^{2}·d) [g·mm/(m^2·d)], 100°F (38°C), 90% rh			16 [6.3]	10 [3.9]	na
COF, face-to-face	D 1894				0.2
back-to-back			1.0	1.0	0.2
test conditions: 73°F (23°C), 50% rh					

consider in any application would be DOP (DEHP), because it is the most widely used plasticizer and low in cost. At the 40 phr level, it would offer a sufficiently low modulus, and because it is highly solvating, the film would exhibit excellent optics. On the other hand, DOP's low temperature properties are marginal; and of critical importance, the FDA will allow it to be used only in films for packaging high water-content foods.

Several regulated adipates are available that can be used with fatty foods, and offer good low-temperature characteristics along with meeting the permeability and modulus requirements. Included are n-octyl n-decyl adipate (NODA), diisononyl adipate (DINA), 7,9 adipate, and di(2-ethylhexyl) adipate (DOA). Of these, the NODA and 7,9 adipate are the most expensive, and NODA is not quite as compatible as the others. DINA is cost-competitive with DOA, but has an efficiency factor of 105–110% (requires 5 to 10% more loading to achieve the same properties). DOA is clearly the best choice. A common loading in flexible PVC meat wrap is 20–25%.

With a few departures, the use of adipate plasticizers form the basis for all food (and many nonfood) film formulations. A film designed for frozen-food packaging would contain about 10–15% more adipate than fresh-meat wrap, and it would usually be about 50% thicker in order to provide abuse resistance and a satisfactory WVTR. Regarding this latter property, flexible PVC's room temperature properties might appear to indicate that its WVTR is too high for frozen-food protection. This is not the case. As temperature falls, free volume is reduced along with WVTR. A 10°C drop reduces the water permeability coefficient of a 50 phr DOP plasticized film by nearly 30% (12). By reducing the adipate concentration by 50%, modulus increases about 120%. This puts the film in the stiffness-range requirements of an overwrap or shrink-bundling film. Intermediate adipate concentrations find application in dispenser (cutter box) films and oriented meat wraps.

Specialty plasticizers are often used in conjunction with the primary plasticizer in order to impart or enhance a specific physical property. For example, butyl benzyl phthalate (BBP) and acetyl tributyl citrate are both highly solvating plasticizers and are used to improve the surface quality (gloss and cling) of films. Both are FDA regulated. BBP has an extraction limitation. Octyl epoxy tallate can reduce the onset of cold-crack.

Properties of three types of flexible PVC film are shown in Table 2. The properties are for 1-mil (25.4 μm) film, but this is not necessarily the most common thickness in commercial practice.

BIBLIOGRAPHY

1. H. A. Sarvetnick, *Polyvinyl Chloride,* Van Nostrand Reinhold Co., New York, 1969, p. 1

2. *Annual Book of ASTM Standards,* American Society for Testing and Materials, Philadelphia, Pa., Part 35, D 883, 1979

3. Ref. 1, p. 139.

4. A. S. Pazur, *J. Vinyl Technol.* 5(3), 126 (Sept. 1983).

5. J. K. Sears and J. R. Darby, *The Technology of Plasticizers,* John Wiley & Sons, New York, 1982, pp. 18, 26, 257

6. *Ibid.,* pp. 665–672.

7. T. J. Kraus and J. E. Copus, *Plastics Compounding* 3(5), 30 (Sept./Oct. 1980).

8. M. C. McMurrer, *Plastics Compounding* 5(7), 78 (Nov./Dec. 1982).

9. D. Bower, *Plastics Compounding* 2(1), 65 (Jan./Feb. 1979).

10. M. C. McMurrer, *Plastics Compounding,* 5(4), 77 (July/Aug. 1982).

11. A. L. Brody, *Flexible Packaging of Foods,* CRC Press, Cleveland, Ohio, 1970, pp. 74–76.

12. Ref. 5, p. 437.

D. G. James
Borden Chemical

FILM, FLUOROPOLYMER

Fluoropolymers are a family of materials that have a general paraffinic structure with some or all of the hydrogen atoms replaced by fluorine. All members of the fluoropolymer family are available in film form, but only Aclar film is used extensively in specialty packaging applications. Allied Corporation is the sole supplier of Aclar film. Some of its useful properties are: extremely low transmission of moisture vapor and relatively low transmission of other gases, inertness to most chemicals, outstanding resistance to forces that cause weathering (uv radiation and ozone), transparency, and useful mechanical properties from cryogenic temperatures to temperatures as high as 300°F (150°C). The initial interest in Aclar film was for packaging military hardware, but its water-vapor-transmission properties led to interest by pharmaceutical companies for packaging moisture-sensitive products (see Pharmaceutical packaging). Available in several grades, Aclar is more expensive than most thermoplastic films, with current pricing in the $20–25/lb ($44–55/kg) range.

Composition and extrusion. Aclar is an Allied Corporation trademark for film made from Aclon, a modified PCTFE (polychlorotrifluoroethylene) fluoropolymer, that contains greater than 95 weight % chlorotrifluoroethylene (1,2). It is converted to film by melt extrusion (see Extrusion). It is a difficult thermoplastic to extrude because it has a very high melt viscosity and a low, critical shear rate for melt fracture. The commercially available films have different machine-direction and cross-direction properties because some orientation is induced during fabrication. Aclar 22 films have good tear strength and can be used unsupported as well as in thermoformable laminates (see Laminating; Thermoforming). Aclar 33, made from a different copolymer, offers superior dimensional stability and resistance to chemicals and water vapor transmission.

Fabrication. Aclar film can be heat sealed (see Sealing, heat), laminated, printed, thermoformed, metallized (see Metallizing), and sterilized (see Health-care packaging). The unsupported and laminated varieties can be handled and processed on most common converting and packaging machines. Unsupported and laminated films can be heat sealed by machines that employ constant heat, thermal impulse, radio frequency, or ultrasonic energy. In some cases, special precautions must be observed.

Most Aclar film for packaging is converted to some type of laminate. It can be laminated to paper, polyethylene (low and medium density), and to preprimed PVC, aluminum foil, polyester, nylon-6, and cellophane. A typical extrusion lamination (see Extrusion coating; Multilayer flexible packaging) uses molten LDPE as an adhesive between the fluoropolymer film and one of the substrates listed above. An MDPE tie layer produces a laminate with better properties at elevated temperatures. Laminates can also be produced by adhesive lamination. In that case the Aclar film should be preprimed. Preprimed film can be bonded to polymeric substrates with two-component urethane adhesives. The adhesive system is applied in water or organic solvent and the liquid is evaporated in an oven. Because of its low surface energy, Aclar film does not have acceptable bond strength to most substrates unless the film is corona-treated (see Surface modification) to increase its surface energy. With corona treatment, the film can be laminated to aluminum and steel foils by using an epoxy-polyamide curing adhesive.

All grades and gauges can be thermoformed. Typical thermoforming temperatures are 350–400°F (175–205°C). It is done close to, but not above, the melting point of the film. One of the important uses of Aclar film is the use of thermoformed laminated Aclar 22A/PVC in blister packs for ethical drugs. Heavier gauges (5–10 mil or 127–254 μm) of unsupported Aclar film can be thermoformed using ceramic or quartz heating units. Thermoformed film and laminates should be quick-quenched to maintain low crystallinity and prevent brittleness.

Most applications require transparency, but Aclar film can be metallized with aluminum. Corona-treated film can be printed with polyamide-based inks (see Inks). Best results are obtained if the film is corona-treated in-line prior to printing.

Properties. Aclar film can be sterilized with steam and with ethylene oxide (ETO) systems. It is being tested to determine if radiation sterilization of Aclar film will deteriorate its properties in the cobalt-60 dosage range of $2.5-5 \times 10^6$ rad ($2.5-5 \times 10^4$ Gy). Tests at 1×10^6 rad (1×10^4 Gy) (3) indicate that little or no property loss occurs at that dosage (see Radiation, effects of). Aclar films have outstanding barrier properties to water vapor and to other gases (see Barrier polymers). Properties at a wide variety of temperatures can be determined by a method that is described in ref. 4. Table 1 compares transmission rates of Aclar films with a number of other films.

Aclar film is inert to acids, bases, strong oxidizing agents, and most organic chemicals. It exhibits excellent dimensional stability in inorganics, including water, salt solutions, strong acids, and bases. Some polar organics, especially hot polar solvents, diffuse into Aclar film and act as a plasticizer. These solvents cause it to become more flexible and sometimes hazy. There are no known solvents that dissolve the film at temperatures up to 250° (120°C). Mechanical properties are usually not an important factor in packaging applications because the important mechanical properties are those of the laminates. Aclar films are not particularly strong or tough, but they have outstanding abrasion resistance, important in clean-room packaging of military hardware.

Applications. Aclar film is used in military, pharmaceutical, electrical/electronic, and aircraft/aerospace component applications. It serves as a key component of a transparent laminated barrier construction that meets the requirements of the MIL-F-22191, Type I specification (see Military packaging). This laminate is used for packaging moisture-sensitive military hardware. Aclar also passes the LOX compatability impact test under NASA specification MSFC-106A. It is used for packaging components designed for service in liquid oxygen and other oxidizers used in spacecraft applications.

The major commercial applications are in packaging moisture-sensitive drugs. Laminates are used for rigid and semirigid blister packs, unit-dose packages, and aseptic peel-packs. They are also used in medical applications; for example, as an overwrap for plastic containers for biomedical specimens and/or pathology specimens. This application requires a film that is sterilizable by heat and ETO and that does not absorb or denature biological fluids (5).

Safe handling considerations. Aclar film is inert and nontoxic and safe to handle at ordinary temperatures. It is thermally stable for up to one hour at temperatures as high as 446°F (230°C). It can be processed for a few seconds at temperatures as high as 554°F (290°C) provided the machine is ventilated with an exhaust fan. Vacuum forming is typically done at 374°F (190°C) with exhaust-fan ventilation. If Aclar film is continuously processed at temperatures above 446°F (230°C), the processing machinery should be equipped with ventilating equipment or the work should be performed in an exhaust hood. The film should not be disposed of by burning. Exposed to flame, it degrades to fluorochlorocarbon gases, some of which are toxic. In the presence of oxygen and olefins or

Table 1. Approximate Comparative Transmission Rates

Films	Moisture^a vapor, g·mil/(100 in.²·d) [g·mm/(m²·d)]	O₂^b	N₂^b	CO₂^b
		cm³·mil/(100 in.²·d·atm) [cm³·μm/(m²·d·kPa)]		
Aclar 33C	0.025 [0.01]	7 [27.2]		16 [62.2]
Aclar 22C	0.045 [0.018]	15 [58.3]	2.5 [9.7]	40 [155]
Aclar 22A	0.041 [0.016]	12 [46.6]	2.5 [9.7]	30 [117]
PVDC	0.20 [0.78]	0.8–1.0 [3.1–3.9]	0.12–0.16 [0.47–0.62]	3.0–4.6 [11.7–17.9]
polyethylene				
low density	1.0–1.5 [0.4–0.6]	300–700 [1166–2720]	130–260 [505–1010]	1400–2800 [5440–10,879]
medium density	0.4–1.0 [0.16–0.4]	170–500 [661–1943]	100–120 [389–466]	500–1500 [1943–5828]
high density	0.3–0.7 [0.12–0.28]	34–250 [132–971]	40–55 [155–214]	250–720 [971–2798]
Nylon 6	19–20 [7.5–7.9]	2.6 [10.1]	0.9 [3.5]	9.7 [37.7]
fluorinated ethylene				
polypropylene (FEP)	0.4–0.5 [0.16–0.20]	750–1000 [2914–3886]	300–400 [1166–1554]	1600–2000 [6217–7771]
poly(vinyl fluoride) (PVF)	2.0–3.2 [0.79–1.3]	3.0–3.3 [11.7–12.8]	0.25–0.70 [1.0–2.7]	11–15 [42.7–58.3]
polyester	1.0–3.0 [0.4–1.2]	4.0–8.0 [15.5–31]	0.7–1.0 [2.7–3.9]	12–25 [46.6–97.1]

^{a} ASTM E 96 at 100°F (38°C) and 90% rh.
^{b} Dry gas at room temperature.

polyolefins, it may form HF and/or HCl when exposed to flame. These acids are toxic if inhaled and are corrosive to metals.

BIBLIOGRAPHY

1. J. P. Sibilia and A. R. Paterson, *J. Polym. Sci. C* **8**, 41–57 (1965).
2. A. C. West, "Polychlorotrifluoroethylene," in M. Grayson and D. Eckroth, eds., *Kirk-Othmer Encyclopedia of Chemical Technology*, 3rd ed., Vol. 11, John Wiley & Sons, Inc., New York, 1980, pp. 49–54.
3. King, Broadway, and Pallinchak, *The Effect of Nuclear Radiation on Elastomeric and Plastic Components and Materials*," Report 21, Addendum AD 264890, Radiation Effects Information Center (now defunct), Battelle Memorial Institute, National Technical Information Series of the Department of Commerce, Aug. 31, 1964.
4. N. Vanderkooi and M. Ridell, *Materials Engineering*, 58 (March 1977).

A. B. ROBERTSON
K. R. HABERMANN
Allied Fibers and Plastics

FILM, HIGH DENSITY POLYETHYLENE

HDPE film has been used for a variety of specialty applications since the mid-1950s, but it was not until the 1970s that major growth began. United States consumption in 1984 was about 400 million (10^6) lb (1.8×10^5 t) (1). More than half of that amount was for grocery sacks and merchandise bags and a few other nonpackaging applications. The other half was accounted for by a variety of packaging uses that require the unique property profile that HDPE films provide.

In packaging applications, HDPE films compete with LDPE and LDPE/LLDPE blends, cast PP (see also Film, nonoriented polypropylene), and oriented PP (see Film, oriented polypropylene). These polyolefin films have many properties in common, but with some significant differences in degree. Typical HDPE film properties are shown in Table 1.

HDPE, by virtue of its higher density, exhibits higher tensile strength, lower WVTR, and greater stiffness than LDPE or LLDPE/LDPE blends. PP films can offer higher tensile strength because of a higher degree of crystallinity and the ability to be biaxially oriented. PP films also have excellent optical properties due to fine crystal structure and biaxial orientation behavior. LDPE and LLDPE/LDPE blends exhibit good clarity due to their lower degree of crystallinity. HDPE is the most opaque of the three polyolefin types. This is an advantage if opacity is desired, because a relatively low amount of pigment is required to achieve that opacity (see Colorants). LLDPE and LLDPE/LDPE blends exhibit higher tendency to elongation under load because of their lower crystallinity and hence, lower tensile modulus. This property is useful in stretch-wrap applications (see Films, stretch) but it is a disadvantage if low elongation is required (as in handled grocery sacks). It is important to recognize, however, that the properties of all polyolefin films can be tailored during polymerization or through the use of blends or coextrusions.

Packaging applications. Two of the most important applications for HDPE film are retail grocery sacks and merchandise bags, which are not considered packaging applications in this Encyclopedia. Together, they account for almost 200 million (10^6) lb/yr (9×10^4 t). Other important nonpackaging applications are trash-can liners and typewriter ribbons.

In packaging, HDPE films have replaced large quantities of glassine (see Glassine, greaseproof, and parchment) for pack-

Table 1. Typical Properties of High Density Polyethylene Film, 1-mil (25.4 μm)

Property	ASTM test method		HDPE-HMW (blown)	HDPE-MMW (blown)
specific gravity	D 1505		0.950	0.950
yield, in.²/(lb·mil) [m²/(kg·mm)]			29,200 [1635]	29,200 [1635]
haze, %	D 1003		78	78
tensile strength,	D 882	MD/XD, Yield	5,200 (35.9)/5,000 (34.5)	4,600 (31.7)/4,400 (30.3)
psi (MPa)		MD/XD, Failure	7,500 (51.7)/7,000 (48.3)	6,500 (44.8)/6,1000 (43.1)
elongation, %	D 882			
		MD	450	350
		XD	500	550
tensile modulus, 1% secant, psi (MPa)		MD	125,000 (862)	120,000 (828)
		XD	130,000 (897)	125,000 (862)
tear strength, gf/mil (N/mm)	D 1922	propagating, MD/XD	20 (7.7)/150(57.9)	15 (5.8)/100 (38.6)
service temperature °F(°C), range			200–250 (93–121)	200–250 (93–121)
WVTR, g.mil/(100 in.²·d) [cm³·mm/(m²·d)]; 38°C, 98% rh			0.8 [0.3]	0.8 [0.3]
COF, face-to-face	D 1894		0.3	0.3

aging cereals, crackers, and snack foods. In these applications, clarity is not desirable, and the translucency or opacity of HDPE films is an advantage. They are used in conjunction with bag-in-box vertical form/fill/seal equipment (see Bag-in-box, dry product; Form/fill/seal, vertical). Rubber-modified HDPE film is used for medical overwraps. HDPE films are also used to produce shipping sacks. Heavy-gauge coextrusions of HDPE with low density polyethylene are used for all-plastic shipping sacks (see Bags, heavy-duty plastic), and thin-gauge HDPE films are used as moisture-barrier plies in multiwall bags (see Bags, paper). Very thin embossed HDPE film is also as a replacement for tissue paper, and without embossing it is used as an inner wrap for delicatessen products.

Film polymers. Three principal polymer properties can be adjusted to tailor the performance of HDPE films (see Polyethylene, high density). Density plays the most significant role as a measure of the degree of polymer crystallinity. Crystallinity has a direct influence on tensile strength, WVTR, hardness, opacity, and coefficient of friction. HDPE film resins are $0.941–0.965$ g/cm^3, depending on comonomer type and concentration. The most common comonomers are butene and hexene. As density increases, tensile modulus and tensile strength increase as well (see Fig 1). The same is true for film hardness, and to some extent for water-vapor transmission rate (see Fig 2).

The molecular weight (MW) of the polymer has a major influence on toughness and impact strength. For many years, HDPE films were made from medium molecular weight (MMW) or low molecular weight (LMW) polymers, for applications that required only medium-to-low toughness. The development of high molecular weight (HMW) polymers has changed the range of applications available to HDPE films (2). The influence of MW on film impact strength is significant (impact resistance increases with molecular weight), but the processing of the resin is critical as well (3).

Molecular weight distribution (MWD) is the third parameter influencing resin design, because it influences processing. Broad-MWD resins are easier to process than narrow-MWD resins. The ability to process HDPE at economic speeds and at controlled melt temperatures requires careful resin design. High flow LMW film resins can be narrow or medium in MWD

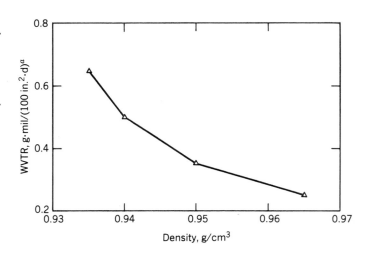

Figure 2. Influence of density on WVTR. To convert g·mil/(100 in.^{2}·d) to g·mm/(m^2·d), multiply by 0.3937.

and still be extruded at high output rates without excessive shear that would degrade the polymer. MMW resins require some careful control for high speed extrusion (4). The use of HMW-HDPE film resins has required extreme MWD control to ensure both acceptable output rates and control of melt temperatures. Thermal degradation must be avoided to maintain the long molecular chains necessary to impart high impact strengths. This MWD has been achieved in some cases by bimodal polymerization technology. The polymer chemist has certain secondary design criteria available to fine-tune the HDPE film polymers. These include special process stabilizers, high-heat stabilizers, antistatic agents (see Additives, plastics), different comonomer types and the physical form in which the resins are supplied (ie, pellets or coarse powders).

Film manufacturing. HDPE films are produced using two principal extrusion techniques: blown and cast (see Extrusion). The most common is blown-film extrusion, in which $1\frac{1}{2}$–6-in. (3.8–15.2-cm) extruders extrude the resin in tubular form. The tube is cooled, collapsed to a flat film, and wound either in tubular form or slit and folded. The properties of

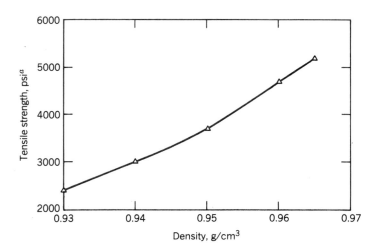

Figure 1. Influence of density on tensile strength. To convert psi to MPa, divide by 145.

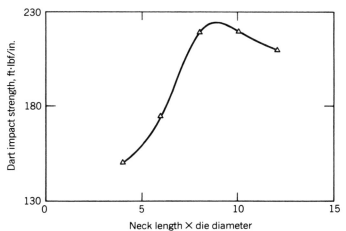

Figure 3. Effect of neck length on film toughness. To convert ft·lbf/in. to J/m, multiply by 53.38 (see ASTM D 256).

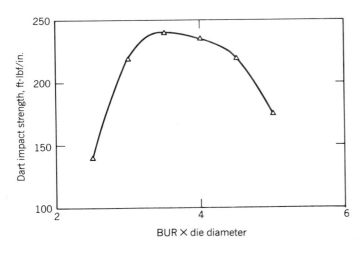

Figure 4. Effect of blow-up ratio (BUR) on film toughness. To convert ft·lbf/in. to J/M, multiply by 53.38.

blown HDPE film are highly dependent on the type of resins used and the design of the extrusion equipment employed (5). MMW-HDPE films, monolayer or coextruded, are generally made with extruders in the 2½–6-in. (6.4–15.2-cm) diameter range operating at relatively low screw speeds yet achieving high output at moderate melt temperatures. The screw L:D ratio is normally 25–30 and the barrel design is normally smooth. These conventional extruders do not allow the full development of the resin capability in terms of optimum toughness, but they serve a useful role in providing high outputs at economic rates for applications which do not require the full strength potential of HDPE.

In contrast, the newer designs for HMW–HDPE film extrusion are based on forced-feed grooved sections. They operate at high screw speed, using relatively short (18–21) L:D ratios (6). These extruders achieve high output with the minimum residence time required to produce a homogeneous melt for the subsequent blowing and orientation processes necessary to develop the optimum toughness and tensile strength. For HMW film it is imperative to use a relatively long neck length (frost height) (6–10 × die diameter) and high blow-up ratio (3–5 × die diameter) to achieve balanced properties. The influence of these processing parameters on the optimum toughness of the film is illustrated in Figures 3 and 4.

LMW-HDPE films are generally made with cast-film technology, in which free-flowing polymer melts are extruded from flat dies on to substrates or chill rolls. These polymers do not have the melt strength necessary for the blown film process and their applications are limited.

BIBLIOGRAPHY

1. J. S. Siebenaller and R. H. Nurse, *Packaging Developments Utilizing HMW-HDPE In the USA Market*, presented at SPI/SPE, June, 1984 Conference, Philadelphia, Pa.

2. A. Brockschmidt, *Plastics Technology*, **30**, 64 (Feb. 1984).

3. H. R. Seintsch and H. E. Braselmann, *"Further Evaluations of HMW-HDPE Film Processing and Effects on Film Properties"*, TAPPI Paper Synthetics Conference, Washington, D.C., 1979.

4. H. E. Braselmann, R. E. Barry, *"HMW/MMW HDPE Extrusion and Fabrication Techniques for Grocery Sacks and Merchandise Bags"*, TAPPI 1984 Polymers, Laminations and Coatings Conference, Boston, Mass.

5. H. R. Seintsch and H. E. Braselmann, *Paper, Film and Foil Convertor*, **55**, 41 (March, 1981).

6. J. Sneller, *Modern Plastics* **59**, 50 (March 1982).

General References

E. F. Giltenan, "The International Flap Over Packaging", *Chemical Business*, **6**(10), 36–40 (Oct. 1984).

D. Sommer, "It's Plastics vs. Paper at Grocery Checkout Counters", *Sales and Marketing Management*, **133**(5), 59 (Oct. 8, 1984).

R. R. MacBride, "HMW-HDPE Extruder Runs Ultra-Thin Film", *Modern Plastics*, **57**, 12 (Sept. 1980).

R. H. NURSE
J. S. SIEBENALLER
American Hoechst Corporation

FILM, IONOMER.   See Ionomers.

FILM, IRRADIATED.   See Films, shrink.

FILM, NONORIENTED POLYPROPYLENE

The combination of high clarity, stiffness, heat resistance, oil resistance, and abrasion resistance at reasonable cost make nonoriented polypropylene films a good choice for a number of packaging applications. They are often called "cast polypropylene" films because they are generally made by the chill-roll cast process, but they can also be made by other methods. They do not, however, go through a biaxial-orientation step (see Film, oriented polypropylene). Annual consumption in the United States is roughly 65 million (10^6) lb (29,500 t). This includes all forms of nonoriented homopolymer, random copolymers (1.4–3.0% ethylene), and block copolymer films (see Polypropylene). Of the dozen-or-so major markets, about two-thirds are food related.

In the food-packaging markets, the use of nonoriented polypropylene is limited owing to its brittleness at below-freezing temperatures, and it is not generally recommended for use with heavy, sharp, or dense products unless laminated to stronger, more puncture-resistant materials. Cheese-wrap accounts now for more than one half of the food market, and its use is growing. In this case, the differential slip characteristics of an LDPE/PP coextrusion are excellent for single-slice cheese wrapping (see Coextrusions for flexible packaging). The market for produce wrap, which once accounted for a considerable amount of nonoriented polypropylene, has made a gradual switch to lower-cost LDPE materials albeit at a loss in clarity, gloss, and scuff resistance. The same thing is true for some bakery applications, especially the packaging of buns in bulk.

Textiles, including hosiery, mens' shirts, blankets, sheets, and draperies, account for a large proportion of the nonfood markets. Its appeal here is strongly related to superior clarity versus LDPE materials and outstanding sealing characteristics with side-weld overwrap machinery. The use of nonoriented PP film for health-care packaging is growing rapidly (see Health-care packaging), largely due to increasing use of medical disposables. The relatively high temperature resis-

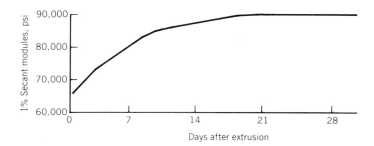

Figure 1. Stiffness vs time for cast random copolymer film, 1 mil (25.4 μm). To convert psi to MPa, divide by 145.

tance of PP permits its use in autoclave equipment where temperatures of 250–275°F (121–135°C) are common. Heat resistance also makes these films ideal candidates for the seal layer in retortable pouches and other applications where the film or lamination is subjected to the stress of high temperatures (see Retortable flexible and semirigid packages).

Film production. Polypropylene film is most commonly made by the chill-roll cast process, although water-quenching methods, tubular or slot-die, are also used (see Extrusion). Polypropylene resins are usually processed with long barrel (28:1 to 32:1 L:D ratio) barrier-type mixing screws.

Flat film dies can be flexible lip design and may incorporate choker bars. Cooling is accomplished on one or more water-cooled chill rolls. The surface of these rolls can be a mirror or matte finish. Rapid quenching, necessary to limit crystal growth, can be accomplished with the aid of an air knife or vacuum box. The quick quenching produces a film of relatively high clarity (low haze) and relatively low stiffness.

Like other members of the polyolefin family, polypropylene must be surface-treated to obtain affinity for printing inks or adhesives. This treatment is usually done with a corona discharge system (see Surface modification). A state-of-the-art beta or infrared scanning system for both MD (machine-direction) and XD (cross-direction) gauge control is a must for high quality polypropylene film, because gauge bands, baggy film, and thin spots are generally not acceptable. Many of the new gauging systems offer computer control of the average thick-

Table 1.   Typical Properties of Nonoriented Polypropylene Films, 1-mil (25.4-μm)

Property	ASTM test		HPP	RCPP
specific gravity	D 1505		0.888	0.887
yield, in²/(lb·mil) [m²/(kg·mm)]			31,185[1746]	31,213[1748]
haze, %	D 1003		3.5	2.0
light transmission, %			92	92
tensile strength, psi (MPa)	D 882	MD	9400(64.8)	9800(67.6)
		XD	6000(41.4)	5200(35.9)
elongation, %	D 882			
		MD	650	550
		XD	750	750
tensile modulus, 1% secant, psi (MPa)		MD	110,000(758.6)	85,000(586.2)
		XD	110,000(758.6)	85,000(586.2)
tear strength, gf/mil (N/mm)				
initial	D 1004		25(9.7)	25(9.7)
propagating	D 1922		400(154)	500(193)
water absorption, 24 h, %	D 570		0.01	0.03
change in linear dimensions at 100°C for 30 min, %	D 1204		1.0	1.0
service temperature, °F(°C), range			35–280 (2–138)	30–230 (−1 to 110)
heat-seal temperature range, °F(°C)			330–370 (166–188)	290–330 (143–166)
oxygen permeability cc·mil/(100 in²·d·atm) [cm³·μm/(m²·d·kPa)] 23°C, 95% rh	D 1434		240[933]	260[1010]
water vapor transmission rate, g·mil/(100 in²·d) [g·m/(m²·d)] 100°F (38°C), 90% rh			0.7[0.28]	0.8[0.31]
COF, back-to-back test conditions, 23°C, 50% rh	D 1894		0.20–0.40	0.10–0.30

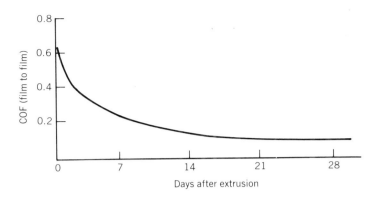

Figure 2. COF vs time for cast random copolymer film, 1 mil (25.4 μm).

ness and XD profile. Some systems offer microprocessor control of the barrel and die temperature profiles to ensure steady-state operation. A well-designed extruder screw and uniform resin feed are also needed to minimize MD thickness variation. Randomization in the cross-machine direction (XD) improves roll conformation. Before winding polypropylene film, other operations such as edge trim slitting, secondary cooling (to remove the heat of surface treating), and static elimination are performed (see Slitting and rewinding machinery). Static can be reduced with ionized air blowers, nuclear sources or tinsel grounded to the machine frames.

An important factor in polypropylene film production is the secondary crystallization that takes place. Polypropylene crystallites are formed during the first quenching on the chill roll. A secondary crystallization continues at a decelerating rate for several weeks. Figures 1 and 2 show typical changes that take place over such a period. Secant modulus increases, whereas coefficient of friction (COF) decreases. Internal slip agents, such as erucamide, are added to give the film its final slip characteristics for the particular end use. It is the continuing crystallization that forces the insoluble slip agent and other additives to the film surface with time.

Film properties. Compared to polyethylene films, nonoriented polypropylene offers great clarity and higher heat resistance. It does not have the strength and barrier properties of oriented polypropylene, but its cost is much lower (see Films, plastic). Typical properties of a homopolymer film and a random copolymer film are shown in Table 1.

The advent of coextruded-film technology has enabled film producers to build a number of synergistic benefits into basic polypropylene films. The technology permits the incorporation of a variety of desirable features which polypropylene alone cannot provide. By combining multiple yet distinct layers of similar or dissimilar materials into one film, coextrusion permits the creation of films with entirely new properties, such as improved heat seal, barrier, stiffness, and toughness characteristics (see Coextrusions for flexible packaging).

BIBLIOGRAPHY

General References

C. R. Oswin, *Plastic Films and Packaging*, Elsevier Applied Science, London, 1975.

C. J. Benning, *Plastic Films for Packaging: Technology Applications & Process Economics*, Technomic Publishing Co, Inc., Lancaster, Pa, 1983.

J. H. Briston and L. L. Katan, *Plastics Films*, Longman, London, 1983.

Ivan Miglaw
Ed Pirog
Crown Advanced Films

FILM, ORIENTED POLYESTER

Biaxially oriented polyester (PET) film is a high-performance film used in a wide range of applications including packaging, magnetic-tape products, floppy disks, microfilm, graphic arts, labels, solar-control window films, pressure-sensitive tape, and photoresist and hot-stamping foils.

In contrast to most packaging films (see Films, plastic) that derive most of their demand from flexible packaging, only about 20% of the demand for polyester film is for flexible packaging. Five domestic producers supply polyester film to the United States merchant market; American Hoechst Corporation (Hostaphan), Bemis Converter Films (Esterfane), DuPont Company (Mylar), ICI Americas (Melinex), and 3M (Scotchpar).

Film Manufacturing Process

To improve physical properties, polyester film is oriented (stretched) in either a blown-bubble process or a cast-tentered process (1). The cast-tenter process (see Fig. 1) is more capital intensive than the blown-bubble process, but because it results in better gauge control and property uniformity, most of the world's biaxially oriented PET film is tenter-oriented. PET is the polycondensation product of ethylene glycol and terephthalic acid (2) (see Polyesters, thermoplastic). The molten resin is extruded onto a casting drum. The polymer molecules are oriented in the machine direction by heating the cast sheet and stretching it to 3 or 4 times its original length. The molecules are then oriented in the transverse direction in the tenter, which is a long oven with several precisely controlled temperature zones. As the film enters the tenter, it is gripped on each side by clips on fast moving chains. In the tenter, at appropriate temperatures, the chains diverge to 3–4 times the original width, orienting the molecules in the cross direction. In the final section of the tenter oven, the film is annealed (heatset).

This stretching process makes PET a biaxially oriented film. For most packaging applications, the film is oriented to the same extent in both directions, which imparts balanced properties. Orienting the film greatly improves tensile strength, clarity, stiffness, chemical resistance, and barrier properties. The annealing step imparts heat stability and controls shrinkage. Film properties can be modified by coextrusion (see Coextrusions for flexible packaging), modifying the polymer, varying processing conditions, or altering the film surface by chemical treatment or corona discharge (see Surface modification).

Film Properties

Properties of the base PET film which are most important to the converting and packaging industries are heat stability, moisture resistance, chemical resistance, toughness, clarity, wide useful-temperature range, stiffness, handleability, and

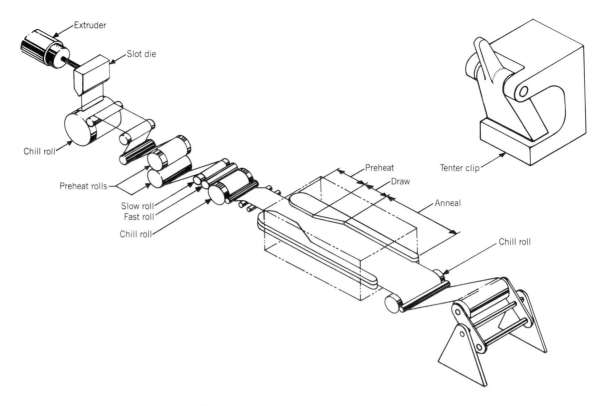

Figure 1.   Tenter orientation process.

barrier properties. The functional and barrier properties of the base film can be substantially improved and tailored for specific applications by further coating, laminating, or metallizing (see Multilayer flexible packaging). Biaxially oriented PET film, with its excellent thermal properties, high mechanical strength, and excellent dimensional stability, is particularly suited for the more demanding and close-tolerance requirements needed in many printing and converting operations that involve aqueous, solvent, extrusion, and solventless coating and laminating. All are basic processes for the production of flexible packaging, labels (see Labels and labeling), and hot-stamping foils (see Decorating).

Because polyester film has excellent thermal properties (see Table 1), it can be processed and used in applications over a wider temperature range than most common packaging films. With a normal temperature range for prolonged use (more than several hours) of -70 to $150°C$, it is an ideal component for boil-in-bag, retort packaging (see Retortable flexible

and semirigid packages), and "dual-ovenable" lidding (see Lidding). The film can withstand considerably higher temperatures for the shorter periods of time that might be experienced in extended drying ovens, in hot-stamping, and on high speed, high temperature form/fill/seal packaging equipment.

Dimensional stability is another important property. Unrestrained shrinkage at $150°C$ for 30 min. is "typically" on the order of 1.5%. Shrinkage levels are designed to meet the specific requirements of tight-register printing for flexible packaging or the high pressures and temperatures of hot-stamping foils. The moisture resistance of the film means it does not change barrier properties, swell, shrink, or otherwise distort in changing humidities. Chemical inertness and good barrier properties are important for many food, pharmaceutical, and medical products. Polyester films are available in grades that comply with FDA (3) and USDA requirements, and can be used in steam, ethylene oxide, and radiation sterilization processes (see Health-care packaging). The film is also highly

Table 1.   Polyester Film Thermal Properties

Property	Value	Test method
melting point	493°(256°C)	ASTM D 2117
service temperature range for extended periods[a]	-100 to 300°F(-73 to 150°C)	
coefficient of thermal conductivity, at 32–180°F (0–80°C)	1.03 Btu·in./(ft²·h·°F) or 3.5 × 10⁻⁴ Cal·cm/(cm²·s·°C) or 0.148 W/(m·K)	
specific heat, at 20°C	0.075 J/(g·K)	

[a] Temperatures up to 204°C (400°F) may be tolerated for 15 to 20 min.

Table 2. Typical Properties of Biaxially Oriented Polyester Film (Plain, Uncoated, 48 gauge)

Property	ASTM test method	Value
thickness, mil (μm)	E 252-78	0.48 (12.2)
specific gravity	D 1505-68	1.4
yield, in.²/lb (m²/kg)		41,700 (59.3)
yield stress, psi (MPa) (F-5)-MD		14,500 (100)
haze, %	D 1003-61	4.0
tensile strength, psi (MPa) MD	D 882-80	30,000 (207)
ultimate elongation, % MD	D 882-80	100
shrinkage, % MD		1.8
(30 min, 150°C) XD		0.4
tear strength, gf (N)	D 1922-67	9 (0.088)
modulus, psi (MPa) (1% secant)	D 882-8	700,000 (4,828)
burst strength, psi (kPa)	D 774	80 (552)
light transmission, %	D 1003-61	88
kinetic coefficient of friction, film/film	D 1894-78	0.45

resistant to a wide range of chemical reagents. It is not dissolved, even when heated, by most common solvents.

To improve the adhesion of various inks (see Inks) and coatings to the polyester surface, producers may corona treat inline or chemically prime the base film. The chemical priming actually changes the surface chemistry of the film and does not decay with time, as may be the case with corona treatment. Coextruded polyester films are produced with an amorphous polyester layer which may be heat-sealed (see Sealing, heat).

Types and Applications

Of the estimated 325 million lb (10⁶) (1.47 × 10⁵ t) of PET film sold into the United States merchant market in 1984, about 50 million lb (2.27 × 10⁴ t) went into flexible packaging applications and about 6–8 million lb (2700–3600 t) into secondary packaging applications including labels and hot-stamping foils. Several types of PET film are used for a wide variety of packaging applications.

Flexible packaging. Most of the PET film used in flexible packaging is 48 gauge (0.48 mil or 12.2 μm). Typical physical properties of a plain uncoated 48 gauge PET are shown in Table 2. It is frequently used as a part of a composite lamination in conjunction with a sealing layer in order to achieve a balance between cost and function (see Multilayer flexible packaging). Because of its strength, printability, stability, and tolerance to high processing and heat-sealing temperatures, PET film is most often the outside and primary support film of laminations (see Laminating) used in many demanding applications such as the flexible retort pouch, boil-in-bag, and "ovenable" lidding. It is used in many pharmaceutical (see Pharmaceutical packaging) and medical-device (see Healthcare packaging) packages. Because of increased packaging speeds requiring higher heat-seal temperatures, PET is increasingly used in snack-food laminates.

PET film may be corona or chemically treated to promote improved adhesion to a variety of inks, coatings, and adhesives. Some of the chemically treated films effectively improve adhesion of water-based ink, adhesive, and cohesive systems (see Adhesives). The barrier properties of polyester film are enhanced by coating with poly(vinylidene chloride) (PVDC) (see Vinylidene chloride copolymers) or metallizing (see Metallizing) with aluminum, as shown in Table 3.

Barrier-type films constitute 50% of current United States PET film consumption, with approximately 20 × 10⁶ lb (9000 t) PVDC-coated by the PET producer and 6 × 10⁶ lb (2700 t) metallized by companies specializing in metallizing substrates. Coextruded/heat-sealable films are frequently metallized and used as the inner ply of snack-food packages, which require barrier properties and opacity to prevent rancidity of the product. PVDC-coated PET is primarily used in transparent packages for such food items as cheese, processed meats, and nuts. The PVDC coating inhibits mold and oxidative degradation. Two-side coated grades provide higher barrier and are available with the coating designed for heat-sealability. A major special application is the single-slice cheese wrapper. Metallized PET is typically used in bag-in-box laminates (see Bag-in-box, liquid product) for wine and condiments and in packaging materials for institutional coffee, candy, cookies, snacks, processed meats, and cheese. For its aesthetic appeal, it is used on cosmetic cartons and for label base. Polyester films are also available for low-draw thermoforming (see Thermoforming), heat-shrink applications (see Films, shrink), and oven use.

Labels. In a typical pressure-sensitive label (see Labels and labeling) or decal (see Decorating), polyester may play one or all of three roles. It can be the base stock to be printed, the over-laminate to protect the printing, or the release liner to carry the label. Polyester films with crystal clarity and chemical primers are highly attractive. The clarity of these films has

Table 3. Typical Barrier Properties of 48 Gauge Polyester Films

Test	Uncoated	PVDC-coated	Metallized
WVTR g/(100 in.²·d) [g/(m²·d)] 100°F (38°C), 90% rh ASTM E 96	2.7 [41.9]	0.5–0.9 [7.8–14]	0.03–0.10 [0.47–1.56]
O² transmission cm³/(100 in.²·d·atm) [cm³/(m²·d·kPa)] 73°F 0% rh ASTM D 1434	6.1 [0.93]	0.3–0.4 [0.045–0.06]	0.02–0.10 [0.003–0.015]

opened new markets to clear pressure-sensitive labeling systems for many products including cosmetic, pharmaceutical, and personal-care products. When metallized these films exhibit mirror-like characteristics. In 1984, approximately 2–4 $\times 10^6$ lb (900–1800 t) of PET film, primarily in 1–2 mil (25–51 μm) thickness, was consumed in packaging related label applications in the United States.

Hot stamping. Approximately 40% of the 10–12 million lb (4500–5400 t) of PET film used in hot-stamping goes into secondary packaging applications such as metallic decoration and coding (see Code marking and imprinting). In the hot-stamping process, PET film, usually plain 48-gauge, serves as the carrier for a preprocessed continuous decoration which is die-stamped under heat and pressure onto the item to be decorated or identified. Hot-stamped metallic foil decorates containers and boxes for personal care and cosmetic products and pigmented stamping foil is used for coding or dating.

Trends

It is estimated that overall PET film demand in packaging will increase at the annual rate of 5–7% through the 1980s, with chemically treated films growing at an even greater rate of 10–12%. A major trend in the use of PET film is the replacement of glass and metal containers with lower-cost plastic laminates. Many of these flexible laminates offer opportunities for current and new types of PET film. The retort pouch for consumer products is not as widely used in the United States as it is in Japan, where a substantial percentage of the estimated 30 million lb (27,000 t) of PET-packaging film is used in retort packaging. The substitution of the no. 10 can (see Cans, steel) by flexible pouches has started to take place in some areas of the United States institutional market, where economic incentives and waste disposal are the motivating factors. Vacuum-packed, bricklike packages of coffee (see Vacuum coffee packaging) are now making inroads into the consumer market which had been dominated by cans. Wines, condiments, soft-drink syrups, and fruits are being packaged in bag-in-box containers ranging in size from 3L to 300 gal (1136 L).

PET film producers, to meet the challenge of new applications and competing films, are focusing on the development of new PET films to meet the demands of the converting/packaging market.

BIBLIOGRAPHY

1. U.S. Pat. 2,823,421 (Feb. 18, 1958) A. C. Scarlett (to E. I. du Pont de Nemours & Co., Inc.).
2. U. S. Pat. 2,465,319 (May 22, 1949) J. R. Whinfield and J. T. Dickson (to E. I. du Pont de Nemours & Co., Inc.).
3. *Code of Federal Regulations*, Title 21—Food and Drugs; Section 177.1630 Polyethylene phthalate polymers; Section 177.1390 High Temperature Laminations.

General References

C. J. Benning, *Plastic Films for Packaging: Technology, Applications and Process Economics* Technomic Publishing Co., Inc., Lancaster, Pa., 1983.

Food and Drug Packaging, (April, 1983); discusses barrier properties of metallized polyester.

R. M. Kimmel, "Metallized Films: The Basics of Polyester Film for Metallizing," *Food and Drug Packaging*, (Feb. 23, 1983).

J. R. Newton, *High Barrier Materials,* 21, ICI Americas, Wilmington, Del., 1984 (presented at the *Future Trends in Vacuum Web Coating* seminar sponsored by Association of Industrial Metallizers, Coaters, and Laminators, Oct. 20, 1983.)

B. L. KINDBERG
R. L. OLDAKER
S. H. ROUNSVILLE
American Hoechst Corporation

FILM, ORIENTED POLYPROPYLENE

Oriented polypropylene film was made commercially for the first time by Montecatini of Italy in the late 1950s. Shortly after, Imperial Chemical Industries of the United Kingdom entered into OPP film production. Kordite (the predecessor of Mobil's Commercial Films Division), Hercules, and DuPont started production in the early 1960s. In 1984, about 900×10^6 lb (408,000 t) of effective annual capacity was in place worldwide. Shipments of OPP film are estimated at 650×10^6 lb (295,000 t). Consumption of this high-performance thermoplastic film is expected to grow 6–8%/yr through 1990.

Oriented polypropylene film is made by over 30 companies around the world. Hercules has plants in the U.S., Canada, U.K., and Brazil. Mobil produces in the U.S., Canada, Belgium, and Italy. Other U.S. manufacturers include Borden, Curphane Division of Bemis, and Norprop, a unit of Internorth. BCL Limited produces OPP in England, France, and Australia. ICI has plants in the UK, Belgium and Australia. Kalle, a part of Hoechst, manufactures the film in the FRG and South Africa. Wolf, a unit of Bayer, is the other major FRG producer. There are 13 Japanese suppliers. These include Futamura-Shansho, Honshu Paper, Toray, Toyobo, Tocello, Gunze, Daicel, Tokuyama Soda, Kojin, Okura Industry, Mitsubishi Rayon, Shinetsu Film, and Totoshina Film. A number of smaller producers are located in Taiwan, South Korea, Venezuela, Indonesia, Mexico, and Israel.

Raw Materials

The basic raw materials for OPP films are four different polypropylene-rich resins: homopolymer, copolymer, terpolymer, and modified resins (see Polypropylene). Homopolymer resins are made by polymerizing propylene monomer with high productivity catalyst. Copolymers have a relatively small percentage of ethylene comonomer (0.1–15%) in addition to the main feedstock. The amount of comonomer used depends on the film's end use. Terpolymer production uses a small amount of butene along with ethylene and propylene feeds. Polypropylene homopolymer resins are also modified with terpenes and styrenes to give them certain heat sealing properties (see Sealing, heat).

Manufacturing Process

When a film is biaxially oriented, it is mechanically stretched in directions which are at right angles to each other. By doing this, the film's molecules are aligned in the cross (transverse) machine direction (XD) as well as in the machine direction (MD). Once biaxial orientation has taken place, the film exhibits a marked improvement in optics, strength, moisture barrier, and low temperature durability over cast polypropylene film (see Film, non-oriented polypropylene). The increase in strength allows OPP film to be made as thin as 0.45

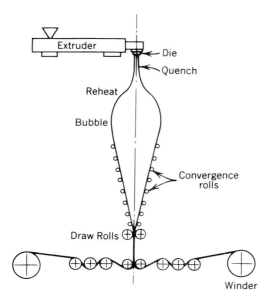

Figure 1. Bubble process.

mils (11 μm) and still function as a laminating substrate for packaging applications (see Laminating; Multilayer flexible packaging). Balanced orientation is obtained when the degree of orientation as expressed by the film's tensile strength is approximately equal in each direction. There are two distinctly different methods of orienting OPP film. They are called the "blown tubular" and "tenter frame" processes.

Blown tubular process. In this process, molten polypropylene resin is extruded through a circular die and is cooled (see Extrusion). At this point, the film is an amorphous tube, many mils ($\times 25.4$-μm) thick. It is then reheated by circular radiant heaters to a temperature above the polymer's T_g (see Polymer properties) but below the melting temperature. The use of air pressure blows the tube into an extended bubble (see Fig. 1). At the same time, the bubble is taken off the orienting unit faster than it is fed. The size of the bubble's circumference controls the amount of XD orientation and the speed differential between the in-feed and the take-off mechanisms controls the degree of MD orientation.

After orientation, the film bubble is collapsed and is either wound up as a shrink film or is annealed for use as a stabilized, heat-set packaging film. In this process, heat stabilization is accomplished by one of two methods. The "double bubble" approach reinflates the bubble while the film's temperature is elevated to a point just below the T_g and then is cooled. The other method passes the collapsed bubble over a series of heated rollers which raises the film's temperature in a similar fashion to that outlined above. In both methods the film is mechanically restrained from shrinking back in the machine direction (MD). The amount of restraint in the transverse direction (XD) varies from a small amount down to essentially none. This heat treatment relaxes most of the high internal stresses and tensions built up in the film during orientation and provides a functionally stabilized product.

Normally, a heat-set OPP film can withstand brief contact with sealing mechanisms that are 300°F (149°C) or slightly above that without surface distortion in the seal areas. Sealing mechanisms above this temperature do create progressively

greater distortion potential until at about 310°F (154°C) the film attempts to return to its unoriented dimensions in a wild and random fashion leaving the affected portions of the sheet a mass of distorted plastic. Because of this reaction to very high temperatures, OPP film's sealing range has an upper limit of 300–305°F (149–152°C).

Heat set OPP film 1.00-mil (25.4-μm) thick, when subjected to a glycol bath for five seconds at 248°F (120°C), exhibits an unrestrained shrinkage of 0.5% in the XD and 2.5% in the MD. As a result, heat-set OPP films will show some shrinkage if they are exposed to very hot environments such as is found in retort packaging (250°F (121°C) at 15 psi (103 kPa)). However, below 230°F (110°C) OPP film shows very little or no shrinkage, and a package wrapped in this material maintains its dimensional integrity.

In the bubble-orientation process, after heat stabilization the sides of the collapsed bubble are slit and the film is wound simultaneously into two mill rolls. During this winding operation the film is frequently corona treated (see Surface modification). This surface treatment consists of a controlled electrical bombardment. The film is slightly oxidized which produces a multitude of sites that are chemically attracted to the polymers used in off-line coatings, printing inks, and laminating bonding agents such as adhesives and polyolefin extrudates.

Tenter frame process. In this process, polypropylene resin is extruded out of a horizontal flat die, sometimes called a slot die. The extrudate, an amorphous sheet many mils ($\times 25.4$-μm) thick, is quickly cooled by a chilled roll and/or in a cold water bath. This heavy sheet is then reheated to a temperature that is above the polymer's T_g but below its melt point. After reheating, it is first oriented in the MD. The first mechanical stretching is accomplished by drawing the sheet off the unit by a set of rollers faster than it is fed on by another set of rollers. On some lines a number of sets of infeed and takeoff rollers are used, stretching and thinning the sheet in sequential steps. The next step is XD orientation, which is done in a "tenter frame." The tenter frame has the configuration of an open-bottomed "V" (see Fig. 2). The MD-stretched film enters the small open end of the "V" and is gripped securely by its sides with clips. The clips are mounted on moving chains which run along tracks on each side of the frames forming the "V". As the chains move forward and outward along these tracks, the film is stretched in the XD.

To achieve good heat stabilization, the film, still securely

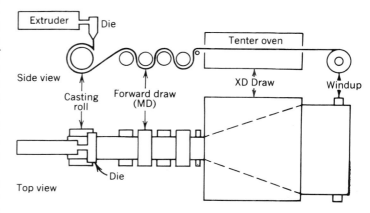

Figure 2. Tenter frame process.

held by the clips, moves on into a heat-setting (annealing) oven section. In this section, the clip/chain systems at each side of the film have passed out of the diverging XD stretching section into a section of the tenter in which the chains run parallel to each other. Here the film is subjected to elevated temperatures for a few seconds while it is completely restrained in both directions. The film is then cooled, released from the clips, has the unstretched portion under the clips removed, is generally corona-treated, and finally is wound up into mill rolls. Some tenter frame users split the full width film into two roughly equal-sized mill rolls by an in-line slitting operation as the OPP film comes off the line (see Slitting and rewinding).

Following orientation and stabilization, regardless of the process used, the mill rolls are usually stored on racks for several days. During this time, remaining internal tensions are relaxed and a further degree of crystallinity builds up in the film. Additives included in the resin, such as those needed for slip, are also given time to bloom to the surface (see Additives, plastics). After the proper time interval, the OPP film is ready for further processing in an off-line coating and/or a finished slitting operation.

In the early days of OPP film production, blown bubbles were 120-in. (3.05-m) in circumference and tenter frames were 84–120-in. (213–3.05-m) wide. Over the years, lines twice as large have been engineered for both processes. Today, 240-in. (6.1-m) circumference bubbles and 240-in. (6.1-m) wide tenter frame lines are in operation in Europe and the United States. The Japanese have OPP film tenter frames that are 320-in. (8.13-m) wide. This movement to much wider lines has greatly increased the OPP film producers' productivity and has also significantly reduced the generation of narrow side rolls in slitting operations. Both factors have contributed to OPP film's reputation of being the most economical of all highly engineered, high performance flexible packaging materials.

The very low density of polypropylene (0.905 g/cm^3) allows OPP film to have the greatest area-coverage yield per unit of weight of any commercially significant thermoplastic or cellulose-based sheet. This high area yield and ability to be made into very thin films are important contributions to its overall favorable economics.

There are relatively few physical property differences between films made by the two processes (see Table 1). Both films possess excellent moisture-barrier properties. This value increases and decreases proportionately with the thickness of the film under consideration. Tensile strength and elongation of blown films tends to be about the same in both directions. XD strength is higher than MD strength in tenter-frame films, and MD elongation is higher than XD elongation. Tenter-frame film has a generally superior heat stability over blown film because of more severe annealing conditions that can be practiced in the tenter-frame operation. Blown-bubble OPP films have somewhat better impact strength. Only in the most demanding and arduous packaging applications would these relatively small differences become significant.

Product Developments

The first two OPP films produced were homopolymer shrink film, and heat-stabilized film that could not be heat sealed. The former had limited application because of its need for hot-wire sealing and tremendous shrink energy which originally precluded it from wrapping high profile products. The development of polypropylene copolymers with relatively high ethylene content during the 1970s has alleviated this problem in many applications.

The latter, heat-set OPP film, gained wide acceptance as a laminating substrate when combined with heat-sealing polymer-coated cellophanes (see Cellophane) and glassine papers (see Glassine). The OPP film contributed strength, moisture barrier, and high suface gloss to the lamination. However, before acceptance was forthcoming from the market place, OPP film producers had to improve the film's slip characteristics over the relatively high coefficient of friction (COF) that is natural to the film. The addition of internal surface-blooming slip agents reduces the film-to-film COF to 0.2–0.4. This COF provides excellent machineability on automatic bag forming machinery used in the snack food and candy industries. Converters developed the hot-melt-adhesive-based, heat-sealing thermal stripe that allows the production of lap back seals on vertical form/fill/seal machines (see Form/fill/seal, vertical). This technology widened the use of OPP film laminations for snack foods, candy, and pasta products. Another application for this non heat-sealing-film is the inner liner of paper bags for cookies and pet foods.

Sealability. The quest for functionality through heat-sealing on conventional sealing mechanisms used on most packaging machines led OPP film producers down four avenues of development. The first was the modified polymer route. This film uses a polypropylene that is modified with terpenes. The film is heat-sealable, although its sealing range is relatively narrow (25°F or ca 14°C), and it does not have good hot-tack-seal strengths needed for form/fill/seal packaging. In overwrapping applications where the product to be packaged is contained in a carton or some other self-supporting structure, this OPP film has found good acceptance. Because this film is only partially heat-stabilized, it can be snugly tightened around the carton or bundle through the use of a heat tunnel. One version is used today as a cigarette-packaging material.

The second method of imparting sealability to OPP film is by addition of an off-line heat-sealable coating based on an acrylic polymer (see Acrylics). This coating adds no barrier properties to the OPP film, but it does provide a relatively wide sealing range of 80°F or ca 45°C, and adds sparkle and good machine slip characteristics. It has a film-to-film COF of 0.2–0.3. It has high-volume use in horizontal form/fill/seal operations (see Form/fill/seal, horizontal) for items such as baked foods (cookies in particular), and candy. As an overwrap film it is used for pet foods and various tobacco products such as cigarettes, cigars, and pipe-tobacco cartons. It's hot-tack-seal strength characteristics are such that it is used extensively for VFFS packaging of light weight products, such as snack foods, but is not generally proposed for similar packaging of heavy products. These acrylic coatings are compatible with many inks (see Inks) and adhesives (see Adhesives) used by converters and are also heat-sealable with the PVDC coatings used on cellophane, glassine and other OPP films (see Vinylidene chloride copolymers).

Oriented polypropylene film in its uncoated form is not considered to be a gas-barrier film (see Table 1). Its oxygen permeability is relatively high at 73°F (23°C) using 1.00-mil (25.4-μm) thick film (see Table 1). The third method to obtain heat sealability is to coat the OPP film off-line with a PVDC coating, which also overcomes the gas-barrier deficiency. PVDC-coated OPP films fall into three categories. The first is readily

Table 1. Properties of Oriented Polypropylene Films, 1 mil (25.4 μm)

Typical film properties	ASTM test	Blown OPP film	Tenter-frame OPP film	Slip-modified film	Acrylic-coated film	PVDC-coated film	Coextruded film	White opaque film	Metallized film
specific gravity	D 1505	0.905	0.905	0.905					
yield in.^{2}/(lb·mil)		30,600	30,600	30,600		27,300	30,600	48,000	
[m^2/g·m]		[1714]	[1714]	[1714]		[1529]	[1714]	[2688]	
haze, %	D 1003	3.0	3.0	3.0	3.0	3.0	3.0		
light transmission %								50–60	
tensile strength, psi (MPa)	D 882								
MD		30,000 (207)	18,000 (124)						
XD		30,000 (207)	34,000 (234)						
elongation, %	D 882								
MD		80	130						
XD		80	50						
tensile modulus 1% secant, psi (MPa)									
MD		350,000 (2414)	350,000 (2414)						
XD		350,000 (2414)	600,000 (4138)						
tear strength, gf/mil (N/mm)									
propagating	D 1922	4–6 (1.5–2.3)	4–6 (1.5–2.3)	4–6 (1.5–2.3)	4–6 (1.5–2.3)				
service temperature, °F (°C), range						70 (39)	50 (28)	40–110 (22–61)	
heat seal temperature °F (°C), range						230–300 (110–149)	250–300 (121–149)	190–300 (88–149)	
oxygen permeability cm^3·mil/(100 in.^{2}·d·atm) [cm^3·μm/(m^2·d·kPa)];	D 1434								
77°F (23°C), 0% rh		160 [622]	160 [622]	160 [622]	150 [583]	1–3 [4–12]	160 [622]	160 [622]	4–18 [16–70]
77°F (23°C), 50% rh						4–6 [16–23]			
water vapor transmission rate (WVTR) g·mil/(100 in.^{2}·d) [g·μm/(m^2·d)]; 100°F (38°C), 90% rh		0.25 [0.1]	0.30 [0.12]	0.30 [0.12]	0.30 [0.12]	0.30 [0.12]	0.30 [0.12]	0.60 [0.24]	0.1–0.3 [0.04–0.12]
COF, face-to-face	D 1894			0.2–0.4	0.25	0.2–0.4	0.2–0.4		
back-to-back		0.4	0.4	0.2–0.4	0.25	0.2–0.4	0.2–0.4		
test conditions, 73°F (23°C), 50% rh									

machineable on heat sealing, automatic packaging equipment. The sealing range in this case is relatively wide (60–80°F or 33–45°C), and the oxygen-transmission rate is improved (see Table 1). These films are available with one or two sides coated. The former is a laminating substrate for use with other OPP films and cellulosic materials, to package snack foods and candies. The latter is used in unsupported form for bags and carton overwraps. In all of these applications, an aroma barrier is generally desired by the end user.

The second category of PVDC-coated OPP film has a superior gas barrier, an oxygen-transmission rate of 8 to 12 cm^2/ (m^2·d). However, because its sealing range is no more than 50°F (28°C), this film is not used for its heat sealability. Rather it is used as a laminating substrate with various other materials for the controlled atmosphere packaging of coffee, natural cheese, and processed meats (see Controlled atmosphere packaging).

The third category combines an acrylic coating on one side and a PVDC coating on the other side. The advantage cited for this film is that it provides a degree of gas and aroma barrier, is lap sealable, machines well, and costs somewhat less than two-sided, heat-sealable, PVDC-coated OPP films.

The fourth avenue of development of heat sealability is through coextrusion. Although coextruded, multi-ply, heat-sealing OPP films have been available since 1965, this type of film has come into large-volume usage only in recent years. Coextrusions are made using two basic approaches. In one approach, two or more extruders feed their individual molten polymer streams to a single extrusion die (see Coextrusion, flat; Coextrusion tubular). In the upstream section of the die is a mechanism that brings these polymer flows together and layers them into a single sheet as it passes out of the die lips. The other method is termed "tandem" coextrusion. With this procedure the film's core material is extruded from its own extruder and die. This extrudate is then extrusion-coated (see Extrusion coating) from one or more extruders and dies, or it is hot-nip laminated to other compatible independently manufactured films. In both cases the multilayered sheets, many mils ($\times 25.4$ μm) thick, are moved into the orientation units of the line. Coextrusions can be made by both the blown bubble and tenter frame processes.

Polypropylene does not anchor well to many other polymers. As a result, great care must be taken in the choice of materials used in OPP film coextrusions or inter-ply bonds may be broken during orientation or later by the stresses of package forming techniques and distribution. Other demands on outer ply materials include hot-slip characteristics that allow easy film movement over hot-packaging machine parts and surface-slip characteristics for good overall trouble-free automatic high speed machining. The ply destined to be printed or laminated must also be corona-treated.

Today most coextruded OPP films consist of two or three layers. The majority of the film's mass is in the core ply which is usually made with virgin homopolymer and some regrind. One or both sides of the core may be covered with a polypropylene-rich copolymer or terpolymer, an ionomer (see Ionomers), or a vinyl acetate-modified polyethylene layer. The choice of polymer depends on the purpose for which the finished film will be used. Polypropylene-rich copolymers of propylene and ethylene are used for their good hot-tack seal strengths and excellent finished seal strengths. Terpolymers of propylene, ethylene, and butene are used to improve the film's breadth of sealing range, but compared to the copolymers, today's terpolymers have a poorer hot-tack strength. Vinyl acetate-modified polyethylene plies are used for their anchoring potential in off-line coating, laminating, and metallizing operations (see Metallizing). Ionomer resin is used for its wide sealing range and ability to repel grease and oils. It is somewhat difficult to handle in the orientation process, but once oriented and heat stabilized, it provides lower level seal strengths of about 300 gf/in. (116 N/m) which can be used advantageously in packages that demand an easy-open feature.

Coextrusion sealing temperature ranges very greatly depending on the sealing ply materials. General categories are listed in Table 2.

Table 2. Coextruded OPP Film Sealing Temperature Ranges

OPP film	Sealing temperature range
copolymer-group A(0.5–3.0% ethylene)	30–50°F (16–28°C)
copolymer-group B(≥4% ethylene)	60–70°F (33–39°C)
terpolymers(ethylene and butene)	85–95°F (47–53°C)
ionomer	95–105° (53–58°C)

Thin coextruded OPP films, 0.57–0.70-mil (14–18-μm) thick, are often laminated to other substrates such as polymer-coated cellophane, glassine, paper, and other OPP films for snack food, candy, processed cheese, and pasta-product packaging. In unsupported form they package lightweight products such as single-service packages of crackers used in restaurants. Medium weight films, 0.80–0.90-mil (20–23μm) thick, are used for carton and tray overwraps by the bakery and candy industries, as well as for form/fill/seal single wall pouches by pet food producers. A specially designed medium weight material is used as a cigarette package overwrap. Heavy, thick coextruded OPP films, over 1.00-mil (25.4-μm) thick, are normally used in unsupported form to package cookies, candies, snack foods in small bags and other products that need the stiffness, strength, excellent moisture barrier and rich feel offered by these films.

Opaque films. White opaque film is one of the fastest growing OPP film developments in the United States and Western Europe. The most widely accepted film products of this type are made by the tenter-frame process. Homopolymer resin is evenly mixed with a small amount of foreign particulate matter. In one product, when the thick filled sheet is oriented, the polypropylene pulls away from each particle creating an air-filled void (closed cell). After heat stabilization the OPP film is similar to a micropore foamed product. In the second product, the material produced as a filled film without voids. The opacity is a direct result of the amount of particulate material included in the film.

In the film with air-filled voids, the imparted opacity and whiteness is created to a small degree by the encapsulated particulate matter. However, the primary opacification is brought about by light rays bouncing off the polypropylene cell walls and the air within each cell. The refractive index of the air is less than for polypropylene. This refraction difference results in a TAPPI opacity of about 55%. The light diffusion gives the film the visual effect of pearlscence. Brightness values are calculated at 65–75% by the GE method. The actual gauges of this white opaque film are deceptively thick when their area yield per unit of weight is considered. A 1.5-mil (38-μm) thick white OPP film has an area yield of 30,000 in.^{2}/lb (427 cm^2/g), which is about the same yield provided by a 1.0-mil (25.4-μm) thick transparent OPP film.

Coextrusion and out-of-line coating techniques have greatly expanded the market acceptability of white opaque OPP film. By using these approaches to film manufacture, the film can be made one- or two-side heat-sealable. These steps can also increase the film's moisture barrier. White opaque OPP films are finding growing volumes in snack-food packaging, candy-bar overwraps, beverage-bottle labels, soup wrappers, and other applications that traditionally have used specialty paper-based packaging materials.

Metallizing. The use of OPP film as a metallizing substrate is another area of new market growth (see Metallizing). In this process, aluminum is vaporized onto one side of a film or paper in a high vacuum environment. The speed of the substrate's passage over the vaporization pots controls the amount of metal deposited on the film. If the metallized surface is to be used primarily for decorative purposes, a relatively light coating of approximately 4 Ω (710 nm) is applied. This gives OPP film, originally in transparent form, a light-transmission rate of about 5%. If the metal coating is used for significant improvement of the finished film's barrier properties, a heavier disposition of 2 Ω (1350 nm) is employed. This level lowers light transmission to about 0.6%. Moisture and oxygen barrier are improved dramatically (see Table 1). In packaging, metalized OPP films are widely used in snack-food laminations and for confectionery wraps.

RANDAL J. HASENAUER
Mobil Chemical Corporation

FILM, RIGID PVC

Poly(vinyl chloride) (PVC) is extraordinarily adaptable to custom compounding for desired performance. This diversity of physical properties at relatively low cost has been the driving force behind its popularity in the packaging industry.

Homopolymer PVC Resin

Commercial PVC resin is a dry free-flowing powder produced by the polymerization of vinyl chloride monomer (see Poly(vinyl chloride).

Molecular weight/viscosity. The fundamental property of PVC is its molecular weight; a measure of its mean polymer chain length and a parameter proportional to the resin's viscosity. In general, higher molecular weight (higher viscosity) resins require higher processing temperatures and yield film or sheet with higher heat distortion temperatures, impact resistance, and stiffness than do resins of lower average molecular weight. The processing equipment for generating film and sheet for the packaging industry demands low- to medium-viscosity PVC resins with relative viscosities of 1.75–2.10 in contrast to extrusion pipe and construction markets that typically use medium- to high-molecular-weight resins.

Heat stability. All PVC resins are subject to thermal degradation during processing and must be compounded with appropriate heat stabilizers to minimize discoloration. These stabilizers serve to scavenge free radicals which perpetuate degradation, as well as hydrogen chloride, the principal degradation product. The highly complex thermal degradation mechanism proceeds by an "unzipping" process whereby liable allylic chlorides act as reaction sites for the liberation of hydrogen chloride and the formation of conjugate-bonding systems. When the conjugation exceeds six bonds in length, color development begins, and if unchecked by ample and appropriate use of heat stabilizers, will progress from a very subtle yellow tint to amber, and finally to black. Commercial processing methods occasionally generate "burned" material which has reached these initial stages of degradation. It is important to recognize that such thermal degradation is both time- and temperature-dependent and that although stabilizers retard the rate of degradation during processing, they do not prevent it. The stabilizers also help protect the film or sheet during subsequent processing (eg, thermoforming), and during the lifetime of the package itself.

Compounding for Properties

Among all polymers used in the packaging industry, PVC is widely regarded as the most versatile and suitable for custom compounding to deliver special properties (see Additives, plastics). It may be compounded for high clarity and sparkle or for maximum opacity and it accepts a full range of custom colorants (see Colorants). Properly compounded, PVC film and sheet are approved for food and drug contact and are available

Table 1. Examples of Custom Compounding

higher heat distortion temperatures for hot-fill packaging
gamma-ray sterilizable film for medical-device packaging
ethylene oxide (ETO) sterilizable film without water-blush for medical devices
improved low temperature impact resistance for drop tests of shipping cartons
improved uv resistance
improved outdoor weatherability
improved sealability (impulse heat, rf, ultrasound)
denesting formulations for machine-fed blisters
static-resistant formulations
optimum performance in laminating to other materials (PVDC, PE, etc)
formulations for vacuum metallizing
improved printability
absence of "white break" or crease-whitening

with residual vinyl chloride monomer (VCM) levels below 10 ppb. Examples of what custom compounding can produce are shown in Table 1.

After identification of such desired properties for the specific packaging application, the compounder selects a suitable resin viscosity. If the film is to be approved for food or drug contact, the resin's VCM level must be low enough before processing to ensure that the resulting film will meet all customer requirements on residual VCM. A heat stabilizer must then be selected. Tin mercaptides are frequently chosen due to their high efficiency, excellent early color, good light stability, and excellent crystal clarity in the product (1). Some of these stabilizers (octyl tins) are cleared for food and drug contact. Uncleared options include lead stabilizers, which are limited to opaque systems, and combinations of barium, cadmium, and zinc. Although there are a few calcium/zinc systems in limited use in food packaging, the tin stabilizers dominate the packaging field. Octyl tins are the principal systems used in food and pharmaceutical packaging (2). Stabilizers are also available that impart improved uv resistance and gamma-ray resistance. Resistance to gamma-ray sterilization is particularly important for packaging medical devices (see Health-care packaging; Radiation effects). With ETO currently under OSHA scrutiny, gamma-sterilizable PVC has become a new growth field now that compounding protection to four megarads can be achieved. All of the stabilizers mentioned above are used in rigid PVC at only very small loadings.

In contrast, impact modifiers may be present up to 15% of the product's weight. As a result, the proper selection and loading of impact modifier is an important compounding decision. Clear packaging films typically contain MBS impact modifiers because of their superior clarity, heat stability, and room-temperature efficiency. ABS modifiers are good for opaque products, and chlorinated polyethylene (CPE) and acrylics (see Acrylics) are often selected for outdoor applications and/or low temperature environments in opaque systems. Pigments may then be added to provide custom color and titanium dioxide is generally used at levels as high as 15% to provide the desired level of opacity. Fillers may be used for cost reduction in opaque systems, and in many cases, to improve such physical properties as impact strength, stiffness, and heat-distortion temperature. Present in very low levels are a variety of proprietary lubricants and processing aids that

Table 2. Typical Physical Properties of Rigid PVC (Clear)

Property	Test method[a]	Units	Values
specific gravity[b]	D 1505		1.30–1.36
yield (1.30 sp gr)	D 1505	in.²/lb(cm²/g)	
7.5 mil (0.19 mm)			2850 (40.5)
10.0 mil (0.25 mm)			2130 (30.3)
12.0 mil (0.30 mm)			1780 (25.3)
15.0 mil (0.38 mm)			1420 (20.2)
20.0 mil (0.51 mm)			1070 (15.2)
tensile strength (yield)	D 882	psi (MPa)	6500–7800 (44.8–53.8)
tensile modulus	D 882	psi (MPa)	$(2.5–4.0 \times 10^5$ (1723–2757)
elongation (break)	D 882	%	180–220
Izod impact (¼ in. or 6.4 mm)	D 256	ft·lbf/in (J/m)	0.5–20.0 (26.7–1068)
gloss, 20°	D 247		120–160
heat-distortion temperature (264 psi or 1.82 MPa)	D 648	°F(°C)	158–169 (70–76)
cold-break temperature	D 1790	°F(°C)	14 to −40 (−10 to −40)
WVTR (38°C, 90% rh)	DIN 53122	g/(100 in.²·24 h) [g/(m²·d)]	
7.5 mil (0.19 mm)			0.30 [4.7]
10.0 mil (0.25 mm)			0.20 [3.1]
surface resistance	DIN 53482	ohm	$10^9–10^{13}$
specific resistance	DIN 40634	ohm·cm	$10^{13}–10^{15}$
dielectric strength	DIN 40634	kV/mm	60–70
specific heat (20°C)		kJ/(kg·K)	0.8
thermal conductivity		W/(m·K)	0.16
linear thermal expansion		K⁻¹	$7.0–8.0 \times 10^{-5}$
Ir absorption[c] (3–18 μm)			various intensities

[a] Ds are ASTM test methods and DINs are German (Deutsche) Industrial Norm test methods.
[b] Indirectly related to amount of impact modifier. Increased opacity may raise to 1.40.
[c] 20 mil (508 μm) unmodified.

are necessary to facilitate processing and to provide desirable properties such as slip, denesting, and improved thermoformability for the film processor/packager. Flame retardants, antioxidants, coupling agents, antistatic agents, phosphite stabilizers, and a host of additional additives may be included if necessary. Because of this tremendous facility for custom compounding, and the variety of products that result from it, Table 2 must be considered only as a general guide to typical rigid PVC properties.

Film and Sheet Production Methods

Extrusion (see Extrusion) and calendering are the principal methods of producing rigid PVC for the packaging industry. Extrusion is used to produce very thin blown films (see Film, flexible PVC) as well as heavy gauge sheeting nearly 1-in. (2.54-cm) thick produced by sheet-die methods. Calendering requires a much greater capital investment, but it offers much greater production rates, superior gauge control (cross-direction and machine-direction) (±5%), superior cosmetic-quality including clarity, and much wider versatility in accommodating gauge and width changes. Calendered film and sheet generally has better dimensional stability, which provides thermoforming consistency throughout a given lot. Rigid calendered PVC is available in thickness of 2–45 mil (51–1143 μm) with gloss, matte, or embossed surfaces, either in rolls or in sheets up to about 60-in. (1.5-m) wide. Calendering is the principal means of processing rigid PVC film for packaging.

Calendering. In a modern rigid-PVC calendering operation, compounding is done by computer-controlled electronic scales that supply precise amounts of each ingredient to a high intensity mixer designed to incorporate all liquids into the resin particles and to secure uniform distribution of all powdered ingredients. Blending is generally done for a specific time period and to a specific temperature. The still-dry, free-flowing blend is then charged to a feed hopper where it is screw-fed into a continuous mixer such as an extruder or kneader. Under the action of this mixer's reciprocating screw in the confined volume of the mixing chamber, the blend begins to flux or masticate into the plastic state. It is then forced out of the barrel of the mixing chamber. The continuous strand may be chopped into small fist-sized buns of hot material or simply exit as a continuous rope. This material may then be directly conveyed to the calender, or it may first pass through a two-roll mill. The calender is a large unit, typically consisting of four or five heated rolls designed to process masticated PVC buns into a continuous web of designated width and thickness (see Fig. 1).

Figure 2 illustrates the typical "L" and "inverted L" configurations generally used for rigid and flexible production respectively (3). The calender rolls have separate temperature and speed controls as well as roll bending and crossing capabilities to control profile across the web. Proper use of these controls, as well as speed and stretch in the take-off train, allow the production of an extremely flat sheet with a profile tolerance less than ±5% across and down the web. Such control is maintained by continuous beta scanning equipment that traverses the web constantly and calls for adjustments in nip openings, skew, and/or roll bending. Such constant in-process

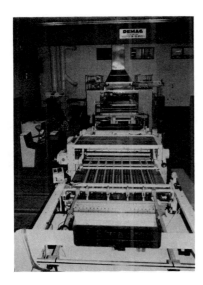

Figure 1. PVC calender.

monitoring and continuous profile adjustments is a significant advantage of calendering over other processing methods. By using special grit-blasting techniques, the third and fourth calender rolls may be custom-surfaced to generate a uniform two-sided matte product. Alternatively, one or more down-stream embossing stations may be utilized to produce a custom surface on one or both sides of the film. Antistatic and/or denesting slip agents may be applied to the surface(s) of the web after separating from the last calender roll. Finally, after the cooling section, the web is cut in-line into finished sheets or wound about a core into a master roll for subsequent custom slitting. Typical slit widths are made to the nearest 1/32 in. (0.8 mm) on 3- or 6-in. (7.6- or 15.2-cm) cores with roll diameters of 14–40 in. (36–102 cm).

Package Production by Thermoforming

Most commercial PVC packages are the result of thermo-forming roll stock into custom blisters. In those cases where further enhancement of PVC's own oxygen and/or moisture barrier properties are required, barrier materials (see Barrier polymers) such as PE, PVDC, or fluoropolymer film (see Film, fluoropolymer) may first be laminated to the PVC web prior to thermoforming. Thermoforming processing conditions are generally dictated by the PVC material itself regardless of lamination. Since PVC is an amorphous material (see Polymer properties), it softens over a large temperature range and has no sharp melting point. There are two temperature ranges in which rigid PVC can most readily be formed (see Fig. 3). It must be emphasized that these temperatures are actual film temperatures which must be measured with thermocouples located directly on the surface of the film (heating element temperatures are very much hotter).

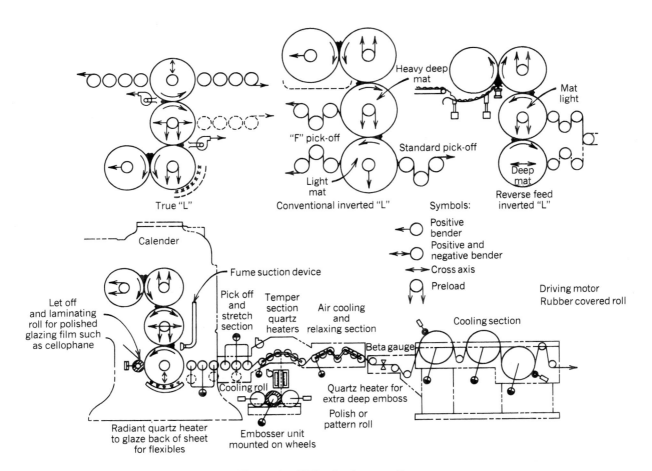

Figure 2. PVC calender operation.

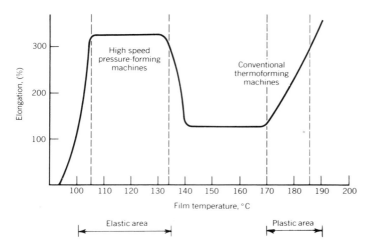

Figure 3. Temperature ranges for PVC thermoforming.

The first plateau at 221–275°F (105–135°C) is the elastic area best for most high speed form/fill/seal pressure-forming machines (see Thermoform/fill/seal). In this area, the film has sufficient hot strength to elongate or stretch to the conformation of the mold. Between 275 and 338°F (135 and 170°C) is a region of inadequate elongation for proper forming and attempts to process in this area may result in blowing holes, tear-offs, and poor definition. Failure to respect this "no man's land" is among the most common reasons for thermoforming problems seen in the field by PVC technical service representatives. The elongation quickly increases in the film at 338°F (170°C) and excellent forming is possible up to about 365°F (185°C). This is the range for optimum performance for conventional commercial thermoforming machines (see Thermoforming). Beyond 365°F (185°C), the material will sag excessively, resulting in webs, wrinkles, holes, or thin areas. In general, rigid PVC film thermoforms best when the film temperature is slightly above 338°F (170°C), with relatively cool molds (as low as 50–60°F or 10–16°C), and when plug assist temperatures are about 194°F (90°C). Mold design should avoid exceeding a 1:1 draw ratio and few extra degress of draft and extra radius on corners will help avoid problems.

Film shrinkage for optimum forming. Since distinct types of equipment are used in each of the two forming regions, it is particularly important that the PVC-film supplier have a thorough understanding of the appropriate shrinkages to be put into the feedstock for each type of machine. PVC destined for pressure-forming over the lower temperature region should have its shrinkage controlled by the producer at 284°F (140°C), a temperature which will release all plastic memory relevant to the processing range of 221–275°F (105–135°C). Typically best results are achieved with PVC film having 284°F (140°C) shrinkages of 2–5% in the machine direction and 0 to 1% in the transverse direction. Such very slight growth in the transverse direction is desirable to compensate for the repeating necking-in which may otherwise occur across each line of blisters such as those used for press-through packaging of ethical drugs.

In the higher temperature range for conventional thermoforming, film shrinkages should be controlled by the film supplier at 350°F (177°C), a temperature which will release all plastic memory relevant to this higher processing range. Best results are typically achieved with PVC film having 350°F (177°C) shrinkages of 4–8% in the machine direction and −1 to +1% in the transverse direction. Excessive shrinkages may cause the film to pull out of the chains, clips, or frame. Insufficient machine direction shrinkage and/or excessive transverse growth may lead to webbing. Thermoforming machine operators often respond to such webbing problems by reducing operating temperatures to prevent excessive sag, but this step can result in blowing holes and poor wall distribution because it takes the film out of the optimum thermoforming temperature range and elongation falls off rapidly. Careful shrinkage control is critical to successful thermoforming of rigid PVC film with all types of forming equipment. Proprietary process controls are available in modern PVC calendering technology to custom produce film with the proper shrinkages for optimum thermoformability within each temperature region and on each design of forming equipment.

Packaging Market

U.S. consumption of rigid PVC film and sheet in 1983 has been estimated at 256 × 10⁶ lb (116,000 t); 75% calendered and 25% extruded (4). In the same year, about 30% of this material was used in packaging (4) (see Table 3). The highest growth rates are in food, pharmaceutical, and medical packaging. Total 1990 consumption of rigid PVC for packaging is projected at nearly 200 × 10⁶ lb (90,700 t). Although the food-packaging market is clearly in a growth mode, with PVC competing with acrylonitrile (see Nitrile polymers) and polyesters (see Polyesters, thermoplastic), ethical-drug packaging is the fastest-growing PVC-packaging market. The popularity of the unit-dose "press-through" package is fueling the demand for both rigid PVC alone and rigid PVC with barrier coating(s) or laminates.

Medical packaging generally utilizes 20–30-mil (5.1–7.6-mm) rigid sheet for thermoformed blisters for packaging diagnostic test kits, disposable surgical instruments, etc. These blisters are often sealed with a permeable Tyvek (DuPont Company) backing for subsequent ETO sterilization. All-vinyl

Table 3. United States Markets for Rigid PVC[a]

	Market 10⁶ lb (10³ t)[b]
packaging (FDA grade)	
food	22 (10.0)
pharmaceuticals	25 (11.3)
medical supplies	27 (12.2)
	74 (33.5)
packaging (general-purpose)	
cosmetics/toiletries	4 (1.8)
pens/lighters	9 (4.1)
razors/blades	5 (2.3)
tools/hardware	23 (10.4)
floppy disk envelopes	25 (11.3)
cassette trays	8 (3.6)
	74 (33.5)
Total packaging (46.5%)	*148 (67.1)*
Total nonpackaging (53.5%)	*170 (77.1)*
Total rigid PVC	*318 (144.2)*

[a] Source: Ref 4 and Klöckner-Pentaplast.
[b] 1985 estimate.

packages, made from specially compounded PVC suitable for gamma-ray sterilization, are increasingly popular. Since these medical supplies are generally not ingested, the PVC does not necessarily have to be an FDA grade. These general-purpose thermoforming grades of PVC are also used in the other market areas cited in Table 3. Particularly noteworthy are the rapid growth rates and high volumes in floppy disk jackets and cassette trays. Rigid PVC dominates each of these areas. In addition, its excellent sparkle, clarity, and printability makes it an excellent choice for clear box lids and now, with the "soft-crease" process, for folding or setup boxes.

BIBLIOGRAPHY

1. A. A. Schoengood, ed., *Plast. Eng.* **32**(3), 25 (March 1976).
2. M. McMurrer, ed., "Update: PVC Heat Stabilizers," *Plastic Compounding*, Resin Publications, Inc., Cleveland, Ohio, July/Aug. 1980, pp. 83–90.
3. L. R. Samuelson, *Plast. Des. Process.* **21**(8), 14 (1981).
4. R. McCarthy, ed., *Film and Sheet Markets, World Analysis of the Vinyl Market*, Vol. XVII, Springborn Laboratories, Inc., Enfield, Conn., 1982, pp. 1–17.

General References

P. Bredereck, *J of Vinyl Technology* **1**(4), 218–220 (Dec., 1979).

Guide to Plastics, McGraw Hill, New York, 1979, p. 27.

M. McMurrer, ed., *Plastic Compounding 1984/85 Redbook*, Vol. 7, no. 6, Resin Publications, Inc., Cleveland, Ohio, 1985.

M. McMurrer, ed., "Update—U.V. Stabilizers," *Plastic Compounding*, Resin Publications, Inc., Cleveland, Ohio, Jan./Feb. 1985, pp. 40–57.

Thermal Stabilization of Vinyl Chloride Polymers, Tech. Rept. 3250, Rev. 7/68, Rohm & Haas, Philadelphia, Pa., 1968.

L. R. Samuelson, *Plastics Design and Processing* **21**(8), 13–15 (1981).

V. Struber, *Theory and Practice of Vinyl Compounding*, Argus Chemical Corp., New York, 1968.

H. J. G. VAN BEEK
R. G. RYDER
Klöckner-Pentaplast of America, Inc.

FILMS, PLASTIC

It is difficult to give a completely satisfactory definition of the difference between film and sheet. At one time anything up to 250 μm (10 mil) was generally recognized by the plastics industry as film (1). Anomalies arise when considering materials like unplasticized PVC or polystyrene, which are appreciably rigid at thicknesses of 150–200 μm (6–8 mil) (1). Recent technical developments have enabled satisfactory performance to be achieved with reduced thicknesses of plastic films and the upper figure of 250 μm now appears to be an academic one. Other proposed upper limits have included 75–150 μm (3–6 mil) depending on the stiffness of the plastic itself (2).

The range of materials now utilized as film formers is extremely large and cannot be covered fully in this article. The most important films are dealt with in other, more specific, articles in this encyclopedia. Other, more expensive high-performance materials are mentioned in the articles on coextrusions. In general, the usage of plastic films can be divided into single and composite films. The latter may be formed by coating a basic film (see Coating equipment), by lamination (bonding of two or more films with an adhesive or by heating) (see Laminating), and by coextrusion (3) (see Coextrusion, flat; Coextrusion, tubular; Coextrusions for flexible packaging; Coextrusions for semirigid packaging).

Packaging Films

Until about 1950 there were no large-tonnage thermoplastic packaging films. In the first half of the century, the dominant transparent packaging film was nonthermoplastic-regenerated cellulose (see Cellophane). In 1950, satisfactory low density polyethylene (LDPE) films were extruded for the first time (see Polyethylene, low density). A brief overview of the strengths and weaknesses of these two important, but very different, films is useful as a background to the developments that have taken place since midcentury.

Cellulose is a long-chain carbohydrate with no cross-linking. It is not thermoplastic, however, because the presence of a large number of hydroxyl groups in each molecule results in a high degree of hydrogen bonding and a consequent strong attraction between the chains (see Polymer properties). Exposure to heat does not soften the material. Charring occurs before the hydrogen bonding is overcome sufficiently to allow the chains to move. This means that cellophane (now a generic term except in the United Kingdom) cannot be heat sealed (see Sealing, heat), without a heat-seal coating. At first, cellophane was used only for decorative purposes. Its use in protective packaging was limited until the development of heat-seal coatings and moisture-proof coatings to improve its barrier properties. Cellophane is as clear as glass, has high tensile and bursting strength, and is a good barrier to oils, greases, and odors. The dry film is almost impermeable to gases but water-vapor permeability is high, and the gas permeability increases with moisture content.

LDPE does not have the tensile strength and stiffness of cellophane, but its flexibility and toughness permitted its use in applications where cellophane was useless (eg, heavy duty shipping bags) (see Bags, heavy-duty plastic). Also, its heat sealability opened up many other bag-packaging applications, including fresh produce, hardware items, toys, and many others (see Bags, plastic). Cellophane and LDPE can no longer be considered as competitors because of the many other films available, but they did compete years ago and much development was necessary before the limp LDPE film could be handled on high speed packaging machinery.

A more challenging competitor to cellophane was polypropylene. The nonoriented film entered the market in 1959 (see Film, nonoriented polypropylene). Polyethylene and polypropylene (PP) are both polyolefins, and both present profiles quite different from cellophane. However, PP differed in a few useful ways from LDPE because it had better transparency and gloss, twice the tensile strength, and higher heat resistance. Over the years, nonoriented PP has been used as an alternative to LDPE because of improved appearance (eg, in textiles packaging), and higher heat resistance (eg, in retortable pouches) (see Retortable flexible and semirigid packages). To put these three films in perspective, annual usage of LDPE film for packaging purposes in the United States is 2–3 × 10⁹ lb/yr (0.9–1.4 × 10⁶ t/yr). The use of cellophane peaked at 200–300 × 10⁶ lb/yr (91–136 × 10³ t/yr). Because nonoriented PP is more expensive than LDPE, so that its property profile is warranted in selected applications only, its use has not exceeded 100 × 10⁶ lb/yr (45,000 t/yr). Biaxial orientation of polypropylene improved properties such as clarity, impact strength, and barrier properties (see Film, oriented polypropylene). By offering comparable properties at lower cost, oriented PP films have replaced cellophane in many of its high volume applications (eg, cigarette wrapping, packaging of snacks, etc).

The assets of other packaging films can now be viewed against this backdrop. The discussion is limited to single (monolayer) films only. Laminations have been used for many years to provide the combined properties of films, paper, and foil (see Laminating; Multilayer flexible packaging), and in recent years, coextrusions have replaced monolayer films and laminations (see Coextrusions for flexible packaging).

Medium density polyethylene (MDPE). MDPE films are used as alternatives to LDPE where greater stiffness is required (eg, wraps for paper products) or higher heat resistance (eg, heat-seal layer of boil-in-bag packages).

High density polyethylene (HDPE). With its higher density and higher crystallinity, the property profile of HDPE film is quite different from that of LDPE (see Film, high density polyethylene). Tensile strength and stiffness are higher, elongation is lower, and haze is so high that the film is translucent to opaque. In its cast form, properties such as tensile strength and tear strength are highly directional, and this limited the use of HDPE for many years. Later, methods were developed for producing tough blown films from high molecular weight resins. Today, HDPE film has replaced LDPE in some applications where great toughness is required (eg, supermarket carryout shopping bags). HDPE has always been a useful alternative to LDPE where clarity is either not needed or not desirable, as in replacement of glassine (see Glassine) for bag-in-box cereals and crackers (see Bag-in-box, dry product), or where heat resistance is necessary, as in hot-fill bag-in-box constructions for chicken broth (see Bag-in-box, liquid product).

Linear low density polyethylene (LLDPE). Compared to conventional LDPE produced under high pressure, LLDPE is polymerized under low pressure (see Polyethylene, low density). Compared to conventional LDPE films, LLDPE films have greater puncture resistance and tear strength. They provide comparable properties in thinner gauges, which has led to large-scale replacement of LDPE by LLDPE (see Bags, heavy-duty plastic).

Oriented/irradiated polyethylene. Specialty films that are oriented and/or irradiated are useful as high clarity shrink films for consumer products (see Films, shrink).

Ethylene–vinyl acetate copolymers (EVA). Compared to LDPE, EVA copolymers have a greater elongation, greater elasticity, and higher impact strength. Barrier properties are lower, however. EVA copolymers have been most useful as stretch films (see Films, stretch).

Poly(methylpentene). This has a particularly high heat resistance but it has found only limited application as a packaging film because of its high gas and water vapor permeability.

Polybutylene. Polybutylene is tougher than LDPE but is more expensive and has found very limited use as a packaging film.

Ionomers. Ionomers are flexible, tough materials with extremely good clarity (see Ionomers). Because ionomers have relatively high melt strength, their drawing characteristics are extremely good. They perform particularly well, therefore, in extrusion coating (see Extrusion coating) and skin-packag-

ing (see Carded packaging) applications. Ionomers are also used in coextruded films because of their grease resistance in thin gauges.

Vinyls. The most common vinyl is poly(vinyl chloride) (see Film, flexible PVC; Film, rigid PVC; Poly(vinyl chloride)). Unplasticized PVC film is stiff and has a high tensile strength. Water vapor permeability is greater than that of LDPE but the gas permeability is lower. Plasticization of PVC film increases the limpness and softness and improves the low temperature properties (see Additives, plastic). Thin plasticized PVC film is widely used for shrink wrapping trays containing cuts of fresh meat and fresh produce.

Polyesters. Polyester films are very tough and strong, and are normally transparent (see Film, oriented polyester; Polyesters, thermoplastic). However, slip characteristics are poor unless slip additives are incorporated, and these make the film slightly hazy. Water-vapor permeability is of the same order as that of LDPE but gas permeability is lower than that of PVC. Uses for polyester film include the vacuum packaging of cooked meat products and boil-in-bag packs (laminated to MDPE), and bags for oven roasting meat and poultry. It is also one of the most useful metallized films (see Metallizing).

Styrene polymers and copolymers. Biaxially oriented PS film is used in some packaging applications that require high water-vapor transmission (see also Foam, extruded polystyrene; Polystyrene; Styrene-butadiene copolymers).

Nylon. In general, nylons are tough materials with high tensile strengths and high softening points (see Nylon). Nylon films have fairly high moisture vapor permeabilities but they are very good gas barriers (of the same order as PVC but not as good as polyester). Nylon films have been used for the vacuum packaging of foodstuffs, for boil-in-bag packs, and the packaging of surgical equipment for steam sterilization (see Healthcare packaging).

Polycarbonate. Polycarbonate has an outstanding combination of high impact strength, high temperature resistance and clarity (see Polycarbonate). The water-vapor permeability of PC film is higher than that of the polyolefins, PVC, and moisture-proof grades of cellophane. PC has been used for boil-in-bag packs and for skin packaging (when coated with LDPE).

Acrylic multipolymer. Acrylic multipolymer has good transparency, especially when in contact with liquids (see Nitrile polymers). It has low gas permeability, comparable to that of PVC, but water-vapor permeability is higher than that of LDPE or PVC.

Acrylonitrile–methyl acrylate copolymer. This material has good clarity and high barrier properties (see Nitrile polymers). In combination with LDPE it has been used for cheese and meat packs.

Fluoropolymer. Poly(chlorotrifluoroethylene) (PCTFE) has the lowest water vapor permeability of any polymer film (see Film, fluoropolymer). Gas-barrier properties are also good, nearly as good as those of nylon. The main use for PCTFE film is blister (strip) packaging of pharmaceutical tablets and capsules where a high degree of protection from water vapor is required (see Pharmaceutical packaging).

Production Methods

Extrusion. Most flexible films are produced by tubular (blown) or slit-die (cast) extrusion (see Extrusion). Tubular extrusion produces tubular film, which can be used in that form or slit to produce a lay-flat film. Casting produces a lay-flat film. The method of production affects film properties (see Extrusion).

Orientation. Some orientation always exists for both blown and cast film, due to the blowup ratio in blown film and directional flow properties in cast film. The deliberate production of a more balanced biaxial orientation, however, can produce marked property improvement (see Film, oriented polyester; Film, oriented polypropylene). Clarity and impact strength are among the properties improved. Particularly in the case of polypropylene, the barrier properties are also improved. A range of films is obtainable: biaxially oriented film with balanced properties (the same draw ratio in each direction); biaxially oriented film with unbalanced properties (different draw ratios in the two directions); and uniaxially oriented film. Uniaxially oriented films tend to fibrillate when stretched at right angles to the direction of orientation and this phenomenon is utilized in the production of film tapes (see Tape, pressure sensitive) and fibers.

In the cast-film process, orientation in the machine direction is achieved by feeding the film through a series of rolls running at gradually increasing speeds. These rolls are heated sufficiently to bring the film to a suitable temperature below the melting point. Transverse orientation is obtained by use of a tenter frame, which consists of two divergent endless belts or chains fitted with clips. The clips grip the film so that as it travels forward, it is drawn transversely at the required draw ratio. The tentering area is also heated, usually by passing the film through an oven. In the tubular process, transverse drawing is obtained by increasing the air pressure in the tube, the draw ratio being adjusted by altering the volume of entrapped air. The air is trapped by pinch rolls at the end of the bubble remote from the extruder which are run at a speed faster than the speed of extrusion, drawing the film in the machine direction.

Calendering. In this process continuous film or sheet is made by passing heat-softened material between two or more rolls. In essence, calendering consists of feeding a plastics mass into the nip between two rolls where it is squeezed into a film that then passes around the remaining rolls. It then emerges as a continuous film, the thickness of which is determined by the gap between the last pair of rolls. After leaving the calender the film is cooled by passing it over cooling rolls, and is then fed through a beta-ray thickness gauge. Compared to extrusion, calendering usually produces film with a better gauge uniformity. It is used mainly for the production of PVC film and sheet.

Solvent casting. In solvent casting, a solution of polymer is deposited on an endless belt, the solvent is driven off by heating, and the film is then stripped off the belt. After being stripped from the casting surface, the film is seasoned in a heated drying cabinet, cooled by passing over chilled rollers, and wound into reels. The solvent coating process is expensive, but cellulose nitrate film was made in this way because of the flammable nature of the polymer. Some cellulose acetate film is produced by solvent casting, and the process has also been used for producing vinyl chloride/vinyl acetate copolymer films.

Measurements

Thickness. In the Imperial system, used in the United States and to some extent in the United Kingdom, film thick-

ness is expressed in terms of mils: 1 mil = 0.001 in. The term "gauge" is sometimes used for very thin films: 100 gauge = 1 mil. For example, 0.00048 in. polyester film is called "48 gauge" film. In the metric system, thickness is expressed in micrometers, ie, 1 mil = 25.4 μm.

Yield. The cost of film cannot usefully be expressed in terms of price/pound because different films have different yields, dependent on specific gravity (density). The formula for calculating in.^{2}/lb is

$$\frac{27.69}{\text{specific gravity} \times \text{thickness (in.)}} = \text{in.}^2/\text{lb (14.2 cm}^2/\text{kg)}$$

Mechanical Properties

Tensile and yield strength, elongation, Young's modulus. These four related properties can be measured on the same equipment. Tensile strength (more properly Ultimate Tensile Strength) is the maximum stress that a material can sustain. Yield stress is the tensile stress at which the first sign of nonelastic deformation occurs. Yield strength is usually more important than ultimate tensile strength, particularly in wrapping or converting equipment where a sudden "snatch" could cause a nonreversible distortion with consequences such as out-of-register printing. Elongation is the elongation at breaking point and is important as a measure of the film's ability to stretch. Young's modulus is the ratio of stress to strain during the period of elastic deformation (ie, up to the yield point). It is a measure of the force required to deform the film by a given amount and is thus a measure of the intrinsic stiffness of the film.

Burst strength. The burst strength of a film is the resistance it offers to a steadily increasing pressure applied at right angles to its surface under certain defined conditions. The burst strength is taken to be the pressure at the moment of failure, and is essentially a measure of the capacity of the film to absorb energy.

Impact strength. Impact strength of a film is a measure of its ability to withstand shock loading. One method of measuring this is the Falling Dart Method. In one variation of this, the dart is dropped from a constant height and its weight is adjusted from a minimum (just too light to rupture the film test piece) up to a maximum (just heavy enough to cause rupture of all the test pieces). The weight at which 50% of the test pieces rupture, multiplied by the height through which the dart falls, is taken to be the impact strength of the film.

Tear strength. The usual tests for measuring tear strength actually measure the energy absorbed by the test specimen in propagating a tear that has already been initiated by cutting a small nick in the test piece with a razor blade. Tear strength requirements may be high or low according to the particular end-use. Where tear tapes are to be incorporated, ease of tear propagation in one direction is desirable; but such propagation must be avoided in shipping sacks, which might be punctured during transit. In all tear tests, test pieces should be cut from both machine and transverse directions because the tear strength can vary widely according to the direction of tear.

Stiffness. Stiffness, which can be considered to be the resistance of the film to distortion, particularly bending, depends on the film thickness and the inherent stiffness of the material. As mentioned earlier, Young's modulus is one measure of the intrinsic stiffness of a film. A more direct method uses the Handle-O-Meter Stiffness Tester. This simple test measures the force required to push a sample of film into a slot of a given width over which the film has been laid. The method is suitable mainly for thin film. For thicker films, stiffness can be measured by treating a film strip as a beam. The test strip is placed on two supports and the resultant beam is loaded at the center. The deflection produced for a given load is then measured. Measurements of stiffness are made in machine and transverse directions. Stiffness is a factor in machine running whenever the film has to bridge a gap during passage through the machine. Stiffness may also be important to the final application. For example, in packaging woolen goods, a limp film is preferred so that the softness of the contents can be appreciated.

Flex resistance. Resistance to repeated flexing or creasing is important in use. Some films are highly resistant, whereas other fail by pinholing or total fracture. Even if failure does not occur, certain properties may be seriously impaired. Resistance to flexing is measured by repeatedly folding the film, backwards and forward, at a given rate. The number of double folds up to the point of failure is recorded. Another way of testing is to subject the film to a given number of cycles in the test equipment and then compare the relevant properties with those of the uncreased film.

Coefficient of friction. The frictional properties of a film are important when the film comes into close contact with metal surfaces such as "ploughs," formers, guides, etc, which are used to preform tubes or to fold the wrapping material round the product. High slip (low coefficient of friction) is desirable in these situations. Friction is also a factor when film passes over free-running rollers. A coefficient of friction that is too low could lead to slippage instead of a positive drive of the rollers; a coefficient that is too high could lead to wrap-around of the roller should the film break.

One method of measuring the coefficient of friction utilizes a sled, made from a metal block wrapped around with a sheet of foam rubber. The film sample is taped to the rubber in order to give the sled base a smooth, wrinkle-free surface. If the film/metal coefficient is required, the sled is placed on a smooth metal table. For measuring the film/film coefficient, another sample of film is taped to the metal table. Either the sled or the table can be motor driven, and the horizontal force on the sled is measured, using a spring balance or a strain gauge. The original deflection of the load-measuring device is noted and used to calculate the static coefficient of friction by dividing by the weight of the sled. The average deflection is also noted and used to calcuate the dynamic (sliding) coefficient of friction.

Blocking. Blocking is the tendency of two adjacent layers of film to stick together, particularly when left under pressure for some time, as when films are stacked in cut sheets. Degree of blocking is determined by the force required to separate the two layers of blocked film, when the force is applied perpendicularly to the surface of the film. Blocking can also be assessed by measuring the force required per inch of film to draw a 1/4-in. diameter rod perpendicular to its axis at a rate of 5 in. per minute between the adhering films, thereby separating them.

Physical and Chemical Properties

Table 1 lists specific properties for many of the plastic films.

Optical properties. Optical properties are particularly important in these days of self-service selling. Light transmission is often quoted, but this gives no indication of image distortion and is, therefore, not particularly relevant for packaging films. "See-through" clarity, haze, and gloss are usually more relevant because of the influence of these properties on visibility of the contents and on the "sparkle" of the film.

Light transmission is the ratio of the intensity of a light source measured with the film interposed, to the intensity without the film, expressed as a percentage. A blurred image can be obtained even through a film with a high light transmission value. Such a film might be suitable as a tight overwrap, but might be unacceptable when the contents are distant from the film, as in a window carton. A better guide to image quality is given by "see-through" clarity, which indicates the degree of distortion of an object when seen through the film. One test compares the optical definition of a well-illuminated standard wire mesh grid, viewed through the test film, with a set of eight standard photographs covering the range of clarities normally encountered in the material under investigation. The test sample is then given a number corresponding to the standard it most closely resembles.

Haze is a measure of the "milkiness" of the film and is caused by light being scattered by surface imperfections or by inhomogeneities in the film. The latter can be caused by large crystallites, incompletely dissolved additives, voids, or cross-linked material.

Gloss is a measure of the ability of the film to reflect incident light in the same way as a mirror (ie, angle of incidence = angle of reflection) A high gloss will, therefore, produce a sharp image of any light source, giving rise to a pleasing sparkle on the film. Gloss is determined by measuring the percentage of the light, incident at an angle to the film surface, that is reflected at the same angle.

Permeability. Very often, one of the prime functions of a packaging film is to act as a barrier to gases and vapors (see Barrier polymers). Cookies and snack items, for example, need to be kept dry; conversely, cigarettes have to be protected from loss of moisture. Many items, particularly fatty foods, have to be protected from oxygen pickup. The measurement of permeability is, therefore, important. One standard test for water-vapor permeability consists of wax-sealing test specimens over the mouths of metal dishes containing a desiccant. The dishes are weighed initially and then placed in a temperature- and humidity-controlled cabinet. Weighings are carried out at regular intervals and the weight gains calculated. Permeability is calculated as weight of water vapor per unit area per unit time, at a given temperature and humidity. (see Table 1).

Measurement of gas permeability is carried out under controlled conditions of pressure, in addition to controlled temperature and relative humidity. The film is used as a partition between a test cell and an evacuated manometer. The pressure across the film is usually 1 atmosphere (101.3 kPa). As the gas passes through the film test piece, the mercury in the capillary leg of the manometer is depressed. After a constant transmission rate is attained, a plot of mercury height against time gives a straight line, the slope of which can be used to calculate the gas-transmission rate (see Testing, permeation and leakage).

Density. The density of a film sample is often measured by means of a density gradient column which is a mixture of two fluids of different density, the proportions of which change uniformly from the top to the bottom of the column. A series of calibrated glass floats covering the required density range is placed in the column. Film samples are cut and then immersed for 30 s in, for example, ethanol or methanol. Then they are placed into the column using tweezers, taking care not to in-

Table 1. Properties of Plastic Films

Property	Regenerated cellulose MS	LDPE	HDPE	Cast PP	OPP	PVC	Polyester	EVA	Ionomer
yield (m²/kg) (25-μm film)	27.4	42.6	41.2	44.0	44.0	28.4	28.4	41.9	42.0
tensile strength (MPa)	48–110	8.6–17	17–35	41.5	165–170	45–55	175	14	17–24
elongation at break (%)	15–25	500	300	300	50–75	120	70–100	650–800	300–400
water-vapor transmission (g/m²·d) (25-μm film at 90% rh and 38°C)	5–15	15–20	5	10–12	7	30–40	25–30	50–60	25–35
oxygen permeability (cm³/m²·d·atm) [cm³/m²·d·kPa] (25-μm film)	670 [6.6] (when dry)	6500–8500 [64–84]	1600–2000 [15.8–19.7]	3700 [36.5]	2000–2500 [19.7–24.7]	150–350 [1.5–3.5]	40–50 [0.4–0.5]	11000–14000 [109–138]	6000–7000 [59–69]
resistance to oils and greases	excellent	some swelling	good	excellent	excellent	excellent	excellent	excellent	very good

troduce air bubbles. After three hours, measurement is made, against a scale, of the heights of the floating samples, and of the markers against their respective densities. The heights of the floats are then plotted against their respective densities and the best-fitting curve drawn through the points. From this curve, the densities of the film samples can be read. Density is important as an indication of the film yield in in.²/lb (cm²/kg).

Heat sealability. The heat sealability of a packaging film is a very important property when considering its use on wrapping or bag-making equipment, and the integrity of the seal is important to the ultimate package. The strength of a heat seal is determined by measuring the force required to pull apart the pieces of film that have been sealed together. Two tests are in common use: dynamic and static. In both tests a 1 in. (25 mm) strip is cut through the heat seals. The dynamic test uses a sensitive tensile-testing machine, with the two free ends of the film strip placed in the machine clamps. The force necessary to peel apart the two pieces of film is then measured. In the static test the strips are hung from a frame with one free end clamped and the other attached to a weight. The seals are examined at intervals for signs of failure. The weight and the length of time to failure are both noted.

Water absorption. The effect of water absorption varies considerably from film to film, ranging from negligible, as with the polyolefins, to solution of the film, as with poly(vinyl alcohol). Water absorption is measured by immersing a sample of film in water for a given time (usually 24 h) at a standard temperature, and noting the change in weight.

Resistance to chemicals. The effect of chemicals on a packaging film is an important factor when assessing its suitability for containing a particular product. The normal test for chemical resistance is total immersion in the chemical under investigation, using specified conditions.

Resistance to light. Prolonged exposure to light may bring about many undesirable effects, such as brittleness, loss of clarity, surface imperfections, and color changes. Films intended for long use in sunlight should contain stabilizers such as uv absorbers. Tests are usually carried out by exposing samples of the film to light of a specified wavelength, or combination of wavelengths, for a given time and noting the effect on various properties such as tensile strength.

Resistance to heat. Resistance to heat may be necessary for a particular use, such as boil-in-the-bag foods. High temperatures may also lead to undesirable changes, especially for plas-

ticized film. Heat can cause loss of plasticizer which results in brittleness of the film.

Resistance to cold. Low temperature behavior of film is also important in many applications, including the packaging of frozen foods. Where low temperature use is envisaged it may be preferable to carry out tests on the actual package because stresses and strains in the film may influence its behavior.

Film Packaging Techniques

There are four basic styles of wrapping: hand-wrapping, assisted hand-wrapping, bagging, and mechanical wrapping (4). The packages produced by these styles of wrapping can be improved by the application of other processes such as shrinking, vacuum packing, etc. The two styles of wrapping that are most important to modern packaging are bagging and mechanical wrapping (see Wrapping machinery; Wrapping machinery, shrink film).

Wrapping. The mechanical wrapping machines introduced around the turn of the century were designed to use paper as the wrapping material. Seals were produced by adhesives and, later, by the use of wax-coated paper in combination with heat sealing (see Sealing, heat). The advent of coated cellulose film produced some complications, including the need for a higher sealing temperature [275°F (135°C) instead of 150°F (65°C)] (5). This meant that better thermostats had to be fitted to the sealing jaws in order to avoid scorching of the cellulose. When thermoplastic films like LDPE were used they added two important problems. Many were too limp to be handled by the type of equipment developed for paper and cellulose film. The problems of heat sealing were also complicated by the fact that the entire thickness melts when heated, whereas coated cellulose film and waxed paper have infusible substrates that support the molten material on the surface until it is cool. These problems have been overcome in a number of different ways. Machines have been designed that pull a limp film through the various stages of the equipment instead of trying to push it, and heat sealers have been developed that incorporate a cooling cycle. During this cooling period the film is supported until is strong enough to withstand the stresses encountered.

Bagging. The advent of plastic films brought changes to the bagging techniques used for paper. Paper bag manufacture involves folding, to a greater or lesser degree depending on the design, and the bags are sealed using adhesives (see Adhesives). The low rigidity of films like LDPE and the initial lack of suitable adhesives countered the use of similar production methods; but the ease of production of LDPE film in tubular form, coupled with hot-wire cutting and sealing, made the production of simple LDPE bags a viable proposition (see Bags, plastic). In addition to the simple bags made by a transverse seal on the tubing, a form of block-bottom bag can be made by using gussetted tubing. Later developments included the production of side-weld bags from flat film. The film is folded over a triangular plate and then intermittently cut and sealed by a heated knife. If the triangular plate is set off-center, the bag so produced has an easy-open flap. The filling of such side-weld bags can be partly mechanized using stacks of bags, drilled through the flaps and fitted over a peg ("wicket"). Opening of the top bag is accomplished by a jet of air and the items to be packaged are inserted down a chute formed by metal arms which swing into the opening. Closing of plastic bags can be achieved by twisting, taping, folding, heat sealing, and with plastic clips or wire ties (see Closures, bag).

Plastic films have also displaced paper in many areas of the heavy-duty bags field, particularly in packaging fertilizers (see Bags, heavy-duty plastic). Original open-mouth bags tended to be pillow shaped when filled, and these were difficult to stack. Some improvement was obtained by the provision of microperforations at the end of the bag, thus allowing internal air to escape during handling and movement on conveyor belts. Later, in the late 1960s, block-bottomed bags were developed, fitted with valves so that they could be filled on the equipment developed for valved-paper sacks. (6) (see Bag-making machinery).

Form/fill/seal. One of the outcomes of rethinking the mechanism of packaging machines to adapt them to plastic films was the development of form-fill-seal machines. A definition of the form-fill-seal operation is "A reel or reels of flexible packaging material is formed into a container, filled and sealed in one series of operations to produce a package, containing a predetermined quantity of product" (6). Form-fill-seal operations can be carried out in three main ways: (a) a web of material is formed into a tube which is filled and sealed at intervals; (b) a web of material is folded along its length, sealed at intervals to form a series of pouches (sachets) which are then filled and closed; and (c) a web of material is thermoformed to give a series of traylike depressions which are filled, and then sealed by means of a second web (7).

Equipment based on the operation described in (a) can be subdivided into horizontal and vertical machines. The horizontal machines are particularly useful for packaging cookies and sweet confectionery (see Form/fill/seal, horizontal). A continuous tube of film is formed around the product, and the longitudinal seal is made by carrying the material past heater blocks, followed by cold rollers. The ends of each pack are sealed and cut simultaneously, usually by rotary crimping sealers. The difference between vertical and horizontal machines is that on the horizontal machines the product acts as the former, whereas on vertical machines the film is formed around a metal tube and the product is delivered independently, through the forming tube (see Form/fill/seal, vertical). After the web of film has been formed into a tube by passing it around a vertically mounted forming tube it is sealed by a vertically mounted sealer bar. Heated crimp jaws, horizontally mounted, close on the formed web of film, making a transverse seal which forms the base seal of the pack. A measured weight or volume of the product is then allowed to fall through the forming tube into the partially formed pack. The crimp heat-

sealing jaws, while still closed, descend, drawing the pack and the web downwards by a predetermined amount. They then open and return to their original position, at which point they close again to form the top seal of the pack and the bottom seal of the one following. A cutoff knife separates the filled pouch from the web and the whole cycle is repeated.

Type (b) sachet-making machines can also be subdivided into horizontal and vertical types. On horizontal machines the sachet (pouch) is usually formed from a single web. This is passed over a triangular-shaped forming shoulder to produce a double thickness. Heat seals are made in the vertical plane and, in some instances, along the folded edge as well to produce a series of pockets along the web which are cut into individual sachets, opened, filled, and finally heat sealed along the top edge to complete the package. A wide range of products can be handled on such machines, including granules, powders, creams, and liquids. Most vertical sachet machines also use a single web of film, folded over a triangular forming shoulder as in horizontal machines. However, the action is different and resembles that of the horizontal machines of type (a). The web is folded to pass on both sides of a small diameter filling tube and sealed on three sides of the sachet (bottom and side seals) by reciprocating platens. The sachet is filled, via the filling tube, and the final top seal is made at the same time as the next set of side seals and bottom seal. The filled sachet is then separated by cutting the horizontal seal along the center line. The chief material requirement is rigidity, which enables the separated sachets to withstand the opening, filling, and sealing operations. Because the small-diameter filling tube must be introduced from the side, this type of machine is more suitable for the filling of liquids and creams which can be pumped (8).

The third main form-fill-seal operation is thermoforming (see Thermoform/fill/seal). In thermoforming, a heat-softened plastic film or sheet is formed either into or around a mold (10) (see Thermoforming). The forming is achieved by air pressure against the heated film or sheet, either directly (pressure forming) or by drawing a vacuum through the mold (vacuum forming). The process is well-adapted to form-fill-seal operations. In one type of machine, blisters are formed to a profile roughly similar to that of the article to be packaged. Articles are fed continuously into the blisters and a reel of fiberboard is sealed over them. The individual packages are then trimmed and separated (see Carded packaging). The other main use is the packaging of a wide range of foodstuffs, including jam, sauces, creams, and cheese. Two reel-fed plastic webs are used, the first being formed into a series of traylike depressions. This formed web is then indexed under a filling head and the filled compartments are lidded by sealing the second web of material on top (see Lidding). The web of lidded containers is then cut and the individual packs separated. The most common plastic films used in form-fill-seal thermoforming are PS and PVC but a thermoplastic-coated paper or aluminum foil can also be used for lidding. A somewhat similar technique is used for the packaging of pills, tablets, or capsules into "push-through" blister packs. (see Pharmaceutical packaging; Tamper-evident packaging).

Skin packaging. A number of other packaging techniques have been developed based on the various properties offered by plastic films. One of them is skin packaging, a vacuum forming process in which the objects to be packaged act as their own mold (9) (see Carded packaging). The articles are placed on a porous board, and the whole assembly is placed in the skin packaging equipment. Film is drawn from a reel and clamped in a frame set above the loaded board. A cutter severs the film from the reel, and a heater moves above the film. After a set period the heater retracts and the frame descends, draping the hot film over the articles on the board. A vacuum is then applied beneath the board and the film is drawn around the articles to be packaged and into intimate contact with the board, forming a permanent bond. Films used for skin packaging should be flexible enough to provide a good contour wrap and have a high melt strength to avoid pinholing and puncturing at sharp corners. LDPE and ionomer films are particularly suitable for skin packaging. PVC is also used, especially for heavy objects and those having sharp edges or corners.

Shrink wrapping. Another technique that utilizes a property specific to plastic films is shrink wrapping, which produces a tight wrap even on awkwardly shaped articles (see Films, shrink; Wrapping, shrink machinery). The principle on which shrink wrapping is based is known as "plastic memory." In other words, a film that has been stretched during manufacture (at a temperature above its softening point) and then cooled to "freeze-in" the consequent orientation of the molecules, tends to return to its original dimensions when reheated. Applications for shrink wrapping range from the collation of cans, bottles, or cartons to the wrapping of whole pallet loads. LDPE and LLDPE films are used most widely because they are cheap, tough, and waterproof. Where a clearer or more sparkling film is required, then PP or PVC are the more likely choices (10). The heating is usually carried out by means of hot air tunnels.

Stretch wrapping. A technique that achieves a similar result, but without the application of heat, is stretch wrapping (see Films, stretch; Wrapping, stretch machinery). Here, the film is first stretched around the article to be wrapped and then heat sealed. The residual tension in the film gives a tight contour wrap. The important stretch films are LDPE, LLDPE, EVA, and PVC, the choice depending on factors such as appearance, the protection required, and the susceptibility to damage by compression. LLDPE is making inroads into the stretch-wrap market because of cost/performance benefits. One advantage of stretch wrapping over shrink wrapping is that no shrink tunnel is necessary. Apart from the savings in capital cost, there are savings in running costs and a direct saving of energy.

Vacuum packing and gas flushing. When it is necessary to reduce oxidation of the package contents, the pack can be evacuated before sealing (see Controlled atmosphere packaging). Certain cooked meats such as bacon are often packaged in this way. The wrapping material must have a high resistance to the passage of oxygen and produce an effective heat seal. In practice a multilayer structure is generally used. One ply is chosen for its oxygen-barrier properties and its strength, the other for its heat-seal properties. The package, almost completely sealed, is evacuated through a small tube or inside a closed chamber before final sealing. In some instances oxidation can be slowed still further by displacing the air in the pack with an inert gas such as nitrogen or carbon dioxide. Among the many products now being packaged using gas flushing techniques are roasted peanuts, potato powder, dried milk products, and coffee (11).

Metallization. New applications for plastic films have been opened up by the technique of vacuum metallization (see Met-

allizing). Aluminum wire is heated in a vacuum to a temperature high enough to vaporize it. The vapor then deposits on the plastic as a very thin film, giving a bright decorative surface (12). The emphasis has now shifted from decoration to barrier performance aspects. According to work carried out in Japan (13), metallized films are less liable to loss of barrier properties when flexed than aluminium foil.

BIBLIOGRAPHY

1. J. H. Briston and L. L. Katan, *Plastics in Contact with Food*, Food Trade Press, London, 1974, p. 331.
2. C. R. Oswin, *Plastic Films and Packaging*, Applied Science Publishers Ltd., London, 1975, p. vi.
3. J. H. Briston, *Plastic Films*, 2nd ed., George Godwin, London, 1983, p. 282.
4. C. R. Oswin, in F. A. Paine, ed., *The Packaging Media*, John Wiley & Sons, Inc., New York, 1977.
5. Ref. 2, p. 138.
6. D. J. Flatman, in ref. 4, p. 3.83.
7. D. J. H. Giles, in F. A. Paine, ed., *Packaging Materials and Containers*, Blackie, London, 1967, p. 311.
8. Ref. 7, p. 319.
9. J. H. Briston, *Converter* **20**(6), (June 1983).
10. D. C. Miles and J. H. Briston, *Polymer Technology*, 2nd ed., Chemical Publishing Co., Inc., New York, 1979, p. 594.
11. W. Guise, *Converter* **20**(8), p. 17 (Sept. 1983).
12. J. H. Briston, *Converter* **20**(3), 10 (Mar. 1983).
13. *Packaging, Japan* **3**(10), (Apr. 1982).

General References

J. H. Briston, *Plastic Films*, 2nd ed., George Godwin, London, 1983.

C. R. Oswin, *Plastic Films and Packaging*, Applied Science Publishers, Ltd., London, 1975.

O. J. Sweeting, ed., *The Science and Technology of Polymer Films*, Interscience Publishers, New York, Vol. 1, 1968, Vol. 2, 1971.

F. A. Paine, ed., *The Packaging Media*, Wiley, New York, 1977.

D. C. Miles and J. H. Briston, *Polymer Technology*, 2nd ed., Chemical Publishing Co. Inc., New York, 1979.

J. H. BRISTON
Consultant

FILMS, SHRINK

Almost all plastic films shrink to some extent (see Films, plastic). In most packaging applications, dimensional stability is prized, and shrinkage is a detriment. In some applications, however, controlled shrinkage is a major asset. Shrink films are used, for example, to package meats and poultry, to produce a tight pilferproof wrap over cartons and boxes, to unitize shipments for distribution packaging. A variety of shrinkable films have been developed to meet the requirements of many different end-use applications.

Heat-shrinkable films are made by stretching (orienting) a conventional film at a temperature close to its softening point (T_g) and then quenching (freezing) the film in the oriented state. What happens is depicted in the following figures. Figure 1 is a schematic representation of a single long-chain polymer molecule in the relaxed and considerably curled state that is characteristic in unoriented film. If the film is stretched in one direction (uniaxially oriented) while it is heated, the ran-

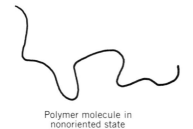

Polymer molecule in
nonoriented state

Figure 1. Polymer molecule in nonoriented state.

domly twisted and intertwined molecules line up roughly as depicted in Figure 2. If the film is stretched not just in the machine direction, but in the transverse direction as well, molecules will line up in both directions, as shown in Figure 3. There will be some orientation of molecules at intermediate angles, but the majority of molecules will line up as depicted.

Orientation is imparted by blown-film extrusion (see Extrusion) or on tenter frames. When the film is quenched after orientation, the molecules are "frozen" in the oriented position; but they still "remember" their original shape, and will return to that shape if the film is reheated to the orienting temperature. When this happens, the film shrinks around the product. The concept can be demonstrated by stretching an ordinary rubber elastic band to about three times its original length and freezing it in this shape with a CO_2 fire extinguisher. The band will remain stretched until it warms up to room temperature, when it will shrink back to its original state. The concept as applied to plastic films is similar, but the shrink temperatures are higher than room temperature.

If, instead of quenching, the film is reheated near its softening temperature and allowed to cool, the molecular "memory" is destroyed, resulting in a biaxially oriented film with no shrinkage properties but with better physical properties than the original film. Most of the BOPP (see Film, oriented polypropylene) and oriented-polyester (see Film, oriented polyester) packaging films are "heat-set" in this way, and their dimensional stability is a major asset. The excellent dimensional stability of BOPP was, for example, one of the reasons why BOPP replaced cellophane for soft cigarette packages in the United States. Cellophane (see Cellophane) has excellent dimensional stability too, but it shrinks about 3–5%, and that was enough to crush soft packs under the influence of temperature and humidity changes. The changeover in Canada was slower because rigid flip-top box packages are more popular and the small amount of cellophane shrink and tight

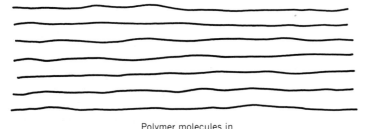

Polymer molecules in
uniaxially oriented film

Figure 2. Polymer molecules in uniaxially oriented film.

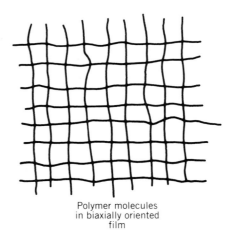

Polymer molecules
in biaxially oriented
film

Figure 3. Polymer molecules in biaxially oriented film.

cling enhanced the appearance of the package. Both uniaxially and biaxially oriented PP and PET films are available in non-heat-set grades that are used in special shrink packaging applications (eg, phonograph records for OPP, some meat and poultry for PET).

Shrinkable films. Blown LDPE film is the least expensive shrink film, used for relatively undemanding short-term applications, but there are other films available for more demanding uses. The development of the shrink-film concept started with the more demanding applications. Natural latex was used in France in the 1930s to shrink wrap perishable foods. The first major commercial breakthrough came in 1948, when Dow Chemical developed Saran vinylidene chloride resins (see Vinylidene chloride copolymers) and the Dewey & Almy (now Cryovac) Division of W. R. Grace used them to make blown shrinkable films that could be used for frozen poultry. The birds were placed in a Saran bag, a light vacuum was drawn, and the bag was closed. Because Saran shrinks at the temperature of boiling water, immersion of the bagged bird in boiling water produced an air-tight package that eliminated the risk of freezer burn.

Later, Cryovac turned its attention to the concept of irradiating polyethylene before blowing into film. Because radiation cross-links the molecules (see Polymer properties), the film can be stretched without becoming fluid at the melting point of a nonirradiated film. Compared to nonirradiated LDPE, tensile strength, shrink tension, and shrink percent are all greatly improved. The concept was broaded to include irradiated versions of other polymers.

DuPont Company took a different approach, blending LDPE and HDPE resins. HDPE film is not used as a shrink film because its very rapid crystallization limits the extent to which it can be stretched (see Film, high density polyethylene). When the resin is blended with LDPE, the rate of crystallization is slowed; but the crystallinity imparts properties similar to those achieved through radiation cross-linking. HDPE/LDPE blends (roughly 70:30) form the basis of DuPont's line of Clysar films. Today, Clysar grades are produced from other polyolefins as well and some are irradiated. Non-heat-set BOPP shrink films on the market today are members of this family. Property differences between conventional and irradiated LDPE are shown in Table 1. Another important

Table 1. Some Typical Shrink Film Properties

Film type	Tensile strength, psi (MPa)	Elongation, %	Tear strength, gf/mil (mN/m)	Maximum shrink, %	Shrink tension, psi (MPa)	Film shrink temp. range, °F (°C)
polyethylene (low density)	9,000 (62)	120	8 (3.1)	80	250–400 (1.7–2.8)	150–250 (65–120)
polyethylene (low density-irradiated)	8,000–13,000 (55–90)	115	5–10 (1.9–3.9)	80	400 (2.8)	170–250 (75–120)
polyethylene (copolymer)	19,000 (131)	130	7 (2.7)	50	450 (3.1)	180–260 (85–125)
polypropylene	26,000 (179)	50–100	5 (1.9)	80	600 (4.1)	250–330 (120–165)
polyester	30,000 (207)	130	10–60 (3.9–23.2)	55	700–1500 (4.8–10.3)	170–300 (75–150)
poly(vinyl chloride)	9,000–14,000 (62–97)	140	Variable	60	150–300 (1–2.1)	150–300 (65–150)

Table 2. Shrink Films, Performance Comparisons

Film type	Advantage	Possible problems
polyethylene (low density)	strong heat seals; low-temperature shrink; medium shrink force for broad application; lowest cost	narrow shrink temperature range; low stiffness; poorer opticals; sealing wire contamination
polypropylene	good optical appearance; high stiffness; high shrink force; no heat sealing fumes; good durability	high shrink temperature; high shrink force, not suitable for delicate or fragile products; brittle seals; high sealing temperature
copolymers	strong heat seals; good optical appearance; high shrink force; no heat sealing vapors	high shrink force, not suitable for fragile products; higher shrink temperature; higher heat-seal temperature; lower film slip-may give machine problems
poly(vinyl chloride)	lowest shrink temperature range; wide shrink temperature range; excellent optical appearance; controlled stiffness by plasticizer-content control; lowest shrink force for wrapping fragile products	weakest heat seals; least durable after plasticizer loss; toxic and corrosive gas emission from heat sealing, good ventilation required; durability problem at low temperature; low shrink force inhibits use as a multiple-unit bundling film; low film slip causes machine wrapping difficulties
multilayer coextrusions	excellent optical appearance; good machineability; low shrink temperature	In coextruded films, one ply compensates for the deficiencies of the other. As a result, they are superior films with no significant performance shortcomings. The wide variability in layer composition and number of layers makes performance analysis difficult.

difference is in creep resistance. Irradiated films retain shrink tension for a longer time. The special polyolefin films are used as food wraps and also as wraps for phonograph records, tapes, housewares, and many other consumer products.

PVC shrink films are used for many of the same applications. As shown in Table 1, the shrink tension of PVC is lower than the special polyolefins, but this can be an asset in packaging fragile products (see Table 2). PVC shrink (and stretch) wraps are also used for in-store packaging of red meats and produce.

Packaging equipment. A shrink film is loosely applied to the product, either as a premade bag or as a heat-sealed overwrap, and then passed through a heated tunnel where hot air impinging on the film shrinks it tightly around the product. The wrapping operation can be done by hand, or by semiautomatic or fully automatic equipment. Applications for premade bags range from the popular poultry bags to very large bags for shrink wrapping palletized loads of heavy durable goods. One of the more popular semiautomatic shrink packaging systems operates by passing the center-folded shrink film through an "L"-type sealer (see Wrapping machinery, shrink film). Shrink films can also be run on fully automatic overwrap machines utilizing folded lap seals. This system requires good heat sealability and good machine processing characteristics, including high stiffness and high slip. Limp, tacky films such as PVC can present problems on overwrap machines.

BIBLIOGRAPHY

General References

C. R. Oswin, *Plastic Films and Packaging,* Elsevier Applied Science, London, 1975.
C. J. Benning, *Plastic Films for Packaging: Technology Applications and Process Economics,* Technomic Publishing Co, London, 1983.
J. H. Briston and L. L. Katan, *Plastics Films,* Longman, London, 1983.

GEOFFREY BIDDLE
Consultant

FILMS, STRETCH

Stretch film, as defined in this article, is a stretchable, elastic, continuous thin plastic film which is stretched and wrapped around one or more items to protect them from the environment or unitize for handling, storage, or shipping. The specific films discussed are those used in materials handling for unitization of pallet loads of goods. This application used about 185 million (10^6) lb (83,900 t) of stretch film in 1985 in the United States. Stretch film is also used to some extent for bundling smaller units and for unitizing long, unwieldy items such as doors, carpeting, and lumber. Stretch films for supermarket-tray wraps are not discussed (see Film, flexible PVC).

Stretch film entered the U.S. market in the early 1970s as a replacement for shrink wrap (see Films, shrink) used to unitize nonreturnable glass bottles. The most common type of stretch film is the cling type, which is easy to use since the wrapping is completed by cutting the film between the load and the film roll, and merely wiping the loose tail of film against the load. The film-to-film adhesion, ie, cling, holds the tail in place. Other, less common, ways of attaching the film tail are adhesives, heat sealing, mechanical fasteners, and tying.

Pallet stretch films are used to unitize or protect a tremendous diversity of goods. They are used to a great extent to unitize products in corrugated boxes, but are also used for bulk shipments of glass and plastic bottles; shipping bags of chemicals and pet foods; building products such as doors, window frames, and foam insulation; automotive parts and garden implements; books and magazines; large rolls of many kinds, eg, carpeting, paper, and aluminum; and a great many other products.

The list shows that load characteristics vary widely, as do the demands placed on the stretch film. The load can be heavy, fragile, unstable, compressible, irregular in shape, or have sharp protruding corners. It can be sensitive to moisture, temperature, dust, uv light, or abrasion, or susceptible to damage by static electrical charges (1). Versatility is one of the chief advantages of stretch film compared with other methods of unitization such as strapping, gluing, or shrink bags. A user can wrap an endless variety of products and load configurations with stretch film stocked in a single size (2). Today's films can perform well at a wide range of tensions, and different load requirements can generally be handled by a simple adjustment to the wrapping-machine controls. The advantages and disadvantages of stretch wrap compared with other unitization methods are summarized below.

Advantages. Compared with shrink film and bags, stretch wrapping affords large energy savings and does not require the availability of fuel. Compared with shrink bags in particular, it simplifies inventory. Compared with strapping, stretch wrapping eliminates the need for corner boards and prevention of cutting or crushing of the load, since the load-holding force is adjustable and distributed over the film width. Stretch wrapping offers better protection from hostile environments, and clear, tinted, or opaque films are available for product identification or pilfer protection. Compared with both shrink films and strapping, stretch films are better able to withstand shock and vibration owing to their elasticity and memory.

Disadvantages. Stretch wraps generally have less moisture resistance than shrink bags, but a top sheet can be dropped over the top of the load for added protection before stretch wrapping. The cling property that makes shrink wrapping possible can also promote load-to-load sticking and abrasion. Stretch film cannot be used to compress a pallet load in the vertical direction since it has its primary holding force in the direction of wrap, ie, horizontal.

Stretch Films

The selection of the proper film for the application is generally based on the answers to the following questions:

Has the film been proven in actual field tests to perform the intended function?
What is the final cost per load, including film, labor, product damage, energy, and changeovers?

Since the film represents a significant cost to the user, the overall cost is the primary consideration in film selection. Other considerations include film consistency and supplier technical support and service level.

The polymers used for stretch wrap include LLDPE and conventional LDPE (see Polyethylene, low density), ethylene–vinyl acetate copolymer (EVA), PVC (see Film, flexible PVC; Poly(vinyl chloride)), and, to some extent, polypropylene (PP) copolymers (3–8). The base materials can be blended or coextruded in various ratios to obtain desired film properties. LLDPE is relatively new, but it is gaining wide acceptance as a base resin because of its excellent tensile strength, elongation, tear resistance, and puncture resistance. Most premium high performance stretch films contain LLDPE. Conventional LDPE, the second-oldest stretch-wrapping resin, is used to only a limited extent. EVA copolymers, which have very good stretch properties, most commonly have VA content of 3–12%. PVC was the first stretch film used in the United States. A disadvantage of PVC compared with the polyolefins relates to stress retention. After 16 h, the polyolefins retain ca 65% of their holding force, and PVC ca 30%.

Additives (see Additives, plastics) include cling agents, ie, tackifiers, such as glyceryl mono-oleate and polyisobutylene (PIB) and antioxidants such as t-butylated hydroxytoluene (BHT) and t-butylated hydroxyethylbenzene (BHEB) to protect against degradation during processing (3,5,7–9). Other additives include antistatic agents, antiblock agents which prevent the layers of film on a roll from becoming permanently bonded together; uv inhibitors to extend the life of film used outdoors, and colorants for load identification or pilferage prevention. Each base resin can be selected from several melt indexes (viscosities), densities, molecular weight distributions, and/or percent comonomer. The final film properties and processing characteristics depend on the combination of base resin and additives.

Important properties (see Films, plastic) for a stretch film are very high elongation, minimal neck-in when stretched, high ultimate tensile strength in the machine direction, high puncture strength, low stress relaxation, high elasticity, high creep and fatigue resistance, high resistance to tear propagation, and good cling properties. Film performance depends in part on the polymer and partly on the manufacturing method.

There are two primary physical differences between cast and blown films (see Extrusion) that affect stretch-film performance. One stems from the method used to cool and solidify the molten plastic. Cast film is cooled rapidly by one or more highly polished rolls containing circulated chilled liquid as the heat-exchange medium. Blown film is cooled less rapidly, generally by chilled air having a high velocity. Cast film has more uniform gauge and higher cling and clarity than film of the same material produced by the blown process.

The other difference is molecular orientation. Because cast film is stretched to some extent in the machine direction during the production process, it has monoaxial orientation in that direction. This means that cast film has a higher ultimate tensile strength in the machine direction (MD) than the transverse direction (XD). Cast films have a low tendency to propagate tears, ie, zipper, in the XD, thus reducing the probability of film breakage during wrapping, storage, and shipment. Blown film is stretched in both directions during manufacture and is more prone to XD tear propagation than the same material produced by the cast process.

Coextrusion is the simultaneous extrusion of two or more layers which are brought into contact while still molten and bond permanently together upon cooling (see Coextrusion machinery flat; Coextrusion machinery, tubular). This process offers tremendous opportunity to use combinations of resins and additives which would not be compatible with each other when used in a simple mechanical blend in a single layer. One

or more of the coextruded layers may be an adhesion-promoting material for resins which do not adhere well to each other. All of today's high performance stretch films are coextruded.

Selection Criteria

Temperature. High ambient temperatures can cause stretch films to relax and lose their original holding force more than they would at standard conditions. The greatest amount of relaxation under standard conditions occurs within 24 h of wrapping. Generally LDPE-, EVA-, and LLDPE-based films retain 60–65% of their original holding force, whereas PVC typically retains ca 25% at ambient temperatures (10). Low temperatures can reduce properties such as cling, toughness, and stretch. Cling levels are reduced by low temperatures in many films, although some sources claim adequate cling performance at −40°F (−40°C) (7). Most successful stretch films are usable between −20 and 130°F (−29 and 54°C).

Humidity. Elevated humidity can sometimes enhance film cling because some cling additives work by attracting moisture from the atmosphere. The same principle that makes it difficult to separate two smooth surfaces having a moisture layer between them is at work in stretch film used in high humidity. Low humidity can result in a higher static cling, but this cannot be depended upon to effectively hold the film in place during storage or shipping because the static charges dissipate with time.

Dirt. Dust and dirt are detrimental to cling. Airborne particles that settle between the film layers and prevent them from contacting one another can dramatically lower cling levels. Stretch films used in this type of environment must sometimes rely on other attachment means, such as glue, heat sealing, or tying.

Stretch-wrapping equipment. The type of application equipment determines to a great degree how a film will work in a specific application (see Wrapping machinery, stretch film). The equipment–film combination is determined primarily by the load characteristics and throughput required, ie, pallets/h or pallets/d. Equipment choices include spiral rotary (about 75–80% of all equipment in use), full-web rotary, hand wrappers, pass-through wrappers, and circular rotating wrappers (11).

Even in well-maintained systems, the variation in stretch due to various types of conventional braking systems can be as high as 16%. This means that equipment can make the difference between success and failure for a film, with properties that are marginal for the application. Certain types of equipment can utilize only a small percentage of the stretch available in a film before the load crushes. Core-lock braking systems, surface-controlled pre-stretch, and electromagnetic- and magnetic-particle brakes all vary in their ability to consistently obtain a particular yield from a film (12,13).

Conventional stretch-wrap equipment does not allow high stretch levels because high stretch force applied to the load can cause crushing. Also, the tension in the film between the brake and the load, ie, about six ft, results in severe neck-in, ie, loss in XD width when stretched in the MD.

Pre-stretch allows the film to be stretched in isolation from the load and then applied to the load. The film drives a pair of rollers which rotate at different speeds and do most of the film stretching; typically, 60–100% stretch is obtained. Stretch varies slightly with film-roll diameter. Some secondary stretch occurs between the rollers and the load. Film choices for this equipment are limited to films providing these stretch levels on a consistent basis.

The newer power-assisted prestretch equipment can obtain very high, ie, up to 250%, elongations. The tension of the film to the load can be set independently of the stretch force, and neck-in is minimal because stretch takes place over ½ in. (13 mm) instead of several feet (ca 1–1.5 m) (12). These machines obtain the highest performance and yield from a given film and generally provide the lowest cost per wrapped load.

Efficient film utilization depends in part on the machine operator, who should be trained to minimize film cost and avoid a tendency to overpackage, ie, use more film than necessary. Different operators may adjust the equipment to their own preference, and this can have a substantial effect on film cost per pallet. The abuse given to the roll of stretch film by the operator during roll changes can result in edge nicks or crushed core ends which can render entire rolls of film unusable.

Load configuration. The shape, stability, fragility, and crushability of the load to be wrapped plays a key role in the performance of the film selected. Lightweight, uniformly shaped loads can be wrapped with less lower cost film than heavy, irregularly shaped loads with sharp protrusions which create a puncture hazard. The latter dictates a premium quality film designed to withstand these extra demands.

Wrapped-product sensitivity. Film formulations containing additives, such as PVC plasticizers, which may migrate onto the wrapped articles or react with them chemically are a real concern (14). Stretch wraps have been known to remove the graphics from beverage cans where there is contact. In general, the wrapped product's sensitivity to chemicals, light, heat, moisture, and dirt can play a role in the selection of the proper film–equipment combination.

Packaging and shipping. The method of packaging the stretch film itself is extremely important to the way in which the film ultimately performs. Because of the high demands placed on the film by today's equipment, any slight scuff or edge nick can result in film breakage during wrapping. Packaging methods vary from individually boxed rolls to use of closed-cell expanded polystyrene foam (EPS) cradles, which provide the ultimate in roll protection and convenience. Cradles have the advantage of providing safe storage method for partially used rolls during roll changes. The operator can easily set a partial roll in the cradles without the scuffing that is unavoidable when a roll is slid back into a closely fit corrugated box.

Film supplier or manufacturer. A key consideration in selecting a stretch film is the manufacturer of that film. A user must be certain that truckload quantities of film will consistently perform as well as the single-roll samples made available for pretesting during supplier or film qualification. All reputable film suppliers have a full-time quality-control staff with modern testing equipment, as well as technical field-support personnel.

Standard tests. The stretch-film user should be aware of the limitations of standard tests for film properties. The basic problem with published laboratory test data of individual film properties is that they are obtained independently of every other property, under laboratory conditions, on equipment which does not necessarily approximate the stresses or rates which the film will undergo in the real world.

Table 1. Cost of Stretch Wrapping

Pallets wrapped per 8-h shift	Film used per 8-h shift, lb (kg)	Film cost per 8-h shift
1–10[a]	1–10 (0.45–4.5)[b,c]	$0.80–$8.00[b]
10–100[d]	5–50 (2.3–23)[c]	$4.75–47.50[c]
100–400[e]	50–200 (23–91)[c]	$47.50–190.00[c]

[a] Manual hand-held wrapper.
[b] Assumes 16-oz (453-g) film per pallet at $0.80/lb.
[c] Assumes 8-oz (227-g) film per pallet at $0.95/lb.
[d] Semiautomatic wrapper with prestretch.
[e] Conveyorized automatic wrapper with prestretch.

Property charts might indicate, for example, that film A breaks at 2000 psi (13.8 MPa), and film B at 4000 psi (27.6 MPa). This information does not necessarily indicate which is the better film. The film with the lower breaking strength might have four times the elongation at break. The point is that all of the test results must be reviewed together; even then, only the most general of conclusions can be drawn about one film versus another. Real-world wrapping and shipping tests should be done on actual loads with actual equipment before a film-purchase decision is made.

Pricing and Economics

The total cost of using stretch film includes not only the cost of the film, but the labor and equipment operating costs less any savings in product spoilage experienced by using the film. Stretch films are typically sold direct or through distributors at $0.80–1.10/lb ($1.76–2.43/kg). The common practice is to sell by the roll and not by weight. The price depends on the film gauge, resins and additives required for the application, and volume. Representative film usage rates and costs are shown in Table 1.

In analyzing stretch-film costs, one must not draw quick conclusions based on price per roll. The lowest cost per roll does not necessarily mean the lowest cost per skid wrapped. High performance (higher cost per roll) films, when used on wrapping equipment that prestretches the film, can result in cost savings of more than 50% over films wrapped conventionally (13). Therefore, film cost should be determined by wrapping the load, conducting shipping and storage tests to verify load integrity and spoilage rates, and cutting the film used off the skid and weighing it to determine actual film usage. Since the cost and weight of a full roll of film are known, it is a simple matter to find the film cost for one skid.

State of the Art

The stretch-film market tends to be segmented into three unofficial categories. A relatively undemanding, low volume Class 1 application uses films with low stretch, ranging from almost none up to ca 75%. This film is used to wrap regular, irregular, or randomly shaped loads using relatively slow wrapping equipment. Just a few years ago, stretch percentages higher than 30% could not be achieved because of equipment limitations, but the situation today is much different. In the intermediate Class 2 area, films are stretched up to 120%. Irregularly or randomly shaped loads, including heavy or shifting products, tend to fall into this category. The equipment is typically semiautomatic conveyorized higher speed equipment that places greater demands on the film. Class 3 is the high performance area. Films are stretched up to 250%, generally at high speeds on conveyorized, automatic equipment. Loads handled are of the most demanding types. Films in this class are usually coextrusions that incorporate LLDPE.

Trends

Lighter gauge films are continuing to replace thicker, lower performance films. In the 1970s, the standard stretch film was 1 mil (25 μm) thick. Today, 0.8 mil (20 μm) is the most common, with some films as low as 0.5 mil (13 μm) thick. Availability of new additives and base resins will result in higher performance films, offering lower cost per load owing to increased stretch levels. New equipment and applications requiring higher stretch levels will result in the development of new films. Coextrusions will continue to dominate the market, but full-web films may replace spiral wraps because they reduce wrapping time. Equipment will dictate film-usage shifts. Automatic equipment is expected to become more prevalent as labor costs increase. This will demand more consistency and performance from stretch films, as downtime caused by film breakage represents a higher number of unwrapped pallets.

BIBLIOGRAPHY

1. G. Schwind, *Mater. Handl. Eng.*, 70 (Oct. 1984).
2. *Packag. Dig.* **19**, 43 (Mar. 1982).
3. U.S. Pat 4,430,457 (Feb. 7, 1984), D. V. Dobreski (to Cities Service Co.).
4. U.S. Pat. 4,222,913 (Sept. 16, 1980), B. A. Cooper (to Bemis Co., Inc.)
5. U.S. Pat. 4,073,782 (Feb. 14, 1978), H. Kishi and co-workers.
6. U.S. Pat. 4,379,197 (Apr. 5, 1983), C. Cipriani and H. J. Boyd (to El Paso Polyolefins Co.).
7. U.S. Pat. 4,311,808 (Jan. 19, 1982), C.-S. Su (to Union Carbide Canada, Ltd.).
8. U.S. Pat. 4,436,788 (Mar. 13, 1984), B. A. Cooper (to Bemis Co., Inc.).
9. U.S. Pat. 3,986,611 (Oct. 19, 1976), D. H. Dreher (to Union Carbide Corp.).
10. D. O. McLeod and R. E. Shaw, *Mater. Handl. Eng.*, 2 (Dec. 1976).
11. U.S. Pat. 4,317,322 (Mar. 2, 1982), P. R. Lancaster and W. G. Lancaster (to Lantech, Inc.).
12. U.S. Pat. 4,387,548 (June 14, 1983), P. R. Lancaster and W. G. Lancaster (to Lantech, Inc.).
13. U.S. Pat. 4,418,510 (Dec. 6, 1983) P. Lancaster III and W. G. Lancaster (to Lantech, Inc.).
14. U.S. Pat. 4,409,776 (Oct. 18, 1983), A. Usui (to Shinwa Kagaku Kogyo Kubushinki Kaisha).

General References

ASTM Task Group D10.14 on Stretch, Shrink & Net Wrap Materials, *Standard Guide For The Selection of Stretch, Shrink, and Net Wrap Materials*, The American Society for Testing and Materials, Philadelphia, Pa., Draft 12, Jan. 31, 1984, 99 pp.

M. Burke, *Mater. Handl. Eng.*, 64 (Nov. 1976).

C. J. Benning, *Plastic Films for Packaging*, Technomic Publishing Co., Inc., Lancaster, Pa., 1983, Chapt. 7, pp. 135–139; Chapt 12, pp. 159–167; Chapt. 13, pp. 168–181.

ROBERT LE CAIRE
Presto Products, Inc.

FOAM CUSHIONING

The packaging industry is about to witness a metamorphosis of its traditional approach to foam molding. With the advent of the new moldable grades of "engineering foams," molders and designers have new opportunities to enhance product protection and reduce costs at the same time. The new foams are relatively expensive, but through innovative design, the cube and material use can be minimized, resulting in smaller cartons, less warehouse space, less shipping weight, reduced product damage, and cost savings. The materials are not in themselves the answer; the quality and degree of imagination and dedication to the testing and design phase will indicate the level of success in unlocking and successfully applying their dynamic properties (see Cushioning, design; Testing, cushion systems).

A list of suppliers and their offerings is presented as Table 1. Comparative properties are shown in Table 2.

Expanded Polystyrene (EPS) Foam

EPS is a moldable lightweight, semirigid, low cost, closed cell, homopolymer polystyrene foam (see also Foam, extruded polystyrene). Available in the United States since 1954, it is the most popular cushioning foam. The basic resin for EPS is in the form of small granules, which are impregnated with less than 8% hydrocarbon blowing agent (pentane). When heated to 185–205°F (85–96.1°C), the blowing agent vaporizes to create the internal pressure that expands the granules into small beads called pre-puff. For most packaging applications the expansion multiple of the bead is 25 to 40 times to give 1.0–1.6 lb/ft³ or 16–26 mg/cm³ density. To shape the molded package, the pre-puffed expanded beads are injected into cast- and machined-aluminum molds that are mounted into molding presses and held together under several tons of pressure while steam is introduced into the mold to heat and reactivate the bead expansion. The heat and pressure within the enclosed area fuse the beads together into a semirigid closed cell foam piece that conforms to the dimensions and contours of the mold. With the mold held in its closed position, the cooling portion of the cycle continues until sufficient Btu (kJ, energy)

Table 1. Suppliers of Foam Cushioning Materials

	EPS	EPE	EPC	SAN
ARCO Co. Newtown Square, Pa.	Dylite		Arcel	Arsan
BASF-Wyandotte Corp. Parsippany, N.J.	Styropor	Neopolene		
Kanlgafuchi Chem. Ind. Co. Japan		Eperan		
Texstyrene Plastics, Inc. Fort Worth, Texas	Spantex			
Dow Chemical U.S.A. Midland, Mich.		Ethafoam		Techmate
Sentinel Foam Products Hyannis, Mass.		Sentinel foam Polyam		

Table 2. Comparative Properties of Foam Cushioning Materials

Properties	Fabricated Extruded:		Molded:				
	EPE	Polyurethane	EPE	SAN	EPC	EPS	Polyurethane
shock absorption below 0.8 psi (5.5 kPa) static load	good	excellent	very good	good	fair	poor	excellent
above 0.8 psi (5.5 kPa) static load	fair	poor	fair	good	fair	excellent	poor
multiple impact properties	excellent	good	excellent	good	excellent	poor	good
vibration damping	poor	poor	good	good		good	fair
cost raw material, per lb (per kg)	$2.50–2.75 ($5.51–6.06)	$2.00–2.50 ($4.41–5.51)	$2.00–2.25 (4.41–4.96)	$2.00–2.25 (4.41–4.96)	$2.00–2.25 (4.41–4.96)	$0.65–0.70/lb ($1.43–1.54)	$2.00–2.25 ($4.41–5.51)
tooling (mold cost)	<$5.00	<$5.00	$5,000–15,000	$5,000–15,000	$5,000–15,000	$4,000–8,000	<$2,000
ability to produce intricate parts	poor	poor	good	good	good	good	good
level of labor rep.	high	high	low	low	low	low	high
tool lead times	2–3 wk	2–3 wk	8–10 wk	8–10 wk	8–10 wk	8–10 wk	4–5 wk
ability to modify design	good	good	poor	poor	poor	poor	fair
repeatability of parts	varies	varies	good	excellent	good	excellent	good
surface skin	fair	none	good	excellent	good	excellent	good
unidirectional performances	varies	good	good	good	good	good	good
tensile strength	good	fair	good	poor	excellent	poor	good
toughness	good	fair	good	fair	excellent	poor	good
resilience	good	good	good	good	good	poor	good
heat sensitivity	high	low	high	medium	high	medium	low
water absorption	low	high	low	low	low	low	high
resistance to static compressive creep	fair	poor	fair	good	fair	excellent	poor
permanent set	medium	high	medium	low	medium	low	high
compressive strength	medium	low	medium	high	med/high	high	low
dynamic set	low	low	low	medium	low	high	low
ESD treatment	available	available	available	available	available	available	available
density, lb/ft³ (mg/cm³)	2.2 (35)	1.5 (24)	1.9 (30)		1.4 (22)	1.2 (19)	
optimum load, psi 2-in. (51-mm) thick at a 30-in. (762-mm) drop—(kPa)		0.5 (3.45)	0.06 (0.41)	0.6 (4.14)	0.8 (5.52)	1.2 (8.28)	
single impact, G	45	29			40	30	

and residual blowing agent are removed to establish and maintain dimensional stability.

Properties. Molded EPS products are semirigid and impact absorbent, with low WVTR and water absorption, and they have excellent insulation properties. The smooth, slightly resilient molded surface will not abrade most product surfaces. In spite of its semirigid classification, the foam can be designed into ribs that will compress as much as 75% of its thickness with a substantial degree of "memory" (ability to return to its original shape), but less than the "resilient" foams. A high weight:strength ratio results in minimum set or creep under loads. EPS raw material is available with low flamma-

bility and/or antistatic impregnation introduced at the expanding station capable of dissipating electrostatic buildup (see Electrostatic discharge protection).

EPS is relatively inert and has FDA clearance for food contact. It is vulnerable to most solvents and mineral spirits, but resistant to acids, alkalies, and aliphatic compounds. Its semirigid properties make it well suited to applications involving heavier static loads, when less-fragile products are being packaged that only require single-impact protection. The economics of EPS production strongly suggest that all avenues of design be exhausted before opting for a more expensive alternative.

Availability. There are many manufacturers of EPS resins in the United States today. The five major suppliers are Arco Chemical, BASF Wyandotte, Texstyrene, Huntsman-Russtek (formerly American Hoechst), and Georgia Pacific. Material cost is approx $0.55/lb ($1.21/kg). Tooling typically costs $4,500–15,000.

"New-generation" EPS. Until the energy crisis of the early 1970s, energy consumption was a low priority matter in the United States. As utility bills escalated, the problem was finally addressed and more energy efficient materials, molds, and equipment were developed. Traditional materials gave way to new lower molecular weight expandable resins with increased sensitivity and tighter tolerance on molding parameters. This resulted in a new generation of equipment with the necessarily increased capability to handle the more efficient material. New molding techniques and tool designs now utilize vacuum-assist, fast cycling times, positive and negative pressure containment, lightweight castings, and generous venting on all molding surfaces. The result has been greatly improved molding economies, and an engineering material that lends itself to more effective designs with better cushion performance characteristics at lower densities for fragile products produced in large quantities. EPS's limited ability to torque, bend, or elongate is somewhat restrictive when designing long narrow ribs for light fragile products that require heavier static loading. Many of those limitations are being overcome with improved testing and "fine-tuned" designs initiated by a small group of progressive EPS molders. EPS is now able to fulfill lower G specifications and many customers are reevaluating multiple-impact requirements (which are often not necessary) in pursuit of the attractive cost savings. Several major personal-computer manufacturers have successfully converted to the new-generation EPS foam instead of more costly resilient foams.

Expanded Polyethylene Foam (EPE)

EPE is a low density semirigid, closed-cell, weather stable, polyethylene homopolymer. It is easier to compress than cellular polystyrene but less compliant than flexible open-celled polyurethane. Although this material has been used extensively throughout Europe and Asia, its introduction into the United States market is very recent. Unlike its predecessor, extruded polyethylene foam, EPE is similar to expanded polystyrene (EPS) foam in its molding process. The material is supplied by the manufacturer in a cross-linked expanded form ready for molding. (Molded EPE planks are also available for fabrication applications.) Conventional EPS molding presses can process EPE with the addition of a filling device, provision for higher molding pressure, and postmolded oven curing. The expanded particles do not contain a "blowing agent" and can be stored for long periods of time at room temperature. Density range is 1.8–7.5 lb/ft^3 (29–120 mg/cm^3). The most commonly used density throughout Europe and Asia is 1.8 lb/ft^3 (29 mg/cm^3).

Properties. Molded EPE cushions are soft and nonabrasive. EPE has superior multiple-impact cushioning properties (low transmission of acceleration to the packaged product upon repeated impact without significant loss of resilience) and high mechanical strength (good tensile strength and insensitivity to notch stress). EPE is highly resistant to chemicals and solvents. It is not damaged by acids or alkalies, but strong oxidizing agents may effect it at higher temperatures.

It has very low water absorption: only about 1% of its total volume after a week of total immersion. Additional properties include very good vibration-damping characteristic, good thermal insulation, and high elasticity with resistance to deformation over the temperature range of −60 to 185°F (−51 to 85°C). In tension (stress/extension), the relative elongation at break is about 50% the tensile strength varies from 25 psi (172 kPa) at a density of 1.87 lb/ft (30 mg/cm^3) to 65 psi (448 kPa) at a density of 5.0 lb/ft^3 (80 mg/cm^3). Creep (thickness loss under constant compressive strain at 73°F (22.8°C)) curves are superior to polyurethane but not as good as EPS. EPE's ability to elongate, torque, compress, return to within 98% of its original thickness after deformation from strain, damp-vibrational input, and transmit low levels of velocity change (deceleration) per given mass, are characteristics that will make it a favorite over other packaging materials when heavier, costlier, fragile products are to be packaged.

Availability. EPE was available only in extruded plank form until the recent introduction of a moldable grade into the United States (Neopolene, BASF). Now engineers have a material which offers a broader range of creative avenues in design and packaging applications. BASF is considering domestic United States production, as are two other companies that would use Japanese technology. ARCO Chemical Company plans to test market a moldable EPE from Japan using Japan Styrene Paper Corporation technology. Kanegafuchi Chemical Industry Company, of Japan, is now shipping small quantities of molded "Eperan" (EPE) product into the United States as well as a moldable "pre-puff" from Belgium to be custom molded. They have also indicated their intent to build a domestic resin producing plant in the near future. Raw material cost is approx $2.25/lb ($5/kg). The cross-linking of EPE propagates static decay, but not as effectively as the extruded types with inherent antistatic properties. Tooling is required for the molding process, designed similar to that of conventional EPS mold construction with modifications to handle higher operating pressures. The cost is slightly higher: $5,500 to 15,000.

Availability of extruded and laminated EPE. There are three leading U.S. suppliers of extruded/fabricated EPE: Dow Chemical (Ethafoam), Packaging Industries (Sentinel Foam), and Richter Manufacturing (Poly Plank). The material is available in specific thicknesses and densities with or without antistatic properties. It comes in extruded plank, rounds, and sheet. Extruded plank is approx $0.50/board ft ($1.18/1000 m^3). Tools and dies are relatively inexpensive ($200.–600.).

Extruded/fabricated EPE vs molded EPE. Until the introduction of moldable EPE foams, fabricated (die-cutting, bonding, routing, saw cutting, and hot-wire cutting) extruded polyethylene foam was the only choice when the packaging application called for PE foam. Inconsistent dimensional repeatability, intensive labor requirements, and extrusion-direction orientation represent obstacles in some applications, yet it is an engineering material that can give a fast "turn-around" with low cost tooling and quick design changes, and it is readily available from many fabricators.

Molded EPE can be designed and molded into elaborate shapes with a high degree of repeatability. The density can be adjusted as required within a broad range to maximize the performance of the package with minimal cube and material. The material is nondirectional and performs consistently in all axes. The use of a mold offers rapid, close tolerance, high volume production. Costly tooling can be a disadvantage to the user of molded EPE. Once the tool is built, it is difficult to make major changes. The performance of the materials are similar with respect to vibration damping and shock absorption (≥20 G where adequate thickness and proper loading is available), but the ability to mold intricate shapes and ledges, without additional labor costs, can give some very significant advantages to the molded grade of EPE over extrusion types in attaining unique dynamic performance and high volume cost-effective applications.

Expanded Polyethylene Copolymer (EPC)

The recently introduced moldable, expanded polyethylene copolymer is a combination of approximately 50% polyethylene and 50% polystyrene resin. By combining the properties of both resins it widens the selection of resilient materials for serious packaging engineers and designers. EPC is a low density, semirigid, closed-cell material that requires refrigerated storage below 40°F (4.4°C) in its raw granular form and has a shelf life of at least one month. The material expansion and conveying of the sensitive pre-puff requires special handling and molding within a short period of time. The molding process and equipment are similar to that of EPS, but with slower molding cycles. Unlike expanded polyethylene, EPC contains and depends on a blowing agent for processing and must be handled accordingly. It is produced and marketed in the United States by ARCO as Arcel.

Properties. EPC features outstanding resilience with exceptional tensile toughness and puncture resistance (tensile strength at 1.5–2.5 lb/ft³ (24–40 mg/cm³) density, 35–47 psi (241–324 kPa)). Elongation is 15–25% at temperatures of up to 180°F (82°C). EPC has a low WVTR, and low water absorption (less than 1% of the volume for densities of 1.5 to 2.5 lb/ft³ (24–40 mg/cm³)). Compressive creep is lower than EPS but better than EPE (2.5% set after 1000 hours at 4 psi (28 kPa) and 73°F (23°C). Dynamic set (set as result of impact) during impact testing measured 7.5% deformation (at 1.0 psi (6.9 kPa) static stress), which is better than EPS, but not as good as EPE. EPC resists common solvents including gasoline, mineral spirits, and aliphatic compounds. The material is available with an antistatic additive to protect against electrostatic discharge. It can be expanded as low as 1.4 lb/ft³ (22 mg/cm³) density by double-passing it through a continuous expander. Most applications are in the 2.0 to 2.5 lb/ft³ (32–40 mg/cm³) density range.

Availability. ARCO is the only producer and supplier of EPC in the United States today. It is available through ARCO licensees and most other EPS molders. Price is in the $2.50/lb ($5.51/kg) range. Aluminum tooling similar to EPS is required. Tooling cost is $5,000–14,000.

EPC vs EPS and EPE. Expanded polyethylene copolymer is a material that falls between EPS and EPE in performance, but exceeds both materials in toughness. Tensile and puncture resistance of EPC is superior to all of the moldable resilient foams available in the United States today. It has good multiple-impact performance characteristics with better memory than EPS, but not as good as EPE. Cushion performance of EPC parallels EPE but at higher levels (7–12 G) even after repeated drops. EPC is especially good for reusable material-handling trays and packaging applications that require a non-abrasive, solvent resistant, impact absorbing material with a superior toughness that elongates, compresses and flexes without material fatigue. EPS molders are capable of molding this material with minor adjustments to their existing equipment.

Expanded SAN Foam

SAN (styrene–acrylonnitrile copolymer) is a recently introduced moldable, lightweight, semirigid, closed cell, highly resilient styrene copolymer resin (see Nitrile polymers). The resin relies on a fluorocarbon blowing agent to preexpand and fuse the material during its molding phase. Refrigeration is not needed for raw material storage, and shelf life of six months is possible with some simple precautions. The processing is much like EPS, but with extended cycles due to the higher level of blowing agent. EPS equipment and standard EPS mold designs can be used without any major changes or adaptations. Postmolding oven-curing is not necessary. Low density of more than 40 times expansion can be attained during the preexpanding phase (as low as 1.0 lb/ft³ (16 mg/cm³) density on the first pass through the expander and 0.8 lb/ft³ (13 mg/cm³) density on the second pass). Antistatic properties are available.

Properties. Molded SAN products are semiresilient and impact absorbent, with low WVTR and low water absorption (less than 0.33% of the volume) at 1.2 lb/ft³ (19 mg/cm³) and limited elongation properties. SAN gives good insulation (k-factor (thermal conductivity) at 1.1–1.2 lb/ft³ (18–19 mg/cm³) density: 0.18–0.23 (Btu·in.)/(h·ft²·°F) [0.026–0.033 W/(m·k)] 75–77 °F (24–25°C) mean. The smooth, highly resilient molded surface will not abrade most product surfaces. The foam can be designed into rib patterns to distribute a predetermined static load. After deformation of the ribs (up to 75%), the memory of the material will return the shape to within 95% of the original configuration (dynamic set of 2-in. (51-mm) thick, 1.15 lb/ft³ (18 mg/cm³) density, at 0.75 psi (5.2 kPa) was a 4.5% deformation). Compressive creep of SAN is similar to EPS at comparable densities (1.5% deformation after 300 h at 4 psi (27.6 kPa) at room temperature). SAN shows good vibration-damping characteristics. SAN has an advantage in cushion performance over EPS when comparisons are made after repeated drops (multiple impact) on the same surface with a static loading of greater than 0.8 psi (5.5 kPa) or on single impacts on surfaces of less than 0.8 psi. These properties make this material suitable for applications where multiple impact, low fragility, high static loads, and high value items are involved or in cases that require low fragiltiy performance with low static loads. The higher cost of SAN over EPS must be incorporated into any comparative evaluation to determine the economics and feasibility of application.

Availability. There are two sources of moldable SAN copolymer resin in the United States: ARCO Chemical Company (Arsan 600) and Dow Chemical U.S.A. (Techmate). The cost of the resin is in the $2.00–2.50/lb ($4.40–5.51/kg) range. The applicable densities are near the 1.0 lb/ft³ (16 mg/cm³) range instead of the 1.7–3.0 lb/ft³ (0.027–0.048 g/cm³) densities that are typical for the other resilient foams and must be considered when making cost comparisons. Several major molders are licensed to mold SAN.

SAN vs other resilient molded foams. SAN is an attempt to take the best from EPS and EPE and combine it into a viable, low density, commercial product for industries that produce high value, high volume, and low fragility products. SAN is very similar to EPE in multiple-impact resiliency and will give impact protection, even after five successive drops, with dynamic cushion curves similar to EPE. It does not have the tensile and elongation properties (ie, toughness) of EPE and is of limited value in "low static-load" requirement design concepts that call for large deflection of thin/long cross-sections where SAN is susceptible to cracking and breaking. SAN can handle heavier static loading, simlar to EPS and not display the creep or compressive set problems of EPE, from prolonged storage or excess heat. Impact protection of low density (0.8–11 lb/ft^3 or 13–18 mg/cm^3) SAN is near that of 2.0 lb/ft^3 (32 mg/cm^3) EPE and could give it a cost advantage over both EPC and EPE.

PETER RODGERS
Foam Fabricators

FOAM, EXTRUDED POLYSTYRENE

Polystyrene foam sheet is a reduced-density sheet made from polystyrene by the extrusion method (see Extrusion). Produced in many forms, is most easily classified by its density. This article deals with sheet that has density of 3–12 lb/ft^3 (0.05–0.19 g/cm^3) and thickness of 0.015–0.150 in. (0.38–3.8 mm).

The first extruded foam sheet was produced in 1958 by extruding expandable polystyrene beads. The direct-gas-injection extrusion technology that is most popular today was developed in the early 1960s. About 15 million (10^6) lb (6800 t) was produced in 1966. The growth of this industry has been substantial, with over 450 million (10^6) lb (204,000 t) produced in 1984 in the United States. Most was used for disposable packages such as meat and produce trays and egg cartons, and as containers and trays for carry-out meals and disposable dinnerware. It is also used for drink cups, bottle labels (see Glass container design; Glass container manufacturing), and miscellaneous cushion-type packages (see Fig. 1). Future growth for this product is expected to exceed that of the plastics industry in general. In addition to normal growth for current products, new applications will be developed by laminating paper or films to foam sheet (see Laminating machinery). Fabrication techniques such as die cutting and scoring will be used to produce folding containers, and ultrasonic and heat welding will expand potential as well.

There are four direct-gas-injection processes in use, all with the same objectives: to melt the polymer, uniformly mix in the blowing agent and nucleator, cool the melt, and expand the mix to form a biaxially oriented sheet. The four processes employ large-diameter or long-L/D single-screw extruders; twin-screw extruders; the Winsted system using a single-screw extruder and a patented cooling system; and two-extruder, ie, tandem, systems. Of these processes, the most successful and widely used is the two-extruder tandem system (see Fig. 2).

The tandem extruder system utilizes two single-screw extruders: a primary extruder for melting, mixing, and feeding extrudate to a secondary extruder for cooling and additional mixing of the extrudate prior to exiting an annular die. Most tandem systems use 4.5-in. (11.4-cm) primary extruders and 6-in. (15.2-cm) secondary extruders with output rates of 500–1200 lb/h (227–544 kg/h), but the 6-in. (15.2-cm) primary and

Figure 1. Extruded polystyrene foam products.

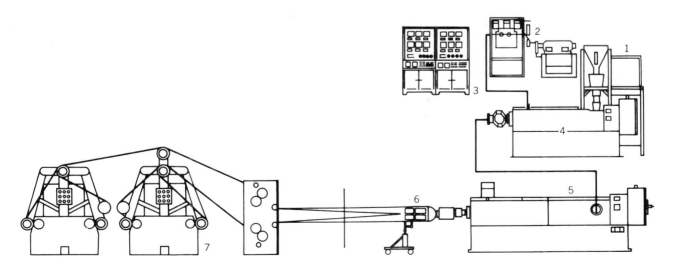

Figure 2. A two-extruder tandem system for the production of foamed sheet: 1, continuous feeding and blending system; 2, volumetric pump system; 3, process control; 4, primary extruder; 5, secondary extruder; 6, annular die and cooling mandrel; 7, draw rolls and winders.

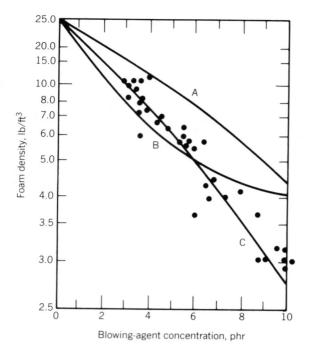

Figure 3. Blowing-agent requirement for polystyrene foam: A, Freon-11 (DuPont): B, pentane; C, Freon-12 (DuPont). To convert lb/ft³ to g/cm³, multiply by 0.0162.

8-in. (20.3-cm) secondary extruder systems are becoming more popular as producers strive for higher output rates, ie, 900–1800 lb/h (408–816 kg/h). Tandem foam lines are supplied by many extruder manufacturers.

In addition to the rather unique extrusion system, the die, mandrel, and takeoff system must be carefully designed. As mentioned above, an annular die is generally used because of the three-dimensional expansion of the foam. The foam is drawn over a cooling–sizing mandrel to cool the foam and provide the desired sheet width. Blowup ratios, ie, die diameter : mandrel, must be calculated based on the density and desired orientation of the foam. Ratios of 3 : 1 to 9 : 2 are common. This tubular sheet is slit into the required widths as it is drawn off the mandrel and wound into rolls (see Slitting and rewinding machinery).

The ingredients needed to produce foam sheet are resin, nucleator, and blowing agent. The resin is normally a high heat general-purpose polystyrene. Nucleators, such as talc or a citric acid–sodium bicarbonate mixture, are added to provide foaming sites to obtain the desired cell size and uniformity. The blowing agent, usually a fluorocarbon or light aliphatic hydrocarbon, is injected as a liquid into the primary extruder. A positive-displacement volumetric pump is used, and the addition port is usually approximately ⅔ of the way up the barrel of the primary extruder. The amount and type of blowing agent control the density of the foam produced. Figure 3 shows these relationships for three of the most commonly used blowing agents.

A large percentage of all foam sheet is thermoformed (see Thermoforming). To achieve good postexpansion in this process, the sheet must be aged for 3–5 d. This aging allows cell gas pressure to reach equilibrium. Matched metal molds are normally used in the forming process. Oven-temperature con-

trol is critical for consistent forming. Scrap from the extrusion and thermoforming processes can be reprocessed by grinding the scrap and densifying it in an extruder. Ecologically, polystyrene foam conforms favorably to current forms of disposal, such as landfill and incineration. In landfill, foam remains inert but packs and crushes easily and there is no pollution of underground water streams by decaying material. In incineration, the chief products of combustion are water, carbon dioxide, and carbon monoxide, typical of organic materials.

P. A. WAGNER
Dow Chemical U.S.A.

FOIL, ALUMINUM

The Material

Aluminum foil is a thin-rolled sheet of pure or alloyed aluminum, varying in thickness from about 0.00017 (4.3 μm) in. to a maximum of 0.0059 (150 μm) in. (1). By industry definition, rolled aluminum becomes foil when it reaches a thickness less than 0.006 in. (152.4 μm) (see Table 1).

Aluminum, from which the foil is made, is a bluish silver-white trivalent metallic element that is very malleable and ductile. Noted for its light weight, good electrical and thermal conductivity, high reflectivity, and resistance to oxidation, aluminum is the third most abundant element in the earth's crust (1).

Aluminum always occurs in combination with other elements in mineral forms such as bauxite, cryolite, corundum,

Table 1. Physical Properties of Aluminum Foil

Property	Value
density	0.0976 lb/in.³ (2.70 g/cm³)
specific gravity	2.7 (approx.)
melting range	1190–1215°F (643–657°C)
electrical conductivity	59° IACS, vol., 200% IACS (approx.), weight
thermal conductivity	53 W/(m · K) at 25°C
thermal coefficient of linear expansion	13.1×10^{-6} per °F, 68–212°F (23.6×10^{-6} per °C, 20–100°C)
reflectivity for white light, tungsten filament lamp	85–88%
reflectivity for radiant heat, from source at 100°F (37.8°C)	95% (approx.)
emissivity, at 100°F (37.8°C)	5% (approx.)
atomic number	13
atomic weight	26.98
valence	3
specific heat at 20°C	0.21–0.23
boiling point	3200°F (1760°C)
temperature coefficient of resistance (representative values per °C)	
at 20°C	0.0040–0.0036
at 100°C	0.0031–0.0028

low temperature properties—aluminum increases in strength and ductility as temperature is lowered, even down to −320°F (−195.6°C) alunite, diaspore, turquoise, spinel, kaolin, feldspar, and mica. Of these, bauxite is the most economical mineral for the production of aluminum. It can contain up to 60% alumina, which is hydrated aluminum oxide. It takes about 4 kg of bauxite to produce 1 kg of aluminum (2).

Alumina is converted into aluminum at a reduction plant or smelter. In the Hall-Héroult process, the alumina is dissolved in a molten salt called cryolite. The action takes place in steel boxes lined with carbon called pots. A carbon electrode or anode is lowered into the solution, and electric current of 50,000–150,000 A flows from the anode through the mixture to the carbon-cathode lining of the steel pot. The electric current reduces, or separates, the alumina molecules into aluminum and oxygen. The oxygen combines with the anode's carbon to form carbon dioxide. The aluminum, heavier than cryolite, settles to the bottom of the pot from which it is siphoned into crucibles. The molten aluminum is eventually processed into products.

Foil

One of aluminum's most common uses is as foil. About 897 million (10^6) lb (4.1×10^5 metric tons) of aluminum foil were shipped in the United States in 1984. Aluminum foil is generally produced by passing heated aluminum-sheet ingot between rolls in a mill under pressure. Ingot is flattened to reroll sheet gauges on sheet and plate mills and finally to foil gauges in specialized foil-rolling mills.

A second method of producing aluminum foil, rapidly gaining popularity, involves continuous casting and cold rolling. This method can eliminate the conventional energy-intensive and costly steps of casting ingot, cooling, transporting to rolling plants, and then reheating and hot rolling to various gauges.

First produced commercially in the United States in 1913, aluminum foil became a highly marketable commodity because of its protective qualities, economic production capability, and attractive appearance. The first aluminum foil laminated on paperboard for folding cartons was produced in 1921. Household foil was marketed in the late 1920s, and the first heat-sealable foil was developed in 1938.

World War II established aluminum as a major packaging material. During the war, aluminum foil was used to protect products against moisture, vermin, and heat damage. It was also used in electrical capacitors, for insulation, and as a radar shield.

After the war, large quantities of aluminum foil became available for commercial use. Its applications boomed with the postwar economy. The first formed or semirigid containers appeared on the market in 1948. Large-scale promotion and distribution of food service foil in 1949 quickly expanded the market (1).

Aluminum foil's compatibility with foods and health products contributes greatly to its utility as a packaging material. A concise guide to the behavior of aluminum with a wide variety of foods and chemicals is a reference entitled *Guidelines for the Use of Aluminum with Food and Chemicals*, published by the Aluminum Association (3).

Standard aluminum foil alloying elements are silicon, iron, copper, manganese, magnesium, chromium, nickel, zinc, and titanium (see Table 2). These elements constitute only a small percent (in most cases, no more than 4%) of aluminum-foil alloy composition.

Table 2. Principal Aluminum Foil Alloys (nonheat-treatable)

Alloy and temper (Aluminum Association Number)	Aluminum, %	Principal other elements,[a] %
1100—H19	99.00	0.12 Cu
1145—H19	99.45	
1235—H19	99.35	
1350—H19	99.50	
3003—H19	97.00	0.12 Cu, 1.2 Mn
5052—H19	96.00	2.5 Mg, 0.25 Cr
5056—H19	93.6	0.12 Mn,
5056—H39		5.0 Mg, 0.12 Cr
heat-treatable		
2024—T4	91.8	4.4 Cu, 0.6 Mn, 1.5 Mg

[a] Nominal compositions.

Properties

Chemical resistance. Resistance of aluminum foil to chemical attack depends on the specific compound or agent. However, with most compounds, foil has excellent to good compatibility.

Aluminum has high resistance to most fats, petroleum greases, and organic solvents. Intermittent contact with water generally has no visible effect on aluminum otherwise exposed to clean air. Standing water in the presence of certain salts and caustics can be corrosive. (3).

Aluminum resists mildly acidic products better than it does mildly alkaline compounds, such as soaps and detergents. Use with stronger concentrations of mineral acids is not recommended without proper protection because of possible severe corrosion. Weak organic acids, such as those found in foods, generally have little or no effect on aluminum. A clear vinyl coating, however, is recommended for use with tomato sauce and other acetic foods.

Temperature resistance. Since aluminum foil is unaffected by heat and moisture, it is easily sterilizable and is actually sterile when heat treated in production. Unlike many packaging materials, aluminum foil increases in strength and ductility at lower temperatures. Its opacity protects products that would otherwise deteriorate from exposure to light (see Table 3).

Mechanical properties. The addition of certain alloying elements strengthens aluminum. The alloys produced from these compositions can be further strengthened by mechanical and thermal treatments of varying degree and combinations. For this reason, the mechanical properties of aluminum foil are significant to an understanding of its versatility.

The lowest, or basic, strength of aluminum and each of its alloys is determined when the metal is in the annealed, or soft, condition. Annealing consists of heating the metal and slowly cooling it for a predetermined period of time. Reroll stock from which foil gauges are produced is annealed (a process of heat-

Table 3. Functional Properties of Aluminum Foil

form	continuous rolls and sheets	hygienic	sterile when heat-treated in
thickness	0.00017–0.0059 in. (4.3–150		production; smooth
	μm)		metallic surface sheds
			most contaminants and
			moisture of sterilization
maximum width	68 in. (1.7 m) for pack-rolled	sterilizable	metal unaffected by heat and
	lighter gauges; 72 in. (1.8 m)		moisture of sterilization
	for gauges		(except for staining in
	0.001 in. (25.4 μm) and		some cases)
	heavier, single-web rolled		
impermeability	(WVTR)^a 0.001 in. (25.4 μm)	nontoxic	inert to or forms no harmful
	and thicker is impermeable;		compounds with most food,
	0.00035 in. (8.9 μm) has a		drug, cosmetic, chemical or
	WVTR of $\leq$ 0.02 g/100 in.^{2}		other industrial products
	(0.065 m^2); 24 h at 100°F	tasteless, odorless	imparts no detectable taste
	(37.8°C)/100th—WVTR		or odor to products
	drops to practically zero		
	when 0.00035-in. (8.9-μm)		
	foil is laminated to		
	appropriate film		
corrosion resistance	aluminum's natural oxide	opacity	solid metal, transmits no
	shielding, which is	permanence	light
	maintained in the presence		highly corrosion resistant in
	of air, renders it		most environments
	substantially corrosion		
	resistant		
compatibility with food,	nontoxic; corrosion resistant to	sealability	excellent dead fold and
drugs and cosmetics	many compounds in solution		adhesion to a wide variety
			of compounds
formability	dead fold	insignificantly magnetic	provides excellent electrical,
			nonmagnetic shielding
nonabsorptivity	proof against water and wide	nonsparking	the leading metallic material
	variety of liquids		for applications with
greaseproof	nonabsorbent		volatile, flammable
			compounds

^{a} WVTR = water vapor transmission rate.

ing and cooling) prior to the foil-rolling operations to make it softer and less brittle.

All alloys are strain-hardened and strengthened when cold worked, as in foil rolling. When the product is wanted in the soft condition, it is given a final anneal (1).

Converting

Aluminum foil is converted into a multitude of shapes and products (see Table 4). Processes involved may include converting mill rolls of plain foil by rewinding into short rolls, cutting into sheets, forming, laminating, coloring, printing, coating, and the like (4) (see Slitting and rewinding; Laminating; Printing; Coating equipment).

Packaging products account for about 75% of the market for aluminum foil (5). Aluminum foil is also used in households and institutions as a protective wrap, and for decorative and construction purposes.

Among packaging end uses, aluminum foil is formed into semirigid containers produced from unlaminated metal for frozen and nonfrozen foods, as well as caps, cap liners, and closures for beverages, milk, and other liquid foods (see Closure liners; Closures). It is also formed into composite containers with films and plastics to package powdered drinks, citrus and other juices, and motor oil and other auto supplies (see Cans, composite).

Aluminum foil combined with other materials such as paper or plastic film can be used to package a host of basic nonfood products, including tobacco, soaps and detergents, photographic films, drugs, and cosmetics.

When water-vapor or gas-barrier qualities are critical to the success of a packaging material, aluminum foil is usually considered. Aluminum foil containers are odorless, moistureproof, and stable in hot and cold temperatures. They were originally developed to supply bakers with cost-cutting disposable pie plates and bake pans.

A major reason for the wide use of aluminum foil in packaging is its versatility. It is adaptable to practically all converting processes and can be used plain or in combination with other materials. Aluminum foil can be laminated to papers, paperboards, and plastic films. It can be cut by any method and can be wrapped and die-formed into virtually any shape. Foil can be printed, embossed, etched, or anodized.

Foil lamination. In many laminations, light-gauge foil is the primary barrier against water vapor transfer. While creasing can create pinholes or breaks in this barrier, problems can be minimized by proper lamination (see Laminating machinery).

Laminations of foil and waxed paper have been popular as overwraps and liners for cereal packages for more than 40 yr. Snack foods that once presented a problem because of their

Table 4. Classifications for Converted and Nonconverted Aluminum Foil End Uses[a]

Packaging end uses
semirigid foil containers (including formed foil lids) produced from
    unlaminated metal for
      bakery goods
        frozen
        nonfrozen
      frozen foods, other than bakery
caps, cap-liners and packaging closures for
    beverages and milk
    foods
composite cans and canisters (including labels and liners for
    composite cans) for powdered drinks; auto supplies; food snacks;
    citrus and other juices; refrigerated dough; other refrigerated
    and frozen products
flexible packaging end uses (including labels, cartons, overwraps,
    wrappers, capsules, bags, pouches, seal hoods and overlays for
    semirigid foil containers)
    food products
      dairy products—cheese, butter, milk, milk powder, ice cream
      dried and dehydrated food products—fruit, vegetables, potato
        products, soup mixes, yeast
      baked goods—bread, cookies, crackers
      cereals and baking mixes—cereals, rice, cake mixes, frosting
        mixes, macaroni products
      powdered goods—coffee, tea, gelatins, dessert mixes, drink
        powders, cocoa, dry concentrates, sugar, salt
      meat, poultry, and seafoods (fresh, frozen, irradiated, dried,
        retorted)
      frozen prepared foods
      confections—candy, mints, chewing gum, chocolate bars
        (converted foil only)
      dry snack foods—potato chips, popcorn, including coated
        popcorn
      beverages—soft drinks, beer, distilled liquors, wines
      food products, n.e.c.[b] (including pet foods)
    nonfood products
      tobacco—cigars, cigarettes
      soaps and detergents
      photographic film and supplies
      drugs, pharmaceuticals, cosmetics, toiletries, kindred products
      nonfood products, n.e.c.[b]
      military specification packaging[c]
Selected nonconverted foil products[d]
packaging end uses (unmounted, unconverted foil stock sold to end
    users for candy and gum wraps)

[a] U.S. Department of Commerce.
[b] Not elsewhere classified.
[c] On direct government orders only.
[d] Reported by foil producers (rollers) only.

high oil content are now wrapped by specially formulated foil–paper and foil–film laminates.

Aluminum foil laminates have been designed to meet exacting requirements of drugs used in transdermal medication systems that deliver medication through the skin at a constant rate over a specific period of time.

Peelable foil-laminated pouches protect transdermal drugs. Space-shuttle astronauts wore a U.S.-quarter-sized transdermal patch behind the ear to help prevent motion sickness.

Printing. Either side of foil may be printed directly, or the foil may be laminated to a reverse-printed clear film to provide attractive designs with accents (6). Use of transparent inks

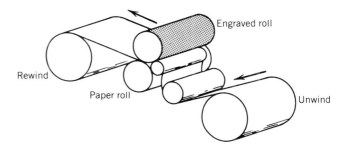

Figure 1. Foil embossing unit. An engraved steel roll, usually the top roll, carries the design, and a matching paper roll becomes the matrix. This matrix roll is constructed of layers of paper (wool-rag) rings compressed solidly into one continuous mass and mounted on an appropriate core.

through which the foil can be seen produces pleasing, metallic colors with no loss in brightness or sparkle.

The same presses and the same types of processes used for printing paper and plastic films are used for printing on foil and foil–paper laminations. Processes include rotogravure, flexography, letterpress, lithography, and silk screen (1) (see Printing; Decorating). Inks for all of these processes are readily available in formulations expressly made for printing on aluminum (see Inks).

To provide anchorage for the inks, prime or wash coatings are nearly always used on foil to be printed. The coatings also provide a barrier that prevents offsetting of undesirable materials from the paper to the foil surface.

Embossing. In-line or out-of-line printing and embossing units allow designs of limitless combinations of color and form. Any embossing pattern in foil produces two basic visual effects, namely, three-dimensional patterns or illustrations and continual reflective contrasts.

Foil embossing often is performed in continuous roll form by passing the web or sheet through a roll stand equipped with one engraved steel roll and a soft matrix roll of paper. The pressure for embossing may be obtained by maintaining the paper roll with the axis in a fixed position and using only the weight of the steel roll to depress the negative pattern into the paper roll (see Fig. 1).

Flexible Foil Packages

Flexible foil-containing packages are extremely popular for food packaging because they provide superior flavor retention and longer shelf life than packages formed from other flexible materials. Flexible foil packaging is impervious to light, air, water, and most other gases and liquids. The packages protect contents from harmful oxygen, sunlight, and bacteria (see Fig. 2) (see Multilayer flexible packaging).

One of the most important flexible packaging applications of aluminum foil is the form/fill/seal pouch. (1). Pouches are formed continuously from roll-fed laminated material and are filled and sealed immediately as formed (see Form/fill/seal, horizontal; Form/fill/seal, vertical; Pouches). The pouch is one of the oldest automated-package forms. The retort pouch, developed in the early 1950s, represented a significant advance in food packaging. It is a flexible package made from a laminate of three materials: an outer layer of polyester for strength; a middle layer of aluminum foil as a moisture, light,

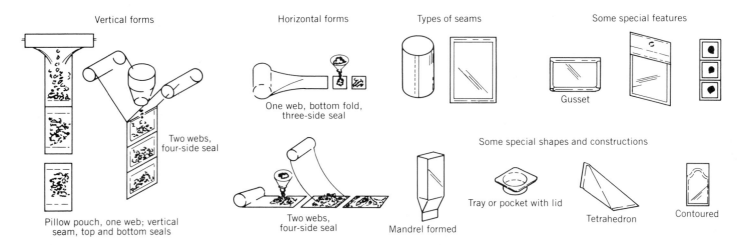

Figure 2. Features and constructions of pouches.

and gas barrier; and an inner layer of polypropylene as the heat seal and food-contact material (7) (see Retortable flexible and semirigid packages).

The Q-Pouch is a new foil-lined paperboard pouch that can be turned into a drinking cup. A user of the package simply rips the top edge along perforations and squeezes the sides of the packet. Scored paperboard opens into a hexagonal-shaped cup (8).

Aluminum foil is also used to some extent for bag-in-box liquid packaging (9) (see Bag-in-box, liquid).

Foil Lidding

Another popular packaging application is the flexible closure or flexible lid. The flexible closure got its start with single-service dairy creamers, yogurt, and cheese dips. It is also widely used in the health-care industry, especially in hospitals where the trend is toward disposable packaging (see Health-care packaging). Liquid medications are also packaged in single-dosage containers, which are convenient and provide easy evidence of tampering (10) (see Pharmaceutical packaging; Tamper-evident packaging).

Flexible lids were introduced in 1966 when a U.S. Health Service regulation required that the pouring lip of dairy containers be covered during shipment and storage. Besides covering the pouring lip of the container, a heat-sealed foil lid offered a more reliable seal, longer shelf life, and greater protection than other lidding materials.

The use of inductive heat-sealing equipment instead of conductive heat sealing is broadening the applications for foil lidding. Inductive sealing does not heat the lid by direct contact. Heat is brought into the lidding stock by a magnetic field through which the container and lid pass. The magnetic field heats the foil in the lid and an effective, hermetic seal forms between the lid and the container. Flexible foil lidding can be used on glass, plastic, or composite cans. This key development enables the aluminum foil industry to provide tamper-evident packaging (see Tamper-evident packages).

Regulated Packages

Aluminum foil is an integral part of many tamper-evident packages. Although the term came into prominence only in the 1980s, this kind of packaging has been on the market before then. As now defined by the *Proprietary Association,* a tamper-evident package is "one which, if breached or missing, can reasonably be expected to provide visible evidence to consumers that the package has been tampered with or opened" (11).

Of 11 methods listed by the Proprietary Association which conformed to the regulation when it was first pronounced, six featured aluminum foil as part of the tamper-evident system.

For a system to be tamper-evident, its package cannot be removed and replaced without leaving evidence. It is extremely difficult to mend or repair a foil-membrane seal after it is broken or to return an aluminum roll-on cap to its original state once it has been removed.

One of the first major laws that required protective packaging was the Poison Prevention Packaging Act of 1970, which essentially said that drugs and dangerous substances must be packed in child-resistant packages. This type of package includes blister packs of plastic–foil–paper combinations; tape seals of paper or foil which adhere to the cap or shoulder of a bottle; and pouches which can be tightly sealed.

Aseptic Packaging

Most aseptic packages are laminations of paperboard, plastic, and aluminum foil. In aseptic packaging, the product is sterilized with high heat and then cooled. The package is sterilized separately, and then the sterile product and package are combined in a sterile chamber (12) (see Aseptic packaging).

Foil provides superior adhesion for the containers which, in the main, hold highly acidic products. In almost all methods of aseptic container manufacturing, aluminum foil is used as a barrier to light and oxygen (12).

Foil and Microwave Ovens

In 1984, Underwriters Laboratories undertook a study on the use of aluminum foil trays to reheat food in microwave ovens. Using various models of microwave ovens and standard types of aluminum trays containing frozen foods, the study concluded that (13):

The power density of microwave radiation emissions did not exceed maximum allowable limits.

There was no significant change in heat going into various liquids and frozen foods in aluminum containers as compared to other containers.

The temperatures in the foods and liquids being tested were generally comparable in the various containers.

The foil trays containing different amounts of water and empty trays produced no fire emissions of flaming or molten metal.

The test trays used for heating frozen foods in accordance with specific instructions did not increase the risk of radiation, fire, or shock hazard.

Semirigid Packaging

Other classifications of aluminum packaging include semirigid and rigid containers. Semirigid containers include those that are die-formed, folding cartons, and collapsible tubes. Die-formed containers are one-way, disposable devices such as pie plates, loaf pans, and dinner trays. Folding cartons come in many sizes and shapes and hold such products as dry cereals, eggs, and milk or other liquids (see Cartons, folding; Cartons, gable top). Collapsible tubes were first made of lead and used by artists more than 100 yr ago. In modern times, a proprietary aluminum foil/film version of the collapsible tube holds toothpaste and products of similar consistency, including hair coloring and depilatories (1) (see Tubes, collapsible).

Aluminum foil provides the barrier to permeation of the oils used in most products and compounds that go into the tubes.

Aluminum tubes have the advantage of providing light weight, high strength, flexibility, and good corrosion resistance. Aluminum tubes also have low permeability and offer quality appearance. Traditionally, aluminum collapsible tubes were impact-extruded, but a new process forms the laminated roll stock into continuous tubing by use of heat and pressure. The tubing is cut into individual sleeves and is automatically headed by injection molding.

As opposed to flexible containers that conform to the shape of the product, semirigid containers have a shape of their own. They can be deformed from their original shape either while they are emptied (as in the collapsible toothpaste tube), or before they are filled (as with the folding carton).

Die-formed aluminum containers are among the most versatile of all packages. They easily withstand all normal extremes of handling and temperature variation. A product in an aluminum container can be frozen, distributed, stocked, purchased, prepared, and served without soiling a single dish.

Bare foil is used for most formed aluminum foil products but protective coatings are used on the containers for some foods and other products. Although the frozen food tray is the most common, the aluminum-formed container is available in scores of shapes and sizes from the half-ounce portion cup to full-size steamtable containers for institutional feeding. Closures for these containers vary from laminated hooding to the hermetically sealed closure according to the amount of protection needed for the product.

Folding cartons offer protective and display characteristics unique in packaging and have some of the advantages of both the flexible and the rigid container. Before use, when it is folded flat, the folding carton offers the storage economy of the flexible bag. When it is filled, it offers much of the protection of the set-up box.

Rigid Containers

Composite cans and drums feature aluminum foil combined with fiber. They are widely used for refrigerated dough products, snack foods, pet foods, and powdered drink mixes.

In some processes, a composite can is made up of paper–polyethylene–aluminum foil–polyethylene laminated stock with a foil-membrane closure (14). Cans are produced from preprinted gravure rolls of bottom aluminum stock. (see Cans, composite).

BIBLIOGRAPHY

1. *Aluminum Foil,* the Aluminum Association, Washington, D.C., 1981.
2. *The Story and Uses of Aluminum,* the Aluminum Association, Washington, D.C., 1984.
3. *Guidelines for the Use of Aluminum with Food and Chemicals,* the Aluminum Association, Washington, D.C., 1984.
4. *Aluminum Foil Converted,* U.S. Department of Commerce, Washington, D.C., 1981.
5. *Aluminum Statistical Review for 1982,* the Aluminum Association, Washington, D.C., 1983.
6. W. C. Simms, ed., *1984 Packaging Encyclopedia,* Cahners Publishing, Boston, Mass.
7. *Food Technology,* Institute of Food Technologists, June 1978.
8. *Food Eng.* 55, 60 (Feb. 1983).
9. *Packaging Newsletter,* the Aluminum Association, Spring 1981.
10. *Packaging Newsletter,* the Aluminum Association, Fall 1983.
11. News release, the Proprietary Association, Washington, D.C., Oct. 14, 1982.
12. *Food Eng. Int.,* (Jan./Feb. 1984).
13. Underwriters Laboratories, letter of July 23, 1984.
14. *Packaging Newsletter,* the Aluminum Association, Fall 1979.

Foil Division of the
Aluminum Association, Inc.

FOLDING CARTONS. See Cartons, folding.

FOOD CANNING. See Canning, food.

FOOD, DRUG, AND COSMETIC PACKAGING REGULATIONS

The history of food, drug, and cosmetic regulation in the United States is essentially one of reaction to a real or perceived crisis. The Pure Food and Drugs Act of 1906(1) was passed in the wake of public outcry after the so-called muckrakers, such as Upton Sinclair in his book, *The Jungle,* exposed the fraudulent and hazardous practices in the food and drug industry, particularly those involving meat packing (2). In keeping with the governmental philosophies most readily employed then, the act creating the FDA established the agency as a policing authority, not as a regulatory body of the type popularized in the thirties.

Under the 1906 Act, it was unlawful to ship or deliver adulterated or misbranded food or drugs in interstate commerce. Although the Agency had seizure and condemnation powers, as well as the general power to make rules and regulations for enforcement purposes, it was not authorized to issue industry-wide regulations or make standards to protect the public health (3). Nor was the Agency given preclearance authority over foods, drugs, or anything else.

Following the elixir sulfanilamide tragedy of 1937 that caused more than 100 deaths (4) and reports that instances of blindness occurred after the use of eyelash dye (5), new legislation was passed. Under a new statute, the Federal Food, Drug, and Cosmetic Act of 1938 (FD&CA or Act) (6), the FDA was given additional authority to promulgate regulations to establish definitions and standards of identity for foods (7). From a precedential viewpoint, it was ordered to preclear all "new drugs" after appropriate safety evaluations. Under the regulatory scheme, if no objections were raised by the FDA, a new drug application became effective once the statutory period for taking action on a new-drug application expired (8). Under the same new statute, cosmetics became subject to the Act's misbranding and adulteration provisions (9).

During the post-World War II period, concern over the growing use of food additives led to a series of Congressional hearings on their safety and the passage of key amendments in 1958 that resulted in a requirement that all food additives—including any substance that might transfer to foods from packaging materials or other food contact surfaces—be subjected to premarket approval treatment unless they fall within specified exemptions from the definition of a food additive (10). The amendments marked another acknowledgment that a technology, namely food manufacturing, had become too complex for direct Congressional control so that more authority had to be delegated to an expert agency. For the same reason, following close on the heels of the food additive amendments was a 1960 amendment requiring preclearance of color additives (11).

In 1962, another major amendment was enacted (12) in the wake of the discovery that the drug thalidomide (ironically never cleared for use in the United States because of the discipline already imposed by the 1938 law) led to severe birth defects (13). This amendment constituted a landmark in the bestowal of power based largely on subjective expert judgment because it ordered the FDA to base the preclearance of drugs on efficacy, as well as safety. In addition, the new provisions strengthened the 1938 preclearance requirement for drugs. The 1938 Act allowed a drug to be considered cleared if the FDA raised no objections to its use before the expiration of a review period; the new law required the Agency to issue an affirmative decision (14).

FOOD REGULATION

Food and Drug Administration

Prior to 1958, food packaging was only regulated in a very oblique way under the adulteration and misbranding provisions of the Act. Thus, a product was deemed adulterated if it were packed under unsanitary conditions or if its container were composed of any poisonous or deleterious substance that might render the product harmful to health (15). A product would be misbranded if the container was "so made, formed or filled as to be misleading" (16).

In 1958, with the advent of the Food Additives Amendment, components of packaging materials became subject to the same premarket approval as direct food additives if and when

they fell within the definition of a food additive. As a result, components classified as food additives cannot be marketed until their use is authorized by a food-additive regulation issued by FDA (17). No statutory distinctions were drawn between direct additives (ie, substances added directly to food) and indirect additives (packaging materials that migrate into food). Both were treated alike in terms of regulatory consequences (18).

When is a packaging material a food additive? The statutory definition of the term "food additive" is the starting point for determining the regulatory status of packaging material components or other food-contact substances. The definition appears in Section 201(s) of the Act, 21 U.S.C. §321(s), and provides that a food additive is

> any substance the intended use of which results or may reasonably be expected to result, directly or indirectly, in its becoming a component or otherwise affecting the characteristics of any food (including any substance intended for use in . . . packaging . . .), if such substance is not generally recognized, among experts qualified by scientific training and experience . . . to be safe under the conditions of its intended use; except that such term does not include. . . .

> (4) any substance used in accordance with a sanction or approval granted prior to the enactment of this paragraph. . . .

Under this definition, therefore, only three categories of substances in packaging material that come into contact with food are not food additives and therefore not subject to FDA regulation. They are substances that (1) may not reasonably be expected to become a component of food (19), (2) are generally recognized as safe (GRAS), or (3) are prior sanctioned. All of these important exceptions are discussed in detail below. In addition, although not mentioned in the Amendments themselves, the legislative history and FDA guidance make clear that housewares (ie, empty food containers, dinnerware, and utensils designed for repeated use by homemakers) are exempt from the definition of a food additive (20). Finally, the Agency itself, with judicial approval, has determined that substances separated from food by a functional barrier are not food additives and can be used without the need for any regulation (21).

Prior sanctions. Before 1958, the FDA and the USDA routinely received voluntary inquiries from manufacturers concerning the appropriateness of using a substance as a food additive or a food-packaging material. Favorable or "no objection" responses to these inquiries attained official status as highly prized "prior sanctions" with the enactment of the 1958 amendments. In a few instances, these pre-1958 clearances are listed in the scanty FDA prior-sanction list in the Code of Federal Regulations (22). There are, however, many more prior sanctions than are listed, and some of these can be controversial because of semantic accidents and differences about interpretation; the Agency takes the position that any prior sanctions are to be interpreted very narrowly.

GRAS. Aside from reliance on the list of substances the FDA told Congress it considered generally recognized as safe (GRAS) during the hearings that led to the Food-Additive Amendment (23), there are three other ways in which a substance may be considered to have GRAS status. First, the FDA publishes a list of GRAS substances (24). If a substance appears on this list, it is regarded as GRAS only if (1) it complies with specifications established by regulation or in the Food Chemicals Codex, a compilation published by the National Academy of Sciences; (2) it performs an appropriate function in the food-contact article; and (3) it is used at a level no higher than necessary to perform its intended function (25). The FDA is currently in the process of evaluating the substances on this list for the purpose of affirming appropriate uses as GRAS.

Second, manufacturers may file a GRAS Affirmation Petition (GRASP) seeking the FDA's concurrence that a substance not on the GRAS list is in fact GRAS (26). Evidence must be submitted to demonstrate that there is a general recognition of safety among experts qualified to evaluate a given chemical (27). For substances used before 1958, GRAS status may be established through experience based on common use in food (28). For substances first used after 1958, GRAS status must be established through scientific procedures (28). The advantage of filing a GRAS Affirmation Petition instead of a conventional Food-Additive Petition is that pending FDA action on a petition, a manufacturer may continue to sell his product once the Petition is accepted for filing without having to wait for FDA action on the GRASP (29).

Finally, if a manufacturer is confident that his product is GRAS even if it does not appear on any recognized GRAS List, he may market the product on his own responsibility without seeking FDA concurrence. The view that a product is GRAS should be based on the strength of sound, generally accepted scientific knowledge as reflected in published articles.

Substances which may reasonably be expected to become components of food. The most significant and most often controversial question concerning the regulatory status of a food-packaging material is whether any uncleared substances therein may reasonably be expected to become a component of food within the meaning of the Act. Section 170.3(e) of the Food-Additive Regulations provides the following explanatory information:

> A material used in the production of containers and packages is subject to the definition if it may reasonably be expected to become a component . . . directly or indirectly, of food packed in the container. . . . If there is no migration of a packaging component from the package to the food, it does not become a component of the food and thus is not a food additive (30).

This definition has been explained to some degree by the United States Court of Appeals in *Monsanto Co. v. Kennedy* (31). The *Monsanto* court stated that migration occurs within the meaning of the statute if a substance's "presence in food can be predicted on the basis of a meaningful projection from reliable data." It rejected the FDA's contention that mere contact between food and its container made the container a food additive. Rather the court held that:

> For the component element of the definition to be satisfied, Congress must have intended the Commissioner to determine with a fair degree of confidence that a substance migrates into food in more than insignificant amounts (32).

In other words, to be a food additive, a substance that comes into contact with food must be expected to become a component of food in more than *de minimis* amounts.

Although the principle of law involved is well established, neither the courts nor FDA have been willing to provide a hard and fast rule for determining what is *de minimis*. This has posed a continuing problem for the packaging industry which has been left without official guidelines for determining whether a particular item is subject to regulation.

The special case of carcinogens. Further complicating the question of what is a *de minimis* amount of a substance is the issue of carcinogenicity. As part of the 1958 amendments, Congress enacted a highly controversial provision known as the Delaney Clause. This clause provides that a food additive is not deemed safe and a regulation is not issued:

> if it is found to induce cancer when ingested by man or animal, or if it is found, after tests which are appropriate for the evaluation of the safety of food additives, to induce cancer in man or animal . . . (33)

Since its enactment, the Agency and industry have been attempting to develop a reasonable interpretation. In 1982, in the wake of *Monsanto v. Kennedy* and with heavy reliance upon its spirit, the FDA published an Advance Notice of Proposed Rulemaking, officially titled "Policy for Regulating Carcinogenic Chemicals in Food and Color Additives" (34). Although the policy has not even been officially proposed, let alone enacted, it is being applied by the Agency and has already passed judicial muster.

In large part, this proposal evolved from the holding in *Monsanto* that the FDA had authority to disregard insignificantly small migration. The proposed policy has three elements. First, it would attempt to clarify the statutory definition of "food additive" to distinguish between the additive itself and unwanted contaminants. Second, it would interpret the Delaney Clause as applying only when the additive itself has been shown to cause cancer. Third, the policy would adopt risk assessment as one of the tools for determining whether the carcinogenic constituent or contaminant is safe under the general safety provisions of the FD&CA.

In *Scott v. FDA* (35), the Sixth Circuit upheld the FDA's application of its carcinogen policy to D&C Green No. 5, a color additive known to be ingested in small amounts. The FDA has also used the policy as the basis for issuing food-additive regulations for acrylonitrile–styrene copolymer beverage bottles (36) and an antioxidant–stabilizer (37). (See Additives, plastics; Carbonated beverage packaging; Nitrile polymers).

The food-additive petition process. If it is determined that a substance is a food additive, and therefore not subject to any of the exclusions or special treatments discussed above, a food-additive petition must be filed with complete data establishing the safety of the additive in light of its maximum expected dietary intake (EDI). A food-additive petition must contain all pertinent information concerning the food additive including its chemical identity and composition; a statement of the conditions of the proposed use of the additives, including directions for use, specimens of labeling, relevant data bearing on the physical or other technical effect the additive is intended to produce, and the quantity of additive required to produce such an effect; a description of the practicable methods for determining the quantity of the additive in or on the food; and full reports of investigations made with respect to safety (38).

With respect to testing for safety, it has become Agency practice to require only an acute oral-toxicity test if the new material has an estimated daily intake (EDI) of less than 50 ppb. If the new material is expected to have an EDI greater than 50 ppb, subchronic toxicity tests are required involving two species with an *in utero* phase (39). In 1984, subchronic tests are estimated to cost in the neighborhood of $125,000. If the EDI is over one or two ppm, chronic feeding studies are required. These include two-year feeding studies that require a minimum of three years to conduct. Currently, chronic studies are said to cost $1.25–1.5 million (106).

Based on the data submitted to it in a food-additive petition, the FDA makes a determination on the safety of the packaging material under its intended conditions of use. The rule making encompasses consideration of the probable consumption of the additive, the cumulative effect of the additive in the diet of man or animals, other safety factors, and a determination as to whether the Delaney Clause is applicable (40).

The statute directs the FDA to issue a regulation authorizing the use of the additive or an order denying the petition within 90 days after the petition is filed. The FDA may extend this period by an additional 90 days if it requires the added time to study the petition. The regulation of an order denying the petition must then be published in the *Federal Register*. Objections may be filed within 30 days of publication, and public hearings may be held upon request. Judicial review may follow (41).

> Although the statutory mandate is clear, the time limits are meaningless in practice. The number of petitions filed each year for indirect food additives is hardly overwhelming. For the calender years 1977 to 1982, the number of petitions filed annually ranged from a low of 13 in 1978 to a high of 47 in 1982. Yet at the end of each year, between 46 and 85 petitions were still pending (42).

In June 1983 over 73% of the pending petitions were pending in excess of the six-month statutory time limit (42).

> Although these figures represent unacceptable delay, they understate the actual amount of time the petitions have been before the Agency. When a manufacturer wishes to obtain FDA approval for an indirect food additive, it is common practice to meet with the appropriate agency personnel to discuss a draft food-additive petition. The draft is then modified in light of informal FDA comment. Because of the difference between the official filing date and initial agency contact or submittal, determining the total actual time a petition has been pending is difficult. Thus, the figures presented here . . . certainly understate the length of time necessary to obtain FDA approval (43).

In a regulation establishing clearance for a substance, the FDA specifies the conditions under which the additive may be safely used, the maximum quantity that may be permitted to be used or permitted as a residue in food, the manner in which the additive may be used, and any required directions or other labeling or packaging requirements for the additive (44). Although a petition may be filed by a single applicant, the resulting rule is generic, applying to all manufacturers. If a manufacturer wishes to deviate from the regulation, he/she must file a petition to amend.

Once a packaging material is cleared, it must still comply with good manufacturing practices (GMPs) for food-contact materials set out in Section 174.5 of the Food Regulations governing indirect additives. These practices include using components of packaging substances that contact food in amounts no more than is reasonably required to accomplish the intended physical or technical effect, utilizing materials of a purity suitable for intended use, and complying with other

sections of the Food, Drug, and Cosmetic Act that are intended to make certain that food is not rendered unfit for consumption (45).

One of the most troubling aspects of food-packaging regulation is the fact that it is so misunderstood by packaging fabricators and food companies, particularly when the packaging material or a component thereof is not a food additive within the meaning of Section 201(s) of the Act. Often companies will buy only a material that has an official imprimatur of some type from the FDA or the USDA. But, official "approval" from FDA is not needed unless a substance is a food additive. Thus, an attorney Peter Barton Hutt stated just prior to becoming FDA General Counsel:

> . . . it is the primary and initial responsibility of the manufacturer of a product to determine the proper classification of his product, and to make certain that it meets all applicable legal requirements. It is in no instance necessary, and in most instances inadvisable, to ask the Food and Drug Administration for its opinion on the proper jurisdiction over the product. . . . [It] will probably seize upon any opportunity to state that the product should be handled as a [food additive]. It is therefore usually preferable for the manufacturer to exercise the obligation of proper classification given to him by the statute, rather than abdicating that responsibility to the Government (46).

The advisability of marketing a product based on one's own scientifically supported determination that a product is not an additive is best illustrated by an FDA chief counsel's memorandum which forms a part of the record in *Monsanto Company v. Kennedy* (47). The memorandum duly notes that the burden of proof of additive status is always on FDA:

> Finally, if any court action is brought, we (FDA) have the burden of proving two things: first, that the ingredient may reasonably be expected to become a component of the food, and second, that the amount of migration involved is not generally recognized as safe. We would need expert testimony on both issues. The fact that extreme conditions produced extraction would not, in my opinion, be sufficient evidence in and of itself to justify a food additive conclusion. We would be required to put on evidence of experts showing that the extraction studies are reasonably related to actual use conditions and thus that the results can be extrapolated to normal use. We would also be required to show that the amount that might reasonably be expected to migrate is not generally recognized as safe, and thus is a food additive (48).

Bureau of Alcohol, Tobacco, and Firearms

Alcohol is considered to be a food within the meaning of the Food, Drug, and Cosmetic Act (49). However, it is also subject to regulation by the Bureau of Alcohol, Tobacco, and Firearms (BATF) (50).

In 1982, the BATF codified its practice of deferring to the FDA on questions of the safety of packaging materials and indicated that it would focus its efforts on determining whether a given container adequately protects the revenue (51). Thus, a packaging material which passes muster under the FD&C Act is a legally satisfactory package for distilled spirits, provided its use does not adversely affect revenue collection (eg, does not result in alcohol content depletion).

U.S. Department of Agriculture

Although the FDA has jurisdiction over food generally, Congress has given USDA primary jurisdiction, among other things, over meat and poultry (52). With respect to packaging, the USDA has quite recently adopted regulations governing the use of packaging materials used in federally inspected meat and poultry plants. Inspectors are instructed (1) to ask for assurances or guarantees from suppliers of packaging materials, stating clearly that their products comply with the Federal Food, Drug, and Cosmetic Act and the applicable food-additive regulations; (2) not to ask for chemical acceptability letters issued by the USDA's Food, Safety, and Inspection Service; and (3) not to rely on such USDA letters unless they are accompanied by the required supplier assurance statement or guarantee (53).

A policy directive addressed to inspectors, circuit supervisors, and area supervisors discusses the types of materials that require supplier letters of assurance or guarantees. Packaging materials that actually contact food require an appropriate form of supplier statement of either type. Packaging materials that do not contact food, such as shipping cartons that are not the immediate containers, and labels applied to cans or other containers after the food is sealed inside, do not require status-assurance documents. Furthermore, no type of guarantee is required for packaging materials containing incoming ingredients that are used in the manufacture of the meat or poultry product; thus, packages with contents such as antioxidants, binders, and seasonings do not require letters of assurance or guarantees (54).

One distressing problem has arisen in connection with the use of colorants in packaging materials (see Colorants). Since the October 1983 FDA rule governing colorants (55) does not provide a formal reference list of acceptable pigments, packagers supplying federally inspected meat and poultry plants must ask color suppliers whether specific pigments have been cleared by the FDA. In many cases, pigments had been "approved" by the USDA but not the FDA, and under the present USDA policy, USDA "approval" by itself is insufficient. The FDA, however, has apparently abandoned its previous policy of providing opinions of "no objection" to colorant manufacturers when "no migration" evidence was supplied, letters which used to take two months to obtain. Instead, in late 1984 the FDA began requiring colorant manufacturers to file formal petitions for approval, a process which can entail a waiting period of 1 ½ to 2 years. Such requests have been made of colorant manufacturers even where they have submitted extraction tests (56) demonstrating no color extraction or migration with methods sensitive to 1 ppb.

DRUG REGULATION

Under the statutory scheme established by both the 1938 Act and its 1962 amendments, drugs can be classified into two categories, "new drugs" and "old drugs." A new drug is one that is not generally recognized as safe and effective (GRASE), or which has been recognized as GRASE, but has not been used "to a material extent or for a material time" or was not used before 1938 for identical uses with identical labeling (57). An old drug is not defined, except as a drug that is not a new drug. Thus, an old drug is one that is generally recognized as safe and effective and has been used "to a material extent or for a material time" or that was used before 1938 for identical uses with identical labeling (See also Pharmaceutical packaging).

A drug is defined as

(A) articles recognized in the official *United States Pharmacopeia*, official *Homeopathic Pharmacopeia* of the United States or official *National Formulary*, or any supplement to any of them; and (B) articles intended for use in the diagnosis, cure, mitigation, treatment, or prevention of disease in man or other animals; and (C) articles (other than food) intended to affect the structure or any function of the body of man or other animals; and (D) articles intended for use as a component of any articles specified in clause (A), (B), or (C); but does not include devices or their components, parts or accessories (58).

The Drug-Approval Process

With respect to new drugs, packaging is considered to be part of the drug itself, and therefore, data on packaging must be submitted as a part of the new-drug application (NDA) which must be approved by the FDA prior to marketing a new drug (59). Unlike a food-additive regulation, approval of an NDA is not generic; it is valid only to cover the party who filed the NDA. Changes in packaging do not, however, result in an old drug becoming a new drug since the distinction between old and new drugs depends on whether the current labeling contains the same representations concerning the conditions of its use as the pre-1938 product or whether the drug, apart from its package, is GRASE.

The approval process usually begins with a "Notice of Claimed Investigational Exemption for a Drug" or IND application (60), which is filed 30 days prior to the first human experimentation with a drug, which gives the Agency an opportunity to evaluate the proposed testing (61). Following testing under the IND, the NDA is filed. The NDA itself is usually voluminous. According to former Health and Human Services Secretary R. Schweiker, "The average application today contains 100,000 pages, filling hundreds of volumes. Applications arrive at FDA, literally in truck loads" (62).

Once the NDA is approved, supplemental NDAs must be filed for certain changes or new indications, including packaging changes that may alter the "safety, effectiveness, identity, strength, quality, or purity of the drug." These changes include those made to bring the package into compliance with child-resistant and tamper-resistant packaging requirements, as well as packaging modifications necessitated by the use of a new component (63) (See Child-resistant packaging; Tamper-evident packaging).

Abbreviated new drug applications (ANDA) may be filed by manufacturers of drug products that are duplicates of products previously granted approval by an NDA. The scope of the FDA's regulations governing ANDAs (64) was drastically changed with the passage of the Drug Price Competition and Patent Restoration Act of 1984 (65). The thrust of the new law permits ANDAs for a wide variety of "me, too" products in exchange for a specified amount of time added to the patent life of a pioneer drug as compensation for the time it takes to go through the regulatory-approval process.

The FDA's regulations governing packaging of drugs provide in pertinent part that each NDA application must include

a full description of the methods used in, and the facilities and controls used for, the manufacture, processing, and packing of the drug. Included in this description should be full information with respect to any new drug substance and to the new-drug dosage form, as follows, in sufficient detail to permit evaluation of the adequacy of the described methods of manufacture, processing, and packing and the described facilities and controls to determine and preserve the identity, strength, quality, and purity of the drug (66).

Information is required "with respect to the characteristics of and the test methods employed for the container, closure, or other components of the drug package to assure their suitability for the intended use" (67). Samples of the finished market packages of each dosage form of the drug must accompany the application (68).

In the event that a drug is intended for use by prescription only, its label must bear a statement directed to the pharmacist specifying the types of containers which may be used for dispensing purposes to maintain the drug's identity, strength, quality, and purity (69).

Drug master files. Since the FDA regulates drugs on a product-by-product basis, its interest in the packaging is limited to that used with a particular drug (70). The FDA does not evaluate individual drug-packaging materials except as a part of its evaluation of specific drug applications. Therefore, information about drug-packaging materials must be supplied with each application; and, since information on the composition of the package is frequently proprietary, as is the case with so-called drug substances sold to drug manufacturers, a "master files" procedure is frequently used by packaging materials sellers to protect their trade secrets. This involves the submission by the packaging manufacturers of relevant information on their materials directly to the FDA on a confidential basis. The FDA can then cross-reference such data in considering drug applications (71).

No fixed rules or requirements exist for this completely voluntary procedure. Packaging manufacturers usually provide the FDA with data relating to the percentage concentrations of components or formulations, relevant details about the manufacturing processes used, and explanations of product designations. Sometimes the master-file information may include references to food-additive regulation clearances, documentation for prior sanction or GRAS status for use as food packaging, and available toxicological or analytical data.

When data for inclusion in a master file are received, the FDA assigns a file number and advises the company that supplies the data of this number for its future use. The FDA does not assess the information submitted; that is, it does not approve master files per se. Rather, master files are a convenient source of confidential information that may be cross referenced by authorized drug manufacturers for use by the FDA. A master file holder must expressly authorize the FDA to refer to its master file in conjunction with a specific drug manufacturer's new-drug application. Individual companies often have several master files since the FDA sets up separate ones for clearly unrelated products or for each manufacturing site. When a company manufactures several different types of basically similar products, they may be consolidated in a single file.

Even though the master file supplies important basic data about a packaging material, generally it does not directly answer the FDA's concern about whether materials used in packaging a specific drug will affect the drug's safety or efficacy. In connection with an NDA or a supplemental NDA, the drug manufacturer must demonstrate by satisfactory data that the use of a proposed package will maintain the quality, purity, safety, identity, and strength of the drug.

The degree of proof the drug manufacturer will have to supply depends on all of the circumstances presented, i.e., the nature of the drug, the potential of the packaging material to affect its safety and efficacy, the possibility of reactivity between the drug and the package, etc. This is why FDA takes the general position that each case must be handled individually with proof being provided that the precise package specified will be safe for the specific drug application proposed (72).

There are several key points that a packager must keep in mind in preparing a drug master file.

1. The file must contain confidential material or the FDA does not afford master-file treatment. If all of the data about a packaging material is nonconfidential, it should simply be provided directly to the drug manufacturer for inclusion in its new-drug application. The information in a master file is kept confidential within the meaning of Section 314.11 of the new-drug regulations. A properly set up drug master file may, of course, contain confidential and nonconfidential material that assists the reviewer in evaluating the suitability of the packaging material.

2. The supplier should notify drug customers and update FDA master files whenever a substantive change is made in the products covered by the existing file. The FDA also requests that annual submissions be made to master files in order to keep them active. Often, updates merely confirm that the information in the file is still current.

3. Suppliers may be asked by drug customers to submit additional information if it is requested by the FDA.

4. Suppliers should avoid submitting information in such a way as to leave themselves with the alternatives of (1) being able to buy components from only a single manufacturer (ie, avoid using trade names for components of packaging materials whenever possible) or (2) having to ask a drug customer to file a supplemental new-drug application before the supplier can switch to a new source of supply (73).

One other factor should be considered with respect to packaging for drugs. There is often a fine line between drugs and devices (74), a point made clear in the regulation of blood bags. Bags that are used to collect blood are considered drug containers and are regulated as part of the NDA process applicable to all drugs because the bags are required to contain an anticoagulant, which is a drug (75). In contrast, bags intended for use in the collection of such blood products as plasma and platelets are shipped without any anticoagulant or other drugs and are therefore regulated as devices (76) (see Health-care packaging).

Good manufacturing practices. Finally, in terms of obtaining drug approval (77) and of avoiding charges of adulteration (78) and misbranding (79), good manufacturing practices (GMPs) must be followed in packaging a drug. In this regard, the primary requirement is that drug-product containers and closures may not be "reactive, additive or absorptive so as to alter the safety, identity, strength, quality or purity of the drug to beyond the official or established requirements (80). If the drug is recognized in an official compendium such as the *United States Pharmacopeia* (USP), it is misbranded unless the manufacturer follows the packaging requirements specified in the relevant compendium or receives permission to deviate from them (81).

The USP also provides test procedures for glass and plastic containers. Significantly, the USP test for high density polyethylene containers for capsules and tablets is generic (82).

Containers that meet this test can generally be used interchangeably for such products without prior approval from the FDA.

COSMETICS REGULATION

Cosmetics are the least regulated products subject to the Food, Drug, and Cosmetic Act. They are defined as "articles intended to be rubbed, poured, sprinkled, or sprayed on, introduced into, or otherwise applied to the human body or any part thereof for cleansing, beautifying, promoting attractiveness, or altering the appearance" of the user (83).

A cosmetic is deemed to be adulterated if it has been packed under insanitary conditions or its container consists of a poisonous or deleterious substance that may render its contents injurious to health (84). There are no premarket approval requirements for cosmetics, nor are there published GMPs.

Other than the proscriptions against adulteration or misbranding, there are no statutory requirements pertaining to cosmetics; however, manufacturers frequently require compliance with food-additive regulations for ingestible cosmetics such as mouthwash and toothpaste. With respect to noningestible cosmetics, there is minimal concern over meeting food-additive regulations, but manufacturers police products to prevent adulteration.

COLOR ADDITIVES AND COLORANTS

In 1960, Congress adopted the Color Additives Amendments to the Federal Food, Drug, and Cosmetic Act (85). The statute now provides that color additives are unsafe unless they comply with a color-additive regulation or are exempt from that requirement. The FDA was given authority to promulgate regulations providing for the separate listing of color additives for use in food, drugs, devices, and cosmetics (86). The term color additive is defined as:

(A) . . . a dye, pigment, or other substance made by a process of synthesis or similar artifice, or extracted, isolated, or otherwise derived, with or without intermediate or final change of identity, from a vegetable, animal, mineral, or other source, and (B) when added or applied to a food, drug, or cosmetic, or to the human body or any part thereof, is capable (alone or through reaction with other substance) of imparting color thereto: except that such term does not include any material which the Secretary, by regulation, determines is used (or intended to be used) solely for a purpose or purposes other than coloring (87).

The Agency has promulgated a food-additive regulation which defines colorants used in food packaging and other food-contact materials. It provides that

The term 'colorant' means a dye, pigment, or other substance that is used to impart color to or to alter the color of a food-contact material, but that does not migrate to food in amounts that will contribute to that food any color apparent to the naked eye. For the purpose of this section, the term 'colorant' includes substances such as optical brighteners and fluorescent whiteners, which may not themselves be colored, but whose use is intended to affect the color of a food-contact material (88).

The definition is troubling and legally defective because it encompasses substances which do not migrate into food.

Therefore, they are not food additives within the meaning of Section 201(s) of the Act, and are not subject to FDA regulation. The FDA's use of rule making to extend its jurisdiction to areas excluded by the statute is a source of continuing vexation.

Colors used in drug packages are dealt with at the time the drug itself is cleared. Of course, if the color migrates from the package to the drug so as to visibly color the drug, it would be a color additive and would require clearance as a color additive to drugs. Similarly, colors used in cosmetic packages are required to meet general standards for cosmetics, (ie, they do not render cosmetics adulterated). Again, if a color should migrate to the cosmetic or impart visible color, it would be a color additive and require clearance as such. Generally, color migration from packages to drugs or cosmetics is not tolerated by the manufacturers of the products so they seldom present a serious problem.

BIBLIOGRAPHY

1. The Pure Food and Drugs Act of 1906, ch. 3915, 34 Stat. 768 (1906).

2. For an in-depth analysis of the events leading to the passage of the 1906 Act, see J. E. Hoffman, *FDA's Administrative Procedures,* in Seventy-Fifth Anniversary Commemorative Volume of Food and Drug Law 1-2 (Food and Drug Law Institute, Washington, D.C. 1984); P. B. Hutt and P. B. Hutt, *A History of Government Regulation of Adulteration and Misbranding of Food,* 39 *Food Drug Cosm. L.J.,* **1**, 47–53 (1984); and J. O'Reilly, Food and Drug Administration § 3 (1983).

3. J. E. Hoffman, ref. 2 at 2-3.

4. *Ibid.,* at §§ 3.04 and 13.04 (citing Report of the Secretary of Agriculture to Congress. For a more in-depth analysis of the events leading to the passage of the 1938 amendments, see I-VI *A Legislative History of the Federal Food, Drug, and Cosmetic Act and its Amendments* (FDA) [hereinafter cited as *Legislative History*].

5. M. Gilhooley, *Cosmetic Regulation: Going Beyond Appearance,* Seventy-Fifth Anniversary Commemorative Volume of Food and Drug Law, Food and Drug Law Institute, Washington, D.C. 1984, pp. 323, 325.

6. Public Law No. 717, 75th Cong., 3d Sess., 52 Stat. 1040 (1938) *Note:* the 1938 Act and all subsequent amendments have been codified at 21 U.S.C. §§ 301 *et. seq.* (1982).

7. *Ibid.,* at § 401.

8. *Ibid.,* at §505.

9. *Ibid.,* at §§ 601 and 602.

10. Food Additives Amendment of 1958, Public Law No. 85-929, 72 Stat. 1784 (1958). For more detail on the legislative history, see XIV *Legislative History,* ref. 4.

11. Color Additive Amendments to the Federal Food, Drug, and Cosmetic Act, Public Law No. 86-618, 74 Stat. 397 (1960). For more detail on the legislative history, see XIV *Legislative History,* ref. 4.

12. Drug Amendments of 1962, Public Law No. 87-781, 76 Stat. 780 (1962). For more detail on the legislative history, see XVII *Legislative History,* ref. 4.

13. O'Reilly, ref. 2, § 13.02. Thalidomide had been used in Europe and was available in the U.S. for investigational use only.

14. FD&CA §§ 201(p), 505, 21 U.S.C. §§ 321, 335 (1982).

15. FD&CA §§ 402(a)(4), (a)(6); 21 U.S.C. §§ 342(a)(4), (a)(6) (1982).

16. FD&CA § 403(d), 21 U.S.C. § 343(d) (1982).

17. FD&CA § 409, 21 U.S.C. § 348 (1982).

18. For a discussion of color additives, see below.

19. Food is defined as "(1) articles used for food or drink for man or other animals, (2) chewing gum, and (3) articles used for components of any such article." FD&CA § 201(f), 21 U.S.C. § 321(f) (1982).

20. 104 Cong. Rec. 17418 (1958) (Statement of Rep. Williams).

21. In *Natick Paperboard Corp. v. Weinberger*, 525 F.2d 1103 (1st Cir. 1975) the court dealt with the food additive status of paper and paperboard that contained excessive levels of polychlorinated biphenyl (PCB) compounds. The court held that:

> If the packager or other claimant can show that the food placed in or to be placed in the paper container is or will be insulated from PCB migration by a barrier impermeable to such migration, so that contamination cannot reasonably be expected to occur, the paperboard would not be a food additive and would not be subject to seizure under the Act.

Ibid. at 1107.

22. 21 CFR § 181.22 (1984).

23. *Food Additives: Hearings on Bills to Amend the Federal Food, Drug and Cosmetic Act Before a Subcomm. of the House Comm. on Interstate and Foreign Commerce,* 85th Cong., 2d Sess. 461–462 (1958).

24. 21 CFR § 182 (1984).

25. 21 CFR § 170.30(h) (1–3) (1984).

26. 21 CFR § 170.35 (1984).

27. 21 CFR § 170.30(a) (1984).

28. 21 CFR § 170.30(c) (1984).

29. *See* 37 *Fed. Reg.* 6207 (1972) (Implies that the food ingredient whose status is being reviewed is present in the food supply while the review is proceeding).

30. 21 CFR § 170.3(e) (1984).

31. 613 F.2d 947 (D.C. Cir. 1979).

32. *Ibid.,* at 955.

33. FD&CA § 409(c)(3)(A), 21 U.S.C. § 348(c)(3)(A) (1982).

34. 47 *Fed. Reg.* 14,464 (1982).

35. 728 F.2d 322 (6th Cir. 1984).

36. 49 *Fed. Reg.* 36,635 (1984).

37. 48 *Fed. Reg.* 37,615 (1983).

38. FD&CA § 409(b), 21 U.S.C. § 348 (b) (1982).

39. *See* Notes from Presentation by M. Van Gemert, FDA, to Society of the Plastics Industry, Inc., Food, Drug and Cosmetic Packaging Materials Committee (May 31, 1983) (available from Keller and Heckman, 1150 17th Street, N.W., Washington, D.C. 20036).

40. FD&CA § 409(c)(5), 21 U.S.C. § 348(c)(5) (1982).

41. *Ibid.,* at §§ 409(c)(2), (e), and (g); 348(c)(2),(e), and (g).

42. Statement of J. H. Heckman on behalf of the Society of the Plastics Industry, Inc. before the United States Senate Committee on Labor and Human Resources concerning the Need for Food Safety Amendments 20 (June 10, 1983).

43. *Ibid.* at 20-21. The damage to industry that the long petition process can involve is well illustrated by the handling of Monsanto's food-additive petition for acrylonitrile copolymer as a packaging material in its Cycle-Safe soft-drink bottle. Despite the *Monsanto* court decision in 1979 establishing the *"de minimis"* concept and remanding the case for agency reconsideration, and despite strong evidence that migration of the component at issue effectively ceased, FDA did not clear the use of acrylonitrile bottles until September 1984. In total, it took more than 7 ½ years for the bottle to be cleared and what at one time might have been a bottle of choice, now has a small possibility of being used for one-trip soft drinks because the market for plastic containers for these beverages has been satisfied during the intervening time period by a competing material.

44. FD&CA § 409(c)(1), 21 U.S.C. § 348(c)(1) (1982).

45. 21 CFR § 174.5 (1984).

46. This statement was made in 1969 in a paper entitled "Proper Classification of Products Under the Federal Food, Drug and Cosmetic Act." The paper dealt primarily with cosmetics and drugs, but the principles enumerated are apt here. With the permission of Mr. Hutt, the bracketed [food additive] has been substituted for the word "drug."

47. FDA docket no. 76N-0070 Ex. M-70.

48. For further discussion of this issue see, *Coping with "Reg-u-cide,"* a paper prepared for presentation at the Society of the Plastic Industry, Inc./Society of Plastics Engineers Plastics Show & Conference East by J. H. Heckman (June 22, 1984) (Available from Keller & Heckman, 1150 17th Street, N.W., Washington, D.C. 20036).

49. FD&CA § 201(f), 21 U.S.C. § 321(f) (1982).

50. 26 U.S.C. § 5301(a) (1980).

51. 47 *Fed. Reg.* 43944, 43946 (1982).

52. The Meat Inspection Act, 21 U.S.C. §§ 601-95; the Poultry Products Inspection Act, 21 U.S.C. §§ 451-70. (1982).

53. 49 *Fed. Reg.* 2230 (1984).

54. USDA Food Safety and Inspection Service Directive 7410.1 (June 29, 1984).

55. *See* discussion of colorants within Section IV of the Oct. 1983 FDA rule.

56. *Chemical Week,* 38–39 (Sept. 26, 1984).

57. FD&CA § 201(p), 21 U.S.C. §321(p) (1982).

58. FD&CA § 201(g)(1), 21 U.S.C. § 321(g)(1) (1982).

59. The statute does not differentiate between over-the-counter (OTC) and prescription (Rx) drugs. Both require NDAs for new drugs. However, because of administrative necessity, OTC drugs are being considered generically by type of drug (eg, analgesics) rather than on a product-by-product basis as has been the case for Rx drugs. OTC drugs will be subject to a definitive classification as old (generally recognized as safe and effective) or new drugs when final monographs have been published. In the interim, OTC drugs that contain active ingredients identical to those in products marketed over the counter pre-1975 for single ingredient drugs and pre-1972 for combination products, may be marketed without adverse regulatory action. *See* 21 CFR § 330.13 (1984); FDA Compliance Policy Guide 7132b.16 (May 1, 1984).

60. *See* FD&CA § 505(i), 21 U.S.C. § 355(i) (1982); 21 CFR § 312.1 (1984).

61. 21 CFR § 312.1.

62. O'Reilly ref. 2, § 13.11 (quoting Address by HHS Secretary R. Schweiker to National Pharmaceutical Council (June 23, 1982). Proposals have been made to reduce the burdensomeness of the IND and NDA phases of the drug approval process. *See* 48 *Fed. Reg.* 26,720 (1983) and 47 *Fed. Reg.* 46,622 (1982).

63. 21 CFR § 314.8 (1984).

64. 21 CFR § 314.2 (1984).

65. Public Law No. 98-417, 98 Stat. 1585 (1984).

66. 21 CFR § 314.1(c)(8) (1984).

67. *Ibid.,* at (c)(8)(i).

68. *Ibid.* at (c)(9)(ii).

69. *Ibid.* at (c)(4)(g).

70. This is significantly different than the Agency's treatment of food packaging, which is cleared for use with any number of types of food.

71. 21 CFR §§ 314.11 (1984).

72. The Society of the Plastics Industry, Inc., *Plastics Packaging for Drug Products—The Regulatory Story* 6 (September 1967) (written by J. H. Heckman).

73. *Ibid.,—The Regulatory Story 7.*

74. The regulation of medical devices is beyond the scope of this article; however, the rules for packaging of devices can be simply stated. If a device is a class I device, ie, one subject to the general controls of registration, record keeping, etc, there is no prior approval of packaging. If the device is considered to be a class II device, which is subject to performance standards, packaging requirements for a device could ultimately be mandated, although at the present time, no such performance standards are in place. Class III devices are subject to premarket approval, and as such, the packaging itself, where relevant, can be part of the approval process, just as with drug packaging. *See* FD&CA §§ 513-515, 21 U.S.C. § 360c, d and e (1982).

75. 21 CFR § 640.4(c) (1984).

76. *Ibid.* at § 864.9100 (1984).

77. *See* FD&CA § 505(b)(4), 21 U.S.C. § 355(b)(4) (1982).

78. FD&CA § 501(a)(2)(B), 21 U.S.C. § 351 (a)(2)(B) (1982).

79. FD&CA § 502(g), 21 U.S.C. § 352 (1982).

80. 21 CFR § 211.94(a) (1984).

81. FD&CA § 502(g), 21 U.S.C. § 352 (1982).

82. *United States Pharmacopeia XX,* (*USPXX–NFXV*), The United States Pharmacopeial Convention, Inc., Rockville, Md, 1980, p. 953.

83. FD&CA § 201(i), 21 U.S.C. § 321(i) (1982).

84. FD&CA § 601(c) and (d), 21 U.S.C. § 361(c) and (d) (1982).

85. Public Law No. 86-618, 74 Stat. 397 (1960); FD&CA § 706, 21 U.S.C. § 376 (1982).

86. FD&CA § 706(b)(1), 21 U.S.C. § 376(b)(1) (1982).

87. FD&CA § 201(t)(1), 21 U.S.C. § 321(t)(1) (1982).

88. 21 CFR § 178.3297 (1984).

J. H. Heckman
I. R. Heller
Keller and Heckman

FOOD PACKAGING

Packaging protects food against a hostile environment. Being biological, food can deteriorate to lose nutritive value; change color, flavor, and masticatory properties; and in some instances can become a toxicological hazard.

Food deteriorates by four vectors: biochemical, enzymatic, microbiological, and physical. The biochemical vector is the result of interaction of food chemicals because of proximity to each other. Enzymatic deterioration is biochemical deterioration catalyzed by enzymes naturally present in food. Microbiological deterioration is the most common food spoilage vector. Microorganisms ubiquitous in food products include yeasts, molds, and bacteria.

Yeasts and molds are generally found in association with high-acid sugary products while bacteria usually are found throughout all foods. Yeast and molds grow best on the surfaces of high-acid, sugary food products but will grow on virtually any food surface exposed to air. Bacteria will grow almost anywhere both within and on the surfaces of foods. Food sterilization is designed to destroy all microorganisms that could grow in or on food products. Water at almost any temperature from 32°F (0°C) to 140°F (60°C) permits the growth of microorganisms. Microbial growth is slowed by reduction in temperature with the rate of reaction changing by a factor of between two and four for every 18°F (10°C) change in temperature.

Bacteria include the largest groups of organisms capable of causing infections and intoxications. Further, low-acid food products with a pH above 4.5, in the absence of oxygen can support the growth of the *Clostridia* organisms capable of producing botulism toxins under anaerobic conditions. Most bacteria, however, do not produce toxic products but rather spoil the food product, reduce its nutritional value, and adversely alter its appearance and flavor.

Damage to food products not associated with biochemical, enzymatic, or microbial spoilage is usually physical, such as gain or loss of water or color. Elevated water activity by dry-food products increases the rate of biochemical reactions. In food products with high water contents such as fresh produce or meat, water loss alters physical characteristics and can lead to conditions for microbiological growth. From an economic standpoint, the water content must be maintained at its original level. Much of the function of food packaging is to ensure against gain or loss of water from the product. Almost all adverse reactions are accelerated by increasing temperatures.

Packaged Food Classification

Approximately half of all packaged products in the United States is foods. Few foods are not packaged in some manner, apart from fresh foods such as produce sold at roadside stands.

Fresh-food products include all fresh meats, vegetables, and fruits that are unprocessed except for removal from the original environment and limited trimming and cleaning. Because of spoilage vectors, fresh foods should be consumed as soon as possible and handled in a manner that retards their deterioration, which is relatively rapid at ambient temperature or above. Thus, meats from freshly killed animals are chilled rapidly to below 50°F (10°C). Most vegetables and fruits are generally reduced to below 40°F (4.4°C) by low temperature air, water, or ice.

Partially processed foods include those which have been altered to help retard deteriorative processes. These include many dairy products which must be refrigerated after pasteurization and cured meats which also must be kept refrigerated to ensure against microbial growth, etc.

Fully processed foods are those intended for long-term shelf life at room temperature, and include almost all heat processed, dried, etc, foods that have traveled through a food-processing plant before packaging.

Fresh Foods

About a quarter of the value of food products in the United States consists of animal meats, including beef, poultry, fish, mutton, veal, and pork, all of which are susceptible to microbiological, enzymatic, and physical changes.

Meat. The color of red meat depends upon the presence of oxygen. The natural color of myoglobin meat pigment is purplish. The basic color of the red meat is oxymyoglobin or oxygenated pigment. Oxidized myoglobin is the brown color seen when meat is exposed to the air for extended time periods.

To preserve red meat, the objective is to retard spoilage, to permit some enzymatic activity to improve tenderness, to retard weight loss, and to ensure an oxymyoglobin or cherry-red color at consumer level.

In distribution, most red meat is packaged under vacuum in high oxygen–water vapor-barrier flexible packaging materials to retard deterioration. Less than half of red meat in the United States is distributed from slaughter to retail use without packaging or with minimum packaging. At the retail level, packaging restores the bright cherry-red oxymyoglobin color. Oxygen-permeable flexible packaging such as poly(vinyl chloride) (PVC) film permits oxygen into the package while retarding the passage of water vapor. All fresh meats are moved from the slaughter to retail at temperatures below 50°F (10°C) in order to retard deteriorative processes.

Poultry. Poultry is extremely susceptible to microbiological deterioration as it is an excellent substrate for the growth of *Salmonella* microorganisms. Thus, it is vital that the temperature be reduced as rapidly as possible. Many times, poultry is immersed in cold water or ice to reduce the temperature to below the optimum for microbiological propagation. Poultry is shipped in wet or dry ice to retail level where packaging most often occurs. Packaging is in soft film such as PVC which retards water vapor loss. Increasing quantities of poultry are centrally packaged in similar systems and maintained under carefully controlled temperature conditions. Turkey and other poultry consumed seasonally are preserved by freezing after packaging in low oxygen–water vapor-permeable heat-shrinkable films similar to those used for refrigerated distribution packaging of primal cuts of fresh beef.

Fish. Fish are generally taken from cold waters and temperature must be reduced immediately. Fish must be kept so that there will be little or no weight loss. Much fish is marketed as fresh from bulk ice. Significant quantities, however, are frozen and packaged (or packaged and frozen). Packaging generally has low water vapor permeability to permit long-term frozen distribution without freezer burn or surface desiccation. Wrapped or coated paperboard cartons and flexible films are employed to package frozen fish.

Produce. Subtropical and tropical fruits and vegetables such as pineapple, banana, and tomato are susceptible to chill damage at temperatures below 50°F (10°C). At temperatures above 50°F (10°C), normal enzymatic and microbiological deteriorations occur. Products susceptible to chill damage must be reduced in temperature to the lowest level at which chill damage does not occur. The main deterioration vectors are enzymatic, retarded by temperature reduction through immersion in chilled water, air blast, or ice.

Fruits and vegetables should be handled relatively gently because of the ubiquitous presence of microorganisms on the surface. Damage to the product surface provides channels through which the microorganisms can enter to initiate spoilage.

Fruit and vegetable packaging is often in bulk in a variety of almost traditional wooden boxes and crates and corrugated fiberboard cases. In Western Europe, expanded polystyrene foam trays are often used. At or near retail level, bulk produce may be repackaged in oxygen-permeable flexible materials, with or without a tray. Access of oxygen, which may be achieved by punching holes in the package, eliminates the possibility of anaerobic respiration which is another spoilage vector. Small quantities of fruit and vegetables are packaged under controlled atmospheres (see Controlled atmosphere packaging).

Partially Processed Food Products

Partially processed food products have received more than minimum processing but still require refrigeration.

Processed meats. Processed meats such as ham, bacon, sausage, bologna, etc, are salted to reduce the water activity, are spiced for flavor, and may also have ingredients to main-

tain the desired red color. Curing agents include sodium nitrite and sodium nitrate. Because the color is fixed, oxygen is not a vector of color loss for extended time under refrigeration. Thus, cured meats maintained with an absence of oxygen will have extended shelf life (see Shelf life). Most processed meats are packaged under vacuum in thermoform/vacuum seal systems and distributed under refrigeration. Small quantities are packaged in pouches under inert atmosphere such as nitrogen.

Dairy products. Dairy products are derived from milk, which, despite best efforts, usually is not sterile. Pasteurization is a low-heat process to destroy disease microorganisms but not to destroy all microorganisms that could cause spoilage. Pasteurized dairy products must be maintained under refrigeration. The packaging of pasteurized milk must be in clean containers. Nonreturnable packages such as blow-molded high density polyethylene bottles or polyethylene-coated paperboard gabletop packages are most often used for packaging and distributing milk under refrigeration in the United States. Returnable glass bottles are still employed to a minor degree in the United States and to a great extent in the United Kingdom. Returnable milk bottles are carefully cleaned on each cycle to remove debris and then sterilized to destroy microorganisms.

Aseptic packaging (see Aseptic packaging) for milk has been used in countries around the world. In aseptic packaging, the milk is sterilized, that is, rendered free of microorganisms. Simultaneously, the high barrier paperboard/foil/plastic-lamination packaging material is sterilized. The two are brought together in a sterile environment and the package is sealed to produce sterile milk in a sterile package. The increased heat required for sterilization of the milk can lead to flavors different from pasteurized refrigerated milk. Aseptically packaged milk may be distributed at ambient temperature.

Both fresh and cured cheeses in the United States employ pasteurized milk into which the enzyme rennin plus pure culture microorganisms are introduced to cause the precipitation of the proteins. With fresh cheese, the protein precipitate is cut and salted slightly for flavor and then packaged in closed containers such as waxed paperboard, thermformed plastic tubs, foil lamination, or plastic wraps to restrict water loss and retard airborne organisms. For practical purposes, such products are solid versions of milk subject to the same microbiological deterioration. Cheese is protected against water loss to ensure retention of physical and flavor integrity.

Hard cheeses, on the other hand, continue to undergo processing to become less susceptible to microbial deterioration. In a few cases, hard cheeses can be distributed without refrigeration. Hard cheeses include parmesan, mozzarella, and Swiss. Mold cheeses include Roquefort, Stilton, and bleu.

In the past, surface-applied wax retarded moisture loss; today, high oxygen-barrier flexible film coupled with vacuum or inert gas packaging greatly retards water loss and helps suppress microbial activity by retaining an oxygen-free interior.

Yogurt is a partially fermented milk product subject to microbial deterioration. Many yogurts also contain jams, jellies, etc, which may contain spoilage microorganisms. Yogurt products must be maintained under refrigeration in closed waxed paperboard or thermoformed or injection-molded plastic cups or tubs used for packaging.

Ice cream is a frozen aerated emulsion of milk fat with sugar. Ice cream is not subject to microbiological deterioration because the distribution and use temperatures are too low for microbial propagation. Packaging is generally minimal: lacquered or polyethylene extrusion-coated paperboard cartons, with a few in molded plastic tubs.

Other products. Fruits and vegetables subjected to natural and/or artificial drying include apples, prunes, raisins, grapes, and apricots. A preservative such as sulfur dioxide retards microorganisms and drying reduces the water content. Dry fruits and vegetables are susceptible to changes in texture from gain or loss of moisture and can be subject to microbial spoilage if the water content increases sufficiently to permit the growth of microorganisms. Although often distributed with minimum packaging, dry fruit and vegetables may be sealed in low water vapor-permeability packaging to maintain the water content.

Fully Processed Foods

Fully processed foods are processed and packaged so that their ambient temperature shelf life can exceed three to six months.

Canned foods. Canning is the process of extending shelf life by thermally destroying all microorganisms and enzymes present and maintaining that sterility by hermetic sealing in oxygen- and water vapor-impermeable packaging that excludes microorganisms. The resulting food product has been fully cooked as a consequence of the heat process (see Canning, food).

Whether we use a metal can or a glass jar, the process begins with treating the food product prior to filling. Initial operations inactivate enzymes so that the enzymes will not degrade the product during processing.

The package is cleaned. The product is introduced into the package, usually hot. In general, air that can cause oxidative damage is removed from the interior. Air removal leads to an anaerobic condition which can foster the growth of *Clostridia* organisms. The package is hermetically sealed (see Can closing) and then subjected to heating. The package must be able to withstand heat up to about 212°F (100°C) for high-acid products and up to 260°F (127°C) for low-acid products which must receive added heat to destroy heat-resistant microbial spores. Packages containing low-acid (above pH 4.5) foods must withstand pressure. In glass-jarred products, there must be external overpressure to ensure that the closure stays on the package.

The thermal process is calculated on the basis of the time required for the most remote portion of the food within the package to achieve a temperature that will destroy *Clostridia* microorganisms. After reaching that temperature, the package must be cooled rapidly to retard further cooking.

Since the products contained are biochemically and chemically aggressive, significant interaction between the food product and the internal can coating can occur. Thus, high temperature-resistant inert organic coatings are employed inside the can.

The packaging must contain the product, exclude air, withstand heat conditions, and also maintain hermetic seal throughout its distribution life and ensure that no microorganisms can reenter the package. It must be easy to open and dispense, inert to contents and the environment, shock and vibration resistant, and cost effective.

Attempts have been made to perform the same thermal retorting in a flexible pouch or tray. The retort pouch (see Retortable flexible and semi-rigid packages), under develop-

ment for many years, has a much higher surface-to-volume ratio and employs a heat seal (see Sealing, heat) rather than a mechanical closure.

The original intent of the retort pouch was to provide a convenient package for military ground forces. This objective was extended to attempting to introduce better quality food by virtue of less heat to effect the thermal sterilization.

Aseptic processing. Aseptic packaging is the independent sterilization of product and package, bringing them together in a sterile environment and sealing so that the contained food may be stored for prolonged periods at ambient temperatures. An important objective is the ability to sterilize the product outside the package, use thin-film heat exchangers, and thus impart less thermal damage to the products. This limited the application of aseptic packaging to liquids such as milk, juices, juice drinks, teas, puddings, etc. Most aseptic packaging is done with laminations of paperboard, aluminum foil, and polyethylene into block shapes; some is being performed in thermoformed cups on thermoform/fill/seal or deposit/fill/seal systems.

Frozen foods. Freezing requires reducing the temperature below the freezing point of the water so that microbiological, enzymatic, and biochemical activities are significantly retarded. In freezing, the product temperature passes through the transition from liquid water to ice rapidly so that ice crystals are relatively small and do not physically disrupt food cells.

The product may be frozen inside or outside of the package. In the early days of frozen foods, most freezing was performed with the product inside a waxed paper-wrapped paperboard carton; plate freezers, with direct contact of the cold surface to the package, were employed widely. More recently, however, freezing processes use high velocity cold air or even liquid nitrogen to remove the heat from the bulk unpackaged or individually quick frozen (IQF) product. Large quantities of cooking vegetables are frozen usually in IQF form.

The frozen product is then packaged in paperboard cartons or polyethylene pouches. Frozen foods are susceptible to sublimation of ice which can lead to freezer burn.

Probably the most important products frozen today in the United States are precooked, processed entrees in meal-size portions, often in aluminum-foil trays with thin-foil closures, overpackaged in printed-paperboard cartons. Soft baked goods such as cakes, pastries, pies, and breads are frozen after packaging in coated paperboard cartons which are grease- and water vapor-resistant.

Dry foods. By removing all water from food, biochemical activity ceases. Biological activities in dry-food products are related to water activity A_w. Water activity is the ratio of water vapor pressure of the food product to the water vapor pressure of pure water under the same conditions. Thus, a food product with an A_w of 0.6 would have an equilibrium relative humidity of 60%. If the relative humidity outside of the product were to exceed 60%, water would enter the product; if the relative humidity outside of the product is below 60%, water leaves the product. Water activity depends not just on the free water but rather on the active water present and capable of being evaporated.

By water activities, food products may be classified into very dry, with a water content of less than 10%; intermediate water activity or a water content between 10–85%; and completely hydrated with water activity of 0.85 or above. Products with water activities of 0.85 and above are essentially fresh or processed products susceptible to microbiological and rapid enzymatic deterioration and so must be processed for extended shelf life.

Dry products include those dried from liquid or engineered mixes of dried components that become dry products. In the first category are instant coffee, tea, and milk. The liquid is spray, drum, or air-dried because water content above 1% can lead to browning or significant product deterioration as well as particle agglomeration that restricts product rehydration. Engineered mixes include beverage mixes such as sugar, citric acid, color, flavor, etc, and soup mixes which include dehydrated soup stock plus noodles and some fat products, all of which become a liquid on rehydration.

Very dry products are often specially dried with porous surfaces to facilitate rehydration and so are very susceptible to moisture in the air. Oils within the product flavor are susceptible to oxidation, and so many such products are sealed under inert atmospheres to ensure against oxidation. Products with relatively high fat such as bakery mixes or soup mixes must be packaged so that the fat does not interact with the packaging materials. For example, low density polyethylene on the interior of the package can adversely interact with fat products. Further, seasoning mixes that contain herbs and volatile flavoring components can interact with plastic packaging materials. Interaction may be scalping or removal of flavor from the product or the product removing components from the packaging material. The package must be hermetically sealed; that is, it must provide a total barrier against access by water vapor, and for products susceptible to oxidation, also exclude oxygen.

Only very dry products with A_w below 10% require the very high barriers. Other dry products are more susceptible to the interaction of the fatty and flavor components of the product with packaging materials and less with the hygroscopicity of the product.

Fats and oils.   In general, fats and oils may be classified into those with and without water. Cooking oils and hydrogenated vegetable shortenings contain no water and are extremely stable. Hydrogenated fats and oils are subject to hydrolytic rancidity through reaction at high temperatures or long-term reactions at low temperatures with small quantities of water.

Unsaturated fats and oils are subject to oxidative rancidity; that is, breaking of the fatty acid chain at the double-bond sites to form peroxides and ultimately low molecular weight odorous aldehydes and ketones. Oils are more subject to oxidative rancidity than fats and both are usually packaged under inert atmosphere such as nitrogen. Hydrogenated vegetable shortenings generally are packaged in metal, composite paperboard, or plastic cans with nitrogen to ensure against oxidative rancidity. Oils are packaged in glass and blow-molded plastic such as high density polyethylene or PVC.

Fats and oils containing water include margarine and butter. Such products contain water-soluble ingredients, eg, salt, as well as milk solids which impart flavor and color to the product. Generally, these products are distributed at refrigerated temperatures to retain their quality. Greaseproof packaging such as coated paperboard, aluminum foil/paper and parchment wraps, and plastic tubs are used to contain butter and margarine.

Grain products.   The largest volume of food product in the world is in the form of grain, ie, rice, wheat, corn, oats, etc.

Water activities are reduced so that the starch and protein contents are stable. Most grain products contain oil subject to rancidity. The presence of water and air renders grains susceptible to biochemical deterioration. Bulk grains also are subject to infestation from insects, rodents, etc. Refined-grain products such as flour have relatively high water activity and so require no special packaging. On the other hand, grain products may contain insect eggs and can be the target of insects and rodents.

Grain products are converted into more edible products. Breakfast cereals contain most of the whole grain in cooked or toasted form plus flavorings, vitamins, and minerals. Cereals have a sufficiently low water content that they are susceptible to moisture absorption. At elevated temperatures, the fat can separate and release liquid fat. Breakfast cereals, therefore, in general require good water vapor- and grease-barrier packaging. Packaging generally should retain the delicate flavors. Breakfast cereals are packaged in glassine, waxed glassine, or polyolefin coextrusions in the form of pouches or bags within paperboard carton outer shells (See Bag-in-box, dry product). Sugared cereals are often packaged in aluminum-foil laminations to retard water vapor transmission (see Multilayer flexible packaging).

Soft baked goods including breads, cakes, and pastries are highly aerated structures containing about 45% water content. Soft baked goods are subject to dehydration and staling. Baked goods are also subject to microbiological deterioration as a result of growth of mold and other microorganisms. Equilibrium relative humidity must be maintained for product distributed at ambient temperature. Because of the short shelf life, fairly good water vapor barriers such as polyethylene-film bags or polyethylene-coated paperboard are used for packaging.

Hard baked goods such as cookies and crackers generally have a relatively low water content and a high fat content. Water can be absorbed, and the product can lose its desirable texture and be subject to rancidity. At elevated temperatures, the contained fat can be exuded and so dry bakery products should be packaged to exclude water vapor and also to eliminate the possibility of fat from the product leaking into the packaging material. Packaging for cookies and crackers includes waxed glassine or polyolefin-coextrusion pouches within paperboard shells, polystyrene trays overwrapped with polyethylene or oriented polypropylene film, etc. Soft cookies are packaged in high water-vapor-barrier laminations containing aluminum foil.

Snacks.   Snacks include dry grain or potato products such as potato and corn chips, and roasted nuts. Snacks usually have low water content and relatively high fat content. Snack packaging problems are compounded by salt, a catalyst for oxidative rancidity and an ingredient which can abrade the packaging material surface. Snacks are most often packaged in pouches that have low water vapor transmission, relying on rapid distribution to obviate any fat oxidation problems. Some snacks are packaged under inert atmospheres in sealed rigid containers such as composite cans to permit long-term distribution. Generally, light harms such products, so opaque packaging is usually employed.

Candy.   The three basic types of candies of concern are chocolate and hard and soft sugar. Chocolates may be either solid, solid with inclusions such as nuts, or molded with centers. Chocolate is a mixture of fat and nonfat components, infrequently subject to flavor or microbiological change. Inclusions and fillings are susceptible to water gain or loss. The chocolate acts as a water vapor barrier that helps protect the inclusions such as nuts, crisped rice, liquid cherry. Chocolates are packaged in greaseproof materials such as glassine or polypropylene plastic.

Hard sugar candies are amorphous sugar with added flavors and are extremely hygroscopic because of their very low moisture content. Further, sugar candies are hard and fragile and should be protected against physical damage.

Soft sugar products include jellies, marshmallows, etc. These contain a matrix of water which can be lost as a consequence of evaporation, and so these products must be protected against water loss.

Sugar candies are sealed in low-water vapor-transmission packaging.

Spreads. Bread and cracker spreads include jams, jellies, preserves, and peanut butter. A serious defect of jams, jellies, and preserves is browning that takes place as a consequence of interaction of oxygen at the product surface which leads to darkening of the color.

Peanut butter is a fatty product containing ground peanuts and is susceptible to hydrolytic rancidity. The package, of course, must be greaseproof.

Most spreads are packaged in glass with reclosable lids.

Beverages. Beverages may be still or carbonated, alcoholic or nonalcoholic. The largest quantity of packaging in the United States is for two carbonated beverages: beer and soft drinks. (see Carbonated beverage packaging).

Beer and other carbonated beverages contain dissolved carbon dioxide which creates internal pressure within the package. Thus, the package must be capable of withstanding the internal pressure of carbon dioxide up to 90 psi (620 kPa) in some instances. Carbonated beverages are subject to oxidative changes in flavor and so should be protected against oxygen. Further, the packaging material should not contain components that can be scalped or removed by the product. Organically lined metal cans and glass bottles are the most used for packaging carbonated beverages. Since the 1970s, polyester plastic bottles have been used for soft drinks.

Beer is more sensitive than other carbonated beverages to oxygen, to loss of carbon dioxide, to off flavor, and to light. Further, most U.S. beer undergoes thermal pasteurization. Most commercial pasteurization is done after sealing in the package, and the internal pressure within the package can build up to well over 100 psi (689 kPa) at 145°F (63°C), the usual pasteurization temperature. Relatively few beers in the United States, except for draft keg beers, are not exposed to heat for stabilization. Beer is sensitive to flavors from packaging materials in which it is stored. Beer and other carbonated beverages are generally packaged at relatively high speeds, which means that the packaging materials must be extremely uniform, free of defects, and dimensionally stable.

About 10% of beer and a quarter of the soft drinks in the United States are sold in returnable glass bottles. Reusable bottles are cleaned and inspected before each use to ensure against chips, cracks, stones, and foreign objects.

Other liquid beverages packaged in the United States include wine, most of it in glass bottles. Very small quantities of wine are packed in metal cans, and very slight quantities are packed in paperboard–foil composites such as currently used for aseptic packaging of juices.

Liquor packaging requirements are subject to regulatory restrictions. The packages must be resistant to both water and alcohol and must not alter the product flavor or the proof or alcohol content. Alcohol extraction of components from coatings is not unknown. Distilled alcoholic spirits have traditionally been packaged in glass although now liquor is being packaged in lightweight nonbreakable plastic packages.

BIBLIOGRAPHY

General References

Packaging of Horticultural Produce, symposium proceedings, Sprenger Institute, Wageningen, the Netherlands, 1981.

A Processors Guide to Establishment, Registration and Process Filing for Acidified and Low Acid Canned Foods, FDA, HHS Publication 80-2126, U.S. Department of Health & Human Services, 1980.

Retort Pouch Technology Seminar II—Proceedings, Pouch Technology, Inc. Oak Brook, Ill., 1982.

Proceedings of Conference *Aseptic Processing and the Bulk Storage and Distribution of Food,* Food Science Institute, Purdue University, Lafayette, Ind., 1978.

Proceedings *International Conference on UHT Processing and Aseptic Packaging of Milk and Milk Products,* Dept. of Food Science, North Carolina State University, Raleigh, N.C., 1979.

Proceedings of First International Conference on Aseptic Packaging, Schotland Business Research, Princeton, N.J., 1983; Second Conference, 1984.

The Science of Meat and Meat Products, W. H. Freeman & Co., San Francisco, Calif., 1960.

H. M. Broderick, *Beer Packaging,* Master Brewers Association of America, Madison, Wisc., 1982.

A. L. Brody, *Flexible Packaging of Foods,* CRC, Cleveland, Ohio, 1970.

A. L. Brody, "Food Canning in Rigid and Flexible Packages," *Crit. Rev. in Food Tech.* **2**(2), (July 1971).

A. L. Brody, and E. P. Schertz, *New Developments in Meat and Meat Packaging Technology,* Iowa Development Commission, Des Moines, Iowa, 1968.

A. L. Brody and E. P. Schertz, *Convenience Foods: Products, Packaging, Markets,* Iowa Development Commission, Des Moines, Iowa, 1970.

D. Carracher, *Seminar Proceedings Aseptic Packaging,* Campden Food Preservation Research Assoc., Chipping Campden, Glos., UK, April 1983.

C. M. Christensen, *Storage of Cereal Grains and Their Products,* American Association of Cereal Chemists, St. Paul, Minn., 1982.

R. B. Duckworth, *Fruit and Vegetables,* Pergamon Press, Oxford, UK, 1966.

A. Griff, *Plastic Containers for Soft Drinks, Beer and Liquor,* Edison Technical Services, Bethesda, Md., 1981.

R. Griffin and S. Sacharow, *Food Packaging,* 2nd ed., AVI, Westport, Conn., 1982.

R. S. Harris and E. Karmas, *Nutritional Evaluation of Food Processing,* AVI, Westport, Conn., 1975.

A. C. Hersom, and E. D. Hulland, *Canned Foods. An Introduction to Their Microbiology,* Chemical Publishing Co., New York, 1964.

D. S. Hsu, *Ultra High Temperature Processing and Aseptic Packaging of Dairy Products,* Damana Tech, New York, 1979.

T. P. LaBuza, "Moisture Gain and Loss in Packaged Foods," *Food Tech.* **36,** (April 1982).

R. Lampi, "Retort Pouch. The Development of a Basic Packaging Concept in Today's High Technology Era," *J. Food Proc. Eng.* **4**(1), (1981).

R. A. Lawrie, *Meat Science,* Pergamon Press, Oxford, UK, 1966.

A. Lopez, *A Complete Course in Canning,* The Canning Trade, Baltimore, Md., 1981.

B. J. McKernan, "Developments in Rigid Metal Containers for Food," *Food Tech.* **37,** (April 1983).

C. R. Oswin, "Isotherms and Package Life: Some Advances in Emballistics," *Food Chem.* **12,** 3 (1983).

C. R. Oswin, "The Selection of Plastic Films for Food Packaging," *Food Chem.,* **8**(2), 121–127 (1982).

F. A. Paine and H. Y. Paine, *A Handbook of Food Packaging,* Leonard Hill, London, 1983.

N. Potter, *Food Science,* AVI, Westport, Conn., 1978.

Aaron L. Brody
Schotland Business Research, Inc.

FORM/FILL/SEAL, HORIZONTAL

A variety of packages can be made on horizontal form/fill/seal equipment. This article deals primarily with pouch making, but similar concepts are applied to thermoform/fill/seal (see Thermoform/fill/seal) and bag-in-box packages (see Bag-in-box, dry product). Pouch styles include the following: three-sided fin seal; four-sided fin seal; single gusset; double gusset; pillow pouch (lap seal/fin seal); and shaped seal. The package is usually made from rolls of film (see Films, plastic) or other "webs" (see Multilayer flexible packaging). The sides are normally heat-sealed (see Sealing, heat), but other methods such as ultrasonic, laser, or radio-frequency welding can be used to meet specific requirements.

Filling can be done in a number of ways, depending on the characteristics of the product. Fillers include liquid fillers (see Filling machinery, still liquid), paste fillers, augers, pocket fillers, vibratory, orifice-type, and gravimetric units (see Filling machinery, dry product). Accessories can be provided with most form/fill/seal equipment to provide: registration (feed-to-the-mark or stretch hardware); web splicing ("flying splice" automatic or manual equipment); in-line printing (see Printing); coding (noncontact, hot-leaf stamping, or printing units) (see Code marking and imprinting); embossing; perforating; notching; vacuum or inert gas packaging (see Controlled atmosphere packaging); aseptic packaging (see Aseptic packaging); tear string; cartoning or bagging.

Some machines are small and relatively portable; others are massive and fixed. Some versatile machines can be readily changed over within limits. Other equipment is dedicated to a given size, and any change requires extensive and costly modifications. Cost of equipment varies widely depending on the output required and the extent of the system purchased, and on accessories specified and design requirements (eg, sanitary criteria and environmental considerations).

This discussion covers pouch, thermoform, and horizontal bag/box machinery according to the outline shown in Table 1. The equipment is classified by type and functional sequence, and further subdivided by design parameters. These are all horizontal form/fill/seal machines, even though the pouch may be horizontal or vertical (see Form/fill/seal, vertical). It would be impossible to mention all the machinery available. Representative equipment is mentioned for clarity, not as endorsement.

Table 1.   Horizontal Form/Fill/Seal Equipment

A. pouch form/cut/fill/seal, pouch vertical
  1. in-line equipment
    a. single-lane, intermittent, and continuous motion
  2. rotary equipment
    a. single-lane, intermittent motion
B. pouch form/fill/seal/cut, pouch vertical
  1. single-lane, continuous motion
  2. multilane, continuous motion
  3. single-lane, intermittent motion
C. pouch form/fill/seal/cut, pouch horizontal
  1. single-lane, intermittent, and continuous motion
  2. multilane, intermittent, and continuous motion
D. thermoform/fill/seal equipment
  1. multilane, intermittent, and continuous motion
E. horizontal form/fill/seal bag-in-box equipment
  1. single-lane, intermittent motion

A. Pouch Form/Cut/Fill/Seal, Pouch Vertical

In-line, single-lane, intermittent motion. The basic "workhorse" of the pouch machinery group is the single-lane, in-line vertical pouch form/cut/fill/seal intermittent motion machine. It requires some web stiffness to transfer from the cutoff knife to the bag clamps. Size changes, within limits, can be accomplished by adjustments to the machine. Its flexibility includes the ability to handle a wide variety of pouch materials, including most self-supported heat sealable materials like polyesters, polyethylenes, foils, cellophanes, and paper. Liquids, creams, pastes, granular materials, pills, tablets and small hard goods, or a combination of these may be packaged on this type of equipment. Fillers include piston fillers, auger fillers, vibratory, volumetric, gravimetric, count, and timed cutoff units. Combination of fillers can be used to permit formulation in the pouch. Gassing and sterilization systems can be provided. The major factor in the selection of this equipment is versatility.

Bartelt Machinery Division of Rexham builds equipment of this type (see Fig. 1). The machinery produces up to 200 packages per minute. Pouch styles that can be produced on this equipment are three- and four-sided fin seal, bottom gusset pouches for greater volume, wraparound pouches, multiple compartment pouches, and die-cut pouches as well as some combinations of these. Bartelt has been a pioneer in the production of retort pouches and they can supply automated retort systems from roll stock through cartoning.

Costs for equipment of this type vary widely. For instance, N and W Packaging Systems of Kansas City, Missouri, makes a relatively simple unit that produces 55 packages per minute. Bartelt makes complete systems that take packages from several machines with an "on-demand" feature and cartons them on high speed equipment.

Hassia, a division of IWK, West Germany (FRG), builds a flat pouch unit that operates at rates up to 120 packages per minute. Another German firm that produces in-line intermittent motion pouch equipment is Hamac-Hoeller, a division of the Bosch Packaging Group.

In-line, single-lane, continuous motion. The sequence of form/cut/fill/seal on a continuous motion machine entails running a cut pouch into clips attached to a continuously moving

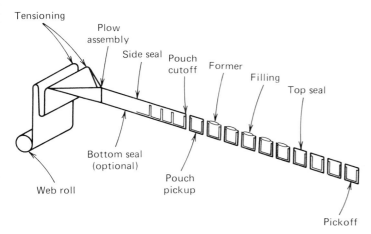

Figure 1. Material-flow diagram of typical Bartelt intermittent bottom horizontal form/fill/seal/machine. Filling of individual pouches allows maximum versatility for multiple component fills and easy pouch-size change. The machine primarily utilizes supported paper, foil or laminated film structures.

chain. Filling is usually accomplished by funnels that move with the pouch to allow time for the fill cycle. Sealing is done either with moving heat sealing bars (see Sealing, heat) or band or contact heaters with squeeze rolls at the discharge of the heater elements. Continuous motion equipment is more complex and expensive than intermittent motion machines. They are relatively fixed in the pouch size and do not provide the flexibility of intermittent units, but they operate at significantly higher rates.

Delamere and Williams, a subsidiary of Pneumatic Scale Corporation, builds a Packetron Pouch Maker that will run at speeds up to 500 pouches per minute. This machine can package products like instant cocoa mix, oatmeal, coffee, seasonings, dairy mix, powdered beverages, and most free-flowing food products. Sealing can be accomplished on three or four sides. A gusseted bottom can be supplied to provide an increased pouch volume.

Rotary single-lane, intermittent motion. In this design concept the pouch sides are sealed, then the pouch is cut and transferred to an indexing wheel. Pouch opening, filling, and top sealing take place at various stations around the circumference of the indexing assembly.

Hamac-Hoeller manufactures a unit of this type called the BMR-100. This machine can run up to 120 pouches per minute and produce three- and four-sided fin seals, bottom gusset, standup, and shaped seal pouches. The BMR-200 machine handles two pouches at a time. This unit has a top rate of 200 packages per minute and makes the same style pouches as the BMR-100 plus a twin pouch with a central seal. These machines can be fitted with various type fillers and auxiliaries to meet specific needs.

The Wrap-Ade Machine Company of Clifton, New Jersey, makes a more economical model of a single-lane rotary pouch machine. This unit produces pouches at a rate of 15–60 pouches per minute. It is a 12-station indexing system with seven possible loading stations.

A somewhat related system is available from Jenco (FRG). This machinery produces a stand-up bag of relatively large volume. The system can be provided with gas flushing, post-pasteurization, or sterilization. Jenco offers complete systems from bag-making to cartoning. System rates are up to 300 bags per minute.

The machines described above produce relatively small pouches. Horizontal form/fill/seal intermittent motion equipment to make and fill large bags is described below.

B. Pouch Form/Fill/Seal/Cut, Pouch Vertical

This area of pouch making has advanced significantly in the past decade. This concept takes the web continuously from a roll, runs it over a forming plough, makes the side seals on a rotary drum (and bottom seal if required), then forms and fills the pouch while it moves around a rotary filling hopper. The top of the web is stretched and the top seal is usually accomplished by running the web through contact heaters and squeeze rolls. A rotary knife is used to accomplish the pouch cutoff. This type of machine can produce packages at very high rates, up to 1300 packages per minute on a single-lane machine. It is an inflexible system and is normally dedicated to one size.

Single-lane, continuous motion. R. A. Jones & Co., Inc., Cincinnati, Ohio, makes a machine of this type called a Pouch King (see Fig. 2). A Japanese firm, Showa Boeki Company Ltd., makes a slower unit called the Toyo type R-10.

The Jones unit was originally developed by the Cloud Machinery Company of Chicago for use on sugar pouches with a paper-poly film (see Multilayer flexible packaging). The current capability of the equipment has been greatly expanded. The machine runs a wide range of sizes and flexible heat sealable structures. These include glassine, cellophane, foil, polypropylene film, and other materials with a variety of sealant materials.

The original machine could only fill relatively small product volumes relative to pouch size but Jones has developed a system called the "tucked-up-bottom" that holds a significantly larger product volume.

Various types of volumetric fillers that handle free-flowing

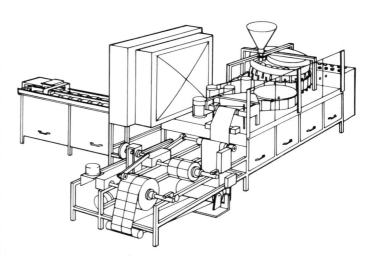

Figure 2. High speed (700–1200 pouches per minute)—horizontal/fill/seal machine is all rotary motion to provide long filling and sealing time. Pouches are automatically collated in the count and pattern required. Courtesy of R. A. Jones & Co., Inc.

products can be used on this equipment, including augers, orifice metering, vibratory, and pocket fillers.

Jones now has six models that run a wide range of products including salt, pepper, sugar, beverage mixes, instant coffee, roasted coffee, cocoa mix, and many other free-flowing products. Output are 750–1300 packages per minute.

Multilane continuous motion. A relatively recent introduction into this category is the Matthews Industries (Decatur, Alabama) "Ropak" machine. This equipment was also originally designed to run sugar pouches at rates up to 2000 packages/min. This rate is obtained by running two lanes at 1000 packages/min per lane. Until recently, the only structure run on this unit was a paper-poly combination, but a paper-poly-foil-poly web is now being successfully packaged. Machines are currently running sugar, pepper, and instant coffee. The majority of the product is bulk packed in drums or cases for institutional use. The machine is equipped with a predetermined counter-and-swing spout for bulk packing by count.

Single-lane, intermittent motion. Simplicity, economics, versatility, and space saving were the criteria for this approach to form/fill/seal pouches. Semirigid packaging material is drawn from roll stock across a vertically mounted tension control. It then passes over an adjustable guide roller into a horizontal and flat position. At this point, the packaging material is folded in half vertically by passing through two round guide bars. In its horizontal direction of travel, the package material passes the bottom seal station (not required for a three-sided seal pouch). A pouch-opening device separates the folded packaging material for entry of the filling funnels before the side seals are completed. After the product is filled, the packaging material passes the top sealing station, followed by the pouch-cutting device. The final station is the packaging material transport station. The machine is adjustable to a maximum pouch of 5 1/8 × 5 1/8 in. (130 × 130 mm) at output speeds up to 120 pouches/min (240 for duplex operations). The machine is manufactured by Kloeckner Wolkogon, a division of Otto Haensel (FRG).

C. Pouch Form/Fill/Seal/Cut, Pouch Horizontal

In this wrapper-type equipment the product is fed horizontally into a web that is wrapped around it and sealed. The product can be fed onto the moving web or the web can be formed around the product being carried on an indexing conveyor. The web is longitudinally sealed to form a tube around the product and the ends are sealed and cut off. The machinery is normally single lane but dual lanes can be run on some equipment. Machines are available that utilize automatic product feed, manual product feed, run registered web, and they can be equipped for inert gas packaging. A wide range of films and sizes can be run on this equipment. Products such as cookies, candy bars, cheese, and other rigid-type products can be packaged. The pouch is basically a pillow pouch style with a lap- or fin-longitudinal seal. Gussets can be added to handle increased product volumes. Four-sided fin-seal packages can also be produced on this equipment. Machinery can be supplied that operates on either an intermittent or a continuous basis. Speeds up to 300 packages per minute are available. The equipment is relatively flexible and changeovers are normally made by adjustments. An interesting innovation recently offered by Weldotron Corporation on its wrapper is a computer-controlled changeover system. Simply by pressing a button,

the machine will adjust itself to any one of six preset sizes. This system was developed by Omori Machine Co. of Tokyo and is distributed in the United States by Weldotron's OMC Packaging Division.

Machinery of this type is supplied by Hayssen Package Machinery, Sig, Bosch, Doboy and Oliver, to name a few. Many of these companies specialize in wrapping a given type of product.

Another category of horizontal form/fill/seal equipment is the pouch strip-packaging unit. This machine is basically used to package low profile products such as flat candy bars, tablets, hardware, medical, and novelty items. The machinery is capable of running two different webs and can run multiple rows of the same or different products.

D. Thermoform/Fill/Seal Equipment

Thermoform equipment takes the horizontal-packaging concept further. In this machine concept a thermoplastic web is heated and formed. The cavity is filled, lidded, and cut from the web. In some machines, a pressure-forming die can be substituted for the heat-form station to form a soft aluminum tray (see Thermoform/fill/seal).

A wide variety of packages can be made on intermittent and continuous motion machines of this type. Liquids, solid foods, pharmaceuticals, medical devices, hardware, and beauty aids can all be packaged on thermoform machines. Packages can range from small blisterpacks to deep drawn cups. The product can be gas, vacuum, or aseptically packaged. Films are available today that provide excellent forming characteristics and barrier protection for sensitive food products.

There is a wide range of thermoforming machines available. The method of transporting the web through the machine depends on the characteristics of the package and product requirements. There are die machines where the film is formed into a die and the die train moves through the machine, supporting the web at every station. By far the most prevalent is the dieless clip machine where the film is carried on its sides by means of clips. Some machines do not use clips but rely on the strength of the web to pull the packages through the machine. The film is heated, prior to forming, by contact or radiant heaters, then formed at the same station or indexed to a separate forming station. Accessories for film forming such as plug or pressure assists can be supplied to obtain deep draws with a uniform wall thickness.

E. Horizontal Bag-in-box Form/Fill/Seal Equipment

There are several types of equipment that produce this type of package. One is a system that uses a rotary indexing mandrel to form the inner and outer container. The inner container is usually a flexible film with barrier requirements to meet the needs of the product. The outer packaging material is usually for package appearance or structural requirements and can be either a printed web or a box. Another system forms the box in-line with the rest of the system, then the inner liner is formed and inserted in the box and the container is filled. A variation on this is that the container is filled while the inner liner is being inserted in the box (see Bag-in-box, dry product).

The major manufacturers of this type of equipment are Pneumatic Scale Corp., Hesser, and Sig Industrial Company. Only Sig and Hesser provide a vacuum system. They also provide a package with a valve system that will allow a product such as coffee to outgas without causing the package to rupture or balloon (see Vacuum coffee packaging).

Hesser also makes a system that produces a container similar to a composite can. This machine takes a laminate from rollstock and forms a rectangular body. It then attaches one end, fills the container and seals a lid to it. This equipment has the capability to make an aseptic package.

Horizontal form/fill/seal is an extremely dynamic segment of the packaging industry, for materials and equipment are continually improving, presenting new opportunities. New coextruded structures offer barrier and machining possibilities that are expanding the range of food products that can be packaged in flexible film (see Coextrusions for flexible packaging; Coextrusions for semirigid packaging). Aseptic packaging is another growth area that will add new dimensions to food packaging. In conjunction with the advance in these technologies are equipment developments that will provide a basis for expansion and cost reduction.

General References

Packaging Encyclopedia and Yearbook 1985, Cahners Publishing.

PMMI Packaging Machinery Directory, Annual, Packaging Machinery Manufacturers Institute, Washington, D.C.

J. H. Briston, L. L. Katan, and G. Godwin, *Plastics Films,* Longman, New York, 1983.

F. A. Paine and H. Y. Paine, *A Handbook of Food Packaging,* Blackie and Son Ltd., Glasgow, 1983.

R. F. Bardsley
Bard Associates

FORM/FILL/SEAL, VERTICAL

The term form/fill/seal means producing a bag or pouch from a flexible packaging material, inserting a measured amount of product, and closing the bag top. Two distinct principles are utilized for form/fill/seal packaging; horizontal (HFFS) (see Form/fill/seal, horizontal) and vertical (VFFS). Generally, the type of product dictates which machine category applies. This article deals specifically with VFFS equipment, which forms and fills vertically. It is used to produce single-service pouches for condiments, sugar, etc, as well as bags for retail sale and institutional use. The range of products and sizes is very large.

Package Styles

VFFS machines can make a number of different bag styles (see Fig. 1):

A pillow-style bag with conventional seals on the top and bottom, and a long (vertical) seal in the center of the back panel from top to bottom. The long seal can be a fin seal or a lap seal (see Fig. 1 **a** and **b**)

A gusseted bag with tucks on both sides to make more space for more product and maintain the generally rectangular shape of the filled bag (see Fig. 1 **c**). This style is used inside folding cartons for cereal and other dry products (see Bag-in-box, dry product)

A three- or four-sided seal package is similar to those made on HFFS machinery (see Fig. 1 **d**).

A stand-up bag (flat bottom, gabletop) of the type that used to be common for packaging coffee.

Other special designs such as tetrahedrons, parallelograms, and chubs (see Chub packages).

A flat-bottom bag needs a relatively stiff material to hold its desired shape, but any type of machinable material can be used to make a pillow-style bag. Various options are available, such as a hole punch for peg-board display, header labels which are an extension of a standard top of a bag, carry-handles for large consumer-type packages, and special sealing tools for hermetic seal integrity.

Materials

Two types of packaging materials are suitable for VFFS: thermoplastic and "heat-sealable" materials. Polyethylenes (thermoplastics) require a special bag-sealing technique. Polyethylene films must be melted under controlled conditions until the areas to be attached to each other are fused. The operation is analogous to welding metals. Heat is applied to fuse the materials and then a cooling process allows the seal to set. The sequence for making good seals requires careful control in order to get quality-seal integrity. Impulse sealing is used to seal thermoplastics on VFFS machines. A charge of electricity is put into a Nichrome wire which heats to a pre-established temperature (governed by material thickness) that will melt and fuse the materials. Since thermoplastics become sticky when melted, the Nichrome wire is covered by a Teflon (DuPont Company) sheath. The principle of impulse sealing does not require any specific tooling pressure.

Thermoplastic materials are generally used when a high degree of product protection is not required and low material cost is important. Polyethylene materials have some porosity and are not ideal for applications where hermetic seals are

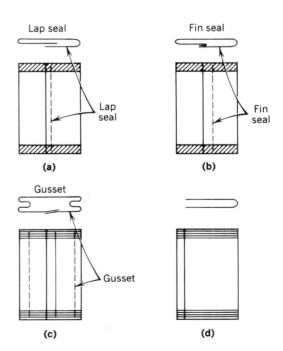

Figure 1. Selected package styles on VFFS machinery. (**a**) and (**b**), Pillow style; (**c**), gusseted style; (**d**) three-sided seal.

necessary for good shelf life, product freshness, gas flushing, etc. They are used, for example, for frozen foods, chemicals, confectionary items, fertilizers, and peat moss.

The class of "heat-sealable" materials or "resistance seal films" includes paper and cellophane as well as some coextrusions and laminations. Because these materials do not melt at sealing temperatures, or do not melt at all, they require a heat-seal layer that provides a seal with the right combination of time, temperature, and pressure. The sealant layer can be on one or two sides of the web, depending on the desired package configuration (see Multilayer flexible packaging).

A fin seal (see Fig. 1) can be made of materials with sealing properties on one side only, because the "heat-sealable" surface seals to itself. This seal is effective for powder products that need the seal to eliminate sifting. It is also a good seal if hermetic-seal integrity is important, as in gas-flush packaging. A lap seal uses slightly less material, but it requires sealing properties on both sides because the lap is made by sealing the inner ply of one edge to the outer ply of the other edge.

Machine Operation

A VFFS machine produces a flexible bag from flat roll stock. Material from a roll of a given web dimension is fed through a series of rollers to a bag-forming collar/tube, where the finished bag is formed (see Fig. 2). The roller arrangement maintains minimum tension and controls the material as it passes through the machine, preventing overfeed or whipping action. The higher the linear speed of the film, the more critical this handling capability becomes.

The bag-forming collar is a precision-engineered component that receives the film web from the rollers and changes the film travel from a flat plane and shapes it around a bag-forming tube. The design of the bag-forming collar can be engineered to get the optimum efficiency from metallized materials, heavy paper laminates, etc. As the wrapping material moves down around the forming tube, the film is overlapped for either the fin or lap seal. At this point, with the material wrapped around the tube, the actual sealing functions start. The overlapped material moving down (vertically) along the bag-forming tube will be sealed. The packaging material/film advances a predetermined distance that equals the desired bag-length dimension. The bag length is the extent of the material hanging down from the bottom of the tube. The bag width is equal to ½ of the outside circumference dimension of

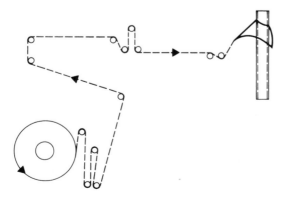

Figure 2. Typical film feed path through a vertical form/fill/seal machine.

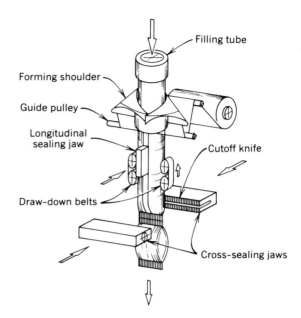

Figure 3. Typical VFFS configuration.

the tube. After the film advance is completed, the bag-sealing and -filling completes the remainder of one cycle (film advance/fill/seal). There are two sets of tooling on the front of the machine. One of the sealing tools, the vertical (longitudinal or back) seal bar, is mounted adjacent to the face of the forming tube. Its function is to seal the fin- or lap-longitudinal seal which makes the package material into a tube.

The other set of tooling, the cross (end) seal, consists of a front and rear cross-sealing jaw that combines top- and bottom-sealing sections with a bag cutoff device in between. The top-sealing portion seals the bottom of an empty bag suspended down from the tube, and the bottom portion seals the top of a filled bag. The cutoff device, which can be a knife or a hot wire, operates during the jaw closing/sealing operation. This means that when the jaws open, the filled bag is released from the machine. All vertical bag machines utilize this principle to make a bag (see Fig. 3).

Machine Variations

Film transport. Two distinct machine designs are used for transporting the packaging material/film through the machine. The traditional design clamps the material with the cross-seal jaws and advances the material by moving the cross-seal jaws down. This is called a "draw bar" (reciprocating up-down cross-seal jaws). The other is a drive-belt principle for film advance, which leaves the cross-seal bars in a fixed horizontal position with only open-close motion. The belt-drive film-advance principle has been shown to be the most versatile design for high speed packaging and simplicity of operation, and a number of companies have converted to this principle.

Power. There are several approaches to providing power for material/film transport and the filling and sealing operations: all electromechanical; electromechanical/pneumatic; and electromechanical/pneumatic/vacuum.

The electromechanical vertical-bag machine incorporates a cam shaft with a series of cams to operate the various functions. The package material/film drive motion works off a motor/reducer/clutch/brake arrangement. The long-seal and the cross-seal tooling are operated by cams. This allows cycle-to-cycle repeatability of machine settings. This is a very basic principle that has fixed timing on all components and is generally accepted as a heavy-duty, low maintenance design.

The most common VFFS design incorporates electromechanical power and pneumatics. This combination offers a manufacturing cost advantage in a highly competitive industry; but the trade off is the need to control air supplies carefully in order to keep the performance of the machine up to its top efficiency. The use of solenoid valves, pneumatic valves, flow controls, and air lines increases the maintenance requirements somewhat.

The electromechanical/pneumatic/vacuum principle is quite unique. The design is similar to the types utilizing air, with the addition of vacuum material/film transport belts. This principle locks the film to perforated drive belts by means of a vacuum pump. It works quite well, but generally imposes limitations on speed and minimum bag-width dimensions due to the design requirements for utilizing vacuum draw-down belts. Also, the vacuum pump adds another power requirement and noise factor.

Bag-Filling Factors

The product being packaged is generally the limiting factor regarding production rate capabilities on any of the machine designs. Machine operation is affected by product characteristics such as dust, fines, and stickiness, as well as by piece size, piece weight, and product volume. Some products create a piston effect when dropped down inside a bag-forming tube, by pushing air down into the sealed end of the packaging material. This air must escape somewhere. There are various controls such as inner fill tubes, snorkel tubes, etc, to release air pressure before the product drops down the tube.

All of these factors affect the end result. Too often, cycle capabilities of the bag machine and of the product measuring system are calculated independently, without considering what happens when the product moves from the measuring system down through the tube and into the bag. Achievable production rates are based on the compatibility of the three components: bag machine, filler, and product(s). Users of VFFS machinery should supply complete information concerning the products to be packaged to the manufacturers of the equipment so they can factually evaluate the achievable speed, weight accuracy, and efficiency capabilities.

Product Fillers

There are several different choices of measuring/filling equipment: net weigh scales, auger fillers, volumetric fillers, counters, bucket elevators, and liquid fillers.

Net weight scales are the most accurate means of measuring products for packaging. The invention and marketing of the multiple-head computer scale system in the past ten years has literally revolutionized the product-weighing industry. Package weight controls can be held to ± 1 g regardless of the size of the piece or particle being packaged.

Next in line for accuracy is the *auger filler*. This is applicable to products that are powdery in form and can be handled through a screw contained inside a tube. Most chemicals, baking products, and other powdery forms use an auger filler. The accuracy is dependent on bulk density control of the product throughout the augering system as well as the cycle repeatability of the chosen auger filler.

A volumetric cup filler fills by volume and is generally used for inexpensive products where high production rates are desirable and product overweight giveaway is unimportant. Counters (see Filling, by count) apply to applications such as hardware, confectionary items, and other items that must be packaged by count. The counter(s) can be mounted directly over the bag machine, or they can work in conjunction with a bucket elevator.

The *bucket elevator* can be an intermediate between any of the other product fillers, but it should be utilized only where space restraints or other impractical reasons dictate needs. The shortest distance between two points is the rule of thumb on VFFS systems, so mounting the filler directly above the vertical bag machine is the best and should be first choice where applicable.

The major VFFS equipment suppliers to the United States market are General Packaging Equipment Corporation, Hayssen Manufacturing Company, Package Machinery Company, Pneumatic Scale Corporation, Rovema Packaging Machines, Inc., Triangle Package Machinery Company, The Woodman Company, Inc., and Wright Machinery Division.

BIBLIOGRAPHY

General References

E. Dierking, F. Klien Schmidt, and J. Hesse, *Packaging on Form, Fill, and Seal Flexible Bag Machines*, Massuers Wolff Walsorde, FRG, 1975; discusses the importance of the bag-forming collar with regard to package quality as well as the relationship of material friction on machines utilizing the belt-drive principle.

"Extending the Shelf Life of Foodstuffs by Using the Aroma Perm Protective Gas Flushing System for Flexible Bags, "*Verpack. Rundesch.* **4**, 412–414 (1979).

The PMMI Packaging Machinery Directory, The Package Machinery Manufacturers Institute, Washington, D. C., annual.

GEORGE "ROCKY" MOYER
Rovema Packaging Machines

G

GAS PACKAGING. See Controlled-atmosphere packaging.

GABLETOP CARTONS. See Cartons, gabletop.

GLASS CONTAINER DESIGN

Glass has many advantages as a packaging material. It is rigid, transparent, inert, impermeable, and odorless, and a wide variety of shapes, sizes, and colors provide customer appeal. An abundant supply of inexpensive raw materials makes glass a natural packaging material. An effective glass container design considers many factors that relate to shape, manufacturing, marketing, and strength.

Container Shape and Dimensions

The basic container shape is determined primarily by the type and quantity of the product. Each product group, such as beverages, wine, liquor, food, pharmaceuticals, cosmetics, and chemicals, has a few characteristic container shapes. A sampling of these shapes is shown in Figure 1. In general, liquid products have small-diameter finishes for easy pouring. Some food containers need larger finishes for filling and removing the contents. A specific product may be available in a family of similar shapes to provide a range of desired capacities. Certain products have practical size limits; for example, perfume containers are much smaller than wine bottles.

The overall container shape can be further defined by three methods. First, the shape is defined by the Glass Packaging Institute (GPI). If several glass manufacturers supply containers to a few large-volume fillers, certain critical container dimensions are agreed upon to ensure that the container will be compatible with labeling, filling, packaging, and shipping operations. Second, the shape can be specified from the manufacturer's drawing of a stock job. Container manufacturers have a catalog of stock jobs for each product type. Third, a distinctive shape can be specially designed for a customer.

Finishes. After the general container shape has been defined, some very specific dimensions are required to ensure finish compatability with the closure (see Fig. 2 for container nomenclature). Finishes can be broadly classified by size, sealing method, and special features. The size, expressed in millimeters, refers to the nominal finish diameter. A closure size of approximately 1.38 in. (35 mm) is the dividing line between narrow-neck (blow-and-blow process) and wide-mouth (press-and-blow process) finishes (see Glass container manufacturing). Common sealing methods are continuous-thread, cork, crown, threaded crown, roll-on, and lug (see Closures). Special features include handle, pour-out, sprinkler, and snap caps. Selected examples are shown in Figure 3.

Head space. The product capacity and head space determine the container's interior volume. Headspace is needed to accommodate any thermal expansion of the product and to assist the filling operation.

Container stability. The center of gravity and bearing surface diameter determine a container's stability. Stability con-

tributes to forming and filling-line efficiency and minimizes customer-handling problems.

Manufacturing Conditions

Dimensions. Forming-machine requirements also affect container shape and dimensions. Bottle-forming machines have inherent height and diameter restrictions. The popular IS machine (see Glass container manufacturing) can make heights of roughly 1–12 in. (25.4–305 mm). IS-machine diameter limits depend on whether 1, 2, 3, or 4 bottles are made on a single section. The widest diameter range is approximately 0.5–6 in. (12.7–152 mm). Other glass-forming machines may have different limitations.

Contours. Hot viscous glass flows easier into a mold with a smooth, rounded shape. Containers with sharp corners or square heels are more difficult to manufacture than those with rounded profiles.

Tolerances. Acceptable limits or tolerances on container dimensions are necessary because forming-process variables prevent exact container specifications. The Glass Packaging Institute standard tolerances apply to capacity, weight,

Figure 1. Typical glass container shapes. (**a**) Laboratory: (left to right) sample oil bottle, amber gallon jug, amber Boston round; (**b**) beverage: (left to right) liquor round, wine round, single-serving juice bottle; (**c**) food: (left to right) dressing/sauce bottle, peanut butter jar, handled round sauce/oil decanter, spice bottle, specialty mustard mug; (**d**) drug and pharmaceutical: (left to right) amber shelf-pack, amber wide-mouth packer round; amber Boston round, amber iodine dropper bottle; (**e**) toiletries and cosmetics: (left to right) round nail-polish bottle, oblong cologne bottle, opal cream jar, and crystal-cut perfume square. Courtesy of Continental Glass & Plastic, Inc., Chicago, Illinois.

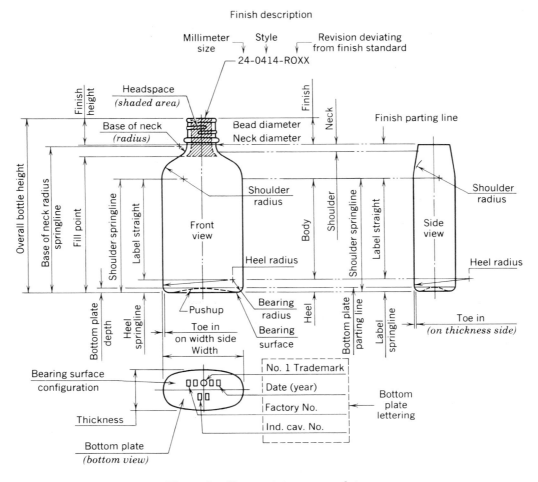

Figure 2. Glass container nomenclature.

height, and diameter. The capacity tolerance varies from 15% for very small containers to less than 1% for large bottles. The weight tolerance is approximately 5% of the specified weight. Height limits range from 0.5 to 0.8% of the overall height. Diameter tolerances vary from 3% near the minimum 1-in. (25-mm) diameter limits to 1.5% at the maximum 8-in. (200 mm) diameter size.

Marketing and Sales Factors

Glass containers enhance product customer appeal with distinctive shapes, attractive colors, graphic labels, and special decorations.

Shape. Distinctive container shapes have instant manufacturer or brand recognition. Customers can identify certain brands of beer, wine, or food by the container shape. An attractive shape, especially for cosmetic items, promotes increased sales. Sometimes the container has more sales appeal than the product.

Color. Flint glass (colorless) provides the customer with a clear view of the container's contents. In addition to their aesthetic appeal, amber and green glass protect certain foods and chemicals by limiting the transmission of ultraviolet light (see Colorants). Blue and opal glass can be used for decorative effects. Many shades of each glass color can be produced to suit the customer's preference.

Labels. Container labels are both attractive and informative (see Labels and labeling). They can be applied by firing an enamel, gluing on paper, or shrinking a preprinted plastic foam. Foam labels also provide cushioning and abrasion protection. Labels and labeling require consideration during container design. Flat areas or areas with only slight curvature must be provided when a label is to be glued or wrapped on the container. Designs often include an indented label panel to protect the label from abrasion.

Decoration. Decorating techniques include cut design, frosting, decal application, and fired enamels. Geometric patterns machined in blow molds provide a high quality image at a low cost. A frosted appearance can be achieved by sandblasting or hydrofluoric acid etching the exterior surface. Both organic and inorganic enamels can be applied by spraying or screening, and then they are heated to fuse the coating on the glass.

Container Strength Factors

A glass container's strength is controlled by shape, surface condition, applied stresses, and glass weight. The following general statements on container strength assume that all other strength factors are constant. This is valid for discussion purposes, but the strength of a specific container is governed by a combination of these factors.

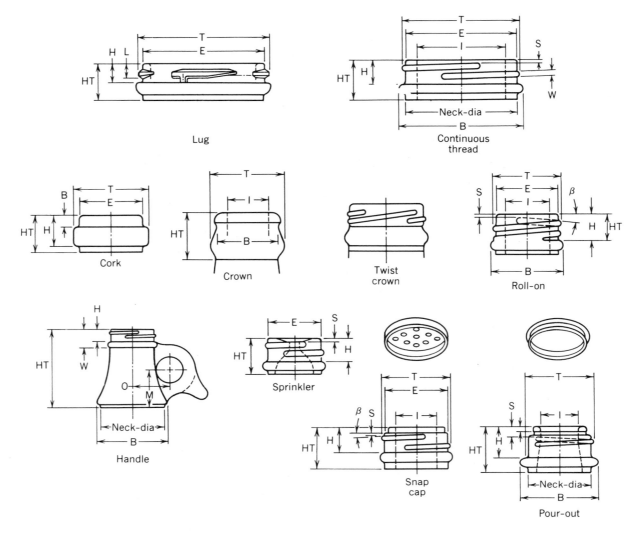

Figure 3. Typical glass container finishes. B = Bead diameter; β (beta) = angle of thread helix; E = wall diameter; H = intersection of "T" with bead or shoulder; HT = finish height; I = inside opening through finish; S = start of thread; T = diameter of thread; W = width of thread. (Thread must make a minimum of one complete turn from center line to center line of cutter.)

Shape. A container shape with a smooth vertical profile is usually stronger than one with sharp transitions. Similarly, a round-cross-section container is usually stronger than a rectangular one. High stress concentraions in sharp transition regions tend to influence container strength. Computer stress analysis of many container shapes shows that a balanced design of the shoulder, heel, and bottom regions can contribute to a container's strength.

Surface condition. A glass surface normally contains small surface imperfections. These imperfections are inherently formed during the manufacturing process and subsequent handling operations as a result of surface contact. The number, magnitude, and effect of these imperfections are controlled by good melting, forming, and handling procedures. Container design, surface coatings, or wraps also reduce the opportunity for formation of imperfections due to surface contact.

Containers can be designed with specific contact areas that concentrate abrasions where they will have minimal effect on glass strength. One example is the use of knurling on the bearing surface. The knurls (small protrusions) are the contact points between the containers and conveyors or other surfaces. Abrasions are concentrated on the tips of the knurls. Since the tips of the knurls are under less stress than the underlying glass, the abrasions have less influence on the container strength than they would otherwise.

Very thin, invisible coatings or surface treatments are often applied to a newly formed container. These treatments lubricate the glass surface and reduce surface abrasions. Bottles with a combination hot- and cold-end coating can be several times stronger than uncoated containers. Hot-end coatings, applied before annealing, include the oxides of tin and titanium. Possible cold-end coatings, applied after annealing, include polyethylene, stearates, oleic acid, silicones, and waxes.

Applied stresses. Container design must consider the type and magnitude of forces applied during its intended use. The most common forces applied to a container are internal pressure, vertical-load, impact, and thermal shock. Tension stress, created by these forces, influences glass strength more than compression stress.

Internal pressure.   Internal pressure stresses are developed from a carbonated liquid, or a vacuum-packed food product. The internal pressure of a soft-drink container may reach 50 psi (0.34 MPa). During pasteurization, internal pressure in a beer bottle may be as high as 120 psi (0.83 MPa) (see Carbonated beverage packaging).

Internal pressure in a closed container generates predominately circumferential and longitudinal stresses. In the cylindrical part of a typical bottle, the circmferential stress S depends on the bottle diameter d, the glass thickness t, and pressure p as approximated by the following equation:

$$S = \frac{pd}{2t}$$

The longitudinal stress in this part of the bottle is one half of this circumferential value. In the noncylindrical parts of a bottle, this equation does not accurately describe the stress. The rapidly changing curvature and walll thickness in the shoulder, heel, or bottom make it necessary to use much more complicated equations or numerical methods to calculate stress. Finite-element computer programs can be used to calculate the stresses in these regions. These programs make it possible to evaluate the stress conditions in alternative designs without actually manufacturing bottles. Figure 4 shows the highly magnified deformation of a container, caused by internal pressure.

Vertical-load stresses.   Vertical-load stresses are generated by stacking containers on top of each other or by applying a closure. These compressive forces produce tensile components in the shoulder and heel region, which may approach 1000 psi

Figure 4.   Magnified bottle deformation from internal pressure. ——, Displaced profile; _ _ _ _, original profile.

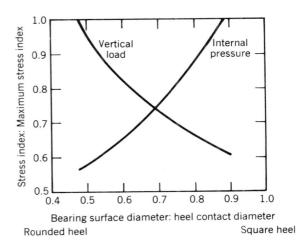

Figure 5.   The effect of heel shape on bottle stress for vertical loading and internal pressure.

(6.9 MPa). Vertical-load stresses can be lowered by reducing the compressive force or modifying the container design to withstand these forces. Computer stress analyses show that vertical-load stresses can be lowered by decreasing the diameter difference between the neck and body, increasing the shoulder radius, and reducing the diameter difference between the body and bearing surface. The vertical-load stress reduction from the latter case is shown in Figure 5.

This graph also shows the associated higher internal pressure stress for this heel-shape change. It also illustrates that a particular shape may be better for one type of loading and less favorable for another type. The designer must balance the design to accommodate all the requirements.

Impact stresses.   Impact stresses are produced when a container contacts another object. These contacts can occur during manufacturing, filling, transporting, and consumer handling. Figure 6 shows the three main stresses created during a container impact: contact, flexure, and hinge. The contact and flexure stresses are relatively high compared to the hinge stress, but localized to a small area. The lower hinge stress is important because it acts on a larger area on the outer surface.

Impacts produce a very complex and dynamic event. The stresses are related to the impact location on the container and the impact material. Also, the stresses vary depending on whether the container is empty or full, stationary or moving,

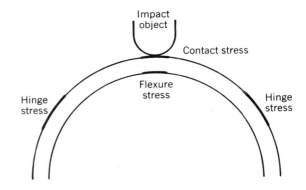

Figure 6.   Stresses produced during impact (1).

or supported or free-standing. All these factors make a dynamic impact analysis exceedingly difficult, but a steady-state finite-element model of a side load can provide useful design guides. These analyses show that higher hinge stresses are developed by impacts in the midbody than in the shoulder and heel regions.

Thermal shock. Thermal shock stresses are developed from rapid temperature changes during pasteurizing or filling a hot or cold product. The thermal-expansion differential between the hot and cold surfaces generates these thermal stresses. Tension stresses are produced on the cold side, and compression stresses are developed on the hot surface. This stress pattern is complicated by bending stresses generated by expansions and contractions in the bottle. Thermal stresses can be reduced by minimizing the temperature gradient from the hot to cold side, decreasing the glass thickness, and avoiding sharp corners, especially in the heel. A finite-element analysis can be used to evaluate a container design for thermal shock.

Glass weight. The intended use and expected stress conditions are important factors governing the glass weight. For example, a single-trip beverage container is subject to much less abuse than a multiple-trip container, so the single-service bottle need not be as heavy. Conversely, a high-internal-pressure champagne bottle must be heavier than a noncarbonated juice container. Use of surface treatments or coatings on the external surface of glass containers can reduce the amount of glass required to make the containers. Special-shape cosmetic containers are a category where the lightest weight may not be important. In this case, the aesthetic appeal may override glass-weight factors. For a new container design, the glass weight is determined by a combination of computer strength analyses and a comparison to similar jobs that have performed well. For a long-running job, weight reduction may be possible because the design and forming process can be fine-tuned.

Computer Uses in Glass-Container Design

State-of-the-art computer-aided design (CAD), computer-aided manufacturing (CAM), and computer-aided engineering (CAE) systems are used for container development. Computers have reduced design cost and improved container performance. Container shape definition can be modified on a computer to achieve the desired image. Tedious volume calculations that previously required hours are now completed in seconds. Clearer and more accurate bottle and mold-part drawings are plotted from computer-defined containers. Exact container geometry is supplied for numerical-control mold-part manufacturing. Computer-forming process simulation programs assist mold designers to achieve better parison shapes. Finite-element analysis of containers provides quantitative stress predictions that were previously unavailable. All these factors have played an important part in lightweighting glass containers.

BIBLIOGRAPHY

1. R. E. Mould, *J. Am. Ceram. Soc.* **35**, 230 (1952).

General References

B. E. Moody, *Packaging in Glass,* Hutchinson & Co. Ltd., London, 1963.

G. W. McLellan and E. B. Shand, *Glass Engineering Handbook,* McGraw-Hill Book Co., New York, 1984.

J. F. Hanlon, *Handbook of Package Engineering,* McGraw-Hill Book Co., New York, 1971; 2nd ed., 1984.

F. A. Paine, *Packaging Materials and Containers,* Blackie & Son Ltd., London, 1967.

F. V. Tooley, *Handbook of Glass Manufacture,* Ogden Publishing Co., New York, 1961.

S. M. Budd and W. F. Cornelius, *Glass Technol.* **17**, 54 (1976).

P. W. L. Graham, "Lightweighting, Strengthening, and Coatings," *Glass Technol.* **25**, 7 (1984).

D. L. HAMBLEY
Owens-Illinois

GLASS CONTAINER MANUFACTURING

This article covers the production of soda–lime glass containers directly from raw materials. It does not cover the production of ampuls and vials from either soda–lime or borosilicate glass tubing (see Ampuls and vials). Soda–lime glass derives its name from the two main constituents of the glass (other than silica): soda ash and limestone. The largest use for these containers is in the bottling of alcoholic and nonalcoholic beverages, foods, health supplies, and toiletries. Plastics have made major inroads into all of these areas during the last decade, but the most severe erosion has been in the latter two markets.

The topics in this article are covered in the same order that material flows through a glass container plant. The first topic is raw materials, followed by mixing and melting of the raw materials, forming the molten glass into a container, annealing, surface treating, inspection, and packing of the containers for shipment to customers. Finally, some statistical data concerning the number of glass containers produced for various applications are presented. Design considerations are discussed in a separate article (see Glass container design). The discussions in this article focus on the most modern techniques currently employed in the glass-container industry rather than techniques employed in the past. The Appendix contains a list of definitions of terms that are unique to the glass-container industry.

Plant Overview

The production area of a typical modern glass container plant occupies more than 15,000 m², and the adjacent warehousing space generally exceeds 20,000 m². New glass-container plants are frequently located near large customers and operate from one to three furnaces. In a modern high-productivity plant geared to the production of similar jobs, each furnace feeds only one or two production lines. In older plants and in those where several types of different containers are produced, each furnace might feed as many as six production lines. A line consists of one feeder, one high-productivity forming machine, an annealing lehr, and inspection and packaging areas.

Glassmaking raw materials received at the plant by rail or truck are unloaded under the direction of a computerized materials-handling system and directed into large silos near the furnace end of the plant for storage. This same computerized raw-materials-handling system controls the weighing and mixing of the glassmaking materials and delivery to the furnaces. Here the raw materials are melted by the 2640°F (1450°C) temperatures in the interior of the furnace. The molten glass leaves the furnace and is distributed to the forming machines through the forehearth and feeder. Gobs formed at the end of the feeder are delivered to the forming machine that makes (forms) the container. The formed container then passes through the annealing lehr. A surface treatment can be applied to the container just before or just after annealing, or both. The container is then inspected and packed, either for delivery to a decorating or labeling department in the plant or for delivery to a warehouse prior to shipment to the customer.

Melting

Raw materials. The oxide basis composition of typical soda–lime container glasses are shown in Table 1. Most of the raw materials used to achieve these oxide compositions are found in abundance in many locations throughout the United States. The major exception is soda ash, which comes primarily from deposits in Wyoming.

The major constituent in soda–lime glass is silica (SiO_2), which is the major constituent in sand; sand is normally over 99% SiO_2. Depending upon the impurities present in the sand deposits and the color of the glass to be produced, the sand is purified, perhaps crushed, and then screened to remove particles that are too large or too small (1).

The second-largest constituent in glass batches is cullet or recycled glass. One source of cullet is offware produced by the factory itself, but recycled glass has now become the principal source (see Recycling). Recycled glass has increased the amount of cullet in the batch to the point that in some furnaces cullet is the main constituent of the glass batch. The economic advantage of using clean recycled glass as a raw material does not lie in the material costs, since the cost of cullet is, depending upon location, comparable to the cost of the raw materials in the batch. The savings are associated with the lower energy costs (remelting cullet is easier than melting new raw materials (see Energy utilization)) and in improved furnace life. The quality of recycled glass cannot normally be controlled as tightly as the quality of the other raw materials. As a result, many plants have installed expensive equipment to process the cullet in order to remove items from the glass, eg, metals, ceramic materials, paper, labels, etc, which could potentially cause problems in the furnace. Poor separation of the cullet by color can also cause problems for some types of glass.

The next two largest constituents in glass are soda ash (Na_2CO_3) and limestone ($CaCO_3$ or $CaCO_3 \cdot MgCO_3$). During the melting process, large amounts of CO_2 (up to 185 times the volume of glass produced) are evolved from these two raw materials. Alumina (Al_2O_3) is an important but not large constituent in the batch. Depending upon the geographical location, feldspar, aplite, nepheline syenite, or feldspathic sand is used as a source of Al_2O_3. The major function of alumina in a soda–lime container batch is to enhance the chemical durability of the glass.

Refining agents are added to the batch to aid in melting reactions and in removing gas from the glass. Without the use of refining agents, much higher temperatures and longer melting times would be required in order to obtain glass free of bubbles (2). These refining agents typically contain sulfides and sulfates. Blast-furnace slag has been a popular refining agent for a number of years, but it is decreasing in popularity now because of its increasing cost.

Finally, colorants (see Colorants) such as Cr_2O_3 (for green

Table 1. Typical Container Glass Composition

Oxide	Wt %	Mol %
silica, SiO_2	68–73	68–73
calcia, CaO	10–13	10–13
magnesia, MgO	0.3–3	0.4–5
soda, Na_2O	12–15	12–15
alumina, Al_2O_3	1.5–2	0.9–1.2
ferric oxide, Fe_2O_3	0.05–0.25	0.02–0.13
sulfur trioxide, SO_3	0.05–0.2	0.04–0.15

glass), Fe and S (for amber glass), and Co_3O_4 (for blue glass) are added to the glass in small amounts to achieve a particular color. Decolorizers are added to flint (clear) glass batches to mask the color imparted to the glass by trace amounts of transition-metal impurities such as iron (3,4).

Charging. The raw materials are removed from their storage silos, weighed, mixed, and delivered to the furnace under control of a computer. Water is frequently added to the batch in the mixer or in the batch charger to reduce the number of fine particles blown into the regenerators by the strong combustion gas currents present in the furnace. The materials are typically "charged" into the furnace either by an auger-type screw charger or a pusher-bar charger.

Charging of the furnace occurs when a level gauge (5) senses that the level of molten glass in the furnace has dropped below a minimum set point. Charging continues until the glass level reaches the maximum set point. The glass level in a furnace is typically held constant to within one millimeter. Because of the intermittent nature of the charging process, the screw-type charger places "piles" of batch on top of the molten glass in the furnace. The pusher-type charger places "logs" of batch on the molten glass. The initial decomposition of carbonates and chemical reactions between the various raw materials begin in the batch piles or logs floating on the molten glass in the furnace.

The function of a glass-melting furnace is to convert the raw materials into molten glass that is chemically homogeneous and virtually free of gaseous inclusions. The chemical reactions occurring between the raw materials in a glass-melting furnace are complex, involving decomposition, solid-state reactions, and liquid–solid reactions as well as melting. Since they occur at very high temperatures, they are relatively hard to study experimentally (6).

The furnace. Glass-melting furnaces are constructed from refractory bricks. Different types of bricks are used in different parts of the furnace. For example, the bricks in contact with the molten glass are chosen for their high resistance to chemical attack by the glass. These bricks are quite dense and have a relatively high thermal conductivity. They typically contain large amounts of Al_2O_3 or ZrO_2 (zirconium oxide), or both. Outer courses of brick are chosen mainly for their insulating properties. The bricks are held in place by a steel framework surrounding the furnace (7).

There are two principal types of furnaces used for melting container glass: end-port and side-port (see Figs. 1 and 2). Either type of furnace can be fossil-fuel fired, but electric boosting is frequently employed to increase the glass-melting capacity. All-electric furnaces are used in some special situations to melt container glass. The names end-port and side-port refer to the manner in which the fuel is introduced into

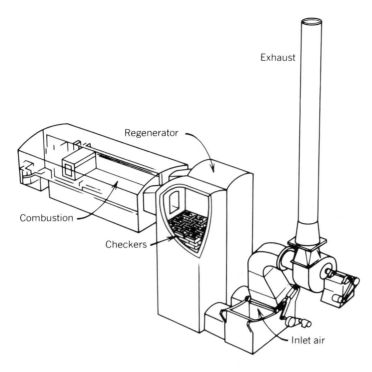

Figure 1. Side-port furnace with natural draft stack.

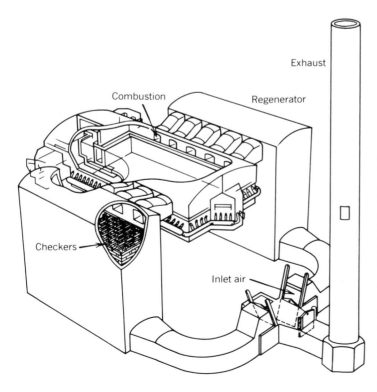

Figure 2. End-port furnace with forced draft stack.

the furnace. In an end-port furnace, there are two large openings (ports) in one end of the furnace. Combustion air is fed to the combustion space of the furnace through one of these ports while combustion gases leave the furnace through the other port. Fuel is introduced into the combustion air by various types of injectors located near the mouth of the port.

Side-port furnaces have several smaller ports along each side of the furnace. Combustion air and fuel are introduced along one side of the furnace and the combustion gases are exhausted through the other side of the furnace.

The size of container furnaces is measured by the surface area of the glass bath. End-port furnaces typically contain a glass bath with a surface area less than 700 ft^2 (65 m^2). Side-port furnaces typically have glass bath surfaces larger than 600 ft^2 (55 m^2). The glass depth typically varies from 3.5 ft (1.1 m) for colored glasses to as much as 6 ft (1.8 m) for flint glasses. A container furnace normally produces 3500–8000 kg of molten glass per hour.

In order to improve fuel efficiency, most container furnaces have either regenerators or recuperators to heat the incoming combustion air. Regenerators are simply large stacks of refractory bricks (30–45 ft or 9–14 m high) through which cool combustion air and hot combustion gases are alternately passed (see Figs. 1 and 2). For example, in a side-port furnace, combustion air enters through the regenerators on the left side of the furnace and hot combustion gases exhaust through the regenerators located on the right side of the furnace. The exhaust gases heat the refractory bricks they pass around. After a period of time, typically 15–30 min, the air flow through the furnace is reversed: combustion air enters around the hot refractory bricks in the right-side regenerator; and combustion gases exit around the cool refractory bricks in the left-side regenerator. In a recuperative furnace, the hot combustion gases and the cool combustion air pass through a network of passages separated from one another by thermally thin walls. In a modern furnace, the amount of fuel supplied to the combustion space is controlled by a computer that monitors several thermocouples located at various locations in the furnace. The air:fuel ratio is also controlled by this same computer, which monitors oxygen sensors in the exhaust gas stream.

Distribution and Conditioning

Function. Distribution of the glass involves providing a constant flow of glass to each of the furnace's container-forming machines. Conditioning of the glass involves cooling the glass from the melting temperature (1250–1350°C) to the forming temperature (~1150°C) and removing any thermal gradients from the glass stream. If the glass supplied to the forming machine is not of constant quantity, is thermally inhomogeneous, or is not at the proper temperature, severe problems are encountered in the forming process. Therefore, distribution and conditioning are very important in achieving good container production.

Distribution. The glass moves from the furnace through a narrow passage called a throat and into a refiner where the cooling and thermal homogenization of the glass begins. The glass continues its motion from the refiner into either a shallow chamber called an alcove and then into the forehearth, or directly into the forehearth. A forehearth is a channel-like structure that carries the glass to a forming machine. Two views of a forehearth are shown in Figure 3.

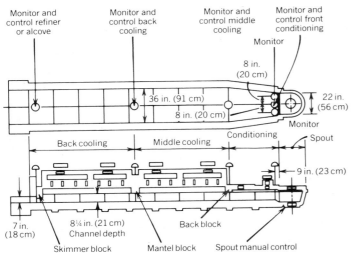

Figure 3. Top and side view of a forehearth.

At the end of the forehearth is a hemispherically shaped container called the bowl. The glass exits the bottom of this bowl in one to four continuous streams depending upon whether the forming machine works with one, two, three, or four gobs simultaneously. Immediately after exit from the bowl, the glass streams are cut (sheared) by rapidly moving steel blades. Each of the resulting cylindrically shaped gobs must contain the same volume of glass if consistency is to be obtained in both the forming of the containers and in the capacity of the finished container. Therefore, it is important that a constant head of glass be maintained by the distribution system at the bowl.

Forehearth. The forehearth is an element of both the glass-distribution system and the glass-conditioning system. The final adjustment to the average glass temperature and thermal homogenization of the glass occurs in the forehearth. The stream of glass in the forehearth is relatively wide (up to 4.3 ft or 1.3 m) and shallow (4–6 in. or 10–15 cm). Forehearths generally range in length from 16 to 33 ft (5 to 10 m). The reason for the wide, shallow configuration is that controlled heat exchange with the glass takes place through the top surface of the glass. Because glass is a rather poor thermal conductor, it is very difficult to obtain good vertical thermal homogeneity simply by exchanging heat at the top surface. By working with a shallow stream of glass with a large surface through which heat can be exchanged, better control over both average glass temperature and thermal gradients can be obtained.

Blenders are sometimes used near the downstream end of a forehearth in order to improve the thermal homogeneity of the glass leaving the forehearth. Blenders are frequently auger-shaped and constructed from ceramic materials. The blenders are rotated in the glass and thus impart a vertical and horizontal velocity component to the glass in the vicinity. The result is a stirring and mixing action that improves the temperature homogeneity of the glass.

Heat is normally removed from the glass by air blown over its top surface, and normally added by radiation from fossil-fuel flames directly over the glass. Some forehearths now uti-

lize electrodes introduced through the side of the forehearth to add energy instead of fossil fuels. These electric forehearths are much more energy efficient than fossil-fuel-fired forehearths. However, special steps must be taken for some types of glass to ensure that the electrodes do not introduce bubbles into the glass through electrochemical reactions.

Controlled addition of heat to or removal of heat from the glass is based upon readings from thermocouples immersed at different depths in the glass and located at different positions in the forehearth. Thermocouples at different depths are required because of the large vertical temperature gradients that can be easily set up in the glass. Information at different locations is necessary because the heat exchange in different zones of the forehearth must be controlled independently in order to supply glass to the forming machine at the proper temperature and with the proper thermal homogeneity.

Colorant forehearth. In addition to their distribution and thermal homogenization function, forehearths can be used to change the color of a glass. This is done when the demand for a specific glass color is not sufficient to justify running an entire furnace with that color. In such situations, special coloring materials can be added in the forehearth to, for example, change an otherwise clear glass into a colored glass. The coloring materials (frits) are relatively low-melting materials with relatively high concentrations of typical glass-coloring agents. The frits must be low melting so that they can be easily dissolved and mixed in the relatively cool glass in the forehearth. Forehearths used to change the color of the glass have a battery of stirrers (blenders) located just after the frit is dropped onto the surface of the glass. These stirrers mix the glass and frit so that the glass exiting the forehearth is of uniform color. Colorant forehearths are normally longer than standard forehearths to allow space for the addition of the frit and the stirrers and still maintain adequate thermal control over the glass.

Delivery

The function of the delivery system is to move gobs from the point where they are sheared into the blank mold. A diagram of the delivery system for an Individual-Section (IS) container-forming machine is shown in Figure 4. The vertical drop of the gob ranges 13–18 ft (4–5.5 m), depending upon the installation. The horizontal displacement of the gob through the delivery system can be as much as 10 ft (3 m). The terminal velocity of the gob upon exiting the deflector is 5–7 m/s.

The delivery system must be capable of consistently placing each of the gobs in the center of the appropriate blank mold at precisely the correct time. If either the placement of the gob in the blank or the time of arrival of the gob at the blank is erratic, problems in the formation of the container will result. In addition, the delivery system must minimize the amount of heat it removes from the gob so as to minimize any resulting thermal inhomogeneities in the gob.

Delivery systems are normally made from iron. The glass-contact surfaces are periodically coated, generally with a graphite-based compound, to reduce the coefficient of friction between the glass and the delivery system. The low coefficient of friction (typically 0.1) and the insulating nature of the coating serve to reduce the heat transfer coefficient between the glass and delivery system. This minimizes the amount of heat transferred from the glass to the delivery equipment.

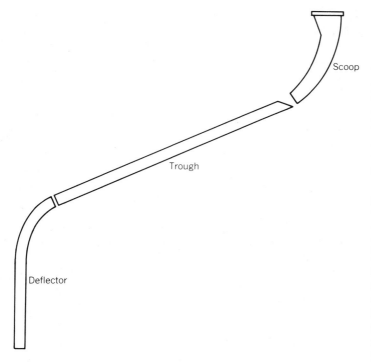

Figure 4.   Delivery system for Individual-Section (IS) machine.

Forming

The process of converting a cylindrically shaped gob of glass into a bottle or jar is called forming. There are various types of forming machines used throughout the world for making glass containers; but because the predominant type, by far, is the IS machine, most of this discussion pertains to the production of glass containers on an IS machine. The physical principles of forming with the IS machine are the same for all types of container-forming machines, but there are differences in the mechanics of loading the blank mold, transferring the parison between molds, and/or the transfer of the container out of the machine.

Function.   The forming machine performs two basic functions. It shapes the solid cylindrically shaped gob into a hollow container. Simultaneously, it removes heat from the gob, which has the consistency of thick molasses when sheared, to prevent the container from deforming significantly under its own weight. The objective of the forming process is to carry out heat removal and shaping functions as rapidly as possible without introducing defects into the container (see Fig. 5).

Forming processes.   There are two basic types of processes used to make containers on the IS machine. These are called blow-and-blow (B&B) and press-and-blow (P&B). In the B&B process, both the preform and bottle are made by blowing air (at 7–30 psig or 150–300 kPa) into the glass. In the P&B process, the preform is made by pressing the glass into the appropriate shape, and the final container is blown by air.

Blow-and-blow. In the blow-and-blow process (see Fig. 6), the glass enters the blank from the top ("loads the blank") after it has free fallen about 20 in. (50 cm) from the end of the deflector (see Fig. 4). As mentioned above, the gob is traveling at a speed of 16–23 ft/s (5–7 m/s) as it enters the blank. This high speed helps the gob to completely fill the lower portion of

the blank mold, which is normally smaller than the gob. The funnel helps to ensure that the gob is properly aligned with the blank cavity.

After the gob has been loaded into the blank mold, the baffle is seated on top of the funnel and settle-blow air is applied. The settle blow forces the glass down into the finish of the container, which is formed by the neck ring. Vacuum applied from the bottom can also be used in place of or in addition to the settle blow to force the glass into the neck ring. Before application of settle blow, a small plunger assembly is inserted into the neck ring. This plunger assembly forms the top and inside surface of the finish.

After settle blow, the plunger is retracted, the funnel is removed, and the baffle is seated directly on the blank. When the baffle is seated, counterblow air is applied to the gob from the bottom. The hole left in the gob by the retraction of the plunger serves as the starting point for the "bubble" that is blown in the gob by the counterblow air. (Counterblow pressure is typically 4–12 psig or 130–180 kPa.) The counterblow air enlarges the size of this bubble until the glass is pressed out against the blank mold and the baffle. The resulting hollow preform is called the parison. About 35% of the energy removed by the blow-and-blow process is removed during the formation of the parison. As a result of this heat loss, the parison (especially its outer surface) is now much stiffer than the original gob.

There is a large radial thermal gradient (as much as 2300°F/in. or 500°C/cm) set up in the parison as it is being formed. The reason for this is that heat cannot flow from the interior of the glass to the surface as rapidly as it can be removed from the outer glass surface by the blank mold. As a result, the interior-surface glass temperature has not dropped significantly when the blank opens, but the outer surface temperature of the parison has dropped more than 540°F (300°C).

After the blank opens, the parison is transferred by the invert mechanism to the blow mold. During transfer (invert), the parison is held in the neck ring by the finish. From the time the blank mold starts to open and continuing during

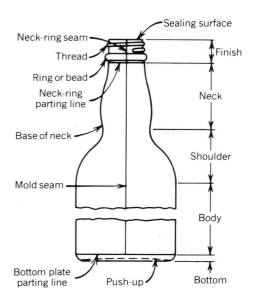

Figure 5.   Parts of a bottle.

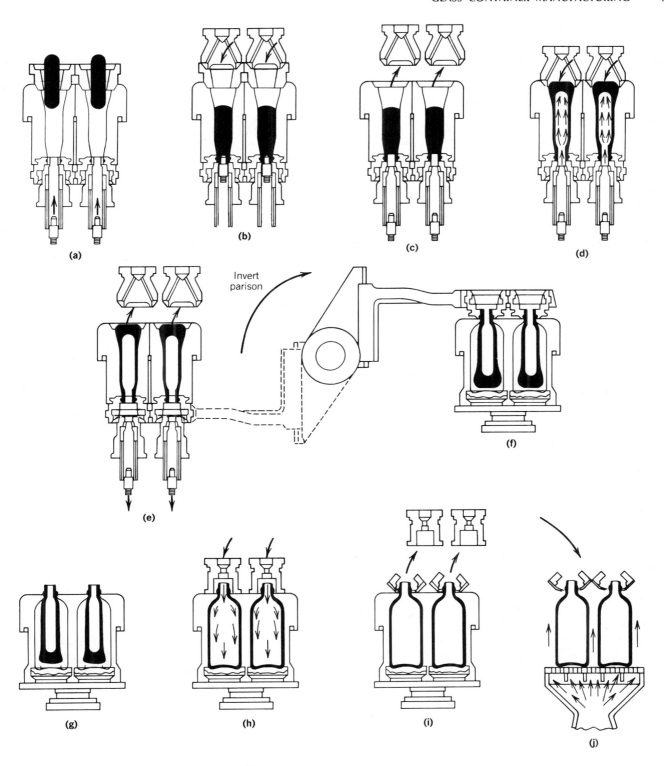

Figure 6. A functional diagram of parison and bottle-forming by the blow-and-blow process. (**a**) Gob delivery through funnels into closed blank molds. (**b**) Baffle on, settle blow pushes glass downward onto plunger and into finish. (**c**) Corkage reheat, plungers down, baffles and funnels off. Bubble inside finish reheats. (**d**) Baffles down, counterblow air forms parison. (**e**) Baffles off, thimbles down, blanks open, parison reheat starts. (**f**) Blow molds close and neck rings open, parison released into blow molds. (**g**) Reheat continues as parison elongates. (**h**) Blow heads down and blow air on. Bottles formed to blow mold. (**i**) Blow heads off and blow molds open. Bottles taken out with takeouts. (**j**) Bottles held in takeout jaws over deadplates then released and swept onto conveyor.

invert, the heat flow from the interior of the glass to the outer surface is larger than the radiative and conductive loss from the parison surface. As a result, the surface temperature of the glass increases or "reheats" and the parison becomes more fluid.

Upon completion of invert but before the final blow air blows the parison out against the walls of the blow mold, the parison elongates (runs) under its own weight. Without the reheat step just prior to run, there would be very little elongation of the gob before final blow. The reheat and run steps have a very large influence upon the distribution of the glass throughout the final product.

After the parison runs the proper amount, the final blow air is applied and once again the glass is pressed against a metal surface. (Final blow pressure is typically 9–22 psig or 160–250 kPa.) Heat is rapidly removed from the surface of the glass and large radial thermal gradients are again set up. However, the temperature differential between the interior and surface glass is not as extreme as it was in the parison for three reasons. First, the glass in the final container is much thinner than it was in the parison. Second, the average temperature differential between the glass prior to blowing and the mold surface is less in the parison than it was in the gob. Finally, unlike the counterblow air, which is virtually stagnant, fresh final blow air is continually circulating throughout the interior of the container. This air flow reduces the interior temperature of the container. As a result of all these factors, the reheat of the outer glass surface is much less after the blow mold opens than after the blank opens. Consequently, after the blow mold opens the container will no longer deform significantly unless an external force is applied to it.

The takeout mechanism is used to transfer the container from the mold to the deadplate. The takeout holds the container over the deadplate for a brief time to allow air to flow up through the deadplate and around the container to further cool it. Finally the takeout jaws open, allowing the container to drop about 2.54 in. (1 cm) onto the deadplate. It is then pushed onto the machine conveyor and transported to the annealing lehr.

Press-and-blow. In the press-and-blow process (see Fig. 7), the gob loads into the blank through a funnel just as in the B&B process. However, unlike the B&B process, the baffle never seats on the funnel. Instead, the funnel is immediately removed, and the baffle seats directly on the blank mold. The plunger then moves up and presses the glass out against the baffle and blank mold walls. When the cavity is filled, glass then is pushed down into the neck ring and the finish is formed. The rest of the steps in the P&B process are identical to the steps in the B&B process. The amount of heat removed in the blank mold is greater (by ~20%) in the wide-mouth P&B process than in the B&B process. This is because in the P&B process, heat is being extracted both from the interior glass surface by the plunger as well as the exterior glass surface by the blank mold.

Process comparison. The main difference between the P&B and B&B processes is the method by which the parison is formed. In the B&B process, the finish is formed first by air blowing on the top of the gob. The rest of the parison is then formed by blowing a bubble in the gob from below through the newly formed finish. In the P&B process, the parison is formed by the plunger moving up and forcing the glass against the blank mold. The finish is the last part of the parison to be formed in the P&B process.

The B&B process has traditionally been used for narrow-neck containers (less than 1.5-in. or 38-mm diameter finish), and the P&B process for wide-mouth containers (jars). This is changing somewhat because the P&B process is now being adapted to narrow-neck containers. Making narrow-neck containers by the P&B method is difficult because the largest diameter of the plunger must not be larger than the size of the inside of the finish. This results in a relatively long and narrow plunger that is more susceptible to alignment problems, scouring, and sticking, and makes plunger penetration into the glass more susceptible to variations in glass volume (gob weight).

The main advantage of the P&B process is that it produces a parison that has the same wall thickness at any position around the parison. The B&B parison has wall-thickness variations that result from the bubble-formation process. The bubble is formed by air pushing against the fluid glass, causing it to flow. Hotter glass flows more easily, since its viscosity is lower. As a result, the bubble is normally not centered exactly in the parison nor exactly cylindrically shaped because there are virtually always small thermal inhomogeneities present in the gob. Since a P&B container has more uniform wall thickness than a B&B container, it can be made with less glass. This is because the more uniform wall thickness requires less of a safety margin to assure that the thinnest portion of the container wall is above minimum wall thickness specifications.

The time required to make a container by the B&B and P&B processes is about the same for many containers. The total time from the gob loading into the blank mold until the container is released onto the deadplate is typically about 12 s for a container with a capacity of 12 oz (0.35 L). This is not, however, the cycle time of the IS machine. One section of the machine can simultaneously be working on the gob in the blank mold and the parison in the blow mold and be holding a container in the takeout tongs over the deadplate. As a result of this overlapping, new gobs are typically supplied to the blank mold less than 5 s apart for a 12-oz (0.35-L) container weighing 6.7–7 oz (190–200 g). Heavier-weight containers require a longer machine cycle time, lighter containers can be made faster.

Forming machines. The IS machine is made up of a variable number of sections arranged in a straight line. Each section is, in effect, a container-making machine. It contains one set of blank molds, one set of blow molds, and associated transfer equipment. The U.S. glass industry is moving toward the use of eight- and ten-section machines, although six-section machines are still used on some lines. Each section of an IS machine can produce more than one container each machine cycle. A double-gob machine (each section contains two cavities for receiving gobs as illustrated in Figures 6 and 7) produces two containers from each section every machine cycle. Double- and triple-gob machines are the most common in the industry. Single-gob machines are used to produce large-volume (more than 2.1-qt or 2-L) containers. Quadruple-gob machines can produce containers up to a capacity of about 12-oz or 0.35 L.

A picture of a ten-section triple-gob machine producing containers is shown in Figure 8. This machine is about 18-ft

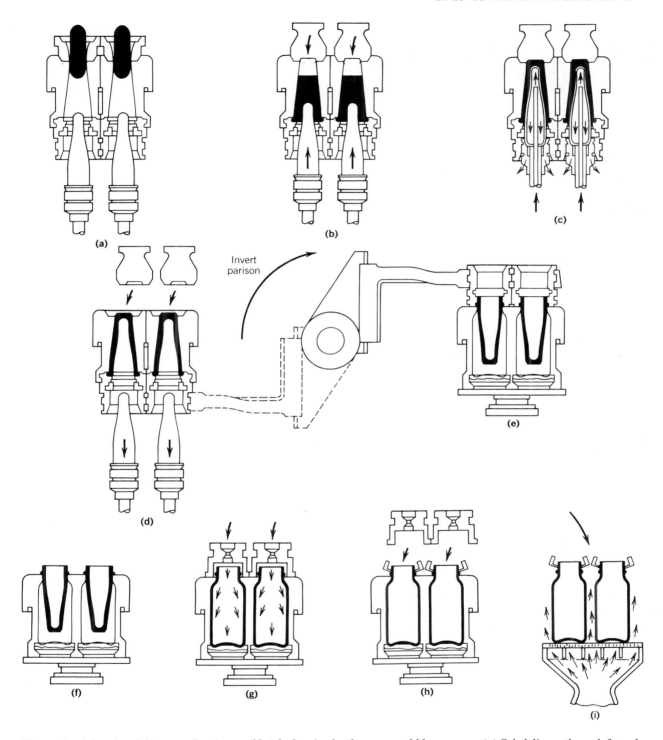

Figure 7. A functional diagram of parison and bottle-forming by the press-and-blow process. (**a**) Gob delivery through funnels into closed blank molds. Plungers in position for loading. (**b**) Baffles on and plungers move up starting to press glass. (**c**) Pressing of glass into parison shape and into finish is completed. (**d**) Plungers down, baffles off and blanks open. Parison reheat begins. (**e**) Blow molds close and neck rings open, parison released into blow molds. (**f**) Reheat continues as parison elongates. (**g**) Blowheads down and blow air on. Bottles (jars) formed to blow mold. (**h**) Blowheads off and molds open. Bottles taken out with takeouts. (**i**) Bottles held in takeout jaws over deadplates then released and swept onto conveyor.

Figure 8. AUTO-MOT controlled ten-section triple-gob IS machine.

(5.5-m) long, more than 3.3 ft (1 m) deep, about 10-ft (3-m) high, and weighs more than 55,000 lb (25,000 kg). It is capable of producing about 400 12-fl oz (0.35-L) capacity bottles per minute (bpm). Machine speeds in the industry run typically in the range of 150–400 bpm depending upon the size and shape of the container, the number of sections in the machine, and the number of cavities in each section. Ten-section quadruple-gob machines are capable of running at speeds in excess of 500 bpm for 10-oz (0.3-L) capacity containers. The highest-speed jobs (in terms of number of containers produced from each cavity per minute) are baby-food jars. An eight-section triple-gob machine making 4-oz (0.12-L) capacity baby-food jars by a P&B process can produce more than 400 containers per minute.

The IS machine is an "in-line" machine, but most of its competitors over the past 60 years have been rotary machines. Rotary machines have their molds arranged on one or more circular platforms that rotate in order to receive gobs and discharge finished containers at one position. These types normally have no delivery system. The gob drops directly into the blank mold after it is sheared. Currently, the most widely used rotary machine is the Roirant, used primarily in Europe. Roirant machines have been used in glass container production nearly as long as IS machines have been used. The Heye 1-2 machine is an example of a new type of rotary machine. This machine uses the P&B process and has two blow molds for each blank mold (8).

A number of machines have been designed, such as the Heye 1-2 machine, which contain two blow molds and one blank mold. This is because an acceptable parison can be formed in the blank mold in a much shorter time than is required for the formation of the container in the blow mold (only about 35% of the heat extracted in the B&B process is removed by the blank molds). Thus, the speed of the machine is normally limited by the amount of time required to extract the necessary amount of heat in the blow mold. Two other recent examples of such a machine are the hybrid machine developed by Owens-Illinois in the United States and the RIS machine developed by Veba Glas and Emhart Machinery Co. in Switzerland. Neither of these machines is in commercial production yet.

IS machine mechanisms and control methods. The various mechanisms on IS machines are currently powered by com-

pressed-air-driven pistons. The machine mechanisms that basically move in a vertical direction are coupled directly to a piston. Those mechanisms that have an arcuate motion are coupled to drive pistons via special linkages or with rack and pinion type gearing, or both.

One consequence of the mechanisms being driven by air is that the speed of some of the mechanisms is sensitive to ambient air temperatures. As a result, operational problems can occur when there are large variations in day and night air temperature. Electric drives have been considered for some mechanisms, as a means to overcome these variations in motions. Unfortunately, electric motors with sufficient torque are not available in sizes that can be fit into existing IS machines. Hydraulic mechanisms have been used periodically on glass-making machines for many years. Many of the earlier problems associated with the use of hydraulics on glassmaking machines, eg, flammable oils, varying response as hydraulic-fluid temperatures change, have been solved in recent years. As a result, hydraulic mechanisms have been installed in a number of experimental machines, such as the Hybrid and RIS machines mentioned above, and are being tested on standard IS machines. Widespread use of these mechanisms might make small increases in machine speed possible and at the same time reduce some types of defects that can be caused by erratic or "jerky" transfer mechanisms.

The various mechanisms on an IS machine operate in a timed sequence. The start of this timed sequence for existing machines is related to the shearing of the gob, not the actual time that the gob arrives in the blank mold. For this reason it is imperative that the delivery system loads the gob into the blank mold at a precise time. In the past, the relative timing of the various mechanism motions was determined by "buttons" on a rotating drum tripping levers, which in turn operated air valves. Directly or indirectly, these air valves allowed air to enter the appropriate cylinder and move the piston. Within the last ten years, computer-based electronic systems have been developed that are taking the place of the mechanical timing drums. These new systems are much easier and more efficient to use and allow much more accurate control over the relative timing of the various mechanism motions.

The advent of electronic timing on the IS machine opens the door to much more active control of the IS machine by computers (9). One such system is in commercial use presently. This system uses sensors on the machine to determine when various mechanisms on the machine actually move and also the precise time when the gob is loaded into each blank. This information is used by other computer programs in the control system to maintain a more stable mechanical operation and also to increase machine speeds slightly by reducing the amount of tolerance required between potentially conflicting mechanism motions.

Future control systems will undoubtedly take into account the thermal- and heat-transfer processes occurring as the container is being formed. However, these control systems will not be developed easily or quickly. The reason is the complexity of the forming process itself, which makes it difficult to model. Heat transfer is occurring by conduction, convection, and radiation simultaneously. The glass flows under gravitational forces as well as under mechanically applied forces. Because the viscosity of the glass is very dependent upon glass temperature, the stress required to produce a given strain is continually changing during the forming process. The resulting induced thermal and mechanical stresses can cause defects in the finished container. Some progress is being made in mathematical modeling of the forming process not only for glass-container forming but for other glass-forming processes as well (10).

In summary, the IS machine has proven to be a very versatile machine for making glass containers. It has been a principal factor in glass-container production since the 1920s, and it will be a major factor for the foreseeable future. The current IS machine is the product of more than 60 years of refinement and evolutionary improvements. The increase in the rate of container production (more than a factor of 10), the decrease in container weights, and improved container quality have, in large part, been made possible by these improvements. In the future, improvements will continue to be made in mechanism performance and in control systems. Computerized control systems are becoming an important tool of human operators in controlling the process and making the adjustments required to maintain high-quality production in the complicated high-speed modern IS machinery. With improved mechanisms and control systems, the weight of glass containers should continue to decrease as their quality and performance improve.

Mold materials. The molds used in the forming of glass containers are normally made from gray or ductile cast iron, although compacted graphite holds promise for larger and hotter-running molds (11). A copper–nickel–aluminum alloy is sometimes used to fabricate neck rings because of its high thermal conductivity. Baffles and plungers are sometimes made from a nickel–boron alloy because of its high-temperature hardness and glass-release properties. A related nickel–boron alloy is often flame sprayed onto metal-to-metal contact surfaces, eg, seams, of molds to increase their hardness and thus the useful lifetime of the molds (11).

Blank molds are coated with a graphite-based compound to improve release properties of the glass from the mold (12). Historically, graphite-based liquid "swabbing" compounds have been applied every 10–15 min by the machine operator using a swab. After a few hours of operation, carbon deposits have built up on portions of the blank mold. These deposits on the mold act as an insulator, thus inhibiting the flow of heat from the glass into the mold metal and interfering with the heat-removal function of the IS machine. As a result, the blank molds must be taken off the machine and the carbon deposits removed, generally by sandblasting.

Recently, compounds have been developed that can be applied to the blank molds before they are placed in service on the machine (12). This solid-film-lubricant (SFL) coating reduces the need for swabbing and increases the time that a blank mold can operate on the machine before it has to be cleaned. The drawback to the SFL is that is has to be carefully applied each time the blank mold is cleaned. Improperly applied SFL can be more of a detriment to the forming process than an aid.

Mold cooling. Molds are cooled by blowing air on or through them, or both. For many years, air has simply been blown onto the outside of the blank molds and blow molds from very crude "wind stacks" placed next to them. Large volumes of low-pressure air must be supplied for the cooling. This in turn contributes to the noise levels present in the vicinity of the machine. As machine speeds have increased and container weights have decreased, the wind stack method of cooling is proving to be inadequate. Recent developments have

been aimed at applying the cooling air more efficiently and more controllably to the molds. The simplest form this has taken is in better-designed wind stacks (13). Monitoring of the cooling capacity of the air and subsequent control of the cooling air supply has been used (14). A third method has been developed in which the cooling air is passed through holes drilled vertically in the blank and blow molds (15). These axial cooling systems allow for much more uniform and efficient cooling of the molds, which in turn leads to more uniform distribution of glass in the final container, higher production speeds, and lower cooling-energy requirements.

Water cooling of blank and blow molds has been commercially tested (16). Two of its main advantages are its lower energy requirements and its insensitivity to changes in ambient air temperatures. Both of these advantages arise because of the great cooling capacity of water flowing through molds. In fact, it is so great that insulating materials must be placed between the water-filled cooling tubes and the glass-contact surface. This results in the cooling capacity of the system being dependent upon this insulation barrier. It is rather insensitive to water-flow rates and temperatures, as long as the cooling water does not boil. Consequently, the machine operator has very little control over the cooling of the molds, and this means that the cooling is very dependent upon the manufacture of both the molds and the insulating materials.

Hot-Bottle Handling

Hot-bottle handling involves transferring the newly formed containers from the deadplate on each section of the IS machine into the annealing lehr. In a typical plant layout, this involves three 90° turns for the container (see Fig. 9). The most critical mechanisms involved in hot-bottle handling (the sweep out, the curved-chain transfer, and the lehr loader) are associated with these 90° turns. In addition, a hot-end treatment is frequently applied to the container at some point between the deadplate and the lehr loader.

Materials. During transfer of the containers from the deadplate to the annealing lehr, they are susceptible to deformation by moderate forces. In addition, they can be thermally shocked, causing small cracks, if they come into contact with cold materials. As a result, those parts that contact the sidewall of the container must not exert large forces against the container, and they should be made of low thermal conductivity materials. Low thermal conductivity ensures that the glass-contact points will run fairly warm and also that they will not conduct too much heat away from the glass at the contact points. Normally, glass-contact parts are fabricated from carbon-based material or from metal coated with carbon-based materials (17). Use of carbon-based materials also reduces the possibility of the surface of the container being scratched at the contact point, thus producing a defective container. This is a difficult application for carbon-based materials, since container temperatures during transfer are 975–1110°F (525–600°C).

Mechanisms. The sweep-out mechanism consists of a retractable, rotatable arm with fingers in the end of the arm. These fingers and the arm serve to make three-sided pockets for the containers. After the container has been released from the takeout mechanism, the sweep-out arm extends out inserting the fingers between each of the containers now resting on the deadplate. The arm then begins to rotate through approximately 90°, dragging the containers along with it. The sweep

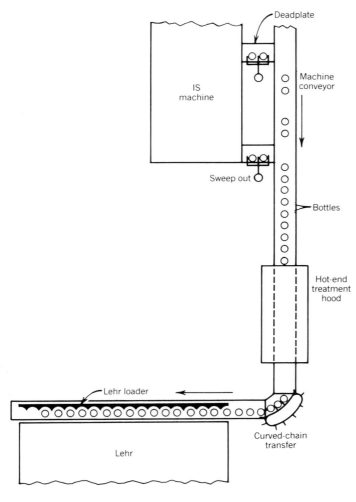

Figure 9. Hot-end handling process.

out moves the container from the stationary deadplate to the moving machine-conveyor belt. The angular acceleration of the sweep out is critical. The containers must be pressed lightly enough against the fingers to avoid deformation of the container. At the same time, the sweep-out arm must accelerate the containers to make them move at almost the same speed as the machine conveyor (also called flight conveyor) when they are swept out onto it. The timing of the sweep-out motion is also critical because this determines the spacing on the conveyor between containers from different sections. Most IS machines currently use a mechanical cam to determine the angular velocity of the sweep-out arm. However, several electrically driven systems are now being marketed and in the future most sweep-out mechanisms will probably be electrically driven and electronically controlled.

The machine conveyor is a metallic belt moving at a precisely controlled speed, as much as 165 ft/min or 50 m/min. This conveyor moves the containers past the end of the machine to a position just to one side of the annealing lehr. There may be facilities for applying a hot-end treatment to the container while it is still on the machine conveyor. These treatments are discussed below.

At the end of the machine conveyor is a device called a curved-chain transfer. On the machine conveyor, the contain-

ers are moving parallel to the lehr. They must now be placed across the front of the lehr and loaded into it. The curved-chain transfer moves the containers from the machine conveyor, accurately spaces the containers with respect to one another, and loads them onto the cross conveyor. The cross conveyor is moving perpendicularly to both the machine conveyor and the lehr. This conveyor is very similar in construction to the machine conveyor. Its function is to move the containers in a single line in front of the annealing lehr.

The lehr loader pushes the containers off the cross conveyor in groups so that lines of containers are placed across the lehr mat at the entrance to the lehr. The lehr loader is simply a cam-driven bar containing semicircular pockets that fit loosely around the containers. This bar, as long as 12 ft or 3.5 m, can push more than 30 containers onto the lehr mat at one time.

During transfer from deadplate to lehr, heat is lost from the containers by radiation and by conduction to the ambient air. As a result, during this time (typically 10–30 s) the temperature of the container normally drops 90–180°F (50–100°C). The temperature of the containers is generally just below the annealing point when they enter the lehr.

Annealing

The first function of the annealing lehr is to remove any residual stresses that may be in the container resulting from nonuniform cooling rates during forming and hot-end handling. It does this by raising the temperature of the container to a point just above the annealing temperature and holding it there for a few minutes. It then slowly cools the container until its temperature is several degrees below the annealing point. After this, it cools the containers as rapidly as possible so that any cold-end treatment may be applied to the container before inspection and packing. The lehr forms the dividing line in the plant between hot-end operations and cold-end operations. A schematic of an annealing lehr is shown in Figure 10.

The lehr is normally 80–115 ft (25–35 m) long and up to 10–13 ft (3–4 m) wide. The lehr mat is a continuous metallic belt the width of the lehr and more than twice as long. The containers are placed in rows on the lehr mat by the lehr loader. The lehr mat with its load of containers moves slowly but continuously from the hot end of the lehr to the cold end; it typically requires 30–45 min. After the containers are discharged from the lehr mat at the cold end, the mat returns inside the lehr to the hot end.

Some heat, often in the form of fossil-fuel fires, is normally supplied to the lehr at the hot end to increase the temperature of the containers to slightly above the annealing temperature. After that, cool air is introduced to lower the container temperatures. As a result, large-capacity fans are a principal component of lehrs. Much effort goes into the design of the baffles and ducts that direct the air flows within the lehr (18). These air flows assure that uniform temperatures are maintained at all points across the lehr and that a proper thermal gradient is maintained along the length of the lehr. Currently, the capacity of the lehr is limited by its ability to remove heat from the containers after they pass through the annealing zone rather than by the amount of heat the annealer must supply to anneal the containers properly.

Surface Treatment

There are three general types of surface treatments that are applied to containers. The first type, concerned with the

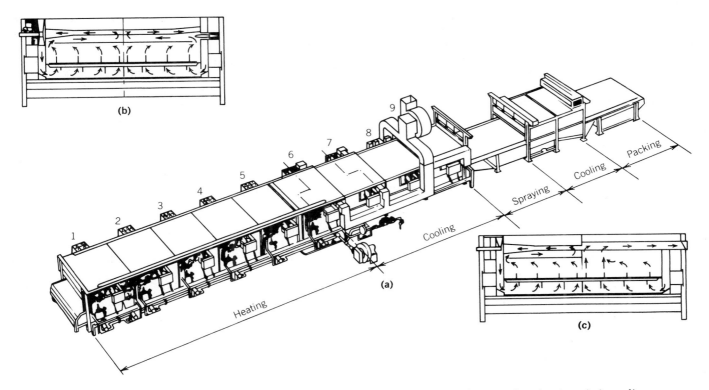

Figure 10. (a) Annealing lehr. (b) Cross-section showing air flow in heating section. (c) Cross-section showing air in cooling section.

inner surface, enhances the chemical durability of the container, eg, packaging parenteral solutions. The second type is concerned with protecting the outer surface from scratches. Outer surface treatments are designed to increase the lubricity (slipperiness) of the container so that the containers will slide easily across other containers and across bars or plates, without scratching the container sidewall, either in the container plant's cold-end handling equipment or in the customer's handling lines. The third type of treatment is concerned with strengthening the container. If the container is stronger, less glass is required for the same service requirements, resulting in a lighter-weight container.

Inner-surface treatments. Two types of treatments are used to improve the chemical durability of interior surfaces of glass containers. The first method uses SO_2 gas injected into the container while it is on the machine conveyor. The gas contacts the glass surface at a temperature of about 1000–1100°F (550–600°C) and reacts with the Na atoms in the outer few atomic layers of the glass. The result is a depletion of Na atoms in these outer layers of the glass surface and the formation of a hazy deposit of Na_2SO_4 on the surface of the glass (19). This hazy layer is subsequently washed off by the customer. The resulting glass surface is extremely resistant to chemical attack.

The second method involves the use of a fluorohydrocarbon gas in place of the SO_2 (20). The F in the gas reacts with the hot glass. The exact mechanism by which the reactivity of the glass is decreased is unclear. One theory suggests that NaF is formed, which then volatilizes and escapes. The other theory suggests that the F atoms enter the glass structure and block the mobility of the Na ions, thus improving the durability of the glass. The result is that the chemical durability of the interior surface of the glass is greatly improved, and there is no residual film on the surface, as there is with the SO_2 treatment.

Protective outer-surface treatments. Protective outer-surface treatments can be one-step or two-step processes. In the two-step process, both a hot-end and a cold-end treatment are applied. In the one-step process, only a cold-end treatment is applied.

The hot-end treatment in the two-step process involves contacting a gas containing tin (Sn) or titanium (Ti) (generally $SnCl_4$ or $TiCl_4$) with the outside surface of the container in a hood covering the machine conveyor. The Sn or Ti reacts at the glass surface to form a thin metal-oxide layer. This layer apparently provides a surface that allows the cold-end coating to adhere better to it than it would to a pristine glass surface.

The cold-end treatment usually involves spraying an organic material in an aqueous base onto the outside of the container. This spraying is done near the end of the annealing lehr when the container temperature is still above about 100°C. The coatings utilized include waxes, stearates, silicones, oleic acid, and polyethylene (21). The main function of this organic layer is to provide a surface with a low coefficient of friction.

Strengthening surface treatments. Surface treatments that increase the strength of the container rely upon chemical reactions that alter the container's surface. It is extremely difficult to break glass with compressional forces but tensile forces on the glass tend to make any surface flaws already present grow until fracture occurs. Containers can be strengthened, therefore, by surface treatments that place the outer surface of the containers in compression. Any tensile stress applied to the container will have to overcome the compressional stress present before it can begin to increase the size of any surface flaws.

The surface layer of the glass can be placed in compression by ion-exchange techniques. The most common method of strengthening containers with ion exchange involves exchanging K ions for Na at the outer surface. This is usually carried out at an elevated temperature by placing a molten K salt in contact with the container surface. An exchanging of K for Na ions takes place. The rate of the exchange is a function of the reaction temperatures, the salts used, and the composition of the glass (22). Processes suitable for use on container production lines have been developed (23,24), but the process has not been demonstrated to be commercially feasible in the United States. Chemical strengthening by ion exchange is currently being conducted on some container lines in Japan (25).

Inspection

The traditional purpose of glass-container inspection is to locate those containers that do not meet specifications and remove them from the ware that is shipped to the customer. In addition, the information obtained in cold-end inspection is being utilized more and more by hot-end personnel and hot-end process computers to modify the forming process in order to eliminate the production of off-specification containers. Inspection is done to find such things as stones, blisters, checks, thin glass, blemishes in or on the container surface, and variations in any one of a number of dimensional specifications. Because of the large variety of criteria being evaluated, no single inspection device has been developed that can make all the necessary measurements adequately. A series of microprocessor-based inspection devices, each looking at specific criteria, are normally employed to detect off-specification containers. These inspection devices cannot normally run fast enough to handle all the output of one high-speed IS machine. Thus the cold-end handling system divides the containers coming off the lehr into as many as four single-file lines. Each of these lines (legs) has a complete set of the inspection devices required for the particular job.

All (100%) of the glass containers packed are inspected by the microprocessor-based devices. In addition, supplemental manual inspection at a light station is often used as a last check before the ware is packed. Finally, statistical sampling methods are employed for those test methods, eg, pressure strength to destruction, which might damage the ware.

Optical inspection devices.   Optical devices are used to locate stones, blisters, blemishes, and checks; however, the same device is not used for all these characteristics. Stone, blister, and blemish devices normally work with the amount of light transmitted through the container. The container is rotated in front of a light source and the amount of light transmitted is detected by a photocell array. The output of each photocell is compared with a reference value. If the output is too low, then the detector signals a device that will eject the container from the cold-end handling equipment and onto a belt that conveys the container to the plant cullet-processing system for remelting. Checks are detected by monitoring light reflected from the container. The crack surface reflects more light than the unbroken container surface.

Thickness gauges.   The thickness of the glass in a container is determined by using a capacitance-type device. The capacitance measured by a sensing head is a function of the thickness of the glass near the head. With proper calibration, the

actual wall thickness can be read in inches or millimeters. As used in a production line, there are normally several sensing heads located in a vertical line with respect to each other. The number of heads is determined by the type of container being produced. The container is rotated in front of the sensing heads and the minimum thickness is determined at each of the heads. If the thickness is less than some preset minimum value, the container is ejected to the cullet belt.

Dimensional gauges.   The dimensional tolerances are normally checked by a variety of devices that make physical contact with the container. Typical dimensional tolerances checked include container height, leaners, unfilled finish, choked neck, and out-of-round or ovality. A leaner is a container with sidewalls not at a right angle to the base. An unfilled finish has a top surface that is not flat. A choked neck is a container (generally narrow-neck) with an opening through and below the finish into the container that is less than specifications. An out-of-round container is a container with either sides or seams flattened with respect to a circle.

Miscellaneous.   Two other types of devices are frequently found on the cold-end handling lines of modern container plants. One of these is an impact stimulator. This is normally the first tester in the series of testers. Each container is "squeezed" between rollers. This action causes stresses in the container that simulate the stresses caused by an impact.

The other type of device, which is being found on more and more production lines, is a cavity-identification device or CID. The CID can operate faster than the other inspection devices. As a result, one CID can inspect all the containers produced by one IS machine. A CID reads a code of rings, bars, or dots located in the bottom or lower sidewall of the container. The rings, bars, or dots are produced by engraving these shapes into the bottom plate or the heel area in the blow mold. Each bottom plate or blow-mold cavity has a unique combination of rings, bars, or dots engraved in it. By "reading" the shapes (using reflected light), the CID is able to determine which cavity produced the container. One CID checking all of the containers produced can be used to sample ware from each cavity periodically or reject all containers produced by one or more cavities on the IS machine. The sampling function is useful for collecting statistical quality-control samples from each cavity. Items checked in these periodic samplings include capacity, color, pressure strength, and additional dimensional tolerances, especially in the area of the finish. If any problems specific to one or more cavities are discovered by these samples, that cavity number is "placed on the CID" and all the containers produced by that cavity are rejected to the cullet belt.

Data logging and reporting.   Because each of the inspection devices is microprocessor-based, it is possible for the device to keep a tally of the number of containers rejected by it and to pass this information on to another "supervisory" computer. If a CID is placed on each leg of the inspection system, it can be coupled with the inspection devices to count the number of and log the characteristic quality measurements of containers being produced by each cavity on the IS machine. These data can then be displayed on monitors at both the cold and hot ends as well as stored for summary tables produced each hour, shift, and day. By observing the hot-end monitor, the IS machine operator is able to take appropriate action to maintain the quality of container being produced by a specific cavity.

Post-Processing

Packing. After the containers have been inspected, they are normally packed automatically in reusable corrugated boxes or on bulk pallets and placed in a warehouse. If the containers are packed in cartons, automatic equipment is also available for assembling the cartons themselves as well as placing the filled cartons on pallets and wrapping them. (Occasionally the containers may be conveyed directly to a labeling or decorating operation.) The pallets are then moved to a warehouse. The containers in the warehouse may be delivered directly to a customer, or they may be moved to another post-processing operation in the plant, such as decorating or labeling.

Decorating. Although there are a number of materials and processes used in decorating glass containers, they basically fall into two categories. One category involves applying a material onto the surface of the glass; the other involves etching away a portion of the glass surface.

A number of materials and techniques can be employed to apply decorating materials to the surface of the container. The oldest method involves applying a glass or ceramic frit (powder) to the surface by screen-printing methods (see Decorating), then firing (heating) the container to the point when the frit will react with the glass and flow together to make a smooth label, which is an integral part of the container. Multiple colors require multiple passes through screen-printing apparatus. Newer methods employ, among other things, the application of decals, either ceramic or organic, various types of thermoplastic inks and enamels, uv-curable inks (see Inks), or noble metals or metallo–organic compounds (26). Discussions of the latest techniques and processes can be found in the papers presented at the annual seminar of the Society of Glass Decorators.

Etching involves first protecting the surface of the glass in a negative pattern; that is, the portion of the container that is not to be etched must be protected. This is normally done by applying a waxlike coating to the container by a screen-printing or other process. The container is then exposed to an etching solution, normally containing HF acid, for a controlled length of time. The container is then washed to remove the protective film before the container is packed for shipping.

Decorating is losing its importance because of the shift to nonreturnable containers in the beer and soft-drink industries and the move to alternative materials in the drug and cosmetic industries. Decorating adds appreciably to the the cost of the container because of added material costs (glazes, inks etc), the screen-printing or other application equipment, the relatively slow application processes, heat-treating equipment, energy requirements, and space required for the entire process.

Labeling. Labeling of containers before shipment to customers is becoming more popular, especially in the beer and soft-drink industries. Labeling can be done relatively inexpensively in the container plant, normally by applying a paper label over part or all of the container sidewall or applying a polystyrene shrink-wrap (27), or other plastic or composite materials, around the lower portion of the bottle. Both the paper and polystyrene labels can be produced in vivid colors, thus supplying advertising impact for the product. The labeling can be done at high speeds and material costs are relatively low. In addition, the polystyrene wrap provides a cushioning effect and the containers require less protection during shipment, thus reducing the cost of the shipping containers.

Uses of Glass Containers

Advantages of glass. Probably the greatest advantage that glass has over competing packaging materials is its inertness to a wide variety of food and nonfood products. For example, glass is very inert to acidic foods, which can attack metal fast enough that the taste of the food is altered in a relatively short time. Also, foods will not leach materials out of the glass, which can alter the taste of the foods, as is the case with some plastics.

A second and related advantage has to do with the impermeability of glass to gasses. Because the diffusion rate for O_2 and CO_2 through glass containers is, for all practical purposes, nil, the integrity of the product is maintained over long periods of time. This is especially valuable in packaging beer and soft drinks (see Carbonated beverage packaging).

The clarity of glass is also cited by many as an advantage, for product visibility. The glass container is also very rigid. Thus when it is picked up or when the pressure of the contents varies, the container will not deform. The cost of the glass container is low, often lower than that of competing materials. In addition, the raw materials of glass are in abundant supply throughout much of the world.

Disadvantages of glass. The two main disadvantages cited for glass containers are weight and breakability. Both of these are associated with the inherent brittle behavior of glass itself. Glass can break in tension as a result of surface damage. The surface may be damaged by abuse or misuse such as that during handling, packaging, shipping, filling, distribution, or use. Glass containers are designed to withstand normal handling and use. This is the reason that glass containers are often heavier than containers made from competing materials in spite of its relatively low density (2.5 g/cm^3, about the same as Al).

Glass-container production. Total United States glass-container production for 1983 totaled about 295 million (10^6) gross (one gross equals 144 containers), or more than 185 containers per capita. The market price of these containers totaled about $5.5 billion ($10^9$).

There are a number of ways in which glass-container production can be categorized. One method used by the U.S. Department of Commerce (28) categorizes glass containers by use. These uses include beer, food, beverages, liquor, wine, medicinal and health supplies, toiletries and cosmetics, and household and industrial chemicals. The largest single usage (by number of containers made) is for beer; the smallest category is chemicals. About four times more narrow-neck containers are used than wide-mouth containers. Beer and beverage uses consume about 55% of domestic glass-container production; food about 30%; and the other five categories, about 15%. Only about 1% of this production is for reusable containers, used almost exclusively for beer and beverage containers.

Since 1980, when shipments totaled 326 million gross, the total U.S. market for glass containers has been decreasing. At the same time, imports of glass containers, primarily from Mexico and Canada, have been increasing. In 1983, imported containers represented about 4% of domestic sales, more than double the 1981 percentage. The only segment of the container market that grew in 1984 was the beverage segment. The driving force in this market is the 0.3- and 0.5-L (10- and 16-oz) nonreturnable soft-drink container.

The greatest variety of container shapes and sizes is found in the toiletries and cosmetics market. In this market, distinctive shapes and sizes are considered to be a principal marketing advantage. Some of the containers made for this market have become collector items, with the current value of the container being far more than the original cost of the container and product.

Appendix

annealing temperature (or annealing point)	The temperature at which stresses in the glass are relieved in a few minutes (ASTM C 336-71)
blank mold	The mold used to shape the parison.
blister	Bubble in glass with a diameter greater than 0.06 in. (1.5 mm).
blow mold	The mold used to shape the container.
cavity	The blank-mold and blow-mold combination used to produce one container.
check	Small crack in the container.
cullet	Crushed glass.
finish	The top portion of the container over which the closure is applied.
flint glass	Colorless glass.
forehearth	The channel-like system used to distribute glass from the furnace to the forming machine.
forming implements	The removable equipment on an IS machine which comes into contact with the glass.
gob	Quantity of molten glass used to make one container.
IS machine	Individual Section machine. The machine most widely used for producing glass containers.
line	The equipment associated with one forming machine, which is required to produce, inspect, and pack containers.
offware	Containers that do not meet all manufacturing criteria.
parison	The preform from which the container is blown.
refractory brick	A ceramic material in the shape of a brick that is used as the lining for or the actual construction material for high-temperature reaction chambers.
seed	Bubble in glass with a diameter less than 0.06 in. (1.5 mm).
stone	Particle of unmelted batch or refractory material in a container.
ware	Another name for glass containers.

BIBLIOGRAPHY

1. E. J. Pryor, *Mineral Processing,* Elsevier, Amsterdam, 1965.
2. M. C. Weinberg, *Glastechn. Ber.* **56,** 60 (1983).
3. H. N. Mills in S. J. Lefond, ed., *Industrial Minerals and Rocks,* American Institute of Mining, Metallurgical, and Petroleum Engineering, New York, 1975, pp. 327–334.
4. H. Moore, *Glass Ind.* **64,** 18, 31 (Jan. 1983).
5. P. J. Finch & C. Wickstead, *Glass Technol.* **25,** 177 (Aug. 1984).
6. E. L. Fords, L. D. Pye, H. J. Stevens, and W. C. LaCourse, *Introduction to Glass Science,* Plenum Press, New York, 1972, pp. 273–327.
7. J. D. Gilchrist, *Fuels, Furnaces and Refractories,* Pergammon Press, Oxford, UK, 1977.
8. R. D. Heather, "Forming Machines," *Proceedings of the International Congress on Glass,* Prague, 1977.
9. N. T. Huff and D. M. Shetterly, *Glastechn. Ber* **56,** 271 (1983).
10. H. P. Wang and R. T. McLay, *Glastechn. Ber.* **56,** 289 (1983).
11. "Cast Iron," *Machine Design* **54** (8), 10 (Apr. 15, 1982).
12. A. F. Beare, *Glass Technol.* **10,** 114 (Aug. 1969).
13. Brit. Pat. 1,370,192 (Oct. 16, 1974), R. Kent, F. Shaw, and P. Grayhurst (to British Glass Industry Research Association).
14. Brit. Pat. 2,011,128 (May 19, 1982), C. Wickstead (to Rockware Glass).
15. S. P. Jones and J. H. Williams, *Glastechn. Ber.* **56,** 277 (1983).
16. U.S. Pat. 3,887,350 (Feb. 11, 1974), C. W. Jenkins (to Owens-Illinois Inc.).
17. B. W. Spear, *Glass Ind.* **63,** 22, 34 (July 1982).
18. G. E. Walker, *Glass Int.* **82,** 20 (July 1982).
19. F. R. Bacon, *Glass Ind.* **49,** 494 (Sept. 1968).
20. U.S. Pat. 3,314,772 (Apr. 18, 1967), J. P. Poole, H. C. Snyder and R. H Ryder (to Brockway Glass Co.).
21. A. S. Sanyal and J. Murkerji, *Glass Ind.,* **28,** 31 (Nov. 1982).
22. H. M. Garfinkel in A. G. Pincus and T. R. Holmes, eds., "Strengthening Glass by Ion Exchange" *Annealing and Strengthening in the Glass Industry,* Magazines for Industry, New York, 1977, pp. 301-307.
23. U.S. Pat. 3,607,172 (Sept. 21, 1971), J. P. Poole, H. C. Snyder and M. A. Boschini (to Brockway Glass Co.).
24. J. P. Poole and H. C. Snyder, *Glass Technol.* **16,** 109 (Oct. 1975).
25. H. Ono, *Glass Technol.* **22,** 173 (1981).
26. A. G. Pincus and S. H. Chang, eds., *Decorating in the Glass Industry,* Magazines for Industry, New York, 1977.
27. U.S. Pat 3,760,968 (Sept. 25, 1973), Stephen W. Amberg and co-workers (to Owens Illinois Inc.).
28. Current Industrial Reports, U.S. Department of Commerce, Washington, D.C., Series M32G.

General References

F. V. Tooley, *The Handbook of Glass Manufacture,* Ashlee Publishing, New York, 1984. Third edition of a standard two volume reference. Broad coverage of many glassmaking areas in addition to container manufacture. Much of the material retained from the first two editions is dated but the updated work in this edition is timely.

D. R. Uhlman and N. J. Kreidl, ed., *Glass Science and Technology,* Academic Press, Inc., New York, 1980, 1983, 1984. Three volumes of this multivolume work have been published. Good timely reference work in many areas of glass technology and production.

Proceedings of the Annual Seminar of the Society of Glass Decorators, Port Jefferson, N.Y. These proceedings contain discussions of the latest techniques and equipment in the decorating of all types of glass and ceramics.

W. Giegreich and W. Trier, *Glass Machines,* Springer-Verlag, New York, 1969. Somewhat dated by still a good reference work for the theory and operation of many types of glassworking machines.

N. T. HUFF
Owens-Illinois Inc.

GLASSINE, GREASEPROOF PAPER, AND PARCHMENT

Glassine, greaseproof paper, and parchment, considered to be packaging films, are produced from cellulose fibers. These films are special in that they are very dense and have an inherent grease resistance not normally found in uncoated cellulosic paper-based materials (see Paper). All three have printable and coatable surfaces and can be used alone or in combination with other mateials in both rigid and flexible packaging applications. When treated with wax, for example, these sheets also become water resistant. Except for glassine, which is transparent, they are naturally translucent. Opacity can be achieved by adding fillers such as titanium dioxide.

The properties of these materials are similar, but the manufacture of parchment is quite different from that of greaseproof paper and its supercalendered counterpart, glassine. The latter two rely upon extensive refining or beating of carefully selected pulps to obtain the hydration of fibers which are formed into grease-resistant papers. In the manufacture of parchment, the hydration or gelation of the cellulose fibers is done chemically as the paper web is passed through a sulfuric acid bath.

Greaseproof Paper

Greaseproof is a term accepted for use in the paper industry as representing a translucent machine-finished paper manufactured from wood pulp and hydrated to give grease and oil resistance. It should not be confused with other types of grease- and oil-resistant papers, such as vegetable parchment, which is produced by treating paper with sulfuric acid. Nor does it refer to papers surface-coated with grease- and oil-resistant films that are not an integral part of the base-sheet structure.

Most paper is made from cellulose fibers obtained from wood. Some kinds of paper are made from a wide variety of species, but this is not true of greaseproof sheet as defined above. Most suitable are woods of great density and long fiber, such as slow-growing northern spruce. The cooking of the wood chips, which gives it the special characteristics required to make it greaseproof, is a particularly slow and exacting process.

For most papers, the beating of pulp is a relatively simple mechanical procedure whereby clumps and bundles of fiber are separated by a pounding action in order to obtain uniform dispersion of individual fibers throughout the slurry of water and cellulose. This process also wets the fibers to make them cohere and form on the machine wire into a structure that will become paper. The preparation of pulp for the manufacture of greaseproof paper is far more subtle and delicate. The beating process, known as fibrillation, is both gentle and severe in transforming bundles of fiber into hairlike fibers, ie, fibrillae. This must be accomplished with a minimum of splitting or shortening of the fibrillae. As the tiny fibers of cellulose pass through a small space between revolving lava stone rolls and a stone bed plate continuously for more than an hour, their physical appearance changes completely. Looking like frayed ropes, the fibrillae absorb so much water that they become superficially gelatinized and sticky. This physical phenomenon is called hydration.

The fibers, in their hydrated condition, are then passed over the wire of the paper machine. They felt and amalgamate into a continuous and homogeneous sheet or membrane free from pinholes and resistant to grease. So intertwined and closely grouped are these fibers that there are very few interfiber spaces for grease to penetrate. In addition, the water absorbed by these fibers becomes trapped within their dense cellular structures and cannot escape. It remains there, rendering the fibers hostile to grease.

Greaseproof papers can therefore be characterized as papers that, in addition to their obvious grease resistance, are highly refined and have excellent formation. They are also slow running, expensive to produce, and tend to be weaker and more brittle than other types of paper. Blends of various types of pulps are being used to improve the paper's strength. Surface treatments are being applied to the paper on the machine to enhance grease resistance while permitting faster running speeds and reductions in refining time. Greaseproof papers can be used as produced or can be further converted into glassine. Greaseproof papers are used in a variety of packaging applications, including margarine wrap, french-fry bags, and as the inner liner for multiwall bags (see Bags, paper) containing greasy products such as dry dog and cat food. They are also used as a liner in spiral cans (see Cans, composite) and a corrugating medium in protecting and cushioning bakery products and prepackaged cookies.

Glassine

Glassine derives its name from its glassy smooth surface, its high density, and its transparency. Glassine is made by densifying machine-finished greaseproof paper. This is accomplished by running the paper through a glassine supercalender. A calender is a machine that presses material as it passes between two rollers. Paper machines frequently have calenders consisting of two or more metal rolls. The freshly made paper is run through them to obtain smoothness, finish, and uniformity. A supercalender is a separate machine for doing what its name implies: a super or extra job of calendering. It contains a vertical stack of heavy rolls, eg, 14, alternately made of steel and fiber (see Fig. 1).

The paper is carefully dampened with water and run through a battery of steam-heated rolls. The rolls are under pressure, which, varying with the grade of glassine to be made, may be as high as 3000 psi (20.7 MPa) at the bottom nip. The effect is similar to ironing, as the steel rolls slightly indent the softer fiber rolls. The paper fibers, lubricated by moistening, are relocated by the pressure and compacted into a still denser formation. The interfiber air spaces are largely eliminated, increasing the index of light refraction which results in high transparency. The paper has now become somewhat more resistant to grease and is known as glassine. The transparency can vary widely depending upon the degree of hydration of the pulp and the basis weight of the paper, and it can also be made opaque with a pigment such as titanium

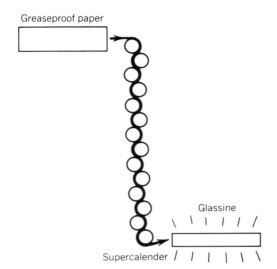

Figure 1. Glassine is made by densifying greaseproof through supercalendering.

dioxide. In the latter, gloss and finish would be enhanced by calendering.

Glassine and greaseproof papers are used in many different packaging applications. Their inherent grease resistance permits them to be used alone or in combination with other materials and coatings. Glassine and greaseproof papers may be tailor-made to meet specific property requirements or to be run on special packaging or converting machines. By varying plasticizer levels, pliability or stiffness can be controlled. The same denseness of paper that prevents the migration of grease also enhances gas-barrier properties because it restricts the passage of air. The papers are sterilizable, scuff resistant, and will accept wet-strength and moldproof treatments. The smooth surface of glassine is particularly well suited for coating and laminating and provides a particularly fine printing surface.

Glassine has long been used as a single-ply structure for wrapping candy bars, single-serving pies, and other bakery novelties. In multiple-serving snack packaging, glassine is combined with polyolefin films to add strength and stiffness as well as grease resistance to the structure. Chocolate-colored glassine also acts as a barrier to uv radiation and helps keep potato chips fresh. Waxed glassine is also a functional, moderately priced material used in cracker and cereal packaging (see Waxes). It provides an excellent moisture barrier, and the packages are durable and easy to open and reclose. For presweetened cereals, glassine coated with PVDC (see Vinylidene chloride copolymers) and then overwaxed provides a moisture barrier comparable to aluminum foil. Other properties available in glassine include a bacterial barrier for sterile medical packaging (see Health-care packaging) and high transparency, which provides good legibility for reverse printing and for window-envelope patching material.

Parchment

Parchment, commonly called vegetable parchment, is produced by chemically treating the paper. Unlike greaseproof sheet, the parchment-based sheet is unsized and highly water absorbent. It is called waterleaf. There is little refining of the cellulose fibers during the papermaking process. Instead, the grease resistance of this sheet is due to the parchmentizing process. The waterleaf web is passed through a sulfuric acid bath. The fibers partially dissolve and become gelatinous as they react with the acid. These sticky fibers permanently fuse together to make grease-resistant parchment. It is then necessary to stop the chemical reaction before the fibers are completely dissolved. This is done by passing the web through a dilution bath and then through water and acid-neutralizing baths before drying.

As with greaseproof paper, if no fillers have been added to the base sheet, the parchment is translucent. Fillers and dyes (see Colorants) can be added to the paper-machine beater during pulping for opacity or color. The dyes must be acid resistant to withstand parchmentizing. Either bleached or unbleached base stock may be used depending upon the end use.

As a result of parchmentizing, vegetable parchment has several distinctive properties. It is partially or completely grease-resistant depending upon its thickness and degree of parchmentizing. It has a very high wet strength, retaining as much as 60% of its dry bursting strength. In light weights, it permits the passage of air, but the heavier weights are nonporous. It is both odorless and tasteless. Since fibers cannot be picked from the surface, it is lint-free. Parchment can be treated with mold inhibitors, waxed for water resistance, or crêped for added strength, and it is printable.

Because of its grease resistance and wet strength, parchment strips away from any food material without defibering. This makes it an excellent material for packaging butter and shortening and as an interleaver between slices of prepared meats. Labels and inserts in products with high oil or grease content, eg, cake mixes and coffee, are frequently made from parchment (see Inserts). Treated with mold inhibitors, parchment is used to wrap foods such as cheese. It is also used to wrap surgical instruments, since it does not disintegrate during steam sterilization. In addition to these packaging applications, parchment is used in numerous printing, duplicating, and copying applications.

BIBLIOGRAPHY

General References

W. H. Bureau, *Paper—From Pulping to Printing*, Graphic Arts Publishing Co., Chicago, Ill., 1968, pp. 273–276.

E. E. Libby, *Paper*, Vol. 2 of *Pulp and Paper Science of Technology*, McGraw-Hill, Inc., New York, 1962, pp. 5, 6, 124.

J. N. Stephenson, *Manufacture and Testing of Paper and Board*, Vol. 3, McGraw-Hill, Inc., New York, 1953, pp. 716 and 717.

R. Wirtzfeld, *The Paper Yearbook*, Harcourt Brace Jovanovich, Inc., New York, 1984, pp. 262–266.

R. K. Steindorf
Nicolet Paper Company

GLUE.   See Adhesives.

GRAPHICS.   See Decorating; Printing.

GUMMED TAPE.   See Tape, gummed.

H

HEALTH-CARE PACKAGING

Health-care packaging (HCP) is done by hospitals, and by manufacturers of devices utilized in medical or surgical procedures. Most of these items must be delivered sterile to the ultimate user. HCP provides a protective envelope for devices that shields them during sterilization, transportation, storage, and delivery to the end user. The package may gather and hold in readiness several items needed for a given surgical procedure.

The term device encompasses a broad range of products. A ball of cotton, a cotton swab, and a tongue depressor are all examples of "disposables." They are used once and then discarded. Disposability is not necessarily synonymous with inexpensive. Some very expensive devices, such as dialysis filters, may be too fragile to clean and sterilize for reuse. For the sake of cost containment, reuse is desirable. The greater portion of HCP materials goes to hospitals, where scissors, scalpels, and many other items are frequently washed, repackaged, and resterilized. Class II devices (eg, iv sets, syringes, scalpels), usually relatively expensive, can be disposable or reuse items. These products must meet certain performance standards. The package, in addition to preserving sterility, is often called upon to protect the integrity of the product (see Regulatory considerations, below). Class III devices (eg, pacemakers) are of a critical nature. Malfunction could be life threatening.

The package must be designed with the product and its function in mind. This article does not deal with the packaging of drugs and other medications (see Pharmaceutical packaging), nor with consumer health-care devices normally sold as "nonsterile" (eg, toothbrushes, cotton swabs).

Package Types

A large variety of packages is utilized in health-care packaging, but they can be discussed in terms of five groups (see Table 1).

Flat pouches. Small- to medium-sized devices packaged in relatively small quantities are prime candidates for this type of package (see Fig. 1). The pouch usually consists of paper or Tyvek (DuPont Company) (see Nonwovens) on one side and a transparent plastic composite on the other. If the product is steam sterilized, paper is a must because of its porosity and PP is prescribed as the sealant member of the plastic composite because of its heat resistance. Radiation sterilization permits the use of an all-plastic pouch. Peelability must be considered

in the pouch design. A possible self-seal feature is discussed below under "Closure". Flat pouches are generally not suitable for very bulky items.

Vent bags. A plain LDPE bag would be less expensive than a flat pouch with chevron seal and peelable opening, but total package cost depends on more than cost of materials. Inexpensive PE bags were used at one time with ETO sterilization, but because the gas penetrates PE slowly, long sterilization cycles were necessary. Eventually the industry realized that the cheap bag was really very costly. Sterilization time is expensive. Furthermore, the thin 1.5 mil (38 μm) bags failed too frequently, and repacking and resterilizing added to the overall cost. A new, stronger, and more rapidly sterilizable bag evolved. The bag consists of LDPE at least 3-mils (76 μm) thick, with a patch of Tyvek in its wall or mouth (1). This type of breathable bag is well suited for bulky but lightweight products, such as baby pajamas (2).

Gusset pouch. For somewhat heavier bulky products, an expandable pouch is available (3) (see Fig. 2). This package is similar to the flat pouch in composition, but the plastic portion can be opened wider. The closing of this type of pouch can present a problem, but it is available with a self-seal feature (4). Failure during sterilization is common in very large gussets. Air is often entrapped in the large-size bag and during heating the expanded volume of air brings appreciable pressure to bear on the seals. It is important to expel as much of the air as possible prior to sterilization. Opening, too, can be

Table 1. Package Types in HCP

Type of package	Example of item packaged	Advantages
flat pouch	surgical gloves	inexpensive
vent bag	vinyl tubing	very inexpensive
gusset pouch	surgical tray	for bulky items
preformed tray/lid	catheter	maximum protection
thermoform/fill/seal	syringe	suitable for mass production

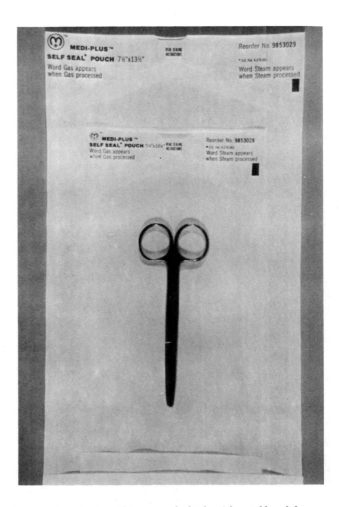

Figure 1. Pouch within a pouch, both with a self-seal feature.

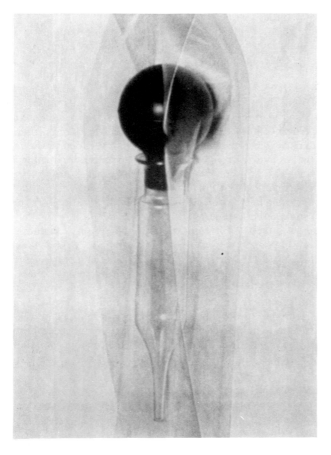

Figure 2. Gusset pouch

difficult. The size of the gusset determines the ease of handling.

Preformed tray/lid.   There are many reasons for turning to a more expensive tray package (see Fig. 3). The item might need better protection than a flexible pouch affords. Often the tray contains several items, all needed for a single surgical procedure. The trays are made from any of the rigid plastics discussed below. The gauge of plastic depends on the fabrica-

tion method, the depth and other dimensions of the tray, and the size, shape, and weight of product(s) packaged. The trays should be designed to permit nesting, which reduces storage space requirements for empty trays. The denested trays are filled, and a lid, generally coated Tyvek, is heat sealed onto the flange of the tray (see Sealing, heat). The sealing of the lid can be tricky at times. The seals are tested with pressure and visually inspected after peel. Increased temperature and/or pressure can harm the seals. As the seal temperature increases, the tray flange may distort, diminishing the contact area. Increased pressure can drive the liquefied heat seal coating into the paper or Tyvek, causing a starved joint. Adjustments should be made with care.

Thermoform/fill/seal.   Large volume products cannot rely on any of the aforementioned types of packages. Productivity would be too low and the cost too high. The form/fill/seal (see Thermoform/fill/seal) operation starts with rolls of material. The machine sets the pace. The machine forms a flexible or rigid tray from thermoplastic laminates (see Multilayer flexible packaging; Laminating) or coextrusions (see Coextrusions for semirigid packaging). Placement of the device(s) in the tray is usually manual, but the operation can be automated as well. A lid is applied and the finished packages are separated, sometimes die cut, and packed into cartons. The number of packages produced per minute per operator depends on the package size as well as the complexity of the thermoforming operation, but rates of several hundred packages per minute are common.

Thermoform/fill/seal involves an enormous investment in machinery. The justification for such capital outlay requires at least one million (10^6) packages per year of one type and size. Many of the modern machines have fast changeover features, but it is nevertheless desirable to have an assured minimum volume of one size. Additional size packages utilize some of the downtime of the equipment and help amortize it sooner, but that should not be part of the original utilization formula (5).

Thermoform/fill/seal saves space and, reduces labor cost. It may also reduce materials cost. Compared to a pouch, a blister pack is much more compact (see Fig. 4). Some of the forming webs may be more costly on a unit area basis than the equivalent pouch materials, but by cutting the area requirements in half or less, the thermoformed package is actually less expensive.

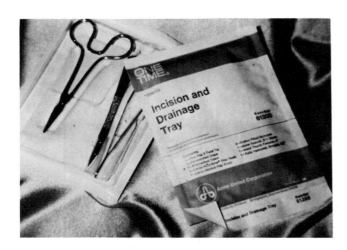

Figure 3. Tray with printed Tyvek lid.

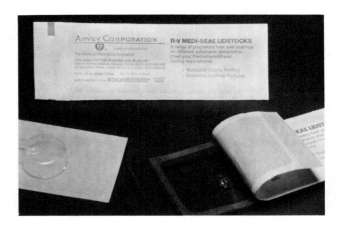

Figure 4. Thermoform/fill/seal package. Polyester blister and Tyvek lid (note tell-tale seal transfer).

Sterilization

Most health-care products are delivered sterile to the physician. The package plays an important role in maintaining sterility of the device until use. The mode of sterilization is thus an important factor in package selection. The primary sterilization processes are steam, gas, and radiation. Hospital sterilization is about 80% steam, 20% gas. In the United States today, manufacturers use gas and radiation in a ratio of about 80:20 (less than 1% steam).

Steam sterilization. The oldest and most reliable method of sterilization involves exposure of the product to heat/steam for a measured time interval. It is important to emphasize that devices must be clean and relatively free of bacteria prior to sterilization. Exposure to steam at 250°F (121°C) for about 30 min or to 270°F (132°C) for about 3 min reduces the bacterial population to almost zero. For the successful steam sterilization of a packaged product, the package must be heat resistant and porous. There can be no package failure due to 270°F (132°C) exposure or due to reaction at the high moisture level, and the package must admit steam to contact the product during sterilization and permit withdrawal of the excess moisture thereafter. Steam sterilization is the most prevalent sterilization process in hospitals. It is relatively quick, reliable, and relatively inexpensive. However, it does distort most packaging materials. This negative appearance factor precludes the use of steam sterilization by most device manufacturers.

Gas sterilization. Several gases have been used as sterilants. In the UK, and to some extent on the European continent, formaldehyde mixed with steam is in wide use. In the United States, the most common gas is ethylene oxide (ETO) (6). Packages must still be porous to permit the free flow of gas, but there is no requirement for high temperature stability, because the ETO cycles are run at about 140°F (60°C). Moisture content is still an issue, because steam is injected into the sterilizer to assure ≥50% rh. (7). Both ETO and formaldehyde have potential hazards attached to their use (8). Such sterilization equipment should be entrusted to trained operators only. Lengthy aeration cycles are required to expel the gas from the package. Specialized equipment is available to evacuate packages for this purpose. Because some materials absorb ETO, the aeration cycle may proceed for up to 24 hours, thus increasing the cost of ETO sterilization.

Radiation sterilization. Health-care products can be sterilized by exposure to low doses of gamma radiation. The radiation source is primarily cobalt, but some electron beam radiation is used as well. The need for porosity in the package, an important consideration in steam or gas sterilization, is eliminated by radiation. Irradiation is run at room temperature under ambient humidity conditions. Thus less stress is placed on the package in general and the seals in particular. Impingement of gamma rays may alter the appearance or performance of several plastics, however (9). PVC, for example, turns brown, and polypropylene brittle upon exposure to low doses [<5 Mrad (50 kGy)] of radiation. Some of the deleterious effects are not noted immediately, but after several weeks or months (10) (see Radiation, effects on packaging materials). Nevertheless, radiation sterilization is gaining wider acceptance in industry. It is cost-effective for large-volume users and the only sterilization method amenable to dosimetry (measurement of the quantity of radiation administered). Other methods require monitoring of each cycle via biological indicators (11). This is costly and delays product release.

Sterilization indicators. The HCP industry seeks sterility assurance. To promote this effort, many redundancies are introduced into an overcontrolled process. Biological indicators are widely used to simulate pathogenic bacterial termination. Chemical indicators on the outside of the package do not attest to the sterility of the content. These surface indicators are used to distinguish between "processed" and yet-to-be processed items. Bioequivalent chemical indicators are available (12) to mimic the action of spores under given sterilization conditions (see Indicating devices).

Packaging Materials

Almost all materials are suitable for HCP (13), if they have adequate biological barrier, have the required porosity (depending on sterilization mode), survive sterilization procedure unaltered, and age well.

The package's primary function is the maintenance of sterility of the sterile product within. For this purpose the material must provide a biological barrier that prevents microorganisms in the environment from entering the package. The porosity required for steam and gas sterilization must be balanced against limiting pore size to prevent bacterial penetration.

Materials can be classified as porous (eg, paper, woven, and nonwoven textiles) or impervious (eg, foil and plastic films) (14).

Textiles. Muslin is one of the oldest HCP materials. Some devices are still wrapped in several layers of muslin in hospitals. The "package" is taped to maintain sterility for at least 30 days.

The use of muslin has been reduced by other materials that provide longer shelf life, visibility of product, easier opening of the package, and other improvements in package performance. However, muslin is still used widely throughout the world. It offers a long history of successful use, a relatively low unit price, and ready availability.

Nonwovens. One must be selective in the choice of nonwovens. Only select grades of these materials are suitable for HCP. The synthetic fibers must be heat- and moisture-resistant and the materials dense enough to constitute a biological barrier. Tyvek, a spun-bonded polyolefin (15,16) made by DuPont, has found wide acceptance in HCP. Its white color, smooth appearance, water repellence, good porosity, and high tear strength are all beneficial performance characteristics. It has a number of limitations as well. It is relatively expensive, and its print quality is poor. Ink adhesion is imperfect, and the web is nonuniform in thickness and density. This in turn raises questions regarding its biological barrier properties. There are several grades of Tyvek available. The most widely used HCP grade is Tyvek 1073B. It has been compared to surgical kraft and found to be as good or better than kraft in HCP.

Because Tyvek may not be used at temperaures much above 212°F (100°C), it is unsuitable for steam sterilization. Seal temperatures should be kept as low as possible, preferably below 250°F (121°C), depending on dwell time and pressure. When Tyvek is heated it may melt or char. Prior to disintegration it will "transparentize." The white opaque Tyvek turns translucent and this is an indication that the properties of the material may have been compromised and it is not suitable for use in HCP.

Surgical kraft. Paper is widely used in HCP, especially for packaging reusable items in the hospital. Industry also utilizes paper for packaging large-volume inexpensive products such as surgical gloves.

Paper has many features to recommend it. It is inexpensive and in good supply, and it provides good porosity and printability. It also has disadvantages. Paper is nonuniform in gauge, density, and porosity. It is moisture sensitive, and the physical strength of the paper may change with humidity conditions. Furthermore, its biological barrier properties could be destroyed by contact with water. Its physical strength is rather limited and paper can be easily torn or punctured, exposing the packaged product to bacterial contamination. The internal strength (cohesiveness) of the paper is usually weak, and surgical kraft is subject to "fiber tear." This causes problems in opening, as discussed below. Not just any kraft paper can be used. There are many factors to be considered in selecting a "surgical" kraft. A number of paper mills specialize in this field. Surgical kraft is available in a range of basis weights, from 20–90 lb/ream (9–41 kg/ream) (see Paper). Most frequently, a basis weight of 40–60 lb (18–27 kg) is employed in HCP. The paper must exhibit some wet strength in order to survive steam sterilization or coating with aqueous heat seal adhesives. Some concern has recently been raised regarding the presence of formaldehyde resins in the paper. Formaldehyde has been identified as a carcinogen and its release during steam sterilization may constitute a hazard to exposed employees. Some "formaldehyde-free" surgical kraft grades are now available. Paper mills have attempted to address the dual problem of fiber tear and moisture sensitivity. The attempt to reduce both of these problems with resin impregnation has met with limited success. The paper is dipped in a resin solution and the solvent, mostly water-evaporated. The resin is expected to coat the paper fibers, making them less water sensitive and improve the adhesion of the fibers to one another. Too high a resin content would reduce the porosity of the paper and interfere with sterilization. Too low a resin pickup, on the other hand, would achieve little improvement in either moisture repellency or internal tear resistance. Resin-impregnated ("reinforced") paper has not fulfilled its advertised promises. Fiber orientation is another property of concern. Every paper has a smooth and a rough side. The fibers have some degree of machine-direction orientation as well. These points must be considered in printing and package formation.

Foil. Aluminum foil, in gauges from 0.0003 to ~0.001 in. (7.6-~25.4 μm), is sometimes employed in HCP (17). It is normally avoided because of its high cost, lack of transparency, and lack of porosity. A compelling reason for the use of foil is the moisture sensitivity of the product packaged. Nothing equals foil in moisture-barrier properties. Foil is usually combined with other materials to provide a functional package. Because aluminum is relatively expensive, it is used in very thin gauges. Foil thinner than 0.001-in. (7.6-μm) thickness is prone to tear and puncture. To forestall foil failure, aluminum foil is often laminated to thin plastic films. The composite can have good abrasion, tear, and puncture resistance as well as many other advantages over the nascent foil. Aluminum foil, because it lacks porosity, poses a sterilization challenge. Conventional steam and gas sterilizations are out of question. Radiation sterilization can be utilized wherever the device can survive such exposure. Dry-heat sterilization, normally at about 356°F (180°C) for about two hours can be substituted, but because no seals survive such treatment, the method is impractical. One trick devised to sterilize devices in a foil pouch by either the steam or gas methods involves the sterilization of the devices in open pouches. The sterilant enters the pouch through its open mouth, and the pouches are removed to a "clean room" and sealed under virtually sterile conditions.

Plastics. HCP has embraced plastic materials for many reasons (18). Some plastics are very well suited for HCP, but not all are equally suitable. Desirable features of plastics are transparency, thermal stability, physical strength, formability, sealability, biological barrier, radiation resistance, and disposability. Plastic materials offer these properties in varying degrees. To attain all the desired performance characteristics, two or more plastics must be combined into a composite packaging material. One component provides strength and thermal stability, and another contributes sealability or some other feature (see Table 2).

Polyester. A thin PET film (see Film, oriented polyester) contributes a high degree of physical strength to the package (19). Normally, the film is 0.0005–0.001-in. (12.5–25-μm) thick. The film has excellent optical properties and very good thermal stability. This film can withstand all the major modes of sterilization. One point of caution: the use of PVDC-coated polyester should be avoided in HCP. This film is very widely used in food packaging, but the PVDC may react unfavorably under sterilization procedures (see Vinylidene chloride copolymers). Polyester film must be combined with a sealant (see below). Heavier gauge semirigid unoriented polyester film [4–25 mil (102–635 μm)] is used in blister packs (see Polyesters, thermoplastic). This material thermoforms well, but some grades are difficult to die cut.

Nylon. This polyamide film (see Nylon) is available in thickness of 0.00075-~0.005 in. (19-~127 μm). Nylon has wide usage in form/fill/seal food packaging applications. The film forms well, and it has good strength characteristics, and excellent transparency. However, its sterilization survival is not good. It has been reported that nylon degrades when exposed to radiation sterilization. The film is also heat- and moisture-sensitive and wrinkles badly during steam sterilization. It has thus failed to make any impact on HCP.

PVC. Heavy-gauge [5–25 mil (127–635 μm)] PVC (see Poly(vinyl chloride)) is widely used in food and pharmaceutical packaging. The film has almost glasslike clarity and is reasonably priced (see Film, rigid PVC). It makes a very appealing blister pack. In HCP, PVC has severe limitations owing to its performance under sterilization conditions. PVC should not be used in conjunction with radiation sterilization. The plastic turns red-brown when exposed to gamma rays. This is probably due to a breakdown of the polymer and purports more serious physical deterioration than the unpleasant appearance change might indicate. PVC cannot be used in conjunction with steam sterilization, because it is too heat-sensitive to survive exposure to temperatures above 250°F (121°C). This leaves ETO gas as the only permissible mode of sterilization. However, PVC absorbs this gas and holds on to it tenaciously. Thus degassing procedures must be extended to assure full desorption of the sterilant.

Polystyrene. This inexpensive plastic (see Polystyrene) has very limited use in HCP. It suffers from two unforgivable deficiencies: lack of impact strength and poor optical properties in impact-modified grades. In the packaging of very lightweight products, such as cotton balls, polystyrene may provide a very

Table 2. Plastic Films for HCP

Film type	Package type	Gauge	Pros	Cons	Sterilization		
					Steam	Gas	Radiation
Tyvek	pouch or lid	0.008 in. (203 μm)	good porosity physical strength	expensive heat sensitive	no	yes	yes
polyester, oriented	pouch	0.0005–0.001 in. (13–25 μm)	strength, optics, thermal stability		yes[a]	yes[a]	yes
nylon	form/fill/seal	0.001–0.005 in. (25–127 μm)	good thermoform	moisture sensitive	?	yes	limited
PVC	blister form/fill/seal	0.005–0.025 in. (127–635 μm)	crystal clear good forming	sterilization	no	yes longer degassing	no changes color
AN and PET copolymers	blister form/fill/seal	0.005–0.025 in. (125–635 μm)	good replacement for PVC		no	yes	yes
acrylic multipolymer	blister	0.005–0.020 in. (127–508 μm)		cloudy flex cracking	no	yes	yes
PS	blister	0.005–0.012 in. (127–305 μm)	inexpensive	optics impact sensitive	no	yes	yes
LDPE	pouch	0.001–0.006 in. (25–152 μm)	good sealant	high temperature sensitive	no	yes[a]	yes
polypropylene	pouch	0.001–0.006 in. (25–152 μm)	withstands steam sterilization	embrittles by radiation	yes[a]	yes[a]	no

[a] In conjunction with proper sealant and/or porous web.

suitable plastic container. Its high degree of porosity may reduce ETO sterilization cycles.

Polyethylene. This is the most widely used plastic in the world (see Polyethylene high density; Polyethylene low density). LDPE has a lower melting point than HDPE and it is easier to heat seal, but HDPE, with more tightly packed molecules, has better moisture-barrier properties and less elongation (better tensile strength). The materials can also be differentiated by molecular weight or melt index (MI). The basic PE is often modified by the addition of small quantities of co-monomers. For example, the addition of 3–5% of ethylene vinyl acetate (EVA) enhances the performance of PE appreciably. The EVA copolymer has better sealing characteristics and improved adhesion to several substrates. Other comonomers are acrylic acid, methacrylic acid, and methyl acrylate.

In HCP, polyethylene and copolymers are normally used in gauges of 0.001–0.006 in. (25–152 μm). They are used as primary packing films (eg, as vent bags); but the use of these materials as sealants is more important. Through the application of heat and pressure over a relatively short time interval, they can be made to adhere to a wide range of other materials. This process is utilized to close health-care packages. PE is suitable for either ETO or radiation sterilization. It is not advisable to use LDPE for steam sterilization because of its low melting point, but some grades of HDPE have successfully survived this process.

Polypropylene (PP). This plastic, too, represents a family of products of homo- and copolymers, of cast and blown films, with and without orientation (see Film, oriented polypropylene; Film, nonoriented polypropylene). In food packaging the biaxially oriented homopolymer is the high volume wrap. In HCP it is the nonoriented cast copolymer that is of most interest. The sensitivity of PP to radiation has been mentioned. PE, at a lower price, is adequate for ETO sterilization. Thus PP is a sealant of limited interest, used primarily for steam sterilization.

Other plastics. Acrylonitrile, acrylic multipolymers, and polycarbonate, find use in heavy gauge blister packs and in form/fill/seal applications. Each offers some advantages, and some shortcomings. Acrylonitrile (see Nitrile polymers) is in many respects similar to PVC, but it can be radiation sterilized. Acrylic multipolymers (see Nitrile polymers) are more widely used in drug packaging. They are cloudy and present potential flex cracking problems. Polycarbonate (see Polycarbonate) is expensive and is limited to specialty applications where high temperature or high impact resistance are of great value.

Heat-seal coatings. Several plastic films have been briefly reviewed above. It has been mentioned that two or more layers of different plastics may be required to achieve all the necessary performance characteristics. Thus PE film, for example, may be the sealant component in a multilayer composite (see Multilayer flexible packaging). Heat-seal coatings afford an alternative approach to furnishing a sealant layer. Plastic film cannot be handled in exceedingly thin gauges, but coatings can be applied in 0.0001-in. (2.5-μm) thickness. The coating can even be applied in patterns, covering but a segment of the substrate. By applying a very thin, or pattern, coating onto paper or Tyvek, one can achieve a very high degree of porosity that could never be duplicated by a PE lamination (see Laminating) or coextrusion. The coating has the further advantage of adhering to a wider range of substrates. Formulation of heat-seal coatings is as much an art as a science and is proprietary. Environmental considerations have forced many formulations to change from solvent to aqueous coatings. One must evaluate the heat seals for "sterilizer creep." Under highly humid conditions and tension (generated in the sterilizer evacuation cycle) the seals may fail partially.

Laminations. Very often no one plastic material can offer all of the desired properties. The package therefore is composed of two or more plastics, each bringing to the task some of the required performance characteristics. Several layers of different plastic materials may be combined with the aid of an adhesive to form a laminate. Oriented polyester film adhesive laminated to polypropylene film (OPET(A)PP) is frequently used for steam sterilization.

OPET(A)PE provides better seals at lower temperature. Still lower seal temperatures are permissible with EVA or ionomer (see Ionomers), but the heat sensitivity of the sealant excludes steam sterilization. With OPET(A)AF(A)LDPE, the foil provides moisture barrier; the polyester (or paper) contributes puncture, tear, and burst strength; and the sealant layer (which can also be PVC, EVA, or EAA) assures closure. The choice of adhesive is important (see Adhesives). Not just any adhesive will do. First of all, the adhesive must be suitable to join the respective webs to one another and maintain a strong bond over several years. The adhesive must retain its properties over a broad temperature range. It must have FDA clearance and may not embrittle, soften, or change color with age.

Coextrusions. In the lamination process, two or more web materials are joined. The webs need not be plastic. One can, for example, laminate paper to foil. Coextrusion is limited to plastics (see Coextrusion, flat; Coextrusion, tubular).

The advantages of the coextrusion process are quite apparent. It has reduced the multiprocess lamination to a single, much less expensive, step. The coextrusion, however, has limitations that must be recognized. It is not suitable for webs, such as paper, which cannot be made by the extrusion process. Not all plastics can be coextruded successfully. There must be a degree of compatibility between the various layers in order to achieve interply bond. Another serious shortcoming of coextrusion relates to printing. Laminations can be preprinted on either web prior to the combining process (see Printing). Thus the printing ink (see Inks) can be "trapped" between plies. In coextrusion the ink must be on the surface, where it is subject to abrasion and scuffing, or if on the inside surface, may contaminate the device (20,21).

Blends. Some years ago, in attempt to reduce the cost of ionomer resin, a mixture of PE/ionomer was extruded. The resulting blend was not just cheaper, but actually offered unique properties. Several such blends are currently in use as thermoformable webs. Most have sacrificed optical clarity but have gained good thermoformability plus sealability (and peelability) to a range of materials. Above all, these blends are relatively inexpensive.

Static control. With respect to materials generally utilized in HCP, it is necessary to consider "static control" (22–24). Static electricity generated by the package is to be avoided. Some fear the potential fire hazard of electrical discharge in the operating room. Others point out that this fear is unfounded in modern times because ignition-prone ether, the traditional anesthetic, has been replaced by modern anesthetics that are no longer poured from a bottle, but administered intravenously or at least in very controlled fashion. In any case, static electricity is a hazard to electronic equipment and devices. Furthermore, dirt and dust are attracted by static to the surface of the package and thus increase the chance of contamination. Many of the plastic materials are available with built-in or surface-coated antistat chemicals (see Additives, plastics).

Closure

To maintain sterility, the HC package must be hermetically sealed. This means that the closure must prevent ingress or egress of any substance. In the old-fashioned overwrap

method the closure is a piece of tape. There is no hermetic seal. The wrap is folded around the device in such a fashion as to prevent ingress of bacteria. The tape merely acts to hold the wrap in place. In the modern package heat and/or pressure are applied to mating surfaces to achieve a lasting marriage of two surfaces.

The closure must withstand the stresses generated during sterilization, shipping, and storage of the package, but because it is normally also the port of entry utilized to extract the device from the package, it cannot be a "weld," unless some other entry mechanism has been provided (25,26). The closure cannot be of a pressure-sensitive nature, because it would be absolutely unacceptable for a package to be "reclosable." A package once opened must be presumed to be "contaminated." To reclose a package and return it to stock would place in question the sterility of all items in storage. Cold seals are generally not suitable for HCP for this reason. When peeled open, these adhesives often permit resealing.

Assurance of closure perfection is most desirable, but the industry has not instituted any procedure to examine every closure for defects. Some imperfections are obvious and can be culled by the casual observer. Voids, tunnels, and foreign matter in the closure area are more easily detected if the closure is colored (27). Destructive tests to measure the physical strength of the closure are discussed below. They can provide statistical data based on proper sampling of a large mass of individuals. However, such destructive testing cannot afford assurances regarding the infallibility of any one item.

Automated inspection of the closure, or better yet the entire package, would be most desirable. Attempts to utilize spectroscopy, lasers, ultrasonics, and other means have met with little success. A method that has been evaluated in the food industry with some degree of success could be adapted to HCP. A minute quantity of an indicator gas (freon, xenon, carbon monoxide) is injected into the package just prior to closure. The finished package passes through a tunnel where slight vacuum and a sniffing device combine to detect traces of the indicator gas. If indicator gas is found, the package is rejected and the item goes back for repacking. This method cannot be employed with porous packaging materials such as paper or Tyvek, but for nonporous, radiation-sterilized HCP this seems a sterility assurance method deserving serious consideration.

Self seal (4,28,29) is one of the more recent innovations in HCP (see Fig. 1). A pouch is provided with an open end and a pressure sensitive tape attached thereto. After the device is placed in the pouch, the open end is closed by means of the tape. This mode of packaging eliminates the need for special equipment, such as heat sealers, to effect a closure of the open end. It was mentioned above that pressure-sensitive tape is not acceptable for package closure. In this case, however, the pressure-sensitive adhesive is extraordinary in that it "cures" during sterilization and becomes a permanent seal.

Opening

Many health-care packages are opened under stress conditions requiring rapid access to the item packaged. Most items require sterile presentation. The opening mechanism must be such as to facilitate both of these functions.

Easy opening is best attained with peelable seals (see Fig. 5). What constitutes "easy" is ill-defined. The industry would like the seals to be strong enough to withstand the usual sterilization and handling rigors and yet weak enough to open easily without material failure. This normally means a seal strength of 0.75–2.64 lbf/in. (0.13–0.46 N/mm). Both the device and the packaging materials must be considered in selecting the seal-strength range. If the weight and shape of the device tend to exert more pressure on the seals, the minimum allowable seal strength must be adjusted upward (30). Conversely, for weaker packaging materials, one must lower the maximum allowable seal strength. To improve the peel, packages are often designed with chevron or corner starts to further promote quick opening.

The package should not be cut open, even with sterile scissors. The exterior of the package is contaminated and care must be exercised not to compromise the sterility of its content.

Rapid access through peelability calls for ever-weaker closures and this may result in package failure (pop open) during sterilization, shipping, handling, or at any other time of its life cycle. Overcompensation for this potential problem could give rise to a new set of difficulties. Package failure in a futile opening attempt can result in device contamination. In evaluating the opening mechanism of a package, one must choose the desirable (or permissible) opening feature(s). For example, a 60 in. (1.52 m) long catheter pack should not be opened entirely, because the device would probably touch the floor. It is therefore of no consequence if this package starts tearing after having been peeled down to about 36 in. (915 mm). Encapsulation (the inability to extract the device from the package) is absolutely unacceptable. If opening of the package does not permit the ready removal of its contents, then the package needs redesigning. Particulate matter, either in the package or on its closure, is unacceptable. At one time some suppliers of health-care devices insisted that their packages show "fibers" in the opened closure in order to guarantee proper seals. Such "fibers" are most undesirable. The particulate matter may contaminate the devices and have serious health consequences. The idea of seal inspection after opening has been refined. The "telltale" seal, which can be examined as the product is removed, testifies to the proper performance of the closure (see Fig. 4).

In testing closures, one must consider the performance requirements under the highest stress conditions. Many seals diminish in strength during sterilization (31) and often remain at a lower level therafter. Nevertheless, seals are evaluated almost exclusively prior to sterilization.

Shelf Life

How long can a health-care package be used? There is no suitable answer to this wrong question. The survival of any package is event-related, not time-related. It has been reported that some surgical cotton wrapped in relatively poor paper survived sterile between the two world wars.

Suppliers of devices have for many years marked their packages as "sterile until opened." Recently, the demand for expiration dates has created competitive pressure to mark packages with unconfirmed claims of five years or longer acceptance. It is very difficult to simulate extended abuse conditions (32). Packages can be stored at high temperatures, low and/or high humidity conditions, they can be placed on shake tables to simulate overland truck shipments, and they can be dropped from measured heights to induce failure. It has been shown that it is not a simple matter to induce failure in a well-designed health-care package. Yet, actual recall of packages is testimony that failure is real.

Testing

A great many tests are conducted in HCP. Many of these tests are meaningless. Thought must be given to the test method, the meaning of the results, and the relation of the test conditions to actual performance. Above all, one must determine the necessity of the test (33).

Several societies have established standard test procedures that are applicable to HCP. Some of these tests procedures can be obtained from:

ASTM—American Society for Testing, and Materials 1916 Race St., Philadelphia, PA 19103

AAMI—Association for the Advancement of Medical Instrumentation 1901 N. Fort Myer Drive, Arlington, VA 22209

HIMA—Health Industry Manufacturers Association 1030-15th St. N.W. Washington, DC 20005

TAPPI—Technical Association of the Pulp & Paper Industry 15 Technology Parkway South, Norcross, GA 30092

All of these societies are concerned primarily with the physical performance of packaging materials. Attention must also be given to the evaluation of the safety of packaging materials with respect to possible transfer of chemicals onto the device. Potential toxic hazards must be assessed (34).

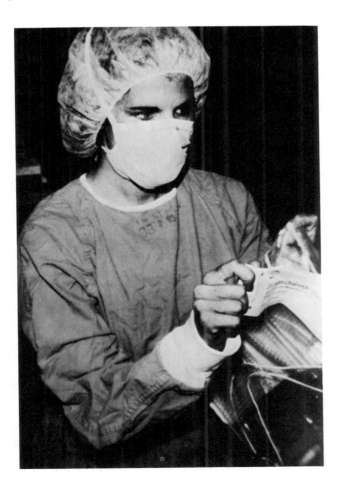

Figure 5. Proper package opening (peel).

Possible nondestructive tests were mentioned above. HIMA suggests nine microbiological test methods to assess both packaging materials and the finished package (35). Unfortunately, all of these methods are of a destructive nature. Furthermore, no single test has universal acceptance. The results are open to challenge as either too mild or too severe test conditions are used and thus not representative of normal occurrences.

Regulatory Considerations

HCP is subject to a wide range of government regulations. In the United States, most of these are promulgated by the FDA. Labeling restrictions are very strictly enforced. The government insists that very specific information be prominently and legibly displayed on the package (36). This requirement leads to a series of problems and different approaches to resolving them. Manufacturers of limited-volume items often resort to pressure-sensitive or heat-seal labels that are applied to the unprinted package (see Labels and labeling). As the quantity of the device produced increases, the label is often replaced by preprinted packages. As the complexity of the device increases, so does the need for longer instructions. The surface of the package is often inadequate to carry the lengthy message. One may insert an instruction sheet or print some or all the message on the inside of the package. The ability to print both sides of a packaging material is thus an important asset.

Multidevice manufacturers often color-code the packages. Either all of the print is in a distinct color for each product, or the background or even a patch on each package has a specific color corresponding to each product. This avoids potential product mix up.

The packaging operation is also subject to FDA controls known as Good Manufacturing Practices (GMPs) (37). These regulations are very voluminous but simple. Devices are classified by expert panels into three classes. The most critical devices, Class III, have very stringent conditions attached. Classes I and II are supervised very liberally. GMPs are based on the 1976 amendments to the Food & Drug Act. The FDA generally will not second guess the manufacturer as to the number and types of quality control procedures needed. A GMP, however, requires a formal organization, written procedures, and adherence to the procedures. What is important is "documentation" (37,38). For example, if a manufacturer has in a written procedure a periodic calibration of an instrument, then he must keep a log of such calibration.

It is important to keep "complaint" files. The reason for the complaint, the action taken to investigate its validity, and corrective actions (if any) must all be documented. In short, GMPs require the manufacturers to be in control of the process.

The EPA and OSHA are also concerned with HCP, especially with respect to sterilization and disposability of packaging materials.

BIBLIOGRAPHY

1. U.S. Pat. 4,057,144 (Nov. 8, 1977), S. J. Schuster.
2. C. D. Marotta, *Medical Device and Diagnostic Industry* 3(9), 33 (Sept. 1981).
3. U.S. Pat. 3,851,814 (Dec. 3, 1974) L. Stage (to Arvey Corp.).
4. U.S. Pat. 4,318,506 (March 9, 1982), A. Hirsch (to Arvey Corp.).
5. J. A. Goodman, *Packaging Technology,* 11(6), 11 (Dec. 1981).
6. L. Oster, *Medical Device and Diagnostic Industry,* 2(1), 11 (Jan. 1980).
7. J. Murtaugh and G. Whitaker, *Medical Instrumentation,* 17(3), 211 (May–June 1983).
8. "Occupational Exposure to ETO" (OSHA), *Fed. Regist.,* 49(122) (June 22, 1984).
9. *Radiation Compatible Materials, HIMA Report No. 78-4.9,* June 1980.
10. A. Hirsch and S. Manne, *Packaging Digest,* 17(1), 40 (Jan. 1980).
11. R. A. Caputo and C. C. Mascoli, *Medical Device and Diagnostic Industry,* 2(8), 22 (Aug. 1980).
12. A. Hirsch, *Bioequivalent Chemical Steam Sterilization Indicators,* Arvey Corp. Cedar Grove, NJ (presented at AAMI 18th National Meeting, Dallas, Texas, May 22–25, 1983).
13. *Guidelines for Evaluating the Safety of Materials Used in Medical Devices, HIMA Report No. 78-7,* June 1978.
14. "Guide to Medical Device Packagers and Packaging Suppliers," *Medical Device and Diagnostic Industry,* 5(8), 33 (Aug 1983).
15. A. Tallentire and C. S. Sinclair, *Medical Device and Diagnostic Industry,* 6(7), (July 1984).
16. C. D. Marotta, *Medical Device and Diagnostic Industry,* 3(5), 18 May 1981.
17. C. D. Marotta, *Medical Device and Diagnostic Industry,* 5(6), 36 (June 1983).
18. C. D. Marotta and A. M. Friedman, *Medical Device and Diagnostic Industry,* 6(4), 22 (April 1984).
19. B. V. Harris, *Medical Device and Diagnostic Industry,* 5(6), 72 (June 1983).
20. R. C. Griffin, Jr., *Paper, Film, and Foil Converter,* 57(1), 42 (Jan. 1983).
21. S. Anthony, Jr., *Medical Device and Diagnostic Industry,* 5(7), 18 (July 1983).
22. S. A. Halperin, *Medical Device and Diagnostic Industry,* 4(5), 35 (May 1982).
23. S. A. Halperin, *Medical Device and Diagnostic Industry,* 4(6), 45 (June 1982).
24. U.S. Pat. 4,098,406 (July 4, 1978), N. J. Otten and E. J. Presnell (to Tower Products Inc.).
25. A. Hirsch and S. Manne, *Medical Device and Diagnostic Industry,* 4(2), 23 (Feb. 1982).
26. C. D. Marotta, *Packaging Technology,* 11(6), 12 (Dec. 1981).
27. U.S. Pat. 3,533,548 (Oct. 13, 1970), M. Teterka (to C. R. Bard, Inc.)
28. U.S. Pat. 4,276,982 (July 7, 1981), J. S. Sibrava and F. E. Caroselli (to Arvey Corp.)
29. U.S. Pat. 4,358,015 (Nov. 9, 1982), A. Hirsch (to Arvey Corp.).
30. W. Haltberg, *Medical Device and Diagnostic Industry,* 3(10), 33 (Oct. 1981).
31. C. D. Marotta, *Medical Device and Diagnostic Industry,* 5(11), 24 (Nov. 1982).
32. A. Hirsch and S. Manne, *Packaging Engineering,* 27(6), 74 (June 1982).
33. A. Hirsch, *Medical Device and Diagnostic Industry,* 3(4), 14 (April 1981).
34. *Biocompatibility, HIMA Report No. 80-1,* Proceedings from the Medical and Scientific Section Meeting, May 1980.
35. *Microbiological Methods for Assessment of Packaging Integrity, HIMA Report No. 78-4.11,* June 1979.

36. T. Riggs, *Medical Device and Diagnostic Industry,* 1(6), 20 (June 1979).

37. V. Goetz, *Medical Device and Diagnostic Industry,* 1(10), 36 (Oct. 1979).

38. Food and Drug Administration, "Classification of General Hospital and Personal Use Devices," *Fed. Regist.,* (Aug. 24, 1979).

A. A. Hirsch
Arvey Corporation

HEAT SEALING.   See Sealing, heat.

HEAT TRANSFER LABELS.   See Decorating; Labels and labeling.

HEAVY-DUTY PLASTIC BAGS.   See Bags, heavy-duty plastic.

HOT STAMPING.   See Decorating.

I

INDICATING DEVICES

One summer in the 1930s, Eastman Kodak Co. received a telegram from the Kitt Observatory in Arizona stating that a shipment of photographic plates had no light sensitivity. Each of these plates was developed completely and was uniformly black. A new set of plates was carefully packed and shipped to the remote observatory. Again, all of the plates were black when developed. A well-respected Pinkerton guard accompanied the next shipment. He never let the box of plates out of his sight, and the mystery was solved at a small siding in Arizona. The rail clerk received the box clearly marked "FRAGILE GLASS PLATES," and methodically opened the box, took out each plate in the bright afternoon sun, removed the black paper cover, and examined each one. "All in perfect shape," he said to the guard as he rewrapped the plates and returned them to the carton.

A shipper must know what happens to a package after it leaves the shipping room, but it is impossible to send a security guard with every carton. Instead, a variety of indicating devices have become the security guards for the packages.

Perishable products may speak for themselves by their physical condition or their odor. Damage to other products from inappropriate handling may not be evident until actual use. As an example, whole blood stored below 50°F (10°C) is

Table 1. Biologics, Shipping Temperatures[a]

Product	Maximum temperature, °F (°C)
cryoprecipitated antihemophilic factor[b]	<0 (−18)
measles, mumps, and rubella virus vaccine, live	<50 (10)
measles and rubella virus vaccine, live	<50 (10)
measles–smallpox vaccine, live	<50 (10)
measles virus vaccine, live, attenuated	<50 (10)
mumps virus vaccine, live	<50 (10)
poliovirus vaccine, live, oral, type 1	<32 (0)
poliovirus vaccine, live, oral, type 2	<32 (0)
poliovirus vaccine, live, oral, type 3	<32 (0)
poliovirus vaccine, live, oral, trivalent	<32 (0)
red blood cells[b], frozen	<−85 (−65)
red blood cells[b], liquid	34–50 (1–10)
rubella and mumps virus, live	<50 (10)
rubella virus vaccine, live	<50 (10)
single-donor plasma[b], frozen	<0 (−18)
smallpox vaccine, liquid	<32 (0)
source plasma[b]	23 (−5)
whole blood[b]	34–50 (1–10)
yellow fever vaccine	<32 (0)

[a] Ref. 1.
[b] Human.

stable for months. If the temperature rises above that point for 20 min, however, enzymatic changes could occur that would make it life-threatening if transfused. The potency of vaccines is lost by exceeding critical temperatures; an even greater hazard may result by presuming that it is effective. Because microcircuits are sensitive to mechanical shock and electrical potential, individual components and wired circuits must be monitored.

Time and Temperature Indicators

The fully integrating monitoring device can indicate the temperature gradient to which it has been exposed, as well as the amount of time that it has been at that temperature. Many inventors and their companies have focused on developing such devices, which operate on the following principles: physical change; chemical change; electrochemical indication; electromechanical indication; electronic readout; and others. The devices monitor perishable foods such as fish, fruit, and vegetables, as well as pharmaceuticals and vaccines. Table 1 lists the FDA temperature limits for various biological preparations.

Indicators that show time only are also available in many styles. These are used primarily on shipping containers, not on consumer packages. Life-dated products, such as photographic film and some foods, can have greatly extended salable time by refrigerated storage. A time indicator would benefit the manufacturer and retailer as well as the consumer.

The Andover Laboratories manufactures a time–temperature integrator called the Tempchron, formerly known as the Ambitemp (see Fig. 1), which functions with a fluid that has a specific melt temperature for the product to be monitored. This device can be described as an integrator because it provides information on the multiple of the time and temperature, giving a readout in degree minutes that can be interpreted from a chart. A wide range of temperatures can be monitored by selection of the liquid that is to be frozen in the tube.

Figure 2 shows an interesting concept of a time–temperature integrator that was at one time manufactured by Honeywell Corp. This device, the TTI, comprised an electrolytic battery that was activated by breaking an ampul of electrolyte between copper and cadmium strips. An electrochemical color reaction that was accelerated at higher temperatures ensued.

Kokum Chemical in Malmo, Sweden, manufacturers a time–temperature monitor that utilizes an enzyme to bring about a color change. This product (see Fig. 3), the I-Point

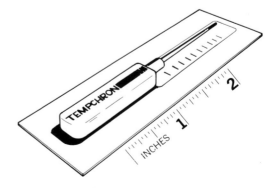

Figure 1. Integrated time–temperature indicator. The Tempchron (Andover Laboratories) utilizes the melt temperature of the product to be measured. To convert in. to cm, multiply by 2.54.

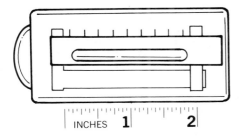

Figure 2. Integrated time–temperature indicator. The TTI (Honeywell Corp.) used an ampul of electrolyte between metal plates. To convert in. to cm, multiply by 2.54.

TTM, is intended to match the enzyme reactions of the product it is monitoring. A progression of color changes occurs.

Another time–temperature integrator functions by the use of a selective gas-absorption plastic membrane and leuco (colorless) dyes that are sensitive to oxygen (2). The leuco dye is converted to the colored form inside the pouch in proportion to the rate of oxygen penetration through the membrane and the time (see Fig. 4). The rate of penetration of the oxygen through the membrane is also related to the temperature. The storage life of foods for emergency survival has been studied as a function of the temperatures at which they were stored. With the accurately calibrated barrier films (see Barrier polymers) now available, this may prove a valuable area of investigation.

Allied Chemical Co. has a patent (3) on work in which combinations of conjugated acetylene compounds irreversibly change color when heated. After finalizing the above design, the Program for Appropriate Technology in Health (PATH) in Seattle, Wash., has marketed a product called the PATHmarker, which indicates the change in color. Figure 5 illustrates the relationship between the color change of this product and the thermal degradation of measles vaccine.

There are two patents (4,5) for time–temperature integrators under the name Monitor Mark (see Fig. 6a). In this product, a dyed meltable solid travels down a porous wick when heated above its melting point. The time interval is a function of the travel of the color down the wick. The device is activated by removal of a barrier film that is positioned between the wick and the meltable solid (see Fig. 6b). The product is available in several response temperature ranges.

The Therma-Gard recorder (Impact-O-Graph Corp.) monitors temperatures on a cassette and functions for 30 d. This unit is depicted in Figure 7. The TSI International Corp. has

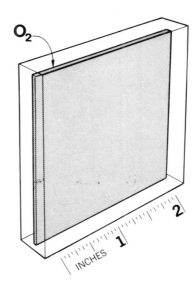

Figure 4. An integrating time–temperature indicator using the penetration of oxygen through a membrane to trigger the formation of a color from a leuco dye (2). To convert in. to cm, multiply by 2.54.

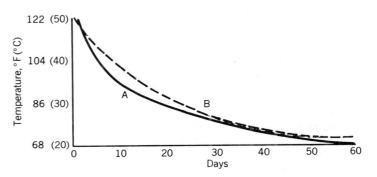

Figure 5. The relationship between measles-vaccine degradation and the color change of the PATHmarker: A, indicator color change to black; B, typical measles-vaccine degradation, based on 7-d exposure to 98.6°F (37°C), the criterion used in current WHO (World Health Organization) recommendations for measles-vaccine stability.

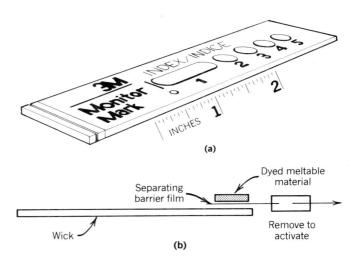

Figure 6. (a) The Monitor Mark Time Temperature Tag (3M Co.), an integrator available in many temperature ranges. To convert in. to cm, multiply by 2.54. (b) Cross section.

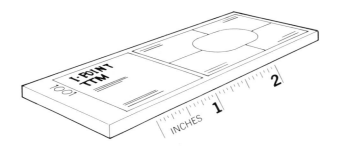

Figure 3. The I-Point TTM (Kokum Chemical) is an integrating time–temperature monitor utilizing enzyme-produced color change. To convert in. to cm, multiply by 2.54.

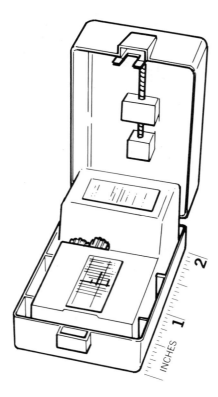

Figure 7. Therma-Gard recorder (Impact-O-Graph Corp.). To convert in. to cm, multiply by 2.54.

introduced a recorder that operates for up to 90 d in the indicating range of −20 to 100°F (−29 to 30°C) (see Fig. 8).

Two workers (6), following the lead of another (7), have fabricated two styles of electronic recorders with built-in memories (see Fig. 9); however, they are not commercially available. Workers at 3M Co. have invented a Thermal History Indicating Device (THID) (8). Another product that interfaces with a computer for its readout is the Tattletale thermograph (Onset Computer Co.) (see Fig. 10). This thermograph reports temperature transients to a memory and can be programmed to record at 5-s intervals for 5.5–6-h intervals for 32 mo. The functional temperature range is −41 to 185°F (−41 to 85°C).

Temperature Indicators

Temperature-measuring devices are numerous. In the simplest form, a sphere of ice is frozen with one half clear and the

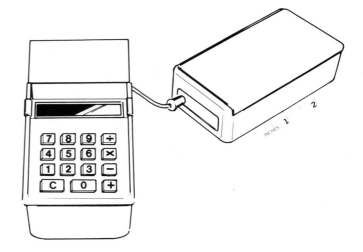

Figure 9. Electronic recorder (6,7). To convert in. to cm, multiply by 2.54.

other red (see Fig. 11). If the product thaws, the sphere becomes pink. High technology temperature indicators include thermistors, circuits, and liquid-crystal displays (LCD).

Most temperature indicators depend upon a chemical color change. These compact units provide direct, reliable information on the temperature to which the product was exposed. They are widely used for materials with critical temperatures such as pharmaceutical and biological preparations (Table 1) and fresh and frozen fish. Table 2 lists some of these indicators; Figure 12 is representative of these devices.

There are two other types of indicators not listed in Table 2: the Criti-Temp (Schobl Enterprise, Inc.), which is a bimetallic spring-loaded monitor (see Fig. 13), and the Precision Digital Thermometer (TSI International Corp.), which functions from a thermistor to give a temperature display (see Fig. 14).

Freeze and Thaw Indicators

The main purpose of the indicators listed in Table 2 is to show a response to a rising temperature well above room temperature. Devices to indicate temperatures near the freezing point of water require special design since they must be unbroken before use and function at low temperatures.

Akzo N.V. of the Netherlands has exerted a strong influence in the monitor area through its subsidiary, Organon, heir to the BMS disposable-thermometer technology, and through

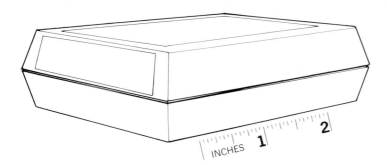

Figure 8. TSI recorder (TSI International Corp.). To convert in. to cm, multiply by 2.54.

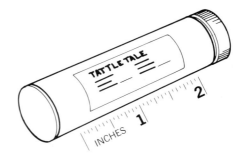

Figure 10. Tattletale thermograph electronic recorder (Onset Computer Corp.). To convert in. to cm, multiply by 2.54.

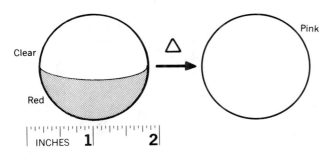

Figure 11. Thaw indicator. Half-colored sphere of ice becomes homogeneous when thawed. To convert in. to cm, multiply by 2.54.

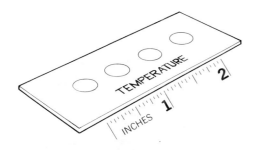

Figure 12. Temperature indicator representative of devices listed in Table 2. To convert in. to cm, multiply by 2.54.

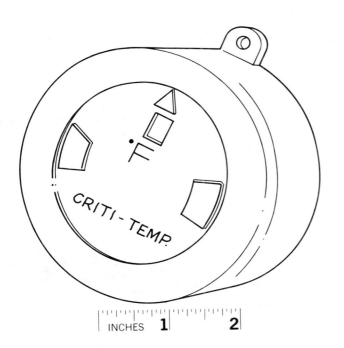

Figure 13. Criti-Temp monitor (Schobl Enterprise, Inc.). To convert in. to cm, multiply by 2.54.

Table 2.   Color-change Temperature Indicators

Product name	Company and location	Temperature range, °F (°C)
Celcistrip	Solder Absorbing Technology (Agawan, Mass.)	104–465 (40–240) 105–500 (40–260)
Thermomarkers	W. H. Brady Co. (Milwaukee, Wisc.)	100–500 (40–260)
Tesa-Temperatur Indikatoren, irreversible	Biersdorf AG (Hamburg, FRG)	175–360 (80–180)
Thermolabel	Paper Thermometer Co. (Greenfield, N.H.)	140–180 (60–81)
Tattletherm	Everest Interscience (Tustin, Calif.)	100–500 (40–260)
Reatec	MRC Corp. (Wayne, Pa.)	100–420 (40–213)
Template	Wahl Corp. (Los Angeles, Calif.)	100–1000 (40–590)
Telatemp	Telatemp Corp. (Fullerton, Calif.)	100–350 (40–155)
Thermo Strip	Archie Soloman and Associates (Roswell, Calif.)	219–435 (103–220)
Thermindex	Thermindex Chemicals and Coatings, Ltd. (Missisauga, Ontario, Canada)	100–500 (38–260)
OmegaLabels	Omega Engineering (Los Angeles, Calif.)	100–400 (40–202)
Hermet	Markal Co. (Chicago, Ill.)	100–500 (40–260)
Temp Tabs	Jardine Engineering (Hong Kong)	100–900 (40–500)
Celsipoint	Signalarm, Inc. (Springfield, Mass.)	100–500 (40–260)
Templilabel Temp-Alarm Templistik	Big Three Industries, Inc. (South Plainfield, N.J.)	125–750 (50–395)
Thermax	Thermagraphics Measurements, Ltd. (Chicago, Ill.)	150–400 (65–200)
T-Dot	Westemp Instruments Co. (Cardiff, Calif.)	100–400 (40–200)

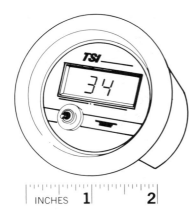

Figure 14. Precision Digital Thermometer (TSI International Corp.). To convert in. to cm, multiply by 2.54.

Figure 15. Freeze-Watch (Info-Chem Protective Products).

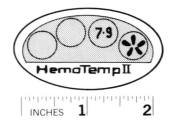

Figure 16. HemoTemp II (Biosynergy, Inc.). To convert in. to cm, multiply by 2.54.

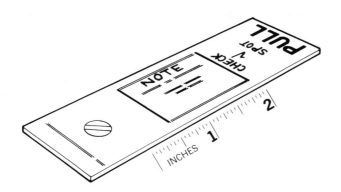

Figure 17. Check-Spot (Check-Spot, Inc.). To convert in. to cm, multiply by 2.54.

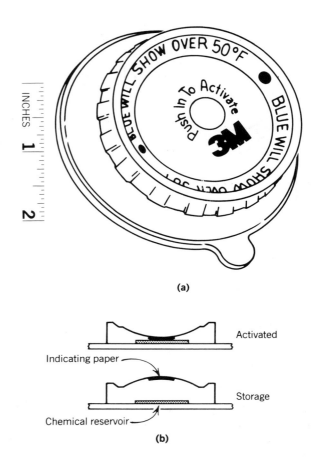

Figure 18. (**a**) Monitor-Mark Button (3M Co.). To convert in. to cm, multiply by 2.54. (**b**) Cross section.

its Info-Chem Protective Products Division. Info-Chem manufactures the Thaw-Watch and the Freeze-Watch. These products utilize an ampul filled with a colored fluid. Freezing breaks the ampul, which spills the contents onto a paper indicator (see Fig. 15).

Biosynergy, Inc. manufacturers a liquid crystal HemoTemp II for use on blood-collection bags. This device, after being frozen, indicates the temperature of the blood (see Fig. 16).

The Check-Spot (Check-Spot, Inc.) (9) functions at 3–32°F (−16 to 0°C) and is based upon a solid emulsion (see Fig. 17).

The Monitor-Mark Button (3M Co.) (see Fig. 18**a**) operates by means of a meltable, dyed compound contained in a porous reservoir. In the inactivated form, a domed indicator paper is

Figure 19. Monitor-Mark Freeze Indicator 32F (3M Co.). To convert in. to cm, multiply by 2.54.

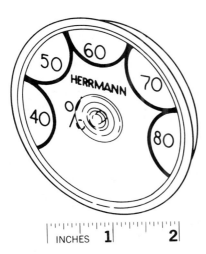

Figure 20. Humidity indicator (Herrmann Chemie and Packmittel of Sod-Chemie AG). To convert in. to cm, multiply by 2.54.

Figure 22. Shockwatch gravitational indicator (Media Recovery Co.). To convert in. to cm, multiply by 2.54.

separated from the reservoir by a small distance. When the dome is pressed, the two materials come in contact, allowing wicking to occur when the melt temperature is reached (see Fig. 18b). This product has many different applications, including irreversible monitoring of bags of whole blood. The specifications for blood banking have been outlined (10).

Other styles of freeze indicators are manufactured by 3M Co. (11–13), including the Monitor-Mark Cold Side Indicator, model 32F (see Fig. 19). When temperatures fall below 32°F (0°C), there is an irreversible color change. In the past, the IWI, or irreversible warm-up indicator, was manufactured by Artech Corp. (14).

Humidity Indicators

Many items are sensitive to moisture in ways that are irreversible. Simply drying-out or humidifying does not restore the original function, and the damage that may have occurred during shipping or storage may not be evident to the recipient. To protect the manufacturer and the customer, humdity changes must be monitored. A humidity monitor based on a color change is available from Herrmann Chemie and Packmittel (see Fig. 20).

A U.S. patent has been granted for a time/history humidity indicator (15). This device comprises a salt that absorbs moisture from the air, a water-soluble dye, and an absorbent material. After activation by removal of a barrier film, the dissolved dye migrates into the absorbent material as the humidity increases. This product resembles the product shown in Figure 6.

Gravitational-force Indicators

Gravity is a useful force in packaging because it keeps containers in place. Many times the normal force of gravity may be applied to a product in shipping and handling. This abusive treatment can cause hidden damage. The Hump-Gard (Impacto-Graph Corp.) is a 30-day monitor for the high *G* forces encountered in railroad shipping. It resembles the Thermo-Gard (see Fig. 7) in external appearance.

The Shockwatch is a compact device for measuring high gravitational forces (16). The active element is a capillary tube with surface-energy differences in each end (see Fig. 21). A colored fluid is placed in the -philic end, and when forces exceed a predetermined value, the liquid is forced into the -phobic end where it is visible (see Fig. 22). Once distributed by 3M Co., it is now available from Media Recovery Co.

This article has discussed but a few of the hundreds of indicating devices available (17). A very low cost indicator is still needed, preferably one that could be imprinted as part of the label of consumer products such as frozen foods.

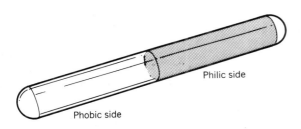

Figure 21. Capillary tube of Shockwatch.

BIBLIOGRAPHY

1. *Fed. Reg.* **39**(219) (Nov. 12, 1974).
2. K. H. Hu, *Food Technol.* **26**, 56 (Aug. 1972).
3. U.S. Pat. 3,999,946 (Dec. 28, 1976), G. N. Patel (to Allied Chemical Co.).
4. U.S. Pat. 3,954,011 (May 4, 1976), W. Manske (to 3M Co.).
5. U.S. Pat 3,962,920 (June 15, 1976), W. Manske (to 3M Co.).
6. A. J. C. Carselidine and R. R. Weste, *CSIRO Food Res. Q.* **36**, 41 (1973).
7. M. J. M. Van't Root, *Proceedings of the 13th Congress on Refrigeration*, Washington, D.C., Vol. 4, pp. 445–451.

8. The author and F. R. Parham, 3M Co., Saint Paul, Minn., have been active in microprocessor monitors.

9. U.S. Pat. 2,971,852 (Feb. 14, 1961), J. Schulein (to Check-Spot, Inc.).

10. B. A. Myhre, *Quality Control in Blood Banking*, John Wiley & Sons, Inc., New York, 1976, pp. 168 and 169.

11. U.S. Pat. 4,132,186 (Jan. 7, 1979), W. Manske (to 3M Co.).

12. U.S. Pat. 4, 457, 252 (July 3, 1984), W. Manske (to 3M Co.).

13. U.S. Pat. 4,457,253 (July 3, 1984), W. Manske (to 3M Co.).

14. Product literature, Artech Corp., Falls Church, Va.

15. U.S. Pat. 4,098,120 (July 4, 1978), W. Manske (to 3M Co.).

16. U.S. Pat. 4,068,613 (Jan. 17, 1978), U.R. Rubey (to Detectors, Inc.).

17. W. Manske, 3M Co., Saint Paul, Minn., has a working file of over 255 U.S. patents on Indicating Devices.

DEE LYNN JOHNSON
3M Company

INJECTION MOLDING

Injection molding, one of the principal methods of forming thermoplastic materials, involves feeding plastic resin to a rotating screw in a heated barrel. There the plastic is melted and mixed. The resulting hot plastic is then injected at high pressure into a closed mold that has one or more cavities in the shape of the desired part. The mold is cooled, and when the plastic solidifies, the mold is opened and the parts are ejected. The process can then repeat itself.

Typical injection-molded products (see Fig. 1) for packaging are thin-walled yogurt and margarine containers, 5-gal (19-L) shipping pails (see Pails, plastic), PET preforms (see Blow molding), and base cups, and threaded closures (see Closures, bottle and jar).

Several developments in the last 15 years have transformed injection molding for the packging industry. Resin costs increased drastically with the rise in oil prices. Competition from paper packaging as well as other processes (see Blow molding; Thermoforming) has intensified. To effectively compete under these conditions, the cost of injection-molded parts

Figure 1. A selection of injection-molded containers and components, including a preform and base cups for blow-molded poly(ethylene terephthalate) (PET) soft-drink bottles.

had to be reduced without compromising the strength or quality of the part. Resin producers helped by developing high flow polyethylene (see Polyethylene, low density; Polyethylene, high density) and polypropylene (see Polypropylene) resins to allow the molding of parts with thinner walls and 10–15% less weight, and equipment suppliers developed new breeds of high performance injection-molding machines for the packaging industry.

The Injection Mold

The single most important component of the injection-molding system is the mold. Mold design has tremendous impact on system productivity and product quality, and therefore on the overall economics of the injection-molding operation. The design of a mold is heavily influenced by the characteristics of the part. Most injection-molded parts for the packaging industry have relatively thin walls and relatively large length-to-thickness (L: T) ratios. Length is the maximum flow length in the cavity, and thickness is the average wall thickness of the part. Ten years ago, L: T ratios ranged up to 200:1; now they can be as high as 400:1. For such thin-walled parts, design emphasis must be placed on the hot-runner system, part ejection, mold-cooling, alignment, and mold-material selection. A typical mold for thin-walled containers is shown in Figure 2.

Operation. As the mold closes, the cores and cavities are aligned by the corresponding tapers on the lock rings and the cavities. Plastic melt, ie, melted resin, is injected by the injection unit of the molding machine into the sprue bushing and forced through a network of flow channels called the runner system. In Figure 2, the runner system is heated by electrical heaters and the term hot-runner system applies. The hot-runner system maintains and controls the temperature of the melt right up to the gates by use of a sprue heater, manifold heaters, and nozzle heaters. Past the gates, the melt flows into the cavities. After the melt has cooled and solidified, the mold opens, and the parts are blown off the cores by blasts of air and dropped between the mold halves.

Runner system. There are basically two types of runner systems: hot and cold. In cold-runner systems, the melt in the flow channels between the sprue bushing and the gates is allowed to cool and solidify during each cycle and must be ejected before the next cycle begins. Cold-runner systems are simpler and less expensive than hot-runner systems. A disadvantage is that the plastic of the cold runner must be reprocessed or scrapped after each cycle. Also, the cold runner can significantly slow the speed at which the mold operates, since the cold runner must be cooled before ejection. For these reasons, hot-runner systems are generally favored in packaging applications.

To ensure consistent quality, the hot-runner system should be balanced so that it supplies melt to each cavity at the same pressure and temperature. Flow balance is achieved by ensuring that each flow channel from the sprue bushing to the cavity gate has an equal length and an equal number of turns. Temperature balance is achieved through uniform heating and insulation. Hot-runner molds with up to 64 cavities operate reliably with this balanced approach. The hot-runner system should also have channels with smooth corners to reduce friction and pressure drop. There should be no dead spots where plastic can be trapped and degraded.

Cooling. Cooling time is a large part of the total molding-cycle time. To decrease cooling time and thus increase produc-

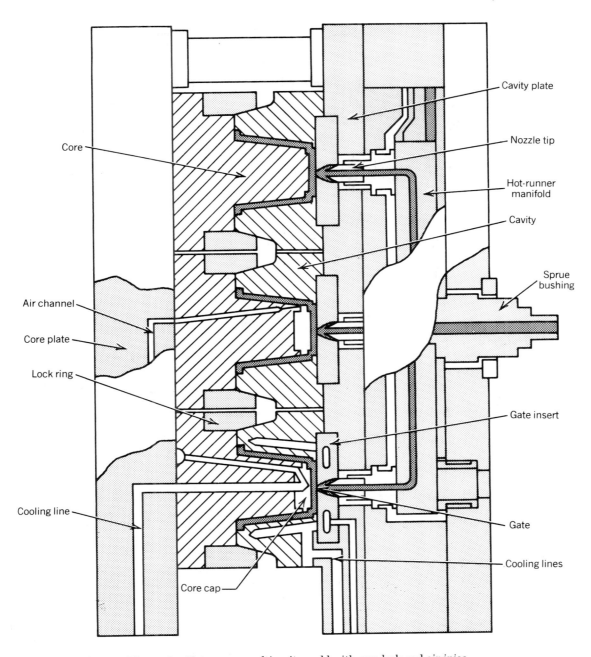

Figure 2. Hot-runner multicavity mold with core lock and air injection.

tivity, cooling channels should be near the molding surface. High coolant-flow rates are necessary to achieve high heat-transfer rates. Beryllium copper inserts can be used to improve heat transfer where high heat is generated, such as around the gates. Beryllium copper has approximately 10 times the coefficient of thermal conductivity of steel. With a well-designed cooling system, typical thin-walled parts can be cooled and ejected only a few seconds after injection. Cycle times as short as 2 s are possible. It is also important for the cooling system to keep the cores and cavities at approximately the same temperature. This prevents excessive wear between the interlocking faces due to differences in thermal expansion between the mold halves.

Part ejection.   There are two basic methods of part ejection: mechanical and air. Mechanical ejection commonly uses strip-

per rings surrounding each mold core to physically push the parts off the core. The stripper rings can be activated by hydraulic cylinders attached to the machine platen, air cylinders in the mold, or a mechanical linkage tied to the motion of the machine platen. Alternatively, air ejection uses blasts of air to loosen and blow the parts off the cores. Air ejection is the more popular method because it involves fewer moving parts and thus less maintenance, the mold can be more compact. Also, parts can be made thinner and ejected earlier in the cycle because the sidewall strength required for mechanical ejection is not needed for air ejection. Mechanical ejection is preferable for polystyrene parts, however, because the force of the air blast can easily crack this stiff material.

Alignment.   The proper alignment of core and cavity is critical in thin-walled molding. Slight misalignment leads to

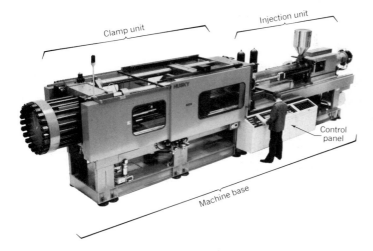

Figure 3. Injection-molding machine.

preferential filling of the thicker part of the cavity, which causes a pressure imbalance around the core and cavity, thus further shifting the core. The overall result is uneven wall thickness in the molded part. In extreme cases, it can result in incomplete filling of the cavity, ie, a short shot, and rejected parts. To achieve satisfactory part quality, center-to-center alignment of the core and cavity within ± 0.4 mil (± 0.01 mm) may be necessary.

There are three main methods of aligning the cores and cavities of packaging molds: the stripper-ring, core-lock, and floating-core methods. The *stripper-ring method* uses tapers on the stripper ring to align the core and cavity. This popular method of alignment has disadvantages for modern thin-walled molding, where high alignment forces cause relatively rapid wear of the stripper rings. Since the stripper rings form part of the molding surface, excessive wear in this area reduces part quality. To avoid a drop in part quality, frequent mold maintenance is required. Wear is not critical with the *core-lock method,* since the core-locking rings are not part of the molding surface. Either air or mechanical ejection can be used with core-lock alignment. The *floating-core method* is a recent design that appears to be the most suitable for most packaging applications. The cores and locking rings are not rigidly fixed to the core plate, but are allowed to float to more easily align themselves with the cavities. Easier alignment

results in less wear on the aligning tapers. This method can also be used with either air or mechanical ejection.

Mold materials.   High quality mold material is of the utmost importance. Most mold components are made of through-hardened high quality tool steel. Hardness ranges from RC 50 (Rockwell C scale) for the cores and cavities to RC 30 for the mold plates. Molds of high quality steel usually become obsolete before they wear out. Only periodic rebuilding of wear items, such as stripper rings and leader-pin bushings, is required.

Stack molds.   A stack mold has two identical sets of cores and cavities which are stacked together back to back. Use of a stack mold almost doubles the productivity of a machine and results in approximately 20% less energy consumption per part produced. For these reasons, stack molds are increasingly popular.

The Injection-molding Machine

An injection-molding machine consists of three main components: injection unit, clamp unit, and machine base (see Fig. 3). The injection unit plasticizes, ie, melts, and injects the resin at high pressure into the closed mold. The clamp unit supports the mold halves, closes and clamps them together during injection, and opens them for ejection. The machine base supports the clamp and injection units and houses a variety of electric, electronic, pneumatic, and hydraulic equipment needed to power and control the machine.

Injection unit.   There are two types of injection units used in packaging applications: reciprocating-screw and two-stage.

Figure 4 shows a current-design reciprocating-screw unit. As the motor rotates the screw, the shearing action on the resin generates most of the heat needed to plasticize the resin. Only about 20% of the heat requirement is from the conductive heaters around the barrel. As the screw rotates, it retracts and feeds the melt to the front of the screw. When the required volume of melt has accumulated at the front of the screw, the screw stops rotating and moves forward to inject the resin into the mold. Injection takes place at a predetermined speed and pressure to ensure complete filling of the mold. Once the melt has been injected, the screw continues to hold the resin in the mold under pressure for a set period of time to compensate for shrinkage of the part as it cools in the mold.

A two-stage unit is shown in Figure 5. This type of unit has separate components for plasticizing and injection. The extruder screw feeds the melt to a separate shooting pot, from which the melt is injected into the mold by an injection piston.

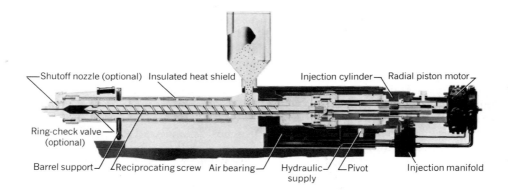

Figure 4.   Reciprocating-screw injection unit.

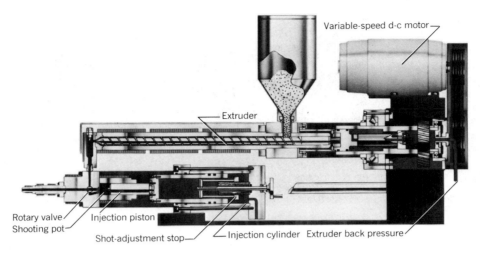

Figure 5. Two-stage injection unit.

In thin-walled molding, injection-unit design emphasis must be placed on fast injection speeds, high injection pressures and plasticizing rates, and shot-to-shot repeatability. Injection speed is critical. Molten resin moving through a thin cavity over a long distance (high L : T ratio) tends to cool quickly and freeze off before completely filling the mold. To prevent this, injection speed must be very rapid. With today's multicavity molds, this requires high injection rates. Injection rates of 3.3 lb/s (1.5 kg/s) or more are not uncommon in thin-walled molding.

High injection pressures are mandatory to achieve the high injection rates into thin-walled cavities. Injection pressures of 29,000 psi (200 MPa) are generally required to fill with viscous resins. Hydraulic accumulators can provide the instantaneous power necessary for injection without oversized hydraulic pumps and motors. High plasticizing rates are needed with the short cycle times characteristic of packaging applications. An advantage of two-stage injection is that the extruder screw can continue to turn without interruption during both the injection and hold phases.

Shot-to-shot repeatability is essential for consistently high quality parts and for keeping part weight at the desired minimum. The two-stage design shown in Figure 5 can offer ±0.1% shot-to-shot repeatability in injection volume through mechanical shot-volume control. Generally, the accuracy of the reciprocating-screw unit is less than that of a two-stage unit because of the time delay for the ring-check valve to seal off

during injection. Repeatability can be enhanced by the use of fast-acting low leakage hydraulic valves and accurate screw or injection-piston sensors. Also, an accumulator dedicated solely to injection can ensure that the same level of hydraulic energy is stored and ready for every shot.

Clamp unit. There are two broad categories of clamps: hydraulic and toggle. Hydraulic clamps use hydraulic cylinders to open and close the mold, and toggle clamps use mechanical linkages.

One type of hydraulic clamp is shown in Figure 6. Two large platens are provided for mounting the mold halves. One platen is firmly attached to the machine base and the other can be moved to open and close the mold. A small-diameter clamp stroke cylinder moves the moving platen at high speeds with minimum oil consumption. A hydromechanical cushion in the clamp cylinder provides consistent slowdown before mold faces contact. After the mold faces are brought together and the mold is closed, shutters block the clamp column, and high clamping force is then applied by the large clamping piston. The large clamping piston also provides high initial opening force for mold breakaway. Tie bars, which join the stationary platen and the clamp mechanism, absorb the reaction forces during application of clamp tonnage and guide the moving platen to ensure alignment with the stationary platen. An ejection mechanism supplies the force necessary to strip the parts off the mold core.

A high speed toggle clamp is shown in Figure 7. Power from

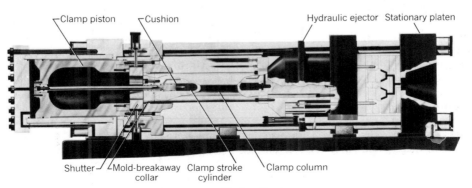

Figure 6. Hydraulic clamp.

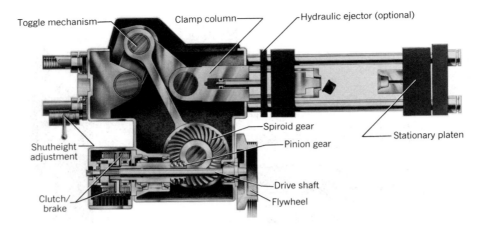

Figure 7. Toggle clamp.

an electric a-c motor is transmitted through a clutch/brake mechanism, then through a pinion and bevel gear to a crank plate. From there, power is transmitted through a connecting rod to the toggle mechanism and the clamp column. Toggle linkages inherently provide high speed and low mechanical advantage at the middle of the stroke, and low speed and high mechanical advantage at the end of the stroke for clamp-up.

To design a clamp unit suitable for a packaging application, several factors must be considered. The fast molding cycles possible with thin-walled containers demand high clamp speed. The need for consistent wall thickness demands clamp rigidity and alignment, and the high cost of modern molds requires reliable mold-protection mechanisms.

Fast speed can be achieved with a hydraulic clamp by using a relatively small clamp stroke cylinder. However, the oil required to rapidly open or close the clamp still places a peak demand on the hydraulic system. To avoid oversizing the hydraulic pumps to meet this demand, hydraulic accumulators can be used. With accumulators, speeds of 5 ft/s (1.5 m/s) are economically feasible. Toggle clamps are generally capable of higher speeds than hydraulic clamps and provide smooth acceleration and deceleration.

To ensure clamp rigidity, massive platens are necessary to limit deflection to under 5 mil (0.13 mm) at full clamping pressures. Also, a large centrally positioned platen backup is important to transmit the clamping force evenly from the column to the moving platen. Excessive platen deflection must be eliminated to avoid flashing (ie, separation of mold parting line during injection and resultant penetration of melt) and uneven wall thickness of the molded part.

Center-to-center alignment and parallelism of the platens is critical. For example, the lower tie bars that guide the moving platen (see Fig. 6) should be supported by the machine base to prevent tie-bar droop and resultant misalignment. Also, the clamp unit should be designed and built with close tolerances to ensure parallelism between platens.

Mold-protection mechanisms prevent the mold halves from slamming together during closing of the clamp. They also prevent damage to cores and cavities when parts or other foreign objects become accidentally jammed between the mold faces. Mold protection can be provided by a variety of machine features. In addition to a hydromechanical cushion, the pressure in the clamp stroke cylinder can be automatically reduced before closing the mold. The shutter design can also prevent application of full locking force unless the mold is fully closed and the shutters have moved in.

Controls. Injection-molding-machine controls can be divided into two categories: sequence controls and process controls.

Sequence controls provide the logic signals to lead the machine through its normal operating sequence, including steps such as closing the mold, injection, starting of screw rotation, etc. Current control design is based on programmable logic, which allows the flexibility of changing the operating sequence and timing to suit different molding jobs and to incorporate downstream parts-handling equipment into the machine-operating sequence.

Process controls control the variables affecting the molding process and therefore the quality of the molded parts. Examples of process-control variables are injection speed, injection pressure, screw speed, and resin temperature. Process controls can be divided into two main types: open-loop and closed-loop. In open-loop control, no feedback is available to determine if a process variable has reached the desired setting. It is assumed that it has. In closed-loop control, feedback is provided and automatic corrections are made to ensure that a process variable reaches and remains at the desired setting.

Although the concept of closed-loop control is attractive, it is not yet practical for high speed packaging operations. With injection times of approximately 0.3 s, at the rate of 3.3 lb/s (1.5 kg/s), current hydraulic valves are simply not fast enough to successfully respond to correction commands within the cycle. Thus, the only practical approach is to design and build inherently accurate components capable of repeatability with open-loop control.

Machine controls must provide the accuracy and repeatability required by the injection and clamp units. Accuracy and repeatability are achieved by a well-matched combination of electronics, hydraulics, and transducers. A suitable electronic package should include digital controls for accuracy and solid-state circuitry for reliability. A microprocessor can store set-up parameters, monitor system performance, and provide management information. The hydraulic package should include accurate and fast-response hydraulic valves. Valve leakage, which can destabilize machine functions, must be mini-

Figure 8. Positive take-out robot. (**a**) An electrical signal from the platen switch box activates an air piston that drives the robot's arm between the open mold halves; parts are ejected into the takeoff plate. (**b**) When the arm reaches the "out" position, a hydraulic cylinder moves the plate 90° down; transfer occurs during the mold-closed cycle. (**c**) As the plate moves down, a gear arrangement turns the plate 90° for parts transfer and ejection; here, the plate is in position for parts ejection. (**d**) After ejection, the plate rotates and tilts once again back into a vertical position for the next cycle.

mized. Cartridge valves are most suitable for these requirements. The transducer package must include accurate sensors for clamp and screw position. Noncontact digital transducers are available today with an accuracy of ±0.4 mil (±0.01 mm).

Product handling. Manual product handling can account for up to 40% of the total processing cost of the product (excluding resin cost). Adding to this problem is the changing attitude of a labor force that is reluctant to work at dull and repetitive jobs. As a result, many molders are reducing costs and increasing product quality with automated product-handling methods.

Product handling begins with ejection of the parts from the mold. The three basic categories of automated handling are as follows:

1. Free drop and unscrambling—after parts drop from the mold, they are unscrambled and reoriented to meet downstream-operation requirements, such as stacking, assembly, printing, and packaging.

2. Controlled drop—guide pins or guide rails are built into the mold to control part orientation as the parts fall by gravity into chutes located beneath the mold; the orientation of the part established in the mold is never lost.

3. Positive take-out robots—parts are mechanically removed from the mold and transferred to product-handling equip-

ment located beside the molding machine without losing the orientation of the part.

For thin-walled containers, the unscrambling and orientation method is the most realistic alternative today (see Unscrambling). This is because controlled-drop methods are generally unsuitable because of the depth of the part, and positive take-out robots are not yet capable of handling the fast cycle times involved. There are, however, problems with the unscrambling and orientation approach. Parts can be damaged during the free fall from the mold, square containers are difficult to handle, and even the best-engineered unscrambling units have less reliability than other product-handling methods.

For parts with relatively thick walls and longer cycles, positive take-out robots are becoming increasingly popular. A typical robot application is shown in Figure 8, handling PET bottle preforms. Robots are highly reliable, can handle deep parts with various shapes, and virtually eliminate parts-handling damage. As a result of these advantages, considerable research is being devoted to producing a high speed positive take-out robot for thin-walled containers.

BIBLIOGRAPHY

General References

Injection Molding Operations—A Manufacturing Plan, Husky Injection Molding Systems Ltd., Bolton, Ontario, Canada, 1980.

C. Kirkland, "Injection Machine Controls: Understanding 'State-of-the-Art'," *Plast. Technol.* **30**(4), 65–72 (Apr. 1984).

J. A. Sneller, "Digital Process Controls: A New Force in Injection Molding Productivity," *Mod. Plast.* **61**(3), 48–50 (Mar. 1984).

G. R. Smolok, "Robotics Applications in the Plastics Industry," *Mod. Plast.* **61**(11), 48–51 (Nov. 1984).

A. Kuhn
Husky Injection Molding Systems, Inc.

INKS

Printing inks are used to decorate the exterior of virtually every package and substrate used in packaging. Since there is a variety of substrates, a number of different printing methods are required to satisfy the needs of this entire market (1). In general, the inks required may be subdivided into two classes: liquid inks and paste inks. The liquid inks include those printed by the flexographic, rotogravure, and screen process methods; the paste inks include those printed by lithographic offset, letterset, and letterpress (see Decorating; Printing).

Basically, all printing inks consist of a colorant (see Colorants), which is usually a pigment but may be a dye, and a vehicle that acts as a binder for the colorant and a film former. Vehicles generally consist of a resin or polymer and a liquid dispersant, which may be a solvent, oil, or monomer. Many other additives are used to provide some specific property or function to the ink or ink film. The ink constituents are discussed in detail in the following sections. Approximate viscosities and film thicknesses used in different printing processes are given in Table 1.

Liquid Inks

Flexographic ink. Flexographic printing utilizes a rubber or plastic printing plate and is essentially a typographic pro-

Table 1. Ink Viscosity and Film Thickness by Process

Process	Viscosity, poise (Pa · s)	Printed film thickness, μm
lithography	50–500 (5–50)	1–2
letterpress	20–200 (2–20)	3–5
flexography	0.1–1 (0.01–0.1)	6–8
gravure	0.1–0.5 (0.01–0.05)	8–12
letterset	30–300 (3–30)	1.5–3
screen	1–100 (0.1–10)	20–100

cess, printing from a raised area. The inks used are very low in viscosity (see Table 1) and dry rapidly because of their relatively high volatility. Since the printing press used has a very simple ink distribution system, these volatile inks do not cause problems in drying on the inking rollers. The thickness of the ink film is controlled by the depth of engraving used on the anilox inking roller. The commercial speeds used for this type of printing range from 500 to 1000 ft/min (152–305 m/min). Flexographic ink accounts for more than half of the ink utilized in decorative packaging, as shown in the last section of this article.

Flexographic inks may be subdivided into two general classes: solvent-based and water-based. The volatile solvents selected for the formulation must be chosen with care, since they are in constant contact with the plate elastomer. It is important to screen these ink solvents with the plate and roller elastomers to ensure that no swelling or attack of the surface takes place. In most cases, alcohols are the material of choice for solvent-based flexographic inks, with additions of lower esters and small amounts of hydrocarbons. These are used as needed to achieve solubility of the vehicle resin and proper drying of the ink film at press speed. Solvent blends are required to achieve the best balance of viscosity, volatility, and substrate wetting, and must also take into account EPA (Environmental Protection Agency) requirements.

Water-based inks are increasing in popularity because of air pollution concerns and are being used in increasing amounts for both absorbent and nonabsorbent substrates. The solvent used in these inks is usually not 100% water. They may contain as much as 20% of an alcohol to increase drying speed, supress foaming, and increase resin compatibility. The wetting of plastic substrates is also greatly aided by the lower surface tension of the alcohol–water mixture.

The resins and polymers used in the flexo-ink vehicles cover a wide range of chemistry and are selected to achieve adhesion to various substrates or to confer resistance properties to the dried ink film. A number of resin classes and the substrates upon which they are generally used are shown in Table 2. Note that the water-based vehicles are listed in a separate portion of the table, since they are generally emulsions or colloidal dispersions rather than true solutions.

A typical formulation for a white solvent-based flexographic ink used in high-speed printing of polyethylene film for bread bags illustrates the types of constituents present:

Titanium dioxide pigment	40%
Alcohol-soluble polyamide resin	20%
Nitrocellulose varnish	5%
Normal propyl acetate	5%
Ethanol	26.7%

Table 2. Liquid Ink Applications for Packaging: Flexographic and Gravure

	Paper/ paperboard	Foil	PE	PP	Vinyl	PET	Cellophane
Solvent ink systems							
NC–maleic	X	X	X	X		X	
NC–polyamide	X	X	X	X		X	
NC–acrylic		X		X			X
NC–urethane				X		X	
NC–melamine	X	X					
chlorinated rubber	X	X					
vinyl		X			X		
acrylic	X	X	X		X		X
ASP–acrylic			X	X			
CAP–acrylic			X	X			
styrene	X						
Water ink systems							
acrylic emulsion	X	X	X	X	X	X	X
maleic resin dispersion	X	X					
styrene–maleic anhydride resins	X	X	X	X			

Slip additive	0.3%
Wax	1%
Plasticizer	2%

Gravure ink. Gravure printing utilizes an etched or engraved cylinder to transfer ink directly onto the substrate. Because of the mechanics of filling the very small cells, the viscosity of gravure inks must be relatively low (see Table 1). This is also required in order to transfer ink from the engraved cells at high speeds. The rotogravure press uses a simple inking system, where ink is applied directly to the gravure cylinder and excess ink is scraped off with a doctor blade. The ink-film thickness is determined by both the depth of engraving and the area of the cells, as these two factors determine the volume of ink transferred. Commercial speeds vary from 800 to 1500 ft/min (244–457 m/min) or more. Gravure ink is the second-largest category of packaging inks. Together with flexographic ink, these two liquid inks account for more than 80% of all packaging ink.

Rotogravure inks can also be classified as solvent-based or water-based types. The solvent-based inks are not as restricted by plate compatibility problems as are the flexographic inks because the plate is metal and resistant to nearly all solvents. Since a wider variety of solvents can be utilized in manufacturing a rotogravure ink, a much wider range of resin types and resin molecular weights can be utilized in achieving the desired ink properties. This means that gravure inks for packaging can be tailored for good adhesion to the widest selection of substrates. The classes of solvents that may be found in these inks include the following types: aromatic hydrocarbons; aliphatic hydrocarbons; alcohols; esters; ketones; chlorinated solvents; nitroparaffins; and glycol ethers. Even uncommon solvents such as tetrahydrofuran can be utilized in the manufacture of inks for certain speciality applications.

The resins and polymers used in packaging gravure vehicles also cover a wide range of chemistries and are chosen to achieve adhesion to various substrates or to confer specific properties to the finished ink film. These are listed in Table 2.

Water-based gravure inks are used widely for printing paper and paperboard substrates, and their use for nonpaper substrates is growing because of concern about air pollution. The formulation of these inks is very similar to those used in water-based flexography and the polymers are also very similar. These are listed in a separate section of Table 2. A significant problem that can occur with water-based inks in gravure is the drying of the vehicle polymer or resin in the engraved cells. This can occur during shutdowns of the press. Some of these materials are not readily resoluble once they have dried because they are usually emulsions or colloidal dispersions in water. To assist in the drying of water-based inks, the gravure cylinders are usually engraved or etched with shallower cells. A thinner ink film is applied, which contains less water to evaporate. This also means, however, that the amount of pigment in the press-ready ink must be higher in concentration to achieve the same relative printing density as the solvent-based ink. A typical formulation for a water-based packaging ink such as used for printing of paperboard cartons (see Cartons, folding) is given below:

Organic pigment	16.5%
Clay extender (wiping aid)	5%
Acrylic emulsion	40%
Plasticizer	2%
Isopropyl alcohol	7%
Wax	2.5%
Morpholine (pH controller)	1%
Water	26%

Screen ink. Screen printing, which accounts for a relatively small percentage of ink used in the packaging market, is used on low volume specialty items or where very thick films are desired (see Table 1). Screen ink is included in the liquid-ink section because of its paintlike rheology and chemistry.

Screen inks are applied by squeegeeing the ink through a stencil screen with a rubber blade. The inks must have adequate flow to pass through the screen, but must also have enough body to resist dripping and stringing when the screen is lifted. The two largest classifications of screen inks are solvent types and plastisol types, although water-based, radiation-curing, and two-part catalytic systems are also available. A solvent-based formulation for printing on vinyl is given below:

Organic pigment	10%
Titanium dioxide	20%
Vinyl resin	18%
Acrylic resin	8%
Glycol ether solvent	29%
Ketone solvent	15%

Paste Inks

Offset lithographic ink. Lithographic printing uses a planographic plate in which the ink-receptive image is chemically differentiated from the nonimage area. Since these plates are constantly wetted with a dampening solution, the inks must resist the chemicals contained in these solutions without changing in their printing characteristics. The thickness of the ink film in this process is 1–2 μm which is the thinnest film in any commercial process. Because of this, the colorant concentrations in lithographic inks are generally higher than those found in inks for other processes. Lithographic inks generally have relatively high viscosities due to the ink-distribution systems used on the press equipment for this process. It is also common to find a gelled consistency in the body of these inks because of the need to obtain high printing resolution and faithful reproduction of the plate image. Image quality of halftone reproduction is extremely high.

Sheetfed offset lithography. This process is widely used for printing on packaging board, paper, metal, and plastic sheets. The use of the offset blanket permits excellent reproduction even on surfaces that are not entirely flat owing to the compressibility of the blanket material. The inks used for most conventional sheetfed printing are dried by an oxidative process and may also be accelerated by the use of infrared radiation. These inks set rapidly in a matter of seconds but are not

truly dry for several hours. In the case of metal, the inks are usually reactive only at high temperatures of $\geq 300°F$ ($\geq 149°C$), which are achieved by passing the metal sheets through a heated oven after removal from the press.

A newer technique for the drying of sheetfed lithographic inks is the use of ultraviolet radiation, which is applied by means of high-powered mercury arc lamps immediately after printing. This process is widely used in the production of packaging that must be die cut and finished in-line, such as cartons for cosmetics and alcoholic beverages. Ultraviolet drying of the ink offers significant energy savings in metal decorating where it replaces long, energy-consuming ovens.

The formula given below is for a typical sheetfed offset ink for paperboard:

Organic pigment	20%
Phenolic–hydrocarbon resin varnish	40%
Drying oil alkyd	10%
Hydrocarbon solvent	25.5%
(500–600°F or 260–316°C)	
Wax	2.5%
Cobalt drier	1%
Manganese drier	1%

A formula for a simple radiation-curing (see Curing) metal-decorating ink is

Organic pigment	15%
Acrylate oligomer	40%
Acrylate monomer	30%
Photoinitiator and sensitizer	8%
Wax	3%
Tack reducer	4%

Web offset lithography. In web offset, the same lithographic principles that are used in sheetfed lithography are applied to

Table 3. Estimated 1982 Consumption of Inks by Process (for United States and Canada), 10^6 lb (kg) and %

	Flexo	Gravure	Litho	Letterpress	Letterset	Screen	Total
corrugated	69.8(31.7)			8.1(3.7)			77.9(35.3)
boxes	89.6%			10.4%			100.0%
flexible	104.7(47.5)	62(28.1)	8.6(3.9)	1.0(0.45)		2.2(1.0)	178.5(81.0)
packaging	58.7%	34.7%	4.8%	0.6%		1.2%	100.0%
folding	7.5(3.4)	27.0(12.2)	14.0(6.4)	1.0(0.45)		1.2(0.54)	50.7(23.0)
carton	14.8%	53.3%	27.6%	1.9%		2.4%	100.0%
food	19.3(8.8)	0.9(0.4)					20.2(9.2)
container	95.5	4.5%					100.0%
household	10.5(4.8)	2.0(0.9)					12.5(5.7)
paper	84%	16%					100.0%
multiwall	30(13.6)	1.5(0.7)		1.1(0.5)			32.6(14.8)
bag	92%	4.6%		3.4%			100.0%
plastic	3(1.36)				1.05(0.48)	1.4(0.63)	5.45(2.47)
container	55%				19.3%	25.7%	100.0%
gift wrap	1.0(0.45)	26.3(11.9)					27.3(12.4)
	3.7%	96.3%					100.0%
wall	8.7(3.9)	7.4(3.4)				1.0(0.45)	17.1(7.8)
covering	50.9%	43.3%				5.8%	100.0%
vinyl		20(9.1)				3.0(1.4)	23.0(10.4)
substrates		87%				13%	100.0%
metal cans			5(2.3)		5(2.3)		10.0(4.5)
			50%		50%		100.0%
Total wt	*254.5(115.4)*	*147.1(66.7)*	*27.6(12.5)*	*11.2(5.1)*	*6.05(2.74)*	*8.8(4.0)*	*455.3(206.5)*
Total %	*56.0*	*32.2*	*6.1*	*2.5*	*1.3*	*1.9*	*100.0*

the printing of the substrate in the form of a web. The web of the substrate is fed into the printing press for decoration from a large roll. The printed substrate, upon exiting the press, must be dried immediately so that the finished product can either be rewound, sheeted, or finished in-line. The most common method to dry the printed ink immediately is the utilization of a high-temperature oven that employs recirculated hot air at a temperature of about 250–350°F (121–177°C). For this reason, the handling of plastic substrates is generally not possible because of the distortion of the substrate at these high temperatures. Again, the newer technology of drying inks with ultraviolet or electron beam radiation is being applied to web offset printing, which allows the printing of temperature-sensitive substrates. Generally, web offset printing in packaging is confined to paper and board printing. However, the new drying technology utilizing radiation curing promises to offer the packaging market the advantages of high-speed printing, the quality of offset lithography, and the lower cost of offset plates compared to gravure cylinders. Typical speeds for web offset printing are 800–1200 ft/min (244–366 m/min).

A formula for an electron-beam-curable web offset ink is given below:

Organic pigment	16%
Acrylate oligomers	40%
Acrylate monomers	30%
Extender resins	10%
Wax	2.5%
Stabilizer	1.5%

Letterset ink. This printing process, formerly known as dry offset, is a combination of letterpress and offset in that the printing plate uses raised images, but the printing on the substrate is accomplished with a rubber blanket. Therefore, the inks used in letterset generally have the viscosity and body of a letterpress ink. The predominant use of letterset printing in packaging is for the decoration of two-piece metal cans (see Cans, fabrication). Plastic preformed tubs and containers are also printed utilizing the letterset process on a mandrel press. Two-piece can printing is accomplished on special presses that produce at the rate of 1200 cans per minute with up to five colors and a clear varnish. The inks for beverage cans are generally dried by heat. Ultraviolet drying is used by several metal-decorating printers, primarily for beer cans. Ultraviolet is widely used for curing plastic containers where thermal sensitivity is a serious problem and heat curing cannot be used. The metal-decorating ovens used for thermal curing have recently gone to short cycles that use temperatures of 600°F (316°C) for only a few seconds to cure the inks. A formula for a typical heat-curing two-piece can ink for short cycle ovens is given below:

Titanium dioxide pigment	55%
Melamine resin varnish	13%
Polyester resin varnish	26%
Acid catalyst	1%
Slip additives	2%
Tack reducer	3%

Letterpress inks. These inks are used primarily for printing corrugated packaging (see Boxes, corrugated), although small amounts are still used for folding-carton and multiwall-bag (see Bags, paper) printing. Since letterpress uses raised images, the inks are fairly heavy in body (Table 1), and the

primary mechanism for drying is oxidative or absorptive. The inks are generally formulated in a manner similar to sheetfed offset inks but with a slightly lower viscosity. This printing method has been losing market share to the other printing processes for a number of years because of high preparatory and labor costs. This is particularly true in corrugated printing, which is going primarily to flexography.

Ink Consumption

Table 3 shows the 1982 estimated consumption of packaging inks by process by weight and percentage.

BIBLIOGRAPHY

1. *The Printing Ink Manual,* 3rd ed., Northwood Books, London, 1979.

General References

E. Apps, *Printing Ink Technology,* Leonard Hill, London, U.K., 1958.
L. Larsen, *Industrial Printing Inks,* Reinhold Publishing Corp., New York, 1962.
H. Wolfe, *Printing and Litho Inks,* MacNair-Dorland Co., New York, 1967.
The Printing Ink Handbook, National Association of Printing Ink Manufacturers, Elmsford, N.Y., 1980.

R. W. Bassemir
A. J. Bean
Sun Chemical Corporation

IN-MOLD LABELING. See Decorating.

INSERTS AND OUTSERTS

The three basic functions of product literature are to inform, instruct, and promote. When used effectively the literature can become an important informational link between the product manufacturer and the user of the product. This is particularly important with products that could be dangerous if used incorrectly, such as pharmaceuticals.

Product literature can be either an insert or outsert. The difference between the two is the placement of the enclosure as a component of the product package. An insert is placed "in" the package; for example, a folding carton containing a labeled bottle and its insert. An outsert is attached to the "outside" of the product container. Many prescription drug packages utilize the outsert style because it eliminates the folding carton without jeopardizing the safe delivery of the product and its literature as a single, complete unit to the pharmacist or patient.

Product literature is designed, written, and produced according to the physical, legal, and informational requirements set by the manufacturer and in some cases regulatory agencies. Some products, such as prescription drugs, require governmental approval of copy that must be submitted and approved prior to use. These agencies regulate copy content and certain physical aspects of the literature such as type size (see Food, drug, and cosmetic regulations; Laws and regulations, EEC; Pharmaceutical packaging).

Preproduction of the literature includes a review of the copy requirements, writing the copy, setting the type, and proofing for submission to various departments within the manufacturing company (marketing, legal) and to outside regulatory agencies for approval. A technical evaluation is also conducted to determine the final size of the enclosure based upon the manufacturing capabilities of the printer and the packing capabilities of the product manufacturer.

Production of the literature begins with the printing that includes such processes as flexography, lithography, letterpress, and gravure (see Printing). The literature may be printed in sheets which are stacked on pallets for further finishing or on a continuous web which is left in roll form. The finished press sheets or web must then be cut to the proper folder-sheet-feed size. There may be more than one piece of literature on each sheet. Multiple inserts on a single sheet can be folded as one, then slit apart into separate pieces. This multiple-folding operation can significantly reduce the cost of the folding operation.

Fold Styles

Fold styles can be categorized into two basic groups. If all folds are parallel to each other the literature is a straight-line fold. If the fold lines intersect at right angles the literature is a right-angle fold. Generally the straight-line fold is faster than the right-angle style because the sheet does not have to change its direction of travel to accommodate the right-angle fold. A straight-line fold travels in a straight line on the folder from beginning to end. The right-angle fold requires the sheet to take a 90° change in direction. The folder itself is "L"-shaped. As the literature hits the 90° corner it changes direction for the final folds.

The fold is accomplished by either a buckle- or knife-folding operation. Buckle folding is based upon the "buckle" created in a sheet when the sheet comes to a sudden stop. By placing revolving rollers in a predesignated position the sheet buckles into the rollers which pull the sheet between them creating the fold. Knife folders do not require the sheet to buckle. Instead they use a blade that pushes the sheet between two revolving rollers that create the fold. This method is used to fold literature that has already been folded many times and can not be expected to buckle into the rollers. Most folders utilize both folding operations, with the buckle-folding method performing the first few folds and the knife style located near the end of the folder where the piece has grown in thickness. A correct setup is critical to a high quality folding operation. If the fold plates are not parallel or the knife action is either premature or late the result is substandard.

Outsert literature requires an additional finishing operation on the folder. Prior to the last fold a spot of adhesive is placed on the final fold flap or body of the literature. When the final fold is made the adhesive spot binds the final flap to the body of the enclosure. This converts the literature into an outsert that cannot be opened unless the adhesive seal is broken. The outsert can now be handled and applied to the product container without fear that it will open or unfold prior to its delivery to the consumer.

Folding Cartons

Pharmaceutical companies are the primary users of the outsert style enclosure. The conversion from glass to plastic containers eliminated the need for case dividers or folding cartons which separated the glass containers to reduce breakage. The only other function of the folding carton was to keep the product literature and product together so that they would be delivered as a single unit (see Cartoning machinery, end-load). This requirement was fulfilled by the outsert concept which mechanically affixed the literature to the plastic container. This is accomplished by many materials such as double-sided tape, adhesive, and banding. The result is a cost savings with no depreciation in product or instructional integrity.

Folding cartons are important for advertising and product recognition however, and their elimination requires careful study. The outsert is normally used with products that are not taken off the shelf by the consumer. Prescription drugs are chosen by the consumer's physician and filled by the pharmacist. Shelf recognition is not relevant.

Roll-fed inserts are delivered to the packaging line in a continuous web or roll. The individual inserts are then cut from the web with sheeters. At this point the insert can be either folded or left flat and inserted into the product container on the packaging line. This method is most commonly used on high volume lines that are dedicated to the production of a few products.

Regulated industries (eg, the pharmaceutical industry) are extremely sensitive to printing errors such as missing, incorrect, and poorly printed copy. Printers for the pharmaceutical industry must be willing to accept high quality and control demands. Recent electronic advances in laser- and optical-barcode technology (see Barcoding; Code marking and imprinting) have dramatically increased the ability of both printers and their customers to check printed package components for incorrect or missing copy. Other advances have led to near-perfect count accuracies, which can be vitally important to the pharmaceutical companies during package component and product count reconciliation after the packaging line has completed a particular batch or lot. The printer must maintain strict control of the manufacturing facility and environment. Each job must be completely traceable throughout each phase of production with accurate written records and quality checks.

Donald J. Jones
Pharmagraphics, Inc.

INSTRUMENTATION/CONTROLS

Today's packaging machines are marvels of mechanical operation. The manufacturers of packaging machinery look to the control system as the most likely method of improving machine performance.

Packaging machines utilize a number of control circuits in their operation. These operating functions have traditionally been controlled by electromechanical devices such as timers and relays. Machinery manufacturers are now turning to the microprocessor (MP) for control of these functions. As recently as 1980, the majority of packaging machinery was operated by electromechanical-control systems. In 1984, MP control in the form of the programmable controller (PC) and the microcomputer has become the norm. The use of the MP in packaging machinery has increased from only 5% in 1976 to over 75% in 1984 (see Fig. 1) (1).

A basic reason for this turn of events can be traced to the electronics industry and its rapid advances in large-scale integration (LSI). LSI resulted in the MP: a computer on a chip.

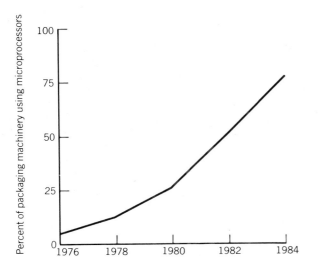

Figure 1. Growth of packaging machinery use of microprocessors.

This low-cost LSI chip, containing arithmetic-, control-, and logic-capabilities, is the basic component of both the microcomputer and the PC. When the MP was developed, packaging machinery manufacturers saw its possibilities for application with their products. When it became cost effective, these same manufacturers quickly applied it to their machine designs. The first applications were tentative. The MPs simply replaced more traditional control methods, but the machines themselves did not change, and no new capabilities were added. More recently, the designers have been reaching out and making better use of the MP, designing machines to take advantage of its tremendous power and adding functions and capabilities that previously were impossible.

Programmable Controllers

The PC is a commercially available, general-purpose control package designed for use in industrial environments. It can operate in the hostile environs of a production area without unusual protection. PCs are primarily used to replace complex control circuitry and timing and sequencing controls such as relays, counters, timers and limit switches. PCs are sequential-logic devices that control machines through input/output (I/O) ports. The components of a PC include an MP, a power supply, and I/O ports. The processor contains the control program, memory, and I/O connections. PCs have many advantages over conventional relay controls, such as (2):

The PC is reprogrammable. Relay controls require rewiring and, possibly, hardware changes.

PCs are more compact than conventional relay systems.

PCs have greater reliability than relay controls and maintenance is easier.

Interfacing with computers is markedly easier than it is with relay systems.

They are modular in design and can be used in a building-block fashion.

Instruction sets for PCs are generally written in relay-ladder logic (3); a knowledge of computer programming and its arcane languages is not required. Electrical technicians can program these controllers, and skilled electricians can make most programming changes (4). But relay-ladder-logic programming creates some limitations in the application of PCs because it does not have the flexibility of more powerful (and more complex) program languages.

The capacity level of a PC system is an important consideration to the user whether the PC is designed as part of a particular packaging machine or the prime control method for a totally integrated line. Even though PCs are expandable, the amount of that expansion is generally governed by the initial capacity of the system (5). It is important to take both present and future needs into consideration when making the capacity decision. Careful analysis will assure that a system is neither too basic nor too elaborate for the application.

Microcomputers

Two different types of microcomputers are being used in packaging machinery today. The first type is the machine-specific computer, designed to operate a particular packaging machine. The second type is the general-use computer, which through its software and I/O ports can be used to perform a variety of machine-monitoring and control functions. The general-use computer, often a popular desktop model, is more likely to be used for monitoring, analyzing, or coordinating overall packaging-line performance. Essentially the two computer types are the same. They both use an MP as their central processing unit (CPU), both have memory, and both have I/O ports (see Fig. 2) (6). The differences are in design philosophy rather than in basic concept.

Machine-specific computers generally contain only the hardware necessary for the operational tasks for which they were designed. They are preprogrammed with the necessary instructions for the machine(s) they are designed to operate (7). The readout is normally a two- or four-line alpha-numeric display using liquid–crystal (LCD) or light-emitting-diode (LED) technology. Operator input is usually through a keypad designed especially for the machine's instruction needs. Printers that provide permanent recordkeeping are often available as options. In a typical application, the microcomputer prompts the operator for setup instructions, stores the responses in its memory, and runs the machine according to the instructions it has received. If so designed, the microcomputer will monitor certain key machine functions and will either make the proper adjustments or shut down production if the machine is operating outside preset parameters.

General-purpose computers almost always utilize the fa-

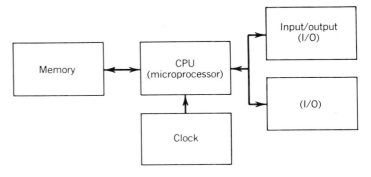

Figure 2. Basic microcomputer architecture.

miliar cathode-ray-tube (CRT) display and full typewriter-style keyboard. Although they can be preprogrammed, the programming is usually on a floppy disk and loaded into the machine by the operator. A printer is normally a standard feature with these systems. In a typical packaging-line-monitoring application the general-purpose computer might collect and analyze such data as production rate, actual throughput, rate of flow, reject counts, machine down-time, total downtime, efficiency against a standard, and inventory usage. Status is continually updated on the CRT, often in graphic form for instant recognition. If the individual packaging machines in a line are controlled by their own microcomputers, then two-way communication is possible, and the operating parameters of the individual machines might be adjusted and balanced for optimum line efficiency by the monitoring general-purpose unit. In this instance the general-purpose computer, in effect, works in a supervisory capacity.

Benefits of Microprocessor Control Systems

Both programmable controllers and microcomputers have a number of significant benefits over traditional control methods. These benefits are important to both the machinery manufacturer and the packager. Because they are solid-state and have no moving parts, MP controls are faster and more accurate than electromechanical devices. The operation time of a typical relay is 20–50 ± 8 ms. A comparable PC control will do the same task in 8–9 ms, and a microcomputer control will perform it in 0.5–0.6 ms (8). This kind of increase in control speed has very positive effects on machine-operating speed and accuracy.

The improved reliability of MP controls is another benefit. A relay has a normal operating life of approximately 50 million (10^6) cycles. The functional life of an LSI logic circuit is billions (10^9) of cycles. When it is considered that the failure of one component in one machine often shuts down an entire packaging line, the improved reliability of MP circuitry is a very important factor.

In addition to improved speed, accuracy, and reliability, microprocessor-control systems provide many new features:

Display screens with instruction prompts are used to guide operators during setup, changeover, and operation.

Memory capabilities can retain the variable operating settings required for running different products.

On-line calculations can monitor machine and production performance and provide instant status reports.

Microprocessors can communicate with each other either directly or through a host computer.

Self-diagnostic tests can identify failure areas and recommend corrective actions in both the control system and the machine.

Plug-in circuit boards and modules provide fast and easy expansion or repair.

Machine operations can be changed through programming alterations rather than hardware and wiring changes.

The features outlined above can be classified as convenience benefits. The more tangible benefits to the packager come in operating results. Here the MP can really come into

Table 1. Comparison Accuracy Test Results. Microprocessor-Based Auger Filler Vs. Conventionally Controlled Fillers[a]

Test description	Accuracy, % (at 2 sigma[b])				MP filler better than next best filler by:
	MP filler	Filler A	Filler B	Filler C	
ground coffee					
60-g fill	0.68	1.49	1.46	2.19	0.78
200-g fill	0.32	0.57	0.72	1.84	0.25
average accuracy	0.50	1.03	1.09	2.02	0.53
best accuracy	0.32	0.57	0.72	1.84	0.25
milk powder					
60-g fill	1.60	1.73	2.23	2.60	0.13
200-g fill	0.72	0.68	1.42	12.50	− 0.04
average accuracy	1.16	1.20	1.83	7.55	0.05
best accuracy	0.72	0.68	1.42	2.60	− 0.04
foundry sand					
60-g fill	0.69	1.14	1.16	3.78	0.45
200-g fill	0.45	0.65	1.07	2.15	0.20
average accuracy	0.57	0.89	1.12	2.96	0.32
best accuracy	0.45	0.65	1.07	2.15	0.20
Overall average accuracy	0.74	1.04	1.34	4.18	0.30
Accuracy gain	0.00	40.36	80.72	461.88	40.36
Best attained accuracy	0.32	0.57	0.72	1.84	0.25
Accuracy gain vs. others	0.00	78.12	125.00	475.00	78.12

Courtesy of Mateer-Burt Co., Inc.
[a] Ref. 10.
[b] The term "2 sigma" indicates accuracy of results 95% of the time, a generally accepted norm in the packaging industry.

its own. When used in such a way that it not only controls the packaging machine, but also monitors the machine performance and corrects for operating variations, microprocessor control can result in substantial speed and accuracy improvements (9). The example in Table 1 shows the accuracy of an MP-controlled auger-filling machine compared to similar machines with traditional electromechanical controls. In a series of six tests, the MP-controlled filler improved on the overall accuracy of the next best performer by 40% (11). This translates into an annual savings of over 6500 kg of product when 500-g containers are filled at a rate of 35 containers/min.

Control Opportunities with Microprocessors

Central-control systems have always posed a problem for the packaging-machinery user. Because a typical packaging line consists of a number of machines, each made by a different manufacturer, central control has been difficult and expensive. Consequently, there has been little integration of entire packaging lines. Before the use of MPs, what was done was usually limited to on/off controls, line guards, and the like. Individual packaging machines controlled by MPs can communicate with each other as well as with a host computer. This allows a packaging line composed of machines from a variety of manufacturers to be controlled from a central supervisory computer, where everything from initial setup through performance analysis to trouble-shooting diagnostics can take place.

Once individual packaging lines become fully integrated, the next logical step is controlling several complete lines with the supervisory computer. Then it is realistic to consider integration of the packaging function with upstream and downstream activities. At this level, a central plant computer will communicate with the supervisory computers which, in turn, will act as communication-relay stations, distributing instructions to the individual machines throughout the production process (12). This kind of highly integrated control can be installed in a step-by-step process (12); it does not have to be accomplished at one time. Careful planning and the MP make it possible.

BIBLIOGRAPHY

1. T. M. Oleksy, "Microprocessors and Packaging Machinery Requirements for Performance," Paper presented at *The Pack Expo 84 Conference*, Chicago, Ill., October 31, 1984, p. 1, Mateer-Burt Co., Inc., Wayne, Pa.

2. M. P. Groover, *Automation, Production Systems, and Computer-Aided Manufacturing*, Prentice-Hall, Englewood Cliffs, N.J., 1980, p. 317.

3. *Packaging*, **29**(13), 43 (Nov. 1984).

4. Ref. 2, p. 316.

5. *Packaging*, **29**(13), 45 (Nov. 1984).

6. J. Hall, "New Applications of Microprocessors with Packaging Machinery and Microprocessors in Auger Filling Systems", *Packag. Technol.*, **13**(6), 36 (Nov. 1983).

7. Ref. 2, pp. 290–291.

8. Ref. 1, p. 5.

9. B. Holmgren, "Line's Peak Output Rests on Interruption-Free Filling," *Packaging*, **29**(9), 33 (Aug. 1984).

10. D. T. Simmons, *Microprocessor Control of Auger Filling*, Mateer-Burt Co., Inc., Wayne, Pa., p. 9.

11. *Ibid.*, p. 5.

12. Ref. 2, pp. 325–328.

General References

Ref. 2 is a good general reference. It is a fine guide to understanding the basics of computerized production methods.

W. H. Buchsbaum, *Encyclopedia of Integrated Circuits: A Practical Handbook of Essential Reference Data*, Prentice-Hall, Englewood Cliffs, N.J., 1981.

DONALD T. SIMMONS
The Marketing Link

INSULATION

The most important aspect of insulative package design is knowledge of the product and understanding its distribution requirements. The best packaging system is one which protects the product in a cost-effective manner. An extensive value analysis should include the cost of raw materials, labor, setup, filling, and distribution.

Design Principles

The volume, density, and mass of the product requiring insulative protection must be known in order to determine the effect of temperature. The temperature of the product may differ from the environment. When the product and the environment are brought into contact, they achieve thermal equilibrium (1). The basis for insulative packaging is to extend the time before thermal equilibrium takes place. Consideration of the product needing protection includes study of the environment (the surroundings) to which it is typically exposed. The packaging (a boundary), if properly selected and designed, can dramatically extend the time to thermal equilibrium. Some popular diathermal packaging boundaries, those through which a system may be influenced thermally, are listed below in order of decreasing resistance to heat flow (2):

Rigid polyurethane foam
Rigid expandable polystyrene foam
Shredded paper
Flexible polyurethane foam
Built-up corrugated fiber
Excelsior
Plywood
Wood

The difference in thermal protection from wood to rigid polyurethane foam is over 400%. It is important to compare the same thickness of alternative materials in identical environments. This allows an "apple-to-apple" comparison of how well a material insulates. Two commonly used measurement criteria are the K factor and the R value. A K factor describes the ability of a material to conduct heat (thermal conductivity). The lower the K factor, the better are the insulative properties. An R value is a measurement of a material's ability to resist heat flow. The higher the R value, the better are the insulative properties of that material.

The analytical science associated with thermodynamics is the basis for an understanding of how products, packages, and environments relate. Unfortunately, as products are shipped through distribution, they are subjected to constantly changing systems. As these systems change and come to equilibrium, the heat flow in and out of the package eventually changes the temperature of the product (3).

An insulated package can be referred to as a closed system. A closed system allows an exchange of energy but not matter (4). Knowledge of the range of variables in the environment to which a product is exposed permits the selection of the material that can best resist the flow of energy through the package. The ideal package resists the flow of energy (heat) from the environment to the product and vice versa (5).

Without an ideal insulative package, the heat flow can affect the product. The dimensions of the product may change (6). Addition of coolants (eg, dry ice, wet ice, gel packs) can dramatically affect volume. If coolants are used, the internal dimensions of the package should incorporate the coolant and the stability this coolant gives to the volume of the product. This becomes an important element in the design of the cost-effective insulated package. For example, as a frozen product with certain cubic dimensions is allowed to thaw, thermal expansion occurs as equilibrium is approached, and the initial cube increases. The packaging designer has two options: (1) design a package allowing for this expansion, or (2) design an insulative package which eliminates the chance for thermal expansion. The first option affects economics by increased raw-material consumption, higher storage-space costs, the potential for lost product owing to stacking damage, and increased shipping charges due to a larger cube in the shipping vehicle. Designing an insulative package may cost more per unit, but an overall value analysis may exhibit cost savings.

Applications

Aquaculture. This is a rapidly growing business segment which raises edible seafood and ornamental fish, such as catfish, clams, oysters, salmon, shrimp, and tropical fish. The majority of these products are shipped fresh. Packages currently being used are corrugated boxes shipped in reefer trucks; 1.5-in. (38-mm) EPS (expanded polystyrene foam) coolers shipped via parcel delivery services; polyurethane-lined corrugated; and extruded polystyrene foam-paper laminate containers shipped airfreight (see Foam cushioning).

Chemicals. Many acids and thermally reactive resins need temperature protection. Some products must not freeze, and others begin to react when they reach certain temperatures. The most important function of the package is to protect these products from undesirable thermal shock. Packages include EPS coolers in corrugated boxes; polyurethane filling between two corrugated containers; EPS panel liners inside corrugated containers; and extruded foam-paper laminate containers. Most of these products are transported by truck.

Gifts. Many high priced cuts of meat, smoked seafood, prime fruits, cheeses, and sausages are commonly shipped via parcel delivery services. These items need protection against adverse environmental elements. The most common choice is the EPS cooler inside a corrugated box.

Horticulture. Bulbs, freshly cut flowers, and plants are shipped by air and truck. These products must be kept from freezing or wilting. The acceptable thermal range of 34–74°F (1–23°C) is most desirable, depending on the product. EPS coolers, polyurethane-coated corrugated, and foam-paper laminates protect the products from going outside that range.

Medical products. This market has a wide range of expensive products, all with very different requirements. A frozen (−40°F or −40°C) product might melt at room temperature within one hour. Another product cannot be frozen. Some products are shipped in individual motorized refrigerated shipping containers that are returned to the pharmaceutical company once a product is unloaded. The other types of insulative packaging described above are widely used.

Seafoods. This is one of the most diverse markets with numerous species shipped fresh and/or frozen. Containers range from corrugated boxes to large molded polycarbonate (see Polycarbonate) insulated shipping cases. It depends on the species amount shipped, mode of transportation, and time in transit. Fresh products need to be held at 34–40°F (1–5°C) for the entire transit time. Frozen products must be held below 32°F (0°C). Products are shipped via truck and air. Spoiled seafood becomes a human health problem. Insulative packaging has become a very important factor in the seafood business because it protects the product and extends the shelf life.

Frozen novelties. Ice-cream products, juice bars, and premium "all natural" items are prime candidates for insulative packaging. Most of these products melt at temperatures well below freezing (eg, 12°F or −11°C) because of their high sugar content. If adequate protective packaging is not used, these products will melt during shipment at 32°F (0°C) in a refrigerated truck. Many companies ship these very sensitive products at 0°F (−18°C) in ice-cream trucks.

Summary

To design the best insulative packing system one needs to know the product, how various environments affect the product, and which packaging materials afford the longest time before the product and its environment come to thermal equilibrium.

BIBLIOGRAPHY

1. T. J. Quinn, *Temperature,* Academic Press, New York, 1983, pp. 3–5.
2. C. J. Adkins, *Equilibrium Thermodynamics,* McGraw-Hill, Inc., New York, 1968, pp. 3–5.
3. L. J. Stavish, *Pharm. Manuf.,* 23 (June 1984).
4. I. Prigogine, *Thermodynamics of Irreversible Processes,* Interscience Publishers, New York, 1961, pp. 3, 4, 8, 16.
5. A. L. King, *Thermophysics,* W. H. Freeman and Company, 1962, pp. 2, 4, 5, 50, 51, 55.
6. B. Yates *Thermal Expansion,* Plenum Press, New York, 1978, pp. 74, 76.

R. M. BELICK
Cherokee Corruguard

INTERMEDIATE BULK CONTAINERS

Intermediate bulk containers (IBCs), also called semibulk containers, are larger than conventional drums and heavy-duty shipping bags but smaller than a bulk vehicle (eg, a tanker truck). One of its major distinguishing features is the absence of protrusions (eg, bilges) for manual handling. It is, by definition, designed for mechanical handling. International agreement has been reached on that point, as well as its size range: water capacity of not less than 0.25 m³ (250 L) and not more than 3.0 m³ (3000 L).

Most IBCs are used for granular and powdered solids, viscous materials, and free-flowing liquids with a vapor pressure not exceeding 16 psi (1.1 bar or 110 kPa) at 50°C, and which are filled and/or discharged without the application of pressure. IBCs contain the product directly, unlike box pallets (see Pallets, wood) which generally contain packaged products.

IBCs are available in many different materials. Those that carry solids are generally made of rigid or collapsible metal; rigid plastics; corrugated paperboard; wood or reconsituted wood; heavy-duty coated fabric; or plastic-tape fabric. Those that carry liquids are generally made of rigid metal; rigid plastic; corrugated paperboard; or flexible materials (eg, rubber-coated fabric, or polyurethane film). These containers can be used for the transport of hazardous materials (dangerous goods) if they comply with standards set forth by the United Nations (still in development) and several countries (see Standards and practices, national and international).

BIBLIOGRAPHY

General References

C. Swinbank, *Intermediate Bulk Containers,* The Institute of Packaging, Melton Mowbray, Leics, UK, 1983.

IONOMERS

Ionomers were discovered by the Du Pont Company in the early 1960s (1). The materials were derived from ethylene copolymers with methacrylic acid, and the copolymer was neutralized to varying degrees with either sodium or zinc. It was soon realized that the neutralized acid groups conferred quite distinctive properties on the copolymer.

These properties were a consequence of the ionic groups segregating into large clusters. Wide-angle x-ray and neutron diffraction and electron microscopy revealed that these clusters are about 100 Å (0.01 μm) in extent. Thus, whereas ordinary polyethylene consists of crystalline and amorphous phases, the ionomers additionally contain an ionic phase. All these phases are molecularly interconnected (2,3). The chemical formula is:

$$+CH_2CH_2\rightarrow_n CH_2 \overset{\overset{\displaystyle CH_3}{|}}{\underset{\underset{\displaystyle COO^-\ (Na^+\ or\ \frac{1}{2}\ Zn^{2+})}{|}}{C}}+CH_2CH_2\rightarrow_m+CH_2 \overset{\overset{\displaystyle CH_3}{|}}{\underset{\underset{\displaystyle COOH}{|}}{C}}+$$

What is perhaps more revealing than the symbolic formula is a schematic representation of the crystalline structure (see Fig. 1) (4), which shows the relationship between amorphous hydrocarbon, fringed micellar crystals of polyethylene and layered, structured ionic clusters.

This structure is called the "sheath-core model" (3). The ionic core appears to have a rather specific structure consisting of two opposing layers of carboxylate ions with anions in between. The sheath is less organized and contains a higher proportion of carboxylate and carboxyl groups than the bulk of the substantially hydrocarbon fraction. These clusters persist to high temperatures, at least up to 572°F (300°C), and are the reason for the high melt viscosity of ionomers compared to ethylene copolymers of similar molecular weight. In packaging applications, they are responsible for the excellent hot tack and high melt strength of ionomers.

The term ionomer in recent years has come to be applied to any polymer that consists of a copolymer of a carboxylic or sulfonic acid wherein the fraction of acid is not greater than 20% and has been neutralized to varying degrees, mostly with a metal or quaternary ammonium ion. Generally, the ionomers retain their organic, thermoplastic nature; the amount of combined metal is not so great as to confer water solubility, although their water absorption is generally appreciably greater than the unneutralized copolymer.

The only significant commercial production is of the ethylene-based ionomers produced by the DuPont Company under the trade name Surlyn resins (5). Ionomers derived from backbones using styrene, butadiene, and fluorocarbons have all been described, but these are not of importance in the present context, and further discussion will be confined to the Surlyn resins.

The Surlyn resins are derived from copolymers of ethylene and methacrylic acid produced by the high pressure process for the manufacture of polyethylene (LDPE). The commercial conversion of the copolymer to the ionomer is a proprietary process.

Properties and Applications

In the simplest terms, ionomers resemble LDPE (see Polyethylene, low density), but the differences determine their applicability and ability to command a higher price. Qualitatively, these differences include higher tensile strength, higher modulus (room temperature), higher toughness, lower softening point, greater clarity, lower haze, greater abrasion resistance, and better oil resistance. In thermoforming, they offer faster heat absorption and dissipation, lower forming temperature, and better hot draw strength (see Thermoforming). There are features where ionomers are at a disadvantage; as films, ionomers have outstanding clarity and low haze but have noticeably poorer slip and block characteristics. Zinc ionomers adhere well to foils and to other polyolefins and nylon in coextrusion. Selected properties of a typical ionomer are given in Table 1.

The most important applications for ionomers are in the manufacture of films and extrusion coatings used primarily for various kinds of food packaging (see Coextrusions; Extrusion coating; Multilayer flexible packaging). All grades sold for packaging applications have been cleared by the FDA for use in contact with foodstuffs. In film applications, the properties which distinguish ionomer films from other varieties of flexi-

Table 1. Selected Properties of Surlyn 1601

Property	ASTM test method	Value
flow (melt index)	D 1238	1.3 g/10 min
density	D 792	0.940 g/cm³
tensile strength	D 638	4800 psi (33 MPa)
% elongation to yield	D 638	400%
defl. temp. at 66 psi (455 kPa)	D 648	111°F (44°C)
Vicat soft point	D 1525	163°F (73°C)
Chemical resistance		
weak acids	D 543	unaffected
strong acids	D 543	slow attack
weak alkalis	D 543	unaffected
strong alkalis	D 543	unaffected
hydrocarbons	D 543	slight swell
alcohols	D 543	slight swell

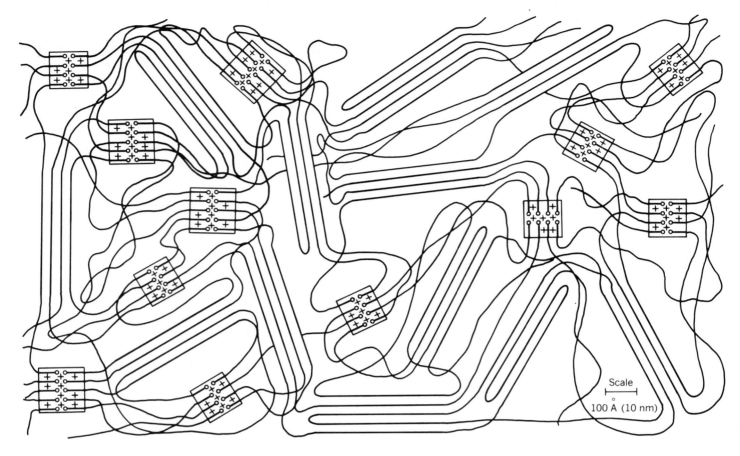

Figure 1.   Schematic structure of an ionomer, showing polyethylene crystals, lamellar ionic clusters, and amorphous hydrocarbon.

ble packaging materials are heat sealability, ease of thermoforming, toughness, and excellent optical properties (see Films, plastic). Also important are resistance to food oils and, in coextrusions with nylon, excellent adhesion. Laminated films with nylon and polyesters are widely used in packaging processed meats and cheese (see Laminating).

Processing Techniques

Ionomers are readily processed by machinery suitable for low density polyethylene. Thus, extrusion into film, sheeting, tubing, and pipes can readily be accomplished. For packaging applications, this includes blown and cast film and coextrusion. The major difference from LDPE is in the higher power requirements (see Extrusion; Coextrusion machinery, flat; Coextrusion machinery, tubular).

In contrast to polyethylene, ionomers are more sensitive to temperature changes. Processing temperatures in excess of 610°F (321°C) should be avoided, and at comparable melt flow index, ionomers have a higher viscosity than LDPE at lower temperatures.

The techniques of manufacture of blown and cast film from LDPE can be applied to ionomers (see Polyethylene, low density). The optical properties of certain ionomers are outstandingly good. Gloss increases and haze decreases as melt temperatures are increased from 360°F (182°C) up to a maximum of 450°F (232°C). In contrast to LDPE, where fast cooling inhibits the growth of spherulitic crystallinity, relaxation processes result in ionomers having better optics with slow cooling.

This factor means that thicker films have surprisingly good opticals because of the slower cooling. With ionomer films there is the interesting effect that the stiffness increases by about 50% on aging for about 10 days after extrusion, if kept dry. The reason for this effect is not clear, but it appears that there is a molecular rearrangement of the carboxylate-containing noncrystalline hydrocarbon segments in the vicinity of the ionic cores of the clusters.

Ionomers are hygroscopic and absorb moisture when exposed to humid environments. This affects processability; the higher the moisture content, the lower the maximum extrusion temperature. If this problem is encountered, surface defects caused by smeared bubbles may appear, and if excessive, bubbles within the body of the film or sheet may appear. Also, the melt flow index increases markedly with water content. As packaged, the water content is less than 800 ppm (10^6), and with care these problems will not be serious. If the opened packages are left exposed to humid atmosphere then the moisture level can rise to as high as 6000 ppm, and the material will become unusable.

Resin Types

Selected properties of the most commonly used ionomer, Surlyn 1601, are given in Table 1. This resin is a sodium-neutralized material with a melt index of 1.3 g/10 min. About a dozen grades have been formulated with specific property improvements for specific packaging applications. The principal variables are molecular weight, ion type, and content. The melt index depends on all three of these variables and varies in this series from 0.7 to 14.0 g/10 min. The ion may be either sodium or zinc. There is not a unique relationship between ion content and either melt index or modulus. Using zinc instead of sodium gives improved low temperature properties. Other grades provide higher heat-seal strength or higher modulus (stiffness). Specific additives have been incorporated in certain grades for better slip and/or antiblocking characteristics.

Acidic Copolymers of Ethylene

Copolymers of ethylene with either acrylic or methacrylic acids have been introduced recently. The acrylic acid copolymers (EAA) are offered by Dow Chemical Co. (Primacor) and the methacrylic acid copolymers (EMA) by the DuPont Co. (Nucrel). In Europe, acidic copolymers are supplied by BASF (Lupolen), CDF (Lotader), and Esso Germany (Escorene).

As with the ionomer resins, comonomer levels are high enough to confer properties distinctly different from low density polyethylene, even though LDPE is the benchmark for comparison.

These resins are somewhat more expensive than LDPE. As films, they have better clarity, toughness, and stress-crack resistance (partly as a consequence of the inevitable decrease in crystallinity). However, the most important advantages for packaging applications, compared to LDPE, are improved adhesion to metal foils (eg, in extrusion coating) and superior hot tack and heat-seal strength. These resins are generally infe-rior to ionomers in hot-tack, and heat-seal strength. In processing, ionomers can be drawn down to thinner gauges and can be oriented more easily in blown film manufacture.

In other properties, such as tensile strength and rheological behavior, these materials are quite similar to LDPE.

BIBLIOGRAPHY

1. U.S. Pat. 3,204,272, R. W. Rees.
2. R. Longworth and D. J. Vaughan, *Nature* **218**, 85 (1968).
3. E. J. Roche, R. S. Stein, T. P. Russell, and W. J. MacKnight, *J. Polym. Sci. Polym. Phys.* **18**, 1497 (1980).
4. T. R. Earnest and R. Longworth, private communication.
5. For further information about Surlyn resins, consult the Marketing Communications Division, E. I. du Pont de Nemours & Co., Inc. Wilmington, Del. 19898.

General References

R. Longworth in L. Holliday, ed., *Ion-Containing Polymers,* John Wiley & Sons, Inc., New York, 1973, Chapt. 2.

T. R. Earnest and W. J. MacKnight, *J. Polym. Sci. Macromol. Rev.* **16**, 41 (1981).

R. Longworth in A. D. Wilson and H. J. Prosser, eds., *Developments in Ion-Containing Polymers,* Vol. I, Applied Science Publishers, Ltd., Barking, UK, 1983, Chapt. 3.

RUSKIN LONGWORTH
E. I. du Pont de Nemours & Co., Inc.

L

LABELS AND LABELING MACHINERY

Labels can be affixed to almost anything to indicate its contents, nature, ownership, or destination. Labeling is the art of applying or attaching the label to a particular surface, item, or product (see also Tags).

Originally, labels were used merely to identify a product or to supply information about the properties, nature, or purpose of the items labeled. Today their use is often required by legislation (1), and they are seen as sales and marketing aids in the total design of a package or product. In some cases, the labels have become products in themselves.

Types of Labels and Materials

The range and variety of labels used, and the markets and applications for them, are increasingly diverse. Materials include paperboard, laminates, metallic foils, paper, plastics, fabric, and synthetic substrates. These in turn may be adhesive or nonadhesive. Even then, the range of label materials can be further subdivided into a variety of different types: coated or uncoated; pressure-sensitive or heat-sensitive; conventional gummed or particle gummed (2). Figure 1 shows the main types of labels and materials.

Of all the different types of labels, nonadhesive plain paper labels applied by wet gluing tend to dominate the total user market throughout the world. In recent years, however, there has been a marked trend towards self-adhesive pressure-sensitive labels. Newer methods of labeling such as shrink sleeve, in-mold, and heat transfer have also gained market acceptance (see Bands, shrink; Decorating).

In 1984 the total label market in the United States for all forms of labels was estimated to be in excess of $2 billion ($10^9$). Estimated market shares for different types of labels are shown in Table 1.

Overall, the U.S. pressure-sensitive (self-adhesive) roll label market is growing at about 10–15% per year, with some printers and market sectors showing growth of 20% or more. The two main markets showing above average growth are electronic data processing (EDP) labels and prime labels.

Plain paper labels. The conventional wet-glue plain paper label is widely used, particularly for large volume items such as beer, soft drinks, wines, and canned foods, where high label application speeds, sometimes in excess of 80–100,000 labels per hour, are required.

Most of the papers used in this type of labeling are one-sided coated grades, which make up the familiar can and bottle labels, but a fair volume of uncoated grades is still used.

Table 1. Estimated Market Share for Different Types of Labels in the United States, 1984

Type of label	% of market
plain paper and wet glue	50
pressure-sensitive/self-adhesive	33
heat-seal and transfer	10
gummed paper	7

These tend to be limited to special effects, such as colored papers, embossed papers, or antique laids for wine labels, where particular characteristics are required (3).

Gummed paper labels. Gummed paper labels have been declining in usage in recent years and are now limited mainly to applications such as address labeling for cartons, point-of-sale displays, and any labeling applications where automation is either difficult or unnecessary. Two main types of gummed labels are in use: conventional gummed and particle gummed (4).

Conventional gummed labels are those which are made up of a base support of paper coated on the reverse side with a film of water-moistenable gum. On particle gummed papers, the adhesive is applied in the form of minute granules. This avoids problems of curling often associated with conventional gummed papers. Most of the newer applications for gummed papers require the properties of high tack particle gummed papers, which offer many advantages in terms of processing efficiency and security of application.

Self-adhesive labels. Self-adhesive label materials (see Fig. 2) today range from permanent to removable adhesive types, as well as a whole range of special adhesive materials (see Adhesives).

Permanent adhesives are those that are required to stay in position for a long time or for application to surfaces which are round, irregular, or flexible. Because of the strong adhesive 'grab', the label normally becomes damaged or defaced if attempts are made to remove it.

Removable adhesives are those that *can* be removed after a specified time without damage to the surface on which they have been applied. Uses include short-term food-packaging applications, point-of-sale or advertising stickers, china, and ovenware, where the label should be easy to remove before use.

Within the broad categories of permanent and removable self-adhesive labels there is a wide variety of special adhesive types to cover requirements such as water removability, low or high temperature adhesion, ultraviolet light resistance, and high and low tack properties. Adhesives, which at one time were largely solvent based, have moved rapidly in recent years to hot melt or water-based acrylic adhesives (5) (see Acrylics).

For all normal self-adhesive labeling requirements (food and supermarket labels, retail/price marking labels, etc.) surface papers in a variety of weights and finishes, colors, and radiants are available. For special applications, such as outdoor labels, instruction and name plates for appliances, underwater labels, ampules and vials, and luxury-labeling of cosmetics and toiletries, there is a wide range of nonpaper materials. These include metallic foils and plastic films.

One of the most recent developments in self-adhesive materials is a range of thermal imaging label facestocks which have become popular for in-store supermarket labeling of meat and fish and fresh produce (fruit and vegetables). These materials contain a surface coating of chemicals which darken under the action of a heated print head (see Fig. 3) (6), thereby providing clear price and weight information and accurate scannable bar codes (7) (see Bar Code; Code marking and imprinting).

Almost all self-adhesive label materials are available in both sheet form, from paper merchants and suppliers, or, for reel-fed conversion, from manufacturers and specialist suppliers in roll form.

The main markets for self-adhesive labels in the United

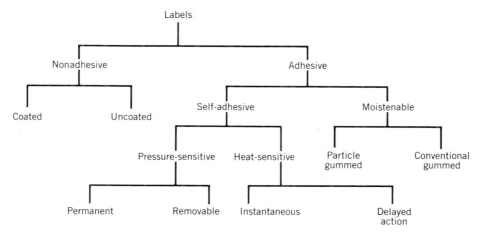

Figure 1. The main types of labels and materials.

States are basically segmented into three areas: industrial labels (or primary and secondary labels used in retail, whole-sale, and industrial packaged goods) which make up about 80% of the U.S. self-adhesive label market; EDP labels (those used for marking/imprinting variable information); and price-marking and imprinting (which addresses label applications through hand-held dispensers and the whole range of imprint technology). Split into some of the main market application areas, this includes promotional labels, price labels, functional labels, office/retail labels, nameplates, stickers, EDP labels, and prime labels.

Heat-sensitive labels. There are two basic types of heat-sensitive or heat-seal labels: instantaneous and delayed action. In the former, heat and pressure are applied to the label to fix it directly on to the product, whereas with delayed action applied heat turns the product into a pressure-sensitive item. No direct heat is applied to the goods and this clearly is vital to some products, such as food.

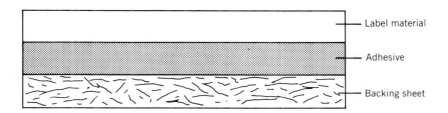

Figure 2. Self-adhesive labels consist of three main layers: the label-face material itself, the pressure-sensitive adhesive coating, and a release coating on a backing material to prevent the adhesive from sticking to the face material in the reel or sheets.

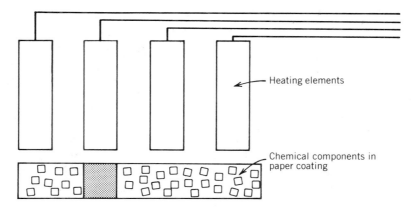

Figure 3. The chemical components in the thermal label paper coating combine and darken when activated by heat.

Typical market applications for delayed action heat-sensitive label papers include pharmaceutical, price/weigh, some glass-bottle, and rigid or semirigid plastic container labeling. Instantaneous materials are used for applications such as end seals on biscuit or toilet tissue packs, labels on "pleat wrapped" articles (eg, disinfectant pads, pies, cakes), and for some banding applications (8).

Label printing

The main printing processes in use for label production are flexography, letterpress, gravure, lithography, silkscreen, and to a lesser extent, hot foil stamping. This applies for most types of labels whether printed flat or from a reel (see Decoration; Printing).

Although the principles of each process remain the same for reel- or sheet-fed printing, the selection of the particular process and the cost factors in that decision are different for each particular section of the label industry. For example, printing of sheet labels is done primarily by lithography or gravure; but letterpress and flexography are major processes for the production of reel-fed self-adhesive labels.

The fastest-growing technique of label printing is the narrow-web roll label conversion of self-adhesive materials using presses which may be printing and converting webs as narrow as 2 in. (50 mm), ranging up to 250, 300, or 400 mm wide (9). These presses not only print substrates up to 6, 8, 10, or more colors in one press pass, but also undertake flat-bed or rotary die cutting of the labels, and possibly uv varnishing, overlaminating with clear synthetic films, removing matrix waste, slitting, fan-folding, sprocket-hole punching, etc, and finally, rewinding of the printed reel of labels ready for automatic application (10). A schematic diagram of a typical roll label printing and converting press is shown in Figure 4.

Printing Methods

Flexography. The main advantage of flexography, the most widely used process for reel-fed label printing, is its output speed, which is due to full rotary printing and machine die cutting. In addition, spirit-based inks evaporate quickly, allowing printing on a wide variety of substrates other than paper (see Inks).

Owing to speed of output, flexo labels generally cost less than labels printed by the other major processes. Because presses can be purchased at a relatively low capital cost, many of the smaller label printers use the process. The presses do, however, vary enormously in complexity. Equally important, the quality of the labels produced depends greatly on the skill of the operator.

Letterpress. Although letterpress has declined in usage in the general printing industry, it is a major process in the label industry and gaining in popularity. It is the second most popular process in the roll label industry after flexography, and in some countries is almost the only process used.

The process has adapted well to the needs of the label industry, giving good quality reproduction and relatively low-cost origination. Presses are available in flat (platen) form, semirotary, or full-rotary letterpress (9).

The main advantage of letterpress is superior quality and ability to hold accurate print definition throughout the run. Solid and fine line printing are no problem, the inks being fully pigmented with excellent light fastness and color matching. Printing plates can be metal, or photopolymer.

The development of uv-cured inks has meant that previously difficult or even impossible to print substances such as vinyls and polyesters can now be "cured" using polymer inks and uv lamps.

Gravure. The gravure process gives excellent quality and print definition. With a very fast throughput it is best suited to long runs where the high cost of originating printing cylinders can be recovered.

Silkscreen. The unique method of applying ink through a screen gives screen process an ink coverage that cannot be completely matched by any other printing method (9). It will also print on virtually any substrate.

The presses are made as complete screen printing units, or a series of screen printing heads can be added to suitable letterpress machines, giving a combination of benefits.

Hot-foil stamping. Hot-foil stamping is generally done in one of two ways: by complete hot-foil presses in their own right, and hot-foil units set in line on to an existing label press. The process gives striking effects when combined with other processes and is very much a designer's medium (9).

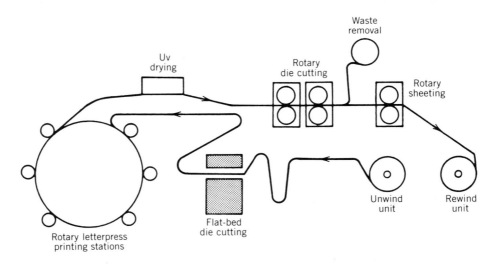

Figure 4. Rotary letterpress roll label press.

Labeling

Once a label has been produced it must be applied in the correct position to a particular surface or product. The label must be applied securely enough for it to remain fixed in that position throughout the useful life of the product or container. It must stand up to whatever conditions it is likely to be exposed to, and should also maintain a good appearance. In the case of returnable bottles, however, the label must be able to be removed easily during the washing operation.

The labeling operation must also be able to meet production requirements. On packaging lines, the labeling machine must be able to keep pace with the line, often at high speeds, having minimum stoppages and down time (11).

Some labels are applied to products or containers with a simple hand-held gun or dispenser; others are applied by a semiautomatic or fully automatic labeling machine suited to the particular type of label: wet glue, pressure-sensitive, or heat-seal (heat-sensitive).

Wet-glue labeling machinery. Wet-glue labeling is the least expensive labeling system in terms of label costs. There are many machines on the market from the simple semiautomatic to the high speed advanced models for speeds up to 600 containers per minute. Straight line or rotary, vacuum transfer of the label or transfer by picker plate, and all-over gumming or strip gumming all have a place and are used in many industries, particularly in food, wine, and spirits plants.

New types of adhesives make glue labeling important for plastic containers as well as glass. Hot-melt adhesive can be used. Large labels are probably cheaper to apply with wet glue than with a heat-seal or pressure-sensitive system, and wet-glue labels are probably superior for applying multiple labels to a range of containers, such as liquor bottles.

Labels may be applied by hand or by semiautomatic or automatic machines (see Fig. 5). There are many different types of wet-glue labeling machines, but they all have to perform the following functions:

1. Feed labels one at a time from a magazine.
2. Coat the labels with adhesive.
3. Feed the glued labels on to the articles to be labeled in the correct position.

4. Ensure that the articles are held in the correct position to be labeled.
5. Apply pressure to smooth the label on to the article and press it into good contact.
6. Remove the article when it has been labeled.

If any of these operations are not performed at the right times, with everything positioned accurately, the results will be unlabeled articles, badly positioned, skewed, or torn labels, and machine stoppages. Machines vary mainly in the way they perform the various functions. For example, some machines may use suction to remove labels from the magazine, and others use the tack of the adhesive itself (11).

The adhesives used for wet labeling fall into five main classes: dextrin-based, casein-based, starch-based, synthetic-resin dispersions, and hot melts.

Apart from the hot melts, all these adhesives are waterborne. The speed at which they set depends, therefore, on the rate at which the water phase can be removed by the absorbency of the label stock. If the water cannot get away, they will not set.

Synthetic resin dispersions, of which the most widely used are those based on poly(vinyl acetate) (PVA), have the major advantage of faster setting, owing to the ability of the polymer particles to draw together to form a continuous film for the loss of much less water than is the case with those adhesives based on natural materials. They are used particularly in the labeling of plastic bottles or coated glass, where normal dextrin, casein, and starch adhesives have difficulty in producing a permanent bond. They are limited to nonreturnable bottles, however, as the dried film has considerable resistance to water and cleaning fluids. One advantage is a low initial tack, which restricts their application to certain types of machines with sufficient brushing-on capacity.

Hot melts are 100% solids, melting when heated and setting almost instantaneously on cooling. They have high initial tack, thus labels adhere to surfaces such as PVC and polyethylene (PE) at high speeds, but are not suitable for wet bottles. To select the correct adhesive for a particular application every factor of the operation must be taken into account: the operating conditions, type, and condition of the articles to be labeled, nature of the label papers, transport

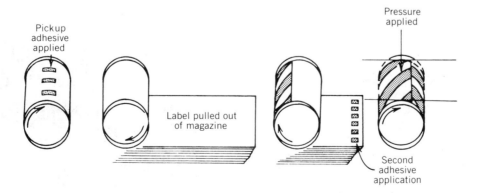

Figure 5. Stages in the labeling of cans. Adhesive is first applied to the container. As the container rotates over the label magazine it picks up its own label, which is then rolled around it. A second adhesive application to the trailing edge of the labels, followed by pressure, completes the labeling operation.

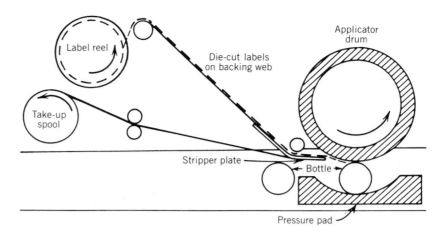

Figure 6. A common type of self-adhesive label applicator.

and storage conditions, and any particular usage requirements (12).

Pressure-sensitive labeling machines. A pressure-sensitive adhesive is one that remains permanently and aggressively tacky in the dry form and has the ability to bond instantaneously to a wide variety of materials solely by the application of light pressure. No water, solvent, or heat is needed to activate these materials.

Because the adhesive is permanently tacky, it would stick to anything it contacted, so a backing sheet is required to protect the adhesive layer until it contacts the article to be labeled. The backing sheet is coated with a release coating to prevent the label from sticking too firmly to it (13).

There are many different types of applicators, but they all have one thing in common: a means of peeling the labels away from the backing (see Fig. 6). This is usually accomplished by unwinding a reel of die-cut labels, and pulling them under tension around a stripper plate. As the backing is bent around a sharp angle, the front edge of the label peels away. Once the labels have been detached from the backing, there are various ways of feeding them forward and pressing them on to the containers in the correct position. Containers may be fed forward to an applicator drum where the label is transferred to it under light pressure created by an applicator drum and pressure pad. Alternatively, the label may be held in position on a vacuum box or drum and released on to the article when it is in the correct position, or it may be blown on to the container by releasing the vacuum and applying air pressure. The backing paper is then rewound on a take-up spool (12).

Comparatively simple systems have been devised to apply the label to the product. The choice ranges from a dispenser which releases a label for hand application up to high speed automatic labelers (see Fig. 7). Labeling machines for applying pressure-sensitive labels vary in complexity according to the nature of the package to be labeled, the number of labels to be applied at one time, and the speed of application. There are, of course, semiautomatic labelers for single-product medium-volume application. Although pressure-sensitive labelers are available as standard machines, they are usually readily customized to meet special requirements. Generally, they are only about one-third to one-half the cost of most competitive labeling systems.

The advantages of pressure-sensitive labels are said to be numerous. They are cleaner than other methods, less wasteful than wet glue, and more easily controlled because they come in roll form so labels cannot be mixed up. A wide variety of adhesives for different surfaces is available, and clear acetate, vinyl, or polyester film can be used for "see-through" labels.

Heat-seal labeling machines. Heat-seal labeling has most of the advantages of pressure-sensitive labels but generally offers a lower label cost. Heat-seal describes a plasticized paper which, when heated by the machine, becomes sticky on its underside. Labeling equipment for heat-seal ranges from simple aids to automatic high speed machines, giving the unique advantage of being able to apply to many varied types of surface labels which can be precision-placed, give all-over adhesion, and be perfectly clean when applied.

The machines free the user from mixing and adding glue, selecting grades, controlling viscosity, and so on. The absence of glue eliminates cleaning the machine, and these savings give substantial increases in productivity and reductions in maintenance expenses. The labels offer a high degree of adhe-

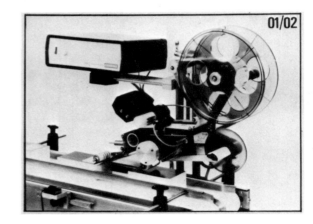

Figure 7. An automatic self-adhesive labeling machine which can either be mounted on existing production lines or used with its own conveyor for any combination of top, side, or under labeling.

Figure 8. A small hand-operated overprinter with two wheelsets and two type channels to carry short text information.

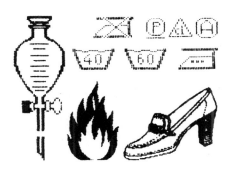

Figure 10. Electronically produced graphics now becoming available on electronic label overprinters.

sion security that is useful, for example, for bottles subjected to high humidity, or steam of water saturation (14).

Of the two types of heat-seal paper, the delayed-action type is best suited for machine operation. The ability of the adhesive to remain tacky after removal of the heat source allows the machine designer to separate the heat source from the pressure application of the label. Thus the heating plates can be simple whereas the pressure-applying devices can be of complex shapes, enabling a diverse range of articles to be labeled, varying from flat to round or irregular shapes (10).

Label Overprinting Machinery

For certain label applications there is a requirement to have a plain paper or preprinted label or labels overprinted with price, price/weight, bar code, or other variable information just prior to or at the point of application. A variety of mechanical, electronic, or computer-based overprinting systems are available for such operations. These are now widely used by department stores, the retail and wholesale trade, industry, and even some label producers (see Fig. 8).

Mechanical overprinting systems range from small hand-operated machines with one or two wheelsets and two type channels, to hand-operated and electronic versions used for price/description labeling with either conventional or EDP-readable fonts (see Fig. 9) (10).

The current trend in overprinting is away from mechanical to totally electronic overprinters which can now print labels and tags without any form of marginal or sprocket hole punching. Free formating and graphics, as well as alphanumeric information and bar codes are now possible using such overprinting technology, the graphics being keyboard entered and edited. Graphics suitable for reproduction include product outlines, such as shoes, laboratory equipment, small hardware items, safety hazard designs, and laundry symbols. Figure 10 shows examples of such electronically produced and overprinted graphics.

BIBLIOGRAPHY

1. *A Guide to Legislation and Standards on Labelling* (UK), Labels & Labelling Data and Consultancy Services Ltd, Potters Bar, Herts, UK, 1984.

2. D. G. N. Alder, "Label Papers and Their Uses," *Labels & Labelling*, **14**, 16 (Jan. 1979).

3. "The Most Important Types of Paper for Beverage Labels," *Krones Manual of Labelling Technology*, Hermann Kronseder Maschinefabrik, Neutraubling, FRG, 1978, pp. 118–121, 1978.

4. A. Hildrup, "Review of the (UK) Gummed Paper Market," *Labels & Labelling Yearbook*, 23 (1981).

5. M. Fairley, "The challenge for water-borne adhesives," *Labels & Labelling*, 29 (Sept. 1984).

6. M. Fairley and R. Brown, *Thermal Labelling*, Labels & Labelling Data and Consultancy Services Ltd., Potters Bar, Herts, UK, 1984.

7. M. Fairley, *Bar Coding*, Labels & Labelling Data and Consultancy Services Ltd., Potters Bar, Herts, UK, 1983.

8. N. Henderson, "Trends and Developments in Heatseal Label Materials" *Labels & Labelling Yearbook*, 31 (1981).

9. M. Fairley, *Label Printing Processes and Techniques*, Labels & Labelling Data and Consultancy Services Ltd., Potters Bar, Herts, UK, 1984.

10. *Labels—A Product Knowledge Book*, National Business Forms Association, Alexandria, Va., 1983.

11. *Labelling Operations*, Pira, The Research Association for the Paper and Board, Printing and Packaging Industries, Leatherhead, Surrey, UK., 1981.

Figure 9. One of the latest types of electronic label overprinters used in retail labeling applications to overprint a large amount of highly variable data.

12. E. Pritchard, "Labelling and Labelling Systems," *Labels & Labelling,* 14, 16 (Mar. 1982).

13. *Labelling—Basic Principles,* Pira, The Research Association for the Paper and Board, Printing and Packaging Industries, Leatherhead, Surrey, UK., 1980.

14. D. Miles, "Pre-adhesed Labelling," *Labels & Labelling,* 20 (Jan. 1981).

M. C. FAIRLEY
Labels and Labelling Data
and Consultancy Services
Limited

LAMINATING MACHINERY

Flexible-web materials are combined on laminating machinery to optimize physical and barrier properties. Laminates produced on machinery using a laminating process are found in many industries, but when applied to flexible packaging, the term laminator nearly always refers to machines that combine two or more plies of materials on rolls. After processing, the combined structure is rewound or sheeted (see Slitting/rewinding). Much flexible packaging laminating today includes an extrusion process, which is not covered here (see Extrusion coating; Multilayer flexible packaging).

In general, laminating is defined in three ways: wet-bond, dry-bond, and thermal. Recent developments have necessitated an expansion of these fundamental concepts, but the machine configurations have not changed significantly. The concept of bonding is central to any laminating operation. One might intuitively assume that a properly laminated structure is inseparable. This is rarely true, either by design or unintentionally, with real values varying from a slight "cling" to a bond that exceeds the strength of the materials. There is also no fundamental relationship between the three types of laminating and adhesion. Of equal importance to actual bonding is the created structure's resistance to attack by chemicals in the product to be packaged, and adhesives are now being modified to provide barrier qualities. These factors affect the laminator and the techniques used to operate it. Each bonding type has a set of unique mechanical requirements which makes a universal laminating machine impractical.

Bonding Variables

All of the laminating techniques described here involve bringing a receptive web into intimate contact with a second material by use of an elastomer-covered steel roll "nip" (see Fig. 1). Receptivity is determined by the presence of some adhesion-promoting element. This may be as simple as first exposing the surface to corona-discharge treatment (see Surface modification) or as complex as one or more chemical pretreatments followed by the application of a cross-linking adhesive (see Adhesives). A previously applied thermoplastic adhesive or an adhesive in web form are two more variations. Combining variables are heat, pressure, and time (see Table 1).

Time and pressure. The time under process pressure is only a few milliseconds at most with typical "footprint" widths of less than 10 mm and line speeds commonly exceeding several hundred meters per minute. Extending this time is impractical and often creates new problems of wrinkle control. Elastomer hardness (measured in the Shore A range) and deflection

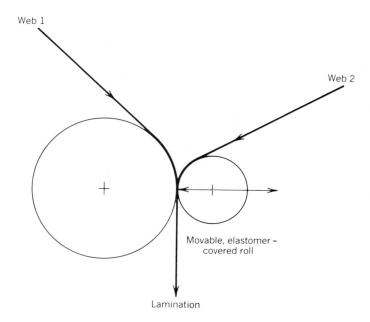

Figure 1. Typical laminator with movable elastomer-covered pressure roll acting against a driven steel roll to form a "nip".

or bowing of the laminator pressure roll are complicating factors. Deflection results in less contact width in the center of the web (see Fig. 2). Such nonuniform conditions produce undesirable effects which are discussed below.

Different techniques are used to minimize deflection, including a variety of patented support structures, but there are two other solutions to this problem. The first is to grind the elastomer roll with a "crown" or larger diameter in the center in such a way that when the roll deflects, the rubber deforms to create an equal-width footprint (see Fig. 3). Note that the rubber "deforms;" it does not "compress". The other correction is to skew the axis of the elastomer roll so that it is not parallel to that of the heated roll (see Fig. 4).

If the nip roll is covered with a "soft" material (eg, 60–70 Shore A), the footprint will be wider than that of a "hard" (90–100 Shore A) covering. For a given nip actuation force, the pressure per unit area of contact will be higher for the harder roll, but the time under maximum process pressure will be less. A varying-width footprint is also detrimental because of an increased tendency for wrinkles to form. Deflection produces nonuniform roll diameters across the web. This results in a varying instantaneous velocity of the elastomer, with the speed of rotation being constant. Therefore, operation and setup tend to become more difficult as the hardness decreases and/or the nip pressure increases. On the other end of the scale, a very hard roll will be unforgiving of side-to-side pressure misadjustment and will tend to bounce at higher speeds.

Table 1. Typical Bonding Variables

Method	Time	Temperature	Pressure
dry-bond	short	medium to high	medium to high
wet-bond	short	low or ambient	low to medium
thermal	short to medium	high	high

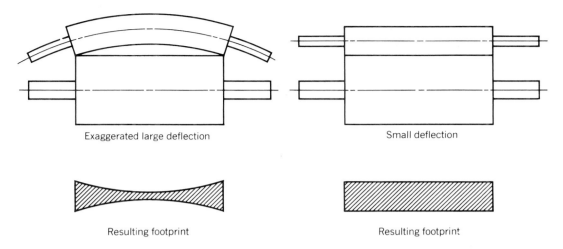

Exaggerated large deflection   Small deflection

Resulting footprint   Resulting footprint

Figure 2.   Deflection of pressure roll with resulting footprint.

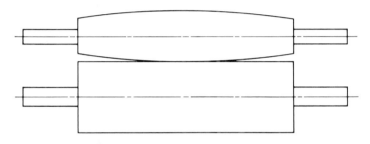

Figure 3.   Minimizing pressure-roll deflection by grinding the outer edges resulting in a larger diameter or crown in the center.

Temperature. As a process variable, temperature is the most important and the most controllable. Although little can be done to manipulate pressure, several avenues exist to change temperature: by variable-position pre- and post-web contact with the hot roll; through the use of auxillary pre- and post-heat rolls; or by using infrared radiant heat boosters. Provisions must be made to protect the thermally sensitive webs during periods of machine shutdown. These include automatic control of heat sources and the actuation of web lifting yokes to raise webs from heated surfaces.

Heat sources. Radiant heaters and heat-transfer media are used to provide energy to the laminator. Hot-oil, hot-water, and steam systems have applications. Each has advantages, with maximum temperature, speed of response, and nontechnical reasons all being significant factors in selection. Product uniformity can be assured only if the process conditions across the web (in the cross direction) are uniform.

Heating a laminating roll uniformly is not a simple matter. Careful design is needed to ensure the required 2–3° accuracy. These techniques include building a hot roll consisting of a roll within a roll: the so-called "double-wall" roll. The heat-transfer medium is channeled between these walls, which tends to give a controlled flow that minimizes stagnant areas and enhances heating uniformity. This solution, though, is still susceptible to "layering" and nonuniform flow, so it is often necessary to weld rods between the walls to give precise controlled fluid flow. Several fluid feed points and high volume flow ensure constant roll-face temperature.

Wet-Bond Laminating

Wet bonding refers to any process where a liquid adhesive is applied to a substrate that is then immediately combined with a second ply to create a laminate. One of the first applications was laminating light-gauge aluminum foil to tissue, us-

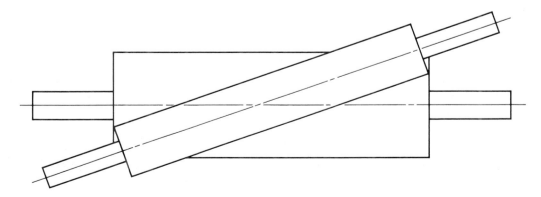

Figure 4.   Minimizing pressure-roll deflection by skewing its axis in relation to the drive steel roll so that more contact area is provided at the center of the roll.

ing either casein or sodium silicate adhesives. The waterborne adhesives endowed the process with the name "wet" laminating since drying occurred after the combining point. Solvent-free or 100% solids adhesives, although not wet, still combine immediately after adhesive application. The wet-bond laminator usually contains a low pressure unheated nip. If the adhesive is solvent-borne, one of the webs must be permeable to the solvent vapor.

Dry-Bond Laminating

In dry-bond laminating, a solvent-borne adhesive is applied to a substrate that passes through a dryer to evaporate the carrier solvent. The web is then combined with a second substrate in a heated pressure nip. Actual applied pressure is typically 10–50 kg/cm of machine width with temperatures of 50–90°C.

Thermal or Pressure Laminating

The thermal laminating "adhesive" takes one or more of the following forms:

A previously applied thermoplastic adhesive which is activated by the heat and pressure of the thermal laminator;

A special sealing layer on one or two sides of a structured film;

Adhesive in dry-film form;

A relatively low melting point film, treated and introduced partially as a sealant;

A "coated" adhesive on a nonreactive (pressure-sensitive) carrier. Some of these adhesives function at room temperature as well.

It is extremely difficult to separate some thermal bonding applications from dry bonding. It could be reasonably argued that the coated-adhesive technique is quite similar in practice to 100% solids wet-bond laminating. Similarly, a PVDC barrier adhesive (see Vinylidene chloride copolymers) requires the combining characteristics of a thermal laminator even though it also has the attributes of a dry-bond lamination. Categorizing processes into historically accepted types is becoming difficult. In this article, thermal laminating is defined as the bonding of two or more flexible webs with one or more of the adhesive types listed above under heat and pressure, without the requirement of a simultaneous application of solvent- or solvent-free coating. It is understood that a thermoplastic film, layered coating, or hot melt is heated to provide a bonding medium and that the bonding is done under quite high pressures.

Process Technique

A variety of web defects can be created during bonding. These include: inadequate bond strength; poor clarity (bubbles, haze, etc); curl in the machine direction; curl in the cross direction; diagonal curl or "twist"; wrinkles; and tunneling.

Bond strength. Ultimate bond strength is determined by the adhesive's specific adhesion to each web surface, and the inherent strength of each adhesive and substrate. This is further complicated by effects of contaminants on or in the plies or the adhesive. Fortunately, few packaging applications require bond strengths equal to the packaging-film components.

Many techniques are used for bond enhancement: selection of adhesive type and quantity; surface preparation and primers; temperature; pressure; corona-discharge treatment; and time or dwell under process. Of these, only temperature, pressure, and time are laminator-related.

Clarity. A transparent laminate, unless designed for translucency, should have clarity and freedom from visible imperfections, at least to the degree of the individual components; but this is a standard quite difficult to achieve in practice. The most common degradation in laminate clarity is caused by incomplete contact across the adhesive/secondary-web interface. This is caused by: insufficient specific adhesion to the secondary web; residual adhesive patterns resulting in high and low points which are not flattened in the combining nip; particulate or other contaminants; and machine-related problems such as elastomer roll bounce, dirty or contaminated roll faces, and inadequate combining pressures and temperatures.

Machine-direction curl. Machine-direction curl is caused by an incorrect ratio of primary to secondary web tension, or by the differential heating of the laminate components inherent in a combining nip together with different coefficients of expansion. Most of these effects can be eliminated by adjusting the tension ratio. This is an extremely difficult condition to achieve and maintain without instrumentation for both web tensions. Maintaining minimum tension into the laminator is also good practice.

Cross-direction curl. Cross direction (transverse) curl and diagonal curl is more difficult to control. Efforts to improve optical quality usually lead to ever-increasing combining temperatures. Transverse curl then becomes a frequent problem. As pressures are increased, the nonuniform width footprint leads to the development of diagonal curl. Web preheat and the ability to skew the elastomer-covered roll are used to eliminate this curl.

Wrinkles. Wrinkling in the combining nip is the result of insufficient precontact wrap on the elastomer roll, incorrect secondary web tension, or nonuniform side-to-side pressure. The actual cause is easily determined by inspection. A wrinkle which maintains an essentially constant position is probably related to high tension. Those that move erratically accompanied by web steering indicate insufficient tension.

Tunneling. Tunneling is merely a manifestation of insufficient bond to overcome the elastic forces of incorrect web tensioning. Examination of the web will show which is under highest tension since the tunnels will be in the low-tension web.

Elastomer Roll Coverings

No single roll covering offers uncompromised properties to withstand all combining-nip stress. Various coverings may have one or more of the following: temperature and pressure resistance; outstanding elasticity without danger of assuming a permanent "set" if the roll is accidentally left closed; resistance to common solvents to permit cleanup after accidental contamination, and toughness for resistance to mechanical abuse. Some applications permit the use of replaceable sleeves, but these trade elasticity for ease of cleaning and release of coatings or softened web materials. Roll hardness, as measured with a tester using the "A" scale adapter, falls between 45 and 105. Most work today is done using rolls between

75 and 85 Shore A. Hardness selection is often based on tradition. Because elastomers harden with usage, many operators are unaware of the hardness of the roll they happen to be using.

Process Development Department
Faustel, Inc.

LAWS AND REGULATIONS, EEC

Although there is no separate branch of law that may be conveniently classified as the "law on packaging," statutes in all the European countries affect packaging. Legislation on the sale of goods, trade descriptions, transport, weights and measures, food and drugs, and environmental issues, are all concerned with packaging. The sale of goods is now an advanced business, and just as retailing has become more complex, so as a consequence has the range of legislation affecting packaging been vastly extended. The remarkable growth of packaging for food, pharmaceuticals, and other goods, and the developments in packaging technology to meet self-service requirements have resulted in consumers needing information about their purchases.

The Influence of the EEC

The European Economic Community has had a great effect upon both the packaging user and the packaging manufacturer. As progress continues towards the creation of a true Common Market in the European Community more and more national laws and regulations are either being adapted to a European pattern or are being replaced by EEC law. Industry is now regularly confronted with new regulations from the EEC designed to harmonize national laws throughout the Community. It is therefore essential to understand the working of the Community, in order to anticipate the future.

The founders of the EEC realized that a closer union among the peoples of Europe could be a means of removing many of the economic and social ills within the member states. Continuing differences in the legislative provisions in member states governing such details as the nature, composition, manufacturing conditions, handling, packaging, and labeling of, for example, foodstuffs hampered efforts to create this union. The Treaty of Rome recognizes this problem and requires that the Community "shall remove obstacles to the free movement of goods" by working towards the "approximation of the laws of the member states"; in other words, by harmonization.

Directives and regulations. The approximation of laws in accordance with Article 100 of the Treaty of Rome has resulted in the emergence of a framework of common legislation. The measures adopted under Article 100 by the Community are "directives." They do not constitute supranational legislation, but they are binding on member states, with the mode and means left to the discretion of each national authority. They must be, however, unanimously agreed by the Council of Ministers of the EEC and adapted into national legislation within specified periods. Failure to do so results in an action brought before the European Court of Justice.

Legislation may also be accomplished by means of an EEC regulation made under Article 43 of the Treaty of Rome, which is concerned with agricultural products and the Common Agricultural Policy. Once passed, an EEC regulation is binding on all member states from the day it comes into force, and it overrules national laws. A directive, on the other hand, has no force in a member state until it is incorporated into national legislation. A time limit always accompanies a directive, giving member states time to amend their national legislation.

The main bodies of the EEC. The main bodies involved in the preparation and passing of directives are the Council of Ministers and the Commission. Other institutions of the EEC are the European Parliament, the Economic and Social Committee (ECOSOC), and the European Court of Justice.

The Council of Ministers consists of ten members, one from the government of each member state. On food legislation matters, this is usually a minister of agriculture or of health. The Council actually promulgates the legislation, and this usually requires unanimous voting. The Commission is responsible for initiating legislation and ensuring that it is correctly applied. The Commission consists of 14 members or commissioners who are appointed by the ten member states for a four-year renewable term. They act only in the interests of the Community and do not receive instructions from member governments. Both bodies are assisted by other Community institutions, the most important of which are the European Parliament and the Economic and Social Committee (ECOSOC) (see Fig. 1).

The European Parliament consists of members elected in and by the member states. It must be consulted on Commission proposals before the Council can take a decision on them (see Fig. 2). The Economic and Social Committee consists of representatives from employers' organizations, trade unions,

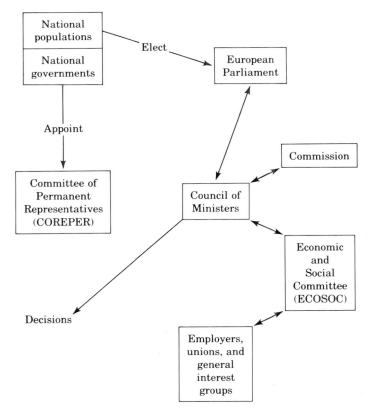

Figure 1. EEC institutional relationships.

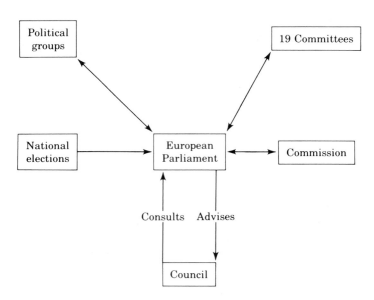

Figure 2.   The European Parliament.

and other trade interests; it also has an advisory capacity only. The role of the Court of Justice is discussed below.

How does the system work?   This can be illustrated by using the food industry as an example. Harmonization of food products is usually accomplished by means of a horizontal or vertical directive. *Horizontal directives* affect all food products and deal across the board with subjects such as labeling, additives, packaging materials, and packing by weight or volume.

Examples of directives in the pipeline at the moment are those on claims; acids, bases, and salts; flavorings; and food packaging materials. *Vertical directives* deal with particular subjects. Existing vertical directives include those concerned with cocoa and chocolate products, honey, sugar, preserves, coffee, and fruit juices. Other examples in the pipeline are soft drinks, milks for babies, tomato products, caseins, and caseinates.

Permanent or temporary derogations may be included in all directives to allow continued marketing of a particular national product or extra time for a country to adapt its laws. Most derogations have to be negotiated and are subject to review. They are eventually either permanently incorporated into law or removed.

When the Commission draws up a proposal it must be well informed about the subject in question. To this end it consults with national authorities who have formed working parties and subgroups to discuss the subjects being treated. In the food area, the Commission consults with the EEC Scientific Committee for Food, and it also consults professional groups such as trade associations, consumer organizations, etc. Because the opinions of these organizations vary, they are represented on an Advisory Committee for Foodstuffs, where they can compare and discuss their different views and approaches (see Fig. 3).

The Commission decides when to submit a proposed directive to the Council, always after discussions and consultations and after the opinions of the European Parliament and the Economic and Social Committee are taken into acount. It is also referred to the Committee of Permanent Representatives (COREPER) (see Fig. 1) which plays an important role in deliberations and can sometimes settle previously unresolved difficulties. Once a directive is adopted, it is published in the

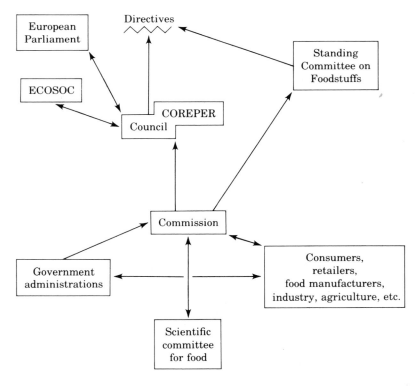

Figure 3.   Directive-making in the food industry.

Official Journal of the European Communities and sent to each member state with the request to amend their laws within the time prescribed, which is usually about two years.

This procedure can be very lengthy and cumbersome. Therefore if it is obvious that technical subjects need not be discussed by nontechnical committees such as the already overworked ECOSOC or by the European Parliament, directives can be adopted by a procedure that requires only the unanimous approval by the Council of the proposal submitted by the Commission. Directives on purity criteria for food additives, emulsifiers, stabilizers, thickeners and gelling agents, and directives on criteria of purity for antioxidants have been adopted in this way.

The Scientific Committee for Food (SCF) is one of the more important bodies of the EEC. It advises the Commission on problems relating to health and safety of persons arising from the consumption of food, in particular its composition, processes that may modify food, the use of food additives and processing aids, as well as the presence of contaminants. The SCF consists of 15 independent members chosen from the various member states who are highly qualified in medicine, toxicology, nutrition, or chemistry. Subcommittees can also be formed to look at specialized topics. Thus there is a wide range of expertise brought to bear in formulating decisions. Not everyone admires the SCF or its decisions, which often mean more expense for industry in the form of funding more toxicological work, but increasing demands from consumers for reassurances on foodstuffs and food additives will probably push the Commission into making more use of it. Here too the consultation process is very lengthy, but sometimes it can be sim-

plified; for example, it is not necessary for ECOSOC or the European Parliament to discuss very technical subjects such as purity criteria for additives.

In 1969 the Standing Committee on Foodstuffs was created, which can also simplify and accelerate the program. This committee consists of delegates from member states and is chaired by a representative from the Commission. It gives opinions by majority voting, and if the opinion is in accordance with the Commission's views, the proposal can be adopted without reference to the Council. This procedure can only be used where the Council has delegated its responsibility to the Commission. Progress on harmonization has been very slow, and where directives have been passed, the derogations and differing interpretations on them sometimes renders harmonization ineffective. To export any product to another country in the EEC it is still necessary to check national laws. It is not enough to comply with directives.

The Cassis de Dijon Case.   A serious challenge came in 1979 from the celebrated Cassis de Dijon Case when a West German importer tried to import the French blackcurrant liqueur "Cassis de Dijon," which has a low-alcohol content of 15–20%. FRG regulations prohibit the sale of potable spirits with a wine-spirit content of under 32%. The case came before the Court of Justice in Luxembourg, the body that interprets Community law and sits as a Court of Reference in disputes of this kind (see Fig. 4). Its decisions are binding on all parties.

The Court ruled that in this case the obstacle created by the FRG regulations was incompatible with Article 30 of the Treaty of Rome, which states: "Quantitative restrictions on imports and all measures having equivalent effect shall, with-

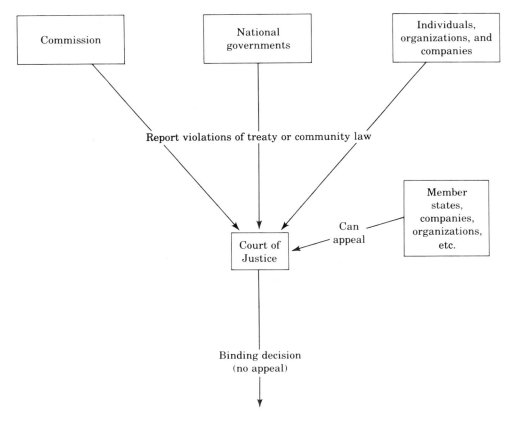

Figure 4.   The Court of Justice.

out prejudice to the following provisions be prohibited as between Member States." Indeed, the Court stated that "an obstacle to trade of this kind was only acceptable when satisfying mandatory requirements relating to what is termed the effectiveness of fiscal supervision, protection of public health, fairness of commercial transactions, and defense of the consumer" as laid down in Article 36 of the Treaty of Rome. In so doing the Court seemed to put forward a general principle regarding the trade in food between member states; that is, provided a product complies with regulations in the country of manufacture, it may be admitted to the market of any other member state. This judgment was upheld by the Commission in a statement published in 1980, but since then work on some of the commodity directives has dropped significantly.

Nevertheless, barriers to trade that cannot take refuge under Article 36 still exist and can be serious obstacles to the free movement of goods. The European Court of Justice has pronounced on some of these nontariff barriers, others have yet to be examined. Problems of this kind can be attributed to protectionism or simply different interpretations of the original directive. Different definitions within and different philosophies behind national food law can contribute to the confusion; for example, UK food law is based upon general principles and case law, whereas most European law is based on Roman codified law (see Table 2). In effect, this means that in continental Europe, what is not specified is not allowed, whereas in the UK what is not specifically banned may well be permitted. Harmonization therefore has a long way to go before all the difficulties are ironed out.

Food/Package Compatibility

Few materials suitable for food packaging are completely inert towards food, and those materials that are inert often have to be used in conjunction with others that are not. The care taken to produce wholesome and attractive foods must be matched by the care taken in the production of the packaging used to contain and protect them. Recognizing this, responsible suppliers of packaging materials and food manufacturers have for many years worked together for the benefit and protection of the consumer. Until joining the EEC, UK food laws were such that packaging materials were not subject to regulations banning the use of specific materials, or insisting that the only materials used were those on a permitted list. The fitness of packaging materials for their purpose was governed by the requirement that food shall be "of the substance and quality" demanded. Any packaging material that did not adversely affect the flavor and the wholesomeness of the food was therefore allowable.

Since we receive almost all our food in a packaged condition, we need to know the extent of any interactions between foods and the materials or containers in which they are packaged. Clearly, any interactions must be small. If they were large the container would cease to be efficient; for example, wet foods are not packed in unprotected paper, nor acidic foods in unprotected metal cans, for obvious reasons. Small interactions can be detected by the senses of taste and smell or the sensitive instruments available to the analytical chemist, or both.

Interactions may result from the movement of volatile or nonvolatile substances. The movement may be in either direction; that is, from packaging to food, or the reverse. Volatile substances can move (migrate) from food and the packaging, but for the movement of nonvolatile substances to occur, there must be contact.

The basis of safety evaluation. The basic equation for all toxic hazards is

$$\text{toxic hazard} = Q \times I$$

where Q is the quantity of material ingested, and I is its intrinsic toxicity.

For the toxic hazard to be acceptable, either the intrinsic toxicity must be so low that any level of ingestion can be tolerated (a situation that does not apply to the majority of plastics components) or, alternatively, the quantity of migrant ingested must be known as well as its intrinsic toxicity.

Originally, toxicology was the study of poisons and their effects. Anything ingested was either a poison or not. The poisons gave severe adverse effects with very small amounts. Nonpoisons gave effects at levels so high as not to be normally encountered. Today, although highly toxic materials are still studied, much of the interest lies in the effects of materials of low toxicity over long periods.

Several interesting viewpoints have been put forward in recent years:

1. The present approach to toxicity testing, based on the concept of a "no-effect level," has proved useful in the past but is now out of date for four reasons: analytical techniques are now very sophisticated; animal toxicity studies require too many specimens; experimental animals are limited in value as predictors of human reactions; and as knowledge of the effects of food components increases, the cost of conventional methods will become impossibly expensive. As a result, the evaluation of safety should be related to the expected exposure. Risk assessment should replace the "no-effect level" concept (1,2,3).

2. Risk assessment implies that some risks are acceptable.

3. Risk assessment must also take account of the hazards of having no food and adverse microbiology.

4. Research in food toxicology has made little progress since the 1960s. One reason for this is the virtually exclusive reliance on experiments with animals. The fallacies of following the established rituals of *ad lib* feeding, maximum tolerable doses, and the neglect of dietary effects have hindered the growth of understanding.

Table 2. Types of Packaging Legislation in EEC Countries Prior to EEC Legislation

Country	Statute law	Common law	Code of practice	Draft law
Belgium	X			
Denmark		X		
France and Luxembourg	X			
Germany			X	
Greece	X			
Ireland		X		
Italy	X			
Netherlands				X
United Kingdom		X	X	

In short, a new approach is necessary based on the integrated use of animal studies, *in vitro* studies, and clinical toxicity studies in man, and where packaging is concerned, a procedure such as the following would be preferable (4):

1. Make a very primitive assessment of toxicity and exposure (the latter being related to migration). The former need not use animals, the latter can be based on theory and calculation. Allocate categories 1 (low), 2 (medium), and 3 (high) to each.

    For toxicity, the categories could have roughly the following meanings: 1 = little or no concern; tolerable daily intake estimated in the order of g/person per day; 3 = high concern, eg, suspect carcinogens, cumulative poisons, radioactive materials, physiologically active materials, analogues of known materials of high toxicity, etc. Estimated tolerable daily intake well below 1 mg/person per day; and 2 = medium concern; all those between.

    For exposure, the numbers would mean something similar to the following: 1 = very low; unmeasurable to 10 ppb (1×10^{-8}) in the diet; 2 = moderate; 10 ppb to 10 ppm (1×10^{-5}) in the diet; and 3 = high; above 10 ppm in the diet.

2. Sum the numbers allocated to the two areas to produce overall classifications 2, 3, 4, 5, or 6.

3. Treat these as follows: 2 = no further testing required—accept risk; 3 = probably no further testing required, possible confirmation of preliminary allocations; 4 = further testing required; 5 = in-depth testing required; and 6 = banned.

Following the first evaluation, some systems may change their classification. Evaluation is terminated when data are considered adequate (based on depth of testing mentioned) to allocate use levels. It is the principle that is important; the above figures illustrate possible levels of concern, but could be the subject of much further discussion.

It is recognized that some toxicologists would flinch at such an approach, but epidemiology must be considered in the context of the real world. No one has been shown to have suffered harm, let alone fatal injury, from packaging migration, but tens of thousands or more have died of starvation and many hundreds from microbiological food contaminations. Yet some minor components of packaging are tested every bit as fully as direct food additives, although the majority are not tested at all.

Food-contact directives. This is the category of main direct relevance. Seven directives are now in force in the EEC countries.

1. Directive 76/893 is a general, basic, directive, rather like a "frame" law. Its main provisions are that it applies to materials and articles in contact with, or intended to come into contact with, food; these should not endanger human health; there should be no unacceptable organoleptic changes in the food; further directives of a specific nature are authorized; and there must be some indication that containers are for food contact. This may be obvious from their nature, eg, domestic holloware, by labeling, eg, "for food use" in English, or by use of a symbol. Where the packaging is not sold directly for retail use, this may be accompanying documentation.

2. Directive 78/142 controls the limits of VCM (vinyl chloride monomer) in plastics and in food. Essentially the requirments are a maximum of 1 ppm in containers and 10 ppb in food.

3. **and 4.** Directives 80/766 and 80/432 prescribe the analytical methods for VCM in plastics and food, respectively. These are needed to implement 78/142.

5. Directive 80/590 describes the symbol to be used on food packages.

6. Directive 82/711 is on plastics.

7. Directive 83/229 on regenerated cellulose film (RCF).

The last two are important "specific directives" on particular materials, both foreshadowed in 76/893.

Plastics. Directive 82/711 is the most controversial of all the EEC food-packaging directives. It was originally planned to cover all relevant aspects; positive lists, specific migration, and organolepsis. It became clear that this was a monumental task (compare FDA legislation), and an attempt at simplification was made in 1978 by introducing an all-embracing "global" migration limitation (now referred to as overall migration) already in use in Italy and France. There are scientific and other objections to the global migration concept of the testing method however, and it was opposed by Germany, Ireland, and the UK. A modified version was issued in 1980, which was more acceptable, since it proposed the introduction of global migration in parallel with specific migration limits. Neither would be implemented before a classification of foods, and their corresponding test liquids, had been agreed. However, sufficient opposition existed for this version also to be rejected, and the final version issued in October 1982 omitted all reference to global migration. It provides a framework for methodology, devoid of specifications or concrete requirements beyond those already made in Directive 76/893.

It had no immediate impact, and the Commission is now concentrating on preparing positive lists. The first two, one for monomers and other starting substances and the second for additives, have been issued and amended several times. Considerable use is made of previous studies and conclusions by the Council of Europe.

Regenerated cellulose film (RCF). Directive 83/229 was issued in April 1983. It includes two positive lists, one for coated film and one for uncoated film. Most of the materials approved in the lists are given with no specific limitations, although there are overall limitations dictated by limiting film and coating thickness (see Cellophane).

Further proposals on food-contact directives. Work is in hand on proposed directives for stainless steel, glass, paper, etc. In particular, a proposal for ceramic ware was first issued in 1974. It included all ceramic and glass-coated articles for food contact, ie, cookware as well as packaging. Development was slow until 1980, when active work was resumed. Its main concern is to limit heavy metals, especially lead and cadmium.

Working document III/808/82 has been issued by the Commission. It concerns draft proposals for a directive harmonizing legislation on food additives, and suggests replacing existing directives on preservatives, additives, emulsifiers, and enlarging the scope to cover many groups not yet dealt with, eg, sweeteners. It also proposed that labeling regulations be extended to processing aids.

There are many directives and amendments on food additives, eg, colorants, antioxidants, emulsifiers, stabilizers, etc.

They are not complete in coverage of member states' national legislations, nor do they align with it. The proposed new directive would update EEC legislation and give comprehensive coverage.

Most of the implications in these are for the material itself, but the following are related to its packaging:

Indirect implications. Migration can infringe legislation controlling food quality (contamination). This is usually just a theoretical possibility, but sometimes it merits consideration; for example, antioxidants, such as BHT, can migrate from plastics into a food in which the antioxidant is forbidden or limited.

Direct implications. The regulations on labeling.

Significant implications. Packaging migrants are not proposed for inclusion. In some countries migrants are treated legally as indirect or accidental food additives. Clearly the EEC is not going down this road, but prefers to treat packaging migrants quite separately under legislation dealing exclusively with food packaging and contact.

Labeling

Although the labeling directive, EEC Directive 79/112: *Labeling, Presentation, and Advertising of Foodstuffs for Sale to the Ultimate Consumer,* was proposed in 1974, it was not adopted until December 18, 1979. It was then incumbent on member states to permit trade in goods labeled according to the directive two years later (December 18, 1981) and to prohibit goods not labeled according to the directive four years later (December 18, 1983). Most member states have now implemented the directive to this time table. Also, in order to accommodate nationally established practices, the final directive allowed member states to be selective sometimes, and also permitted more detailed national labeling controls to be imposed, providing a notification procedure was followed. There are about 30 options or derogations in the directive, from which member states can choose. Those chosen differ from one country to another so that the 10 sets of regulations stemming from the same directive each differ in some way. Harmonization, therefore, is only partially achieved.

Basic principle. Judging by experience of the UK, member states have faced problems in including the provisions of the labeling directive into national legislation. The directive is based on the principle that labeling methods used must not mislead purchasers as to the nature, identity, properties, composition, quantity, durability, origin or provenance, method of manufacture, or production of a food; by attributing to it properties that it does not possess; or by suggesting that it has special characteristics when all similar foodstuffs also have them.

Requirements of the labeling directive. Leaving the exemptions and derogations aside, the directive requires the following particulars (5) to be shown on food labels: the name under which the product is sold; a list of ingredients; the net quantity; the date of minimum durability; any special storage conditions or conditions of use; the name or business name and address of the manufacturer or packager, or of a seller established within the Community; the place of origin where failure to give such information could mislead the consumer as to where the foodstuff came from; and instructions for use when it would be impossible to make proper use of the foodstuff if these were not supplied.

Member states may retain national requirements for an indication of the factory or packaging center where domestic production is involved; this is so in France. Member states may also apply more extensive provisions regarding weights and measures, and most EEC countries have such a requirement.

Product name. Product naming must follow a set pattern. First, if the name is prescribed by community of national law or by administrative rules, then that is the name which must be used. If a prescribed name does not exist, a customary name may be used, or the product must be described precisely to inform the purchaser of its true nature and to distinguish it from products with which it could be confused. A customary name is a name that is customary in the member state in which the product is to be sold. Some customary names are used in many countries, eg, spaghetti and frankfurter, but generally speaking, they are more local (UK examples are haggis, Pontefract cakes, and Cornish pasties).

The directive prohibits the substitution of a trade mark, brand name, or fancy name for the product name. Hence, although brand names such as Pepsi Cola and Coca Cola are internationally recognized, they must be more fully described.

A product name must also include or be accompanied by details as to the physical condition of the foodstuff, or of a treatment it has undergone, if the omission of such information could create confusion for the consumer. If it is not readily apparent that a food is powdered, freeze-dried, concentrated, cured, etc, then this must be made clear in the product name. By this, it is presumed that a product name should reflect, for example, the polyphosphate treatment of chilled or frozen chicken, as well as mentioning whether the product is chilled or frozen.

Ingredients. "Ingredients" means any substance, including additives, used in the manufacture or preparation of a food and still present in the finished product, even if it has altered in form. Additives must be listed, together with either their chemical name or the EEC serial number. However, additives used only as processing aids need not be declared, nor those additives which are there because they were present in one or more of the ingredients, as long as the levels of carryover are insufficient for them to have a technological function in the finished product.

Ingredients must be listed in descending order of weight, determined at the mixing stage. The only exceptions are volatile products and added water, which must be listed according to the amount present in the finished product. This allows for cooking losses. Less than 5% of added water need not be declared.

The names of ingredients must be those used when the ingredients are sold separately. Some can be described generically, eg, fish, poultry, meat, and cheese. Whether oils and fats are of animal or vegetable origin must be stated, but there is no provision to permit manufacturers to indicate the type of oil and fat *likely* to be present, in order to allow more flexiblity in the composition of the fat blend used and to accomodate fluctuations in the supply and world market prices of specific oils and fats.

The foregoing is not a comprehensive account of the provisions in the directive covering ingredient listing, but it does identify the main points. Where the labeling of a food places emphasis on the presence or low content of one or more of its ingredients, or where the description of the food has the same effect, the minimum or maximum percentage, as appropriate, used in its manufacture must be stated.

Quantity. When considering the quantity marking of goods, whether or not these are foodstuffs, other European community legislation must be taken into account. For example, there is a directive concerned with the packing of goods in prescribed quantities (80/232); another directive specifies the prepacking of solids on a qualified average-quantity basis (76/211); and similar control is imposed on prepackaged liquids (75/106).

Foods prepackaged in quantities greater than 5 g or 5 mL must have the quantity marked. There is some freedom for member states to increase the 5 g or 5 mL threshold in exceptional cases and also to make national provisions in specific circumstances, to derogate from the need to quantity mark at all.

Briefly, goods packed according to EEC weights and measures legislation must contain, on average, at least the stated quantity. Variation from the average must be within specified tolerances, which are related to the quantity in the pack. Packers must keep adequate, statistically based, quantity-control records, and inspectors must carry out reference tests in a prescribed manner. The symbol "e" alongside a quantity mark on a label means that it conforms to EEC quantity-control standards and has been checked by the member state concerned. Control of the volumes of liquids packed in bottles is covered by a directive (75/107) that defines the accuracy to which the volume of bottles must conform if the bottles are used as measuring containers. The checking of fill levels is the basis for quantity measurement.

Open date marking. The principle on which EEC date marking is based is that of minimum durability. This is defined as the date until which a foodstuff retains its specific properties when properly stored. The mode of marking is specified in the directive. In most instances it takes the form of "Best before" followed by the minimum durability date. In the case of foodstuffs which are highly perishable from a microbiological point of view, member states can adopt the words "Use before," or some equivalent phrase, instead of "Best before." The equivalent chosen by the UK is "Sell by (date): Best eaten within (number) days of purchase."

A number of foodstuffs are not required to have open date marking. Fresh fruits and vegetables, wines, beverages with more than 10% alcohol, vinegar, and cooking salt are examples. Also, member states may exempt long shelf-life foods (those remaining in good conditions for longer than 18 months). The UK has done this.

To reduce practical problems in applying date marks, the "Best before" statement can be followed either by the actual date or an indication of where the date is placed on the pack, eg, Best before: see date on can end). Remember, however, that if particular storage conditions are needed to keep the product for the specified time, then these particulars must follow the "Best before" indication; for example, Best before (date): Keep in refrigerator.

Instructions for use. The directive specifies that instructions as to the method of using a foodstuff shall be provided to ensure it is used appropriately. UK labeling regulations have always required that if additions to the food are needed, this must be clearly stated. The directive adopts the same principle so that if, for example, a prepacked cake mix requires the addition of an egg or other ingredient, this is made clear on the label in close proximity to the product name.

Manner of marking. The directive does not demand an information panel, nor does it specify type-size requirements to determine how the required particulars must be marked on the label. It only specifies that they be easily understood, be placed conspicuously, and be easily visible, clearly legible, and indelible. Furthermore, they must not be hidden, obscured, or interrupted by other written or pictorial matter. The product name, the quantity, and the date must all appear in the same field of vision.

Nutritional labeling. Nutritional labeling is not a mandatory requirement in any of the EEC member states, although when such claims are made, nearly all countries require them to be capable of substantiation and to be supported by the appropriate details on the label. The nearest approach to nutritional labeling is provided in the *Directive on Food for Particular Nutritional Uses (77/94)*. This is concerned specifically with foods that are nutritionally balanced, eg, slimming foods and infant and baby foods, and other foods that are designed to meet a specific need, eg, for diabetics. The directive requires that these foods be suitable for their use, the claims can be substantiated, and certain labeling provisions be met. Specifically the manufacturer may not claim that the food can prevent, treat, or cure human diseases, except in exceptional and clearly defined cases provided for under national legislation. This directive is solely concerned with claims related to dietary and dietetic foods, and they are still subject to the wider requirements of the labeling directive and any national labeling rules permitted by it.

Summary of position on labeling. The EEC labeling directive provides only a starting point for determining labeling requirements within the countries of the EEC. Member states use the permissible exemptions and derogations in the directive extensively, and this means that national regulations based on the directive vary in a number of respects. Labeling requirements in any specific country are not confined to those of the EEC directive. Hence, exporters, particularly those operating from non-EEC countries, wishing to label products to be sold in one or more EEC countries correctly, are advised to consult a labeling expert in the country of sale. Alternatively, consult an expert who has the necessary contacts in EEC countries and is skilled in finding out the precise requirements.

Environmental Directives

There are directives on several environmental aspects that affect packaging indirectly, mostly in manufacturing or waste disposal. Concerning manufacturing, directives on air pollution, water quality, toxic waste, and waste disposal are relevant. Waste disposal is affected by directives on water quality, eg, limits on cadmium that may arise from leaching from cadmium pigments, and other aspects of solid waste disposal.

The most directly relevant Environmental Directive relates to energy and recycling: in particular, there is a proposed directive on beverage containers. This would encourage returnable "multi-trip" containers, partly by facilitating consumer return, but also possibly by penalties, eg, deposits or taxes, on one-trip containers. The effects of this directive would change current market patterns. Although the ultimate effects would be complex, there is little doubt that it would favor glass containers as against paper, plastics, or metal packagings. This is one area where U.S. legislation has influenced the EEC considerably, especially legislation in Minnesota, Oregon, and Washington.

Proposed directive on beverage containers. The European Parliament spent 1982 discussing the EEC Commission's proposals on beverage containers without getting very far. The issue has also become a political football with voting along party lines. The Left broadly supports the directive, and the Right opposes it. Votes swing either way, depending on who is present at the meeting in question. The Council of Ministers discussed the directive in June 1983, when seven member states voted in favor of it, two against (UK and Ireland), and the Italians were undecided.

Document 1-1187/82 drawn up on behalf of the committee on the Environment, Public Health, and Consumer Protection of the European Parliament calls on the Parliament to forward a new resolution to the Council and Commission proposing that the directive be replaced by a recommendation. The draft directive is faulted on several points, including failure to provide a basis or evidence for the proposed measures and for being unclear and badly drafted.

Well-informed people connected with the beverage industry think that a diluted version of the directive will eventually emerge without too restrictive an effect on the industries concerned, nor causing distortion of trade among member states. Legislation is also often threatened in the European countries that would control packaging to prevent "waste."

The prime function of packaging is to enable consumers to receive products in good condition at the lowest reasonable price. Any manufacturer, distributor, or retailer concerned with design or use of packaging has a responsibility to ensure that there is a regular review of packaging having regard to the economics of the total manufacturing/distribution chain and to consideration of reuse and disposal. Marketing and commercial considerations should be reconciled as far as possible with economy in the use of materials and energy and the environment.

1. Packaging must comply with all legal requirements.
2. In containing a product the package must be designed to use materials as economically as practicable, while at the same time having due regard to protection, preservation, and the presentation of the product.
3. Packaging must adequately protect the contents under the normal foreseeable conditions of distribution and retailing and also in the home.
4. The package must be constructed of materials that have no adverse effects on the contents.
5. The package must not contain any unnecessary void volume nor mislead as to the amount, character, or nature of the product it contains.
6. The package should be convenient for the consumer to handle and use. Opening (and reclosure where required) should either be obvious or indicated and convenient and appropriate for the particular product and its use.
7. All relevant information about the product should be presented concisely and clearly on the package.
8. The package should be designed with due regard to its possible effect on the environment, its ultimate disposal, and to possible recycling and reuse where appropriate.

Figure 5. UK Code for the Packaging of Consumer Goods

Excessive packaging. The principle criticism is probably the accusation that "packaging is excessive." These accusations are usually made where the selling and convenience factors are concerned. It must be remembered, however, that the decision by the manufacturer of a product to put conveneience or selling into his packaging results from the decision that this will provide an advantage over a rival product produced by a competitive company. Most manufacturers are very concerned to keep their packaging costs to a minimum, and the objective of any convenience or sales appeal in the packaging must be better sales for the product.

Moreover, the social implications of excessive packaging may be measured against a code for good retail packaging first suggested by the Japanese Packaging Institute. A code of practice along the lines of the Japanese suggestion has been produced by the United Kingdom Packaging Council (see Fig. 5), responsible for considering complaints referred to it by any interested parties who feel that a particular package breaks the guidelines set out in the Consumer Goods Packaging Code.

Industry on both sides, user and maker, in the UK and many parts of Europe would favor the self-regulating approach suggested here, rather than legislation, and any proposals for regulations in this field will be resisted. The Japanese and UK codes are working reasonably well, although a second tier of the codes, detailing the means for determining compliance is needed.

BIBLIOGRAPHY

1. D. M. Conning, *International Symposium on Food Toxicology—Real or Imaginary Problems,* University of Surrey, July 1983, British Industrial Biological Research Association (BIBRA), UK.
2. P. S. Elias in Ref. 1.
3. L. Golberg in Ref. 1.
4. L. L. Katan in Ref. 1
5. A. Turner, "Labelling Legislation," *Symposium on Trends in Food Legislation,* Nov. 14–16, 1983 Hyatt Regency Hotel, Chicago, Leatherhead Food RA, UK and Grocery Manufacturers of America, Inc.

F. A. Paine
Packaging Consultant and Secretary General, IAPRI

LIDDING

Lidding is a very specialized aspect of flexible-packaging technology. The advent of portion packaging and dispensing packages created a need for flexible-packaging lidding materials, and liddings are frequently used to seal other types of packages, including semirigid containers. Lidding materials are rarely composed of just one layer. One or more layers provide physical properties, and other layers provide sealability. Generally, the ideal lid is one that is easily peelable, leaves no traces of sealant residue, and is tamper-evident (see Tamper-evident packaging). The sealant should melt at a relatively low temperature unless heat resistance is necessary for sterilization of the contents or reheating of a food product. Typical examples of lidding applications are shown in Figure 1.

The first considerations in the choice of a lidding must be

the intended use. Some of the questions that should be asked follow:

Must the lid prevent contamination or aid in dispensing the product by being peelable or having push-through properties?

What are the requirements for gas, moisture-barrier and light protection?

Should the lid be fusion-sealed or peelable?

What temperature resistance is required of the lidding? Is the product to be packaged hot? Will retained heat be a problem?

Should the lidding be heat-sealable or pressure-sensitive? Will the use of cold-seal adhesives be advantageous?

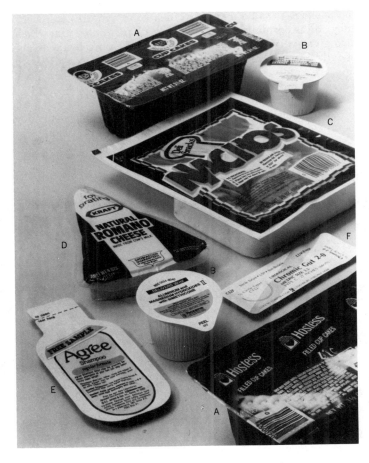

Figure 1. Typical examples of liddings used in packaging applications. A, Transparent lidding sealed to opaque preformed cupcake tray; because of rapid turnover, this type of product requires only the minimal barrier properties of a peelable lidding. B, Peelable lidding sealed to a vinyl-coated aluminum cup for liquid unit-dose drug applications; this type of lidding should be easily peelable, but also requires tamper evidence. C, Snack package utilizing a flexible, peelable lidding with good gas- and moisture-barrier properties. D, Peelable cheese lidding with good gas- and moisture-barrier properties. E, Sample package of shampoo utilizing a flexible lidding sealed to a tray formed from polyethylene-coated PVC sheet; this is an example of a fusion seal. F, Lidding for a PVC preformed tray containing sutures packed in alcohol.

Should the lidding be tamper-resistant?

What type of container will the lidding cover, and how will it be sealed to the container?

Will the lidding be left on during a temperature cycle, (eg, sterilization)?

Must it also be resistant to electron-beam or nuclear sterilization? If a lidding is used on a "cook-in-tray," will it be left on during heating in a microwave or conventional oven? Liddings for cook-in-trays must also have good adhesion to the tray while the product is in a refrigerated or frozen state.

If the lidding is for an industrial application, it might be very easy to find an appropriate material. A typical example would be a polyethylene-lined cannister filled with lubricating oil. An excellent lidding would be nylon/LLDPE (see Polyethylene, low density). However, if the application is for a food product, the parameters of the barrier properties, compatibility of food and container, storage conditions, and when and how the lid is to be removed must be evaluated. Meeting applicable FDA regulations is another important consideration. With medical products, the problems of contamination of the packaged item also have to be very carefully considered.

How a lid is removed is of prime importance on many packages. A tab, or something to hold onto, is highly desirable. Ideally, the lid should peel off in one piece. A residue of sealant should be avoided if possible, but sometimes the peelability of the lid is designed to come about by separating the coating from the base stock. Examples of good bases for lidding stocks are paper/polymer/foil (see Multilayer flexible packaging) or oriented polyester film (OPET), which also might have a barrier coating for improved protection.

In a paper/polymer/foil base stock, the polymer is generally polyethylene, but it can also be an ionomer (see Ionomers) or ethylene–acrylic acid (EAA) to meet specific requirements. An ionomer might be used to improve toughness or tear resistance. The next consideration is the protection of that side of the foil that will face the container from the product being packaged. If the product is inert to foil, protection is not necessary, but because few products are chemically neutral, a barrier layer next to the foil is generally essential. This can be as simple as a vinyl coating, a vinyl film, a polyolefin film, or a polyester film, depending on the product being packaged and the type of protection needed. The heat-seal material is then applied over the protective layer (see Sealing, heat).

One of the simplest overwraps is nothing more than a corona-treated (see Surface modification) OPET film (see Film, oriented polyester). Polyester film is normally not heat-sealable, but corona treatment changes the surface so that it can be heat-sealed to itself. The seal must occur at a temperature near the polymer's melting point with high seal pressures. The seal is a fusion-type seal, but it is brittle and tears open easily. The chief use of this type of material is to overwrap school lunch trays or sandwiches. The physical protection is minimal, but it does act as a cover to prevent direct contamination. Another use is as a wrapper around frozen pizza. It is one of the most inexpensive liddings that can be sealed to a polyester-coated tray, but it is normally not used in "ovenable" applications.

Another important group of materials for tray liddings is

OPET film coated with ethylene–vinyl acetate (EVA) applied from a solvent system or as an extrusion coating (see Extrusion coating). Normally, lower coating weights can be applied from a solvent system. The coating makes the material not just sealable to itself, but to a variety of other materials. It also seals at relatively low temperatures. EVA coatings do not form fusion seals, but the seals are peelable and usually removable in one piece. This type of lidding is very popular for polystyrene sandwich trays or other food trays. Tray packs of cheese and luncheon meat often use a PVDC-coated OPET film with an EVA coating.

Ovenable liddings are usually solvent-based polyester coatings applied to a polyester base film. The coating is used to provide heat sealability, and by proper selection of polyester resins used in the coating formulation, different seal ranges can be obtained and the degree of peelability regulated. Another route to obtaining peelability is to incorporate inert fillers into a coating which normally makes fusion seals. Polyester coatings, in addition to sealing to polyester materials, usually also seal to vinyl materials. For example, a polyester-coated OPET lid seals very well to a semirigid PVC blister.

PVC (see Poly(vinyl chloride)) films and solvent coatings are used as sealants on liddings where fusion seals to PVC semirigid stocks are required. The coating can be modified to provide peelability. Liddings for orange-juice portion packs have traditionally been aluminum foil with vinyl-type coatings which seal to a vinyl cup. The vinyl coating is inert to the acidic juice and is also good film-former to protect the foil from corrosion. Newer-type liddings are foil/film laminations (see Laminating machinery) coated with a peelable heat-sealed coating. Similar liddings are used for yogurt, but they also require good barrier properties to extend the shelf life of the product.

Medicinal products in pill form are sometimes packed in PVC trays with a push-through-type lidding for ease in dispensing (see Pharmaceutical packaging). The tray is designed with wide flanges around each pill so that every pill is fusion-sealed in its own compartment and kept free of contamination. The lidding is usually a vinyl-coated aluminum foil, at least 0.001 in. (25.4 μm) thick for good barrier properties, which is sealed to a tray formed from semirigid PVC. The other popular pill package is the strip package. The strip package generally incorporates a peelable lidding having several plies. The outer layer is usually paper (to provide a good printing surface) which is then mounted to foil either by extrusion coating or adhesive lamination. The sealant side of the lidding can be a film or coating or a combination. The actual construction depends on the barrier requirements. If an extremely good barrier is necessary for very long shelf life, the structure can contain Aclar (Allied Corp.), which has exceptional barrier properties (see Film, fluoropolymer).

Other medical uses for liddings are safety seals on bottles to prevent tampering. These are combinations of materials that form fusion seals and are destroyed, ie, delaminated, when opened so that resealing is difficult. A safety seal is often fabricated with aluminum foil and mounted to a bottle-cap liner stock (see Closure liners) with wax. The combination of safety seal and cap liner is die-cut and placed in the bottle cap. The filled and capped bottle is passed through an induction sealer that fuses the safety seal to the bottle and melts the wax

adhesive layer so the two parts separate when the bottle is opened.

For medical devices (see Health-care packaging) that are to be ethylene oxide (ETO) sterilized, a popular packaging technique is to use a PVC tray with a Tyvek-coated lid. Tyvek (DuPont) medical-grade materials are porous to ETO gas, but not to bacteria. The sealant requires only limited heat resistance, but the web must be porous to allow the ETO to penetrate the Tyvek. Tyvek can be coated with a sealant in a pattern that does not change the porosity of the lidding. Another variation is to use a coating that is heat-sealable but not fused in drying so that it does not form a continuous film and therefore maintains porosity. Yet another option is to put the sealant material on the forming web so the Tyvek does not have to be coated.

Aluminum foil is normally used in a lidding if excellent barrier properties and protection from light are required. If the package is to be microwave-heated or requires transparency, foil is normally replaced with a PVDC-coated film which also gives very good barrier properties. Sealing methods are usually conductive-type heat seals. Important exceptions to this are safety seals in conjunction with a bottle-cap liner which use induction-type sealing equipment.

Further refinements of lidding technology can be expected as part of the current focus on semirigid replacements for metal cans (see Retortable flexible and semirigid packages).

BIBLIOGRAPHY

General References

C. J. Benning, *Plastic Films For Packaging,* Technomic Publishing Co., Inc., Lancaster, Pa., 1983.

R. C. Griffin, Jr., S. Sacharow, and A. L. Brody, *Principles of Package Development,* 2nd ed., AVI Publishing Co., Inc., Westport, Conn., 1985.

R. C. Griffin, Jr. and S. Sacharow, *Food Packaging,* AVI Publishing Co., Inc., Westport, Conn., 1981.

A. R. Endress, "Heat-Seal Coatings—Water and Solvent Based For Paper, Foil and Film," *Paper Synthetics Conference, Atlanta, Ga., Sept. 27–29, 1976,* TAPPI, Atlanta, Ga., 1976, pp. 55–57.

A. R. Endress, "Water Base and High Solids Coatings For Complying With E.P.A. Regulations," *Paper Synthetics Conference, Cincinnati, Ohio, Sept. 15–17, 1980,* TAPPI, Atlanta, Ga., 1980, pp. 205–207.

R. B. Schultz, "Heat Seal Coatings For Disposable Medical Device Packaging," *Paper Synthetics Conference, Atlanta, Ga., Sept. 13–15, 1982,* TAPPI, Atlanta, Ga., 1982, pp. 13–15.

F. R. Solenberger, "Health Care Packaging With Ionomers and EVA," *Paper Synthetics Conference, Atlanta, Ga., Sept. 13–15, 1982,* TAPPI, Atlanta, Ga., 1982, pp. 65–69.

R. Pilchik, "Lidding Materials for Formed-Filled-Sealed Packages of Medical Devices," *Paper Synthetics Conference, Lake Buena Vista, Fla., Sept. 26–28, 1983,* TAPPI, Atlanta, Ga., 1983, (1), pp. 221–224.

W. R. SIBBACH
Jefferson Smurfit Corporation

LITERATURE.   See Networks.

LITHOGRAPHY.   See Printing.

METALLIZING

Metallized products are somewhat ordinary materials such as paper or plastic films that are treated in a special way to enhance metal adhesion and then plated with a thin coating of metal (generally aluminum). The process and its products are viewed in several different ways, and there has been no consensus. For example,

1. The metal is a coating, like any other coating (eg, PVDC), applied to achieve specific properties. In this case, the coating material is aluminum.

2. The metal is a very thin foil (300 times thinner than the thinnest commercial foil) with appropriate properties, increased flex–crack resistance, and somewhat poorer unstressed barrier properties.

3. This is an entirely new technology producing a unique product.

In any case, aluminum, 30-nm thick, provides barrier properties for the package that are difficult to achieve in any other way. Moisture- and gas-barrier properties are improved, and the metal keeps out light that would cause oxidative rancidity of fatty and fried foods, such as snacks, nuts, and coffee. Since the consumer equates the "foil look" with freshness, a metal coating offers an added bonus in that respect.

The Process

In most ways, a metallizing machine is similar to a conventional coater (see Coating equipment) A roll of film or paper is passed over a spray head, the coating is applied, and the web is wound on a core (see Fig. 1). The roll of film or paper is loaded on an unwind stand (A), and threaded over rollers to a windup stand (C). In this case, the coater (F) that it passes on the way consists of evaporator pots or boats containing the melted metal, opposite a chill roll (B). The metal is replenished through the wire-feed system (E). The only unusual thing

about this is that the whole process takes place in a vacuum chamber. In the process, called distillation, the aluminum is melted, vaporized, and condensed on the substrate just as steam condenses on a bathroom mirror after a shower. Aluminum will not vaporize except in a high vacuum, in the neighborhood of 0.9 in. of mercury (3.1 kPa). Simple mechanical pumps cannot achieve sufficient vacuum, and diffusion pumps are needed to bring the vacuum to the levels needed. The amount of aluminum is measured by percent light transmission, electrical resistance (ohm/sq), and/or optical density, which has become the most popular specification method. Normal plating levels are 1.3–2.3 (see Table 1).

The amount of metal on the substrate is controlled by: temperature of the aluminum (hotter = more metal); running speed (slower = more metal); and number of plating stations (more = more metal).

In packaging, the need for controlled plating is generally related to barrier properties, but in some cases (eg, solar control or see-through electric-eye cutoff) it is required for functional properties. And of course, there are economic considerations. Thicker plating costs more, because more metal is required and also because the running speed is slow.

The plating level is monitored in several ways. Most metallizing machines have viewing ports. The film passes over a fluorescent tube, and the experienced operator can often assess the quality of the plating by eye, or at least the uniformity of the plate. Newer machines have TV cameras to observe the plating process and its results. The very latest have electric-eye monitors on each of the evaporator banks which monitor the plating level continuously. Some have automatic feedback that adjusts the temperature of the evaporator boats to make the deposit more uniform.

Applications and Properties

Metallizing for packaging is presently a $40–60 million dollar business in the United States (see Table 2). Most of the applications for metallized film have not involved replacement

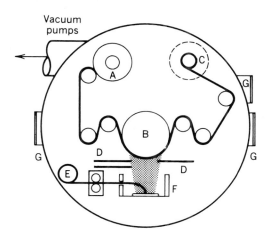

Figure 1. A metallizing machine. A, Payoff mandrel; B, cooled drum; C, take-up mandrel; D, aperature and shutter assembly; E, wire-feed system; F, source; and G, viewing ports.

Table 1. Conversion Table for Metallized Film (Aluminum)

Metal thickness		Macbeth optical density	Light transmission, %	Resistance (Ω/sq)
nm	10⁻⁶ in.			
0	0	0	100.00	0
3.05	0.12	1.00	10.00	6.70
5.08	0.20	1.10	7.943	5.35
6.35	0.25	1.20	6.310	4.55
7.11	0.28	1.30	5.012	3.98
8.13	0.32	1.40	3.981	3.50
8.4	0.33	1.50	3.162	3.32
9.65	0.38	1.60	2.512	3.06
10.2	0.40	1.70	1.995	2.86
10.7	0.42	1.80	1.585	2.68
11.4	0.45	1.90	1.259	2.50
12.2	0.48	2.00	1.000	2.35
12.7	0.50	2.10	0.7943	2.2
13.5	0.53	2.20	0.6310	2.05
14.2	0.56	2.30	0.5012	1.93
15.5	0.61	2.40	0.3981	1.80
17.3	0.68	2.50	0.3162	1.66
25.4	1.00	3.00	0.1000	1.12
36.3	1.43	3.50	0.0316	0.78

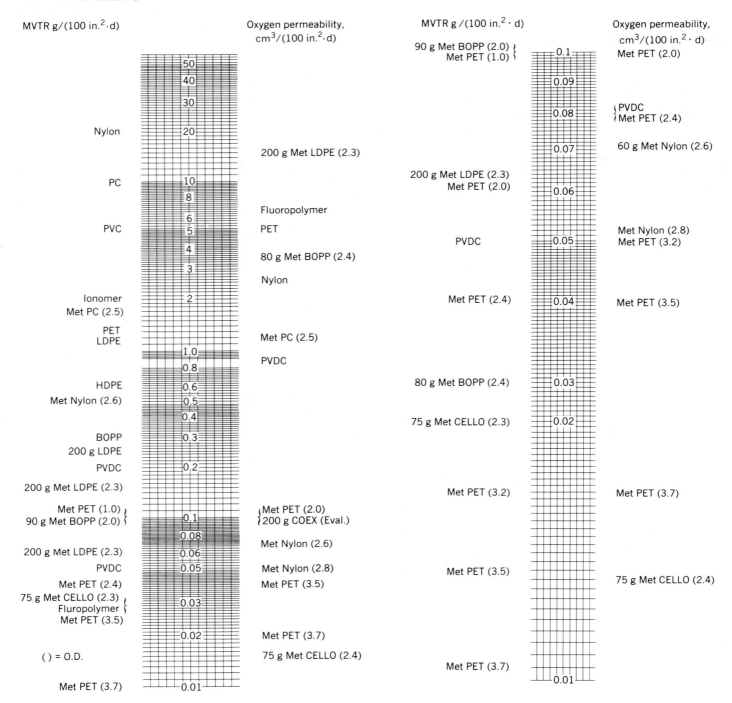

Figure 2. Barrier properties of unmetallized and metallized films. Unless otherwise stipulated, all materials (base substrates) are 0.001-in. (25.4-μm) thick. MET Substrates are 48 ga (0.00048 in. or 12.2 μm) unless noted otherwise. To convert g/(100 in.² · d) to g/(m² · d), multiply by 0.0645. To convert cm²/(100 in.² · d) to cm³/(m² · d), multiply by 0.0645. Courtesy of ICI Americas, Inc.

Figure 3. Barrier properties of metallized films. See Figure 2. To convert g/(100 in.² · d) to g/(m² · d), multiply by 0.0645. To convert cm²/(100 in.² · d) to cm³/(m² · d), multiply by 0.0645. Courtesy of ICI Americas, Inc.

Table 2. Sales of Metallized Packaging Films, million (10^6) lb (t)

Product	1980	1984	1985[a]	1986[b]
PET	2.5 (1134)	3.5 (1588)	6.0 (2722)	5.5 (2495)
OPP	2.8 (1270)	4.5 (2041)	6.0 (2722)	8.0 (3629)
ONF	1.4 (635)	1.2 (544)	0.8 (363)	1.5 (680)
PE	3.1 (1406)	2.3 (1043)	2.2 (998)	2.3 (1043)
PVC	0.1 (45)	0.1 (45)	0.1 (45)	0.1 (45)
other		0.1 (45)	0.3 (136)	0.6 (272)
Total	*9.9 (445)*	*11.7 (5307)*	*15.4 (6985)*	*18.0 (8165)*

[a] Estimated.
[b] Projected.

of aluminum foil, but rather the improvement of the barrier properties of clear products (see Films, plastic). Today, with refined technologies for the attainment of barrier properties, the potential for replacing foil is greater than it was before.

In packaging, particularly with films, companies strive to get the proper balance of properties: specific barriers, cost, aesthetic appeal, and machine performance. This is done in two ways; by selecting the best plate level and the appropriate base film. The plating can generally be regarded as an enhancement of the natural properties of the films. For instance, unmetallized biaxially oriented (BON) nylon film (see Nylon) has good oxygen resistance but poor barrier to moisture so it is not used in applications where products must remain either crisp or moist. Metallized BON is an excellent oxygen barrier. Biaxially oriented polypropylene film (BOPP) (see Film, oriented polypropylene) has good moisture-barrier properties, with or without metallizing, and it can be used to package crisp snack foods; but metallization can improve its moisture barrier by as much as 20 times.

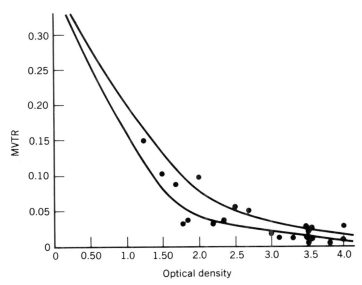

Figure 5. Metallized BOPP film (80 ga), optical density vs WVTR.

The second factor that base film contributes is smoothness of substrate surface. The smoother the surface, the higher the barrier properties. If the surface is micro-rough, the barrier properties achieved are somewhat lower than if the surface is micro-smooth. For that reason, brightness and barrier go hand-in-hand. The rule of thumb is: "the brighter the film, the better the barrier." Cast films generally have better barrier than blown films (see Extrusion; Films, plastic).

The third contributing factor is the level of plate. The more metal, the more the barrier; but the relationship is not linear (see Figs. 2–5). The greatest contribution to barrier is made by the first metal laid down. For the same reason, scratches are less important than one might expect, since the scratches rarely reach the base film. In general, the level of plate is less important than the base film properties, until "ULTRAPLATE" levels are reached. In this newest of developments, metal in excess of 3.0 OD is put on films. OPET and BOPP films plated to the "ULTRA" level have barriers equal to uncreased foil and vastly superior to creased foil. Problems of metal cohesion, adhesion of metal to film, and spitting (pinholes) have been overcome.

Typical substrates and applications are listed below:

Application	Substrate
Snack foods	BOPP, OPET, or PE
Coffee	BON or OPET
Candy	BOPP, OPET, or Cello
Hosiery	LDPE
Processed meat	OPET
Decorative material	OPET, PVC, or Cello

Nonheat-seal films are usually used on the outside of laminations (see Multilayer flexible packaging). When heat-seal films (eg, coextruded BOPP or heat-sealable OPET) are used as inner webs with reverse-printed outer webs, printing is locked in. Typical structures are shown in Figure 6.

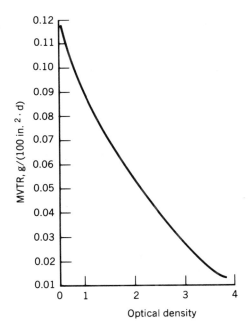

Figure 4. Metallized OPET film (48 ga), optical density vs WVTR. To convert g/(100 in.^{2} · d) to g/(m^2 · d), multiply by 0.0645.

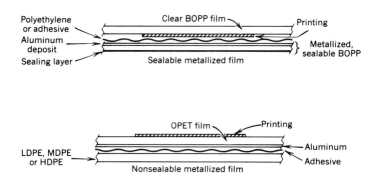

Figure 6. Typical structures containing metallized film.

Variations include stripe metallization, which creates clear areas between the metal, all in the machine direction. The demetallization process, which removes metal in the printing process, is a later development and more costly. It does offer more decorative aspects than the stripe. One of the newest packaging innovations has now spilled over into metallizing: the use of latex coating for "cold-seal" or "heat-seal" closures. A combination of the correct substrate and the proper overall construction provides aspects of packaging (ease of use, economies due to lower waste and higher speeds, easier opening qualities) not available from other laminations. Metallized films provide the barrier needed to compete with and replace foils.

Developments in film technology are reducing costs and expanding potential for metallized films. BOPP, available now as thin as 32 ga (0.32 mil or 8.1 μm), competes directly with foil. Coextrusions can be cost-effective alternatives to laminations, with barriers that are further enhanced by metallization. New films with ultra-smooth surfaces are yielding high barriers with conventional plate levels. Cost-reducing machinery developments are expected as well: wider machines (now being developed at 96 in. (244 cm) and high-speed units running up to 1200 ft/min (366 m/min). These developments will lower the cost of materials and make metallized films and papers more competitive.

PHILIP E. PRINCE
Inovpack

MILITARY PACKAGING

Military packaging has long been considered a separate and distinct entity, even though the materials, equipment, and processes in the military are the same as in commercial packaging. Even graphic packaging advertising considerations are important to military packaging when shipping to commissaries, PXs, and other similar government organizations that purchase brand-name products for resale. To understand why it has historically been considered a separate category in packaging technology, it is first important to understand what the term *military packaging* means and all that it encompasses.

There are hundreds of military documents that include definitions of the term *packaging*. One publication states that "The word packaging is used in the general sense to indicate all of the operations involved in preservation and packing. . ." (1). A more recent document provides a more detailed definition of packaging: "The processes and procedures used to protect material from deterioration and damage. It includes cleaning, drying, preserving, packing, marking, and unitization" (2). Government civil agencies, along with industry, also often reference commercial publications providing definitions of the term packaging (3,4). In addition, both military and civil agencies frequently use the term "Preparation for Delivery" to describe the preservation, packing, and marking requirements applicable to a particular item or group of items.

Because of the number of different government definitions and terms, one must take a comprehensive approach toward packaging to avoid problems when interpreting military- or civil-agency specifications or contract requirements. As a general rule, it is best to consider government- or military-packaging requirements as encompassing the complete protection of the item from the environment and the environment from the item, from the point the manufacturing process begins to the point where the item is placed into end use.

U.S. Military-Packaging Requirements

Military packaging derives its uniqueness from the special nature of the overall government-supply system. The shipping and storage conditions encountered in the supply pipeline, the large number of items purchased by the government, and the very nature of the government as an organizational entity all contribute significantly to the military-packaging mystique.

Although much of the government's supply system is similar to a commercial supply operation, there are also unusual supply conditions caused by the worldwide presence and the often highly complex requirements of many agencies. The U.S. Government must satisfy its essential supply needs, anywhere in the world, at any point in time, and in the face of potentially severe environmental conditions of all kinds. The material must arrive in a usable condition and there is often no "acceptable" level of damage. Since the government must, as a result, engineer requirements in anticipation of this worst-case supply situation, many government-packaging requirements automatically exceed those of a normal commercial operation. In many cases, this makes the government requirement appear to be a case of overdesign, but commercial shipments would be packed the same way if similar shipping conditions had to be anticipated.

The number of items purchased (the government buys almost every known commodity) also contributes to the confusion and apprehension over military packaging. Since so many items are purchased, standardization is employed wherever possible. The inevitable failure to consider individual-item characteristics and specific shipping environments on a case-by-case basis results in different interpretations and frequent differences of opinion. Adding to this problem are the increasingly complex and expensive weapon systems and high tech items being acquired. This equipment must perform with an unusually high degree of reliability during end use, placing great demands on the packaging systems used for protection. These characteristics of the government-supply system lead to the specification of unusually high cost, difficult-to-meet packaging requirements, erring on the side of safety and reliability rather than minimizing costs at the risk of the item. In most commercial operations, an acceptable level of risk is routinely considered in the development of the packaging requirements. Frequently, there is no acceptable level of risk when developing requirements for military packaging.

Lastly, the government is a large bureaucracy, as are many large corporations. Most businessmen recognize there is a price to pay in dealing with large corporations compared with dealing with small businesses, and this applies to the government as well. There is less flexibility, more red tape, and a lengthier and seemingly overly complex process. This is particularly true of government packaging. There are costs incurred in military packaging that far exceed those associated with typical commercial operations. There are even problems in interpreting the details of the government's packaging requirements because of the way the requirements are described in government contracts and specifications.

All of these features have caused military packaging to evolve into a separate packaging category. Most companies try to understand and cope with the idiosyncrasies of the military's packaging system, since most simply cannot afford to overlook the possibility of doing business with a customer as large as the government.

Types of Requirements

Military-packaging requirements are presented in many different formats. There is an order of precedence that must be known, ie, which requirements take precedence over others, and additional information is needed in order to interpret the applicable requirements. This is especially critical in government contracting, since it is the contractor who is responsible for ensuring compliance with the specifications and standards cited in the contract. Unless specifically instructed otherwise, the contractor is also responsible for ensuring that the latest edition of any referenced specification or standard is followed. The government publishes two indexes to help in this effort, one listing federal specifications and standards, and one covering both federal and military specifications and standards (5,6).

When preparing items for delivery under a government-supply contract, contractors must always meet certain minimum requirements. This minimum degree of protection is often called commercial packaging, standard industry practice, normal commercial practice, etc. These terms mean the same thing: a contractor must provide adequate preservation and packaging to prevent damage to the item under normal transportation and storage conditions and comply with the rules and regulations of the applicable transportation carrier (7,8).

The military extensively uses coded packaging data to describe packaging requirements above this minimum level. One recent Navy solicitation, issued to establish a supply contract for circuit-card assemblies, typically included the following information to describe the packaging requirements applicable under the proposed contract:

MIL-STD-726
3P/1/1/00/AA/AA/Y/XX/5/D3/A/AG/AB TP 3

Obviously, without a copy of the standard (9), there is no way to know the type and amount of packaging the contractor must provide for each circuit board. Without this knowledge, it is impossible to bid intelligently or perform the contract responsibly. The government solicitation or specification always references any base document that must be used to interpret the various packaging codes included therein. Contractors must obtain the specific document and edition referenced by the government to interpret packaging requirements accurately.

Although the circuit-board solicitation referenced MIL-STD-726 (9) as the applicable document for interpreting the packaging codes listed, Department of Defense (DOD) packaging personnel now expect MIL-STD-2073-1A (2) and MIL-STD-2073-2A (10) to become the principal government standards for coding military-packaging information. The classifications and codes provided in these standards are not only essential to interpreting specific packaging requirements contained in a solicitation or contract, but also necessary when a contract requires the contractor to develop packaging requirements for the item being supplied. Under such government contracts, packaging data describing those requirements must be provided by the contractor in the proper coded format.

In the example cited above, the code reveals that the basic level of protection established for the proposed contract is Level A. Specifically, the government contractor is required to provide water-vapor-proof protection for each circuit-board assembly by providing preservation protection in accordance with submethod IA-15 of military specification MIL-P-116 (11). This submethod requires that the item, preserved, wrapped, and cushioned as required, be enclosed in a snug-fitting carton or box, as applicable. The box must then be enclosed in a sealed bag conforming to MIL-B-117 (12). Any applicable process of MIL-P-116 (11) can be used to clean the item. No preservative material application is needed. The general requirements of MIL-P-116 (11) apply to wrapping and cushioning the item. The cushioning thickness is left to the discretion of the contractor, as long as other requirements are met. The contractor must then place five of the unit packages into an intermediate container, which must be a folding, metal-stayed, setup, or fiberboard box conforming to Federal Specifications PPP-B-566 (13), PPP-B-665 (14), PPP-B-676 (15), or PPP-B-636 (16), as appropriate. The maximum allowable weight of the unit pack is 0.23 lb (0.1 kg). The maximum cube is 5 in.^{3} (82 cm^3).

All the foregoing information can be obtained directly from translation of the codes provided in the solicitation. Final details of the requirements are in the applicable specifications or other referenced documents.

Government-Packaging Levels

The above example illustrates the need for a thorough understanding of the government's "Levels of Packaging." This is critical since the government's estimated level of protection determines the extent of the packaging requirements specified. "Levels" actually refer to a set of performance criteria specifying the minimum performance required of the packaging under certain anticipated shipping and storage conditions (17). They are used to describe degrees of protection relating to preservation, unit packaging, and packing and are sometimes used to differentiate between varying degrees of marking and unitization requirements.

Level A protection. Level A is the maximum degree of protection provided under a government contract. It is the degree of preservation, packaging, and packing required to protect materiel against the most severe conditions known or anticipated by the government to be encountered during shipment, handling, and storage. Since the government often does not know the conditions to which items will be exposed or even the ultimate destination, Level A protection often means providing adequate protection under the most extreme conditions imaginable, which may include worldwide distribution, long-term storage, and multiple reshipments via all transportation modes (see Distribution hazards; Export packaging). Protection designated as Level A is specifically designed to protect shipments against direct exposure to extremes of climate and terrain and severe operational and transportation environments, without any protection other than that provided by the pack. This is by far the most complex and costly level of protection required by the government. Water-vapor-proof barriers, wood containers (see Boxes, wood), humidity indicators (see Indicating devices), and cushioning (see Cushioning, design) are commonly required when Level A protection is specified.

Level B protection. Level B is the degree of protection required under known favorable conditions during shipment, handling, and storage. The key difference between Level B and Level A is the word *known*. Level B is specified when the government can anticipate, with reasonable certainty, the relative distribution conditions to which the pack will be exposed. These known conditions are usually less severe than the worst-case conditions which must be considered under Level A.

Level C protection. This level of protection is defined as that degree of protection required for conditions known to be less severe than Level B. It is designed to protect materiel under known favorable conditions of shipment, limited handling, and storage. The protection provided must as a minimum meet referenced applicable public laws, codes of federal regulations, or a specific military or federal specification, standard, or instruction.

Commercial or industrial packaging. Civil agencies refer to this level as commercial packaging, meaning the degree of protection employed by commercial firms to satisfy the packaging requirements of the commercial retail distribution system. Commercial packaging must protect the item against physical and environmental damage during shipment, handling, and storage (18). The word *retail* is used to distinguish between wholesale or bulk practices and the government's commercial packaging. The military, preferring to accept no packaging below Level C, recently defined this level of packaging as "Industrial Packaging" and stated that it is acceptable for any level of protection "whenever the technical design details of the package meet all of the conditions of the level of protection specified" (2). "Industrial Packaging" must offer the same protection against physical and environmental damage as the military package. Bulk-type practices such as those used for inter- and intraplant shipments or shipments to jobbers are not acceptable by the military unless they are the usual trade practices for selected commodities such as petroleum or coal. A close reading shows that the military's definition of industrial packaging is not synonymous with commercial packaging, and that the military has no formal level below Level C. Rather, contractors are simply permitted to meet the designated level in a way other than that specifically described by the government.

Examples of Military Requirements

The impact and meaning of the government's levels of packaging is best understood by reviewing a few examples of how the government describes detailed packaging requirements for specific commodities. If more specific data are required than those provided by the codes and the basic interpretive documents, the codes or the solicitation will make reference to detailed specifications, standards, drawings, or other documents applicable to the contract or commodity.

Mounted magnetic compasses. Military specification MIL-C-38214B (19) covering mounted magnetic compasses specifies that these compasses shall be packaged as follows, depending upon the level specified by the procuring activity:

Preservation and Packaging:

Level A. Compasses shall be preserved in accordance with MIL-P-116, Method IA-8 and packaged one each in a container conforming to PPP-B-636, PPP-B-665, or PPP-B-676. (There is no Level B requirement.)

Level C. Compasses shall be provided sufficient protection against physical and mechanical damage to assure safe delivery without degradation of equipment reliability from the supply source to the first receiving activity for immediate use.

Packing:

Level A. Compasses shall be packaged . . . and shall be packed for shipment in exterior-type shipping containers conforming to PPP-B-636, water-resistant type.

Level B. Compasses shall be packaged and shall be packed for shipment in exterior-type shipping containers conforming to PPP-B-636.

Level C. Packages that require overpacking for acceptance by the carrier shall be packed in exterior shipping containers in a manner that will insure safe transportation at the lowest rate to the point of delivery. Containers shall meet the consolidated freight classification rules or regulations of other common carriers as applicable to the mode of transportation.

These requirements illustrate the fact that government-packaging requirements become less specific as the level is lowered from Level A, always the most stringent requirements, to Level C and below.

Pitot static tubes. Pitot static tubes conforming to MIL-T-5421B (20) are avionic instruments used on aircraft to provide an accurate source of impact pressure and are used with airspeed indicators under extreme environmental conditions. The specification requires that they be packaged as follows:

Packaging: Preservation and packaging shall be Level A or C, as specified.

Level A.

Cleaning. Pitot static tubes shall be cleaned in accordance with C-1 of MIL-P-116.

Drying. Pitot static tubes shall be dried in accordance with process D-4 of MIL-P-116.

Preservation application. Not applicable.

Unit package. Unless otherwise specified by the procuring activity, each pitot static tube shall be packaged in quantity unit packs of one each in accordance with Method IC-1 of MIL-P-116. Each pitot static tube shall be overboxed in a PPP-B-636 carton. The carton shall be large enough for the application of sufficient cushioning material between the container and bag, of a type, density and thickness, to insure shock transmission does not exceed peak values in G's established for the item when completed packs are subjected to the rough handling drop tests of MIL-P-116.

Level C. Pitot static tubes shall be clean, dry, and individually packaged in a manner that affords adequate protection against corrosion, deterioration, and physical damage during shipment from supply source to first receiving activity.

Packing:

Level A. Pitot static tubes packaged as specified . . . shall be packed in shipping containers conforming to PPP-B-601, Style A or B, Class overseas, unless otherwise specified by the procuring activity. Insofar as practical, exterior shipping containers shall be of uniform shape, size, and of minimum tare weight and cube, consistent with the protection required.

Level B. Pitot static tubes packaged as specified . . . shall be packed in weather-resistant shipping containers conforming to PPP-B-636. Other requirements as specified for Level A above are applicable.

Level C. Packing shall be applied which affords adequate protection during shipment from the supply source to the first domestic receiving activity for immediate use. This level shall conform to applicable carrier rules and regulations and may be the contractor's commercial practice provided the latter meets the requirements of this level.

The PPP-B-601 (21) container, Level A packing, is a cleated-plywood, wood box. The PPP-B-636 (16) container, Level B packing, is a fiberboard box. This typifies another characteristic of military-packaging levels. The higher the level, the more costly the materials and procedures involved. A wood box costs more to buy or build and is more difficult to work with than a fiberboard shipping container. Level C does not reference a specific container, but leaves the choice to the contractor, making this level the easiest and cheapest with which to comply.

Floor mops. Floor mops conforming to Federal Specification T-M-580 (22) must be packaged as follows:

Packaging: Packaging shall be Level A or C, as specified.

Level A. Mops shall be bundled in units of 10, with 5 swab heads at each end of the bundle. The swab heads shall be tightly bound with not less than four turns of marline.

Level C. Mops shall be packaged in accordance with the manufacturer's standard commercial practice.

Packing:

Level A. Mops packaged as specified . . . shall have the swab ends wrapped in not less than 60 pound kraft paper conforming to UU-P-268. The wrap shall be secured in place with either twine or gummed tape. Four wrapped packages shall be overwrapped with osnaburg or burlap cloth conforming to CCC-C-429 or CCC-C-467 respectively, and securely tied with twine.

Level B. Mops, packaged as specified . . . shall have the mop heads wrapped with not less than 60 pound kraft paper conforming to UU-P-268 and secured with 3-inch wide gummed kraft tape. Two bundles shall be securely tied with twine.

Level C. Mops, packaged as specified . . . shall be packed in a manner to insure carrier acceptance and safe delivery at destination. Containers shall be in accordance with the rules or regulations of carriers applicable to the mode of transportation.

This example illustrates that the packaging requirements in federal specifications are presented in the same way as in military specifications. Level A is, again, much more specific than for the lower levels. The level of packaging indicates the potential cost and difficulty of complying with the government's requirements and is also the key to determining the specifics of the stated requirements.

The Elements of Government Packaging

Given the system's definition of packaging, government-packaging requirements consist of seven essential elements (23), each of which must be considered when evaluating or developing government-packaging requirements. Although a particular element may not apply to a specific item or contract, evaluating the requirements by considering each element minimizes overlooking or misunderstanding the government's needs relative to that item or contract. The seven elements follow.

Preliminary item preparation. The four functions covered by this element of government packaging are cleaning, drying, item marking, and preserving items. As a minimum, government contractors must provide preliminary item preparation to the extent necessary to ensure that the item is functional and undamaged when needed. The government currently lists 14 cleaning processes in MIL-P-116 (11), which may be specified in the contract or chosen by a contractor. MIL-P-116 (11) also provides specific tests to verify that items are cleaned properly and five authorized drying procedures to remove any cleaning solution or residual moisture. Although the "US" property marking is sometimes required, it is only one of many that may be required on each item. Item-marking requirements identify the specific item and manufacturer, not just the owner (24). After items are thoroughly dried, appropriate preservatives are applied as a barrier against moisture, air, or other agents of corrosion (11), or volatile corrosion inhibitors (VCIs) are used to deposit an invisible protective film on the items (25).

Unit packaging. This element consists of what most people think of as packaging. It involves the actual placement of the item into the unit package, including any cushioning or blocking and bracing that may be required. The unit pack, or unit package, is the first tie, wrap, or container applied to a single item or group of identical items, which provides a complete and individually identifiable pack, or package (1). The military defines various unit-packaging techniques as standard "Methods of Preservation" (11), which are really descriptions of various procedures, or combinations of procedures, combining unit-packaging materials and containers to provide different degrees and types of item protection. The methods not only include the unit-packaging requirements, but all the preparation-for-delivery requirements up to, but not including, placement of the item into the shipping container and beyond. A military method of preservation is a specific combination of cleaning, drying, preservative application, and unit-packaging requirements, which together protect an item from an anticipated level of deterioration or damage. The item's characteristics, susceptibility to various contaminants, potential hazards, and the anticipated shipping environment all dictate the specific method or submethod to be used.

Packing. This element covers the placement of the item or unit package into an intermediate pack or shipping container. If the unit container serves as an intermediate or shipping container, placement of the item into the unit package may also be considered a packing function (26).

Unitizing and palletizing. This consists of the procedures to follow when several shipping containers are consolidated for storage, handling, or transportation purposes (27).

Marking and labeling. Unit packs, shipping containers, and unitized loads must be legibly marked for identification and transportation purposes. The government's marking requirements often far exceed those normally encountered in industry, making this element potentially the most unique of all the packaging requirements placed upon the contractor. Despite the significance of this potential cost, the additional expenses associated with complying with the requirements of this element are frequently overlooked. In military packaging, the primary specification for marking is MIL-STD-129 (28). In civil agencies, it is Federal Standard No. 123 (29). Bar-code marking, increasingly required in military packaging, is covered by MIL-STD-1189 (30).

Special-situation packaging requirements. For any item, contractors must determine whether special requirements are applicable. These may apply owing to the nature of the item, its destination, or the transportation mode to be used. This is an extremely important element to consider. Not only are special protection and shipping requirements applicable to the item, but potentially hazardous situations to personnel or the environment are also included. This category includes special programs such as hazardous-material packaging (31); the DOD Container Design Retrieval System, an automated system to help design specialized containers (32); static-free packaging materials (33); multiapplication containers (2); and package-design testing (2).

Packaging quality assurance. When packaging military shipments or designing military-packaging requirements, contractors must ensure that required procedures are followed. This category consists of the who, what, and when of controlling the quality of a contractor's packaging operations in accordance with government quality-assurance requirements (34,35).

The Future of Military Packaging

Unfortunately, military packaging appears to be in for a rather schizophrenic future. Two vastly differing forces are impacting military-packaging requirements. First, the Federal Acquisition Regulations (36) make it clear that commercial packaging and distribution resources should be used whenever possible. Although this requirement alone would not force the military to adopt this as normal policy, when it is coupled with the ever-increasing pressure by Congress and industry to reduce waste and lower procurement costs, it represents a significant driving force pushing the military to accept more true commercial packaging.

The other principal force facing military packaging is the continuing need to increase the automation and reliability of its supply system, increasing the standardization of packaging units to the point where the system may control the packaging requirements more than the item itself. Improvement is necessary owing to decreasing civil-service employment levels in the military-supply area; the need to improve the military's readiness posture as evidenced by the emergence of rapid-deployment units; and the increasingly technological nature of the equipment and supplies used at all organizational levels within the military.

In the meantime, one substantial change has already occurred. The latest high-tech advances in the military have significantly affected related packaging requirements, forcing the government to develop intricate materials and processes to accommodate state-of-the-art equipment and distribution techniques. Such advances may lead to military-packaging materials, equipment, and processes truly becoming unique in the field of packaging. As it stands today, military packaging has become one of the more interesting and dynamic of the packaging-technology categories.

BIBLIOGRAPHY

The government specifications and standards listed below can be obtained from the U.S. Naval Publication and Forms Center, 5801 Tabor Ave., Philadelphia, Pa., 19120.

1. *DSAM 4145.2, Defense Supply Agency Manual, Packaging of Materiel, Preservation* (Vol. 1), Sept. 1976 (ie, *TM 38-230-1, NAVSUP PUB 502, AFP 1–15, MCO P4030.31B*).

2. *MIL-STD-2073-1A, DOD Materiel Procedures For Development And Application Of Packaging Requirements*, July 16, 1984.

3. *ASTM D 996-83a, Definitions of Terms Relating To Packaging and Distribution Environments*, American Society For Testing and Materials, Philadelphia, Pa., Oct. 26, 1983.

4. *ANSI MH 15.1-79, Glossary of Packaging Terms*, American National Standards Institute, New York, 1979.

5. *FPMR 101-29.1, The Index of Federal Specifications and Standards*, Apr. 1983.

6. *Department of Defense Index Of Specifications and Standards (DODISS)*, DOD, Washington, D.C., July 1, 1983.

7. *Uniform Freight Classifications*, Western Railroad Association, Chicago, Ill.

8. *National Motor Freight Classification*, American Trucking Associations, Inc., Washington, D.C.

9. *MIL-STD-726G, Packaging Requirement Codes*, Mar. 5, 1984.

10. *MIL-STD-2073-2A, Packaging Requirement Codes*, July 16, 1984.

11. *MIL-P-116H, Methods of Preservation*, Dec. 1, 1980.

12. *MIL-B-117E, Bags, Sleeves, and Tubing—Interior Packaging*, July 15, 1975.

13. *PPP-B-566, Boxes, Folding Paperboard*, Nov. 28, 1980.

14. *PPP-B-665D, Boxes, Paperboard, Metal Edged and Components*, Nov. 1, 1972.

15. *PPP-B-676D, Boxes, Setup*, Aug. 10, 1977.

16. *PPP-B-636J, Boxes, Shipping, Fiberboard*, Jan. 20, 1982.

17. *Federal Standard 102B, Preservation, Packaging and Packing Levels*, Jan. 29, 1983.

18. *MIL-STD-1188, Commercial Packaging of Supplies and Equipment*, cancelled Jan. 31, 1984 and replaced by *ASTM D 3951-82*, The American Society for Testing and Materials, Philadelphia, Pa., 1982.

19. *MIL-C-38214B, Compass, Magnetic, Mounted*, Dec. 1, 1980.

20. *MIL-T-5421B, Tubes, Pitot, Electrically Heated, Aircraft*, June 19, 1985.

21. *PPP-B-601G, Boxes, Wood, Cleated-Plywood*, Mar. 29, 1985.

22. *T-M-580, Mop, Wet, Cotton*, July 28, 1965.

23. D. K. Eary, *The Commercial Guide To Government Packaging—Contractual Obligations and Technical Requirements*, Global Engineering Documents, Santa Ana, Calif., 1985, pp. 3–15.

24. *MIL-STD-130F, Identification Marking Of US Military Property*, July 2, 1984.

25. *MIL-I-8574E, Utilization of Volatile Corrosion Inhibitors*, Apr. 2, 1985.

26. *DLAM 4145.2, Defense Logistics Agency Manual, Packaging of Materiel, Packing (Vol. II)*, June 1977 (ie, *TM 38-230-2, NAVSUP PUB 503*, Vol. II, AFP 71–176, MCO P4030.21C).

27. *MIL-STD-147C, Palletized Unit Loads*, Sept. 30, 1981.

28. *MIL-STD-129J, Marking For Shipment and Storage*, Sept. 25, 1984.

29. *Federal Standard 123D, Marking For Shipment (Civil Agencies)*, July 20, 1982.

30. *MIL-STD-1189A, Standard Department of Defense Bar Code Symbology*, Sept. 4, 1984.

31. *AFR 71-4, Preparation of Hazardous Material For Air Shipment*, Jan. 25, 1982 (ie, *TM 38-250, NAVSUP PUB 505, MCO P4030.19, DLAM 4145.3*)

32. *MIL-STD-1510A, Procedures For Use of the Container Design Retrieval System*, June 15, 1978.

33. *DOD-STD-1186, Electrostatic Discharge Control Program For Protection of Electrical and Electronic Parts, Assemblies and Equipment*, May 2, 1980.

34. *MIL-I-45208A, Inspection System Requirements*, July 24, 1981.

35. *MIL-Q-9858, Quality Program Requirements*, Aug. 7, 1981.

36. Federal Acquisition Regulations Title *48 Code of Federal Regulations*, U.S. Government Printing Office, Washington, D.C., Apr. 1, 1984, Chapt. 1.

D. K. Eary
Government Contract Services Company

MULTILAYER FLEXIBLE PACKAGING

Converters of flexible packaging materials can combine the properties of many different substrates into composite structures. Years ago, potato chips were sold in small paper bags, but they had to be sold quickly, before moisture made them soft, oxygen made them rancid, and the oil in the product penetrated the paper. If the chips were made in the back of the store, they had to be sold in the front of the store because a paper bag could not provide physical protection in transport. Waxed paper, one of the first multilayer structures, imparted some moisture protection, and glassine imparted oil protection, but improvements attainable with any one substrate are very limited. Today, potato chips and innumerable other products can be shipped long distances and stored for months because the properties of many materials can be combined into multilayer structures that provide all of the key characteristics of flexible packaging to the right degree: protection from dust and dirt; moisture barrier; gas barrier; protection against flavor and odor loss or gain; transparency or opacity; protection against ultraviolet light; flexibility or stiffness; sealability; machineability; aesthetics; and durability.

Sealability

Flexible packages are sealed by the packager on the packaging line as the last step in the production chain. The subject warrants some preliminary treatment here, however, because the method of closure must be known at the earliest stage of package design. It influences the nature and placement of some of the components of the structure and the production sequence. Mechanical wrapping machines (see Wrapping machinery) were introduced around the turn of the century for paper wraps. Seals were produced with glue applied by the packager, or with wax that was applied by the converter and heat-sealed by the packager. Other heat-seal coatings were developed later for other materials. The term *heat-seal coating* has traditionally pertained to coatings applied from solvent solutions or water emulsions by the converter (see Coating equipment) and heat-sealed by the packager. There is a clear distinction between the adhesives used to bond the webs of a lamination together (see Laminating) and the heat-seal coatings that are used later to seal the package (see Sealing, heat).

With the advent of extrusion coating and extrusion lamination, the distinction is blurred because the same material can be used as an adhesive and as a coating. Low density polyethylene, for example, can be used as a laminating extrudate, an extruded heat-seal coating, and a structural layer. The development of effective heat seals led to the development of a variety of packaging concepts and configurations.

The heat-seal coating is positioned on the inside of the composite if the structure is to be used to form a pouch. The heat-seal coating is positioned on the inside and outside of the composite if the structure is to be used to produce lap seals on horizontal or vertical FFS packaging machinery (see Form–fill–seal, horizontal; Form–fill–seal, vertical). The heat-seal coating or heat-sealable component of a laminated structure must be positioned so that it will be brought in intimate contact with itself or another sealable component on the packaging machine. In most overwrapping operations, the back seal is overlapped and the end seals are folded, so both sides of the material must be sealable. There are HFFS and VFFS machines that form either a lap or fin seal for the back seam. In these cases, the end seals are always fin seals. Where lap-seal equipment is used, a heat-sealable inner component seals to a heat-sealable outer component. The outer component can be an overall coating, or it can be a strip of material applied only in the area of overlap. If the inner sealant is PVDC (see Vinylidene chloride copolymers), the outer coating can be PVDC, acrylic (see Acrylics), or a copolymer heat-seal strip that usually contains waxes (see Waxes). If the equipment forms a fin-back seal, no additional coating is required.

Converting Equipment

The various types of converting equipment are discussed under separate headings. One of the major converting operations is printing (see Inks; Printing). Coating (see Coating equipment) and laminating (see Laminating) are others. In a converting plant, all of these steps can be combined to accomplish several steps in one pass. By adding a laminator to a flexopress, for example, one side of a film can be printed and adhesive (see Adhesives) can be applied. A second web can be introduced and adhesively bonded to the first at that point. The result is a two-layer lamination with printed film on one side and another film on the other. The relatively thin adhesive layer serves no structural function. Many variations are possible, and the structure may be complete at this point.

If the structure is not complete at that stage, the roll of laminated material (see Roll handling) may be moved to a line that includes one or more extruders (see Extrusion). On this line, the lamination is fed under an extruder that deposits melted thermoplastic material (generally polyethylene or ethylene copolymer) on one side of the web. If no second web is fed in at that point, the thermoplastic material is a coating, and the process is extrusion coating (see Extrusion coating). If a second web is fed in at that point, the thermoplastic material acts as a bonding layer, and the process is extrusion lamination (see Laminating). In this case, the relatively heavy bonding layer serves a structural function as well.

Some converting operations are performed by the web suppliers. For example, converters can purchase PVDC-coated balanced oriented polypropylene film (BOPP) or nitrocellulose- or PVDC-coated cellophane. Film is generally purchased from a merchant film supplier, but some converters produce their own film. The increasing use of coextruded films (see Coextrusion, flat; Coextrusion, tubular; Coextrusions for flexible packaging) has encouraged integration because coextrusions enhance the productivity and versatility of the converting operation.

Because there are so many ways to combine layers, it is difficult to describe all of the variables in one system of notation. The system used in this article is an attempt to provide maximum clarity in minimum space. It is not a reflection of widespread industry practice, but it incorporates some of the concepts proposed recently by DuPont Company. The abbreviations and symbols used are shown below. Structures are read left to right from the outside to the inside of the package.

Abbreviation or symbol	Converting materials and operations
ADH	solvent or emulsion adhesive
AF	aluminum foil
AM	aluminum metallization
BK	bleached kraft
BON	biaxially oriented nylon
BOPP	balanced oriented polypropylene film
CELLO	cellophane
EAA	ethylene acrylic acid
EMA	ethylene methacrylate
EMAA	ethylene methacrylic acid
EVA	ethylene–vinyl acetate
gauge	cellophane, in.^{2}/lb × 100 [eg, 250 ga = 25,000 in.^{2} (16 m^2)]
gauge	films, foil 100 ga = 0.001 in. = 1 mil (25 μm)
HFFS	horizontal form–fill–seal
MG	machine glazed
NK	natural kraft
OL	overlacquer
ON	oriented nylon
PET	poly(ethylene terephthalate)
PR	printing
VFFS	vertical form–fill–seal
/	lamination
(---)	coextrusion
-	inked, metallized, or other coated surface
#	pounds/ream = lb/3000 ft^2 (454 g/278.7 m^2)

Substrates

Paper. The papers used in multilayer flexible packaging include light 8# tissues to 70# natural krafts; groundwood to highly refined glassine (see Glassine, greaseproof, parchment); and 26–50# clay-coated papers (see Paper). The paper industry has done an excellent job of developing special "laminating-grade" papers for the converting industry. The laminations of many of the tissues (lightweight papers of any kind) are described below in the section on aluminum foil. Additional uses include waxed twisting-grade tissue laminated to strips of aluminum foil for individual candy packaging.

The MG bleached kraft paper used for converting is generally a mixture of sulfate and sulfite pulps blended to combine physical strength with surface smoothness for line printing (see Printing). The blend can be tailored to physical strength requirements. Tissues and bleached grades can be colored by vat dyeing or the application of a colored coating (see Colorants). Special radiation-free vat-dyed tissue is used as a component of a tissue/PE/AF/PE construction for photographic film wrap. Fluorocarbons can be incorporated to prevent staining by oils and greases, eg, shortening in flour mixes and baked products.

The basis weight of the paper is chosen for functionality, which includes packaging-machinery performance and economics. For example, individual sugar or artificial sweetener pouches are 16–18# tissue, extrusion coated with a very thin layer of polyethylene; cake-mix liners are 25–30# MG bleached kraft with a 1–1.5 mil (25–38 μm) layer of polyethylene.

MG bleached kraft paper with a holdout coating is used for paper–PVDC/PE structures in which the basis weight of the paper, the amount of PVDC, and the thickness of polyethylene can be varied to achieve the characteristics needed in the final packaging material. In order to obtain the desired barrier properties, the PVDC is applied in two or more applications. The heavier the PVDC, the greater the barrier to moisture vapor and oxygen.

Supercalendered pouch paper is often used in laminations. It is made from highly refined pulp that often contains plasticizers for property enhancement. This paper provides a printing surface somewhat better than MG bleached kraft, and its flexibility is useful in laminations subjected to flexing, eg, over

Table 1. Typical Paper Structures

Structure	Application
16–18# tissue/10–15# LDPE	individual sugar and sweetener packets
25–30# MGBK/LDPE	cake-mix liners
wax–glassine/wax/glassine–wax	cereal carton liners, for moisture protection
wax–glassine–wax	cereal carton liners, for less moisture protection
transparent film–PR/ADH/ glassine–PVDC	snack foods
BOPP/ADH/AF/PE/glassine–PVDC	large snack packages

the forming collar of VFFS equipment. Glassine, widely used in laminations for flexible packaging, is produced from a very highly refined pulp that is supercalendered to provide a very dense paper that has excellent grease resistance and very low porosity. Examples of structures are shown in Table 1.

Clay-coated papers, which provide the ultimate printing surface, are used where multicolor printing is required for high-quality pictorials. These papers can then be laminated to a wide number of different films or foils to produce the finished packaging material. Through reverse printing (printing on the side of plastic films that will be buried in a lamination) high-quality graphics can be achieved with less expensive paper. For example, a reverse-printed BOPP or PET film, extrusion laminated to a 30# MG bleached kraft, provides high gloss, excellent scuff resistance, and excellent graphics. The film also protects the paper surface from the rigors of distribution. The applications mentioned above are but a few of the opportunities to use paper as a component of multilayer flexible packaging.

Aluminum foil. The primary reasons for using aluminum foil (see Foil, aluminum) in laminations are barrier properties and aesthetics. In many laminations where foil is chosen for its protective properties, the aesthetic advantages can also be attained by graphic design. Aluminum foil for multilayer flexible packaging is available in thickness from 0.25–5.9 mil (6.4–150 μm). Most laminating foils are 25–150 gauge, but some special laminations for pharmaceutical packaging (see Pharmaceutical packaging) incorporate 150 and 200 gauge foils, and some formed container applications use 300–400 gauge foils.

Aluminum foil used in multilayer flexible packaging has been annealed under special conditions to the maximum degree to a condition called "dead soft." The annealing process also removes the residual rolling oils from the surface and produces a "0"-wettable surface that allows excellent coverage by solvent or emulsion adhesives. Thin-gauge foil cannot function alone in a package, but it can be laminated with most adhesive systems. Foil is readily combined with any of the other materials to form composites that meet specific end-use requirements. Some examples are shown in Table 2.

Cellophane. Cellophane was the first transparent flexible material (see Cellophane; Films, plastic). For some products and package sizes a single ply of cellophane was not sufficient, so two layers were used, initially as two separate webs placed

on the packaging machinery. Two webs of heat-seal-coated cellophane could be combined at the seal areas for package integrity. The duplex cellophane package led to the lamination of the two sheets of cellophane, and ultimately to other cellophane laminations. Uncoated cellophane is very moisture sensitive, coming to moisture equilibrium with the atmosphere. If it picks up moisture, it becomes somewhat soft; if it loses moisture, it becomes brittle. The uncoated film does not provide good consistent barrier properties, nor is it heat-sealable. Cellophane used in multilayer flexible packaging is coated to provide these features, generally with PVDC on one or both sides. Nitrocellulose-coated cellophanes are seldom used.

The thickness of cellophane is expressed in terms of yield. The thinnest gauge generally used in laminations is 250 ga (25,000 in²/lb or 355.6 cm²/g), which is about 0.8 mil (20 μm) thick. An adhesive lamination of two webs of 250 ga PVDC–CELLO-PVDC is more durable than a single web with the same overall thickness. This lamination also allows printing between the film webs, which provides greater gloss and protection of the ink in transit.

PVDC–CELLO-PVDC/ADH/PVDC–CELLO–PVDC laminations have been used for VFFS candy packages and for cookie-tray overwrap formed on horizontal equipment, using the lap-seal capability of the two-side coating. In addition to barrier properties, these laminations provide stiffness, gloss,

Table 2. Typical Paper/Foil Structures

Structure	Application
28.5–35 ga AF/wax/tissue	Chewing gum wrappers. Materials usually combined on the packaging machine. Wrap is not sealed, but provides enough protection against moisture loss for required shelf life.
28.5 ga AF/ADH/BK	Cigarette wrappers. Aqueous adhesive lamination provides some moisture protection without seal.
35 ga AF/ADH/tissue	Margarine and butter wraps. Adhesive prevents corrosion of the aluminum by ingredients in the product. Wrap protects product from external odors and flavors.
PR–28.5–35 ga AF/ADH/MGBK–modified wax	Soap wrappers. Modified wax is heat-seal layer. Package protects against moisture and fragrance loss.
PR–28.5–35 ga AF/ADH/25# bond paper–modified wax or PR–28.5–35 ga AF/ADH/12–15# tissue/wax/8# open porous tissue	Carton overwrap. When heat is applied to wax-laminated structure, wax bleeds through for heat seal. Products packaged include cookies, confections, hygroscopic detergents and water softeners.
30–35 ga AF/wax/12–15# tissue/wax/8# open porous tissue	Carton liner for sugar-coated cereals. Products are hygroscopic, requiring protection from moisture gain. This structure works well on "double package makers" (see Bag-in-box, dry product).

and transparency; all desirable in the marketing of cookies. The crispness of the wrap implies crispness in the cookies.

PVDC–CELLO–PVDC/PE/PVDC–CELLO–PVDC–PE provides a new dimension in a pouch, with high heat-seal strength and much greater durability than the adhesive lamination. The moisture and gas barrier are basically the same as those of the cellophane, but the ability to form high-strength hermetic seals allows the use of these structures for a wide range of products (see Table 3). If transparency is not desired or if graphic considerations dictate a white package, white-pigmented PE film can be used in the lamination. For different requirements (lower temperature heat seal, hot tack, seal strength), PE is frequently replaced by ethylene copolymers. Cellophane is also used in structures such as PVDC–CELLO–PVDC–/PE/AF–PE. The extrudate can be LDPE, EVA, EMAA, EAA, or ionomer (see Ionomers), depending on the specific properties desired, and the gauge can be custom-tailored to the application. The foil gauge can be 28.5–100, but it is generally in the 28.5–35 range (see Table 3). These structures provide maximum barriers to moisture vapor and oxygen, as well as barriers to essential oils and light.

Laminations of cellophane with a variety of BOPP films provide the stiffness and barrier properties of cellophane, but with better durability. Lap seals on HFFS and VFFS equipment can be made with PVDC–BOPP/ADH/CELLO–PVDC laminations. This structure has replaced CELLO/CELLO laminations for cookie-tray overwraps. If fin-seal back seals are

required, eg, VFFS snack products, cellophane is laminated to coextruded heat-sealable BOPP.

Biaxially oriented heat-set polypropylene (BOPP). BOPP has become one of the most widely used families of plastic films in multilayer flexible packaging (see Film, oriented polypropylene). This material provides unmatched versatility. Like cellophane, it can be coated for property enhancement, but unlike cellophane, it can also be coextruded with other resins to produce composite structures with heat-seal properties to meet many different specific requirements. Uncoated monolayer homopolymer BOPP film is generally used as the printed external component of laminations. It is available in gauges ranging from 0.45 to 2 mil (11.4 to 51 μm). The printing is generally done on the reverse side, with the film providing gloss and scuff protection for the inks. The gauge is chosen based on the package size and other components in the laminated structure to optimize product protection, machineability, and economics. Examples of structures utilizing the nonsealable film are included in Table 4. They are a few of the many combinations that are routinely used in flexible packaging for a wide range of applications including cookie-tray overwraps, dump candy bags, snack foods, and dry food mixes.

BOPP is also available with PVDC coatings on one or two sides. The PVDC coatings vary in barrier properties and heat sealability. The heat-sealable PVDC coatings provide less oxygen barrier than the nonsealable coatings. Heat-sealable PVDC-coated BOPP can be used as the inner component of a

Table 3. Typical Cellophane Structures

Structures	Applications
PVDC–CELLO–PVDC/ADH/PVDC–CELLO–PVDC	Early structure used for VFFS candy packages and HFFS cookie-tray overwrap.
PVDC–CELLO–PVDC/ADH/PVDC–CELLO–PVDC	Current structure used for HFFS cookie-tray overwrap.
PVDC–CELLO–PVDC/PE/PVDC–CELLO–PVDC–PE	Frozen ice; gas-flush packaging of nuts, processed sliced meat and poultry products; atmospheric packages of dry-powdered products, nut meats, sunflower and other snack seeds; pharmaceutical tablets; prophylactics
250 ga PVDC–CELLO–PVDC/10# PE/35 ga AF–22.5# PE	Small gas-flushed nut packages; dehydrated soup and sauce mixes; dry chemicals; pharmaceutical powders
PVDC–CELLO–PVDC/ADH/BOPP–PVDC	VFFS snack products

Table 4. Typical BOPP Structures

Structures	Applications
Uncoated film	
70 ga BOPP–PR/200 ga LDPE	candy, dump cookie bags
120 ga BOPP–PR/7# EAA/35 ga AF/7# EAA/350 ga LLDPE	high-barrier cookie packaging
75 ga BOPP–PR/7# LDPE/30 ga AF–15# LDPE	cookies, snacks
50 ga BOPP–PR/ADH or PE/34# clay–paper/LDPE	pouch for dry foods
50 ga BOPP–PR/ADH or PE/150 ga white opaque BOPP	candy bars
45 ga BOPP–PR/ADH or PE/30# glassine–PVDC	snack foods
75 ga BOPP–PR/ADH or PE/30# glassine–PVDC	snack foods
Coated film	
120 ga PVDC–BOPP–PR/ADH/250 ga PVDC–CELLO–PVDC	cookie-tray overwraps
75 ga PVDC–BOPP–PR/ADH/250 ga PVDC–CELLO–PVDC	cookie-tray overwraps
45 ga PVDC–BOPP–PR/ADH/250 ga PVDC–CELLO–PVDC	cookie-tray overwraps
75 ga BOPP/15# LDPE/70 ga BOPP–PVDC (heat-seal PVDC inside)	
45 ga BOPP/ADH/70 ga BOPP–PVDC (heat-seal PVDC inside)	
100 ga PVDC–BOPP/ADH/200 ga LDPE (high-barrier PVDC outside)	
100 ga BOPP/ADH/70 ga BOPP–PVDC	
75 ga BOPP/ADH/70 ga BOPP–ionomer	
Coextruded BOPP[a]	
80 ga (*–BOPP–*)/ADH or 7# LDPE/80 ga (*–BOPP–*)	
75 ga (*–BOPP–*)/10# LDPE/70 ga (*–BOPP–*)	
75 ga BOPP/ADH or LDPE/(80 ga *–BOPP–*)	
Other	
50 ga BOPP/ADH/opaque BOPP	soft-drink bottle wraparound
50 ga BOPP/ADH/150 ga opaque BOPP–PVDC–cohesive	candy bar wraps

[a] * = coextruded-polymer heat sealant.

lamination for use on VFFS packaging equipment. Care must be exercised that the weight of product drop does not exceed the hot tack or seal strength of the PVDC coating. These films can also be used as the external ply of a lamination where the high barrier available with the nonsealable films is not required. In these structures, the PVDC is buried in the lamination and not used for sealing.

The versatility of BOPP films has been expanded by coextrusions, which fall into three categories: (*1*) two layers, one side heat-sealable (AB); (*2*) three layers with identical outer layers (ABA); and (*3*) three layers with different outer layers (ABC). Coextrusions for multilayer flexible packaging have been designed to provide specific heat-seal properties on one side and optimum conditions for ink or laminating adhesive on the other. It is more difficult to obtain consistent bond

strength with homopolymer BOPP than with other laminating materials, and the best bonds attainable with these films are far inferior to those obtainable with other films, foils, and paper. The ABC coextrusion also provides the opportunity to incorporate differential slip characteristics to meet packaging machinery and package-line requirements. Variable slip remains a problem with BOPP. In a variation of the coextrusion process, BOPP films are given an ionomer surface for low-temperature heat seals with good hot tack.

The variety of coextruded films permits the design of specific laminated products to meet a wide range of end-use requirements. Extreme care and testing must be part of the development and evaluation process to be sure that the structure will perform satisfactorily on the particular packaging equipment at the required speeds. Sometimes the heat required to effect a seal may exceed the heat-set temperature of the film (300°F or 149°C), resulting in distorted, and sometimes nonfunctional, seals.

There are two methods of producing white opaque BOPP films. One method, which utilizes pigmentation, produces a film with slightly higher effective density and physical and barrier properties similar to unpigmented films of the same thickness. The other, which creates opacity by introducing small gas pockets in the film, results in a film with about half the density and half the moisture barrier of conventional BOPP and significantly lower physical properties. Both can be coated or coextruded (see Table 4).

Biaxially oriented heat-set polyester (PET). Biaxially oriented PET films have excellent dimensional stability and heat resistance down to 48 ga (see Film, oriented polyester). They can be printed by flexography or rotogravure in multicolor designs. These characteristics make PET films a natural choice for the external component of laminated structures if exposure to elevated sealing temperatures will be encountered.

A very common lamination is PET–PVDC–PR/PE or ionomer. The polyolefin provides the heat seal. This structure is used as the printed component of processed meat and frankfurter packages and as liquid-filled pouches for freezer pops. Uncoated PET in the same construction is used as a pouch for boil-in-bag products, using MDPE or LLDPE for sealing. Heat-resistant adhesives are used for boil-in-bag and microwave-reconstitution applications.

PET films are also used in more complex laminations, eg, in PET–PR/PE/AF/PE, where the high barrier properties and heat resistance of the PET allows the use of high heat-seal-bar temperatures to attain high packaging machine speeds. Uncoated PET films are available now with a surface modification to provide uniform anchorage of inks and adhesives, as an alternative to the traditional corona discharge (see Surface modification). A coextruded PET film with a heat-sealable copolymer surface can be used as the inner ply of a laminated structure. This film can be metallized to improve the barrier properties and provide an attractive surface for graphics. An excellent snack food structure is BOPP–PR/AM–(PET–*) (see Metallized Films).

The greatest moisture and gas barrier available with PET is provided by the structure PVDC–AM–PET–PVDC. It is being used for gas-flush-packaged snacks and nut products in the combination BOPP–PR/PE/PVDC–AM–PET–PVDC/PE. Since the PVDC coating in this structure is heat sealable, it can be used as the inner ply of such laminations as BOPP–PR/

Table 5. Typical PET Structures

Structure	Application
PET–PVDC/LDPE	processed meat packaging pouches for liquids
PET/heat-resistant ADH/MDPE or LLDPE	boil-in-bag frozen vegetables
PET/LDPE/AF/LDPE	pouches for high-speed packaging machines

Table 6. Typical Nylon Film Structures

Structure	Application
60 ga nylon/ADH/PE or ionomer	forming structure for bacon
48 or 60 ga nylon–PVDC/ADH/200 ga ionomer	forming structure for luncheon meat and frankfurters
PR-48 ga AM–BON/ADH/200 ga LDPE	institutional coffee packaging, VFFS
48 ga BON–PR/PE/28.5 ga AF/PE	high-barrier cookie packaging

50 ga PVDC–AM–PET–PVDC. Uncoated PET films serve a unique function in child-resistant packaging of pills. A closure for blister packages uses the structure 48 ga PET/pressure-sensitive ADH/50–150 ga AF-heat-seal coating. This structure is very difficult to tear or remove from the blister, but if the PET is peeled away from the foil, the pill can be dispensed by pushing it through the foil. A sampling of PET structures is shown in Table 5.

Nylon. Although nylon does not provide good moisture barrier, it does provide some oxygen barrier (see Barrier polymers; Nylon). Nylon is a hygroscopic film that comes to moisture equilibrium at ambient temperature and rh. As moisture content increases, barrier properties decrease. The major nylons used for films in the United States are nylon-6 and nylon-6,6. They are used mainly in packages in which one component is thermoformed, in combination with LDPE, EVA, or ionomer. Thickness of the nylon and sealant layer vary, depending on the draw depth of the package. The thickness of the composite must be maintained in the corner areas, where the greatest draw takes place, in order to ensure package integrity. For vacuum-packaged oxygen-sensitive processed meats, one of the films is coated with PVDC for added oxygen barrier.

These structures can be produced by lamination, and also by extrusion coating one material onto the other. Most frequently, the nylon is extrusion coated onto the material that serves as the moisture barrier and sealant, and the sealant is coated with PVDC for oxygen barrier. Bacon packers who do not require the oxygen-barrier properties use laminations without PVDC. Laminations of nylon/sealant are also used as pouches for primal and subprimal cuts of fresh meat, for block cheese, and as the thermoformed component of a natural cheese package. A PE, EVA/nylon/PE, or EVA lamination is used as a bag-in-box material for liquid soap used in dispensers (see Bag-in-box, liquid product). Nylon films have been used as the heat-sealing component of specialty packages that required the chemical resistance of nylon. The relatively narrow melting range of nylon requires very close sealing-temperature controls.

Oriented nylon (ON). Orientation of nylon film increases tensile strength and oxygen barrier, and decreases elongation and thermoformability. When heat-set, it has enough resistance to deformation to permit printing by flexography or rotogravure and lamination by dry bonding or extrusion lamination. ON also provides excellent puncture and flex-crack resistance. Biaxially oriented nylon film (BON), with or without PVDC coating, has been available only from Japan, but one domestic supplier has announced plans to produce the film

in the United States. It is used in the United States in laminations such as BON–PVDC–PR/LDPE for institutional pre-measured packages of ground roast coffee. Improved-barrier metallized BON can be used as well, ie, PR–AM–BON/LDPE. A more complex structure is used for vacuum packaged coffee: PET–PR/ADH/AF/BON/LDPE. The BON is used in this structure for puncture and flex-crack resistance. The structure used for the nonforming component of packages for processed meat and frankfurters is BON–PVDC–PR/PE or ionomer.

Monoaxially oriented nylon film is being produced in North America, which can generally be used as an alternative to BON. It is oriented in the machine direction, the direction in which dimensional stability is important for printing and laminating operations. These films are also available with and without PVDC coating. Typical nylon film structures are given in Table 6.

Coextrusions. In coextrusion, two or more resin layers can be combined in a single extrusion operation (see Coextrusion, flat; Coextrusion, tubular; Coextrusions for flexible packaging). In many cases, coextrusions are replacing laminations of the same or similar films. For example, bacon wraps were formerly laminations of PET/PE or PET/ionomer for the nonforming side, and nylon/PE or nylon/ionomer for the forming side. Today, many are made of coextruded nylon/PE or nylon/ionomer for both webs.

There is a growing use of coextrusions as components of laminations to enhance economics and functional characteristics. The simplest of these is the replacement of a monolayer film with a coextruded film of the same thickness in which half of the expensive resin is replaced by a much less expensive material. Blown and cast coextrusions are candidates for laminations. The many combinations available today provide unlimited opportunities for designing complete structures with very specific barriers, heat seal, flexibility or stiffness, opacity or transparency, and color or combinations of colors. The development of resins to promote adhesion between different polymers is opening additional opportunities for tailor-made combinations through research in adhesive resins using terpolymers, quadrapolymers, and more advanced polymer chemistry.

Another way to use the concept of coextrusion in conventional converting processes is coextrusion coating, in which two or more extruders feed an extrusion-coating die. The primary use of coextrusion coating has been to replace a significant percentage of a relatively expensive resin with a lower-cost resin without sacrificing the sealing characteristics. Another approach is to use one of the components to promote adhesion between the other component and the substrate. The use of coextruded films as components of laminations is just beginning. It is expected to become a very important technique in the future. Typical structures with coextruded films are nylon/(PE-ionomer); BOPP–PR/ADH/(white HDPE-chocolate HDPE-heat seal); and BOPP–PR/ADH/(white HDPE-heat seal).

Metallized films. The application of a very thin layer of aluminum to plastic films by vacuum deposition provides a material with the appearance of bright aluminum foil, permitting eye-catching graphic opportunities. Advances in metallizing technology and the application of heavier metal deposits have increased the functionality of metallized films (see Metallizing). The base film has a significant effect on the brightness of the metal and on the barrier properties at a given level of metal deposition. PET currently provides the best surface for metallizing, providing substantially better barrier properties than the other films.

Metal deposits on films for flexible packaging are described in terms of light transmission through the film. Four percent light transmission provides a good metal surface appearance but does not significantly improve the moisture- and gas-barrier properties. This level of metallization does permit the material to be used on packaging equipment that uses transmission electric eyes to control cutoff. It is now possible to specify light-transmission values of 1% and 2% for good barrier properties and most suppliers are now offering ultrametallized films that have a light transmission well below 1% with excellent barrier properties. The measured barrier properties of these films approach the limits of available testing equipment, so the results would appear to compare favorably with foil structures and may be suitable replacements for foil in some applications. However, for products that are highly hygroscopic or require maximum protection from oxygen, product storage tests should be conducted before making a change.

Two-side PVDC-coated metallized PET provides better barrier properties than any of the other metallized films. The following films are metallized for use in multilayer flexible packaging: PET, coextruded heat-sealable PET, BOPP, coextruded BOPP, cellophane, BON, and polyethylene. Typical structures are shown in Table 7.

Laminating Adhesives

The primary function of an adhesive is to bond two materials together, but because many have properties that contribute additional characteristics to a lamination, converters use a number of different adhesives (see Adhesives). Adhesion to many flexible packaging materials depends on factors other than the mechanical bonding often associated with laminating paper or chemical bonding, in which a chemical reaction at the interface forms a new material that adheres other materials together. Adhesion may also depend on surface energy (electrostatic forces) to provide surface attraction.

An extremely important aspect of joining materials is the ability of the materials to come in intimate and complete contact. This is described as the "wettability" of the surface. Wet solvent or water solutions, or water emulsions, must flow uniformly over the surface of the substrate. Voids in the adhesive layer result in surface defects in the lamination, and if the structure is transparent, the transparency is affected adversely because of light refraction through the small entrapped air pockets. The degree to which an adhesive wets the surface depends on the surface tension of the liquid and the surface energy of the substrate. Surface energy is measured by applying liquids with different surface tension until one is found that does not bead up. The result is reported in dynes (10^{-5} N). The surface energy of plastic materials can be altered by electrostatic (corona) discharge or by flame treating (see Surface modification). Paper and aluminum foil can be wet by most adhesives from solvent or water systems. The cellulose molecule is polar, as is the aluminum oxide surface of aluminum foil. Polyethylene and polypropylene are nonpolar, and the surface of these films does not wet out readily. Corona-discharge treatment is required in lamination or printing.

Solution and emulsion adhesives. Porous papers are laminated to aluminum foil with water-based adhesive systems by a wet-bonding process. The adhesive is generally applied to the foil, and the paper is immediately brought in contact with the wet adhesive under pressure. The laminate then passes through a drying tunnel, where the water is removed. The primary adhesives used are dextrins, sodium silicates, and casein–rubber latex. In laminating two impermeable materials, the adhesive is applied to one of the substrates, and the solvent is removed in a drying tunnel before the materials are combined by heat and pressure. The adhesives used for this type of lamination fall into two categories: thermoplastic and thermosetting.

Thermoplastic adhesives. These soften by heat, resolidify upon removal of heat, and resoften if heat is applied again. For this reason, laminations with thermoplastic adhesives lack heat resistance. They cannot withstand high-temperature heat sealing. The most widely used thermoplastic adhesives are formulated plasticized vinyl acetate–vinyl chloride copolymers. Others have been used for specialty applications. The vinyl copolymers also provide a protective coating on aluminum foil to prevent dissolution of the metal by product ingredients. Variation in the type of bonding can be achieved by varying the plasticizer content, which also has an effect on stiffness.

Thermosetting adhesives. These are preferred in flexible packaging. A chemical reaction is initiated after the two substrates are combined, creating a new chemical compound that holds the lamination together. The most common types are polyurethane and polyester–urethane. The chemistry of these

Table 7. Typical Structures With Metallized Films

Structure[a]	Application
48 ga BOPP–PR/ADH or PE/80 ga AM–(BOPP–*)	snack package
PR–AM–BON/ADH/PE	coffee package
48 ga BOPP–PR/PE/60 ga AM–(PET–*)	pouches for alcohol wipes
48 ga BOPP–PR/PE/AF/PE/60 ga AM–(PET–*)	pouches for medicinal wipes

[a] Key: * = coextruded copolymer heat sealant (see Table 4).

materials provides a wide degree of formulating flexibility to achieve adhesion to different substrates, heat resistance, chemical resistance, and barrier to migration of components of some products. Because of this formulating flexibility, less demanding applications can be served by lower-cost formulations.

Waterborne acrylic emulsion adhesives are being used in greater quantities than before to achieve compliance with EPA regulations (see Acrylics). Waterborne PVDC adhesives are also used to some extent, although efforts to develop PVDC adhesives that provide both good adhesion and good oxygen barrier have not been successful. These two characteristics reflect direct opposites in the chemistry of these materials. One can expect good, but not excellent, adhesion and good oxygen barrier with selected PVDC adhesives.

Extruded adhesives. Polyethylene and ethylene copolymers are widely used as laminating adhesives. They provide adhesion between substrates, add bulk to the lamination, and contribute to moisture barrier in the absence of foil. Polyethylene, EAA, EMA, EMAA, and ionomers are all used, depending on the substrate, bond strengths required, and end use. Polyethylene often requires the use of an adhesion promoter on one or both substrates. Nonpolar PE does not adhere to the polar surfaces of most traditional laminating materials. PE extrusion lamination is accomplished by using high extrusion temperatures to oxidize the surface molecules to some extent, which introduces a polar configuration. The polar ethylene copolymers adhere well to a variety of substrates. EAA, EMA, and ionomers have particular affinity for foil.

Waxes and wax blends. Paraffin or microcrystalline waxes, or both, are used for special applications where high bond strengths and heat resistance are not required (see Waxes). These materials can be modified by the addition of low molecular weight PE, EVA, or other resins and resin additives to increase adhesion, hot tack, and stiffness or softness of the lamination. They are generally applied to one of the substrates by gravure–cylinder application (see Coating equipment) at temperatures high enough to reduce the viscosity of the material within a range that can be adequately controlled by gravure. The second substrate is brought in contact with the hot wax under pressure, and the composite is chilled. If one of the substrates is very porous, the wax is chilled on the primary substrate before the combining operation. Examples of the use of wax as an adhesive include AF/wax/tissue chewing gum wraps; CELLO/wax/AF–modified wax for cheese packaging; and glassine/modified wax/AF/PE for a tobacco pouch.

Miscellaneous adhesives. A small number of laminations utilize the thermoplastic nature of a coating, which is a component of one or both substrates. For example, two films that have a heat-sealable PVDC coating can be joined with heat and pressure, or aluminum foil can be laminated to a prewaxed tissue by the application of heat and pressure. Printing inks can also serve as adhesives if the bond-strength requirements are limited. This technique is used in some PE–PR–PE laminations for chub packs (see Chub packages).

Package Closure

The methods of package closure are heat sealing, cold sealing, and glue sealing.

Heat sealing. The primary method of closing multilayer packages is heat sealing (see Sealing, heat). The inner component of the structure is a thermoplastic material that softens with application of heat and solidifies when the source of heat is removed.

Cold sealing. Heat is not necessary if the coating is a modified-rubber-based material that seals to itself with the application of pressure, but not to other materials. These are called cohesives, or cold-seal coatings (see Adhesives). They can be applied over the entire material surface, but they are most commonly applied as a peripheral coating at the edges of the material in register with a printed design on the package face. This is done to reduce materials cost, but also because cohesive materials have very high coefficients of friction. Reducing the covered surface area facilitates movement over the packaging equipment. Lap-seal applications are not feasible with cold seals because the cohesive would have to be coated on both sides of the material, which would result in sealing in the rolls. Care must be exercised in the processing of cold seal materials to reduce the possibility of off-odors affecting the product. Cold-seal materials are used to package ice cream, confections, and candy bars that would be damaged by heat.

Glue sealing. This involves the application of an adhesive (glue) to specific areas of the packaging material on the machine that forms the package (see Adhesive applicators). Glue sealing is not widely used for multilayer flexible packaging, except for products that do not require continuous seals and only on paper-containing packaging materials. The adhesive is partly dried by absorption of the water from the glue into the paper.

Heat-Sealable Materials

The success of multilayer flexible packaging is directly related to the use of heat sealing, which involves positioning the faces of the materials so they will be combined by melted sealant upon application of sufficient heat and pressure. This is dependent on the solidification of the sealant material when the heat is removed.

Heat-seal coatings. Heat-seal coatings are generally defined as coatings applied from solvent solutions or water emulsions. The coating weight is generally between 1 and 5 lb/3000 ft² (0.45 and 2.3 kg 278.7 m²). Coatings on films and paper can be vinyl acetate–vinyl chloride copolymers, nitrocellulose, acrylics, or PVDC. PVDC coatings can provide barrier as well as heat seal (see Vinylidene chloride copolymers), but those formulated for maximum barrier properties are not heat-sealable. They are used as inner components of a lamination or on the outside surface. Vinyl chloride–vinyl acetate copolymers have been used for some time as heat-seal coatings on films and foil. They also protect aluminum from corrosive agents present in many products. Structures of this type are generally used as lid stocks (see Lidding) for rigid containers or as the closure portion of pharmaceutical pill packages. Emulsion coatings of low molecular weight EVA copolymers are used as the heat-seal component of lid stock for cups. Other heat seal coatings include waxes, nitrocellulose, eg, on cellophane, and acrylics, eg, on BOPP.

Waxes. Wax heat-seal coatings on paper also provide moisture barrier (see Waxes). Paraffin and microcrystalline waxes are used for sealing, as well as blends of the two. Waxes are available with a range of melting points, and care must be taken to choose a wax or blend that will achieve the desired sealing characteristics without blocking or sticking in the rolls. Waxes and wax blends can be modified with low molecular weight LDPE to harden them slightly. This makes them less susceptible to blocking, and it also strengthens the heat seal. Low molecular weight EVA added to waxes tends to improve the hot tack, seal strength, and resistance to blocking in roll form. Other tackifiers can be added to make the seals more tenacious. Wax coatings are sometimes applied on the packaging line. Some carton liners are heat sealed through the use of a wax stencil applied on the packaging machine to supplement the wax contained in the packaging material or to areas where no wax sealant is applied by the materials producer. This is primarily done on the foil surface of AF/wax/tissue laminations.

Hot melts. Hot melts, applied in molten form, utilize a number of thermoplastic resins, waxes, and modifiers. They are applied as coatings over the entire packaging material surface or in a pattern, registered with printing, to effect positive package closure. Hot melts can be formulated with one of a number of resins as the principal component, eg, ethyl cellulose, nitrocellulose, EVA, or polyamide.

Heat-sealant films. Most heat-sealant films are polyolefins.

Low density polyethylene. The most common heat sealant is low density polyethylene. This is not a single product, but a family of materials that vary in density, melt index, and molecular weight distribution (see Polyethylene, low density). As density increases, sealing temperature, heat resistance, strength, and stiffness increase, and sealing range, clarity, and barrier decrease. The differences are relatively minor within the range of low density polyethylenes.

Melt index has a more significant effect. As melt index increases (molecular weight decreases), sealing range and clarity increase, and sealing temperature, heat resistance, strength, stiffness, and chemical resistance decrease. Barrier properties are not affected. In general, polymers with melt index from 0.24 to 5.0 are used for manufacturing blown films, and those from 2.0 to 30.0 for cast films and extrusion coating. Molecular weight distribution (MWD) is a function of the polymerization process. Narrow MWD resins have a narrower sealing range than broad-MWD resins.

The range of polyethylenes provides great latitude in tailoring structures to meet package/product requirements. They are incorporated through film lamination, extrusion coating, or as components of coextrusions. The density of linear LDPE is within the range of conventional LDPE, but some of the properties tend to be more like HDPE. Compared to LDPE, for example, the heat-seal initiation temperature and physical strength of LLDPE are much higher. In many cases, LLDPE film can be thinner than an alternative LDPE film. LLDPE resins can be blended with conventional LDPE in any ratio. The heat-sealing and physical characteristics of films made from blends range between the properties of either material. Processing of blends is much less demanding than processing straight LLDPE. LLDPE can also be blended with EVA copolymers to further enhance sealing properties.

Medium density polyethylene. MDPE is sometimes used as a component of multilayer structures where slightly higher heat resistance or barrier properties are desirable. MDPE is used as the inner component of boil-in-bag material and for some medical devices subjected to retort sterilization. LLDPE blends are replacing MDPE in many of these applications.

High density polyethylene. HDPE (see Film, high density polyethylene; Polyethylene, high density) is rarely used as the sealing medium in flexible packaging because of its high seal-initiation temperature and narrow seal range. Where heat resistance and high strength are required in special applications, eg, medical products, a rubber-modified HDPE is used.

Polypropylene. PP (see Polypropylene) is not widely used as a heat sealant except as a component of coextrusions used in special market areas. Polypropylene copolymers are used in coextruded BOPP films as the heat-sealant component and in retort pouch structures (see Retortable packages).

Ethylene copolymers. Copolymers of ethylene and vinyl acetate (EVA), acrylic acid (EAA), or methacrylic acid (EMAA) have properties that are very different from LDPE–LLDPE (see discussion under Ionomers). Each of the copolymers is available in a range of comonomer percentages. EVA is available with 4–30% vinyl acetate, for example. The acid copolymers, EAA and EMAA, provide excellent adhesion to metals, as do ionomer-modified acid copolymers. With increasing comonomer content, clarity and heat-seal range increase; crystallinity, stiffness, and heat-seal temperature decrease. They can all be used in multilayer structures as films that are laminated, as extrusion coatings, or as components of coextrusions. The copolymers cost more than LDPE and are used only if their specific properties justify the additional expense.

In summary, the variables considered when choosing a polyolefin sealant film are density, melt index, molecular weight distribution, homopolymer or copolymer, comonomer and percent comonomer, blend of resins, and film thickness. All of these materials can be modified with slip and antiblock additives, and they can be pigmented. Care must be exercised when using some of the copolymers if the packaged product is susceptible to picking up off-odors or flavors.

Sealing Properties of Heat Sealants

Seal-initiation temperature. This is the lowest temperature at which a seal can be achieved. Activation temperatures range from relatively low for waxes to relatively high for MDPE and HDPE. In multilayer structures, this involves not only the melting point of the sealant, but the conductivity of the other components of the structure. Most flexible materials are poor thermal conductors, which means that heat transfer is not rapid. In sealing operations in which the heat is applied to the outer surface of the material, more often from only one side, the amount of heat required is increased because of the insulation effect of the components. Aluminum foil is an excellent conductor, but because it transmits the heat laterally and takes the heat away from the seal area, more energy is necessary to effect a seal. Heat-seal initiation temperatures are determined by establishing a standard dwell time and pressure, and then progressively increasing the temperature until a seal is obtained. Information obtained in this way provides a starting point for establishing packaging machine conditions.

Seal range. This is the range of temperatures at set conditions of pressure and dwell time in which effective seals can be obtained. The seal-initiation temperature is at the bottom of the range; the top is the highest temperature at which a satisfactory seal can be obtained without deterioration of the seal or the structure.

Hot tack. Hot tack is the resistance to separation of a seal immediately after removal of the pressure and temperature of sealing. Waxes generally have poor hot tack; ethylene copolymer films, particularly ionomers, have good hot tack. This feature is very important in VFFS operations, where the product is dropped into the formed tube while the seal jaws are closed. The full weight of the product is then on the hot seal when the jaws open. If the hot tack is not sufficient to hold the weight of the product, the seal fails.

Strength Properties of Heat Sealants

Impact, tensile, and tear strength. As the internal ply of laminations, heat-sealant films can add strength to the total structure. These characteristics are important in providing durability to the packaged product during distribution.

Seal strength. This parameter is easily defined as the force necessary to separate a seal, reported in pound-force (N) per inch (cm) width. Unfortunately, the numbers do not provide more than a happenstance relationship to the effectiveness of a seal, nor do they predict with any accuracy the performance of a package. The term seal strength implies the force necessary to separate the seal at the interface of the two layers of the sealant material; however, because properly sealed sealant materials fuse together, the interface is lost in the homogeneous mixture. Thus, the failure is somewhere else in the material.

If the sealant material has low cohesive strength, eg, wax, the failure leaves a portion of the sealant on both interfaces. Sometimes the sealant material fractures adjacent to the seal area on one side of the material and peels away from the balance of the structure; in that case, the force necessary to fracture the sealant material is relevant, as well as the bond strength of the lamination of the sealant material. Sometimes the total structure fails in a tensile mode with the application of force. In that case, the failure is not due to limited seal strength; on the contrary, the seal strength exceeds the tensile strength of the composite.

Seal strength is not a measure of seal continuity. A continuous seal can be obtained with a wax or heat-seal coating, which has very low seal strength, if the coating fills any voids that might have appeared. Peelable seals on portion-control packages are continuous, but they have low seal strength. Conversely, it is possible to have high seal strength, but very poor continuity and compromised product protection.

Seal integrity. The role of package sealing in product protection is extremely important. A high-barrier material is of little value without a continuous seal. Channels in the seal would allow the atmosphere to penetrate the headspace of the package. Packaging professionals have not yet agreed on the best method for testing seal integrity. As noted above, when testing for seal strength the mode of failure must be recorded. The material usually fails in an area adjacent to the seal, where the materials have been weakened slightly by the heat-seal jaws. If the sealant film fractures, the product can still be contained, but the barrier properties contributed by the outer components are lost.

Seal continuity can be evaluated by placing the sealed pouch in a vacuum jar under water, and then removing the air from the jar with a vacuum pump. This effectively increases the pressure inside the pouch and a stream of bubbles will indicate a point of failure. This method also reveals fractures or pinholes in the packaging material. For packages in which paper is the external component, glycerine or light oil can be used instead of water. In a similar test, a needle attached to a source of low pressure is inserted through the wall of the submerged pouch. The needle can be forced through a small rubber stopper or gasket pressed against the pouch to prevent air from leaking through the injection hole.

The seal strength provided by heat-seal coatings and waxes is generally much lower than that achieved with the thermoplastic films or relatively heavy extrusion coatings, but seal continuity can still be achieved and product protection assured. The greater the amount of sealant medium, the greater is the tolerance for packaging machine deficiencies. A heat-sealable coating of 3 lb/3000 ft^2 (1.36 kg/278.7 m^2), which is about 0.2 mil thick (5/μm), does not tolerate significant misalignment of the heat seal jaws, but a 2-mil (51-μm) polyolefin film provides enough material to flow into areas where pressure is reduced because of misalignment.

Resistance to seal contamination. Sealants vary significantly in their ability to seal through or around surface contamination. Surface contamination can be in the form of dust, or powders from flour and similar products; coffee grounds; shortening; grease from meat products; peanut husks; sauces; or moisture. Some of the contaminants are attracted to the seal surface by static; therefore, static elimination can minimize contamination. The choice of sealant must take into account any potential source of contamination.

Barrier properties. Sealant materials have very specific barrier properties with respect to a broad range of the components of the atmosphere or packaged products. Polyethylene provides a good barrier to moisture, but not to oils, greases, or essential oils. Care must be exercised in bonding the polyethylene to other substances so that there will be sufficient resistance to delamination of the plies during distribution. In some cases, if acidic components of a product in the presence of essential oils will permeate the sealant, and the sealant is next to aluminum foil, a protective coating is required on the foil for corrosion protection.

Absorption of product components. Some product components, frequently essential oils, are absorbed by the heat-sealant layer. This can result in a change of the product flavor. Sealants must be chosen to preclude any adverse effects on the product.

The choice of the sealing medium depends on product weight and bulk density, packaging machine speed (heat and dwell times available), desired product shelf life, susceptibility of product to deterioration from atmospheric exposure, stress encountered in distribution, and ingredients of product that could affect the sealing medium adversely.

Protective Coatings

Coatings are used in multilayer flexible packaging to provide scuff resistance to the printed surface of laminations; heat resistance; release properties to prevent sticking of the material in rolls; gloss; control of friction coefficient (slip); heat seal; barrier to oxygen, moisture, essential oils, or active ingredients in a product; or combinations of the above.

Coatings used over printing inks can be formulated from a number of different resins and resin combinations, including nitrocellulose, ethylcellulose, acrylics, polyamides, and or mixtures of the above with the addition of modifiers to achieve specific properties. Waxes can be added to these formulations to provide low friction coefficient and to improve scuff resistance. Where maximum heat resistance and high gloss are desired, thermosetting-type coatings are used that react chemically to provide a coating that is not affected further by heat. These coatings are used where the laminations may be subjected to very high heat-sealing temperatures.

Printing

The primary printing processes for multilayer flexible packaging are rotogravure and flexography (see Printing). The preponderance of packaging rotogravure presses in the United States have eight printing stations, allowing any combination of eight colors, overprint varnishes, or adhesion promoters. Many of these presses are equipped to use up to two of these stations for printing or coating the opposite side of the web of material. This feature is frequently used for the printing of coupons, the application of a secondary message, application of a heat- or cold-seal coating, or a colored coating. In multilayer flexible packaging the flexographic presses are mainly central-impression types with six-color capability. Rotogravure printing is generally used where excellent pictorial reproduction is desired and generally where there are long runs or multiple runs. The cost of engraved cylinders and the setup time for a gravure press make it uneconomical for short runs.

There have been significant developments in flexographic process printing in recent years that now permit printing pictorials flexographically. Because the initial setup charges for flexography are much less than rotogravure, flexography is widely used for short-run items, ie, items where copy changes are made frequently.

The use of transparent colors on highly reflective aluminum foil or metallized films provides eye catching appeal to a flexible package. Printing the reverse side of transparent films, used as the outer component of a lamination, offers high gloss and excellent resistance to abrasion, which is difficult to achieve in single-ply structures. In many instances, this is sufficient reason to choose a lamination over a single-ply structure.

Properties of Multilayer Structures

Flexible packaging is expected to do many things: protect dry products from atmospheric moisture and protect high-moisture products from moisture loss; protect products susceptible to oxidation from atmospheric oxygen; contain the CO_2 or N_2, or both, used to replace atmospheric oxygen in gas flush packages (see Controlled atmosphere packaging); protect products from odorous substances in the atmosphere; prevent the loss of volatile flavoring components from a packaged product; etc. The barrier properties of individual materials are discussed in other articles (see also Barrier polymers; Testing, permeation and leakage), and discussion here is limited to particular considerations that apply to multilayer structures. The properties of many different structures are shown in Table 8.

Homogeneity. Because permeability in cellulosic and plastic films is homogeneous, it can be measured and expressed in terms of surface area. Composite structures containing metal foils are not homogeneously permeable, since the undisrupted metal surface provides a complete barrier. Where discontinuities occur, permeation can occur as well, but when the foil is covered by other materials, the opportunities are minimal. The introduction of aluminum foil into a lamination changes all the rules of moisture vapor transmission. Much has been said and written about the effect of pinholes in foil on moisture and gas permeability. The only true test of performance of a foil-containing structure is actual package performance tests. The traditional moisture-vapor transmission tests are so inaccurate for high-barrier materials that they are useless.

Property correlation. Correlation between barrier properties cannot be assumed. There is no correlation, for example, between the passage of liquid water and water vapor through a material. The water (moisture) vapor reacts in the gas phase, and in many ways it can be likened to gas permeability. It has been common practice to relate the oxygen transmission rate of plastic materials to permeability of flavorants, but these properties have been proven to lack correlation.

Barrier placement. It is important in the design of packaging structures to place the barrier component as close to the product as possible. For example, if the product is very hygroscopic or susceptible to moisture loss, nothing more than a heat-seal layer should separate the foil from the product. A paper/PE/AF/PE structure provides significantly better protection than a AF/PE/paper/PE structure. The paper component is susceptible to gain or loss of moisture and the PE coating cannot prevent it. As the product lowers the rh in the headspace of the package, there is a driving force for the moisture in the paper to permeate the polyethylene. As the paper loses moisture, there is a new driving force to attract moisture from the atmosphere at the exposed edges to bring the paper to its equilibrium moisture content, which perpetuates the transfer to the product. This activity cannot be seen when laboratory moisture-vapor transmission rates are conducted, because the edges of the material are not exposed. It is sometimes very surprising when actual product storage tests do not correlate with laboratory data.

Tailored properties. It is very important in the choice of materials that the product definition include as much information as possible concerning the specific degree of protection needed in each category. For example, there is no need to provide a great moisture barrier if the product is not hygroscopic, ie, does not remove moisture from the atmosphere. Likewise, there is no need to provide an excellent barrier to oxygen if the product does not react with oxygen.

Flavor barrier. Flavors and fragrances, desirable and undesirable, constitute a wide range of chemical compounds, and it is not possible to generalize about the flavor-barrier properties of packaging materials. Studies are underway at Rutgers University to provide a reproducible method of establishing permeability characteristics of various packaging materials to essential oils. The solubility of certain flavorants in some plastics can result in loss of product flavor even if the plastic provides a good barrier to the flavor. For example, polyolefins absorb many essential oils. The passage of essential oils through one component of a multilayer structure can also create problems in maintaining packaging material integrity.

Where the inner components of a structure are aluminum foil and a polyolefin, permeation of an essential oil can cause a bond failure, resulting in delamination. If the product also contains organic acids, they can go through the polyolefin along with the oil and react with the aluminum. This problem is encountered in packaging ketchup. It is overcome by providing a protective coating between the foil and the polyolefin.

Light transmission. Multilayer flexible packaging can provide the utmost extremes of transparency or opacity, or a controlled level of light transmission between these extremes. Transparency may be chosen for product visibility; opacity may be chosen for light protection. Aluminum foil provides the maximum degree of opacity and light barrier. The use of pigmented polymers (see Colorants) allows choices of degree of

Table 8. Properties of Selected Structures[a]

Structure[b]	Yield, in.²/lb[c]	Basis wt, lb/3000 ft²[d]	Thickness, mil[e]	Tensile strength, gf/in.[f]		Tear strength, gf[g], Elmendorf		Seal strength, gf/in.[f]	Water vapor transmission, g/(100 in.² · d)[h]	Oxygen transmission, cm³/(100 in.² · d)[i]
				MD	XD	MD	XD			
100 ga high-barrier PVDC–BOPP/ ADH/200 ga LDPE	9,600	45	2.9	12,500	9,900	66	230	4,500	0.3	0.9
100 ga BOPP/ADH/70 ga BOPP–PVDC	16,500	26.2	1.8	31,900	18,500	20	11	150	0.2	4.0
80 ga (*–BOPP–*) ADH/80 ga (*–BOPP–*)	18,000	24	1.7	12,900	24,000	24	23	700	0.3	>50
100 ga BOPP/ADH/70 ga (*–BOPP–*)	16,900	25.6	1.8	15,000	17,000	20	18	200	0.3	>50
48 ga PET/200 ga LDPE	10,600	40.7	2.58	3,200	3,800	200	150	4,000	1.0	6.0
48 ga PET/7 LDPE/AF–22 LDPE	8,400	51.4	2.8	7,400	10,600	32	32	2,800	20.1	<0.1
45 ga uniax ON/200 ga (LDPE-ionomer)	11,200	38.6	2.6	40,000	9,000	40	60	4,000	0.8	9.0
100 ga PVDC–nylon/200 ga LLDPE	8,500	50.8	3.2	8,700	5,500	80	136	4,500	0.34	0.6
48 ga PET/150 ga AM–LDPE	13,000	33	2.1	6,600	8,000	34	28	1,900	0.1	15
48 ga PET/ADH/48 ga AM–PET–200 ga LDPE	8,000	54	3.0	14,900	18,100	26	32	2,500	0.05	0.08

[a] At 65% rh.
[b] Key: * = coextruded copolymer heat sealant.
[c] To convert in.²/lb to cm²/kg, multiply by 14.22.
[d] To convert lb/3000 ft² to kg/278.7 m², multiply by 0.4536.
[e] To convert mil to μm, multiply by 25.4.
[f] To convert gf/in. to N/m, multiply by 0.3861.
[g] To convert gf to N, multiply by 0.0098.
[h] To convert g/(100 in.² · d) to g/(m² · d), multiply by 15.5.
[i] To convert cm³/(100 in.² · d) to cm³/(m² · d), multiply by 15.5.

light transmission as long as the pigment does not affect polymer properties. The degree depends on the color and the film thickness. This technique cannot reduce light transmission to 0%, but light barrier to the range of 5–10% can be achieved by combining opaque or colored papers with pigmented polymers.

Light transmission can also be controlled by vacuum metallization, which enhances the moisture and gas barrier of the film at the same time. Heavy deposits provide the lowest level of light transmission and the highest barrier, but cost increases along with the amount of metal. When the packaging equipment is equipped with electric eyes that depend on transmission of light through the packaging material to control package cutoff, that transmission must be at least 4%. This

level can be achieved in metallized films, but it must be specified. Normal metallization is 1–2%. Electric eyes that function by reflection of the light beam must be used for opaque materials such as foil, highly metallized films, or highly opaque paper structures.

The metal deposit can be applied in lanes (strip metallization) and placed in a position relative to printing that permits product visibility as well as a highly reflective metallic appearance that enhances graphics. A process has been developed to remove the metal from metallized films in a pattern in register with the printed design. Demetallization provides designers with exciting possibilities of combining the aesthetics of a highly reflective package with a window for product display. The barrier of the demetallized material is between that of the unmetallized structure and the fully metallized structure, and directly proportional to the surface area in which the metal has been removed.

Products susceptible to deterioration from exposure to uv radiation can be protected, even in transparent packages, if one of the components absorbs uv radiation. Processed meats, for example, can be protected from uv-induced color change by including PVDC, which absorbs the uv light.

Flexibility versus stiffness. The desired degree of stiffness or softness is defined by the product, packaging machinery, and marketing considerations. Crisp dry products such as cookies and crackers need crisp (stiff) packaging materials. When high-moisture-content pipe tobacco is in a soft package, the consumer feels that the product is soft, and therefore fresh. Some packaging machines function only with relatively stiff materials. In carton overwrapping equipment, for example, the sheet is cut to size and pushed into place over the product elevator. The material must be stiff enough to be pushed. Many other kinds of machines are not sensitive to stiffness, and the choice of materials can be based on marketing considerations.

Stiffness or flexibility can be designed into a package by the choice of components and adhesive. Modified waxes are frequently used as adhesives in very soft, flaccid laminations. The adhesive itself has a soft feel, and in addition, the type of bond between the adhesive and the substrate allows some freedom of movement, which also contributes to a soft hand. Thermosetting adhesives would provide a much stiffer feel with the same components. By choice of specific adhesive formulations and catalyst-to-resin ratio, the stiffness can be varied within limits. The use of extruded resins as bonding adhesives provides another level of stiffness and body, which can be altered over significant ranges by choice of resin and the amount of extrudate applied.

Durability. This is the ability of the package to retain its protective properties through the distribution cycle until the product is consumed. Materials that become brittle because of moisture loss, eg, cellophane and glassine, lose durability when exposed to an atmosphere with low relative humidity. These otherwise excellent materials show a high rate of package failure during the winter in colder climates. Flexible packages must also resist fracturing during the flexing that is normally encountered in shipment. In laminations, materials subject to fracture can be encased between fracture-resistant materials.

Composite properties. The attainment of optimal properties requires great care in the choice of materials and combining methods. Physical properties are not the sum of the individual components. In many cases, a desirable property is lost in the lamination. For example, PE has excellent tear resistance and elongation, but if it is tightly bonded to paper, the combined material has the tear and elongation properties of the paper. If the bond is rather loose, the PE retains its properties and the lamination has greater durability.

Selected Applications

The following miscellaneous applications illustrate some of the points made above.

Chewing gum wrappers: 28.5–35 ga AF/wax/tissue, materials combined on the packaging machine. The wrap is not sealed, but it provides sufficient protection against moisture loss for the required shelf life.

Cigarette packages: 28.5 ga AF/aqueous ADH/lightweight bleached sulfite–sulfate blend paper. The wrap is not sealed, but it provides some protection against moisture loss.

Margarine and butter wraps: 35 ga AF/special ADH/tissue. Special adhesive contains a protective coating to prevent corrosion of the aluminum by product ingredients. Wrap keeps out external odors and flavors.

Soap wraps: 28.5–35 ga AF/aqueous ADH/bleached MG kraft–modified wax blend. Wax coating provides heat seal; package prevents loss of moisture and fragrance; foil is printed for product identification and consumer appeal.

Carton overwraps: 28.5–35 ga AF/ADH/20–25# bond paper–modified wax; or 28.5–35 ga AF/ADH wax/12–15# tissue/wax/8# open and porous tissue. When heat is applied, wax in the latter structure bleeds through the tissue to form a seal. Foil is printed. Products include cookies, confections, hygroscopic detergents, and water softeners.

Carton liners: 30–35 ga AF/wax/12–15# tissue/wax 8# open and porous tissue used as carton liner for hygroscopic sugar-coated cereals. Structure works well on double package makers (see Bag-in-box, dry product).

Lidding: Structures used for lids of thermoformed portion-control packages are generally paper or film/ADH or PE/AF/ heat-seal coating. The choice of heat-seal coating, which depends on the composition of the formed component, is determined also on the basis of easy-peel opening. The most common heat-seal coatings are vinyl acetate–vinyl chloride copolymers, EVA, and LDPE.

Drug tablets: For child resistance, 100–150 ga AF/soft or pressure-sensitive adhesive/PET. In use, the PET is removed and the tablet can be pushed through the foil membrane.

Effervescent tablets: 20–30# MG bleached kraft/PE/50–70 ga AF/150–200 ga LDPE. These products require the maximum protection against moisture gain.

Labels: AF/ADH/paper has been used extensively for beverage-bottle labels. The graphic opportunities using reflective foil coupled with embossing has been widely used for beer labels. The choice of the proper inks and paper produces a label that will disintegrate completely in the bottle-washing operation.

Retort pouch: PET/heat-resistant ADH/AF/heat-resistant ADH/PP copolymer film (see Retortable flexible and semirigid packages).

Bulk packaging: 40# natural kraft/PE/35 ga AF/PE as inner ply of multiwall bags (see Bags, paper) for hygroscopic products or as liner for bulk boxes.

Composite can liners: Paper/AF laminations are used as the inner ply of spiral-wound composite cans for dry hygroscopic products and liquid products. Slip coatings, heat-seal coatings, or protective coatings, or all three, are applied to the foil surface, depending on the product requirements. When liquids are packaged, a strip can be sealed over the seam to prevent product seepage into the paper constituent. Uses include refrigerated biscuits, snacks, and motor oil.

Composite can labels: AF/paper laminations are used as labels for special and convolute wound cans. In many cases, the foil prevents surface staining that would be apparent if the exterior were paper.

Pouches. Pouches warrant some special attention here because their development and proliferation depended on multilayer flexible packaging technology. Most pouches contain aluminum foil, and the use of foil was made possible by the

Table 9. Typical Pouch Structures

Structure	Applications
PR–25# MGBK/7# LDPE/30 ga AF/22# LDPE	dry soup mix
PR–25# MGBK/30# LDPE/30 ga AF/22# LDPE	dry soup mix, sharp components
PR–25# MGBK/7# LDPE/30 ga AF/22# ionomer	dry soup mix, sharp components
OL–PR–clay–paper/7# LDPE/30–50 ga AF/22–30# LDPE	drink mix
25# pouch paper/7# LDPE/30–50 ga AF/22–30# LDPE	drink mix
48 ga PET–PR/7# LDPE/30 ga AF/22# LDPE	drink mix
25# transparent glassine–PR/7# LDPE/30 ga AF/15–22# LDPE	tobacco
PR–25# MGBK/7# LDPE/35 ga AF/22# ionomer	cereal bars
25# MGBK/7# LDPE/35–50 ga AF/22–30# LDPE	instant potatoes, gas flush
OL–PR–18# tissue/7# LDPE/28.5–35 ga AF/15–22# EAA	moist towelettes
OL–PR–20# MGBK/7# LDPE/30 ga AF/22# EAA	
BOPP–PR/22# LDPE	candy
250 ga cello–PR/22# LDPE	
PR–BON–AM/LDPE	coffee
LDPE–PR/LDPE	rice

development of extrusion coating and extrusion lamination. The most common structures are generally referred to as paper/poly/foil/poly, film/poly/foil/poly, or foil/poly/paper/poly. ("Poly" can stand for any of the polyolefins). The widespread adoption of these structures resulted from the choice available from the broad generic classes of films, foils, papers, and extrudates to tailor the products for the specific end use.

As a component of pouch material, aluminum foil provides the ultimate moisture and oxygen barrier and protection against loss or gain of flavors and fragrances. Gauges from 25–70 are commonly used, depending on the absolute barrier required and the abuse expected in distribution. Almost any type of paper can be used, depending on strength requirements, packaging machine function, graphics or aesthetics, and economic considerations. The papers include 15–18# tissues, 20–40# BK and NK, 20–30# pouch papers, 25–55# clay-coated papers, and some 20–25# glassines. Special oil and puncture-resistant papers are available. When a tissue, pouch paper, or MG bleached kraft paper is the external component of a pouch, it is generally printed by flexography or rotogravure using line printing. Clay-coated papers, printed by process rotogravure or process flexo, are used where product vignettes are desired. High-gloss overlacquers are generally used to enhance the appearance of the pouch.

Extrusion lamination generally utilizes approximately 7# of extrudate, although heavier applications can be used for functional purposes. LDPE is most often used, but EAA, EMA, EMAA, and ionomers are also used. Where tight bonds to aluminum or improved puncture resistance are required, the acid copolymers perform very well. Heat-seal extrusion coatings are generally 7–30#. Depending on end-use requirements, the resins can be LDPE, LLDPE, EVA, EMA, EMAA, or ionomer.

If a transparent film is used as the exterior component, instead of paper, the foil shows through. The foil can then be used not just for its barrier properties, but also for its aesthetic appeal. The film in this case is not used for gas or moisture barrier, but only as a protective carrier web for the foil. The films used are cellophane, BOPP, PET, and ON. The specific film characteristics that govern the choice are flex-crack resistance, heat resistance, the need to seal the external component (as in gussets), economics, and packaging machinery perfor-

mance. Cellophane provides good heat resistance and external sealing; PET has excellent heat resistance but is difficult to cut on some packaging equipment. BOPP and ON have excellent resistance to fracturing (sometimes called flex cracking but more accurately described as durability), but these films lack heat resistance and will shrink if exposed to high heat-sealing temperatures. The durability of cellophane is significantly lowered by exposure to dry conditions, eg, cold dry weather. The uses of these structures is similar to those of the paper/poly/foil/poly structures. The same opportunities exist to use varying gauges of PE or any of the copolymer materials as the adhesive or coating medium.

End-use requirements for pouches depend on the characteristics of the product, the type of packaging machine that will be used, shelf-life demands, and secondary packaging. Some examples of typical pouch structures are shown in Table 9.

E. L. MARTIN
Printpack, Inc.

MULTILAYER PLASTIC BOTTLES

The concept of combining different materials into one package has been used for many years in the production of flexible bags and pouches (see Coextrusion machinery, tubular; Coextrusions for flexible packaging; Laminating machinery; Multilayer flexible packaging) and thermoformed containers (see Coextrusion machinery, flat; Coextrusions for semirigid packaging) (1). In countries outside the United States, particularly Japan, the concept has been applied to blow-molded bottles (see Blow molding) as well (2), but it has only recently attained commercial significance in the United States. Commercial development here had to await the availability of suitable barrier materials (see Barrier polymers), as well as the refinement of processing techniques.

For many applications, monolayer bottles are not adequate, because no one polymer has all the properties that many packaging applications require at prices competitive with other packaging materials, such as glass and metal. Multilayer structures are called for whenever a combination of properties not found in a single polymer is needed. For example, food containers may call for high barrier properties, resistance to high temperatures during hot filling or retorting, and chemical resistance to certain agents, all at a cost that is competitive with other packaging media. Typically, a bottle made from a single high barrier, multifunctional plastic would be far too expensive. In multilayered bottles, a strong but low cost plastic is used to satisfy the mechanical requirements, and the usually expensive high barrier plastics are used in very thin layers.

At present, the chief driving force for the development of multilayer bottles is the need to improve the gas barrier of the commodity resins. This, however, is not the only reason. Some of the earliest coextruded bottles and tubes in Japan were three-layer structures with nylon on the outside for abrasion resistance and aesthetic effects. Coextrusion can also be used to combine pigmented and nonpigmented layers. Resistance to nonpolar hydrocarbons (see Surface and hydrocarbon-barrier modification) can be imparted to polyolefins by coextruding a resistant layer of nylon (see Nylon) in contact with the solvent. In a proprietary process from DuPont, blending and processing nylon and HDPE (see Polyethylene, high density) produces a solvent-barrier container. The nylon forms plates in the structure which retard the solvent loss. Another method of producing a solvent container is to use a fluorination process to change the HDPE structure (3). Multilayer containers have entered the market produced by coinjection blow molding. Two companies, Nissei and American Can Corp., are active in this area. The advantage of this process over extrusion blow molding is that no scrap is produced. The Nissei machine injects three layers consecutively, and the American Can process injects five layers simultaneously. It is difficult to get all layers to move at the same rate, but American Can has solved this problem and is marketing a retortable plastic "can" (see Cans, plastic; Retortable flexible and semirigid packages). Another approach combines blow molding with thermoforming. The Co-Pak bottles and jars made by National Can Corp. are made by a three-stage process: a barrier liner made of coextruded sheet is thermoformed into a shape corresponding to that of a preform. A structural layer is injection-molded around that preform, which is then blow-molded (4).

Work is under way in many countries to improve the gas barrier of PET (see Polyesters, thermoplastic) bottles to make them suitable for packaging oxygen-sensitive beer and food products. This is being done to some extent by coating the finished bottles with PVDC (see Vinylidene chloride copolymers) (5). Since PET requires biaxial orientation for clarity, production of multilayer bottles containing PET will require techniques for producing coextruded or coinjected parisons for subsequent stretch blow molding (6–9).

Multilayer blow molding is still in its infancy, and this article is an introduction to the many new developments that can be expected in the near future. The discussion below pertains to coextrusion blow molding only.

Materials

When choosing the resins that are to make up a given multilayer container, the following properties of the desired bottle must be accounted for: oxygen-barrier requirement; water-vapor transmission requirement; carbon dioxide transmission requirement; aroma and flavor retention; stability at elevated temperature, eg, for hot filling, autoclaving, retorting; clarity, opacity, or decoration; mechanical strength, ie, rigidity and flexibility; surface characteristics, eg, scratch resistance, printability, gloss; effect upon content, solvent resistance extractability, and absorption; and size and shape of the container.

Barrier resins. Good to excellent oxygen and moisture barriers are required in most multilayer applications (see Barrier polymers). If the product to be contained is an aqueous product, an internal layer of a good water-vapor barrier is needed, eg, polyethylene (PE), polypropylene (PP), or PVDC. If the product is a relatively nonhygroscopic material, a nylon, PET or PVC can be used. The currently available oxygen barriers are EVOH (see Ethylene–vinyl alcohol), PVDC, acrylonitriles (see Nitrile polymers), PET, and rigid PVC (see Poly(vinyl chloride)). At present, EVOH is the most economical resin for oxygen barrier, even though it is the most expensive resin per unit weight. PVDC must be 6–10 times thicker to obtain the same barrier as dry EVOH. Nitriles must be 5 times thicker, and PET 14 times thicker. Nylon would have to be 20 times as thick, and PVC 38 times.

The barrier layer in coextrusion does not necessarily have to be thin, so any of the above resins can be used in a coextruded container if the economics are right. EVOH is much easier to process than PVDC, which requires special materials in the extruders, adapters, and die head. EVOH does lose some barrier owing to moisture, but recent shelf-life tests indicate that the barrier loss does not seem to be significant even after autoclaving. The EVOH must be protected from direct contact with moisture by an adequate moisture barrier.

Structural resins. To satisfy the required mechanical properties, the bulk of the container can normally be made of any of the common plastics, such as LDPE, HDPE, PP, PVC, PS, or PC. In monolayer form, HDPE is the most widely used blow-molding resin and is one of the best water-barrier resins; LDPE is used for improved flexibility with slightly lower water barrier. PP has better heat resistance than HDPE, and in monolayer form is used for containers that must be hot-filled, eg, syrups, or sterilized, eg, parenteral solutions. In addition, it provides better contact clarity than HDPE. If contact clarity is not sufficient, one of the transparent resins must be used as the structural layer. PVC is used for its clarity and resistance to certain chemicals to which polyolefins are sensitive. Polystyrene (PS) is used for clarity, rigidity, and low cost, but PS containers tend to be brittle. PC has clarity and outstanding heat resistance, but it is expensive and a poor barrier to both water and oxygen.

To create a container that can withstand high temperatures is a complex problem. A successful container depends on the heat resistance of the materials, the design of the container and closure, the heat content, headspace, and the processing temperature. Heat-resistant plastics such as PP or PC may be used in multilayer containers to support heat-sensitive plastics during hot filling, autoclaving, or retorting (10).

Adhesives and tie materials. The resins chosen for the desired combinations of their individual properties must be bonded together during container manufacture to prevent separation in use. Most polymers cannot be bonded without adhesives. The effectiveness of the adhesives varies with the type of resin to be bonded, and even with the specific nature of resins within the same family. It changes from brand to brand: an adhesive that bonds to PP from one manufacturer may not bond well to PP from another. Also, processing parameters have a definite effect on adhesion (11). For this reason, adhesives should be selected very carefully. Information about adhesives is best obtained directly from the manufacturers. The principal producers of adhesives for coextrusion are Northern Petrochemical (Plexar), DuPont (CXA), Mitsui (Admer), and Mitsubishi (Modic). Some companies, eg, Shell Chemical Co. and Eastman Chemical Products, Inc., make selected special adhesives.

Coextrusion Blow Molding

Coextrusion blow molding has been in existence since the early 1970s when Toyo Seikan Kaisha and Toppan Printing in Japan independently developed their own coextrusion die heads (12). Extrusion blow molding is the same, whether applied to a monolayer or a coextruded parison. Container designs can be identical, except that the coextruded containers can be made thinner owing to the extra strength of a multilayer container (the principle is the same as in double-bagging groceries). An optimum coextrusion machine can be designed for any given container. The main differences between a conventional blow-molding machine and one equipped for coextrusion are in the die head, the extruders, and the controls.

Heads. The die head is a critical part of the coextrusion system, since it is here that the different resins are brought together to create a single parison, which is then blown in the same manner as a conventional monolayer tube. The consistency and integrity of each layer in the finished product largely depend upon the design and operation of the head (13).

There are two basic ways for the head to combine the resins. The feedblock approach brings all the layers together at the same point. This allows for a shorter head, but requires that the viscosities of the resins be closely matched. This limits the combinations possible with such a head. Bottles with inner and outer layers of different materials would be difficult to produce (14). The multimanifold design permits each layer to form fully before adding the next. This overcomes many of the viscosity problems, but usually results in a longer head (14). Heads can be built for any number of layers, ie, from two to seven or more. Most firms build their own heads. The only commercial head available in the United States today is the Mac Head, imported from Japan.

Extruders. After the head is chosen, it is necessary to choose the type of extruders (see Extrusion), at least one for each type of resin. If the inner and outer layers are to be of the same material, a five-layer container can be made by using only three or four extruders, by splitting the output of the extruder to two ports of the head (13).

The type of extruder used depends on what plastic is to be extruded. The extruders for the adhesive and barrier layers have a very low flow rate, but it is generally not feasible to use an extruder smaller than 1.25 in. (32 mm) because of the torsional strength of the screw. At present, no extruder manufacturer is making extruders specifically for coextrusion blow molding.

The head must be connected to the extruders through melt piping. A $40:1$ $L:D$ ratio would probably not cause an excessive pressure drop, and even a $60:1$ ratio might be acceptable, but this must be engineered for each application, taking into account the specific plastic used and the rate of flow (13). Ideally, the melt piping should have only gradual bends, if any.

Controls. When extruding only a single layer of plastic, only a few trial setups are required to optimize the process.

When coextruding multiple layers, however, the viscosity of the material is critical for control of each layer's thickness (13). It is therefore necessary to measure the temperature of each material as it is extruded. Accordingly, the number of heaters and controls increases in coextrusion along with the number of extruders.

Present methods of temperature control rely on measuring the temperature of the extruder barrel for control of the heater elements. This is only indirectly related to the temperature of the plastic itself. Changes in room temperature, in material flow rate, and in the rate of heating by the heating elements all influence the temperature of the material, although the temperature of the barrel remains unchanged (13).

Because pressure also has an effect on the viscosity of various materials, if screen packs are used, pressure transducers are needed on both sides of the pack to control the pressure in the respective locations. Suitable means should be used to measure the rate of extruder-screw rotation, as related to flow rate.

With a five-layer container, it might be necessary to have 15 zones to control heat as well as cooling and possibly as many as 16 heat-only zones. At least five pressure transducers would also be required (13). This is very different from single-layer blow molding, where three temperature controllers for one extruder and two die-head temperature controllers are common, without any pressure transducers. With the use of microprocessors thermocouples will change to resistive temperature transducers. Several commercial systems using microprocessors are available for controlling blow-molding machines, including the Maco V from Barber Coleman, EM1 from Eurotherm, and the Harrel CP 670. Individual control units are also available from companies such as Love, LFE, Athena, and Eurotherm. The memories of microprocessors can retain the operating parameters in order to run the same job automatically again and again (15).

Economics

There are several blow-molding systems for coextruded parisons on the market; more are being developed. They vary in price, capacity, and capability. Just as with conventional blow-molding equipment, both wheel and shuttle machines are available. For high volume production, the wheel is more economical (16). Amortized equipment cost can be less than $0.01/container.

After the initial capital investment, the chief economic consideration in multilayer bottle manufacturing is the cost of resins. Many of the resins used in multilayer packaging are expensive, but their function is to reduce costs by providing more effective properties. For example, a layer of EVOH that costs $0.0058/container provides the same oxygen barrier as a (much thicker) bottle made of rigid PVC costing $0.064/bottle (11). Another example is the use of coextrusion to sandwich nonpigmented plastic between two thin layers of pigmented plastic, saving most of the cost of the expensive colorant (see Colorants). Substantial savings can be realized by using different layers for barriers, strength, and aesthetics, instead of looking for one material that meets all of these needs. The relative costs of coextruded containers compared with glass and metal cannot yet be expressed with precision, but some studies point to considerable savings (17).

Scrap. An important element in the cost of coextruded containers is the multilayer scrap generated in the production process. Cost-effective in-plant recycling is seldom, if ever, possible. If clarity is not essential, scrap can be incorporated in one of the bottle layers. Another solution is to use the scrap for another container, eg, as the inside layer of an opaque plastic product. If the scrap is simply thrown away, it can easily add 20% to resin costs. Most scrap consists of the ends of the parison that must be trimmed. Since the amount of this necessary scrap is generally constant for any given machine and independent of size, larger bottles have proportionally less scrap. In the production of small bottles, scrap can be reduced by 50% by blowing bottles in pairs, mouth-to-mouth, and then cutting them apart. On a wheel machine, scrap can also be reduced by adjusting the diameter of the wheel to minimize trim scrap (16).

Coextrusion blow molding with accumulators. For coextruding containers larger than 1 gal (3.785 L), it is necessary to use an accumulator. The Ishikawajima-Harima Heavy Industry (IHI) Co. in Japan developed a three-layer coextrusion blow-molding machine using an accumulator head. The extruders feed the resins intermittently into separate concentric chambers within the head. Concentric rams inject the materials simultaneously against a mandrel to form a multilayer parison. Speed and travel of each piston are individually controlled, with feedback to the extruders for automatic speed and pressure adjustments. IHI claims that the machine can stop the barrier layer before it gets into the trim, thus making scrap easier to recycle. Containers of 2–30-gal (7.6–113.6-L) capacity can be made on this equipment.

BIBLIOGRAPHY

1. F. R. Nissel, "Sheet Coextrusion After 15 Years: A 1981 Status Report," *Coextrusion Comes of Age,* Rosemont, Ill., June 15–17, 1981, The Society of Plastics Engineers, Inc., Brookfield Center, Conn., 1981, pp. 173–182.

2. M. F. X. Gigliotti, "State-of-the-Art Report: Multilayer Blown Containers, Coating vs. Coextrusion vs. Coinjection," *Coex '83,* Oct. 3–4, 1983, Düsseldorf, FRG, Schotland Business Research, Princeton, N.J. pp. 141–178.

3. K. J. Kallish, B. D. Bauman, and K. A. Goebel, "Direct In-Line Fluorination of HDPE Fuel Tanks," 43rd ANTEC, Washington, D.C., Apr. 29–May 2, 1985, The Society of Plastics Engineers, Inc., Brookfield Center, Conn., 1985 pp. 876–877.

4. *Package Eng.* **27**(3), 57 (Mar. 1982).

5. G. C. Dodson, "Coating: A Cost-effective and Quality Alternative to High-barrier Plastics," *9th International Conference on Oriented Plastic Containers, Atlanta, Ga.,* Mar. 25–27, 1985, Ryder Associates, Whippany, N.J., 1985.

6. R. B. McFall, "New High Barrier PET Resins For Coextrusion Applications," *Coex '84,* Düsseldorf, FRG, Sept. 19–21, 1984, Schotland Business Research, Princeton, N.J., FRG, 1984, pp. 277–287.

7. K. H. Siefert, "Available Equipment for Orientation Blow Molding with Specific Consideration of Extrusion Systems, *5th International Conference on Biaxially Oriented Containers, Amsterdam,* Oct. 16–17, 1984, The Society of Plastics Engineers, Inc., Brookfield Center, Conn., 1984.

8. R. W. Seymour and S. Weinhold, "High-barrier PET Containers," in Ref. 5.

9. T. Noble, "Heat-set, Hot-fill, and Other Topics," in Ref. 7.

10. A. J. Dambrauskas, "Polycarbonates in Coextruded Film, Semi-Rigid and Rigid Packaging," in Ref. 2, pp. 31–59.

11. A. M. Adur, J. Machonis, and M. Shida, "What's New In Plexar Resins," *1984 TAPPI Polymers, Laminations and Coatings Conference,* Boston, Mass., Sept. 24–26, 1984, TAPPI, Atlanta, Ga., 1984, pp. 647–652.

12. E. M. Naureckas, "Coextrusion Blow Molding: Developments Through Today and Beyond," *1985 TAPPI High Barrier Packaging Seminar,* Hilton Head, S.C., Apr. 1, 1985, TAPPI, Atlanta, Ga., 1985, pp. 7–15.

13. E. M. Naureckas, "Some Technical Aspects of Retrofitting Blow Molding Equipment for Coextrusion," in Ref. 3, pp. 896–897.

14. E. M. Naureckas, "Non-Symmetrical Layer Structure in Coextrusion Blow Molded Containers," in Ref. 6, pp. 361–372.

15. J. E. Roberts, "Anatomy of a Coextruder," in Ref. 1, pp. 79–88.

16. W. J. Fudakowski, "Selecting Technically and Commercially Appropriate Coextrusion Blow Molding Equipment," *Future-Pak '84,* Atlanta, Ga., Dec. 3–5, 1984, Ryder Associates, Whippany, N.J., 1984, pp. 207–248.

17. R. Foster, "Comparative Economics of Composite Containers," in Ref. 16, pp. 175–205.

General References

K. Ikari, "Oxygen Barrier Properties and Applications of Kurraray Eval Resins," *Coex '82,* Düsseldorf, FRG, Nov. 3–5, 1982, Schotland Business Research, Princeton, N.J., 1982, pp. 1–42.

G. A. Kruder, F. R. Bush, and D. L. Melvin, "Is Coextrusion the Best Answer?" *Coextrusion Comes of Age,* Rosemont, Ill., June 15–17, 1981, The Society of Plastics Engineers, Inc., Brookfield Center, Conn., 1981, pp. 37–52.

"Coextrusion Blow Molding for Gas Tanks and Industrial Parts," *Mod. Plast.* **10**(2), 27–37 (Mar. 1977) Vol. 10, No. 2, pp. 27–37.

A. I. Fetell, "Unique Process For Laminar Barrier Containers," pp. 275–288; J. A. Wachtel, B. C. Tsai, and C. J. Farrell, "Retorted EVOH Multilayer Cans With Excellent Oxygen Barrier Properties," pp. 5–33, *Future-Pak '84,* Atlanta, Ga., Dec. 3–5, 1984, Ryder Associates, Whippany, N.J., 1984.

E. M. NAURECKAS
J. K. NAURECKAS
Coex Engineering, Inc.

N

NETTING, PLASTIC

Plastic netting is produced through counterrotating die lips as a feed stock with strands at a nominal 45° to machine direction and then oriented into a lightweight tubular netting. A variety of mesh sizes, diameters, and colors are available. The primary resin used for plastic netting in the United States is high density polyethylene (HDPE) (see Polyethylene, high density).

Most extruded plastic netting used for flexible packaging is produced by a process invented in the mid-1950s and marketed throughout the world as Netlon (1).

In the United States and most other countries in North and South America the Du Pont Company introduced Netlon through a licensing agreement under the trademark Vexar in 1959.

Plastic netting is available in the following forms from manufacturers and converters:

tubular netting on traverse-wound rolls (commonly called "rope" by the industry);

header-label bags converted from rope, heat-set into open tubes (usually gussetted) and sealed by the manufacturer with a folded sewn printed label;

"G bags" which are lengths of material cut with a hot knife to produce a gathered heat seal on one end; and

"sleeving" or "cartridges" which is netting shirred onto a collapsed corrugated board. The board can be opened into a square or rectangle for transfer onto a tube or funnel.

The largest market for plastic netting is consumer-sized packaging in the produce industry. Nearly complete ventilation reduces spoilage of many fruits and vegetables such as citrus and onions. Automatic packaging equipment using rope shirred onto tubes offer labor and material costs competitive with other consumer-size packages (see Fig. 1). Premade header-label bags are also widely used by produce packers and by supermarkets that package in store.

In the United States plastic netting is used to overwrap frozen poultry, primarily turkeys. Netting is applied over the film bag with a stapled loop at one end as a handling convenience to the shopper. The netting also offers some protection to the film package.

Netting is used to bale or compress Christmas trees to prevent limb damage, increase density for shipment, and provide more convenient transportation and handling for the consumer.

Plastic netting is also used to package toys, plastic housewares, and various other consumers products. The variety of colors to complement the product or label and the conformability of netting to irregular shapes are advantages for these applications.

Other applications include:

"stockinettes" or carriers for curing or smoking meat, poultry, or cheese products;

consumer-sized packages of shellfish and fish-bait bags;

and unitization of various products as an alternative to polyethylene film or corrugated cartons for outdoor storage. Black netting is usually used to prevent deterioration from prolonged exposure to sunlight.

Plastic mesh is available as an alternative to plastic film for stretch-wrap containment of pallet loads or skids. It is used with spiral stretch-wrap machines, usually in 20-in. (51-cm) web widths. There are two concepts and various materials commercially available. One product is extruded from a resin or resin blend that is elastic and is stretched upon application (2). It continues to apply constant pressure on the wrapped load. The other product is a high tensile fully oriented material that relies upon tight initial application but offers little compensation for settling or compression of the wrapped product during shipment and storage (3).

In general, mesh is more expensive than film. Primary applications include:

products requiring ventilation to retard spoilage (eg, fresh fruits and vegetables);

hot-packed products (eg, flour and pet foods) since dissipation of heat is important to minimize condensation;

food products blast-frozen after loads are unitized (eg, dairy products, meat, and poultry);

![Figure 1. Oriented plastic netting tubing ("rope").]

Figure 1. Oriented plastic netting tubing ("rope").

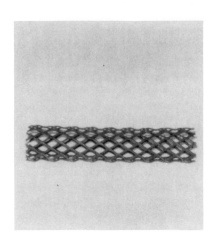

Figure 2. Protective shaft cover.

extremely irregular loads, with or without sharp projections, difficult to contain by other methods and requiring ventilation; and

highly combustible or flammable products.

Unoriented mesh tubing is used to protect machined or polished tubes, shafts, gears, and other metal parts against scuffing, abrasion or other damage (see Fig. 2). Sleeving is available in color-coded sizes. Each size is somewhat elastic and fits a range of diameters.

Protective sleeving is extruded of low density polyethylene (LDPE) or other relatively soft resins. This product can be applied without need for hot-air guns or shrink tunnels and does not entrap moisture or solvents.

BIBLIOGRAPHY

1. U.S. Pat. 2,919,467 (Jan. 5, 1960), F. B. Mercer (to Plastic Textiles Accessories Ltd.).
2. Stretchnet manufactured by Bemis Company, Inc., Multi-Net Division, St. Louis, Mo.
3. Tensionet and Tensionet 2 manufactured by Conwed Corporation, Plastics Division, Minneapolis, Minn.

K. F. OEKEL
Bemis Company, Inc.

NETWORKS

"Networking" has become such a popular term that some may believe it is also a new concept. It is not. Popularized recently in any number of contemporary articles and books, and particularly prevalent in the computer community, the term simply describes a communications process.

In fact, business and professional people have always depended upon their own information networks in order to keep current in their respective fields. In packaging, networks are particularly important because of the diversity of interests often required to plan and coordinate the design and implementation of a successful package.

In this process, a mechanical engineer may be required to design, build, or otherwise tinker with a packaging machine or machinery system. An industrial engineer may be involved with packaging material supply or plant layout; a manufacturing engineer, with productivity and integration with other elements in the factory; a chemical engineer, with migration; a quality-control engineer, with finished product integrity; a distribution engineer, with warehousing and safe transportation systems. Each may belong to his or her own respective professional society, but that society may only peripherally be involved in packaging issues and information.

Meeting grounds to address packaging concerns specifically had to be created.

Trade and Business Media

The first to serve the growing need for up-to-date knowledge and information was a publication. Many publications now serve the field. Some focus on packaging alone; others treat packaging as an important adjunct to their primary editorial coverage.

A list of the North American publications currently devoted to packaging interests is presented here as Table 1.

Importantly, most of these publications are distributed to "qualified readers" free of charge. In many large companies, a library is established to house a large number of publishing titles to assure broad access to information by employees.

Anyone who spends a considerable amount of time on packaging matters benefits from access to one or more of the packaging publications.

Shows and Conferences

Trade shows and conferences are meeting grounds in a literal sense. There, individuals from different industries and with different job responsibilities—engineering, production, marketing and sales, and management—meet to address common problems. Shows can often spark new ideas, because packaging is often readily adaptable from one application or industry to another.

Thus, although packaging may be represented at trade shows organized for specific end-use industries, those who have a substantial responsibility in packaging should also attend the principal international packaging expositions. In the United States, this is Pack Expo (see Exhibitions).

Associations and Societies

Organizations provide opportunities for direct, personal contact, and they often have the added advantage of incorporating published information as a member service. The Society of Packaging and Handling Engineers, for example, publishes a directory, a regular newsletter, and the *SPHE Technical Journal* twice a year. SPHE and The Packaging Institute, U.S.A., are both individual membership societies serving the broad interests of packaging suppliers and users.

Other professional societies listed in Table 2 (eg, the American Society of Mechanical Engineers and the American Society for Quality Control) may have sections or committees that deal with packaging matters. For international concerns, the World Packaging Organization is now located in Paris. North American voting members are the SPHE and the Packaging Association of Canada.

Table 2 also contains a listing of the various trade associations, groups comprising company memberships from specific supply or use industries.

Again, some of these are concerned only with specific packaging applications or materials. The Glass Packaging Institute, the Flexible Packaging Association, Paperboard Packaging Council, and the Packaging Machinery Manufacturers Institute are examples. Others, such as the Aluminum Association, The Society of the Plastics Industry, and the American Iron and Steel Institute have broader concerns but are nonetheless very much involved in certain packaging matters.

Many of these associations and societies provide opportunities for direct involvement through committee work and various education seminars offered periodically on specific interest topics.

Education

For many people involved in packaging, seminars and conferences are a key means of meeting leading experts in a particular field, interchanging information and experiences with peers, and brushing up on technical information.

A wealth of packaging-related information is available through online databases. Since most databases are online versions of published literature, the distribution of packaging-

Table 1.   North American Publications

Beverage Digest P.O. Box 238 Old Greenwich, CT 06870 (203) 629-8423	*Boxboard Containers* 300 W. Adams Chicago, IL 60606 (312) 726-2802
Canadian Packaging 481 University Avenue Toronto, Ontario, M5W 1A7, CANADA (416) 596-5744	*Converting Product News* 400 N. Michigan Avenue Chicago, IL 60611 (312) 222-2000
Distribution Chilton Way Radnor, PA 19089 (215) 964-4383	*Food & Drug Packaging* 7500 Old Oak Boulevard Cleveland, OH 44130 (216) 243-8100
Food Engineering Chilton Way Radnor, PA 19089 (215) 964-4447	*Good Packaging* 1315 E. Julian St. San Jose, CA 95116 (408) 286-1661
Handling & Shipping Management 1111 Chester Avenue Cleveland, OH 44114 (216) 696-7000	*Material Handling Engineering* 1111 Chester Avenue Cleveland, OH 44114 (216) 696-7000
Modern Materials Handling 221 Columbus Avenue Boston, MA 02116 (617) 536-7780	*Packaging Printing & Diecutting* 401 N. Broad St. Philadelphia, PA 19108 (215) 238-5300
Packaging 1350 Touhy Ave. Box 5080 Des Plaines, IL 60018 (312) 635-8800	*Packaging Digest* 400 N. Michigan Ave. Chicago, IL 60611 (312) 222-2000
Packaging Newsletter 818 Connecticut Avenue., NW Washington, DC 20006 (202) 862-5100	*Packaging Technology* 101 West Street Hillsdale, NJ 07642 (201) 664-7525
Paper, Film & Foil Converter 300 W. Adams Street Chicago, IL 60606 (312) 726-2802	*Paperboard Packaging* 7500 Old Oak Boulevard Cleveland, OH 44130 (218) 723-9465
SPHE Technical Journal (Society of Packaging and Handling Engineers) Reston International Center Reston, VA 22091 (703) 620-9380	*Powder Bulk Solids* 15 S. 9th St., Ste 320 Minneapolis, MN 55402 (612) 370-0413

related information is as at least as wide as its distribution in the trade literature. There are articles about metal cans cited, for example, in the Metals Abstracts/Alloys Index (Metadex) database maintained by the American Society for Metals and The Metals Society, London. The subject of food packaging is covered in the food-related databases (eg, Foods Adlibra), plastics in the plastics databases (eg, Rapra), paper in the paper databases (eg, Paperchem), etc.

In addition, there are several specialized packaging databases. One is PIRA, produced by PIRA, Leatherhead, Surrey, UK, which contains abstracts of scientific and technical literature on all aspects of papermaking, boardmaking, packaging, and printing. Another is Packaging Science and Technology Abstracts (PSTA), produced by the International Food Information Service, Frankfurt, FRG.

Increasingly, the nation's colleges and universities are also playing a role in this, especially in helping prepare talented young people for productive careers in packaging. Rutgers, the State University of New Jersey, maintains the Ira Gottscho Library, a specialized packaging library, as a part of its packaging-education facilities.

Several are now offering degrees which provide a minor or option in packaging engineering, and others are integrating packaging courses into more established disciplines such as mechanical engineering and food science. Some of the universities have opted for offering these courses during evening hours or in concentrated summer programs so that participation by those already in industry is encouraged.

Table 3 presents a list of U.S. institutions offering packaging courses or, in a few instances, options in major degree paths. For example, the following disciplines are represented among the students and faculty at Michigan State University School of Packaging in 1985: agricultural chemistry, agricultural engineering, applied mechanics, botany, chemical engi-

Table 2. North American Associations and Organizations

The Adhesive and Sealant Council
1600 Wilson Boulevard, Suite 910
Rooslyn, VA 22209
(703) 841-1112

Aluminum Foil Container Manufacturers Association
P.O. Box 7
Walworth, WI 53184
(414) 275-6838

American Defense Preparedness Association
1700 North Moore Street, Suite 900
Arlington, VA 22209
(703) 522-1820

American Frozen Food Institute
1700 Old Meadow Rd., Suite 100
McLean, VA 22102
(703) 821-0770

American Institute of Food Distribution, Inc.
28-12 Broadway
Fair Lawn, NJ 07410
(201) 791-5570

American Management Association
135 West 50th Street
New York, NY 10020
(212) 586-8100

American National Standards Institute (ANSI)
1430 Broadway
New York, NY 10018
(212) 354-3300

American Paper Institute, Specialty Packaging and
  Industrial Div.
260 Madison Avenue
New York, NY 10016
(212) 340-0600

American Society for Quality Control
230 W. Wells Ave.
Milwaukee, WI 53203
(414) 272-8575

American Society for Testing and Materials (ASTM)
1916 Race Street
Philadelphia, PA 19103-1187
(215) 299-5400

American Society of Mechanical Engineers (ASME)
345 East 47th Street
New York, NY 10017
(212) 705-7470

American Trucking Association
1616 P Street, NW
Washington, DC 20036
(202) 835-9100

Associated Cooperage Industries of America
2100 Gardiner Lane
Louisville, KY 40205
(502) 459-6113

Association of American Railroads
1920 L Street, NW
Washington, DC 20036
(202) 484-6400

Association of Independent Corrugated Converters
414 N. Washington St.
Alexandria, VA 22314
(703) 836-2422

Association of Professional Material-Handling
  Consultants
1548 Tower Road
Winnetka, IL 60093
(312) 441-5920

Can Manufacturers Institute
1625 Massachusetts Avenue, NW
Suite 500
Washington, DC 20036
(202) 232-4677

Carded Packaging Institute
P.O. Box 1333
Stamford, CT 06904
(203) 323-3143

Closure Manufacturers Association
6845 Elm St.
Suite 208
McLean, VA 22101
(703) 821-1118

Composite Can and Tube Institute
1742 N Street, NW
Washington, DC 20036
(202) 223-4840

Containerization Institute
299 Madison Avenue
Suite 1000
New York, NY 10017
(212) 697-3120

Dairy and Food Industries Supply Associations, Inc.
6245 Executive Boulevard
Rockville, MD 20852
(301) 984-1444

Fibre Box Association
5725 N. East River Road
Chicago, IL 60631
(312) 693-9600

Flexible Packaging Association
1090 Vermont Avenue, NW
Suite 500
Washington, DC 20005
(202) 842-3880

Flexographic Technical Association
95 West 19th Street
Huntington Station, NY 11746
(516) 271-4224

Food Processing Machinery and Supplies Association
1828 L Street, NW
Suite 700
Washington, DC 20036
(202) 833-1790

Food Processors' Institute
1401 New York Ave. NW
Suite 400
Washington, DC 20005
(202) 393-0890

Glass Packaging Institute
6845 Elm Street, Suite 209
McClean, VA 22101
(703) 790-0800

Grocery Manufacturers of America
1010 Wisconsin Avenue, Suite 800
Washington, DC 20007
(202) 337-9400

Gummed Industries Association
380 North Broadway
Jericho, LI, NY 11753
(516) 822-8948

INDA, Association of the Nonwoven Fabrics Industry
1700 Broadway
New York, NY 10019
(212) 582-8401

Table 2. (*Continued*)

Institute of Food Technologists
221 North LaSalle Street
Chicago, IL 60601
(312) 782-8424

International Material Management Society
650 East Higgins Road
Schaumberg, IL 60195
(312) 310-9570

Material Handling Institute
1326 Freeport Road
Pittsburgh, PA 15238
(412) 782-1624

Metal Tube Packaging Council of North America
118 East 61st Street
New York, NY 10021
(212) 935-1290

National Association of Container Distributors
1227 Hennepin Ave.
E. Minneapolis, MN 55414
(612) 331-8880

National Barrel and Drum Association
910 17th Street, NW—Suite 912
Washington, DC 20006
(202) 296-8028

National Classification Board
1616 P Street, NW
Washington, DC 20036
(202) 797-5308

National Council of Physical Distribution Management
2803 Butterfield Road #380
Oak Brook, IL 60521
(312) 644-6610

National Food Processors Assn.
111 E. Wacker Drive
Chicago, IL 60601
(312) 644-6610

National Food Processors Association
1401 New York Ave, NW
Washington, DC 20005
(202) 639-5900

National Frozen Food Association
P.O. Box 398
Arlington, VA 22202

National Paper Box Association
231 Kings Highway East
Haddonfield, NJ 08033
(609) 429-7377

National Wooden Pallet and Container Association
1619 Massachusetts Avenue, NW
Washington, DC 20036
(202) 667-3670

Package Designers' Council
P.O. Box 3753
Grand Central Station
New York, NY 10017
(212) 682-1980

Packaging Association of Canada
111 Merton Street
Toronto, Ontario, Canada M4S 3A7
(416) 929-3194

Packaging Education Foundation
Reston International Center
Reston, VA 22091
(703) 620-2155

The Packaging Institute, USA
200 Summer St.
Stamford, CT 06905
(203) 325-9010

Packaging Machinery Manufacturers Institute (PMMI)
1343 L Street, NW
Washington, DC 20005
(202) 347-3838

Paperboard Packaging Council
1101 Vermont Avenue, NW, Suite 411
Washington, DC 20005
(202) 289-4100

Paper Converters' Association
1133 15th Street, NW
Washington, DC 20005
(202) 429-9440

Paper Shipping Sack Manufacturers' Association
2 Overhill Road
Scarsdale, NY 10583
(914) 723-6440

Single Service Institute
1025 Connecticut Avenue, NW
Suite 513
Washington, DC 20036
(202) 347-0020

Society of Packaging and Handling Engineers
Reston International Center
Reston, VA 22091
(703) 620-9380

Society of the Plastics Industry
355 Lexington Avenue
New York, NY 10017
(212) 503-0600

Steel Shipping Container Institute
2204 Morris Avenue
Union, NJ 07083
(201) 688-8750

Tag and Label Manufacturers Institute
196 North Street
Stamford, CT 06901
203-323-3143

Technical Association of the Pulp and
   Paper Industry (TAPPI)
Technology Park Atlanta
P.O. Box 105113
Atlanta, GA 30348
(404) 446-1400

Textile Bag Manufacturers Association
P.O. Box 2145
Northbrook, IL 60062
(312) 272-3930

Warehousing Education and Research Council
4022 Countryview Drive
Sarasota, FL 33583
(813) 924-3544

Western Packaging Association
309 Matthew St.
Santa Clara, CA 95050
(408) 986-8212

Wirebound Box Manufacturers' Association
380 W. Palatine Road
Wheeling, IL 60090
(312) 520-3280

Table 3. U.S. Institutions Offering Education in Packaging

California Polytechnic State University San Luis Obispo, California 93407	Rochester Institute of Technology One Lomb Memorial Drive Rochester, New York 14623
Chapman College 333 N. Glassell Street Orange, California 92666	Rutgers University P.O. Box 909 Piscataway, New Jersey, 08854
Clemson University Clemson, South Carolina 29631	San Jose State University San Jose, California 95192
Cornell University Ithaca, New York 14853	Sinclair Community College 444 W. Third Street Dayton, Ohio 45402
Indiana State University 6th and Cherry Streets Terre Haute, Indiana 48709	Spring Garden College 102 E. Mermaid Lane Chestnut Hill, Pennsylvania 19118
Joint Military Packaging Center Aberdeen Proving Ground, Maryland 21005	Texas A&M University College Station, Texas 77843-3123
Michigan State University East Lansing, Michigan 48824-1223	University of Detroit 4001 W. McNichols Road Detroit, MI 48221
New York University 80 Washington Square East, Rm. 53 New York, New York 10003	University of Missouri-Rolla 301 Harris Hall Rolla, Missouri 65401
North Carolina State University at Raleigh Box 5992 Raleigh, NC 27650	University of Wisconsin-Stout Menomonie, Wisconsin 54751
Pratt Institute Pratt Studios 24 Brooklyn, New York 11205	

neering, chemistry, education, electrical engineering, fermentation science, food science, food science and human nutrition, food science packaging, food technology, forest products, forestry, general science, geology, horticulture, industrial engineering, international trade, mathematics, mechanical engineering, organic chemistry, packaging, pharmaceutical chemistry, photography and printing, physics and crystallography, plant pathology, printing management, resource development, social work, textile engineering, and wood technology.

International Research Institutes

One prominent international network is the International Association of Packaging Research Institutes. It was formed "to establish and advance professional personal relations between research workers in packaging in various parts of the

Table 4. Members of the International Association for Packaging Research Institutes (IAPRI)

Australia: National Materials Handling Bureau, NSW
Austria: Packaging Laboratory for Foodstuffs and Beverages,
  Vienna
Belgium: Belgium Packaging Institute, Brussels
Brasil: CETEA, Campinas, Sao Paulo
Czechoslovakia: Institute of Handling, Transport, Packaging and
  Storage Systems, Prague
Denmark: Danish Packaging Research Institute, Copenhagen
Federal Republic of Germany: Advice and Research Unit for
  Maritime Transport Packaging Institute for Export Packaging,
  Hamburg
  Institute of Food Technology and Packaging (ILV), Munich Federal
    Institute for Materials Testing, Berlin
Finland: The Finnish Pulp and Paper Research Institute
  The Finnish Packaging Research Institute, Helsinki
France: French Packaging Institute (IFEC), Paris
  National Testing Laboratories (LNE), Trappes
Israel: Technion, Israel Institute of Technology
Japan: Japan Packaging Research Institute, Tokyo
	Mexico: National Laboratory for Industrial Development,
	  Mexico City
	Morocco: Moroccan Packaging Institute, Casablanca
	The Netherlands: Institute TNO for Packaging Research
	  (IVV), Delft
	Norway: Norwegian Agricultural Institute for Research in
	  Food Packaging, Vollebekk
	Poland: Polish Packaging Research and Development
	  Centre, Warsaw
	Sweden: Swedish Packaging Research Institute, Stockholm
	Switzerland: EMPA. Swiss Federal Laboratories for Testing
	  Materials and Research, St. Gallen
	United Kingdom: Pira. The research association for the
	  paper and board, printing and packaging industries,
	  Leatherhead
	United States of America: Rutgers University, New Brunswick,
	  New Jersey, and Michigan State University, School of Packaging,
	  East Lansing, Michigan

world and, as far as possible, to avoid duplication of work between institutes."

Its members are both universities and professional societies or institutes established within a number of countries. A list of members is presented in Table 4.

W. C. Pflaum
W. C. Pflaum Co., Inc.

NITRILE POLYMERS

Nitrile polymers are generically those which contain the cyano ($C\equiv N$) functional group, also called the nitrile group. The commercial development of these materials was due in large part to the 1957 discovery by Sohio of a low-cost one-step process for acrylonitrile (AN) production (1). The pure nitrile polymer, polyacrylonitrile (PAN), is 49% nitrile. It is an amorphous transparent polymer with a relatively low glass-transition temperature ($T_g = 87°C$) (see Polymer properties) that provides an outstanding barrier to gas permeation and exceptional resistance to a wide range of reagents. Unfortunately, its combination of properties is not of commercial value in packaging. Its primary deficiency is that it is not melt processable. Degradation occurs at 428°F (220°C), below that required for melt processing. To overcome this handicap, nitrile polymers are produced using acrylonitrile ($CH_2{=}CHCN$) as a comonomer with other monomers that impart melt processability. Through copolymerization, the desirable properties can be retained and the undesirable properties can be suppressed.

Copolymers

Styrene–acrylonitrile (SAN). Typical SAN polymers are made using a 3:1 ratio by weight of styrene to acrylonitrile. The copolymer has a combination of properties which reflect the processability of the styrene component and the chemical resistance of the AN; but gas-barrier properties are low because of the relatively low AN concentration. SAN is of relatively minor significance in packaging, used in applications where PS would suffice with an added measure of chemical resistance.

Acrylonitrile–butadiene–styrene (ABS). ABS is a graft copolymer of SAN onto a polybutadiene backbone. The SAN forms a matrix phase; the polybutadiene a discrete (dispersed) phase. A tough impact-resistant thermoplastic is produced by using the grafting mechanism to compatibilize the two phases. Although the polybutadiene is an excellent impact modifier with a low T_g (ca −85 °C), its refractive index is different from the SAN matrix polymer. In contrast to PS (see Polystyrene), PAN, and SAN, therefore, it is not transparent. ABS is a major commercial thermoplastic, but it is rarely used in packaging because of its opacity, lack of gas-barrier properties, and economics relative to other commodity resins. Thermoformed ABS (see Thermoforming) has been used, for example, to produce margarine tubs. This is an application in which PS would suffice if not for its limited resistance to stress cracking. A wide range of SAN ratios is used to achieve properties of value in nonpackaging applications.

Acrylic multipolymers. Some acrylic multipolymers are produced using AN as a comonomer with methyl methacrylate (MMA); for example, XT Polymer (Cy/Ro Industries). Refractive-index-matched rubber modifiers are incorporated to combine toughness with transparency. These materials are used

in health-care packaging (2) (see Health-care packaging) and in some food packaging applications. Because the AN concentration is low, these materials do not have exceptional gas-barrier [18.7 cm³ · mil/(100 in.² d · atm) [72.6 cm³ · μm/(m² · d · kPa)] at 73°F (23°C)] properties.

High-nitrile resins (HNR). As noted above, SAN is a styrene–acrylonitrile copolymer. Its gas-barrier properties are limited, however, by the low nitrile content (eg, 12%). High barrier melt-processable polymers can be produced by raising the nitrile content above 25%. Over this threshold, copolymer properties begin to resemble those of PAN, particularly with respect to gas barrier and chemical resistance.

Acrylonitrile–styrene copolymers (ANS). High-nitrile copolymers can be produced by combining acrylonitrile and styrene in a 70:30 ratio. This was the approach taken by Monsanto Company (Lopac) and Borg-Warner (Cycopac) during the 1960s and early 1970s in their development of resins for carbonated-beverage packaging. As major commercial development began, toxicological problems with AN surfaced (3). The FDA banned the use of HNRs in beverage packaging because of concern for potential AN-extraction from the bottle into the beverage. Since beverages are a major component of the diet, their treatment by FDA was most severe. The Agency continued to permit the use of HNRs for direct and continuous nonbeverage-food-contact applications (4) with filling and storage temperatures less than 150°F (65.6°C); but because the major commercial significance of ANS polymers was in carbonated beverage packaging, all commercial production of those polymers was discontinued.

In 1984, the FDA amended its position on HNR-beverage applications (5). Monsanto Company had petitioned on behalf of an ANS high nitrile resin. As described in their process patent (6), the bottle preform is irradiated with an electron beam prior to blowing of the bottle. The company claimed that this process resulted in a bottle that, because of the thermodynamics of the extraction process, would have essentially no extraction of AN by the contained beverage. The FDA did not accept this claim, but it did decide that the extraction would be below the detection limit of 0.16 ppb. The Agency ruled that AN concentrations at or below this detection limit would be considered acceptable, and limited the residual AN content of the finished container to 0.1 ppm.

Rubber-modified acrylonitrile–methacrylate copolymers (AN/MA). High barrier properties can also be achieved by copolymerizing acrylonitrile and methlacrylate in a 75:25 ratio onto a nitrile rubber backbone. This is the approach taken by Sohio Chemical Company (formerly Vistron Corporation) in the production of Barex 210 resins. ANS and rubber-modified AN/MA copolymers (RM-HNRs) have similar barrier properties, but the AN/MA copolymer is tougher because of the rubber content. It can be blown into bottles by all conventional blow-molding methods (see Blow molding). In contrast, ANS copolymers required stretching to compensate for their toughness deficiency. Transparency is retained through the use of a refractive-index-matched rubber modifier.

Properties

Among the commerical packaging polymers that have the physical properties required for monolayer structures. HNRs offer the greatest gas barrier [0.8 cm³ · mil/(100 in.² · d · atm) or 3/cm² · μm/(m² · d · kPa)] and chemical resistance. Their gas-barrier properties (see Barrier polymers) are surpassed only by EVOH (see Ethylene-vinyl alcohol) and PVDC (see Vinylidene chloride copolymers), which are used as components of multilayer structures (see Coextrusions for flexible packaging; Coextrusions for semi-rigid packaging; Multilayer plastic bottles). Because of the polarity that the nitrile group imparts to the molecule, HNRs show an affinity for water. Water-vapor barrier is lower than that of the nonpolar polyolefins (eg, polyethylene, polypropylene); but that same polarity imparts resistance to nonpolar solvents. The relatively high flexural modulus means that for moldings with identical geometry, the HNR parts can be designed with thinner walls for equivalent stiffness compared to the polyolefins, PVC, and thermoplastic polyesters.

Applications

RM-HNR is used in packaging in a variety of physical forms. These include: film, semirigid sheet, and injection-molded (see Injection molding) and blow-molded containers. Blown film (see Extrusion) is used in polyolefin-container structures to provide formability, chemical resistance, and gas barrier. Spices, medical devices, and household chemical products are example of such applications. Laminations with aluminum foil are used in applications ranging from food packaging to oil-drilling core wraps (7). These structures have exceptional barrier properties, as well as sealability and chemical resistance. Oriented RM-HNR film, now commercialized in Japan, is also used as a safety shrink wrap around glass beverage bottles (see Table 1).

The semirigid sheet market for RM-HNR is primarily meat and cheese packaging in thermoformed (see Thermoforming) blister packages. With its excellent gas barrier, clarity, and rigidity, the RM-HNR is the premium packaging material. Of increasing importance in semirigid applications, however, is disposable medical device packaging. The RM-HNR can be sterilized by either ethylene oxide (ETO) or gamma radiation and is unaffected by plasticizers present in many devices. As a result of its combination of properties, usage has been increasing in this market area.

Blow-molding applications for RM-HNR are dominated by chemical-resistance requirements. Injection blow molding is the most widely used method for manufacturing small containers; for example, bottles for correction fluid, nail enamel, and other cosmetics. Larger bottles are generally extrusion-blow-molded. Some important applications include: pesticides and other agricultural chemicals, fuel additives, and hard-to-hold household chemicals. Extrusion-stretch-blow-molding is gaining increasing acceptance for bottles 16 ounces (473 mL) or larger. The orientation achieved during stretching greatly increases the drop-impact performance, and the walls can be relatively thin.

RM-HNR has been coextruded with many different polymers, but the polyolefins have been of greatest commercial significance. Coextrusions are available in sheet, film, and bottle form. They typically gain gas barrier or chemical resistance from the HNR and water-vapor barrier and economics from the polyolefin. In structures with polypropylene, the

Table 1. Properties of Rubber-Modified HNR Films

	ASTM test	Units	1 mil (25.4) μm) film Blown	1 mil (25.4) μm) film Biaxially oriented
specific gravity	D 1505		1.15	1.15
yield		in.²/(lb·mil)[m²/(kg·mm)]	24,080 [1,348]	24,080 [1,348]
haze	D 1003	%	5–7	4–6
light transmission		%	92.5	92.5
tensile strength	D 882	psi (MPa)	8,500–11,000 (58.6–75.9)	16,000–24,000 (110.3–165.5)
elongation	yield	%	4	4
	break	%	100–150	50–100
tear strength				
initial	D 1004	gf/mil (N/mm)		456 (176)
propagating	D 1922	gf/mil (N/m)	9.0 (3474)	4.0 (1544)
tensile impact	D 1822	lbf/in² or psi (kPa)	75–200 (517–1380)	475–650 (3280–4480)
bursting strength	D 774	1 mil (25.4μm), psi (kPa)	30–50 (207–349)	50–75 (349–517)
water absorption, 24 hr	D 570	%	0.25	0.25
folding endurance (tension of 1.5 kgf or 14.7N)	D 2176	cycles	425	425
change in linear dimensions at 100°C for 30 min.	D 1204	%	40	60
service temperature range		°F (°C)	−50–165 (−46 to 74)	−50–165 (−46 to 74)
heat-seal temperature		°F (°C)	250–375 (121–191)	250–375 (121–191)
oxygen permeability at 23°C, 90% rh	D 1434	cm³·mil/(100 in.²·d·atm) [cm³·μm/(m²·d·kPa)]	0.8 [3.1]	0.35 [1.36]
water vapor trans. rate, 38C, 90% rh		g.mil/(100 in.²·d)[g·mm/(m²·d)]	5.0 [1.97]	3.5 [1.38]
COF, face-to-face back-to-back	D 1894		0.45 static 0.35 kinetic	0.45 static 0.35 kinetic

heat-deflection temperature (HDT) of the structure is increased by the higher HDT of the polypropylene. This permits the use of RM-HNR in high-temperature environments such as microwave ovens.

The adhesive used to combine the layers in an HNR coextrusion are typically styrene–isoprene or styrene–butadiene block copolymers (see Multilayer flexible packaging). Scrap is reusable in the polyolefin layer if the nonolefin percentage in that layer is well dispersed and of lower concentration than about 15%.

Chemical-resistant coextruded bottles containing RM-HNR are now being commercialized (see Multilayer plastic bottles). RM-HNR is the inner contact layer enclosed by adhesive and polyolefin, typically HDPE. Use of a three-layer structure limits the cost and complexity of the machinery. It also places the solvent-resistant polymer in contact with the chemicals. Five-layer or laminar structures (see Surface modification; Nylon) using other barrier resins place the polyolefin in direct contact with the aggressive contents of the container. The three-layer structure also allows visual inspection of the barrier layer and maximizes the sealing area of the barrier layer at the pinchoff of the bottle. RM-HNR coextruded containers offer high performance with economics superior to the other packaging alternatives (8).

The HNR family offers a unique combination of properties to the packaging industry. Developing layer-combining technologies, increasing consumer acceptance of plastic packaging, and the changing FDA status bode well for HNRs. Displacement of metal and glass in packaging applications should be accelerated by the new processing technologies in blowmolding. Sheet and film applications are expected to show growth in both medical disposable and food market areas.

BIBLIOGRAPHY

1. U.S Pat. 2,904,580 (Sept. 15, 1959), J. D. Idol (to Standard Oil, Ohio).
2. J. M. Lasito, "Acrylic Multipolymers in Medical Packaging," *Proceedings of the Technical Association of the Pulp and Paper Industry* **71,** 74 (Sept. 15, 1982).
3. Report to the FDA by the Manufacturing Chemists Association, Jan. 14, 1977.
4. *Fed. Regist.* **41,** 23,940. Title 21, Part 177.1480 (June 14, 1976).
5. *Fed. Regist.* **49,** 36,637, Title 21, Part 177 (Sept. 19, 1984).
6. U.S. Pat. 4,174,043 (Nov. 13, 1979), M. Salome and S. Steingher (to Monsanto).
7. U.S Pat 4,505,161 (Mar. 19, 1985), P. K. Hunt and F. J. Waisala (to Standard Oil, Ohio).
8. J. P. McCaul, "The Economics of Coextrusion," *Proceedings of the High Technology Plastic Container Conference,* Society of Plastics Engineers, Nov. 11, 1985.

JOSEPH P. MCCAUL
SOHIO Chemical Co.

NONWOVENS

Some forms of nonwovens have been in existence for a hundred years or more (1), but it has only been the last 35 years that their importance has gained recognition. A nonwoven is a fabriclike structure composed of fibers that are bonded to give the structure sufficient integrity.

Web Manufacture

The fibrous webs can be produced by several methods. These include carding, air-laying, wet-laying, and the spunbonded method.

Carding. The process of carding was initially used in the manufacture of textile yarns. The carding machine uses rolls that are covered with either metallic or fillet clothing. Metallic clothing resembles a band-saw blade which is spirally wound around the rolls. Fillet clothing incorporates staplelike wires which are punched through a ¼ in. (6 mm) thick rubberized fabric base with a width of 1–3 in. (25–76 mm). The carding rolls, including main cylinder, workers, strippers, etc, are set close together. During carding either the action of working or stripping can take place. When working, the direction of the wire points, and the speed and direction that the wire travels, is such that the clothing operates point-against-point to clean, open, disentangle, straighten, and distribute the fibers in a uniform manner. Stripping takes place when the same variables are such that the fibers are raked from one roll by another. It was recognized over 100 years ago that this process could be used to make certain types of nonwovens if the web of fibers is kept at their full width and then bonded using some form of a binder or adhesive (2).

In carding, the fibers tend to be oriented in the machine direction as the web exits the card. If bonded in this state, the nonwoven is called a parallel-laid nonwoven, which has high strength and low elongation in the machine direction (MD), and low strength and high elongation in the cross-machine direction (CD).

The web exiting the card may also be folded back and forth on a moving apron or conveyor. In this case, the product produced is called a cross-laid nonwoven. Because the fibers are oriented in a crossing pattern, the strength and elongation properties are more uniform in all directions. Sometimes a composite-carded type nonwoven is made where both parallel and cross-laid webs are layered together to combine the beneficial properties of each web.

A recent trend in carding technology is the manufacture of machines that produce a randomized carded web. This is accomplished by having the doffer and the surface of the cylinder move in opposite directions. Randomized cards combine the high production rate of parallel-laid carding with the uniform tensile and elongation properties characteristic of the slower cross-laid process.

Air-laying. In the late 1940s, the air-laying technique for making nonwovens was developed. In this process the fibers are opened, disentangled, and straightened using carding elements similar to those used in carding. Then the fibers are stripped off a roll and conveyed using an air stream. A rotating condensing cylinder or moving screen is used to collect the fibers to form the web. Upon exiting the machine, the web is then bonded by one of several methods. Air-laid products tend to be soft and lofty and are often used where this property is desirable.

Wet-laying. Wet-laid nonwovens are produced using papermaking equipment. Initially, conventional Fourdrinier paper machines were used. However, an improved product can be made by using inclined-wire paper machines. This permits handling a larger volume of water and the sheet is formed from a pool of water. As a result, longer fibers may be used and sheet formation and properties are superior to that produced on a flat tabled Fourdrinier machine.

Spunbonded methods. Spunbonded nonwovens are produced as the fibers are extruded from the spinnerettes. The fibers are layed randomly on a moving screen and then bonded by one of several methods. The thermoplastic fibers such as polypropylene, polyethylene, polyester, and nylon are the most common, but rayon spunbonded nonwovens have also been made by this method. The spunbonded nonwovens are especially known for their high strength and light weight. In addition to fibrous-type spunbondeds, there are modified films, nets, and scrims in this category.

Bonding

The web must be bonded in order to have sufficient integrity.

Inherently bonded nonwovens are bonded by their own extruded interconnected structure. Examples are Delnet, Sharnet, and Conwed.

Thermally bonded nonwovens are bonded at elevated temperature. Thermoplastic fibers, scrims, and powders are commonly used to bond the fibrous web together using ovens or heated calender rolls.

Mechanical bonding is accomplished by interlocking the fibers in the web using barbed or forked needles, hydraulic needling, or by stitch-through methods.

Chemical bonding generally uses binders or adhesive-like materials applied by saturation, spraying or painting. Sometimes solvents are incorporated to partially dissolve and fuse the fibers together.

Packaging Applications

Because nonwovens can be made from virtually any fiber (3), diverse properties can be obtained. Fibers vary considerably in strength, elongation, toughness, absorbency, warmth, chemical and abrasion resistance, light resistance, heat sealability, and numerous other properties. The range of properties offered by a given fiber can also be broadened by varying the web production, bonding and finishing methods.

Carded and air-layed nonwovens. Because of their soft protective characteristics, carded and air-laid nonwovens are used for packaging home and office electronic equipment, delicate instruments, glass separators, silverware, electric razors, thermometers, and pen and pencil sets. New automobiles can utilize temporary seat covers to prevent soiling by mechanics during shipment. Nonwovens made from reclaimed fiber (shoddy) is often used to hold pillows in place and protect furniture during shipment. In horticulture, properly spaced seeds are packaged in strips ready for planting. The entire strip is buried and the cotton or rayon fibers decompose leaving only the remaining plants. Nurseries also use nonwovens to wrap the root ball of plants, bushes, and trees.

Wet-laid nonwovens. Wet-laid nonwovens are used in the production of tea bags. A portion of the sheet consists of a thermoplastic fiber that acts as a binder fiber and enables the bag of tea to be heat-sealed. Houses are wrapped in nonwovens for insulation purposes and lumber is shipped in nonwovens to prevent weathering and pilferage.

Spunbonded nonwovens. These are excellent packaging materials, offering exceptionally high strength, toughness, tear and puncture resistance, and light weight. They are used for mailing envelopes, baling, bags for small hardware, and sand. High strength, spunbonded tape is used to bundle glass insulation. Most hang tags for furniture are made with Tyvek (Du Pont Company) spunbonded nonwovens. Disposable tarpaulins, automobile enclosures, and carpet wrapping are excellent applications. Nonwovens are used extensively for sterile microbe-barrier medical packaging (see Health-care packaging). Other applications include book covers, fruit and vegetable bags.

Nonwovens are versatile packaging materials. They are soft and inexpensive, combining high strength and toughness with exceptional aesthetics. They can be printed and sewn, and if the nonwoven is produced with thermoplastic fibers, it can be thermoformed or heat-sealed. There are numerous types of nonwoven materials having diverse and often opposing properties that can be selected for particular packaging objectives.

BIBLIOGRAPHY

1. J. R. Wagner, *TAPPI J.* **66**(4), 41 (1983).
2. Brit. Pat. 114 (1853) Bellford.
3. M. L. Joseph, *Introductory Textile Science,* 3rd ed., Holt, Rinehart, and Winston, New York, 1977.

J. R. WAGNER
Philadelphia College
of Textiles and Science

NYLON

Nylons are selected for applications in the packaging industry mainly for their functional contributions. In general, they offer clarity, thermoformability, high strength and toughness over a broad temperature range, chemical resistance, and barrier to gases, oils, fats, and aromas. For most packaging applications, nylons are used in film form, as single components in multilayer structures (see Films, plastic; Multilayer flexible packaging).

Nylons, or polyamides, are thermoplastics characterized by repeating amide groups (—CONH—) in the main polymer chain. The various types of nylon differ structurally by the chain length of the aliphatic segments separating adjacent amide groups. The combination of hydrogen bonding of amide groups and crystallinity yields tough, high-melting thermoplastic materials. The flexibility of aliphatic chains permits film orientation to further enhance strength (1). As the length of the aliphatic segment increases, there is a reduction in melting point, tensile strength, water absorption, and an increase in elongation and impact strength. Copolymerization also tends to inhibit crystallization by breaking up the regular polymer chain structure and likewise yields lower melting points than the corresponding homopolymers (2). The selection of a particular nylon for an application involves consideration of specific physical requirements (mechanical properties, barrier properties, dimensional stability), processability (melting point), cost, etc. Table 1 lists comparative properties of commercial nylon films.

As noted above, polyamides can be described as long-chain molecules with amide functionalities (—CONH—) as an inte-

Table 1. Properties of 1-mil (25.4-μm) Nylon Films

Property	ASTM test	Nylon-6	Nylon-6,6	Nylon-6/6,6	Nylon-6/12	Nylon-6, biax. oriented
				Value		
melting point, °F (°C)		424–428 (218–220)	510 (266)	388–395 (198–202)	386–392 (197–200)	424–428 (218–220)
specific gravity	D 1505	1.13	1.14	1.11	1.10	1.15
yield, in.^{2}/(lb·mil) [m^2/(kg·mm)]		24,500 [1,372]	24,300 [1,361]	25,000 [1,400]	25,200 [1,411]	24,000 [1,344]
haze, %	D 1003	1.5–4.5	1.5	2.0		1.3
light transmission, %	D 882					
tensile strength, psi (MPa) MD	D 882	12,000 (82.8) 9,000 (62.1)	16,000 (110.3)	4,640 (32.0)	32,000 (220.7)	
XD		10,000 (69.0)	9,000 (62.1)	16,000 (110.3)	4,500 (31.0)	32,000 (220.7)
elongation, %	D 882					
MD		400	300	400	400	90
XD		500	300	400	500	90
tensile modulus, 1% secant, psi (MPa) MD		100,000 (689.7)	100,000 (689.7)	100,000 (689.7)		250,000 (1724.1)
XD		115,000 (793.1)	100,000 (689.7)	100,000 (689.7)		250,000 (1724.1)
tear strength, gf/mil (N/mm)						
initial	D 1004	500 (193)	600 (232)	500 (193)		200 (77.2)
propagating	D 1922	35 (13.5)	35 (13.5)	70 (27)		10 (3.9)
bursting strength, 1 mil (25.4 μm), psi (kPa)	D 774	does not burst, 10–18 (69–124)	18 (124)			
water absorption, 24 h, %	D 570	9	8	9	3	7–9
folding endurance cycles	D 2176	>250,000				
change in linear dimensions at 212°F (100°C) for 30 min, %	D 1204	less than 2		1		less than 2.5
service temperature, °F (°C) range		−50–250 (−46–121)	−100–450 (−73–232)	−50–240 (−46–116)		−76–266 (−60–130)
heat-seal temperature, °F (°C)range		410–420 (210–216)	490–500 (254–260)	375–385 (191–196)	360–375 (182–191)	410–420 (210–216)
oxygen permeability cm^3·mil/(100 in.^{2}·d·atm) [cm^3·μm/(m^2·d·kPa)]	D 1434					
23°C, 0% rh		2.6 [10.1]	3.5 [13.6]		2.0 [7.8]	1.2 [4.7]
23°C, 50% rh				5.0 [19.4]		
23°C, 95% rh		3.5 [13.6]	16.0 [62.2]		2.2 [8.5]	3.5 [13.6]
water vapor transmission rate, g·mil/100 in^2·d) [g·mm/(m^2·d)], 100°F (38°C), 90% rh		18 [7.1]	19 [7.5]	31 [12.2]	7 [2.8]	17 [6.7]
COF, face-to-face back-to-back	D 1894	0.25–0.65	0.45			

gral part of the repeat unit. Film-forming nylons are usually linear and conform to either of two general structures.

(1)

(2)

Examples of the first-type polymers are nylon-6 (R=(CH$_2$)$_5$), nylon-11 (R=(CH$_2$)$_{10}$) and nylon-12 (R=(CH$_2$)$_{11}$). Here, the

Table 2. Typical Temperature Profiles for Nylon Extrusion, °F(°C)

	Nylon-6	Nylon-6,6	Nylon-6/6,6[a]
feed zone	446–482 (230–250)	500–554 (260–290)	401–464 (205–240)
transition zone	437–500 (225–260)	500–545 (260–285)	401–482 (205–250)
metering zone	396–527 (220–275)	500–545 (260–285)	392–491 (200–255)
head	437–518 (225–270)	500–545 (260–285)	401–482 (205–250)
die	419–518 (215–270)	494–563 (255–295)	392–482 (200–250)
melt temperature	437–518 (225–270)	500–512 (260–300)	401–482 (205–250)

[a] Nylon 6/6,6 temperature profile reported by Allied Corporation for XTRAFORM resin.

nylon type corresponds to the total number of carbon atoms in the repeat unit. Examples of the second-type nylons are nylon-6,6 ($R' = +CH_2+_6$, $R'' = +CH_2+_4$) and nylon-6, 10 ($R' = +CH_2+_6$, $R'' = +CH_2+_8$). In the case of the second-type polymers, R' refers to the number of methylene (CH_2) groups or carbon atoms between the nitrogen atoms and the R'' refers to the number of methylene groups or carbon atoms between the CO groups. The n in the formula is the degree of polymerization and its value determines the molecular weight of the polymer.

Nylons can also be prepared as copolymers. For example a nylon 6/6, 6-copolymer might have the formula

$$NH_2-R-\underset{\substack{\| \\ O}}{C}+N-R'-\underset{\substack{\| \\ H}}{N}\underset{\substack{\| \\ O}}{C}-R-\underset{\substack{\| \\ H}}{N}\underset{\substack{\| \\ O}}{C}-R''-\underset{\substack{\| \\ O}}{C}+_n N-R'-NH_2$$

$$R = +CH_2+_5$$
$$R' = +CH_2+_6, \qquad R'' = +CH_2+_4$$

If the amounts of the nylon-6 and nylon-6,6 monomers are varied, many combinations of the comonomers are possible. This would give rise to a series of nylon-6/6,6 copolymers with different properties (3).

Processing Methods

Extrusion. Nylons are melt-processable via conventional extrusion, but some parameters differ from those used in extruding other resins (see Extrusion.) Since the quality of extruded film is sensitive to raw material defects nylon resins should be clean and free of gel particles. Selection of resin viscosity depends on use. Low viscosity resins are used in extrusion coating to allow rapid drawdown rates, whereas medium- to high-viscosity resins are preferred for film production. Because nylons are hygroscopic, special care is required to assure sufficiently low water content (<0.10 wt %) for the extrusion process. Unless properly packaged and stored, nylon resins absorb moisture from the atmosphere and inevitably pose processing difficulties (4). In any extrusion operation, continuous production of uniform quality film as well as maintenance of a safe work environment is best achieved by monitoring and tightly controlling all machine variables (ie, temperatures, head pressure, extruder drive load, screw rpm, etc) (5,6).

Many film converters have been successful in producing nylon film on conventional polyethylene extruders, making only small modifications. The most important factors to consider are temperature control and screw design. The extruder must be equipped with adequate heaters for the required processing temperatures (5,7). In addition to heating capability, temperature control should permit little fluctuation (±3°C) to assure delivery of a proper and consistent melt. Typical temperature profiles are shown in Table 2.

Although a variety of screws have been used successfully for processing nylon resins, not all screws are optimum for the nylon being processed. In general, a nylon screw is thought of as a rapid-transition metering-type screw. A compression ratio (volume of a feed flight relative to the volume of a metering flight) of 3:1 to 4:1 is acceptable for most nylons. Most nylon screws are 40% metering, 3–4 turns transition, and the remainder feed zone. Length to diameter (L:D) ratios of 20:1 and 24:1 are common and acceptable. In designing a screw for a specific nylon, factors that must be considered include melting point of the resin, melt viscosity characteristics, machine type, extrusion rate, etc (4).

Film manufacture. Nylon film can be produced by either the cast-film process or the blown-film process. Semicrystalline polymers like the nylons can be made "amorphous" by rapidly quenching the melt via the cast-film process. That is, the polymer chains are prevented from aligning and organizing into regular, three-dimensional "crystalline" structures. Thus, by varying quenching rates, nylons are capable of existing in a low-order or amorphous state and in a highly crystalline state. The properties of the final film are highly dependent on the crystalline state of the polymer. By rapidly quenching the melt, favorable properties of transparency and thermoformability are induced.

In producing nylon film by the blown-film process, the cooling rate is much slower than for the cast-film process: the film is allowed to crystallize and film clarity is generally sacrificed. Blown films are used in applications requiring tubular film, or films of higher strength or better gas barrier properties than yielded by the cast process. A chrome-plated bottom-fed die with a spiral mandrel is generally recommended for blown-film processing to minimize weld lines and polymer degradation due to stagnant hold-up of melt in the die. Because of their stiffness, nylons pose wrinkling problems in the bubble-collapsing phase of production. Special care is required to properly align the collapsing frame and nip rolls, minimize gauge variations, and limit drag force in the collapsing assembly. It has been suggested that the lower-density nylons show less-severe wrinkling problems (4).

Oriented nylon films. The high strength and toughness properties of nylon films can be further enhanced by orientation. The increased alignment and tighter packing of polymer chains resulting from the process also yields improved barrier

Table 3. Comparative Properties of Unoriented, Uniaxially Oriented (MD), and Biaxially Oriented Nylon-6 Films (1-mil or 25.4-μm Films)

Property		Nylon-6 film, unoriented	Nylon-6 film, MD oriented	Nylon-6 film, biax. oriented
			Values	
specific gravity		1.13	1.14	1.15
tensile strength, psi	MD	12,000 (82.8)	50,000 (344.8)	32,000 (220.7)
(MPa)	XD	10,000 (69.0)	10,000 (69.0)	32,000 (220.7)
elongation, %	MD	400	60	90
	XD	500	450	90
tensile modulus	MD	100,000 (689.7)	300,000 (2069)	250,000 (1724.1)
psi (MPa)	XD	115,000 (793.1)	100,000 (689.7)	250,000 (1724.1)
initial tear strength, gf/mil (N/mm)	MD	500 (193)	650 (251)	200 (77.2)
	XD	500 (193)	1,300 (502)	200 (77.2)
propagating tear strength, gf/mil (N/mm)	MD	35 (13.5)	40 (15.4)	10 (3.9)
	XD	35 (13.5)	100 (38.6)	10 (3.9)
oxygen permeability, $[cm^3 \cdot \mu m/(m^2 \cdot d \cdot kPa)]$ $cm^3 \cdot \mu m/(100\ in.^2 \cdot d \cdot atm)$		2.6 [10.1]	2.4 [9.3]	1.2 [4.7]
WVTR, $g \cdot mil/(100\ in.^2 \cdot d)$ $[g \cdot mm/(m^2 \cdot d)]$		18 [7.1]	18 [7.1]	17 [6.7]

properties. Table 3 lists property comparisons for unoriented vs biaxially oriented films. Preferential orientation, typically machine-direction, improves strength and toughness in the direction of orientation. Biaxial orientation yields films with balanced properties. Market development efforts in this area are concentrated on critical packaging applications requiring soft films that permit tight package conformation and offer improved impact strength and reduced pinholing, superior burst strength, and flex crack resistance. As with most other flexible substrates, oriented films permit further conversion, (eg printing, lamination, metallization, etc).

Three processes are used to manufacture biaxially oriented nylon film:

One-step tenter frame. This process simultaneously draws cast nylon film in both the machine and transverse directions (8).

Two-step tenter frame. This is a two-step orientation process in which a nylon film that has been modified with a plasticizer is first drawn in the machine direction and then drawn in the transverse direction (9).

Blown bubble. Nylon film extruded from a circular die is oriented in the transverse direction by controlled internal air pressure, and oriented in the machine direction by regulating the bubble takeoff speed (10).

Coextrusion. Nylons used in film extrusion are often combined with other plastic materials via coextrusion. In most cases, polyolefins are used as coextrusion partners for nylon to provide heat sealability, moisture barrier, and good economics. As in single-layer-film extrusion, nylons can be processed by cast-film coextrusion and by blown-film coextrusion (see Coextrusion, tubular). The combining-adapter technology is generally the preferred method for joining layers in cast film coextrusion of nylon, although multimanifold-die systems are also used. Special care in matching resin viscosities is required

when using the combining-adapter system in order to produce films of uniform layer profile. For both systems, nylon-6 is the most common and preferred polyamide used in cast film coextrusion.

Because the blown-film coextrusion process employs air as the cooling medium, the melt is cooled slowly, permitting spherulite formation in semicrystalline nylon homopolymers (ie, nylon-6, nylon-6,6) and film transparency is sacrificed. For this reason, less-crystalline nylon copolymers such as nylon-6/12 and nylon-6/6,6 are gaining acceptance in blown film coextrusion.

Applications for nylon/polyolefin coextruded films include vacuum packaging of meat products, cheese-ripening pouches, consumer packaging of cheese and fish products, and several nonfood packaging applications including containers for chemicals, fertilizers, and animal foods.

Extrusion coating. Nylons with lower viscosity permit rapid drawdown rates for extrusion coating (see Extrusion coating.) Typical substrates range from heavy-duty paperboard to intermediate aluminum foils and papers, to thin polyethylene films. Published literature describes the nylon extrusion coating process in detail (4,11). As in cast-film production, annealing the nylon coating is necessary to impart dimensional stability when a flexible substrate is used.

Blow molding. Nylons of ultrahigh viscosity are commonly used in blow molding (see Blow molding). Because of their excellent impact strength, chemical resistance, toughness, and wide temperature use properties, nylons are ideally suited for several blow-molded applications, including industrial containers, moped fuel tanks, and automotive oil reservoirs. A recent development for nylon in blow molding has been in the area of blow-molded hydrocarbon-barrier containers. The process employs a blend of a modified nylon barrier resin and a polyolefin, extrusion blow molded under controlled mixing conditions to produce a barrier container competitive with surface-treated HDPE bottles (1,2,11). Packaging applications include containers for charcoal lighter fluids, general purpose cleaners, waxes, polishes (see Surface modification) and those described in several patents (12).

Secondary Conversion

Thermoforming. Nylon films are readily thermoformed by conventional methods (see Thermoforming.) Ease of formability is affected by nylon type (melting point), molecular weight, degree of crystallinity, and machine variables. In general, nylon films offer excellent thermoformability due to their high elongation. Further, the high elongation facilitates deep draw, and flex- and stress-crack resistance of nylons minimize film breaks during and after forming. Current applications include vacuum and gas packaging of meats and cheeses, and thermoform/fill/seal packaging of disposable medical devices (see Controlled Atmosphere Packaging; Health care packaging; Thermoform/fill/seal.)

Heat sealing. Because of their high melting points, nylons are typically coextruded with, laminated to, or coated with a polyolefin heat seal layer (eg PE, EVA, EAA, ionomer, etc). However, unsupported nylon films are heat sealed for applications that require heat-seal integrity under high temperature exposure (eg, oven "cook-in" bags). By properly balancing the variables of time, temperature, and pressure, unsupported nylons can be heat sealed at relatively low temperatures. For most commercial applications, however, it is necessary to heat the films to temperatures that closely approach their melting points. This factor, compounded by the relatively narrow melting range of nylons, necessitates precise temperature ($\pm3°C$) and pressure control. By making necessary modifications to conventional machinery, nylons are successfully heat sealed by thermal impulse and constant-heat techniques.

Adhesive lamination. Nylon films are combined with other flexible materials via adhesive lamination to produce multiple structures, each ply contributing to the requirements of the end product (see Laminating; Multilayer flexible packaging). Typical substrates include sealant webs (PE, EVA, ionomer, etc), and aluminum foils. Converters may choose to laminate rather than coextrude nylon to minimize scrap losses for short runs. Lamination is also necessary when combining non-coextrudable or incompatible plastics. The lamination process is described in detail in published literature (13). The adhesive is generally applied to the nylon web which has been corona treated (see Surface modification) to assist wettability, as the high melting point of nylon provides suitable stability in solvent-drying ovens. Adhesive systems are typically two-component types that vary for nylon types and substrates. Nylon film suppliers recommend adhesives for specific substrates (see Adhesives.)

Vacuum metallizing. Applications for metallized oriented nylon films are expanding in the packaging industry. Metallization offers functional contributions of improved moisture, oxygen, and light barriers and unique aesthetic appeal at the consumer level (14). The resultant film offers excellent flexibility, oxygen barrier, flex-crack resistance, antistatic properties, and printability. These properties meet the necessary requirements for use in such applications as institutional coffee pouches, metallized balloons, and liquid-box containers (see Bag-in-box, liquid product; Metallizing).

Packaging Applications

Although nylons are not generally considered commodity-packaging resins, the added material cost is easily justified in specific demanding applications where the nylons' physical properties provide added protection, extended shelf life, or reduced losses of expensive contents. The combination of excellent thermoformability, flex-crack resistance, abrasion resistance, gas-, grease-, and odor-barrier, and tensile-, burst- and impact-strength over a broad temperature range make nylons well suited for many packaging applications. For most applications, nylons are combined with other materials that add moisture barrier and heat sealability, such as low density polyethylene, ionomer, EVA, and EAA.

Most nylon-containing packaging films are used in food packaging, principally in vacuum-packing bacon, cheese, bologna, hot dogs and other processed meats (15). A variation of this package includes a carbon dioxide flush prior to heat sealing to remove traces of oxygen, thus prolonging shelf life for foods such as poultry, fish and fresh meat (16). Two recent developments are nylon composites used in vacuum-packing cooked whole lobster, boasting a two-year shelf life and canless "canned" ham, where the product is vacuum-packed and cooked in its package (17). Poly(vinylidene chloride) (PVDC) coatings are offered for improved oxygen-, moisture vapor-, or uv light-barrier properties (see Vinylidene chloride copolymers).

Nylon-6 is the nylon resin used most frequently for packaging applications because of the balance of cost, physical properties, and process adaptability. For blown or cast extrusion as well as cast coextrusion, nylon-6 resins are favored by most converters, while lower-melting nylon copolymers (nylon-6/6,6 or nylon-6/12) have been developed primarily to aid blown-film coextruders (lower melting points permit lower process temperatures for faster melt quenching). In addition to lower melting points, the nylon copolymers are less crystalline than their corresponding homopolymers and provide better clarity and thermoformability. On the high end of the melting point scale, nylon-6 and nylon-6,6 resins are appropriate for use in oven-cooking bags, where high-temperature tolerance is a key requirement.

Medical-packaging applications, such as packaging of hypodermics and other medical devices, are a relatively new and expanding area for the nylons. The combination of toughness, puncture resistance, impact strength, abrasion resistance, and temperature stability make nylons appropriate for protecting sterile devices during shipping and storage. Although ethylene oxide and steam have always been appropriate means of sterilization for nylons, modified-nylon resins have recently been introduced that permit radiation sterilization as well (18).

Biaxial orientation of nylon films provides improved flex-crack resistance, mechanical properties and barrier properties. These films have new applications in packaging foods such as processed and natural cheese, fresh and processed meats, condiments, and frozen foods. They are used in pouches and in bag-in-box structures (see Bag-in-box, dry product; Bag-in-box, liquid product) (19). In other areas (eg, cooked meats, roasted peanuts, smoked fish) the nylons compete with biaxially oriented polyester (see Film, oriented polyester). Although oriented nylons offer better gas barrier, softness and puncture resistance, oriented polyester offers better rigidity and moisture barrier.

Some recent packaging developments for nylon film include a composite pouch for wine (20) a nylon-6 shrink film (see Films, shrink) for meat and fresh-vegetable packaging (21), a nylon composite film used in a system to produce greaseless fried chicken (27), a uniaxially oriented nylon-6 film for food packaging (23) and a nylon-6 film with improved thermo-formability (24).

BIBLIOGRAPHY

1. R. D. Deanin, *Polymer Structure, Properties and Applications,* Cahners Books, Division of Cahners Publishing Company, Inc., Boston, Mass., 1972, pp. 455–456.

2. K. J. Saunders, *Organic Polymer Chemistry,* Chapman and Hall Ltd., London, 1976, pp. 175–202.

3. U.S. Pat. 4,417,032 (Nov. 22, 1983), Y. P. Khanna, E. A. Turi, S. M. Aharoni and T. Largman (to Allied Corporation).

4. R. M. Bonner, "Extrusion of Nylons", M. I. Kohan, ed., *Nylon Plastics,* John Wiley & Sons, Inc. New York, 1973.

5. E. C. Bernhardt, *Processing of Thermoplastic Materials,* Reinhold Publishing Co., New York 1959.

6. J. M. McKelvey, *Polymer Processing,* John Wiley & Sons, Inc., New York, 1962.

7. R. J. Welgos, "Polyamide Plastics" in M. Grayson and D. Eckroth ed, *Kirk-Othmer Encyclopedia of Chemical Technology,* Vol. 18 3rd ed., Wiley Interscience, New York, 1982, 18 pp. 406–425.

8. U.S. Pat. 3,794,547 (Feb. 26, 1974), M. Kuga and co-workers (to Unitika, Ltd.)

9. U.S. Pat. 29,340 (reissued Aug. 2, 1977), K. Khisha (Orig); U.S. Pat. 3,843,479 (Oct. 24, 1974), I. Hayashi and K. Matsunumi (to Toyo Boseki, Ltd.)

10. U.S. Pat. 3,499,064 (March 3, 1970) K. Tsuboshima and co-workers (to Kohjin Co., Ltd.)

11. S. M. Weiss, in J. Abranoff, ed., *Modern Plastics Encyclopedia,* Vol. 16, No. 10, McGraw Hill Publications, New York, 1984, pp. 199–202.

12. U.S. Pat. 4,410,482 (Oct. 18, 1983) P. M. Subramanian; U.S. Pat. 4,416,942 (Nov. 22, 1983) R. C. Di Luccio; U.S. Pat. 4,444,817 (Apr. 24, 1984) P. M. Subramanian,

13. J. A. Pasquale, J. Agranoff, ed., *Modern Plastics Encyclopedia,* Vol. 61, No. 10, McGraw Hill Publications Co., New York 1984. pp. 284–286.

14. W. Goldie, *Metallic Coating of Plastics,* Electrochemical Publications Ltd., Hatch End, Middlesex, 1969.

15. E. C. Lupton, *Mod. Packag.* 52(12), 26 (Dec. 1979).

16. M. Gilbert, *Plast. Eng.* 34(10), 41 (Oct. 1978).

17. *Mod. Plast. International* 12(7), 21 (July 1982).

18. Allied Corporation, *Mod. Plast.* 60(9), 114, 119 (1983).

19. D. May, *Pap. Film Foil Converter* 56(1), 46 (Jan. 1982).

20. *Mod. Plast.* 59(12), 28 (1982).

21. *Plast. Ind. News* 29(10), 148 (1983).

22. *Food & Drug Packag.* 46(12), 30 (1982).

23. *J of Commer.* 353(25,263), 22B (July 20, 1982); *Plastics World* 40(8), 76 (1982).

24. *J. of Commer.* 345(24,747), 5 (July 2, 1980).

M. F. Tubridy
J. P. Sibilia
Allied Corporation

OFFSET PRINTING.   See Printing; Decorating.

P

PACKAGE HANDLING SYSTEMS.   See Conveying

PAD PRINTING.   See Decorating.

PAILS, PLASTIC

Plastic pails can be either open-head containers with removable lids or containers produced as a single unit called tight-head containers. The container top or lid can be manufactured with one or more openings for filling and dispensing. The openings in turn are designed to be used with a variety of closures. In this article, only the range of container sizes and designs that are normally considered to be shipping containers are discussed in detail. These can be defined as single-wall, heavy-duty containers with a handle. These containers range in capacity from 1 gal (3.785 L) to 6 gal (22.7 L) (see also Drums, plastic).

In the open-head configuration, the containers are usually supplied with a removable lid designed to be used with either liquids or solids. The lids are available with pour fittings of various types. The most commonly used pour fitting for liquids is the flexible polyethylene spout, which either snaps into position in a preformed hole or is a spout that incorporates a metal collar that is crimped onto a formed, ridged opening in the lid. These fitting designs feature the ability to be recessed into themselves and not protrude above the lid surface when in a nonuse position. When positioned for use, however, they extend up and out to form a convenient pour spout. Open-head containers are, with few exceptions, of tapered design, which allows the individual pails to nest into each other for efficient storage and transportation, prior to being filled and sealed. These pails are universally injection-molded, as are the matching lids (see Injection molding).

In the closed-head configuration, no nesting advantage exists, but this product has certain advantages in the area of structural integrity arising from the one-piece construction. This construction meets particular government specifications for packaging of hazardous products (dangerous goods). Most closed-head plastic containers manufactured and sold in the North American market are integral one-piece units produced by the blow-molding process (see Blow molding).

The use of plastic pails for the shipment of industrial products in the United States and Canada was estimated to have reached a level of 160 million (10^6) units in 1984 (1). Of the total, less than 25% are closed-head (also called tight-head containers). The uses of these pails and the approximate breakdown of the principal categories of use are shown in Table I.

Within the subject capacity range, the 5-gal (18.9 L) open-head pail is the predominant package, comprising 60–70% of the total number of pails consumed (2). Other popular packages in the United States include a 4-gal (15.1 L) "food package" and the 3½-gal (13.25 L) open-head pail used in the chemicals industry.

In Canada, since the government-mandated conversion from Imperial units to Metric standards in the early 1980s, the 20-L container size has become the predominant pail size (2). The relative closeness in capacity between the 5-gal (18.9 L) and the 20-L size would seem to indicate that a future U.S. conversion to metric would favor the adoption of the 20-L size in the United States as well.

Raw Material Considerations

Open-head and closed-head pails are made of high density polyethylene. The following discussion pertains to the predominant open-head injection-molded pails only (3).

Four basic parameters must be considered in the design of a polyethylene resin for injection molding industrial containers. These are melt index, density, molecular-weight distribution and type of comonomer incorporated (see Polyethylene, high-density). Adjusting each of these parameters results in a tradeoff between the various designed properties. Although such tradeoffs make resin design difficult, there seems to be a fairly small range in each of the above four parameters that gives optimum properties. The injection-molding pail resins from all resin suppliers appear to fit into the ranges listed below.

Melt index is one of the two most important factors. The higher the melt index, the easier the processability of the resin, allowing faster molding cycle times and lower production costs. On the other hand, increasing the melt index reduces all the physical properties, including drop impact strength, top-load capacity, and environmental-stress crack resistance (ESCR). In general, a melt index of 4–8 g/10 min gives an optimum balance of these properties. The smaller the volume of the pail and the thinner the wall, the higher the optimum melt index of the resin should be.

Density is the other most important factor: It is controlled by the amount of comonomer that is polymerized with the ethylene. Lowering the density increases the drop impact strength and the ESCR, but decreases the top-load capacity of the container. Processability is virtually unaffected by density. Most injection-molding pail resins have densities of 0.950–0.955 g/cm^3.

Another factor is the molecular-weight distribution (MWD), which is determined by the type of polymerization process, as well as the type of catalyst used. A narrow MWD means that most of the polymer chains are of similar length, whereas a broad MWD implies that the polymer chains vary in length. A broader MWD aids in processability and generally increases ESCR, but at the expense of some drop-impact strength and top-load capacity. An intermediate MWD seems to be optimum for pail applications.

Table 1. Industrial Uses of Plastic Pails[a]

Type	Quantity, $\times 10^6$
printing inks	5
janitorial products	9
chemicals	22
petroleum products	11
paints	27
building products (adhesives, cements, joint compounds, etc)	52
foods	30
miscellaneous	4
Total	*160*

[a] Statistics are provided for comparative purposes only.

The type of comonomer also affects the properties of the injection-molded part. A comonomer such as butene or octene is used in the polymerization process to add branches to the polymer chain and thus lower the density. The shortest-chain comonomer available is propylene (C_3), whereas the longest comonomer generally used for polyethylene is octene (C_8). Propylene copolymers possess very poor ESCR. A comonomer with a long-chain, such as octene, results in an increase in drop-impact strength and a large increase in ESCR. However, in the melt index range under discussion, the superior properties of octene can only be realized if the density of the copolymer is below 0.950 g/cm^3. Above this density, so little comonomer is incorporated into the polymer that the octene acts solely as a density modifier and not as a physical property enhancer. As a result, injection-molding pail resins are generally ethylene–butene (C_4) copolymers.

Performance Requirements

Plastic pails are used to package many different products, and many of these products are likely to be shipped across state or national boundaries. As a result of these shipments, the containers themselves are subject to government regulations and specifications. The specifications and regulations that apply to the pails are determined by the nature of the product being transported. Products are categorized as either "Hazardous" or "Nonhazardous." In terms of packaging performance requirements, hazardous products are further split into liquids and solids. All hazardous products shipped in interstate commerce in the United States are regulated by the Department of Transportation (DOT), which defines hazardous materials as "a substance or material in quantity and form which may pose an unreasonable risk to health and safety or property when transported in commerce" (4).

Pails used for the packaging of nonhazardous products are required to meet specifications that are established by the Uniform Classification Committee (UCC) of the Association of American Railroads (for rail transport) and the National Motor Freight Classification Committee (NMFC) (for highway transport).

There are additional regulations for food products. In the United States they are as follows:

1. All containers used for food products must meet the requirements of the Food Additives Law.
2. All containers used for meats and poultry must be approved by the United States Department of Agriculture (USDA).

As a general rule all ingredients used in the manufacture of pails destined for food packaging are expected to comply with FDA regulations. Users of pails for food-packaging applications typically request certification from their suppliers that this is the case (see Food, drug, and cosmetic regulations).

Specifications

Neither the above review of applicable regulations nor the specifications that follow are intended to be a comprehensive and up-to-date list of all specifications and regulations. A number of these regulations and specifications are currently under review. The interested reader is urged to obtain up-to-date information from the appropriate authority. Sources of information can be found in Ref. 5. The primary purpose of this

Table 2. Compression Load Requirements

Marked (rated) capacity, gal (L)	Compression test, lbf (N)
2.5–6.5 (9.5–25)	600 (2669)
15 (56.8)	1200 (5338)
30 (113.6)	1800 (8007)

discussion of specifications is to examine the basic parameters for the design of plastic pails that are on the market and in use at this time in North America.

DOT Specifications for hazardous commodities. *Specification 34-Liquids* (6) covers reusable polyethylene containers for use without overpack, which range in size up to 30 gal (114 L) and have no removable heads. Until recently, in addition to limiting the container size, this specification listed in detail allowable limits of melt index, density, tensile strength, and elongation. All of the above are for polyethylene packagings only. The most recent direction is to eliminate specific raw material criteria, extend the range of packaging sizes, and in general to increase the emphasis on performance requirements. The test procedures described below have provided the basis for design of current products.

Tests. Space limitations do not allow for a detailed listing of the test procedures; however, they are summarized below.

1) A container filled with liquid conditioned to 0°F (−18°C) is required to withstand a 4-ft (1.22 m) drop on the top diagonal edge, or whichever part is weakest. The container must not leak after the test.

2) A container must withstand a hydrostatic pressure of 15 psi (103.4 kPa) for 5 min without pressure drop.

3) A container must not deflect in excess of 1 in. when filled with water and subjected to a compression load as per Table 2. The load is held constant for 48 h.

Records of test results must be maintained in current status and retained by each manufacturer at each producing plant. The preceding performance requirements of *Specification 34* are likely to be changed to add a requirement to establish permissible rates of permeation for various hazardous commodities. Also to be included is a test procedure involving time, temperature combinations, and both product weight loss and material testing.

Specification 35-Solids (7) is for nonreusable molded polyethylene drums for use without overpack with removable head required. As in Specification 34, specific parameters of polyethylene are included, together with construction and capacity details. The specifications are likely to change in the future in line with the trends described above, but the existing basic tests are the design parameters used for current DOT specification containers.

Performance requirements are summarized below:

1. In a drop test a container is filled with dry powder, topped with sodium bicarbonate, and closed with lid. Container and product conditioned to 0°F (−18°C) must survive without spillage each of the following drops, with no container being required to submit to more than one test: 4-ft (1.22-m) drop onto concrete flat on bottom, 4-ft (1.22-m) drop onto concrete on top edge, and 4-ft (1.22-m) drop onto concrete on bail ear (side drop).

2. A vibration test is required, although vibration is not normally a cause of failure.

3. A compression test is required, where container and dry product are conditioned to 130°F (54°C) and are required to withstand without deflection in excess of 1 in. (2.54 cm) a 600-lb (272-kg) top load.

Transport specifications for products classified by DOT as nonhazardous commodities. Rail shipments of products packaged in plastic pails are regulated by the UCC of American Railroads. The most-current applicable regulation is covered under *UFC 6000C, Rule 40, Sect. 7¼.* This specification covers open-head pails specifically and in addition to material requirements and construction specifications, there is a brief listing of performance requirements.

Performance requirements. Pails, filled with commodity to marked capacity, must meet the following performance standards without failure. Failure is defined as leakage or spillage of contents. Each test must be performed on a minimum of three sample pails, but no single pail will be required to withstand more than one test.

1) Pails must be conditioned to 0°F (−18°C) for a minimum of 4 h. Drop tests must be performed with the pail flat on its side and also at a 45° angle on the bottom chime onto solid concrete from a height of 48 in. (1.22 m).

2) Pails must be conditioned to 130°F (54°C) for 4 h, stacked three high, and vibrated for 1 h at 1 g to a vertical linear motion.

3) Pails filled with commodity to marked capacity must withstand a static load to 600 lb (272 kg) for a period of 48 h without defect or damage.

Tests may be performed using water when the viscosity of the commodity does not exceed 5000 cP (5 Pa·s), or sand when the viscosity of the commodity exceeds 5000 cP (5 Pa·s).

Highway shipments of nonhazardous commodities are regulated by the NMFC. The most current ruling affecting open-head plastic pails is *Classification 100J.* These regulations are similar to those of the railroad regulatory group.

A review of the performance standards listed shows that packaging of nonhazardous commodities is required to undergo much less rigorous testing than that required of pails to be used for hazardous commodities, particularly for liquids. Because the number of open-head pails used for hazardous commodities is relatively small and the emphasis on higher performance requirements for these pails is increasing, some manufacturers of open-head pails do not offer their products for these applications. The market is served by manufacturers of closed-head plastic pails and those with a specialized heavy-duty-design open-head pail package. Recent emphasis by DOT on compatibility testing is a further barrier to be overcome by the HDPE packages that are aimed at the hazardous-commodities packaging markets. Although a number of plastic pail designs currently on the market meet all of the requirements established by the DOT, the majority of product being sold is aimed toward meeting the less stringent nonhazardous-packaging regulations.

Design

An examination of patents relating to plastic pail design reveals more than a dozen spanning the past 15 years (8). Actually, most plastic pails are currently being manufactured

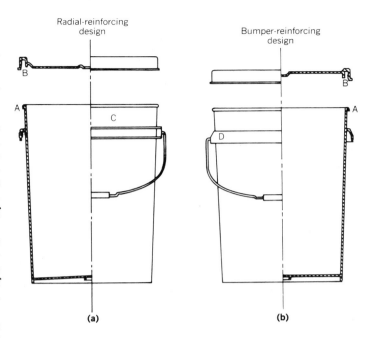

Figure 1. Plastic pail designs; (**a**) radial-reinforcing design and (**b**) bumper-reinforcing design. Parts A-D are described in text.

and sold without benefit or concern for patent protection. In order to meet previously described performance requirements, there are a number of design features that tend to be common to all plastic pails.

These features are related to the ability to withstand impact, without leakage; the ability to withstand compressive loads at elevated temperatures; the maximization of the raw material component of the package and resin processability; and convenience and aesthetic features.

Design features review. The features listed in the previous section are examined with reference to the designs illustrated in Figure 1.

Impact strength. The fit of lid to pail body and the ability of the combination to withstand deformation is key to maintaining package integrity. Where severe impact inevitably results in deformation, that deformation must occur in such a controlled fashion as to maintain a water-tight seal. As illustrated in Figure 1, pail bead size and design (A), corresponding cover fit (B), together with an adequate reinforcing structure of either radial ribs (C), or a bumper configuration (D) are all contributing factors to the ultimate success or failure of the design to withstand impact.

Compression capability. The following are generally considered important factors in maximizing top-load strength (9): wall thickness and uniformity—No other factor has a greater effect given specific material characteristics of stiffness; number and location of either radial or vertical stiffening structures; lid design; and angle of taper and ratio of container diameter to height.

Processing factors. Design features that relate to processability and effective use of the polymer. These features may include variable wall cross sections of pail walls and lids, and variations in gate sizes and designs.

Convenience and aesthetics. Other areas of design interest are taper or nesting capability; size and texture of the printing-surface area of the container; chime, base ribbing configuration, and ear design for bail handle; design of cover for removal and installation ease, pouring, or ability to avoid surface-water pooling; and general ease of cleaning after filling. Ability to be handled with automatic stacking and palletizing equipment.

Mold design is often tailored to product design and *vice versa*. A successful marriage of the two is essential for a satisfactory outcome.

Processing Equipment

The injection-molding process is generally well known, and information regarding the process is readily available from a variety of sources, such as the Society of Plastics Industry (SPI), New York, N.Y., and the Plastics Shipping Container Institution (PSCI), Chicago, Illinois (see Injection molding). The predominant type of machinery used for pails is the all-hydraulic, horizontal injection-molding press, with 500–700 tons (4.45–6.23 MN) of clamp capability. Over the past decade, development of processing machinery and expansion of the plastic-pail market have resulted in equipment built and targeted for specific uses. These tailored specifications have contributed to substantial improvements in both product quality and output.

The handling, processing, and use of polyethylene is generally considered to be a relatively safe activity. Polyethylene is used widely as a raw material of choice for industrial packaging, and as evidenced by the widespread acceptance of the product in the food industry, can be considered as a safe product in its finished form. Much has been discussed and written with regard to the toxicity of the products of combustion of plastics. Available research (10) indicates that the products of combustion of a typical plastic pail are not greatly different from those of a pine log.

BIBLIOGRAPHY

1. Sales Statistics, Vulcan Industrial Packaging, Ltd., Toronto, Canada, 1982.

2. *Vulcan Industrial Packaging Statistics 1983/1984,* Vulcan Industrial Packaging, Ltd., Toronto, Canada, 1984.

3. R. Scott, research data, Dow Chemical Canada, Ltd., Canada, 1984.

4. *Title 49, Hazardous Materials Transportation Act,* U.S. Department of Transportation, Washington, D.C.

5. DOT, Materials Transportation Bureau, Hazardous Materials Regulations, Research and Special Programs Administration-Standards for Polyethylene Packaging, U.S. Department of Transportation, U.S. Government Printing Offices, Washington, D.C.; NMFC-100, Classes & Rules, National Motor Freight Traffic Association Inc., Alexandria, Va.; UFC-6000, Ratings, Rules, and Regulations, Uniform Classification Committee, Association of American Railroads, Chicago, Ill.

6. Hazardous materials regulations, *Specification 34, DOT Requirements, Title 49, Code of Federal Regulations* (49CFR), Sect. 178.19, U.S. Department of Transportation, Washington, D.C., 1982.

7. Hazardous materials regulations, Specification 35, *DOT Requirements, Title 49, Code of Federal Regulations* (49CFR), Sect. 178.16, U.S. Department of Transportation, Washington, D.C., 1982.

8. U.S. Pat. 3,510,023 (May 5, 1970), F. E. Ullman, W. Klygis, and M. J. Klygis (to Inland Steel); U.S. Pat. 3,516,571 (June 23, 1970), W. Roper, R. E. Roper, R. Roper, C. R. Roper, F. Roper and R. A. Miller; U.S. Pat. 3,804,289 (Apr. 16, 1974), R. G. Churan (to Vulcan Plastics Inc.).

9. J. E. Boyd and S. B. Falk, *Strength of Materials,* McGraw-Hill, Inc., New York, 1950.

10. W. J. Potts, T. S. Lederer, and J. F. Quast, *Combust. Toxicology* 5, 412–433 (Nov. 1978).

PETER KIRKIS
Vulcan Industrial Packaging, Ltd.

PALLETIZING

A pallet is a low platform used to stack or accumulate a number of smaller units of product so that they may be conveyed by mechanical means. Pallets may be made of wood (see Pallets, wood), plastic (see Pallets, plastic), or corrugated kraft board (see Pallets, expendable corrugated), and are usually rectangular rather than square. The arrangement of product on the pallet (pallet pattern) is critical to an orderly, compact loading, and distribution system. Patterns correspond to the pallet shape and are designed to afford the maximum load in the minimum space without forfeiting structural strength. Pattern design may be restricted by the primary and secondary product packages, the shipping method and size of conveyance, and warehouse space.

The quantities of pallets and tiers are also determined by the shipping container, shipping method, and distributor space allotment. The second layer, or tier, may be exactly like the first, or rotated 180°, or it may be an entirely different pattern. Unless the cases are heavy and/or large, the pattern of the second layer is changed to form a load which will not break apart. Rotation by 180° is a common solution. The stacking strength of the load must be judged against the space limitations of the pattern. A column stack (each layer identical) with corners one above the other is strongest, but it is most likely to topple since the tiers are not interlocked. Many different pallet patterns are shown in Figure 1.

Palletizing

Once the pattern and number of tiers have been designed, the method of product-to-pallet transfer is determined. Hand palletizing is the most versatile method and is effective when loading is slow. Single-product, single-pattern high level palletization developed from hand palletization. Recent technological advances permit automated palletization of almost every product, but this may not always be economical. The selection of a palletizer demands consideration of performance requirements and space limitations. The palletizer is an element in the total packaging/distribution system, and it should never be a limiting factor. The palletizer should be capable of speeds exceeding normal line speed by 5–10% and change over time from product to product should be minimal. A fully automatic palletizer does not require an operator in attendance full time.

High level palletizers. High level palletizers (also called "moving-pallet palletizers) pick up the product at a level above the height of a full pallet and form the tier pattern on a bed. The pallet is raised to the bed level, and the product is transferred (by a sweep or rake-off system) to the pallet. Then, the pallet is lowered one layer, and the next tier is formed in the same manner and placed atop the first. This process continues

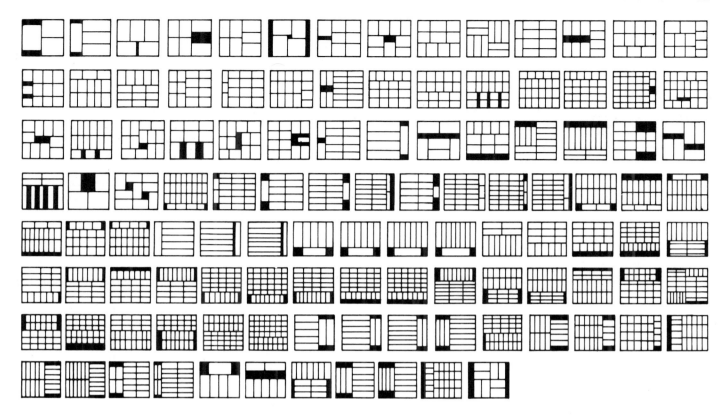

Figure 1. Pallet patterns.

until the pallet reaches the prescribed height. It is then transferred out to be replaced by an empty pallet which is raised to the bed for a repetition of the process. Numerically controlled tape programs provide simple multipattern capability and are adaptable for a variety of users.

Low level palletizers. Low level palletizers (also called "fixed-pallet" palletizers) were developed more slowly than high level palletizers, but with the introduction of programmable logic controllers in the last decade, they have become highly competitive with high-level units in cost and speed. The low level palletizer operates at floor level (see Fig. 2). The pallet is neither raised nor lowered, and the tier is formed at low level. The transfer bed or apron is raised or lowered to the appropriate level and the tier is transferred to the pallet. The bed returns to the tier-forming position, and the next layer is formed. This is repeated until the pallet is full. After the full pallet is discharged, an empty one is transferred into the load station, and the process is repeated.

Robotic palletizers. In the past, "pick-and place" systems picked up a product at one point and placed it at a second point repeatedly. At present, robotic palletizers may be thought of as intelligent, discriminating systems capable of picking up several different products and placing them at several different points. These units are usually capable of movement in two or three planes, often on rotational coordinates operating from a fixed point at the center of a circle. They are capable of picking up a variety of products from one point on the circumference of the circle and placing it on one of several pallets located at other points around the circumference. Robotic palletizers are most applicable when a variety of products require different pallet patterns and relatively slow speeds are acceptable.

Bulk palletizers. The shift to bulk handling has reduced the demand for case depalletization of empty containers. There are numerous applications for palletizing without benefit of a case; for example, the transfer of empty cans from the can manufacturer to the filler. Cans are bulk palletized successfully because they are strong enough to withstand relatively rough treatment. Improved handling methods permit bulk palletization of glass containers, and many plastic containers now are being bulk palletized as they gain a greater place in the market.

Nested container-to-container bulk pallets can save up to 15% space for the same number of containers compared to column-stacked cases. That means 15% more containers can be shipped in the same truckload to the user, cutting time and expense. The compact loads also reduce warehouse requirements. In addition, bulk palletizing eliminates the labor and expense associated with uncasing equipment and additional conveyors for empty cases.

Bulk palletization is essentially an adaptation of high- or low-level palletization. The product is accumulated on a table or bed, then transferred by lifting a tier or clamping the perimeter and sweeping the tier onto a pallet. A corrugated sheet the size of the pallet, called a tier sheet, is generally used between layers to provide stability to the load. Further stability can be gained by use of an inverted tray instead of a tier sheet.

Miscellaneous palletizers. Other types of palletizers include drum palletizers, keg palletizers, and bag palletizers. In most cases, these machines address a particular end-use and are concerned only with a narrow portion of the market place.

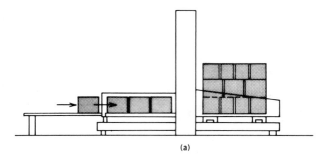

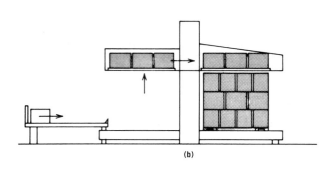

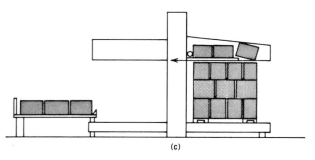

Figure 2. Operation of a low-level palletizer. (**a**) Operation 1. Sealed shipping cases feed in and are oriented to the preprogrammed pattern. When one row is formed, the cases move forward. The next row forms and moves forward, continuing until the layer is complete and the loading plate is filled. (**b**) Operation 2. The layer is lifted to the height of the existing pallet stack. The filled loading plate moves into position just above the pallet stack. (**c**) Operation 3. The loading plate retracts, allowing the cases to settle, row by row onto the top of the pallet stack. The pallet is squared by a squaring bar which also assures complete unloading. The loading plate returns to starting position where another accumulated load is ready.

Depalletizing

The removal of product from pallet depends upon the conformation of the product. Bulk depalletizers remove tiers of product from the pallet in much the same manner as bulk palletizers in reverse. In one approach, the tops of the containers are gripped mechanically, pneumatically, or with a vacuum, and the tier is lifted onto a discharge table. In another, the tier is swept onto the discharge table. Removal of the tier sheet or inverted tray is as critical here as in bulk palletization. Product stability is a key factor in all bulk handling operations and the primary determinant of method.

Depalletization of plastic cases or crates may require modified bulk depalletizers or specialized robotic depalletizers. Plastic crates usually have an interlocking feature which requires a tier to be lifted clear of the one below before transfer to the discharge table, thus precluding a sweep system. Most pail depalletizers must handle the products individually in addition to lifting clear of the pail below.

Depalletization of corrugated cases is more difficult than palletization. The flaps on the cases get caught on one another, preventing consistent sweep-off. Corrugated cases do not interlock, so clamping the perimeter causes the center cases in the tier to slip down. The most reliable way to remove a whole tier of corrugated cases is to use tier sheets, but the additional cost discourages wide acceptance. The next-best method is a combination clamping and vacuum system. Some automated warehousing systems remove cases from pallets one at a time in a type of "order-picking" operation. Little effort is being expended today on finding better case-depalletization methods because the shift to bulk handling has shifted research and development work in that direction as well.

S. D. ALLEY
ABC-KCM Technical Industries, Incorporated

PALLETS, EXPENDABLE CORRUGATED

Corrugated board can be combined with other materials to produce pallets for lightweight loads. Several manufacturers provide expendable pallets that use corrugated materials for the deck and various other products for the support structure. The deck can be supported by plastic legs, cut-down paper cores, or corrugated buildup material. All provide the features required by materials handling systems. Most expendable corrugated pallets have a load limitation of 1500 lb (680 kg). They can be custom manufactured to fit the exact dimensions of the load placed on them. Typical users include manufacturers of foam products, electrical components, and plastics.

PALLETS, PLASTIC

Plastic pallet construction began during the late 1960s when the low cost of commodity resins such as polystyrene and polyethylene encouraged scores of molders to enter this promising market. In the early 1970s, HDPE was priced at about $0.16/lb ($0.35/kg). Still in its infancy, the plastic pallet market was severely curtailed when prices of commodity resins more than doubled at the time of the 1973 oil embargo. A 2:1 price differential between wood and plastic pallets quickly became a 4:1 price disadvantage. In the United States today, plastic pallets still represent only about 1–2% of a $950 million ($10^6$) pallet market (1).

A number of changes taking place in the 1980s has encouraged the use of plastic pallets:

packaging is now often considered part of direct production cost instead of fixed overhead. This highlights the savings generated by reusable pallets;

Figure 1. Typical single-faced plastic pallet.

adoption of the Just-In-Time inventory concept (2) including reusable packaging, inventory reduction, and higher quality;

greater use of robots and automated palletizers which require uniform size/weight pallets; and

increased awareness and regulation of plant sanitation.

Typical single- and double-faced pallets are shown in Figures 1 and 2.

Materials. Most plastic pallets are manufactured from HDPE (see Polyethylene, high density). Materials such as polystyrene, fiberglass-reinforced plastics (FRP), and polypropylene are used occasionally. Heavy pallet loads and unsupported pallet racking may dictate the use of stiffer polystyrene (see Polystyrene).

FRP (see Thermosets) are used for low-volume custom pallet requirements or prototype pallets. In this situation, low-cost wooden tooling is used with the hand lay-up fiberglass technique. Polypropylene (see Polypropylene) has been used to construct structural foam plastic pallets by companies with excess virgin or regrind polypropylene. Polypropylene is not normally used in pallet construction because it requires relatively expensive impact modifiers for cold-weather performance.

Polyethylene is favored for a number of reasons; commodity status (ie, low cost, uniform performance, readily available, wide acceptance); excellent resistance to impact; good performance under a wide range of operating conditions (ie, temperatures of −30 to 150 °F (−34 to 66°C), indoor or outdoor applications, light- to heavy-weight loading); outstanding chemical resistance to most acids and bases; USDA and FDA clearance for use in food and pharmaceutical plants; easy cleaning; and outstanding molding and design flexibility.

Polyethylene's one glaring weakness is its inability to resist deflection (bending) under load. This deflection problem is especially serious in pallet-racking applications. Unsupported racks do not have center supports or decking. In these racks, the pallet must span an open space while maintaining the load. With loads of over 2000 lb (907 kg), plastic pallets are prone to bending (deflection). In addition to the initial deflection, the plastic pallet will continue to bend or creep for up to two weeks. Over time, it may become difficult to reenter the pallet with the forks of a lift. Most standard pallet rack, drive-through-rack, and gravity-flow rack is "unsupported."

In situations where heavyweight racking is a must, steel-reinforced plastic or stiffer polystyrene (PS) are frequently used. Steel reinforcements add expense, and compared to HDPE, PS costs more and offers less chemical and impact resistance. One solution to the racking problem is in the design of rackable pallets. Two-way-entry pallets can rack over 3000 lb (1360 kg) in an unsupported rack (see Fig 3). Experimental plastic resins are also being tried in an attempt to solve the racking dilemma.

Design and construction. Pallet design and the method of construction greatly influence pallet performance, price, and acceptance. Today plastic pallets are designed and built using several different processing techniques (see Table 1).

Structural foam molding. Most plastic pallets made today are made by structural foam molding. (3,4) This low pressure injection molding process produces parts with a solid skin surrounding a foamed core. Compared to high-pressure injection molding, structural foam molding allows the economic production of heavy wall sections and helps reduce stress points throughout the pallet. The structural foam process provides outstanding design flexibility. Wall thickness of 3/16 to 3/4

Figure 2. Typical double-faced plastic pallet.

Figure 3. Two-way entry racking pallet.

in. can be molded to produce pallets ranging from lightweight single-faced units to super-duty racking pallets. Another benefit is high-speed production, with cycle times as low as 2–3 minutes. Good resistance to impact, high strength per pound (kilogram), and good deflection strength are all positive characteristics that make structural foam a good choice for large scale production of both custom and proprietary pallets. The chief limitation of structural foam is that relatively high volume (3000 total units minimum) is required to amortize the relatively high tooling cost. When compared to high-pressure injection molding, the tooling for structural foam may be less costly. Most low-pressure foam tools may be built from machined aluminum or Kirksite, which reduces tooling costs by up to 50%. When structural foam tools are built from steel, the cost savings are negligible.

Injection molding. High-pressure injection molding (see Injection molding), also offers design flexibility. It is used for the production of pallets that range from very lightweight disposables to heavy-duty 60-lb (27-kg) reusables. Injection molded parts generally have narrower wall sections than structural foam, ≤ 0.300 in. (≤ 7.62 mm), and rely on their rib design for structural integrity (5). Injection molding excels in lightweight large-volume production. Cycle times for 1/8 in. (3.2 mm) injection-molded pallets can be under one minute. Heavy-duty parts with wall sections $\geq 1/8$ in. (3.2 mm) offer high strength and excellent durability. Because high molding pressures require expensive equipment and hardened-steel tooling, high-volume production runs ($\geq 10,000$) are generally required to amortize tooling and press costs.

Rotational molding. Rotational molding (3) (see Rotational molding) uses a heated tool into which solid or liquid polymer is placed. This process offers the most economical tooling costs. Myriad sizes and designs of relatively low-quantity (1000–2000 units) can be economically justified. Design innovation, including the molding of steel-encapsulated, smooth-skinned pallets is a feature of rotational molding. Rotationally molded parts offer good resistance to blunt impact and the repair of small puncture damage is possible. Its drawbacks include relatively long cycle times (as high as five min) and relatively narrow 3/16-in. (4.8-mm) wall thickness, (6) limiting rotationally molded pallets to medium-duty applications. Some rotationally molded designs can accomodate the addition of steel reinforcement for heavier loads and pallet racking operations.

Thermoforming. Thermoformed plastic pallets (see Thermoforming) are offered in dozens of low-cost lightweight designs. Inexpensive tooling allows faster amortization of low-volume custom pallets. For example, custom reusable dunnage trays are often thermoformed. Thermoforming too, has its disadvantages. With cycle times averaging five minutes (5), high-volume projects are sometimes impractical. In addition, relatively narrow wall thicknesses limit these pallets to lighter loads, usually under 3000 lb (1360 kg). They are not often found in heavy-duty racking applications. Twin-sheet vacuum forming allows heavier loads with reduced deflection, but it lengthens cycle times and adds cost.

Reaction injection molding (RIM). RIM polyurethane pallets are starting to enter the market now. RIM utilizes two or more liquid components (polyol and isocyanate) which are mixed, then injected into a closed mold. These two components react to form a finished polymer taking on the shape of the tool. The chief advantages of RIM are lower cost equipment and tooling especially in building large parts such as pallets. The chief disadvantage of RIM pallets is the lower resistance to deflection. For this reason, many large RIM parts are steel reinforced or manufactured with fiberglass or mineral fillers. These stiffening techniques add cost.

Advantages. Plastic pallets are used primarily in the food, pharmaceutical, textile, high-technology, and automotive industries. Due to the higher cost of plastic pallets, most purchasers use their pallets in-plant, or in a closed-loop shipping system. Plastic pallets are almost always found in applications where the user can retrieve most of the pallets after each trip.

Plastic pallets of all types offer certain generic advantages which make them attractive alternatives to other pallet materials. Listed below are some of the plastic pallet's chief benefits.

Long pallet life. The relatively expensive plastic pallet must offer a long service life. Many customers experience plastic pallet life of five to nine years and more (7).

Reduced load damage. Smooth molded plastic helps eliminate product damage. There are no broken boards or protruding nails to damage sensitive loads (7).

Easy cleanup. Plastic pallets are easy to clean and keep clean (8).

USDA and FDA clearance. Both polyethylene and polystyrene are acceptable in food and pharmaceutical plants. Pallets made from these materials can be approved on a case by case basis by the on-site inspectors.

Reduced worker injury. Smooth construction and consistent weights help to eliminate minor cuts and back strain (7).

Chemically inert. Polyethylene plastic pallets are highly resistant to acids and bases, and at ambient temperatures, hydrocarbon solvents.

Moisture proof. Plastic pallets will not absorb moisture and soak loads. Plastic pallets will not rust or break down in wet conditions (8).

No harbor for pests. Plastic pallets will not harbor or support the growth of worms, eggs, molds, or mildew.

Design advantages of plastic pallets can include:

nestability (single-faced pallets can nest with each other when unloaded). This feature can save over 50% of valuable truck or dock space; and

interstacking (the ability to positively locate one loaded pallet on top of another loaded pallet).

Table 1. Plastic Pallets, Production Methods

Molding process	Plastic pellets advantages	Plastic pellets disadvantages	Ideal pallet application	Secondary applications	Tooling options	Average cycle	Wall thickness
structural-foam molding	economic production of heavy wall sections; short cycle times; good impact resistance; good deflection strength; high strength per pound (kilogram); good weight and dimensional tolerance; allows complex shapes	high cost tooling and processing equipment	large volume custom or proprietary pallets with runs of 1000 pallets or more; minimum custom order quantity 3000 units	manufacture of heavy duty racking pallets is possible by using filled polyethylene pallets or polystyrene wall thicknesses of up to 1 in. (2.54 cm) can be used when necessary	Kirksite; aluminum; steel	2–4 min	3/16–1.0 in. (4.8–25.4 mm)
injection molding	flexible process allows production of light weight disposable as well as heavy duty returnable pallets; allows complex geometry	highest tooling cost; highest equipment costs; high energy costs	largest volume custom and proprietary pallets	lightweight disposable pallets can be inexpensively produced by keeping wall sections narrow and cycle times short; heavy duty racking pallets can be manufactured by using heavier wall sections and a well integrated rib design	hardened steel	30 s to 3 min	1/32–3/8 in. (0.8–9.5 mm)
rotational molding	low equipment cost; low tooling cost; production of double walled parts	relatively long cycle times; limited weight and dimensional stability; limited to simpler design (geometry)	low volume production of large pallets; custom pallet projects of 1000 units or more are feasible	manufacture of heavy duty racking pallet is possible by encapsulating steel reinforcements into the pallets	cast aluminum; fabricated; metal plated nickel	3–6 min	1/8–1/4 in. (3.2–6.4 mm)
vacuum forming	low cost equipment; low cost tooling	relatively long cycle times; limited wall thickness; limited depth of draw; limited design complexity	lower volume, low cost, lightweight pallets; pallet projects of 500 units and above are feasible	heavier loads up to 3000 lb (1361 kg) can be accomodated by using twin sheet vacuum forming; vacuum-formed pallets are not generally used for heavy duty racking applications	metal; plaster; epoxy; wood	3–6 min	1/8–1/4 in. (3.2–6.4 mm)
reaction-injection molding	lighter weight tooling and equipment costs less than injection molding processes; allows complex designs; lower pressures and temperatures afford significant savings (70%) over injection-molding processes	limited deflection strength; slightly longer cycle times than injection-molding processes; limited dimensional stability	lighter duty custom and proprietary pallets; pallet projects of 1000 units and above should be justifiable	fiberglass-reinforced reaction injection molding is used to increase deflection strength for heavier applications; steel reinforcements can be encapsulated for additional strength	lightweight; steel; aluminum; Kirksite; sprayed metal	2–4 min	1/8–2.0 in. (3.2–51 mm)

BIBLIOGRAPHY

1. *U.S. Industrial Outlook,* United States Department of Commerce, Jan. 1984.
2. J. M. Callahan, "Just-In-Time A Winner," *Automotive Industries Magazine,* **65**(3), 78 (March 1985).
3. *Modern Plastics Encyclopedia,* 1983–1984 ed., McGraw-Hill Publications Co., New York, N.Y.
4. "What's Available in Plastics?", *Warehouse Supervisor's Bulletin,* National Foreman's Institute, Waterford, Conn., June 25, 1984.
5. "Pallets Take Off in All Directions," *Modern Plastic Magazine,* 64–66 (March 1971).
6. "Fitting Plastic Pallets to the Job," *Plastic Design Forum,* **9**(3), 57 (May/June 1984).
7. R. F. Ellis, "Plastic Pallets Eliminate Product Damage in Storage," *Modern Materials Handling Magazine,* **39**(8), 87 (June 8, 1984).
8. "Molded Plastic Pallets Solve Odor Transfer Problem," *Food Processing Magazine,* **46**(3), 108 (March 1985).

L. T. Luft
Menasha Corporation

PALLETS, WOOD

A pallet is a fabricated platform used as a base for assembling, storing, handling, and transporting materials and products in a unit load. A pallet container, or bin pallet, is a pallet having a superstructure of at least two sides (fixed, removable, or collapsible), with or without a lid (1). Pallets and containers are constructed of a variety of structural forms of wood, metals, plastics (see Pallets, plastic), paperboard (see Pallets, expendable corrugated), and various combinations of these materials.

Pallets were introduced as a materials handling tool in the 1930s after the forklift truck manufacturers developed small highly maneuverable lift trucks and hand-jacks (2). Wooden pallets were first employed on a large scale during World War II by the military services. They purchased more than 50 million (10^6) pallets from 1941 to 1945, and annual production increased from 11 to 32 million (10^6).

In 1946, the food-processing industry, together with transportation companies, terminal warehouse companies, and the pallet-manufacturing industry, recommended the adoption of 40×32 in. and 40×48 in. pallet sizes. These recommendations were incorporated in the Department of Commerce Simplified Practice Recommendation No. R228-47, "Pallets for the Handling of Groceries and Packaged Merchandise."

In 1947, the National Wooden Pallet Manufacturers' Association was formed as a part-time division of the National Wooden Box Association, and its first set of specifications was issued in 1949. The first Federal specification for pallets was issued in 1947, and by 1952, there were at least three military specifications in force.

During the period from 1952 to 1965, pallet production expanded from 33 million (10^6) to 88 million per year. The National Wooden Pallet Manufacturers' Association (NWPMA) separated from the National Wooden Box Association in 1953. The NWPMA published new specifications for hardwood pallets and for pallets produced from West Coast woods in 1962. The scope of the association was expanded to include wooden containers, and the name was changed to National Wooden Pallet and Container Association (NWPCA) in 1967.

The first industry pallet pool in the United States was formed in 1945 in the brick industry, and a national interindustry pool was started in Sweden in 1947 (3). By 1962, 13 national interindustry pallet pools were in operation in Europe, and 10 company-operated national pools were in existence in the United States. Intercompany pallet pools have operated primarily within groups of related industries since the early 1960s in the United States. These pools are labeled as pallet-exchange or pallet-interchange pools to denote the transfer of ownership of the pallets when the ownership of the goods on the pallet is transferred. The pallet-exchange pool operating among firms in the food and related industries is variously labeled as the "food pallet pool" or the "GMA pool" initiated by firms within the Grocery Manufacturers of America (GMA).

After 1965, research focused on solving problems of equality of value and quality of pallets exchanged in pool operations of the food industry. A procedure for estimating the strength of wooden pallet deckboards was published in 1959 (4). Procedures for estimating strength and stiffness of deckboards and stringers under a variety of load and support conditions were published in 1976 (5). These procedures were developed as a computer program in 1978 (6). The William H. Sardo, Jr., Pallet and Container Research Laboratory was established on the campus of Virginia Polytechnic Institute and State University at Blacksburg, Va., in 1976.

In 1984, the NWPCA copyrighted an improved computerized standard design procedure developed cooperatively by NWPCA, the Pallet and Container Research Laboratory, and the U.S. Forest Service, entitled, "Pallet Design System" (7). For location and purpose of principal research laboratories, see Table 1.

Standards and Specifications

A list of U.S. pallet standards is contained in Table 2. The American Society of Mechanical Engineers in conjunction with the American National Standards Institute has published standards on pallet definitions and terminology, pallet

Table 1. Principal Pallet Research Laboratories

Universities
Virginia Polytechnic Institute and State University Blacksburg, Va. conducts research into engineering design characteristics of pallets, pallet materials, and pallet fasteners

United States Government
United States Department of Agriculture, Forest Service
Northeastern Forest Experiment Station
Forestry Sciences Laboratory
Princeton, W.V.
   conducts research on supply and demand for pallets and pallet materials

Forest Products Laboratory
Madison, Wisc.
   conducts research into engineering properties of wood and wood fastenings

United States Department of Defense, Department of the Army
Mobility Equipment Research and Development Command
Fort Belvoir, Va.
   conducts research and testing and publishes specifications for DOD pallets

Table 2. U.S. Pallet Standards

American National Standards Institute (ANSI)

ANSI/ASME MH1.2.2-1975	pallet sizes
ANSI/ASME MH1.4.1-1985	procedures for testing pallets
ANSI/ASME MH1.1.2-1978	pallet definitions and terminology
ANSI/ASME MH1.5M-1980	slip sheets

American Society for Testing and Materials (ASTM)

ASTM D 1761-77(84)	standard methods of testing metal fasteners in wood
ASTM F 547-77(84)	standard definitions of terms relating to nails for use with wood and wood-based materials
ASTM F 592-80(84)	standard definitions of terms relating to collated and cohered fasteners and their application tools
ASTM F 680-80(84)	standard methods of testing nails
ASTM D 2555-81(84)	standard methods for establishing clear wood strength values
ASTM D 1185-73(85)	standard methods of test for pallets and related structures employed in materials handling and shipping

National Wooden Pallet and Container Association (NWPCA)

1980	specifications and grades for warehouse, permanent, or returnable pallets of southern pine
1981	specifications for Douglas-fir and western softwood plywood pallets
1982	logo-mark hardwood pallet standards
1982	logo-mark West Coast pallet standards
1983	logo-mark white woods pallet standards
1985	logo-mark pallet repair standards

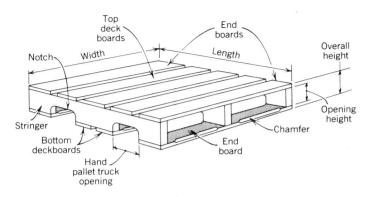

Figure 1. Principal parts of stringer-type pallets.

height is the overall vertical dimension including decks, stringers, stringer boards, and blocks. Pallet opening, or inside height, is the vertical dimension between the decks or between the top deck and the support base to accommodate entry of materials handling equipment (see Figs. 1–3).

The National Wooden Pallet and Container Association in 1981 estimated that the most common sizes used in the United States were as follows:

	Percent of total production
48-in. (1219-mm) length	
40-in. (1016-mm) width	28.5
42-in. (1067-mm) width	3.2
36-in. (914-mm) width	1.3
40-in. (1616-mm) length	
48-in. (1219-mm) width	4.8
36-in. (914-mm) length	
48-in. (1219-mm) width	2.4
Square pallets	
48 in. (1219 mm)	4.2
44 in. (1118 mm)	1.3
42 in. (1067 mm)	5.4
40 in. (1616 mm)	2.9
36 in. (914 mm)	2.2

The remaining 43.8% includes a variety of other sizes, each representing less than 1% of total production.

Pallets are constructed of one or two decks secured to deck spacers, which may be stringers or blocks. These are often referred to as stringer-type (Fig. 1) or block-type (Fig. 2) pallets. Pallets of both types are produced in a variety of designs

sizes, procedures for testing pallets, and slipsheets (see Slipsheets). The American Society for Testing and Materials (ASTM) has published standards on methods of test for pallets, definition of terms for fasteners, testing nails, and testing fasteners in wood.

The NWPCA has published standards for pallets produced from hardwoods, from West Coast woods, from white woods, from plywood, and from Southern pine. Standard specifications for pallet fasteners (8) are currently under review. The NWPCA also offers a logo-mark pallet inspection and certification service to pallet purchasers to ensure that pallets purchased conform to the specifications.

The International Organization for Standardization (ISO) is currently developing international standards for pallets for unit load methods of materials handling: on performance requirements and on methods of test, under their technical committee, ISO/TC 51 (see Standards and practices).

Designs of Pallets

Pallet size is defined by the dimensional length and width of the top deck (1). Conventional designation is to state the length first and then the width. The length is determined by the length of the stringer or the stringer board, except for plywood-deck block pallets in which case the length is the dimension perpendicular to the face grain of the plywood. Width is the dimension perpendicular to the length. Outside

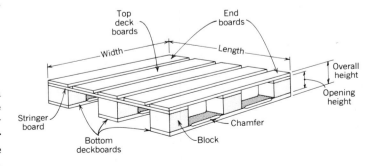

Figure 2. Principal parts of block-type pallets.

to suit any application. Single-deck pallets or skids have only a top deck; and double-deck pallets have top and bottom decks.

Reversible pallets have similar top and bottom decks capable of carrying a load; nonreversible pallets have dissimilar top and bottom decks with only the top deck capable of carrying a load. Wing pallets are pallets constructed with the decks overhanging the stringers to permit the pallets to be handled with slings. The edge stringers are set in from the ends of the deckboards. Single-wing pallets have overhang only on the top deck, and double-wing pallets have overhang on both decks. Flush pallets have no wings.

Two-way entry pallets are stringer pallets with solid stringers so that fork-truck tines or other equipment can enter the pallet only from the two ends of the pallet. Four-way entry pallets are block-type with the blocks spaced so the fork trucks and hand trucks can enter the pallet from any of four directions. Partial four-way pallets are stringer-type with notches cut out of the bottom of the stringer to permit entry of fork-truck tines from four directions, but hand trucks may enter from two directions.

Bin pallets (Fig. 3) may have any design of pallet as a base, and the bin may be constructed of plywood, lumber, wire, mesh, corrugated paper, plastic, or metal. The bin may be attached permanently to the pallet base or it may be removed and folded, or collapsed, to become a collapsible bin.

Pallet Fasteners

Pallet fasteners are normally steel nails or staples, although bolts are used in special cases. A wide variety of metal or plastic clamp devices is also used in the construction of bins either in combination with or in lieu of nails and/or staples. Pallet nails vary in length from 1 to 3.5 in. (2.5–9 cm). The wire sizes vary from 0.085 to 0.135 in. (0.2–0.35 cm) diameter. Nail length is the length of the nail shank, measured from the underside of the head to the tip of the point.

Plain-shank nails are produced from smooth wire. Deformed-shank nails may be helically threaded, helically fluted,

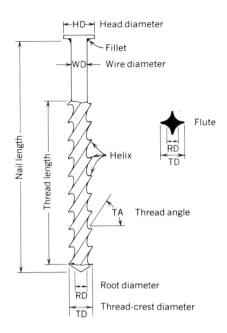

Figure 4. Nomenclature of helically threaded pallet nails.

or annular ring. Helically threaded nails (Fig. 4) are used in most cases. The thread is formed by passing the plain shank through a die. Only the lower portion of the shank is threaded, leaving about 0.75 inch (1.9 cm) of the plain shank under the head. Helically fluted nails are formed by twisting a squared wire, and are usually threaded the full length of the shank. Annular ring nails are formed by passing the plain shank through a die which forms usually 16 to 20 rings per inch (6–8/cm) of thread length. The thread rings are discrete threads perpendicular to the nail-shank axis.

Thread angle is the angle formed on the helically deformed nails with the axis of the shank. The thread angles vary from 60 to 75 degrees. Thread-crest diameter is the diameter of the shank measured across the deformed portion of the shank. Flutes are the number of helical threads formed around the shank. Four flutes are most common. Helix is the number of helical thread crossings counted along the threaded shank.

Pallet structures are unique as compared to most other structures in that they must be able to absorb stresses and strains over a relatively broad range. Consequently, the joints must be able to give or flex. Rigid joints such as are obtained with adhesives have been found to result in excessive damage to the structure. Flexible adhesives have also been tried and were found to be inadequate.

The influence of the fasteners on pallet performance is of major importance. The fasteners must be suited to the properties of the woods employed. Recommended standards for nailing pallets were published in 1974 (9). Recommended performance rating criteria for pallet fasteners, based on scales of relative performance for withdrawal resistance and for shear resistance were published in 1982 (10, 11). Recommend standard specifications for nails and staples for pallets were prepared in 1984 (8).

The NWPCA specifies helically threaded nails with wire diameters from 0.105 to 0.120 in. (0.27–0.305 cm) and permits

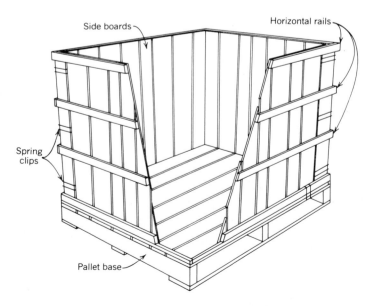

Figure 3. Principal parts of bin pallets.

other fasteners that provide equivalent performance. Quality of nails varies widely depending on thread characteristics and on the quality of wire used to produce them. Wire quality is defined in terms of its resistance to bending when samples of nails are subjected to a standard impact bending test. Terms commonly employed are soft-wire fasteners that may be clinched, stiff-stock fasteners that are not heat treated and tempered, and hardened-steel fasteners that are heat treated and tempered.

Staples are widely used in lieu of nails for short-life pallets. Staples are usually produced from 15-gauge flattened soft-steel wire with a 0.375-inch (0.95-cm) crown, although both 16- and 14-gauge wire are also used (see Staples). A few pallet manufacturers also use a round wire staple produced from 0.080-in. (0.20-cm), 0.10-in. (0.25-cm), and 0.125-in. (0.318-cm) diameter wire. The length of staples, as for nails, varies depending on the thickness of the deckboards and the amount of penetration desired or needed.

Most pallets are produced with two fasteners per joint in boards ≤4 in. (≤10.2 cm) in width and with three fasteners per joint in boards ≥5 in. (≥12.7 cm) in width. In general, the trend of using hardened-steel nails has permitted use of smaller diameter nails, thus reducing danger of splitting without reducing pallet performance. The helically threaded nails are used almost universally in preference to plain-shank or smooth-wire nails because the helically threaded nails retain their holding power after the wood dries in the pallet. Since pallets are produced primarily with green wood, bolts are seldom used due to loosening of the joint as the wood dries.

Pallet Construction Materials

About 95% of all pallets are constructed of lumber and plywood employing almost all species of wood grown in North America. In other parts of the world, native and imported species are used. The only restrictions on wood used for pallets are the very light and low strength species and the very heavy and hard species found in some of the tropical forest areas of the world. Thus, almost any species of wood may be used in pallets if proper design is exercised in the construction. With the exception of some of the western coniferous species, and for special uses, essentially all pallets are produced from green lumber.

In the western regions, where pallet requirements are small relative to total lumber production, the pallet manufacturers use primarily the lowest grades of construction lumber such as economy grade and some utility grade. For pallet manufacturers employing Southern pine, grades No. 4 and No. 3 are used.

Historically, the pallet users prepared detailed specifications for their pallets, which the pallet manufacturers followed. The pallet-manufacturing industry produced whatever the users requested. In recent years, development of the computerized design procedures has prompted pallet users to rely on the pallet manufacturers for their specifications.

Prior to development of the design procedures, the pallet materials were grouped in three classes or four groups of species based primarily on specific gravity. These groupings are still widely used but are inadequate for the new design procedures. The primary properties needed to design safe pallets are bending strength (modulus of rupture in bending, or MOR), stiffness (modulus of elasticity in bending, or MOE), and spe-

cific gravity. Furthermore, the variance associated with the averages for these three properties is also required. The properties of the commercial species in the United States and Canada, based on tests of small clear specimens, are contained in ASTM D 2555-81 (12).

The National Wooden Pallet and Container Association standards for hardwood pallets (13) recommends a combined B&C Class and an A Class:

Class B&C		Class A
ash	madrone	alder
beech	magnolia	aspen
birch	maple	basswood
cherry	myrtle	buckeye
elm	oak	chestnut
hackberry	pecan	cottonwood
hickory	persimmon	poplar
gum	sycamore	willow
locust	tupelo	

The West Coast standard (14) recommends Douglas-fir, hemlock, and larch for warehouse pallets and any western softwood for shipping pallets. The white woods standard (15) recommends white woods (spruce, pine, and fir) for both warehouse and shipping pallets. The NWPCA standards do not include properties information for design.

Control of the quality of the individual pallet parts (pallet shook) is almost universally accomplished by specifying the minimum quality allowed in a pallet under the assumption that all of the higher quality shook obtained from the source lumber, cant, or log is included in the mix. The grade criteria reflect industry practice in regard to quality classes. Some of the grade criteria are related to defects that affect the bending strength and some are related to defects that affect nailing and to appearance. Four grade classes ranging from No. 1 for essentially clear defect-free shook to No. 4 for shook with large knots and/or steep cross grain have been defined for use in the computerized design procedures.

Increasing emphasis is being placed on designing pallets to specified performance requirements. Also, NWPCA standards now require specification of actual dimensions for length, width, and thickness of the shook used for assembly. This is a significant departure from the specification of nominal dimensions. The design procedures employ weighted average properties based on the mix of species used in the pallets. These procedures provide estimates of load-carrying capacity and deflection under load for pallets placed in stacks and in racks. They also provide estimates of the expected life of the pallets and cost per trip to use the pallets. They are based on results of testing in normal use in the field and on laboratory testing conducted during the past 20 years. The introduction of the logo-mark inspection and certification service by NWPCA provides a means for ensuring that the pallets do in fact meet the specifications.

Pallet Production and Use

The pallet and container industry represents a major market for lumber and is the largest single market for hardwood lumber. It consumes nearly 50% of hardwood lumber production and 8% of softwood lumber production, plus hardwood roundwood equal to 10% of lumber production. Pallets and containers are produced, repaired, and distributed by about 2500 firms (16). Current annual production is over 250 million (10^6). Pallets are widely used by essentially all industries (17), with food and similar products accounting for about 30%.

Pallet demand is expected to increase to 500–600 million(10^6)/year by 1995. In order to satisfy the demand projections, with timber supplies increasing at a 1–2% rate, the efficiency of use of pallets must increase at a rate of about 5%. This means that each board foot of lumber used in pallets needs to provide increased service to the pallet user. About 85% of the annual production of pallets is for replacement of damaged and discarded pallets (16). Most of this is related to the quality of pallets used. Industrial pallet users have long recognized the potential cost savings inherent in using pallets designed to provide longer life, lower repair cost, and lower cost of use. In order to realize these benefits, a national pallet-exchange system is needed.

Pallet Pools

A national pallet pool for the United States has been the subject of numerous investigations and research since the 1960s (3, 18). In 1964 and 1965, Butterick Management Services conducted a series of studies of pallet-pool operations in the food industry to identify problems and solutions (19–21). These early efforts to create interindustry pools were frustrated by a lack of uniform standards for pallet construction and for repair of pallets. Consequently, the exchanges of pallets were not equal in value. Since then, the NWPCA has revised its pallet standards to permit more precise specification of pallets, and it has initiated a logo-mark inspection and certification service for pallet purchasers. Most recently, it has copyrighted a computerized standard design procedure that enables the pallet users to obtain equal pallets in terms of strength and durability. They are presently preparing standard pallet repair procedures also backed by logo-mark inspection and certification. The pallet purchasers now have the means of specifying and ensuring that the pallets are equal, regardless of species of wood used to manufacture the pallets.

The other principal problem of pallet exchange, which has not yet been resolved, is management of a national system (22). Many companies operate limited pools among their establishments that are national in character. National Pallet Leasing Systems, Inc. operates a limited national leased-pallet pool. They own the pallets, maintain them, and replace worn-out pallets. A similar system is in operation in Australia, England, and Canada. National pools in Europe are managed by the national railroads. A national pool for the United States is generally recognized as a means of providing significant cost savings. The continued growth in pallet use is preparing the foundation for a national pool insofar as sufficient volumes of goods are being shipped on pallets to permit pallets to move loaded in all directions with limited backhauls of empty pallets.

BIBLIOGRAPHY

1. *Pallet Definitions and Terminology,* MH1.1.2-1978, American National Standards Institute, The American Society of Mechanical Engineers, New York, 1979, 9 pp.

2. R. K. Day, *The Wood Pallet Industry, Its Development and Progress toward Standardization,* FPL Rep. R1957, U.S. Department of Agriculture, Forest Service, Forest Products Laboratory, Madison, Wisc., 1953, 20 pp.

3. G. C. Thomas, ed., *Mod. Mater. Handl.* **18**(10), 42 (1963).

4. T. B. Heebink, *Load-Carrying Capacity of Deck Boards for General-Purpose Pallets,* FPL Rep. 2153, U.S. Department of Agriculture, Forest Service, Forest Products Laboratory, Madison, Wisc., 1959, 10 pp.

5. W. B. Wallin, E. G. Stern, and J. A. Johnson, *Determination of Flexural Behavior of Stringer-Type Pallets and Skids,* Bull. No. 146, Virginia Polytechnic Institute and State University, Wood Research and Wood Construction Laboratory, Blacksburg, Va., 1976, 34 pp.

6. W. B. Wallin, and K. R. Whitenack, *Pallet Analysis Program for Strength and Durability,* U.S. Department of Agriculture, Forest Service, Forestry Sciences Laboratory, Princeton, West Virginia, 1978 (rev. 1984). Unpublished computer program.

7. M. S. White, Pallet Enterprise 4(2), 2 (1985).

8. E. G. Stern, and W. B. Wallin, *Standard Specification for Pallet Nails and Staples—Hammer, Tool, and Machine Driven,* National Wooden Pallet and Container Association, Washington, D.C., Sept. 1984. Working draft, unpublished.

9. W. B. Wallin, and E. G. Stern, *Tentative Nailing Standards for Warehouse and Exchange Pallets,* Bull. No. 129, Virginia Polytechnic Institute and State University, Wood Research and Wood Construction Laboratory; Blacksburg, Va., 1974, 16 pp.

10. W. B. Wallin, and K. R. Whitenack, *Pallet Enterprise* **1**(3), 21 (1982).

11. W. B. Wallin, and K. R. Whitenack, *Pallet Enterprise* **1**(4), 20 (1982).

12. *Standard Methods for Establishing Clear Wood Strength Values,* ASTM D2555-81 (84), American Society for Testing and Materials, Philadelphia, Pa., 1984, pp. 513–534.

13. *NWPCA Logo-Mark Hardwood Pallet Standards,* National Wooden Pallet and Container Association, Washington, D.C., 1982, 5 pp.

14. *NWPCA Logo-Mark West Coast Pallet Standards,* National Wooden Pallet and Container Association, Washington, D.C., 1982, 5 pp.

15. *NWPCA Logo-Mark White Woods Pallet Standards,* National Wooden Pallet and Container Association, Washington, D.C., 1983, 5 pp.

16. R. E. Buckman, *Research and Development of Pallet Industry,* keynote address to NWPCA Manufacturing and Sales Promotion Clinic, Hot Springs, Va., June 25–27, 1984, 12 pp.

17. R. S. Bond, and P. E. Sendak, *The Structure of the Wood-Platform Industry of the Northeast,* Exp. Sta. Bull. No. 586, Agricultural Experiment Station, College of Agriculture, University of Massachusetts, Amherst, Mass., 1970, 70 pp.

18. J. A. Eaton, ed., *Mod. Mater. Handl.* **18**(11), 42 (1963).

19. G. C. Thomas, ed., *Mod. Mater. Handl.* **20**(3), 36 (1965). (Source: Butterick Management Services, John J. Strobel, General Manager.)

20. G. C. Thomas, ed., *Mod. Mater. Handl.* **20**(4), 60 (1965). (Source: Butterick Management Services, John J. Strobel, General Manager.)

21. J. J. Strobel, and W. B. Wallin, *The Unit-Load Explosion in the Food Industry,* Res. Pap. NE-121, U.S. Department of Agriculture, Forest Service, Northeastern Forest Experiment Station, Broomall, Pa., 1969, 60 pp.

22. W. B. Wallin, *Pallet Enterprise* 4(1), 24 (1984).

W. B. WALLIN
United States Department of Agriculture, Forest Service

PAPER

In the United States today, approximately 1.2×10^{10} lb (5.5×10^6 t) of paper are consumed annually for packaging purposes (1). This represents about 15% of total domestic paper production (2). There is no strict distinction between paper and paperboard, particularly in view of the tremendous variations in density possible with current technology. Generally, structures less than 0.012-in. thick (12 "points" or 305-μm) are considered paper regardless of weight per unit area. Except for the use of paper as an overwrap for folding cartons, most packaging papers are used in flexible applications.

The primary intermediate product used to make paper is wood pulp. The properties of an individual paper or paperboard are extremely dependent on the properties of the pulps used. Pulp preparation from deciduous (hardwood) or conifer (softwood) species may be done by mechanical, chemical, or hybrid processes (3–5). These hardwood or softwood pulps may be used unbleached, or they can be bleached to varying degrees by a diversity of techniques.

Mechanical pulps produce papers that are characterized by relatively high bulk, low strength, and moderate to low cost (6). Their use in packaging is very limited.

The kraft (sulfate) pulping process, introduced about 100 yr ago, dominates the chemical pulping industry: yields are higher, pulps are stronger, and process chemicals more completely and economically recovered than with any other process. Although the sulfite processes were extremely prominent 75–125 yr ago, their chemicals are difficult to recover, the resultant pulps are significantly weaker than those produced by the kraft process, and they produce no unique paper properties. Unbleached pulps are generally stronger, stiffer, and more coarse than their bleached counterparts, but papers made of white, conformable fibers are used in many more applications. General treatments of pulping and bleaching are provided in references 7–10.

The standard ream basis for packaging papers in the United States is 500 sheets cut to a size of 24×36 in. (61×91.5 cm) (3000 ft^2 or 279 m^2). On this basis, packaging papers normally weigh 18–90 lb/ream (8.2–40.8 kg/ream), but some specialty applications require weights as low as 10 lb (4.5 kg) or as high as 200 lb (90.7 kg)/ream. At any given basis weight, density may typically vary from 0.08–0.16 lb/in.^{3} (2.2–4.4 g/cm^3), providing a very wide range of thickness and strength properties.

The two most general classifications of packaging papers are coarse and fine. Coarse (kraft) packaging papers are almost always made of unbleached kraft softwood pulps. Fine papers, generally made of bleached pulp, are typically used in applications demanding printing, writing, and special functional properties such as barriers to liquid and/or gaseous penetrants.

Kraft Papers

Kraft papers, produced by the Kraft process, derive their name from the German word for "strong." Kraft paper is made from at least 80% sulfate wood pulp. It is typically a coarse paper with exceptional strength. Sometimes made with a rough finish to keep bags from sliding off piles, these papers are often made on a fourdrinier machine, and then either machine finished with a calender stack or machine glazed by using a Yankee dryer (7,8). The surface of these papers is acceptable for printing by letterpress, flexography, and offset processes (see Printing). In addition to wrapping applications, kraft papers are used for multiwall bags and shipping sacks (see Bags, paper), grocers' sacks, envelopes, gummed sealing tape (see Tape, gummed), butcher wraps, freezer wraps, tire wraps, and specialty bags and wrappings that require economy and strength. Many papers formerly manufactured from sulfite pulps, especially those of tissue weight, are now manufactured with kraft pulps. Unbleached "sulfite" papers are used for products such as oil cans (intermediate liner) and single-service food packages.

Extensible kraft papers have satisfied a special niche in the packaging industry. Although creped papers have long served both decorative and functional roles, other papers capable of absorbing energy at sudden rates of strain have become increasingly important in uses such as shipping sacks. Conventional creping is performed either at the wet press section of a paper machine (wet crepe) or on a Yankee dryer (dry crepe). Dry creping is most commonly used to generate qualities such as softness and absorbency; wet creping is a technique for making tough, flexible papers capable of absorbing tensile energy. A secondary creping operation rewets a dry sheet, and may be done in-line on the paper machine or as an independent manufacturing process. Whereas a standard kraft paper might have a stretch (before breaking) of 3–6%, creped papers generally may be stretched 35–200% of their original length before breaking. Manufacturers of creped, extensible, and other coarse (kraft) papers may be found in references 11 and 12.

Bleached Papers

Packaging applications that place a higher priority on printing, writing, and special functional properties than on economy and strength generally utilize bleached papers. The pulps used to manufacture bleached papers are relatively white, bright, and soft, and they are also receptive to special chemicals necessary to develop many functional properties. Although generically not as strong as unbleached kraft papers, bleached papers can be manufactured to meet simultaneous requirements of both strength and printability. Their whiteness enhances print quality and generates a perception of cleanliness and quality. The aesthetic appeal of bleached packaging papers may be augmented by clay coating one side (C1S) or both sides (C2S). The increasing demand for a combination of functional performance and top quality graphics favors the C1S manufacturer who can satisfy the variety of challenges of this market.

Vegetable Parchment

The process for producing parchment paper was developed in the 1850s, making it one of the grandfathers of special packaging papers. By soaking an absorbent paper in concentrated sulfuric acid, the cellulosic fibers are swollen tremendously and partially dissolved (see Glassine, greaseproof, and parchment). In this state the plasticized fibers close their pores, fill in voids in the fiber network, and thus produce intimate contact for extensive hydrogen bonding. Rinsing with water causes reprecipitation and network consolidation, resulting in a paper that is stronger wet than dry, lint free, odor free, taste free, and resistant to grease and oils. By combining parchment's natural tensile toughness with extensibility imparted by wet creping, paper with great shock-absorbing capability can be produced. Special finishing processes provide qualities ranging from rough to smooth, brittle to soft, sticky to releasable. Parchment was first used for wrapping fatty substances like butter, but this versatile paper is now also used whenever food is prepared, frozen, packaged, or displayed, and when tough, lint-free, chemically pure surfaces are needed for special packages.

Greaseproof and Glassine

Because cellulose is hydrophilic, it is a good substrate to use for resisting penetration of hydrophobic liquids. As noted above, vegetable parchment performs well as a greaseproof paper because it is essentially pore-free and composed of a hydrophilic material. "Greaseproof" paper is a substrate manufactured to also have an essentially pore-free consolidation; but mechanical refining ("buffing" or cutting) is used in its production instead of swelling with concentrated sulfuric acid. Refining fibrillates, breaks, and swells the cellulose fibers to permit consolidation of a web with many interstitial spaces filled in. Glassine paper is produced by further treating "greaseproof" paper with a supercalender operation. The supercalender step involves moist high temperatures (steam), pressure (several hundred pounds per lineal inch or ca 100 kg/cm), and differential hardness (one roll typically cotton or soft rubber, the other roll hard rubber or metal) to polish the surface. Supercalendering a greaseproof paper generates such intimate interfiber hydrogen bonding that the refractive index of the glassine paper approaches the 1.02 value of amorphous cellulose. This indicates that very few pores or other fiber/air interfaces exist for scattering light or allowing liquid penetration.

Greaseproof and glassine papers are frequently plasticized to further increase their toughness. They have a reputation of running well on high speed packaging lines, and have served well for odor- and aroma-barriers. Like other flexible packaging papers, they can be chemically modified to enhance functional properties (eg, wet strength, adhesion, release). When waxed, they are standard materials for primary food pouches used to package dry cereals, potato chips, dehydrated soups, cake and frosting mixes, bakery goods, candy and ice cream confections, coffee, sugar, pet food, etc. In addition to their protective functions, these papers heat-seal easily when waxed and reclose well.

Water, Grease, and Oil-Resistant Papers

The distinction between greaseproof papers and grease-resistant papers is a fairly subtle one, involving an understanding of the methods of penetration of liquids into surfaces. Because the primary mechanisms involve capillary penetration and/or wetting, it is appropriate to consider the severity of the packaging requirement. The requirement may be for minimal staining by grease, oil, or water under negligible pressure; or it may be for absolute resistance to any penetration of the liquid over long periods of time and/or under substantial pressure; or it may lie between these extremes. Parchment, glassine, and greaseproof papers offer decreasing protection from grease and oil at the more restrictive end of the spectrum. As noted above, their resistance is generated by the lack of capillaries and the oleophobic nature of cellulose.

A consideration in designing primary packages is the economics of using the various barriers available for the job. A bag for a single-service consumable (eg, french-fried potatoes) may require resistance to staining for only several minutes. A lubricating oil package, on the other hand, may require a stain-free barrier for several months. Imparting sufficient resistance to liquid penetration to meet the requirements of the less demanding applications can be done very economically with chemical treatments. If grease and/or oil penetration is the only concern, moderate resistance (package life of minutes to days) can be developed using waxes (see Waxes) and other low surface-energy materials such as fluorocarbons. Fluorocarbon technology in papermaking has expanded from multiwall and consumer bags to labels, coupons, carry-out food packaging, perishable bakery goods packaging, candy and confection packages, and form-and-fill packages where edge wicking may be an important consideration.

In many applications (eg, carry-out food packaging) resistance to both water and oils must be developed for adequate performance. The use of rosin-based chemicals for developing water repellency in paper requires the use of alum ($Al_2(SO_4)_3$) or other multivalent cations which destroy the grease-resisting properties of fluorocarbons. The simultaneous development of both water and oil resistance requires the use of oleophobic chemicals which react with hydroxyl and carboxyl groups on the cellulose fibers, and the use of hydrophobic chemicals such as fluorocarbons which have also been modified to react with the same cellulosic functional groups. Typical oleophobic chemicals currently in widespread use are alkylsuccinic anhydrides and alkylketene dimers. The most successful economic choices are made through close consultation between the paper manufacturer and the user, which allows the careful selection of designs for meeting specific performance requirements. Grease and oil-resistant papers can be made to run well in most any converting and printing processes.

Waxed Papers

Waxed papers are age-old papers that have served the packaging industry well in applications requiring direct contact with food for barrier against penetration of liquids and vapors, as well as heat sealability, lamination, and even printing. Waxing can be performed in-line with the paper-manufacturing process, in-line with printing, converting or lamination processes, or as a discrete process. A great many base papers are suitable for waxing processes, including greaseproof and glassine papers, and water-resistant papers. There are two fundamentally different waxing processes, generating different characteristics for the finished sheet (see Waxes). Wet waxing is an operation in which the wax coating is applied to the surface of the sheet. Surface wax is desirable for heat sealing and lamination, and essential for vapor-barrier development. Dry waxing is performed to absorb wax into the sheet, leaving a surface that often does not look or feel waxy. Penetration of wax allows additional surface treatments for special release applications, or for further lamination. Absence of the continuous film of wax on the surface characteristic of wet waxed paper also allows the dry waxed paper to "breathe" moisture, carbon dioxide, and oxygen.

Waxed papers provide an economical choice for primary food packaging not only because of their versatility, but also because of their safety as tasteless, odorless, nontoxic, and relatively inert materials. Their widespread use in conventional packaging applications includes delicatessen pick-up sheets, box liners, cover, scale, and utility sheets, patty papers, sandwich wraps and bags, laminations to other papers and paperboard for food trays, locker papers, carry-out cartons, food pails, baking cups, folding cartons, cereal liners, and folding carton overwrap.

Specialty-Treated Papers

Many packaging applications require barrier to substances other than water, grease, or oil. Most food packaging applications are well-served with water- and/or oil-penetration resistance, but a number of products require more elaborate barriers. Meat-wrapping paper demands exceptional strength, resistance to grease, moisture, and blood, easy release with no residual taste or odor, plus "bloom" retention. Freezer paper must remain pliable at low temperatures, and offer puncture resistance, moisture-vapor barrier, exceptional seal integrity, and easy release from frozen or thawed meats. Other products such as chemicals, drugs, cosmetics, personal-care items, and industrial products require package functions such as acid resistance, alkali resistance, alkali solubility, mold resistance, flame retardation, solvent resistance, organic polymer adsorption/absorption resistance or affinity, adhesion, release, tarnish or rust inhibition, heat stability, sterilizability, specific-ion adsorption, conductivity, resistivity, stiffness, or flexibility. To address the general manufacturing techniques or even the functional property classification of such a variety is beyond the scope of this publication. Reference 13 provides a general treatment of some of these products which have transcended proprietary technology. Because many of the manufacturing techniques employed to generate these functional properties are considered trade secrets, the reader is referred to directories of the specialty-packaging paper manufactures (11,12).

Wet-Strength Papers

Conventional papers are not strong when wet. The most predominant fraction of paper's strength is the result of hydrogen bonding between hydroxyl and carboxyl groups on adjacent fibers. The removal of water during the papermaking process generates these bonds, and the process is reversible. The two strategies for manufacturing strong-when-wet papers are (1) keep the water out of the paper and (2) introduce additional chemical bonds between fibers which are not influenced by the introduction of water.

As mentioned above, paper that has been parchmentized is actually stronger wet than dry, principally because of the loss of individual fiber identity during the gelatinized stage of the process. Through advances in chemical technology, several more economical alternatives exist for generating wet-strength papers. In general, the chemicals used to augment the natural hydrogen bonding are cross-linked during the manufacturing process. Common chemicals for producing wet-strength papers include protein, urea, melamine, resorcinol, and other phenolic or amino resins cross-linked with formaldehyde, and condensation products of polyalkylene polyamines with dicarboxylic acids cross-linked with epichlorohydrin (14).

The conventional tests of wet-strength papers are for tensile, tear, and burst strength, as these are the most useful indicators of use requirements. The degree of wet strength is expressed as a percentage of original dry strength, and is referred to as the percent of strength retention. The typical range for sack or pouch papers is 15–30%. Wet strength may also be generated in a variety of permanence levels, so that the product will either remain tough-when-wet, or eventually disintegrate with soaking time or application of force.

Absorbent Papers

Papers designed for absorption of specific fluids are an important part of the packaging industry. Although they are a member of the class of "Specialty-treated papers," they are distinctive enough to warrant separate discussion. Typically at the low end of the strength spectrum, absorbent papers must not only be exceptionally porous, they must have surface modifications to render affinity to the target liquid. When that liquid is aqueous they must generally also have a definitive level of wet strength. Providing the special affinity to a given liquid is often a proprietary technology involving chemicals that are substantive to cellulose and to the target liquid. This special class of papers currently services industries ranging from fresh-food packaging to industrial-products packaging.

Tissue Papers

Tissues form a special group of fine packaging papers because of their versatile performance. Always fairly thin, tissues range from semitransparent to totally opaque. They can be waxed or treated with any of the specialty treatments (eg, edible oils for fruit wrapping, antitarnish metal protection), or they can be used "as is" for applications ranging from intermediate lamination steps in composite-container construction (see Cans, composite) to gift wrapping. They may be made exceptionally weak for softness, or surprisingly tough in all directions. Tissue papers are generally either machine-finished (MF) or machine-glazed (MG). Machine finishing involves calendering between rolls which are usually constructed of highly polished steel which are hydraulically loaded to squeeze and polish both sides of the paper to similar levels of smoothness. Machine glazing of papers produces a smooth glazed side and a rough back side. The special finish is the result of drying the paper with 60–70% moisture on a Yankee dryer, 6–18-ft (1.8–5.5-m) dia with a mirrorlike surface. This glazing process produces a very smooth surface for printing, adhesion, release, or wet waxing. A physical fusion takes place with the surface fibers, producing a physical barrier similar to a cast film. MG papers may also be machine finished to improve the smoothness on both sides or to produce intermediate rolls which process better through subsequent converting steps. A great many applications for tissue paper today utilize special treatments for adhesion or release in an intermediate package-manufacturing process where the paper's light weight and thinness make it an economical carrier for more costly substances (see Multilayer-flexible packaging).

Coated Papers

Coated papers is a term generally reserved for papers acting as a base for aqueous mixtures of clay and/or other mineral pigments with natural and/or synthetic polymers as pigment binders. The term does not typically refer to papers which have been extrusion- or solvent-coated with organics or plastics, or surface-treated with specialty treatments listed above for functional improvements. Aqueous coatings of paper are performed primarily for market appeal where graphics are important. Several hundred thousand metric tons of coated papers are consumed annually for packaging purposes, not only for labels and multiwall bags, but also for lamination or combination with other functional materials in composite structures. Coated papers can be designed for printing by any process from letterpress to ink jet (see Code marking and imprinting). The most valued grades are typically produced in discrete manufacturing processes which vary widely (7,8,10). Detail of coating technology state-of-the-art are contained in annual TAPPI Coating Conference Proceedings (15).

Nonwovens

Nonwovens are materials used as cloth substitutes, made entirely or partially from cellulosic fibers. As an industry, nonwovens is dynamic and growing tremendously in medical, health-care, industrial, food-processing, consumer- and household-products areas. Differentiated from classical paper, which is formed in water and consolidated with interfiber hydrogen bonds, nonwoven-manufacturing technologies include resin and thermally-bonded carded web process, meltblown process, and an air-laid process. Because the nonwovens industry is less than two decades old, much of the technology is proprietary (see Nonwovens).

BIBLIOGRAPHY

1. *Paper, Paperboard, and Woodpulp Capacity,* American Paper Institute, New York, 1984.
2. *Statistics of Paper, Paperboard, and Woodpulp,* API, New York, 1984.
3. S. A. Rydholm, *Pulping Processes,* Wiley-Interscience, New York, 1965.
4. M. G. Halpern, ed., *Pulp Mill Processes, Developments Since 1977,* Noyes Data Corp., Park Ridge, N.J. 1981.
5. *Proceedings of the Alkaline Pulping Conference, 1981 and 1985,* TAPPI Press, Atlanta, Ga.
6. D. R. Allan, ed., *Uncoated Groundwood Papers,* Miller Freeman Publications, San Francisco, Calif., 1984.
7. J. P. Casey, ed., *Pulp and Paper Chemistry and Technology,* 3rd ed., Vols. I–IV, Wiley-Interscience, New York, 1980.
8. R. G. MacDonald, ed., Pulp and Paper Manufacture, 2nd ed., Vols. I–III, ed., McGraw-Hill New York, 1969.
9. *Pulp Technology and Treatment for Paper,* Miller Freeman Publications, Inc, San Francisco, Calif., 1979.
10. M. J. Kocurek and C. F. B. Stevens, eds., Pulp and Paper Manufacture, 3rd Ed., Joint Textbook Committee of The Paper Industry, Atlanta, Ga. 1984.
11. H. Dyer, ed., *Lockwood's Directory of the Paper and Allied Trades,* 108th ed., Vance Publishing Corp., New York, 1985.
12. *Post's Pulp and Paper Directory,* Miller Freeman Publications, Inc., San Francisco, Calif., 1985.
13. R. H. Mosher and D. S. Davis, Industrial & Specialty Papers, Vol 1–4, Chemical Publishing Co., 1970–1974, 1969–1973, Vol. 1 (*Technology*) 1968, Vol. 2 (*Manufacture*) 1968, Vol. 3 (*Applications*) 1969, Vol. 4 (*Product Development*) 1970.
14. U.S. Pat 2,926,154, (Feb. 23, 1960), G. A. Keim, (to Hercules Powder Company).
15. *Proceedings of the Polymers, Laminations, and Coatings Conference,* annual publication of TAPPI Press, Atlanta, Ga.

General References

B. Wirtzfeld, ed., *The Paper Yearbook,* H. B. Jovanovich, New York, comprehensive listing of typical uses.

The Competitive Grade Finder, 18th ed., Grade Finders Inc., Pub. Bala-C., vvd Penn., 1984.

J. Hube, ed. *Kline Guide to the Paper Industry,* 4th ed., C. H. Kline Pub. Fairfield, N.J., 1980.

API annual *Statistics of Paper and Paperboard* for relative shipment and sales volume.

Pulp and Paper North American Industry Fact Book, Miller Freeman Publications, Inc., San Francisco, Calif., 1982.

The Dictionary of Paper, 4th ed., American Paper Institute, Inc., Pub. New York.

B. Toale, *The Art of Papermaking,* Davis Publications, Inc., Worcester, Mass., 1986; concise but fairly comprehensive treatment of historical techniques for making contemporary decorative papers.

I. P. Leif, *An International Sourcebook of Paper History,* Archer Dawson, Hamden, Conn., 1978.

H. F. Rance, ed., *Handbook of Paper Science,* Elsevier Scientific Publishing Co., New York, 1982, Vol. 1, *Raw Materials and Processing,* Vol. 2, *Structure and Properties.*

R. P. Singh, ed., *The Bleaching of Pulp,* 3rd ed., TAPPI Press, Atlanta, Ga., 1979.

G. R. Hutton, ed., *Phillips Paper Trade Directory-Mills of the World,* Derek G. Muggleton Publishers, Kent, UK, 1985.

W. S. Dempsey, "The Manufacture of Dry Crepe Paper" *Pulp Paper Mag. Can.* **60**(5), T1 39–40, (1959).

A. Koebig, "Manufacture of Creped Paper," *Pulp Paper Mag. Can.* **39**(3), 237–238, 1938.

B. J. Diaz, "Clupak Paper-a New Type of High Stretch Paper, Its Manufacture and Its Performance, *Symposium on Paper and Paper Products, ASTM Special Tech. Publ. No. 241,* Philadelphia, Pa., pp. 51–63 1959.

C. T. Waldie, The Manufacture of Wet Creped Paper: *Pulp Paper Mag., Can.* **60**(7), 224–226 (1959).

M. SIKORA
James River Corporation, Specialty Packaging Papers Group

PAPERBOARD

Paperboard, often called simply board, is one of the major raw materials used in packaging. The term includes boxboard, chipboard, containerboard, and solid fiber. Its application covers a wide range of uses from simple cartons (see Cartons, folding) to complex containers used for liquids. It can also be converted into drums for bulk packaging of chemicals or combined with other materials to produce containers large and strong enough for the protection of large, heavy, and often fragile items during transport. In addition to product-protection requirements, paperboard must often have at least one smooth surface capable of accepting high-quality print (see Printing).

Terminology

Paper is the general term for a wide range of matted or felted webs of vegetable fiber (mostly wood) that have been formed on a screen from a water suspension. The general term can be subdivided into paper (see Paper) and paperboard. There is no rigid line of demarcation between paper and board, and board is often defined as a stiff and thick paper. ISO standards state that paper with a basis weight (grammage) generally above 250 g/m² (~51 lb/1000 ft²) shall be known as paper-board, or "board." The definition becomes less clear due to the fact that in some parts of the world board is classed as such when its caliper (thickness) exceeds 300 μm (~12 mil) [in the United Kingdom, 250 μm (~10 mil)] There are exceptions to the above: blotting papers and drawing papers thicker than 300 μm are classified as paper, and corrugating medium, liner-board, and chipboard less than 300 μm are classified as paper-board.

It is also not possible to strictly define paperboard by its structure or by the type of machine used to produce it. For example, paperboard can have either a single- or multi-ply structure and can be formed on a Fourdrinier-wire part, a single or a series of cylinder molds, or a series of modern form-ers, or sometimes by means of a combination of one or more of the above (see diagrams and related text). For a small part of the market, paperboard is produced by laminating sheets of paper together. In that case, the product is solid fiber (see Boxes, solid fiber).

The following terms are in general use in paperboard man-ufacturing and associated converting industries:

Basis weight. This is the weight of a known area; for exam-ple, g/m² (grammage) or (lb/1000 ft²), the weight in pounds of a ream (usually 500 sheets) of paperboard cut to its "basic size."

Caliper (thickness). The thickness of the sheet expressed in μm or thousandths of an inch (mil or points) (mil = 25.4 μm).

Size of a sheet. The width and length of a sheet of paper-board. The width, always expressed first, is the dimension cut at right angles to the direction of the sheet. Length is the dimension cut in the machine direction. For example, 20 × 30 in. means 20 in. (50.8 cm) cut across the machine by 30 in. (76.2 cm) cut in the direction of the machine. The first dimen-sion is often termed the cross-direction and the second dimen-sion the machine direction.

Structure. The composition of the web (see Fig. 1).

Ply. A fibrous layer of homogeneous composition.

Topside. The side of the web opposite to the wire side is the normal paper definition for this term. In the case of paper-board, topside can also mean liner side, generally the better-quality face of the web. Some grades of paperboard are known as double-lined, in which case the two faces of the web are both of high quality.

Liner. A ply of good-quality fiber (usually white), on the topside.

Underliner. The layer (ply) of fiber between the external layer (topside/liner) and the middle.

Middle. The layer (ply) of fiber between the two external layers or between the underliner and an external layer.

Backs. The outside layer (ply) of fiber directly opposite to the liner layer.

Duplex (or biplex). This is a board consisting essentially of two layers of different furnish.

Triplex. This is a board consisting essentially of three dif-ferent furnish layers (external furnish layers may have the same composition).

Multiplex. This is a board with more than three furnish layers. Two or more of the layers can have the same composi-tion. Also known as a multilayer board.

Furnish. The constitution of the various materials that are blended in the stock suspension from which the paperboard plies are made. The main constituents are the fibrous material (pulp or secondary fiber), sizing agent, fillers, and dyes.

Structure and Properties

The structure of a typical multi-ply paperboard (see Fig. 1) consists mainly of cellulose fibers. The most common source is mechanical and chemical pulps derived from wood. Secondary fibers are used widely in the cheaper grades. Other sources of cellulose fiber are occasionally used such as straw and esparto.

Examination of the structure shows that it consists of a compact network of the fibers bonded together by mechanical entanglement and chemical links. This structure can be a thick single homogeneous ply or, as is more often the case, of two to eight thinner plies. The multi-ply construction allows different types of fibers to be used for the different plies. Im-proved fiber economy is achieved by selected distribution of the fiber types through the web (eg, by using cheaper fibers for the inner layers). Improved characteristics are obtained by using the correct selection of fiber for the individual layers (eg, stronger fibers for the outer plies). A typical example of a multi-ply construction is a type of paperboard known as white-lined chip. The top ply (liner) is made of bleached chemical wood pulp, which provides surface strength, good appearance, and printing properties. For the other plies, secondary fiber is used.

It is of extreme importance that the individual plies of mul-tilayer paperboard are bonded together. If the bonding is poor, the structure can break up, with resultant deterioration in strength during subsequent processing and use. The level of bonding achieved, which must meet certain requirements, is dependent on having the right balance of mechanical entan-glement and chemical bonding. If, for example, a paperboard sheet is subjected to continuous folding (eg, the hinged lid of a cigarette carton) and the bonding level is too high, cracking at the hinge will occur.

Physical Characteristics

Mechanical properties. Because paperboard has the same type of fibrous structure as paper, the strength-to-weight rela-tionship is of the same order as paper. Because of its extra thickness and the bonding between the plies it normally has

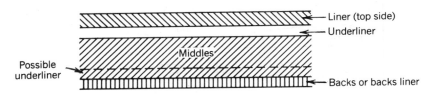

Figure 1. Structure of a typical multi-ply paperboard.

considerably greater flexural rigidity (stiffness). The choice of furnish for the individual plies influences the stiffness characteristics: increased stiffness is obtained by increasing the strength of the outer plies by using a stronger pulp. During the converting of paperboard, it is often necessary to fold the sheet, and because of the thickness of the paperboard, large internal forces are generated. These can cause structural damage to the outer layers (cracking). In order to carry out the folding operation with minimum damage to the outer layers, the ply bonding is broken down locally by a creasing operation before the paperboard is folded. A paperboard sheet has grain characteristics similar to those of paper, with the sheet being strongest in the machine direction.

Optical properties. Paperboard is normally opaque by virtue of its thickness and only the color (whiteness, etc.) and occasionally the gloss of its outside plies are important. For a multi-ply paperboard with a white liner on darker under plies (eg, when waste paper is used) liner-ply opacity is important in order to maintain the white appearance of the liner surface. If a dark waste paper is used, an underply (underliner), consisting either of a lighter colored waste paper or a mechanical pulp furnish, is applied. This reduces the show-through of the dark waste layer and improves the whiteness of the liner surface.

Absorptive properties. The surface of paperboard is often required to have characteristics suitable for printing. To obtain these characteristics the surface layers are sized in ways similar to those used for paper. Paperboard, however, must often be glued during the manufacture of cartons, and for this the surfaces must be absorbent. In a typical case, the back surface has higher absorption characteristics than the liner ply, thus allowing the back ply to absorb some moisture during the gluing operation. Like paper, paperboard changes dimensions if it absorbs moisture. The degree of change depends on many factors such as type of fiber, condition of fiber, structure of the fibrous network, etc. This effect of dimensional change with change of moisture content presents many problems during converting operations such as printing. For example, in lithography and laminating (see Laminating) serious dimensional change can cause print misregister or curling of the paperboard web.

Paperboard Manufacture

The methods of fiber treatment (beating, refining, cleaning, etc) are essentially the same as those used in manufacturing paper grades. After the sheet has been formed, the methods used for removing excess water and finishing the web (pressing, drying, calendering, etc) are also essentially the same as those in the manufacture of paper. The main difference is found at the forming section of the machine, where the web is formed.

The following are the main forming methods used in the manufacture of paperboard:

Single-ply paperboard: mainly fourdrinier.

Multi-ply paperboard:
1. Fourdrinier with secondary head boxes.
2. Rotary formers (multiples and using multilayer formers).
3. Twin wire formers (multiples and using multilayer flow boxes).
4. Combination of the above.

Fourdrinier machine. The forming section of a Fourdrinier machine is made up of two essential parts: the flow box and the drainage table. The operation of both parts can influence the structure of the resulting paperboard web. It is normally the aim to deliver the fiber suspension, well dispersed, to the moving screen at approximately the same velocity as the screen. The concentration going to the screen for paperboard generally ranges between 0.4–1.2% depending on the furnish and product requirements (Fig. 2).

On early machines and some still used today, preliminary dewatering takes place at the table rolls. The action of each table roll can be considered as a pump, drainage being induced by the suction on the downstream side of each table roll, so that the drainage flow is intermittent. A positive-pressure pulse exists on the upstream side of each table roll, so water is forced up through the screen and the deposited mat. This mechanism has considerable influence on web structure. The magnitude of the positive and negative pulses depends on table roll diameter and screen speed.

Modern practice is to replace the table rolls by stationary foils. These units generate less intense suction pulses than table rolls, and the length of the drainage zone can be considerably extended. These foils are often closely spaced along the Fourdrinier forming table so that there is a gain in available drainage. Further down the Fourdrinier table, when concentration is between 3–4%, suction flat boxes are used to continue the drainage at a rate controlled by the level of vacuum in the box. It should be noted that flat vacuum boxes can be used at the initial drainage zone, and today a unit is often used

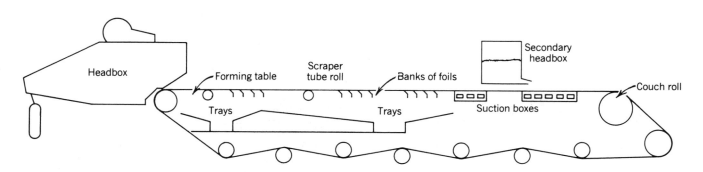

Figure 2. Fourdrinier forming.

that in effect is a combination of foil sections in a vacuum enclosure (wet suction box). When producing heavy-weight papers and boards, because drainage is more difficult, it is usual to use additional vacuum boxes. This can lead to excessive drag on the screen, resulting in higher power requirements, excessive wear of suction box tops and screen, and may even lead to screen-stalling. For the production of multi-ply paperboard, such as linerboard, a secondary flow box is often used. A base ply is formed first and after sufficient dewatering has taken place a second ply is applied by means of the secondary flow box. The associated fibrous suspension is then drained through the base ply and dewatering is carried out essentially by flat vacuum boxes with thickening as the main mechanism.

Rotary forming devices. In these units the forming screen is in the form of a drum and the fibrous suspension is fed to the screen by various methods ranging from simply immersing the rotating screen in a chamber (vat) containing the fibrous suspension, to the use of a type of flow box.

Cylinder mold machines. There are two basic types: Uniflow and Contra Flow. In the Uniflow machine, the fibrous suspension is fed into the vat at the ingoing face of the cylinder mold (forming screen); in the Contra Flow, the suspension enters at the emerging face (see Fig. 3).

The draining forces are low, typically 1–5 in. of water static head difference. The mechanism of forming is complex, and because of the low drainage forces considerable wash-off of the fragile newly formed web followed by redeposition takes place in the forming area. The forming zone is obviously too long with continual washing off and repositioning taking place so that the overall mat deposition is inefficient. Various modifications have taken place in attempts to overcome this problem. Further development has led to a variation known as the rotary former.

Roll formers. Fig 4 shows a typical rotary former. It is easy to see that it consists essentially of a cylinder-forming screen with an associated type of flow box. The forming length has been considerably decreased, and the drainage forces can be far higher than is the case with the previously described cylinder mold units (Fig. 4).

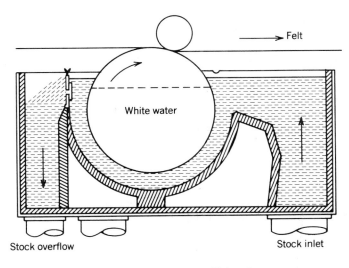

Figure 3. Cylinder-mold forming.

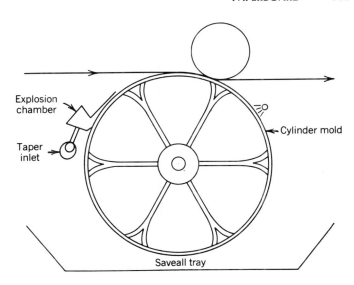

Figure 4. A cylinder former.

The cylindrical screen can be of relatively simple construction relying on a pressure force in the forming zone to assist drainage or it can be in the form of a suction roll using a series of vacuum boxes to further assist dewatering. Compared to cylinder molds, roll formers have several advantages:

1. They can develop and tolerate higher levels of turbulence in the initial forming zone because the drainage zone is enclosed, whereas the initial forming zone in a cylinder-mold machine starts at a free surface.

2. High pressure in the free suspension and the possibility of using suction on the underside of the forming screen permit a much higher rate of drainage.

3. The rotary former has a more uniform metering of the fibrous suspension onto the forming screen.

Although both types of rotary forming devices can be used to produce paper, the greatest application by far is the manufacture of multi-ply board. A number of the units are operated in series, progressively building up a multi-ply web.

Twin-wire formers. The third basic method used for paper and paperboard forming is a relative newcomer, invented in the 1950s and called the twin-wire method. In this technique the paper web is formed between two forming screens. The idea dates back to the nineteenth century, but it was only in the 1950s that serious development took place, taking advantage of improved ancillary equipment. The process became a commercial reality and a viable contender in some applications. Figure 5 shows a typical twin-wire forming unit, in this case the basic Inverform concept, a U.K. invention and the first commercial twin-wire system.

In all twin-wire formers, the fibrous suspension is fed into the gap between two converging forming screens by means of a flow box where dewatering and associated web forming takes place. The actions related to web structuring in the forming zone are still not fully understood but with regard to dewatering, the twin-wire concept offers the opportunity to carry out symmetrical drainage of the fibrous suspension from both sides. This allows a symmetrically structured web to be formed and provides the opportunity to greatly increase the dewa-

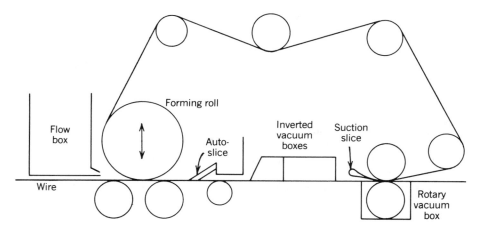

Figure 5. An Inverform (Beloit) twin-wire forming section.

tering potential. Dewatering is assisted by the use of deflectors, which press into the forming screens, and/or vacuum boxes operating on one or both forming screens. A further benefit of the twin wire concept is the absence of a free surface in the forming zone.

In general terms, multi-ply webs are produced by: (1) Separately forming the individual plies and then combining them together; (2) Forming onto an existing ply or plies, to form more plies; (3) Using multilayer (stratified) flow boxes; or (4) A combination of 1–3.

Examples of multi-ply arrangements in use today are shown in Figures 6 and 7.

Machine finishing. The surface of the paperboard web can be treated during the manufacturing process by various means according to the characteristics required of the finished product. It is not unusual to use as part of the drying process an M.G. (Yankee) cylinder, which imparts a smooth surface to one side of the web without too much densification of the web taking place. In the drier section, there is often a size press where chemicals can be added to the surfaces of the web in order to impart certain characteristics (eg, hard sizing or barrier properties). Paperboard machines often have coaters "in line" with the operation at which one or more layers of coating medium can be applied. At the end of the manufacturing process, right before the windup, one or more stacks of calenders are installed. A stack of calenders consists of a number of horizontal cast iron rolls set one above each other. As the web of paperboard passes through the nips, the calenders increase the smoothness and gloss of the surface of the web.

Types of Paperboard

The simplest types of paperboard are single-ply thick papers used for many nonpackaging purposes (eg, index board for card-filing systems and display mounting). They have a degree of stiffness, an acceptable appearance (eg, a uniform surface). This type of paperboard can also be made from waste paper. The product will have good stiffness characteristics, but not necessarily good appearance. Its main use is as package inserts and envelopes. Paperboard with similar structure, but with improved printing surfaces, has a wide range of uses (eg, high-quality display). The greatest volume of paperboard is used, however, in packaging.

Single-ply paperboard made from 100% bleached-chemical wood pulp is used for food packaging where purity and clean appearance is required together with a degree of strength and a surface of sufficient quality to accept good-quality print. Materials of this type are often coated or laminated with a plastic film to improve barrier properties. Compared to single-ply structures, multi-ply paperboard can be used over a much wider range of applications because virgin-pulp outer layers

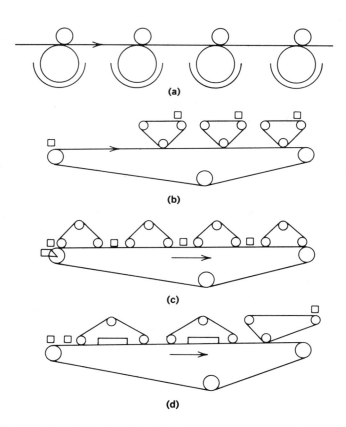

Figure 6. Typical multi-ply forming arrangements. (**a**) Typical cylinder M/C. Common throughout the world. Can be Uniflow, Contra Flow etc, including mixtures. (**b**) Fourdrinier with "on top" mini Fourdriniers. Examples in Europe and the U.S. (**c**) Suction breast roll with Inverform units. Example in the U.S. (**d**) Fourdrinier with bel bond units and mini Fourdrinier. Examples in UK and Europe.

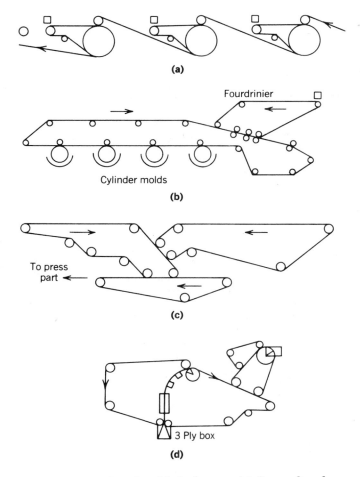

Figure 7. Examples of multi-ply formers. **(a)** Super ultra former. Examples Japan, the U.S., Canada, etc. **(b)** Typical combination M/C. Examples mainly in Europe. **(c)** Typical multiwire M/C. Example Finland. **(d)** Commercial application of contro flow former. Example Finland. Contro flow former combined with an arcu (rotary) former.

with good appearance, strength, and printing properties can be combined with lower-grade middle plies. These paperboards are often combined with other materials such as plastic film or high-quality paper in order to extend the range of application. The general term for this range of paperboard is folding box board.

Another type of paperboard, known as fiberboard, is used to produce large and strong cases. The materials used for the construction of these containers are made from several layers of paperboard. There are two main types: solid board (two or more boards are laminated together) and corrugated board. In the production of corrugated board, two facings are glued to both sides of the corrugating medium to produce a single-wall corrugated board. For stronger and larger boxes three facings and two media are used (double-wall corrugated). Triple-wall corrugated uses four facings and three media (see Boxes, corrugated).

The types of paperboard mentioned have, in general, a basis weight range from 200–600 g/m² (~41–123 lb/1000 ft²). Thicker and heavier paperboards are used in many applications ranging from building materials to suitcases. They are often produced by laminating many layers of thinner paperboards.

Table 1 shows a typical range of paperboard grades with comments on their furnish, requirements and end usage. The following definitions describe the most common types of paperboard used in packaging:

Body or base board. This is a board that is ultimately treated by, for example, a coating or a surface application.

Lined board. A multi-ply board with a liner ply, usually of high-grade material (eg, white-lined chipboard).

White-lined board (duplex). A board with a bleached pulp liner and the remainder of the board made up of, for example, a mixture of chemical pulp and mechanical pulp. Often used for food packaging.

Kraft-lined chip board. An unbleached kraft liner on a waste-paper base used for packaging products such as electrical and mechanical components.

Test jute liner. This is a type of kraft-lined chip board. Sometimes the liner is made from a strong kraft waste furnish. This material is combined with fluting medium to produce corrugated board.

Double-lined board. A board lined on both sides. For example, the outer surfaces (liner) can consist of bleached pulp and the middle mechanical pulp. It is used for high-quality packaging of foods and cosmetics.

Carton board (folding-box board). Paperboard of various compositions used for the manufacture of folding cartons and set up boxes.

Food board. Single- or multi-ply paperboard used for food packaging. It is hard-sized for water resistance.

Liquid packaging board. Also called special-food board, or milk-carton board, this strong board is usually 100% chemical pulp, often plastic-coated. It is formed into containers for a wide range of liquids (eg, milk, other beverages).

Frozen food board. Single- or multi-ply paperboard with high moisture- and water-vapor-resistance. It is often single-ply, made from bleached wood pulp with a surface coating for high-quality print.

Kraft liner board. A strong packaging paperboard with a two-ply construction made essentially from virgin kraft pulp and produced on a Fourdrinier machine. The top ply is added by means of a secondary flowbox. It is used in the manufacture of large containers, combined with fluting medium to form corrugated board.

Fluting medium (corrugating medium). A board usually with a basis weight of 100–125 g/m² (20–25 lb/1000 ft²) and a caliper of 225 μm (9 mil) made from semichemical hardwood pulp or waste paper. The material is fluted in a corrugating machine and combined with linerboard to produce corrugated board.

Chipboard. A paperboard made from waste paper, used in low grade packaging, solid fiber, and bookboard.

Machine-glazed board. Paperboard that in its manufacture has had one face made smooth and glossy by drying on a large polished steam-heated drying cylinder (Yankee cylinder).

Tube board. A paperboard generally unsized and smooth finished. It is slit into narrow widths for winding and pasting into spiral or convoluted mailing tubes, cores etc.

Can board. Paperboard used for the manufacture of composite cans and fiber drums. The cans are used for packaging a wide range of materials including liquids and powders.

Coated boards. Paperboards of various grades that have been coated on one or both faces to make the surfaces suitable for high-quality printing.

Table 1.   Examples of Typical Paperboard Grades

Board grade	Grammage range g/m² (lb/1000 ft²)[a]	Typical furnish	Special requirements (physical)	Typical usage
white-lined folding-box board	200–800 (40–160)	liner, virgin pulp under-liner, mech. pulp middles, waste backs, mixture (mech. and chem. pulp)	bending printing plybond stiffness	general packaging cartons
chipboard	200–800 (40–160)	100% waste	bending plybond	packaging cartons, tubes, and stiffeners
gypsum board	300–800 (60–160)	100% waste	plybond	outer component of plaster board
test liner	150–300 (30–60)	liner, virgin pulp (kraft): rest, waste	bending bursting strength plybond crush resistant	outer components of corrugated container board
liner board	150–300 (30–60)	100% virgin pulp	bursting strength	outer components of corrugated board
food board	200–600 (40–120)	100% virgin pulp (single or multi-ply)	bending printing plybond stiffness	All foods, especially frozen foods (high quality containers)
liquid packaging	200–400 (40–80)	100% virgin pulp (single or multi-ply) with barrier (coating and/or laminate)	bending printing plybond stiffness	containers for wide range of liquids including milk
fluting medium	90–200 (18–40)	Typical—100% waste or semichemical hardwood pulp	crush resistance	inner components of corrugated container board

[a] g/m² = 0.2 lb/1000 ft².

BIBLIOGRAPHY

General References

American Society for Testing and Materials, Philadelphia, Pa. (ASTM Special Publication No 60 B), 1963.

British Standard Institution—Glossary of Paper Terms (B.S. 3203), London, 1964.

R. Higham, *A Handbook of Paper and Board Manufacture* Vol. 1 (Business Books Ltd.) London, 1968.

C. Klass, Cylinder Board Manufacture (Lockwood Trade Journal Co. Ltd.) London, 1968.

Dictionary of Paper, American Paper and Pulp Association, New York, 1965.

E. Labarre, *Dictionary and Encyclopedia of Paper and Papermaking*, Oxford, 1952 and 1967 supplements.

Pulp and Paper Manufacture, Vol. III, 2nd ed., McGraw-Hill, New York, 1970.

Paper and Board Manufacture—Technical Division, British Paper and Board Industry Federation, London, 1978.

J. Grant, *A Laboratory Handbook of Pulp and Paper Manufacture*, Arnold, London, 1961.

J. P. Casey, *Pulp and Paper*, 3rd ed., Vols. I, II, III, and IV, (John Wiley and Sons, New York, 1980–1983.

B. Attwood, "Inverform Developments." *Paper Technology*, 13(4), 253–258 (August 1972).

B. Attwood, *Multi-Ply Web Forming—Past, Present, and Future*, Proceedings of the Tappi Annual Meeting, Atlanta, Ga, February 1980, pp 229–241.

B. W. ATTWOOD
St. Annes Paper and Paperboard Developments Ltd.

PERMEABILITY.   See Barrier polymers; Testing, permeability and leakage.

PHARMACEUTICAL PACKAGING

There are two major categories of pharmaceutical products: ethical products and OTC (over-the-counter) consumer products. Because pharmaceutical companies usually manufacture and package in the same facility, packaging of ethical and

OTC drugs is consistent with recognized standards of the pharmaceutical industry.

Ethical Pharmaceutical Packages

The word *ethical* is used to define a product that is sold only on prescription. Federal law prohibits the sale of this drug class in any other way in the United States. (In certain other countries, the ethical or "legend" drug can sometimes be legally obtained without a prescription.) The principles of package design are, therefore, somewhat different from those used in the design of consumer packages. The physician, dentist, nurse, pharmacist and medical technician are the users of prescription packages.

With the exception of unit-of-use packages and selected injectable products, prescription tablets, capsules, oral liquids, some ointments and creams, and suppositories, are generally purchased by the pharmacist in bulk packages and dispensed according to the physician's directions. The typical prescription is, therefore, not usually sold in the original manufacturer's package, but is transferred to a container supplied and labeled by the pharmacist.

The ethical-prescription package supplied by the manufacturer requires special attention because of the role it plays in the normal distribution process. The primary container and closure (the items in direct contact with the drug) must protect the product both chemically and physically. The selection of primary package material and size is determined by exhaustive scientific study to confirm the product-package compatibility. Testing, as it relates to light protection, water-vapor transmission, gas permeation, potency stability, and product durability under simulated and actual shipping conditions are all important evaluations before the necessary approvals can be requested (see Testing, consumer packages; Testing, packaging materials). The results of development testing also determine the required expiration date of the product. They may rule out certain desirable package features; for example, clarity, shatter resistance, snap cap, plastic syringe, large container, light weight, and low cost.

Product-package testing requires scientific data to substantiate the package choice, and to confirm the requested expiration dating. United States regulations require that these data be submitted to the FDA as part of the New Drug Application (NDA) or an NDA supplement. Written FDA approval of the NDA is then required prior to interstate distribution (see Food, drug, and cosmetic regulations). NDA documentation contains a complete description of the primary package, including desired package sizes. The description of the package material may include manufacture of the material composition (if a compound), the Drug Master File (DMF) number, as well as the type of closure (see Closures), and specifications of cap liner (see Closure liners). After the original NDA has been submitted, one may expect a request for more detail from the FDA examiner. Such demands usually involve primary package data.

All package copy and labeling is also submitted as part of the NDA. This includes the physician's insert or leaflet (see Inserts and outserts). This information folder must be included with every prescription package. Where and how to include the insert with each package is often a challenge. The usual methods include placing the insert and primary container in a folding carton (see Cartoning machinery); automatically locating the insert under the label of the bottle in a way that allows its removal without destruction of the label; placing it under the closure; and taping or banding it to the container.

The Poison Prevention Packaging Act of 1970 (1) and subsequent supplements created new and significant packaging changes for pharmaceuticals. This law meant all oral prescription products sold in containers designed and intended to be used by the patient or ultimate consumer must have "special packaging", the term in the law to mean child-resistant packages, more commonly referred to as child-resistant safety closures (see Child-resistant packaging; Closures). For those prescription products packaged in containers for bulk dispensing by the pharmacist, the law holds the pharmacist responsible to provide the child-resistant package at the time of the sale to the customer. The law provides that the prescription customer may request a conventional closure, thereby assuming the safety-measurement risk, and the pharmacist is then allowed to sell the product in a package that does not have a safety cap or child-resistant packaging (2).

There are three other types of special-purpose packaging for ethical pharmaceuticals: unit dose; unit-of-use; and parenteral packaging.

Parenterals. Parenteral is a medical term for "outside of the intestine." The word is commonly used to mean sterile products injected into the body, either into the muscle (IM) or into the vein (IV). In some cases, the physician can decide whether the product will be given IM or IV. The package design accommodates this choice in many cases and contributes to more convenient and accurate administration. Parenteral or sterile pharmaceutical products are further subdivided into small-volume parenterals (SVPs) and large-volume parenterals (LVPs). The SVPs contain less than 100 mL in a single package. The LVPs contain 100–1000 mL.

Until the late 1970s, parenteral products were always packaged in glass. Such packages took the form of glass ampuls, rubber-stoppered or sealed vials, and rubber-stoppered bottles. The ampul is glass tubing that is product filled and heat sealed (see Ampuls and vials). The constricted neck of the container is snapped off to enable withdrawal of the sterile product with a syringe. Glass vials, usually SVPs, permit multidose packaging by allowing a number of syringe withdrawals from the same package. The rubber compounds used to seal and stopper the vials are chosen for their resealing ability, in addition to other physical and chemical demands.

The LVP glass bottle normally contains intravenous (IV) medication. Popular sizes include 250 mL, 500 mL, and the most-common 1000 mL bottle. The medication is administered to the patient through an IV needle attached to flexible plastic tubing. The other end of the tubing is connected to a large metal or plastic needle that pierces the rubber-stoppered bottle. The flow rate to the patient is controlled by a pinch clamp on the flexible tubing.

Advancement in plastic packaging is now allowing a limited number of sterile products to be approved in plastic containers. The LVP products are leaders in this early trend. Both rigid and flexible containers store the widely used electrolyte and sugar supplements, and a few pure intravenous drug compounds are now available in "ready-to-use" bottles or bags. Such packaging eliminates the need to add the drug product to the larger infusion (eg, 5% dextrose) container. With a "piggyback" tubing attachment, the drug product is administered through the same tube that carries the infusion solution. The "ready-to-use" IV packaging system is expected to be widely accepted when FDA clearance is in place.

Unit dose.   Another package innovation evolved from hospital use of pharmaceuticals: hospital unit-dose packaging. The product is contained in a package that allows and controls the dispensing and the administration of a prescribed single dosage at the right time with the right product. Hospital unit-dose packaging has significantly reduced hospital medication errors since their introduction, in the form of prefilled disposable syringes, over 30 years ago. In addition to single-dose prefilled syringes, many parenteral drug products became available in single-dose vials.

The most significant advancement for hospital packaging was the introduction of unit-dose oral products. During the early 1950s some tablets and capsules were available in foil-strip packages. Tablets or capsules were packaged in individual pouches that were attached to each other and separable by means of a perforation in the foil strip of 100 tablets or capsules. Each small packet or pouch was labeled to include the product trade name, the generic chemical name, expiration date, lot number, and name of the manufacturer. From this early concept, the present hospital unit-dose blister package evolved.

Blister packages (see Fig 1) are recognized as an improvement over the early foil-strip package. The tablet or capsule is visible through the blister side of the unit-dose package. This provides one more safety measure in the hospital dispensing process. Recognition of the product prior to opening the package reduces medication errors, as well as waste of products opened by mistake that must be destroyed. The transparent blister can be made of one of several thermoformable (see Thermoforming) polymers or combinations of polymers that provide improved barrier properties, or the heat-seal capabilities needed to seal the blister side to the lidding stock (see Lidding; Sealing, heat). The selection depends on the chemical

and physical barrier demands of the pharmaceutical product. The more moisture sensitive the product, the better the moisture-vapor transmission barrier properties must be. Available materials range from relatively inexpensive PVC to the expensive PVC/chlorotrifluoroethylene (see Films, fluoropolymer). The choice of film thickness affects both material costs and barrier properties. Other considerations are machineability, production rates, depth of the blister, wall thickness and uniformity of the blister, and sealing properties to the lidding stock. It is obvious, therefore, that expensive and time-consuming testing must take place to verify the package-product acceptance demands.

The reverse side of the hospital unit-dose package is the lidding stock. This usually takes one of two forms: a lamination of aluminum foil/paper (see Multilayer flexible packaging) or preprinted aluminum. The more popular type is the paper/foil combination. The foil side is sealed to the blister containing the product with a heat-seal coating material. The paper side is printed on line with the required label copy. The foil component of unit dose must perform the same protection and sealing functions as the blister. For this reason, a specification of 0.001 in. (25 μm) is the usual standard (see Foil, aluminum). This thickness is considered to be pinhole free and provides the optimum ratio of cost : product protection.

As with all prescription products submitted to the FDA for approval, complete identification and specifications for hospital unit-dose packaging are included. To assist inventory requirements, a group of products, known as controlled substances, are sometimes numbered sequentially on each blister. These products include the narcotic compounds, the barbiturates, and other habit-forming substances specified by the FDA (3).

Many hospitals have converted all, or almost all, of their dispensed prescriptions to the unit-dose system for in-patient care. Because unit-dose packages are not commercially available for all prescription pharmaceuticals, many hospitals have invested in automatic or semiautomatic blister packaging machines. The products are purchased in bulk containers from the pharmaceutical manufacturer and then repackaged into the blister form.

Unit-of-use.   Unit-of-use packaging is not the same as hospital unit-dose packaging: unit dose is prescribed for hospital inpatients; unit-of-use is intended for the ambulatory or prescription customer. This package is designed to contain that amount of drug to satisfy the patient's therapeutic requirements for a period of not more than 30 days. The product is intended to be dispensed by the pharmacist in the original package, as supplied by the pharmaceutical manufacturer. The typical pharmacist's label, indicating the dosage prescribed by the doctor, is also attached to the original package. The early unit-of-use packages were probably ointment tubes and oral contraceptive products. Oral pediatric liquid antibiotics and eye-drop preparations are also provided in unit-of-use packaging. It is difficult to package all ethical drugs in this form because of the wide range of dosage requirements due to patient needs, physician judgment, and economics. But the popularity of unit-of-use is steadily increasing in the United States and is now the preferred method in many European areas.

Unit-of-use assures that the product is packaged in an approved container that is labeled by the manufacturer and carries the approved expiration date, as well as the manufactur-

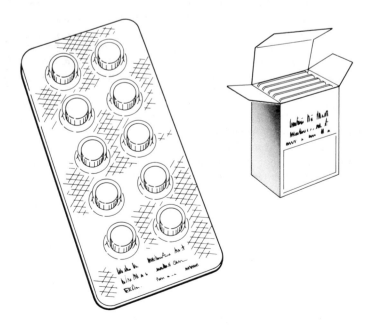

Figure 1.   Blister packaging. A push-through vinyl blister with foil back. Either vinyl or foil can be printed.

er's lot or batch number (see Code marking and imprinting). Such measures assure the quality and uniformity of the product as backed by the manufacturer. Like hospital unit-dose packaging, unit-of-use packaging is popular for the hospital out-patient. The out-patient is one that is being discharged to go home or visits the hospital for care and medication without being admitted to the hospital as an in-patient. The convenience of unit-of-use packaging is obvious from the pharmacist's viewpoint, and it gives the doctor an easier way to write complicated prescription directions. It also places responsibility for regulatory compliance in the hands of the manufacturer. Proof of child-resistant closure effectiveness is also the responsibility of the drug packager. This is done according to a Federal protocol testing procedure (4,5).

The term *secondary packaging* describes the remaining packaging components, which are not in direct contact with the drug product. Folding cartons, paperboard sleeves, corrugated boxes, corrugated dividers, thermoformed trays, and plastic foams are examples of secondary package components used to provide shipping protection. Laboratory testing and actual shipping trials determine the optimum cost-performance relationship between materials-design and product damage.

OTC Packages

Unlike most ethical product packages, OTC packages are researched, designed, and produced for the consumer. They are sold not only in prescription outlets, but also in supermarkets, department stores, small neighborhood convenience stores, and vending machines. The structural design of the package, and its graphics, must appeal to the consumer, since the package frequently sells the product. Focus interviews and other market-research techniques are used to measure this appeal.

Safety-closure regulations apply to selected OTC pharmaceuticals (5). The law allows the buyer to choose one package size without the child-resistant feature. If only one size is available, the safety cap must be part of the package supplied by the drug manufacturer. The pharmacist supplies a conventional cap or package only at the request of the customer. Manufacturers must also comply with the regulation promulgated in 1982 that called for tamper-evident packaging (see Tamper-evident packaging).

Secondary packaging for OTC pharmaceuticals includes display packers that hold 12–36 individual packages, corrugated stands that contain several dozen packages of different sizes, and packages that are designed to be displayed on a wire stand. Trial size packages are recent additions to OTC packaging. The amount of product is intended to acquaint the prospective customer with a sample quantity for taste, size, short-term effect, and price. Trial-size packages are not routine production packages either in terms of volume or machine demand, and on a unit-cost basis, they are relatively expensive.

The FDA governs the use and acceptance of consumer-drug products. A system called the drug monograph regulates those products that are safe and effective and available without prescription. The drug monograph is to OTC products what the NDA is to the ethical products. The FDA is continuing research to determine which prescription products can be authorized for consumer-product status. A number of ethicals have already been changed to OTCs, based on long-term prescription usage and safety. This trend is important in the packaging of OTC products. Because the requirements for ethical and OTC packages are so different, a change in the status of a product presents interesting new package-design challenges.

Production Activities

Pharmaceutical manufacturers use a relatively high number of dissimilar materials and packages, and packaging operations form an intrinsic part of pharmaceutical manufacturing. The Federal government controls manufacturing activities in United States plants, and its guidelines are often used in other countries. The regulations are known as CGMPs or Current Good Manufacturing Practices (6). The purpose of the regulations is to ensure that all pharmaceutical products are produced and packaged under specified conditions to assure safety and effectiveness. These rules include periodic inspection of the pharmaceutical-manufacturing facility.

Due to the sometimes large number of package sizes of the same product, and the varied package designs of different products, packaging operations are generally not standardized in large and medium-sized pharmaceutical companies. Long production packaging runs are not routine and frequent machine changeovers are common. This makes it especially important that packaging materials comply with written and issued specifications (see Specifications and Quality Assurance). Successful suppliers of packaging materials and components recognize the importance of consistently providing the pharmaceutical packager with products of the highest quality. Careful incoming inspections are made to ensure that the parts meet the detailed specifications. In most cases, the primary packaging components are examined analytically before production approval. The health industry and medical profession recognizes the importance of pharmaceutical packaging. The quality of the product, the protection and potency of the medication and the accurate means of identification all contribute to modern medical-pharmaceutical success.

BIBLIOGRAPHY

1. *Public Law 91-60/5.2162,* Dec. 30, 1970.
2. *Fed. Regist.* **37**(32) (Feb. 1972).
3. 21 USC 801, Comprehensive Drug Abuse Prevention and Control Act, 1970.
4. *Fed. Regist.* **36**(225) ¶295.10 (Nov. 20, 1971).
5. *Fed. Regist.* **38**(151) (Aug. 7, 1973).
6. CFR, part 211.

H. C. WELCH
PharmPack Company

PIGMENTS.   See Colorants.

PLASTIC PAPER

There is no strict definition for "plastic paper." The term has been applied to many kinds of synthetic products. In the early 1970s, a great deal of research and development work was directed toward plastic paper on the assumption that resin prices would continue a downward trend that would make plastic papers relatively inexpensive substitutes for wood-based paper. The oil crisis put an end to that assumption, but the design and development of synthetic papers for specialty applications continued.

Most synthetic papers are film-based, but brief mention of synthetic pulp is warranted. Synthetic pulp is made in a polymerization reactor by vigorous stirring that forms fibers. The process has been applied to HDPE and other resins, but the applications are outside of the packaging area. Some special corrugated boards have been made in Japan by combining wood pulp and synthetic pulp, but the economics do not favor large-scale use of the concept. Also the synthetic component would interfere with recycling. Fibrous spunbonded polymers (eg, Du Pont's Tyvek) are used much like paper in some applications (eg, envelopes, lidding), and can be considered plastic papers.

The term is generally applied, however, to nonfibrous plastic films that are "paperized" by internal or external methods. The internal methods involve cavitation during orientation, generally with fine-particle fillers as the sites of cavitation. This concept has been applied to HDPE/calcium carbonate and PP/clay/titanium dioxide. Another method of obtaining cavitation is by combining two incompatible polymers. One dispersed polymer acts as a filler in the continuous phase of the other. Externally paperized products include certain coated films, and some plastic papers are made by combining fillers and coatings. Much of the development efforts today are directed toward the use of synthetic papers for printing and writing and not for packaging; but there are some heavy-gauge pigmented films in the market that have potential for replacing high quality paperboard. The packaging film that best fits the definition of a synthetic paper is opaque BOPP (see Film, oriented polypropylene; Multilayer flexible packaging).

BIBLIOGRAPHY

General References

Proceedings, TAPPI Paper Synthetics Conference, TAPPI Press, Atlanta, Ga., 1977.

Proceedings, TAPPI International Synthetic Paper and Pulp Symposium, Tappi Press, Atlanta, Ga., 1980.

L. Leese, Current Status of Film-based Synthetic Papers, *Proceedings, TAPPI Polymers, Laminations, and Coatings Conference,* Tappi Press, Atlanta, Ga., 1984.

PLASTICIZERS.   See Additives, plastic.

PLASTICS, PROPERTIES.   See Polymer properties.

POLYAMIDES.   See Nylon.

POLYCARBONATE

Polycarbonate (PC) is an amorphous resin that does not require orientation to achieve its full property profile. The molten resin can be extruded by the blown or cast processes (see Extrusion), injection molded (see Injection molding), or blow molded by the extrusion-blow or injection-blow techniques (see Blow molding). As the polycarbonate cools into film, sheet, or containers, it exhibits excellent dimensional stability, rigidity, impact resistance, and transparency over a wide range of temperatures and loading rates (see Table 1). Because PC is amorphous, its wide softening range and added strength in thermoforming (see Thermoforming) operations provides deep-draw capabilities.

Polycarbonate provides an excellent combination of tensile strength and flexural modulus at high temperatures (see Table 2). Its heat resistance, combined with superior impact resistance at both high and low temperatures, makes it an excellent structural layer in coextruded or laminated packaging for hot fill at 180–210°F (82–99°C), retorting at 250°F (121°C), autoclaving at 270–280°F (132–138°C), and frozen-food packaging. In addition, PC can be sterilized with both gamma and electron-beam irradiation with good stability.

Polycarbonate has light-transmittance values of 88–91% as compared with 92% for clear plate glass. It has a haze factor of less than 1%, and maintains these values throughout the temperature scale. Its high gloss and easy colorability and printing contribute to distinctive package design. Polycarbonate has high resistance to staining by tea, coffee, fruit juices, and tomato sauces, as well as lipstick, ink, soaps, detergents, and many other common household materials. Its relatively dense composition makes it immune to odors, and its hard, smooth surface facilitates easy removal of foodstuffs.

Some of the most significant attributes of polycarbonate stem from its very low water absorption. Added weight increase after 24 h immersion at room temperature is only 0.15%. This low absorption level helps account for the resin's excellent dimensional stability and stain resistance. It also indicates that the resin, itself tasteless and odorless, is unlikely to pick up food odors. PC is available in grades that meet FDA and USDA regulations and is recognized as safe for food-contact applications.

Packaging Applications

Refillable bottles. Polycarbonate is the material of choice for use in reusable bottles, particularly 5-gal (19-L) water bottles, which represent the resin's chief packaging application. These bottles take advantage of polycarbonate's toughness (to resist breakage) and clarity (to see the contents). The fact that PC is much lighter than glass provides fuel savings as well as productivity improvements, since several bottles can be carried at once. Systems have been developed to wash polycarbonate and provide clean bottles for reuse with minimum impact on trippage.

Medical-device packaging. Polycarbonate meets many requirements of medical-device packaging (see Health-care packaging). It is clear, tough, and can be sterilized by commercial sterilization techniques: ethylene oxide (ETO), radiation, and autoclave sterilization (see Radiation, effects of). The development of coextrusion technology has afforded opportunities in all sterilization systems. In thin films, PC can be coextruded with polyolefin heat-seal layers (see Coextrusions for flexible packaging; Coextrusions for semirigid packaging) to produce a cost-effective alternative to laminations based on oriented films (see Films, plastic). Because it is amorphous,

Table 1. Typical Property Values for Polycarbonate[a]

| Property | ASTM test method | Melt flow indexes | | | | PC copolymer |
		22	16	10	6	
specific gravity	D 792	1.20	1.20	1.20	1.20	1.20
light transmittance, 0.125 in. (3.2 mm), %	D 1003	89	89	89	89	85
haze, 0.125 in. (3.2 mm), %	D 1003	1	1	1	1	1–2
deflection temp at 264 psi (1.8 MPa) °F(°C)	D 648	260 (127)	265 (129)	270 (132)	270 (132)	325 (163)
flammability rating[b], UL 94, at 0.060 in. (1.5 mm)		V-2	V-2	V-2	V-2	HB
tensile strength, yield, psi (MPa)	D 638	9,000 (62)	9,000 (62)	9,000 (62)	9,000 (62)	9,500 (65.5)
tensile strength, ultimate, psi (MPa)	D 638	9,500 (65.5)	10,000 (68.9)	10,000 (68.9)	10,500 (72.4)	11,300 (77.9)
elongation, rupture, %	D 638	120	125	130	135	78
flexural strength, psi (MPa)	D 790	13,500 (93.1)	14,000 (96.5)	14,000 (96.5)	14,200 (97.9)	14,100 (97.2)
flexural modulus, psi (MPa)	D 790	335,000 (2,310)	340,000 (2,340)	340,000 (2,340)	340,000 (2,340)	338,000 (2,330)
Izod impact strength, notched, ⅛-in. (3.2-mm) thick, ft·lbf/in. (kJ/m)	D 256	12 (0.64)	13 (0.69)	15 (0.80)	17 (0.94)	10 (0.53)
tensile impact, ft·lbf/in.^{2} (J/cm^2)	D 1822	180 (37.8)	225 (47.3)	275 (57.8)	300 (63.0)	275 (57.8)

[a] Properties shown are average values that can be expected from typical manufacturing lots and are not intended for specification purposes. These values are for natural color only. Addition of pigments and other additives may alter some of the properties.
[b] This rating is not intended to reflect hazards of this or any other material under actual fire conditions.

heat sealing does not shrink or embrittle the film (see Sealing, heat). This virtually eliminates puckering, which can lead to hairline cracks and shattering upon opening. A soft blister package with good puncture resistance can be produced by thermoforming heavier-gauge film. For increased stability in the autoclave, a polycarbonate copolymer, poly(phthalate carbonate), can be incorporated into the structure. Properties of monolayer PC films are listed in Table 3.

Food packaging. Coextrusions of polycarbonate are being evaluated for use in several segments of the food-packaging market. The snack-food industry, and other users of flexible-packaging materials, are discovering ways to use PC in coextrusions to replace laminated films in a portion of a structure or to replace the entire structure (see Multilayer flexible packaging). It is polycarbonate's toughness without orientation that makes it a good candidate for coextrusion.

For frozen foods, the resin's low temperature impact strength adds durability to dual "ovenable" trays. In coextrusions, it can add toughness to crystallized polyester or polyetherimide trays. In cases where the wall thickness is deter-

mined by impact-strength requirements, gauge reductions up to 33% are possible.

In high barrier multilayer containers, polycarbonate offers the rare combination of dimensional stability at retort and hot-fill temperatures, along with crystal clarity (see Retortable flexible and semirigid packages; Multilayer plastic bottles). To produce a 1% distortion in PC requires 3000 psi (20.7 MPa) stress. A similar distortion in polypropylene occurs at less than 500 psi (3.4 MPa) stress. Polycarbonate is about three times as expensive, but dimensional stability can be translated into value by making lighter containers with thinner walls by faster and more reliable closing at hot-fill temperatures, higher retort temperatures with less critical overpressure control, and enhanced container-design flexibility.

An unexpected benefit from the use of PC in retort applications is an increase in the effectiveness of the barrier material EVOH (see Ethylene–vinyl alcohol). In a retort, moisture is driven into all layers of a plastic container. When the outside skin layer is polypropylene, the polypropylene prevents moisture from entering the structure, but is also prevents moisture trapped in the EVOH layer from escaping. Because polycarbonate has poorer moisture-barrier properties than polypropylene, it allows much more rapid drying of the EVOH layer and longer shelf life. As coextruded plastic containers become more readily available and begin commercial penetration, polycarbonate will become a more important material in food packaging.

Table 2. Tensile Strength and Modulus of Polycarbonate Over a Temperature Range

Temperature, °F(°C)	Tensile strength, psi (MPa)	Flexural modulus, psi (MPa)
73 (23)	10,000 (68.9)	320,000 (2,206)
212 (100)	5,800 (40)	233,600 (1,610)
270 (132)	4,000 (27.6)	211,200 (1,456)

J. M. MIHALICH
L. E. BACCARO
General Electric Plastics

Table 3. Polycarbonate Film Properties, 1 mil (25.4 μm)

Property	ASTM test		Value
specific gravity	D 792		1.20
yield, in.²/(lb·mil) (m²/(kg·mm))			23,100 (1,294)
haze, %	D 1003		0.5
optical clarity	D 1746		86–88
tensile strength, psi (MPa)		MD	10,735 (74)
		XD	10,009 (69)
elongation, %	D 882	MD	91
		XD	92
secant modulus, psi (MPa)		MD	185,000 (1,275)
		XD	196,000 (1,351)
tear strength, gf/mil (N/mm)			
initial[a]	D 1004		454 (175)
propagating[b]	D 1922		16 (6.2)
tensile impact, S type, ft·lbf/in.² (J/cm²)	D 1822		225–300 (47.3–63.0)
bursting strength, 1 mil, psi (kPa)	D 774		27.4 (189)
water absorption at 24 h and 73°F (23°C), %	D 570		0.15
folding endurance[c], cycles	D 2176		11,000
heat-seal temp at 40 psi (276 kPa) for 3 s, °F (°C)			400–420 (204–216)
oxygen permeability at 77°F (23°C) and 0% rh, cm³·mil/(100 in.²·d·atm) (cm³·μm/(m²·d·kPa))	D 1434		240 (933)
water-vapor transmission rate at 100°F (38°C) and 90% rh, g·mil/(100 in.²·d) (g·mm/(m²·d))	E 96-66		6.5 (2.6)
coefficient of friction			
static	D 1894		0.570
kinetic			0.542

[a] Graves.
[b] Elmendorf.
[c] Double folds.

POLYESTER FILM.   See Film, oriented polyester.

POLYESTERS, THERMOPLASTIC

Poly(ethylene terephthalate) (PET) was first developed by a British company, Calico Printers, in 1941 for use in synthetic fibers. The patent rights were then acquired by DuPont and Imperial Chemical Industries (ICI), which in turn sold regional rights to many other companies. Polyester fibers have since made a considerable impact on the textile industry. The second principal application of PET was film. In 1966, PET became available for the manufacture of injection-molded and extruded parts.

The amazing growth of PET in beverage packaging began in the early 1970s with the technical development of biaxially oriented PET bottles (see Blow molding) and with the introduction of the first 2-L PET beverage bottle in 1976. Since then, the U.S. PET beverage-container market has grown from 200 million (10^6) units in 1977 to 5 billion (10^9) units in 1985.

PET has the following formula:

$$\text{HOCH}_2\text{CH}_2\left(\!\!\text{OC}-\!\!\!\bigcirc\!\!\!-\text{COCH}_2\text{CH}_2\!\!\right)_n\!\!\text{OH}$$

where $n = 100–200$.

Manufacture of PET

Raw materials. The raw materials for PET (see Fig. 1) are derived from crude oil. *Para*-xylene, one of the two starting materials, is part of the naphtha feedstock which used to be fully available for chemicals because it was a by-product of limited value to the oil refiner. Now it has become the source of additives that replace lead in unleaded gasoline in many countries. Of the mixed xylenes, as they come from the reformer, only *p*-xylene is suitable for building straight polymer chains, and the straight configuration is necessary to give the polymer its fiber- and film- (or bottle-wall-) forming characteristics and ultimately its high tensile strength. The other raw material for polyester is ethylene, which is contained in the crude oil's gas fraction. It is converted to ethylene glycol by oxidation and hydrolysis. All of these raw materials are now derived from oil, but it is technically feasible to produce them from coal-tar distillation.

Intermediate products. Two different routes are used to manufacture PET: one by way of dimethyl terephthalate (DMT), and the other by way of terephthalic acid (TPA). Both are dibasic acids. Figure 1 shows both routes. The plants for intermediate and end products are shown together, but they are normally physically separated. The process of making DMT is relatively simple: one end of *p*-xylene is first oxidized with air and esterified with methanol to yield a half-ester which is subsequently converted at the other end. The resulting DMT is purified by distillation and repeated crystallization to remove isomers and other impurities. The TPA route is similar. TPA can be produced by oxidation of *p*-xylene in solution and purified by solvent extraction.

Melt polycondensation. Both batch- and continuous-polycondensation processes are used. The continuous process inherently allows better product uniformity; the batch process is preferred for small quantities of specialty resins. On the DMT route, DMT and ethylene glycol (EG) are continuously metered into the ester interchanger, where the methyl end groups of the DMT are replaced by ethyl end groups to form diethylene glycol terephthalate (DGT), the monomer of PET. In this step, EG is consumed while methanol is evaporated and collected to be returned to the DMT plant. On the other route, TPA is esterified with EG to DGT, and water is removed as a by-product. Subsequent polycondensation is the same for both routes. In EG takeoff, excess EG is removed and sent to the distillation plant for recovery. Polycondensation takes place in vacuum reactors designed to evaporate EG, the condensation by-product, thereby shifting the equilibrium toward long polymer chains. The final step includes extrusion of the melt as strands or ribbon, quenching in water, and cutting to the desired chip size.

Solid-state polycondensation. Melt polycondensation produces amorphous PET as used in most fiber and film applications. Unfortunately, this product is not suitable for the injection molding of food containers because the inherently high acetaldehyde level would affect the taste of some foods; there are other impurities that might promote degradation during the injection-molding process; and amorphous resin tends to fuse and form lumps in the drying hopper. The polymer must be upgraded by the solid-state polycondensation process. The chips are crystallized to avoid later sticking and then dried to reduce hydrolysis at high processing temperatures. Solid state polycondensation takes place in the reactor, where the chips are subjected to high temperatures under vacuum (batch process) or in a nitrogen or dry-air stream (continuous process). The product's intrinsic viscosity (IV) is normally between 0.70 and 1.0. (Intrinsic viscosity is a method for the characterization of the average length of the molecule chains in PET.) High viscosity resin is relatively expensive because its production is lengthy. High viscosity PET, ie, having longer molecule chains, offers better mechanical properties than the average-viscosity resin. These properties compensate for certain deficiencies of molded articles, eg, excessive volume expansion of beverage bottles can be limited by higher viscosity resins that creep less under load. The final step is cooling, since the resin should not be exposed to moist air while it is hot.

Homopolymers and Copolymers

PET is a homopolymer made from one part dibasic acid, ie, TPA or DMT, and one part EG. A copolyester (copolymer) is made from more than one dibasic acid and/or glycol. Copolymers remove processing limitations and provide increased physical properties at elevated temperatures. In addition to DMT or TPA, isophthalic acid (IPA) can be used as a comonomer to reduce the rate and degree of crystallization to an extent that depends on its dosage. This broadens the processing parameters of food-container manufacturing machines.

Glycols offer several opportunities for modification. During polycondensation, EG reacts with itself to some extent to form diethylene glycol (DEG). Higher amounts of DEG affect many polymer properties. There are other glycols available as partial substitutes for EG, eg, neopentyl glycol, cyclohexane dimethanol, etc. All these modifications lead to desired polymer property changes, ie, reduction of the crystallization rate, melting point, etc. Cyclohexane dimethanol can react with a mixture of terephthalic and isophthalic acids in order to increase the melt strength of the polymer for extrusion processes (1).

On the other hand, some injection-molding (see Injection molding) and thermoforming (see Thermoforming) applications require accelerated crystallization rates in order to set up crystallization in the article which prevents physical deformation at elevated temperatures. This objective can be achieved by nucleation, which involves the addition of other ingredients to the polymer. Inert, nonsoluble substances (eg, mica, talc, etc), organic substances (eg, aromatic alcohols), and certain polymers (eg, PP, PE, etc) can be used as nucleation ingredients to increase crystallization rates.

The use of PET and optional added substances for food-packaging applications is governed by FDA Regulation No. 177-1630 of March 16, 1977. Homopolymers and copolymers and additives must conform to this regulation.

Packaging Applications

Homopolymers. By strict definition, most PET resins are modified homopolymers. These homopolymers are used to manufacture containers, ie, bottles, by injection blow molding or injection-stretch blow molding (see Blow molding) (2). About 70–80% of the resin consumption is used for soft-drink bottles (see Carbonated-beverage packaging). PET bottles are

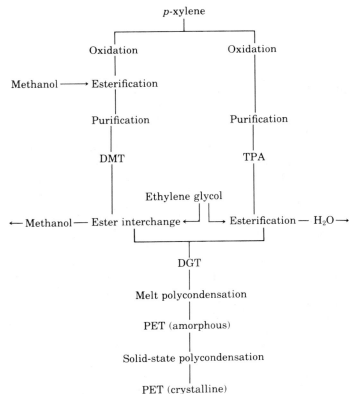

Figure 1. Manufacture of PET.

Table 1. PET Polymer Resin Options for Different Manufacturing Processes and Applications

Manufacturing process or application (PET resin type)	Standard homopolymer	Standard bottle resin	Slow-crystallizing copolymer	Fast-crystallizing homopolymer
injection blow-molded bottles (nonoriented)		X		
injection-stretch blow-molded bottles (biaxially oriented)		X		
extrusion blow-molded bottles			X	
sheet extrusion[a]	X		X	X
film casting[a] (nonoriented)	X		X	X
biaxially oriented film	X			
heat-set film				X
crystallizable PET trays				X
PET coating for paperboard	X			

[a] Resin option depends on desired application.

also used for liquor, wine, food, toiletries, and pharmaceuticals, and for beer in some countries. Homopolymers cannot be processed by extrusion blow molding because of insufficient melt strength.

Biaxially oriented PET film (see Film, oriented polyester) is usually manufactured by polycondensation and subsequent continuous casting of the film plus direct biaxial orientation, ie, molecule chains of the resin become biaxially oriented. Nonoriented PET film and sheet are manufactured by melting PET resin in an extruder and casting the melt through a flat die with subsequent calendering.

A fast-growing application for PET is "ovenable" trays for frozen food and prepared meals. These trays are thermoformed from cast PET film and crystallized. Crystallization heat-sets the article to prevent deformation during cooking and serving. The main advantages of PET for this application include suitability for both conventional and microwave ovens, light weight, and superior aesthetics (as compared with foil trays).

Copolymers. Commercially available copolymers offer improved melt strength for the vertically positioned extrusion blow-molding process. They are used to some extent for bottles, and to a greater extent as extruded sheet for blister-pack applications. Since the FDA regulation permits certain copolymers for food-contact use, future applications may include packages for noncarbonated drinks, cooking oil, and vitamin preparations.

Polymer options, manufacturing processes, and applications are summarized in Table 1.

The manufacturing process of PBT is very similar to that of PET. Instead of EG, 1,4-butanediol (HO $+CH_2\rightarrow_4$ OH) is used to react with either TPA or DMT. PBT is rarely used in packaging applications.

BIBLIOGRAPHY

1. P. Aspy and co-workers, "Controlled Crystallization—PET Copolyester," *Paper presented at the Seventh Annual International Conference on Oriented Plastic Containers,* Atlanta, Ga., Mar. 1983, Ryder Associates, Whippany, N.J., 1983.

2. U.S. Pat. 3,733,309 (May 15, 1973), N. Wyeth (to E. I. du Pont de Nemours & Co., Inc.).

General References

General Information about Hoechst Thermoplastic Polyester Resin, Technical Bulletin 2, American Hoechst Corp., Somerville, N.J., 1981.

K. D. Asmus, "Polyalkylenterephthalate," *Kunststoffe Handbuch,* Carl Hauser Verlag, FRG, 1972.

Chemie-Kompendium für das Selbststudium, Hoechst AG, Frankfurt, FRG, 1972.

E. H. NEUMANN
Hoechst AG

POLYETHYLENE, HIGH DENSITY

High density polyethylene (HDPE) is a milky-white, odorless, and tasteless nonpolar thermoplastic. Each molecule is made by addition polymerization, which covalently bonds many thousands of ethylene molecules together end-to-end by way of a catalyst to form a long linear chain of carbon atoms.

HDPE is one of the most versatile polymers in the packaging industry (see Table 1). Blow molders (see Blow molding) use it to make 55-gal (208-L) industrial-chemical drums (see Drums, plastic), bleach and detergent bottles, and containers for bottling milk, juice, and water. Injection molders (see Injection molding) rely on HDPE to make, for example, pails and buckets (see Pails, plastic), thin-walled dairy containers, and closures (see Closures, bottles and jars). Injection blow molding is used to make cosmetics containers, pharmaceutical bottles, and containers for personal-care products such as shampoo and deodorant. Blown or cast HDPE film (see Film, high density polyethylene) is utilized in flexible-packaging applications such as deli wrap, snack-food wrappers, breakfast-cereal liners, merchant bags, produce bags, and multiwall bag liners. Some rotational molders (see Rotational molding) use HDPE to make drums, crates, and various other shipping containers. In 1984, an estimated 5.3 billion (10^9) lb (2.4×10^6 metric

Table 1. U.S. Consumption of HDPE for Packaging

Applications	1983, 10^6 lb (10^3 t)	1984, 10^6 lb (10^3 t)
bottles, jars, vials	1118 (507)	1280 (581)
food containers, including cups	552 (250)	580 (263)
flexible packaging		
household and institutional		
bags and film	90 (41)	110 (50)
other	160 (73)	180 (82)
all other packaging	671 (304)	805 (365)
Total HDPE for packaging	*2591 (1175)*	*2955 (1340)*
Total domestic HDPE produced	*4706 (2135)*	*5315 (2411)*
Percent packaging	*55.1*	*55.6*

tons) of HDPE was produced in the United States alone, and of that, nearly 3 billion lb (1.36×10^6 t) was for packaging applications (see Table 1).

History

In the 1950s, 20 years after ICI pioneered low density polyethylene (LDPE) in the UK using high pressure technology (see Polyethylene, low density), three separate research groups almost simultaneously but completely independent of each other, developed low pressure processes for making HDPE with properties very different from LDPE. The catalysts used in two of the processes, ie, chromium oxide on a silica base discovered by J. P. Hogan and R. L. Banks of Phillips Petroleum Co. (1) and metal halide–aluminum alkyl discovered by Professor Karl Ziegler of the FRG (2,3), now form the bases for the two leading low pressure processes for producing HDPE worldwide. Since that time, many subtle but commercially significant varieties of HDPE have been made using either new catalyst types, new processes, or both.

Originally, many manufacturers formed HDPE in solution at relatively low pressures and high temperatures. Later, slurry processes formed HDPE as discrete particles in hydrocarbon diluent. Less energy was needed to separate and dry these discrete HDPE particles than in the solution process. A more recent process utilizes gas-phase polymerization. After particles of HDPE leave the reactor and are dried, they resemble coarse white laundry detergent. This powder can be sold as is, usually with thermal stabilizer added, but it is generally sold as pelletized resin.

Molecular Structure and Packaging Properties

The chemical and mechanical properties of HDPE packaging materials and components stem from the polymer's molecular structure. The four principal molecular factors affecting the properties of HDPE are short-chain branching, molecular weight, molecular weight distribution, and long-chain branching. Processing conditions and additive changes can also drastically affect chemical and mechanical properties.

Short-chain branching/crystallinity. When HDPE is composed of only ethylene molecules linked together, it is called an ethylene homopolymer. If other olefins, such as 1-butene, 1-hexene, or 1-octene, are incorporated during the reaction, the

resulting resins are called ethylene–butene, ethylene–hexene, or ethylene–octene copolymers, respectively. When ethylene is copolymerized with any of these comonomers, they form short-chain branches (SCB). For example, the use of 1-butene produces an ethylene–butene copolymer with ethyl branches; use of 1-hexene produces an ethylene–hexene copolymer with butyl branches.

HDPE forms a large amount of ordered, tightly packed crystalline regions as it cools to the solid state. This tight packing contributes to HDPE's good moisture-barrier properties and chemical resistance. These crystalline regions form when the long PE molecules fold back and forth on themselves, approximately every 50 carbon atoms (see Fig. 1). Compared with most other polymers, ethylene homopolymers have an extremely high percentage of their structure made up of these rigid, tightly packed crystalline regions. The amount, type, and distribution of SCB used in making PE determines crystallinity. Short-chain branches disrupt the tight packing of the HDPE molecules, thereby reducing the resin's crystallinity and lowering its density, which results in lower barrier and chemical resistance, among other things. Just two or three branches per 1000 carbon atoms suppress the density of HDPE from >0.960 to 0.940 g/cm³. PE density is measured by compression-molding a slab under controlled conditions, then measuring where a punch-out from this slab settles in a graduated density column (ASTM D 1505). ASTM has designated that HDPE be split into two types: Type III (0.941–0.959 g/cm³) and Type IV (0.960 g/cm³). This breakdown serves to differentiate between homopolymers and copolymers (see Fig. 2).

As density increases, stiffness, melting point, barrier, hardness, chemical resistance, and tensile yield strength increase;

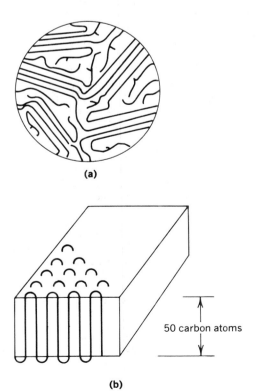

50 carbon atoms

Figure 1. Two models for PE crystal structure: (**a**) extended-chain crystallites; (**b**) folded-chain crystallites.

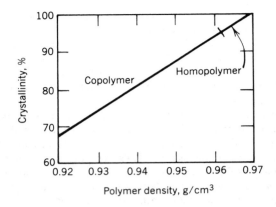

Figure 2.   Relationship between crystallinity and density.

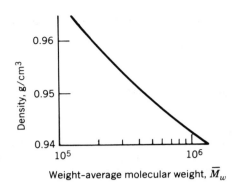

Figure 3.   Effect of molecular weight on homopolymer density.

however, properties such as environmental stress-crack resistance (ESCR) and impact strength decrease (see Table 2). Most dairies, for example, use the highest density homopolymer for maximum stiffness to give acceptable crush resistance in filling, capping, and stacking since they make their bottles as lightweight as possible. On the other hand, packages of industrial and household chemicals are more concerned about ESCR, so they use lower density copolymers and accept the stiffness reduction. Such trade-offs are common when selecting the right density. Since direct measurements of SCB and crystallinity are expensive and require much time and expertise, density is the preferred method to characterize these properties, which it does reasonably well. As the length of PE chains increase, their density is suppressed (see Fig. 3), because the longer polymer chains inhibit crystallization.

Molecular weight.   HDPE resins are available in a wide range of melt indexes. Melt index (MI) is a function of each resin's molecular weight. MI is simply a measurement of the rate of flow of the molten HDPE at 374°F (190°C) through a small orifice at the end of a barrel. In the case of ASTM D 1238

Condition-E melt index, a 4.76-lb (2.16-kg) weight is used to push the extrudate out of the barrel. Condition F (HLMI) uses a weight 10 times heavier: 47.6 lb (21.6 kg). This heavier weight is used for more viscous melts. High MI corresponds to low molecular weight, since short molecules flow more easily than long ones. The molecular weight of HDPE used in packaging is usually between 100,000 and 250,000, or roughly 7,000 to 18,000 carbon atoms per HDPE chain. The term weight is somewhat a misnomer since the most common R&D-type evaluations do not directly measure molecular "weight." Rather, these tests measure the molecular volume (eg, gel-permeation chromatography (GPC) and size-exclusion chromatography (SEC)), the resin's melt viscosity (eg, melt index, capillary rheometry, disk rheometry), or the resin's solution viscosity (eg, intrinsic viscosity). Molecular weight is then inferred from each of these indirect measurements. Membrane osmometry and light scattering, the only true measurements of molecular weight, are rarely used to characterize HDPE because they require time, expense, and expertise.

As the molecular weight of HDPE increases, melt elasticity, toughness, ESCR, elongation, and tensile break strength

Table 2.   Effects of Density, Molecular Weight, and Molecular Weight Distribution (MWD)

Property	As density increases	As molecular weight increases	As MWD broadens
melting point	increases	decreases slightly	decreases slightly
softening point	increases	decreases slightly	decreases slightly
melt viscosity		increases	
flow		decreases	increases
stiffness	increases	increases slightly	decreases slightly
tensile strength	increases	increases	decreases
impact strength	decreases	increases	decreases
low temperature brittleness	increases	decreases	decreases
abrasion resistance		increases	decreases slightly
hardness	increases	increases slightly	decreases slightly
chemical resistance	increases	increases slightly	
ESCR	decreases	increases	
permeability	decreases	decreases slightly	
gloss	increases	decreases	decreases
haze	increases	increases	
shrinkage	increases	increases	decreases

increase; brittleness temperature, glass-transition temperature (Tg) and processability decrease. A balance of all these properties must be found for each application. A company manufacturing large blow-molded drums would use a high molecular weight (HMW) resin to ensure good impact strength and ESCR, accepting the sacrifice in processibility. Injection molders typically use lower molecular weight resins which do a good job of flowing long distances in the mold, filling nooks and crannies, and maximizing production rate. HDPE resins may be labeled according to their molecular weight as follows:

Type of resin	Molecular weight range
Medium molecular weight	<110,000
High molecular weight	110,000–250,000
Very or extra high molecular weight	250,000–3,500,000
Ultrahigh molecular weight	>3,500,000

Packaging processes are organized according to MI (ASTM Condition E) and HLMI (ASTM Condition F) in Figure 4. The high flow 100-MI resin at the top is used to fill extremely thin walls in injection molds; the zero HLMI resin at the bottom does not flow enough for processing by typical thermoplastic methods.

Molecular weight distribution. Just as all trees in a forest do not grow to be the same size, neither do all molecules within a given HDPE-resin type grow to be exactly the same length (molecular weight). SEC and GPC can graphically depict the variation in size and amount of these HDPE molecules (see Fig. 5). The distribution of molecular weights is bell-shaped, with most molecules clustered around a certain molecular weight called the number-average molecular weight, $\overline{M}_n$. The weight-average molecular weight, $\overline{M}_w$, is a molecular weight value taken from the higher molecular weight end of the resin's curve.

The $\overline{M}_w/\overline{M}_n$ ratio, sometimes called the heterogeneity index or polydispersity index, is a measure of the breadth of the resin's molecular weight distribution (MWD). The larger the ratio, the broader the molecular weight distribution. Narrow-MWD resins have more molecules of the same length, and broad-MWD resins have more molecules that are both shorter and longer than the number-average molecular weight (see Fig. 6). HDPE with a $\overline{M}_w/\overline{M}_n$ of 6 or less is considered narrow MWD; a value of 12 or greater indicates broad MWD. HDPE with values between these two is considered medium MWD.

Another method for characterizing a resin's molecular weight distribution takes advantage of the fact that, when molten, HDPE flows faster as shear stress increases. Two HDPE resins might have the same MI, but one can have much higher HLMI (MI Cond. F) than the other at this higher shear stress (see Fig. 7). This might be the case if the resin with higher flow had more low molecular weight ends than the other, ie, broader MWD. These shorter HDPE chains would act as a "lubricant" at higher shear rates, thus providing higher flow.

As the MWD of a HDPE resin broadens, impact strength, melt strength and warpage resistance decrease and processibility and ESCR generally improve (see Table 2). An injection molder, for example, can select a resin with relatively narrow MWD to compensate for the relatively low impact strength of the LMW resin chosen for flowability. On the other hand, since blow molders use higher molecular weight resins to obtain impact strength and ESCR, they do not need narrow MWD;

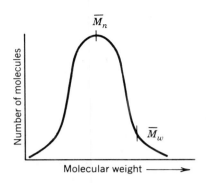

Figure 5. Typical molecular weight distribution curve for HDPE: $\overline{M}_n$ = number-average molecular weight; $\overline{M}_w$ = weight-average molecular weight.

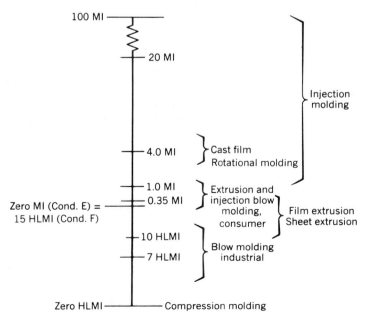

Figure 4. Packaging processes organized according to melt index.

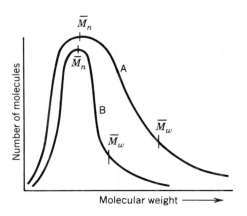

Figure 6. Molecular weight distribution curves: A, broad MWD; B, narrow MWD.

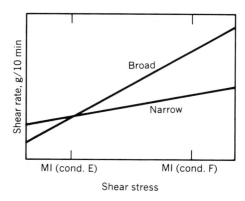

Figure 7. Effect of molecular weight distribution on processibility.

rather, they would prefer to use a broad MWD to maximize processibility and production rate. All of these properties are discussed in more detail below.

Long-chain branching. Short-chain branches are formed intentionally by adding a comonomer. Long-chain branching usually occurs accidentally, either in the reactor or during subsequent extrusion. Long-chain branches (LCB) can be defined as any branches of eight or more carbons. There is no quantitative means by which to measure LCB; it can only be inferred. As the amount of LCB increases in HDPE, melt appearance, ESCR, and high shear melt viscosity decrease, with the only potential benefit being that melt-strength increases. Most HDPE resin suppliers intentionally make injection, extrusion, and blow-molding resins with little or no LCB, so it is not a factor in resin selection. Long-chain branching can intentionally be carried to an extreme to form cross-linked HDPE for nonfood contact applications such as wire coating or rotationally molded tanks and containers that do not require

FDA clearance. This is done by using agents such as organic peroxides (chemical) or beta or gamma rays (radiation). Each advanced LCB provides a site for visbreaking and ultimately cross-linking where one HDPE polymer molecule covalently bonds with one or more of its neighboring molecules. This cross-linking makes a part that is essentially thermoset (see Polymer properties), with excellent impact and abrasion resistance, but no scrap-regrind capability during processing.

Chemical Properties

Chemical resistance. In general, HDPE is very good for use in the packaging of water-based products, medium-to-high molecular weight aliphatic hydrocarbons, alcohols, ketones, aldehydes, as well as dilute acids and bases. If HDPE is used to package aromatic hydrocarbons such as benzene or halogenated hydrocarbons (eg, carbon tetrachloride), the package, or at least the interior, should first be treated to resist excessive content permeation and subsequent part deformation. There are four chief ways to do this: fluorination (4), sulfonation (5), and the use of a mixture of nylon and HDPE compounded and processed using special techniques to obtain enhanced barrier properties (6) (see Surface and hydrocarbon-barrier modification). The fourth method, coextrusion is still probably the most common method packaging manufacturers use to take advantage of the good barrier properties associated with HDPE in coordination with the good properties associated with other resin types.

The chemical resistance of HDPE stems largely from its density compared with its low density counterparts because HDPE has a higher percentage of tightly packed crystal lamellae which allow less chemical permeation that could lead to swelling, staining, and softening (or embrittlement, depending on the chemical). Chemical-resistance data for a few select chemicals are given in Table 3 (7,8). Such data can be used only as preliminary guidelines because chemical compatibility

Table 3. Resistance of HDPE to Selected Chemicals

Product	Chemical resistance[a]	Permeability, loss/yr, %	Can cause stress cracking	Comments
Acids				
acetic	E	<3	yes	
hydrochloric	E	<3	no	slight staining at elevated temperatures;
sulfuric, 80%	G	<3	no	stiffening and embrittlement occur at elevated temperatures
Bases				
ammonium hydroxide, 30%	E	<3	no	
Industrial chemical				
benzaldehyde	E	<3	no	
benzene	G	high	no	
chloroform	P	high	yes	softening and part deformation
ethyl alcohol	E	<3	yes	
ethyl ether	F	140	no	softening and part deformation
Foods				
ketchup	E	<3	no	slight staining
cider	E	<3	yes	
margarine	G	<3	yes	

[a] Key: E = excellent; G = good; F = fair; P = poor.

strongly depends on wall thickness, part size and geometry, package shelf life, storage temperature, and actual use conditions. Furthermore, many products are a combination of chemicals, so the ultimate compatibility with the packaging material involves testing the product in its proposed container.

Environmental stress-crack resistance (ESCR). Unlike other mechanical tests for polymers, which are usually conducted at constant elongation (eg, tensile strength) or constant stress (eg, creep), most ESCR tests are conducted under constant strain in some sort of chemical that accelerates environmental stress cracking (ESC) at an elevated temperature. The most common ESCR test immerses bent HDPE strips that have a partial notch on their face in a surfactant, ie, soap, or a surfactant–water mixture. Sometimes called "Bell ESCR," this test is described as ESCR (Cond. A) by ASTM D 1693. The bend induces a stress on the crystalline network of the solid HDPE strips, the notch gives the stress a place to focus, and the surfactant acts as a plasticizing agent which, theoretically, allows the tie molecules between crystals and the crystal lamellae network to slip out of these crystals, thereby causing failure. A low ESCR value means that all of the samples failed quickly. A HMW resin has relatively high ESCR, since it has more and longer tie molecules to disentangle from the crystalline network. Another factor controlling ESCR is density: as density decreases, ESCR increases. This is true because lower densities reflect less lamellar orientation and certain lamellar orientation hurts ESCR. The SCB which decrease density by disrupting lamellar orientation may sometimes act as tie molecules to cement crystals together and further increase ESCR. When selecting a resin for packaging, it must be remembered that increasing SCB affects other properties.

As the amount of surfactant in the ESC solution decreases, ie, to about 7%, ESCR decreases. That is why a resin takes less time to fail using ESCR (Cond. B, 10%) than using ESCR (Cond. B), with a default condition of 100% surfactant. The surfactant has a hydrophilic "tail" and a hydrophobic "head." The more dilute the solution, the more easily surfactant molecules can align their tails into the water and allow the head to align itself with the HDPE and thereby increase the rate of plasticizing and, subsequently, accelerate ESC failure. This one complex interaction illustrates the dynamic interactions packaging companies must contend with when using HDPE to package hundreds and even thousands of different pure chemicals as well as chemical mixtures. Sometimes containers are tested for ESCR by filling them with a surfactant–water solution and placing them in a room with a controlled elevated temperature. The containers may be capped with the cap connected to a constant-pressure manifold to maintain a certain internal bottle pressure. This accelerated test attempts to simulate real-world conditions, eg, a hot warehouse in the summer. Such internal pressures could be generated if the liquid within these containers had components that could volatilize. Some fabricators avoid the need for a constant-pressure manifold by putting a measured amount of alcohol in the surfactant–water mixture that generates internal pressure.

Fungus resistance. HDPE showed no evidence of deterioration or corrosion due to fungus exposure when tested according to Military Specification MIL-E-527A. This involved a 28-d test exposing tensile bars to a sprayed suspension of five groups of fungi at 95 ± 5% rh at 86 ± 2°F (30 ± 1°C).

Decorating and bonding. Inherently, PE has a chemically inert, nonpolar surface. Oxidation of this surface is required

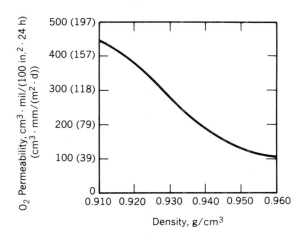

Figure 8. Effect of resin density on oxygen (O$_2$) permeability.

for acceptable adhesion of decorations, coatings, and adhesives. The oxidation of polyolefin surfaces may be accomplished by flame, chemical, or electrical processes (corona discharge, suppressed spark, or plasma method) or the USM (United States Machinery) process (9) to produce a receptive polar surface (see Surface modification). After the surface is properly treated, parts made from HDPE may be decorated by any number of the conventional methods of decorating plastics (see Decorating). Adhesive bonding may be accomplished by any adhesive that "wets" the treated surface (see Adhesives). The degree of treatment can be measured by the polymer's surface tension. Untreated HDPE has a critical surface tension of ca 29 dyn/cm (= mN/m). A minimum of 36 dyn/cm (= mN/m) is necessary for ink bonding (see Inks) and a minimum of 40 dyn/cm (= mN/m) for adhesive receptivity.

Gas permeability. As the crystallinity, ie, density, of PE increases, its gas permeability decreases (see Fig. 8). Gas permeability is measured by ASTM D 1434 (see Testing, permeation and leakage). HDPE can sometimes prevent oxygen and other gases from spoiling food, but its gas-barrier properties are relatively poor (see Barrier polymers). It may be coextruded with other resins to enhance the overall gas-barrier properties of the package. Coextrusion is now commonly done in many extrusion processes and some blow-molding processes (see Coextrusion, flat; Coextrusion, tubular; Multilayer plastic bottles). In ASTM D 1434, the PE sample acts as a membrane between two pressure vessels. The vessel on one side of the sample is charged with a test gas and maintained at a specific temperature and one atmosphere pressure. The vessel on the other side of the sample is evacuated and maintained under vacuum. The volume of test gas passing through the sample is recorded periodically until a steady state is established (see Table 4).

Moisture (water) permeability. The standard method for measuring moisture-barrier properties is ASTM D 96 (Desiccant Methods). The preferred procedure employs a desiccant, such as anhydrous calcium sulfate, sealed in an aluminum dish by a film sample with one surface exposed to the environment. The assembly is conditioned in a humidity cabinet which circulates air maintained at 100°F (37.8°C), 90% rh. Water-vapor transmission rate (WVTR or MVTR) is determined by measuring the weight gained, ie, water capture, over a period of time. Increasing the density of PE significantly

Table 4. Gas Permeability of Homopolymer HDPE

Gas	Rate, cm³·mil/(100 in.²·24 h) (cm³·mm/(m²·d))
carbon dioxide	345(136)
hydrogen	321(126)
oxygen	111(44)
helium	247(97)
ethane	236(93)
natural gas	113(44)
freon 12	95(37)
nitrogen	53(21)

reduces its WVTR, but not linearly (see Fig. 9). The WVTR of PE decreases exponentially as the thickness of the sample increases (see Fig. 10).

Physical Properties

Impact resistance. Many different impact tests are utilized for HDPE. Whether impacting compression-molded sample bars using an Izod or Charpy impact testing device (see Testing, packaging materials) or actually impact testing the final package (see Testing, consumer packages), impact strength is a function of processing conditions, part design, MW, MWD, density, and the testing environment. A general rule of thumb for selecting a resin with acceptable impact resistance is that impact strength increases as MW increases, MWD narrows, and density decreases.

In injection molding, narrow-MWD resins are generally used to boost impact strength, since lower molecular weight HDPE is needed for flow. A broader-MWD resin in most injection-molding applications would produce differential flow of the melt into the mold; it would build in internal stresses, causing poor impact strength in the part as well as warpage problems due to subsequent differential cooling.

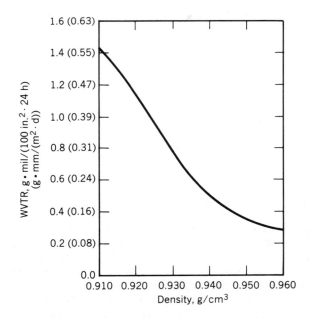

Figure 9. Effect of resin density on WVTR.

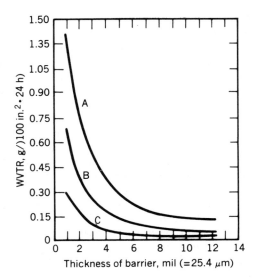

Figure 10. Effect of PE film thickness on WVTR: A, 0.920 density; B, 0.940 density; C, 0.960 density. To convert g/(100 in.²·24 h) to g/(m²·d), multiply by 15.5.

In film extrusion, sheet extrusion, blow molding, and other applications where medium-to-broad-MWD resins are used to obtain good processability and part appearance, molecular weight and density are used to control impact resistance. HDPE, like other polymers, is "notch-sensitive." In other words, significantly less energy is needed to break a bar or part if it has been notched or an imperfection has been introduced that gives the impact energy a place to concentrate.

As the testing temperature decreases, impact strength decreases. A typical range for tensile impact (ASTM D 1822) for HDPE is 50–80 ft·lbf/in.² (10.5–16.8 J/cm²), with some resins as low as 25 ft·lbf/in.² (5.25 J/cm²) and some as high as 200 ft·lbf/in.² (42.0 J/cm²). These values do not reflect UHMW HDPE values, which are higher. Some HDPE samples do not break even when notched and tested on the Izod machine; others have Izod impact values as low as 0.4 ft·lbf/in.² (0.084 J/cm²). Parts may be impact-tested by dropping a heavy object onto them, eg, Gardner impact or actually dropping the finished package, usually with something in it, eg, water, and sometimes by chilling to a lower temperature to accentuate impact failure.

Load deformation, constant loading (stackability). It is important for packaging manufacturers to know how well dairy tubs, shampoo bottles, acid drums, etc, will withstand loading when stacked one on another in the store shelves or warehouse, sometimes at elevated temperatures. For one-time-use packages containing chemically "bland" substances in relatively small amounts, higher density HDPE can be used to maximize the stiffness–thickness ratio of the part wall with little worry for ESCR, impact resistance, etc. As density is decreased to meet higher impact and ESCR demands, deformation resistance decreases. Long-term stackability begins to depend more and more on MW as the container size and content weight increase. The higher the molecular weight, the better the long-term stackability. Longer molecules do a better job of holding the crystalline network together under long-term stress. This phenomenon is somewhat analogous to tensile creep.

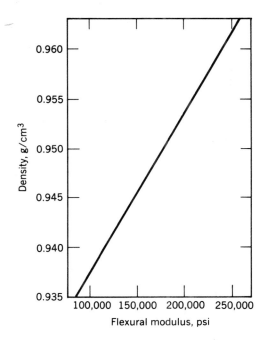

Figure 11. Effect of density on flexural modulus of HDPE. To convert psi, to MPa, divide by 145.1.

Load deformation, momentary loading (top-load). In in-plant dairy-bottle production, for example, it is important to have good immediate development of top-load strength, ie, resistance to bottle crush as measured by various machines. Without this, bottles coming to the filler from the blow molder would deflect and deform too much when the filler mechanism or capper mechanism descended onto them. MWD can influence this rate of crystallization, ie, rate of top-load development.

Stiffness. Whether stiffness is measured by flexural modulus, ie, force measured on a bar being deflected in its center while being supported at each end (ASTM D 790) (see Fig. 11), tensile modulus of elasticity (ASTM D 638) (see Fig. 12), or by

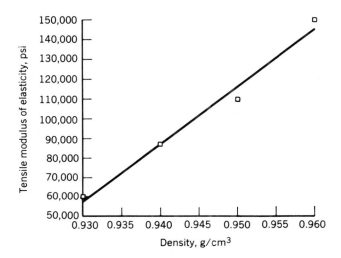

Figure 12. Tensile modulus of elasticity for HDPE vs density. To convert psi to MPa, divide by 145.1.

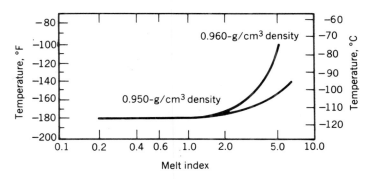

Figure 13. Melt index vs brittleness temperature. Limit of test is −180°F (−118°C).

the cantilever-beam method (ASTM D 747), stiffness increases as density increases.

Within the PE family, 260,000 psi (1,792 MPa) flexural modulus represents maximum homopolymer stiffness, compared with a branched-LDPE stiffness of 40,000 psi (276 MPa). Mineral-filled HDPE can meet and even surpass the stiffness of polymers such as polystyrene at 500,000 psi (3,446 MPa) flexural modulus (10). As resin stiffness increases, wall thickness of a package can be reduced for cost savings, as long as no other important physical properties are compromised. By properly designing the part and selecting the appropriate density, HDPE can be used in applications requiring either stiffness, eg, industrial pallets that must not bend when supporting heavy loads (see Pallets, plastic), or flexibility, eg, squeeze bottles.

HDPE has extremely good low temperature flexibility as measured by brittleness temperature (see Fig. 13). This is the temperature at which the sample gives a brittle fracture at a fixed rate of loading. This makes HDPE ideal for freezer packages, cryogenic packaging, and winterproof packaging. As molecular weight increases and density decreases, brittleness temperature decreases; thus, below 1.0 MI, all HDPE resins have a brittleness temperature at or below −180°F (−118°C), which is the lower limit of the test. This correlates well with HDPE T_g of −193°F (−125°C), at which large-scale molecular motion does not occur and the polymer assumes the properties of glass, including brittleness.

Melt strength. To obtain maximum melt strength, a resin with relatively high MW, ie, lower MI, should be selected if a decrease in production rate can be tolerated. Typical MI ranges for various applications are presented in Table 5.

Melting point. The melting point of HDPE is primarily a function of its density. As density increases, melting point increases (see Fig. 14). Vicat softening temperature (ASTM D 1525) quantitatively determines a resin's softening temperature, which, like melting point, is influenced by density. Vicat softening temperature for HDPE varies anywhere from 270°F (132°C) for homopolymers to 250°F (121°C) for a resin with a 0.94-g/cm³ density. Another applications-oriented measure of melt temperature is the heat-distortion (heat-deflection) temperature (HDT), specified by ASTM D 648 as the temperature at which an injection-molded sample bar ½ × ½ × 5 in. (13 × 13 × 127 mm) begins to deflect by 0.01 in. (0.25 mm) under a flexural load of 66 psi or 264 psi (455 or 1820 kPa, respectively) placed at its center. HDPE varies in HDT at 66 psi (455 kPa) from 160 to 210°F (71–99°C) and at 264 psi (1820 kPa) from

Table 5. Sampling of HDPE Resins Used in Packaging

Process	Blow molding			Injection blow molding	Injection molding				Cast film	Blown film		
density, ASTM D 1505, g/cm³	0.950	0.955	0.964	0.957	0.945	0.950	0.955	0.967	0.960	0.947	0.956	0.955
MI, ASTM D 1238[b], g/10 min	10 HLMI[a]	0.35	0.75	1.0	1.5	6.0	18.0	30.0	5.0	0.28	1.0	9 HLMI[a]
MWD	medium	medium	medium	narrow	medium	narrow	narrow	narrow	narrow	medium	medium	broad
flexural modulus, ASTM D 790, psi (MPa)	175,000 (1,206)	200,000 (1,379)	240,000 (1,654)	225,000 (1,551)	150,000 (1,034)	175,000 (1,206)	200,000 (1,379)	260,000 (1,792)	240,000 (1,654)	128,000 (882)	216,000 (1,489)	200,000 (1,379)
ESCR, ASTM D 1693[b], F_{50}, h	800	45	15–20	30	100	25	<1	<1	6	>1,000	50	>1,000
tensile strength at yield, ASTM D 638, Type-IV specimen, 2 in. or 5 cm/min., psi (MPa)	3,800 (26.2)	4,000 (27.6)	4,400 (30.3)	4,100 (28.3)	3,400 (23.4)	3,700 (25.5)	4,200 (29.0)	4,600 (31.7)	4,200 (29.0)	2,800 (29.0)	3,000 (20.7)	4,000 (27.6)
typical applications	55-gal (208-L) drums, gasoline and chemical tanks	household and industrial-chemical containers, ice chests	milk, juice, and water bottles	cosmetic packages and pharmaceutical bottles	closures	industrial containers, closures	milk cases, housewares	thin-walled containers, overcaps	frozen-food bags, snack-food packaging medical packages	merchant and produce bags, multiwall bag liners	coextruded snack-food packaging, cereal-box liners, cracker bags	grocery sacks and merchant bags

[a] Cond. F.
[b] Cond. A.

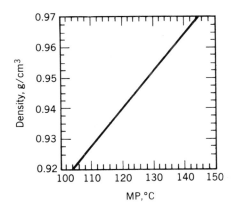

Figure 14. Melting point vs density of PE

120 to 127°F (49–53°C). Such information might be helpful in determining which HDPE would make the best drum or pail for long-term stackability in a hot warehouse.

Flammability. The relative flammability of PE is rated as being low (11). Although HDPE melts at temperatures as low as 248°F (120°C), it is difficult to ignite at 660°F (349°C). If an ignition source is present, HDPE can ignite at 645°F (341°C). Both self-ignition and assisted-ignition temperatures were determined by ASTM D 1929. Once PE ignites, it burns at a rate of 1 in/min (2.54 cm/min) as measured by ASTM D 635 and MVSS 302. When HDPE is burned using the ASTM E 84 tunnel test, it has a flame-spread rating of 97.8, where a red-oak control rates at 100. HDPE received a fuel-contribution rating of 30 and smoke-density rating of 350, also by ASTM E 84. The maximum continuous-service temperature for HDPE is 250°F (121°C) (11). Should better flammability ratings be desired, three families of flame retardant chemicals can be added. Halogenated aliphatics and brominated aromatics suppress rapid oxidation and decrease the heat as it evaporates, thereby keeping the polymer cool and preventing ignition. Certain organophosphorus compounds represent the third family of flame retardants, which decompose to form a char during rapid oxidation rather than carbonaceous gases which would contribute to combustion.

Packaging Regulations

HDPE easily meets the FDA specifications for use in food-contact service. Some additives used by HDPE manufacturers are not FDA-approved for food contact or may have guidelines restricting their use to certain conditions, eg, packaging temperature, types of foods to be packaged, etc. HDPE manufacturers will know what these restrictions are if there are any questions concerning suitability. Many HDPE resins are listed in the Drug Master File, which is a prerequisite for use in packaging drugs (see Food, drug, and cosmetic packaging regulations).

BIBLIOGRAPHY

1. Belg. Pat. 560,617 (Jan. 24, 1955) and U.S. Pat. 2,825,721 (Mar. 4, 1958), J. P. Hogan and R. L. Banks (to Phillips Petroleum Co.).
2. K. Ziegler, E. Holzkamp, H. Breil, and H. Martin, *Angew. Chem.* **67**, 426, 521 (1955).
3. V. L. Folt, "Physical Properties of Ethylene-1-Olefin Copolymers," *18th ANTEC*, Vol. VIII, Pittsburgh, Pa., Jan. 30–Feb. 2, 1962, The Society of Plastics Engineers, Inc., Brookfield Center, Conn., 1962.
4. U.S. Pat. 3,862,284 (Jan. 21, 1975), D. D. Dixon, D. G. Manly, and G. W. Recktenwald to Air Products and Chemicals, Inc.).
5. K.-D. Johnke and P. Behr, "Status Report on HDPE Fuel Tanks in European Automobiles: Characteristics of Service Life on Performance," *Passenger Car Meeting,* Troy, Mich., June 7–10, 1982, The Society of Automotive Engineers, Warrendale, Pa., 1982, SAE Technical Paper 820800.
6. U.S. Pat. 4,410,482 (Oct. 18, 1983); 4,416,942 (Nov. 22, 1983); and 4,444,817 (Apr. 24, 1984); P. M. Subramanian, (E. I. Du Pont de Nemours & Co., Inc.).
7. *Technical Information Bulletin on Packaging Properties,* Plastics Technical Center, Phillips Chemical Co., Bartlesville, Okla., 1986.
8. *Technical Service Memorandm No. 243,* Plastics Technical Center, Phillips Chemical Co., Bartlesville, Okla., June 1983.
9. U.S. Pat. 3,607,536 (Nov. 7, 1968), R. A. Bragole (to USM Corp.).
10. D. L. Peters and R. C. Kowalski, "Automotive Blow Molding Opportunities from Fuel Tanks to Load Floors," *SPI 1985 National Plastics Exposition Conference,* Chicago, Ill., June 20, 1985, The Society of the Plastics Industry, New York, 1985.
11. D. V. Rosato and J. R. Lawrence, eds., Plastics Industry Safety Handbook, The Society of the Plastics Industry, New York, 1973, p. 81.

M. A. SMITH
Phillips Chemical Co.

POLYETHYLENE, LOW DENSITY

Polyethylene is a thermoplastic polymer formed from the polymerization of ethylene. It is available in a variety of molecular weights and densities tailored to specific end-use markets. ASTM has divided polyethylene into four general categories according to density:

Type	Nominal density, g/cm³
I	0.910–0.925
II	0.926–0.940
III	0.941–0.959
IV	≥0.960

In general, the polyethylene industry does not always follow these designations, but has broken polyethylene into two broader categories: high density polyethylene (HDPE), ≥0.940 g/cm³, and low density polyethylene (LDPE), 0.915–0.939 g/cm³. In addition, there is now a new type of polyethylene, very low density polyethylene (VLDPE). The Dow Chemical Co. markets linear low density polyethylene at 0.912 g/cm³ (1), and Union Carbide Corp. has introduced polyethylene with densities of 0.900 and 0.905 (2).

Low density polyethylene was first produced in the UK in 1933 by Imperial Chemical Industries (ICI, Ltd.) when ethylene gas was compressed to high pressures and heated to high temperatures (3). ICI's development of a commerical process for the manufacture of LDPE was closely followed by the wartime use of the product in critical areas such as high-frequency cables for ground and airborne radar equipment.

In the early 1950s, another type of polyethylene, HDPE (see Polyethylene, high density), was commercially introduced by several companies that had developed new low pressure processes for its production (4). HDPE is a linear polymer, without any of the long-chain branching characteristics of LDPE.

Table 1. HP-LDPE Homopolymer Film Resins

type	HP-LDPE	HP-LDPE	HP-LDPE	HP-LDPE	HP-LDPE	HP-LDPE
comonomer	none	none	none	none	none	none
melt index	0.2–0.8	1.5–2.0	1.5–2.0	1.2–2.0	1.0 –2.0	5.0 –10.0
density, g/cm^3	0.919–0.923	0.922–0.925	0.930–0.935	0.918–0.924	0.923–0.927	0.917–0.930
molecular weight distribution	broad	broad	broad	broad	broad	broad
process	blown	blown	blown and/or cast	blown	cast	cast
applications	shipping sacks; heavy-duty applications	bread and bakery; general-purpose packaging	overwrap	general-purpose packaging	bread and bakery; general-purpose packaging	extrusion coating
critical properties	toughness	clarity	clarity; stiffness	extrudability; toughness	extrudability; good tear	drawdown adhesion pinhole resistance

Because of its different structure, HDPE possessed different and complementary properties to LDPE and was quickly utilized in many new packaging applications. These two polyethylenes, differentiated by density, properties, and manufacturing processes, coexisted and grew until the early 1960s, when a third type of polyethylene, LLDPE, was introduced. Because the molecular structure of this new type of LDPE resembled that of HDPE, the term linear low density polyethylene (LLDPE) was coined.

Therefore, the polyethylene industry today is composed of three types of polyethylene: HDPE, LDPE (sometime also referred to as high pressure LDPE or HP-LDPE), and LLDPE. Some confusion exists in the terminology used for LDPE and LLDPE. In some articles and publications, LDPE is used as a generic term for polyethylene below 0.935 g/cm^3, thus covering both HP-LDPE and LLDPE. In other cases, LDPE is used to cover only HP-LDPE, and LLDPE is a separate category of polymer. In this article, the second type of categorization is used. LLDPE is treated as a separate type of polyethylene and

referred to only as LLDPE; LDPE implies only low density polyethylene produced by the high pressure process. When written out in words, (low density polyethylene), both LLDPE and HP-LDPE are included (see Tables 1 and 2). LLDPE has made inroads in many of the markets currently served by LDPE and also competes with HDPE in some new applications.

Characterization of Polyethylene

The properties of products made from polyethylene depend on some basic characteristics of the polyethylene itself. Several of the terms commonly used to describe polyethylene are listed below.

Melt index. The melt index (MI) of a polymer is used as an empirical measure of its molecular weight. To measure melt index according to ASTM D 1238, a polymer sample is melted and forced through a small orifice of fixed size under a fixed pressure. The weight of polymer that is extruded in 10 min under 44 psi (303 kPa) pressure is called the melt index. When

Table 2. HP-LDPE Copolymer Film Resins

type	HP-LDPE	HP-LDPE	HP-LDPE	HP-LDPE
comonomer	2–5% VA[a]	3–5% VA[a]	7% VA[a]	15–18% EA[b]
melt index	1.5–2.0	0.2–0.4	0.2–4.0	2.0–6.0
density, g/cm^3	0.925–0.930	0.923–0.927	0.927–0.945	0.927–0.940
molecular weight distribution	broad	broad	broad	broad
process	blown	blown	blown; extrusion coating	blown; extrusion coating
applications	frozen food	ice bags	sealing layer; liquid packaging	disposable gloves; ID cards
critical properties	clarity; low temperature properties	low temperature properties	adhesion; low temperature properties	low stiffness; adhesion to polar substrates

[a] Vinyl acetate.
[b] Ethyl acrylate.

the pressure is increased to 440 psi (3034 kPa), the weight of polymer extruded in 10 min is the flow index (FI). Since a polymer with very high molecular weight is very viscous and resistant to flow, it does not pass through the small orifice quickly and the weight obtained in 10 min (or melt index) is low. Melt index is thus inversely proportional to molecular weight. Typical LDPE melt indexes range from 0.2 to >100 g/ 10 min. In general, products in the lower melt-index range are used for film extrusion and the higher melt-index products for injection molding and extrusion coating. Within any fabrication process, use of a lower melt-index resin results in a stronger product, although usually with some sacrifice in ease or rate of production.

Melt–flow ratio. The melt–flow ratio (MFR) is a rough estimate of the molecular weight distribution (MWD) of a resin. All polymer chains in a given resin are not exactly the same length, and a measurement of molecular weight distribution describes how dissimilar the chains are from each other. Melt–flow ratio is the ratio of the flow index to the melt index. The higher the ratio, the broader the molecular weight distribution and the more dissimilar the chains are from each other. A polymer with every chain exactly the same length would have a very narrow MWD and a very low MFR. Melt–flow ratios of commercial LDPE vary from about 20 (very narrow) to about 100 (very broad). Narrow-MWD polymers give stronger products but are more difficult to extrude than broad-MWD polymers.

Density. The density of a polymer is a measure of its crystallinity. Density measurement according to ASTM D 1505 consists of taking a small sample of polymer which has been molded in a carefully prescribed manner and dropping it into columns with solutions of different viscosities. The position of the unknown polymer in the column is then compared with standard samples of known density. The density of a film or molded article is only partly controlled by the density of the resin used to make the product. The rate at which the product is cooled also plays an important role. The faster a film or molded article is cooled, the less time there is for the polymer chains to crystallize and the lower the density of the final product. For example, the density of a sample cut from a blow-molded polyethylene bottle was 0.945 g/cm³. The same polyethylene resin, when compression-molded into a plaque and cooled according to ASTM D 1505, measured 0.954 g/cm³. Product properties such as stiffness, rigidity, environmental stress crack resistance (ESCR), and water-vapor transmission rate (WVTR) are affected by density. The lower the density of a product, the more limp and flexible it is. A 0.918-g/cm³ polyethylene product has better ESCR and higher WVTR than one with a density of 0.930 g/cm³.

LDPE Process

LDPE is made by the high-pressure polymerization of ethylene (4,5). In either a tubular or autoclave reactor, ethylene is pressurized to more than 20,000 psi (138 MPa) and heated to above 300°F (150°C). Small amounts of an initiator, typically oxygen or peroxide, are added to start the polymerization process. Comonomers such as vinyl acetate or ethyl acrylate can be added to make EVA and EEA copolymers, respectively. Critical molecular characteristics such as molecular weight, molecular weight distribution, and density are controlled by reaction temperature, ethylene pressure, and the concentration of chain-transfer agents. Constraints on the viscosity of the polymer solution in the reactor limit the rate at which products with high molecular weight or high density, or both, can be made in the high-pressure process. Products made by the tubular high pressure process differ subtly in the type and degree of branching from those made in the autoclave process.

Density depression occurs in all types of polyethylene because of the presence of short branches along the backbone of the polyethylene chain. These short branches, typically 1–5 carbon atoms long, prevent the long polyethylene chains from folding together and forming crystals. However, early in the development of LDPE, it was found that the presence of these short branches did not completely explain some of the rheological properties of LDPE. It was then hypothesized and proven that LDPE also contains low levels of long-chain branching (6). These long chains, which can be over 1000 carbon atoms in length, have a minor effect on the density of LDPE but a significant impact on the processing and properties of LDPE (see Fig. 1).

The tubular reactor makes LDPE with a large number of long-chain branches. The branches are of relatively short length, however. The autoclave reactor, on the other hand, yields products with low levels of long-chain branches of extremely long length. Because of the difference in long-chain branch type and frequency, some specialization in product applications has occurred. The autoclave reactor produces products that are especially useful in high speed extrusion coating and in film applications requiring toughness. The tubular reactor gives products with the best clarity and processing characteristics.

LLDPE Process

Low pressure processes for the polymerization of ethylene were developed in the 1950s (4, 5). These processes used organometallic catalysts to polymerize ethylene to form HDPE at moderate pressures, ie, ca 300 psi (2 MPa), and tempera-

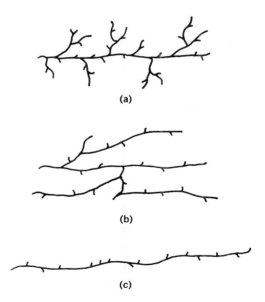

(a)

(b)

(c)

Figure 1. Molecular structures for LLDPE and HP-LDPE: (**a**) HP-tubular process; (**b**) HP-autoclave process; (**c**) LLDPE.

tures of ca 212–392°F (100–200°C). The three basic types of low pressure systems are solution (polymer completely dissolved in a solvent at high temperatures); slurry, (solid polymer particles physically suspended in a solvent at lower temperatures); and gas-phase (solid polymer in contact with only the polymerization gases). Again, because of process constraints in the reactor, molecular weights and densities were thought to be limited, especially in the solution and slurry systems. The density limitations of the low pressure process and high pressure process were thought to form incompatible above 0.935 g/cm^3 and the high pressure process making polyethylenes below 0.935 g/cm^3.

The first commercial production of LLDPE was made in a solution process by DuPont Canada in 1960 (7). A large market for this new polymer did not develop until 1977, when Union Carbide Corp. began licensing its gas-phase process for the manufacture of linear low density polyethylene. Since that time, several other polyethylene manufacturers have announced the conversion of low pressure HDPE processes or even high pressure LDPE processes to make LLDPE, and some of these processes are available for licensing.

The structure of these new linear low density polymers is very different from the LDPE made by the high pressure process. In LLDPE, there is no long-chain branching. Density is controlled by the addition of comonomers such as butene, hexene, or octene to the ethylene (see Table 3). These comonomers give rise to short-chain branches of different lengths: 2 carbon atoms for butene, 4 for hexene, and 6 for octene. The length of the short-chain branches determines some of the strength characteristics of LLDPE. The absence of long-chain branches in LLDPE plays a significant role in the difference in extrusion characteristics between LLDPE and LDPE (see below).

Low Density Polyethylene Properties

Low density polyethylene is one of the most widely used commerical plastics. Its utility in a variety of different applications is not due to some single outstanding property or characteristic, but usually to a combination thereof. The low price of polyethylene compared with that of wood, metal, and other polymers has accelerated its penetration into many applications. In addition its low cost, the excellent toughness, flexibility, moisture barrier, chemical resistance, electrical insula-

tion, and light weight of polyethylene films, bottles, pipes, cables, and other articles make them superior to articles made from conventional materials. The superior properties of LLDPE have led to its use in new applications for polyethylene as well as the replacement of LDPE and HDPE in some areas. Compared with LDPE, LLDPE at the same melt index and density offers better toughness, rigidity, stress crack resistance, elongation, and moisture barrier, as well as a higher melting point.

Low Density Polyethylene Markets

United States. Low density polyethylene is used in a wide variety of applications, ranging from thin-gauge film to 500-gal (1893-L) water-storage tanks. The total 1984 production of LDPE, ie, defined as below 0.940 g/cm^3, in the United States is estimated at ca 8.5×10^9 lb (3.86×10^6 metric tons) (8). Of this, ca 2.3×10^9 lb (1.04×10^6 t) was LLDPE (9). Production capacity is increasingly difficult to measure because of the versatility of the various low pressure processes. It is now relatively easy for a polyethylene producer to switch production from HDPE to LLDPE, depending on market situations. As more producers convert existing high or low density plants or build new plants capable of manufacturing both HDPE and LLDPE, the distinction between production capacities of the two polymers will blur and there will be a unified polyethylene market.

In 1984, ca 750×10^6 lb of LDPE (340,200 t) and LLDPE was exported from the United States (8). A very small quantity, less than 100×10^6 lb (45,360 t), was imported. Imports will be influenced in the next few years by the start-up of some large petrochemical complexes in Saudi Arabia, Canada, and other nations (10). The effect of this new capacity on the pricing stability of polyethylene in the United States is unclear at this time. It will certainly affect the export of LDPE and might change the import situation as well. Pricing of LDPE and LLDPE is extremely variable, and discounting off list price is common. In 1984, the list price of LDPE and LLDPE in the United States varied from $0.31 to 0.46/lb ($0.68–1.01/kg).

In the next few years, the growth rate for LDPE is expected to decrease as LLDPE penetrates the traditional markets for LDPE (see Fig. 2). It is expected that by 1991, all LDPE markets will be penetrated to some extent by LLDPE. Growth of

Table 3. LLDPE Film Resins

type	LLDPE	LLDPE	LLDPE	LLDPE	LLDPE
comonomer	butene	hexene; octene	hexene; octene	hexene; octene	hexene; octene
melt index	0.8–2.5	2.0–4.0	2.0–5.0	0.7–1.5	0.8–1.5
density, g/cm^3	0.917–0.922	0.912–0.919	0.928–0.935	0.924–0.928	0.917–0.923
molecular weight distribution	narrow	narrow	narrow	narrow	narrow
process	blown	cast	cast	blown	blown
applications	general-purpose packaging	stretch wrap	bread and bakery; overwrap	grocery sack	blending; ice bags
critical properties	extrudability; toughness	puncture and tear resistance	stiffness; moisture barrier	stiffness; tear resistance	excellent toughness; low temperature properties

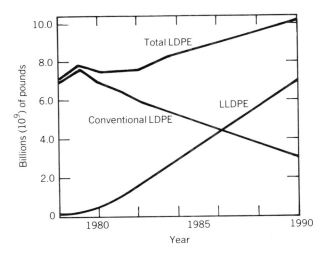

Figure 2. LDPE polyethylene sales and use. To convert 10^9 lb to 10^6 t, multiply by 0.4536.

the total polyethylene market should be low, probably less than 3%/yr through 1990 (11). This low growth reflects not only the mature state of the market, but also the practice of downgauging, ie, using less polyethylene in the final product. Downgauging has become very widespread because of the superior properties of LLDPE in film and molded products. The overall low growth rate reflects a decreasing market for LDPE and a healthy growth rate for LLDPE. By far the largest domestic application for LDPE and LLDPE is in film and sheet. Nearly 65% of all low density polyethylene is used in these applications (8). Other markets, listed in order of size, are injection molding, extrusion coating, wire and cable, and blow molding.

World. The total world demand for LDPE and LLDPE is predicted to be 36.8×10^9 lb (16.7×10^6 t) in 1990 (12). Based on nameplate capacity and announced expansions, the supply of LDPE and LLDPE in 1990 is projected at 43×10^9 lb (19.5×10^6 t). The rest of the world will lag behind the United States

in incorporating LLDPE into the market. The penetration of LLDPE into all LDPE applications in the United States is expected to be about 45% in 1990. In contrast, the penetration in Western Europe will only be ca 24%, and in Japan, 35%. (12). This is due in part to market characteristics and conditions and also to the fact that LLDPE has been selling at a 15–20% premium outside the United States (13,14).

Applications of LDPE and LLDPE

Film and sheet. About 65% of all the low density polyethylene in the world today is used for film, and a relatively small amount for sheet (see Extrusion; Films, plastic). These applications include garbage bags, grocery sacks, garment bags, heavy-duty bags, shrink film, stretch wrap, carrier film, pond liners, construction and agricultural film, and food packaging. Polyethylene film resins range in melt index from 0.2 to 6 and in density from 0.915 to 0.935 g/cm^3. The advantages of LDPE/LLDPE include low cost, flexibility, toughness, chemical resistance, and moisture-barrier properties (see Tables 1–3).

This has been the market most rapidly penetrated by LLDPE in the United States. In 1984, 74% of the LLDPE used, ie, 1.7×10^9 lb (7.71×10^5 t), was in the film area (9). This rapid penetration is due, in part, to the outstanding physical properties of LLDPE film (see Table 4). Dart-drop and Elmendorf tear tests measure the toughness of film in a high speed test and give an indication of the failure behavior of film under catastrophic conditions. Puncture and tensile tests are done at relatively slow speeds and give an indication of intrinsic properties and failure modes that can occur with long-term use.

The differences in molecular structure between LLDPE and LDPE (linear versus branched molecules and molecular weight distribution) affect the rheology of the two materials. LLDPE is more viscous at extrusion shear rates and requires more power to extrude. In addition, it is necessary to use a wide die gap to avoid melt fracture when extruding 100% LLDPE. Therefore, in order to extrude LLDPE on extruders designed for HP-LDPE and obtain optimum film properties, some relatively minor modifications to the screw and die gap

Table 4. Blown Film Properties^a of LLDPE and HP-LDPE

	ASTM test		HP-LDPE	HP-LDPE	LLDPE	LLDPE	LLDPE
melt index			2.5	0.2	1.0	1.0	1.0
density, g/cm^3			0.921	0.923	0.918	0.918	0.920
comonomer			none	none	butene	hexene	octene
Dart drop, gf/mil (N/mm)	D 1709		75 (29)	185 (71.4)	100 (38.6)	200 (77.2)	250 (96.5)
puncture energy, in.·lbf/mil (kJ/m)			6 (26.7)	5 (22.2)	16 (71.2)	17 (75.6)	
Elmendorf tear, gf/mil (N/mm)	D 1922	MD	160 (61.8)	90 (34.7)	140 (54)	340 (131.2)	370 (142.8)
		XD	110 (42.5)	100 (38.6)	340 (131.2)	585 (225.8)	800 (308.8)
tensile strength, psi (MPa)	D 882	MD	2900 (20)	2800 (19.3)	5000 (34.5)	5200 (35.9)	6500 (44.8)
		XD	2700 (18.6)	3000 (20.7)	3800 (26.2)	4700 (32.4)	5100 (35.2)
tensile impact strength, ft·lbf/in. (MPa)		MD	440 (36.4)	500 (41.4)	1200 (99.3)	1930 (159.7)	
		XD	650 (53.8)	1050 (86.9)	900 (74.5)	1760 (145.6)	

^{a} All properties measured on 1.5-mil (38-μm) film produced at 2:1 blow-up ratio.

Table 5. HP-LDPE and LLDPE Molding Resins

type	HP-LDPE	HP-LDPE	HP-LDPE	LLDPE	LLDPE	LLDPE
melt index	6–10	15–25	35–50	0.8–1.2	12–30	50–100
density, g/cm³	0.924–0.926	0.914–0.918	0.923–0.925	0.918–0.922	0.920–0.926	0.926–0.935
molecular weight distribution	broad	narrow	narrow	broad	narrow	narrow
process[a]	IM	IM, BM, PE	IM	BM, IM, PE	IM	IM
applications	bottle closures	drug bottles; aseptic packaging	lids	drum liners; irrigation tubing; spouts	industrial containers; specialty housewares	lids; housewares
critical properties	ESCR[b]; stiffness	ESCR[b]; low modulus	cycle time	ESCR[b]; processibility	ESCR[b]; low temperature impact	cycle time

[a] Key: IM = injection molding; BM = blow molding; PE = profile extrusion.
[b] ESCR = environmental stress crack resistance.

should be made (15). These modifications may not be necessary in the future because equipment manufacturers are designing a new generation of film extruders to handle LLDPE. In addition, there are additives that allow the extrusion of LLDPE through narrow die gaps without melt fracture (16). It is also very likely that new-generation LLDPE resins will be developed that can be extruded through conventional HP-LDPE extruders without modification.

Injection molding. This market includes lids, buckets, wash basins, housewares, toys, freezer containers, and general housewares (see Injection molding). The advantages of low density polyethylene include good low temperature properties, low cost, light weight, and flexibility (see Table 5). Low density polyethylene in the 2–100 melt-index range and the 0.920–0.930-g/cm³ density range is commonly used in injection-molding applications. LLDPE has made significant penetration into markets such as lids and housewares, where property advantages such as higher stiffness, improved ESCR, and heat-distortion resistance, can be coupled with higher melt indexes to result in faster cycle times and downgauged articles (see Table 6) (17). The 1984 consumption of LLDPE for injection molding in the United States is estimated at 330×10^6 lb (149,700 t) (9).

Extrusion coating. In this application, polyethylene is used as a coating on another material, eg, paper, aluminum foil, or paperboard (see Extrusion coating; Multilayer flexible packaging). These coated products are used for packaging liquids such as milk and juice and for the new aseptic packaging of nonrefrigerated juices (see Aseptic packaging). The polyethylene serves as an adhesive, moisture barrier, seal layer, printable surface, or barrier to tear. Because of its molecular structure, ie, the type of long-chain branching, polyethylene made in the high-pressure autoclave reactor is the most successful in this market. Resins with melt indexes of 4–10 and densities of 0.920–0.930 g/cm³ are commonly used. Due to its linear structure, LLDPE has not penetrated the extrusion-coating market, except as blends with LDPE.

Blow molding. High density polyethylene is the preferred material for this market because of its rigidity and barrier properties (see Blow molding). Low density polyethylene is used in those segments where flexibility and excellent ESCR are required, such as squeeze bottles, toys, and drum liners. Fractional-melt-index polyethylene, ie, 0.920–0.935 g/cm³, is typically used in this market. LLDPE offers better stiffness and ESCR compared with LDPE.

Safety and Health

Polyethylene is generally recognized as a safe packaging material by the FDA. Resin suppliers will state which of their resins comply with regulations governing polyethylenes used in food-contact applications. These regulations are covered in the U.S. Food, Drug, and Cosmetic Act as amended under Food Additive Regulation 21 (CFR 177.1520). Polyethylene is one of the most inert polymers and constitutes no hazard in normal

Table 6. Comparison of LLDPE and HP-LDPE Molded Properties

	LLDPE	LLDPE	LLDPE	HP-LDPE	HP-LDPE
melt index	20	50	100	23	49
density, g/cm³	0.924	0.926	0.931	0.924	0.924
dishpan impact, ft·lbf (J) at −20°C	30 (40.7)	24 (32.5)	3 (4.1)	9 (12.2)	3 (4.1)
failure mode	ductile	ductile	ductile	shatter	shatter
ESCR[a] (F_{50} h)[b]	150	3	<2	<1[c]	

[a] Environmental stress crack resistance.
[b] At 50°C, 100% Igepal, no slit, ASTM D 1693.
[c] F_{100}h.

handling. Resin suppliers will provide Material Safety Data Sheets for polyethylene resins on request.

BIBLIOGRAPHY

1. *Plast. World.* **43**, 92 (Jan. 1985).
2. *Chem. Week.* **135**, 66 (Sept. 19, 1984) *Plast. World.* **42**, 8 (Oct. 1984).
3. Brit. Pat. 471,590 (Sept. 6, 1937) and U.S. Pat. 2,153,553 (Apr. 11, 1939), E. W. Fawcett and co-workers (to ICI, Ltd.).
4. A. Renfrew and P. Morgan, eds., *Polyethylene*, Interscience Publishers, Inc., New York, 1960, Chapt. 2
5. R. A. V. Raff and J. B. Allison, *Polyethylene*, Interscience Publishers, Inc., New York, 1956, Chapt. 3
6. J. J. Fox and A. E. Martin, *Proc. R. Soc. London Ser. A* **175** 211, 216 (1940).
7. U.S. Pat. 4,076,698 (Feb. 28, 1978), A. W. Anderson (to E. I. du Pont de Nemours & Co., Inc.).
8. *Chem. Eng. News* **62**, 14 (June 25, 1984).
9. *Plast. Worlds* **42**, 108 (Oct. 1984).
10. *Ind. Week* **213**, 62 (May 3, 1982).
11. *Monthly Petrochemical Analysis,* Chemical Data Systems, Inc., Oxford, Pa., July 1984, p. 41.
12. *World Polyolefin Industry, 1982–83*, Vol. 1, Chem Systems, Inc., Tarrytown, N.Y., June 1983, pp. I-10 and I-13.
13. *Eur. Plast. News* **9**, 13 (Dec. 1982).
14. *Mod. Plast. Int.* **10**, 6, 8, (Oct. 1982).
15. S. J. Kurtz and L. S. Scarola, *Plast. Eng.* **36** (6), 45 (June 1982).
16. *Plast. World* **41**, 52 (Dec. 1983).
17. *Plast. World* **42**, 42 (Mar. 1984); **41**, 37 (Mar. 1983).

General References

J. H. DuBois and F. W. John, *Plastics*, Van Nostrand Reinhold Co., New York, 1981. Compares properties of thermoplastics and thermosets.

T. O. J. Kresser, *Polyethylene*, Reinhold Publishing Corp., New York, 1961. Contains general information on polyethylene applications and a table of the comparative properties of transparent packaging films of different plastics.

R. A. V. Raff and J. B. Allison, *Polyethylene* (1956); A. Renfrew and P. Morgan, eds., *Polyethylene* (1960), both by Interscience Publishers, Inc., New York. Both books contain background information on polyethylene manufacturing and characterization and are good sources for basic molecular descriptions of polyethylene.

World Polyolefin Industry, 1982–83, Chem Systems, Inc., Tarrytown, N.Y., June 1983. Describes the economics of the polyethylene industry and contains marketing forecasts of supply and demand.

Polyolefins Through the 80s—A Time of Change, SRI International, Menlo Park, Calif., Aug. 1983. A four-volume study of the marketing and production aspects of the polyethylene industry.

N. J. Maraschin
Union Carbide Corporation

POLYMER PROPERTIES

Polymers are high molecular weight organic materials produced by combining highly purified simple molecules under controlled heat and pressure, frequently in the presence of catalyst, accelerator, or promoter chemicals. The initial or-

ganic building blocks for polymers are 10–50 times more expensive than ores and silica, the raw materials for metals and glasses. The profound impact of polymers on the packaging industry is due to intelligent application of modern chemistry and engineering principles that has utilized both the versatility and vast array of inherent polymer properties and high speed, low energy automated forming techniques. The result has been the development of cost-effective packaging systems that compete well with conventional packages of glass and metal. There are more than 20 principal classes of polymers today, with new subclasses being created by ingeneous combinations of polymers, additives, fillers, etc. Development of desired properties for a specific packaging application depends upon a good understanding of comparative molecular architectures of candidate polymers. This cursory overview points out some polymer features that influence processing properties and part performance.

THERMOSETTING POLYMERS

There are two general categories of polymers: thermosetting and thermoplastic. The most important types of thermosetting polymers are intrinsically cross-linked resins such as phenol–formaldehyde, epoxies, polyurethanes, and unsaturated polyester resins. In packaging, phenolics are used for closures (see Closures, bottles and jar) epoxies are used as can coatings (see Cans, steel), and polyurethanes are used as cushion and insulation foams (see Foam cushioning; Insulation). Unsaturated polyesters in the form of bulk-molding compounds (BMC) are used for disposable "ovenable" trays and in the form of sheet-molding compounds (SMC) as returnable shipping containers. These materials attain their final form by reaction of relatively simple chemically unsaturated molecules. Unsaturation occurs as isolated regularly spaced double bonds along the carbon–carbon backbone, as $C=C$. A three-dimensional structure is formed by opening the double bonds with catalysts, and, in some cases, heat. The product is a network polymer that takes a permanent shape. In general, thermosets cannot be reused or returned to their original forms.

THERMOPLASTIC POLYMERS

More than two-thirds of all polymers used in the world today are thermoplastic polymers. In their final form, thermoplastics are characterized by extremely long molecules with saturated carbon–carbon backbones, as $-C-C-$. Unlike thermosets, thermoplastics have thermal and chemical stability at typical processing temperatures. As a result, most thermoplastics can be softened or melted, formed into useful articles, and then softened or melted again for reuse. Consider polyethylene, the simplest of thermoplastic polymers. If an ethylene gas molecule ($CH_2=CH_2$) were enlarged 100 million (10^8) times, each $-CH_2-$ unit would be about $3/8$ in. (9.5 mm) long. A single ethylene unit ($-CH_2CH_2-$) in a polyethylene molecule is called a repeat unit. An olefin oil or grease would have molecules about 3 ft (91 cm) long, if the chains could be fully extended. Low density polyethylene (LDPE), commonly used as film wrap, would have 1,000 repeat units and would be about 60 ft (18 m) long. Ultrahigh molecular weight (UHMW) polyethylene, used for friction-and-wear surfaces, would have

Table 1. Comparative Sizes of Polymer Molecules[a]

Material	End-to-end distance, nm	Number-average chain length	Model length of —CH$_2$—groups of 3/8-in. (9.5-mm) dia
phenolic	4.01	8	8.5 in. (21.6 cm)
melamine	3.54	5	7.5 in. (19 cm)
alkyd, unsaturated polyester resin	9.44	19	20 in. (51 cm)
epoxy adhesive	1.59	1	3.3 in. (8.4 cm)
epoxy resin, medium MW	11.09	6.5	24 in. (61 cm)
epoxy resin, high MW	58.06	34	10 ft (3.0 m)
olefin grease	20.0	67	50 in. (127 cm)
LDPE	200	670	42 ft (12.8 m)
polyethylene, UHMW[b]	9,100	30,000	0.4 mi (0.64 km)

[a] Fully extended chains, scaled 100,000,000 : 1 ($10^8 : 1$).

100,000 repeat units and would be about 1 mi (1.6 km) long. In contrast, thermosetting molecules, prior to reaction into three-dimensional networks, are relatively short, bulky molecules that are typically only 10–20 units long. On the same 100 million-to-one scale, phenol–formaldehyde is about 1 in. (25 mm) long and nearly 1 in. (25 mm) in diameter. Other comparative molecular sizes are given in Table 1.

As might be expected, polymeric molecules can become twisted, coiled, folded, or entangled. Resistance to shearing forces is affected by the relative length of the molecule, the flexibility of the main polymer chain, and the relative bulk of any side groups, either as pendant groups or branches. Prior to reaction, typical thermoset resin viscosities are quite low, ie, 100–10,000 ($10^2–10^4$) times that of water. As the three-dimensional networks form, viscosities climb rapidly until the materials gel or form solids. Thermoplastic-resin viscosities are essentially time-independent at processing temperatures. The very long molecular chains, and, in some cases, bulky pendant groups and side chains, cause considerable interference and entanglements as molecules slide over one another during flow. As a result, thermoplastics usually have viscosities $10^5–10^{10}$ times that of water.

Certain thermoplastics such as polyethylene can be cross-linked by chemical (peroxide) or nuclear means (gamma radiation) to produce three-dimensional structures. The primary action is to remove an atom such as hydrogen from the saturated carbon–carbon backbone, thus creating a reactive site. This site can then react with another molecule. Although the number of cross-links per backbone carbon tends to be small, ie, 0.1–10 per thousand, high temperature properties, such as creep resistance, are obtained. The reprocessing of cross-linked thermoplastics usually involves mechanical destruction of some carbon–carbon bonds.

Addition and condensation polymerization. There are two general types of materials within the general category of thermoplastics, as distinguished by the methods of formation from monomers. Addition polymers are usually formed by continuous reaction of an unsaturated carbon–carbon bond monomer, eg, ethylene, to the active end of a preexisting polymer chain. The largest class of addition polymers is based on the generic vinyl structure $R_1R_2C=CR_3R_4$ and includes commodity polymers such as PVC, PP, PS, and PE (see Table 2).

Condensation polymers are usually produced by reaction of one or two (or more) saturated monomers with reactive end groups, ie, hydroxyl, carboxyl, amine. The polymerization reaction is characterized as an equilibrium reaction. A small molecule such as water is liberated as a by-product and as long as it continues to be physically removed, polymerization continues. Nylons, thermoplastic (saturated) polyesters, and polycarbonates are condensation polymers. These and others are summarized in Table 3.

Aliphatic and aromatic polymers. A further way of classifying thermoplastics focuses on the polymer's degree of unsaturation. Polymers such as PE have no unsaturation, with a simple C–C backbone, and are identified as aliphatic polymers. Polymers such as PS have pendant benzene groups distributed along a simple C–C backbone. These are the simplest kind of aromatic polymers. A higher order aromatic polymer exists if the benzene ring is included in the polymer backbone, as is the case with PET. Final properties such as rigidity and chemical and thermal stability strongly relate to the degree of aromaticity in the polymer (see Table 4).

Definition of Molecular Weight

The molecular weight of any polymer molecule is obtained by multiplying the molecular weight of the repeat unit by the number of times the monomer is repeated, then adding the end-group molecular weights. For a polyethylene, the molecular weight of the repeat unit $+CH_2CH_2+$ is 28. The molecular weight of the polymer (having 10,000 units) is thus 280,000. Unfortunately, the polymerization process cannot be controlled precisely enough to produce a polymer with molecular chain lengths that are all the same. All commercial materials, therefore, are made of polymeric chains of various lengths. One way of characterizing the average polymer chain length is to calculate the total weight of all chains and divide by the total number of chains:

$$\overline{M}_n = w/n = \Sigma\, N_i M_i / \Sigma\, N_i$$

This is the number-average molecular weight, which is the value implied in this article by the term molecular weight (MW). The weight-average molecular weight is obtained by multiplying the weight of a chain of a given number of repeat

Table 2. Chemical Structure of Vinyl-type Thermoplastics

$$\left(\begin{array}{cc} R_1 & R_3 \\ | & | \\ C{-}C \\ | & | \\ R_2 & R_4 \end{array}\right)_n$$

Common name	Abbreviation	R₁	R₂	R₃	R₄	Subspecies	Glass-transition temp (T_g), °F (°C)	Crystalline melt temp (T_m), °F (°C)	Melt temp of pure crystal polymer[a], °F (°C)
polyethylene	PE	H	H	H	H	LDPE, branched	−166 (−110)	234 (112)	
						LLDPE, linear low	−166 (−110)	253 (123)	
						HDPE, linear	−166 (−110)	273 (134)	279 (137)
polypropylene	PP	H	H	H	CH₃	atactic	5 (−15)	A[b]	
						syndiotactic	41 (5)	329 (165)	338 (170)
polyisobutylene		H	H	CH₃	CH₃		−85 (−65)	113 (45)	262 (128)
polybutene		H	H	H	CH₂CH₃		−94 (−70)	A[b]	
polybutadiene (divinyl)		H	H	H	HC=CH₂		−67 (−55)	A[b]	
poly(vinyl chloride)	PVC	H	H	H	Cl		194 (90)	A[b]	414 (212)
poly(vinyl fluoride)	PVF	H	H	H	F		−4 (−20)	A[b]	
poly(vinyl dichloride)	PVDC	H	H	Cl	Cl		1.4 (−17)	A[b]	
poly(vinylidene fluoride)	PVF₂	H	H	F	F		−31 (−35)	A[b]	
polytetrafluoroethylene	PTFE	F	F	F	F		257 (125)	619 (326)	
polystyrene	PS	H	H	H	C₆H₅		201 (94)	A[b,c]	
poly(vinyl alcohol)	PVOH	H	H	H	OH		185 (85)	A[b]	
poly(vinyl acetate)	PVAc	H	H	H	$\overset{\text{O}}{\overset{\|}{\text{OCCH}_3}}$		86 (30)	A[b]	
poly(methyl methacrylate)	PMMA	H	H	CH₃	$\overset{\text{O}}{\overset{\|}{\text{COCH}_3}}$		212 (100)	A[b]	
polyacrylonitrile	PAN	H	H	H	CN		219 (104)	527 (275)[d]	621 (327)

[a] Ref. 1.
[b] A = commercially amorphous polymer.
[c] Isotactic melting point = 464°F (240°C).
[d] As a highly oriented fiber.

units by the number of such chains and then dividing by the total weight of the chains:

$$\overline{M}_w = nw/w = \Sigma\, N_i \cdot N_i M_i / \Sigma\, N_i M_i$$

The number-average and weight-average molecular weights, and their ratio, $\overline{M}_n : \overline{M}_w$, known as the dispersity index, help define the characteristic chain length or molecular weight distribution (MWD) of a polymer.

Molecular weight and properties. For a given polymer, low molecular weight permits easier processing, and high molecular weight provides better finished-part properties. The MWD of some polymers, eg, PE and PP, can be substantially altered by polymerization or controlled depolymerization. Broad-MWD polymers, characterized by viscosities that are shear-sensitive over wide processing ranges, are used in extrusion coating (see Extrusion coating) and heat-sealing (see Sealing, heat), which require adequate melt strength over a wide temperature range. Narrow-MWD polymers are used for highly oriented film, which requires high melt strength. Generalizations are difficult, however, owing to immense chain lengths, energy interactions on a molecular level (eg, hydrogen-bonding forces), chain entanglements, pendant groups, and side-chain branching. Many polymers exhibit high elastic strength that imparts such phenomena as melt stability during fiber processing, characteristic radial swelling, and parison wall-thickness control in blow molding.

Morphology and Properties

Polymer morphology or structure dramatically affects processibility and product performance. It is difficult to envision the high degree of order needed to form crystals in these long-chain, highly entangled polymers. Yet, certain polymers, eg, polyolefins, PET, and nylons, crystallize readily upon cooling from the melt state and are called *crystalline* polymers. These materials have rather well-defined melting points. Polymers with many bulky side groups or very stiff backbones, eg, PMMA, PS, PVC, do not crystallize, and so are called *amorphous*. Polymers that are 100% crystalline are rare outside the laboratory. Most crystallite regions in commercial crystalline polymers are mixtures of spherulitic and dendritic crystals in amorphous regions. Crystalline architecture is best observed with conventional x-ray crystallography, although crystalline levels much below 20% are difficult to interpret. Techniques have been devised for determining the extent and direction of orientation in crystalline materials. The crystalline concentration is determined by the intrinsic nature of the polymer

Table 3. Chemical Structures of Typical Condensation Thermoplastics

Common name	Abbreviation	Structure	Glass-transition temp (T_g), °F (°C)	Crystalline melt temp (T_m), °F (°C)	Melting point of pure crystal polymer, °F (°C)
poly(ethylene terephthalate)	PET	$\left(CH_2\right)_2OC-\!\!\bigcirc\!\!-CO\!\!\left.\right)_n$	158 (70)	500 (260)	513 (267)
nylon-6,6	PA-6,6	$\left(N\left(CH_2\right)_6NC\left(CH_2\right)_4C\right)_n$	122 (50)[a]	464 (240)	509 (265)
nylon-6	PA-6	$\left(N\left(CH_2\right)_5C\right)_n$	122 (50)[a]	410 (210)	491 (255)
polycarbonate	PC	$\left(O-\!\!\bigcirc\!\!-\overset{CH_3}{\underset{CH_3}{C}}-\!\!\bigcirc\!\!-OC\right)_n$	302 (150)	A[b]	
polyacetal (poly(oxymethylene))	POM	$\left(\overset{H}{\underset{H}{C}}O\right)_n$	−76 (−60)	356 (180)	
cellulose[c]		(ring structure with CH$_2$—R and R substituents)			
cellulose, R = OH			104 (40)	[d]	
cellulose nitrate, R = NO$_3$			127 (53)		
cellulose triacetate, R = OCCH$_3$ (=O)			158 (70)		581 (305)
cellulose tributyrate, R = OCC$_3$H$_7$ (=O)			248 (120)		365 (185)

[a] Dry; T_g increases when water is absorbed.
[b] A = commercial amorphous polymer.
[c] Natural polymer.
[d] Infusible; degrades before melting.

and the way in which it is processed. High molecular weight, narrow MWD, and linearity in the polymer backbone can yield high crystallinity. Small amounts of nucleants or impurities such as catalyst, filler, or pigment, can enhance the crystallization rate. High shear during processing and rapid cooling inhibit crystallinity, especially spherulitic crystallinity; annealing and orientation enhance it.

For example, highly long-chain-branched LDPE has low crystallinity and a broad melting point (see Polyethylene, low density). LDPE blown films have very low haze and high gloss. HDPE chains are nearly devoid of branching and so can be crystallized to more than 50% (see Polyethylene, high density). HDPE resins have rather sharp melting points, but high MW HDPE films tend to have observable haze and can be semiglossy. Deliberate production of very small spherulites can yield improved clarity and surface in HDPE films (see Film, high density polyethylene).

Amorphous, ie, noncrystalline, polymers have no melting point. They simply soften when heated, in much the same way as glass. The glass-transition temperature (T_g) is the point at which a polymer changes from a brittle, glasslike material to a rubbery one. This temperature defines the energy level where substantial chain-segment mobility can take place along the polymer backbone. Under load, permanent chain motion is possible, as evidenced by gross material deformation. As processing temperatures increase above T_g, amorphous polymers become increasingly easier to process. Crystalline polymers also have T_gs (see Tables 2 and 3), but are restricted in the extent of deformation until the crystalline melt temperature is reached. As a result, crystalline polymers retain much of their integrity under load until just before the melting temperature is reached (see Fig. 1). Crystalline polymers are not ideal for many commercial processes, eg, blow molding, stretch blow molding, thermoforming, foam-sheet extrusion, that require

Table 4. Examples of Aliphatic and Aromatic Polymers

Species	Structure		Glass-transition temp (T_g), °F (°C)	Flexural modulus, 10^3 psi (MPa)	Deflection temp under 66-psi (455-kPa) load, °F (°C)
fully aliphatic polymer, eg, PE	$+CH_2CH_2CH_2CH_2 +_n$	LDPE	−166 (−110)	40 (276)	109 (43)
		HDPE	−166 (−110)	200 (1379)	190 (88)
aliphatic main chain with aromatic pendant groups, eg, PET	$+CH_2CH_2OC$—⬡—$CO+_n$		203 (95)	400 (2759)	180 (82)
polymer with aromaticity in main chain, eg, PS	$+CH_2CHCH_2CH+_n$		158 (70)	400 (2759)	109 (43)
fully aromatic polymer, eg, polydiphenyl[a]			590 (310)[b]	450 (3103)	680 (360)
ladder-type aromatic polymer, eg, polyetherimide			419 (215)	480 (3310)	410 (210)

[a] Not commercial.
[b] Estimated.

polymers to be easily deformed over wide temperature ranges. The ratio of melt temperature to glass-transition temperature for crystalline polymers is 1.4–2.0 (in K).

Orientation and Properties

Useful properties can be obtained from some polymers by stretching the materials above T_g. If a high performance tape is required, the polymer is stretched uniaxially. If a tough and/or transparent film is needed, the polymer is stretched biaxially. Orientation can be achieved for both crystalline and amorphous polymers. Unique combinations of properties can be achieved with crystalline polymers by carefully matching levels of mechanical stress with heating and cooling rates. Crystallites thus formed can be highly oriented to yield dramatic increases in ultimate tensile strength, albeit at reduction in elongation at break. Moderate amounts of uni- and biaxial orientation of amorphous polymers that are inherently brittle, eg, PS or PMMA, yield films with substantially increased ultimate elongation and ductility.

Chain Mobility and Stiffness

Although the intrinsic strength of the polymeric C–C backbone is extremely high, products made of polymers never achieve this strength. Instead, macroscopic mechanical strength derives from the ability of the polymer chains to resist deformation or disentanglement while under continuous load or to absorb energy without backbone cleavage or chain-to-chain separation while under impulse load. The rigidity, hardness, ductility, and impact resistance of a plastic product is related to the nature of the polymeric molecular structure. Certainly, macroscopic polymer flexibility can be directly related to the degree of freedom for rotation about C–C bonds

that make up the backbone of the polymer. Polymer stiffness is increased if double bonds are present in the main chain. Further increase in stiffness occurs if there is conjugation, ie, if every other main chain bond is unsaturated (—C=C—C=C—). Aromaticity in the backbone causes additional stiffness, particularly if benzene rings alternate with adjacent aliphatic carbons, as with PET and PC. If there are no aliphatic carbon bonds, as with polycyclic diphenyls, the polymer backbone is further stiffened. Some of the stiffest poly-

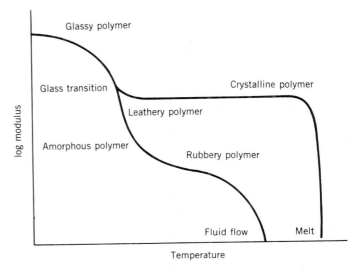

Figure 1. Stiffness profiles of amorphous and crystalline polymers with identical T_gs (1).

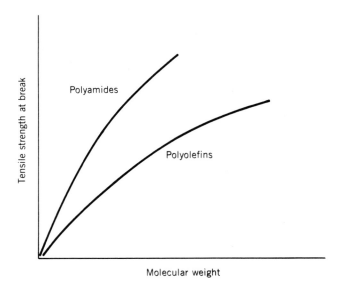

Figure 2. Effect of hydrogen bonding on tensile strength (2).

mers are the polyimides, in which backbone bonding occurs at four rather than two points on the aromatic ring, thus forming ladderlike structures.

Stiffening can occur through steric hindrance of side groups to the main chain. Not all pendent groups cause stiffening, however. Consider long-chain branching on LDPE. The apparent effect of these side groups is to separate main chains, causing high free volume in the solid, as manifested by lower density. Increased branching provides greater flexibility, lower strength, and lower T_g and T_m. The development of LLDPE is predicated on evidence that low levels of short-chain branching can increase stiffness, particularly in films. PP is a limiting case of regular short-chain branching, with methyl side chains occurring on every other backbone carbon. This steric hindrance forces the polymeric chain into a helix. PP T_g and T_m values are therefore higher than those for other polyolefins. Because benzene rings as pendent groups provide even greater steric hindrance and greater polymer stiffening, PS is a brittle, rigid, amorphous polymer. Although not a pendent group per se, the chlorine atom in PVC is much larger than a hydrogen atom. As a result, its substantial steric hindrance significantly stiffens the polymer. Another factor is important here as well. Chlorine and fluorine atoms are highly electronegative, and so tend to repel one another. This repulsion helps stiffen the polymer backbone into a rodlike configuration. As a result, all fluorine- and chlorine-substituted polymers are very difficult to process without flow aids, ie, plasticizers. Intrinsically high hydrogen-bonding levels yield stiffer polymers. The secondary hydrogen bonds to such main-chain groups as amines (NH) and hydroxyls (OH) make the structural unit appear larger and more rodlike, and thus better able to resist mechanical deformation. This is seen in Figure 2, where at the same molecular weight, polyamides are in general stiffer than olefins.

Stress-crack Resistance

The ability of a polymer to withstand an aggressive medium under load is known as environmental stress-crack resis-

Table 5. Extent of Short-chain Branching— Decreases With Increasing Density for Polyethylene[a]

Melt index, g/10 min	Density at 73°F (23°C), g/cm³	Bent strip cracking time, F_{50}, h
0.60	0.9218	>1000
0.76	0.9320	7
0.71	0.9415	>1000
0.61	0.9472	130
0.52	0.9492	200
0.74	0.9502	12
0.69	0.9573	72
0.68	0.9613	12

[a] Materials tested in Igepal according to ASTM D 1693 (3). F_{50} = time when 50% of samples fail.

tance (ESCR). Strong solvents dissolve or swell polymers when the polymer–polymer intermolecular attraction forces are less than the polymer–solvent attraction forces. The solvent molecules thus move between and separate the adjacent polymer main chains. The classicial example is PS and toluene. When a fluid swells or dissolves the polymer only when the polymer chains are under strain, the solvent is said to be weak. Typically, increased ESCR is obtained by increased polymer MW. This dramatic effect is seen when PE is exposed to Igepal, ie, a classic stress-cracking agent (see Tables 5 and 6). With polyethylenes, the degree of crystallinity is another contributing factor to improved ESCR. For long-term, low level strain, ESCR is improved with decreasing crystallinity. For constant applied stress, ESCR is improved with increasing crystallinity. The effect of polyethylene molecular characteristics on stress-crack resistance is examined in Tables 5 and 6.

Water Absorption

Water absorption depends upon polar attraction forces and hydrogen-bonding energies. Condensation polymers produced by water by-product elimination, eg, nylon and PET, tend to have strong affinities for water. Water absorbed at temperatures substantially below normal processing temperature

Table 6. Melt Index—Decreases With Increasing Molecular Weight for Polyethylene[a]

Melt index, g/10 min	Intrinsic viscosity in xylene at 185°F (85°C)	Cracking time, F_{50}, h
0.08	1.35	>500
0.22	1.17	>500
0.35	1.04	>500
0.60	1.10	>500
1.30	0.980	>500
7.8	0.940	<0.5
16.0	0.895	<0.5

[a] Footnote to Table 5 applies here as well.

tends to plasticize the polymer, and that absorbed during processing tends to depolymerize the polymer. On the other hand, polymers with low or no hydrogen-bonding tendencies, eg, polyolefins, are excellent water barriers.

Gas Permeation

As with liquid solvents, gas transmission through polymers strongly depends on the relative orders of magnitude of polymer–polymer and polymer–gas molecule intermolecular forces (see Barrier polymers). Gases chemically similar to the polymeric repeat unit tend to permeate easily, whereas polymers tend to act as barriers to chemically dissimilar gases. Thus, cellulosics transmit water vapor, and LDPE does not; LDPE tends to transmit CO_2, and PVDC does not. Gas-molecule permeability is the product of the solubility of the gas in the plastic and its diffusion rate through the plastic. As noted above, the solubility directly relates to molecular affinity, but the diffusion rate can be controlled in several ways. In highly crystalline materials, the diffusional path is tortuous and so the diffusion rate is low. In partially crystalline materials, such as the polyolefins, an increasing degree of crystallinity directly decreases the permeability of all molecules. Orienting a crystalline polymer substantially increases the already long diffusional path, and so highly oriented polymer films of PET (see Film, oriented polyester) and nylon become efficient gas barriers.

Polymerization and Properties

To this point, emphasis has been on pure polymers made from a single set of monomers. Many polymers must be altered in some way to be processed in conventional equipment or to meet specific end uses. A common method of altering polymer processing properties and solid properties is by copolymerization. Small amounts of reactive monomers are coreacted with the primary polymeric material. By controlling the method of polymerization, the copolymer can be added randomly along the main chain, attached as pendent groups (branched copolymerization), or added as long-chain homopolymer blocks into the main chain (block copolymerization). If the comonomers are randomly copolymerized, the general effects are broadening of melt and glass-transition temperatures, increased flexibility and melt strength, and reduced melt viscosity and crystallinity. Examples include ethylene into PP to improve low temperature impact strength and sheet thermoformability and sodium methacrylate into PE to produce an ionomeric polymer (see Ionomers) with reduced crystallinity and improved transparency and toughness in thin films. Block copolymerization can provide main-chain flexibility in otherwise brittle polymers, as with butadiene in PS, for improved impact strength (see Styrene–butadiene copolymers; Polystyrene). The butadiene blocks are not soluble in PS and so form a separate but chemically linked phase. Many thermosets, such as polyurethane and reactive polyester, can be similarly flexibilized by using a long-chain unsaturated polymer as one of the reactants. An example is thermoset polyester resin with PET between the reactive maleic acid units.

Blends

It is not always necessary to coreact monomers to improve polymer properties. If two polymers are completely soluble in each other, intensive shear-melt mixing of a physical blend of resins yields a single phase. Resulting polymer blend properties are similar to those obtained by copolymerization. Typically, such physical blends should be sought between homologues in each polymer species, such as acrylates, vinyls, or cellulosics. Occasionally, however, unique new polymers are created by mixing nonhomologous but completely soluble polymers such as PVAc and PMMA, or PS and polyphenylene oxide. Insoluble blends yield macroscopic two-phase systems that behave in much the same way as their equivalent block copolymers. The classic example is the physical mixture of polybutadiene rubber and PS.

Plasticizers and Properties

Plasticizers are usually small molecules of a chemical nature similar to the polymer. Plasticizers dissolve or swell the polymer. The most apparent effect is main-chain separation, causing reduced polymer resistance to applied stress. Thus, plasticized polymers have lower processing viscosities; increased flexibility, tear strength, and ultimate elongation; and reduced stiffness, T_g and T_m (if any), and continuous-use temperatures. The extents of these effects greatly depend on the thermodynamic compatibility of polymer and plasticizer, the T_g of the plasticizer, and its concentration in the polymer. In general, the T_g region is broadest when the plasticizer is a poor solvent for the polymer. The classic example is PVC, which is intractable in the unplasticized state. When dioctyl phthalate (DOP), a poor solvent, is added to 40 vol %, T_g is reduced from 212°F (100°C) to about 41°F (5°C) and the peak is broadened from less than 20°F (−7°C) to about 55°F (13°C). Further, to ensure maintenance of long-term properties, plasticizers must have very low vapor pressures at room temperature and be nonmigrating.

Additives and Properties

Many other chemicals are added in small quantities to change neat polymer properties. Processing aids such as silicones and stearates are added to change film surface energies, to reduce blocking in polyethylenes, bacterial growth in vinyls and polyurethane foams, static charge in oriented PS films, or to make fillers more compatible with polymers (see Additives, plastics). Antioxidants are required in some polymers, eg, PP, to minimize yellowing during processing. Tints are dyes (see Colorants) that are used to change transparent polymer colors from natural to perceived "water-white," as in PET. Organic dyes color transparent plastics. Organic and inorganic pigments color opaque plastics. Titanium dioxide is used as a white opacifier. Carbon black is used as an opacifier and as a uv absorber, particularly in PVC and polyolefins. Blowing agents are extremely fine powdered ultrapure chemicals that decompose at processing temperatures to produce gases such as N_2 or CO_2. Materials such as sodium bicarbonate–citric

$$H_2NC\overset{\displaystyle O}{\overset{\|}{}}N{=}NC\overset{\displaystyle O}{\overset{\|}{}}NH_2$$

acid and azodicarbonamide ($H_2NCN{=}NCNH_2$) are used to foam plastics such as PS (see Foam, extruded polystyrene). Frequently, chemical foaming agents are used in combination with physical foaming agents such as fluorocarbons and hydrocarbons to produce low density foams for cushion packaging. The chemical agents act here as nucleating sites for bubbles.

Fillers and Properties

Common fillers are inexpensive inorganic minerals such as talc or calcium carbonate. In addition to acting as extenders when used to levels of 40 wt %, thus reducing overall resin cost, fillers increase polymer stiffness and continuous-use temperature by interfering with polymer-chain-segment mobility. In this same manner, however, fillers restrict bulk-chain straightening and flexing under load, leading to dramatically reduced ultimate elongation, tensile strength, and impact strength. Some property improvement can be gained by adding coupling agents, eg, titanates, to improve the interfacial bond between polymer and filler. Talc is used to 20 wt % in PP for improved part stiffness and to broaden the processing window in thermoforming. Calcium carbonate and talc are used in unsaturated polyester resin for stiffening bulk-molding compounds. Some fillers such as graphite, polytetrafluoroethylene (as very fine powder), or molybdenum disulfide are added to improve polymer part lubricity. Exceptional improvement in polymer stiffness is possible by adding inorganic reinforcing elements such as glass fibers or mica. Applications in packaging are restricted to nonfood-contact structural materials, such as SMC handling and shipping containers.

Multilayer Materials

Today, many polymers are physically combined in multiple layers of thin films or coatings to achieve barrier properties that could otherwise not be achieved by single polymers. There are some special concerns for polymers that comprise these multilayer structures. For example, additives must be nonmigratory to prevent interfacial delamination. During multilayer extrusion, polymer melt viscosities must be carefully matched (see Coextrusion, flat; Coextrusion, tubular; Multilayer plastic bottles). Mismatched viscosities can lead to interlayer thickness variation. Some control can be achieved by proper selection of polymer MW and MWD or through copolymerization, but care must be exercised to avoid compromising desired properties. Additional control is obtained by adjusting individual polymer temperatures prior to extrusion through the multilayer die block. Where needed, plasticizer concentrations must also be carefully controlled.

Care must be taken during biaxial orientation of multilayer sheet to minimize formation of microvoids in inherently weak films such as EVOH (see Ethylene–vinyl alcohol). The level and rate of crystallinity in polymers such as PET and nylon are much more difficult to control, particularly if high clarity is sought, quench cooling is restricted, and films are necessarily oriented to less then optimal levels.

BIBLIOGRAPHY

1. R. D. Deanin, *Polymer Structure, Properties and Applications,* Cahners Books, Denver, Colo., 1972.

2. N. M. Bikales, ed., *Mechanical Properties of Polymers; Encyclopedia Reprints,* Wiley-Interscience, New York, 1971.

3. J. B. Howard, "Fracture; Long-Term Phenomena," in N. M. Bikales, ed, *Encyclopedia of Polymer Technology,* Vol 7, Wiley-Interscience, New York, 1967, pp. 261–291.

General References

F. W. Billmeyer, Jr., *Textbook of Polymer Science,* 3rd ed., John Wiley & Sons, Inc., New York, 1984.

P. Meares, *Polymers: Structure and Bulk Properties,* Van Nostrand, New York, 1965.

M. L. Miller, *The Structure of Polymers,* Reinhold, New York, 1966.

L. E. Nielsen, *Mechanical Properties of Polymers,* Reinhold, New York, 1962.

A. V. Tobolsky, *Properties and Structure of Polymers,* John Wiley & Sons, Inc., New York, 1960.

S. L. Rosen, *Fundamental Principles of Polymeric Materials,* Wiley-Interscience, New York, 1982.

F. Rodriguez, *Principles of Polymer Science,* 2nd ed., McGraw-Hill, Inc., New York, 1981.

Z. Tadmor and C. G. Gogos, *Principles of Polymer Processing,* Wiley-Interscience, New York, 1979.

J. L. Throne, *Plastics Process Engineering,* Marcel Dekker, Inc., New York, 1979.

J. L. Throne
Consultant

POLYPROPYLENE

Polypropylene is an extremely versatile material in the packaging industry. The reason for its adaptability is the ease with which its polymer structure and additive packages can be tailored to meet diverse requirements. Many useful properties are inherent in polypropylene. It has low density (high yield), excellent chemical resistance, a relatively high melting point, good strength, at modest cost.

General Categories and Definitions

Polypropylene is the result of linking a large number (typically 1,000 to over 30,000) of propylene molecules to build long polymer chains (see Fig. 1). Polymers made up only of propylene are called homopolymers. If another monomer is added (typically ethylene), the polymer is called a copolymer. The order and regularity of the monomer units in the polymer control the properties of the product.

One end of a propylene molecule is different from the other. The end with three hydrogens is called the head; a head-to-tail linkage is called stereoregular. If the process links the monomers head-to-head about as often as they are linked head-to-tail, the resulting polymer has little order and does not crystallize (1). At room temperature, this material has a density of about 0.850 g/cm³. It is called atactic or amorphous polypropylene and is soft, tacky, and soluble in many solvents. Such polymers are useful in hot-melt adhesives (see Adhesives) and several other applications.

If monomers are connected head-to-tail almost every time, the polymer is said to be isotactic and crystallizes (see Fig. 2).

Figure 1.   Polypropylene

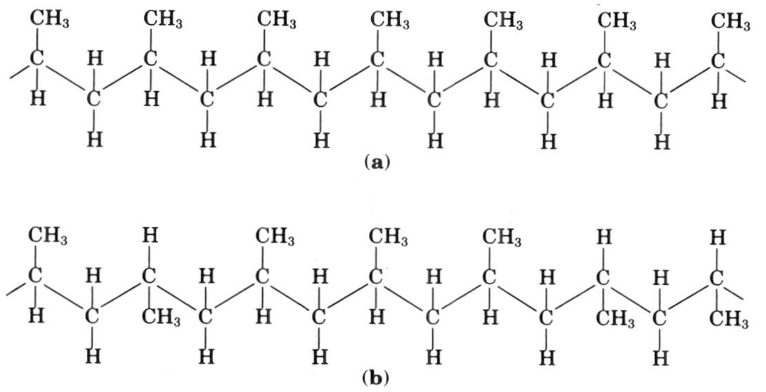

CH₃ CH₃ CH₃ CH₃ CH₃ CH₃ CH₃

(a)

CH₃ H CH₃ CH₃ CH₃ H H

(b)

Figure 2. Comparison of the structure of isotactic and atactic polypropylenes. (**a**) Isotactic PP–methyl groups in orderly alignment. (**b**) Atactic PP–methyl groups in random alignment.

Crystallinity is the reason for the solvent resistance, stiffness, and heat resistance of the commercial plastic material. At normal conditions, isotactic polypropylene is usually about half crystalline and has a density of 0.902 ± 0.005 g/cm³. Normal polypropylene melts at about 329°F (165°C) when heated slowly and contains roughly 5% of atactic material as well as intermediate structures. In this article, the term polypropylene (PP) implies the plastic of commerce.

When discussing copolymers and alloys, an additional level of complexity is added. Alloys, also called blends, are mixtures of polymers. Copolymers are the result of polymerizing two or more monomers together. There are many possible types (structures) of copolymers (2), not all of which are useful.

Random copolymers result when a small amount of comonomer, usually ethylene, is polymerized at random intervals along the PP chain. Typical ethylene levels are from 1 to 5 wt %, and the product has only one phase. These copolymers are relatively clear and have lower and broader melting points than PP homopolymers. The lowering of the melting point is proportional to the randomness of incorporation and the amount of comonomer incorporated. At ethylene levels well above 10%, the product becomes noncrystalline and is called EPR, or ethylene–propylene rubber. One cannot make an alloy (polymer blend) that resembles a random copolymer.

Impact copolymers generally contain larger amounts of ethylene monomer, typically 4 to >25%, and are heterophasic. The ethylene may be polymerized in the form of polyethylene and/or EPR. Usually, such products are made by polymerizing a homopolymer and changing the conditions to add ethylene to the polymer chain. Reasonably comparable materials can be made by mechanically blending polypropylene with EPR and/or polyethylene. If crystalline polyethylene is present, it can be detected by a second melting point at 244–271°F (118–133°C). These products are characterized by lower stiffness, much enhanced toughness at low temperatures, and a relatively opaque appearance (see Table 1).

The molecular weight (related to the average number of monomer units in a chain) and the molecular weight distribution (MWD) are significant polymer characteristics. High molecular weight (HMW) polymers are highly viscous when melted and are difficult to injection mold or push through restrictive dies; but they have high toughness and good "melt strength" (melt elasticity). Melts of low molecular weight (LMW) polymers are more fluid. They have less toughness and lower melt strength. In a narrow-MWD polymer, there is less variation in the length of the chains than in a broad-MWD polymer. Narrow MWD allows retention of the toughness of HMW grades (3) in a more-easily processed material. Narrow MWD materials do not generally have high melt

Table 1. Typical Polypropylene Properties

Property	Values			
	Homopolymer	Random copolymer	Impact copolymer	High impact copolymer
melt flow cond. L, g/10 min.	4	6	4	4
tensile strength (yield), psi (MPa)	5,050 (34.8)	4,000 (27.6)	3,900 (26.9)	2,900 (20)
yield elongation, %	11	14	13	7
flexural modulus, (1% secant), psi (MPa)	260,000 (1,793)	155,000 (1,069)	200,000 (1,379)	155,000 (1,069)
hardness Rockwell R (HRC)	100	83	80	70
heat deflection temp (66 psi or 455 kPa), °F (°C)	216 (102)	167 (75)	178 (81)	171 (77)
notched Izod, 23°C, ft-lbf/in. (J/m)	0.8 (43)	2 (107)	2.5 (133)	7 (374)
application	injection molding	injection molding film	injection molding	injection molding

Table 2. Typical Properties of Filled Copolymers

	Values			
Property	$CaCO_3$ 20%	$CaCO_3$ 40%	$CaCO_3$ 40%	Talc 40%
melt flow cond. L, g/10 min.	0.4	0.3	4.0	4.0
tensile strength (yield), psi (MPa)	3,800 (26.2)	3,600 (24.8)	3,000 (20.7)	3,700 (25.5)
yield elongation, %	15	14	6	4
flexural modulus, (1% secant), psi (MPa)	220,000 (1,517)	310,000 (2,138)	325,000 (2,241)	420,000 (2,897)
hardness Rockwell R (HRR)	78	83	87	85
heat deflection temp (66 psi or 455 kPa), °F (°C)	201 (94)	237 (114)	220 (104)	260 (127)
notched Izod, 23°C, ft-lbf/in. (J/m)	3.3 (176)	2.8 (149)	0.8 (43)	0.6 (32)
density, g/cm³	1.04	1.22	1.22	1.23
application	sheet extrusion thermoforming	sheet extrusion thermoforming	injection molding	injection molding

strength. They offer advantages in fiber spinning and injection molding.

Substantial modification in properties can be achieved by the use of additives (see Additives, plastics) and fillers. Virtually all commercial grades are stabilized to increase resistance to oxidation on aging or at elevated temperatures. Additives can confer resistance to sunlight, reduce the tendency to retain static electric charges, modify the coefficient of friction, and prevent surface tackiness.

The additives should be selected for the application. In food packaging, the stabilization should be chosen to avoid transfer of taste and odor to the package contents. Antistats reduce static electricity, which attracts dust and makes packages appear dirty. In sunlight, the damaging wavelengths are in the uv range. Uv resistance is important for items that may be stored outdoors. Coefficient of friction and antiblocking properties are important in the winding and handling of film (see Slitting/rewinding). Frictional characteristics are also important in threaded closures (see Closures).

Fillers can greatly increase stiffness, improve processing behavior, confer conductivity, and change the appearance of PP. Commonly used fillers are talc, calcium carbonate (usually powdered marble), glass fiber, mica, carbon black, clays, cellulose fibers, lubricants, and pigments. Many exotic fillers have been used for special purposes. The use of fillers in PP usually reduces toughness and raises density and cost per volume (see Table 2).

Processing

Melt flow rate (MFR) is one of the key variables in the processing of PP. The American Society for Testing and Materials (ASTM) specifies (4) that PP is to be tested at 446°F (230°C) under a pressure exerted by a nominal 4.4-lb (2-kg) mass in the apparatus specified under ASTM Standard D 1238. The result is the weight of material extruded through the standard orifice in 10 minutes. This test is a crude measure of the melt viscosity of the plastic under low shear rate. Melt viscosity correlates to the weight-average molecular weight of the polymer. Commercial PP grades of interest in packaging span a MFR range of 0.3–40 g/10 min. The low flow grades (up to about 2.5 g/10 min) are used in sheet extrusion (see Extrusion) and blow molding (see Blow molding). Film is manufactured from intermediate-flow grades (MFR 3–15 g/10 min). Injection-molding processes (see Injection molding) ordinarily use PP with MFR of 3–40 g/10 min (see Table 3).

Applications and Markets

In 1984, consumption of polypropylene in packaging applications in the United States was roughly as follows (5):

Applications	1×10^6 lb (1×10^3 t)
Containers and lids	
Blow molded	107.8 (48.9)
Injection molded	165.0 (74.8)
Thermoformed	55.0 (24.9)
Film	
Oriented	319.0 (144.7)
Nonoriented	81.4 (36.9)
Closures	
Injection molded	200.2 (90.8)

Thermoforming. Sheet extrusion for thermoforming (see Thermoforming) is a small, rapidly growing segment of the PP market. Until the late 1970s, few converters were willing to attempt thermoforming of PP. Since that time, thermoforming techniques and grades have been improved. Random copolymers and impact copolymers have a broader temperature "window" and are more easily formed than homopolymers.

Table 3. Effect of Molecular Weight (Melt Flow) on Homopolymer Properties

Property	Values			
melt flow cond. L g/10 min	0.4	0.8	12	35
tensile strength (yield), psi (MPa)	5,400 (37.2)	5,150 (35.5)	5,050 (34.8)	4,750 (32.8)
yield elongation, %	13	13	11	12
flexural modulus, (1% secant), psi (MPa)	240,000 (1,655)	245,000 (1,690)	260 (1,793)	210,000 (1,448)
hardness Rockwell R (HRR)	95	95	100	98
heat deflection temp (66 psi or 455 kPa), °F (°C)	194 (90)	203 (95)	212 (100)	194 (90)
notched Izod 23°C, ft-lbf/in. (J/m)	3 (160)	2.5 (133)	0.7 (37)	0.6 (32)
application	sheet extrusion thermoforming blow molding	sheet extrusion thermoforming blow molding	injection molding film	injection molding

Several grades now in development promise excellent forming characteristics, even in homopolymers. Mineral-filled grades also form well, probably because they have higher thermal conductivity and lower heat of fusion than other types. The factors that promote the use of PP in thermoforming are low odor and taste transfer, good moisture barrier, chemical resistance, and adequate clarity in thin sections.

Blow molding. In blow molding, about 65% of the resin consumed is for consumer products and about 35% for medical products. High melt strength is required for extrusion blow molding, which is used to produce relatively large containers up to 5.3 gal (20 L) in size. Injection blow molding usually requires MFR of 1–2.5 g/10 min. This process is particularly useful for relatively small containers. The ability to withstand temperatures or aggressive chemicals that would stress crack or attack other materials is usually the reason to blow mold PP today. In detergent exposure tests that crack HDPE in a day or two, PP does not fail even after many weeks. Random and impact copolymers are most often used. The development of high melt-strength resins will assist the growth of PP blow molding. Some blow molded PP bottles are biaxially oriented. If the chains of PP are stretched, they line up to give remarkable strength, clarity, and toughness (even at low temperature). This process uses a preform that is stretched and blown while warm. Homopolymers and random copolymers can be used since low temperature toughness is provided by biaxial orientation. Typical resin MFR is 1–3 g/10 min for these applications.

Film. The use of PP in film is very large. Oriented films typically have high toughness and excellent clarity (see Film, oriented polypropylene). They can be produced by a high expansion bubble process or a tenter process. Product variations are possible based on the amount of transverse and machine direction orientation. Nonoriented cast films are usually made by a chill-roll process but there are also water-quench and water-quenched bubble processes in use. (see Extrusion; Film, nonoriented polypropylene). Oriented films can have a stiff feel or "hand." They sparkle and tend to "crinkle" audibly.

These films are employed as cigarette wrap, candy wrap, and snack-food pouches, often in replacement of cellophane (see Cellophane). Oriented films have been tailored for superior barrier properties (by coating), heat-seal strength (by resin modification or coating), heat-shrink properties (see Films, shrink), printability, and electrical properties. A relatively new addition to the family of oriented films is opaque film. The opacity is produced by a filler and by controlled voiding. A type of decorative ribbon is foamed uniaxially oriented PP with a colorant (see Colorants).

The use of woven slit tape or slit film for heavy-duty agricultural bags is not a major use in the United States, but these bags compete with jute and other natural fiber in carrying much of the world's grain. A related product is strapping, which is made by extruding either sheet or tapes. If sheet is extruded, it is subsequently slit (see Slitting/rewinding). The filaments are then stretched while warm to give tensile strength values of ≥50,000 psi (>345 MPa), ten times that of unoriented PP.

Nonoriented film has a number of growing markets. Compared to oriented film, it is available in thicker gauges and has a softer "hand" at the same thickness. Some applications are release sheets, sanitary products, disposable-diaper layers, bandages, and apparel packaging. The use of PP in composite film and sheet materials is small, but growing rapidly. PP is used in combination with other PP structures, paper, metal foils, fabric (woven and nonwoven), and other plastics. Such composites can be made by coextrusion (see Coextrusion), lamination (see Laminating), or extrusion coating (see Extrusion coating). The motivation for making such structures is often related to barrier properties, temperature resistance, chemical resistance, and cost. The principal market for these products is in food packaging.

Most film processes use PP grades with a MFR in the range of 2–10 g/10 min. Extrusion coating uses materials with flow ranging from 10 to over 60 g/10 min. Selection of the additive package is an important consideration. Printing, winding, static-charge buildup, blocking (sticking together), odor, color, heat-seal strength, and other properties are influenced by the additives.

Injection molding. Injection molding produces many familiar packaging items, including threaded, dispensing, and pump closures (see Closures); aerosol valves and overcaps (see Pressurized packaging); wide-mouthed jars, totes, crates (see Crates, plastic), apparel hangers, snuff boxes, cosmetic containers, drug syringes, barrel bungs, delicatessen tubs, and many others. Most general-purpose PP molding grades are well suited for packaging. In food and drug packaging, one must ensure that the particular grade is suitable for the product to be packaged. Conventional injection-molding grades span a MFR range of 2->40 g/10 min. Generally, the lower-flow materials are tougher and process less rapidly. By using narrow-MWD or "controlled-rheology" resins, high MFR grades with good toughness can be made. These grades are appropriate choices for thin-walled moldings and for large multiple-cavity molds. They offer better dimensional control, but are not as stiff as broader-MWD material.

Normal homopolymers are brittle at refrigerator temperatures. The use of copolymers is recommended when shipping or use expose the part to low temperature impact. A demanding application that requires stiffness at microwave-oven temperatures without sacrificing toughness at freezer temperatures might require a mineral-filled copolymer.

Health and Safety Issues

Except for fire-retardant grades containing antimony, PP is generally a nontoxic material. Many grades are available that comply with FDA requirements for food packaging. PP is used in drug packaging and medical devices. It is fiber-spun for use in undergarments, upholstery, sanitary products, and bandages. It is not soluble at normal temperatures in food and beverages. PP is combustible and burns completely when adequate air is available to the flame. As sold by resin manufacturers, PP is usually in a coarse granular form and presents no unusual fire hazard. If it is finely divided (>200 mesh or 74 μm), however, polypropylene can present a combustible dust hazard as do most organic materials. Like other organic materials (eg, wood, wool, flour, etc), the products of incomplete combustion include carbon monoxide and can include a number of unpleasant, partially oxidized pyrolysis products (aldehydes, ketones, etc).

Under most processing conditions, little hazard exists with the use of PP. One should avoid contact with the molten polymer. If the plastic is exposed to air at temperatures above 500°F (260°C), proper ventilation should be used. The autoignition temperature is 675–700°F (357–371°C) for most grades. In summary, there are no unusual risks associated with the use of PP.

BIBLIOGRAPHY

1. M. R. Schoenberg, J. W. Blieszner, and C. G. Papadopoulos, "Propylene," in M. Grayson and D. Eckroth, eds., *Encyclopedia of Chemical Technology*, 3rd ed., Vol. 19, John Wiley & Sons, Inc., New York, NY 1982, p. 228.

2. J. R. Fried, "Polymer Technology—Part 1: The Polymers of Commercial Plastics," *Plastics Engineering*, **38**, 49–55 (June 1982).

3. J. R. Fried, "Polymer Technology—Part 3: Molecular Weight and its Relation to Properties," *Plast. Eng.* **38**, 27–33 (Aug 1982) (Parts 4, 5, and 6 followed).

4. *1983 Annual Book of ASTM Standards*, Vol. 8.01, American Society for Testing and Materials, Philadelphia, Pa., pp. 569–581.

5. *Mod. Plast.* **62** (1), 69 (Jan. 1985).

General References

J. R. Fried, "Polymer Technology—Part 2: Polymer Properties in the Solid State," *Plast. Eng.* **38**, 27–37 (July 1982).

J. L. Szajna, "A Supplement to 1980 ANTEC Paper, Functions vs. Economics vs. Aesthetics," presented at the *40th Annual Technical Conference and Exhibition of the Society of Plastics Engineers*, May 10–13, 1982.

H. Gross and G. Menges, "Influence of Thermoforming Parameters on the Properties of Thermoformed PP," presented at the 40th Annual Technical Conference and Exhibition of the Society of Plastics Engineers, May 10–13, 1982.

R. C. Miller
Himont U.S.A., Inc.

POLYPROPYLENE FILM.   See Film, oriented polypropylene; Film, nonoriented polypropylene.

POLYSTYRENE

Polystyrene is one of the most versatile packaging resins. Easily extruded and thermoformed, it is commonly used to make tubs for refrigerated dairy products. As a foam, it is one of the major materials for protective cushioning. These are just two examples of the many packaging applications of polystyrene.

Polystyrene resins are generally classified as either crystal or impact grade. Crystal grades are selected whenever clarity is important enough to outweigh their inherent brittleness. Impact grades are characteristically translucent or opaque because of rubber compounds added to enhance impact resistance. Expandable polystyrene (EPS) is a form of crystal polystyrene supplied as a partially expanded bead. Crystal and impact grades are fabricated by conventional molding and extrusion processes. EPS beads contain a low boiling hydrocarbon compound which vaporizes on application of heat to form a low density foam structure. Annual United States consumption of polystyrene is about 4×10^9 lb (1.8×10^6 t). About one-third of the total is accounted for by packaging and disposable cups and plates.

Manufacture

Polystyrene is manufactured by polymerizing styrene monomer in the presence of heat. The reaction can be shown schematically as

styrene   polystyrene

Polystyrene can be polymerized by bulk, suspension, emulsion, and solution processes. Only the bulk and suspension processes are commercially significant today. In the bulk polymerization process, step-wise conversion of the monomer is done in a series of two or three reactors at 248–392°F (120–200°C). Reaction rate is accelerated by addition of initiators (organic peroxides and hydroperoxides). The mixture of monomer (and rubber for impact grades) is pumped through an agitated-reactor series and a devolatilizer to remove residual monomer, extruded, and pelletized. Heat resistance and physical properties are improved by a process variation involving the addition of ethylbenzene solvent and adjustment of reaction conditions. A process developed by BASF also provides resin with enhanced properties. This partial conversion process involves a modified reactor chain and a rapid two-stage devolatilization step to reduce residual volatile material.

The suspension process is generally used to produce EPS. (It can also be used to produce crystal grades.) The styrene monomer is polymerized in water-containing protective colloids. The feed stream also contains 6–8% of a low boiling liquid such as petroleum ether which is dispersed in the bead product. Polymerization is carried out below 122°F (50°C) to avoid premature reaction of the low boiling foaming agent. After separation and washing, beads are resuspended and heated to 194°F (90°C). Bead volume increases about 300%. The foaming reaction is completed during subsequent molding at 230–248°F (110–120°C), filling the mold.

Properties

Polystyrene is generally characterized as a noncrystalline (amorphous) polymer. Melt flow and mechanical properties (tensile strength, flexural strength, impact strength, heat-deflection temperature) depend on molecular weight and molecular weight distribution. Polystyrene exhibits a gradual melting range (194–212°F or 90–100°C), brittle fracture (crystal grades), and tendency to flow under stress. Clarity and impact properties are the primary differentiating features between crystal and impact grades. Resistance to uv-radiation is generally poor. Typical properties of crystal and impact grades are shown in Table 1.

Crystal grades with optimum clarity transmit about 90% of visible light radiation. Izod impact resistance and tensile elongation (2.0–3.5%) are low. Tensile strength up to 8000 psi (55.2 MPa) and flexural modulus approaching 500,000 psi (3,450 MPa) place the resin among the most rigid of unreinforced thermoplastics; therefore, creep-related deformation is not an important design consideration.

Polystyrene impact grades provide resistance to part failure under impact loading. These resins consist of two phases: a continuous glassy phase of pure polystyrene and a discrete rubber phase. This latter phase consists of rubber which has been modified by graft copolymerization with styrene monomer. The rubber addition usually does not exceed an 8–10 wt %. Balance of impact and other significant mechanical properties is the primary control on rubber levels. The mechanism by which the rubber phase influences performance is a matter of some disagreement. In general, the rubber phase enhances impact resistance by restricting propagation of microcracks formed during impact loading. Izod impact values can approach 5 ft·lbf/in. of notch (267 J/m) depending upon the amount of elastomeric compound added. A simultaneous increase in elongation to about 60% is observed, along with tensile strength values of 2500–5000 psi (17.2–34.5 MPa). Commercially available resins provide a broad selection of property combinations within the ranges characteristic of impact grades.

Specialty variations of the classical crystal and impact grades are available now as well. Products with impact and environmental stress crack properties (ESCR) rivaling those of some ABS products have been developed through use of elastomeric additives. An increasing emphasis on product safety in nonpackaging applications has prompted the development of fire-retardant grades.

Processing and Applications

Polystyrene is readily processed by all the popular techniques for fabricating thermoplastics. Injection molding (see Injection molding) and sheet extrusion (see Extrusion) account for the major volume. Bottles are blown by injection-blow molding (see Blow Molding). Industrial applications also use profile extrusion and structural foam molding. Extrusion and subsequent thermoforming (see Thermoforming) are the most common processes in packaging applications (eg, for dairy containers, rigid-box inserts). Extruded foam is used for egg cartons and produce and meat trays (see Foam, extruded polystyrene), and to jacket glass bottles. Injection molding is used to make containers for some refrigerated products, but it is particularly useful in the production of transparent reusable boxes (see Boxes, rigid plastic). Crystal PS is injection-blow molded to some extent to make bottles for pills, tablets, and capsules.

EPS is supplied as a partially expanded bead. Part molding involves a two-step process. Beads are first charged into a preexpander and heated by steam injection. In the second stage, the charge is transferred to the part mold. Expansion is completed by a second injection of steam resulting in bead fusion into a porous solid. EPS is not injection molded or extruded like solid pellet material because the porous structure

Table 1. Polystyrene Properties

Property	Values	
	Crystal grades	Impact grades
tensile strength, psi (MPa)	5000–8000 (34.5–55.2)	2000–6000 (13.8–41.4)
flexural modulus, 10³ (MPa)	500–500 (2759–3448)	250–430 (1724–2966)
impact strength, Izod ft·lbf/in. of notch (J/m)	0.3–0.5 (16–26.7)	1–5 (53.4–266.9)
deflection temperature, °F (°C)	150–230 (66–110)	160–210 (71–99)
hardness, Rockwell (HRM)	65–80	50–60

would be destroyed during the melting stage of these processes. EPS foam has excellent shock-absorbing and insulation characteristics.

New applications are being developed through the use of coextruded sheet stock incorporating polystyrene and barrier resins such as EVOH and PVDC (see Barrier polymers; Ethylene–vinyl alcohol; Vinylidene chloride copolymers). Coextruded thermoformed wide-mouth containers can be used for shelf-stable food products. Multilayer blow-molded bottles (see Multilayer plastic bottles) are also produced using similar combinations.

BIBLIOGRAPHY

General References

J. D. Griffin and J. Y. Glasc, "Polystyrenes," in D. M. Considine, ed, *Chemical and Process Technology Encyclopedia*, McGraw-Hill, Inc., New York, 1974, pp. 914–917.

C. Harper, *Handbook of Plastics and Elastomers*, McGraw-Hill, Inc., New York, 1975.

D. V. Rosato and R. T. Schwartz, *Environmental Effects on Polymeric Materials*, John Wiley & Sons, Inc., New York, 1968.

H. S. Gilmore and A. R. Hoge, "Polystyrene," in *Modern Plastics Encyclopedia 1984–85*, McGraw-Hill, Inc., New York, 1984.

L. E. Nielsen, *Mechanical Properties of Polymers*, Reinhold Publishing Corp., New York, 1962.

C. C. Winding and G. D. Hiatt, *Polymeric Materials*, McGraw-Hill, Inc., New York, 1961.

A. V. Tobolsky, *Properties and Structures of Polymers*, John Wiley & Sons, Inc., New York, 1960.

G. Forger, "Polystyrene New Grades, New Uses," *Plastics World*, 39–44 (June 1979).

J. S. Houston, *Polystyrene – A Versatile Packaging Material*, Chemical Marketing Research Associates meeting, Houston, Texas, February 2–4, 1983 – "Feast or Famine?" "The Future of Chemicals in the Food Industry."

R. D. Deanin, *Polymer Structure, Properties and Applications*, Cahners Publishing Company, Inc., Boston, Mass., 1972.

J. S. HOUSTON
Amoco Chemicals Corporation

POLY(VINYL CHLORIDE)

Poly(vinyl chloride) (PVC) is, by consumption, the largest member of a group of polymers commonly referred to as "vinyls." These polymers are all based on either the vinyl radical (CH_2=CH—) or the vinylidene radical (CH_2=CR—). Included in this versatile group of polymers are poly(vinyl acetate), poly(vinylidene chloride) (PVDC) (see Vinylidene chloride copolymers), poly(vinyl alcohol), poly(vinyl fluoride), polyvinylidene difluoride, and poly(vinyl butyral).

PVC is polymerized from vinyl chloride monomer (CH_2=$CHCl$) to produce the following polymer structure:

$$\text{--}CH_2CHCl\text{--}_n$$

Because of the versatility of the polymerization process, a number of copolymers (eg. vinyl acetate, vinylidene chloride, acrylic esters) and graft copolymers (eg. ethylene–vinyl acetate, ethylene–propylene diene) have been manufactured. Each of these products has its own special characteristics, and some are used in packaging applications, but the thrust of this article is homopolymer PVC and its applicability to packaging applications.

As a result of its toughness, relatively low cost, and the ability to modify its physical properties, PVC has evolved into one of the most versatile polymers. When first developed, it was almost impossible to process PVC resin into usable products without severe thermal degradation due to its tendency to dehydrochlorinate (1). As scientists discovered additives such as stabilizers, plasticizers, etc, they soon found that PVC can easily be compounded to make it rigid or flexible and thus alter most physical properties across a broad spectrum.

The result has been a wide range of applications that include exterior house siding, electrical-wire coatings, credit cards, and medical blood bags. Packaging represents about 5% of annual United States shipments of ca 5.5×10^9 lb (2.5×10^6 t).

Manufacturing

PVC is normally polymerized from its monomer by one of four different processes: suspension, mass, emulsion, and solution. Each process uses peroxide-type initiators to produce free radicals, and the exothermic reaction is normally carried out at 95–167°F (35–75°C) (2,3). That is where the similarity ends. Under different reactor configurations, agitation, and reaction media, these four processes produce PVC resins with uniquely different characteristics. These characteristics play important roles in subsequent processes such as extrusion and calendering.

There are 10 major manufacturers of PVC resins in the United States.

Structure and Properties

Each of the four resin processes produces a resin with a unique structure or characteristic, more commonly referred to as morphology. As a basis for discussing PVC's morphology (4–6), some typical suspension-resin characteristics are shown in Table 1. Suspension resins are used in one of the largest PVC packaging applications: food film (see Film, flexible PVC).

Molecular weight. The molecular weights of PVC resins produced in the United States are typically from 0.50 to 1.20 inherent viscosity. As molecular weight is increased, physical properties such as tensile strength and tear strength increase proportionately (7,8), as well as melt viscosity. The trade-off in selecting a PVC resin is to choose the minimum molecular

Table 1. Typical Characteristics of PVC Suspension Resins

molecular weight	
inherent viscosity	0.88–0.98
weight average molecular weight	142,000–185,000
number average molecular weight	55,000–62,000
particle size	
mean	130–165 μm
bulk density	0.450–0.550 g/cm^3
porosity	
ASTM D 2873	0.23–0.35 cm^3/g
residual vinyl chloride monomer	
ASTM D 3749 or D 3680	1.0 ppm

weight to meet the end-product's physical requirements without exceeding any melt viscosity restrictions.

Particle size. As a result of the suspension polymerization process (agitation, suspending agents, etc) PVC particles are somewhat spherical. The resin particles generally have a size distribution from 70–250 μm in diameter, which results in a mean size of 130–165 μm.

Bulk density. In PVC blown-film production the output rate of the extrusion line is normally directly proportional to the bulk density of the PVC resin. PVC has a specific gravity of approximately 1.40, but the resin's bulk density is significantly less: 0.450–0.550 g/cm^3. Bulk density is directly related to the particle's morphology and specifically the resin's porosity, particle size and distribution, and particle surface characteristics, etc.

Porosity. A PVC resin particle is made of a structure that is not a solid uniform polymer matrix, but rather a structure that contains many openings in the resin surface plus a measurable and accessible void within the particle. The amount of non-PVC volume within a resin particle is referred to as its porosity. Techniques such as mercury intrusion (ASTM D 2873) are commonly used to measure a resin's porosity. The porosity of a PVC resin is used to absorb the various liquids added to a PVC compound. The amount of porosity and its accessibility play an important role when considering the amount and viscosity of the liquids, such as plasticizers, added during compounding.

Residual vinyl chloride monomer. During the polymerization of PVC, less than 100% of the vinyl chloride monomer is converted to polymer. Today the amount of residual monomer remaining in the dried PVC resin is typically less than 1 ppm. This level establishes safe-worker levels and provides acceptable levels in the product (see further discussion below).

Compounding

Without the addition of compounding additives, PVC is a very difficult polymer to process into a useful product. For example, the addition of a liquid plasticizer such as di(2-ethylhexyl) adipate (DOA) permits the production of a flexible film with the oxygen-transmission properties required for meat packaging. The addition of a rubbery polymer such as methacrylate–butadiene–styrene (MBS) significantly affects the toughness or crack propagation characteristics as measured by impact tests (eg, Gardner or Izod). PVC's tremendous versatility results from the ability to tailor these properties to the requirements of the product. With the proper additives, a rigid PVC bottle can be blow-molded (see Blow molding) for edible oil packaging. Other ingredients allow the extrusion of a blown flexible film for produce wrapping. These formulations are generally highly proprietary and very specific to use applications and properties. For information on compounding methods, see the list of references at the end of this article.

Stabilizers. Stabilizers used in food-contact applications must have FDA clearance. These include Ca/Zn salts, epoxidized soybean oil, and octyl-tin mercaptides. In other applications, stabilizers such as butyl or methyl tin mercaptides are used. Stabilizers give PVC the ability to withstand the thermal and shear conditions of processing without polymer degradation (9).

Plasticizers. There are numerous types of plasticizers, and each imparts a specific set of properties to the final product

(10). These liquid or polymeric additives generally reduce the T_g of PVC. (In Europe, the term "additive" is applied only to components added at levels lower than 10%. Plasticizers are not considered additives.) At the same time, they reduce tensile strength and increase elongation. Certain plasticizers, such as di(2-ethylhexyl) phthalate (DOP), improve PVC's water-vapor barrier properties; DOZ (dioctyl azelate) significantly improves its low-temperature impact strength.

Lubricants. Generally, lubricants are added to PVC compounds to reduce the frictional properties between the compound and the metal surface of the processing equipment (11). They are also used to reduce the surface friction of the final product or to reduce the product's surface static properties. Lubricants include such families as paraffinic waxes and metal stearates.

Impact modifiers/processing aids. Many types of impact modifiers and processing aids have been developed for PVC packaging applications. Examples include methylacrylate–butadiene–styrene (MBS), chlorinated polyethylene, and acrylic polymers (see Acrylics). The type used depends on application requirements such as clarity, cost efficiency, low-temperature impact, metal-release properties, etc.

Fillers and pigments. Fillers such as CaCO$_3$ can reduce the raw material costs of a PVC compound. They have little affect on physical properties at low levels. Some fillers can improve properties such as stiffness or abrasion resistance.

PVC Packaging Applications

PVC packaging applications fall into three general categories: food, nonfood, and medical.

Most food-packaging applications (see Table 2) are served by rigid and semirigid sheet (see Film, rigid PVC), film (see Film, flexible PVC), and bottles (see Blow molding). Extrusion-blown food film is used primarily for wrapping meat, poultry, and produce (12). The advantages of PVC are clarity, barrier properties (see Barrier polymers), puncture resistance, and cling for good sealability. The use of PVC bottles for foods and beverages has been held back in the United States by various FDA concerns (see Food, drug, and cosmetic packaging), but bottles are now being used for edible oils, honey, etc. A major potential application in liquor bottles is awaiting regulatory clearance.

In nonfood applications, rigid PVC sheet plays an important role in vacuum-formed packaging applications (13). Because of PVC's clarity, toughness, and excellent vacuum-forming characteristics, it is used for various blister packaging applications (see Carded packaging). Although somewhat limited because of the inroads of LLDPE (see Polyethylene, low density), PVC film is still widely used for applications such as pallet wrap, box wrap, and shirt wraps. PVC bottles are used extensively in nonfood applications. Because PVC's excellent blow molding characteristics can be used for versatile product design, it is used to package toiletries and cosmetics and household detergents.

For over 20 years PVC has been used to package medical devices (14). For example, PVC is extruded into blood or intravenous-solution tubing. It is also calendered or extruded into 0.010–0.020-in. (254–508 μm) thick flexible sheet which is then formed into bags for blood platelets or intravenous solutions (see Health-care packaging).

Table 2. United States Consumption of PVC in Food Packaging in 1983 and 1988(est), 10^6 lb[a]

	1983	1988(est)
milk	0	10
milk products	0	20
cream	0	5
ice cream	0.05	10
butter	1	4
fresh meat	105	132
frozen meat	8	10
cured meat	27	39
fish	10	14
processed meats and fish	3	7
fresh produce	19	25
frozen produce	1	10
processed produce	5	9
edible oil	17	69
shortening	0	2
nonalcoholic beverages	0.2	3
beer	0	10
wine	1	22
alcohol	0	43
shelled nuts	2	8
powdered dry foods	0.5	7
cereal grains	2	4
ready-to-eat cereals	2	6
Total	204	469

[a] To convert 10^6 lb to t, multiply by 453.6.

FDA Status

PVC is used extensively in food-contact applications such as meat wrap. It is prior sanctioned for use in general food-contact applications by virtue of an article published in July 1951 in the *Journal of the Association of Food and Drug Officials of the United States* by A. J. Lehman of the FDA. In addition, PVC resins are listed in a number of specific FDA regulations relating to food-contact substances.

In the mid-1970s the FDA expressed concern about the use of PVC in food or alcoholic beverage packaging due to the discovery of residual vinyl chloride monomer (VCM) within the PVC container. Since VCM does not have FDA clearance for food contact, the concern pertained to residual monomer migrating into the container's contents. The FDA never banned the use of PVC for food products, however, and the PVC industry has responded by producing PVC resins with extremely low reduced levels of residual VCM. The industry has also demonstrated that the maximum potential migration at these new levels of residual VCM is less than the capability of measurement by the most sensitive methods currently available (15). As a result, it is expected that the FDA will favorably clarify its position on the use of PVC for food and alcoholic-beverage packaging.

Editor's note: see Fed. Reg. **51**(22), 4173 (Feb. 3, 1986).

BIBLIOGRAPHY

1. A. Guyot, M. Bert, P. Burille, and co-workers "Trial for Correlation Between Structural Defects and Thermal Stability in PVC," as presented at the Third International Symposium on Polyvinylchloride, Aug. 10–15, 1980, Case Western Reserve University, Cleveland, Ohio).

2. J. Ugelstad, *J. Macromol. Sci. Chem. A* **11**(7), 1281 (1977).

3. L. F. Albright, *Chem. Eng.* 145–152 (June 5, 1967).

4. P. R. Schweagerle, *Plast. Eng.* 42–45 (Jan. 1981).

5. N. Berndstein and G. Manges, *Journal of Pure and Applied Chemistry* **49**, 597–613 (1977).

6. H. Behrens, *Plaste and Kautschuk* **20**(1), 2–6 (Jan. 1973).

7. J. R. Fried, *Plast. Eng.* 27–33 (Aug. 1982).

8. S. Kaufman and M. M. Yocum, *Plastic Compounding* 44–46 (Nov./Dec. 1978).

9. H. O. Wieth and H. Andreas, *Journal Pure and Applied Chemistry,* **49**, 627–648 (1977).

10. H. P. Harris, M. T. Payne, and J. P. Mieure, *Plast. Eng.* 25–26 (Oct. 1982).

11. L. F. King and F. Noel, *Polym. Eng. Sci.* **2**, 112–119 (March 1972).

12. J. Mills, "Blown PVC Film," presented at the *Regional Technical Conference, SPE Mississauga, Ontario, Canada,* Sept. 13, 14, 1983.

13. J. Southus, "PVC Film & Sheet in Packaging," presented at the *Regional Technical Conference, SPE, Mississauga, Ontario, Canada,* Sept. 13, 14, 1983.

14. N. Perry, "Flexible PVC Medical Packaging Applications," presented at the *Regional Technical Conference, SPE, Mississauga, Ontario, Canada,* Sept. 13, 14, 1983.

15. R. Mathis, "PVC Bottles Past, Present and Future," presented at the *Regional Technical Conference, SPE, Mississauga, Ontario, Canada,* Sept. 13, 14, 1983.

General References

L. I. Nass, ed., *Encyclopedia of PVC,* Vol. 1 (1976), Vol. 2 (1977), Vol. 3 (1977), Marcel Dekker, Inc., New York.

J. K. Sears and J. R. Darley, *The Technology of Plasticizers,* John Wiley & Sons, Inc., New York, 1982.

W. S. Penn, *PVC Technology,* MacLaren, London, 1966.

Society of Plastics Engineers Conference Preprint, "Vinyl in Packaging," as presented at the Regional Technical Conference, Mississauga, Ontario, Canada, Sept. 13, 14, 1983.

J. H. Briston and L. L. Katan, *Plastic Films,* Longman, Inc., New York, 1983.

D. A. Cocco
The BF Goodrich Company

PRESSURE-SENSITIVE TAPE.   See Tape, pressure-sensitive.

PRESSURIZED CONTAINERS

Three distinct classes of pressurized containers are used to dispense consumer specialties. Two are self-pressurized: the chemical-propellant sprayer, ie, aerosol, and the nonchemical sprayer, which can take several forms. The third is the mechanical pump, such as the trigger-spray dispenser, in which pressure is generated only within the valve body. The aerosol system is by far the most common. An aerosol consists of a combination of concentrate and propellant, contained in a pressure-resistant dispenser fitted with a valve. The propellant can range from ca 0.5 to 100%, and, at the very high end, the propellant may also function as the concentrate. Aerosols are used to dispense a vast number of products in containers ranging in capacity from 3 to ca 1000 mL.

Liquid sprays are the most common form of delivery for all three dispenser types. There are solid-liquid streams for special purposes, and two rather unusual trigger-type valves are now used to dispense toothpaste, but the aerosol package is undoubtedly the most versatile of the group. About 85% of all aerosols produce sprays, with particle-size distributions ranging from gaslike to extremely coarse. They can also produce foams, liquid streams, pastes, liquid–powder sprays, and even gas–powder sprays. By using special bag-in-can or piston-operated aerosols, fitted with large-orifice valves, liquids with viscosities up to several million (10^6) cP (several thousand Pa·s) can be dispensed. Aerosol-metering systems can provide sprays, liquids, foams, creams, and pastes in dosages ranging from 0.05 to 5.00 g. Sterile foods, pharmaceuticals, and other products can be packaged in aerosols as well. Special filling techniques are used in some cases; in others, the finished dispenser is sterilized with either heat or gamma irradiation.

The total market for pressurized packages stands at slightly over 3 billion (10^9) units per year in the United States alone. The aerosol segment totaled about 2.4 billion (10^9) units in 1984, and pump sprays accounted for nearly all the rest: about 600 million (10^6) units. Sales of the nonchemical sprayer were negligible. Worldwide retail sales of aerosols were estimated at 6.3 billion (10^9) in 1983.

Historical Development

The aerosol concept originated in 1923 when Eric Rotheim of Oslo, Norway, developed a ski-wax spray and several other products using butane and vinyl chloride propellants and heavy brass containers fitted with needle valves. Sample cans still sprayed 50 years later during a commemorative press conference. Nothing was done with the Rotheim development until 1943, when Lyle Goodhue and William Sullivan, working for the USDA, discovered that a 10-μm aerosol spray of insecticide would remain airborne for at least 5 min and was effective against both flying and crawling insects. They patented the discovery, and the patent remains the basis of EPA requirements for preproduction registration of all aerosol pesticides. Heavy steel, brass, and aluminum aerosol insect sprays were shipped to the armed forces in the South Pacific, and a few million (10^6) per year were also marketed domestically, but the costly container and valve had to be made returnable against a deposit, at least for civilian uses, and this greatly inhibited the market. Finally, in about 1947, Harry E. Peterson developed a super-strong beer can which could serve as an aerosol container. He also helped to create a relatively inexpensive valve that was simple to operate. Since then, over 100 billion (10^9) aerosols have been produced and sold worldwide.

Aerosol Containers

About 85% of all aerosols produced in the United States in 1984 were steel (70% tinplate, 14.7% ECCS). Aluminum accounted for 13.0%; glass, including plastic-coated glass, 2.2%; and stainless steel and plastics, ca 0.1%. The "85% rule" seems to apply in several other countries as well, eg, Canada, Mexico, the UK, and Australia, but many other countries use much larger ratios of aluminum cans. They predominate in Italy, and are almost 100% in India, largely as the result of trade barriers against imported tinplated cans.

Steel. Tinplated steel is specified in terms of the weight of metals per base box (see Cans, steel; Tin-mill products). A base box is an area of 31,360 in.^{2} (20.23 m^2). If this area of the untinned steel weighs 55 lb (25 kg), the plate is called 55-lb stock, and it should have an average thickness of 0.0061 in. (0.1549 mm). The steel-mill tolerance is usually $\pm10\%$ on thin plate, but the can companies generally insist on -5 to 10%. A very strong Double-Reduced Temper 8 (DR-8) version of 55-lb (25-kg) plate is now starting to be used for aerosol-can bodies with a diameter of 2.075 in. (52.7 mm). It is the thinnest plate now used for aerosols.

Until recently, most aerosol-can tinplate carried 0.25 lb (113.4 g) of tin per base-box area. Equally thick layers of tin were electroplated on each side, resulting in a plate that could be termed 0.125/0.125-lb (56.7/56.7-g) ETP (electrolytic tin plate), but was always called simply 0.25-lb ETP. Until 1982, this was the thinnest and most economical tin coating available. It was bright, with an almost complete, ie, continuous, layer of tin covering the dark FeSn$_2$ intermetallic compound that adhered to the steel itself. As the price of imported tin rose to an unprecedented ca $8/lb ($17.60/kg), the tin mills reacted by reducing the coating thickness down to the present limit of 0.05-lb (22.7-g) ETP. Considering this as 0.025/0.025-lb (11.3/11.3-g) ETP, each side of the plate carries an average of 3.0×10^{-6} in. (73.5 nm) of pure tetragonal and alloyed tin metal. The tin layer is so thin that nearly all of it is alloyed with iron during flow brightening, and it has a gray appearance. All electrotinplated steel is flow-brightened after tinning to develop a smooth reflective surface and to control the development of a uniform alloy layer. After that, the tinplate is passed through a warm sodium dichromate solution in a process known as cathodic dichromate treatment, which employs an electric current to deposit an exceedingly thin layer of chromium and chrome oxides on top of the tin. This improves the adhesion of can enamels and optimizes oxidation resistance and solderability. Japanese steel companies developed a method to electrodeposit a duplex coating of chromium metal and chromium oxide on steel (ECCS). This coating is very uniform and pore-free even at coating thicknesses of 1×10^{-6} in. (25 μm), and hence inexpensive. Enamel adherence to ECCS is very high. Because of its relatively low cost, it is preferred by many firms that market anhydrous or chemically inert aerosols.

Different thicknesses of tinplate and ECCS are used for aerosol cans. Body plate is the lightest, ranging from 55-lb (25-kg) stock on the smallest diameter cans to 90 lb (41 kg) for the large 3-in. (76.2-mm) dia ones. Plate for the end sections must be considerably heavier in order to hold pressures of at least 160 psig (1.2 MPa) without deforming. The dome of a typical 2.588-in. (65.7-mm) dia aerosol can is made of 128-lb (58 kg) plate, while the base is generally 123-lb (55.8-kg) plate. Domes are frequently manufactured from Type D (Al-killed) steel because of its better forming characteristics. Table 1 provides information on the optimum selection of baseweights and tempers for various aerosol-can components.

As the table indicates, the temper, ie, stiffness, of the plate is an important consideration. A popular plate of intermediate

Table 1. Thickness and Temper for Tinplate Used for Aerosol Cans

Weight per base box, lb (kg)	Average thickness, in. (mm)	Typical temper designation	Typical application	
			Component	Can dia (mm)[a]
55 (25)	0.0061 (0.155)	DR-8	body	202 (54)
65 (29)	0.0072 (0.183)	DR-8	body	207.5 (62.7)
70 (32)	0.0077 (0.196)	DR-8	body	211 (68.3)
70 (32)	0.0077 (0.196)	T-5	body	202 (54)
75 (34)	0.0083 (0.211)	DR-8	body	300 (76.2)
75 (34)	0.0083 (0.211)	DR-8 or T-5	body, DOT Specification 2P	202, 207.5, 211 (54, 62.7, 68.3)
80 (36) 0.0088	(0.224)	DR-8	body	300 (76.2)
85 (39) 0.0094	(0.239)	DR-8	body, DOT Specification 2Q	202, 207.5, 211 (54, 62.7, 68.3)
			body, DOT Specification 2P	300 (76.2)
90 (41)	0.0099 (0.251)	T-5	body, including DOT Specification 2P	300 (76.2)
107 (49)	0.0118 (0.300)	TU or T-5	bottoms	113, 202 (30, 54)
112 (51)	0.0123 (0.312)	TU or T-5	bottoms	207.5 (62.7)
112 (51)	0.0123 (0.312)	T-3	domes	202 (54)
118 (54)	0.0130 (0.330)	DT-2	domes	207.5 (62.7)
123 (56)	0.0135 (0.343)	TU or T-5	bottoms	211 (68.3)
128 (58)	0.0141 (0.358)	T-3	domes	211 (68.3)
128 (58)	0.0141 (0.358)	TU or T-5	bottoms, including DOT Specification 2P	211 (68.3)
			bottoms	300 (76.2)
135 (61)	0.0149 (0.378)	DT-2	domes, including DOT Specification 2P	300 (76.2)
			domes, including DOT Specifications 2P and 2Q	211 (68.3)
135 (61)	0.0149 (0.378)	TU or T-5	bottoms, including DOT Specification 2P	300 (76.2)

[a] 202 = 2 2/16-in. (54-mm) dia, etc.

temper is MR-TU: made of Type MR steel (Minimum Residuals), single (hot) reduced by rolling and annealed by a special continuous process to a Rockwell 30-T of about 57–60. The standard tempers used for aerosols have ranged from T-1 to T-6, plus TU, until recently, when the double (cold) reduced variety began to be used in some applications. Known as DR plate, or, less commonly as 2CR plate, they have apparent hardness values in excess of 75 on the Rockwell 30-T scale. Typical designations of DR-7, DR-8, DR-9, and DR-9-Special are used to indicate relative hardness. Actually, in DR plate, hardness is less meaningful than minimum yield strength as measured in a tensile test or springback tester. The value for the most commonly used type, DR-8, is ca 90,000 psi (621 MPa) at 0.2% offset.

The use of DR-8 plate will undoubtedly increase for can bodies and most can bases. The reason is highly cost related. For example, whereas 85–90-lb (39–41-kg) T-5 plate must be used for 300 (76.2-mm) dia can bodies, the same strength and dent resistance can be obtained with 75–80-lb (34–36-kg) DR-8 plate. This represents a steel reduction of ca 11.4% on a section, which usually amounts to two-thirds of the weight of the entire aerosol can, and the canmakers have acted to reduce their prices accordingly.

One potential problem with the use of DR-8 and similar

thin plate is that they have relatively low implosion, ie, paneling, resistance to vacuums applied to the finished aerosol can during the filling process. After introducing the product concentrate, many aerosol units are vacuum-crimped before adding the propellant. In the vacuum-crimping operation, a machine draws a partial vacuum to ca 18–22 in. (61–74 kPa) of Hg just prior to hermetically sealing on the valve by a crimping process. For a brief period of time, until the propellant is added, the can must withstand this level of evacuation, even under the jostling that normally occurs in production handling. If it caves inward, ie, panels, even at an incidence level of 1 in 200, the filler or marketer must consider taking corrective action. In marginal cases, this can be as simple as reducing the degree of evacuation. Since the vacuum step ties into such things as final can pressure, delivery rate, and, in rare cases, to corrosion aspects, any significant change in the amount of applied vacuum must be approached with caution.

At this time, many fairly aggressive aerosol products, such as starches and disinfectant and deodorant sprays, are being reevaluated to determine their compatibility with DR-8-type cans with either plain or lined 0.05-lb (23-g) ETP or lined ECCS constructions. Test packs are being prepared using regular vacuum crimping (as controls), partial vacuum crimping, and nonvacuum or atmospheric crimping to see if the difference in oxygen content has any significant adverse effect on the product or on the compatibility of product and container. In many instances, the can companies participate in these testing programs. If the containers and product appear satisfactory after 9–12-mo storage at room temperature and 100°F (37.8°C), they may issue a special warranty stating that the dispenser will be acceptable for 1 yr.

Dimensions. As in other industries, the size of steel aerosol containers is described as the diameter across the double seams multiplied by the height from the base to the top of the top double seam (see Figs. 1 and 2). In Europe, the diameter is stipulated as the inside diameter in millimeters. In the United States, the designations almost always have three digits: the first is the inch value, and the next two the number of $\frac{1}{16}$-in.

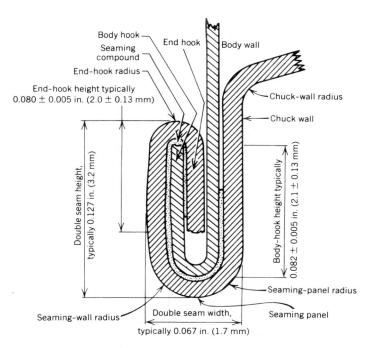

Figure 2. Aerosol tinplate can bottom double seam terminology and dimensions.

increments added. Thus, the over-the-seams diameter of a 202 can should be 2.125 in. (53.98 mm); that of a 211 can should be 2.688 in. (68.28 mm). The actual diameters are slightly different because of a standard correction factor. The height of a typical 211 × 604 can is 6.250 in. (158.8 mm) to the top of the top double seam. The dimensions of several popular aerosol cans are shown in Table 2.

The smallest tinplate aerosol can is the 202 × 200 (54 × 51 mm). The brimfull capacity is only 3.4 fl oz (102 mL), and the largest use is for giveaway samples containing ca 1.5–2.0 oz (42.5–56.7 g) of product. The next smallest is a curious can with a size description of 113 × 313 (30 × 97 mm), representing the smallest height that can be efficiently welded along the side seam. This and two taller versions, ie, 113 × 411 (30 × 119 mm) and 113 × 509 (30 × 141 mm), are also somewhat unique in that the top and bottom double seams are necked in.

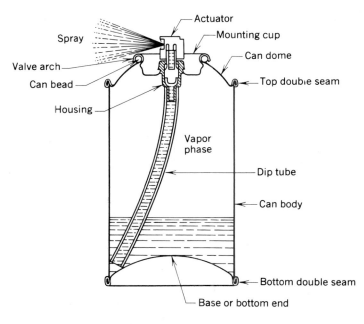

Figure 1. Aerosol dispenser (tinplate can) in operation.

Table 2. Dimensions and Capacities of Typical Aerosol Containers

Nominal size (mm)	Diameter over seams, dimension C, in. (mm)	Top seam height, dimension D, in. (mm)	Total height, dimension E, in. (mm)	Capacity, fl. oz. (mL)
202 × 214 (54 × 73)	2.178 (55.3)	2.865 (72.7)	3.261 (82.8)	5 (147)
202 × 406 (54 × 111)	2.178 (55.3)	4.365 (110.9)	4.761 (120.9)	7.6 (226)
202 × 509 54 × 141)	2.178 (55.3)	5.553 (141.0)	5.949 (151.1)	9.8 (290)
207.5 × 701 (63 × 179)	2.469 (62.7)	7.063 (179.1)	7.844 (199.2)	16.8 (498)
211 × 413 (68 × 122)	2.703 (68.7)	4.803 (122.0)	5.600 (142.2)	13.7 (404)
211 × 604 (68 × 159)	2.703 (68.7)	6.250 (158.8)	7.038 (178.8)	17.7 (522)
300 × 709 (76 × 192)	3.016 (76.6)	7.559 (192.0)	8.351 (212.1)	26.9 (796)

This means that the welded body section is formed to a slightly smaller diameter for the top and bottom, ie, 0.30 in. (7.6 mm) or so, before the flanging operation. A correspondingly smaller diameter dome and bottom are then double-seamed to the necked-in body. Ideally, the base diameter, across the seam, then equals the diameter of the body wall, and the dome diameter, across the seam, is ca 0.040 in. (1.0 mm) smaller. When standard full-diameter plastic caps are applied to such cans, they fit over the top double seam and give the finished unit an elegant profile similar to that of many high priced aluminum impact-extruded containers. Actually, the plastic cap is designed to have a thickness of ca 0.026 in. (0.66 mm) so that the wall protrudes beyond the body-wall diameter by ca 0.006 in. (0.15 mm). This prevents body rubbing in shipment. The practice of necking in bodies has been widely applied in Europe for almost a decade, but the concept is rarely used in the United States.

Linings. In the United States, single, double, and even triple can linings are often applied to increase the compatibility between containers and relatively aggressive products. Linings are used less frequently in the rest of the world, sometimes because marketers will accept shorter shelf lives and partly because linings are often not really essential, eg, for a number of anhydrous formulations. The linings are applied to the body and end sections while they are still "in the flat." This is also the case for exterior coatings, including lithography, which are applied right after the linings have been roller-coated on and thermally cured (1,2).

All single linings are either epoxy–phenolic or urea–formaldehyde–epoxy hybrids in composition. A standard 2 in. × 2 in. (51 × 51 mm) coupon carries ca 9–11 mg of these linings. For double-lined cans, a top coat of modified vinyl is applied at about the same weight per coupon. When extra protection is needed, the vinyl is made more impermeable by adding various beige, tan, gray, or blue–gray pigments; such top coats are called organosol linings. If still more protection is required, the welded seam is striped or the formed body and ends may be organosol spray-lined just before final curing and assembly. If the product contains methylene chloride, acetone, or some other solvent that can dissolve vinyls, then a urea–formaldehyde–epoxy base coat and side stripe might be advised, plus a phenolic top coat. Several can companies have the very resistant organosol linings for ends, but not yet for bodies. A more detailed description of linings is given in Ref. 1.

Pressure resistance. The 70°F (21.1°C) pressure range of aerosols is ca 20–115 psig (239–894 kPa), although most fall in the 32–100-psig (322–791-kPa) area. However, aerosols are regulated according to their pressure at 130°F (54.4°C). The Department of Transportation (DOT) Dangerous Goods Transport Regulations (3) divide regular aerosol dispensers into three categories according to pressure resistance (see Table 3). The DOT regulations also require the hot-tanking of finished aerosols, using methods to ensure that the pressure in the can rises to the 130°F (54.4°C) predetermined value. This is done by drawing the cans through a long trough of water, heated from ca 135 to 160°F (57.2–71.1°C) for total immersion periods of about 40–120 s, depending upon design (4).

Aluminum. Aluminum aerosol cans have been available since about 1948. The present annual requirement of ca 310 million (10^6) units is handled by a domestic production base of ca 180 million (10^6) per year, plus 80 million (10^6) imported empty cans and 70 million (10^6) imported filled cans. The filled cans are nearly all deo-cologne and mousse products made in Europe. Virtually all mousse aerosols are aluminum.

Unlike three-piece welded steel cans, aluminum containers are seamless and described by their diameter and total height. A 2.089 × 6½ in. (53.06 ×165.1 mm) U.S. size, which holds 325 mL brimfull, would be described as 2.052 × 6.500 in. (52.12 × 165.1 mm) in Europe, based on the inside diameter. Aluminum cans (see Cans, fabrication) are available in very small and very large sizes owing to the extreme flexibility of the impact-extrusion process used to produce them. On a worldwide basis, they are available from ⅝ × 2 13/16 in. (16 × 71 mm) 10-mL capacity, to 2.598 × 12 ⅜ in. (66 × 314.3 mm) 993-mL capacity. The upper limit is set only because of EEC regulations that disallow any aerosol can with an overflow capacity of over 1 L (see Laws and regulations, EEC).

The American Can Co. produces two kinds of two-piece aluminum cans: a regular type and a piston type that can be used for products such as cheese spreads, cake icings, toothpaste,

Table 3. Pressure, Thickness, and Inspection Limits for Aerosols

Parameter	Nonspecification	DOT Specification 2P[a]	DOT Specification 2Q[a,b]
DOT product pressure at 130°F (54.5°C), psig (MPa)	140 (1.07)	160 (1.20)	180 (1.34)
DOT plate thickness, min, in. (mm)	none	0.007 (0.178)	0.008 (0.203)
DOT inspection to destruction, min[c]	none	1/25,000	1/25,000
unofficial industry minimum, permanent distortion pressure, psig (MPa)	150 (1.14)	170 (1.27)	195 (1.45)
bursting pressure DOT min, psig (MPa)	210 (1.55)	240 (1.76)	270 (1.96)
unofficial industry minimum, psig (MPa)	225 (1.65)	255 (1.86)	285 (2.07)

[a] The DOT 2P and 2Q plate thickness minimums were designed for steel; aluminum containers must have thicker walls. Also, the 70-lb (31.8-kg) plate for DOT 2P cans and 80-lb (36.3-kg) plate for DOT 2Q cans are selected from coils where over 99.69% (three sigma) of the area is over 0.0070 in. (0.1778 mm) and 0.0080 in. (0.2032 mm), respectively.

[b] The DOT 2Q specification is not offered in the 300 (76.2-mm-) dia can size.

[c] Destructive inspection is generally done by the canmaker, except for aluminum cans, which must be inspected by the marketer or contract filler.

etc. In the latter form, a polyethylene (PE) piston in the shape of an inverted 2 × 1.6 in. (51 × 41 mm) cup is inserted in the finished can dome–body section before the base is attached by standard double seaming. The product is filled into the can, a valve attached, and the can passed to a gasser–plugger machine, where a few grams of food-grade propane–isobutane liquid propellant is added under pressure through a preformed hole in the base. A neoprene rubber cord ca 0.1875 in. (4.76 mm) in diameter is then tightly forced into the hole and cut off, thus sealing the gas–liquid into the bottom portion of the can below the piston, causing it to exert an upward force on the piston that acts to extrude nonaerated product when the valve is actuated. Several piston-type cans are available in Europe, and more will become available in the United States.

Aluminum aerosol cans are often lined and can be decorated in up to about six colors, depending upon available equipment. Interesting brushed effects are possible. Some firms are able to base coat the bottom of the can, whereas others are not. Special linings are required for hair-mousse products, hydroalcoholic–surfactant dispersions that are too aggressive for double-lined tinplate cans, although with large amounts of inhibitor (such as 1.4% Q.A.I., a quaternary ammonium nitrite) or triple coatings, some progress is starting to be made (5).

Aluminum aerosol cans generally have full-diameter metal or plastic caps, giving the unit an elegant cylindrical profile. Gold-anodized aluminum caps are especially preferred for fragrance products: deo-colognes, colognes, sachets, and meter-spray perfumes, whereas the plastic caps find greatest use in the pharmaceutical area. Among the plastic caps are those that are best described as spray caps, such as those used for bronchodilators, where sufficient pressure on the cap actuates a metered amount for inhalation.

Special-feature aerosols. Several special-feature aerosol cans are available. At one time, American Can Co. was marketing a rim-vent release (RVR) can, which had indentations in the top of the double seam at frequent intervals. When pressurized to ca 185–215 psig (1.38–1.58 MPa), depending upon plate selection, the dome would evert, tearing open the work-hardened indentations and creating a series of tiny apertures that would allow the excess pressure to escape before the can could burst. This safety can, as it was called, was offered by other canmakers under license, but in addition to the extra cost there were several disadvantages. No more than about 40 million (10⁶) were ever sold in any one year.

In addition to the piston types, there are other compartmented cans available. The most popular is the Sepro Can (Continental Can Co., U.S.A.), which consists of an accordioned plastic alloy bag within a 202 × 509 (54 × 141 mm) can; the bag is fitted to the aerosol by wrapping partway around the bead at the top of the dome. After filling the bag almost to the top, a standard valve with a slightly undersized mounting cup is fitted into the neck of the bag and crimped outward. The cup is compressed against the bag and the can bead by this expansion process, making an hermetic seal. After this, a device is used to insert a few grams of isobutane gas–liquid into the can (and around the bag) through a loosely rubber-plugged hole in the center of the base, after which an air-operated ram drives the plug tightly into the hole for a seal.

Sepro Cans may be used to dispense nonaerated products, such as pastes, gels, greases, lotions, petroleum jellies, caulks,

syrups, and creams. By including ca 2% of a mixture of isopentane:isobutane (90:10) bp = 79°F (26°C) in certain gel-shaving-cream concentrates and packaging them in the Sepro Can, they may be dispensed as clear gels that magically foam up when placed in the palm of the hand or when stirred slightly with a finger. The postfoaming gels constitute the main commercial application for the present 40 million (10⁶)/yr Sepro Can market (1,6).

A similar can is produced under license from the Presspack Corp., Bronxville, N.Y. Known as the Presspack System, it uses a 202 (54 mm) dia can with inner plastic bag. The bag has a shape designed to fit the can, but with vertical channels to facilitate about a 97% collapse as the product is dispensed. Unlike the Sepro Can unit, the Presspack System can be filled by adding concentrate to the bag, laying the 1-in. (25.4-mm) valve cup on the neck of the bag as it protrudes slightly from the top bead of the can, gassing the "exo-space" between bag and can, using a Kartridg Pak Co. Under-the-Cap (U-t-C) gasser, and then pressing the valve into place and crimp sealing, using the same rotary machine. The Presspack System is still experimental.

In Europe, the Compack unit has been introduced by Boxal/Alusuisse, and the Alucompack, Alupresspack, and Microcompack units by Aerosol Services, SA, Möhlen, Switzerland. In each case, the can is aluminum. In the Alucompack and Microcompack dispensers, the inner container is a very thin-walled aluminum tube. These thin tubes are flanged and gasketed at the top so that a hermetic seal can be made when the valve cup is applied. Because there is no bottom hole in the outer cans, liquid chlorofluorocarbon or hydrocarbon propellant must be introduced atmospherically by pouring in a few grams before inserting the filled inner tube of product and sealing. The extreme flammability of the liquid hydrocarbons has negated use of these dispensers in the United States, where government regulations have banned the use of nonflammable chlorofluorocarbons except in a few products. The Compack and Alupresspack units are plastic inner bags and are similar to the Presspack System just described. All of these systems have found very limited use so far.

A system with good potential is the Enviro-Spray (Enviro-Spray Corp.). A standard aerosol can may be used, eg, a 202 × 509 (54 × 141 mm) size. In the filling process, the product is added to the empty can and a very advanced pouch is inserted. The valve is placed on the can and crimped in place. As a water-soluble poly(vinyl alcohol) film dissolves, the pouch allows a concentrated citric acid solution to contact a compartment containing sodium bicarbonate powder, producing a predetermined amount of carbon dioxide gas. The amount is sufficient to pressurize the container to ca 90–100 psig. (722–791 kPa) by inflating the pouch to a certain extent. As the contents are dispensed, the pressure tends to decrease. Further expansion of the bag then occurs, rupturing an inner wall and releasing another capsule of citric acid to react with the bicarbonate. The process repeats itself about six or seven times during the life of the dispenser, keeping the pressure to within fairly narrow limits. All the gas remains within the bag, so that the product can be delivered nonaerated, unless a low pressure propellant is added to produce a foam, a finer spray pattern, or some other special effect.

Glass aerosols. Glass aerosols make up about 2.2% of the total aerosol market. In the United States in 1983, about 45% were fragrance products and the rest were pharmaceuticals.

Only about 25% of the fragrance bottles were plastic-coated, but at least 70% of the pharmaceutical containers were coated. About 27% contained chlorofluorocarbon propellants, such as P-11, P-12, and (less commonly) a mixture of P-142b:P-22 (60:40). The remaining 73% was pressurized with hydrocarbon propellants or, sometimes, dimethyl ether (DME). Virtually all these bottles have a so-called 20-mm finish, with an orifice size of 0.402 in. (10.2 mm), min.

The bottles are usually made from USP Type III glass of the conventional soda-lime formula (see Glass container manufacturing). If they are to be used without plastic coating or have marginally acceptable geometry from a pressure-resistance standpoint, they are toughened by a "hot and cold end" process that causes titanium(IV) silicate or tin(IV) silicate to be surface-deposited on the exterior. This reduces the chance of scratching, bruising, or abrasion, any of which greatly enhances the possibility of rupture. The silicate film is very lightly sprayed with a lubricant as the ware energes from the cooling end of the annealing lehr. Plain glass aerosols are generally less than 1-fl oz (29.6-ml) capacity to minimize the ill effects of dropping them inadvertently on a hard surface. The smaller bottles have more resistance to bursting, but if they do rupture, the amount and size of flying glass fragments is limited, as well as any hazard from the flammability of the gas–liquid cloud that also forms. The relative danger from fragmentation is further reduced if a relatively low pressure formulation is used, and chemists try to keep pressures below 18 psig at 70°F (226 kPa at 21.1°C).

Glass aerosol coatings in the United States are now all of the bonded type, attached to the glass surface with a clear adhesive. The plain bottles are heated slightly and then passed through a dip machine where they are immersed to the neck in a bath of warmed adhesive. After the excess adhesive is electrostatically drained, the bottle is heated electrically to bring about partial curing the remaining film. In similar fashion, the bottle is then passed through a fairly viscous bath of plastisol: a concentrated colloidal dispersion of high molecular weight PVC in a plasticizer blend. The fluid clings to the bottle as a gel structure of little physical integrity, and the excess is drained off. The bottle, with its 0.02-in. (0.5-mm) coating layer, is heated to fuse the solid resin, bond it to the adhesive, and develop tensile strength, tear resistance, and impact absorption properties. The coated bottle then passes through a final dip machine containing a dispersion of HMW-PVC with a different molecular orientation. Upon curing, the top layer bonds to the "rigisol" lamination below and forms a relatively soft covering that is still high in tensile strength and provides good impact-absorbing properties. The bottles are then trimmed of plastic at the neck and decorated. On bottles up to 4-fl oz (118.3-mL) nominal size, the total coating thickness is 0.035–0.055 in. (0.89–1.4 mm).

Informal industry standards suggest that pressures not exceed ca 40 psig (377 kPa) in coated bottles up to 3-fl oz (88.7-mL) nominal size and 30 psig (308 kPa) in bottles of 3–4 fl oz (88.8-118.3 mL). In still larger bottles, pressures should be kept below 25 psig (274 kPa). The largest coated glass aerosol available in the United States today has a nominal a size of 10 fl oz (295.7 mL) and actual overflow capacity of 12 fl oz (355 mL) (1). There is no regulatory requirement for bottles smaller than 4 fl oz (118.3 mL), but DOT shipping regulations limit larger bottles to pressures of 25.3 psig at 70°F (276 kPa at 21.1°C) and 99.3 psig at 130°F (786 kPa at 54.4°C) unless the content is flammable, in which case the pressure limits drop too low for good aerosol operation (7).

Glass aerosols, being under internal pressure, are of heavier construction and have fewer design options than competitive glass pump sprayers, which are almost indistinguishable from them. Marketers also recognize that the plain glass pump sprayer does not burst apart if the bottle is dropped and cracked nor release a cloud of potentially flammable hydrocarbon or dimethyl ether gas. The mean particle size of the aerosol spray for perfumes and colognes can be readily matched by the pump sprayer. For all of these reasons, there has been a lack of marketer interest in glass aerosol fragrance products, even though relatively nonflammable propellants are available. The situation is different in the pharmaceutical industry, where the particle-size distribution must often be much finer than can be obtained with pump sprayers. For example, the mean particle size of a bronchodilator spray is ca 4 μm, compared with an ethanol-based pump spray formula of ca 60 μm. The dispersing effect of the propellant makes the difference; many bronchodilators have 75–90% chlorofluorocarbon propellant (8).

Aerosol operation. The aerosol dispenser is basically just a container plus a valve mechanism hermetically sealed to it. All steel-based and most aluminum aerosol cans have the so-called 1-in. (25.4-mm) beaded orifice, with a valve cup inserted into the orifice and crimped tightly into it by the expanding action of a crimping collet. The typically 0.970-in. (24.6-mm) dia well of a tinplate mounting cup (with Dewey-Almy GK-45-NVH gasket) is spread to ca 1.070 in. (27.18 mm) at a depth of ca 0.180 in (4.57 mm) below the top arch of the cup and measured to the centerline of the crimp indentations. Either six or eight indentations are made, depending on whether the hardened steel collet has six or eight segments. The eight-segment collet, introduced in 1982, is fast becoming the industry standard.

The integrity of the gasketed seal between valve cup and and bead has been the subject of much research, many articles, and even a fair number of lawsuits. If the seal is grossly defective, the aerosol will leak during the hot-tanking process, as evidenced by gas bubbles of propellant rising through the hot water. However, some immediate leakers lose gas so slowly that they are not normally detected in the hot tank. In rare cases, a dispenser will hold the product for a matter of weeks, even many months, and then leak; these are called latent leakers. The slow leakers and latent leakers often get into the field undetected. About one can in every 2000 will leak sufficiently to become defective or totally inoperative, either in a warehouse, retail store, or in the hands of the consumer. The normal leakage rate for aerosols is 1–2 g/yr at 70°F (21.1°C) and 1.6–3.5 g/yr at 100°F (37.8°C). Strong solvents almost always increase the seepage rate. As a rule, a container leaking at over 3.5%/yr is rejectable (1,9).

A different kind of valve is used for relatively small aluminum containers and all glass bottles. Most of them have what is termed a 0.79-in. (20-mm) ferrule or cap that is clinched inward around the neck finish of the container, much like the cap of a beverage bottle. The very small aluminum and glass containers may use a 0.51-in. (13-mm) ferrule size. A clinching tool with about 16 tines (joined to a tubular upper section) is brought down over the ferrule, after which a mandrel or

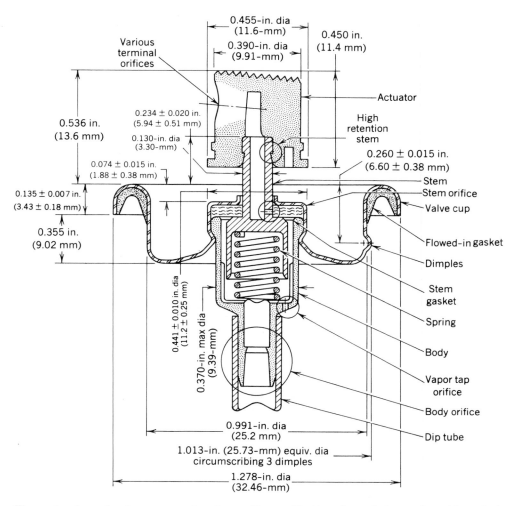

Figure 3.   Aerosol valve component and assembly specifications for male type valve with conical cup and standard unrestricted body tailpiece orifice system.

plunger is forced downward over the tapered tines, squeezing them inward against the ferrule wall so that it is tucked beneath the bottle finish or can curl to effect the seal.

A great many aerosol-valve designs are available, from about 11 U.S. manufacturers, and from at least 20 larger valve makers in other countries. They are vertical- and toggle-acting. Some actuator buttons fit on valve stems; others have integral stems and fit into the stem gasket well within the housing or body section. Most valves have a PE dip tube.

A vertical-acting valve is operated by pressing the button or actuator pad downward with a force of at least 1.6 lb (726 g). This presses the valve stem down, against a coil spring in the housing, bringing a small cross-hole (the stem orifice) to a position below the stem gasket. Product then flows up the dip tube and through the housing, the stem orifice, the stem, the actuator orifice, and into the atmosphere (see Figs. 3 and 4). Some valves have a housing or body orifice; others may have a vapor-tap orifice drilled through the housing. The latter allows propellant gas to flow directly into the valve and out through the actuator, thus making the spray finer and softer, while reducing delivery rate. With larger diameter vapor-tap valves, sufficient propellant may escape during the operation of the

dispenser that, near the end, the spray becomes noticeably coarser and the delivery rate dwindles. Most antiperspirants and air fresheners have vapor-tap valves.

The toggle-acting valve is sprayed by pressing a chamfered button sideways, exposing the stem orifice on the low side of the base to the housing cavity (see Fig. 5). Product flow then commences. Body orifices and vapor taps may be used with this valve design as well. Of some interest is the fact that such valves are highly directional, and there is little doubt of the direction of the spray when actuated.

Valves designed for inverted spraying have either no dip tube or a dummy dip tube (for automatic valve tipping into the container) and an orifice directly into the side of the housing. Valves for both upright and inverted spraying, such as a bathroom-cleaner spray, may have an enlarged diameter "jumbo" dip tube of ca 0.265-in. (6.73-mm) inside diameter, able to hold perhaps 8 g of product for intermittent inverted spraying, or a somewhat more costly "Spray Anyway" (Gulf Oil Co., Seaquist Valve Co.) accessory that operates on a ball valve principle, by which the ball position determines if a dip tube or housing-area orifice will be opened (10).

The internal gasket of an aerosol valve is commonly made

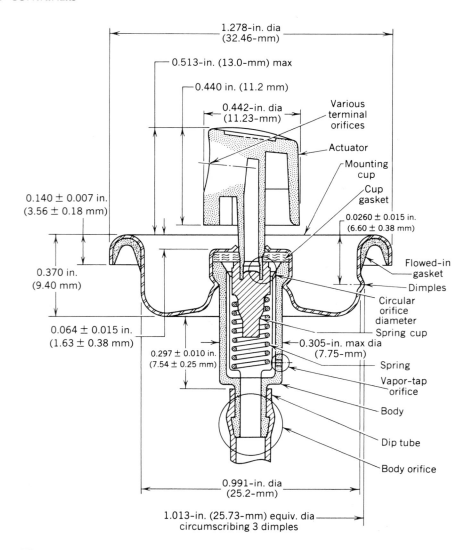

Figure 4. Aerosol valve component and assembly specifications for female type with conical cup and standard unrestricted body tailpiece orifice system.

of either Buna-N (acrylonitrile–butadiene) or Neoprene 759 (polychloroprene). The latter is generally better when large percentages of methylene chloride are included in the formula. The more costly Viton (Du Pont Co.) gasket material is sometimes used because of its resistance to most high solvency formulas. However, it is not resistant to dimethyl ether propellant and diethyl ether solvent. Butyl rubber is good for these ethers and for many other compositions (11).

The aerosol valve routinely includes parts of nylon or acetal plastic for the body and stem, PE or (rarely) polypropylene (PP) for the dip tube and button, stainless steel #302 or #305 for the spring, and tinplate, aluminum, or (rarely) stainless steel #304 for the mounting cup or ferrule. These materials must work in harmony with each other, with a wide variety of products and over temperatures that easily range from 0 to 130°F (−17.7 to 54.4°C). It is a tribute to the valve-making industry that they can produce such valves in a myriad of variations, at speeds of 300/min and price them at less than $40 per thousand. Typical valve assemblies can be seen in the figures.

Aerosol caps. All aerosols have overcaps or spray domes. The spray dome is a combination cap and valve actuator that snaps on to the valve cup just below the cut edge. In the early days of the industry, virtually all caps were of enameled tinplate, but now they are almost exclusively PE or PP. Most are full diameter, ie, the same diameter as the can. Both tamper-proof and tamper-evident caps or related fitments are available (see Tamper-evident packaging). They are often used for paint products. Child-resistant closures are also made for certain insecticides and for some caustic oven-cleaner aerosols (see Child-resistant packages) (1,12).

Nonchemical Sprayers

Some container manufacturers have seized upon the chlorofluorocarbon propellant vs ozone controversy of 1975, as well as the hydrocarbon-propellant flammability issue of the 1980s (13), to suggest that nonchemically powered sprayers are better, safer, and perhaps less costly. The user must operate a pumping-type device in the bottom of the dispenser to build air pressure within it. For instance, Airspray Interna-

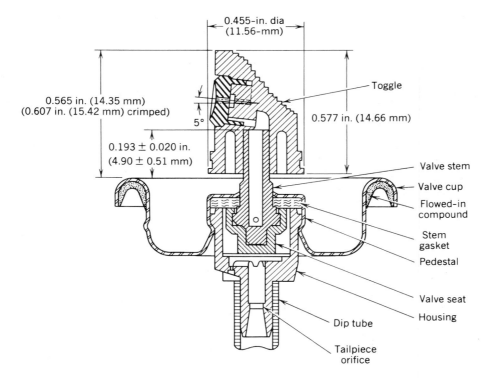

Figure 5. Aerosol valve component and assembly specifications for a toggle action valve with flat cup and standard unrestricted body tailpiece orifice system.

tional BV of the Netherlands has begun worldwide marketing of a spray system based on a cap and valve method, using air only as the propellant. The cap is fitted with a round plunger that is pumped several times within a small plastic chamber, thus building the necessary air pressure within the plastic bottle or can to provide the propelling force to discharge the contents. The pressure can be maintained simply by repeating the pumping action. The system is patented worldwide and is available in refillable and disposable versions. The Airspray system is comparable in cost to aerosols, trigger sprays, and fingertip sprays.

Another product is the Twist-N-Mist, an aerosol-type spray developed by the CIDCO Group, Inc., Aurora, Colo. The device functions much like a syringe. A quarter turn of the twist top draws the liquid inside the can into a rubber bladder. Another quarter turn in the opposite direction builds air pressure around the bladder. When the button is pressed, the air pressure forces the liquid out. At this time, there are still no commercial products in these containers, which can be metal or plastic and refillable or disposable.

As a final example, Container Industries, Inc., Londonderry, N.H., introduced its Exxel elastomer package in 1981, rivaling an earlier composite container marketed in the 1970s by the Selvac Division of Plant Industries, Inc. Both containers consist of a posted heavy rubberized bag mounted within a formed can, plastic bottle, or other rigid dispenser. The bag is pressure-filled with product, such as a lotion, grease, or gel. Upon pressing an actuator, the contents are released as the bag seeks to return to its original deflated state.

All of these innovations, although interesting, have had no noticeable impact upon the aerosol market and are extremely limited in product selection and dispensing formats.

Mechanical Pump Sprayers

The mechanical pump-spray package is a formidable challenger to the aerosol in a number of specialty-product categories. The sprayer design may be either the push-button type or the trigger type (see Fig. 6) (14). The blow-molded container is generally HDPE, but PP, PVC, and PET are used in certain applications. The small sizes, such as cologne and perfume pump sprays, use glass containers. Since the sprayer is relatively costly, many marketers offer refill bottles with a simple screw cap. The refill may be the same size, but is often larger than the original bottle.

The principal product category marketed in pump-spray form is the household specialty cleaner. For example, the 55 million (10^6) unit per year aerosol window-cleaner market of a decade ago has been almost eliminated in favor of the trigger-pump and refill alternative. During the late 1970s, the push-button sprayer made a significant inroad into the troubled aerosol hair-spray market, rising from 4.2% of all hair fixative sales in 1974 to 25.5% in 1980. During 1973, aerosol sales of fragrance products were 169 million (10^6), according to a Commerce Department survey; by 1977, they had fallen to a mere 11 million (10^6) units and pump-spray colognes and perfumes grew to well over 150 million (10^6) units (15).

The only unique thing about the pump-spray package is the mechanical sprayer itself. The sprayer is much like the aerosol meter-spray valve in that it provides a shot or dosage of product. A certain volume of liquid is dispensed for every full actuation once the pump sprayer is primed and full of product. When the pump sprayer is actuated, a mild vacuum is formed in the body of the valve which acts to draw the product up the PE dip tube so that it can fill the metering barrel. Meanwhile,

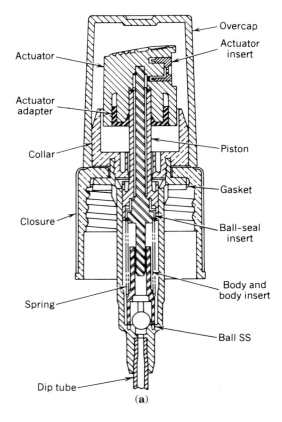

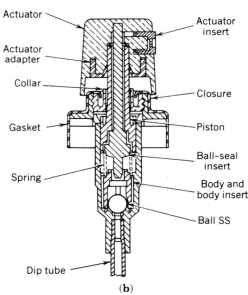

Figure 6. Model SL-200 (**a**) and SL-400 (**b**) Micromist pump sprayers.

strong solvents, dispersions containing solid ingredients, and products containing two liquid phases which would stratify in the metering barrel of the sprayer. Products with viscosities over ca 300 cP (3 Pa·s) should be avoided since the mechanical break-up action of the spray head is insufficient to produce a spray in such cases. Also, refilling of the barrel may be so slow that repeated actuations produce reduced dosages, plus sputtering. Any product that is adversely affected by air (specifically oxygen or carbon dioxide) or microorganisms should not be dispensed by the pump-spray methods. Although pump-sprayer improvements were made during the late 1970s, and to some extent during the 1980s, the mean particle size of the spray is always ca ≥2.36 mil (60 μm). This is because no flashoff action of a propellant is available, as is the case with most aerosol packages. The break-up efficiency of the swirl chamber in the terminal orifice area is the sole determinant of particle size. The efficiency can be more fully realized in the case of low viscosity formulas, volatile products, and for smaller metering valves (the barrels can be as small as 40 μL (0.04 cm³) that allow more air pressure to drive the product through the chamber.

BIBLIOGRAPHY

1. M. A. Johnsen, *The Aerosol Handbook,* 2nd ed., Wayne E. Dorland Co., Mendham, N.J., 1982, p. 65.

2. H. R. Shepard, ed., *Aerosols: Science and Technology,* Interscience Publishers, Inc., New York, 1961, p. 65.

3. T. A. Phemister, *Tariff No BOE-6000-A,* Bureau of Explosives, Association of American Railroads, Washington, D.C., Dec. 18, 1980, Sec. 173.306(a)(3).

4. M. A. Johnsen, *Aerosol Age* **28**(6), 26 (June 1983); also in two subsequent issues.

5. M. A. Johnsen, *Aerosol Age* **29**(8), 42 (Aug. 1984).

6. P. A. Sanders, *Principles of Aerosol Technology,* Van Nostrand Reinhold Co., New York, 1970, p 350.

7. *The Aerosol Guide,* 7th ed., The Chemical Specialties Manufacturer's Association, Inc. (CSMA), Washington, D.C., 1980, p. 42.

8. C. N. Davies, ed., *Aerosol Science,* Academic Press, Inc., New York, 1966, p. 411.

9. J. J. Sciarra and L. Stoller, *The Science and Technology of Aerosol Packaging,* John Wiley & Sons, Inc., New York, 1974, p. 317.

10. P. A. Sanders, *Handbook of Aerosol Technology,* 2nd ed., Van Nostrand Reinhold Co., New York, 1979, p. 103.

11. A. Herzka, ed., *International Encyclopaedia of Pressurized Packaging (Aerosols),* Pergamon Press, Inc., Elmsford, N.Y., 1966, p. 109.

12. Herzka and J. Pickthall, *Pressurized Packaging (Aerosols),* 2nd ed., Butterworth & Co., Ltd., London, 1981, p. 112.

13. *Hydrocarbon Propellants: Considerations for Effective Handling in the Aerosol Plant and Laboratory,* The Chemical Specialties Manufacturer's Association (CSMA), Washington, D.C.

14. M. A. Johnsen, *Aerosol Rep. Switzerland* **21**(2/82), 63 (1982).

15. M. A. Johnsen, *Aerosol Age,* **29**(2) 30 (1984).

MONTFORT A. JOHNSEN
Peterson-Puritan Inc.
Div., Metal Box, p.l.c.

PRINTING

Because the package influences the choice of a product at point of sale, quality, consistency, and integrity in printing the

the material already in the barrel is forced out, through the button or terminal orifice, by the air pressure caused by actuation. The barrel then refills. A stainless-steel (#302) spring restores the button or trigger to its original position. The pump sprayer is made integral with the screw cap and lining.

The pump sprayer is substantially more complex than the standard aerosol valve, and correspondingly less tolerant of

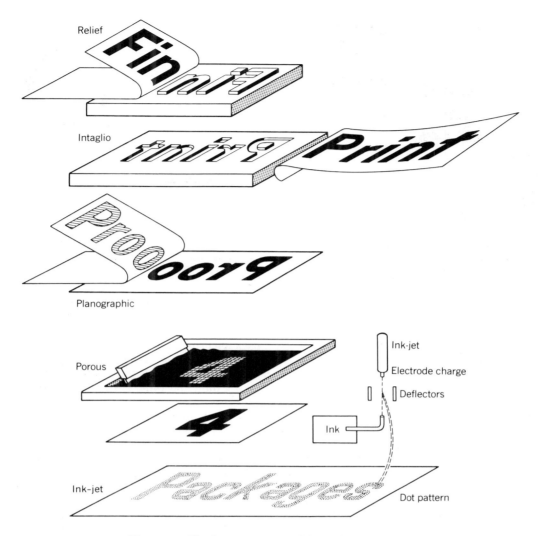

Figure 1. The five processes used for package printing.

package are of paramount importance. All printing production can be roughly divided into four major divisions: (1) art and copy preparation provides original images to be reproduced; (2) the original art is converted into printing image carriers that are inked and used to print an image; (3) presswork produces the final printed image; and (4) finishing operations convert the substrate into its final form. Today, packages are printed by relief, planographic, intaglio, porous, and impactless printing methods (see Fig. 1) each with its own unique characteristics, applications, advantages and disadvantages.

Relief Printing

Relief printing, often called letterpress printing, is a long-established method that originally used cast-or etched-metal type. Today, much of the metal has been replaced by synthetic-rubber or photopolymer printing plates. In relief printing the images (printing areas) are raised above the nonprinting areas. The ink rollers (see Inks) touch only the top surface of the raised areas. The surrounding (nonprinting) areas are lower and do not receive ink. Letterpress printing presses include platen, flatbed (cylinder), and rotary presses. A typical rotary letterpress configuration is shown in Fig. 2. Ink is placed in an

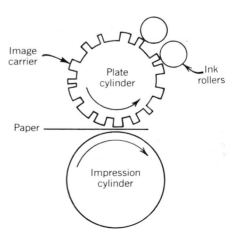

Figure 2. Diagram of a rotary press.

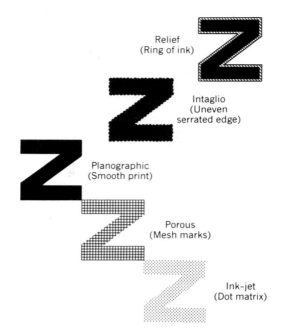

Figure 3. Distinguishing characteristics of printing processes.

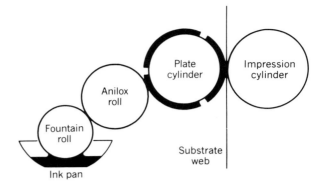

·**Figure 4.** Flexo printing station. The fountain roll (rubber) turns in ink to deliver ink to a steel or ceramic inking (anilox) roll, a plate cylinder, and the impression cylinder. Courtesy of the Flexographic Technical Association.

ink fountain and travels through a series of steel and elastomeric-covered rollers to smooth out the ink and deliver a controlled amount to the printing surface. Plates are fastened to the plate cylinder and receive ink as they contact the inking rollers. An impression cylinder (platen) presses the substrate against the inked plate cylinder (flatbed) and transfers the image. Sheets of paper are printed on sheet-fed presses; rolls of paper on web-fed presses. Letterpress printing applications include the production of corrugated and folding boxes, kraft bags, labels, blister and skin-pack cards, and all types of bags for dry goods, as well as a variety of job-printing applications. A distinctive feature for recognizing letterpress printing is a "ghost-like" image around each character caused by the ink spreading slightly due to the pressure of the plate on the substrate. Sometimes, a slight embossing or denting appears on

the reverse side of the surface, but the letterpress image is usually sharp and crisp (see Fig. 3) (1,2).

Another relief-printing technique with widespread use in package printing is flexography, unique because it was developed primarily for printing packaging materials. Flexo prints from a raised image using flexible elastomeric plates and thin, highly fluid and rapid-drying inks that are both solvent and water reducible. A typical flexo printing unit generally consists of four rollers (see Fig. 4). A rubber fountain roll revolves in an ink reservoir and delivers ink to a steel inking (metering) roll, often known as an anilox roll. The anilox roll is mechanically or laser-engraved, and then chrome or ceramic coated. It is used to deliver a controlled film of ink from the fountain roll to the printing plates. The printing plates are attached to the plate cylinder mechanically, magnetically, or with double-sided tape in such a way that their raised surfaces contact the anilox roll. The substrate passes between the plate surface and an impression cylinder causing the image to be transferred. Flexo press configurations include stack, in-line, and central-impression (CI) types (see Fig. 5). Stack presses have 1–4 individual printing stations mounted one above another on each side of a vertical frame. In-line flexo presses

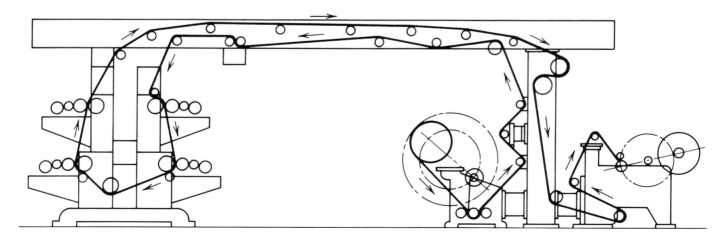

Figure 5. A schematic drawing of a four-color wide-web stack flexo press, type A. The wide-web press can print on papers, films, foils, and high density polymer films in widths of 65–>150 in. (1.65–>3.8 m). Courtesy of the Flexographic Technical Association.

have tandem printing stations placed in a row. Central-impression presses use a common impression cylinder around which 2–7 printing stations are positioned. Flexo has numerous advantages in package printing including the ability to print on a wide variety of substrates. Virtually anything that can go through a web press can be printed by flexography. Its quick-drying fluid inks save energy costs; the increased use of water-based inks reduces air pollution problems; variable repeat (image) lengths and quick makeready and changeover are possible; and equipment costs are less than for other processes. Flexo has made inroads in markets (eg, newspapers) that were traditionally served by other processes. The advent of flexography for printing foils and cellophane was of crucial importance in the development of the flexible-packaging industry. Some limitations of flexo include difficulty in getting a sharp image due to the squeezing action during image transfer, and the tendency for fine type and reverses to fill in due to the low ink viscosity (3).

Planographic Printing

Planographic printing prints from a flat surface. The image and nonimage areas are on essentially the same plane of a thin metal plate (see Fig. 1). The process today is most often known as "lithography" or "offset lithography."

Lithography works on the chemical principle that grease (ink) and water (fountain solution) do not mix. Plates for offset lithography (usually aluminum) are clamped into slots built into the plate cylinder. In operation (see Fig. 6), ink, similar in viscosity to letterpress ink, travels from an ink fountain through a series of rollers that spread the ink evenly and deliver it to the plate. At the same time, water (fountain solution) is traveling through the dampening system and also reaches the plate surface. The plate accepts ink and repels water in the image areas and attracts water and repels ink in the non-image areas. The image on the plate is transferred or "offset" to an intermediate (blanket) cylinder covered with a rubber blanket. The paper picks up the image as it passes between the blanket and the impression cylinder.

Presses for offset lithography are all based on the rotary principle and are available as sheet- or web-fed units. Sheetfed presses are available in many sizes up to six units, which means that six colors can be printed in one pass through the press. Web-fed presses can use eight units. Offset lithography also prints on metal, using special presses designed to handle the heavier metal sheets, in a process known as metal decorating. Many web presses are "perfecting" presses, which can print the substrate on both sides in one pass through the press.

A major advantage of this process is that the soft rubber blanket creates a clear impression on a wide variety of materials, allowing for the extensive use of illustrations. Printing by offset lithography is characterized by smooth, sharp images without the impression or ring of ink characteristic of letterpress. A disadvantage of offset is that for some applications the ink coverage is too light. Problems exist in obtaining consistent colors and special operations are needed for imaging foils and films. Offset is used in producing labels, cartons, and metal containers (4, 5).

A process with characteristics of both relief and planographic printing methods is known as letterset, dry offset, or indirect letterpress. This process, which combines the features of relief printing with indirect image transfer, can be used with conventional sheet- or web-fed presses. Letterset was originally developed to provide letterpress with some of the advantages of offset lithography without the problems encountered with offset's dampening systems. Today, letterset printing typically handles jobs which need an offset process but which cannot be done by lithography. It provides very-low-odor printing for food cartons and confectionary wraps that require the use of water miscible inks. It is also used for printing on plastic bottles, cups, and tubs, and on some metal packages, primarily toothpaste tubes (6) (see Tubes, collapsible).

Intaglio Printing

Intaglio, often called gravure or rotogravure, prints from a sunken or depressed surface and is the only process in which there are no plates to attach to a cylinder. Instead, the plate cylinder is machined to contain the image areas, which consist of cells or wells etched into a copper cylinder. The nonprinting area is the cylinder itself (see Fig. 7). The plate cylinder ro-

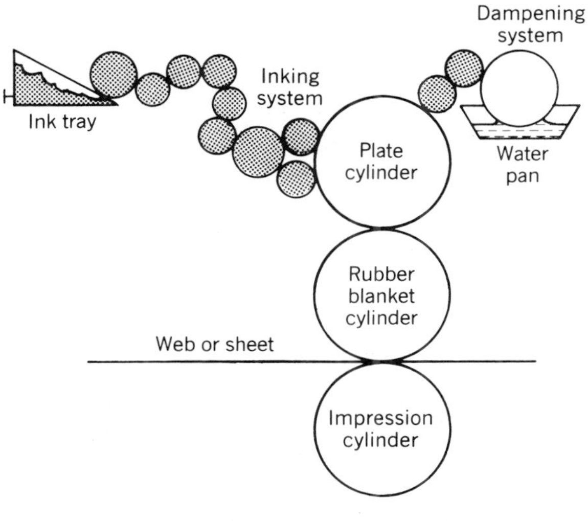

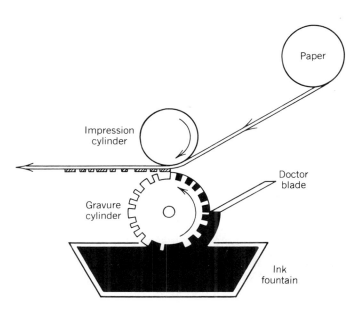

Figure 6. Offset lithography printing station.

Figure 7. Diagram of intaglio process.

tates in a fountain of ink and the individual cells become filled with the thin, fluid ink. As the cylinder rotates and comes out of the ink fountain, the excess ink is removed by the wiping action of a doctor blade which rides on the surface of the gravure cylinder and scrapes away the ink in the nonimage areas. As the cylinder comes in contact with the substrate, an elastomer-covered impression cylinder presses the material against the gravure cylinder and the ink in the cells is transferred to the substrate through capillary action.

Gravure printing can easily be recognized because the entire image area (type and illustrations) is screened to produce the tiny cells in the gravure cylinder (see Fig. 3). Gravure screens usually contain 150 lines per inch (59 lines per cm) but finer screens are often used for high-quality printing. Gravure printing is capable of extremely long press runs because there is little wear on the sunken image areas. However, the preparation of gravure cylinders is a time-consuming and costly process that typically limits gravure printing to long press runs.

Presses for gravure printing are generally of an in-line design with individual stations and dryers arranged in a row. Much like flexography, gravure can print on a wide variety of substrates as the material passes through as many as eight individual printing units. Gravure printing has widespread use in packaging in the printing of cartons, foils, films, cigarette wraps, labels, bags, and laminations. The process is also used to apply special coatings like lacquers and varnishes (7, 8) (see Coating equipment).

Porous Printing

Porous printing, known as screen printing, involves forcing ink through a screen-mesh-supported stencil of silk, synthetic fabric, or stainless steel onto a substrate (see Figs. 1, 8.) A stencil, prepared either by hand or photographically, becomes the image carrier, with the nonprinting areas protected by the stencil. After attaching the stencil to the fabric and frame assembly, printing is accomplished by placing the substrate to be imaged under the screen, applying ink to the inside of the screen, then spreading and forcing ink through the openings in the stencil with a rubber squeegee. Presses for screen print-

ing range from simple, wooden-frame, hand-operated feeding, printing and delivery operations to fully automatic machines that position the item to be printed, pull the squeegee, and dry and stack the final product.

Screen printing is recognized by its heavy layer of ink (often ten times heavier than that of letterpress) and the texture of the screen on the final image. This is a very versatile printing process capable of printing on the widest variety of materials including the rounded and irregular surfaces of tubes, (see Decoration) and other surfaces. There is a wide array of inks available to the screen printer to handle many different substrates. Compared to other processes, equipment is inexpensive well suited for short-run, quality work. It is ideal for imaging glass, preparing promotional pieces for new products, box wraps, folding cartons, labels, gift wraps, and plastic containers. Disadvantages include relatively slow printing speeds, special jigs for holding various objects, and a frequent need for extra drying time (9).

Impactless (Ink-Jet) Printing

Unlike other printing processes that require contact between an inked image and the substrate, ink-jet is a noncontact printing process. Although there are several ink-jet technologies in current use, they all operate on the same basic principles. Jets spray electrically charged drops onto a substrate while dot matrices, housed in the printer's memory, electrostatically direct these drops to form characters. More drops are generated than are needed and those remaining fall into a gutter that returns them to an ink reservoir (see Fig. 1). Available since the mid-1960s, ink-jet printing found an immediate application in coding and dating (see Code marking and imprinting). Ink-jet's ability to project tiny drops of ink into various recesses, the availability of inks which adhere to nonabsorbent surfaces, plus the low cost and high speed compared to labeling have firmly established this process in the packaging industry (10, 11).

BIBLIOGRAPHY

1. V. Strauss, *The Printing Industry,* Printing Industries of America, Washington, D.C.; 1967, gives thorough discussion of printing processes.
2. *Pocket Pal,* International Paper Company, New York, 1983. Good overview of the graphic arts industry.
3. *Flexography Principles and Practices,* Flexographic Technical Association, New York, 1980. Standard reference work on the flexographic printing industry.
4. E. A. Dennis and J. D. Jenkins, *Comprehensive Graphic Arts,* 2nd ed. Bobbs-Merrill Educational Publishing, Indianapolis, Ind., 1983.
5. *The Lithographer's Manual.* Graphic Arts Technical Foundation, Pittsburgh; 1981.
6. "Letterset," in *Inklings,* No. 84, Coates Group of Companies, London UK 1979.
7. J. M. Adams and D. D. Faux, *Printing Technology,* Breton Publishers, North Scituate, Mass. 1982.
8. *The Romance of Roto,* The Gravure Research Institute, Port Washington, N.Y. Very good description of the gravure printing process and the industry.
9. W. P. Spence and D. G. Vequist, *Graphic Reproduction,* Charles A. Bennett Company, Peoria, Ill., 1980.

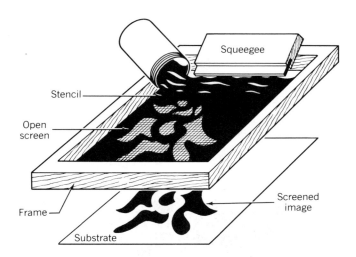

Figure 8. The porous process.

10. J. Lewell, *Graphic Arts Monthly* **54,** 93 (May, 1982).

11. "It's Sunnyside Up for Ink Jet Printing," *American Printer* **52,** 56 (Sept. 1980).

J. Lentz
Webcraft Technologies, Inc.

PULP, MOLDED

Like the paper-making process itself, the origins of pulp molding are pretty well obscured by antiquity. One assumes that artisans in early Oriental, Egyptian, and Greco-Roman civilizations who were involved with screen-felting of vegetable fibers into sheets of paper rapidly discovered benefits that derived from embossing the screening material to obtain contours. It is believed that ornate wall- and ceiling-molding artifacts of those civilizations were fabricated in this manner. With the rapid development of the paper industry (see Paper) around the turn of the century, disposable molded pulp products began to find their way into the mainstream of commercial life as protective vehicles for transport of fragile foods such as pies and eggs. The economics of manufacture of pie transfer plates and egg trays have kept these unglamorous products in the marketplace. Although certainly in the mature part of the viability curve, they remain of considerable commercial interest.

The term "molded pulp" is used to describe three-dimensional packaging and food-service articles that are manufac-

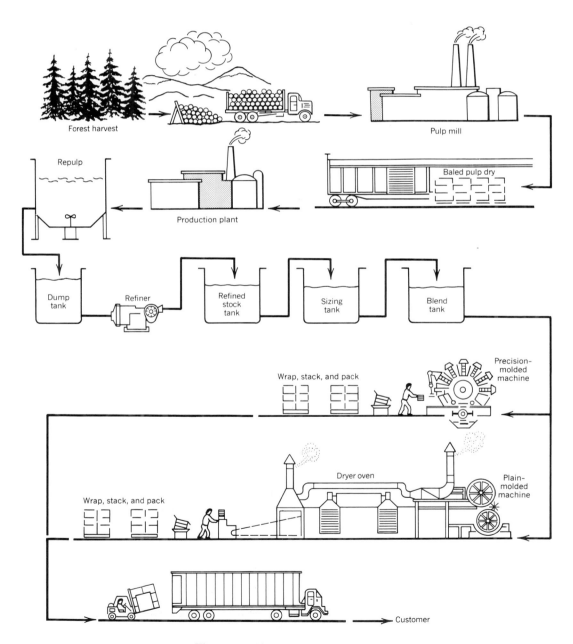

Figure 1. The molded-fiber process.

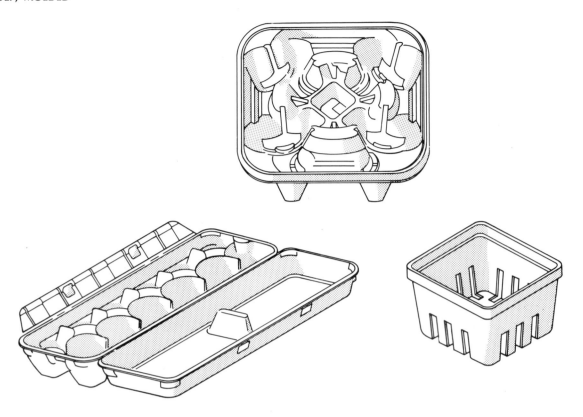

Figure 2. Examples of plain-molded products.

tured by forming from an aqueous slurry of cellulosic fibers into discrete products on screened, formaminated molds in a process analogous to continuous-sheet cylinder-board paper-making. It is not unusual to find the pulp-molding process grouped with other converting processes such as compression-molding of fiber-reinforced resin parts and pressboard converting because of the similarities in finishing steps and common markets, but the molded-pulp process and its products are fundamentally unique. Two basic methods of fabrication are used: plain molding, and precision molding (see Fig. 1).

Products of plain molding are as fundamental as the triangular corner protector pads used in furniture and appliance packing crates, and as complex hinge-lidded, self-locking, one-dozen egg cartons enhanced by hot after-pressing and graphic-

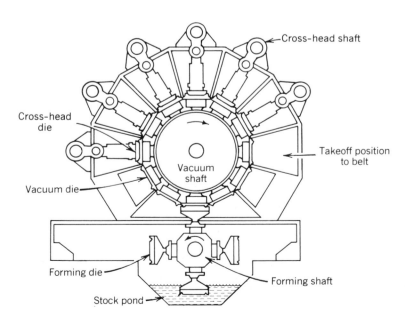

Figure 3. The Chinet precision-molding machine.

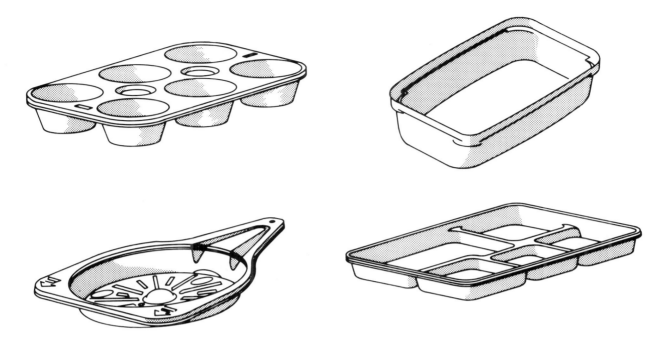

Figure 4. Examples of precision-molded products.

print applications. Between these extremes are products such as berry punnets, peat pots, produce prepackaging trays, and egg and apple-locator trays (see Fig. 2). Such products have commercial relevance because they are produced on highly automated and productive machine modules (>20 tons (>18.1 metric tons) per day), are nestable for efficient transport to users, low in density in comparison to converted paperboard products (see Paperboard), and fabricated from some of the least-costly raw materials available (ie, recycled mechanical cellulosic fibers, groundwood, sphagnum peat fiber, etc).

Precision-molded processing differs from its higher-speed forerunner in drying methodology. Forming of the product and the removal of mechanically bound water by vacuum-assisted compaction is essentially the same; but at that point final drying is accomplished by step- or continuously applied heat between matching mold surfaces (see Fig. 3). Precision-molded products include disposable plates, dishes, bowls, and specialty products such as loudspeaker membranes and cones (see Fig. 4). These products are typically denser, smoother, and more exactly dimensioned and contoured than their free-dried counterparts. Virtually any plain-molded product can also be made in this manner, but they would be more costly because of the complexity of tooling and higher electrical energy required.

Many internal treatments are given to the slurries to render them more valuable to their users. By blending fibers that have received different process and refining treatments, "green" matrix strength can be enhanced; drainage, internal bond strength, and shrinkage can be altered; and biodegradability of products after or during use can be controlled. Colloidal rosin or wax emulsions are commonly added together with paper-making alum to convert hydrophilic fiber into the water-repellent products essential to food handling. Fluorocarbon chemicals in concert with cationic retention aids can be added to impart repellency to low surface-tension oil or greasy liquids. Fertilizers, dyestuffs, flame retardants, and modified starches or wet-strength resins are among other additives that are also commonly added internally where specific end-use effects are desired.

Secondary treatments are sometimes given to nondisposable molded-pulp products. Cafeteria serving trays have been made by after-pressing laminates of molded-fiber preforms loaded with thermosetting polymers. Luggage shells, automobile trunk wells, glove compartments, and door liners are made this way. The hot after pressing and decorative printing of egg cartons also represents secondary treatment that enhances printability and product automation.

A very interesting recent example of secondary treatment is the lamination of a thin thermoplastic film to one surface of molded-pulp trays by vacuum-thermoforming techniques. Products of this type are being used for frozen dinners because of their "dual ovenability" (ie, their suitability for use in microwave and convection ovens). They could be relevant in gas-flush barrier-packaging applications, and where superior decorator graphics are required. Other significant new product opportunities being actively investigated in nonpackaging applications.

BIBLIOGRAPHY

General Reference

U.S. Pat. 4,337,116 (June 29, 1982), P. D. Foster and C. Stowers.

Edwin H. Waldman
Keyes Fibre Company

QUALITY CONTROL.   See Specifications and quality assurance.

R

RADIATION, EFFECTS ON PACKAGING MATERIALS

The medical, pharmaceutical, and food industries use heat, chemicals, and irradiation to sterilize products (see Canning, food; Health-care packaging; Pharmaceutical packaging) (1). Each of these methods has advantages and disadvantages, and each places special demands on packaging materials. The inherent advantages of irradiation sterilization make it a forerunner. The four important advantages are listed below.

1. Sterilization in a completely sealed, impervious package which excludes any possibility of recontamination.
2. Ability to penetrate into the most inaccessible places, bringing the lethal effect to all vegetative and dormant stages of the living organisms within the package.
3. Great reliability and simple control of the process. There are no sterilization conditions to be monitored, and no residuals that must be removed. The only control is the measurement of the radiation dose to assure that it is above the required minimum.
4. Little or no heat is produced. Gamma irradiation produces no heat. Irradiation with accelerated particles might increase the temperature of plastics by 50°F (10°C) and metals by 104°F (40°C).

Two basic processes are utilized: gamma irradiation (^{60}Co, x rays) and accelerated particles (electron-beam accelerators) (2,3). The highest dose is that which extinguishes all forms of vegetative and dormant life, about 3 Mrad (3×10^4 Gy).

Generally, the effect of irradiation upon packaging materials and components is the same for both methods, except for the effect of the small amount of heat generated when accelerated particles are used (4).

The packages consist primarily of the basic packaging materials (glass, metal, paper, and plastics), composite structures of these materials, and packaging components (rubber, adhesives, coatings, and inks).

Irradiation may affect the packages directly or indirectly. It may have direct effects on their physical or chemical properties (5), stability, maintenance of sterility or formation of any toxic, deleterious, or otherwise harmful products. Indirectly, it may affect chemical or biological (6) properties which leads to interaction with the packaged product, changing its efficacy, potency, palatability, or usability.

The direct effect of irradiation may be predicted to some extent; the indirect effect has to be established for each particular product and package.

Effects on Glass, Metals, and Paper

Glass. Glass is not affected by irradiation, except for discoloration. The color formation is proportional to the irradiation dose and may be utilized as a built-in dosimeter. The color fades with time and temperature; therefore, irradiated glass vials, ampuls, or bottles may vary appreciably from batch to batch depending on their aging history (7).

Metals. Aluminum, tin, and steel are used as containers or as foils in composite materials (see Multilayer flexible packag-

ing). The metals are not affected by irradiation, but may react with the product. Reaction with the product is an indirect effect which must be taken into consideration for each packaging application (8).

Paper and paperboard. These, and all other natural products, are affected by irradiation. Owing to the comparatively small dose used for sterilization, this effect, which manifests itself as discoloration, loss of strength, and embrittlement, is not sufficiently significant to prevent the use of these materials in monolayer or multilayer packaging. The dose effect is cumulative, and repeated sterilization of the same package should be avoided.

Effects on Plastics

Plastics form the bulk of modern packaging materials and must be investigated very carefully as to their usefulness for each application. Although general guidelines may be drawn, each material must be viewed for specific applications. There are two basic effects of irradiation on plastics: degradation or cross-linking. There may be simultaneous degradation and cross-linking of the degradation products, and the net result depends upon the comparative rates of the two reactions.

Cross-linking polymers. Table 1 lists polymers which are predominantly cross-linked by irradiation.

There is virtually no effect on polyethylene at the sterilization dose. The effect of irradiation decreases with increasing density: high density polyethylene is less sensitive than medium density polyethylene, which is less sensitive than low density polyethylene (see Polyethylene, high density; Polyethylene, low density). Chemical changes due to irradiation manifest themselves in the changes of infrared spectra, specifically in the type and distribution of unsaturated groups. The presence of oxygen (when irradiated in the air) also has a marked influence on these rearrangements as a result of oxidation. Similarly, spunbonded olefins (see Nonwovens) and ionomers (see Ionomers) are not affected by the sterilization doses.

The effect of irradiation on ethylene copolymers depends greatly on the types of monomers used, but radiation improves impact and tensile strength. Both EVA and EEA are suitable for radiation sterilization (9).

Polystyrene undergoes cross-linking on irradiation; an aromatic ring (phenyl group) attached to alternate backbone carbon atoms adds to the resistance of the plastic to degradation. Mechanical properties are changed very little by irradiation, and only by very high doses. High impact polystyrene, which contains modifiers, is more susceptible to damage, especially to reduction of elongation and impact strength. It can,

Table 1. Cross-linking Polymers

polyethylene (PE)
ethylene-co-vinyl acetate (EVA)
ethylene-co-ethyl acrylate (EEA)
ionomers
polystyrene (PS)
styrene-co-acrylonitrile (SAN)
polyesters
polyurethane
polyphenylene oxide (PPO)
polysulfones, aromatic
poly(vinylidene fluoride) (PVDF)
ethylene-co-tetrafluoroethylene (ETFE)

Table 2. Polymers that Undergo Degradation

cellulose
polytetrafluoroethylene (PTFE)
polychlorotrifluoroethylene (PCTFE)
poly(vinylidene chloride) (PVDC)
polyacetals

however, withstand several radiation sterilizing doses without ill effect.

Styrene–acrylonitrile copolymers are not as resistant to radiation as polystyrene itself, but are still fairly stable and able to resist high irradiation doses.

As in polystyrene, the aromatic ring structure in polyesters absorbs a large part of the radiation energy. Thus, the effect of irradiation on strength and elongation of these materials at sterilization dose is negligible. Similarly, aromatic polyurethanes can be sterilized by irradiation without loss of tensile strength.

Other examples of radiation-resistant plastics containing aromatic ring structures are polyphenylene oxide and aromatic polysulfones (10).

Surprisingly, PVDF and ethylene-*co*-tetrafluoroethylene show improved tensile strength and thermal resistance on irradiation, unlike other fluorine-containing plastics which are degraded drastically (11).

Polymers that undergo degradation.   Table 2 lists polymers which are degraded by irradiation Cellulosics and fluorocarbons are examples of materials which are degraded by irradiation. Cellulose, as mentioned in connection with paper, is quite sensitive to irradiation degradation, which manifests itself in a tensile strength loss of about 25% upon irradiation at 5 Mrad (5×10^4 Gy). The degradation is owing to main-chain scission and steady reduction of molecular weight. After irradiation with 500 Mrad (5×10^6 Gy), cellulose is almost completely degraded to water-soluble fragments.

Fluorine-containing polymers, namely PTFE and PCTFE, show rapid deterioration on irradiation. They are the most radiation-sensitive plastics and preferably should not be used for packaging applications with this type of sterilization. PCTFE is less susceptible to degradation than the tetrafluoropolymer. All these changes occur in the presence of air; in the complete absence of oxygen, the damage is less severe.

PVDC is also not recommended for use with irradiation sterilization. Its mechanical properties start to degrade at 4 Mrad (4×10^4 Gy), and the plastic becomes very discolored.

Polyacetals show a high degree of degradation at low irradiation doses. Copolymers are slightly less sensitive than homopolymers, but both lose most of their tensile strength and become brittle when irradiated with doses above 5 Mrad (5×10^4 Gy) (12).

Intermediate polymers.   A third group of plastics, listed in Table 3, occupies an intermediate position on the scale between cross-linking and degradation.

These materials are generally affected by higher irradiation doses, but if properly stabilized, they withstand sterilization dose with minimal changes which may permit their use where repeated sterilization is not anticipated.

Polypropylene is a good example of such material. When irradiated in the presence of air, it undergoes marked oxidative degradation (13). With proper stabilization, however, this

degradation may be kept to a minimum so as not to impair mechanical properties (14,15).

Polymethylpentene reacts to irradiation in a manner similar to polypropylene. Although it is somewhat more resistant to lower irradiation doses than polypropylene, it is not recommended for repeated exposure.

Indications are that structures with branching alkyl chains are less resistant to irradiation than nonbranched polymer chains. PMMA is an example. Although it can satisfactorily withstand a single radiation sterilization dose, it is not suitable for repeated doses. It degrades in both the main chain and in the ester side chain of the molecule. The plastic becomes brittle and discoloration develops at doses as low as 0.5 Mrad (5×10^3 Gy). Some radiation-grade acrylics are available now for molding medical devices. Polyamides (nylons) react to irradiation in a manner similar to polyethylene insofar as some increase in tensile strength is concerned, but show a much more rapid decrease in impact strength. Therefore, repeated sterilization doses are not recommended.

Esters of cellulose (acetate, propionate, and acetobutyrate), which are of interest in packaging applications, are affected to a lesser degree by irradiation than cellulose itself.

Polycarbonate exhibits some discoloration and becomes brittle at doses considerably above 2.5 Mrad (2.5×10^4 Gy). Although the impact properties are hardly affected, thin irradiated films exhibit some decrease in tensile strength and elongation on aging. Nevertheless, it can be safely exposed to a single sterilization dose (16). Acrylonitrile-*co*-butadiene-*co*-styrene (ABS) is also suitable for a single sterilization, but generally is much less resistant to irradiation than styrene-*co*-acrylonitrile (SAN). PVC does not have a clearly defined position in this scheme. Some investigators report degradation and others cross-linking. Apparently, both statements are true depending on irradiation conditions and the type of additives present in the plastic (17).

Rigid PVC containing 58% chlorine loses HCl upon irradiation and also decomposes into unstable fragments which may then undergo cross-linking and recombination. This dehydrochlorination leads to formation of conjugated double bond structures which manifest themselves in coloration of the polymer (18,19).

This observed change in color may vary from pale yellow to black, depending on irradiation dose and of course the stabilizing system (20,21). Also, evolution of HCl gas is noticeable and can cause metal corrosion and changes in pH. Their spectrum also changes. Tensile strength is affected only slightly, but brittleness increases. Other significant changes in mechanical properties proceed at doses which normally are not used in packaging.

The degree of degradation depends therefore on the type

Table 3. Intermediate Polymers

polypropylene
polymethylpentene
poly(methyl methacrylate) (PMMA)
poly(vinyl chloride) (PVC)
polyamides
cellulosic esters
polycarbonate
acrylonitrile-*co*-butadiene-*co*-styrene (ABS)

Table 4. Radiation Effects on Rubber

Rubber	Stable up to
polyurethane	500 Mrad (5 × 10⁶ Gy)
natural	100 Mrad (1 × 10⁶ Gy)
styrene-*co*-butadiene (SBR)	100 Mrad (1 × 10⁶ Gy)
nitrile	100 Mrad (1 × 10⁶ Gy)
silicone (polydimethylsiloxane)	10 Mrad (1 × 10⁵ Gy)
neoprene	10 Mrad (1 × 10⁵ Gy)
butyl	1 Mrad (1 × 10⁴ Gy)

and amount of the additives (stabilizers, antioxidants, inhibitors, etc.), presence of oxygen during irradiation, aging, and the irradiation dose.

In order to prevent degradation, the plastic must be properly stabilized. When used as packaging material, rigid PVC should retain two important properties: original clarity (be free from discoloration) and low brittleness. There are a multitude of stabilizers to prevent different types of degradation, and combinations of them should be chosen to produce the desired effect. For packaging of foods and drugs, choice of stabilizers is limited owing to toxicity or extractability factors. In fact, a stabilizer should prevent a particular type of degradation, namely, dehydrochlorination, oxidation, destruction of the polymer, or reconstitution (cross-linking) of degraded chains.

As is pointed out in the literature, a combination of several weak stabilizers may provide a very satisfactory synergistic action if used in proper combination. Several firms now offer radiation-resistant PVC compounds which are stabilized to inhibit cross-linking and show low discoloration.

Effects on Rubber

Table 4 shows the effect of irradiation on various types of rubber in an unstressed state. The stability can be influenced by various additives like fillers and antioxidants (22,23).

Effects on Composite Materials

Composite materials may be affected differently than the individual components. For example, the loss of strength of paper or a cellulosic film may not be noticed at all if the film is supported by polyethylene or foil. On the other hand, various types of heat-seal coatings which are not affected by irradiation per se may produce weak seals when used as components of a laminate (see Sealing, heat). This may be caused by the effect of irradiation on the strength of the adhesion between the coating and the substrate rather than on the actual bond between the two sealed surfaces. This effect, if carefully controlled, may be used to produce peelable seals which are so important to ensure sterile delivery of the product.

Aging studies are of utmost importance to determine any progressive degradation which may not have been detected immediately after irradiation. Accelerated aging (high temperature and/or humidity) indicates the trend, but should not be substituted for the actual shelf stability, since it may sometimes cause a change which would not occur under normal conditions (24).

Inks. Commercial printing inks are developed to be light stable; that is, they contain pigments and dyes which do not change color under the influence of visible and ultraviolet light, and they generally resist radiation.

Specially developed inks which are used as radiation dosimeters change color under the influence of radiation due to pH change of the ink system rather than that of the basic colorant (25).

Adhesives and coatings. Most adhesives and coatings are based on polymers and plastics. Their resistance to radiation, therefore, can be predicted from the data developed for the basic components. Generally, thermosetting types are more resistant to radiation than thermoplastic types. Fillers, pigments, plasticizers, and other additives affect radiation stability (26).

Testing. The dose required for sterilization is not usually a satisfactory dose to use for radiation-effects determination. This is because in both gamma and electron-beam sterilization, delivering a specified sterilizing dose to the center of a bulk unit of product involves exposure of the outer regions to a higher dose. The ratio of these two doses is a function of the size and bulk density of the product unit involved. For gamma radiation, this ratio can range from about 1.25 : 1 to 3 : 1; for electron beam it is many times higher. Any exposures for radiation-effects testing should take this into account.

The biological aspects of the irradiation-sterilized package must not be overlooked. Apart from the physical and chemical changes of the packaging materials, their toxicity and any interaction with the product must be determined (27).

Although much work has been done on investigation of the toxicity of various packaging materials, this knowledge is only a starting point for a packaging engineer, as it is obvious that the interaction of the packaging material with the packaged drug or food cannot always be theoretically predicted.

It must first be determined that the product itself as well as the package can withstand irradiation sterilization. The product should be irradiated in an inert medium; a glass container may be used to determine this. The product must then be sterilized in the final package and tested for purity, efficacy, toxicity, palatability, and shelf stability.

BIBLIOGRAPHY

1. R. D. McCormick, *Prepared Foods,* **153**(4), 133 (Apr. 1984).
2. W. J. Maher, "Critical Assessment of Gamma and Electron Radiation Sterilization," presented at *1979 Disposable Medical Packaging–Government Compliance Seminar,* sponsored by TAPPI, Lincolnshire, Ill., Oct. 15–16, 1979, pp. 22–26.
3. J. H. Bly, *Radiat. Phys. Chem.* **14**, 403 (1979).
4. D. A. Vroom, *MD & DI,* 45 (Nov. 1980).
5. J. J. Killoran, *Radiat. Res. Rev.* **3**, 369 (1972).
6. R. K. O'Leary and co-workers, "The Effect of Cobalt-60 Irradiation Upon the Biological Properties of Polymeric Materials," presented at *Sterile Disposable Devices: Updated 1973 Technical Symposium,* sponsored by HIA Sterile Disposable Device Committee, Washington, D.C., Oct. 17–18, 1973, pp. 160–166.
7. H. J. Zehnder, *Alimenta* **23**(2), 47 (1984).
8. P. S. Elias, *Chem. Ind.* **10**, 336 (1979).
9. A. Barlow, J. Biggs, and M. Maringer, *Radiat. Phys. Chem.* **9**, 685 (1977).
10. J. R. Brown and J. H. O'Donnell, *J. Appl. Polym. Sci.* **19**, 405 (1975).
11. G. G. A. Bohm, W. F. Oliver, and D. M. Pearson, *SPE J.* **27**, 21 (July 1971).
12. M. Možišek, *Plaste Kautsch.* **17**(3), 177 (1970).
13. J. L. Williams and co-workers, *Radiat. Phys. Chem.* **9**, 445 (1977).

14. T. S. Dunn and J. L. Williams, *J. Ind. Irradiat. Technol.*, **1**(1), 33 (1983).

15. P. Hornig and P. Klemchuk, *Plast. Eng.*, 35 (April 1984).

16. B. A. Rohn, "Polycarbonate Coextrusions for Medical Device Thermoform, Fill & Seal (TFFS) Packaging" in *Proceedings of the First International Conference on Medica and Pharmaceutical Packaging, Healthpak '84*, May 8–9, 1984, Düsseldorf, FRG, Schotland Business Research, Inc., Princeton, N.J., pp. 65–84.

17. C. Wippler, *Nucleonics* **18**(8), 68 (1960).

18. A. A. Miller, *J. Phys. Chem.* **63**, 1755 (1959).

19. S. Ohnishi, Y. Nakajima, and I. Nitta, *J. Appl. Polym. Sci.* **6**(24), 629 (1962).

20. R. V. Albarino and E. P. Otocka, *J. Appl. Polym. Sci.* **16**, 61 (1972).

21. M. Foure and P. Rakita, *MD & DI*, 57 (Nov. 1983); 33 (Dec. 1983).

22. D. W. Plester, "The Effects of Radiation Sterilization on Plastics" in G. Briggs Phillips and W. S. Miller, eds., *Industrial Sterilization,* International Symposium, Amsterdam, 1972, Duke University Press, Durham, N.C., pp. 149–150.

23. R. W. King and co-workers, "Polymers" in J. F. Kircher and R. E. Bowman eds., *Effects of Radiation on Materials and Components,* Reinhold Publishing Corp., New York, 1964, pp. 110–139.

24. J. P. Ferrua, "New Multilayer Materials Specifically Designed for Radiation Sterilization," in ref. 16, pp. 53–63.

25. D. B. Powell, "Packaging Requirements for Radiation-Sterilized Items," in *Ionizing Radiation and the Sterilization of Medical Products,* Proceedings of the First International Symposium organized by the Panel on Gamma and Electron Irradiation, Dec. 6–9, 1964, Taylor and Francis Ltd., London, pp. 95–103.

26. Ref. 23, pp. 139–144.

27. K. Figge and W. Freytag, *Dtsch. Lebensm. Rundsch.* **73**(7), 205 (1977).

General References

G. O. Payne, Jr., C. J. Spiegl, and F. E. Long, *Study of Extractable Substances and Microbial Penetration of Polymeric Packaging Materials to Develop Flexible Plastic Containers for Radiation Sterilized Foods,* Technical Report 69-57-FL, U.S. Army Natick Laboratories, Food Laboratory FL-65, Natick, Mass., Jan. 1969.

A. F. Readdy, Jr., *Applications of Ionizing Radiations in Plastics and Polymer Technology,* Plastic Report R41, Plastics Technical Evaluation Center, 1971.

A. Chapiro, *Radiation Chemistry of Polymeric Systems,* Vol. XV of *High Polymers,* Interscience Publishers, New York, 1962.

R. O. Bolt and J. G. Carroll, *Radiation Effects on Organic Materials,* Academic Press, New York, 1963.

A. Charlesby, *Atomic Radiation and Polymers,* Pergamon Press, New York, 1960.

M. Dole, ed., *The Radiation Chemistry of Macromolecules,* Academic Press, New York, 1972.

J. Silverman and A. R. Van Dyken, eds., *Radiation Processing,* Transactions of the First International Meeting on Radiation Processing, Vols. I and II, Puerto Rico, May 9–13, 1976, Pergamon Press, New York, 1976.

H. K. Mann, ed., "Radiation Sterilization of Plastic Medical Devices," *Radiat. Phys. Chem.* **15**(1) (1980).

O. Sisman and C. D. Bopp, *Physical Properties of Irradiated Plastics,* ORNL-928 Report, TIS, U.S. Atomic Energy Commission, Oak Ridge, Tenn., June 29, 1951.

T. S. Nikitina, E. V. Zhuravskaya, and A. S. Kuzminsky, *Effect of Ionizing Radiation on High Polymers,* Gordon and Breach Science Publishers, Inc., New York, 1963.

J. E. Wilson, *Radiation Chemistry of Monomers, Polymers and Plastics,* Marcel Dekker, Inc., New York, 1974.

W. E. Skiens, "Sterilizing Radiation Effects on Selected Polymers", Symposium on Radiation Sterilization of Plastic Medical Products, Cambridge, Mass., March 28, 1979, Report No.: CONF-7903108-1, Dept. of Energy, Washington, D.C., PNL-SA-7640.

B. J. Lyons and V. L. Lanza, "Protection Against Ionizing Radiation" in W. L. Hawkins, ed., *Polymer Stabilization,* Wiley-Interscience, New York, 1972, pp. 250–311.

G. de Hollain, *Plast. Rubber: Mater. Applications,* 103–108 (Aug. 1980).

I. W. TURIANSKY
Ethicon, Inc.

RECYCLING

Recycling is a critical issue today in all highly developed countries, as the volumes of solid waste continue to increase and the population centers become more concentrated. Different countries take different approaches which stem in part from government philosophies of regulation and also from the wide variations in the extent to which industries and consumers participate in the recycling effort (1). This article deals with developments in the United States since the first Earth Day in 1970 focused national attention on the growing problem of waste disposal and the need for energy conservation. Many communities are trying to reduce the volume of solid waste through outright bans, deposits or auxiliary taxes, or tax credits for promoting recycling. Many urban centers are running out of disposal sites and are considering either costly incineration or long-distance hauling.

Recycling is the most attractive approach for a number of reasons. It saves energy, to an extent that depends on the energy expended for retrieval and the type of material recycled (see Energy utilization). Recycling aluminum can save 95% of the energy that would otherwise be used in producing aluminum from bauxite; recycling steel and glass scrap saves about 50% of the energy otherwise used in making these materials from ore and silica. The reuse of plastic wastes can also save energy, depending on the application. Fabricating a part or package from plastic scrap saves about 85–90% of the energy that would be used in fabricating from virgin resin, including the energy of the petroleum feedstocks used to manufacture the resin. Simply burning plastic wastes can save energy too, but recycling and reuse frequently double the energy savings. For example, 100 lb (45.4 kg) of high density polyethylene (HDPE) has a fuel value of 20 million (10^6) Btu (19 kJ), but recycling and reuse save 40 million (10^6) Btu (38 kJ).

Recycling also conserves valuable resources. Recycled metals are indistinguishable from virgin metals and are therefore equally valuable. Recycled plastics and paper are generally less valuable than the corresponding virgin material because they normally suffer some degradation during reprocessing, but recycled plastics and paper are nevertheless useful and are less costly. Apart from the role of packaging wastes at disposal sites, they also create litter. This is another reason to promote recycling.

Packaging wastes represent about one-third of the 150 million (10^6) tons (136×10^6 metric tons) of municipal wastes generated annually in the United States. Plastic of all kinds account for about 6 wt %, and plastic-packaging waste alone for about 6 million (10^6) tons (5.4×10^6 t) (2), or 4 wt % of the total, according to recent analyses of refuse collected in 11 U.S. cities (3).

Paper accounts for about 50% of packaging wastes, and glass about 25%, (4), but plastics in solid wastes are frequently the concern of environmentalists and city engineers. This is partly because plastics are relatively new and little is known about their long-term behavior in landfill sites. It is also because plastics have been rapidly displacing conventional packaging materials. The volume of plastics used in packaging has essentially doubled in the past ten years (2). There is also concern about the behavior of chlorinated plastics in incinerators. The plastic content of solid wastes is small on a weight basis but is nevertheless significant on a volume basis. This is particularly true for bulky and resilient plastic bottles which take space in collection trucks and landfills. The cost of solid waste disposal is that of space.

Criteria for Successful Recycling

Society is generally not concerned with the disposal of scrap generated in the manufacture of packaging materials because they are usually homogeneous and readily recycled by the manufacturer. However, society is concerned with postconsumer packaging wastes. Recycling postconsumer wastes includes the following steps: collection; reprocessing; preparation of a product for sale; and sale of a product derived in whole or in part from the reprocessed material.

Thus, the four criteria for successful recycling are (1) a continuing source of scrap; (2) viable recycling and reprocessing technology; (3) applications and markets for products derived from the wastes; and (4) good economics. All of these criteria must be met if recycling is to succeed; the failure to meet any one guarantees failure. Factors affecting economics include convenience of collection and the cost of the disposal alternative. Unfortunately, the selling prices of reprocessed materials of scrap are volatile, and efforts to recycle sometimes fail because the price of the reprocessed or recovered scrap plunges, and the economics of recycling shifts from profit to loss.

Reprocessing Technology and Applications

Paper. Recycling used paper and board products in the manufacture of new paper and board has long been practiced in the pulp and paper industry (see Paper; Paperboard). In the United States, approximately 20% of the total volume of paper and board is customarily manufactured from secondary fiber. This includes both manufacturing and postconsumer wastes. Secondary fiber is derived from a variety of industrial and municipal sources that includes paper-finishing operations (eg, sheeting, cutting, and packaging), corrugated-container manufacture, and municipal refuse. A chief issue in determining the value of these waste fibers in competition with virgin materials is the difficulty of classifying them according to quality. For example, waste corrugated clippings are routinely recycled from container manufacturers to the paper mill, and many kinds of low quality folding-carton wastes are recycled to the boxboard-making industry (see Cartons, folding). In between these quality opposites lies a wide range of secondary fiber derived from paper grades ranging from printing papers to tissue.

The feasibility of recycling papermaking fibers is principally tied to the complexities and costs of fiber collection, classification, purification, and transportation to the papermaking operation. Technical feasibility would permit a substantial increase in the amount of secondary fiber recycled, but the economics of secondary-fiber recycling in the United States, like the economics of all recycled material, is controlled by the cost of competitive virgin materials. There are some technical limitations, eg, recycling through repulping shortens the paper fibers. If these fibers are used in container manufacture, bursting strength is reduced, and this can necessitate the use of heavier boxes to meet the same tear-strength levels. Various contaminants, eg, plastics, glues, inks, adhesives, and laminations with foil, also impede the recycling of paper.

Corrugated containers (see Boxes, corrugated) are easy to collect, because most used corrugated board is recovered from supermarkets. Since these containers are bulky, they are usually flattened and frequently baled in the store, moved to the central warehouse, and finally delivered to the local scrap dealer. Bulky wastes are costly to landfill and of relatively low value if recycled. Nevertheless, because corrugated board is convenient to collect, it is an important source of secondary fiber. Once clean, there is a ready market for secondary fiber in such products as corrugated containers, combination boxboard (eg, soap, cereal, and shoe boxes), construction paper, board for roofing insulation and sound absorbency, toweling and tissues, and decorative and writing papers.

Once waste paper and paperboard become part of municipal solid wastes (MSW), they are very difficult to isolate. Normally, paper wastes are burned to produce energy, either as steam or as refuse-derived fuel (RDF). In 1984, there were 14 large RDF plants in the United States with a capacity of 26 million (10^6) lb (11.8×10^6 kg) of MSW per day (5). Total U.S. capacity of all energy-recovery systems based on municipal solid wastes is about 200 million (10^6) lb (91×10^6 kg) per day (operating, under construction, or under negotiation in early 1985) (6).

Ferrous metal. There has always been an active secondary ferrous-metals industry in the United States. In 1984, domestic iron and steel scrap accounted for 30.5% of the ferrous metal consumed (7), compared with 23.6% in 1979. Most of this is manufacturing scrap. The primary method for recycling steel involves melting the scrap together with other ingredients to make raw steel ingots. Metals do not lose strength when recycled repeatedly, but recycling tinplate (see Tin-mill products) scrap can be a problem because small amounts of tin, ie, as low as 0.01%, can form hard spots in the steel and create subsequent difficulties in rolling. Today, therefore, only low concentrations of can scrap (see Cans, steel) can be used directly in the furnace.

One way to use scrap with tin impurities is as a precipitant in copper leaching. This is an application that is useful only in the Western United States, however, and because the copper market has been depressed for many years, it is not a significant market. The other alternative is detinning, a chemical process that removes tin from tinplated steel by treating the scrap with a hot solution of caustic soda in the presence of an oxidizing agent. This process produces high quality tin and high quality steel. Aluminum residues from bimetallic beverage cans represent a serious economic barrier in detinning because caustic solutions attack aluminum preferentially to tin. Consequently, detinners would rather not purchase bimetallic beverage cans (ie, steel body, aluminum end). In recent years, the steel industry has subsidized the recycling of bimetallic beverage cans recovered from states having deposit laws by offering a premium to beverage-can manufacturers who make steel cans and collect the empty cans. The steel industry uses a nuggetizing process, ie a special shredder, to remove the aluminum end from the steel body; the separated aluminum and the steel are used in the steel-making process. The significance of tin in the waste stream has decreased in recent years, partly because few beverage cans are made of steel now, and also because more tin-free steel is being used. Only a few plants operate detinning lines today.

The steel cans collected today come primarily from cities that operate various types of resource-recovery facilities. Magnetic separation is relatively simple. About 30 cities now operate facilities that separate a ferrous fraction. In addition to the steel beverage cans collected in the deposit states, there are a number of volunteer collection programs. Most of these ferrous packaging wastes are recycled and used in steel making. In addition, some ferrous can scrap is used in the production of ferroalloys and in foundries.

Glass. The glass industry manufactures containers and flat-glass products. Bottles and jars account for more than half of all glass produced, and because bottles are a large factor in the solid-waste stream, interest in glass recycling and reuse has centered around the container business, especially the nonreturnable, oneway bottle. The glass-container industry now uses about 20% cullet, ie, crushed waste glass, in glass making to promote melting and mixing of the sand, limestone, and soda-ash raw materials. About 75% is generated in-house, and about 25% is postconsumer container glass. Adding cullet aids the glass-manufacturing process because cullet melts at a lower temperature than the rest of the raw materials. Thus, less heat is required and furnace wear is reduced. Research has shown that cullet content of 30% or more is feasible; therefore, more recycled glass can be used (see Glass container manufacturing).

Postconsumer container glass is mostly collected from the deposit states, some voluntary programs, and a few resource-recovery facilities. In the latter case, reprocessing is complex and costly. Even when the used glass containers are source separated, some reprocessing is required because glass-container manufacture uses high purity materials. For example, if metal closures are included in the mix, oxides that interfere with the operation of the furnace can form. Ceramic or other foreign particles form defects in the container. Color also can be a problem because colored glass is essentially unusable in making colorless (flint) glass. Green glass can tolerate as much as 15% amber and/or flint glass, and amber can tolerate as much as 10% green and/or flint glass. Consequently, postconsumer cullet must be color-sorted, either by hand or by machine. Cullet used as is, without color sorting, can be used to produce a light-green glass container which is sometimes called Eco-glass, but no beverage producers have yet chosen to use it.

Most postconsumer cullet is used in glass-container manufacture, but, unlike metals, glass can be reused in other industries without refining, particularly as a raw material for construction. For example, glass scrap can be used as aggregate for roadways, building bricks, and for glass-wool insulation and honeycomb structural materials. Crushed container glass has also been used in the production of reflector materials and costume jewelry. A mixture of glass and plastic wastes has been molded into synthetic slate products, and waste glass has even been used to produce sewer pipe.

Aluminum. The entrance of the aluminum beverage can into the market in the late 1960s and early 1970s went hand in hand with recycling. The aluminum industry opened buy-back centers soon after Earth Day 1970, and since then it has done an impressive job of promoting the recycling concept. As a result, the aluminum can has virtually displaced the steel beverage can in the United States. Aluminum-can recovery is a business involving millions (10^6) of dollars, and no subsidies are needed today. Aluminum cans are lightweight, easily crushed, and easily transported to buy-back centers, where they are redeemed for about $0.01 per container. Thus, large quantities of aluminum cans are returned at these centers, particularly in the deposit states. Only a small percentage of the postconsumer aluminum scrap is recovered via resource-recovery facilities operated by municipalities.

In 1979, the aluminum industry mounted an even more aggressive recycling effort to retrieve the metal, which was in short supply and high-priced at the time. At that time, the supermarkets in the nondeposit states also became buy-back centers. Today, more than 5000 retail outlets in nondeposit states buy back aluminum cans. The reverse vending machine was also introduced in 1979. Over 1000 stores use these machines, which either refund the deposit (in the deposit states) or return a smaller amount for the value of the metal.

Used aluminum cans are relatively easy to reprocess. They are shipped to the aluminum producer in bales, briquettes, or shreds, where they are remelted and delacquered. Organic coatings must be removed before the scrap can be used to make new can sheet stock. Delacquering takes place in special furnaces. About 90% of the recovered cans is remelted into the alloy used to produce sheet for the aluminum-can body; the balance is used for other aluminum products.

Plastics. Like other materials, plastics are recycled routinely during the manufacturing process, but the concept of using scrap recovered from urban wastes is quite recent. The process of reclaiming plastics is similar to the compounding or mixing step in the manufacture of plastics, where various additives are mixed with the plastic resin to achieve desired properties. One of the principal differences is that reprocessing usually involves removing contaminants as well as blending. Generally, individual plastic resins must be sorted out before they can be used. In a typical reclamation operation, a chopper or guillotine reduces large chunks of a homogeneous scrap plastic to small pieces that can be fed to a granulating machine. The granulated scrap is then mixed with other additives or fed directly to an extruder. After extrusion, it is pelletized or diced. If the scrap is in film form, it is often made into confetti and fed directly into an extruder. Plastics lose some of their physical properties during reprocessing, but reprocessed material has been used in relatively undemanding applications.

To recover postconsumer plastic wastes, interception of these wastes before they reach final disposal is almost essential, because postconsumer plastic wastes are normally present as mixtures of resins and of components made of other materials. Because discarded plastic bottles are the most easily collected, recycling programs were instituted as far back as 1970. The focus then was on HDPE milk bottles. At that time, a number of dairies began collection programs, but they soon ran out of steam as the reality of recycling economics became apparent. It was not until 1977, when the 2-L PET container for carbonated beverages (see Carbonated beverage packaging; Polyesters, thermoplastic) rapidly took over the family-size beverage-container market, that serious recycling programs began. They were instituted primarily in the deposit states, with the bottles collected at the bottler's facility. Soon entrepreneurs developed a number of systems to process the used PET bottles. Ultimately, reprocessing must separate paper labels, aluminum caps, and polyethylene base cups from the PET. Today, there are several systems available for purchase, franchise, or license that produce a reprocessed PET having virtually the same properties as virgin PET.

Simply producing reprocessed colorless or green PET in granulate or pellet form does not guarantee its sale. In contrast to other materials which are recycled into new containers, reprocessed PET is difficult to use in producing new PET bottles for carbonated beverages. The problems are not technical: they are matters of economics and psychology. Few beverage producers wish to sell their products in PET bottles made from reprocessed PET, and few producers of recycled PET wish to undergo the lengthy period of testing required for FDA clearance (see Food, drug, and cosmetic packaging regulations).

Therefore, recycled PET is now used in a myriad of nonpackaging applications including fiberfill, plastic alloys, fiber-reinforced resins, new injection-moldable compounds, plastic sheet, foamed polyurethanes, and unsaturated polyesters. Many more applications are in the development stage. At this time, fiberfill is the chief use market.

Effectiveness of Recycling Packaging Wastes

The recycling of the aluminum beverage can is the shining example of effective recycling. It meets all of the criteria of success. For the past few years, the recycling of aluminum cans in the United States has achieved a recovery rate of over 50%. In 1983, the recovery rate was 52.9%, based on the statistics of the Aluminum Association. The number of nondeposit states now achieving recovery rates above 60% is truly remarkable. These states include Colorado, Arizona, California, Washington, New Mexico, Arkansas, Texas, North Carolina, Florida, and Georgia. Normally, the recovery rate in a deposit state is 80–90%.

The same cannot be said about the recycling of steel beverage cans. It is not possible to distinguish between steel beverage cans and food cans in the recovery statistics generated by the industry, but probably no more than 2% of the steel beverage and food cans shipped in 1970 was recovered and recycled. Today, the percentage may be as high as 5%, or about 300×10^6 lb (136,000 t) based on shipments of 6.43×10^9 lb (2.92×10^6 t) of steel cans. The problem is twofold: reprocessing steel cans is complex and therefore costly, and, once recovered, steel scrap has relatively low value, ie, about \$0.02/lb (\$0.04/kg) vs \$0.30/lb (\$0.66/kg) for aluminum.

Glass-container recycling is much like steel-can recycling. It is estimated that about 1.75×10^9 (7.9×10^5 t) of postconsumer glass was recycled in 1984 (8). This is glass used for both food and beverage containers. With glass usage for containers at 23×10^9 lb (10.4×10^6 t) that year, recycling accounted for only 7.5% of shipments. In this case, transportation costs for recovered glass are high, but used, color-sorted glass is valued at only \$0.02–0.03/lb (\$0.04–0.07/kg). Again, the economics are poor, and effective recycling suffers.

In the case of paper packaging, the best example is corrugated containers. Over the past several years, these have been recovered and recycled at a rate of 25–32%. The driving force is not economics, but the convenience of collection and the cost of the disposal alternative.

The only type of postconsumer plastic packaging waste being recovered and recycled is the PET beverage bottle, and this is not due to the efforts of the industry, but rather to regulatory pressures. PET bottles are currently recovered primarily in the nine states that have deposit laws: Oregon, Vermont, Maine, Michigan, Connecticut, Massachusetts, Iowa, Delaware, and New York. In nondeposit states, perhaps 1–2% of the PET bottles are recovered.

In 1984, an estimated 100×10^6 lb (45,000 t) of PET was recovered in the United States as used carbonated beverage bottles for recycling. This represents a recovery rate of about 18%, up from only 5% in 1979, and 10% in 1982 (9). Though 18% represents an encouraging recovery rate, it represents less than 4% of all plastic bottles produced in 1984. Furthermore, unless new states adopt deposit laws or industry mounts and aggressive campaign to retrieve PET bottles from nondeposit states, the recovery rate in coming years might increase only slightly as the market penetration of PET increases or levels off. This is the challenge facing the plastics industry. The search is on for plastic products based on recycled PET that can command higher selling prices.

BIBLIOGRAPHY

1. I. Boustead and K. Lidgren, *Problems in Packaging: The Environmental Issue*, John Wiley & Sons, Inc., New York, 1984.
2. *1985 Edition of Facts and Figures of the U.S. Plastics Industry*, The Society of the Plastics Industry, New York, 1985.
3. R. DeCesare, *Pilot-Scale Studies of the Combustion Characteristics of Urban Refuse*, U.S. Bureau of Mines, Washington, D.C., 1980, R.I. 8429.
4. J. Milgrom, "Recycling Packaging Wastes: The Challenge of The Plastic Beverage Bottle," *Future-Pak '83*, Ryder Associates, Atlanta, Ga., 1983.
5. H. Alter, "Refuse-Derived Fuel Production and Combustion in the United States, *Paper presented at the EEC, (European Economic Community) seminar on Sorting of Household Waste*, Luxembourg, 1984, in M. P. Ferranti and G. L. Ferrero, eds., *Sorting of Household Waste and Thermal Treatment of Waste*, Elsevier Applied Science Publishers, New York, 1985.
6. D. L. Klass, *Resource Conserv.* 11(3/4), 157 (1985).
7. Statistical Services, National Association of Recycling Industries, Inc. (NARI), New York.
8. J. C. Cavanagh, "Packaging, Will Plastics Dominate?" *Resource Recyc.* 4(2) 14(1985).
9. J. Milgrom, "Markets for Recovered PET Scrap," *Ninth International Conference on Oriented Plastic Containers*, Ryder Associates, Atlanta, Ga., 1985.

J. Milgrom
Walden Research, Inc.

RETORTABLE FLEXIBLE AND SEMIRIGID PACKAGES

This article pertains to packages made entirely of flexible materials, eg, pouches, as well as semirigid containers that have at least one flexible body wall or lid. These packages, available in many shapes, sizes, and combinations, are used for food and medical products processed by heat and pressure.

The characteristics of "retortable," as related to packages, has been defined by the American Society for Testing and Materials (ASTM) Committee on Flexible Barrier Materials and its Subcommittee on Retortable Pouches and Related Packages. The standard definition relating to retortable flexible barrier materials is that they are "capable of withstanding specified thermal processing in a closed retort at temperatures above 100°C (1). The legal requirements for retort processing of foods in metal cans, aluminum cans, glass jars, and retort pouches in the United States are clearly defined in the *Code of Federal Regulations* (CFR) (2).

Retortable packages must maintain their material integrity as well as their required barrier properties for their designated end-product use during product-to-package handling, thermal processing, and subsequent shipping abuse. For shelf-stable food packages, the materials used must be retortable and still maintain extended barrier characteristics against such effects as light, moisture, oxygen, and microbial penetration.

Steel cans (see Cans, steel) perform very well in a retort environment. Glass jars and lightweight aluminum rigid containers also are well-established containers which can be thermostabilized in this manner, except that they both normally require some special handling in the form of overriding process pressures and critical control of pressure changes. Special handling is required to prevent closure lids on jars from releasing their vacuum seal, easy-open scored lids on aluminum cans from fracturing, and aluminum can bodies from paneling owing to excessive internal or external pressure.

Flexible and semirigid packages also have critical retort process-handling requirements. A variety of high temperature process and equipment systems other than standard retort pressure vessels (see Canning, food) can be utilized as long as proper pressure control and uniform heat distribution are maintained. Most medical-type autoclaves (see Health-care packaging) lack the necessary controls, but high temperature short-time (HTST) systems, vertical or horizontal hydrostatic cookers, microwave chambers, and many variations thereof have been adapted for thermostabilization of flexible and semirigid packages.

Retort processing of these packages may cause degrading effects such as weakened package-seal integrity, flex-crack or flex-stress leading to film delamination or leakage, semirigid container distortion or denting, and even loss of hermetic-seal integrity because of pressure changes during heat processing or between the cooking and chilling processes.

Since flexible and semirigid containers are more sensitive than rigid metal cans to degrading effects during the thermal process, they cannot be retorted in the same way. Rigid metal cans are usually processed in a saturated steam atmosphere, typically at ca 250°F (121°C). The external pressure of this process on the container is therefore 15 psig (205 kPa). The internal pressure of the heated product in a rigid metal can can be different from the external pressure of the steam, within limits, and the rigid container supports this pressure difference through its body structure. Flexible and semirigid packages, however, are susceptible to rupture or seal separation if the internal pressure exceeds the external process pressure excessively. Consequently, the thermal processes for these containers must include an overriding pressure beyond that of the saturated steam. This is accomplished by process-

ing in superheated water with overriding air or steam pressure, or by processing in a combination steam–air atmosphere. Application of an overpressure of 10–15 psig (170–205 kPa) is typical, yielding a processing pressure of 25–30 psig (274–308 kPa) for a retort temperature of 250°F (121°C).

The primary reason for specifying a package to be retortable is for thermal sterilization or at least minimal destruction of pathogenic microbial contamination which could cause spoilage of contained products such as foods and prevent chemical reactions and resulting degradation of other packaged products. Numerous studies of these effects of retort processing on retortable packages have been published (3,4). The most positive effects on foods packaged in the retortable flexible and semirigid packages are the resultant shortened process times for commercial sterilization and improved product quality.

Flexible Packages

The descriptive packaging term *flexible* has been defined by the ASTM Committee on Flexible Barrier Materials as "easily hand-folded, flexed, twisted, and bent" (1). The United States Flexible Packaging Association defines *flexible* packages as "packaging utilizing flexible materials (papers, films, foils, and metallized films) which are normally printed or laminated as roll products and which may conform to the shape of their contents."

A wide range of monolayer materials and multilayer structures (see Multilayer flexible packaging) can be considered as flexible packaging material structures (see Fig. 1). The limitations of each package are determined by their fabrication characteristics, ie, heat sealability, printability, flexibility, and retortability, and by their product-to-package integrity, ie, moisture-, light-, and oxygen-barrier properties, flex-crack resistance, odor, and food-contact compatibility.

The term flexible in packaging is meaningful both as a package characteristic and as it relates to product application. Although savings can often result from the use of one flexible packaging material instead of another or rather than rigid containers, flexible packages must meet the specifications of compatibility with the products contained and also offer convenience, aesthetic values, quality benefits, extended shelf life, or other advantages. The application of flexible packages to foods requires development and evaluation of all package and product characteristics, production operations required, and economics. These steps involve the sciences of food and packaging technology (5). Nonfood products in flexible packaging require multiple technical and marketing disciplines as well.

Retortable Pouch

All sides of retortable pouches are flexible and are constructed of one or more layers of plastic and/or foil layers, each having its package functionality. The choice of barrier layers, sealant layers, and food-contact layers depends on the processing and product application. Flexible retortable pouches collapse or compress tightly around the contained products when a vacuum is applied to the package before sealing.

The retortable-pouch concept dates back to the 1940s, when university and container-industry suppliers evaluated various flexible barrier film materials as a means of improving the economics and quality of shelf-stable foods. Researchers in the

Figure 1. Retortable flexible packages.

United States and around the world soon realized the potential advantages of the thin-profile package concept, and development was rapid. An early pioneer in proving the integrity of this package and its product-quality improvement for military field rations was the U.S. Army Natick Research and Development Center (6). Military acceptance worldwide has led to the gradual replacement of metal cans with flexible retortable pouches for individual portions of shelf-stable foods. The pouches offer product quality; carrying, reheating, serving convenience; and weight and cubage savings. The U.S. military terms this packaged-food concept as "Meal, Ready-To-Eat" (MRE).

A chief technical problem resolved to the satisfaction of many producers and users, but unsolved by normal commercial standards, is the state of development of equipment, manufacturing techniques, and quality-control programs for the forming, filling, vacuumization, and sealing of retortable-pouch packaged foods in comparison with the high volume production-line speeds and economies of other food containers. Achieving package-seal integrity without seal contamination during filling and sealing remains a limitation in the manufacturing process.

The retortable pouch continues to find new applications, new markets, new advantages, and even new equipment and film materials. The impression forming of multiple pouches

across a web stock, followed by filling and sealing with another lid layer of film (see Thermoform/fill/seal), produces another package configuration that offers cost-saving potential. Recent developments in multilayer plastic structures (see Coextrusions for flexible packaging) have produced packages that can be microwave-processed and reheated for serving convenience without sacrificing the long shelf-life previously achieved with foil barrier layers. Improved package integrity has also resulted from film laminations offering more pliability and resistance to flex-cracking. Food and nonfood uses are being developed in small sizes for individual-consumer servings and in larger-volume institutional sizes.

Retortable pouches are also widely used in the sanitary packaging required by the medical and pharmaceutical industries. Products sterilized within such a package include surgical instruments, sterile water, bandages, intravenous solutions, medicated ointments, and liquid-diet tube-feeding foods. A typical retort pouch for packaged liquid-diet food for enteral feeding (see Fig. 2) features foil and PVDC (see Vinylidene chloride copolymers) barrier layers, spike-and-hang and dispensing features, and extended shelf-life without refrigeration.

Retortable Trays

The retortable tray usually has a rigid or semirigid structural supporting body and a sealable flexible lid. Composition may be of foil or plastic combinations which are retortable and have barrier properties. The tray package is a thin-profile container that offers all the advantages of the retortable pouch, ie, sterilization and reheating, and can also be used as a serving dish. The advantages of a thin-profile package, ie, reduced heat-sterilization requirements, improved product quality, and usage conveniences, have led to widespread development of semirigid packages. Improved techniques of thermoforming (see Thermoforming) or (cold) impact forming and coextrusion (see Coextrusions for semirigid packaging) have recently advanced the state-of-the-art of this package. Thermoforming is a relatively low-cost technique of heating a thermoplastic sheet to an optimum temperature, then forcing it against con-

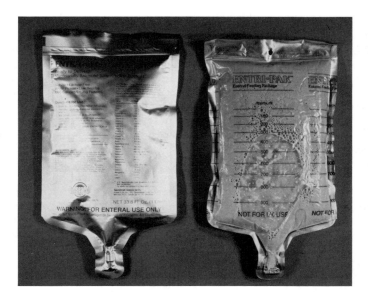

Figure 2. Retortable flexible pouch package for liquid-diet food.

Figure 3. Forming multiple units of flexible packages.

toured molds to achieve desired shapes, and finally cooling to a setting temperature. Forming may utilize either vacuum, pressure, or both. This technique has been applied to medical devices for many years, and more recently to foods.

Advantages of semirigid packages include ease of filling or top-loading of products (most are formed and filled in a horizontal mode) and potential increases in production-line speeds and volume due to multiple pocket forming on a wide-web style operation (see Fig. 3).

Completely nonmetallic varieties of this package are now commercially available for processing or reheating in microwave units. Economic advantages of these materials will soon become apparent. This lightweight, retortable plastic package is available in a wide range of sizes and shapes: tubs, trays, and cans with double-seamed, heat-sealed, or induction-bonded lids. They can be supplied to users as roll-stock for in-line form/fill/seal operations or as preformed packages.

The concept of retorting semirigid nonmetallic packages is relatively new, and all of the technologies are not yet in place. This article has mentioned the use of thermoforming, but retortable plastic containers made by injection blow molding (see Blow molding) have recently been introduced (7). These new developments will necessitate new definitions and standards.

BIBLIOGRAPHY

1. "Paper; Packaging; Flexible Barrier Materials; Business Copy Products"; *1984 Annual Book of ASTM Standards (Sect. 15), General Products, Chemical Specialties, and End Use Products* Vol. 15.09, American Society for Testing of Materials, Philadelphia, Pa., 1984, pp 831–832.
2. *Code of Federal Regulations,* Title 21, Office of the Federal Register, General Services Administration, Washington, D.C., 1982, Pt. 113.
3. R. A. Lampi, *Adv. Food Res.* **23,** 305 (1977).
4. J. P. Adams, W. R. Peterson, and W. S. Ottwell, *Food Technol.,* 123–142 (Apr. 1983).
5. F. A. Paine and H. Y. Paine, *A Handbook of Food Packaging,* Leonard Hill, Blackie & Son Ltd., London, 1984.
6. *Proceedings of the Symposium on Flexible Packaging for Heat-Processed Foods,* National Research Council, National Academy of Sciences, Washington, D.C., Sept. 1973.
7. J. A. Wachtel, B. C. Tsai, and C. J. Farrell, "Retorted EVOH Multilayer Cans with Excellent Oxygen Barrier Properties," *Proceedings of EUROPAK '85, First Ryder European Conference on Plastics and Packaging,* Sept. 17–18, 1985, Ryder Conferences, Ltd., St. Helier, Jersey, UK, 1985.

D. D. Duxbury
Pouch Technology, Inc.

ROLL HANDLING

In web processes, such as slitting and rewinding, coating and laminating, printing, and other applications, rolls must be loaded from the rewind station after the process. Because of weight and size, handling of rolls is done by general-purpose equipment such as cranes and forklifts. For greater productivity, roll-handling equipment is engineered to suit the characteristics of the web-process equipment. An important fringe benefit of engineered handling systems is the reduction of safety hazards associated with general-purpose equipment.

Slitter/rewinder overview. A slitter/rewinder (see Slitting-rewinding machinery) converts master rolls into high quality, narrow-width, rewound rolls which may or may not be the final product. Variables in the web characteristics and in the final form of the product determine which of the several types of slitting and rewinding techniques is used. The slitter/rewinder machine is typically composed of unwind, slitting, and rewind stations. A turret used for unwinding allows the loading of new master rolls while the machine is slitting and rewinding. Continuous unwinding can be attained by automatic splicing equipment.

The slitting methods used depend on the material characteristics and quality of the slit edge desired. Razor slitting is used primarily on films and score slitting on many types of fabrics, nonwovens and textiles. Shear slitting is excellent for most materials, but requires a relatively high initial investment and operator expertise. Locked-core rewinding is successful on good caliper webs and pressure-sensitive tapes with relatively small roll diameters. Differential rewinding is required for off-caliper materials and large-diameter rolls.

An effective method of rewinding is to alternate the slit webs on two rewinding shafts. This method, termed duplex rewinding, is particularly suited for narrow-width slit material. Winding on a single shaft, called simplex winding, is generally effective for slit and unslit paper, nonwovens, and textiles.

Shaft extraction/insertion system. On duplex and simplex slitter/rewinders, webs are slit and rewound on cores supported by rewind mandrels. Rewind mandrels may be extremely heavy, ie, several thousand pounds (kilograms), or cumbersome to handle manually. Each machine cycle requires that a rewind mandrel be loaded up with cores, put into the rewind station, removed from the rewind station, and then extracted from the rewound rolls. This might take place many times a day, and an operator might have to lift and maneuver many tons of rewind mandrels per shift. Apart from the burden on the operator, manual shaft handling hinders produc-

Figure 1. A duplex slitter/rewinder with roll-handling system. Turret carriages are in full roll downward position at slitter/rewind station. New mandrels are in an upward position awaiting turret index.

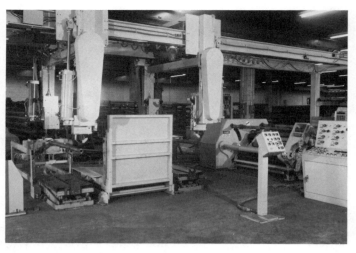

Figure 2. A slitter/rewinder roll-handling system with a shaft type unwind stand. This two-motor drive system has an adjustable vacuum winding capability and an overhead mandrel lift/transfer mechanism.

tion because the cycle time for loading and unloading is relatively long.

The use of a shaft extraction/insertion system minimizes cycle time as well as operator fatigue. With this concept, the finished rolls and rewind mandrels are transferred to a lift table where shaft extractors pull the rewind mandrel from the rolls. The shaft is supported in the extracted position until the rolls are removed from the extractor table. Fixtures called core

boxes are sometimes engineered into the extractor table to locate cores in their proper positions. The extractor inserts the shaft back into the cores. The cored shafts are then brought back into the rewind stations.

Shaft extractors are easily interfaced in close proximity to turret winders, unwinders, and other simplex winders, which often require the handling of steel shafts 6–10 in. (15–25 cm) in diameter by several hundred inches (several meters) long.

Handling systems for duplex rewinders. A duplex slitter/rewinder (see Fig. 1), with the rewind stations located above the main machine, interfaces well with an overhead roll-handling system. After the rewind cycle, the shafts with the rewound rolls are lifted out and the new shafts with cores must be inserted into the machine.

A typical system (see Fig. 2) may have an overhead-powered carriage which supports dual vertically adjustable shaft-clamp assemblies. With two clamps for each vertical lifting

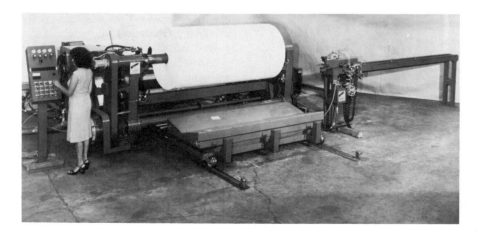

Figure 3. A surface center slitter/rewinder with hydraulically pivoting rewind arms, preset constant surfacing pressing in the center surface rewinding mode, and an automatic minimum gap for center winding.

and rewound rolls become free of the shafts and may be upended on pallets for conveyor or forklift removal into storage.

Floor loading and unloading. Floor loading and unloading features are available as an integral part of many unwinds and simplex rewinds. These units use hydraulically pivoting arms to raise and lower rolls to the floor. The need for auxiliary equipment is eliminated. Lift tables, shaft extractors, and upenders are designed to work with simplex (single-position) rewinders which produce large-diameter rolls of paper, laminates, and nonwovens (see Fig. 3). Driven "V"-trough conveyors can then move the rolls from the rewinders to weighing stations, stretch wrapping, and storage conveyors.

Robotic roll handling. An example of robotics in special-purpose roll handling is shown in Figure 4. A simplex slitter/rewinder with an automatic roll pusher pushes the slit and wound rolls onto the robot shaft. The robot shaft then expands to lock the cores to the shaft. The robot is programmed to shift the axes of the rolls 90° and to automatically position them over a pallet. A rotary table supports the pallet and will index every 90° to allow unloading of finished stacks of rolls in the four quadrants of the pallet.

Total automation. The ultimate in roll handling is shown in Figure 5. This slitter/rewinder for duct tape automatically cuts cores, loads cores at their proper location on dual rewind shafts, feeds the loaded shafts into rewind stations, winds the slit tapes, unloads the rewound rolls from the machine, and feeds the empty shafts back in for core loading. This completely "hands-off" machine is microprocessor-controlled. Cycle times are better than 30 s.

R. E. MASTRIANI
Arrow Converting Equipment, Inc.

Figure 4. A roll unloading and stacking robot with hydraulically controlled cantilever mounted boom with 90° rotational capability on vertical and horizontal axes for pallet offload.

mechanism, new shafts with cores can be loaded into the machine immediately after the shafts with the finished rolls are removed. In other words, a machine with two rewind stations has four shafts circulating through the system simultaneously. As the new rewind cycle begins, the shafts with rolls are conveyed overhead to the shaft extractors, where the slit

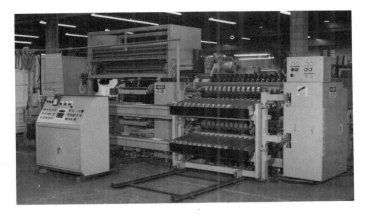

Figure 5. A fully automatic pressure sensitive tape turret slitter/rewinder with automatic core cutter and loader.

ROTATIONAL MOLDING

Rotational molding, also called rotomolding, is seldom used for packaging applications, but it is used to some extent for the production of plastic drums (see Drums, plastic) and plastic pallets (see Pallets, plastic). It is a process that produces hollow plastic parts that can be very large and oddly shaped. Outside of packaging, it is particularly useful for products such as agricultural storage tanks. The resins are generally polyethylenes, sometimes cross-linked for extra heat and chemical resistance. The process starts with the insertion of powder resin in a heated mold that is rotating on two axes. The mold is cooled after the part has formed on the mold surface. The machinery and molds cost relatively little, and the process is well suited for relatively short runs.

BIBLIOGRAPHY

General References

R. E. Ducan, D. R. Ellis, and R. A. McCord, "Rotational Molding," in N. M. Bihales, ed. *Encyclopedia of Polymer Science and Technology*, Vol 9, Wiley Interscience, New York, 1963, p. 118.

S

SEALING, HEAT

Most of the materials used in flexible packaging use thermoplastics in their construction. When heat is applied, the thermoplastics melt and act like a glue in effecting a seal. This article describes how heat is applied in both flexible and semirigid packaging in order to achieve acceptable seals.

Flexible packaging materials fall into two categories: laminated packaging materials (see Laminating; Multilayer flexible packaging) and unsupported films (see Films, plastic). Unsupported films consist of just one or more thermoplastic materials; laminations consist of nonthermoplastic materials with thermoplastic layers for sealing purposes. Even though unsupported films with more than one layer can be laminations, the term lamination generally refers to constructions with a portion that does not melt during sealing.

When laminated materials are to be sealed, the outer face of the material is generally nonthermoplastic. This makes it possible to apply a heated bar directly to the outer face in order to get the heat to the sealing interfaces and join the two package members. The outer face may be a thermoplastic if its melting point is sufficiently higher than that of the sealant so that seals can be made with a hot bar.

Heat cannot usually be applied directly to unsupported films because they melt and stick to the surface of the sealing bar. The seal area is destroyed in the process. For this reason, such materials are sealed by impulse sealing (see below). If the package members are too thick and too insulating, other means must be used to get the heat to the sealing interfaces. Since bar sealing uses the least expensive equipment, it is more widely used for sealing than any other method. The other methods described below are used in applications where bar sealing is not suitable.

When sealing thin materials together, it is generally sufficient to introduce heat from one side of the construction. When using thicker materials, or if higher speeds are required with thinner materials, heat may be introduced from both sides. The fundamental principles in heat sealing are to provide heat at the interfaces, pressure to bring them intimately in contact, and complete a weld, all within an acceptable time period. The only exception to this is in radiant sealing, which relies on film orientation and surface tension as a substitute for applied pressure. All of the important sealing methods are listed in Table 1.

Bar sealing. Bar sealing is the most widely used method for sealing. It is used both to make and seal pouches. It is also used in most form/fill/seal equipment as well. When very long seals are to be made, it is essential that the seal bars be designed to avoid any deflection in order to assure uniform pressure throughout the length of the seal. Since it is important to avoid wrinkles in seals, means are frequently provided to stretch out the seal area of the packaging materials in order to assure that they are flat when sealed. Another approach is to have mating serrations incorporated into the seal bar faces, which will stretch the packaging materials, and hopefully, remove any wrinkles. Care must be taken to see that the serrations do not puncture the films during sealing. Serrated bars

are used where good mechanical strength is required and some tiny leaks in the seal can be tolerated.

When hermetic and/or liquid-resistant seals are desired, they are best made by a flat-faced heater bar, opposed by a bar that has a resilient surface. Silicone rubber is generally used for this purpose. It is best if this resilient surface is curved when viewed from the end of the seal bar. When the bars come together, they first create a line of pressure throughout the length of the seal bar. As the bars close further, this line broadens into a band. The pressure along the initial line is at a maximum and at a minimum along the edges of this band. This assures that the optimum sealing pressures are present somewhere in between. In the event that drops of liquid are in the seal area, this pressure profile expanding from a line will tend to push these droplets out of the seal area. If the droplets are water and they are not pushed out of the seal area, they become steam and rupture the seal.

As an alternative to curving the section of the resilient bar, the heated metal bar can be curved. This is generally not done since it is more difficult to maintain a curved section than a flat section on a metal seal bar. It is also important that the edges of metal seal bars be gently rounded to avoid puncturing the packaging materials. Figures 1 and 2 show the best constructions used for straight seal bars.

Bar sealing is also used for applying covers to cups and trays. The upper heated bar is shaped to match the shape of the rim of the container being covered. The bottom bar is shaped to fit under the rim and support the container. In order to assure good seals, it is essential that uniform seal pressure be effected around the entire rim of the container. Factors contributing to nonuniform pressure are variations in the thickness of the rim of the container and warping of the opposed sealing bars. Using a resilient sealing surface under the rim of the container generally corrects these deficiencies. Another approach is to design the rim of the container to be curved upward in section. When the seal is made, this curved rim deflects much like a resilient backup member.

For some applications, continuous seals can be made by passing thin materials between heated rollers. Unfortunately, the dwell time for actual sealing under pressure between rotating rollers is extremely short. Good seals can be effected only if the rollers are moving very slowly or if the materials are adequately preheated before passing through the heated rollers.

Band sealing. In band sealing, a pouch mouth is introduced between two moving bands, which are pressed together by

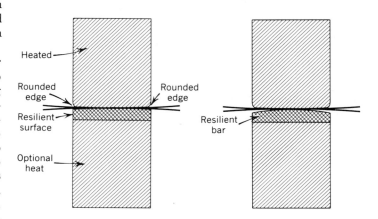

Figure 1. Bar sealer   **Figure 2.** Rounded resilient bar

Table 1. Sealing Methods

bar	*Concept*: Jaw sealer with one or two heated opposed bars. *Applications*: Manufacture and closing of laminated pouches, cup lidding, form fill seal packaging (eg, potato chip bags).	hot melt	*Concept*: Continuous strip or dots of hot, molten thermoplastic sealant are applied between two surfaces just before pressing them together. The hot melt contains sufficient heat to effect seal, and surfaces absorb heat to cool melt rapidly. *Applications*: Paperboard containers, peelable seals, case packers.
band	*Concept*: Two moving bands backed by heated and cooled metal jaws. *Applications*: Sealing filled pouches including those made from unsupported materials.	pneumatic	*Concept*: Heated film is sealed to another surface by application of air pressure. *Applications*: Skin packaging.
impulse	*Concept*: Jaw sealer backed with resilient silicone rubber. Electric current flows through Nichrome ribbon stretched over one or both surfaces and covered with high temperature release film or fabric. *Applications*: Sealing tacky materials, unsupported thermoplastic films (eg, frozen vegetable bags).	dielectric	*Concept*: High frequency electrical field melts materials held under pressure. *Applications*: Unsupported PVC, PVC-coated paper (not for polyolefins).
wire or knife	*Concept*: Hot wire or knife seals and cuts film. *Applications*: Bag making and closing; overwrap for toys, records.	magnetic	*Concept*: Gasket with iron-containing compound pressed between surfaces, assembly placed in magnetic field. *Applications*: Heavy-gauge polyolefins (eg, cap liners or lids).
ultrasonic	*Concept*: Tooling hammers or rubs materials together at high frequency, generating heat for sealing. *Applications*: Sealing biaxially oriented films, thick webs, aluminum foil, rigid container components.	induction	*Concept*: Alternating current is induced in metallic foil, usually aluminum, which heats and melts surfaces pressed against the foil. *Applications*: Tamper-evident closure liners; lids.
friction	*Concept*: Frictional heat generated by rubbing components together generates heat for sealing. *Applications*: Assembly of round containers, sealing ends of strapping.	radiant	*Concept*: Infrared radiation sealing without pressure. *Applications*: Sealing uncoated highly oriented films and nonwovens, including polyester, nylon, polyolefins.
gas	*Concept*: Hot air or gas flame applied to both surfaces and removed; molten surfaces are then pressed together. *Applications*: Manufacture and closing of polyethylene-coated paperboard milk containers.	solvent	*Concept*: Solvent liquefies surfaces which are then pressed together. *Applications*: Sealing configurations where heat may degrade the thermoplastic or is not practical to apply.
contact	*Concept*: Plate is placed between surfaces to be sealed, withdrawn, and the molten surfaces pressed together. *Applications*: Sealing ends of strapping, tubing.		

heated bars (see Fig. 3). The heat passes through the bands and into the pouch material, softening it for sealing. As the pouch continues along between the bands, the bands are next pressed together by chilled bars that withdraw heat from the pouch seal through the bands. The bands then progress to release the pouch. Band sealing is fast and widely used for closing pouches filled with product. It is important that the pouch mouth be flattened before entering the bands if wrinkles in the seals are to be avoided. Band sealing provides a continuous method for sealing, avoiding the problems found in sealing between rotating hot bars.

Impulse sealing. Impulse sealers have the same general configuration and mechanical construction used for bar sealers; the difference lies in the sealing jaws (see Fig. 4). Each of the opposed jaws is generally covered with a resilient surface, such as silicone rubber. A taut Nichrome ribbon is then laid over the resilient jaw and covered with an electrically insulating layer of thin heat-resistant material, such as silicone-rubber-coated fiberglass, Teflon-coated fiberglass, or Teflon-coated Kapton (DuPont). A pouch mouth is placed between the jaws, and the jaws are closed. An electric current passes through the Nichrome ribbon for a brief period of time and is then turned

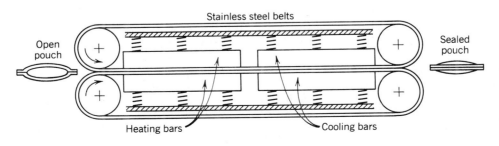

Figure 3. Band sealer

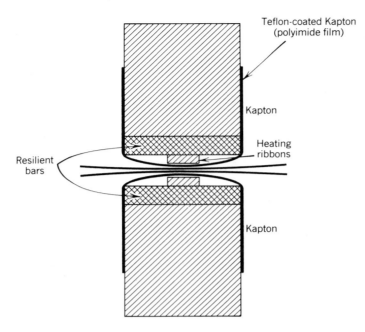

Figure 4. Impulse sealer

off. Heat is withdrawn from the pouch mouth and the ribbon through the resilient jaw surfaces, and through the jaws. The jaws are then opened and the sealed pouch removed.

The advantage of impulse sealing over bar sealing is that the seal is cooled to achieve adequate strength before the jaws are opened. This permits the sealing of constructions that are insufficiently tacky when hot to hold together when a bar sealer is opened. It also permits the sealing of unsupported thermoplastic materials, which would stick to heated bars and fall apart when heated bars are opened. The material which covers the Nichrome ribbon has release properties, so that it does not stick to unsupported thermoplastics, or to many of the coatings that are used on the outside of packaging laminations. Such coatings include PVDC, ink, and varnish. Shaped impulse-sealing ribbons are used for securing lids to cups and trays. The dominant drawback of impulse sealing is high maintenance cost. The impulse ribbons slowly deteriorate and the release coverings over the ribbons degrade requiring frequent replacement. On a production basis, impulse sealing should be used only where bar sealing does not do the job properly.

Hot-wire or knife sealing. Hot-wire or knife sealing is adaptable to a very high speed operation and is used for both sealing and cutting apart polyethylene bags on bag-making equipment (see Bags, plastic). Unsupported films, when trim sealed by this method, tend to form a strong bead in their seal areas due to surface tension and orientation. This method is also used to a limited degree with laminated constructions. Hot-wire cutoff is also used in film-dispensing equipment and with L sealers.

Ultrasonic sealing. In ultrasonic sealing, the sealing heat is produced by mechanically hammering or rubbing the packaging materials together at a high frequency. It is valuable with highly oriented films because sufficient heat is generated to melt the interface, but not enough to heat the rest of the material to the point of degradation. It is also very useful to seal materials that are much too thick to permit heat transfer through them for sealing. It is the only method used for welding aluminum foil in its production.

Friction sealing. In friction sealing, the top and bottom halves of cylindrical containers may be easily joined. The bottom member is held against rotation by a friction brake; the top member is pressed against the bottom member and rotated. The heat generated by friction at the interface between the two halves melts their surfaces and the viscous interface causes the bottom half to rotate in unison with the upper half. This effects the seal. Thermoplastics with surfaces that become slippery when heated are best suited to friction welding; those that become doughlike present problems. Sealing is also effected between the ends of package strapping, which are rubbed together at high speed until the interface melts, causing them to weld. Ultrasonic welding is actually a method of friction welding.

Gas sealing. In gas sealing, where paperboard and thicker thermoplastic materials offer too much resistance to the passage of sealing heat, heat is applied directly to the surfaces to be sealed by a gas. The gas may be either hot air or a gas flame. The most important application of this method is the manufacture of paper containers for dairy products. A gas flame is played on the surfaces about to be sealed, after which they are clamped together between chilled bars. Sufficient sealant must be used on the surfaces in order to fill all the fissures formed at the joints in the paperboard.

Contact sealing. In contact sealing, an alternative to gas, the sealing interfaces are each contacted by a heated plate before being pressed together. This is also used for strapping as an alternative to friction sealing.

Hot-melt sealing. Hot-melt sealing is effected by depositing either a continuous strip, or dots, of a thermoplastic material on a packaging member, just before the next member is pressed against it. It is used as an alternative to glue, since the packaging members quickly remove heat from the melt, making it tacky enough to hold the members together quickly even though spring-back may be tending to open them up (in sealing the covers of a shipping container). With glue, the members must be held together until enough of the glue's solvent is absorbed by the packaging material to make it tacky. Hot melts are also used where applied heat might damage a packaging member or where peelability is desired (see Adhesives; Waxes).

Pneumatic sealing. Pneumatic sealing is used for skin packaging, (see Carded packaging) where pressure from the atmosphere pushes a hot, tacky film into close conformity with an object and seals the film to a substrate on which the object is placed. It is also used in hermetic skin packaging for fresh red meat.

Dielectric sealing. In dielectric sealing, used principally with PVC materials, a high frequency electric field generates heat for sealing. The members are pressed together between a cold bar, generally made of brass, and a cold flat metal surface. The field is generated between the bar and the lower surface, and these members are kept cold to withdraw any residual heat from the materials that have been sealed. The polyolefins, unfortunately, do not respond adequately to a dielectric field.

Magnetic sealing. Magnetic sealing is used where heavy polyolefin package members are to be sealed. A gasket, shaped to fit between the sealing faces of the two package members, is made from the same thermoplastic as the package, but has milled into it an iron-containing compound that has a high hysteresis loss. The package portions are pressed together with the gasket between their sealing surfaces and placed in a magnetic field oscillating at a frequency high enough to melt the gasket. This in turn melts the sealing surfaces and causes them to weld together. As an alternative to a gasket, an iron-containing material can be preapplied to one or both of the sealing surfaces, producing the same result.

Induction sealing. In induction sealing an alternating electrical field heats up a metal, eg, aluminum foil, placed between two members to be sealed. Its most common use is to produce a tamper-evident seal over the top of a bottle. The foil member is generally incorporated under the closure liner (see Closure liners). The assembly is placed in the alternating field, and a current is induced in the foil; this current heats the foil. Foil sticks to polyethylene bottles; with bottles made from other materials, a suitable coating is put on the foil, which will stick the foil to the bottle when the foil is heated.

Radiant sealing. Radiant sealing makes it possible to seal many materials used in packaging that have hitherto been considered either difficult or impossible to seal without coatings (see Fig. 5). Polyester film (see Film, oriented polyester) long used in packaging constructions and many nonwoven (see Nonwovens) materials can now be sealed by this method. These include Tyvek (DuPont), a nonwoven polyolefin used in packaging and for protective garments; Reemay (DuPont), a nonwoven polyester, used in filters; Cerex (Monsanto), a nonwoven nylon, used as a carpet backing and in filters; and nonwovens made from polypropylene and other base materials. This method is used for sealing other oriented films, such as polypropylene and polyethylene. OPET pouches are commercially made for document preservation. OPET is the only material approved by the Library of Congress for use in contact with valuable documents. In addition to making pouches, this sealing method can also be used to produce continuous seals to edge-join rolls of materials at high speed.

Solvent sealing. Solvent sealing is used to join together package configurations that don't lend themselves to heat sealing. It is also used for joining materials that are either not susceptible to heat sealing, or may be damaged by the application of heat (eg, rigid containers made of acrylics or polystyrene).

Selecting a Sealing Method

For laminations, bar sealing is the least expensive of all sealing methods and it should be tried first before considering other methods. If sticking to or contamination of the seal bar is a problem, the following should be tried out before considering impulse sealing:

1. Coating or impregnating the seal bar with Teflon.
2. Periodically wiping the seal bar with a silicone grease.
3. Mounting a release material, such as Kapton (DuPont), between the seal bar and the packaging material.

If all of these fail, impulse sealing should be considered next. For unsupported films, hot-wire, knife, or radiant sealing should be investigated. Although the equipment is more expensive, ultrasonic should also be considered. There are many applications where ultrasonic will be the method of choice when other methods fail. For heavier packaging materials, friction, contact, or magnetic sealing should be considered. As a last resort, solvent sealing should be considered.

When packaging materials are sealed, the package and its function dictate the properties to be achieved. Mechanical strength is important, since the seals should not fail under normal handling. Hermetic integrity is important in food, pharmaceutical, medical, and chemical packaging. Although small wrinkles in a seal area may have slight effect on its mechanical strength, they are completely unacceptable if a hermetic-type package is required. The best approach to avoid seal wrinkles is to hold the seal region of the construction under tension during sealing. If this is impractical, increasing the thickness of the sealant portion of a lamination may assure good seals, even though small wrinkles may be present. Incorporating EVA in polyethylene, or substituting ionomer (see Ionomers) as a sealant, results in improved flow properties, and better tolerance of small wrinkles.

The most reliable seals for hermetic packaging consist of true fusion welds between the opposed sealant members of a lamination. A true weld can be identified by pulling apart the lamination, and noting where seal failure occurs. If it occurs along the interface between the materials, a true weld was not attained. If failure migrates from the interface, through the sealant, and to the outer portion of the lamination, a homogeneous seal has been achieved (see Multilayer flexible packaging).

If a peelable seal is desired, one should not rely on control of temperature or pressure to achieve peelability. Seals made in this way may be peelable, but they can come apart by themselves with handling. The only way to achieve dependable peelability is to have an interface formulated to be peelable. An example of true peelability is something that behaves very much like mending tape: the seal should be a tacky seal. Peelable seal laminations have been developed recently that survive steam sterilization and are now being used in sterile food packaging for market testing.

Seal Testing

The American Society of Testing and Materials (ASTM) has set up a standard for the testing of the seal strength of flexible barrier materials. Their designation for the testing procedure

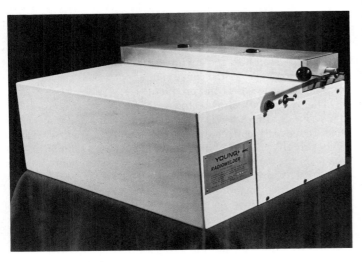

Figure 5. A radiant sealer

is ASTM F 88-68. This standard describes how the specimens are to be prepared, along with the equipment and methods to be employed for testing. Tests in accordance with these standards are extremely useful in comparing the seal strength achieved with various materials under consideration for a particular application.

Seal strength, however, is not the only criterion to be evaluated for many applications. Although some seals may test well in the laboratory, they may fail in service. Reasons for this failure include absence of a true fusion weld of the interfaces of the materials being sealed. Occasionally, unwelded seals perform well under mechanical testing, but fail in the actual application. When there is not a true fusion weld, portions of the packaged product may find their way out of the package through the seal. For seals that must be reliable, particularly where sterility must be assured in a package (see Health-care packaging), sections of the seals should be examined under the microscope. Unless examination shows that the interface between the sealant portions of the laminations has been completely obliterated, the seals should be suspect; with a true fusion weld, there is no visible interface remaining. The exception to this is peelable seals. The only way to assure that these are acceptable for the long term is to subject them to rough-handling testing for extended times, exceeding the environmental extremes likely to be encountered. If heat processing is involved, good sense dictates that the tests be carried out for at least twice the period of time that the product is heat processed. Tests should also be conducted to simulate environmental pressure fluctuations at something in excess of the maximum temperatures expected to be used in processing. For the packaging of chemicals, or other corrosive products, testing should be carried out with filled packages, subjected to rough handling, elevated temperatures, and any other severe conditions that might be experienced.

When dealing with flexible material constructions, even small wrinkles in the seal area produce a statistically significant percentage of leakers. Aside from visual inspection, no satisfactory automatic system has yet been devised to isolate packages with small seal wrinkles, although a great deal of time and money has been spent in this area. When developing a packaging system where seal integrity is paramount, everything should be done in the design of the system and its equipment to eliminate chances for seal wrinkles. The best way is to keep the packaging materials under tension in two directions while they are being sealed. This can do more to achieve success in eliminating seal wrinkles than any other single effort that might be made. Frequently, design of the package can be improved for the sole purpose of reducing the possibility of wrinkles in the seals.

BIBLIOGRAPHY

General References

R. D. Farkas, *Heat Sealing,* Reinhold, New York, 1964.

W. E. Young, "Sealing" in W. C. Simms, ed., *Packaging Encyclopedia,* Cahners Publishing, Boston, Mass., 1984.

W. E. YOUNG
William E. Young Company, Inc.

SHELF LIFE

Shelf life is the time between the production and packaging of a product and the point at which the product first becomes unacceptable under defined environmental conditions. It is a function of the product, the package, and the environment through which the product is transported, stored, and sold. The principles of shelf-life testing hold for virtually any product/package combination. Products packaged in impermeable materials, such as glass or metal, degrade primarily from mechanisms inherent in the chemistry of the product. Since these mechanisms are essentially product-dependent, evaluations of the package would be unproductive. Notable exceptions to this are situations in which the package allows or contributes to deterioration of the product. For example, light transmission through glass can promote oxidation reactions, but the problem can be alleviated by tinting the glass. Products in metal cans can react with the metal substrate itself (eg, pinholes in tinplate), or with components of the can coating. Improvements in can and coating technology are designed to preclude these interactions (see Cans, aluminum; Cans, steel). Since the reactions of these impermeable containers are specific to the containers, the focus of this article is on semipermeable and permeable materials.

Factors that Influence Shelf Life

Product. Products differ greatly in their susceptibility to degradation by various agents. Some products become unacceptable from a change in moisture content. For example, a moisture gain in ready-to-eat breakfast cereals destroys the crisp texture; a moisture loss in tobacco products alters their burning characteristics. Snack items can become rancid due to oxidation of the oils absorbed during frying. Snacks also can be sensitive to moisture change. The mode of failure of the product has a direct influence on the type of protection that must be provided by the packaging material. The match between product susceptibilities and package protection influences shelf life.

Once the mode of deterioration of the product is determined, acceptance criteria must be defined. Acceptance criteria are based on a critical component of the product. This is often done by a sensory panel. If sensory scores can be correlated to analytical data, analysis can provide the index for product quality. The acceptability criteria for any product have a direct influence on the measurement of shelf life. Some products have a clear point of acceptability; others have complex deteriorations which make it difficult to identify a critical point. If the product has multiple or interacting modes of failure, subjective decisions must be made to set the acceptance criteria. Because the consumer is the final judge, concordance between a standardized definition of acceptability (as used within the company) and consumer acceptance is crucial to the commercial success of the product.

The product which is tested for shelf life must be specified. This sounds almost trivial, but reformulations and ingredient-cost reductions can change product characteristics without changing the product name. The result is that assumptions are made on the basis of past performance, which may not hold for the present product. Therefore, the test label should include product name, formulation (or reference to specific formulation in a laboratory notebook), date of manufacture, date of test, conditions of test, and principal researcher. The packaging system must also be identified in terms of key factors described below.

Package. The package protects the product against an agent which degrades the product. For moisture-sensitive products, the degree of protection is measured in terms of the water-vapor transmission rate (WVTR); for oxygen-sensitive products, in terms of oxygen transmission rate (OTR), etc. The critical sensitivity is determined in part by the package. For example, a snack product which is sensitive to moisture gain (loses crispness) and oxygen (becomes rancid) can be viewed as "moisture sensitive" if the texture degrades before the rancidity becomes objectionable. The same product, if packaged in a sufficient moisture barrier, would be identified as oxygen sensitive in terms of package criteria. WVTR and OTR are measurable by a variety of standard procedures (see Testing, permeability and leakage). Permeation rates for flavors and other vapors can be determined by gas-chromatographic, mass-spectral, infrared, and other techniques. There is no standard technique at this time, but efforts are underway to provide such a standard.

The size of the package also influences shelf life. As the package size increases, the surface-to-volume ratio decreases. The result is that the amount of permeant which comes through the package increases in a square function, but the volume of product which absorbs that permeant increases as a cube function. If all other factors remain equal, barrier requirements decrease as package size increases. In summary, the packaging parameters which must be specified for shelf-life testing are the material designation and source, appropriate transmission rate, surface area, and net weight of the enclosed product.

Environment. Product distribution causes stress on the product as a function of product sensitivities and all conditions experienced by the product as it is carried through various distribution networks. Actual conditions vary with where it is shipped, how it is shipped, the seasons, warehouse conditions, etc. It is impossible to account for this variety of conditions by using a single storage condition. The manufacturer of sensitive products (food, pharmaceuticals, cosmetics, tobacco, chemicals), must define the criteria which will be used for the shelf-life determinations. Pharmaceutical manufacturers must package for the worst-case situation to ensure survival under the most severe conditions in the distribution chain. Food products are often packaged for less-than-worst-case scenarios. Products which are not rotated on the supermarket shelf may fail, but this loss is considered more acceptable than increasing packaging costs for the entire line to accomodate this situation. It is common to aim for the survival of 80% of the product through distribution. Once the representative distribution mode is chosen, useful statistical indices can be derived which represent this mode.

The shelf life of a product plays a major role in the structure of a distribution system, and vice versa. Products with short shelf lives require a relatively rapid distribution system. One manufacturer of moisture- and oxygen-sensitive snack items ensures freshness in flexible pouches by shipping from many production facilities through hundreds of distribution centers. Another can ship nationwide from just one production facility (1,2) by using a multilayer composite can (see Cans, composite).

Refrigerated distribution is costly, but it provides additional product protection. Heat processing, sometimes combined with aseptic packaging, extends shelf life and eliminates the need for refrigeration. In Europe, where refrigerated and frozen distribution systems are less extensive than they are in the United States, aseptic packaging was developed years ago to fill the gap that the United States fills with refrigeration (see Aseptic packaging).

Open dating. The measurement of shelf life is becoming increasingly important because intense competition and high energy costs have resulted in renewed interest in reducing packaging costs. An abundance of new packaging materials provides many choices, but also adds to the complexity of the choice. In addition, government requirements for open dating would necessitate shelf life testing. Open dating (3,4) is the inclusion of a date on the package which indicates the date by which the product should be consumed. Dating of products on line is a long-established practice of companies producing food, pharamaceuticals, and other sensitive products (see Code marking and imprinting). Each company developed its own proprietary coding system to identify production dates and production facility when necessary. However, these codes were designed to be unreadable to consumers and competing producers. Open dating would change this.

Various types of open dating are possible. A direct conversion of production codes into consumer readable form gives the date of manufacture. A freshness date ("use by") indicates the date by which the product should be consumed to assure maximum freshness. A "pull date" is the date after which the product should no longer be sold. Pull dates are used on milk and other dairy products.

All of these dates assume storage under specific conditions. For example, dairy products are assumed to be refrigerated. If conditions are not held as anticipated, product quality will not be reflected in the dates. Dating of products will benefit the consumer, and it can also benefit the manufacturer and retailer. It would improve stock rotation, which could ultimately reduce the packaging requirement necessitated by existence of old product.

Traditional Approaches to Shelf-Life Testing

One common procedure for testing shelf life is a shipping test. Shipping tests are often used as an adjunct to other shelf-life testing, but they are sometimes considered a replacement. This is a mistake, because shipping is only one component of the testing environment. This type of testing is characterized by uncontrolled and often unknowable storage conditions. A product is shipped to a warehouse, stored for a specified time, and then returned to headquarters (or R&D) and examined for acceptability. Shipping tests are also used to check for product deterioration from shock and vibration during shipment. At least one company uses shipment through the Chicago post office as the ultimate test for product endurance. (Shock and vibration effects, important for assessing product quality, are abuse dependent, not time dependent.) Shipping tests show product-quality changes with time in real-world situations; but they rarely, if ever, represent all conditions which the product will experience. The test appears to be a representative real-world test, but the changing conditions, (eg, differences between shipment in carton vs pallet quantities) necessitate careful evaluation of the results.

Storage tests. Storage tests consist of maintaining product in a static facility, and evaluating product quality with time. Storage conditions can be uncontrolled or controlled. Warehouse storage is usually characterized by uncontrolled conditions. Storage conditions can be controlled through the use of storage cabinets which maintain temperature (and often humidity) at specified values. Controlled conditions provide more reproducible data. Various conditions can be tested, including accelerated conditions that provide more-rapid results. A typical ambient condition (for the United States as defined by

TAPPI) is 73°F (23°C)/50% rh. Accelerated conditions vary, but a typical condition would be 95°F (35°C)/80% rh. Accelerated storage speeds product degradation. However, the factor that defines this increase is related to the kinetics of the chemical reactions that define the degradation. This varies with product and with mode of degradation. Therefore, a standard multiplication factor is meaningless.

Accelerated studies are useful if an initial evaluation is performed at both ambient and accelerated conditions, and compared. This comparison can be through actual kinetic studies (such as Q_{10} analysis described below) or by shelf-life comparison. It is imperative to assure that discontinuities do not exist between ambient conditions and accelerated conditions. If either the product or packaging passes through a transition, the results of the accelerated test will have little bearing on comparisons with ambient. Transitions which effect packaging include the glass-transition temperature (T_g) of polymers. For example, permeation values for polypropylene cannot be extrapolated to frozen conditions due to a T_g at 14°F (-10°C). Product transitions include melting/freezing of any ingredient (usually water or fats), recrystallizations, etc. In addition, activation energies must be considered, because certain reactions will not occur below certain temperatures. Extrapolation below these temperatures is therefore not recommended.

Q_{10} analysis involves testing products at varying temperatures and defining the difference in reaction rate for a 10°C (18°F) temperature increase. A Q_{10} of 2 means that the reaction rate will double for each 10°C increase in temperature. This is based on the Arrhenius relationship which is discussed in any text on kinetics or physical chemistry. The Arrehenius relationship is useful even though it oversimplifies the complex chemistries of most products. A straight-line relationship with a minimum of three points (over a limited temperature range) indicates that Q_{10} can provide a useful indicator for storage duration. Caution must be exercised with respect to transitions. It also should be noted that the Q_{10} itself is temperature-dependent. However, products are usually studied under a limited range of temperatures so that a constant Q_{10} can be assumed.

Once the Q_{10} is determined (and shown valid through a linear Arrhenius relationship), the value can be used to predict the time necessary to test a product to be equivalent to the shelf life at ambient temperature. For example, a product with a Q_{10} of 2.4 can be stored for 10 weeks at 91°F (33°C) to obtain results similar to 24 weeks storage at 73°F (23°C) (5). (One can view these studies as a test in which the conditions extended to the package are exaggerated in an attempt to simulate the accelerated passage of time. The underlying assumption is not always valid, and the technique is only as good as that underlying assumption.)

An alternative type of "accelerated" test is a study in which the permeation rate of the packaging material is "accelerated." A product which is being tested for sensitivity to moisture is packaged in a material that has a relatively high oxygen barrier, but a relatively low moisture barrier (eg, an acrylonitrile copolymer). With a known permeability, area, and net weight, the amount of moisture permeating the package can be determined. Product can be sampled for acceptability until failure is reached. This defines the amount of moisture transfer that can be tolerated. Calculations with higher moisture-barrier materials can provide performance criteria for other materials. A similar study can be performed for oxy-gen-sensitive products by using a low oxygen/high moisture barrier material (eg, a polyolefin). In essence, this test purposefully introduces a known amount of the critical agent to accelerate the time to degradation without changing the conditions experienced by the product. A cautionary note is important. The accelerated test described above includes the inherent assumption that the changes in the product occur faster than the transmission across the package. If product changes are slower than the influx of permeant, the results of an accelerated test will indicate a shelf-life which is longer than the extrapolation to a reasonable barrier material. This is due to a critical permeant level being obtained faster than the evidence of product degradation.

Computer Models for Shelf-Life Prediction

Shipping and storage tests require that the choice of packaging be made before the study. Chosen materials are then submitted for evaluation with time. Successful completion of the test verifies that the chosen package will last at least as long as the chosen test duration under the conditions of the test. Failure to survive the test conditions demonstrates insufficient protection. The deficiencies of this procedure are that failure is not known until demonstrated and overprotection is not shown at all because testing usually terminates after desired time is reached.

The use of a computer to simulate a storage test provides a means to pretest materials before actual storage testing. By performing these calculations rapidly, as only a computer can do, many iterations are possible within a reasonable time, providing a powerful support tool for packaging development. The simulations are accomplished by separating the product and packaging characteristics. The product is analyzed for changes which occur upon exposure to the shelf-life-limiting parameter which is identified as the mode of failure for the product. The packaging material is analyzed for its barrier properties against that parameter. The computer is then used to combine the protective aspects of the package with the sensitive properties of the product. Since packaging, product, and storage parameters are entered independently, it is possible to test the effect of different packages or conditions simply by entering the new variables.

Computer simulation yields the most cost-effective packaging design. Later real-time storage studies serve the purposes of legal requirements and confirmation of predicted results. Differences between predicted and actual results can be used as indices to check initial assumptions. For example, poor seals will shorten actual shelf life achieved with an otherwise-adequate barrier material. In this example, the false assumption is that the material barrier value is equivalent to the package permeation (see Testing, permeation and leakage).

Simulated studies extend beyond the one package/one condition study. Parameters can be changed to effect another study. A survey of potential new packaging materials can be accomplished without need to retest the product. Only the changes in material parameters need be entered to test performance. This is especially important when limited amounts of product material are available, as in the product-development stage. Changing storage conditions follow a similar procedure, but the caution for passing transitions remains.

Product evaluation. The mode of failure for the product must be known prior to using any simulated approach. This could be moisture gain or loss, oxidation, CO_2 loss, flavor loss, chemical degradation, etc. Once the critical factor is known, testing is conducted to show product changes with change in the critical factor. For example, a moisture-sensitive product is evaluated for change in moisture content as the rh in the storage environment is varied. This test is the moisture isotherm. The isotherm can usually be completed in one to four weeks, depending on the speed at which the samples equilibrate. For oxygen-sensitive products, the testing consists of evaluating product changes with oxygen uptake, and can be performed by inverse gas chromatography (6), respirometry, Warburg, or a variety of tests for degree of oxidation. For carbonated beverages packaged in permeable bottles, the test would measure carbonation loss (soft drinks), or oxidation (beer), or both.

Product testing should be performed at various temperatures if temperature effects are to be modeled. The evaluation of differences in reaction rates vs temperature supplies important information concerning the kinetics of the reaction, and can be used to investigate performance for different distribution environments. A minimum of three tests is recommended. The caution concerning transitions remains valid.

Package evaluation.   For computer-modeling purposes, the chief packaging characteristics are barrier properties. Materials that do not change significantly with machining and flexing can be tested in flat-sheet form for WVTR, OTR, or any other critical permeant. For materials which change with handling (waxed glassine, foil, etc) full package testing is recommended. The computer model measures the change of permeant level in the package as a function of the transmission rate. If the flat-sheet value is used in a system which is poorly sealed, or has developed fractures in handling, the predicted shelf life will be erroneously long.

The package material characteristics of sealability, printability, compatability, cost, mechanical strength, etc, remain the province of the packaging engineer and are not usually incorporated into computer models. Non-barrier parameters are not included in this discussion of shelf life. Those parameters which do effect shelf life, but are not in the model (eg, light transmission for oxygen-sensitive products), must be considered in evaluating any computer printouts.

Model fundamentals.   The internal environment of a package changes as permeants enter or leave the package. In terms of the mathematics, components which enter the package from the outside environment, and those which permeate from the inside to the outside, are identical. Therefore, any discussion or examples of permeation rates are valid for either circumstance.

The transmission rate is standardized for package area, film thickness, time of measurement, and difference of partial pressure of permeant across the package. In general the terms are amount of permeant multiplied by film thickness per package area per time times the unit of partial pressure differential of permeant across the package. The measurement is made at a temperature which is implied but not usually stated in the units. In use, the standardized transmission rate is combined with area, thickness, and time in the model to define how much permeant can enter or leave the package. The effect on the product is a function of the net weight of the product upon which the permeant acts. The partial pressure differential across the package becomes difficult to measure, and changes as the internal environment of the package changes.

The partial-pressure-differential change with permeation is the major reason for the use of computers in shelf-life modeling. If this change did not occur, one could simply multiply transmission rate by time and area, and divide by the thickness to obtain amount of permeation for any time. This assumed straight-line relationship overestimates transmission because the permeation process itself changes the differential of partial pressure across the film and therefore reduces the amount. In addition to the driving-force change due to the transmission of permeant, some of the permeant is absorbed by the product and further complicates the system. In engineering terms, as the permeation process procedes, the driving force of that process (the partial pressure differential) decreases. This change can be described by a differential equation. The computer models incorporate these effects and yield a representative manifestation of the process.

Computer models have been published by various authors (7–13). Many models have been developed which have not been reported in a usable form in the literature. These include models on oxidation-sensitive products, beverages, and products in which product characteristics result in product-dependent models.

The models are based on a standard differential equation which describes mass transfer across a permeable barrier. This is shown in the following equation (14)

$$dW/dt = k/\ell \times A \times (P_{out} - P_{in})$$

where: t = time; W = change in weight of the critical component; k = permeability coefficient for the packaging material; ℓ = thickness of the packaging material; A = area of the package; P_{out} = partial pressure of permeant outside the package; and P_{in} = partial pressure of permeant inside the package.

This equation can be converted into a usable form by substituting measurable values for the W and P_{in} terms. The P_{out} term is readily obtainable: for oxygen it is 0.21 atm (21.3 kPa); for moisture it is the saturated vapor pressure (15) times the rh of storage; for flavors and volatile components it is essentially zero.

The P_{in} term requires knowledge of the reaction of the permeant with the product. In the case of moisture change, this is expressed as the moisture isotherm, which describes change in product moisture vs storage rh. An expression which describes the water activity over the product for any moisture value is used to define the P_{in} term. (Water activity (A_w) is the partial pressure of water above the product divided by the saturated vapor pressure of water at the same temperature.) First the isotherm is determined. Second an expression is derived which describes A_w as a function of moisture content. Finally, P_{in} at any moisture content can be described as the expression for A_w times the saturated vapor pressure. (Note: Water activity is commonly used, but it is valid only at the temperature used in the isotherm study. The preferred procedure would study moisture content as a function of vapor pressure directly, instead of as a function of A_w. The computer model converts A_w into pressure terms.) Models for permeants other than moisture require analogous expressions to relate internal pressure to product changes.

With a substituted expression for the P_{in} term in concentration (rather than weight) terms, the W term must be expressed in concentration terms. An expression is therefore needed which describes the change in concentration as a function of weight change.

The above expressions are substituted into the equation and rearranged to group variables on one side of the equation, and the new equation is integrated to form the model. The limits of integration are permeant concentration from initial to critical values, and time from initial to shelf life (time to reach critical amount of permeant).

The various models described in the literature use approaches related to the above procedure. Simplifying assumptions (such as linearity in the isotherm) are used to develop simpler models, and these models can be used to advantage if the assumptions are kept in mind. The fewer restrictions used in deriving the expressions, the more universal the model; but the mathematics become increasingly complex.

The models do not replace actual storage studies. Storage is still necessary to verify product changes, find unusual effects, and in some cases, meet compliance criteria with government regulations. The use of the models serves to reduce the time and effort necessary to identify optimal packaging protection, and eliminate storage studies on materials with insufficient barrier properties to be reasonable candidates for the product.

BIBLIOGRAPHY

1. *Food Engineering's 1983 Directory of Food and Beverage Plants,* Food Engineering, Chilton Co., Radnor, Pa., 1983.

2. C. E. Morris, *Food Engineering* **56**(9), 65 (Sept. 1984).

3. *Open Shelf-Life Dating of Food,* Office of Technology Assessment, Congress of the U.S., Supt. of Documents Stock No. 052-003-00694-4, 1979.

4. "Open Shelf-life Dating of Food," *Food Technology* **35**(2), 89 (Feb. 1981).

5. T. P. Labuza, "Reaction Kinetics and Accelerated Tests Simulation as a Function of Temperature," in I. Saguy, ed., *Applications of Computers in Food Research and Food Industry,* Marcel Dekker Inc., New York, 1983.

6. S. G. Gilbert, in J. C. Giddings, ed., *Advances in Chromatography,* Marcel Dekker Inc., New York, 1984, Chapt. 6 pp. 199–228.

7. S. Mizrahi and M. Karel, *Journal of Food Processing and Preservation* **1**, 225 (1977).

8. K. S. Marsh, "Computer-Aided Shelf Life Prediction," *1984 Polymers, Laminations and Coatings Conference-Book 1,* TAPPI Press, Atlanta, Ga., 1984.

9. S. Gyeszly, *Package Engineering* **25**(6), 70, (June 1980).

10. R. Heiss, *Mod. Packag.* **3**(8), 119, 172 (Aug. 1958).

11. M. Karel, S. Mizrahi, and T. P. Labuza, *Mod. Packag.* **44**(8), 54 (1971).

12. S. S. Rizvi, "Requirements for Foods Packaged in Polymeric Films," *CRC Critical Reviews,* 111 (Feb. 1981).

13. W. H. Clifford, S. W. Gyeszley, and V. Manathunya, *Package Development & Systems* **7**, 29, (Sept./Oct. 1977).

14. J. Crank and G. S. Park, eds., *Diffusion in Polymers,* Academic Press, London, 1968, p. 42.

15. R. C. Weast, ed., *Handbook of Chemistry and Physics,* 65th ed., CRC Press, Inc, Boca Raton, Fla., 1984.

K. S. MARSH
Kenneth S. Marsh Associates

SHRINK BANDS.   See Bands, shrink.

SHRINK FILMS.   See Films, shrink.

SILK SCREENING.   See Decorating.

SLIPSHEETS

Small shipping containers have been combined into unit loads since the 1940s, when the wooden pallet was developed (see Pallets). The weight of wooden pallets, however, and the space they take in transport equipment precluded economical unitized shipping of many lightweight products, such as frozen foods, cereal products, and cotton balls. Also, as palletized shipping programs expanded the difficulties of controlling pallet quality and costs increased substantially. These shortcomings led to the development of the palletless unit-load system that utilizes a slipsheet as the carrying base. The use of slipsheets lowers freight costs by increasing the payload on transport vehicles and eliminates the need to return pallets to the shipper. Their use requires front-end attachments on forklift trucks.

Materials.   Slipsheets are manufactured from solid fiber, corrugated board, or plastic. Solid fiber is a lamination of plies of kraft linerboard and cylinder board designed to provide adequate tensile strength and other properties. The thickness of a solid-fiber slipsheet is expressed as caliper of board and is measured in points. One point (mil) is 0.001 in. (25.4 μm). Corrugated board (see Boxes, corrugated) is described in terms of flute size and board test, eg, B flute, 250-lb test (linerboards are 69 and 42 lb/1000 ft^2, ie, 337 and 205 kg/m^2). Plastic shipsheets are made of polyethylene, polypropylene, or other polymers. Thickness is expressed in terms of mils.

Configurations.   A slipsheet is a flat sheet of material that is used as the platform base upon which goods and materials may be assembled, stored, and transported. It may have one or more tabs along its side that extend beyond the load to facilitate mechanical handling. The platform base is the area upon which product is placed and stacked to form a unit load. The dimensions of the platform base, excluding the tabs, must be at least as large as the base dimensions of the unit load in order to prevent overhang. Any overhang results in dragging of the product or container, which can break up the unit load (see Fig. 1).

The number of tabs, located on one or more sides, is usually determined by the handling and shipping requirements of the user. The recommended width of tabs is 3 in. (76 mm) min to 4 in. (102 mm) max. These dimensions are compatible with the gripper bar of handling trucks. The most common tab configurations are shown in Figure 2.

Overall size.   The combination of base and tab dimensions determines the overall slipsheet size. The most common size is

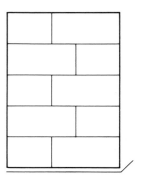

Figure 1.   Slipsheet with side tab.

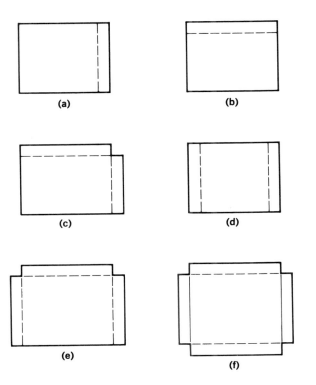

Figure 2. (**a**) One tab on the short side is excellent for unit loads that are positioned in one direction and loaded and unloaded on the short side of the slip sheet. (**b**) One tab on the long side is excellent for unit loads that are positioned in one direction and where it is desirous to load and unload a unitized load on the long side of the slipsheet. (**c**) Two adjoining tabs are perfect for unit loads that are positioned in a pinwheel pattern, that is, one unit load facing in the short-side direction and one unit load facing in the long-side direction, side by side. This results in greater cube utilization of transportation vehicles. (**d**) Two tabs on opposite ends are excellent for those situations where handling requirements demand loading or unloading from opposite ends of the unit load. (**e**) Three tabs, two on the short side and one on the long side, are excellent for those situations where handling requirements demand entry from three sides. It is particularly useful in the doorway area of railcars. (**f**) Four tabs, on all four sides, are excellent for those situations where handling requirements demand entry from all four sides, particularly in the doorway area of railcars.

48 in. × 40 in. (1.219 m × 1.016 m), which corresponds to the standard Grocery Manufacturers Association (GMA) pallet. With two adjoining 4-in. (102-mm) tabs, these dimensions would be described as 48 in./4 in. × 40 in./4 in. (1219 mm/102 mm × 1016 mm/102 mm). Sizes in countries using the metric system correspond to standard metric pallets (eg, 1200 mm × 1000 mm).

The base and tabs should be manufactured within a tolerance of ±¼ in. (±6.4 mm) in the length and width dimensions. Common tab corner configurations are shown in Figure 3.

Properties. *Tensile strength* is the maximum tensile stress developed before failure of the slipsheet. The slipsheet must contain sufficient strength to avoid rupture when the tab is properly pulled by the gripper bar. The gripper bar is that part of the front-end attachment of forklifts and walkie–riders that clamps the tab of the slipsheet uniformly along its length. Tensile strength of paper is tested by either TAPPI D 494 or ASTM D 828. Tensile strength of plastic is tested by ASTM D 638-76.

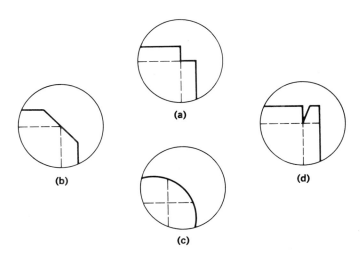

Figure 3. Common tab corner configurations. (**a**) 90° Corner cutout; (**b**) diagonal corner cutout; (**c**) rounded corners; (**d**) slit corner.

Score lines. Slipsheets with tabs are tested through the score lines after bending scores 90°. A score is an impression or crease that is provided to locate and facilitate folding.

Stiffness. Adequate stiffness is required for handling certain products such as bagged or irregularly shaped materials to avoid excessive sagging of the slipsheet tab, which may prevent proper clamping by the gripper bar of the front-end attachment. Additional stiffness is normally obtained by increasing caliper. Stiffness testing is described in TAPPI D 489 and in ASTM D 790-71.

BIBLIOGRAPHY

General References

J. C. Bouma and P. F. Shaffer, *USDA Marketing Research Report, No. 1075,* United States Department of Agriculture, Washington, D.C., 1978.

P. J. Mann, *Food Eng. Int.* **2**(10), 32 (1977).

Quick Frozen Foods **38**(7), 52 (1976).

E. M. O'MARA
Union Camp Corporation

SLITTING AND REWINDING MACHINERY

Flexible plastic packaging materials generally do not have perfectly flat surfaces. The material in a parent roll contains stretch lanes and varies in gauge by about 1–5% across the web width. Slitting and rewinding such rolls requires equipment that can control the rewinding speed of each slit roll independently (1). The rewinding process is further complicated by air entrainment into the coils of the rewinding rolls, which separates the outer convolutions and allows them to slip laterally. Winding difficulty caused by this phenomenon increases when handling high-slip films or films with very smooth surfaces. The primary aim of every slitter design is to provide apparatus to deal selectively or generally with one or more of the diverse products presented to it. All designs are based on center winding (1), surface winding (2), or a combination of both.

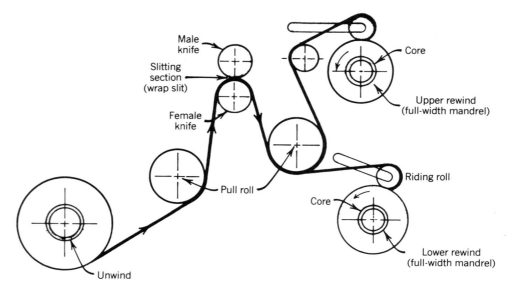

Figure 1. Center winder (duplex winder) illustrating stagger differential winding principle.

Center Winding

Center winder (duplex winder). The most widely used and versatile slitter is a duplex stagger center winder or two-bar winder (see Fig. 1). The web tension at the unwind of the machine is isolated from the rewind section by driven rollers in the slitting section of the machine. These rollers positively control web speed, opposing any overdraw by the rewind. For most applications, the rewinding is done on differential-rewind (slip-core) mandrels in order to maintain suitable rewind tension on each slit roll despite stretch lanes in the web and unequal diameter buildup due to gauge bands commonly found in plastic webs (1). Individual riding rollers forced against each of the rewinding rolls minimize air entrainment into the rewinding roll. For winding paper products, the riding roller also has a surface-winding effect which increases roll hardness without using excessive rewind tension (3).

These machines are suitable for handling web widths up to about 80 in. (2032 mm), rewind diameters up to 24 in. (610 mm), and web speeds up to 1500 ft/min (ca 460 m/min). They are especially suited for narrow-width slitting. The productivity of machines with removable rewind mandrels can be increased by using two pairs of mandrels so that the slitting operation can continue while the alternatve pair is being serviced. Removable mandrels commonly require two operators or one operator and a hoist for handling into and out of the machine. Machines equipped with nonremovable cantilevered mandrels permit one operator to remove the slit rolls, recore, and restart winding quickly without mandrel handling (see Fig. 2). The choice of systems depends on the size and number of slit rolls per setup. Maximum production is achieved by mounting cantilevered mandrels into two-station turrets (see Fig. 3). With automatic cutoff and restart of web around the core, these machines can slit and rewind a web emerging from an extruder on a continuous basis.

Individual rewind arm (IRA) surface-center winders. IRA-winder design is based on both center- and surface-winding principles (see Fig. 4). The design stresses the ability to control the surface-contact force and rewind torque of each rewinding

roll individually. Each rewind core is supported by chucks (one or both driven by a torque-controlled motor) which are supported on pivoted arms arranged in stagger-wind fashion on each side of a winding drum. Handling of rewind mandrels is eliminated (1). The contact force between the rewind roll and the winding drum is exerted by air cylinders controlled manually or by microprocessors. For better control of roll hardness, this force can be programmed as a function of rewind diameter and web speed.

These machines are suited for processing webs in excess of 80 in. (2032 mm) in width and for winding in excess of 24-in. (610-mm) diameter. Maximum design web speed is about 2000 ft/min (610 m/min). Actual production operating speed is frequently limited to ≤1200 ft/min (365 m/min) because of web properties, vibration of the rewound roll (4), parent-roll eccen-

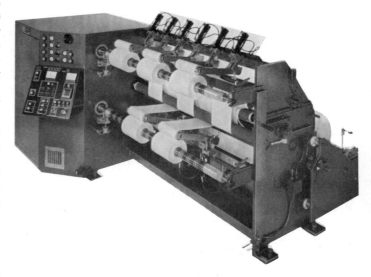

Figure 2. Center winder with cantilevered rewind mandrels and integral unwind station.

Figure 3. Center winder with cantilevered rewind mandrels mounted in turrets for continuous differential winding.

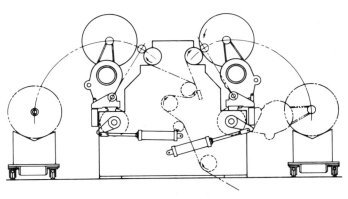

Figure 5. Individual-rewind arm (IRA) surface-center winder with doffing to floor.

tricity, or off-machine roll-handling backup (see Roll handling). Doffing of the rewound rolls can be done in sets with overhead hoist and special "grabbing" devices. Recent designs feature rotating rewind arm support beams, allowing the rewound rolls to be lowered and released directly onto low-bed, wheeled dollies. This is called doffing to the floor (see Fig. 5).

Individual rewind arm (IRA) center winders. This winder is similar to the IRA discussed above except that an individual contact roller, supported by a pair of pivoted arms, contacts the rewound roll in place of a full-width winding drum (see Fig. 6). The contact roller is urged against the rewound roll by air cylinders. The winding tension is developed by a rewind motor driving each core. The primary function of the contact roller is to minimize air entrainment. Sensors detect the buildup of the rewind roll, causing the rewind arms to move outward and maintain a relatively fixed position of the contact roller. This

winding method is more responsive to control of stretch lanes in the web and significantly reduces the roll vibration sometimes associated with high speed surface winders (4). For best results, the contact force and rewind torque must be programmed by microprocessors as a function of web speed and rewind diameter, thereby giving the desired hardness and uniformity in the rewound roll at web speeds upward of 2000 ft/min (610 m/min). These machines also feature doffing to the floor.

Surface Winders

Many versions and sizes of slitting equipment employ the surface-winding principle (2,5). For detailed discussion of larger-size rewinders, see Ref. 6. The machine most widely used for general converting is the two-drum winder with riding roll (see Fig. 7). It is commonly used for slitting plain paper and paper board, as well as many printed, coated, and laminated paper products.

Good-quality paper has the following physical properties conducive to surface winding: uniformity of gauge, freedom from bags, relatively rough surface with medium friction coefficient, high modulus of elasticity, and high degree of compressibility. The density of the finished roll is determined by a combination of constant unwind tension and down force on the riding roll (7).

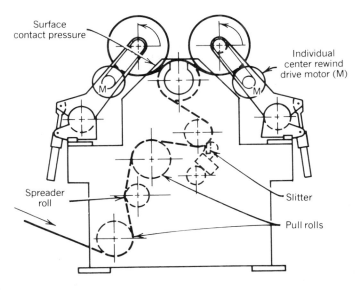

Surface-Center MIR Winder

Figure 4. Individual-rewind arm (IRA) surface-center winder.

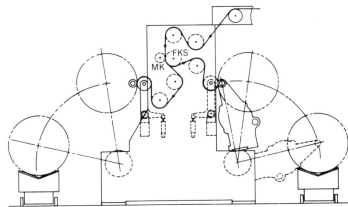

Figure 6. Individual-rewind arm (IRA) center winder with riding roll and doffing to floor.

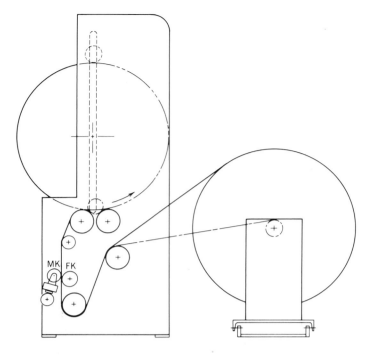

Figure 7. Two-drum surface winder with kiss-shear slitting.

The medium-size, general-purpose converting slitter has the following advantages:

1. The controls and construction are relatively simple.
2. It can wind to large-diameter buildup ratios.
3. It can wind coreless or on almost any diameter core.
4. It winds on one mandrel or without mandrel.
5. The slit-width changeover is accomplished primarily by relocation of knives.
6. It runs at relatively high web speeds, commonly ≥2000 ft/min (610 m/min).

It also has the following disadvantages:

1. The risk of slit rolls interleaving make it difficult or impossible to separate the slit rolls. Bowed spreader rollers can alleviate roll-separation problems (8,9).
2. It lacks versatility for winding baggy or off-caliper webs.
3. It is unsuitable for winding paper with slippery, eg, silicone, coatings.
4. Some grades of paper develop a washboard surface on the periphery of the rewound roll during winding, thereby causing vibrations and restricting operating speed (4).
5. Operator skill is necessary to keep scrap rate low on some products.

Unwinding

Many types of unwind stands are available, from simple shaft-type units with manual side shift to large shaftless turret unwinders (10) with automatic web guiding (11,12). Friction brakes usually provide the holdback torque (13). Regenerative d-c drives are occasionally required for parent-roll acceleration, as well as holdback (14). The choice of automatic tension controllers for maintaining constant unwind tension during acceleration and deceleration, as well as compensating

for decreasing unwind diameter, includes load-cell sensing rollers (15); dancer roller sensors, recommended for use with eccentric parent rolls (16); and open-loop microprocessor control based on real-time calculations of diameter and weight of parent roll. It has been common practice to use overhead hoists for loading parent rolls into operating position, but hydraulic lifts integral with the unwinder are frequently preferred (17).

Tension Control—Rewind

Tension control for surface winders simply requires constant tension control from the unwind stand and has a secondary effect on rewind roll hardness. Tension control of center winders is much more complex because tension is the primary control factor. Rewind buildup ratios of less than 5:1 may require only constant torque (decreasing tension) windup for many products. However, large buildup ratios and tension-sensitive products require either constant tension or programmed tension (18). Depending on the design of the machine, this may require automatic control of pressures to differential-rewind mandrels or air-operated slip clutches, or control of reference voltages to d-c rewind motors (19). When differential winding at high speeds and large buildup ratios, it is necessary to automatically reduce the rpm of the rewind mandrels as a function of rewind diameter. The profile of the tension pattern as a function of diameter must be determined empirically. Microprocessor control is ideal for exploratory profile studies and for maintaining repeatability once the desired results are achieved.

Slitting

Rotary shear knives can be used to slit virtually all packaging-web products. For simplicity of setup and sometimes for cleaner cutting, razor blades are used for plastics thinner than about 0.005 in. (127 μm). For better control, especially when taking narrow trim cuts, the web is wrapped around and is positively supported by a grooved roll at the point of slitting. Similarly, shear slitting is done with the web wrapped around the so-called lower or female knife shaft, known as wrap slitting. If frequent changes of slit width are required when shear cutting, it is desirable to run the web tangentially over the female knife (called kiss slitting) (1) (see Fig. 7). The prime consideration in the slitting section of the machine is ease of changing slit widths. Equipment is being introduced to accomplish this automatically, but it is not yet in widespread use on the average converter-type slitter primarily because of cost and space requirements.

Other Considerations

In addition to the basic features discussed above, many others require consideration in a slitter specification. The more common items include automatic web guiders, spreader rolls, predetermining rewind footage counters, multiple rewind and main drives, special roller coverings, static eliminators, expanding unwind and rewind mandrels, core-locking differential-rewind mandrels, roll-unloading devices, provision for web inspection, web-splicing devices, safety guards, emergency trip cables, and trim-removal equipment. Last, but not least, increasing effort is being made to automate nearly every operational and setup function associated with slitter operation. About half of all slitters sold requires some design modification or accessory to meet the special needs of the user.

New models are constantly being introduced to handle new products, reduce costs, increase productivity, and enhance the quality of slit rolls (20).

The performance of a slitter depends not only on the basic physical properties of the web, but also on the quality consistency of parent rolls. It is difficult to measure some of the more important characteristics, let alone predict how they will affect the slitting operations. Two supposedly identical generic products, each from a different manufacturing source, can exhibit surprising performance differences. Furthermore, problems can be expected when attempting to handle webs that deviate significantly from the norm in physical properties or when web speed and roll size exceed state-of-the-art limits. Consequently, it is always wise to check, if possible, performance of actual parent rolls on existing demonstration equipment to be assured that performance requirements can be achieved.

Nomenclature

Bags	Undulations in a web which otherwise should be a flat surface, usually occurring in lanes of various widths in the machine direction of web. These bands frequently occur where a thickness variation is above average relative to the rest of the web (1).
Buildup ratio	The diameter of a rewound roll divided by the outside diameter of its core.
Doffing	Process of removing or unloading wound rolls from a winder.
Draw	The distance a web must travel unsupported between two web transport rollers.
Gauge	Thickness or caliper of a web. Sometimes expressed in units of points (mil, 0.001 in. or 254 μm); gauge (0.00001 in. or 0.25 μm); or micrometers (0.001 mm or 10^{-6} m).
Gauge band	A machine-direction strip or band of above-average thickness in a web evidenced by a peripheral circumferential bulge on the periphery of a web roll (1).
Guide, edge	Automatic web guide using the edge of a web as reference. Those using an air nozzle as a sensing device are sometimes called air guiders (12).
Guide, line	Automatic web guide using photo cells to track a reference line on a printed web (12).
PLI	Pound force per lineal inch (1bf/in. = 175.1 N/m). Unit of measure of web tension or line contact force of riding roll against a rewinding roll.
Roll	A coiled spool of web material usually wound on a paper core.
Roller	General term for any type of rotating cylinder serving as a web-transport device to support and guide a web through a slitter.
Roll, parent	A large roll from which smaller rolls are slit and rewound. Also called mill roll, master roll, mill reel, bundle roll, unwind roll, and stock reel.
Roll, rewind	Roll resulting from a slitting or trimming operation. Also called coil, spool, and bobbin.
Roller, riding	An idler roller that maintains contact with the surface of a rewinding roll. Also called touch roller, top-riding roller, contact roller, lay-on roller, ironing roller, and squeeze roller. Its purpose is to minimize air entrainment into a rewinding roll.
Slitter	Short term for slitter/rewinder. It is generally understood to include an unwinder.

Winding, differential	A method of stagger winding on two-bar (duplex) center winders whereby the rewinding cores are allowed to slip with controlled torque between keyed spacer sleeves on an overrunning rewind mandrel with the aim of winding each slit strip with equal tension regardless of parent-web defects. Also called slip-core winding (1).
Winding, stagger	Winding alternate slit strips on each of two rewind mandrels so that adjacent slit strips are not wound side by side on the same axis or mandrel (1).
Winding, taper-tension	A reduction of winding tension in a controlled manner from the center of a rewinding roll outward with the aim of giving the desired hardness and uniformity in the roll (18).

BIBLIOGRAPHY

1. J. R. Rienau, *Techniques of Slitting and Rewinding,* John Dusenbery Co., Inc., Randolph, N.J., 1979.
2. J. D. Pfeiffer, "Mechanics of a Rolling Nip on Paper Webs," *Tappi* **51**(8), 774 (Aug. 1968).
3. J. D. Pfeiffer, "Nip Forces and Their Effect on Wound-in Tension," *Tappi* **60**(2), 115 (Feb. 1977).
4. D. A. Daly, "How Paper Rolls on a Winder Generate Vibration and Bouncing," *Pap. Trade. J.* 48 (Dec. 11, 1967).
5. D. Satas, *Web Processing and Converting Technology and Equipment,* Van Nostrand Reinhold Co., New York, 1984, p. 383.
6. L. Rockstrom, *Control of Residual Strain and Roll Density By Three Winding Methods,* Cameron Machine Co., New Brunswick, N.J, 1964.
7. J. Colley, A. J. Keeley, and P. J. Schnackenberg, *Appita* **36**(4), 288 (Jan. 1983).
8. R. G. Lucas, *Pap. Age,* 9 (Sept. 1972 and Nov. 1972).
9. Ref. 5, p. 414.
10. H. L. Weiss, *Coating and Laminating Machines,* Converting Technology Co., Milwaukee, Wisc., 1983, p. 326.
11. Ref. 5, p. 404.
12. H. L. Weiss, *Control Systems for Web-fed Machinery,* Converting Technology Co., Milwaukee, Wisc. 1983, p. 277.
13. *Ibid.,* p. 105.
14. Ref. 12, p. 207.
15. Ref. 12, p. 61.
16. Ref. 5, p. 400.
17. Ref. 10, p. 336.
18. S. E. Amos, "Winding Webs: A Case of Constant Tension versus Constant Torque," *Pap. Film Foil Converter* 56 (Sept. 1970) and 62 (Oct. 1970).
19. Ref. 12, p. 239.
20. R. Aylott, *Pap. Film Foil Converter* **58**(10), 128 (Oct. 1984).

General References

D. Satas, *Web Processing and Converting Technology and Equipment,* Van Nostrand Reinhold Co., New York, 1984, 587 pp. Contains an exhaustive bibliography and comprehensive review of converting machinery.

H. L. Weiss, *Coating and Laminating Machines* (441 pp.) and *Control Systems for Web-fed Machinery* (357 pp.), Converting Technology Co., Milwaukee, Wisc., 1983. Two exceptionally comprehensive reference sources.

R. W. Young
John Dusenbery Company, Inc.

SPECIFICATIONS AND QUALITY ASSURANCE

A specification is a technical definition and description of an artifact or a process. In packaging, both materials and process are involved, which are defined by four kinds of specifications, commonly called "specs". Each of the four kinds of specs is composed of the following elements:

1. Packaging materials specs: descriptions of structure—material(s) of construction, dimensions, and drawings, if necessary; performance—how the material(s) must behave to exercise its intended functions of product protection, communication, and possibly product application; and surface treatment—texture and graphics, if any.

2. Packaging process specs: enumeration of steps in manufacture from receipt of the packaging materials through assembly of the product-in-package, sealed and ready for distribution. Specs include process conditions and limits.

3. Finished-goods specs: definitions of structure—the assembled product/packaging combination, including dimensions of the primary packages, their arrangement and orientation in shipping containers, and the arrangement of shipping containers in unit loads, if so distributed; performance—physical properties of the assemblies in relation to distribution stresses and end-use presentation of the product; and appearance—the condition of the surface treatments on primary packages and shipping containers and unit loads as to identification, distribution, and end-use information.

4. Quality-assurance (QA) specs: enumeration of procedures for ascertaining conformance to the three kinds of specs above, covering sampling, inspection against specified criteria of quality, the test methods by which inspection is conducted, and the actions to be taken on determination of nonconformance.

Conformance is the responsibility of several groups of people, and the quality-assurance specs are the coordinating bond. The finished-goods specs are always those of the product manufacturer, who is responsible for delivering an acceptable and marketable product to customers. The package is the means of delivery. The product manufacturer usually buys packaging materials from vendors, who are responsible for meeting the buyer's needs. Thus, packaging materials specs are those of the buyer (the product manufacturer), who knows the product best and can choose the packaging that meets the product's needs, not only from the standpoint of physical protection, but also the manner of presenting the product in the marketplace.

On the other hand, no one knows the capabilities and limitations of packaging better than the makers and vendors thereof. Therefore, the packaging materials specs are always finally determined between vendor and buyer in conference. Some large corporations manufacture their own packaging materials, but the relationship of vendor and buyer is present, even though both parties are internal to the corporation. The needs are those of the product division, and the package-materials division supplies to the client division's specs. Conformance to packing process specs is the responsibility of the product manufacturer's packing operations department(s); and conformance to finished-goods specs is the net result of product quality, packaging materials quality, and packing operations quality control.

A distinction exists between quality control (QC) and quality assurance. QC refers to actions simultaneous to the production of an artifact, which maintain the production in conformance to specs. QA is a much broader activity that includes quality control plus the issuance and maintenance of all specs, standards, test methods, and quality audits, the relation of quality criteria to customer needs, and the conveyance of the packer's quality needs to vendors. A QC department is usually

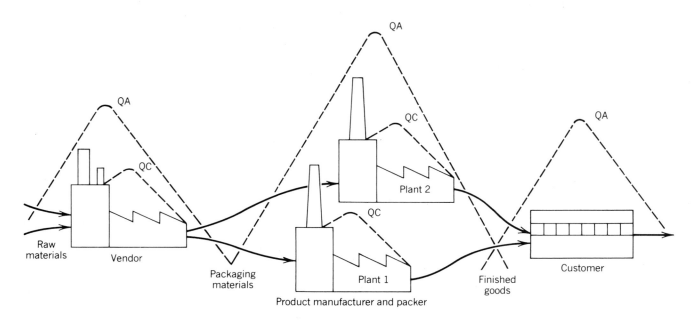

Figure 1.   Quality control (QC) and quality assurance (QA): span of oversight responsibility.

SPECIFICATION FOR: 1-lb Bread Crumb Carton

PROPERTY OF: SANSNOMEN CORPORATION

CODE: CT-F-2050
DATE: Aug. 1, 1985

I. SCOPE: This specification describes a fiberboard carton to contain 16 oz (454 g) of flavored bread crumb mixes. The product is made to Spec. No. P-F-205, dated Mar. 12, 1985.

II. RELATED SPEC: General Specs, CT-1, Printed Cartons

III. CONSTRUCTION:
A. The carton shall be made of 0.024-in (0.62-mm) white machine *clay-coated chipboard,* medium density.
B. Blank shall be dimensioned in accordance with Print No. 2050 of above date, which is part of this spec. Carton shall have full flaps, seal ends, with *Van Buren ears* to prevent sifting, and a perforation pattern at the top of one side panel for easy opening and pouring.
C. 1000 Cartons shall weigh 143 ± 7 lb (65 ± 3 kg) at standard conditions.
D. Manufacturer's joint shall be glued, 0.63 in. (1.6 cm) wide.
E. Printing shall be 4 colors plus scuff-resistant varnish, in accordance with mechanical art specified separately. Grain direction of the paperboard shall be horizontal to the printing.

IV. PERFORMANCE:
A. Cartons shall operate on Brand Z cartoning machines at 200/min.
B. Break-open force of k.d. cartons shall not exceed 6.86 N (700 gf), and springback shall not exceed 2.94 N (300 gf).
C. Filled and sealed cartons shall not allow product sifting when exposed to 1 h of vibration on an L.A.B. vibrator in cases at 1 g acceleration.

V. PACKING FOR DELIVERY
Cartons shall be stacked with uniform facing in layers of 500 in corrugated boxes, 18 boxes per unit load on *slip sheets* or *pallets,* returnable. Each tier shall contain 6 boxes, separated by chipboard pads. Boxes shall be stenciled with supplier's name and plant number, this spec number, and lot number, with date of manufacture.

VI. CLASSIFICATION OF DEFECTS
Class A: Faults that prevent the cartons from containing the product or from identifying it.
A. Excessive opening or springback force.
B. Dimensions outside of tolerances shown on Print No. 2050.
C. Tears, holes, or scuffs that mutilate or make illegible any of the copy.
D. Missing color or colors.
E. Misregistration of colors that makes any copy illegible.
F. Misregistration of scores that causes production difficulty in setup, filling, or sealing.
Class B: Faults that cause marginal function or low-quality appearance.
A. Smudged printing or scuffed surface showing board through the clay coating.
B. Uneven scores making carton setup difficult on the packing line, thus causing loss of packing efficiency.
C. Inadequate perforation for glue penetration on seal flaps or Van Buren ears.
D. Inadequate perforation for easy-opening feature on carton side.
Class C: Minor faults affecting appearance only.
A. Rough printed surface or low gloss from varnish.
B. Slight off-color match.

VII. ALLOWABLE DEFECT LEVELS
A. Class A: 0.4%
B. Class B: 1.0%
C. Class C: 2.5%

VIII. INSPECTION
The supplier is expected to conduct quality control and inspection sufficient to assure compliance with this specification. Should it become apparent because of poor carton performance that a lot (a truckload, ca 500,000 cartons) may not comply, the Plant Quality Manager may call for an audit.

Figure 2.   Representative packaging materials spec in text form.

a plant function, which inspects and accepts or rejects incoming materials, monitors in-plant processing, and rates finished goods from that plant. QA, on the other hand, is always a headquarters function, which is responsible for the specs and standards used by QC at all plants and for auditing the uniformity of applying the specs and standards across all manufacturing units.

Thus, a vendor of packaging material has its own quality-control and quality-assurance functions. The latter is responsible to the vendor's management for clearances to deliver to the packer's plant. The packer's QA function is responsible to its management for clearances to deliver finished goods to the packer's customers. The respective QC functions of vendor and packer are responsible for compliance within their individual plants. Figure 1 shows the relationship of QA and QC in diagrammatic form.

The instruments of specifications and quality assurance are company policies and laws or regulations where applicable; organization, ie, internal relationships, authority, accountability, and mission of QA–QC in relation to operations, purchasing, marketing, and research functions; facilities, usually laboratories in each plant, with testing and storage space; and skilled personnel, who are qualified to perform in the fields of packaging technology, statistical sampling and inspection, interpretation of data, development of specifications, standards and test methods, and communication with vendors, customers, their own management personnel and other functions, and the technical/professional societies in their specialities.

Packaging Materials Specs

The meaning of quality.   The basic function of a packaging spec is to communicate enough information so that a supplier

SANSNOMEN CORPORATION
FOOD DIVISION
PACKAGING MATERIAL SPEC

CODE: CTF2050
EFF. DATE: 080185

SCOPE: Sealed-End Carton for 16-oz (454 g) Bread Crumb Mixes
RELATED SPECS: PF205, GS-CT-1, Drawing 2050

APPLICABLE AT: Plants 2, 3, 5

MATERIAL PROPERTIES

Identity	Specification	Test method	Defect class
clay coated chipboard	0.62 ± 0.026 mm. gauge	PM–01	B
	4.55 ± 0.17 Pa	PM–02	B
brightness, unvarnished	80 ± 2 units	PM–45	B
stiffness, Taber units			
machine direction	280 min	PM–19	B
cross direction	150		
tear strength			
machine direction	195 gmin	PM–08	B
cross direction	320 gmin		
scorebend resistance	700 gf max (6.86 N)	PM–14	A
springback force	300 gf max (2.94 N)	PM–14	A
kinetic coefficient of friction	0.27 max	PM–05	A

CONSTRUCTION

Identity	Specification	Test method	Defect class
style	siftproof, full-flap, sealed ends		A
dimensions, cm			
height	22.2 ± 0.3	PM–03	A
width	15.3 ± 0.2		A
depth	5.5 ± 0.2		B
board use per carton	1402 cm^2		
weight per 100 cartons	65 ± 3 kg	PM–06	B
moisture content	$6 \pm 1\%$	PM–03	C
warp or curl as received	negligible	visual	A
scuff resistance, varnished	50 strokes, 2-kg weight	PM–07	C

IN-BOUND SHIPMENT

Identity	Specification
case, RSC, paper tape closure, returnable	
count per case	500
weight per case	34 kg
cases per pallet or slip sheet	18
unit-load weight	615 kg
markings	spec code, supplier and plant, lot and date of manufacture

AQL	Class	%
	A	0.4
	B	1.0
	C	2.5

Figure 3. Representative packaging materials spec in tabular form.

SANSNOMEN CORPORATION
GENERAL SPECIFICATIONS, PACKAGING MATERIALS

CT-1 PRINTED CARTONS

Printed folding cartons used by the Sansnomen Corp. shall conform to the following general specifications:

A. Side Seam Gluing

Cartons shall be glued so that there will be complete fiber tear along the entire length of the glue seam. Cartons shall not be glued more than 0.03 in. (0.8 mm) out of square. There shall be no glue visible on the outside surface.

B. Scoring

Scoring must be of such depth and uniformity that a straight, well-defined fold will be produced. Scores must break with well-defined beads with minimum cracking when folded and ironed 180°. The flap cut of the cartons should be aligned with the center of the panel scores. All die cuts shall be cut clean and even. All working scores must be prebroken to maximum machine capacity.

C. Flatness

Cartons, upon receipt at the plant, shall be flat. Warp or curl in cartons is a major defect.

D. Blocking

Cartons, upon receipt at the plant, shall be free of blocking.

E. Brushing

Cartons shall be brushed and blown to remove all carton cutting dust and shreds from surfaces before packing.

F. Printing

Printing on cartons shall conform to approved color standards and artwork.

G. Food and Drug Administration Conformity

The materials used in the fabrication shall not contain any packaging migrants in amounts exceeding the tolerance limits set forth in the Food Additive Amendment to the Federal Food, Drug, and Cosmetic Act, nor shall it contain any packaging migrants unapproved by the Food and Drug Administration and all regulation promulagated thereunder. This material shall be manufactured under modern sanitary conditions in accordance with the best commercial practices. All deliveries shall conform to the provisions of the Federal Food, Drug, and Cosmetic Act and all regulation promulated thereunder.

H. Packing and Marketing

Printed cartons will be packed in the prescribed manner either in a corrugated carton sealed with tape or stacked on pallets suitably covered with protective material. Each unit will bear the identification of the supplier, quantity of cartons per container, style of carton and dimensions, date of manufacture, and order number.

I. Storage and Handling

Cartons should be stored in a clean area in which the temperature does not exceed 80°F, and the relative humidity is between 40% and 60%. Cases of cartons should rest always on their flap surfaces and should not be stored near radiators or other heating units, nor should they be stored directly on floors that can become damp. The axiom "first in, first out" should be followed in removing cartons from the storage area. Containers of cartons should be kept closed until they are opened at the production line when ready for use. Only enough containers should be brought into the production area from the storage area for a reasonable production time. The practice of emptying one container before opening the next should be employed at all times.

Figure 4. General specifications for printed cartons.

can make the material, a buyer can order it, a manufacturing organization can use it in a packing department, and a QC department can verify that it does or does not conform to the spec. The most difficult part of the communication is to specify quality, since without further clarification "quality" is an abstract concept, subject to many different interpretations. When thousands and millions of dollars are involved in the selling and buying of packaging materials, the two parties involved in a transaction must have prior common understanding regarding acceptable quality and rejectable lack of quality. It is imperative, therefore, to translate "quality" from the abstract to the specific (1).

Lord Kelvin's comment on science is nowhere more appropriate than to the field of quality: "If you can measure that of which you speak and can express it by a number, you know something of your subject; but if you cannot measure it, your knowledge is meagre and unsatisfactory." Fortunately, much

intellectual effort has been applied to the subject of quality, by mathematicians, engineers, physicists, chemists, and biologists. The net result is that the means for defining, controlling, and auditing quality in most applications is available. It must be stated, however, that quality-control technology is in a continual state of refinement and development in response to the ongoing development of new kinds of products and processes. Quality control and quality assurance are not static arts.

The steps in translating quality of packaging materials from abstract to concrete are

1. Identify a list of quality criteria that are individually necessary and important factors to be incorporated in a given packaging material.

2. Identify a list of defects that are important to avoid in the packaging material.

PACKAGING MATERIAL SPECIFICATION

PROPERTY OF SANSNOMEN CORP.

DATE 080185   PRINT NO. 2050

ITEM 1-lb (0.454-kg) BREAD CRUMB CARTON, KEYLINE, PRINTED SIDE

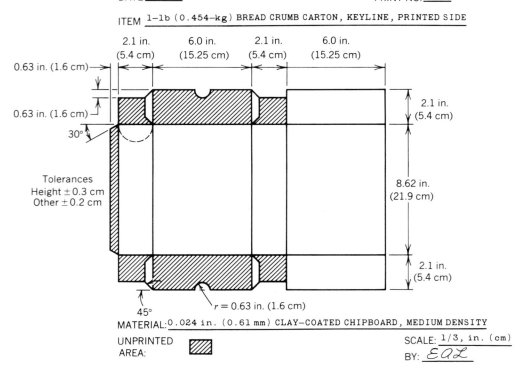

MATERIAL: 0.024 in. (0.61 mm) CLAY-COATED CHIPBOARD, MEDIUM DENSITY

UNPRINTED AREA: SCALE: 1/3, in. (cm)
                                                   BY: EQL

Figure 5. Print of carton blank for sample spec.

3. Categorize the items in the list of defects according to seriousness, or level of severity, usually in three levels, called critical, major, and minor.

4. Determine the number of such defects that can be tolerated per thousand units of material delivered, unless there are established commercial standards for the material that are acceptable. This is called Allowable Quality Level (AQL).

5. Identify the sampling and inspection procedures to be used in determining conformance to the quality criteria in any given lot of the packaging material.

In the aggregate, the quality criteria, the defects and their severity, and the AQL define the quality of the material in concrete terms. The sampling and inspection procedures specify how to measure conformance. This information is usually included in a concise manner in the spec for the packaging material so that seller and buyer have a mutually clear understanding of the quality required and how it is determined. Thus, the content of packaging materials specs includes a description of structure, performance, surface treatment, and concise references to the QA specs, based on a categorized listing of quality criteria.

Representative packaging materials specs. Format is optional and can vary widely, depending on issuing source, from mainly textual to tabular. Figure 2 (textual) and Figure 3 (tabular) are representative packaging materials specs for bread crumb mix packed in folding cartons (see Cartons, folding). The tabular approach is becoming more popular with the

use of word processors and computers, because it is faster to edit or modify. Both examples contain the same elements (2):

Identity of issuer/ownership.

Identifying number or code and date.

Title.

Statement of scope.

Construction, with tolerances.

Performance requirements with references to test methods.

List of or reference to categorized defects and AQLs.

Identity of related prints, drawings, graphics, and color standards.

Statement of how (manner of packing) lots are to be delivered.

Ownership. A materials spec is usually issued by the purchaser (the packer) because of the need for structure and performance. The graphics spec is almost always the purchaser's, but there are exceptions; for instance, a fruit and vegetable canner packs for inventory during harvest season and sells to a number of retailers during the subsequent months until the next harvest. The canner buys a can-maker's standard spec, which is sold to many packers, and stores the packed cans without labels. Retailers supply labels when they purchase from the inventory of packed cans or provide the graphics specs for the canner to purchase and apply labels. In most situations, however, the packer sells under his or her own

SPECIFICATION FOR: Shipper, 24/1-lb Bread Crumbs
PROPERTY OF: THE SANSNOMEN CORP.

NUMBER: CC-F-2050
EFF. DATE: Aug. 1, 1985

I. SCOPE:
This specification covers construction and performance requirements for a corrugated box used to distribute 24 cartons of Spec CTF2050. The related product spec is No. PF205, dated Mar. 12, 1985.

II. CONSTRUCTION:
The box shall be RSC style, $3 \times 8 \times 1$ arrangement, made of 200T board (both liners 42-lb natural kraft board), with C flutes, vertical direction. The manufacturer's joint shall be glued. Top and bottom surfaces of the box shall be treated with nonskid coating to improve pallet stability. Inside dimensions shall be 18.1 in. (45.9 cm) × 17.3 in. (44.0 cm) × 8.7 in. (22.2 cm) tolerances plus or minus 0.12 in. (0.3 cm). See Print No. 2050S, 080185

III. PERFORMANCE:
Boxes shall be delivered with consistent quality for efficient erection and can loading on automatic end-loading case equipment.

IV. DELIVERY:
Boxes shall be fabricated by suppliers with manufacturer's seam glued, edges scored, and KDF. The flat cases shall be stacked in bundles of 50 and tied. Stacks shall be made with the same side up on all cases, and the same edge of each at one side of the stack for automatic erection and proper placement of the date-making stamp.

V. INSPECTION:
Suppliers shall maintain sufficient quality control to assure compliance with this specification, including responsibility for disposition of damage by the carrier in making deliveries to Sansnomen Corp. Quality defects found by Sansnomen shall be reported through the Plant Quality Control Manager to the Plant Purchasing Agent, who will arrange appropriate action with the supplier.

If visual observation and performance of cases on a packing line indicate that a lot may be out of compliance with this specification, the Plant Quality Control Manager may sample and inspect the lot according to the schedule under VII, below.

VI. CLASSIFICATION OF DEFECTS
A. Class A Defects: Faults that preclude shippers from protecting or identifying the product contained through the distribution to retailers and institutional buyers.
   1. Loose manufacturer's joint.
   2. Dimensions outside of limits.
   3. Break or cut-through a liner at a score.
   4. Tears, punctures, holes, or ragged flaps with adhering pieces of excess board.
   5. Wrong, missing, or illegible copy, or wrong color graphics.
   6. Contamination with any foreign matter other than fiberboard dust.
B. Class B Defects: Faults that make function borderline or questionable.
   1. Incomplete or partially illegible printing.
   2. Off-square gluing of manufacturer's joint.
   3. Bends other than on score lines.
   4. Liner incompletely glued to corrugated medium.
   5. Gap greater than 0.12 in. (0.3 cm) when outer flaps are butted at the ends in case sealing.
   6. Nonskid treatment missing.
C. Class C Defects: Faults that impair appearance, but not function.
   1. Light or blotchy printing.
   2. Stains, scratches, or scuff marks.

VII. ACCEPTABLE QUALITY LEVELS
For purposes of inspection and acceptance, a *lot* of corrugated cases shall be a *truckload* as delivered. An average delivery will be 40 pallets of 400, or 16,000 cases. Inspection shall be made when necessary or advisable by sampling in accordance with Mil Std-105D. *Single Sampling, Normal Inspection.* In this case, the sampling shall consist of 320 specimens, taken 4 from each stack*, with inspection for the defects listed under VI, and action taken as indicated by the following table:

Class of defects	AQL	Reject level
Class A	1.0	6 or more defectives
Class B	4.0	15 or more defectives
Class C	6.5	22 or more defectives

VIII. RELATED SPECS
The following two documents are part of this specification: Print No. 2050S, dated 1 August 1985, and CS-1, *General Specifications for Corrugated.*

* Two stacks per pallet

Figure 6. Representative corrugated shipper spec in text form.

label and develops an exclusive spec to differentiate his product(s) in the consumer market. The packer may buy from two or more vendors, to supply multiplant packing operations, and must have the same spec across his entire market area.

Identifying code and date. Every spec must be unique and distinguishable from every other spec, including earlier editions of specs for the same package. Materials specs are surprisingly impermanent. Most packaging specs are modified for some reason within a year of issuance: to reduce cost, to process better on new or altered packing machinery, to improve consumer function, or in response to legal and regulatory initiatives. When such changes are made, it is especially important to avoid ordering an obsoleted version of the specified package or material.

When specs are issued in the "traditional" manner, which is by typing, reproduction, and distribution of "hard" copies,

SANSNOMEN CORPORATION   CODE: CSF 2050
FOOD DIVISION   EFF DATE: 080185
PACKAGING MATERIAL SPEC

SCOPE: Top-Loading RSC (Regular Slotted Case) for 24 Cartons, 16-oz (454 g) Flavored Bread Crumbs

RELATED SPECS: PF-205, CTF2050, Drawing 2050S, General Spec GS-CS-1

APPLICABLE AT: Plants 2,3,5

CONSTRUCTION			Reference
style	RSC, top load, vertical flutes		Print 2050S
mfr's joint glue, inside			Print 2050S
dimensions, cm, ± 0.2	Inside	Outside	
height	22.2	23.0	PM-03
length	46.0	47.0	
width	44.0	45.0	
board area	1.225 m^2		
case cube	0.486 m^3		
case weight, filled	13.0 kg		
corrugation	C-flute, 14 $\pm$ 1 flute per 10 cm		

PROPERTIES		Reference
burst strength, Mullen	91 kg min	PM-04
compression strength	960 kg min	PM-09
moisture content	6 $\pm$ 2%	PM-03
caliper, printed	0.37 cm min	PM-01
weight, each liner	204 gm/m^2 min	PM-02
weight, medium, flat	126 gm/m^2 min	PM-02

DISTRIBUTION INFORMATION		Reference
platform and size	pallet or slip sheet, 120 × 100 cm	
unit-load dimensions	113 × 94 × 91 cm	Print 2050S
case count	20:5 per tier, 4 tiers	Print 2050S
carton count	480	
unit-load weight	260 kg, plus pallet or slip sheet	
clampable	yes	
stabilization	50-μm stretch film, single wrap	

IN-BOUND SHIPMENT OF CASES	
form	KDF, 2 stacks on 120 × 100-cm pallet
count	400 per pallet: 200 per stack
stabilization	polypropylene strapping

AQL	Class	%
	A	1.0
	B	4.0
	C	6.5

Figure 7. Representative corrugated shipper spec in tabular form.

there is a risk of confusion between current and superseded specs, unless spec files are carefully maintained by all recipients. The identifying code and date are the keys to accuracy and currency in the application of packaging specs. Except for small businesses, where less than 100 specs are issued per year, and all from one point in the organization, the structure of coding systems for easy access to a current active spec for a given packaging material is not easy. The design of such systems is beyond the scope of this article. See Refs. 2 and 3 for further information on this subject.

When specs are recorded in electronic form (computerized), it is easy to prevent confusion about the currency of a spec because the issuer deletes the superseded spec when a modification is made; thus any terminal in the hardware network has access to only the current issue. In this system, the maintenance of hard copies in spec files is discouraged.

Title and scope. The title clarifies that it is a packaging materials spec, distinct from a processing spec, finished-goods spec, QA spec, or any other kind, eg, ingredients, raw materials, or product. The scope usually relates the packaging spec to a product or products and delimits its coverage. Most packaging materials specs cover all three levels of packaging: primary, secondary, and tertiary (4). The *primary package* contains the product, eg, can, bottle, carton; the *secondary package* holds a group of primary packages (generally a corrugated case); the *tertiary package* is the shipping unit, eg, pallet or slip sheet stabilized with strapping or film. All three are normally included in the packaging materials spec because all are determinants of safe delivery to point-of-sale and are cost factors to the product. The tertiary package is the unit of inventory and sales accounting, so its structure, and especially the number of secondary packages per unit load, must be standardized across all plants and warehouses.

Some products do not require three levels of packaging. Large furniture items and major appliances have only one; each table or refrigerator is individually boxed and shipped. Other products, such as cement, have two levels of packaging; the primary packages are multiwall bags that are strapped onto a pallet in fixed numbers, then given a secondary packaging of film overwrap. There are products that are sold in both two-level and three-level packaging. An example is dry dog food, which has primary packages from 5 to 50 lb (2.3–22.7 kg) in net weight: 5- and 10-lb bags are grouped into shipping containers, which in turn are unit-loaded; 25-lb (11.3-kg) bags are banded or bagged in twos and unit-loaded; 50-lb (22.7-kg) bags are simply unit-loaded. The scope statement in each spec must clarify what the total packaging materials includes.

Construction with tolerances. The packaging media, ie, materials of construction, are paper, paperboard, glass, metal, plastics, and to a minor extent, wood and textiles in a variety of compositions, combinations, and forms. Glues and waxes are important auxiliary materials of construction. The specification of structure is akin to the specifications and plans for a house: The materials of construction are listed, dimensions of the package or component specified are stated and included on a drawing, and the allowable tolerances (+ and −) are shown. Thickness (gauge), weight, or both are included, also with tolerances. Special attention is given to dimensioning parts that must fit together, such as the thread dimensions on the finish (opening) of a bottle and those of the matching threads on a closure (cap).

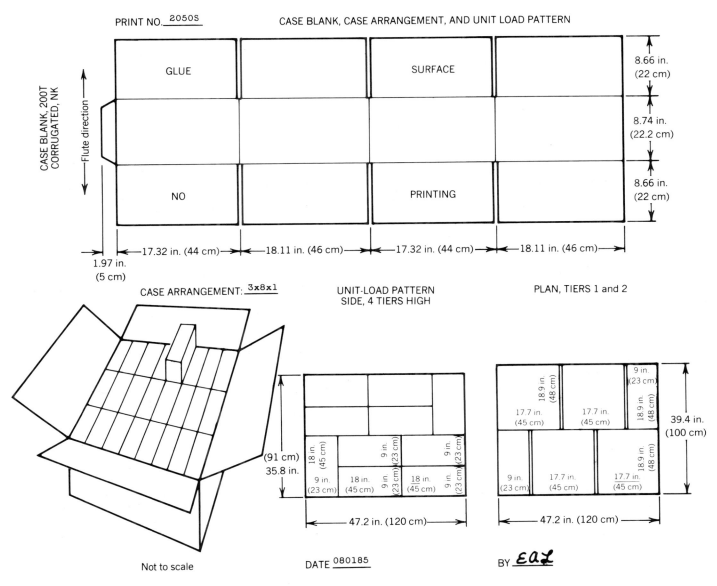

Figure 8. Print No. 2050S. Case blank, case arrangement, and unit-load pattern.

The purpose of careful dimensioning is fourfold: to control material usage and therefore cost; to permit components to fit (caps on bottles, labels on cans, cartons into corrugated boxes); to assure uniform flow through packing machinery; and to present a uniform appearance in the market place.

Performance requirements. Three critical functions that packages must perform are protection of product from physical damage; protection of product from climatic damage or contamination; and communication to the user of product identity, source, and use. Depending on the nature of the product contained, the balance of importance among these three critical factors will vary, but they are dependent on the package structure. For each performance requirement, a reference is made to a test method by which the performance can be measured or estimated. Certain packages are required to perform functions beyond delivery of product to the end user. In such cases, those performance requirements are included in the spec, with appropriate reference to verifying test methods. Ex-

amples are spray cans for applying paint, boil-in-bags for preparing foods, roll-on packages for applying deodorants, and prepacked hypodermics for injection of medicines.

Whereas the function of communication to the product user is designed for execution by graphics and surface treatment, the scuff resistance of the structure is vital to preventing rub-off in distribution and handling.

Categorized defects and AQLs. For each product/package combination there are specific package criteria that must be monitored for conformance in order to meet the package performance requirements. These criteria may be of two kinds: attributes and variables. An attribute is a "yes-or-no" type of quality criterion, eg, a glass bottle is or is not cracked; a label on a can is or is not torn; a cap on a bottle is or is not cocked (misthreaded). A variable is a measurable criterion, eg, height, pressure, thickness, or weight, which can be enumerated as within or outside of specified tolerance. All variables can be measured by some type of instrument to determine

SANSNOMEN CORPORATION

GENERAL SPECIFICATIONS, PACKAGING MATERIALS

CS-1   *CORRUGATED*

Corrugated containers and pads used in the Sansnomen Corp. shall conform to the following general specifications:

A. Material
   Corrugated shipping containers shall consist of 100% natural kraft corrugating medium or semichemical equivalent of not less than 82 g/m² (0.23 mm) with 100% natural kraft liners (facings) of weights as specified, properly glued to both sides of the corrugating medium.

B. Conformance to Regulations
   The corrugated board and shipping cases shall comply with the provisions of Rule 41 of the Uniform Freight Classification and such compliance stated by a box maker's certificate stamped on each case.

C. Printing
   Printing of corrugated materials shall conform to approved color standards and artwork.

D. Case Performance
   Shipping cases shall be so constructed that they will consistantly perform well on high-speed packing lines at each plant. Since some plants may use material from more than one source of supply, the shipping cases at a particular plant shall be completely interchangeable so as to avoid stoppages when switching from one supplier's case to another.

E. Manufacturer's Joint
   Corrugated cases must fold squarely, and the gap at the manufacturer's joint should be uniform along the entire length but shall not vary more than 0.3 cm when the gaps are measured at the flap scorelines. Glued joints must have 100% fiber tear after 72 h at −20°C. Glued joints will always be used unless otherwise specified.

F. Scoring and Slotting
   Scorelines must be of sufficient depth to permit opening the flaps easily, in a straight line, and at the specified location. Slots must be clean cut with no frayed edges or pieces of scrap attached. Flap slots must be within 0.5 cm of the flap scoreline. Scorelines must meet at the manufacturer's joint with an allowable variation of 0.3 cm. Outer liners should show no evidence of washboarding. No finger line can be evident.

G. Sanitation of Materials
   Corrugated materials shall be manufactured under modern sanitary conditions in accordance with the best commercial practices and shall be properly protected to arrive at our plant free from contamination from dirt, insects, chemicals, or other foreign matter.

H. Size Specification
   The ID shall not vary more than ±0.3 cm from the dimensions specified.

I. Flutes and Style
   All corrugated containers will be RSC and will have vertical flutes unless otherwise specified.

J. Packing and Marking
   Corrugated materials will be bundled together in a reasonable number of units and palletized or unitized to aid ease of handling and storage. Each palletized or unitized load will bear the identification of the supplier, quantity per unit, material description and size, proper product code numbers and name, date of manufacture, and vendor's lot number.

K. Storage Conditions and Handling
   Corrugated materials should be stored in a clean area that is neither too dry or too humid. Stacks of corrugated materials should not be so high that flute crushing results. Bundles should not be opened until they are to be used. Inventories should be rotated so that the oldest materials are used first.

Figure 9. Representative general spec for corrugated cases.

conformance, whereas the test method for attributes usually depends on visual or automated machine evaluation and may further require judgmental decision.

Experience with both the product and the packaging medium is necessary to develop a list of categorized defects and AQLs. For an inexperienced packer with a new product in mind, the package supplier is the first resource of experience to set up the criteria and limits of acceptance. Technical committee members of professional packaging societies (see Networks), such as The Packaging Institute, Stamford, Conn.; the Society of Packaging and Handling Engineers, Reston, Va.; the *American Society for Testing and Materials* (ASTM), Philadelphia, Pa.; and the *American Society for Quality Control*

(ASQC), Milwaukee, Wisc. are sources for literature, training, references, and consultants on developing QC and QA programs. Trade associations of packaging materials manufacturers can also provide information on industry standards or guidelines and test methods. The Department of Defense and Department of Commerce (National Bureau of Standards) make publications available on statistical sampling of materials and products for inspection of compliance to specs (5, 6, and General References) (see Military packaging; Standards and practices).

The moment that the AQL is determined for a given defect or category of defects, the sampling plan for judging compliance is fixed and can be found in sampling tables developed by

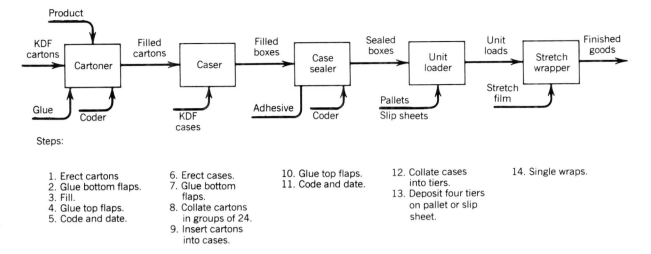

Figure 10. Diagram of a packing process for loose product in cartons and cases.

Steps:

1. Erect cartons
2. Glue bottom flaps.
3. Fill.
4. Glue top flaps.
5. Code and date.

6. Erect cases.
7. Glue bottom flaps.
8. Collate cartons in groups of 24.
9. Insert cartons into cases.

10. Glue top flaps.
11. Code and date.

12. Collate cases into tiers.
13. Deposit four tiers on pallet or slip sheet.

14. Single wraps.

the Department of Defense for lot sizes up to an unlimited number. Sampling for inspection is never included in a packaging material spec because the lot sizes in which goods are manufactured are variable, so a fixed sampling plan would have no value in measuring compliance to spec.

Prints, drawings, graphics, and color standards. To completely understand how a package or packaging material should be made, a dimensioned print is nearly always needed, and a package with several components must have a spec, with print, for each component. Further, most packages must be printed with some copy to identify the product contained at the very least, so a print of the graphics must be supplied to the vendor who will manufacturer a printed component. When color is used, which is true of almost all packages, color specs must be supplied. This is accomplished most simply by attaching visual standards in the form of swatches on paper or thin paperboard, available from a matching-system source (7) in thousands of colors, with ink formula (see Inks) included to facilitate the printer's work. Illustrations are provided in the form of photographic color transparencies. For commercial reproduction, black and three standard primary ink colors are commonly used. If solid-colored package components are required, such as plastic bottles, they are specified by the same color swatches as for printing and matched by blending pigments into the plastic resin at the time of molding. The control of color reproduction is discussed under Quality-assurance Specs.

Delivery of packaging materials. The final part of a materials spec states how the buyer wants the material presented to the receiving department. This is important to the cost of operating, since the material must be kept clean, yet easily accessible with minimum labor to unpack and deliver to the packaging department without generating waste to be disposed. Vendors of packaging materials have standardized a few options for the major packaging media (see Table 1), which have proven satisfactory for most buyers. Other creative variations have been developed by buyers who can minimize their operating costs by calling for nonstandard delivery systems.

Comments on representative specs. Since cartons are the most commonly used of all packaging media, a paperboard carton pack is used to illustrate the materials-spec format op-

tions. Figures 2 and 3 represent text and tabular format, respectively, for the carton itself, that is, the primary package. It is obvious that they communicate in very different ways, and neither is complete in itself. The text form of the materials spec (Fig. 2) is amplified by a general spec (Fig. 4) that applies to all carton specs purchased by the Sansnomen Corp. (a ficti-

Table 1. Normal Delivery Options for Major Packaging Media

Package type	Normal delivery system
glass bottles and jars	(1) in corrugated cases with partitions, on returnable pallets or nonreturnable slip sheets
	(2) bulk-palletized, in layers with tier pads and film overwrap, on returnable pallets and pads[a]
plastic bottles	in bulk, in collapsible, returnable corrugated boxes with film-bag liner
metal cans	bulk-palletized, as described under glass (2); loose ends stacked in thousands, paper wrapped
cartons	(1) folded flat, stacked and film wrapped; stacks bundled in stretch film on slip sheets or pallets
	(2) as flat blanks, otherwise as in (1)
closures	bulk, in corrugated boxes with film-bag liners; boxes strapped to form unit load or palletized
rolls of flexible packaging materials	(1) each roll of film wrapped, eight or ten rolls strapped onto a pallet
	(2) two rolls per fiber drum, four or five drums per pallet; drums and pallets returnable
plastic thermoforms	(1) stacked in corrugated boxes, on slip sheets or pallets
	(2) stacked and film-wrapped, on pallet with tier pads and stretch film overwrap
corrugated boxes and multiwall bags	flat, banded in tens or twenties, strapped on returnable pallets.

[a] Ref. 8.

```
SANSNOMEN CORPORATION          CODE: PP205
FOOD DIVISION                  EFF. DATE: 080185
PACKING PROCESS SPEC
```

SCOPE: Packing of 16-oz (454 g) Flavored Bread Crumbs

RELATED SPECS: PF205, CTF2050

APPLICABLE AT: Plants 2,3,5

PROCEDURE

Activity	Specification	Test method	Defect class
adjust product moisture content	5.0 ± 0.1%	P-101	B
cool product for packing	25°C max	P-105	A
deliver product at steady rate to cartoner	91 kg/min		
fill cartons at 200/min	net weight 456 ± 2 g	P-102	A
seal carton flaps, odorless PVA emulsion glue	siftproof	PM-17	A
	odor level 3	PM-31	B
date and code legibly	day, shift, lot, plant, line number	visual	A
seal cases, hot-melt adhesive	carrier regulations	P-107	B
date and code cases legibly	same as cartons	visual	A
unit load and stretch wrap	case copy legible through wrap	visual	B

QUALITY CONTROL INFORMATION

lot size	1 shift; 100,000 cartons maximum
sample size, cartoning	500; 1 per minute of packing

	Class	%
AQLs, product and cartoning	A	0.65
	B	1.0
AQLs, casing and unit loading	A	1.0
	B	2.5
sample size, casing and unit loading	200 cases; 1 per unit load before wrapping	

Figure 11. Representative packing process spec.

tious company). The separate statement of these generalities permits the abbreviation of the particular spec for the bread crumb product.

The tabular form (Fig. 3) is amplified by the test methods, which are referenced rather than appended. An advantage of the tabular format, in addition to brevity and ease of modification, is that essential data can be seen almost at a glance, and when stored in a data-processing network, can be related easily to cost information.

Figure 5, which is part of both forms of the carton spec, conveys to graphic designers the areas that are available for treatment. In practice, cartons are either handmade or stamped out on a sample press with a temporary die to provide designers with three-dimensional models, in addition to the keyline drawing, to facilitate their work.

Secondary packaging and unit loading are described in Figures 6 and 7, which are comparable. Figure 8 is a part of both, and Figure 9 amplifies Figure 6 in the same relation as Figure 4 to Figure 2. Two terms require explanation. An RSC (Regular Slotted Case) is a box construction in which the long top and bottom flaps are dimensioned to meet at the centerline when the box is closed, leaving little or no gap (see Boxes, corrugated). KDF is an acronym for "Knocked Down Flat", describing the configuration in which cartons and boxes are delivered: two adjacent sides are folded against the opposite sides for compact stacking.

Of the two formats, the textual form is more explicit in its description of defects, whereas the tabular form gives more information of value for inventory of finished goods. Either format could be made to provide the same data as the other, but brevity would suffer in both. Experience builds confidence with either format or an intermediate.

Packing Process Specs

Reduced to the barest essence, a packing process consists of receiving product and packaging materials and combining them in a specified manner to produce marketable finished goods, which are protected for distribution and presentable for sale and use.

The first item in a packing spec is containment, that is, the amount of product per package. If a medicine bottle is to contain 50 pills, it may not be 49 nor 51. A lesser quantity may violate a law or regulation; a larger amount is a cost penalty. The second item in the spec is product condition, such as its temperature, physical form, or concentration. The third is the packaging materials delivery to the packing operation, and possibly its preparation or assembly of components prior to filling with product. The fourth is a specification of the packing rate—a factor of great import to control of cost and quality. The fifth item is a description of the packing steps, which is a function of the machinery that constitutes a given line of hardware wherein the operation is conducted. The last item specifies the QC requirements.

Figure 10 is a diagrammatic illustration of the machinery and steps in packing a loose product (such as bread crumbs) in cartons and corrugated boxes. Figure 11 is a statement of the packing spec, with limits on critical criteria. The QA specs, discussed below, cover the sampling and inspection process. The major responsibilities of the operating personnel are

Maintain product conditions and flow at the specified levels.

Maintain the packing-line rate uniformly at the specified level.

Maintain control of net weight within specified limits (9).

Assure siftproof quality in carton flap seals.

Date and code cartons and corrugated boxes (See Code marking and imprinting) accurately and legibly.

Avoid contamination of product or package with foreign matter.

Finished-Goods Specs

One might be tempted to assume that there should be no need for finished-goods specs if product, packaging materials, and packing process specs are all met within specified limits. Stated another way, if 3.0 is a perfect score, based on a cumulation of 1.0 for each of the three kinds of specs thus far described, there should be no point in reviewing the sum. Unfortunately, in the field of quality, $1 + 1 + 1 = 3$ max or $1 + 1 + 1 \leq 3$. There are several reasons:

Nothing is perfect. Components that have met specs still have slight defects.

Packaging protects the values (quality variables and attributes) in products, but does not add value.

Small separate defects in packaging materials, product, and process can combine to produce a major defect in the finished goods.

Specs cannot cover every conceivable defect without becoming too long and detailed, thus rendering quality control unaffordable.

Therefore, the process of quality control begins with a definition of quality required in the finished product and backs up to define the parameters in product, package, and process that will yield the desired end product. The finished-goods spec is an absolute necessity to prevent a situation in which "all the manufacturing specs were met, but the customer rejected the goods" (10, 11). The example in Figures 12 and 13 pertain to the flavored bread crumbs product, wherein the finished-goods spec looks at the combined product/package from the customer's viewpoint, stressing the criteria that are important for user acceptance, as well as those important to the seller.

To the greatest possible extent, a finished-goods audit should be nondestructive, since all factory-door costs are built into each package by the time it reaches this stage. For example, the check on case count (the number of cartons in a case) can be done simply by weighing a sealed case. If the count is short by one carton, the case will be lighter than standard by 1 lb 2 oz (510 g); ie, 454 g for product and 56 g for the carton. An automated packing line is usually equipped with a scale over which each packed case moves before it enters the unit loader, and an underweight case will be rejected by a signal or sweep bar that is actuated by the scale. Likewise, barcode scanners (see Barcoding) can be built into the packing line, and a vibrator table between cartoner and caser can be used to detect cartons that are not siftproof.

The physical properties of resistance to stacking, handling, impact, and vibration, which are experienced by all products during their distribution and sale, are not required to be tested on routine production, since the construction specs are based on testing for those properties in the package development process. ASTM test methods (3) provide for evaluation of damage resistance up to and including unit loads, using equipment available at numerous contract testing laboratories, universities, and service departments of shipping-container vendors.

Quality-Assurance Specs

The material considered thus far concentrates on the QC function: the activities that oversee the compliance of materials, process, and product with specifications. By contrast, the QA function is responsible for the existence of a system of specifications that in fact identify and cover the quality attributes and variables that relate to legal requirements and company policies and for the organization and procedures of qual-

---

SANSNOMEN CORPORATION	CODE: FG2050
FOOD DIVISION	EFF. DATE: 080185
FINISHED-GOODS SPEC	

SCOPE: 16-oz (454 g) Flavored Bread Crumbs, Packed, Ready for Shipment

RELATED SPECS: PF205, CTF2050, CSF2050, PP205
APPLICABLE AT: Plants 2,3,5

PROCEDURE FOR FINISHED GOODS QUALITY AUDIT

Audit shall be made on each shift of production, which shall be identified as a lot and assigned a recallable code, consisting of date, plant, line, and shift data, to be imprinted on each carton and case. Each lot shall consist of a maximum of 96,000 cartons in 4000 cases, as 200 unit loads, per Process Spec PP205. Acceptance for distribution requires meeting the following criteria:

Criterion	Specification	Defect class	AQL
case count	24 cartons	A	0
net weight	1 carton max in 5 cases below label weight	A	0.1
integrity	no sifting cartons	A	0.1
carton appearance	all copy legible, UPC scannable, no scuff marks	A	0.1
contamination	clean cartons and cases, no dust, foreign matter, or machine oil on surfaces	B	1.0
flap closure	no unglued flaps on cartons or cases	A	0.4
case appearance	ink colors within control limits, all copy legible, bar code scannable	B	0.65

Audit shall be conducted on sample taken per Spec PP205. A lot found that is not in compliance with the above acceptance criteria may not be forwarded for distribution and must be embargoed for sorting out defectives.

Figure 12. Representative finished-goods spec.

---

ity control to measure compliance adequately to the specs and communicate needs for corrective action in a timely and effective manner.

In the broad sense, the QA function translates legal necessity and company objectives regarding finished-goods quality into a technical road map that shows checkpoints along the paths of purchasing and receiving materials, manufacturing and packaging products, and preparing finished goods for distribution, and oversees that the checkpoint activities are appropriate to the quality objectives. In executing these responsibilities, the QA organization is part of the corporate staff in a multidivisional or nondivisional company, whereas QC is either a divisional or plant organization, reporting to operations management. The QA group sets policy as to kinds of specs, classification of defects, lot identification, production sample retention, test methods and standards, and the training of QC personnel. QA also acts as a resource to QC in the setup of sampling plans, the handling of unusual problems, and audits on QC activities. Particular specification writing is usually done by the research function, because development of prod-

SANSNOMEN CORPORATION, FOOD DIVISION
Finished-Goods Specifications

I. Subject: Specification for Shipping Graphics

II. Scope: This spec covers requirements for product, company and content identification; the criteria for open dating; rotation coding; and quality control coding.

III. General: All shippers shall be constructed of natural kraft and printed one or two colors. Exceptions such as a third color for promotional packs or bleached kraft for introductory programs must be specifically approved by a Division Officer.

IV. Universal Code and Product Code:
   A. All four sides of every shipping container shall show the product code at the upper right corner when viewed palletized, with copy right side up. When a case is printed with a tray-cut section and is shipped with the tray at the top of the case, the product code shall be printed at the upper right corner of the panel just below the tray-cut line and must read right side up as palletized.
   B. Product codes shall be printed in numerals not less than ¾ in. (1.9 cm) high nor more than 2 in. (5.1 cm) high.
   C. Wherever shown, the product code shall be preceded by the Sansnomen Corp. UPC: 00000—in numerals ½ in. (1.3 cm) high, and in the same color as the product code.
   D. Top and bottom panels of shippers shall be printed with the Universal code followed by product code at an upper right corner, parallel to a long side. Numeral sizes shall be same as above.

V. Quantity and Pack Size
   Contents of the shipper shall be stated at the upper left of each panel in the same color and numeral size as the product code, on the same horizontal level. Descriptive words, such as jars, cans, or bags, may be in smaller type than the numerals, but not less than ¼ in. (0.64 cm) high.

VI. Product Name and Brand Name
   A. These shall be centered in the four side panels and on the top and bottom panels below the quantity, pack size, universal code, and product code.
   B. On the long side panels, the spaces below the quantity/pack size and below the product code must be left blank for imprinted open date, QC data, sequential code, and rotation code.

C. A clearance of 1 in. shall be maintained for imprinting between the top of the product/brand name lettering and the bottom of the quantity/pack size and universal/product code.

VII. Pallet Pattern and Company Address
   A. Pallet Pattern: A rectangle 3 × 2½ in. (7.6 × 6.4 cm), shall illustrate the arrangement of shippers on one tier of a standard 48 × 40 in. (122 × 101.6 cm) pallet. The pattern shall be printed on the top panel, below the quantity/pack size numerals at upper left. Below the pallet pattern shall be noted the number of tiers per pallet, using the format: _X Tiers_ in characters ½ in. (1.3 cm) high.
   B. The Sansnomen Corp. and Division name, address, and zip code shall be printed in letters ¼ in. (0.64 cm) high along the bottom of the upper flap on cases and at the bottom of the top panel on bag shippers.
   C. Other unused space on the top panel may be used for brand name and opening instructions. Copy on both case flaps shall read the same side up.

VIII. Other Copy
   A. Supplier Information: box-maker's stamp and supplier specification numbers shall be placed on the bottom panel, lower right.
   B. Other Information: corporate seal and address may be repeated, as on top panel. Opening instructions, if any, shall be stated on the bottom panel. Brand and product names may be repeated on unused space on the bottom.

IX. Imprinted Data:
   A. Rotation Code: code must appear on the two long sides of each shipper, just below the product code at upper right. Numerals imprinted shall not be shorter than ¾ in. (1.9 cm), nor taller than 1 in. (2.5 cm).
   B. Open Date and QC Code: these must appear on one long side of each shipper, just below the quantity/pack size at the upper left, and preferably on the same horizontal level as the rotation code, applied by the same imprinter. The open date shall state in numerals the month–day–year (last 2 digits) packed, followed by a space equal to one digit in width and a code starting plant–line–part of shift or a sequential case number. Like the rotation code, the open date and code must be clearly legible and of characters equal in size. When sequential case numbers are used, they shall be at least ¼ in. (0.64 cm) high, below the open date.

Figure 13. Representative graphics spec for shipping containers.

ucts and packages is their responsibility, and the technical expertise on materials and processes resides there. Identification of quality criteria related to a given product/package combination is commonly fixed in committee by QA and the research function. The specs are drafted by the research function, then combined with sampling/testing plans and issued by QA. The latter organization then trains plant QC personnel in new procedures and verifies the adequacy of plant testing facilities and skills. QA policies and procedures are written and issued by the QA function. They tie together the entire system of specs and quality control, coordinating that system with regulatory requirements and company policy on quality. Figure 14 illustrates a summary statement of QA policy and procedures.

Color Control—A Separate Specialty

The research function is the source of structural and functional packaging specs, whereas the graphics are developed elsewhere in a typical company or outside the company. Color standards and the quality control of graphic reproduction are subject, however, to the same disciplines of sampling, inspection, and classification of defects as the structural attributes

SANSNOMEN CORPORATION
SPECIFICATIONS POLICY

PURPOSE:

To assure the existence of uniform comprehensive specifications at all facilities producing products for the Sansnomen Corp. by identifying the types required, defining their content, and assigning the responsibility for their preparation and maintenance.

POLICY:

In order to assure the manufacture and distribution of all products in accordance with management-approved design quality and good manufacturing practice, all product shall be produced by operating divisions in accordance with approved formulas and specifications for:

Raw and Package Materials: to facilitate purchase, receiving acceptance, and storage at plant locations by identifying classified levels of defects and mandatory test requirements.

Process: to identify process control points, classification of process defects, and mandatory test requirements to assure conformance to design quality standards.

Finished Product and Package: to define management-approved design quality characteristics and defect classification.

RESPONSIBILITY:

All specifications shall be prepared by the Research Department and issued by Quality Assurance Department.

APPROVAL AND SIGNATURES

Raw Materials: Technical Research Laboratory Manager
Quality Control Manager(s), producing plant(s)
Purchasing Materials Management Manager

Package Materials: Technical Research Laboratory Manager
Quality Control Manager(s), producing plant(s)
Purchasing Materials Management Manager

Process: Technical Research Laboratory Manager
Plant Manager(s), producing plant(s)

Finished Package and Product: Division Technical Research Manager
Operations Manager
Marketing Manager

SANSNOMEN CORPORATION
CLASSIFYING DEFECT SEVERITY

Some quality characteristics and defects are critically important to fitness for use; others are not. For the purposes of this procedure this limited definition of "Quality" is "the absence of defects." A defect is an identified out-of-limits condition for any specified product quality characteristic. Acceptable Quality Levels (AQLs) are established in recognition of the practical impossibility to achieve zero defects on a continuing basis. AQLs are used in selecting sampling plans to measure performance against specs. Continued generation of defects is not considered to be acceptable performance, even though the total number is within the assigned AQLs.

All finished-product quality defects shall be classified into one of three severity classes, based on the potential impact of the defect upon: (a) consumer health and safety, (b) product acceptability, (c) liability risk, and (d) regulatory compliance. They are called: critical (Class A), major (Class B), and minor (Class C).

All defects falling within a given classification are additive in their impact upon the franchise.

Figure 14.   Representative corporate specs policy.

and variables in packages. The test facilities, instrumentation, and skills required for graphic quality control are quite separate, however, from the usual QC laboratory equipment. First and foremost, the technicians who judge graphic reproduction quality must have good color vision, since they visually compare production samples against color standards composed of three parts: an ideal target, approved by the marketing function; a limit of acceptability on the light side (of the ideal); and a limit of acceptability on the dark side. Next, since visual color matching is affected by the light in which it is seen, a "standard daylight" illumination is required at all plants where inspection is to be made. This is provided in a viewing booth lighted by a blend of fluorescent and incandescent sources in its ceiling, balanced to the same spectrum as average outdoor "north light." Each plant must also have a set of identical standards and limits, and they must be kept in prime condition (12).

It is obvious that some degree of subjectivity is an inevitable risk with visual color matching, and in response to that shortcoming, reflectance densitometers have been developed to the level of limited utility in the present state-of-the-art of

color matching. Visual inspection remains preeminent in dealing with the match of multicolor halftone printing, where for instance the skin color in the illustration of a face is composed of four colors closely registered as minute dots. Without being too light or too dark overall, the flesh tones can vary to be either too green or too red if the corresponding two inks vary light and dark in opposite directions. Densitometer control is most effective in areas of solid color, and combination of visual and densitometer inspection is commonly practiced where halftone graphics are printed on solid background colors (13).

BIBLIOGRAPHY

1. M. G. Baldwin, *New Directions in Packaging*, American Management Association, New York, 1970, pp. 36–43.

2. E. A. Leonard, *Packaging Specifications, Purchasing and Quality Control*, 2nd ed., E. A. Leonard, Yonkers, N.Y., 1976, pp. 6–14.

3. N. S. Gardner, *Package Dev. Mag.* (now *Packag. Technol*), **2**, 26 (Nov.–Dec. 1972).

4. *AMA Management Handbook*, 2nd ed., American Management Association, New York, 1983, p. **13**–4.

5. *Mil Std-105,* Department of Defense, U.S. Government Printing Office, Washington, D.C., 105 D, 1974.

6. *Voluntary Specifications Guideline for Paperboard and Corrugated Board Packaging Systems,* National Soft Drink Association, Washington, D.C., 1979.

7. *GTA Standard Color Charts,* Gravure Technical Association, New York.

8. *Bulk Glass Handling, Technical Publication #TP-649,* Standard-Knapp Division, Emhart Corp., Portland, Conn.

9. E. F. Daigler, *Package Dev. Mag.* (now *Packag. Technol.*), **6,** 19 (May–June 1976).

10. R. E. Seely, *Package Dev. Mag.* (now *Packag. Technol.*), **5,** 14 (Nov.–Dec. 1975).

11. R. E. Schulze, "Graph Simplifies Calculation of Sample Size," Paper presented at *Quality Progress,* Oct. 1978, American Society for Quality Control, Milwaukee, Wisc., 1978.

12. A. W. Harckham, *Package Dev. Mag.* (now *Packag. Technol.*), **8,** 17 (May–June 1978).

13. E. A. Leonard, *Package Eng.,* **19,** 86 (Apr. 1974).

General References

A. F. Cowan, *Quality Control for the Manager,* 2nd ed., 1964, Pergamon Press, Inc., Elmsford, N.Y.

National Bureau of Standards Handbook #130, annual reissue U.S. Government Printing Office, Washington, D.C.

J. W. Leek and R. S. Cowles, *AMA Management Handbook,* 2nd ed., American Management Association, New York, 1983, pp. 4-107 to 4-117.

Glass Packaging Institute, McLean, Va. 22101

Can Manufacturers Institute, Washington, D.C.

Technical Association of the Pulp and Paper Industry (TAPPI), Atlanta, Ga.

American Defense Preparedness Association, Arlington, Va.

E. A. LEONARD
Cornell University

STABILIZERS.   See Additives, plastics.

STANDARDS AND PRACTICES: NATIONAL AND INTERNATIONAL

Standards are developed in a variety of ways. Formal preparation by the official national and international standards organizations is the most obvious route, but regulatory requirements often create standards, directly or indirectly. The purchasing policy of a large corporation, Government agency, or even international bodies such as NATO, frequently involves compliance with standard specifications. Trade practices, often following a market leader, become *de facto* standards. Associations of package or package material manufacturers, and on occasion package users, develop useful trade standards, and learned societies may agree on the best way to carry out an operation, and that may result in a standard test method.

What Are Standards?

A standard has been officially defined as:
"A technical specification or other document available to the public, drawn up with the cooperation and consensus or general approval of all interests affected by it, based on the consolidated results of science, technology and experience, aimed at the promotion of optimium community benefits and approved by a body recognized on the national, regional, or international level" (1).

In contrast, a regulation is
"A binding document which contains legislative, regulatory, or administrative rules and which is adopted and published by an authority legally vested with the necessary power" (1).

A voluntary standard can be incorporated in a regulation and its provisions thus become mandatory.

The aims, or purpose, of standardization have been usefully summarized (2):

1. Provision of means of communication among all interested parties.
2. Promotion of economy in human effort, materials, and energy in the production and exchange of goods.
3. Protection of consumer interests through adequate and consistent quality of goods and services.
4. Promotion of the quality of life; safety, health, and the protection of the environment.
5. Promotion of trade by the removal of barriers caused by differences in national practices.

In short they are communication, variety reduction, quality, safety, and free trade. The word *practice* refers to what industries or other bodies do, but without the official adoption of a standard.

Units of Measurement

Measurement standards of mass, length, time, etc, are relevant to package standardization. New standards are likely to be drafted in terms of Standard International (SI) units, but certain standards and regulations are expressed in non-SI units. This is the case if, for example, the nonmetric measurement is enshrined in legislation, as are capacity and thickness in certain specifications in the United States Code of Federal Regulations; the standard was first expressed in nonmetric terms and its provisions are more memorable than an arithmetic conversion to SI units, which can be shown in parenthesis. For example, the International Organization for Standardization (ISO) freight container with external dimensions of 8 ft × 8 ft × 40 ft, and a maximum permissible gross mass of 30 long tons is expressed that way rather than as 2,438 × 2,438 × 12,192 mm, with a rating of 30,480 kg; and certain components used in packaging, such as a ¾-in. B.S.P. threaded flange and screw bung (which is not interchangeable with a ¾-in. U.S. screw bung) can not be expressed intelligently in metric terms, and recourse has to be made to the use of a code (eg, G¾).

When studying the provisions of any standard, regulation, or specification which are not expressed in SI units, care must be taken to understand what units of measurement are used. The U.S. gallon, for example, is fundamentally different from the Imperial (British) gallon, and in expressions of thickness of sheet steel and plastics film, or weight per unit area of paper there are differences between North American and European practice (3). Virtually all developing countries have adopted metric practices.

Types of Standards

Standards which are relevant to the *Encyclopedia of Packaging Technology* may be grouped as follows (4):

Type A. Standards relating to the style, construction, capacity, dimensions, and quality of manufactured packages or containers which may be:

A.1, for general use (eg, a metal drum or fiberboard case).

A.2, for a specific use (eg, an egg box).

Type B. Standards relating to materials used in packaging, usually covering terminology but some covering matters of quality or purity. It is necessary to distinguish between:

B.1, materials designed for use in the packaging operation, such as speciality films and laminates.

B.2, materials which have many other uses (eg paper, fiberboard, wood, steel, and plastics).

Type C. Standards relating to test methods and/or test requirements. There are three distinct groups:

C.1, tests on complete, filled transport packages.

C.2, tests on empty containers for quality assurance.

C.3, tests on materials.

Type D. Other standards related to the distribution of goods:

D.1, marking and labeling.

D.2, dimensional coordination (package and unit load dimensions).

D.3, pallets.

D.4, freight containers.

Types B.2 and C.3 standards are not normally the responsibility of packaging standards committees nor of persons engaged in the packaging industry. There are other standards, specifications, and practices relating to the packaging methods to be used for specific commodities that are beyond the scope of this article.

International Organization for Standardization (ISO)

Following a meeting in London in 1946, delegates from 25 countries decided to create a new international organization "whose object shall be to facilitate international co-ordination and unification of industrial standards" (5). The new organization, ISO, which began to function officially on February 23, 1947, is based in Switzerland (1, rue de Varembé, Case postale 56, CH-1211 Genève 20).

There are now about 74 ISO member bodies, plus 15 correspondent members. A member body is the national body "most representative of standardization in its country," and thus only one such body for each country is accepted for membership in ISO. Member bodies are entitled to participate and exercise full voting rights on any technical committee of ISO. A correspondent member is normally an organization in a developing country that does not have its own national standards body. Correspondent members do not take on active part in the technical work, but are kept fully informed about the work. Table 1 lists the acronym of each member body, its country, and date of entry into ISO (5). It shows that the American National Standards Institute (ANSI), representing the United States, was a founder member of ISO.

The work of ISO is carried out through more than 150 technical committees (TC) and their subcommittees (SC) and working groups (WG). Each TC and SC has a secretariat assigned to an ISO member body. Details of each TC, its secre-

Table 1. ISO Member Bodies[a]

Member body	Country	Year of entry
ABNT	Brazil	1947
AFNOR	France	1947
ANSI	United States of America	1947
BCS	Sri Lanka	1967
BDS	Bulgaria	1955
BDSI	Bangladesh	1974
BSA	Albania	1974
BSI	United Kingdom	1947
CAS	China	1978
COSQC	Iraq	1964
COVENIN	Venezuela	1959
CSK	Democratic People's Republic of Korea	1963
CSN	Czechoslovakia	1947
CYS	Cyprus	1979
DGN	Mexico	1947
DGQ	Portugal	1949
DIN	Federal Republic of Germany	1951
DINT	Ivory Coast	1978
DS	Denmark	1947
ELOT	Greece	1955
EOS	Arab Republic of Egypt	1957
ESI	Ethiopia	1972
GOST	Union of Soviet Socialist Republics	1947
GSB	Ghana	1966
IBN	Belgium	1947
ICONTEC	Colombia	1960
IIRS	Ireland	1951
INAPI	Algeria	1976
INN	Chile	1947
INNORPI	Tunisia	1984
IRAM	Argentina	1983
IRANOR	Spain	1951
IRS	Romania	1950
ISI	India	1947
ISIRI	Iran	1960
ITINTEC	Peru	1962
JBS	Jamaica	1974
JISC	Japan	1952
KBS	Republic of Korea	1963
KEBS	Kenya	1976
LYSSO	Libyan Arab Jamahiriya	1978
MSC	Mongolia	1979
MSZH	Hungary	1947
NC	Cuba	1962
NNI	Netherlands	1947
NSF	Norway	1947
NSO	Nigeria	1972
ON	Austria	1947
PKNMiJ	Poland	1947
PSA	Philippines	1968
PSI	Pakistan	1951
SAA	Australia	1947
SABS	Republic of South Africa	1947
SANZ	New Zealand	1947
SASMO	Syria	1981
SASO	Saudi Arabia	1974
SCC	Canada	1947
SFS	Finland	1947
SII	Israel	1947
SIRIM	Malaysia	1969
SIS	Sweden	1947
SISIR	Singapore	1966

Table 1.   (Continued)

Member body	Country	Year of entry
SNIMA	Morocco	1963
SNV	Switzerland	1947
SSD	Sudan	1973
SZS	Yugoslavia	1950
TBS	Tanzania	1979
TCVN	Socialist Republic of Viet-Nam	1977
TISI	Thailand	1966
TSE	Turkey	1956
TTBS	Trinidad and Tobago	1980
UNI	Italy	1947
YDNI	Indonesia	1954
ZABS	Zambia	1984

[a] Full names, postal and telegraphic addresses, and telephone and telex numbers are contained in refs. 5 and 6.

tariat, and scope (determined by the ISO Council) are set out in the annual ISO Memento. The ISO technical committees directly concerned with packaging and the distribution of goods are shown in Table 2. It is important to note, however, that the work of other TCs often relates to packaging; for example, TC 6 "Paper, board and pulps," and TC 61, "Plastics," for materials standards and associated test methods. The scope of TC 34, "Agricultural food products," and TC 126, "Tobacco and tobacco products," includes packaging, storage, and transport of such products but is, in practice, limited to general, practical advice.

Because international standardization is a lengthy and expensive process, proposals for new work must be justified by identifying the need, the aim(s) of the proposed standard, and the interests that may be affected. Once work is put in hand it is listed under the relevant Technical Committee in the annual *ISO Technical Programme,* (7) as draft proposals (DP) or draft international standards (DIS) in increasing degrees of formality.

When a technical committee agrees on a draft standard, it is proposed for approval by all ISO members. If 75% of the votes cast are in favor of the DIS, it is sent for acceptance to the ISO Council. This provides an assurance that no important objections have been overlooked. All ISO standards are listed in the *ISO Catalogue,* (6) published each February with three quarterly supplements. The standards are classified by technical committee and then listed in numerical order. Alternative methods of access to the large body of ISO standards are by handbooks covering specific technical fields, or by the ISO

Table 2.   ISO Technical Committees Concerned with Packaging and Distribution[a]

TC 51, Pallets for unit load method of materials handling
Secretariat: British Standards Institution (BSI),
  2 Park Street,
  London W1A 2BS, UK
Scope: Standardization of pallets in general use in the form of platforms or trays on which goods may be packed to form unit loads for handling by mechanical devices.
Committee structure:
  No subcommittees, but specialized working groups.

Table 2.   (Continued)

TC 52, Light-gauge metal containers
Secretariat: Association française de normalisation (AFNOR),
  Tour Europe,
  Cedex 7,
  92080 Paris La Defénse, France
Scope: Standardization in the field of light-gauge metal containers with a nominal material thickness up to or equal to 0.49 mm.
Committee structure:
  SC 4, Open top containers. (Secretariat—AFNOR.)
  SC 5, General use containers. (Secretariat—AFNOR.)
  SC 6, Aerosol containers. (Secretariat—Institut belge de normalisation (IBN), Avenue de la Brabanconne, 29, B-1040 Bruxelles.)

TC 63, Glass containers
Secretariat: Uřad pro normalizaci a měřeni (CSN),
  Václavské náměsti 19,
  113 47 Praha 1, Czechoslovakia
Scope: Standardization of glass containers used as a means of packaging. Excluded: containers made from tubular glass.
Committee structure:
  SC 1, Terminology. (Secretariat—BSI.)
  SC 2, Test methods. (Secretariat—CSN.)
  SC 3, Dimensions. (Secretariat—DIN Deutsches Institut fur Normung, Burggrafenstrasse 4–10, Postfach 1107, D-1000 Berlin 30.)

TC 104, Freight containers
Secretariat: American National Standards Institute (ANSI),
  1430 Broadway,
  New York, N.Y. 10018
Scope: Standardization of freight containers, having an external volume of one cubic meter (35.3 cubic feet) and greater, as regards terminology, classification, dimensions, specifications, test methods, and marking.
Committee structure:
  SC 1, Dimensions, specifications and testing. General purpose containers, series 1. (Secretariat—AFNOR).
  SC 2, Dimensions, specifications and testing. Specific purpose containers, series 1. (Secretariat—BSI).
  plus several working groups.

TC 122 Packaging
Secretariat: Standards Council of Canada (SCC),
  2000 Argentia Road, Suite 2-401,
  Mississauga, Ontario,
  Canada L5N 1V8
Scope: Standardization in the field of packaging with regard to terminology and definitions, packaging dimensions, performance requirements, and tests. Excluded: matters falling within the scopes of particular committees (eg, TC 6, 52, and 104).
Committee structure:
  SC 1, Packaging dimensions. (Secretariat—SCC.)
  SC 2, Sacks. (Secretariat—Suomen Standardisoimisliitto r.y. (SFS), P.O. Box 205, SF-00121 Helsinki 12.)
  SC 3, Performance requirements and tests for means of packaging, packages and unit loads (as required by ISO/TC 122). (Secretariat—BSI.)
  plus several working groups.

[a] Ref. 5.

Table 3. ISO Standards for Light-Gauge Metal Containers[a]

Reference	Subject
	Light-gauge metal containers; definitions and determination methods for dimensions and capacities
ISO 90/1	Part 1, Open-top cans
ISO 90/2	Part 2, General use containers
ISO 90/3	Part 3, Aerosol cans
ISO 1361	Light-gauge metal containers; open-top cans, round cans, internal diameters
	Light-gauge metal containers; open-top cans
ISO 3004/1	Part 1, Round general purpose food cans
ISO 3004/2	Part 2, Food cans for meat and products containing meat for human consumption
ISO 3004/3	Part 3, Cans for drinks
ISO 3004/4	Part 4, Cans for edible oil
ISO 3004/6	Part 6, Round cans for milk
	Hermetically sealed metal containers for food and drinks; food cans for fish and other fishery products
ISO/TR 7423	Sections for nonround cans
ISO/TR 7670	Capacities of round and nonround cans and associated diameters of round cans
ISO/TR 8610	Light-gauge metal containers; round vent-hole cans with soldered ends for milk and milk products; capacities and related diameters

[a] Refs. 6 and 7.

KWIC Index of international standards which is prepared in key-word-in-context format.

There is no requirement for ISO member bodies to publish an ISO standard as a national standard, although it is by such action that ISO standards are put into use throughout the world. The British Standards Institution (BSI), for example, has a policy to publish an ISO standard without change as a dual-numbered British standard or, where minor changes of text but not of substance have been necessary, as a British standard clearly marked as being "technically equivalent" to the identified ISO standard. On occasion, publication of an ISO standard as a national standard is not considered to be appropriate.

ISO Packaging Standards

Three groups of ISO standards on packaging and related activities merit detailed comment.

Freight containers. The 12 standards relating to freight containers (TC 104) are perhaps the best possible example of international standardization, because without them the "container revolution" could never have taken place. Standardized dimensions, maximum gross masses, and above all, standardized methods of lifting (twist-locks and corner fittings) have enabled rapid, safe transfer of the containers from one transport mode to another. These D.4 type standards relate not just to box containers but to tank containers for liquids, gases, and solids, as well.

Light-gauge metal containers. Through the work of TC 52, the number of different sizes of open-top general-purpose food cans has been reduced from over 2000 when work commenced

in 1947 to 35 in the latest revision of ISO 3004/1. This is still too many, but it represents the best possible compromise at the present time. The range of standards for light-gauge metal containers, following the current program of work, is likely to be as listed in Table 3 although some are only at the DIS stage and others as Technical Reports (TR). All these are type A.1 or A.2 standards.

Testing filled transport packages. ISO/TC 122/SC 3 has the responsibilities for these standards. All are type C.1 standards concerned with the testing of filled transport packages ready for dispatch. They have been welcomed in many countries as national standards. They are listed in Table 4.

Other ISO packaging standards are listed in Table 5. These include several DIS standards currently being processed by TC 63 (Glass containers) and TC 122/SC 2 (Sacks).

ISO standards and other publications can frequently be purchased from the national standards body of the country concerned.

Regional Standards Organizations

In addition to ISO there are other international standards organizations in different regions of the world. These include:

ARSO. African Regional Organization for Standardization, which was founded in January 1977 under the auspices of the UN Economic Commission for Africa (ECA). Membership is open to the national standards bodies of African countries who are members of ECA and Organization of African Unity.

ASMO. Arab Organization for Standardization and Metrology, which was founded in 1965 to serve as a specialized technical body for the League of Arab States in the fields of standardization, metrology and quality control.

CEN. European Committee for Standardization (see below).

CMEA. Council for Mutual Economic Assistance. This was founded in 1949 to promote coordination between the USSR

Table 4. ISO Standards for Testing Filled Transport Packages[a]

Reference	Subject
	Packaging; complete, filled transport packages
ISO 2206	Part 1, Identification of parts when testing
ISO 2233	Part 2, Conditioning for testing
ISO 2234	Part 3, Stacking test
ISO 2248	Part 4, Vertical impact test by dropping
ISO 2244	Part 5, Horizontal impacts tests (inclined plane test, pendulum test)
ISO 2247	Part 6, Vibration test
ISO 2872	Part 7, Compression test
ISO 2873	Part 8, Low pressure test
ISO 2874	Part 9, Stacking test using compression tester
ISO 2875	Part 10, Water spray test
ISO 2876	Part 11, Rolling test
ISO 4178	Complete, filled transport packages; distribution trials, information to be recorded
	Complete, filled transport packages; general rules for the compilation of performance test schedules
ISO 4180/1	Part 1, General principles
ISO 4180/2	Part 2, Quantitative data

[a] Ref. 7.

Table 5. Other ISO Standards on Packaging and Distribution

Reference	Subject	Type
DIS 445	Pallets; vocabulary	D.3
ISO 780	Packaging; pictorial marking for handling of goods (see Fig. 1)	D.1
ISO 3394	Dimensions of rigid rectangular packages; transport packages	D.2
ISO 3676	Packaging; unit load sizes, dimensions	D.2
	Packaging; sacks, vocabulary and types	
ISO 6590/1	Part 1, Paper sacks	A.1
DIS 6590/2	Part 2, Sacks made from thermoplastic film	A.1
	Packaging; sacks, description and method of measurement	
DIS 6591/1	Part 1, Paper sacks	A.1
DIS 6591/2	Part 2, Sacks made from thermoplastic film	A.1
	Packaging; sacks, conditioning for testing	
ISO 6599/1	Part 1, Paper sacks	C.1/2
ISO 7023	Packaging; sacks, method of sampling empty sacks for testing	C.1/2
DIS 7458	Glass containers; internal pressure resistance test, test methods	C.2
DIS 7459	Glass containers; thermal shock resistance and endurance, test methods	C.2
	Packaging; sacks, drop test	
DIS 7965/1	Part 1, Paper sacks	C.1
DIS 8106	Glass containers; determination of capacity by gravimetric method; test method	C.2
DIS 8113	Glass containers; resistance to vertical impact; test method	C.2
DIS 8162	Glass containers; tall crown finishes; dimensions	A.1
DIS 8163	Glass containers; shallow crown finishes; dimensions	A.1
DIS 8164	Glass containers; 52 cl 520-mL beer bottles (Euroform); dimensions	A.2
	Packaging; estimating the filled volume using the flat dimensions	
ISO/TR 8281/1	Part 1, Paper sacks	D.2

a Refs. 6 and 7.

ISO 7000 No 0626

Figure 1.   Pictorial marking. KEEP DRY. One of the 12 pictograms in ISO 780 1984 (8) to convey handling or storage instructions. Courtesy of ISO.

PASC. Pacific Area Standards Congress. Founded in 1973 to help the Pacific countries participate in international standards activities and promote closer cooperation between its members.

European Committee Standardization (CEN).   CEN has possibly the greatest impact on world trade because of its close association with the European Economic Community (EEC) and the European Free Trade Area (EFTA). CEN was founded in 1961 and its members comprise the national standards bodies of the countries listed in Table 6, all of whom are members of ISO.

The Councils of EEC and EFTA have accepted the principle of specifying technical requirements in legislation by reference to standards, where possible. CEN provides, where appropriate, the European standards needed to support the EEC program of legislation for elimination of technical barriers to trade under Article 100 of the Treaty of Rome. Once a CEN standard is called up in an EEC Directive the terms of the Directive will determine the extent to which that standard is mandatory in each of the EEC member countries. A further requirement of CEN is that, unlike ISO, European Standards are to be published without variation of text as national standards in the countries approving them.

Activity on standards for packages has been limited to the Technical Committees listed in Table 7 and the standards

Table 6. CEN Member Bodies

Member body	Country	Member of
ON	Austria	EFTA
IBN	Belgium	EEC
SNV	Switzerland	EFTA
DIN	Federal Republic of Germany	EEC
DS	Denmark	EEC
AFNOR	France	EEC
IRANOR	Spain	EEC
SFS	Finland	EFTA
BSI	United Kingdom	EEC
ELOT	Greece	EEC
IIRS	Ireland	EEC
UNI	Italy	EEC
NNI	Netherlands	EEC
NSF	Norway	EFTA
DCQ	Portugal	EEC
SIS	Sweden	EFTA

Note: The twelfth member state of the EEC, Luxembourg, does not have an independent standards body but tends to use Belgian standards.

and East European national standards bodies and to assist the production of common or harmonized standards. In recent years membership has been extended to Cuba, Mongolia, and Vietnam.

COPANT. Pan American Standards Commission, which was founded in 1961 and comprises the national standards bodies of the United States and 11 Latin American countries. It is a coordinating organization concerned with the regional implementaion of ISO standards and recommendations.

Table 7. CEN Technical Committees[a]

TC 63,	packages for washing and cleaning powders
TC 78,	capacities of glass jars for preserved fruit, vegetables and similar products
TC 81,	capacities of metal cans for fruit, vegetables and similar products
TC 101,	steel drums
In each case DIN holds the secretariat	

[a] Ref. 9.

listed in Table 8. The standard EN 23, Part 1, "Packages for washing and cleaning powders" is in use throughout the EEC in support of Council Directive 80/232/EEC (see Table 12) and consists of ten capacities of carton which have fixed dimensions and bear the appropriate "E" number (E.1 to E.30), and six capacities of drum, also of fixed dimensions and bearing the appropriate "ET" number (ET.5 to ET.30). By this means the consumer is not misled by a variety of package sizes which may be similar in appearance but not in capacity. EN 76, "Packages for certain prepacked foodstuffs," was prepared at the request of the EEC Commission for use in Directive 80/232/EEC. It is notable that only 15 capacities of package are specified for general use, one of which is exclusively a glass container. All of the 14 capacities of metal can appear in ISO 3004 which lists 35 capacities. The CEN standard is thus more restrictive than the ISO standard. The two provisional standards relating to metal drums are not likely to be called up in an EEC Directive.

National Standards Organizations

As indicated in Tables 1 and 6, there are a large number of national standards bodies in developed and developing nations. The majority of such organizations are governmental institutions, or incorporated by public law, or, as in the case of BSI, operate under a Royal Charter. All have close links with the public administration in their respective countries. In the United States, some of the ANSI packaging standards are national versions of ISO standards, and others are ASTM methods which have been formally adopted as ANSI standards.

A 1978 report on standards coordination (4) identified over 1500 national standards concerned with packaging which were divided into two broad groupings: those relating to selected package types and packaging materials (see Table 9), and standards relating to the packaging of specific products, or groups of products (see Table 10).

Table 8. CEN Standards on Packaging

Number	Title	Type
	Packages for washing and cleaning powders	
EN23-1	Part 1 Dimensions and volumes of cartons and drums from fiberboard	A.2
EN 76	Packages for certain prepacked foodstuffs; capacities of glass and metal containers	A.2
pr EN 209	Steel drums; removable (open) head drums with a total capacity of 213L.	A.1
pr EN 210	Steel drums; tight head drums with a total capacity of 216.5 L.	A.1

[a] Ref. 9.

Table 9. Principal National Standards on Package Types

No. of standards	Package type	No. of countries
265	glass bottles	32
100	metal drums	18
99	wooden boxes and crates	19
76	metal cans	18
56	plastics packages (excluding boxes/crates, bags and sacks)	18
53	glass jars	16
43	fiberboard boxes	18
plus lesser numbers of 13 other types of package or packaging material		

Identification of an Applicable Standard

At national level the standards body usually publishes a complete list of current standards which may be exclusively domestic or ISO/CEN standards published as dual numbered or technically equivalent national standards. The *BSI Catalogue* (10), which is published annually, is a typical example. It contains some thousands of standards listed by number, title and including a short summary, for example;

BS 6117 : 1981
Specification for glass bottles for light wine. Covers design, dimensions, capacity, finish, verticality, thermal shock, and markings for wine bottles of capacity 350 mL to 2 L. 20 pages.

Despite comprehensive indexing of such publications many national standards bodies also publish sectional lists. These lists, when read in conjunction with the catalogue afford considerable help in the identification of an applicable standard within one's own country.

The problem of identifying an applicable standard in another country is manifestly more difficult, and finding information about the standards, technical regulations, or related testing and certification activities can be a heavy task. ISONET, the ISO Information Network, has been developed to meet this need. Individuals needing information about any standard or governmental technical regulation in another

Table 10. National Standards Relating to the Packaging of Specific Products or Product Groups[a]

No. of standards	Product(s) to be packed	No. of countries
198	fruit and vegetables	24
175	preserved foods	31
82	textiles	12
39	petroleum products	17
32	pharmaceuticals	11
29	milk	12
29	fish	10
29	food, general	9
27	sheet glass	6
25	glass holloware	8

[a] Over 200 product groups were identified: only the 10 most numerous are listed in the table

Table 11. Examples of Other Standards-Making Bodies[a]

European Solid Fibreboard Case Manufacturers' Association (ASSCO), Sutherland House, 5-6, Argyll Street, London W1V 1AD, United Kingdom	International Case Code; test methods for boards; test methods for fiberboard cases
European Corrugated Fibreboard Case Manufacturers' Association (FEFCO), 37, Rue d'Amsterdam, F-75008 Paris, France	
Fiber Box Association (FBA), 224, South Michigan Avenue, Chicago, Illinois 60604	standards for fiber board cases
Federation of European Aerosol Associations (FEA), Waisenhausstrasse 2, CH-8001 Zurich, Switzerland	standards relating to the construction of metal and glass aerosols and their fitments
International Technical Centre of Bottling (CETIE), 7, rue de la Boetie, F-75008 Paris, France	standards for glass bottles for beer, wines, etc
Technical Association of the Paper and Pulp Industry (TAPPI), One Dunwoody Park, Atlanta, Georgia 30341	test methods for materials and standards for paper-based packages
American Society for Testing and Materials (ASTM), 1916, Race Street, Philadelphia, Penn. 19103	test methods for packages and packaging materials

[a] A useful list of names and addresses of more than 300 intergovernmental, international, and national bodies is contained in ref. 11.

country have only to contact the ISONET member in their own country to receive information and assistance from more than 50 other ISONET members, each of whom are responsible for documents issued within their own territory.

Other Standards and Trade Practices

In addition to the official international and national standards there are many which have been developed by professional bodies, national and international associations of package manufacturers, government agencies, or to meet the needs of specific user groups. Examples of the development of such standards are shown in Table 11. *De facto* commercial standards have also been established by long periods of use, or the transfer of technology by multinational companies either by license or investment. A leading example of such a *de facto* international standard is the 210 L/55 U.S. gal/45 Imp gal steel drum for liquids. A variety of diameters was in use prior to the 1950s when the oil industry, in consultation with the chemical industry, embarked on a standardization program which resulted in the adoption of an internal diameter of 22½ in. (571.5 mm). This diameter has been incorporated in certain national standards; for example, those of ANSI, and in the two provisional standards of CEN shown in Table 8.

The conditions of carriage of nondangerous goods by various modes of transport frequently impose packaging standards for the goods to be carried. In European rail transport the UIC Code published by the International Union of Railways, which is binding on the various railway administrations, includes

"Packing conditions for goods in international traffic" (12) and "Loading recommendations to prevent damage to goods" (13), plus four pamphlets relating to pallet size and quality (14). Similarly, in the United States the provisions of the Uniform Freight Classification (15) include in Rule 40 the entire area of shipping containers (drums, pails, crates, carboys, bags, etc.) and in Rule 41 detailed specifications and test requirements for acceptable fiberboard boxes. Rule 222 of the National Motor Freight Classification (16) contains standards for fiberboard boxes which are virtually identical to those of Rule 41 (see Boxes, corrugated).

National and International Regulations

The regulations and recommendations of national governments, intergovernmental agencies, and international bodies exert a powerful influence on package standards, directly and indirectly. Such influences can be summarized broadly as:

Regulations which are designed to ensure that there is no harmful interaction between the package or packaging material and the products intended for human consumption.

Recommendations relating to the packaging and/or distribution of specific products or groups of products.

Regulations which control the weights or volumes in which certain prepackaged products may be sold.

Regulations which are designed to ensure that there is no harmful interaction between the package or packaging material and the products intended for human consumption.

Packaging materials in contact with food. Standards for packages, and particularly packaging materials, may be affected by legislation in this area. The most widely known regulations are those administered by the FDA under the authority of the Food, Drug and Cosmetic Act, 1958 (17) (see Food, Drug and Cosmetics regulations). The Agricultural Stabilization and Conservation Service (ASCS) of the United States Department of Agriculture has published detailed specifications (CMO-1) for the packaging of dairy products, processed grains, salad oil, and shortening which include the materials permitted by the FDA and outer transport packages which meet the appropriate Federal Specification or the requirements of the Uniform Freight Classification (15). Similar but not identical legislation has been in existence for some years in other countries such as the Federal Republic of Germany and the United Kingdom. More recently the European Economic Community (EEC) has published, or has in preparation, a family of Council Directives relating to "Materials and articles intended to come into contact with foodstuffs" (18).

Recommended standards for the packaging and distribution of specific products. Various international organizations have published standards or recommendations relating to a wide range of products, principally fresh or preserved foods. These include:

ISO/TC 34, Agricultural food products, which has published more than 150 standards, which are listed in the *ISO Catalogue* (7). Some of these include guides to the packing, storage, and transport of produce.

The *Codex Alimentarius Commission* (19), which has published a number of standards chiefly related to preserved fruits and vegetables, canned or frozen.

The *UN Economic Commission for Europe (ECE)*. The ECE Committee on Agricultural Problems has published more than 50 standards or recommendations primarily related to fresh, perishable produce for which the packaging requirements are covered in a general, rather than a specific manner. Of more significance is Resolution No. 222 of the ECE's Inland Transport Committee (20), relating to the "standardization of packaging for the international transport of fresh or refrigerated fruit and vegetables", which establishes the dimensions, mechanical strength characteristics, and performance test requirements of rectangular wood, fiberboard and plastics packages to be carried on 800 × 1200 mm and 1000 × 1200 mm pallets.

The *Organization for Economic Co-operation and Development (OECD)* whose international standards relating to fruit and vegetables under the OECD Scheme are closely related to those of the ECE. By Council resolution (21) the dimensional, strength, and test requirements of the ECE Resolution No. 222 have been adopted by OECD.

The principal requirements of the ECE/OECD resolution on fruits and vegetables may be summarized as:
Classification (according to stacking capability)
    Group A, 2.5 m
    Group B, 8 m
External dimensions (maximum)
    600 × 400 mm
    500 × 300 mm
    400 × 300 mm
Compression test using (ISO 2872 and ISO 2874).
    Compressive resistance to be calculated from a formula incorporating weight of filled package, overall height of package and height of stack (2.5 or 8 m). Vertical deflection of 5% or greater not allowed.
Vibration test (using ISO 2247).
    Three loaded packagings placed on top of one another, plus a superimposed load, vibrated for 2 periods of 10 min. No visible permanent deformation permitted.
Drop test (using ISO 2248).
    Two drops per package, one on a bottom-end edge and one on a bottom-side edge, from a height in cm of 70-P, where P is the weight of the filled package in kg. No visible, permanent deformation is permitted.
Approval certificate, including details of package specification and test results.
Marking. Every package to show manufacturer's name or mark; group to which it belongs; and prescribed stamp incorporating the approval certificate number.

Permitted quantities of prepackaged goods. Although "prepackaged" is a common term, it may not be generally realized that within the European Economic Community (EEC) it has a very precise meaning, ie,

    "A product is prepacked when it is placed in a package of whatever nature without the purchaser being present and the quantity of the product contained in the package has a predetermined value and cannot be altered without the package either being opened or undergoing a perceptible modification."

For many years, some countries have had consumer protection legislation providing that certain essential foodstuffs such as butter and sugar may only be sold in specified quantities, the increments in size (weight) being such that there is no

Table 12. Council Directives of the European Economic Community (EEC)

Directive	Title (abbreviated)	Ref.[a]
75/107/EEC	bottles used as measuring containers	L **42** (2/15/75)
75/106/EEC	making up by volume of certain prepackaged liquids	L **42** (2/15/75)
78/891/EEC	-do- (adapting to technical progress)	L **311** (11/4/78)
79/1005/EEC	-do- (amendment)	L **308** (12/4/79)
76/211/EEC	making up by weight or volume of certain prepackaged products	L **46** (1/21/76)
78/891/EEC	-do- (adapting to technical progress)	L **311** (11/4/78)
80/232/EEC	ranges of nominal quantities and nominal capacities permitted for certain prepackaged products	L **51** (2/25/80)
75/324/EEC	aerosol dispensers	L **147** (5/20/75)

[a] From *Official Journal of the European Communities* (22). The dates given are (month/day/year).

possibility of the consumer being deceived. It is this fundamental principle which has been adopted by the EEC in their series of Council Directives on the prepacking of all forms of drink and foodstuffs sold by weight and by volume, and nonfood products which are sold by volume, and nonfood products which are sold by weight (such as cleaning products) or by volume (eg, paints and lubricating oil).

The principal Council Directives are listed in Table 12. In addition to the parent Directive on each subject there are, or may be in the future, further Directives which *amend* the text of the Directive or *adapt* its Annex or Annexes. Member states of the EEC are required to implement the provisions of each Directive within the time scale prescribed in that Directive.

It is not possible to summarize briefly the detailed requirements of these various Directives, but only to refer to the principal features. These have been incorporated into the national legislation of the Member States of the European Community and have influenced the packaging standards developed in those countries.

Directive 75/107/EEC applies to bottles of glass, or materials of similar qualities, between 50-mL and 5-L capacity which are used as measuring instruments. Such bottles may only be marked with the EEC sign (a reversed epsilon) if they conform with the requirements of the Annexes to the Directive. The *Liquids Directive* (75/106/EEC) is a specific Directive which applies to packages between 50-mL and 5-L capacity for wine, beer, spirits, vinegar, edible oils, milk, soft drinks and fruit juices. Annex III of the Liquids Directive, as adapted and amended, sets out the nominal volumes in which such products may only be sold in prepackaged form. Certain sizes may continue to be used for a transitional period. Examples of the definitive range are set out in Table 13. The undermentioned British standards comply with the requirements of 75/107/

Table 13. Nominal Volumes Permitted in Council Directive 75/106/EEC (Examples Only)

Still wines	Beer made from malt	Milk and milk-based beverages
100 mL		
		200 mL
250 mL	250 mL	250 mL
	330 mL	
375 mL		
500 mL	500 mL	500 mL
750 mL	750 mL	750 mL
1.0 L	1.0 L	1.0 L
1.5 L		
2.0 L	2.0 L	2.0 L
3.0 L	3.0 L	
	4.0 L	
5.0 L	5.0 L	

EEC, provide "shape" and "dimensions" for some of the capacities which are permitted in 75/106/EEC, and are based on the work of CETIE (see Table 11):

BS 6117, Specification for glass bottles for light wine.

BS 6118, Specification for multi-trip glass bottles for beer and cider.

BS 6119, Pt. 1, Glass bottles for carbonated soft drinks.

Directive 76/211/EEC (the Solids Directive) applies to packages between 5 g/5 mL and 10 kg/10 L for products not covered by the Liquids Directive. Directive 80/232/EEC extends the Solids Directive and prescribes nominal quantities for prepackaged goods in three groups. Annex I covers food aold by weight (butter, cheese, salt, sugar, cereals, pasta products, rice, dried vegetables, coffee, frozen products such as fruit, vegetables, and fish); foodstuffs sold by volume (ice cream above 250 mL); dry pet food; paints and varnishes; cleaning products; cosmetics and toiletries; washing products; solvents; and lubricating oils. Typical examples are shown in Table 14. Annex II of 80/232/EEC lists the range of permissible container capacities for preserved foods for human con-

Table 14. Range of Nominal Quantities of Contents Permitted in Council Directive 80/232/EEC (Examples Only)

Butter margarine, etc	Sugars	Paint and varnishes	Lubricating oils
		25 mL	
		50 mL	
125 g (0.28 lb)	125 g (0.28 lb)	125 mL	125 mL
250 g (0.55 lb)	250 g (0.55 lb)	250 mL	250 mL
		375 mL	
500 g (1.1 lb)	500 g (1.1 lb)	500 mL	500 mL
		750 mL	
1.0 kg (2.2 lb)	1.0 kg (2.2 lb)	1.0 L	1.0 L
1.5 kg (3.3 lb)	1.5 kg (3.3 lb)		
2.0 kg (4.4 lb)	2.0 kg (4.4 lb)	2.0 L	2.0 L
2.5 kg (5.5 lb)	2.5 kg (5.5 lb)	2.5 L	2.5 L
		3.0 kg (6.6 lb)	3.0 L
		4.0 kg (8.8 lb)	4.0 L
5.0 kg (11.0 lb)	5.0 kg (11.0 lb)	5.0 L	5.0 L
		10.0 L	10.0 L

sumption; moist pet food; and washing and cleaning products in powder form. The Annex incorporates the capacities of tin and glass containers set out in EN 76, and cartons and drums of fiberboard listed in EN 23-1 (See Table 8). Annex III covers products sold in aerosol containers. It lists 13 capacities of metal containers for products propelled by liquid gas and the appropriate volume of the liquid phase for each size. Also specified are six volumes of the liquid phase for products sold in glass or plastic containers. Test pressures and filling requirements for aerosols are covered in Directive 75/324/EEC. The standards developed by FEA (see Table 11) are closely related to these two EEC Directives.

Hazardous materials, transport of dangerous goods. National and international regulations regarding the transport of dangerous goods create "standards" of construction, design, and performance test requirements for the packages used to contain such substances. Some countries incorporate the mandatory requirements in national standards; in other countries the international or national legislation is considered to be sufficient.

There are two approaches to such standards. One is exemplified in the current United States Code of Federal Regulations (23) and the obsolescent Restricted Articles Regulations of IATA, the International Air Transport Association (24). Another approach is that of the United Nations Committee of Experts on the Transport of Dangerous Goods.

CFR 49 contains detailed requirements covering the construction, testing and certification marking of a large number of DOT specification packages. For example, Specification 17E covers the 55 U.S.-gal (208 L) tight-head steel drum used for liquids in all parts of the world: its provisions include minimum and maximum capacity; gauge of metal for body and ends; rolling hoops; closures; marking; type testing (drop and hydrostatic pressure); closures; markings; type testing (drop and hydrostatic pressure); and leakage testing of all drums. It does not specify dimensions, this being left to the industry standard developed by the Steel Shipping Container Institute and adopted as an ANSI standard. The ANSI standard MH.2.1 thus gives shape, dimensions and tolerance to the 55 U.S.-gal-tight head drum for dangerous goods (DOT 17E) and similar drums for nondangerous goods meeting the requirements of UFC Rule 40 (15) and NMFC 260 (16).

In contrast, the UN Recommendations on the Transport of Dangerous Goods (25) do not specify how a package is to be made, but what it is required to do. Chapter 9, "General Recommendations on Packing," lists some 55 types of transport package; allocates to each an alpha-numeric code to denote package type and material of construction; states the maximum capacity and/or maximum net mass permitted for each package type; stipulates performance tests which are to be carried out on each packaging design type and the intensities of such tests; and requires the use on the package of a certification mark indicating that it corresponds to the successfully tested design type. Chapter 9 applies to packages for all classes of dangerous goods other than radioactives and infectious substances, for which separate provision is made.

UN Chapter 9 has been adopted by all the four modes of international transport and its provisions incorporated in:

"Technical Instructions for the safe transport of dangerous goods by air" (26) which are mandatory on all states which are signatories to the Convention on International Civil Aviation (the Chicago Convention).

"The International Maritime Dangerous Goods Code" (27). This has only the status of "recommendations" and requires adoption by signatories to the International Convention for the Safety of Life at Sea (SOLAS), but has been so adopted by most of the important maritime nations.

"ANNEX 1 (RID) to the Convention concerning International Carriage by Rail (COTIF)" (28), which is binding on the 28 states in Europe, the Middle East, and North Africa which are parties to the Convention.

"ANNEX A to the European Agreement concerning the International Carriage of Dangerous Goods by Road (ADR) (29), which is binding on the 19 European states which are parties to the Agreement.

Except for certain transitional provisions, the use of UN tested and certified packagings will be mandatory for all international transport of dangerous goods, and thus the UN has standardized the test methods and levels of performance for all

of the 55 types of transport package listed in Table 15. There are no tests for inner packagings but the ICAO TIs (26) contain certain specifications for such inners which are generally similar to those of the former IATA RAR.

The UN test requirements, too detailed to reproduce *in toto*, are summarized in Tables 16 and 17, the latter making specific reference to a 55 U.S.-gal/45-Imp gal/~210 L tight-head steel drum (similar to that in ANSI standard MH.2.1, or British Standard BS.814, or the provisional CEN standard pr-EN 210). One of the recommendations of the 1978 report (4), which was adopted by the ITC/WPO International Consultation on Packaging (30), was that ISO and its member bodies should give consideration to the development of standards for specific types of container which meet the UN requirements, such standards providing for the use of the UN identification and certification markings. It is expected that the coupling of dimensional and constructional specifications with the UN performance test requirements to form standards for individual

Table 15. Summary of UN Transport Packagings

Package type	Material	Category	Code	Maximum[a] Capacity, L (U.S. gal)	Maximum[a] Net mass, kg (lb)
1. drum	A. steel	nonremovable head	1A1	450 (118.9)	400 (882)
		removable head	1A2	450	400
	B. aluminum	nonremovable head	1B1	450	400
		removable head	1B2	450	400
	D. plywood		1D	250 (66.0)	400
	G. fiber		1G	450	400
	H. plastics	nonremovable head	1H1	450	400
		removable head	1H2	450	400
2. barrel	C. wood	bung type	2C1	250	400
		slack type (removable head)	2C2	250	400
3. jerrican	A. steel	nonremovable head	3A1	60 (15.8)	120 (265)
		removable head	3A2	60	120
	H. plastics	nonremovable head	3H1	60	120
		removable head	3H2	60	120 (165)
4. box	A. steel	unlined	4A1		400
		with liner	4A2		400
	B. aluminum	unlined	4B1		400
		with liner	4B2		400
	C. wood	ordinary	4C1		400
		with siftproof walls	4C2		400
	D. plywood		4D		400
	F. reconstituted wood		4F		400
	G. fiberboard		4G		400
	H. plastics	expanded	4H1		60 (132)
		solid	4H2		400
5. bag	H. woven plastics	without inner lining or coating	5H1		50 (110)
		siftproof	5H2		50
		water resistant	5H3		50
	H. plastics film		5H4		50
	L. textile	without inner lining or coating	5L1		50
		siftproof	5L2		50
		water resistant	5L3		50
	M. paper	multiwall	5M1		50
		multiwall, water resistant	5M2		50
		in steel drum	6HA1	250	400

Table 15. (*Continued*)

Package type	Material	Category	Code	Maximum[a] Capacity, L (U.S. gal)	Maximum[a] Net mass, kg (lb)
6. composite packaging	H. plastics receptacle	in steel crate or box	6HA2	60	75 (165)
		in aluminum drum	6HB1	250	400
		in aluminum crate or box	6HB2	60	75
		in wooden box	6HC	60	75
		in plywood drum	6HD1	250	400
		in plywood box	6HD2	60	75
		in fiber drum	6HG1	250	400
		in fiberboard box	6HG2	60	75
		in plastics drum	6HH	250	400
	P. glass, porcelain or stoneware receptacle	in steel drum	6PA1	60	75
		in steel crate or box	6PA2	60	75
		in aluminum drum	6PB1	60	75
		in aluminum crate or box	6PB2	60	75
		in wooden box	6PC	60	75
		in plywood drum	6PD1	60	75
		in wickerwork hamper	6PD2	60	75
		in fiber drum	6PG1	60	75
		in fiberboard box	6PG2	60	75
		in expanded-plastics box	6PH1	60	75
		in solid-plastics box	6PH2	60	75

[a] The metric values are absolute, the gal and lb conversions are only approximate.

package types will occur at national, rather than international level.

For larger packages up to 3000-L (792.6-U.S. gal) capacity, known officially as "Intermediate Bulk Containers" (IBCs), which are not portable tanks as defined in chapter 13 of the General Introduction to the IMDG Code (27) or tank containers (which are similar pressure vessels) according to the requirements of Appendix X of RID (28) or Appendix B.1b of ADR (29), the UN has in preparation similar design-type tests. Requirements for metallic prismatic IBCs and for flexible

Table 16. **UN Design Type Tests for Transport Packagings**[a]

Test	Applicability
drop	all types of packaging
stack	all types of packagings other than bags (sacks)
air leakage	all packagings for liquids (also to be applied to all such packagings before they are used for the first time, and before reuse).
hydraulic pressure	all packagings for liquids, the minimum being the vapor pressure of the liquid to be carried at a given temperature, multiplied by a safety factor.

[a] Notes:
Conditioning of packagings of paper and fiber-board for at least 24 h, and the drop test to be carried out immediately after removal from the conditioning environment.

Conditioning of all plastics packagings (excluding bags and expanded polystyrene boxes, but including composite packagings with inner plastics receptacle and combination packagings, eg, a box, with inner plastics packagings, eg, a bottle) down to at least −18°C (ca 0°F) prior to the drop test: the test must be conducted with the contents at that temperature which usually necessitates the use of an antifreeze solution as the test contents.

The stacking test for plastics drums, plastics jerricans, and composite packagings (code 6HH) is to be conducted at 40° C (104°F) for a period of 28 d: this compares with 24 h at ambient conditions for other types of packaging.

Table 17. **Testing of Steel Drums, Code 1A1**[a]

Test	No. of samples	Procedure
drop	6	3 drums, filled with water to 98% of their capacity, dropped from a height of 1.2 m[b] diagonally on the climb with the center of gravity immediately above the point of impact. The further 3 drums to be dropped from the same height striking the target on the weakest point not tested in the first drop. Criteria: no leakage *after the external and external pressures have been equalized* (eg, by loosening a bung).
stack	3	3 drums, filled as for the drop test, to be subjected to a superimposed load equivalent to the total weight of identical packages which can be stacked on it up to a height of 3 m, period of test, 24 h. Criteria: no leakage, nor any deterioration or distortion likely to cause instability in stacking or transport.

Table 17. *(Continued)*

Test	No. of samples	Procedure
		Note: the superimposed load for a 210-L/(ca 55-U.S. gal) drum to ANSI MH.2.1 containing a liquid of 1.2 relative density would be approximately: 270 kg × 2.39 = 645 kg (gross mass) (no. in 3 m stack, less 1)
air leakage	3	Each drum to be submitted to an air pressure of not less than 20 kPa[c] (0.197 atm). Criteria: no leakage.
hydraulic pressure	3	Each drum, including its closure(s), to withstand for 5 min a gauge pressure determined by one of the following methods: *(1)* Not less than the total gauge pressure measured in the packaging (ie, the vapor pressure of the filling substance and the partial pressure of the air or other inert gases, minus 100 kPa or ~1 atm) at 55°C (131°F), multiplied by a safety factor of 1.5. *(2)* Not less than 1.75 times the vapor pressure at 50°C (122°F) of the substance to be transported, minus 100 kPa (~ 1 atm) but with a minimum test pressure of 100 kPa (~ 1 atm). *(3)* Not less than 1.5 times the vapor pressure at 55°C (131°F) of the substance to be transported, minus 100 kPa (~1 atm) but with a minimum test pressure of 100 kPa (~ 1 atm). Criteria: No leakage. Note: depending on the gauge of metal and the construction of the end seams 210-L (ca 55-U.S. gal) drums have been certified to pressures of at least 270–400 kPa (2.66–3.95 atm).

[a] Assumptions: relative density of intended contents does not exceed 1.2, for products of medium danger, ie, Packaging Group II
[b] This height is multiplied by a factor of 1.5 for products of great danger (Packing Group I) or divided by a similar factor for products of minor danger (Packing Group III).
[c] For Packing Group I products not less than 30 kPa (0.296 atm) is required.

IBCs have now been incorporated in a new chapter 16 of the UN Recommendations (25), and doubtless will be used in a number of national standards for IBCs. It is expected that requirements for other types of IBC will be added to chapter 16 in due course. There are very few national standards on IBCs, of which the following are examples:

USA DOT Specification 56, Metal portable tanks (for solids). DOT Specification 57, Metal portable tanks (for liquids).

FRG Guidelines for the construction, testing, approval, marking and use of cubical tank containers (KTC) made of metallic materials, TR/KTC 001 (31).

FRG Guidelines for the construction, testing, and approval of transport containers made of plastics materials for the transportation of dangerous goods, RTK 001 (32).

UK BS 6382, Part 1, specification for flexible intermediate bulk containers designed to be lifted from above by integral or detachable devices.

Safety in use of dangerous goods. In addition to the regulations concerning the *conveyance* of packages containing dangerous goods, other regulations (or the same regulations) make provisions designed to ensure *safety* of those goods *in use*. These provisions impinge on packaging standards to a considerable degree. Legislation relating to the size, layout, and content of precautionary labels or markings is outside the scope of this section, but two related subjects directly affect packaging standards: child-resistant closures, and tactile-danger warnings.

In the United States, under the Poison-Prevention Packaging Act, child-resistant closures are required on a number of products (see Child-resistant packages). Enforcement is the responsibility of the Consumer Product Safety Commission (CPSC), whose requirements are in the Code of Federal Regulations. The testing methods for the child-resistant packages (CRCs) have been developed by the American Society for Materials and Testing (ASTM). The ASTM standards have been adopted by ANSI in MH12.20 to MH12.27 inclusive.

The European Economic Community (EEC) has a family of Directives based on 67/548/EEC, "Classification, packaging and labeling of dangerous substances," usually known as the Substances Directive. The Sixth Amendment (33) to the Substances Directive, in the new Article 15 (2) provides that Member States *may* prescribe (i) the use of seals which are irreparably damaged when opened, (ii) child-resistant fastenings, and (iii) tactile warnings of danger. As a consequence a number of EEC member states such as France, Federal Republic of Germany, and the United Kingdom have national standards relating to the testing of CRCs, although not necessarily similar to the ASTM/ANSI standards. Other countries also have standards in this area and a proposal is before ISO with a view to harmonizing the provisions.

On tactile-danger warnings, work is being carried out within CEN to develop a suitable European standard. If and when agreement is reached, the EEC Commission and/or the governments of its member states may adopt the standard, particularly where a national government has exercised its option under Article 15 (2) to introduce a requirement for tactile warnings of danger to be incorporated in the design of the package or its closure.

Dimensional Coordination

In recent years considerable attention has been paid to the reduction of transport costs by determining the optimum way of assembling small packages into transport packages, transport packages into unit loads, and unit loads into vehicle loads. Numerous computer programs have been written to deal with this complex problem, which has also engaged the attention of standards bodies, official and unofficial, without an agreed view emerging. The differences relate not only to perceived objectives, but to the major variations in size of transport vehicles, rail and road tunnels, and other infrastructure in the various regions of the world which are apparently irreconcilable.

The problems were discussed at length in the 1978 report (4) but the basic differences of opinion concern the choice of module; whether it is to be based on a pallet of given dimensions or derived from the ISO series 1 freight container.

The choice of 800 × 1200 mm (31.5 × 47.24 in.) as the deck size for the flat pallet exchange pools of the various European state railways (14) led to the European Packaging Federation's publication in April 1960, "Standardization of packaging dimensions according to a modulus system based on the ISO standard pallet sizes" (see Pallets, wood). The so-called Golden Module of 400 × 600 mm (15.75 × 3.66 in.), now widely used in the grocery trade in both Western and Eastern Europe, will fit four on a European rail pool pallet and five on the 1000 × 1200 mm (39.37 × 47.24 in.) pallet which is used extensively in road transport. It is this module which is used in ISO 3394 (see Table 5). The very detailed pallet standard for use in the European pallet-exchange pools is set out in UIC leaflet 435-2 (14).

In Australia a pallet-hire pool has been in use since World War II and accounts for the majority of movements within Australia. The deck size is 1165 × 1165 mm (45.87 × 45.87 in.) and the pallets conform to the SAA standard 2068. For exports and coastwise shipments by freight container, however, the alternative module (based on the ISO series 1 freight container) is used; the deck size is 1100 × 1100 mm (43.31 × 43.31 in.) and the pallets conform to the SAA standard 1899.

The desirability of reconciling the established practice in the European grocery and consumer goods trade with the needs of nations such as Australia, Japan, the United Kingdom, and the United States, which rely heavily on the ISO container for long distance shipments, was recognized in ISO TC122/SC1. The resultant standard ISO 3676 (see Table 5) is inevitably a compromise. Unfortunately ISO TC.51 has not yet been able to agree on a similar compromise on deck sizes of pallet for an ISO standard which is necessary to replace the ISO Recommendations R.198 and R.329 of 1961 and 1963 respectively, which have been withdrawn.

Possibly the most detailed type D standards relating to dimensional coordination are those of ANSI, eg,

ANSI MH10.1, Unit load sizes for dimensioning transport package sizes.
ANSI MH10.2, Transport package sizes for ANSI MH10.1, unit load sizes.

Future Needs

Although the requirements of the market place are of paramount importance to successful commercial and industrial activity, the Helsinki Consultation (30) expressed particular concern over the position of developing countries as is evident from the undermentioned resolutions:

1. The Consultation is of the opinion that establishing standards for packaging of specific products is of considerable importance to all countries because: it should ensure optimum use of vital resources; it should protect the consumers' interests; and it should contribute to an upgrading of technology within the country.

   Many developing countries do not possess the specialized knowledge and financial resources to develop standards related to packaging of specific products. The Consultation therefore recommends the initiation of a project to identify such products, having special importance for developing

countries, eg, fresh fruit and vegetables, processed foods, etc.

3. Considering the relatively small number or total absence of packaging standards in developing countries the Consultation considers it important to establish the guidelines to serve as a basis for the formulation of national standards for packaging in developing countries for the most important sectors. These guidelines should also facilitate coordination of efforts at an early stage, thus minimizing the need for subsequent requirements for harmonization of such standards between different countries.

5. The Consultation expresses serious concern about national or regional packaging standards which might be considered as nontariff barriers to exports. It recommends that utmost caution should be exercised in creating such standards and that particular consideration should be taken of the practical possibilities for developing countries to comply with such standards in world trade.

The International Trade Centre (UNCTAD/GATT), the United Nations Industrial Development Organization, the various UN Regional Economic Commissions and, of course, the official standards bodies themselves all do valuable work but it is difficult to see how these desirable objectives can be achieved without a massive increase in resources and good will from all parties.

BIBLIOGRAPHY

1. *ISO Guide 2. General Terms and Their Definitions Concerning Standardization and Certification*, International Organization for Standardization, Geneva, Switzerland.

2. *A Standard for Standards. BSO: Part 1. General Principles of Standardization*, British Standards Institution, London, UK.

3. C. Swinbank, *Packaging of Chemicals*, Newnes-Butterworths, London, UK, annual publication.

4. C. Swinbank, *Co-ordination of National, Regional, and International Standardization in Packaging, Report ITC/CONF/P/4*, International Trade Centre (UNCTAD/GATT), Geneva, Switzerland, 1978.

5. *ISO Memento*, International Organization for Standardization, Geneva, Switzerland, annual publication.

6. *ISO Catalogue*, International Organization for Standardization, Geneva, Switzerland, annual publication.

7. *ISO Technical Programme*, International Organization for Standardization, Geneva, Switzerland, annual publication.

8. *ISO 780, 1983, Packaging, Pictorial Marking for Handling of Goods*, International Organization for Standardization, Geneva, Switzerland.

9. *CEN Annual Report*, European Committee for Standardization, Brussels, Belgium.

10. *BSI Catalogue*, British Standards Institution, London, UK, annual publication.

11. *Information Sources on the Packaging Industry*, United Nations Industrial Development Organization, Vienna, Austria.

12. *Packaging Conditions for Goods in International Traffic, U.I.C. Code 260-1*, International Union of Railways, Switzerland.

13. *Loading Recommendations to Prevent Damage to Goods, U.I.C. Code 426 R*, International Union of Railways, Switzerland.

14. *Flat Pallets and Box Pallets, Tariff conditions, etc, U.I.C. Leaflet 277 OR; Characteristics of Loading Pallets Used in International Traffic, U.I.C. Leaflet 435-1; Standard for European Flat Pallet, Made of Wood, 800 × 1200 mm, U.I.C. Leaflet 435-2; Standard for European Box Pallet, Made of Steel, 800 × 1200 mm, U.I.C. Leaflet 435-3; International Union of Railways*, Switzerland.

15. *Uniform Freight Classification*, Railroad Freight Classifications, Chicago, Ill.

16. *National Motor Freight Classification*, American Trucking Associations, Washington D.C.

17. *Code of Federal Regulations, Title 21, Food and Drugs.*

18. *Council Directive 76/893/EEC*, Commission of the European Communities, Brussels, Belgium.

19. *Codex Alimentarius Commission* (Joint FAO/WHO Standards Program), Rome, Italy.

20. *Resolution No. 222, TRANS/GE. 11/3 (AGRI/WP.1/8) Rev. 1*, Economic Commission for Europe, Geneva, Switzerland, 1977.

21. Recommendation of the Council, Organization for Economic Cooperation and Development, Paris, France, 1976.

22. *Official Journal of the European Communities*, Brussels, Belgium.

23. *Code of Federal Regulations, 49 CFR.*

24. *IATA Restricted Articles Regulations, 23rd ed.*, International Air Transport Association, Geneva, Switzerland.

25. *Transport of Dangerous Goods. Recommendations of the Committee of Experts on the Transport of Dangerous Goods. (ST/SG/ AC.10/1/Rev. 3, 1984)*, United Nations, New York.

26. *Technical Instructions for the Safe Transport of Dangerous by Air*, International Civil Aviation Organization, Quebec, Canada.

27. *International Maritime Dangerous Goods Code*, International Maritime Organization, London, UK.

28. *International Regulations Concerning the Carriage of Dangerous Goods by Rail (RID)*, l'Office Central des Transports Internationaux par Chemins de Fer (OCTI), Berne, Switzerland.

29. *European Agreement Concerning the International Carriage of Dangerous Goods by Road (ADR)*, UN Economic Commission for Europe, Geneva, Switzerland.

30. *Report of the ITC/WPO International Consultation on Packaging, Helsinki, (ITC/CONF/P/8)*, International Trade Centre (UNCTAD/GATT), Geneva, Switzerland, 1978.

31. *TR KTC 001*, No. 144, Verkehrsblatt, Vol. 12, 1978.

32. *RTK 001*, No. 89, Verkehrsblatt, Vol. 6, 1976.

33. *Council Directive 79/831/EEC*, (O. J. No. L 259) Commission of the European Communities, Brussels, Belgium.

COLIN SWINBANK
Consultant

STAPLES

In packaging, staples are used to assemble wooden pallets (see Pallets, wood), skids, and boxes and crates (see Boxes, wood). In fastening wood-to-wood, the relatively heavy shank of a nail provides greater shear strength, but staples provide greater holding power because their two legs provide more wood-to-metal contact surface area. Staples are used as a replacement for nails in many applications including furniture, housing, fencing, etc, but the following discussion focuses on packaging applications only.

The staple has three basic parts: the crown, two legs, and two points (see Fig. 1). In selecting the proper staple, four factors must be considered: wire size (gauge), crown width, leg length, and type of point.

Wire size. Heavy-duty staples are designated by the gauge of the wire prior to manufacturing the staples.

$$16 \text{ ga} = 0.062 \text{ in.} = 1.587 \text{ mm}$$
$$15 \text{ ga} = 0.072 \text{ in.} = 1.829 \text{ mm}$$
$$14 \text{ ga} = 0.080 \text{ in.} = 2.032 \text{ mm}$$

In packaging applications, 16 gauge is the most common. Heavier gauge staples are used if a particularly strong combi-

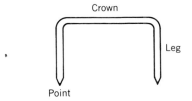

Figure 1. Three basic parts of the staple.

nation of holding power and shear strength is required, as in the assembling of some heavy-duty skids. Lightwire staples are designated by the width of the wire after manufacturing; for example, "30" is equivalent to 0.030 in. (0.762 mm) finished width; "50" is equivalent to 0.050 in. (1.270 mm) finished width. A "50" staple is heavier and more power is required to drive it.

Crown width. The widths most common in packaging are shown in Figure 2.

The staple used most frequently for packaging applications is the heavy-duty ½-in. (1.3-cm) crown. The wide-crown 15/16-in. (2.4-cm) heavy-duty staple is used to prevent pull-through on attachment of porous materials; for example, fastening a corrugated box to a wooden skid. The wide-crown staple is also used over steel strapping to secure it in place. Lightwire staples are used to attach identification tags to the outside of wooden boxes. Plastic liners are attached to boxes with lightwire staples as well. These are just a few of the numerous applications for heavy-duty and lightwire staples in packaging.

Legs. A staple has two legs of equal length. The staple must be long enough to penetrate the receiving member with sufficient length to adequately hold two pieces of material together. The stress of the application must be considered in

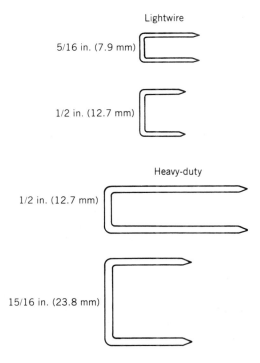

Figure 2. Common crown widths.

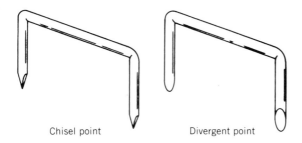

Chisel point   Divergent point

Figure 3. Staple points.

Table 1. Strapping Materials, Comparative Strength

Strapping material	Break strength, lbf (N)	Tensile strength, psi (MPa)
steel	1170 (S204)	117,000 (806)[a]
polyester	600–800 (2669–3559)[b]	60,000–80,000 (4 ± 4–551)[b]
nylon	630 (2802)	63,000 (434)
polypropylene	500–600 (2224–2669)[b]	50,000–60,000 (345–414)[b]

[a] There is also a premium-grade steel material with a tensile strength of 145,000 psi (1000 MPa).
[b] Range results from different characteristics of materials from different suppliers.

order to determine the minimum allowable penetration of the staple in order to do the job.

Points. Staple points are cut at different angles to meet various application requirements (see Fig. 3). A divergent point causes the staple legs to divert in opposite directions when it is driven into the wood. This type of point is advantageous when clinching is required. Clinching is the process of bending the staple points over the opposite side of the work piece. The staple legs penetrate the wood members then strike a steel plate diverting the points back into the wood. The chisel point, which is designed to cut through the wood without diverging, is the most common in the industry.

BIBLIOGRAPHY

General References

Products 1, Paslode Company, Lincolnshire, Ill. 1979.
Paslode Products, Paslode Company, Lincolnshire, Ill. 1983.

J. R. Charters
Paslode Corporation

STRAPPING

Strapping is a material used throughout industry to preserve the integrity of a package or load during handling, shipping, and/or storage. Strapping is used to close cartons, bundle or bale items, unitize pallet loads of materials, or brace shipments of goods during transit.

There are four primary types of strapping materials being manufactured today: steel, nylon, polypropylene, and polyester. A choice of the best strapping material for a particular application requires not only knowledge of the characteristics of the types of strapping, but also the characteristics of the packages or loads to be strapped.

Strapping Materials

Several characteristics highlight the essential differences among steel, nylon (see Nylon), polypropylene (see Polypropylene), and polyester (see Polyesters, thermoplastic) strapping. These characteristics affect the packaging performance of strapping most directly: (1) strength; (2) working range; (3) retained tension; and (4) elongation and recovery.

For the sake of uniformity, the following comparisons are based on strapping samples with single cross sections of $1/2 \times 0.020$ in. (13×0.5 mm). The performance characteristics were evaluated under the typical environmental conditions of 72°F (22°C) and 50% rh.

Strength. The strength of strapping materials is measured in two ways: break strength and tensile strength. Table 1 illustrates that most steel strapping has much higher break and tensile strength than the strongest plastic strapping material. If a package or load is heavy enough to require a strapping with great strength, and if it is strong enough to withstand the application of strapping at high tension, steel strapping is often the best selection. On the other hand, if a package is not heavy and strong, one of the plastic strapping materials will probably render effective protection. However, given the relatively small strength difference among the plastic materials, other criteria must be examined before selecting the proper plastic strapping.

Working range. Working range can be defined as the range of applied tension. The minimum may be virtually zero on a fragile package and the maximum is the highest tension level at which strapping is applied in actual situations. It is within the working range that strapping does nearly all its work securing a package or load. Although impact during handling or shipping may sometimes subject strapping to tensions above the working range, the package itself, not the strapping, usually fails under such conditions.

Figure 1 shows that steel strapping is applied at upper-level tensions of only 700 lbf (3114 N), although its break strength is much higher, 1170 lbf (5204 N). Polyester, nylon, and polypropylene strappings of the same dimension may exhibit break strengths of 600 lbf (2669 N) or more, but they are applied at tensions with typical upper levels of only 300, 250, and 200 lbf (1334, 1112, and 890 N), respectively. Strapping application equipment limitations help explain the wide difference between each material's break strength and its applied tension level. Often, the characteristics of the package or load also restrict applied tension to lower levels.

Retained tension. Strapping is nearly always applied under tension, and retaining that tension over a period of time is important. Figure 2 shows that the ability to retain applied tension on a rigid (noncompressible) package differs with each type of strapping material. Not shown in Figure 2 is a supplementary 4-h test at 140°F (60°C). The test shows no effect on the retained tension of steel strapping, but it reduces the retained tension of polyester and nylon strapping by nearly 50%. The polypropylene strapping retains about 25% of its tension under the same conditions. In addition, if high humidity is present it has an adverse effect on the retained tension of nylon strapping, but not on that of steel, polyester, or polypropylene.

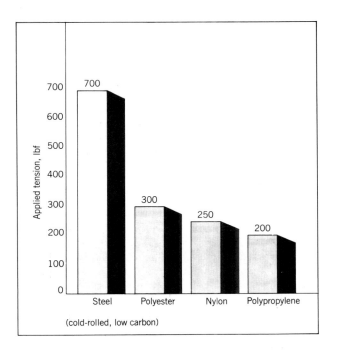

Figure 1. Typical working ranges of $1/2 \times 0.020$ in. (13–0.5 mm) strapping. To convert lbf to N, multiply by 4.448.

If a rigid (noncompressible) package or load requires maximum retained tension to provide sufficient protection, steel strapping is the best choice, followed by polyester and nylon. However, most lightweight packages require only limited amounts of retained tension to preserve their integrity, and polypropylene strapping is often adequate.

Elongation and recovery. Vibration, handling, elapsed time, temperature, and humidity cause many types of packages and loads to shrink or settle. All three plastic strapping materials elongate under tension and seek to recover a part of that elongation as stress is relieved. Elongation recovery enables plastic strapping to contract as the package or load shrinks. In contrast to the three plastic strapping materials, steel strapping elongates a negligible amount, but exhibits 100% recovery.

Plastic strapping varies widely in its capacity to elongate and recover. The data in Figure 3 show the inches of recovery in 10-ft (3-m) samples of various plastic strapping materials. Nylon is by far the best performer. In a 10-ft (3-m) sample, it elongates slightly more than 10 in. (0.25 m) under tension and recovers nearly 9 in. during the five days after removal. Polyester elongates 2.3–4 in. (5.84–10.2 cm) and recovers 2–3.6 in. (5.1–9.1 cm). Polypropylene elongates 4.2–7.2 in. (10.7–18.3 cm) and recovers 3.5–5.4 in. (8.9–13.7 cm)—less than nylon but more than polyester. Figure 3 shows the relationship that occurs when each material is tensioned to the upper level of its own working range. At the more common tension level of 200 lbf (890 N), the relationship remains similar.

Before making a choice among plastic strapping materials on the basis of elongation recovery, two things should be determined: the amount of elongation recovery required; and how much tension can be applied to the strapping without damaging the package or load.

If a large amount of elongation recovery is required and the package can withstand up to about 250 lbf (1112 N) of applied tension, nylon is the clear choice unless high humidity is present. If only a modest amount of elongation recovery is needed and the package can withstand 300 lbf (1334 N) of

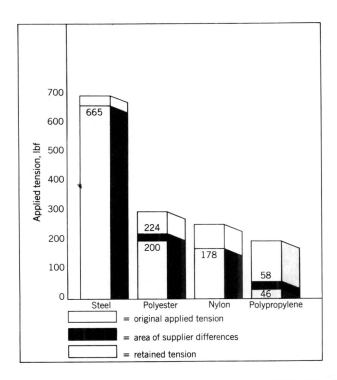

Figure 2. Retained tension after 5 days at 72°F (22°C), 50% rh. To convert lbf to N, multiply by 4.448.

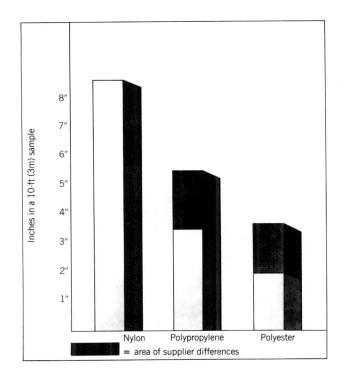

Figure 3. Elongation recovery of plastic strapping 5 days after removal from tension at upper level of individual working ranges, at 72°F (22°C), 50% rh. To convert in. to mm, multiply by 25.4.

Table 2. Comparison of Strapping Qualities[a]

Strapping material	Break strength ($\frac{1}{2} \times 0.020$ in. or 13×0.5 mm)	Working range tension	Retained tension[b]	Elongation recovery	Heat resistance	Humidity resistance	Ease of disposal
polypropylene	moderate	lowest	fair	high	fair	high	excellent
polyester	moderate	moderate	good	fair	good	high	excellent
nylon	moderate	moderate	good	highest	good	low	excellent
steel	highest	highest	highest	negligible	excellent	high[c]	fair

[a] Disregarding supplier differences.
[b] Within the working range, rigid package.
[c] Except for rusting.

tension, polyester is a more economical alternative. If a package can withstand only 200 lbf (890 N) of tension, but elongation recovery is important, nylon ranks first; polypropylene and polyester rank second and third, respectively, with considerable difference among brands.

When a package or load simply cannot bear much tension, polypropylene offers the most cost-effective elongation recovery. Humidity has no effect on the elongation recovery of polyester or polypropylene strapping, but does have an adverse effect on nylon. Finally, temperature has an adverse effect on all plastic strapping, but to the greatest degree on polypropylene. (For a complete comparison of all primary strapping materials, see Table 2.)

Characteristics of Packages and Loads

Knowledge of the characteristics of a given package or load is essential in the selection of the proper strapping material. Loads and packages can be divided into five basic categories:

Rigid. Rigid (noncompressible) loads undergo few dimensional changes during strapping, handling, shipping, and storage. Unitized steel plates and bricks are good examples.

Expanding. The circumference of an expanding load decreases when strapping is applied under tension. Cotton bales are an example.

Shrinking. Shrinking loads settle or shift after strapping. Green lumber, because it dries after strapping, is an example.

Compressible. Bundles of newspapers and telescoping cartons filled with apparel are strapped while under compression. Compressible loads rebound somewhat when compression is removed.

Combination. A combination load compresses when strapped and expands following compression. The load shrinks when other loads are stacked on top of it and rebounds after the weight is removed. Corrugated sheets are an example.

In addition, other factors must be considered before choosing a strapping material. All loads and packages must be evaluated in terms of shape, contour, weight, stability, and method(s) of handling and shipping. When all the variables are determined, a strapping selection can be made on the basis of the aforementioned characteristics.

A choice of the right type of strapping for a particular load requires specialized knowledge. That choice can best be made by consulting with companies that manufacture or distribute the nearly 200 types and sizes of strapping material available today and have the necessary technical-systems evaluation expertise.

Applications

Although careful, professional examination of a particular application is always recommended, some general guidelines can help define the most common uses of the four types of strapping material.

Steel. Steel strapping is commonly used for unitizing very heavy loads or bracing loads inside railcars, trailers, and overseas containers. Steel strapping is seldom used for shrinking loads, but it has the strength to hold highly compressed loads and is a frequent choice for unitizing rigid loads. In general, steel strapping is used where high strength and high retained tension are required. It has the highest tensile strength of all the strapping materials, the least tension decay, and for all practical purposes does not elongate.

Nylon. Nylon strapping has high retained-tension capabilities and is commonly used to unitize shrinking loads of heavy items and loads that can withstand high initial tension. It also elongates and recovers more tension than polyester or polypropylene strapping. Nylon strapping is also the most expensive plastic strapping material.

Polypropylene. Polypropylene is the least expensive strapping material. It is generally used for lighter-duty unitizing, bundling, and carton closure. Polypropylene strapping has high elongation and elongation recovery, but does not retain tension as well as other plastic strapping materials.

Polyester. Polyester offers the greatest strength and highest retained tension of the plastic strapping materials. It remains tight on rigid loads that require strapping tension throughout handling, shipping, and storage. Often, polyester performs as well as lighter sizes of steel strapping. Polyester strapping is commonly used in many applications where high tensile strength, high retained tension, and elongation are required.

Another consideration is the cost of strapping materials. In equal sizes, polypropylene is the least costly, followed by polyester, nylon, and steel. Simple material cost, however, does not take performance into consideration. In the end, the most economical strapping material is the one that produces the highest cost:benefit ratio, that is, the best performance at the lowest cost.

BIBLIOGRAPHY

General References

Standard Specification for Flat Steel Strap and Connectors (ASTM D 3953-83), American Society for Testing and Materials, Philadelphia, Pa., 1983.

Federal Specification for Strapping, Steel and Seals (QQ-S-7814), Federal Supply Service, General Service Administration, Washington, D.C., 1974.

Standard Specification for Strapping, Nonmetallic and Connectors (ASTM D 3950-80), American Society for Testing and Materials, Philadelphia, Pa., 1983.

Federal Specification for Strapping, Nonmetallic and Connectors (PPP-S-760B), Federal Supply Service, General Service Administration, Washington, D.C., 1973.

HERMAN WEIMER
Signode Corporation

STERILIZATION METHODS.   See Health-care packaging; Canning, food.

STOPPERS.   See Closures.

STRETCH FILMS.   See Films, stretch.

STYRENE–BUTADIENE COPOLYMERS

The styrene–butadiene copolymers (SB) that are suitable for packaging applications are those resinous block copolymers that typically contain a greater proportion of styrene than butadiene and that are polymodal with respect to molecular weight distribution. These copolymers, produced by solution polymerization processes using sequential multiple additions of an organolithium initiator, styrene, and butadiene (1,2), are amorphous in nature and transparent. Compared to other transparent styrenic polymers (e.g., general-purpose polystyrene and styrene–acrylonitrile copolymers), the styrene–butadiene copolymers offer outstanding shatter resistance (3). Physical properties are dependent on molecular weight distribution, styrene:butadiene ratio, and the nature of the block structure. Since all three of these parameters are a function of the polymerization recipe, grades having different properties can be produced: Grades tailored for injection molding (see Injection molding) and for extrusion (see Extrusion) are commercially available (see Table 1).

One of the major applications for SB is disposable cups and lids, but a combination of good mechanical and visual properties, low density compared to other transparent polymers, and ease of fabrication by the typical molding and extrusion techniques make styrene–butadiene copolymers a good choice for many packaging applications. These include bottles and jars, hinge boxes (ball-and-socket or integral-living hinges) (see Boxes, rigid plastic), tubs and lids, blisters and skin packaging (see Carded packaging), overwrap and shrink film (see Films, plastic).

Styrene–butadiene copolymers are often blended with other polymers (4). If transparency is a requirement, the blending polymer must have a refractive index very near that of the styrene–butadiene copolymer. This limits the choice to styrenic polymers such as general-purpose polystyrene (GPPS) (see Polystyrene), styrene–acrylonitrile copolymers (SAN) (see Nitrile polymers), and styrene–methyl methacrylate copolymers (SMMA). GPPS is used extensively as a blending polymer for applications requiring sheet extrusion and thermoforming (see Thermoforming) (eg, carded packaging). Blends with SAN or SMMA are mainly used for injection

molded applications (e.g., hinged boxes, video cassette cases). The impact strength of SB can be either enhanced or reduced when blended with other styrenic polymers. For example, a significant loss of impact strength occurs with the addition of even low levels of GPPS. In contrast, certain blends of SB and SMMA have synergistic impact strength. Therefore, the suitable blend ratio for a given application depends on the required combination of stiffness, toughness, and economics. Blend ratios typically range from 40:60 to 80:20 SB:styrenic polymer. If clarity is not a requirement, blends with polymers such as high impact polystyrene (HIPS), polypropylene (PP), and polycarbonate (PC) can provide synergistic enhancement of certain physical properties, unusual aesthetics such as gloss and pearlescence, or reduced cost. Bottles and jars can be produced from nonblended SB or from blends of SB with GPPS or PP by extrusion or injection blow molding (see Blow molding). Blown and cast films can be produced from nonblended SB or blends of SB/GPPS. SB is also suitable for coextrusion (see Coextrusion, flat; Coextrusion, tubular).

U.S. produced styrene–butadiene resinous copolymers meet FDA specifications (CFR 177.1640). Certain grades are fully documented to qualify for U.S. Pharmacopoeia Class VI-50 medical applications and blood contact uses, and have been shown to be nonmutagenic, nonirritants to sensitive tissues, and nondestructive to living cells. In view of these qualifications and because they can be sterilized by both gamma irradiation and ethylene oxide (ETO) procedures, these copolymers are widely used in medical packaging applications.

There are currently five producers of styrene–butadiene copolymers throughout the world (see Table 2). The 1985 selling price in the United States was $0.70–0.75/lb ($1.54–1.65/kg).

Styrene–butadiene copolymers have no demonstrated toxicity in laboratory studies, and the level of machine-side toxic fumes detected during typical fabrication processes is well below OSHA standards. As with most hydrocarbon based polymers, SBs burn under the right conditions of heat and oxygen supply. Combustion products of any hydrocarbon based material should be considered toxic.

Table 1. Styrene–Butadiene Copolymer Physical Properties

Process	Injection molding and extrusion	Injection molding
specific gravity	1.01	1.01
flow rate, condition G, g/10 min	8.0	8.0
tensile yield strength, psi (MPa)	4,000 (28)	4,400 (30)
elongation, %	190	20
flexural yield strength, psi (MPa)	5,500 (38)	6,400 (44)
flexural modulus, psi (MPa)	191,000 (1317)	215,000 (1483)
falling dart impact, gf/20-mil sheet (N/mm)	375 (7.2)	100 (1.9)
hardness, shore D	63	75
heat deflection temperature, 264 psi (1.8 MPa), °F(°C)	157 (68)	170 (77)
vicat softening point, °F(°C)	185 (85)	200 (93)
light transmission, %	90–91	90–91
moisture absorption, 24 h, %	0.09	0.08

Table 2. Producers of Styrene–Butadiene Resinous Copolymers

Company	Location	Tradename
Phillips 66 Company	U.S.	K-Resin
Firestone	U.S.	Stereon
BASF	Federal Republic of Germany	Styrolux
Denka-Kaguku	Japan	Clearene
Asahi Chemical	Japan	Asaflex

BIBLIOGRAPHY

1. U.S. Pat. 3,639,517 (Feb. 1, 1972), A. G. Kitchen and F. J. Szalla (to Phillips Petroleum Company).
2. U.S. Pat 4,091,053 (May 23, 1978), A. G. Kitchen (to Phillips Petroleum Company).
3. L. M. Fodor, A. G. Kitchen, and C. C. Baird, "K-Resin BDS Polymer: A New Clear Impact-Resistant Polystyrene," in D. Deanin, ed., *New Industrial Polymers, ACS Symposium Series 4,* American Chemical Society, Washington, D.C., 1972, pp. 37–48.
4. G. M. Swisher and R. D. Mathis, "A Close-up of Blends Based on Butadiene-Styrene Copolymer," *Plastics Engineering,* **40**(6), 53–56 (June 1984).

R. D. Mathis
Phillips 66 Company

SURFACE AND HYDROCARBON-BARRIER MODIFICATION

The surfaces of materials used in packaging applications are often modified to improve the properties of the container. The reasons for modification and the methods used vary widely. Black plate to be used in steel cans is electrolytically coated with a thin layer of tin, which aids in soldering the side seam and helps protect the outside of the can from corrosion (see Cans, steel). Tin-free steel has a thin coating of chromates, phosphates, or aluminum to protect the exterior surface from corrosion (see Tin-mill products). Aluminum cans do not need an exterior coating, since a thin layer of aluminum oxide forms automatically on the surface, protecting the bulk of the material (see Cans, aluminum).

Thin coatings are applied to glass containers just before annealing (hot-end coating) and just after annealing (cold-end coating) (see Glass container manufacturing). The hot-end coating is usually tin chloride or titanium chloride, which improves adhesion of the cold-end coating and hardens the surface to make it more difficult to scratch. The cold-end coating may be silicone, stearates, polyethylene, or other compounds that lubricate the surface to decrease friction, which in turn improves scratch resistance. Because of the surface-sensitive nature of glass, any process that will decrease the number of scratches on a container will increase its strength.

Laquers and varnishes are applied to the surface of folding cartons (see Cartons, folding) and paper labels (see Labels and labeling) to give a glossy appearance and to improve abrasion resistance. The top and bottom surfaces of multiwall bags (see Bags, paper) and corrugated shippers (see Boxes, corrugated) can be coated to increase their coefficient of friction, thereby helping to maintain unity of pallet loads. Fiber surface treatments that cause covalent and ionic cross-linking (fiber-to-fiber primary bonding, instead of the normal weak secondary bonding) can increase the wet strength of materials manufactured from cellulose fibers, such as kraft paper (see Paper) (1). Adhesion of labels and printing inks (see Inks) to plastic containers (especially polyolefins) can be improved by pretreating, and sometimes posttreating, the containers using flame, corona discharge, or gas plasma techniques.

Barrier properties of materials or containers can be improved with surface coatings or surface treatments which vary in thickness from $4 \times (10^{-3}–10^{-7})$ in. [$1 \times (10^{-4}–10^{-8})$m]. Nonselective barriers, such as metallizing (see Metallizing) decrease transmission rates of all gases and vapors. Metallizing also provides a conductive surface that is easily grounded to dissipate static charges. Other barrier materials are selective with respect to which gas and vapor transmission rates will be decreased. Vinylidene-chloride copolymer (see Vinylidene-chloride copolymers), generally called PVDC, is applied as a coating primarily for its oxygen-barrier properties, although it is also a good barrier to water vapor and to many other gases and vapors (see Barrier polymers). Nitrocellulose coatings are often used on cellulosic materials (see Cellophane) to reduce their water-vapor transmission rates. Fluorination or sulfonation of the surface of a polyethylene container improves its hydrocarbon-barrier properties.

Although not a surface treatment, another means by which hydrocarbon transmission rates can be decreased is by molding a container from a blend of polyethylene and nylon. Films, sheets, and bottles can be coextruded to provide a hydrocarbon-barrier layer. It is apparent that the reasons for surface modification are extremely varied. This article deals only with methods for improving adhesion and hydrocarbon-barrier properties. The methods discussed are summarized in Table 1.

Adhesion Improvement

In order to improve the adhesion of printing inks, coatings, and labels to plastic films and containers, the surface can be cleaned by oxidation. For example, corona-discharge techniques improve the adhesion of uv varnishes to inks, which often have a high wax content, on folding cartons, so the varnish can be applied evenly. This prevents separation during

Table 1. Summary of Methods Used for Improvement of Adhesion and Hydrocarbon-Barrier Properties

Method	Improves	Comments
corona discharge	adhesion	sophisticated, high treat levels possible
gas plasma	adhesion	sophisticated, high treat levels possible
flaming	adhesion	commonly used for plastic bottles
fluorination	hydrocarbon-barrier	post-treat or in-mold treat possible
sulfonation	hydrocarbon-barrier	post-treat, some surface yellowing
polymer blends	hydrocarbon-barrier	polyethylene–nylon, water barrier decreases
coextrusion	hydrocarbon-barrier	films, sheets, bottles, generally containing EVOH, PVDC, or nylon
coating	hydrocarbon-barrier	films, sheets, bottles, generally coated with PVDC

folding and gluing; the varnish will adhere to the adhesive used for the carton flaps and side seams. Also, the varnish will not separate when a pressure-sensitive price label is applied in a retail store (2).

In addition to removing dust, oils, greases, processing aids, mold-release agents, etc, the surface is activated. Its energy is increased, so its wettability is enhanced. The formation of carbon–carbon double bonds and carbonyl ($>$C=O) and hydroxyl—OH) groups produces a surface with a higher electron density (polarity) and thus better bondability. In general, two polar surfaces form a stronger bond than one polar and one nonpolar surface or two nonpolar surfaces, since dipole interactions are involved. Although the mechanism of adhesion improvement is not completely understood, experimental evidence points to changes in the chemical nature of the surface (oxidation and chain scission), in addition to morphological changes at the surface (increased amorphous fraction) and in layers below the surface (increased percent crystallinity) (3).

More sophisticated surface-treating techniques are being used because fire and safety hazards are associated with the flame treatment of containers (4) and because the newer water-based inks and coatings require higher treating levels than can be achieved with flaming (5). If too high a treat level is used, problems can be encountered; for example, in heat-sealing films (see Sealing, heat). Optimization of the treat level is required.

Corona-discharge techniques. A plastic film (web) or sheet moves between an electrically grounded roller and an electrode maintained at high voltage. The air between the two surfaces is ionized and a continuous arc discharge (corona) is generated at the surface of the film or sheet, which cleans, oxidizes, and activates the surface. The process is performed in air at atmospheric pressure and elevated temperatures (6,7). Ozone is a by-product of the corona-discharge method, so its removal from the workplace is required. This can be accomplished with fans that supply fresh air and remove the ozone. Catalysts can be incorporated to destroy the ozone chemically (8). The air in the treating area must be maintained at a low relative humidity, or the oxidation that occurs will result in excessive formation of low molecular weight polymer chains, exhibited as a frosted appearance on the surface (9). Multidischarge corona techniques are also being used to remove the rolling oils from aluminum foil (10).

Gas-plasma techniques. This is an automated batch process in which the containers to be treated are placed in a reaction chamber, which is then evacuated. The chamber is subsequently charged with oxygen, argon, helium, or nitrogen while a radio-frequency field ionizes the gas. A glow discharge is produced, which affects the surface of the containers in the way described for corona discharge. In this case, oxygen cleans and oxidizes the surface, and the other gases activate the surface. The process is performed at low pressures and relatively low temperatures (11).

Flame techniques. Containers are passed by a bank of flame jets. The source of energy is usually natural gas. The flame oxidizes the surface and burns off surface contaminants, such as mold-release agents, by a mechanism similar to that of corona discharge.

Hydrocarbon-Barrier Improvement

Polyolefins are widely used as packaging materials (see Polyethylene, high density; Polyethylene, low density; Polypropylene). Their low densities result in lightweight pack-

ages. Polyolefins are relatively tough materials, able to survive relatively high container drops. Because they are relatively inert chemically, resistant to most acids and bases and many organic solvents, product–package interactions are minimal. Polyolefins are low-cost materials that are easy to blow mold (see Blow molding), and regrind can be added without significant loss in properties. They also have potential for recycling. Polyolefins are good water-vapor barriers; however, they are poor barriers to gases such as oxygen and to vapors such as nonpolar hydrocarbons. The transmission rates of aromas, flavors, fragrances, and hydrocarbon solvents are important to the shelf life of many products.

By treating the surface of polyethylene containers with fluorine or sulfur trioxide, a very thin polar surface can be created on a nonpolar polymer, decreasing the transmission rates of nonpolar hydrocarbons, eg, gasoline, motor oil, propellants, hexane, carbon tetrachloride, and benzene, through the walls of the container. Generally, nonpolar penetrants have high transmission rates through nonpolar polymers and low transmission rates through polar polymers. The reverse is true for polar penetrants. An added benefit of increasing surface polarity is that it facilitates printing, wetting, and labeling, eg, heat-transfer labeling, (see Decorating). However, heat sealing is made more difficult because of cross-linking of the surface.

Surface fluorination. There are two basic methods of treating the surface of polyolefin containers with fluorine: after fabrication and during fabrication (blow molding) (12–14). The Surface-Modified Plastics (SMP) Fluorination System was developed by the Linde Division of the Union Carbide Corporation, which offers the technology under license. It is a batch process in which the manufactured containers are placed in a heated reaction chamber prior to evacuating it. A mixture containing a low percentage of fluorine in nitrogen is allowed to flow into the chamber. The fluorine reacts with the surface of the container, oxidizing it (fluorine is one of the best oxidants available) and creating a polar, cross-linked surface. This eliminates the need for flame, corona discharge, of gas-plasma techniques to improve printability or label-adhesion properties. The chamber is again evacuated to remove the unreacted fluorine and the containers are removed. Both the inside and the outside surface of the container are fluorine-treated in this method. Since it is a posttreat method, there is no possible corrosion of expensive tooling. The process has been cleared by the FDA for use with containers for foods and pharmaceuticals.

The other fluorination method was developed by Air Products and Chemicals, Inc., which offers the technology under license (15). It involves the use of a mixture containing a low percentage of fluorine in nitrogen in the blow gas during blow molding. The parison is prepurged with nitrogen, the blow gas is introduced to form the container and treat the inside surface with fluorine, and the container is purged to remove the fluorine mixture. Since fluorine is such a reactive molecule, fluorination of the container surface takes place at normal extrusion and blow-molding temperatures. Containers that are manufactured by in-mold fluorination have not yet been cleared by the FDA for use with foods and pharmaceuticals.

In both types of fluorine treatment, a polar fluorocarbon layer is formed that is 0.08–0.16 mil (2–4 μm) thick (15). Even this very thin layer decreases transmission rates of nonpolar hydrocarbons by factors of 10–100 and even higher, depending on the exact chemical nature of the permeating molecule (15,16). This hydrocarbon-barrier improvement is useful for fuel tanks and containers for home, automotive, agricultural, and industrial chemicals. If the vapor pressure of the contained solvent is too high, the container may balloon, and rupture, especially at elevated temperatures. On a smaller scale, container creep may thin out the barrier layer, which is already extremely thin, leading to significantly higher transmission rates than would be expected. If this is the case, a change in the structural design of the container may be necessary in order to minimize expansion and creep.

The barrier properties achieved with fluorine-treated pigmented containers are sometimes inferior to those attainable without pigment. This can be overcome by using a coextruded bottle (see Multilayer plastic bottles) with a clear layer inside and a pigmented layer outside (17,18). The fluorination of the clear plastic on the inside will provide a good barrier, even if the outer surface is not treated at all (in-mold treat) or if its fluorine-treat is not optimal (posttreat).

Surface sulfonation. A posttreat process with a mixture of sulfur trioxide in an inert gas has been developed for HDPE containers by the Dow Chemical Co., which offers the technology under license. In this case, polar sulfonic acid groups are bonded to the surface of the polyethylene. These groups are subsequently neutralized with ammonia or sodium hydroxide. Ammonia is preferred, since it can be used as a gas. The barrier properties, ie, the depth of treat, developed by this process depend on the concentration of sulfur trioxide and the time of exposure. This process has been used commercially for treating the inside surface of fuel tanks. It has potential for organic-solvent containers. Some surface yellowing occurs with this process, but this has not been apparent in most experimental containers, which contained carbon black. Permeation rates of gasoline through treated containers were less than 1% of those for identical untreated containers (19). Containers that are manufactured by this process have not yet been cleared by the FDA for use with foods and pharmaceuticals.

Polymer blends. Polyethylene is a good water-vapor and polar-hydrocarbon barrier, but a poor barrier to nonpolar hydrocarbons. Nylon is a good barrier to oxygen and nonpolar hydrocarbons, but a poor barrier to water vapor (see Nylon). Selar (DuPont) resin is a blend of a special nylon resin (5–18 wt%) and polyethylene (12–14,20). Bottles can be blow molded from Selar resin with only minor modification of the extrusion screw in a conventional blow-molding machine. Up to 80% regrind can be used. The nylon forms a maze of overlapping, discontinuous barrier plates integral to the end product (21). The blend provides the good properties of both resins without the need for coextrusion or surface treatment with toxic or corrosive chemicals. Compared to polyethylene, the improvement in solvent-, gas-, and odor-barrier properties is comparable to that provided by the fluorination processes (21,22). Strength, toughness, and heat resistance are also improved, but WVTR is higher than that of a polyethylene container of the same configuration. Containers can be pigmented with no adverse effect on their properties.

Potential package forms for this resin are bottles, pails and drums, injection-molded containers, aerosols, and films. All of these package forms except films can be easily fluorine-treated by the posttreat process; only those which are blow molded can be treated by the in-mold process. Potential markets are the same as those targeted for the fluorine and sulfur trioxide treatment processes: agricultural, industrial, and household

chemicals; oil-based paints and cosmetics; waxes and polishes; insecticides; fertilizers; etc. These containers have not yet been cleared by the FDA for use with foods and pharmaceuticals.

Other polymer blends are being used in packaging today, and growth in this area is predicted (23). Most of the widely used blends, eg, HDPE–LDPE used for grocery bags, have been selected because of the improvement in mechanical properties. However, Continental Can has been producing an accordion-pleated bag from a blend of nylon and LDPE for use in an aerosol can. This bag must be a good barrier to hydrocarbons in order to keep the propellant separate from the product (24) (see Pressurized containers). Increased use of polymer blends is anticipated in packaging that requires a flavor or fragrance barrier, or both.

Coextrusion. Since ethylene vinyl alcohol (EVOH) (see Ethylene vinyl alcohol) and nylon (see Nylon) have good hydrocarbon-barrier properties (25–27), coextrusion with polyethylene or other plastics can be done to manufacture film, sheet, or bottles (28,29) that have low oxygen and water-vapor transmission rates (16,30,31) in addition to low transmission of hydrocarbons, eg, solvents such as toluene, xylene, and methyl ethyl ketone, (13) (see Coextrusions for flexible packaging; Multilayer plastic bottles).

Drying agents are being added to some of the layers of plastic cans (see Cans, plastic) that are being developed for retort applications (see Retortable flexible and semirigid containers) in order to maintain a low moisture content in the EVOH layer, since the permeability of EVOH increases as its moisture content is increased (32). The choice of the tie layers between the layers is usually critical to the performance of the package (33).

Coating. Vinylidene chloride copolymers (generally called PVDC) are good barriers to hydrocarbons, as well as to oxygen and water vapor. PVDC has been widely used as a barrier layer in multilayer films and sheets and is now being used as a coating on plastic bottles (24,34,35). It is applied by spray coating, roll coating, or dip coating in layers as thin as 0.08 mil (2 μm) (36,37). Other materials such as epoxies can also be used to coat containers in order to reduce hydrocarbon transmission rates, but their commercial use has been very limited (24).

BIBLIOGRAPHY

1. A. N. Neogi and J. R. Jensen, *TAPPI* **63**(8), 86 (1980).
2. J. P. Nixon, *Package Print.* **30**(8), 16 (Aug. 1983).
3. B. Catoire, P. Bouriot, O. Demuth, A. Baszkin, and M. Chevrier, *Polymer* **25**, 771 (1984).
4. R. N. Gidwani, *Am. Lab.* **15**(11), 81 (Nov. 1983).
5. T. W. Sprecher, *Pap. Film Foil Converter,* **57**(11), 114 (Nov. 1983).
6. *Flexo* **10**(4), 54, 62 (1985).
7. R. N. Gidwani, *Am. Lab.* **16**(11), 84 (Nov. 1983).
8. *Pap. Film Foil Converter* **57**(7), 50 (July 1983).
9. D. Briggs, C. R. Kendall, A. R. Blythe, and A. B. Wooton, *Polymer* **24**(1), 51 (1983).
10. B. P. Sherman, *Converting World U.K.* **11**, 6 (1985).
11. R. N. Gidwani, *Am. Lab.* **16**(11), 85, 87 (Nov. 1983).
12. *Chem. Week* **134**(11), 50 (1984).
13. *Chem. Week* **134**(11), 52 (1984).
14. *Chem. Week* **134**(11), 55 (1984).

15. *Package Eng.* **26**,(12) 64 (Nov. 1981).

16. M. A. Amini, *Packag. Technol.* **14**(2), 24 (Oct. 1984).

17. R. D. Sackett, *Food Drug Packag.* **47**(10), 34 (Oct. 1983).

18. *Mod. Plast.* **60**(7) 30 (July 1983).

19. *Dow Surface Treatment, Gasoline Barrier Improvement for HDPE Fuel Tanks,* The Dow Chemical Company, Midland, Mich., 1974, Form No. 308-184-74.

20. *Mod. Plast.* **61**(12), 96 (Dec. 1984).

21. *Mod. Plast.* **61**(1), 12 (Jan. 1984).

22. *Packag. Letter* **28**(25), 1 (1983).

23. P. R. Lantos, *Packaging* **30**(6), 59 (May 1985).

24. R. D. Sackett, *Food Drug Packag.* **47**(10), 30 (Oct. 1983).

25. H. L. Allison, *Packaging* **30**(3), 25 (Mar. 1985).

26. R. D. Sackett, *Food Drug Packag.* **47**(10), 28, 30, 33 (Oct. 1983).

27. *Mod. Plast.* 94, **61**(12), 94, 96 (Dec. 1984).

28. G. R. Smoluk, *Mod. Plast.,* **60**(12), 44 (Dec. 1983).

29. J. W. Peters and R. Heuer, *Packaging* **29**(1), 31 (Jan. 1984).

30. F. Labell and J. Rice, *Food Process.* **46**(3), 46 (1985).

31. M. A. Amini, *Packag. Technol.* **14**(2), 22 (Oct. 1984).

32. L. H. Doar, *Food Eng.* **57**(2), 59 (Feb. 1985).

33. D. P. Dempster, *Food Can.* **45**(2), 21, 37 (1985).

34. J. R. Newton, *Plast. Eng.,* **41**(5), 83 (May 1985).

35. H. L. Allison, *Packaging,* **30**(3), 26 (Mar. 1985).

36. J. W. Peters and R. Heuer, *Packaging* **29**(1), 35 (Jan. 1984).

37. G. R. Smoluk, *Mod. Plast.* **60**(12), 46 (Dec. 1983).

Mary A. Amini
Center for Packaging Engineering
Rutgers, The State University of New Jersey

T

TAGS

Tags have been used in all areas of business and industry for centuries. The more progressive merchants of the 1800s used tags in the same manner and for the same reasons as merchants today: to inform, instruct, identify, and inventory (see Fig. 1). Traditional cloth and paper materials have been joined by nonwoven and film-based synthetics (see Nonwovens; Plastic paper).

Tags used in retail environments are generally produced from cast-coated paperboard stocks ranging in thickness from 8 to 10 points (0.008–0.010 in., 203–254 μm). Type stands out and colors are brilliant on the smooth surface of cast-coated stocks. Cloth tags are frequently called law labels. This term describes labels generally sewn onto mattresses and furniture to comply with government regulations regarding the identification of stuffing materials. Tags manufactured of synthetics such as nonwoven Tyvek (DuPont Company) are gradually replacing cloth tags because they exhibit an extremely high tear strength, are insensitive to moisture, and have a smoother, more uniform printing surface than cloth. Plastic tags are generally used for advertising or instructional purposes in the retail industry.

Many fastening methods are available, but the most common in the retail industry for product identification is a knotted string (see Fig. 2(a)). Plastic barbs are generally used for price marking in retail garment applications. Manufactured from an ultrathin plastic, these barbs affix price tags to ready-to-wear merchandise to prevent price-tag switching. Plastic barbs can also be combined with a hook (see Fig. 2(b)) to hang merchandise such as socks.

In manufacturing, processors, converters, and fabricators have a variety of uses for tags. Tags play an important part in tracking serially numbered goods such as electronic components. The simple tag shown in Figure 3(a) uses transfer tape on two of the three stubs to enable the tag to travel with the component throughout the entire packaging cycle without losing track of the serial numbers. Color coding minimizes the need for processing instructions. The entire tag is affixed to the component at the end of the assembly line, using the transfer tape on the gray stub first. Prior to boxing the product, stubs two and three (white and off-white) are detached and affixed to the outside of the box. Upon receipt of shipping instructions, the off-white portion is detached and forwarded with a copy of the shipping document to inventory control.

Tags and pressure-sensitive labels are frequently combined to create unusual tag products. The tag in Figure 3(b) includes four pressure-sensitive labels, each carrying the serial number of the component, in addition to the nameplate, which in this case is affixed to the product after delivery by the installer. The tag is attached to the component with string or transfer tape, depending on the surface or characteristics of the component. The shipping department lifts up part one to reveal part two of the set which consists of a small-face slit label containing the serial number of the component and the instructions "put on shipping document." Part two is torn at the perforation along the top and reveals three additional labels with instructions for distribution (warehouse copy, distributor copy, installer copy). This unusual tag and label combination provides a simple solution to the often difficult task of tracking serially

(a)

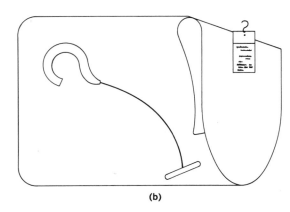

(b)

Figure 2. (**a**) Knotted-string fastener. (**b**) Plastic barb with hook.

Figure 1. Price marking tags from the 1800s.

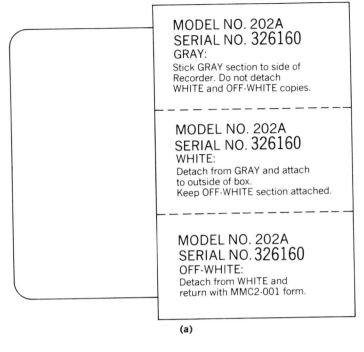

MODEL NO. 202A
SERIAL NO. 326160
GRAY:
Stick GRAY section to side of
Recorder. Do not detach
WHITE and OFF-WHITE copies.

MODEL NO. 202A
SERIAL NO. 326160
WHITE:
Detach from GRAY and attach
to outside of box.
Keep OFF-WHITE section attached.

MODEL NO. 202A
SERIAL NO. 326160
OFF-WHITE:
Detach from WHITE and
return with MMC2-001 form.

(a)

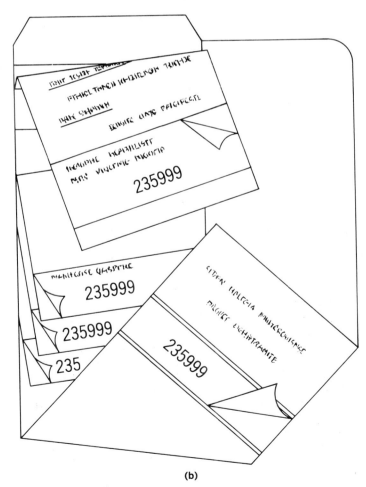

(b)

Figure 3. (a) A three-stat tag. (b) Tag and label combination.

numbered products and components from assembly through shipping to their ultimate destination.

Tag applications in the retail packaging industry include, but are not limited to product description, content identification (ie, materials used in production), care and use instructions (ie, operating and care directions, identification of potential hazards), guarantee and warranty information, decoration and promotion, inventory control, and price marking (frequently combined with inventory control when using magnetic stripes, OCR, or bar codes) (see Bar code; Code marking and imprinting).

Product marking tags, also called hang tags, are printed in up to eight colors on a variety of stocks, and include many features designed to get the attention of the potential consumer. Sizes range from small jewelry marking tags (see Fig. 4(a) to oversized tags used to identify major appliances.

Hang tags are generally shear cut, die cut, or continuous, and offer maximum versatility for product identification, sales promotion, and advertising. The three basic tag styles found in retail tagging applications are shear-cut, die-cut, and continuous tags.

Shear-cut tags. Shear-cut tags, also called shipping tags, are the oldest and most common of all tag styles. Shear-cut tags are simply sheared off the end of the web from the bindery section of the press after all manufacturing processes are completed. These tags may be simple, printed on one side with black ink, with a standard ³⁄₁₆-in. (4.8-mm) punched hole. They may also be very complex, with such features as multicolor printing on both sides. Eyelets in tags used for product identification are rarely reinforced because the life expectancy of the tag is short.

Die-cut tags. Die-cut tags are cut out of the web at the end of the bindery section of the printing press using either stock or custom manufactured dies. Die cutting produces tags that are round, have rounded or scalloped corners, or have unusual shapes as shown (see Fig. 4(b)). Die-cut tags may have all or most of the features of shear-cut tags, but usually do not require features such as large patches or jumbo numbers. They are used primarily as promotional pieces or to convey information regarding a specific product, and are usually not subjected to a great deal of rough handling.

Continuous tags. These are used in applications where variable information is imprinted by a computer printer. Many possibilities exist for the number of perforations and form depths since stop and go presses are used in their manufacture. Continuous tags are generally used for retail price marking (see Fig. 4c). Special features are frequently incorporated to create hang tags that are used effectively as promotional pieces.

Booklet tags are shear-cut tags, scored vertically to form a fold, providing four surfaces for copy instead of two (Fig. 4(d)). Special effects can be added to further enhance the image of these versatile tags. Die cuts on one of the pages, generally the first, can be used to highlight the company logo or trademark or provide a representation of the product (Fig. 5(a)). Die cuts are also used on the front cover to create a window on the cover panel or to create a three-dimensional effect. The die cut is generally backed up with a solid color block or printed design which shows through the die cut. (Fig. 5(b)). *Gatefold tags* are a special variety of booklet tag featuring two vertical scores to form three separate double-sided surfaces for a total of six surfaces available for printing (Fig. 5(c)). *Accordion-fold tags*

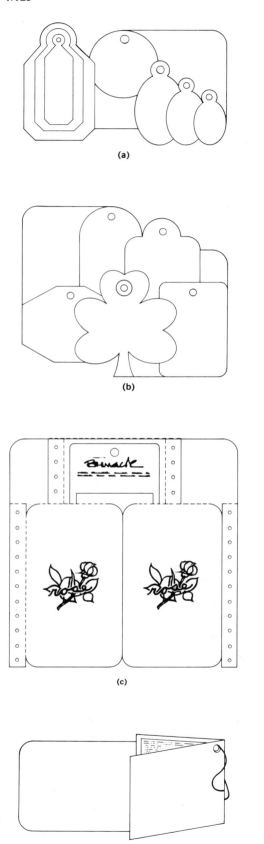

Figure 4.   (**a**) Jewelry tags. (**b**) Die-cut tags. (**c**) Continuous tags. (**d**) Booklet tag.

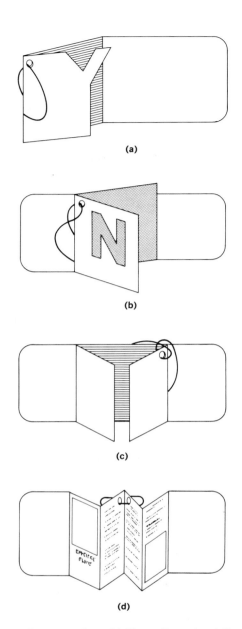

Figure 5.   (**a**) Die-cut outline. (**b**) Three-dimensional die-cut tag. (**c**) Gatefold tag. (**d**) Accordion-fold tags.

have three vertical scores that create a total of four "pages" with eight surfaces for printing. Accordion-fold tags also may be produced with additional scores to add pages and sides (Fig. 5(**d**)).

Many applications require that tags not only carry a message regarding the product but serve as a carrier for the product itself. In instances where merchandise (eg, jewelry is hung on racks for display or when merchandise is seasonal and only temporarily displayed, this type of tag is very effective. Figure 6(**a**) shows a tag with die cuts and additional punches to hold stick pins in a retail jewelry application. The center punch is used to hang the merchandise on a display rack. The tag shown in Figure 6(**b**) uses two die-cut triangles to hang gloves on display racks in order to conserve space because of the product's seasonal nature. The tag is folded at the score over the cuff and stapled to fasten it to the merchandise. The die-

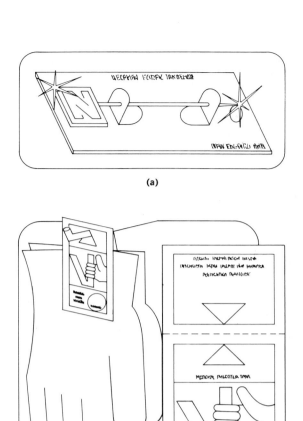

(a)

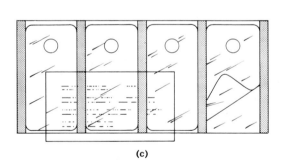

(c)

Figure 6. (**a**) Jewelry display tag. (**b**) Display-rack tag. (**c**) Transparent hang tag.

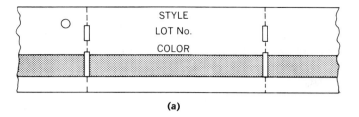

(a)

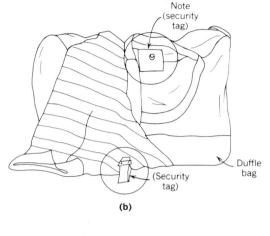

(b)

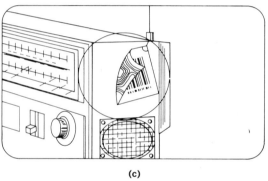

(c)

Figure 7. (**a**) Marking tag with magnetic stripe. (**b**) Plastic electronic security tags. (**c**) Security tag with miniaturized radio circuit.

cut triangle is then used to hang the gloves on the display rack. Figure 6(**c**) shows a hang tag of clear plastic partially coated with a pressure-sensitive adhesive. The coated portion of the tag is applied to the merchandise and the die-cut hole is used to hang the merchandise on the rack.

Tags used for price marking are generally produced as continuous tags. These tags satisfy the special needs of large retail organizations by affording them the opportunity to ship preprinted price tags along with stock shipments to their various distribution locations. The tags come in a variety of sizes

and shapes and are generally affixed with plastic barbs by tagging guns to prevent price switching. The accelerated use of computers in the retail industry has had a dramatic affect on the appearance and use of tags. Price-marking tags are now being produced with magnetic stripes (Fig. 7(**a**)) which can contain information such as style, lot number, and manufacturer. In addition, bar code and OCR (optical character recognition) technologies have gained acceptance in the retail tag market.

Tags can also aid in theft prevention which is one of the biggest concerns in the retail industry. Hard plastic tags such as those shown in Figure 7(**b**) are affixed to articles of clothing in many large retail stores. The tags must be removed at the time of sale to prevent triggering an alarm as the garment passes through a special gate which contains a scanner that is sensitive to the tag. Because of their bulk, these tags are used primarily for clothing. Recent innovations in miniaturization have produced tags with what appears to be a strip of aluminum foil (Fig. 7(**c**)) but which is actually a built-in security device. If not deactivated by another strip of foil affixed at the time of sale, these circuits interrupt radiowaves being transmitted from the exit gate and set off an alarm. These tagging systems are currently marketed by companies that sell retail-security systems.

Industry indicators predict that the future of the tag industry in retail operations will not be adversely affected by advances in computer technology. As long as there are products to be identified, inventoried, and priced there will be a market for tags of all kinds in the retail industry.

LINDA L. BUTLER
National Business Forms Association

TAMPER-EVIDENT PACKAGING

Industry has adopted the term *tamper-evident* packaging, but there is no officially recognized definition for it. The United States Food and Drug Administration (FDA) published the regulation 21 CFR 211.132 "Tamper-Resistant Packaging Requirements for Over-the-Counter Human Drug Products" on November 5, 1982, in the Federal Register (1). In this regulation, FDA defines a *tamper-resistant* package as a package "having an indicator or barrier to entry which, if breached or missing, can reasonably be expected to provide visible evidence to consumers that tampering has occurred." Common industry usage substitutes the term *tamper-evident* for the term *tamper-resistant*. The FDA has not objected to this substitution, nor has it made any official recognition of it as a synonomous term. The issue is still being discussed among practitioners of packaging worldwide and it seems likely that use of both terms will continue for some time. All practitioners and the FDA are agreed that the term *tamper-proof* is not appropriate and must not be used.

Development of the United States Regulation

On October 4, 1982, the Proprietary Association (PA) established the Committee on Product Security, which appointed an Expert Technical Committee on Tamper-Resistant Packaging with members from government and industry (2). The Expert Technical Committee reported its conclusions on October 14, 1982 to the Board of Directors of the PA in New York. On October 20, this Board transmitted recommendations for action to the FDA. On November 5, the FDA published the resultant regulations in the Federal Register. The publication included an extensive statement of background, concept of tamper-resistance, provisions of the rules, legal authority, economic considerations, and environmental impact. The basic regulation is 21 CFR 211.132 "Tamper-Resistant Packaging Requirements for Over-the-Counter Human Drug Products". Regulation 21 CFR 314.8 "Supplemental Applications (to New Drug Applications)" was also affected and 21 CFR 700.25 "Tamper-Resistant Packaging Requirements for Cosmetic Products" was added. In addition, 21 CFR 200.50 "Ophthalmic Preparations and Dispensers" was revised to include reference to the new 21 CFR 211.132, and 21 CFR 800 was changed to add 21 CFR 800.12 "Contact Lens Solutions and Tablets; Tamper-Resistant Packaging".

Placement of the regulation in 21 CFR 211 "Current Good Manufacturing Practice for Finished Pharmaceuticals" is significant. It is under the provisions of this part that manufac-

turers of pharmaceuticals are expected to keep up-to-date (current) in improved technologies for manufacture (and packaging) of pharmaceuticals (see Pharmaceutical packaging). The implication is clear that FDA will expect manufacturers to upgrade tamper-resistant packaging as improvements in it are made. Most discussions of regulatory activity end with the November 5, 1982, Federal Register. However, several subsequent publications have corrected, clarified and amplified the original. These publications are

1. *Federal Register* **48**(2) 334–335 (Jan. 4, 1983), "Tamper-Resistant Packaging Requirements for Certain Over-the-Counter Human Drug and Cosmetic Products; Advisory Opinion".
2. *Federal Register* **48**(10), 1706 (Jan. 14, 1983), "Tamper-Resistant Packaging Requirements for Contact Lens Solutions and Tablets; Correction" and "Tamper-Resistant Packaging Requirements for Certain Over-the-Counter Human Drug and Cosmetic Products; Correction".
3. *Federal Register* **48**(39), 8137 and 8138 (Feb. 25, 1983), "Tamper-Resistant Packaging Requirements for Certain Over-the-Counter Human Drug and Cosmetic Products; Availability of Advisory Opinions".
    a. Interpretation of Retail Sale details public access as part of retail sale and gives examples.
    b. Lozenges exempted from the regulation.
    c. Ammonia inhalant ampuls exempted from labeling requirements.
    d. Aerosol products and compressed medical oxygen exempted from labeling requirements.
    e. One-piece soft gelatin capsules granted extension to May 5, 1983, to comply with regulation.
4. *Federal Register* **48**(76), 16658–16667 (April 19, 1983), "Tamper-Resistant Packaging Requirements for Certain Over-the-Counter Human Drug and Cosmetic Products" and "Tamper-Resistant Packaging Requirements for Contact Lens Solutions and Tablets".
    These are both Final Rules and they list certain amendments and clarifications of sections of the preamble. They also specify products exempted from the regulations. This publication is the FDA response to public comments received regarding the original November 5, 1982, publication of the rules. It contains substantial information detailing the comments received and the agency response to them.
5. *Federal Register* **48**(162) 37624 (Aug. 19, 1983), "Tamper-Resistant Packaging Requirements for Certain Over-the-counter Human Drug and Cosmetic Products and Contact Lens Solutions and Tablets; Stay of Effective Date."
    This publication issues a stay of effective date for requirement of a specific identifying characteristic that is incorporated into the tamper-resistant feature of any product. The new date is Feb. 6, 1984.
6. *Federal Register* **48**(181), 41578 (Sept 16, 1983), "Tamper-Resistant Packaging Requirements; Interim Stay of Retail Level Effective Date" (with three subsequent editorial corrections).
7. *Federal Register* **48**(181), 41601 (Sept 16, 1983), "Tamper-Resistant Packaging Requirements; Proposed Stay of Retail Level Effective Date and Request for Comments, Data, and Information" (with three subsequent editorial corrections).

Table 1. Products Required to be in Tamper-Resistant Packaging (If Accessible to the Public While Being Held for Sale)

Product	Applicable section of regulation
OTC drug products oral (except dentifrices and insulin); nasal; otic; ophthalmic, rectal; vaginal	21 CFR 211.132
Cosmetic products oral liquids such as mouthwashes, gargles, breath fresheners; vaginal	21 CFR 700.25
Contact lens solutions and tablets	21 CFR 800.12

Provisions of the Regulations

A tamper-resistant package is one having an indicator or barrier to entry which, if breached or missing, can reasonably be expected to provide visible evidence to consumers that tampering has occurred. To reduce the likelihood of substitution of a tamper-resistant feature after tampering, the indicator or barrier is required to be distinctive by design (ie, an aerosol product container) or have an identifying characteristic (eg, a pattern, name, registered trademark, logo, or picture) (3). "Distinctive by design" means the packaging cannot be duplicated with commonly available materials or through commonly available processes.

Each retail package of an over-the-counter (OTC) nonprescription drug product covered by the regulation is required to bear a statement that is prominently placed so that consumers are alerted to the specific tamper-resistant feature of the package. This statement is also required to be so placed that it will be unaffected if the tamper-resistant feature of the package is breached or missing (3). The products covered (see Table 1) are required to be in tamper-resistant packaging if they are accessible to the public while held for sale. There are certain conditions under which products are not required to be in tamper-resistant packaging (see Table 2). In general, FDA holds that products distributed in a manner that does not afford the public access to them while they are held for sale were never intended to come within the scope of the regulations. Table 3 lists the three exemptions from the regulation that have been granted by FDA.

Types of Tamper-Resistant Packaging

The FDA has not specified any particular kind of tamper-resistant package. Indeed, the agency emphasizes that any

Table 2. Some Circumstances Under Which Products are not Covered by the U.S. Regulations

Products sold directly to a hospital, an institution, a medical or first-aid unit or a health professional for dispensing directly to patient.

First-aid kits sold directly to organizations and that are not accessible to the general public.

Products sold in vending machines.

Products sold through the mail directly from a firm representative to an individual user.

Products sold door-to-door directly by a firm representative to an individual user.

Table 3. Exemptions Granted

Product	Exemption Published
lozenges	exempted from both packaging and labeling requirements, Ref. 3, p. 16663.
ammonia inhalent ampules	exempted from labeling requirements, Ref. 3, p. 16663.
aerosol products and compressed medical oxygen	exempted from labeling requirements, Ref. 3, p. 16663.

package form that satisfies the definition of tamper-resistant given in 21 CFR 211.132 is acceptable. The agency did, on November 5, 1982, publish a list of package forms (see Table 4) which at that time were deemed to fulfill the requirements of the definition (1). It was emphasized that the list was a listing of examples then available, and it was not intended to exclude any package form which might be developed in the future.

In publishing the list, the FDA was careful to point out that use of one of the listed technologies does not necessarily guarantee compliance. The following statement appears on p.

Table 4. Food and Drug Administration Examples of Tamper-Resistant Package Forms as of November 1982[a]

1. *Film Wrappers.* A transparent film with distinctive design is wrapped securely around a product or product container. The film must be cut or torn to open the container and remove the product.

2. *Blister or Strip Packs.* Dosage units (eg, capsules or tablets) are individually sealed in clear plastic or foil. The individual compartment must be torn or broken to obtain the product.

3. *Bubble Packs.* The product and container are sealed in plastic and mounted in or on a display card. The plastic must be torn or broken to remove the product.

4. *Shrink Seals and Bands.* Bands or wrappers with distinctive design are shrunk by heat or drying to seal the union of the cap and container. The seal must be cut or torn to open the container and remove the product.

5. *Foil, Paper, or Plastic Pouches.* The product is enclosed in an individual pouch that must be torn or broken to obtain the product.

6. *Bottle Seals.* Paper or foil with a distinctive design is sealed to the mouth of a container under the cap. The seal must be torn or broken to open the container and remove the product.

7. *Tape Seals.* Paper or foil with a distinctive design is sealed over all carton flaps or a bottle cap. The seal must be torn or broken to open the container and remove the product.

8. *Breakable Caps.* The container is sealed by a plastic or metal cap that either breaks away completely when removed from the container or leaves part of the cap attached to the container. The cap must be broken to open the container and remove the product.

9. *Sealed Tubes.* The mouth of a tube is sealed and the seal must be punctured to obtain the product.

10. *Sealed Carton.* All flaps of a carton are securely sealed and the carton must be visibly damaged when opened to remove the product.

11. *Aerosol Containers.* Aerosol containers are inherently tamper resistant.

[a] Ref. 1.

50445 of Ref 1: "Conversely, use of one of the identified technologies does not, by itself, constitute compliance with the requirement for the use of a tamper-resistant packaging system if the application of the technology in a particular case does not meet the standard established in this final rule (eg, if the system is inappropriate to the product or is faulty in design)."

Since publication of the list, two of the package forms have required special attention. Seal-end folding cartons (see Cartons, folding) require special attention to the amount and kind of adhesive used to ensure suitable evidence of opening. Several carton designs have been presented to FDA for evaluation as improvements to the system. It was also found that cellulose shrink bands (see Bands, shrink) could easily be soaked for removal and subsequent replacement without detection and these were removed from the list. The FDA has indicated that sealed tubes should be sealed at both ends; that is, the blind-end tube should also be sealed at the crimp end. A simple crimp without seal does not provide sufficient barrier to entry.

Since publication of the original listing, additional designs have been offered as tamper-resistant package forms. Among them are threaded bottle closures (see Closures) designed so that, when rotated to open, a membrane is torn to reveal a characteristic color under a transparent overcap. A patent has been granted for a complex closure design that has a combination of tamper-resistant and child-resistant properties (4). To date, these new designs have not appeared in a widely available commercial form. The familiar depressed-button closure design used for many years on baby food is available now for other vacuum-packed products. When the cap is opened, the release of vacuum allows the concave cap crown to return to flat or convex shape. Another widely available variety is a plastic container/cap system with a tear-away strip that must be removed to open the package.

Given the rapid development of designs, the best way to keep informed about the kinds of tamper-resistant packaging available is to consult the annual buyers'-guide issues of packaging magazines and to contact the FDA for currently acceptable package forms.

Current Status and Trends for the Future

The transition to tamper-resistant packaging for OTC drugs and other covered products was accomplished in a timely manner as required. This was possible because the packaging required was based on existing technologies and the magnitude of the change could be managed by the suppliers and users involved. On September 16, 1983, the FDA issued an interim stay of retail effective date for tamper-resistant packaging, saying in effect that it would not be mandatory that nontamper-resistant packages be removed from retail shelves after February 6, 1984. This stay was issued pending completion of a survey by the FDA to determine whether the original retail effective date was realistic or if it should be modified in some way (5). The stay remains in effect as of December 1985.

Food, another class of ingestibles, has not been regulated. The FDA responding to public comment, has said that it believes the scope of the regulation should not be broadened to other products generally. The implication seems to be that the warning is there and prudent packagers of ingestibles of all kinds will take appropriate steps. Food packagers are, in fact, moving quietly to provide tamper-resistant features in their packaging. Informal surveys by the author have shown a rapid increase in the number of food products now being distributed in tamper-resistant packaging. A formal survey conducted by the Cornell University Department of Food Science in January 1984 (6) revealed that as many as one half of current food packages are already tamper-resistant and would need only the addition of a label statement to be in compliance with the regulation for OTC drugs. The study identified nine categories of foods for which substantial package changes would be required to achieve compliance.

An area of considerable discussion and controversy appeared when charges were made that tamper-resistant packages were too difficult for elderly and manually or visually impaired people to open. In answer, the FDA, together with the Consumer Product Safety Commission (CPSC), initiated a study of the difficulty of opening of tamper-resistant packages. From the results of the study (6) the FDA determined that the elderly can gain access to tamper-resistant packages, that it does take longer to open tamper-resistant packages, and that it is not necessary to establish a standard of tamper-resistant packaging access (7).

As investigators and regulators continue working with tamper-evident packages, nomenclature, terminology, and classification will become a problem. When the FDA published the list of package forms they deliberately avoided specificity to allow industry maximum flexibility in design development. The Cornell study utilized a very specific classification of tamper-resistant packages developed for their purposes. In June 1983 investigators at the Michigan State University School of Packaging published yet another classification (8). Beginning in November 1982, the ASTM Committee D 10 on Packaging initiated development of a Standard Classification for Tamper-Evident Packages. The working group preparing the classification is composed of representatives from the FDA the PA, university researchers, and industry suppliers and users of packaging. The classification has been through several drafts and two letter ballots, with several questions to be resolved. The issue of "tamper-evident" vs "tamper-resistant" is being debated as part of the classification writing process.

Studies conducted at Michigan State University (8,9) indicate a complex involvement of consumer perception, consumer knowledge, and consumer awareness in the whole process of consumer ability to detect tampered packages. The unpublished data are still being analyzed, but results obtained so far indicate that consumers have a limited ability to detect malicious tampering, coupled with a tendency to think that some untampered packages have been tampered.

A few good and bad design features have been identified. With respect to *neck bands*, effectiveness is enhanced by vertical perforations, shrinking over the transfer bead, and extension of the band far down the package. Shrink bands over some child-resistant closures (see Child-resistant packaging) should be avoided, because the large-diameter closure hinders a tight fit. In some instances, transparent bands foiled effective repair; in others pigmented bands prevented tampering/repair and transparent bands did not. In *closures*, horizontal scores on metal breakaway caps are better than vertical scores. Induction-sealed thin-foil membrane seals are effective, but there should be no overhang over the finish circumference. Distinctive logos may hide tampering.

Wraps benefit from transparency, perforations or tear strips, and tight shrink. In *carton sealing*, thin tapes are better than thick tapes, and delaminating tapes are good; but adhesive seal is stronger than tape. Adhesive on seal-end carton should cover a large surface area. Cartons need tape seal on manufacturer's joint (see Cartons, folding); high-gloss carton surfaces should be avoided. *Unit-dose blisters* are difficult to defeat, particularly with embossed seal. Push-through backings are better than peel backings; thin foil backings are best. Any *sound* that must be present to indicate an unopened container is desirable, such as the breaking bridges of breakaway caps, or the sound of air entering or leaving a pressurized or evacuated package.

The data from the studies have been sufficiently complex that they defy simple statistical treatment. Most recently, a clinical medical statistic has been adapted which takes into account all of the four possible responses: (*1*) correct judgment of no tampering, (*2*) incorrect judgment that untampered package was tampered, (*3*) correct judgment of tampering, and (*4*) incorrect judgment that tampered package was not tampered. Extensive computation with this statistic is now underway in an attempt to achieve usable statements of statistically significant differences among consumer judgments about various forms of tamper-evident packages.

The engineering principles and human factors involved in tamper-evident packaging are not yet well defined. Tamper-resistance in packaging is probably not an issue that will fade into oblivion. The fact of its regulation under Current Good Manufacturing Practices just about guarantees that it will not. Furthermore, in a Delphi Study recently done for the Office of Technology Assessment of the United States Congress (10), the 59 experts surveyed agreed that tamper-resistant packaging will increase in importance to packaging and the economy over the next 15 years. All of the evidence available points to more applications and further refinements of tamper-evident packaging between now and the year 2000.

Organizations

Several organizations have primary and continuing involvement with tamper-evident packaging. The following list is representative, but not necessarily all-inclusive.

American Society for Testing and Materials (ASTM), Committee D 10 on Packaging, 1916 Race St., Philadelphia, PA 19103.

Cornell University, Department of Food Science, Ithaca, NY 14853.

U.S. Food and Drug Administration (FDA), Regulations and Policy Branch, 5600 Fishers Lane, Rockville, MD 20857.

Proprietary Association (PA), 1150 Connecticut Avenue, N.W., Washington, DC 20036.

School of Packaging, Michigan State University, East Lansing, MI 48824.

BIBLIOGRAPHY

1. *Fed. Regist.* **47**, 50442–50456 (November 5, 1982).
2. Statement of the Proprietary Association on Securing the Contents of OTC Medicine Packages before The Subcommittee on Health and the Environment Committee on Energy and Commerce, United States House of Representatives, October 15, 1982, Washington, D.C.
3. *Fed. Regist.* **48**(76), 16664 (April 19, 1983).
4. U.S. Pat. 4,526,283 (July 2, 1985), R. A. Skinner.
5. *Fed. Regist.* **48**(181), 41578 (Sept. 16, 1983).
6. J. H. Hotchkiss and R. B. Gravani, *The Current Status of Tamper-Evident Packaging for Foods,* Department of Food Science, Cornell University, Ithaca, N.Y.
7. H. S. Spungen, Division of Drug Quality Evaluation, and D. J. Schuirmann, Division of Biometrics, Center for Drugs and Biologics, Food and Drug Administration, *Accessibility of Tamper Resistant Packaging to the Elderly.*
8. H. E. Lockhart, M. Richmond, and J. Sneden, "Tamper Resistant Packaging: Is It Really?" *Package Eng.* **28**(6), 96 (1983).
9. Tamper-Resistant Packaging Surveys, unpublished data, School of Packaging, Michigan State University, June, Aug. 1983, April 1984.
10. C. J. Mackson, Director, *The Impact of New Technologies on the Food Packaging and Preservation Industries,* School of Packaging, Michigan State University, East Lansing, Mich., Jan 1985; prepared for Congress of the United States, Office of Technology Assessment.

General References

Package Eng. (now *Packaging*), **28**(7) (June 1983).
Packaging Technology, **13**(2) (June 1983).
J. Sneden, *Testing of Tamper-Resistant Packaging.* M. S. Thesis, 1983, University Microfilms, Ann Arbor, Mich.

Hugh E. Lockhart
Michigan State University

TAPED, GUMMED

Water-activated gummed paper tapes used to seal the center and/or end seams of corrugated boxes are very different from what they were at the time of their inception, approximately 100 years ago. Technological advancements in paper, laminates, adhesives, reinforcements, and dispensing and application equipment provide today's user with a scientifically produced high quality product. Depending on the packager's need, gummed tape can be obtained with a wide range of tailor-made special features.

There are two basic forms of gummed tapes; single-ply non reinforced "paper" sealing tape, and double-ply fiber glass "reinforced" sealing tape. Both varieties begin with what is commercially known as "gumming kraft," which differs from ordinary kraft paper in that it is sized to prevent the adhesive from penetrating too deeply into the paper. Following application of a vegetable-based remoistenable adhesive (see Adhesives), single-ply paper sealing tape is slit into roll lengths of 375–6000 ft (114–1829 m). Depending upon the basis weight of the paper (35#, 60# or 90#) (see Paper) the product is categorized for light, medium, and heavy-duty application.

Double-ply reinforced gummed sealing tapes are produced by sandwiching fiber glass yarns between two sheets of kraft paper with either a hot melt or aqueous based adhesive. Fiber glass yarns generally run in three directions (machine direction and both transverse directions) forming a diamond or "Three Way" pattern. As in single-ply paper tapes, a water-remoistenable adhesive of vegetable and/or animal glue formulation is applied to the bottom sheet before or after lamination. Depending upon the user's needs, paper basis weight, yarn spacing and denier, and laminate- and adhesive-coating weight can be varied to produce products for light, medium and heavy-duty application. Finally, the parent or "jumbo" roll is slit into smaller rolls ranging in size from 360 to 4500 ft (110–1372 m).

Both paper and reinforced sealing tapes are usually wound gummed-side-in on a cardboard core. Most manufacturers also supply coreless rolls. Tape widths generally range from 1 to 3 in. (25–76 mm) in ½-in. (13-mm) increments. Gummed tapes are available in a range of colors, widths, and lengths. They can also be custom printed and can be obtained in strippable grades or with special antitheft, tamper-evident, or inventory-control features.

When properly purchased and applied, both paper and reinforced sealing tape fully meet the requirements and specifications for rail (UCC Rule 41), truck (NCB Rule 222), plane, parcel post, UPS, and other parcel delivery service shipments. General Services Administration Commercial Descriptions (CIDs) A-A-1492A, A-A-1671A, and A-A-1672A govern the purchase and use of gummed tapes by the federal government.

The product does not lend itself to differentiation by appearance, but tape users should be aware of the fact that all gummed tape is not the same. Because of diverse manufacturing techniques, adhesive and laminate formulations, and quality-control guidelines, tape-performance differences between producers exist.

Gummed tape dispensing and application equipment fall within several categories. These are (1) hand-operated dispensers; (2) electrically operated dispensers; (3) automatic taping machinery for fixed-size boxes, and (4) automatic taping machinery for random-size boxes.

General References

Gumming Industry Voluntary Product Standard for Paper Sealing Tape (Single Ply), The Gummed Industries Association, Inc., Jericho, N.Y. 1983

Gumming Industry Voluntary Product Standard for Fiberglass Reinforced Paper Sealing Tape (Double Ply), The Gummed Industries Association, Inc., Jericho, N.Y. 1983

How to Use the New GIA Sealing Tape Standards: A Guide to the Proper Selection and Use of Gummed Sealing Tapes, The Gummed Industries Association, Jericho, N.Y. 1983

R. W. McKellar
The Gummed Industries Association

TAPE, PRESSURE-SENSITIVE

Pressure-sensitive tape is used in packaging for closing boxes, combining packages, attaching packing lists, color coding, pallet unitizing, adding carrying handles, splicing, providing ease of package opening, securing the package contents, protecting labels, reinforcing critical package components, and holding documents.

The first pressure-sensitive tape was developed in 1925 for paint masking; it had a paper backing and a glue–glycerol adhesive. Today, there are hundreds of specialty tapes available for specific applications in packaging, general industry, the office, and the home. The common theme is a backing material coated with an adhesive which adheres with a light touch without a need for an activating solvent or heat (see Adhesives).

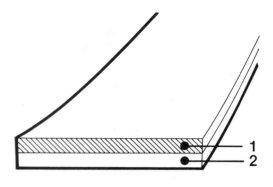

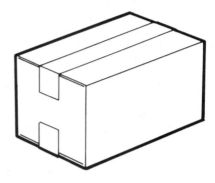

Figure 1. A typical box-sealing tape consists of a backing film (**1**) and a layer of pressure sensitive adhesive (**2**). It is used most often as a closure for regular slotted containers.

Box-Sealing Tape

A major use of pressure sensitive tape in packaging is the closure of regular slotted containers (see Boxes, corrugated). Figure 1 depicts a typical construction of a box-sealing tape; a plastic film is coated on one side with pressure-sensitive adhesive. The film may have a release treatment on the backing to allow easy removal of the tape from the roll during dispensing. Some film backings also are treated or coated on the other side to increase the bond of the adhesive to the backing, although for a box-sealing tape, this is usually not a critical factor. The standard application of a tape to a regular slotted container (RSC) is for a 2-in. wide (or 48-mm wide) strip of tape to be applied over the center seam, extending 2 to 2½ in. (48–65 mm) onto the end panels of the box. This seals the center seam and helps keep dust and dirt out of the box. If a total seal is needed, cross strips of tape can be added at the end edges of the box. This "H-Seal" or "six-strip seal" is specified for some export and military packages. (see Export packaging; Military packaging).

The choice of tape for box closure is very important and affects the performance of the entire package during storage and distribution. The most common backing films are biaxially oriented polypropylene (see Film, oriented polypropylene) or polyester (see Film, oriented polyester) although some unplasticized PVC film is also used. One of the critical backing properties is machine-direction tensile strength (ASTM D 3759). During rough handling, the tape can be stressed severely at the end edge of the box, and adequate tensile strength and elongation are needed to resist a tape tear. A

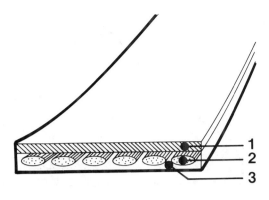

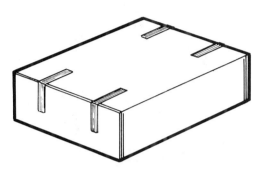

Figure 2. A filament tape has a backing film (**1**) and filaments (**2**) embedded in a pressure-sensitive adhesive (**3**). One of the uses is the L-clip closure of full overlap boxes.

property at least as important is the cross-direction tensile strength (ASTM D 3759). This is needed to keep the tape from splitting down the center during handling and from rupturing when the tape is impacted by another package. The tape's caliper (ASTM D 3652) and rigidity relate to the ease of handling with dispensers and to machinability on some automatic equipment.

A proper backing for a box-sealing tape is a good start but the tape must have an aggressive adhesive if the backing strength is to be realized. The adhesive must be able to adhere to a corrugated box with only a light rub down under unpredictable factory conditions, yet it must withstand the constant recovery force of the box flaps and hold tight during rough handling. Many box-sealing tapes use an adhesive based on rubber and a tackifying resin, but acrylics (see Acrylics) and other synthetic adhesive systems are also used. The most frequently referenced adhesive property is the peel adhesion to steel (ASTM D 3330) but it has dubious value in comparing tapes for box sealing. The shear adhesion to fiberboard (ASTM D 3654, A) correlates much better to the tape performance on boxes because it uses a standard kraft linerboard as a substrate and it applies stress in a shear mode, similar to the stress on the end legs of the tape during use. The literature on pressure-sensitive tape includes several methods of measuring the initial tack or "wet grab" of an adhesive. These are valuable research methods but, because most of them use glass or steel as a test surface, it is very difficult to correlate the results with the performance of tape on a fiberboard box.

A good pressure-sensitive box-sealing tape offers an excel-

lent closure for corrugated boxes. Laboratory tests, field trials, and production shipping have demonstrated its performance and it has been accepted by shippers, carriers, and consignees. The cost effectiveness of pressure-sensitive tape is apparent when the simplicity of labor-saving equipment and the low operating and maintenance costs are considered.

Filament Tape

A second broad category of tape is pressure-sensitive filament tape, sometimes known as "strapping tape". Figure 2 shows that it is made of a film backing (usually polyester) with reinforcing filaments embedded in the pressure-sensitive adhesive. The most common reinforcement is fiberglass, tape reinforcement which provides a high tensile strength with very little stretch. A few tapes have polyester or rayon filaments for extra impact and cut resistance. The tensile strength (ASTM D 3759) is controlled by the number and type of filaments used in the tape construction: Some have as much as 600 lbf of tensile strength per inch of width (105 N/mm). The adhesive requirements of filament tape are as critical as those of box-sealing tape. Care should be taken to choose a tape with a balance of initial adhesion and shear-holding power.

Box closure can be accomplished with filament tape on a variety of box styles. Filament tape is usually recommended

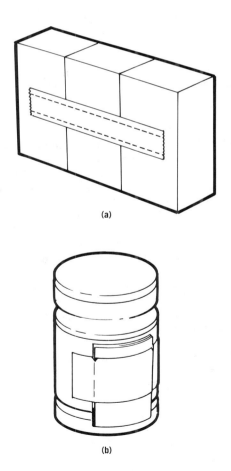

Figure 3. (**a**) Pressure-sensitive tape used to combine three small packages as a unit. Perforations allow easy break-away. (**b**) Tape used to attach instruction sheet to bottle. Tape is nicked to facilitate a clean tear in the tape when the paper is removed.

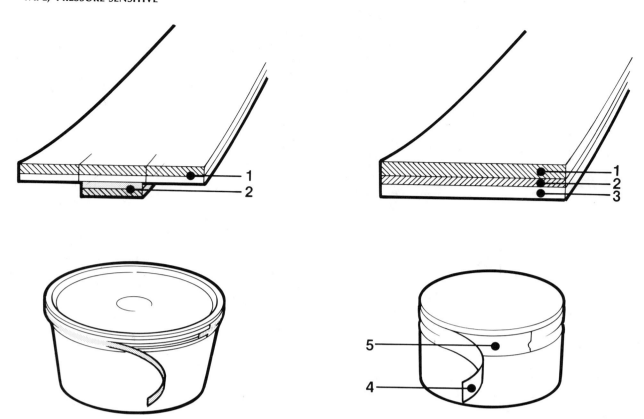

Figure 4. A heat-shrinkable tape (**1**) is laminated to a tape tear strip (**2**) to form a freshness seal on a tub. Pulling the tab opens the tub neatly and cleanly.

Figure 5. Unique tape constructed of (**1**) a rigid heat-shrinkable backing, (**2**) a polyethylene layer, and (**3**) pressure sensitive adhesive. The rigid backing comes off first (**4**), leaving the weak polyethylene (**5**) adhered to the container as an indication of tampering.

for use with boxes with fully overlapping flaps such as a five-panel folder, full-overlap box, or full-telescope box. A high performance filament tape may be applied in L-shaped strips as small as ½-in. (12.7-mm) wide by 4¼-in. (108-mm) long; 2 in. (51 mm) on each leg of the L-clip. A tape with lower tensile or adhesive properties may have to be used in ¾ in. (19 mm) by 6¼ in. (159 mm) L-clips for equivalent performance on a box.

Filament tape is used in dozens of other packaging applications where high strength is required. These include general purpose holding, bundling, recooperage, and pallet unitization. If two packages are being sent to the same destination, the shipping costs often can be reduced by bundling them together with filament tape for the shipment. Convenience handles can be formed easily on boxes with loops of filament tape which have strips of paper or film to matte out the adhesive in the center of the loops. Improved handles can incorporate the tape into the design of the box to reinforce critical areas or to form integral tape/box handles.

Specialty Tapes

With a full choice of films, papers, and foils for backings and a wide range of adhesives available, it has been possible to develop special pressure-sensitive tapes for many packaging uses. For example, Figure 3 shows three small packages combined for a unit sale. The transparent tape used to combine them is perforated to weaken the backing so the end package can be pulled off with a twisting motion. Figure 3 also shows a

bottle with an instruction sheet attached to it by the use of tape. This sometimes can eliminate the need for a folding carton and generate considerable savings in packaging costs.

Tape can be used to form a band around most container sizes and shapes to form an attractive tamper-evident secondary closure. The seals help protect product freshness and are printable to communicate messages to the consumer. Figure 4 shows one method of sealing a lid on a tub. A heat-shrinkable PVC or polypropylene tape (see Films, shrink) is laminated to a tear-strip tape with similar shrink properties. The tape is applied around the circumference of the tub with two nicks applied to the outer tape at the overlap. The tape is shrunk to form a tight seal and a tab is formed, which can be used to pull the tear tape. One of the several other methods of providing a tamper-evident freshness seal is shown in Figure 5. The tape's rigid heat-shrinkable backing is separated from the aggressive pressure-sensitive adhesive by a relatively weak layer of polyethylene film. This tape is applied around a closure and allowed to shrink into a tight band. When the tape is removed, the weak bond between backings separates and the thin layer of polyethylene is left on the container. This film, which may be printed, fractures easily when the cap is removed and provides a means of seeing whether the closure has been opened. It is also possible to perforate the outer shrink band of tape to allow opening the container without removing the tape (see also Bands, shrink).

One more example of a specialty tape system is shown in Figure 6. A tape is used as a peel-open closure for a can end

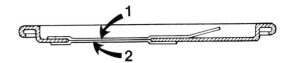

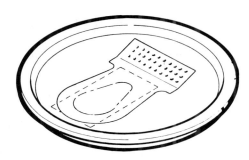

Figure 6. Scotchtab (3 M) peel-open closure system. A top-closure tape (**1**) is peeled off a can end for easy access to juice. A raw edge protection tape (**2**) is sometimes used to isolate the cut can end from the juice.

used for single-strength fruit and vegetable juices. The backing is a vapor-coated polyester film which gives the tape a metallic appearance. The special adhesive is designed to withstand the hot-fill process and to remain secure until it is peeled off the can. With some juice products, such as grape or cranberry juice, a barrier is needed to isolate the cut edge of the pour hole to prevent color or flavor changes. The inner "raw-edge protection tape" acts as the barrier between the cut edge of the pour hole and the contents of the can and gives a clean pull out of the hole upon removal of the top closure tape.

Many other tapes are available with the backing and adhesive engineered for specific purposes such as providing resealable features for flexible bags, attaching packing lists to a package with a tape which is coated with adhesive only at the perimeter, and attaching products to cards. The versatility of pressure-sensitive tape is leading to new applications.

BIBLIOGRAPHY

General References

L. F. Martin, *Pressure Sensitive Adhesives, Formulations and Technology*, Noyes Data Corp., Park Ridge N.J., 1974.

D. Satas, *Handbook of Pressure-Sensitive Adhesive Technology*, Van-Nostrand Reinhold, New York, 1982.

Book of Standards, Vol. 15.09, American Society for Testing and Materials, Philadelphia, Pa., 1984

Test Methods for Pressure-Sensitive Tapes, Pressure-Sensitive Tape Council, Glenview, Ill., 1976.

J. A. Fries, *Paper Film and Foil Converter*, 50 (Feb. 1979).

C. A. Wangman, *Adhesives Age*, 33 (May 1976).

J. Johnston, "Tack", *Proceedings of the Industry Technical Conference*, Pressure Sensitive Tape Council, Glenview, Ill., 1983.

R. L. SHEEHAN
3M Corporation

TESTING, CONSUMER PACKAGES

Consumer packages are tested to verify their ability to meet the requirements of regulatory agencies and trade associations for safety, integrity, and ability to be safely handled in the distribution environment. They are also tested to ensure their fitness to meet the requirements of production and filling equipment and to warrant consumer satisfaction to the greatest extent possible. The reasons for performing certain tests often overlap. Companies that package consumer products usually require at least some of the tests described in this article.

Preproduction testing is especially important for firms engaged in high-speed production and filling. As production speeds increase so does the need to be sure that all the components can be accommodated by the production equipment. Components that are outsized can either jam production equipment, fail to fill or seal properly, or be improperly assembled. Any of these conditions can slow or halt a production line. Thus, in this environment it is important that all the components meet the requirements of the automated handling equipment. As the name implies, preproduction tests are performed on packaging components proposed for use with either a new or revised product or package. These tests are usually performed before any decision is made to procure production quantities of the components. Based upon the results of these tests, decisions can be made regarding the continuation of the project as proposed or with changes to either the packaging components or the production equipment. In the latter case, it is usually consideration of scheduling and cost constraints that dictate which of the elements is to be changed.

To maintain ongoing production, it is customary to perform quality-control audits of the packaging components as they are received (see Specifications and quality assurance). Such audits are often performed on a statistically selected sample of the components. ANSI/ASQC Z1.4-1981 contains useful guidelines for selecting a statistical sample (1). Quality-control audits are usually designed to evaluate only the most critical parameters of the item under consideration. The determination of the most important criteria is usually done during the preproduction testing. In addition to preproduction tests and quality-control audits, some firms also perform forensic evaluations of packaging components. These tests are performed to determine the cause of premature failure or unsatisfactory performance.

Government regulatory agencies such as the Consumer Product Safety Commission (CPSC), the Environmental Protection Agency (EPA), and others also require testing. Such testing is designed to ensure that the packaging components meet specific safety requirements. The child-resistant features of child-resistant closures (see Child-resistant packages) and the tamper-indicating features of tamper-evident packaging (see Tamper-evident packaging) are two examples of such safety requirements. For the packaging of edible products, the Food and Drug Administration (FDA) requires extensive testing of the pigments and resins to certify that harmful elements are not released into the product by the package (see Colorants; Food, drug, and cosmetic packaging regulations). Testing to qualify the ability of packages to meet these requirements is usually performed during the preproduction phase. However, if a component is changed, then these tests may have to be repeated during ongoing production.

The FDA also requires that packages for certain categories of over-the-counter drugs have features to indicate whether the packages have been opened. No formal test procedures have yet been developed. The specific requirements of the FDA can be found in Part 21 of the Code of Federal Regulations (CFR).

Package testing is also performed to ensure consumer satisfaction. The Department of Defense has a myriad of stringent tests (2) for packaging materials and components. Although there are no codified batteries of tests for the individual consumer, the individual consumer is probably just as demanding as the government. To meet the consumer's needs, the package must protect the product, reliably dispense the product with each operation, not fail prematurely, and for certain categories of products at least, look aesthetically pleasing.

Laboratories and Environmental Conditions

Laboratories that perform the testing described herein fall into three principal categories:

1. *Captive laboratories* owned and operated by and for the manufacturing firm, (ie, manufacturers of either packaging materials or the packager of the finished product).
2. *Independent laboratories* (independently owned and operated) perform tests for all client companies that wish to retain their services. Such firms often serve as arbiters of disputes or provide certification to regulatory agencies that certain packages meet the requirements of the law.
3. *Government laboratories* owned and operated by and for regulatory agencies. Such labs are used to monitor compliance with the agencies' requirements.

Although government laboratories are not ordinarily available to perform developmental testing, they often do provide a valuable service to industry by rendering opinions regarding proposed packaging components. Because of the need to get incoming production materials into the manufacturing schedule as quickly as possible, almost all quality-control testing is performed in captive laboratories.

The packaging industry relies very heavily on plastics and paper products. Because the physical characteristics of these materials are substantially affected by ambient environmental conditions, conditioning of these materials prior to testing is extremely important. Conditioning is the exposure of materials to be tested to a controlled environment for a prescribed period of time or until conditions of equilibrium have been reached. Per the requirements of ASTM D 685 (3), paper products are usually conditioned for a period of 24 h in an environment maintained at $73 \pm 2°F$ ($23 \pm 1°C$) and $50\% \pm 2\%$ rh. In addition, the specimens should not be exposed to direct sunlight. Standard conditioning of plastics is performed in accordance with the requirements of ASTM D 618 (4). Specimens less than 0.28 in. (7 mm) thick should be maintained at 73°F (23°C) and 50% rh for a period of 40 h. Although ASTM does not specify standard conditioning atmospheres for glass and metal products, it is strongly recommended that these materials also be maintained at 73°F (23°C) for a period of time prior to testing. This is recommended for the purpose of eliminating spurious effects that might be caused by unanticipated thermal growth or shrinkage. When they are being conditioned, test specimens should be supported in such a manner that all surfaces are exposed to air flow. There are times when other conditioning environments are required. At these times, the conditioning environment should be adjusted to meet the particular application. Whenever possible, testing should be performed in the same environment as that used for conditioning. When this is not possible, it is recommended that specimens be withdrawn from the conditioning environment singly, and tested as soon as possible.

Types of Tests

Testing to determine the ability of packaging systems to meet regulatory requirements often involves panels comprised of people of various age groups who are generally not involved in the packaging industry. For example, in evaluating such components as child-resistant closures, it is necessary for a certain number of the panelists to open and resecure the closure within a specified time period. In the testing of child-resistant closures, harmless placebos are utilized instead of the actual product. One set of pass/fail criteria apply to panels comprised of children and another set of criteria apply to panels of adults. The CPSC is currently reviewing the effects of tamper-evident packaging on the adult population as it relates to the time required for closure removal. Requirements for the compliance of a packaging system can be found by referring to Part 21 of the Code of Federal Regulations. Industry groups are trying to develop their own test protocol to determine the efficacy of tamper-evident packaging. For the evaluation of materials to be utilized in the packaging of food and drug products, tests specified by the FDA are employed. In these tests, potentially harmful chemicals are extracted from the packaging components with one or more solvents. The resulting solutions are then analyzed by one of the techniques of instrumental chemistry. For plastic materials, both resins and pigments must be evaluated in this manner.

In assessing the ability of a packaging component to be satisfactorily filled or assembled in production, one of the most important evaluations is often dimensional analysis. As the name implies, this involves a measurement of the size of the packaging component or a feature on the component. This analysis is important because it helps verify that mating parts will fit together properly, even when the worst tolerance case is encountered. The worst tolerance case occurs when one component has a dimension at one limit of the tolerance range and the mating dimension on the mating part is at the other extreme of its range. Thus the fit of the assembly could tend to be either too loose or too tight. Dimensional analysis is also important for the purpose of ensuring that packaging components are of such a size that they can be accommodated by the production or filling equipment. Similarly, the measurement of the forces or torques required to assemble packaging components is critical. As with the measurement of dimensions, the importance of force and torque measurements increases as the production speed increases. Excessive force or torque requirement can result in either improperly assembled components, reduced production speeds, damaged packages, or jammed equipment. The determination of acceptable forces or torques is dependent upon each application and the type of assembly to be performed. Obviously the requirements of hand assembly differ from those for machine assembly. Also of critical importance in the food and pharmaceutical industries is the cleanliness of the packaging components. Aseptic packaging may be required to preclude the growth of bacteria or spores that could be harmful (see Aseptic packaging). Maintaining the freshness and taste of the product are also concerns in this application.

Probably most tests are performed to warrant consumer satisfaction with the package. In many cases the consumer considers the package and the product as a single entity. Thus, a superb product in a poor package is as likely to be considered unsatisfactory as a poor product in a good package. Consequently many manufacturers go to great lengths to make sure that their packaging performs in a manner consistent with the expectations of the consumer. Among the criteria evaluated in ensuring consumer satisfaction are the following: ease of use; reliability of the packaging system; ability to maintain an aesthetically pleasing appearance during its useful life; satisfactory payoff; ability to retain fragrance and flavor; and the ability to withstand accidental impact. *Payoff* is the quantity of product dispensed with each operation. In tests to evaluate these parameters, the packages are subjected to many more operations than they might normally have to withstand. Some packages are dropped from a prescribed height to determine their ability to survive such impacts (5). Some classes of materials are tested for their ability to resist the buildup of a static electrical charge (6). This is an indication of the ability of the package to withstand the agglomeration of dust and dirt. Long-term storage tests with the product (7) are performed to ensure that there is no significant loss of color, fragrance, flavor, volume, or potency (pharmaceuticals) (see Pharmaceutical packaging). Preliminary indications of the suitability of the protective characteristics of the container are often determined through Water Vapor Transmission Rate Testing (8).

Determination of the ability of packages to withstand the rigors of the distribution environment is as important as any of the tests mentioned above. Many of the tests performed to determine these characteristics involve exposure to shock and vibration. Because these subjects are covered in other articles no further mention of this type of testing will be made here (see Testing, cushion systems; Testing, product fragility; Testing, shipping containers). However, tests of secondary packaging materials such as fiberboard are discussed in this article. In this context, primary packaging is considered to be any packaging material that is in direct contact with the product. Secondary packaging contains the primary package with the product. Packages that contain the primary and secondary packaging, as well as the media exposed to the distribution environment are referred to as shippers.

Test Procedures and Equipment

The following describes some of the detailed considerations of the tests as well as some of the instrumentation that might be used.

Some of the tests performed to warrant the effectiveness of child-resistant packaging systems are performed with panelists who are usually unskilled in packaging technology. Other than a stopwatch and a means of recording data, no specialized instrumentation is required for this testing. Ordinarily this testing is performed under the supervision of one or more observers who provide the necessary instructions to the panelists and monitor their performance. For the testing of the suitability of plastic and glass containers in contact with food and pharmaceuticals, some sophisticated instrumentation is employed. For these applications, the required testing is described in the USP-NF *U.S. Pharmacopeia and National Formulary* (9). In these tests solvents prescribed by the U.S.

Pharmacopeia are used to wash the packaging material. After a prescribed exposure period, the solution is analyzed by one of the techniques of instrumental chemistry. Often infrared spectrometry is utilized for this procedure (10). The solution is analyzed for traces of elements leached from the packaging material and considered to be hazardous to humans. Other techniques utilize chromatography, atomic absorption, or x-ray diffraction.

Dimensional analysis. The most basic tool for performing dimensional analysis is a simple measuring device: a scale or ruler. Increased accuracy is obtained by using micrometers, calipers, optical-measuring equipment or coordinate-measuring machines. For ease of measuring complex shapes such as thread profiles or tapered elements, the optical devices or the coordinate-measuring machines are the most versatile. ASTM D 2911 (11) gives detailed information on the measurement of thread profiles on plastic bottles. Coordinate-measuring machines or optical comparators are extremely accurate and can be used more rapidly than scales, calipers, or micrometers. In optical-measuring devices, the profile of the feature to be measured is projected onto a viewing screen while the component is mounted on a translating table. The table is moved by a precision screw thread controlled either manually by a micrometer drive or by an electric motor. With either type of drive, the amount of table movement is accurately measured. By moving the table so that first one side of the image and then the other is aligned with a cursor on the screen, an accurate measurement of the dimension of the feature can be obtained. Coordinate measuring machines are similar to optical comparators. With a coordinate-measuring machine, however, the specimen is held stationary and a measuring probe mounted on a movable head is employed. The measuring head is connected through variable resistors on each axis so that movement along any axis changes the voltage across the resistor. These voltage changes are calibrated in units of length. Thus, any movement of the measuring head is displayed as a unit of length. Other devices utilize the interruption of a laser beam or video scanners to enable dimensional measurements. In selecting the instrument to be used, it is recommended that the requirements of the application be evaluated against the capabilities of the instrument.

Force measurement. Just as dimensional instruments vary in complexity and accuracy, so do force-measuring instruments. Force-measuring instruments range from the relatively inexpensive and simple "fish scale" to sophisticated devices that are computer-controlled, and provide data that are automatically analyzed by a computer. All the force-measuring instruments with an electrical signal have a load cell as their basic element. The load cell is basically a strain gauge connected in a bridge circuit. Changes in the length of the cell caused by either tensile or compressive forces cause the resistance of the bridge to change. Changes in resistance in turn cause changes in the voltage at the output terminal of the bridge. When calibrated in units of force, the voltage changes are interpreted as forces. Force-testing instruments are often connected to x–y plotters, which can be used to measure force versus deflection. An example of the usefulness of such a plotter would be in the determination of the compressive force required to seat a wiper in a bottle. Wipers are commonly used in the packaging of cosmetics. They are fittings with small orifices that are inserted in the necks of bottles. Their function is to "wipe" excessive product off the applicator as it is withdrawn from the bottle. Excessively high insertion forces could result in unacceptably slow production speeds or in improperly assembled packages. If the interference is too loose, then the wiper could be withdrawn from the bottle along with the applicator. If this were to occur the package would be useless. In measuring the force required to seat the wiper, the bottle is mounted on the moving crosshead of the testing device. A flat platen attached to the load cell is operated in a manner that presses the wiper into the bottle. The point at which the x–y plotter indicates no more crosshead movement as the force increases rapidly, is the point at which the wiper has seated.

Brittleness measurement. The x–y plotter is also useful for determining the brittleness of packaging materials. The brittleness of a packaging material can be affected by exposure to either the product or to some other environmental condition. Through either tensile or compression testing it is possible to determine whether changes in this property have occurred. Simply stated, brittleness is a function of the force and deflection at the yield point of the material. After exposure to a test environment, changes in either the force level or the amount of deflection at the yield point indicate that there have been changes in either strength or ductility or both.

Torque. For packages that employ screw closures, torque testers are important. Torque testers are manually operated. Basically these devices are similar to torque wrenches. For purposes of package testing, however, they incorporate a means of holding a bottle or closure as the other component is manually rotated. Torque testers have internal springs to resist rotation so that the torque required to tighten increases with increasing angular deflection. As the bottle or closure is turned, a pointer attached to the table holding the component moves along a calibrated scale. With this device packages can be sealed with closures at predetermined torque levels. The torque required to remove a closure can also be determined (12). When performing water vapor transmission tests, it is often helpful to have closures sealed at different levels of torque. In evaluating the ability of a package to form a satisfactory mechanical seal, specimens closed at various levels of torque are tested to determine the efficacy of the seal. This type of testing is described below.

Cleanliness. In industries where cleanliness of the packaging components is important, there are several tests that are employed. The simplest of these tests involves visual observation only (2). This entails the use of a bright light and sometimes a magnifying glass or microscope. In this test, the packaging component is examined under the light and turned so that different facets are illuminated. Visual observations of the object are made to determine if any foreign material is present. Another approach is the wipe test (2). A clean, lint-free cloth is utilized to wipe the test specimen. After wiping the entire specimen, the cloth is examined to determine if it has picked up any foreign material. A third method is the water break test (2). In this test, the specimen is dipped into distilled water. When the specimen is withdrawn from the water bath, it is allowed to drain, and the manner of water flow is observed. For a clean specimen the water must drain without breaking the film. Formation of small droplets or breaking of the film indicates the presence of foreign matter.

Closure efficacy. Of critical importance in packaging is the sealing efficacy of the bottle and closure. Tests to evaluate this characteristic are usually performed by exposing the package to either a partial vacuum or to pressure and observing for any evidence of leakage. The simplest form of this test is that in which the bottle is partially filled with dye, the closure applied, and the assembly is inverted in a beaker of clear water. This system is exposed to a partial vacuum for a predetermined period of time. Evidence of the dye in the beaker is an indication of leakage. Selection of the time period for the exposure and the level of partial vacuum should be based upon the requirements of the individual application. Another version of this test employs the use of bottles or closures that have been modified to incorporate a pressure fitting and gauge. In this test the package is pressurized, the source of pressure is removed, and the gauge is observed for changes in pressure. Changes in the gauge reading are indications of leakage. The level of pressure and the duration of the exposure are dependent upon the application. Another source of concern is the fact that once sealed, the packages can undergo some "relaxation" or regression of torque. This is due partially to molecular "slip" and partially to the vibration and temperature changes inherent in the distribution environment. The effect of these phenomena can be accelerated by exposing bottles sealed at known levels of torque to a simulation of the distribution environment, ie, vibration, shock, and changes in temperature. After these exposures, the removal torques are determined. Pass/fail criteria should be related to the long-term storage stability of the product.

Stability tests. In long-term stability tests, the product is stored in the package at elevated and ambient temperature conditions. Packages with screw threads should be sealed at different levels of torque. Periodic weighings of the packages are used to calculate the rate of weight change. The rate of weight change determined when the plot has either stabilized or become linear is used to predict the shelf life of the product. Of course if the resulting weight change is too severe, the package may not be suitable for long-term storage of the product. But if a torque level is found at which the product maintains a satisfactory rate of weight change, then this is the minimum torque that should be employed to determine the pass/fail criteria of the torque regression test.

Testing of blisters. Depending upon the application, blister films can be considered as part of the primary or secondary packaging (see Carded packaging). In either case, there are many tests that are performed to determine the suitability of the blister film. Because it is customary to nest blisters for production, the ability of the blisters to be denested or separated is of concern. One test to evaluate this characteristic of the film involves the determination of the coefficient of friction (13). In this test a sheet of the film is attached to a flat plate, and a second film specimen is attached to a movable sled. The sled in turn is attached to a movable head with a force-measuring device, eg, load cell or fish scale. As the head moves it moves the sled, thereby causing the film to slip over itself. Forces that are developed are measured by the load cell and displayed on a read-out device. From the data generated in this test, the coefficients of static and kinetic friction can be calculated. Of course the lower the value, the more easily the film slides over itself. Another area of concern in plastic films is brittleness. This is especially important during the winter months or when shipping in cold climates. A brittle film is more prone to cold weather cracking than a ductile one. Testing for brittleness is usually performed in accordance with ASTM D 1790 (14). This test requires that specimens of the plastic film be formed into small loops, which are affixed to 3 × 5-in. (8 × 13-cm) file cards. After conditioning at a reduced temperature, the loops are struck with the arm of a hammerlike device. By performing these tests at different temperature levels, it is possible to determine the temperature at which 50% of the specimens would break. The lower this temperature level, the less brittle the film. Because the quality of the bond to a blister card is often dependent upon the quality of the blister, it has been found useful to perform photoelastic evaluations of the formed blisters (15). If the film is not hot enough when the blister is formed, residual stresses can be developed in the blisters. When reheated during application to the blister card, the blister can shrink, thereby giving rise to the potential for an unsatisfactory bond. Thus, each new blister configuration should be evaluated with a polariscope. With the polarizing filter crossed, the ideal blister should be monochromatic. That is, there should be no contrasting light and dark areas. The polariscope consists of two polarizing light filters with the specimen held between. When viewed with light from the back, dark and light areas can be seen. The dark areas correspond to areas of residual stress. Another indication of the strength of the proposed film is its resistance to impact by a free-falling dart. Based upon ASTM D 1709 this test involves clamping a specimen of the film over an annular opening (16). A dart of adjustable weight is allowed to fall freely from a known height onto the film. The energy level at which the film is punctured is an indication of the film's strength.

Testing shipping containers.   In evaluating the suitability of shipping containers, the strength of the corrugated board is an important consideration (see Boxes, corrugated; Testing, shipping containers). To date the most widely used test in evaluating the strength of corrugated board has been the Mullen Burst Test. The detailed procedure for this test is listed in ASTM D 2529 (17). Although the Mullen Burst Test has been widely used for a number of years, many segments of the packaging industry now believe that the Short Column Crush Test of Corrugated Fiberboard provides a more realistic indication of the corrugate's strength (18). This test is described in ASTM D 2808 (19). In essence, this test requires that the corrugated fiberboard specimens with the flutes in the vertical plane be subjected to compressive forces. Other tests subject the entire shipper to compressive forces. These tests have been found useful in determining the maximum stack height of a pallet as well as the maximum number of pallets that can be stacked. Many tests have also been developed to evaluate pressure-sensitive tape that is used to assemble the shipping containers. Such tests as peel adhesion, shear strength, and the ability of a pressure sensitive tape to instantaneously adhere to a surface have been developed (20) (see Tape, pressure-sensitive).

Even though performance testing of packaging components may easily require the most rigorous and imaginative of testing procedures, this subject has received very little attention from organizations that promulgate testing standards. Performance testing can include such evaluations as the number of times a bottle can be squeezed before it will crack or the number of times an applicator can be used before it becomes deformed and useless. For the most part, no special instrumentation is required for performance testing. Usually a balance, an event counter, and graduated cylinders or beakers are the only instruments necessary. In performance testing it is customary to accelerate the test to complete the objective in a timely manner. In most cases a test is accelerated by increasing the intensity of the environment. In general, in the linear relationship between the intensity of the environment and sample life, the life decreases as the intensity is increased (21). In planning a performance test, the severity of the environment should be realistic (22). To ensure credibility and to permit reproducibility to the greatest extent, it is recommended that whenever possible, published test procedures be followed.

BIBLIOGRAPHY

1. *Mil Std-105D, Sampling Procedures and Tables for Inspection by Attributes,* Department of Defense, U.S. Government Printing Office, Washington, D.C., Mar. 20, 1964.

2. *Federal Test Method Standard 101C, Test Procedures for Packaging Materials,* U.S. Government Printing Office, Washington, D.C., Mar. 13, 1980.

3. *ASTM D 685, Method for Conditioning Paper and Paper Products for Testing,* American Society for Testing and Materials, Philadelphia, Pa., 1980.

4. *ASTM D 618, Method for Conditioning Plastics and Electrical Materials for Testing,* American Society for Testing and Materials, Philadelphia, Pa., 1981.

5. The Plastic Bottle Institute, *Test for Drop Impact Resistance of Plastic Bottles, Technical Bulletin PBI 4-1968, Rev. 1-1978,* The Society of the Plastics Industry, Inc., New York, 1978.

6. The Plastic Bottle Institute, *Method of Test for Soot Accumulation on Polyolefin Bottles, Technical Bulletin PBI 9-1978,* The Society of the Plastics Industry, Inc., New York, 1978.

7. The Plastic Bottle Institute, *Method of Test for Polyolefin Bottle Permeability and Compatibility of Packaged Products, Technical Bulletin PBI 5, Rev. 1-1978,* The Society of the Plastics Industry, Inc., New York, 1978.

8. *ASTM D 1251, Test Method for Water Vapor Permeability of Packages by Cycle Method,* American Society for Testing and Materials, Philadelphia, Pa., 1979.

9. *The United States Pharmacopeia XX, (USPXX-NFXV),* U.S. Pharmacopeial Convention, Inc., Rockville, Md. 1980.

10. R. L. Pecsok, L. D. Shields, T. Cairns, and I. G. McWilliam, *Modern Methods of Chemical Analysis,* 2nd ed., John Wiley & Sons, Inc., New York, 1976, p. 165.

11. *ASTM D 2911, Specification for Dimensions and Tolerances for Plastic Bottles,* American Society for Testing and Materials, Philadelphia, Pa., 1982.

12. *ASTM D 3198, Test Method for Measuring Application and Removal Torque of Threaded or Lug-Style Closures,* American Society for Testing and Materials, Philadelphia, Pa., 1979.

13. *ASTM D 1894, Test Method for Static and Kinetic Coefficients of Friction of Plastic Film and Sheeting,* 1978, and *ASTM D 3028, Tet Method for Kinetic Coefficients of Friction of Plastic Solids and Sheeting,* American Society for Testing and Materials, Philadelphia, Pa., 1983.

14. *ASTM D 1790, Test Methods for Brittleness Temperature of Plastic Film by Impact,* American Society for Testing and Materials, Philadelphia, Pa., 1983.

15. *ASTM D 4093, Method for Photoelastic Measurements of Birefringence and Residual Strains in Transparent or Translucent Plastic Materials,* American Society for Testing and Materials, Philadelphia, Pa., 1982.

16. *ASTM D 1709, Test Method for Impact Resistance of Polyethylene Film by the Free Falling Dart Method,* American Society for Testing and Materials, Philadelphia, Pa., 1980.

17. *ASTM D 2529, Test Method for Bursting Strength of Paperboard and Linerboard,* American Society for Testing and Materials, Philadelphia, Pa., 1974.

18. W. D. Godshall, "Mullen or Short Column Test—How Should Corrugated be Graded?", *Packag. Technol.* **14**(2), p. 44 (Oct. 1984).

19. *ASTM D 2808, Test Method for Compressive Strength of Corrugated Fiberboard (Short Column Test),* American Society for Testing and Materials, Philadelphia, Pa., 1976.

20. *PSTC-1, Peel Adhesion for Single Coated Tapes 180° Angle,* 1975; PSTC-12, Shear Strength After Solvent Immersion, 1966; and PSTC-5 *Quick Stick,* 1970, Pressure Sensitive Tape Council, Glenview, Ill.

21. C. Lipson and N. J. Sheth, *Statistical Design and Analysis of Engineering Experiments,* McGraw-Hill, Inc., New York, 1973, p. 163.

22. V. Suben, "Testing to Validate Advertising Claims," *Test Eng. Management* **45**(5), p. 10 (Oct./Nov. 1983).

R. G. McManus
Victor Suben
Noxell Corporation

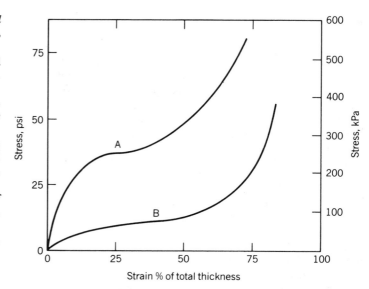

Figure 2. Typical load/deflection results for A, expanded polystyrene; B, poly(ether-urethane).

TESTING, CUSHION SYSTEMS

Cushion materials can be described as mechanical isolators used to mitigate the effects of shock and vibration on a product. "A cushion is anything interposed between one object and another to mitigate the effects of shock and vibration on the first object" is the definition given in Ref. 17.

Cushions deform in response to induced forces. They transform the relatively high G short-duration shock pulse experienced when two rigid surfaces collide, such as a product dropping on the floor, into a lower G longer-duration pulse, such as depicted in Figure 1. Package cushioning normally involves the use of materials rather than devices such as metal springs or shock-absorbing struts. Because package cushioning materials are generally not secured to the product that they surround, they work in compression or flexure rather than in shear or tensile.

Most cushion materials have certain common characteristics that we can study, measure, and use in package design.

These include

1. Compressive stress vs strain characteristics.
2. Compressive creep characteristics.
3. Compressive set.
4. Shock absorption.
5. Vibration transmissibility.

Test Methods

The characteristics of cushion materials can be quantified by the test methods described below (see Foam cushioning).

Stress–strain characteristics. A stress–strain curve is a plot showing the stress (in psi or kPa) on the vertical axis and strain or unit deformation (percent of total thickness) on the horizontal axis (see Fig. 2). The energy absorbed at any particular point is determined by calculating the area under the curve up to that point. The information is generated by measuring the force necessary to compress the material at a uniform rate. Stress–strain curves were formerly the primary influence on cushion design. A quantity known as the "cushion factor" was obtained by plotting static stress vs the ratio of the static stress to energy absorbed at that static stress (2). Because of the static nature of this test, the data are of limited value, and the test is not used extensively by packaging engineers today. However, the stress–strain curve does show the degree of cushion material linearity and an indication of cushion efficiency (available deflection in a cushion as a percent of total cushion thickness).

Compressive creep. Most flexible cushion materials settle or slowly lose their thickness when subjected to a long term compressive load. This characteristic is called "creep" and may be defined as the change in thickness of a cushion under a static compressive load over a long period of time. The procedure used to measure creep is ASTM D 2221 (3). Data obtained by this method (see Fig. 3) are applicable to the cushion under the conditions tested and are not necessarily the same as that obtained in a complete package in the distribution environ-

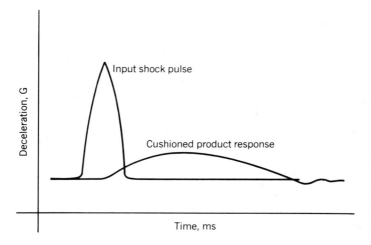

Figure 1. The effect of cushions on shock pulse.

ment. Factors that influence creep include the nature of the package itself, temperature, humidity, shock, vibration, and the static loading on the cushion material. Many resilient and semiresilient cushion materials show high rates of creep when the cushions are loaded for optimum shock and/or vibration characteristics. Yet, the creep properties are rarely investigated when package designs are in the development process. Typical values for creep range from 10% to as high as 40% or more during a 30-day period. These numbers can be higher in elevated temperature environments. This loss of thickness means that a cushion that performed adequately when packed may not provide sufficient protection after an extended period of storage. In addition, vibration characteristics of the package can be altered significantly because of the effects of compressive creep.

Compressive set.   This is a measure of the nonrecoverable loss in thickness of a cushion material after a static load is removed. This permanent set is determined using ASTM D 2221 (3). It is expressed as a percentage of the difference between the initial thickness and the thickness of the material after load removal. This information is important for material-handling applications, such as an interplant container, or where a cushion is designed for multiple use. Permanent set of 10–25% can be anticipated from many resilient and semiresilient cushions when loaded for optimum shock and vibration characteristics.

Shock absorption.   The ability of materials to mitigate shock is an important characteristic to the packaging engineer. It is necessary to know exactly what to expect when using cushion materials in a design situation. Shock performance is measured using instrumented impacts resulting in a cushion curve such as that shown in Figure 4. The cushion curve describes the level of deceleration transmitted through a given thickness of material as a function of the static stress (loading) on the material and the drop height. The procedure for generating the cushion curve is covered by various standards including ASTM D 1596, ASTM D 4168, MIL-P-26514E, MIL-C-26861B, and others (4,5).

Most commercially available cushion curves are based on ASTM D 1596. This procedure involves dropping a guided platen of predetermined mass onto a cushion of known thickness and area from a given drop height. The level of deceleration transmitted through the cushion is measured by an accelerometer mounted on the platen. The results are displayed on an oscilloscope or similar readout device. The resulting information is plotted on a graph with deceleration on the vertical axis and static stress loading on the horizontal axis (static stress equals weight divided by bearing area). Each cushion curve is drawn from a minimum of five test points (static loading lev-

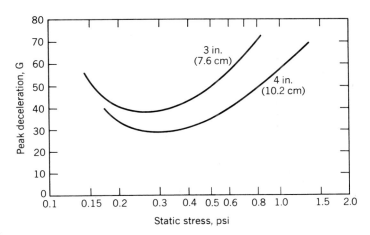

Figure 4.   Cushion curve. Cushion response data, transmitted deceleration vs static stress. Thickness, 3- and 4-in. (7.6- and 10.2-cm); drop height, 1 m (3.28 ft); test method ASTM D 4168. To convert psi to kPa, multiply by 6.893.

els), and each test point is the average of the last four of five deceleration readings from the cushion material.

Most cushion curves have the general shape of those shown in Figure 4. The left portion of the curve shows a relatively high deceleration level transmitted through the cushion. The center portion of the curve represents a more optimum loading where there is sufficient force to deflect the cushion and cause the deceleration to be spread over a longer period of time. The result is a lower deceleration level. On the right portion of the curve the material is being overloaded (it bottoms out), thus it approaches using no cushion at all, resulting in high deceleration levels.

Producers and users of cushion materials should be familiar with dynamic cushion curves in terms of the data and how they are obtained. This is important. For example, information gained from ASTM D 1596 tests will probably be different from information obtained through the use of the Enclosed Test-Block Method described in ASTM D 4168 for testing foam-in-place materials. Similarly, information generated using many of the MIL standards is intended primarily as material qualification data and the information is not necessarily applicable to cushion design. The engineer must be cognizant of how information is generated and the effect of various test procedures on a package design.

Vibration performance.   The vibration characteristics of cushion materials are determined by subjecting them to vibration inputs over the frequency range typical of the distribution environment. The procedure involves placing a test block on top of a cushion to form a spring–mass system. The resonant frequency of this system is determined by placing it on the table of a vibration machine programmed to produce a constant acceleration input (normally ½ G) while the frequency is slowly changed (or swept) from low to high, typically 3–300 Hz. The response:input ratio is plotted as a function of frequency, producing the transmissibility plot shown in Figure 5.

Resonance is that characteristic of all spring–mass systems where, at a given frequency, response acceleration is greater than the input. The resonant frequency is the point where maximum response of a spring–mass system to forced input

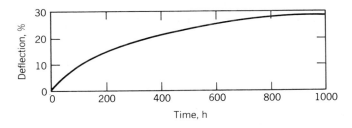

Figure 3.   Compressive creep data. Test method ASTM D 2221.

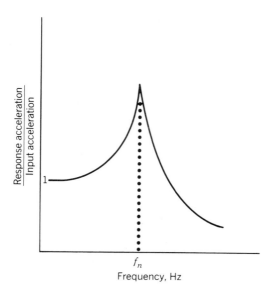

Figure 5. Cushion material transmissibility plot for one static stress loading and one thickness.

occurs. The test setup to produce these data is shown schematically in Figure 6.

The mass of the test block is changed in order to vary the static loading on the cushion, and the test is repeated. Different plots are obtained in this fashion, as shown in Figure 7.

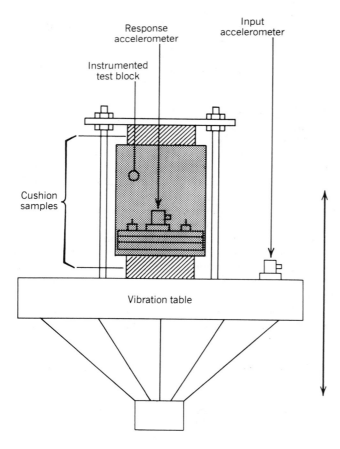

Figure 6. Vibration test setup.

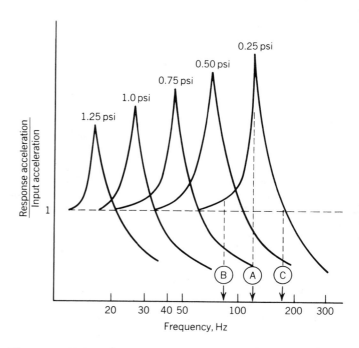

Figure 7. Multiple transmissibility plot. To convert psi to kPa, multiply by 6.893.

A series of five sweeps is recommended to plot the amplification/attenuation (A/A) plot shown in Figure 8.

The A/A plot shows frequency on the vertical axis and static stress loading on the horizontal axis. The center portion is that combination of frequency and loading that results in amplification of vibrational input. This is called the amplification zone. At lower static stress loadings and frequencies, there is a zone where the response : input ratio is approximately 1. This is the unity zone where the cushion material neither amplifies nor attenuates the input. At higher frequencies and static loading levels, the cushion material will attenuate (reduce) vibration input. This area is called the attenuation zone.

The plot in Figure 8 may be interpreted as follows. For a given frequency, the lower static stress loadings result in the same acceleration transmitted to the product as the input. In other words, the response : input ratio is approximately 1. As the static loading level increases, there is a range in which the cushion material amplifies the vibration input. At higher static loading levels, the cushion material attenuates vibration input and the response : input ratio is less than 1.

At the present time there is no recognized standard governing the procedure for running cushion vibration tests. Much of the data currently published, such as *MIL Handbook 304B* (6), were produced using the Fixture Method. This fixture is a device for restraining a test block in two axes while allowing it free movement in the vertical axis. A cushion sample is placed above and below the block and the entire fixture is mounted on the table of a vibration test machine. The remainder of the procedure is identical to that described earlier. Other procedures, including the use of the Enclosed Test Block as described in ASTM D 4168 (5), are also commonly used. It bears repeating that the packaging engineer must be knowledgeable in these various test procedures and understand the effect that test procedures have on data presentation.

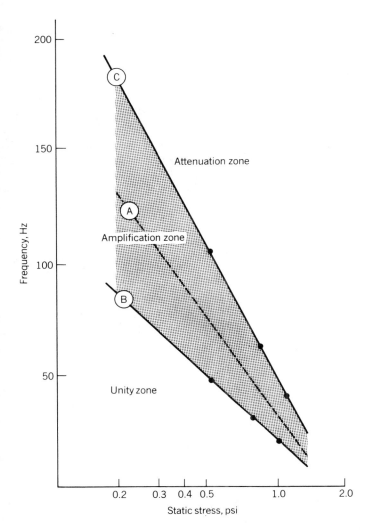

Figure 8. Amplification/attenuation plot. To convert psi to kPa, multiply by 6.893.

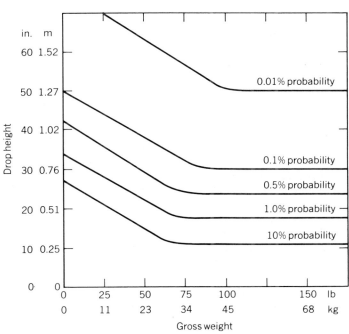

Figure 9. Typical design drop height (7).

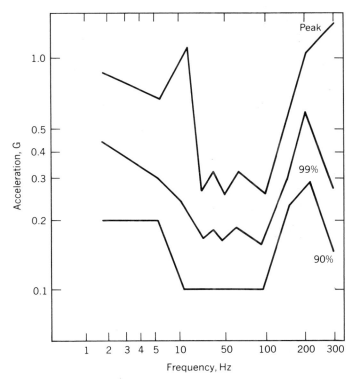

Figure 10. Typical environment data (7).

Cushioned Package Design

There are three vital pieces of information necessary for rational cushion design (see Cushioning, design). These are quantification of the probable environmental shock and vibration input, product fragility data, and cushion performance characteristics.

A necessary prerequisite for package development is a precise definition of the distribution environment (see Distribution hazards). It is during this part of the product's life cycle that the package must perform its job. The distribution environment is known to contain many potentially harmful inputs other than shock and vibration. Some of these others include temperature and humidity extremes, atmospheric pressure changes, compression, electromagnetic fields, etc. This discussion is limited to the effects of shock and vibration. The engineer must be aware of all likely hazards and quantify them when applicable in terms of their ability to cause damage to the product.

Most shock inputs occur during physical handling, especially the loading and unloading of transport vehicles. Defining the distribution shock environment amounts to quantify-

ing the drop height to which the packaged product is likely to be subjected. A number of studies have attempted to define drop height as a function of the size and weight of a package system (7). An example of these data is shown in Figure 9.

Defining the vibration environment is a more difficult task

owing to the complex and random nature of the input. It involves quantifying the acceleration vs frequency profile of vehicles in which the product is likely to be shipped. The information is presented in the form of acceleration vs frequency plots (see Figure 10). These data change with the type of vehicle, location monitored, loading of the vehicle, speed at which it travels, condition of the roadbed, and numerous other factors. A wealth of information is available to help the engineer properly quantify the distribution environment (2,6–9). The result of this process should be a design drop height for shock input and a vibration profile for vibration input.

Product fragility is determined by means of a Damage Boundary Test and a Resonance Search Test (see Testing, product-fragility). The result of product-fragility testing should be a Damage Boundary plot for each axis of the product and resonant-frequency vibration information for all critical components in all axes of the product.

To begin the protective-package-design process, the engineer assembles cushion curves and amplification/attenuation plots for the selected cushion materials. A horizontal line is drawn across the cushion curves tangent with the critical deceleration determined from Damage Boundary testing. Any

portion of the cushion curve that lies below this line will define an acceptable loading for the drop height selected. It may be economically desirable to load the cushion at the highest static stress allowed by the cushion curves, resulting in less material usage and lower cost. However, this procedure increases the potentially harmful effects of creep and compressive set.

Next the engineer draws a horizontal line across the amplification/attenuation plot tangent with the product natural frequencies determined in product vibration testing. The minimum static stress loading is determined from the intersection of the attenuation boundary and the lowest product resonant frequency (see Figure 11). Note that the designer must select the thickness and static loading levels that satisfy both shock and vibration requirements of the product. This process is facilitated by having both cushion curves and A/A plots on the same form. On both plots the horizontal axis is static stress loading, the major variable the package designer has for optimum performance and fine tuning of the package system.

The process outlined above works best for resilient and semiresilient cushions (see Foam cushioning). More rigid materials such as expanded polystyrene (EPS) tend to work best

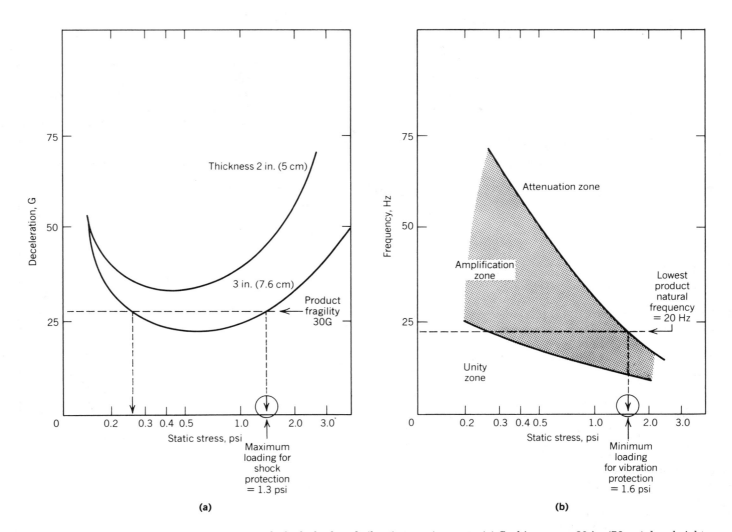

Figure 11. Determining optimum static stress for both shock and vibration requirements. (**a**) Cushion curve, 30-in. (76-cm) drop height. (**b**) Amplification/attenuation plot, 3-in. (7.6-cm) thick cushion material. To convert psi to kPa, multiply by 6.893.

with lighter static-stress loadings, especially in vibration. Because of their relatively stiff nature, these materials tend to have higher natural frequencies than resilient materials for a given loading. This situation results in a high loading necessary for traditional vibration attenuation. The solution is a relatively light loading where the shock protection from the cushion curves is adequate and the material is still in the unity zone in vibration. In this way, the shock requirements are satisfied and the material should not amplify vibration input at product critical frequencies. Note that the highest product critical frequency should be used to make this determination.

Other factors enter into the process of package design. These include fabrication cost, end user requirements, ecological considerations, flammability, electrostatic discharge protection requirements, etc (see Electrostatic discharge protective packaging). The important requirement for dynamics is a static stress-loading and cushion thickness that satisfies the product shock sensitivity and does not result in vibration amplification at product critical frequencies.

Package System Test

Verify package performance by subjecting it to the design drop height and the chosen vibration spectra. This is necessary because the design procedure involves a number of simplifying assumptions that can only be evaluated by test. Some of these are listed below:

1. Cushion response data were generated on materials that may have had an entirely different shape than those used in the final package system. Fabricated cushions are likely to show different response characteristics than flat material.

2. The product is likely to move in all three axes in response to a dynamic input, whereas the cushion material was forced to respond in only one axis during the test to generate cushion curves.

3. The effect of side-pad friction during vibration testing is neglected. Similarly, dealing with the effect of preloading is difficult. These may have a pronounced effect on vibration transmissibility at resonance.

Test procedures are available for verification of package performance and integrity. A relatively new standard, ASTM D 4169, has gained wide acceptance since its release and is recommended for this purpose (10). Performance refers to the ability of the package system to protect the product from shock and vibration. Integrity refers to the ability of the package system itself to withstand the rigors involved during a normal distribution cycle.

The process described above is applicable to designs using a wide variety of cushion materials, many of which were not available ten years ago. Today the engineer has a choice of fabricated materials, including polystyrenes, polyethylenes, polyurethanes, and others. New families of moldable materials have been added to the market, including foam-in-place polyurethanes, beaded polyethylene, copolymer moldable materials, as well as the familiar expanded polystyrene. In order to evaluate economic and performance tradeoffs the engineer must have adequate information on the materials and be able to properly interpret that information as it relates to optimum package design.

BIBLIOGRAPHY

1. G. S. Mustin, *Theory and Practice of Cushion Design*, Shock and Vibration Center, U.S. Department of Defense, Washington, D.C., 1968, p. 1.

2. *Cushion Packaging with Ethafoam*, Dow Chemical USA, Midland, Mich., 1981.

3. *ASTM D 2221-68, Standard Test Method for Creep Properties of Package Cushioning Materials*, American Society for Testing and Materials, Philadelphia, Pa., 1982.

4. *ASTM D 1596-78a, Method of Test for Dynamic Properties of Package Cushioning Materials*, American Society for Testing and Materials, Philadelphia, Pa., 1982.

5. *ASTM D 4168-82, Standard Test Methods for Transmitted Shock Characteristics of Foam-In-Place Cushioning Materials*, American Society for Testing and Materials, Philadelphia, Pa., 1982.

6. *MIL-HDBK-304B, Package Cushioning Design*, Air Force Packaging Evaluation Agency (AFALD/PTPT), Wright-Patterson Air Force Base, Ohio.

7. F. E. Ostrem, and B. Libovica, *A Survey of Environmental Conditions Incident to the Transportation of Materials*, Report PB-204 442, General American Transportation Corp., Niles, Il. 1971.

8. J. P. Phillips, "Package Design Consideration for the Distribution Environment" in *Proceedings of the 1979 International Packaging Week Assembly*, October 1979, Packaging Institute, Stamford, Conn., 1979.

9. R. J. Winne, "What Really Happens to Your Package in Trucks and Trailers," in *Proceedings of Western Regional Forum of the Packaging Institute*, Stamford, Conn., 1977.

10. *ASTM D 4169-82, Standard Practice for Performance Testing of Shipping Containers and Systems*, American Society for Testing and Materials, Philadelphia, Pa., 1982.

General References

ASTM D 775-80, Drop Test for Shipping Containers, American Society for Testing and Materials, Philadelphia, Pa., 1982.

ASTM D 999-75, Standard Methods for Vibration Testing of Shipping Containers, American Society for Testing and Materials, Philadelphia, Pa., 1982.

C. M. Harris and C. E. Crede, *Shock and Vibration Handbook*, Vols., 1, 2 and 3, McGraw-Hill Book Co., New York, 1961.

M. T. Kerr, "The Importance of Package Testing in Today's Data Processing Industry," *Proceedings of Western Regional Forum Packaging Institute*, U.S.A., 1981.

J. H. Mazzei, *Confined State Testing of Dynamic Properties of Cushioning Materials*, Technical Report RFL-TR-45, Feltman Research and Engineering Laboratories, Picatinny Arsenal, Dover, N.J., 1961.

R. D. Newton, *Fragility Assessment Theory and Test Procedure*, Monterey Research Laboratory, Inc., Monterey, Calif., 1968.

M. Schiowitz, *Package Development*, 4(1), 18 (Jan/Feb. 1974).

R. K. Stern, *FPL Dynamic Compression Testing Equipment for Testing Package Cushioning Materials*, U.S. Forest Products Laboratory Report No. 2120, Madison, Wisc., 1958.

H. H. Schueneman
Schueneman Design Associates

TESTING, PACKAGING MATERIALS

The testing of packaging materials covers the evaluation of a wide range of component parts of containers and packages. There are thousands of products and substances shipped via commercial carriers that must be protected in relation to the atmospheric, transportation, handling, and warehouse environments. Many standards and test methods pertinent to particular materials are in existence to meet the necessary requirements specified by the various agencies and classifiation committees. Sources of U.S. standards and tests are the American Society for Testing and Materials (ASTM), the Technical Association of the Pulp and Paper Industry (TAPPI), the Uniform Freight Classification Committee (UFC), the National Motor Freight Classification Committee (NMFC), the Department of Transportation (DOT), the United Nations (UN), and the International Civil Aviation Organization (ICAO) (see General References for further information).

Bags

A bag is a preformed container of tubular construction made of flexible material, generally enclosed on all sides except one that forms an opening that may or may not be sealed after filling (1). It may be made of any flexible material, or multiple plies of the same, or combination of various flexible materials (see Bags, paper; Bags, plastic; Bags, heavy-duty plastic). Since the contents of bags must have considerable resistance to crushing hazards, the outer bag materials must have the flexibility and toughness to withstand the forces induced.

Paper and multiwall bags. Five basic tests are employed to evaluate paper and paper combinations. They are basis weight, tearing resistance, bursting strength, tensile strength, and water vapor transmission.

Basis weight. Normally, this term is expressed in lb/1000 ft^2 or g/m^2. After proper conditioning, the material is cut to an appropriate size and weighed on an accurate scale or balance. Using the results of the mass and the square area, the basis weight can be calculated (2).

Tearing resistance–internal. This test method is designed to determine the average force in gram-force (gf or N) needed to tear a single sheet of paper after the tear has been originated. An elmendorf-type tester, employs the principle of a pendulum, making a single swing and causing the tearing of one or more paper sheets at one time through a fixed distance. The work required for this tearing is measured in gram-force (actually gf · m or N · m = J) as the loss in potential energy of the pendulum. This test is performed, after proper conditioning, in two directions, one with the tear parallel to the machine direction and the other with the tear perpendicular to the machine direction (3).

Bursting strength. This test method is used to determine the amount of hydrostatic pressure in lb/in^2 (kPa) required to obtain a rupture of the material. The pressure is applied at a controlled increasing rate through a rubber diaphragm. A fixed lower plate and an upper adjustable plate clamp the material in position around the circumference with sufficient pressure to prevent slippage during the test but to allow the center to bulge under pressure during the test (4,5).

Tensile strength. This test method determines the resistance of paper to direct tensile stress. The paper, cut to a specific width, is held in place by two clamps, one fixed and one movable, which are aligned to hold the material in the designated plane without slippage. The load is applied at a rate that induces failure within a specified time period. After proper conditioning, the test is performed on specimens cut from both principal directions of the paper (6,7). Certain bag materials are subjected to stress when wet. A wet-strength device is attached to the lower clamp so that the paper remains saturated for the appropriate period of time. The load is applied again at a rate that involves failure within a specified time period. This test is also performed on specimens cut from both principal directions of the material (8,9).

Water vapor transmission. There are test methods applicable to the sheet material, such as paper, multiwall sacks, and plastic films. The purpose of these methods is to determine a rate of water vapor transmission, which is designated as the time rate of water vapor flow normal to two specified parallel surfaces, under steady conditions, through unit area, under the conditions of test. The accepted unit is one gram per 24 hours per square meter (g/(m^2 · d). Open-mouth test dishes are used, with the specimens attached to the top of the dishes by means of a sealant. The dishes contain either a desiccant or water, depending on the method. The complete assemblies are inserted in the appropriate atmosphere, and after specific periods of time, are weighed accurately to determine gain or loss. The test is conducted until a nominal steady state exists (10,11) (see Testing, permeation and leakage).

Plastic film bags. In addition to being evaluated for weight, tearing resistance, bursting strength, tensile strength, and water vapor transmission, plastic bag and sack materials are also tested for heat-seam strength and impact resistance when applicable (see Films, plastic; Sealing, heat).

Heat-seam strength. 1-in. (25.4-mm)-wide bands are cut perpendicular to the seam, with the seam located in the center of the test length. After proper conditioning, the individual specimen is placed in the grips of the testing machine, which are then tightened evenly to prevent slippage during the test. The movable grip pulls the specimen apart at the seam at a specified rate, and the maximum load is recorded. Specimens are tested from seams, both parallel and perpendicular to the machine direction of the material (12).

Impact resistance, free-falling dart. This method determines the energy that causes polyethylene film to fail under specified conditions of impact of a free-fall dart having a hemispheric head. The method specifies polyethylene film, but it applies to other films as well. The dart mass (g), dropped from a specific height, is adjusted to give a 50% failure rate of the specimen. After proper conditioning, a large enough perfect specimen is clamped in place by an air-operated horizontal upper-gasketed steel ring. Sufficient pressure must be employed to prevent specimen slippage. An adjustable bracket holds the dart head, which is positioned vertically above the center of the film and released from a predetermined height to impact the specimen. The specimen is then inspected to determine if failure has occurred. The dart mass is adjusted accordingly to arrive at a point where 50% of the film specimens fail (13).

Textile bags. Two principal textiles, burlap and cotton, are fabricated into sacks. These textiles are laminated with different combinations of papers or films to provide the necessary protection for the application desired. Tensile strength, tearing resistance, and waterproof characteristics are important factors in determining a suitable material.

Bottles

A bottle is designated as a hollow vessel of glass, earthenware, plastic, or similar substance, with a narrow neck or mouth and without handles. It must have a suitable closure to prevent leakage.

Glass. The primary ingredients of glass are soda ash, limestone, and sand. Coloring agents are added to obtain a wide variety of color combinations. Various tests are performed on the completed container in relation to size, thickness, weight, capacity, impact resistance, internal pressure resistance, and thermal shock resistance.

Plastic. Many thermoplastic materials are employed to produce plastic bottles. The type of product being packaged determines which thermoplastic is most applicable. Clarity and oxygen and moisture barriers are important considerations. In addition to tests performed on completed containers, such as those specified for glass bottles, evaluations are conducted in relation to heat, cold, sunlight, stiffness, and resistance to acids, alkalis, oils, and solvents.

Boxes

A box is a rigid container having faces and completely enclosing the contents (14). When this term is used in the classifications, its signifies that if fiber boxes (corrugated or solid fiber) are used, such fiber boxes must comply with all the requirements of NMFC Item 222 and UFC Rule 41. NMFC Item 220 also states that boxes are containers with solid or closely fitted sides, ends, bottoms, and tops. They must be made of wood, metal, plastic, fiberboard, or paperboard and foamed or cellular plastic combined, and completely enclose their contents (14).

Corrugated and solid fiberboard boxes. Corrugated boxes are made of corrugated board, a structure formed from two or more paperboard facings and one or more corrugated members.

Solid fiberboard is a solid board made by laminating two or more plies of container board (14) (see Boxes, corrugated; Boxes, solid fiber). The following tests are required for evaluating corrugated or solid fiberboard: basis weight; bursting and short column- or edgewise-compression strength of corrugated fiberboard; flat crush of corrugating medium; internal tearing resistance of paper; moisture content of paper board, ply separation (wet); puncture and stiffness; ring crush of paperboard; static bending and thickness (see Testing, shipping containers).

Basis weight. See Ref. 2 for further information.

Bursting strength. See Refs. 4 and 5 for further information.

Short column- or edgewise-compression strength of corrugated fiberboard. This test method determines the edgewise-compressive strength, parallel to the flutes, of combined corrugated fiberboard. It is used to compare different material combinations or different apportions of similar corrugated fiberboard. The specimens are accurately cut, usually by a circular saw blade, to a specific width and height so that the flutes are in the vertical plane during the test. Each loading or long edge is dipped to an approximately ¼-in. (6.4-mm) depth and allowed to dry. After proper conditioning, the individual specimen, held vertically by guide blocks, is placed on a bottom platen of a compression tester. The upper parallel platen applies a load to the top of the specimen, the guide blocks are removed, and the pressure is increased until failure occurs. The force is recorded and the average maximum load per unit width is calculated as lb/in. or kg/cm (15,16).

Flat crush of corrugated fiberboard. This test method, primarily used for single-face or single-wall corrugated fiberboard, determines the resistance of the flutes to crushing when pressure is applied perpendicularly to the board surface. Its

purpose is to give a general level of quality in relation to the corrugated board fabrication and whether or not the flutes have been damaged during the printing process. The specimens are usually cut very carefully into a 10 in.^{2} (65 cm^2) circular size. After proper conditioning, the specimen is placed in a flat position on the bottom of the platen of a compression tester. The upper parallel platen applies a load to the top of the specimen until the side walls of the corrugations collapse and fail. The force is recorded and the load is calculated in lbf/in.^{2} or kPa (17,18).

Internal tearing resistance. See Ref. 3 for further information.

Moisture content of paperboard. This test method determines the amount of water in paper and paperboard materials. The individual specimen is removed from unsealed and unprinted sections of a container, set in a holder with a known tare weight, and weighed on a precision balance. The specimen and holder are then placed in an oven set at 105°C and held at that temperature for a minimum of 2 h. The specimen and holder are set into a desiccator, allowed to cool for 1 h, removed from the desiccator, and reweighed. The difference between the original weight and oven-dried weight is calculated as the percentage of moisture in the specimen (19,20).

Ply separation (wet). This test method determines if solid or corrugated fiberboard fabricated with weather-resistive adhesive separates after being exposed directly in water. The individual specimens are placed in a freshly aerated water tank for 24 h (the corrugated fiberboard is set with the flutes in the vertical plane), then removed and allowed to drain. The specimens are then examined at the edges for adhesion between components. If there is delamination, the amount of the separation is specified in in. or cm from the edge (21).

Puncture and stiffness. This test method determines the resistance of paperboard, corrugated fiberboard, and solid fiberboard to puncture. The specimen to be tested is placed between clamping jaws. A triangular pyramid, attached to a pendulum arm, travels through an arc to strike the underside of the specimen. The resistance to puncture or energy required is measured in units, where one unit equals 0.265 lbf-in. or 2.99 N · cm (0.03 J). In determining stiffness, a modified test procedure is used wherein the material is slit prior to releasing the puncture head (22,23).

Ring crush. This test method determines the resistance of paper having a maximum thickness of 0.036 in. or 0.9 mm to edgewise compression. The specimen, 0.5 in. (12.7 mm) wide and 6 in. (153 mm) long, is placed in a holding block having a circular groove. The protruding specimen and holder are positioned on a bottom platen. An upper parallel platen, moving at a uniform rate, applies pressure to the edge of the specimen until it fails. Specimens are tested from samples in both the machine direction and cross-machine direction of the paper. The maximum load is registered in lbf or N (24,25).

Static bending. This test method determines the modulus of elasticity or stiffness of single-wall or double-wall corrugated board. The higher the modulus, the lower the deflection, and vice versa. After proper conditioning, a rectangular specimen is placed on two round-edge supports, usually 9 in. (229 mm) apart. A dial gauge is placed underneath the specimen halfway from each support. A round-edge loading head travelling downwards applies the force at the center, halfway from each support, in line with the dial gauge. Load and deflection readings are taken concurrently so that curves can be established.

Specimens are tested with corrugations both parallel and perpendicular to the length. Loads are measured in lbf or N. Deflection is measured in 0.001-in. or 0.025-mm increments (26).

Thickness. This test method determines the thickness of paper and paperboard by measuring the perpendicular distance between two principal surfaces. After proper conditioning, the specimen is placed between the contact surface of a dial type micrometer and a pressure foot is lowered to record the thickness. The value is listed in in. or mm (27).

Wood boxes and crates.   Crates are containers constructed of members made of wood or metal with apertures between, or members made of wood or metal combined with fiberboard, securely nailed, bolted, screwed, riveted, welded, dovetailed, or wired and stapled together having sufficient strength to hold the article packed therein so as to protect it from damage when handled or transported with ordinary care (28). A crate is usually a framework or open container. Boxes have closed faces (see Boxes, wirebound; Boxes, wood). Crates must be constructed so as to protect contents on the sides, ends, tops, and bottoms, and in such manner that the crate containing its contents may be taken into or out of the vehicle. Contents must be securely held within crates and no part shall protrude, unless otherwise provided in individual items. Surfaces liable to be damaged must be fully covered and protected. Wood boxes and crates are still in demand when specific requirements are called for, such as high compressive or stacking loads, stiffness, resistance to puncture (as a box) and the ability to maintain their strength characteristics under high humidity or very wet conditions. Certain military specifications (see Military packaging) still call for the use of wooden containers to meet the rigors of transportation, handling, and atmospheric environments under the most adverse conditions. Wood combinations are being used more frequently. This category would include plywood, structural-sandwich construction, wood-base fiber and particle panel materials, and modified woods and paper-base laminates. Component materials are evaluated for the properties given below.

Physical properties: appearance, moisture content, shrinkage, density, working qualities, weathering, thermal and electrical properties, and chemical resistance.

Mechanical properties: elastic strength and vibration characteristics, influence of growth, and effect of manufacturing and service environment.

Plastic boxes.   Plastic boxes constructed of high density polyethylene, self-supporting, rigid construction, not extruded nor expanded, must be molded by either an injection molding, a blow molding, rotational molding, or thermal molding process. Tops or covers must be securely affixed. The basic material-properties tests for plastic boxes are specified as melt index, density, tensile strength, and elongation.

Melt index or flow rates. This test is performed on the resin material, as specified by the UFC. This method determines the rate of extrusion of a molten resin through a die under specific temperature, load, and piston position conditions. A deadweight piston plastometer is used, consisting of a thermostatically controlled heated steel cylinder, a die at the lower end, and a piston inside the cylinder. After the test specimen, in the form of powder, granules, or strips, has been inserted in the cylinder base, a preheat cycle is originated. The weighted piston, as designated by the material, is inserted, and a purge sequence takes place. At the specified test temperature, as designated by the material, additional material is inserted after the preheat cycle, and the rate of flow is measured for a specific time period. The extrudate is cooled and weighed accurately in milligrams. The flow rate is designated in terms of grams per 10 minutes (29).

Density. This test method determines the density of solid plastics for identification purposes, to verify uniformity, and to learn if any physical changes have taken place. A temperature-controlled density-gradient tube is used, with different density liquids inserted. Calibrated glass floats are also inserted to cover the various density ranges. Three test specimens are carefully placed inside the tube and allowed to reach equilibrium. The test specimens' densities are determined by their relation to the position of the calibrated glass floats. The reading is specified in g/cm^3 (30).

Tensile strength and elongation. This test method determines the maximum tensile stress and the increase in length produced in the gauge length of the specimen by a tensile load. Tensile and elongation values are helpful for quality control purposes and for research and development studies. The specimens are prepared in the form of "dogbones" by machining, die cutting, or molding. The center section is narrower to assure breakage in that area. Gauge marks or extension indicators are placed in the middle of the narrower center section to measure elongation. After proper conditioning, the specimen is placed in the grips of the testing machine, which are then tightened evenly to prevent slippage during the test, but not too tight to cause crushing of the material. The movable grip pulls the specimen apart at a specified rate, and the load and elongation are recorded accordingly until ultimate failure occurs. The load value divided by the unit area determines the tensile strength in psi or MPa. The elongation, measured as an increase in length of inches or millimeters, is calculated as a percentage increase in relation to the original gauge length (31).

Bulk Containers

To reduce shipping and handling costs, dry or solid materials such as plastic pellets, sand, flour, and chemicals are transported in bulk containers. Provisions are made in certain corrugated fiberboard containers for rapid product removal. Steel and rigid plastic bins are used to handle automotive parts in bulk. Large plastic bags, capable of transporting loads up to 6000 lb (2722 kg) are employed for both domestic and export shipping (see Intermediate bulk containers). These bulk units are designed with special straps for ease of handling to accommodate commercial land carriers and ship operations. The test methods for the component materials of the various bulk containers have been described above (see Testing, shipping containers).

Cans

A can is a receptacle generally of ≤ 10-gal (38-L) capacity, normally not used as a shipping container. The body is made of lightweight metal, plastic, or is a composite of paperboard and other materials having the ends made of paperboard, metal, plastic, or a combination thereof. There is intense competition today among can suppliers to furnish the optimum container. Different can designs, innovative composite combinations, and special attractive features help make this type of container an ever-changing, progressive part of the packaging industry.

The test methods for the component materials of the various types of cans, such as tensile strength and moisture vapor transmissions, have been described earlier.

Cartons

A folding box is generally made from boxboard for merchandising consumer quantities of products (see Cartons, folding). A carton serves a variety of purposes. It must protect a specific amount of contents against warehouse and transportation environments, the store shelf, and home storage hazards; be appealing and send a powerful selling message; it must be small and compact, handle easily, and be capable of dispensing the contents readily; preserve freshness, taste, odor, appearance, and original form; and prevent contamination. It must be produced efficiently and economically and be capable of traveling through packaging systems at high speeds. The board stock must be material that can be folded and creased without breaking.

Paperboard cartons. Many paperboards (see Paperboard) are now being treated with coatings or laminated with other materials to provide the necessary protection for the product. The basic materials are tested for basis weight, thickness, water vapor transmission, and folding endurance.

Basis weight. See Ref. 2 for further information.

Thickness. See Ref. 27 for further information.

Water vapor transmission. See Refs. 10 and 11 for further information.

Folding endurance. This test determines the resistance of paper to folding. After proper conditioning, the specimen is placed between jaws of a tester. A specified test is applied to the test strip, and the machine is activated to give the necessary number of double folds per minute until it breaks. Specimen strips are cut from each principal direction of the paper (32,33,34).

Additional protection tests can be performed in relation to corrosion, mold resistance, possible migration, antitarnishing, heat-sealing characteristics, and abrasion resistance.

Plastic sheet packages. Although plastic sheet packages are not designated as cartons, their application and usage are quite similar. They must meet the same wide variety of purposes that were described previously for cartons. The majority of units produced are blister packs (see Carded packaging), trays, cups, and boxes (see Boxes, rigid plastic). Chipboard material is often combined with the formed plastic sheet to provide printing surfaces for visual merchandising. Test methods for the plastic sheets have been described in previous paragraphs (10,30,31).

Drums and Pails

Drums are cylindrical shipping containers of 3–165-gal (11.4–625-L) capacity, without bilge, with or without bails or handles, and must be made of wood, fiberboard, metal, plastic, or rubber (35).

Pails are containers (1–7-gal or 4–26.5-L capacity) with heads or covers, with or without bails or handles, with bilge, and must be made of plastic or metal.

Fiberboard drums. Fiberboard drums can be used to hold dry or solid materials. They are fabricated with either fiberboard tops and bottoms, steel tops and bottoms or plastic tops and fiberboard bottoms. The fiberboard side walls, tops, and bottoms are evaluated for bursting strength and thickness using two test methods described previously. The steel tops and bottoms are checked for thickness. The plastic tops are checked for density, melt index, and tensile strength using three test methods described above. In addition, the material is checked for its vicat softening point and stress cracking.

Vicat softening point. This test method determines the temperature at which a needle penetrates the specimen under a specific condition. The specimen is placed on a support so that it is under the needle. The entire assembly is placed in an oil bath with a temperature-measuring device as close as possible to the specimen. A dial indicator measures the amount of needle penetration. A specified load is set on top of the needle assembly, and the oil, under constant stirring, is heated at a specified rate. The temperature is recorded when the indicator shows 1-mm (0.04-in.) penetration (36).

Stress Cracking. This test method determines the susceptibility of polyethylene plastics to failure by cracking when exposed to a surface-active agent such as Igepal CO-630. Individual specimens are cut to size, then notched to a specific length and depth. They are bent into position and placed into a brass-channel specimen holder. The entire assembly is set inside a test tube, the agent is induced, and the test tube is sealed. The sealed test tube is inserted into a controlled-temperature hot-water bath for a specified period, after which time the specimens are examined for failure (37).

Plastic drums and pails. Plastic drums (see Drums, plastic) and plastic pails (see pails, plastic) are either of removable-head or tight-head construction. They are suitable for liquid, dry, or solid materials. The plastic components are subjected to melt index, density, tensile strength, and percent elongation tests, using methods that have been described above. When steel covers are installed on plastic pails, the steel gauge is verified by a precision micrometer.

Aluminum or steel drums. Aluminum or steel drums and pails (see Drums and pails, steel) have their side walls and top and bottom heads checked for proper thickness by means of a precision micrometer. Tensile strength and percent elongation tests described above can also be conducted to verify that quality standards have been maintained.

Because of space limitations, certain tests for packaging materials required for other containers and packages, such as aerosols, ampuls, carboys, cups, envelopes, kegs, pouches, tubes, and vials have not been included in this article. For information on testing of cushioning materials, integral components of many packages, see Testing, cushion systems. ASTM and TAPPI both publish detailed indexes with cross-references as guides for locating the right tests for the right materials.

BIBLIOGRAPHY

1. ASTM D 996-83, *Standard Definition of Terms Relating to Packaging and Distribution Environments,* American Society for Testing and Materials, Philadelphia, Pa. 1983.

2. TAPPI T 410-83, *Grammage of Paper and Paperboard (weight per unit area),* Technical Association of the Pulp and Paper Industry, Atlanta, Ga., 1983.

3. TAPPI T 414-82, *Internal Tearing Resistance of paper,* Technical Association of the Pulp and Paper Industry, Atlanta, Ga., 1982.

4. ASTM D 774-71, *Bursting Strength of Paper,* Vol. 15.09, American Society for Testing and Materials, Philadelphia, Pa, 1971.

5. TAPPI T 403-85, *Bursting Strength of Paper,* Technical Association of the Pulp and Paper Industry, Atlanta, Ga., 1985.

6. ASTM D 828-71, *Tensile Strength of Paper and Paperboard,* Vol. 15.09, American Society for Testing and Materials, Philadelphia, Pa., 1971.

7. TAPPI T 404-82, *Tensile Breaking Strength and Elongation of Paper and Paperboard,* Technical Association of the Pulp and Paper Industry, Atlanta, Ga., 1982.

8. ASTM D 829-76, *Tensile Strength of Paper and Paperboard (Wet),* Vol. 15.09), American Society for Testing and Materials, Philadelphia, Pa., 1976.

9. TAPPI T 456-82, *Wet Tensile Breaking Strength of Paper and Paperboard,* Technical Association of the Pulp and Paper Industry, Atlanta, Ga., 1982.

10. ASTM E 96-80, *Water Vapor Transmission of Sheet Material,* Vol. 15.09, American Society for Testing and Materials, Philadelphia, Pa., 1980.

11. TAPPI T 464-85, *Gravimetric Determination of Water Vapor Transmission Rate of Sheet Materials at High Temperature and Humidity,* Technical Association of the Pulp and Paper Industry, Atlanta, Ga., 1985.

12. L-P-378D, *Para. 4.3.8, Heat Seal Strength, Plastic Sheet and Strip,* Federal Specification, U.S. Government Printing Office, Washington, D.C. 1973.

13. ASTM D 1709-80, *Impact Resistance of Polyethylene Film by the Free Falling Dart,* Vol. 08.02, American Society for Testing and Materials, Philadelphia, Pa., 1980.

14. Item 222-6, National Motor Freight Classification National Motor Freight Classification Committee, Washington, D.C., 1984; Rule 41, Sect. 14, Uniform Freight Classification, Uniform Freight Classification Committee, Chicago, Ill., 1985.

15. ASTM D 2808-76, *Compressive Strength of Corrugated Fiberboard (Short Column Test),* Vol. 15.09, American Society for Testing and Materials, Philadelphia, Pa., 1976.

16. TAPPI T 811-83, *Edgewise Compressive Strength of Corrugated Fiberboard (Short Column Test),* Technical Association of the Pulp and Paper Industry, Atlanta, Ga., 1983.

17. ASTM D 1225-71, *Flat Crush of Corrugated Fiberboard* Vol. 15.09, American Society for Testing and Materials, Philadelphia, Pa., 1971.

18. TAPPI T 808-81, *Flat Crush Test of Corrugated Board,* Technical Association of the Pulp and Paper Industry, Atlanta, Ga., 1981.

19. ASTM D 644-82, *Moisture Content of Paper and Paperboard by Oven Drying,* Vol. 15.09, American Society for Testing and Materials, Philadelphia, Pa., 1982.

20. TAPPI T 412-83, *Moisture in Paper,* Technical Association of the Pulp and Paper Industry, Atlanta, Ga., 1983.

21. TAPPI T 812-85, *Ply Separation of Solid and Corrugated Fiberboard,* Technical Association of the Pulp and Paper Industry, Atlanta, Ga., 1985.

22. ASTM D 781-73, *Puncture and Stiffness of Paperboard and Corrugated and Solid Fiberboard,* Vol. 15.09, American Society for Testing and Materials, Philadelphia, Pa., 1973.

23. TAPPI T 803-85, *Puncture and Stiffness Test of Container Board,* Technical Association of the Pulp and Paper Industry, Atlanta, Ga., 1985.

24. ASTM D 1164-73, *Ring Crush of Paperboard,* Vol. 15.09, American Society for Testing and Materials, Philadelphia, Pa., 1973.

25. TAPPI T 818-82, Compression Resistance of Paperboard (Ring Crush), Technical Association of the Pulp and Paper Industry, Atlanta, Ga., 1982.

26. ASTM D 1098-73, *Static Bending Test for Corrugated Paperboard,* Vol. 15.09, American Society for Testing and Materials, Philadelphia, Pa., 1973.

27. TAPPI T 411-84, *Thickness (Caliper) of Paper, Paperboard and Combined Board,* Technical Association of the Pulp and Paper Industry, Atlanta, Ga., 1984.

28. Item 245, National Motor Freight Classification, National Motor Freight Classification Committee, Washington, D.C., 1984.

29. ASTM D 1238-82, *Flow Rates of Thermoplastics by Extrusion Plastimeter,* Vol. 08.01, American Society for Testing and Materials, Philadelphia, Pa., 1982.

30. ASTM D 1505-79, *Density of Plastics by the Density-Gradient Technique,* Vol. 08.01, American Society for Testing and Materials, Philadelphia, Pa., 1979.

31. ASTM D 638-82(a), *Tensile Properties of Plastics,* Vol. 08.01, American Society for Testing and Materials, Philadelphia, Pa., 1982.

32. ASTM D 2176-82, *Folding Endurance of Paper by the M.I.T. Tester,* Vol. 15.09, American Society for Testing and Materials, Philadelphia, Pa., 1982.

33. TAPPI T 423-84, *Folding Endurance of Paper (Schopper Type Tester),* Technical Association of the Pulp and Paper Industry, Atlanta, Ga., 1984.

34. TAPPI T 511-83, *Folding Endurance of Paper (M.I.T. Tester),* Technical Association of the Pulp and Paper Industry, Atlanta, Ga., 1983.

35. Item 255, National Motor Freight Classification, National Motor Freight Classification Committee, Washington, D.C., 1984; *Rule 40, Sect. 4,* Uniform Freight Classification, Uniform Freight Classification Committee, Chicago, Ill, 1985.

36. ASTM D 1525-82, *Vicat Softening Temperature of Plastics,* Vol. 08.01, American Society for Testing and Materials, Philadelphia, Pa., 1982.

37. ASTM D1693-80, *Condition B, 50°C—Environmental Stress-Cracking of Ethylene Plastics,* Vol. 08.02, American Society for Testing and Materials, Philadelphia, Pa., 1980.

General References

Annual Books of ASTM Standards, Vols. 08.01, 08.02, 15.09, American Society for Testing and Materials, Philadelphia, Pa., 1984.

Test Methods, Technical Association of the Pulp and Paper Industry, Atlanta, Ga., 1984.

ICC-UFC No. 6000C, Uniform Freight Classification Committee, Chicago, Ill., 1985. Standards pertain to shipment by rail.

ICC-NMF No. 100L, National Motor Freight Classification Committee, Washington, D.C., 1984. Standards pertain to shipment by truck.

Code of Federal Regulation, Title No. 49, Department of Transportation, U.S. Government Printing Office, Washington, D.C., 1983, Pts. 178–199. Standards pertain to shipments of hazardous materials.

Transport of Dangerous Goods, United Nations, International Regulations Publishing and Distributing Organization, Chicago, Ill., 1984.

Techical Instruction for the Safe Transport of Dangerous Goods by Air, International Civil Aviation Organization, International Regulations Publishing and Distributing Organization, Chicago, Ill., 1984.

CHESTER GAYNES
Gaynes Testing Laboratories, Inc.

TESTING, PERMEATION AND LEAKAGE

There are two ways by which gases and vapors can enter or leave a package: permeation and leakage (1). Leakage is the passage of gases and vapors through discontinuities in a material, such as cracks, pinholes, and microscopic gaps between surfaces (eg, between a closure and a bottle neck·finish) (see Closures). It is a combination of convection (forced flow caused by a total pressure gradient) and diffusion (molecular motion caused by a partial-pressure gradient). For packaging applications, it is very important to determine whether leakage is occurring and it is also useful to know the leakage rate and location for package design and development work. Permeation is the passage of gases and vapors directly through a material by dissolution into one surface, diffusion through the bulk of the material, and desorption from the other surface, all caused by a partial-pressure gradient. For packaging applications, it is very important to determine the permeation rate through the walls of a production package.

Both mechanisms are operative in all packages, but one is often negligible when compared to the other. For example, in a glass bottle with a plastisol-lined metal closure, permeation through the glass, metal, and plastisol may be negligible when compared to leakage between the surface of the neck finish and the closure liner (see Closure liners). On the other hand, leakage of oxygen through the heat seals (see Sealing, heat) in a low density polyethylene pouch may be negligible when compared to permeation directly through the material (see Barrier polymers). Most materials and package tests measure total transmission rates through the sample which are a combination of the permeation and leakage rates. Leakage through pinholes in a film increases the measured transmission rate above that occurring by permeation alone. Weight loss from an aqueous product packaged in a plastic bottle results from permeation through the walls as well as from leakage around the closure (see Fig. 1). For package design and development, it is important to separate leakage and permeation rates, since different design changes must be made to decrease permeation rates or to decrease leakage rates. For example, for a linerless plastic closure, a change to a material which has a lower oxygen permeability coefficient will decrease permeation through the closure; a change to a softer material can improve the mechanical seal and thus decrease leakage between the closure and the neck finish.

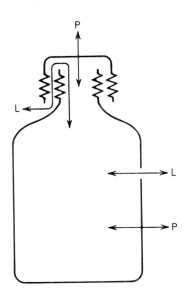

Figure 1. Schematic representation of potential permeation P and leakage L pathways in a plastic bottle with a linerless plastic closure.

Two basic sample types are used in permeation and leakage experiments: materials (films or flat sheets) and packages (see Testing, consumer packages; Testing, packaging materials). The most useful information about end-use performance is obtained by testing a package. Data for a plastic film may differ considerably from data for a sheet cut from a blow-molded bottle of the same resin due to differences in orientation and crystallinity resulting from the different processing methods. Material from a flexible pouch may have a higher transmission rate than the material from the roll before it was formed, filled, and sealed if the material is crease-sensitive (see Form/fill/seal, horizontal; Form/fill/seal, vertical). Also, no indication of the potential for leakage in the heat seals would be obtained from a materials test either before or after the F/F/S operation. Whenever possible, package tests should be performed in preference to materials tests.

Permeation

Gas permeability coefficients are usually reported for use in the packaging industry as $cm^3(STP)\cdot mil/(100\ in.^2\cdot d\cdot atm)$, $cm^3(STP)\cdot mil/(m^2\cdot d\cdot atm)$, or $cm^3(STP)\cdot \mu m/(m^2\cdot d\cdot kPa)$. The temperature and relative humidity used in the test must be reported along with the units. The units represent the volume of the gas which would pass through a piece of the packaging material which is 1 mil (25.4 μm) thick and has a surface area of 100 in^2 (0.065m^2) or 1 m^2 in a 24-h period if the partial pressure differential across the film is 1 atm (101.3 kPa). Permeability coefficients of vapors are reported as $g\cdot mil/(100\ in.^2\cdot d)$, $g\cdot mil/(m^2\cdot d)$, or $g\cdot mm/(m^2\cdot d)$. The temperature and relative humidity must be reported along with the units for water vapor transmission rates. The temperature, relative humidity, and permeant pressure must be reported for vapors other than water vapor. Since there is a factor of 15.5 between 100 in^2 (0.065 m^2) and 1 m^2, it is important to report the units properly. Also, it should be kept in mind that there are a different number of molecules in 1 g and 1 cm^3(STP).

In the initial stages of a permeation experiment, the permeation rates are continually increasing as the gas begins to dissolve into and diffuse through the material. Eventually, a steady-state permeation rate develops which is the maximum possible rate for that sample. Generally, the faster a gas passes through a package, the shorter the shelf life of the product inside (see Shelf life). Therefore, the minimum shelf life will be calculated from the steady-state permeation rate. It is of the utmost importance to be sure that the steady-state rate has been measured in an experiment, so that a conservative estimate can be made of the shelf life. ASTM and TAPPI standard tests (2,3) are summarized in Tables 1 and 2, respectively. A general description of test methods follows.

Weight gain or loss. One of the simplest methods of measuring transmission rates is to package a liquid product (aqueous or nonaqueous) and to store the sample in an environment containing none of the volatile ingredients which could escape from the product. A dry product or desiccant can be packaged and stored in a high-humidity environment to monitor entry of water vapor into the package. In either case, the sample is weighed periodically. Ideally, the steady-state transmission rate can be found from the slope of a weight-versus-time plot (4,5). A common method for measuring water vapor transmission rates of flat-sheet samples is to mount and seal the sample over a metal cup containing a desiccant. The test cell is stored at constant temperature and humidity and weighed periodi-

Table 1. ASTM Test Methods for Permeation and Leakage of Gases and Vapors

Test number	Test title	Sample type	Diffusant	Detection method	Test conditions
D 726-58 (1971)	Standard Test Methods for Resistance of Paper to Passage of Air	material	air	volume change	4.9 and 12.2 in. of water (1.22 and 3.04 kPa) and another undefined pressure
D 814-81	Standard Test Method for Permeability of Vulcanized Rubber to Volatile Liquids	material	volatile liquids	weight loss	volatile liquid on one side; absence of permeant on other side, 77°F (25°C)
D 895-79	Standard Test Method for Water Vapor Permeability of Packages (currently under revision)	package	water vapor	weight gain	usually 90% rh and 100°F (37.8°C)
D 1008-64 (1977)	Standard Test Methods for Water Vapor Transmission of Shipping Containers	package	water vapor	weight gain	usually 90% rh and 100°F (37.8°C)
D 1251-79	Standard Test Method for Water Vapor Permeability of Packages by Cycle Method	package	water vapor	weight gain	1 cycle = 24 h at 0°F (−17.8°C) followed by 6 d at 90% rh and 100°F (37.8°C)
D 1276-68 (1978)	Standard Test Method for Water Vapor Transmission of Shipping Containers by Cycle Method	package	water vapor	weight gain	1 cycle = 1 wk at 0°F (−17.8°C) followed by 3 wk at 90% rh and 100°F (37.8°C)
D 1434-82	Standard Method for Determining Gas Permeability Characteristics of Plastic Film and Sheeting	material	any dry gas	pressure or volume change	partial pressure, temperature, humidity, and hydrostatic pressure not specified, but should be reported
D 2684-73 (1979)	Standard Recommended Practice for Determining Permeability of Thermoplastic Containers	package	product or water vapor	weight gain or loss	50% rh and 73°F (23°C) or 122°F (50°C) at a measured rh
D 3078-72 (1977)	Standard Test Method for Leaks in Heat-Sealed Flexible Packages	package	product or headspace gases	visual observation of bubbles or dyes	5, 10, 15, or 27 in. of mercury (17, 24, 51, or 91.5 kPa) for 30 s
D 3079-72 (1977)	Standard Test Method for Seepage Rate of Aerosol Products	package	water vapor	weight gain	90% rh and 100°F (37.8°C)
D 3094-72 (1981)	Standard Test Method for Seepage Rate of Aerosol Products	package	product or propellant	volume change	80°F (26.7°C) with package submerged in water
D 3199-79	Standard Test Method for Water Vapor Transmission Through Screw-Cap Closure Liners	package	water vapor	weight gain or loss	25% rh and 100°F (37.8°C) when package contains water, 75% rh and 100°F (37.8°C) when package contains desiccant
D 3985-81	Standard Test Method for Oxygen Gas Transmission Rate Through Plastic Film and Sheeting Using a Coulometric Sensor	material	oxygen	coulometric sensor	partial pressure, temperature, and humidity not specified, but should reported
E 96-80	Standard Test Methods for Water Vapor Transmission of Materials	material	water vapor	weight gain	50% rh and 70–90°F (21–32°C) 90°F recommended) or 90% rh and 100°F (37.8°C)
F 372-73 (1978)	Standard Test Method for Water Vapor Transmission Rate of Flexible Barrier Materials Using an Infrared Detection Technique	material	water vapor	infrared detector	100°F (37.8°C) and 81, 90, or 100% rh

Table 2. TAPPI Test Methods for Permeation and Leakage of Gases and Vapors

Test number	Test title	Sample type	Diffusant	Detection method	Test conditions
T 251pm-75	Air Permeability of Porous Papers, Fabrics, and Pulp Handsheets	material	air	pressure drop across orifice	pressure drop determined by apparatus design
T 448su-71	Water Vapor Transmission Rate of Sheet Materials at Normal Temperature	material	water vapor	weight gain	50% rh and 73°F (23°C)
T 460om-83	Air Resistance of Paper (units of Gurley seconds)	material	air	volume change	pressure drop determined by apparatus design
T 464os-79	Gravimetric Determination of Water Vapor Transmission Rate of Sheet Materials at High Temperatures and Humidities	material	water vapor	weight gain	90% rh and 100°F (37.8°C)
T 523om-82	Dynamic Measurement of Water Vapor Transfer Through Sheet Materials	material	water vapor	infrared, electrical resistance, or electrolytic cell	temperature and humidity not specified, but must be reported
T 536pm-79	Resistance of Paper to Passage of Air (H.P. Air Resistance Apparatus)	material	air	volume change	pressure drop determined by apparatus design

cally. Standard methods are available (2,3) for creasing the samples prior to testing (ASTM D 1027 and TAPPI T 465, T 512, and T 533), since the water vapor transmission rate of a material may increase significantly if folded (eg, on a packaging line). Because the sensitivity of weight gain or loss tests depends on the accuracy of balances [which varies from <0.0001 g (<2.2 × 10^{-7} lb) to >1 lb (>453.6 g)] the time required to detect a steady-state increase or decrease in weight may be as long as weeks or months, depending on the transmission rate. However, these tests have the advantage over more sensitive methods in that many samples can be tested concurrently, limited only by the size of the environmental chamber. If a large number of packages are to be tested, consecutive measurements using more sensitive detection methods may take longer than the concurrent testing using less sensitive detection methods.

Pressure and volume changes. The internal pressure in packages containing pressurized products [pressures >1 atm (>101.3 kPa)], such as aerosols and carbonated beverages, or vacuum-packaged products (pressures <1 atm) can change as a function of time due to the transmission of gases into or out of the package. The internal pressure can be measured periodically (often a destructive test), or other methods can be used to detect compositional changes (see below) in a test chamber surrounding the package (an indirect measure of pressure change inside the package). A commonly used method for measuring the transmission rates of gases other than water vapor through flat-sheet samples utilizes a test cell (Dow Cell, Linde Cell, or Volumetric Cell) separated into two chambers by the sample. A constant supply of the permeant gas is provided to the top chamber by connecting it to a compressed-gas cylinder. Initially, a vacuum pump is used to evacuate the lower chamber. The increase in pressure or volume in the lower chamber

is measured with a manometer as the permeant passes through the sample.

Compositional changes. If the test system (materials or package test) is set up so that there is a permeant-free gas on one side of the sample and a known partial pressure of the permeant on the other side, then the composition of the gas on the low concentration side can be measured as a function of time as the permeant passes through the sample and builds up on the low concentration side. Any analytical technique can be used which is specific and quantitative for the permeant under investigation (5–7). The MOCON (Modern Controls, Inc., Minneapolis, Minn.) OX-TRAN (for oxygen transmission rates of films and packages) uses a coulometric detector; the MOCON Permatran C (for carbon dioxide transmission rates of films and packages) and Permatran W (for water vapor transmission rates of films and packages) use a pressure-modulated infrared detector; the MOCON IRD (for water vapor transmission rates of films) uses a mechanically-modulated infrared detector; and the Honeywell Water Vapor Transmission Rate Tester (manufactured by the Thwing-Albert Instrument Company, Philadelphia, Penn. for the testing of films) uses a gold-grid bridge circuit.

For organic vapors, such as solvents, flavors, and fragrances, the packaged product can be placed in a glass or metal test cell whose headspace composition can be periodically analyzed using a gas chromatograph. Other test cells can be used for flat-sheet samples, in which one side of the sample is exposed to the organic vapor or liquid and the other side initially contains none of the permeant. The gas chromatograph, an instrument found in most analytical laboratories, is especially useful, since the amount of almost any gas can be measured on the low concentration side of the sample with the proper choice of detector, carrier gas, flow rate, column, and temperatures. Depending on the detector and methods used, sensitivities will be from 1 part per thousand to below 1 ppm. This is a useful tool for flavors and fragrances, since they are present in such small amounts that their entire loss may not be detectable as a weight loss in the product, and yet even a small percentage loss may result in an unacceptable product. The gas chromatograph can also be used for measuring water-vapor, oxygen, carbon dioxide, and nitrogen transmission rates.

Leakage

If permeation is negligible in a package or packaging material, and leakage is the only operative mechanism, then the tests described previously can be used to measure leakage rates. Otherwise, the most frequently used tests increase the total pressure differential across the sample; for example, by creating a partial vacuum around the package or increasing the pressure on one side of a porous sheet. These tests detect leakage, and they can also locate leakage pathways (8) (see Multilayer flexible packaging). Detection can be visual (looking for air bubbles or dye solutions emanating from the package) or instrumental (using infrared spectroscopy for carbon dioxide leakage or mass spectrometry for helium leakage). The greatest problem with this type of test is that the package is changed physically when the total pressure differential is changed. If a bottle is placed in a vacuum, the closure is "pulled" upward, away from the neck finish. This may cause leaks that would never exist in actual usage. It is extremely difficult to determine how much to accelerate the test; that is, what vacuum or pressure level to use. Optimum test conditions can be determined only by extensive side-by-side testing on a given package utilizing a wide range of pressures (including the pressure differential encountered in end-use) and time periods (including the desired shelf life).

The existence of leakage through pinholes in a film or seams in a package, for example, can be detected in many standard permeation experiments if, after steady-state has been reached, the total pressure on the low concentration side of the package or film being tested can be decreased considerably for a very short time period. In one method the contents of the test chamber are withdrawn with a large volume syringe and then the contents of the syringe are injected back into the test chamber without removing the syringe needle from the septum port of the test cell. The leakage rate is increased temporarily without affecting the permeation rate. This is exhibited as a step-function change in the plot of concentration of permeant in the test chamber versus time. There is a linear portion of the curve representing the steady-state transmission rate before the pressure drop, a step-function increase, then a return to the slope that existed before the pressure drop.

Separation of Permeation and Leakage

One means of separating the two mechanisms involves first measuring the total transmission rate (the combination of leakage and permeation) by any available method. Then, potential leakage pathways can be selectively sealed off with an appropriate adhesive, such as wax, epoxy, or hot melt, and the transmission rate can be remeasured. The choice of adhesive depends on the chemical composition of the leaking gas and the package components involved (see Adhesives). The new

transmission rate can be attributed to permeation and the difference between the two transmission rates to leakage. A similar, but opposite, approach can be taken by coating all surfaces through which permeation could occur with a sufficiently thick layer of a high barrier material, for example, wax for water vapor (see Waxes) or PVDC (see Vinylidene chloride copolymers) for oxygen or carbon dioxide. One must be sure not to block off leakage pathways during the coating process.

If the diffusion coefficient for permeation through the material is very low, or the materials involved are very thick, the two mechanisms can be separated by utilizing the fact that permeation is slower than leakage. It takes a certain amount of time for the permeant to start to dissolve into a material, to diffuse through it, and to begin to desorb from the other side. Until this occurs, no permeant is observed on the low concentration side. The lower the diffusion coefficient, the longer it is before penetrant is observed, since the time to reach steady-state is related to the reciprocal of the diffusion coefficient. Also, the thicker the material, the longer it takes, since the time to reach steady-state is related to the square of the sample thickness. On the other hand, leakage usually occurs very quickly. Therefore, a plot of the amount of permeant which has passed through the sample versus time often displays two linear portions: the first representative of leakage only, and the second representative of the combination of leakage and permeation. The permeation rate can be calculated from the difference between these two slopes.

Conclusions

Many methods are available for measuring the transmission rates of various gases and vapors through packages and packaging materials by permeation and leakage. Whenever possible, package tests should be performed in preference to materials tests, since the data from package tests give more information about end-use performance and package shelf life. There is a general trend in the industry toward performance testing, as opposed to materials testing.

BIBLIOGRAPHY

1. M. A. Amini and D. R. Morrow, *Packaging Development & Systems,* **9**(3), 20 (May/June 1979).

2. *Annual Books of Standards,* American Society for Testing and Materials (ASTM), Philadelphia, Penn., multivolume set published every year.

3. *Official, Provisional, and Historical Test Methods,* Technical Association of the Pulp and Paper Industry (TAPPI), Atlanta, Ga., looseleaf notebooks updated every year.

4. M. A. Amini, *Pharmaceutical Technology,* **5**(12), 38 (1981)

5. *The United States Pharmacopeia,* 21st Rev., United States Pharmacopeial Convention, Inc., Rockville, Md., 1985, updated periodically, revised every five years.

6. M. A. Amini, *Cosmetic Technology,* **3**(12), 30, 52 (1981).

7. J. R. Giacin, B. Harte, H. E. Lockhart, and M. Richmond, *Packaging (The 1984 Packaging Encyclopedia)* **29**(4), 35 (1984).

8. M. A. Amini, "Methods of Evaluating Closure Integrity," *Proceedings of the Third Wisconsin-Extension Update Conference on Packaging,* University of Wisconsin-Extension, Madison, Wisc., Oct. 1983, pp. 69–79.

Mary A. Amini
Center for Packaging Engineering
Rutgers, The State University of New Jersey

TESTING, PRODUCT FRAGILITY

Most products must undergo a certain amount of handling and transportation from the time they are manufactured until they are ultimately used. Normally, products are enclosed in a protective-package system during the time they are exposed to the distribution environment. This article outlines engineering principles and procedures that optimize the protective function of a package system and help guarantee the product's safe arrival at favorable cost.

The waste of resources caused by improper packaging is enormous. Whether this waste shows up as damaged product or an overly expensive package system, it is still waste that can be avoided. The application of sound engineering principles to optimize the product protection system must receive significant endorsement by product managers, quality assurance (see Specifications and quality assurance), and all those concerned with delivering a quality product to the customer.

To develop a protective-package system, three important pieces of data are necessary: (1) information on the likely inputs (shock and vibration) from the distribution environment; (2) product fragility characteristics and sensitivities in terms of the environmental inputs; and (3) the performance characteristics of commonly available packaging materials.

The engineering process for developing a protective-package system involves a step-by-step problem-solving approach similar to other engineering disciplines. The principal focus is on product characteristics because it is impossible to optimize a package system without first knowing basic engineering data of the product itself.

Once the environment has been defined and product fragility has been determined, the engineer can evaluate the economic feasibility of possible product modifications. For example, if a certain component within the product keeps failing at a relatively low input level, it may be economically more desirable to modify that component and increase its ruggedness rather than design an expensive and elaborate package system for the entire product based on the sensitivity of that one component. If it is economically feasible to modify the product in order to effect a package-cost savings, this option should be studied carefully. Many examples exist where slight product modifications have resulted in substantial cost savings in packaging materials and reduced damage in shipment.

There are many potential hazards in the distribution environment including shock, vibration, temperature or humidity extremes, electrostatic discharge, magnetic fields, and compression. This article deals primarily with shock and vibration inputs; however, it is important to explore carefully those areas where the product is sensitive or a large environmental input is likely.

The steps involved in designing an optimized package system are as follows:

1. *Define the environment* in terms of shock and vibration inputs likely during the product's manufacturing and distribution cycle.
2. *Define product fragility* in terms of shock and vibration.
3. *Obtain product improvement feedback,* ie, examine product improvements in light of the economic trade-offs between packaging costs and product modification costs.

4. *Evaluate cushion material performance,* ie, evaluate the shock and vibration characteristics of available cusion materials.
5. *Design the package system,* ie, select cushion materials and design for optimum shock and vibration performance.
6. *Test the final package,* ie, verify that the package system performs as designed and properly protects the product.

The primary purpose of this article is to examine product fragility testing in detail, and therefore the other steps are not covered in a comprehensive fashion. Refer to the bibliography for excellent background material on the other steps in this process (1–6) (see also Cushioning, design; Distribution hazards).

Defining the Environment

The end result of defining the environment should be the establishment of a design drop height and a resonant-frequency spectrum for a particular product in a given distribution environment. That is, based on the size and weight of a packaged product, the engineer selects a drop height which represents a given probability of input. Also, vibration profiles for likely modes of transportation are selected. Both of these pieces of information are used in the last step, testing of the package system (7–10).

Defining Product Fragility

The term "product fragility" is misunderstood by many people and often conjures up images of totally destroyed products, broken bottles, and the like. In reality, product fragility is simply another product characteristic such as size, weight, and color. Just as other product characteristics are determined by measurement, product fragility (or product ruggedness) can be measured with shock inputs. This measurement takes the form of a damage boundary curve for shock and resonant frequency plots for vibration. In both cases, the importance of determining these characteristics cannot be overemphasized. Most people would not think of buying a pair of shoes based on guessing their foot size. It is just as shortsighted to design a package system by guessing at product fragility.

Shock fragility assessment. The damage boundary is the principal tool used to determine the ruggedness of a product; it takes the general shape shown in Figure 1. This plot defines an area on a graph bounded by peak acceleration on the vertical axis and velocity change on the horizontal axis. Any shock pulse which can be plotted inside this boundary causes damage to the product regardless of whether it is packaged. (This is a *product* test.)

Acceleration is a vector quantity describing the time rate of change of velocity of a body in relation to a fixed reference point, ie, it describes the rate at which velocity is increasing or decreasing (deceleration). The terms acceleration and deceleration are often used interchangeably because most products respond similarly to a rapid start or a rapid stop. Both terms are usually expressed in Gs, multiples of earth's gravitational constant, g.

Velocity change is the difference in a system's velocity magnitude and direction from the start to the end of shock pulse. Velocity change is the integral of the acceleration vs time pulse and is directly related to drop height.

To run a damage boundary, mount the product to be tested

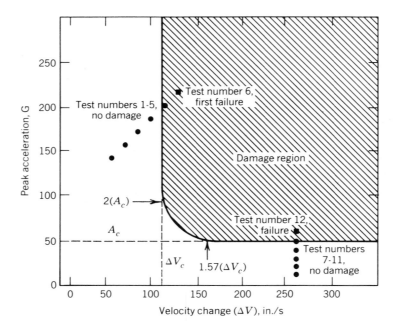

Figure 1. Damage boundary (single orientation). A_c = critical acceleration. V_c = critical velocity. To convert in. to cm., multiply by 2.54.

on the table of a suitable shock-test machine (see Fig. 2). Secure the product with a rigid fixture which lends even support to the product over its entire surface. The fixture must be as rigid as possible so that it does not distort the shock pulse transmitted to the product.

Set the shock machine to produce a low velocity-change shock pulse with a duration of approximately 2 ms (A halfsine waveform is generally used for this test). After the shock pulse is delivered to the product, examine it to determine if damage has occurred. If not, set the shock machine to produce a slightly higher velocity change and repeat the test. Continue this process with small incremental increases in velocity change until damage occurs. The last non failure shock input defines critical velocity change ΔV_c for the product in that orientation.

Next, fasten a new test specimen to the shock table and set the machine to produce a trapezoidal pulse with a low acceleration level and a velocity change at least two times that of the critical velocity determined in the previous test. Program the shock pulse into the product and as before, examine the product to determine if failure occurs. If no failure has occurred, set the shock machine to produce a higher acceleration level at approximately the same velocity change. Repeat this procedure with small increments in acceleration until the failure level is reached. The last nonfailure shock input defines the critical acceleration A_c for the product in that orientation (11).

The damage boundary curve may now be plotted by drawing a vertical line through the critical velocity change point and a horizontal line through the critical acceleration point. The intersection of these two lines (the knee) is a smooth curve as shown in Figure 1. A rectangular corner may be used as a conservative approximation of the damage region (12).

Note that the critical acceleration as determined by a trapezoidal pulse is conservative when compared to that generated

through the use of other waveforms; ie, a trapezoidal pulse is more damaging than other waveforms of the same peak acceleration and duration. This is shown graphically in Figure 3.

Since the shape of the waveform transmitted through various packaging materials during impact is generally not known, the use of a trapezoidal wave during damage boundary testing results in a higher confidence level in a package system and is recommended for this purpose. It should also be pointed out that the use of the trapezoidal waveform results in a nearly linear abscisa on the damage boundary. This means that it is necessary to determine only one point on that axis in order to determine the critical acceleration for the product in that orientation. Other waveforms result in critical accelerations which are a complex function of the natural frequency of components within the product. The procedure described is based on ASTM D 3332 (13).

The damage boundary is a valuable and powerful tool. Critical velocity change is equated to equivalent freefall drop height from the formula $\Delta V = (1 + e) \sqrt{2gh}$ where e is the coefficient of restitution of the impact surfaces, g is acceleration of gravity (386 in./s² at sea level (9.8 m/s²)), h is equivalent freefall drop height in in. (or m). The critical velocity change tells the designer how high the unpackaged product can fall onto a rigid surface before damage occurs in that axis. If this equivalent drop height is likely to be exceeded in the distribution environment, then the product must be cushioned. The performance requirements of the cushion are that no more than the critical acceleration be transmitted to the product.

In theory, the value of e (coefficient of restitution) can vary from zero to one. A value of zero would imply a totally plastic impact with no rebound whatsoever. A value of one implies a perfectly elastic impact where the rebound velocity is exactly equal to the impact velocity. As a practical matter, a range of $e = 0.25-0.75$ produces good accuracy in calculating the range

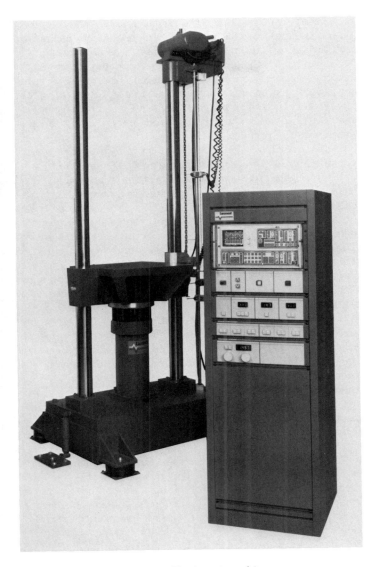

Figure 2. Shock-test machine.

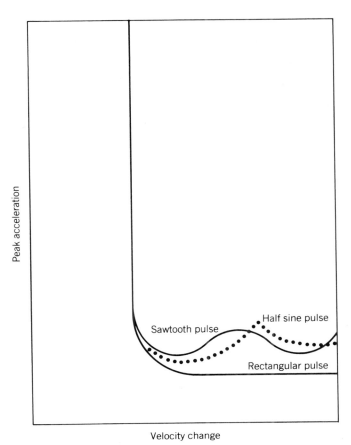

Figure 3. Damage boundary for various pulses.

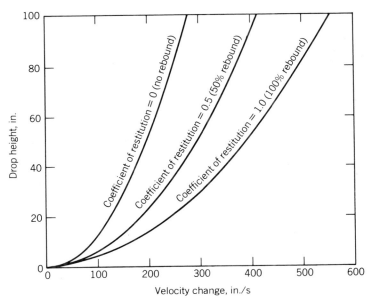

Figure 4. Coefficients of restitution. To convert in. to cm., multiply by 2.54.

of equivalent freefall drop height. The chart in Figure 4 shows the effect of e.

The damage boundary also tells the engineer that at low velocity changes infinite accelerations are possible, and, at low acceleration levels infinite velocity changes are possible without product damage. That means that it is necessary to define both critical acceleration and critical velocity change to properly characterize the fragility of a product.

Before running the damage boundary test, the engineer must define what constitutes damage to the product. On one extreme, damage may be catastrophic failure. However, there are many less severe damage modes which can make a product unacceptable to the customer. In some cases, damage can be determined by observation of the product, at other times it involves running sophisticated functional checks. Once the determination of damage is made, the definition must remain constant throughout the testing and must be consistent with what is deemed unacceptable to the customer.

In general, damage boundary tests must be run for each

axis in each orientation of the product. In the case of a rectangular product such as a television set, this means that a total of 12 specimens are necessary for a rigorous test. However, since this testing should be done in the product prototype stage, this quantity is rarely available for a potentially destructive test. As a practical matter, much information can be gained from a limited number of samples and multiple damage boundaries are often run on the same unit in different orientations.

Vibration-fragility assessment. Determining product-vibration sensitivities involves identifying resonant frequencies of components in the product in each of the principal axes. Resonance is that characteristic displayed by all spring/mass systems wherein at a given frequency, the response acceleration of a component is greater than the input acceleration. This characteristic is shown graphically in Figure 5.

As a general rule, the product will not be damaged due to nonresonant inertial loading caused by vibration in the *distribution environment*. This is because the acceleration levels of most vehicles are relatively low when compared to the critical acceleration of most products. It is only when a component within a product is excited or forced by vibration at its natural frequency that damage is likely to occur.

At frequencies below the resonant frequency, the response of a critical component is roughly equal to the input (the response:input ratio is 1). At frequencies higher than the resonant or natural frequency, the response acceleration is lower than the input. In this region a component acts as its own isolator and results in a condition known as *attenuation*.

At (and near) the product resonant frequency, however, the response acceleration can be very much greater than the input, causing product fatigue and ultimate failure in a relatively short time. The purpose of vibration sensitivity assessment is to identify those critical frequencies likely to cause damage to the product.

A resonant frequency search test is run by attaching a product to the table of a suitable vibration test machine (Fig. 6) and subjecting it to a constant acceleration input at a low level (typically 0.25–0.5 G) over a suitable frequency range, typically 2–300 Hz (cycles per second). An accelerometer is fastened to a critical component within the product in order to determine the component's response to the input acceleration. The response:input ratio is plotted as a function of frequency. This ratio reaches a maximum at the component resonant or natural frequency. The test usually involves monitoring many components in each axis of the product in order to characterize its overall vibration sensitivities (14).

The importance of vibration testing cannot be overemphasized. Any product shipped from point A to point B is subject to vibration because of the transit vehicle it is riding. The probability of this input is 100%. In contrast, the probability of a shock input because of a drop is exactly that, a probability function. In some cases, the drop height experienced by a product may be severe, in other cases, it is hardly measurable. However, any product that is shipped in a vehicle is subject to vibrational input and it should be tested for sensitivity to that input.

Conclusion. At this point, the engineer has sufficient data to make intelligent decisions about tradeoffs between product modifications and package costs. If a fragile component can be ruggedized at minimal cost resulting in substantial package savings, then it makes sense to pursue the product modification.

The package-design process uses environmental data, product fragility information, cushion-performance data (see Foam cushioning; Testing, cushion systems), and a healthy dose of designer creativity. Knowledge of package-fabrication techniques is essential as well as other vital information on flammability restrictions, maximum weight and cube for storage and transportation, recyclability of the package components future cost trends of various key materials, etc (1, 3, 5, 6, 15).

Once the design has been finalized and a prototype fabricated, it must be tested to verify compliance with product requirements. It is important to specify the correct inputs, both their magnitude (or duration) and sequence, in order to closely duplicate the potentially damaging effects of the distribution environment. Recently developed test procedures such as ASTM D 4169 (7) hold great promise for improving the correlation between laboratory tests and field experience (2, 4, 16).

A properly engineered packaged system is now within reach of all manufacturers, distributors, and other package-system users. The wasteful practice of overpackaging can virtually disappear along with damage-in-shipment reports. The tools are available and the technology straightforward. Optimized packaging is indeed an attainable goal.

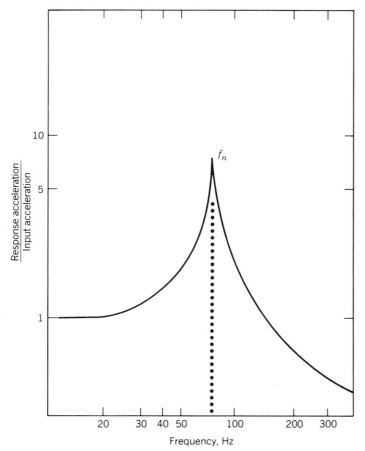

Figure 5.   Resonant frequency plot.

Figure 6. Vibration-test machine.

BIBLIOGRAPHY

1. M. E. Gigliotti, *Design Criteria for Plastic Package Cushioning Materials,* Plastic Technical Evaluation Center, Picatinny Arsenal, Dover, N.J., 1962.

2. T. J. Grabowski, *Design and Evaluation of Packages Containing Cushioned Items Using Peak Acceleration Versus Static Stress Data, Shock, Vibration and Associated Environments Bulletin No. 39,* part II, Office of the Secretary of Defense, Research and Engineering, Washington, D.C., 1962.

3. C. Henny and F. Leslie, *An Approach to the Solution of Shock and Vibration Isolation Problems as Applied to Package Cushioning Materials, Shock, Vibration and Associated Environments Bulletin No. 30,* part II Office of the Secretary of Defense, Research, and Engineering, Washington, D.C., 1962.

4. M. T. Kerr, "The Importance of Package Testing in Today's Data Processing Industry," *Proceedings of Western Regional Forum Packaging Institute,* USA, 1981.

5. R. D. Mindlin, *Bell System Tech. J.* **24**(304), 352 (1945).

6. S. Mustin, *Theory and Practice of Cushion Design,* Shock and Vibration Information Center, U.S. Department of Defense, Washington, D.C., 1968.

7. *ASTM D 4169-82, Standard Practice for Performance Testing of Shipping Containers and Systems,* American Society for Testing and Materials, Philadelphia, Pa., 1982.

8. S. G. Guins, in *Shock and Vibration Handbook,* Part II, McGraw-Hill Book Co., New York, 1961, Chapt. 45.

9. F. E. Ostrem and B. Libovicz, *A Survey of Environmental Conditions Incident to the Transportation of Materials, Report PB-204 442,* General American Transportation Corp., Niles, Ill., 1971.

10. R. J. Winne, "What Really Happens to Your Package in Trucks and Trailers," *Proceedings of Western Regional Forum of the Packaging Institute,* USA, 1977.

11. R. D. Newton, *Fragility Assessment Theory and Test Procedure,* Monterey Research Laboratory, Inc., Monterey, Calif., 1968.

12. D. E. Young and S. R. Pierce, *Development of a Product Protection System, Shock and Vibration Bulletin No. 42,* Office of the Secretary of Defense, Research, and Engineering, Washington, D.C., 1972.

13. *ASTM D 3332-77, Standard Methods for Mechanical-Shock Fra-*

gility of Products Using Shock Machines, American Society for Testing and Materials, 1982.

14. *ASTM D 3580-80, Standard Method of Vibration (Vertical Sinusoidal Motion) Test of Products,* American Society for Testing and Materials, Philadelphia, Pa., 1982.

15. J. P. Phillips, "Package Design Consideration for the Distribution Environment," *Proceedings of the 1979 International Packaging Week Assembly, Packaging Institute, Oct. 1979.*

16. J. F. Perry, *A Brief Summary of Dynamic Test Methods for Shipping Containers,* Del Monte Corporation, Walnut Creek, Calif., March, 1982.

General References

C. M. Harris and C. E. Crede, *Shock and Vibration Handbook,* Vols, 1, 2, and 3, McGraw-Hill Book Co., New York, 1961.

C. E. Crede, *Vibration and Shock Isolation,* John Wiley & Sons, Inc., New York, 1951.

J. F. Perry, "Cost Reduction Through Dynamic Testing of Shipping Containers," *Packaging Technology* (June/July 1982).

W. Silver and E. Szymkowiak, "Recommended Shock and Vibration Test for Loose Cargo Transported by Trucks and Railroads," *Proceedings of the Institute of Environmental Sciences,* (1979).

ASTM D 775-80, Drop Test for Shipping Containers, American Society for Testing and Materials, Philadelphia, Pa., 1982.

ASTM D 1596-78a, Method of Test for Dyanmic Properties of Package Cushioning Materials, American Society for Testing and Materials, Philadelphia, Pa., 1982.

ASTM D 999-75, Standard Methods for Vibration Testing of Shipping Containers, American Society for Testing and Materials, Philadelphia, Pa., 1982.

H. H. SCHUENEMAN
Scheuneman Design Associates

TESTING, SHIPPING CONTAINERS

The purpose of testing shipping containers is to provide the user or designer with information useful in reducing damage, improving costs, or satisfying other objectives in packaging design. Without testing, the design of a shipping container is in question until it has been used for a period of time and shipping results are known. Using meaningful tests provides information on the protective abilities of the container within a very short time and also permits comparison between alternative concepts. Trial shipments provide no fundamental data, are cumbersome, slow, and unduly expensive, and may severely delay the introduction of new products to the marketplace. Laboratory tests are performed much more rapidly and can pinpoint precise causes of failure.

Environmental Factors

The laboratory tests seek to stimulate or reproduce various environmental conditions found in actual handling, storage, and shipping. A detailed discussion of these factors, a report by the Forest Products Laboratory of the USDA, is given in ref. 1. Environmental factors may be divided into two types of hazards (see Distribution hazards). The first is the hazard of natural or atmospheric environment such as temperature and humidity. The second type of hazard is classified as physical or man-made; this includes shock, caused by dropping, railcar switching, accidental impact, etc; vibration, caused by carrier vehicle input during transportation; puncture, caused by dropping, accidental impact, lift truck handling, etc; compression,

caused by warehouse storage and stacking in transportation vehicles; and miscellaneous, such as tension, shear, torsion, and tear.

Sources of Test Standards

A number of groups generate packaging test standards in the United States. The oldest and largest of these is the American Society for Testing and Materials (ASTM) Committee D-10 on Packaging. Operating as a balanced consensus group, the D-10 Committee has generated over 100 packaging standards since its inception in 1914. These are all included in the *ASTM Annual Book of Standards,* Volume 15.09 (2).

The Federal Government is another large producer of packaging test methods. Most of these are included in *Federal Test Methods Standard 101C* (3). Some shipping-container test standards are also developed by the Packaging Institute USA. The National Safe Transit Association (NSTA) developed performance test criteria for shipping packages in the late 1940s which are continually updated in the document entitled *Pre-Shipment Test Procedures* (4).

Other industry groups have generated packaging test standards for their particular product interest, such as the Electronic Industries Association and The Association of Home Appliance Manufacturers. Pallet and slipsheet test methods have been generated by Committee MH-1 of the American Society of Mechanical Engineers (ASME).

Test Methods

Each of the environmental factors has at least one test method to measure the ability of a shipping container in protection of its contents or its durability in rough handling. The following discussion describes many of the test methods available for the various types of shipping containers.

Compression test. The effects of compressive forces from storage in warehouses and stacking in transportation vehicles can be studied with the compression test. ASTM D 642-76 "Method of Compression Test for Shipping Containers" can be used for measuring the ability of the container to resist external compressive load applied to its faces, to diagonally opposite edges, and to corners. This method may be used to compare the characteristic of a given design of container with a standard or to compare the characteristics of containers differing in construction. The method is suitable for testing boxes, crates, barrels, drums, kegs, and pails made of metal, wood, fiberboard, plastics, and combinations of these materials.

The tests are conducted on a machine known as a compression tester, which is calibrated to read compressive loads as well as deflection. The machine may be equipped with an autographic recording device.

The container is usually tested complete with interior packing and product, but testing may be also conducted on empty containers sealed in the proper manner as they are prepared for shipment. Generally a test is continued until the container fails, but in new performance testing sequences, compression tests are often run to a predetermined level and then the load is immediately removed.

The test procedure begins with container placement on the bottom platen of the testing machine and lowering the top platen until it comes in contact with the container. The upper platen can be either fixed or swiveled. An initial pressure or preload is used to ensure definite contact between the test specimen and the platen. The load is then applied in a continuous motion of the movable head of the testing machine at a speed of 0.5 in./min (12.7 mm/min) until failure, maximum load, or both has been reached.

Drop test. Several ASTM test methods are available for measuring the ability of shipping containers to withstand damage caused by sudden shock induced in accidental dropping. Test Method D 775-80 "Drop Test for Loaded Boxes" simulates those shocks that are likely to occur in the handling of packages through a variety of distribution cycles and the ability of the box to protect the contents against these shocks. Drop tests are performed with either of two objectives: measurement of the shipping container's ability to withstand rough handling or measurement of the ability of the total package (shipping container and interior packing materials) to protect its contents. The equipment used in drop testing is an apparatus which permits the box to be positioned accurately, controls the height of drop, and provides a release mechanism that does not impart rotational or sidewise forces to the test package. A rigid and level dropping surface with a mass of at least 50 times of the weight of the dropping container must be used.

In conducting drop tests, one of two types of end points are used: (1) continuation of the test until obvious damage occurs or (2) a specified number of drops followed by examination of resulting damage to contents. The height from which the boxes are dropped depends on the purpose of the test. In a procedure of constant-height drop tests, the package is repeatedly dropped in a specified sequence. In the progressive-height testing, packages are given a specified number of drops from several progressively increasing heights until the end point is reached. A typical test sequence used by many companies and specified in the National Safe Transit Association procedures (4) requires a total of 10 drops from heights that vary according to the weight of the product.

ASTM D 959-80 "Drop Test for Filled Bags" either measures the ability of any type of bag to withstand handling or provides a comparative evaluation of various bag constructions. The equipment used is similar to that for boxes, except a timber approximately 4 in. × 4 in. (102 mm × 102 mm) area and 4 ft (1.22 m) in length is firmly positioned on the dropping surface. This hazard is optional. Bags are dropped in a specified sequence, usually front, back, right side, left side, and optionally bottom and top. Occasionally tests are conducted until failure occurs. This is considered to have occurred when the contents are exposed, contents spill from the bags, or the bag breaks open.

To measure cylindrical containers such as barrels, drums, kegs, or pails, ASTM D 997-80 "Drop Test for Loaded Cylindrical Containers" is used. A rope or belt sling is placed around the container, and a chain hoist or other appropriate lifting device is used in conjunction with a tripping device. A solid surface of concrete and optional timber of 4 in. × 4 in. (102 mm × 102 mm) may be placed on the surface. As with boxes and bags, the test sequence may be for a specified number of drops or until failure. Test surfaces should include edges, sides, top, and bottom. For hazardous materials, test methods are specified by the Department of Transportation (DOT) for domestic shipments and by various international organizations for export.

Incline impact test. Besides accidental dropping, shipping containers receive shocks from other handling and transportation hazards. The incline impact test is used to reproduce these hazards. ASTM D 880-79 "Incline Impact Test for Shipping Containers" includes three different procedures depending on the desired results. Procedure A determines the ability of a container to withstand impact stresses. Procedure B determines the ability of the container, inner packaging, or both, to provide protection to the contents. Procedure C simulates the types of shock pulses experienced by lading in railcar switching. Any of these procedures is suitable for testing various types of containers such as boxes, crates, barrels, drums, kegs, or pails. They are particularly suitable for testing large, heavy, or loaded containers.

The test apparatus consists of a two-rail steel track inclined at 10 degrees from the horizontal, a rolling carriage or dolly, and a rigid backstop. For Procedure C, the carriage is fitted with a bulkhead at a 90 degree angle to the carriage surface, and a programming material or device is fastened to the backstop or carriage to provide a repeatable, controllable shock pulse to the bulkhead and container.

In Procedure A, the container is placed on the carriage, which is pulled up the incline and then released, allowing the container to strike against the rigid backstop. The test is continued until failure of the container occurs. In Procedure B, the same test method is used except a specified number of impacts of prescribed intensity are applied, whereupon the container is opened for inspection. In Procedure C the container is placed against the carriage bulkhead. An impact velocity of the carriage and a programming material which produces the required pulse are used in conjunction with a simulated backload behind the container on the carriage. The backload applies a dynamic compressive force in addition to the shock pulse of the programming material.

Horizontal impact test. Horizontal impact occurs in the distribution environment through rail switching, pallet marshaling, and other sources. ASTM D 4003-81 "Controlled Horizontal Impact Test for Shipping Containers" is a method of applying controlled levels of shock input to obtain optimum design for protecting products and prescribing modes of shipping and handling that do not induce damage to packaged products.

This test method requires a horizontal impact test machine with a guided test carriage on which the test specimen is mounted against an upright bulkhead perpendicular to the specimen. The test machine provides a means of linearly moving the carriage into a programmed impact surface. The programming devices provide shock pulses through hydraulic devices, springs, or cushioning materials. In some tests, such as rail switching, a backload is required and a specially adapted backload fixture may be used. Proper instrumentation and calibration is required for monitoring the acceleration and duration of impact shock pulses.

Test procedures vary according to the desired end results. For instance, the railcar switching impact test may require three or more impacts at several levels of impact velocity. A typical backload equivalent to 3 ft (914 mm) of cargo is used.

Revolving-drum test. This test provides an indication of the ability of a shipping container to withstand a variety of shock and impact stresses. It provides a comparison of different container designs for the same load. The drum test is not intended to simulate any specific environmental hazard and is subject to a high degree of variation. The number of falls which similar containers can withstand in successive test may vary over a wide range. ASTM D 782-82 "Testing Shipping Containers in

Revolving Hexagonal Drums" describes the equipment and procedure.

The test machine consists of a revolving drum in the form of a prism with vertical bases of regular hexagons and lateral bases of rectangles. Two drum sizes are generally used: 7 ft (2.13 m) and 14 ft (4.27 m). Baffles or hazards are fixed on the inside faces of the drum. Normal speed of operation is 2 rpm for the 7-ft drum and 1 rpm for the 14-ft drum.

Two test procedures can be used. In the first, the ability of a container to withstand rough handling is measured as the drum revolves and the container slides, tumbles, and falls in varying positions until the container fails at some point. While the test is in progress, a record is made of the development of damage to the container, that is, punctures, slits, or tears, and failure of fastenings or reinforcement. In the second method, the test is conducted until a predetermined end point has been reached, such as a given number of falls or a particular condition of container.

Vibration test. There are several test procedures available for vibration testing of filled shipping containers. These are all included in ASTM D 999-75 "Vibration Testing of Shipping Containers." The tests are intended to subject filled containers to vibration similar to transportation. They are not intended for determining the response of products to vibration or for product design purposes, nor are they intended for testing products in operational configuration. Three separate methods are included within ASTM D 999-75.

Method A, the Repetitive Shock Test, is suitable for testing individual containers that are transported unrestrained on the bed of a vehicle, and is suitable for testing containers that may be subjected to repetitive shock due to amplification of vibrations in unit loads or stacks. The vibration machine has a horizontal table which operates in either a rotary or vertical linear motion with a double amplitude of 1 in. (25.4 mm) and in a frequency range within 2–5 Hz (cycles per second). In this procedure the test specimen is placed on the platform in its normal shipping orientation with restraining devices to prevent the specimen from moving off the platform, but is not fastened to the platform. The vibration of the platform is steadily increased from 2 Hz until the test specimen leaves the platform or until an acceleration of the platform of 1.1 times the acceleration of gravity is reached. The test is conducted for a specified length of time or until a predetermined amount of damage is detected. A typical test duration is one hour.

Method B, the Single-Container Resonance Test, requires a vibration test machine with a horizontal table and a mechanism capable of producing a sinusoidal vibration in the vertical linear plane. It must control accelerations or displacements, or both, over a controlled, continuously variable range of frequency. Suitable fixtures and attachments are provided to rigidly attach the test specimen to the platform. Accelerometers, signal conditioners, and data display or storage devices are required to measure and control the accelerations of the test surface. Starting at the lowest frequency, the frequency of vibration is swept at a continuous logarithmic rate of approximately ½ octave per minute to the upper frequency range, and then swept back to the lower frequency limits. The cycle is completed twice, recording all resonant responses of the test specimens. A frequency range of 3–100 Hz is typically recommended. After the sweep tests are completed, vibration dwells at each resonant frequency for a specified length of time or until damage to the container is noticed, which ever occurs first.

Method C, the Unitized Load for Vertical Stack Resonance Test, uses the same type apparatus as *Method B.* Full-sized unitized loads of test specimens are placed on the platform (or a vertical column equal in height to that used in the mode of transportation to be utilized). A single vertical column may be used if vertical stacking alignment is used in shipping. The procedure of sweeping twice for the full range and dwelling at each resonant frequency similar to *Method B* is utilized. Typical dwell times are 10–15 min at each resonance point.

Water vapor transmission test. To measure the water vapor transmission rate of bulk shipping containers there are two methods described in ASTM D 4279-83 "Water Vapor Transmission of Shipping Containers—Constant and Cycle Methods." This procedure is designed for larger shipping containers. Smaller shipping containers requiring greater accuracy in weighing should use ASTM D 895 or ASTM D 1251.

Water resistance tests. To determine the water resistance of a shipping container, ASTM D 951-51 "Water Resistance of Shipping Containers by Spray Method" is used. This method is frequently used in conjunction with other tests made prior to or after the spray test, such as drop, impact resistance, and drum.

Conditioning of Shipping Containers

All tests for shipping containers should be conducted in specified atmospheric conditions. A variety of standard conditions are described in ASTM Standard Practice D 4332-84 "Conditioning Containers, Packages, or Packaging Components for Testing." This practice provides special conditions which may be used to stimulate particular field conditions a container may encounter during its distribution cycle. The practice describes procedures for conditioning the containers so that they may reach equilibrium in the atmosphere to which they might be exposed. A standard preconditioning atmosphere of 20–40°C and 10–35% rh is specified. The U.S. standard conditioning atmosphere is listed at 23°C and 50% rh. Special atmospheres are specified as shown in Table 1.

Following the preconditioning cycle the containers are placed within the conditioning chamber and exposed to the required conditions for an amount of time specified, ordinarily at least 72 h or that required to reach equilibrium. Whenever possible, the desired test or measurement should be made at the specified conditioning atmospheres. When this is not possible, the test may be performed as quickly as possible after the specimen is removed from the conditioning atmosphere.

Pallet Testing

Various methods of determining the capability of pallets are described in ASTM D 1185-73 "Methods of testing Pallets." This test method covers the determination of the four most

Table 1. Special Atmospheres

Environment	Temperature, °F (°C)	Relative humidity (rh), %
cryogenic	−67 ± 6 (−55 ± 3)	
frozen food storage	0 ± 4 (−18 ± 2)	
refrigerated storage	41 ± 4 (5 ± 2)	85 ± 5
temperate, high humidity	68 ± 4 (20 ± 2)	85 ± 5
tropical	104 ± 4 (40 ± 2)	85 ± 5
desert	140 ± 6 (60 ± 3)	15 ± 2

important properties of both expendable and reusable pallets, including static-load capacity, shock-load capacity, vibration resistance, and diagonal rigidity. In addition, the following properties of reusable pallets are described only for leading edge impact resistance, impact resistance for stringers or blocks, and pallet stiffness.

Other pallet test methods have been developed by Committee MH-1 of The American Society of Mechanical Engineers (ASME). These are described in ANSI Standard MH 1.4.1 (1977) "Procedures for Testing Pallets." ANSI Standard MH 1.5M (1980) "Slip Sheets" is also a development of Committee MH-1.

Performance testing

Each of the test procedures described previously reproduce within the laboratory a discrete hazard present in the distribution environment. As containers move throughout the environment they usually encounter several of these hazards in various sequences depending on the mode of transportation. Testing to reproduce these series of hazards is called performance testing.

Standardized procedures for preshipment performance testing of containers were developed in the late 1940s by the National Safe Transit Association. Their Preshipment Test Procedures were developed as a means of predetermining probability of safe arrival of packaged products at their destination through the use of tests for simulating shock and stresses normally encountered during handling and transportation. Their procedures are divided into Project I which covers packaged products weighing 100 lb (45.4 kg) and over, and Project IA for packaged products weighing less than 100 lb.

Project IA begins with a vibration test using test methods and equipment as specified in ASTM D 999. Frequency of vibration is determined by the ability to insert a shim at least 4 in. (102 mm) under the bottom of the packaged product between it and the surface of the table. The packaged product is then vibrated for a total of 14,200 impacts. The vibration test is then followed by a drop test or an incline impact test in conformance with ASTM D 775 or ASTM D 880. Each container is given ten drops on various corners, edges, and faces from heights depending on packaged product weight. For 1–21 lb (0.45–9.5 kg), the drop height is 30 in. (76 cm), 21–41 lb (9.5–18.6 kg) is 24 in. (61 cm), and 41–61 lb (18.6–27.7 kg) is 18 in. (46 cm). Above 61 lb (27.7 kg) up to and including 100 lb (45.4 kg), the drop height is 12 in. (30 cm).

For Project I the vibration test procedure is the same except the total number of vibration impacts is reduced to 11,800. This is followed by an incline impact test according to ASTM D 880. The intensity of the impact is determined by velocity of the empty dolly which must be 5.75 ft/s (1.75 m/s) upon impact. The test is conducted on all sides, top, and bottom of the shipping container unless it is not practical to position it for a top impact. Both projects may be followed by an optional static compression test.

In 1982 ASTM published D 4169-82 "Standard Practice for Performance Testing of Shipping Containers and Systems." This procedure recognizes a variety of distribution cycles with varying sequences of hazard elements. Each hazard element corresponds to an existing test method described above (see Table 2). Their sequence of use in various distribution cycles is shown in Table 3.

The tests should be performed sequentially on the same container in the order given. An acceptance criteria must be

Table 2. Hazard Elements and Corresponding Tests

Code	Hazard element	Test simulation of hazard	ASTM designation
A	manual handling up to 200 lb (90.7 kg)	drop	D 775, D 959, D 997
B	mechanical handling over 100 lb (45.4 kg)	drop	D 1083
C	warehouse stacking	compression	D 642
D	vehicle stacking	compression	D 642
E	truck and rail transport, stacked or unitized	vibration	D 999 Method C
F	loose-load vibration	repetitive shock	D 999 Method A
G	vehicle vibration	vibration	D 999 Method B
H	rail switching	longitudinal shock	D 880 Method C, D 4003

established prior to testing and should consider the required conditions of the product as received by the consignee. Each hazard element and test procedure has three assurance levels to choose from. These assurance levels or test intensities should be established prior to testing. Assurance Level II is suggested unless conditions dictate otherwise. Assurance Level I provides a more severe test than II, and Level III provides a less severe test. Assurance levels maybe varied between elements if such variations are known to occur.

Regulatory Agency Use of Container Testing

Until recently the agencies that regulate packaging for transportation have used material specifications as their sole

Table 3. Performance Test Sequence

Distribution cycle (DC)	DC number	Hazard element sequence[a]
general schedule, undefined distribution system	1	A or B,D,E,F,H,A or B
specially controlled environment, user specified	2	user specified
single-package environment, up to 100 lb (45.4 kg)	3	A,D,F,G,A
motor freight, single package over 100 lb (45.4 kg)	4	B,D,F,G,B
motor freight, TL (truckload), not unitized	5	A or B,D,E,G,A or B
motor freight, TL or LTL (less than truckload), unitized	6	B,D,E,B,C
rail only, CL (carload), bulk-loaded	7	A,D,E,H,A
rail only, CL, unitized	8	B,D,E,H,B,C
rail and motor freight, not unitized	9	A or B,D,G,H,F,A or B
rail and motor freight, unitized	10	B,D,E,H,B,C
rail TOFC (trailer on flat car) and COFC (container on flat car)	11	A or B,D,H,E,F,A or B
air and motor freight, over 100 lb (45.4 kg) or unitized	12	A or B,D,E,G,A or B
air and motor freight, single package up to 100 lb (45.4 kg)	13	A,D,F,G,A
warehousing, partial cycle	14	A or B,C

[a] See Table 2

measurement criteria. In the late 1970s, however, the National Classification Board, which develops regulations for the National Motor Freight Classifications, has included performance testing procedures based on ASTM test methods for certain special package numbers for products such as furniture and data processing equipment. The DOT has proposed a change from complex material specifications in the packaging of hazardous materials for transportation to a much simpler performance testing program, based on similar kinds of package testing programs issued by the United Nations. The International Standards Organization (ISO) Technical Committee 122 on packaging has test methods quite similar to those described above for ASTM. Federal and military test methods in the United States follow ASTM procedures where available. Others are listed in Federal Standard 101C (3).

BIBLIOGRAPHY

1. F. E. Ostrem and W. D. Godshall, *An Assessment of the Common Carrier Shipping Environment,* General Technical Report FPL 22, U.S. Forest Products Laboratory, Madison, Wisc., 1979.

2. *Annual Book of Standards, Volume 15.09,* American Society of Testing and Materials, Philadelphia, Pa., 1984.

3. "Test Procedures for Packaging Materials," *Federal Test Method Standard 101C,* Sept. 1983. Supt. of Documents, U.S. Gov't. Printing Office, Washington DC 20402.

4. *Pre-shipment Test Procedures,* National Safe Transit Association, Chicago, Ill., 1984.

A. H. McKINLAY
Consultant, Packaging and Handling

THERMOFORM/FILL/SEAL

Thermoform/fill/seal equipment is used for a variety of food and nonfood packaging applications, Thermoform/fill/seal (tffs) machines use two continous webs or rolls of film (see Fig. 1). Typically the lower web is formed into a cup, which is then

Figure 1. Thermoform/fill/seal machine. Shown with lower web forming station in foreground. Note extensive product-loading area.

filled with product. The upper web becomes the lid. Although over 90% of all tffs machines are run with formed bottom webs and unformed lid stock, it should be mentioned this is not always the case. Sometimes both webs are formed; sometimes the top web is formed, and the bottom one is not.

Packaging Applications

Tffs machines were originally designed as vacuum-packaging machines to package ham, bacon, or sausage. From cured meats, their use spread to other sectors of the meat industry (frozen steaks, certain fresh meats, cook-in-package delicatessen meats), and then to cheese products, baked goods, fresh pasta, and a variety of perishable food products requiring extended shelf life.

Nonfood applications of tffs evolved from the original vacuum-packaging machines for sausage and meat products. These are primarily sterile-medical applications (see Healthcare packaging). Typical medical items packed on tffs equipment are syringes, needles, catheters, kidney-dialysis filters, scrub sponges, surgical drapes and clothing, operating room kits, and many other items. Although medical tffs machines are structurally and conceptually the same as tffs machines used in the food industry, the materials tend to be more difficult to run and the quality standards for cutting and thermoforming are more difficult to meet. Among the tffs applications that are not food or medical, are hardware items, cigarette lighters, cosmetics, and similar applications where packages made on continuous-web machinery are replacing preformed plastic "blisters" (see Carded packaging).

Food packaging. In food applications, the packages are almost always hermetically sealed, and the films have varying degrees of barrier properties. The primary application for vacuum packaging is to improve shelf life of foods that are sensitive to oxygen or susceptible to dehydration. Almost all cured meat products (ie, ham, bacon, sausage, corn beef, beef jerky, etc) are vacuum-packaged, as are many cheeses. An off-shoot of vacuum packaging is controlled gas-atmosphere packaging (see Controlled-atmosphere packaging). A controlled-atmosphere package produced on tffs equipment is a step beyond vacuum packaging. First, the unsealed package is indexed automatically into a vacuum chamber. Then all the air is removed from the package. Once the air is removed, the desired gas mixture is injected into the package, usually until the pressure of gas inside the package reaches about one atmosphere. Table 1 lists some commonly used gas mixtures and their application.

Vacuum and controlled-atmosphere packages are generally used to protect refrigerated perishable foods that would otherwise spoil in less than a week. In a controlled atmosphere, they may stay fresh for four to six weeks or longer. Many frozen foods are also packed on tffs machinery, and a controlled atmosphere can sometimes replace freezing (see Fig. 2). Some nonperishable foods with shelf life of over one year can be packaged on tffs equipment. In Europe, for example, shelf-stable heat-processed vegetables are marketed in retorted thermoformed packages. Thermoforming machines can be modified to mechanically cold-form aluminum foil for retort processing of various meat and nonmeat dishes. The taste, texture, and nutritional quality of food in retort-sterilized aluminum foil pouches is generally higher than that of conventionally canned foods (see Retortable flexible and semirigid packages).

Table 1. Gases Used in Packages Produced on Tffs Equipment

Gas	Application
N_2	As a pressure-relief agent to prevent external atmosphere from crushing the product. N_2 is an inert gas. It does not react with either the food substances or bacteria. Common applications: bulk-pack bacon and sausage, shredded and sliced cheese, and beef jerky.
CO_2	Depending on the application, 25–100% CO_2 may be used. CO_2 lowers the pH of the food product and can exert a powerful slowing effect on the growth of bacteria and molds. Primary application is for baked goods, cookies, cakes, breads, as well as dough and pasta products. CO_2 tends to be absorbed into the actual body of the food product itself. Accordingly, CO_2 is frequently mixed with N_2 to prevent the package from clinging too tightly to the product. On the other hand, some products which are not sensitive to strong pressure or tight cling, but which are susceptible to spoilage by mold growth, are packed in an atmosphere of 100% CO_2. An example of this is chunk cheese. Many CO_2 gas packages have the appearance of a vacuum package, with much of the CO_2 absorbed into the product itself.
O_2	The applications using O_2 are primarily for red meat. Still experimental in the United States, the concept is widely used in Europe for centrally packed retail cuts. The O_2 is used as an oxygenating agent, at levels in the range of 40–80%, to form the bright-red, fresh-meat color called oxymyoglobin. When O_2 is used, it is usually mixed with CO_2 and N_2: CO_2 for its preservation effect; N_2 to provide a bulking agent. The O_2 tends to disappear inside the package. It can be metabolized by the meat to CO_2 which is absorbed in the water phase of the meat as carbonic acid.

Tffs Machines

The essence of tffs machines is their modularity. Within any given machine designation an almost infinite number of configurations is possible. There are six basic operations that make up the production of a package on tffs equipment: film advance, thermoforming, loading, sealing, cutting, and labeling/printing (see Fig. 3).

Film advance. Tffs machines are built like miniature assembly lines with separate stations for different operations. There are several film-advance mechanisms for moving the webs through these stations. The drive mechanisms (ie, AC Digital, DC Digital, Geneva, Elliptical Gear, and Planetary Gear) differ as to accuracy of advance, acceleration curve, or suitability for print registration of stretchable or nonstretchable materials.

Thermoforming. This is the first step in the formation of a package. Thermoforming (see Thermoforming) may involve heating in the forming die or preheating in a separate station. There are two main ways to thermoform: positive (male) and negative (female). Female forming involves using compressed air or vacuum to pull the heated, softened film into a mold. It is this mold or concave surface that is the die that produces package shape and surface detail. A mechanical plug may help push the film down into the cavity. In this case the designation "plug-assist forming" applies.

Positive (male) forming is the reverse. Positive forming usually permits the production of a package with greater sur-

Figure 2. Controlled gas-atmosphere package for perishable foods. Fresh (unfrozen), precooked french fries (10 kg) packed in N_2 atmosphere. Positive (male) formed PVC/PE bottom, polyester/PE cover. Print registration of flexible upper web by means of stretching with a brake.

face detail and more-uniform wall thickness. A male die part operated by a piston provides package shape and surface detail. Air outside the film or vacuum on the inside, or both air and vacuum, force the preheated film to drape itself onto the exterior of the plug. Positive forming requires a separate station or two separate stations to preheat the film. A typical package requires thermoforming of only the lower web, but

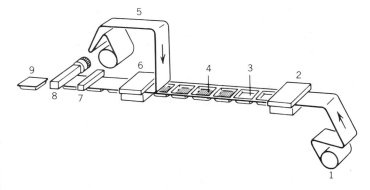

Figure 3. A schematic of the simplest and most basic configuration of a tffs machine as would be used for flexible, not semirigid, upper and lower webs: 1, Lower or forming web; 2, thermoforming die; 3, formed, unfilled pocket; 4, filled pocket; 5, upper or lidding web; 6, vacuum/sealing die; 7, flying knife for across-the-machine direction package cutoff; 8, high speed rotary knives for package cutoff in-the-machine direction; and 9, finished package (rectangular).

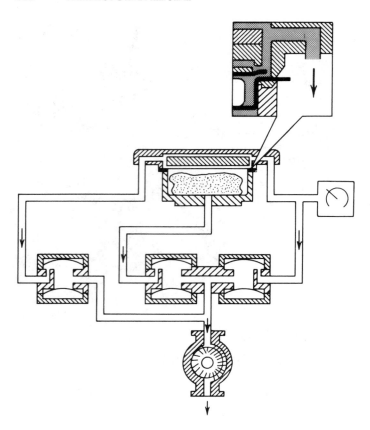

Figure 4. Evacuation. This shows the die after the first or outer box has closed. Vacuum is drawn inside the die chamber. Because vacuum is drawn on the outside of the package both from above and below, air pressure inside the package forces the two webs to separate providing a large opening for air to flow out of the package. During evacuation, the travel of air from the inside of the package is through the gap between the narrow upper web and the wide lower web. Package air exits along with air from above the upper web. In this figure, the seal bar is in its upper, or retracted, position so that air can leave the package.

the machines can also form both lower and upper. Usually each product to be packaged requires its own separate forming die. Major machine manufacturers have designed and built thousands of different forming dies.

Loading.   The lower web emerges from the forming station with the pocket formed and ready to accept product. The load-

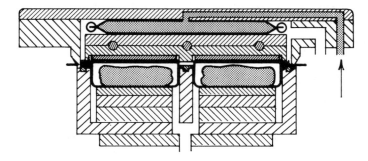

Figure 5.   Heat sealing. After the package is evacuated, an air bladder lowers the sealing bar and the package perimeter is clamped shut and heat sealed. In the next step, air is re-admitted into the die, the die opens and the package is indexed out.

ing area must be long enough to accommodate the people required for loading and inspecting, or automatic loading equipment.

Sealing.   Once the pockets on the web are loaded, the upper web is indexed over them and the two webs are sealed (see Sealing, heat). In food packaging, evacuation and, if needed, gas injection are performed at the sealing station. In order to provide a more attractive package, sometimes all surfaces of the top and bottom webs not in contact with the product are sealed to each other. Sometimes, using a patented process, steam is injected into the sealing die to shrink the lower web (forming web) around the product to form an attractive skintight package.

Cutting.   Cutting systems range from the simple and inexpensive to the complex and costly: paper-cutting-type knives (ie, flying knives and high-speed rotary knives); shear cutting (machine direction) with cross-direction strip removal by means of matched male and female dies; shear cutting (machine direction) and cross-direction cutting by means of steel rule dies; and complete package cutting with matched male and female dies.

Labeling and printing.   Online printing may consist of simply printing, or embossing a code date or lot number (see Code marking and imprinting), or may consist of printing an entire index, actually "bleeding" finished package edges. Labels may be applied to either surface of either web.

Vacuum and Sealing-Die Operation

At this station the package is evacuated and the two webs are sealed together. In order to understand how this die operates, it helps to think of the die as consisting of two nesting, concentric boxes that close in sequence. The outer box closes first (see Fig. 4). It mechanically seals off that portion of the package represented by the lower web. The upper web is narrower than the lower web and is not sealed at this stage.

Figure 6.   Sliced luncheon meat package. Nylon/ionomer flexible forming web, OPET/ionomer top web. Circle cutting by means of flying knife. Print registration by means of stretching film with a brake.

Therefore, until the inner box closes, atmosphere is free to enter or leave the package under the lip of the upper web. The inner box is made up of the seal mechanism. After the entire die chamber, including the interior of the package, is evacuated, the inner box (the seal mechanism) closes on both webs, first sealing them mechanically, so no air can enter or leave the package, then applying a permanent heat-seal (see Fig. 5). The key operation, package evacuation, takes place when air flows out of the package through the gap between the narrow upper web and the outer wall of the die chamber. This occurs during the interval after the outer box has sealed off the lower web, but before the inner box has closed off the package itself.

This is one of the most common methods of package evacuation. If gas injection is required for controlled-atmosphere packaging, the operation takes place in the sealing die.

Materials

The detailed design of a tffs machine depends on the packaging materials to be run. Therefore, any discussion of how tffs machines operate must refer to materials as well. The two webs consist of a bottom web and a top web on the type of equipment generally known as horizontal thermoform/fill/seal machines. In the usual case, the bottom web is thermoformable and is formed by heat into a cup or cavity which forms a receptacle for the articles to be packaged. However, all variations are possible:

Top web	Bottom web
Nonformable	Formable
Nonformable	Nonformable
Formable	Formable
Formable	Nonformable

When food products are packaged, the webs are usually made of multilayer materials selected for their barrier and strength properties. Typically, a barrier material protects

Figure 7. Thermoformed medical-kit package. Typically this type of package would have a semirigid HIPS (high-impact polystyrene) thermoformed bottom and a Tyvek or paper lid. Forming would be positive (male-plug). Note both severe draw ratio and enhanced surface detail. Print registration would be by means of a d-c drive since paper cannot be registered by stretching. Cutting could be a complete 360° cut with matched male/female dies, or shear cut (machine direction) with strip removal by means of matched male/female dies (cross-direction). Package is shown with lid removed.

Table 2. Cutting Methods for Rigid Materials

In-the-machine direction	Across-the-machine direction
squeezing knives shear cut knives with or without strip removal	steel rule dies with rounded corners strip removal for rounded corners by means of matched male and female dies complete 360° cut by means of matched male and female dies

Tables 3 and 4 list some commonly used materials in tffs applications.

against transmission of water and oxygen into or out of the package. In the medical industry, however, products are typically sterilized by ETO gas. In this case, high gas-transmission is desired. As a result, the tffs machines use an upper web (also called lidding material) (see Lidding) selected for high gas-transmission properties as well as strength. Typical high gas-porosity materials include paper and Tyvek (DuPont Company), a tough, spunbonded polyolefin.

Table 3. Common Food-Packaging Webs[a]

Structure	Comments
Nonforming web Polyester/polyethylene or Polyester/ionomer These structures may contain an intermediate layer of PVDC or EVOH as oxygen barrier.	Polyester is used for printability, strength, and general resistance to moisture and abrasion. Normally it is reverse (capture) printed (ie, the polyester is printed on its inside surface for protection of the printing). Printed polyester is normally stretch-registered.
Forming Web *Flexible* Usually a nylon base web is used for strength and formability, combined with PE or ionomer as heat sealant and moisture barrier, and, if needed, PVDC or EVOH as oxygen barrier.	Forming of flexible web by compressed air, vacuum, or compressed air and vacuum possibly has a heated plug for severe draw ratios. Cutting by means of flying knife in the cross-direction (XD), and high-speed rotary knives in the machine direction (MD) for rectangular packages. Circular, oval, or shape cutting by means of flying knife following a pattern or cam.
Rigid Semirigid PVC, acrylonitrile, polyester, or HIPS, combined with PE or ionomer for moisture barrier and if needed, PVDC or EVOH as oxygen barrier.	Forming of rigid web by either negative (female) forming with plug assist or positive (male) forming for more uniform forming and surface definition. Cutting by steel-rule die, or matched male and female die and strip removal across-the-machine direction with shear, cut strip removal in-machine direction. Complete cut by means of matched male and femal dies shaped to the final package size.

[a] See Ionomers; Film, oriented polyester; Nylon; Nitrile polymers; Polyesters, thermoplastic; Polystyrene; Vinylidene chloride copolymers; Ethylene–vinyl alcohol

Flexible materials such as nylon (see Fig. 6) can usually be formed by heating and pressure-forming into a female die part. Cutting is usually be means of a flying knife in the cross-direction and high speed rotary knives in the machine direction. More severe draw ratios call for a preheat station and plug assist. Nonrectangular shapes, such as circles and ovals, can be cut with a flying knife traveling on a cam.

Rigid materials usually require costlier and more-elaborate forming and cutting methods. Rigid films usually call for one or more preheat stations and plug-assist forming. If good surface definition and uniform wall thickness is required, positive (male plug) forming is usually required.

In order to avoid sharp corners, rigid materials must be cut by one of the methods shown in Table 2.

Table 4. Common Medical-Packaging Webs

Structure	Comments
Non-forming web	
Tyvek (spunbonded polyolefin) or paper used for wet strength and resistance to puncture, also exhibits minimal fiber generation when cut. Paper is used for superior printability and lower cost.	Tyvek can be stretched registered (by means of photocell and brake). Paper cannot be stretch registered, so it needs to be registered by controlling the advance with either an a-c or d-c servo-drive.
Both Tyvek and paper are usually heat-seal coated. Used for ETO gas sterilization because of their porosity. Polyester or polyester/aluminum foil for gamma sterilization or gas or moisture barrier. These materials are used in conjunction with appropriate heat seal coatings and laminating adhesives.	Stretch or advance registration.
Forming web	
Flexible forming webs: Polyolefin laminates or blends.	Forming methods are similar to those used in flexible food packaging. Shear cutting is normally used in both directions for superior cleanliness of cut with paper or Tyvek upper webs.
Rigid forming webs: PVC or acrylic multipolymer or high impact polystyrene (HIPS) or copolyester.	Forming is typically with temperature controlled plug or positive forming depending on draw ratio and degree of surface definition required. Cutting, typically, matched male-female in across-the-machine direction. Shear cut in-the-machine direction. Or 100% matched male-female cutting. These cutting systems provide relatively particulate-free packages with radiused corners.

BIBLIOGRAPHY

General References

Modern Plastics Encyclopedia, McGraw Hill, Inc., New York, published annually.

The Packaging Encyclopedia, Cahners Publishing Co., Denver, Colo., published annually.

J. M. Ramsbotton, "Packaging," in J. F. Price and B. S. Schweigert, *The Science of Meat and Meat Products,* 2nd ed., W. H. Freeman and Company, San Francisco, Calif., 1960, pp 513–537.

L. D. STARR
Koch Supplies Inc.

THERMOFORMING

Thermoforming is the means of shaping thermoplastic sheet into a product through the application of heat and pressure. In most cases, the heat-softened plastic assumes the shape by being forced aganst the mold until it cools and sets up. Forming pressure may be developed by vacuum (atmospheric pressure), positive air pressure, or by mating matched molds.

The word sheet as it relates to thermoforming is used to describe flat extruded plastic material that is generally relatively heavy, in contrast to comparatively thin plastic films. It is ambiguous, however, in two ways: thermoforming is also used to shape relatively thin films, and the word sheet is also used to distinguish between thermoforming machines that form cut pieces (sheet-fed) rather than as-extruded webs (web-fed). The distinction between sheet-fed and web-fed machines is similar to that between sheet-fed and coil-fed equipment in the metal-can industry, except that in plastics forming, the wide web can comes from either coiled stock or directly from the extruder.

Thermoforming is an important thermoplastic fabrication process which began in the early 1960s when containers and lids formed from high impact polystyrene (PS) were first used by the dairy industry to package cottage cheese, sour cream, yogurt, and other dairy foods. Currently, roll-fed thermoforming machinery is available that can produce beverage cups at rates of 75,000–100,000 pieces/h while consuming plastic sheet in excess of 1 ton/h (910 kg/h). Sheet consumption for larger and heavier containers can exceed 2 tons/h (1.8 metric ton/h), but unit production rates may be lower because fewer mold cavities can be mounted in the machines and more time is required to cool the thicker walls.

The ability to produce extruded PS foam sheet (see Foam, extruded polystyrene) has provided additional packaging markets for thermoforming. The first of these was meat and produce trays, and later egg cartons. In the United States, approximately 90% of the meat and produce tray market is served by the foam trays and about 60% of consumer-market eggs are packaged in thermoformed foam cartons. Other applications include fast-food carry-out cartons, institutional dinnerware, medical trays and blisters, and inserts for rigid boxes.

Process Steps

The thermoforming process is used to make products from thermoplastic sheet by a sequence of heating, shaping, cooling, and trimming. Trimming is not necessarily an integral part of the forming cycle, but few applications can use the formed web without some kind of trimming.

Heating. Thermoplastic sheet is typically heated to a temperature range adequate for forming, usually 285–325°F (141–163°C), depending on the material used. Temperature control is critical because of plastic's poor thermoconductivity and because temperature affects the forming characteristics, ie, ductility, of the materials: too much heat and the sheet flows without drawing; too little and it ruptures early in the forming process.

Most thermoplastics absorb infrared energy emitted in the 3.0–3.5-μm wavelength range, which makes them ideally suited to heating by radiation. Heating time is governed by the heating process used, surface conditions of the material, and the combination of low thermoconductivity and relatively high specific heat capacity. Any of the common transfer processes such as convection, conduction, or radiation can be used. Convection heating would be ideal because the sheet could be soaked in hot air at just the right temperature with the assurance that the entire sheet would be uniformly heated. However, straight convection systems are not practical because they present material-handling problems relative to transferring the sheet from the oven to the forming station. Therefore, most thermoforming machines are equipped with systems that take advantage of both radiation and convection. Most of these ovens are electrically heated, but there are also large number of cut-sheet machines using gas-fired ovens. Most are capable of heating the sheet from two sides, which is especially advantageous if the sheet is ≥40 mil (≥1 mm). The ovens should also be appropriately zoned and temperature-regulated by instrumentation enabling the operator to maintain good control over the heating process.

Today any number of electrically powered heating system can be specified, including the conventional tubular steel rods, glass or ceramic panels, quartz lamps, emitter strip panels, and small ceramic modules. The most common continues to be the calrod type, but use of rectangular ceramic elements has been steadily increasing. Their relatively small size permits them to be incorporated into elaborate microprocessor-control systems to provide preferential heating in localized areas. Emitter strip panels have been introduced as an alternative to provide temperature uniformity with a less complex installation. They are gaining popularity because they provide a full area heat source and can be mounted closer to the sheet and operated at lower temperatures, thereby conserving energy.

Conduction or contact heating, sometimes called trapped sheet heating, has been used successfully with materials such as oriented polystyrene (OPS) and PVC. Here, the plastic sheet is trapped between the mold and the temperature-controlled hot plate which is perforated with extremely small holes, ie, ca 0.020 in. (0.5 mm) dia. The holes are drilled in a grid pattern on ≤1-in. (2.5-cm) centers. After the sheet has been in contact with the heated plate for a predetermined time, compressed air is injected through the plate, forcing plastic off the heated platen and into the mold. This system is generally reserved for materials ≤0.015 in. (0.38 mm) thick.

Solid-phase and conventional thermoforming. Solid-phase pressure forming (SPPF) means forming below the crystalline melting point, ie, 5–8% lower than melt-phase forming, depending on the material. For example, polypropylene (PP) is formed in the melt phase at 310–315°F (154–157°C) but at 285–295°F (141–146°C) in SPPF. SPPF does not require special thermoforming equipment. A thermoforming machine forms products within the thermoforming window, and prod-

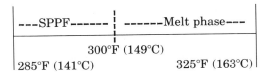

Figure 1. Thermoforming window.

uct-configuration requirements, ie, stress, strength, rigidity, flexibility, determine the proper temperature range.

Figure 1 indicates that SPPF is done in a temperature range within the window, as in melt-phase forming. Compared with products formed in the melt phase, SPPF can produce stiffer parts with less materials. Solid-phase forming improves the sidewall strength of a container and increases the stress factor. When forming cups, trimming in the mold is recommended. Plug assist and high forming pressure, ie, 100 psi (689 kPa), are required, and the plug design is more critical than in conventional melt-phase forming.

Forming. There are two basic forming methods from which all others are derived: drape forming over a positive (male) mold, and forming into a cavity (female) mold. Product configuration, stress and strength requirements, and material specifications all play a part in determining the process technique. Generally, forming into a female mold is used if the draw is relatively deep, eg, cups. Female molds generally provide better material distribution and faster cooling than male molds. Male-mold forming is preferred for certain product configurations, however, particularly if product tolerances on the inside of a part are critical. Male-mold forming produces heavier bottom strength; female-mold forming produces heavier lip or perimeter strength. An advantage of straight forming into a female mold is that parts with vertical sidewalls can be formed and extracted, stress-free, from the molds because of the shrinkage that occurs as the part cools.

In drape forming, when the hot plastic sheet touches the mold as it is being drawn, it chills and start to set up. To successfully drape-form, several variables must be considered. One of the most significant is shrinkage. Since all plastic materials have a high coefficient of thermal expansion and contraction, ie, about 7–10 times that of steel, care must be taken when designing the mold to provide sufficient draft on the sidewalls so the part can be extracted from the mold. It is not unusual for parts to rupture upon cooling on an improperly designed drape mold. Another potential problem is that the part may become so highly stressed during forming and cooling that it loses most of the physical properties that the sheet would otherwise provide.

Natural process evolution has combined the two systems to take advantage of the better parts of each method. The plug-assist process, similar to matched-die forming, involves a male mold (or plug) that ranges in size from ca 60 to 90% of the volume of the cavity. By controlling the geometry and size of the plug and its rate and depth of penetration, material distribution can be improved for a broad range of products. The plus-assist technique is used to manufacture cups, containers, and other deep-draw products.

Many thermoforming techniques have been developed to obtain better material distribution and broaden the applicability of the process. Some of the more popular methods are illustrated and described in Figures 2 through 10. Most of these

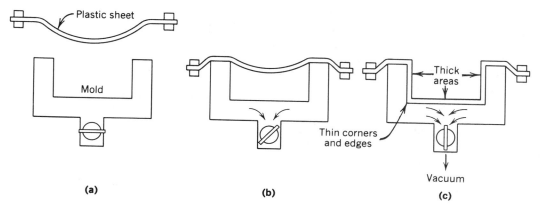

Figure 2. Shallow-draw forming, female-mold bottom. (a) The plastic sheet is clamped and heated to the required forming temperature. (b) A vacuum is applied through the mold, causing the plastic sheet to be pushed down by atmospheric pressure. Contact with the mold cools the newly formed plastic part. (c) Areas of the sheet reaching the mold first are the thickest. Applications: ice-chest lids, tub enclosures, etc.

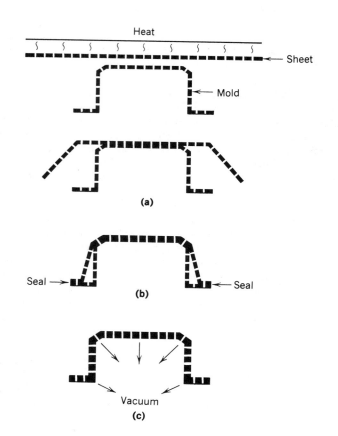

Figure 3. Vacuum (drape) forming. (a) The plastic sheet is clamped and heated to the required forming temperature. (b) The sheet is sealed over the male mold. (c) Vacuum from beneath the mold is applied, forcing the sheet over the male mold and forming the sheet. Material distribution is not uniform throughout the part, as the portion of the sheet touching the mold remains nearly the original thickness. The walls are formed from the plastic sheet between the top edges of the mold and the bottom-seal area at the base. Applications: trays, tubs, etc.

techniques can employ vacuum, pressure, or a combination to apply the force necessary to shape the heat-softened plastic sheet.

Cooling. The time required to cool the heat-softened plastic below its heat-deflection temperature while it is in contact with the mold is often the key to determine the overall forming cycles. Cooling is accomplished by conductive heat loss to the mold and convective heat loss to the surrounding air. Cooling rate depends upon the tooling, because in all methods except matched mold, the plastic is in contact with the mold on one side only. The opposite side is cooled convectively by forced air or ambient air. Water sprays are sometimes used but often pose as many problems, eg, water spotting, as they solve. Pressure forming helps minimize cooling time because the higher air pressure keeps the sheet in more intimate contact with the mold surface.

Trimming. A number of trimming methods are available, including hand cutting, rough shearing of the web, punching parts out from the web, or trimming around a periphery with a

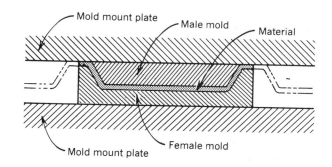

Figure 4. Matched-mold forming. The plastic sheet is clamped and heated to the required forming temperature. The heated sheet can be positioned over the female die, or draped over the male mold. The male and female halves of the mold are closed, forming the sheet, and trapped air is evacuated by vents located inside the mold. Material distribution varies with mold shape. Detailed reproduction and dimensional accuracy are excellent. Applications: foam products, egg cartons, meat trays, etc.

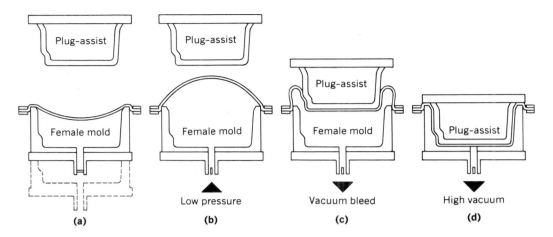

Figure 5. Pressure-bubble/plug-assist vacuum forming. The plastic sheet is clamped and heated to the required forming temperature. (**a**) The heated sheet is positioned over the female mold and sealed. (**b**) Prestretching is accomplished by applying controlled air pressure through the female mold, creating a bubble. (**c**) When the sheet is prestretched to the desired degree, the male plug is forced into the sheet. (**d**) When the male plug is fully engaged, a vacuum is applied through the female mold to form the sheet. Pressure may be applied through the plug to assist forming, depending on the material and forming requirements. Applications: refrigerator liners, bathtubs, etc.

steel rule or sharp-edged die used to penetrate the web. The punch-and-die trim provides the greatest accuracy and longest life, but at the highest cost. These dies are usually designed with a hardened-steel punch and will pass through a slightly softer steel die that can be peened when it dulls. These trim dies are designed so the parts are punched through the die successively and exit from the trim station in a nested fashion. This simplifies the material handling required to pack or prepare the parts for the next post-forming operation, eg, rim rolling, printing, etc.

Steel-rule dies are better suited for the lower volume, thin-sheet operations. Steel-rule trim dies are used predominantly in custom thermoforming operations, where the run size does not justify the cost of matched shearing dies. They are used almost exclusively in the manufacture of blister packages and decorative packaging for the cosmetics industry.

The trimming operation can be done inside or outside the mold. The remote-trim configuration utilizes a separate trim press which punches out 1–3 rows of parts at a time. The parts are discharged in nested horizontal stacks ready for subse-

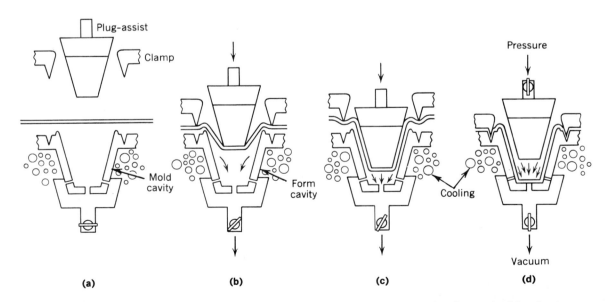

Figure 6. Plug-assist forming sequence. (**a**) The plastic sheet is clamped and heated to the required forming temperature. The sheet is then sealed across the female-mold cavity. (**b**), (**c**) The male plug is forced into the sheet. The depth and speed of penetration, as well as plug size, are the primary factors in material distribution. (**d**) When the plug is fully engaged, vacuum, pressure, or a combination are applied to form the sheet. Applications: cups, food containers, etc.

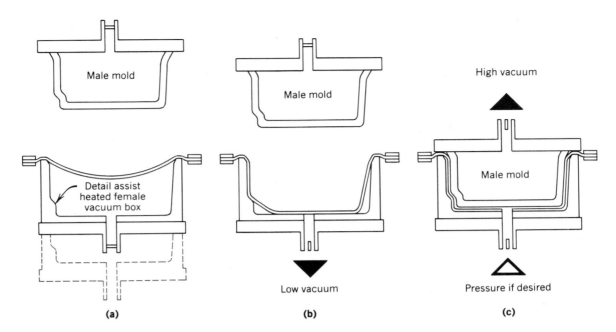

Figure 7. Vacuum snapback. (**a**) The plastic sheet is heated and sealed over the female vacuum box. (**b**) When vacuum is applied to the bottom of the vacuum box, the plastic sheet is prestretched into a concave shape. (**c**) After the plastic sheet is prestretched to the desired degree, the male plug enters the sheet. A vacuum is applied through the male plug while the vacuum box is vented, creating the snapback. Light air pressure may be optionally applied to the vacuum box, depending on the materials used and forming requirements. Applications: deep-draw parts (ie, luggage, auto parts), gun cases, cooler liners, etc.

quent operations or for packing. Trimming within the mold provides the most accurate alignment of the trim cut to the formed part, but the close proximity of the trimming to the mold body is a complex arrangement which may limit cooling capacity near the cutting surface, and after trimming, precise control of the part ceases. Complex mechanisms are required to sort and nest these parts downstream. Trimming is done by sawing or routing in the production of low volume odd-geome-

try industrial parts, eg, boat hulls, recreational vehicle bodies, bathtubs, but not in packaging.

Forming machines. There are basically two types of thermoforming machines: sheet-fed and web-fed. Sheet-fed machines operate from sheet cut into definite lengths and widths for specific applications. The sheet is generally heavy, ie, 0.060–0.5 in. (1.5–13 mm), and the products are industrial. Packaging-related applications include dunnage trays and

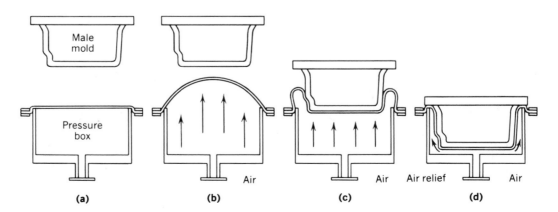

Figure 8. Pressure-bubble vacuum snapback. (**a**) The plastic sheet is clamped and sealed across the pressure box. (**b**) Prestretching is accomplished by applying controlled air pressure under the sheet, through the pressure box, creating a bubble. (**c**) When the sheet is prestretched to the desired degree, ie, 35–40%, the male plug is forced into the sheet while the air pressure beneath the sheet remains constant. (**d**) After the male plug is fully engaged, the lower air pressure is increased. A vacuum is simultaneously applied through the male plug, creating the snapback. Applications: intricate parts where material distribution is critical.

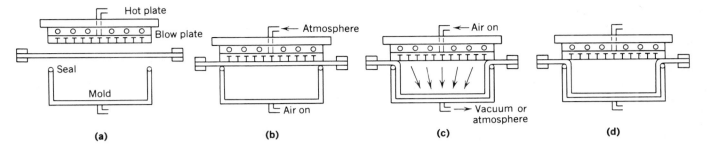

Figure 9. Trapped-sheet, contact-heat, pressure forming. The plastic sheet is clamped between the female-mold cavity and a hot blow plate. (**a**) The hot, porous blow plate allows air or vacuum to be applied through its face. (**b**) The mold cavity seals the sheet against the hot plate. (**c**) Controlled air pressure, applied from the mold, blows the plastic sheet in contact with the hot plate to ensure complete contact with the heating surface. (**d**) When the desired heating has been accomplished, air pressure is applied through the hot plate, forcing the plastic sheet into the mold. Venting of the mold can be simultaneous, depending on materials and forming requirements. Steel knives may be inserted in the mold for sealing and in-place trimming if additional closing pressure is available. Applications: OPS containers, covers, etc.

pallets, large produce bins, shipping-case dividers, box liners, crates, and carrying cases formed to the shape of the product.

Packaging applications generally employ web-fed machines, which use either coil stock or a web that comes directly, ie, in-line, from a sheet extruder. The need to form a coil limits the thickness of the web to ca 0.125 in. (3.2 mm). Web thickness can be increased by operating directly in-line with an extruder.

Tooling

Among the plastic fabricating processes, thermoforming permits the use of the broadest range of mold-tooling materials. Mold materials such as wood, epoxy, polyesters, or combinations thereof can be used if the volume is not sufficient to warrant the cost of advanced, temperature-controlled metal molds. For high volume applications, eg, food packaging,

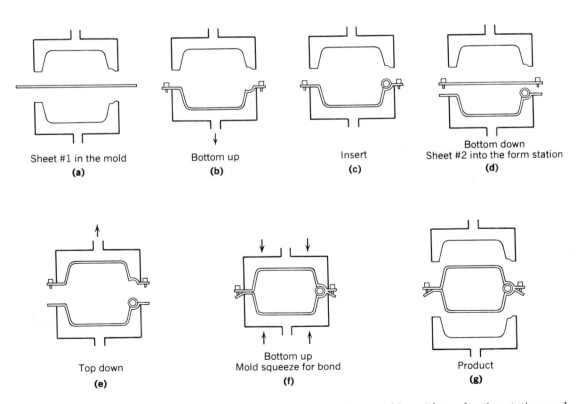

Figure 10. Twin-sheet forming. This technique utilizes a rotary-type machine with two heating stations and bottom-load clamp frames. The first sheet (**a**) is heated and formed into the lower half of the mold (**b, c**) while the second sheet is being heated. After forming the first sheet, the second sheet is indexed into the forming station (**d**) and vacuum is applied to the upper mold half to form the sheet (**e**). A bond is simultaneously made between the parts by pressure applied to the mating surfaces (**f**). After cooling, the molds are removed to give the product (**g**).

molds are normally manufactured from cast or machined aluminum, sometimes with hard coats applied to the wearing surfaces for longer service life. Molds are occasionally made from beryllium copper or brass which is chrome-plated for extended service life.

Vacuum and compressed air are the primary means used to form the heated, thermoplastic sheet to the mold configuration with rigid-type materials. For foam materials, matched metal molds are frequently used to compress the material to the desired thickness without compressed air. Vacuum is transmitted through small holes that are located to allow the sheet to come in intimate contact with the mold surface. The size of these holes ranges from ca 0.013 to 0.030 in. (0.33–0.76 mm) dia, depending on the type of thermoplastic to be formed. Other means of applying vacuum in the mold include the incorporation of narrow grooves or slots, which often are easier to incorporate than vacuum holes and allow greater evacuation of the air in the tool. Polyolefin materials require small vacuum holes and slots; PS materials, which cool faster, can use larger holes. Final product design and material selection determine the exact size of the holes required.

In addition to vacuum holes and slots, molds often require a complete sandblasting of the molding surface to minimize the possibility of trapped air between the sheet and the mold. Eliminating trapped air is important because it prevents imperfections on the part surface and aids in rapid transfer of heat out of the material by providing close contact between the plastic and the cavity. Undercut areas of molds are used for PS are normally not sandblasted because this material is relatively rigid and sandblasting would make part removal difficult. Molds used for polyolefins are normally sandblasted completely over the forming surface. Vacuum is usually pulled through the forming mold, male or female, and the compressed air is introduced from the opposite side of the sheet material. In foam-forming with matched metal molds, vacuum is normally applied only through the female cavity.

Tooling for high production applications is most often aluminum because of its thermal conductivity, light weight, and cost. Most crucial is thermal conductivity, because in any thermoforming process, the residual heat of the plastic must be removed as rapidly as possible for rapid cycling. The temperature control of the mold is accomplished by water channels in the mold and designing them to provide maximum heat removal from the sheet. Ideally, all molds should provide a temperature differential between inlet and exit water temperature of no more than 5°F (3°C).

Many techniques are used to incorporate cooling in the molds for improved cycle times. The most common methods are contact cooling and direct water cooling. Contact cooling consists of coring a mold base and carefully machining cavity inserts so that contact occurs when the cavity insert expands from the heat of the plastic sheet while the mold base contracts from the coolant circulating through it. The most common method today for rigid PS is to circulate water through hollow-shelled cavities. This is more expensive but the most effective way to cool the plastic sheet rapidly.

Tooling for low volume work, eg, box inserts, blisters, and other nonfood packaging, and prototyping can often be simple and effective without any of the special cooling features described above. This is because it normally takes longer to heat the sheet than to cool it, and the work is generally done on slow, low production equipment that does not require the rapid cycle rates. With these molds, cooling is very often augmented by the use of fans to increase the air circulation over the material after it is formed. In such applications, low cost, simple molds of plastic, plaster, wood, or other materials are used effectively. Mold dimensions must allow for the different shrinkage rates of different types of thermoplastics. In the manufacture of a family of products, it is sometimes possible to utilize the same cooling base and interchange only the molding inserts. Depending on the trim dimensions of the products, the same cutting tool might be suitable for more than one product; if not, they can be designed to be easily interchanged from the trim die shoe.

Cut-sheet Thermoforming

Cut sheet is processed on two types of thermoforming machines: the shuttle style and the rotary, or carousel, style. In single-station shuttle machines, the plastic sheet is placed in a clamping frame that shuttles into the oven for heating and back again for forming. Another very popular shuttle machine is the double-ender, which has a common oven with a form station at each end. This arrangement provides 100% utilization of the oven because heating and forming can be done simultaneously.

Rotary or carousel thermoformers are most frequently used for heavier gauge materials, ie, starting at ca 0.050 in. (1.3 mm). Most rotary machines have three working stations: load/unload, heat, and form. In a typical machine, the carousel frame rotates 120° on a time-controlled basis. Generally, the time cycle is dictated by the period required for the plastic sheet to cool below its heat-deflection temperature. When cooling time is faster than the heating time, four-station rotary machines are often used; these have two heating stations. Another type of four-station machine has a separate load and unload for automation and parts handling.

Most cut-sheet thermoformers are designed to operate as vacuum formers, but rotary and single-station (shuttle-type) machines are available with pressure-forming capabilities and sustain forming pressures up to 50 psi (345 kPa), depending on mold area. Most sheet-fed machines can be adapted for any of the standard thermoforming techniques. Cut-sheet machnes have been built to accept sheet 10 ft (3.05 m) wide and 20 ft (6.1 m) long with the capability of making parts up to 4 ft (1.22 m) deep.

Twin-sheet Thermoforming

Twin-sheet thermoforming can produce hollow parts eg, pallets, from cut sheet. These parts can be produced from both cut-sheet and roll-fed machinery. In sheet-fed machinery, the material to be formed can be 0.050–0.500 in. (1.3–13 mm) thick. Roll-fed machines are limited to materials 0.005–0.125 in. (1.3–3.2 mm) thick. With a typical web-fed, twin-sheet system, two rolls of plastic materials are simultaneously fed, one above the other. The webs are transported through the oven on separate sheet-conveyor chains and heated to a formable temperature. At the forming station, a specially designed blow pin enters the space between the two sheets before the mold closes. Air pressure is introduced between the sheets through the blow pin, and, at the same time, vacuum is applied to each mold half. Twin-sheet forming is done by a slightly different method on specially designed rotary thermoformers.

Equipment Improvements

Rapidly expanding markets for thermoformed products are attributable to technological advancements in equipment design. Better process control has enabled converters to increase productivity and improve product quality. Use of microprocessor-control systems allows the operator to enter data, examine existing values in the program, or trouble-shoot problems by viewing a television monitor. The conventional control panel includes a cathode-ray tube (CRT) and keyboard, form-on and emergency stop push buttons, and indicator lights to signal form or automatic modes. Machine functions may be displayed on the CRT upon request and function control is obtained by means of the simplified keyboard.

The microprocessor-control systems feature control of oven-temperature orientation in multiply zoned ovens, interface speed between former and trim press, sheet-index length, machinery functions and timing, and control and graphic display of start and end points of forming functions, storage of production information, and diagnostic capabilities. Use of the microprocessor results in tremendous savings in labor, energy, and downtime. Many of these systems are available as a retrofit to existing equipment. Equipment has also been designed to monitor several lines of similarly equipped machines from one central location. Such data-management systems provide comparative data, setup parameters, production rates, maintenance schedules, alarms and safety signals, and storage of reference information, resulting in a true on-line management-information reporting system.

Design improvements that have contributed to improved productivity include redesigned forming presses that provide front and rear access to the forming station to facilitate tooling installation and periodic maintenance; access doors to oven heaters for quick cleaning; ovens that eject automatically if the sheet overheats; and automatic lubrication systems that increase component life substantially and allow faster mechanical operation due to reduced friction.

Microprocessor heating control allows the operator to control the temperature profile of the sheet from edge to edge or to provide localized heating. Solid-state instruments used in conjunction with mercury or semiconductor-controlled rectifier (SCR) relays have made it possible to maintain extremely close heater temperature control.

Forming presses are now available with precision guidance systems on the platens to permit critical operations to take place in the forming molds that formerly had to be done on auxiliary equipment. An example is the method of incorporating the latching slots which are cut in the sidewalls of egg cartons, fast-food cartons, and trays. Previously, these slots were punched in on a special press located downstream from the forming station. Registration problems relative to the slot location were often encountered because of variations in the index length by the thermoformer web-conveyor system. These problems were overcome by expensive drive systems. On machines with precision-guidance systems, mating punches can be incorporated in the forming molds that engage during the last 5–10% of press closure and form a slot in the heat-softened plastic that has reinforced edges similar to button holes.

High speed trim presses are available with supported moving platens that carry up to 85% of the gross-platen and trim-tool weight. The advantage of this system is that it permits the press tie rods to function primarily as guidance members rather than as both platen transporters and guides, in addition to serving as tie rods. The system reduces trim-tool maintenance by up to 50%.

Significant improvements have been made in product-handling systems, which can now automatically count, stack, and package products at the trim-press discharge. The design of these automated systems varies from relatively simple systems using guide chutes or magazines and transport conveyors to those employing robotics and advanced packaging machinery.

Automation has been extended to the material handling of large plastic sheets into sheet-fed machines. The development of automatic sheet loaders provides automatic loading systems that enable use of the entire machine area. This results in a 20–30% increase in productivity, achieved largely from savings in loading time, since a manually loaded machine normally used only one quarter of the machine effectively.

BIBLIOGRAPHY

General References

"Thermoforming Techniques," in J. Agranoff, ed., *Modern Plastics Encyclopedia 1985–1986*, McGraw-Hill, Inc., New York, 1985.

Lynn McKinney
William Kent
Richard Roe
Brown Machine

TIN-MILL PRODUCTS

Tin-mill products are thin coils or sheets of low-carbon steel. If they have been coated with pure tin, the product is tinplate. If coated with chrome–chrome oxide, the product is electrolytic chrome-coated steel (ECCS). ECCS is often called tin-free steel (TFS). If the steel is left uncoated, it is black plate, also tin-free. Steelmakers have been experimenting with alternatives to chrome, including nickel, but none has been marketed. Apart from its protective functions, tinplate is the only type of steel for can making that can be soldered and is easily welded (see Cans, fabrication; Cans, steel). It is required for the two-piece D&I process, and it is used to some extent for two-piece DRD cans. ECCS is the standard material for two-piece DRD can bodies and is widely used for ends on cans fabricated by all methods. It can be welded, but the chrome coating must first be removed by edge treaters, and special consideration must be given to corrosion resistance. Black plate is not used for food or beverage cans, but it is used to package some less demanding products, such as motor oil and spices.

The metal can or "tin can" has been used since the early 1800s (see Cans, steel). Prior to the mid-1930s, tinplate used for can manufacture was produced by the hot-dip method. Steel sheets were dipped in molten tin. The concept of continuous coating of tinplate began in the late 1930s. Aside from the obvious benefits of speed and uniformity from a continuous line, the chief purpose for this move to continuous electrolytic plating was the savings in tin. Hot-dip tinning had resulted in the use of a standard coating of 1.50 lb/base box (33.63 g/m²), which is roughly 90×10^{-6} in. (2.3 μm) thick. Electrolytic tinning afforded the opportunity to produce a much lighter coating of tinplate. World War II provided the impetus to reduce this tin coating because tin was in short supply.

Production

Iron from a blast furnace is refined to a basic low carbon steel. Today, this is commonly accomplished in a basic oxygen furnace. Upon reaching the proper temperature and chemistry, the steel is poured into an ingot mold or continuously cast into slab form. (Continuous casting is becoming the predominant casting form.) If cast into an ingot, the ingot is reheated to uniform temperature in a soaking pit and then rolled to a slab. The slab is reheated and hot-rolled to what is known as a hot band on a hot strip mill. Here, a slab up to 10 in. (25.4 cm) thick and 15 ft (4.57 m) long is hot-reduced to a thickness of 0.075–0.100 in. (1.9–2.54 mm), depending on the end use. This hot band is the raw material for the tin-mill product. At this point, steel chemistry, steel cleanliness, and, to a large extent, the condition of the surface are fixed. The steps in the manufacture of tin-mill products are shown in Figure 1.

First, the iron oxide (scale) on the surface of the hot band is removed in a pickle line, where the hot-rolled strip, ie, hot band, is flexed to break the brittle scale and passed through a series of tanks containing acid, generally H_2SO_4 or HCl. The base metal is allowed to react with the acid, removing the scale. As the pickled strip is recoiled, it is coated with an oil which serves as a rust preventative and as a lubricant in subsequent cold-rolling operations. The pickled hot band then goes to a multistand (generally five or six) cold-reducing mill. In this operation, the hot band is reduced by 90% at speeds of ≥5000 ft/min (1524 m/min). In the case of a single-reduced product, the final gauge is determined at this point. If the steel is going to be reduced again, ie, double-reduced, it is rolled to an intermediate gauge at this point. The oils used in reduction are removed by a combination of electrolytic action and scrubbing of the strip. The cold band is now a "full hard" product, ie, the material is quite hard and brittle, and for most applications it must be softened by annealing. In annealing, the hard cold-reduced band, ie, coil, is heated to temperatures of 1100–1300°F (593–704°C). This "recovers," ie, recrystallizes, the internal structure of the steel, making it a much more usable product. A full hard product will barely take a 180° bend. After recovery, the product can be drawn, beaded, flanged, etc.

Annealing. Annealing can be done in a batch or continuous operation. In batch annealing, the coils are stacked and covered with an airtight cover in a furnace. They are surrounded with an inert gas and heated to the desired temperature for a predetermined period of time. Batch annealing provides the opportunity to produce softer tempers. Softness is a function of time and temperature, and batch annealing is a slow process providing opportunities for both. In continuous annealing, the coil is passed single-strand through a furnace. The steel is heated to the desired temperature, then cooled, all in an inert atmosphere that prevents oxidation (scaling) of the strip surface. Continuous annealing, done at speeds of up to 1500 ft/min (457 m/min), provides less time for annealing. This results in a grain structure smaller than that obtained from batch annealing, and the manufacture of softer materials is much more difficult to accomplish. It does, however, yield a more uniform product than batch annealing.

Until this point, single- and double-reduced products have followed the same route. If the coil is to be double-reduced, it goes to a second cold-reduction mill after annealing. In this process, a coil that has been cold-rolled to an intermediate gauge on the tandem mill is given a second cold reduction similar to the first, except that the reduction is about 35% instead of 90%. For instance, a 0.0102-in. (0.26-mm) cold band off the tandem mill is given a second cold reduction down to 0.0066 in. (0.168 mm), or what is commonly known as 60-lb basis weight (1345 g/m²). At this point, single- and double-reduced product join the same processing line again.

The as-annealed product has almost the right hardness, but if it were used in this condition it would flute, panel, or crease very severely. To reduce the potential for these problems and to impart the desired surface finish, a skin-passing (tempering) is necessary. This is just a very light cold reduction that changes surface hardness, ie, temper, only slightly but imparts "springiness" to the steel along with the proper surface finish. If one were to bend an as-annealed panel, it would feel quite "dead," in contrast to a tempered product. Reductions accomplished in tempering are generally in the 0.5–2.0% range, which does not amount to a measurable change in thickness.

The temper mills produce two basic finishes. The more common finish in the can industry is the #7 smooth finish, which is created by a shot-blasted roll in the first stand and a ground roll in the second stand. The grinding in the second roll determines the final finish. A #5 rough finish is created by shot-

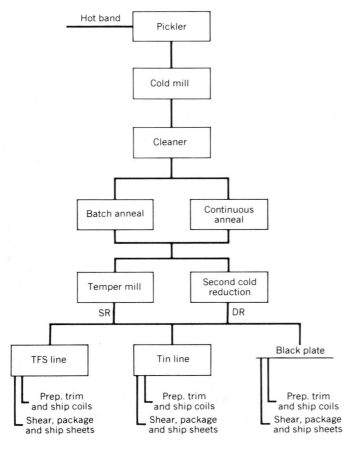

Figure 1. The manufacture of tin-mill products.

blasted rolls in both stands. At this point, the product is called black plate. Black plate, both single- and double-reduced, has a number of uses, eg, motor-oil cans, spice cans, and several industrial applications. In a tin mill, black plate is the raw material for electrolytic tinplate (ETP) and for ECCS, also known as TFS.

Black plate. Tinplate is black plate with an electrolytic tin coating on its surface. Coils are entered into the tin line, which operates at speeds of up to 2000 ft/min (610 m/min). The strip passes through a caustic scrubber cleaning operation and a light acid bath or light pickle operation to prepare the surface for plating. There are a number of plating methods, but the two principal methods are the acid-sulfate process, and the halogen process. Of the large steel companies, some use one, some use the other (see Ref. 1 for explanations of all processes). As-plated tinplate has a dull-gray matte appearance that is acceptable for some applications, but the most common variety is the bright (or "brite") shiny product produced by reflowed tin. This is done by passing the strip through an induction heater, ie, britener, where it is heated to the melting point of tin, ie, 460°F (238°C). Then it is immediately quenched in water. The melting (reflowing) produces the bright reflective appearance normally associated with tinplate.

After this step, the strip is subjected to one of several postbrightening chemical treatments which stabilize the tinplate surface and inhibit oxidation. The most common treatment for normal applications is an electrochemical treatment in sodium dichromate. The strip passes through a solution of sodium dichromate in water and between conductor grids. Cathodic treatment serves to stabilize the tin oxide surface and retard the growth of the oxide. This also plates a very fine film of metallic chrome and chrome oxides on the surface, which allows storage of tinplate for reasonable lengths of time without deterioration of the surface. It also provides a uniform surface for subsequent enameling or lithographing.

After the chemical-treatment section, the strip is dried and lightly oiled with acetyltributyl citrate oil, which provides lubricity to the sheet so that recoiling, shearing, and subsequent handling do not scratch the soft tin coating. This oil is at very low film weights and is compatible with today's enamels and inks (see Inks). The tinplate is then recoiled and either packed for shipment or cut to length.

The manufacture of ECCS involves plating of chrome and chrome oxides rather than tin. Black plate is coated with a very thin layer of chromium, ie, 0.3×10^{-6} in. (0.0076 μm) thick. The main difference between the two plating systems is that a tin line uses pure-tin consumable anodes which maintain the tin level in the electrolyte. On a chrome line, the anodes are nonconsumable, and additions of chromic acid are required to maintain the chromium level in the electrolyte. Modern ECCS lines are two-step lines. In the first step, metallic chrome is plated; in the second step, a cathodic chromic acid treatment deposits chrome oxides on the surface over the metallic chrome. The ECCS is then oiled with butyl stearate and recoiled.

Terminology

The procedures described above produce a thin, strong, and ductile steel-based product covered by layers of iron–tin alloy and free tin, or chrome and chrome oxides, and oil. This sandwich of layers must withstand the rigors of bending, drawing, coating, printing, and soldering or welding and still satisfy

Table 1. Nominal Base Weight and Theoretical Thickness

Base weight, lb/base box (g/m²)[a]	Theoretical thickness	
	in.	mm
50 (1121)	0.0055	0.140
55 (1233)	0.0061	0.155
60 (1345)	0.0066	0.168
65 (1457)	0.0072	0.183
70 (1569)	0.0077	0.196
75 (1681)	0.0083	0.211
80 (1794)	0.0088	0.224
85 (1906)	0.0094	0.239
90 (2018)	0.0099	0.251
95 (2130)	0.0105	0.267
100 (2242)	0.0110	0.279
103 (2309)	0.0113	0.287
107 (2399)	0.0118	0.300
112 (2511)	0.0123	0.312
118 (2645)	0.0130	0.330
128 (2870)	0.0141	0.358
135 (3027)	0.0149	0.378

[a] One lb/base box = 22.42 g/m².

FDA requirements for health and safety. It must look good, lay flat, and be of uniform size and thickness. This unique product has its own terminology. Most steel products are described in terms of short ton (0.907 metric tons). Tin-mill products are described in terms of base boxes and packages. These terms originated early on, when tinplate was sold in units of 112 sheets each, called packages. Each sheet was 14 × 20 in. (35.6 × 50.8 cm); a package of 112 sheets was called a base box. The area it contained, 31,360 in.² (20.2 m²), survives today as the unit area used to sell tin-mill products. Table 1 shows the relationships between lb/base box and thickness (2,3).

Material thickness is expressed as a basis-weight unit, or as pounds per base box. For instance, a 75-lb (34-kg) basis-weight plate weighs 75 lb/base box or 1681 g/m². Basis weight is converted to nominal (in.) decimal thickness by multiplying the basis weight by 0.00011 (eg, 75 lb × 0.00011 = 0.0083 in.) and to mm thickness by multiplying the kg basis weight by 0.00618. Control of basis weight, or gauge, is extremely important. The tin-mill products industry works to very tight tolerances on gauge (2,3).

Tin coating weights are also expressed in pounds per base box or g/m². This refers to the weight of the tin deposited on a base box of black plate, actually both sides of the base box. For instance, "quarter-pound tin" means that a quarter of a pound (0.113 kg) of tin is deposited on both sides of a base box or both sides of 31,360 in.² (20.2 m²), ie, 0.125 lb or 0.057 kg on each side (see Table 2) (3). The thickness of the tin in this case is 0.000015 in. (0.381 μm). This product is designated No. 25.

Tin coating weights are generally available (Table 2) in ranges from 0.05 lb/base box (1.12 g/m²) to 1.35 lb/base box (30.27 g/m²) in practically all variations thereof. Differentially coated tinplate has a heavier coat on one side than the other. Control of tin coating weights are generally controlled by continuous line monitoring. Control of proper tin coating

Table 2.   Electrolytic Tin Plate Coating Weight[a]

Designation no.	Nominal tin coating weight for each surface, lb/base box[b]	Coating for each surface, g/m²	Minimum average coating weight for each surface, test value, lb/base box	Minimum coating for each surface, g/m²
10	0.05/0.05	1.12/1.12	0.04/0.04	0.9/0.9
20	0.10/0.10	2.24/2.24	0.08/0.08	1.8/1.8
25	0.125/0.125	2.8/2.8	0.11/0.11	2.5/2.5
35	0.175/0.175	3.92/3.92	0.16/0.16	3.6/3.6
50	0.25/0.25	5.6/5.6	0.23/0.23	5.15/5.15
75	0.375/0.375	8.4/8.4	0.35/0.35	7.84/7.84
100	0.50/0.50	11.2/11.2	0.45/0.45	10.1/10.1
50/25	0.25/0.125	5.6/2.8	0.23/0.11	5.15/2.5
75/25	0.375/0.125	8.4/2.8	0.35/0.11	7.84/2.5
100/25	0.50/0.125	11.2/2.8	0.45/0.11	11.2/2.5
100/50	0.50/0.25	11.2/5.6	0.45/0.23	11.2/5.15
135/25	0.675/0.125	15.1/2.8	0.62/0.11	13.9/2.5

[a] ASTM A 624.
[b] One lb/base box = 1 lb/31,360 in.² = 22.42 g/m².

Table 3.   Temper Designations

Product type	Designation	Rockwell hardness ranges 30-T[a]	Characteristics	Applications
box annealed product, single-reduced	T-1	52[b]	soft for drawing	drawn requirements, nozzles, spouts, closures
	T-2	50–56	moderate drawing where some stiffness is required	rings and plugs, pie pans, closures, shallow drawn and specialized can parts
	T-3	54–60	shallow drawing; general-purpose with fair degree of stiffness to minimize fluting	can ends and bodies, large diameter closures, crown caps
	T-4	58–64	general-purpose where increased stiffness desired	can ends and bodies, crown caps
continuously annealed product, single-reduced	T-4-CA	58–64	moderate forming fair degree of stiffness	closures, can ends and bodies
	T-5-CA	62–68	increased stiffness to resist buckling	can ends and bodies

[a] These hardness values are only a guide in the ordering and processing of tin-mill products. Values above and beyond the ones shown are encountered in a commercial product. Rockwell values are too varied to permit establishment of ranges. These values are based on the use of the diamond anvil.

Note: Lighter base weight products, ie, nominally 75 lb per base box (0.21 mm) and lighter, are normally tested using the 15-T Rockwell scale and the results converted to the 30-T Rockwell scale.

Based on ASTM A 623 or A 623 M and industry practice.

[b] Max.

Table 4. Typical Mechanical Characteristics of Double-Reduced Product

Designation	Longitudinal tensile strength, psi[a]	Tensile yield strength, MPa, longitudinal L at 0.2% offset	Rockwell hardness, 30-T[b]	Applications
DR-8	80,000	550	73	small-diameter round can bodies and ends
DR-9	90.000	620	76	larger-diameter round can bodies and ends

[a] Approximate.
[b] See footnote[a] of Table 3.

weight is necessary for the protection of the material to be packed in a can.

Performance

Tinplate, particularly brite tinplate, is a combination of steel base and iron–tin alloy, free tin, tin oxide, and oil. The five components all play a role in the performance of the tinplate. The steel base provides the basic strength of the material. The iron–tin alloy, which is formed during brightening, provides some corrosion resistance. The free tin provides anodic protection between the steel base and the product. Free tin refers to the tin that is not alloyed. For instance, the 0.125 lb/base box (2.8 g/m^2) of tin on one side of No. 25 might have 0.035 lb/base box (0.8 g/m^2) as alloy and 0.09 lb/base box (2 g/m^2) as free tin.

Control of the very thin tin oxide and very light oil levels are necessary for enameling. Most cans produced today are enameled (see Cans, fabrication), and oil, oxide, and chrome concentrations must be controlled to provide a suitable substrate. Control of surface chrome levels is also a factor in controlling sulfide staining in selected sanitary packs (see Canning, food; Cans, steel).

Temper of tin-mill products refers to surface hardness, measured as Rockwell 30-T, a measure of resistance of the surface to penetration. As shown in Table 3 (2,3), hardness and strength increase as the temper number increases. Ductility (ability to be formed) generally decreases as temper increases. These physical, ie, mechanical, properties are controlled by combinations of chemistry, hot strip-mill practice, cold-rolling practices, annealing practices, temper practices, or, in the case of double reduction, double-reduced practices. They are guides to the ability of the material to be formed. Double-reduced product, because of the high degree of cold work imparted, has high strength but is very directional and has low ductility. Double-reduced product is generally produced to two temper ranges, as shown in Table 4. Tensile strengths are generally more indicative of the formability characteristics of the product than are hardness values. Tensile elongation values seldom exceed 2% on double-reduced product.

BIBLIOGRAPHY

1. *Guide to Tinplate, Publication No. 622*, International Tin Research Institute, Middlesex, UK, 1984.
2. *Tin Mill Products Manual*, American Iron and Steel Institute, Washington, D.C., 1973.
3. *ASTM Standards A 623, 623M, 624, 625, 626, 630, and 650*, American Society for Testing and Materials, Philadelphia, Pa.

J. M. Crossett
LTV Steel Company

TRAYS, STEAM-TABLE

Reusable steam-table trays are common in institutional food service. These shallow rectangular serving pans are usually made of heavy-gauge stainless steel. Foods prepared in the institutional or commercial kitchen may be transferred to a steam-table tray for holding on a serving line equipped to maintain foods at appropriate serving temperatures. Full-size steam-table trays are dimensionally standardized to fit most serving line equipment. Such trays are also available in one-half and one-third sizes to serve multiple items in the same space required by full-size trays. Identical trays are also used on cold serving lines. In Europe, the standardized food equipment system is called "Gastronorm."

Disposable steam-table trays are now available to food processors for packaging prepared foods such as entrees, vegetables, fruits, desserts, and some baked goods. Foods packaged in steam-table trays are available in the food-service market in frozen, chilled, and shelf-stable (heat-processed) forms. The advantage of all these packaged forms is that the foods are shipped, stored, reheated (or otherwise prepared for serving), and served from the same disposable containers. These types of fully prepared foods are of particular interest in food-service systems where skilled chefs and cooks are not available or affordable. They also eliminate the cost of cleaning reusable trays.

Frozen prepared foods are commonly offered in full- and half-size steam-table trays (see Fig. 1). The trays are usually made of heavy-gauge press-formed aluminum foil. Lids for shipping, made of fiberboard or aluminum, are removed or loosened for reheating in conventional convection ovens. Individual pans are usually contained in disposable outer cartons for protection during shipment and storage. In some cases, foods that are meant to be served cold may be packaged in formed plastic trays with low heat resistance.

In the early 1970s, two companies in the United States (Kraft and Central States Can) began production of steel half-size steam-table tray containers designed for production of thermally stabilized or "canned" food items (see Canning, food) (1,2). These containers are formed (drawn) from tin-free

Figure 1. A half-size steam-table tray. Courtesy of Central States Can Company.

steel (see Tin Mill Products), and shipped to the food processor nested (a space-saving transportation advantage over round cans) with separate steel lids. The processor fills, seals, and retorts trayed food products as in the conventional canning process (see Cans, steel). After processing, these foods are shelf stable, without costly refrigerated transport or storage. Inherent in the retortable steam-table tray design is the potential for shorter heat-processing time to achieve commercial sterility as compared to round (cylindrical) cans. This shorter process time is said to allow processors to produce foods with substantially improved product flavor, texture, and color (3). Central States Can Co., now the sole tray maker in the United States, has licensed half-size steam-table trays for manufacture in Europe (Groupe Carnaud) and Japan (Kojima Press) (1,4).

An aluminum one-third-size steam-table tray package was reported in 1982 for a line of high quality entrees (5). Heat sterilization times were significantly reduced in this case through use of a rotating retort cooker to produce superior flavor and avoid "canned" flavor and texture. This tray and its package assembly, developed by Food America Corp. and called Entree-Pak, won a 1982 Food and Drug Award (6) for innovative design, but it has not yet become a commercial reality.

Since the early 1970s the United States Department of Defense has been actively interested in the advantages to be gained from shelf-stable foods packaged in half-size steam-table trays (7) (see Shelf life). The objective has been to develop new, simplified combat food-service systems that will achieve substantial savings in transportation, storage, and skilled-manpower requirements. A tray-pack-based ("T-Ration") food-service system of special interest to the U.S. Army (8) has been tested by U.S. Army Natick Research and Development Center for use by the other U.S. military services (9) (see Military packaging).

Frozen foods are associated with the aluminum steam-table trays in the minds of many food service operators (10), and will likely continue that way because of long-time usage in the industry. On the other hand, shelf-stable foods in trays have not yet attracted a substantial food-service market (1). A factor that may affect future markets for trayed foods is the successful development of a retortable plastic tray. Plastic composite structures with high oxygen barriers have reportedly

already been used successfully in the United States (11) and Europe (12) to form containers that will withstand elevated food-sterilization temperatures. Although size constraints on a plastic steam-table tray have not yet been adequately explored, the appearance on the market of a practical steam-table tray that can be heated in a microwave oven could have a profound effect on future use of shelf-stable, tray-packed foods by food-service operators.

The future use of the retortable steam-table tray as a packaging system may depend on the success of the U.S. military in developing a total food system based on shelf-stable foods in half steam-table trays (13). The range of products packaged in trays for a successful military system must include a far greater variety of items and higher quality than is currently available commercially. This U.S. military effort, if successful, could make available the overall benefits of a simplified food-service system to civilian food-service operators.

BIBLIOGRAPHY

1. Central States Can Co., Massillon, Ohio, private communication.
2. *Canner/Packer* **146**(8), 40 (1977).
3. *Food Eng. Int.* **4**(5), 38 (1979).
4. G. Booth, *Food Flavour. Ingredients Pack. Process.* **1**(8), 36 (1980).
5. *Nat. Provis.* **27**(2), 33 (1982).
6. *Food Drug Packag.* **46**(6), 1 and **46**(12), 31 (1982).
7. J. Szczeblowski, *Activities Report of the Research and Development Associates for Military Food and Packaging Systems,* **25**(1), 77–84 (1973).
8. R. S. Maize, II, *Activities Report of the Research and Development Associates for Military Food and Packaging Systems,* **35**(2), 46–50 (1983).
9. H. J. Kirejczyk, *The Cold Weather 83 Evaluation of Mobile Food Service Unit and Tray Packs, Technical Report 85/009,* U.S. Army R & D Center, Natick, Mass., August 1984; *Field Feeding Systems to Support U.S. Marine Corps Forces in the 1990s, Technical Report 85/011,* July 1984.
10. K. S. Ferguson, *Activities Report of the Research and Development Associates for Military Food and Packaging Systems,* **32**(2), 20–22 (1980).
11. J. Szczeblowski, *Activities Report of the Research and Development Associates for Military Food and Packaging Systems,* **32**(2), 23–31 (1980).
12. *Int. Z. Lebensmittal Technologie Verjahrenstechnik* **34**(6), 541, 544 (1983).
13. A. Dungan, *Activities Report of the Research and Development Associates for Military Food and Packaging Systems,* **35**(2), 41–45 (1983).

Avalon L. Dungan
L. J. Minor Corporation

TUBE FILLING

The various types of collapsible tubes (see Tubes, collapsible) do not require different types of filling equipment. Metal, plastic, and laminated tubes can all be filled on the same basic machines through the use of optional equipment. The choice of machine and options depends on the product, productivity requirements, and frequency of changeover. Simple manual units with outputs of 5–10 tubes/min are available for laboratories and very small production runs (eg, sample runs). For

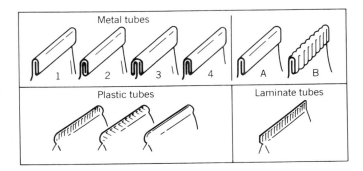

Figure 1. Metal plastic and laminate tube closure types. **1,** single-fold; **2,** double-fold; **3,** saddle-fold; **4,** double-fold; **A,** plain; and **B,** crimped.

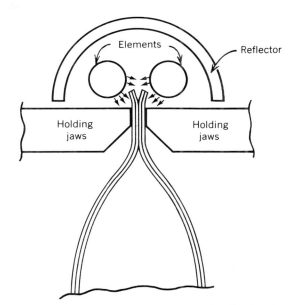

Figure 2. Radiant-heat sealing system for plastic tubes.

very high production rates (eg, for toothpaste) machines are available that fill up to 200 tubes/min. The capacity of a machine depends on the number of nozzles, and on the product and tube size. One nozzle typically fills and seals 40–100 tubes/min; two nozzles, 80–200/min. Within those categories, the lower part of the range applies to viscous products filled into large tubes. The highest speeds are attainable with low viscosity products in small tubes. Modified tube-filling machines are also used to fill nonstandard package configurations (eg, syringes, vials, and mascara cartridges).

Loading. The smallest single-nozzle machines are loaded by hand. In the larger machines the tubes are loaded into tube holders automatically. This is done with a gravity track infeed that loads tubes one by one. This is sometimes done with the aid of a cassette infeed that feeds tubes from cartons to the gravity infeed. It can also be done with a fully automatic system that removes tubes a row at a time from cartons or trays, places them on a powered conveyor, and inserts the tubes auto-matically into the tube holder. In the tube holder the tubes are cleaned by a blow-and-suction device that removes dust and lint.

Cap tightening. Because tubes are filled from the bottom, caps are tightened before filling. This is done with a device that has an adjustable torque slip clutch or by electrical pulsation. Sensors detect tubes that have reached this point without caps. These are ejected automatically or removed by an operator. The cap tightening station cannot handle tubes that have the same outside diameter as the cap. The caps on such tubes cannot be tightened in production.

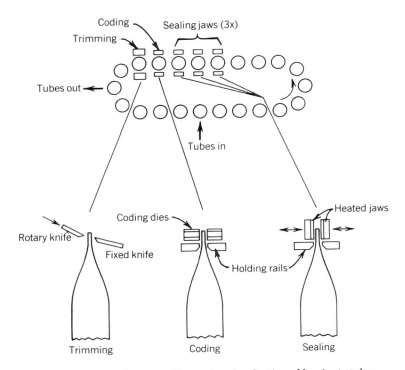

Figure 3. Heated-jaws sealing system for plastic and laminate tubes.

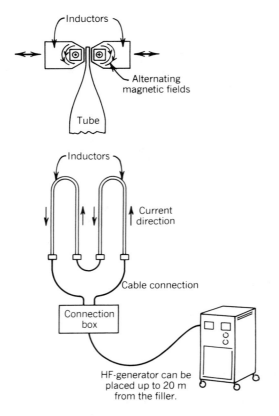

Figure 4. High frequency sealing system for laminate tubes.

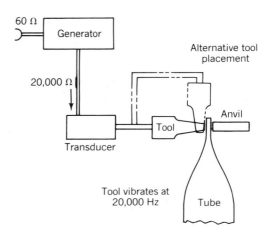

Figure 5. Ultrasonic sealing system for laminate and plastic tubes.

Filling. Tube filling machines must handle many different types of products and accuracy of fill depends on metering devices. The most accurate metering device is the volumetric piston pump. The newest machines incorporate positive product cutoff to prevent stringing or dripping at the end of each cycle. They can also be disassembled quickly for cleaning and/or sterilization. It is important in filling to keep air out by lifting the tube or lowering the nozzle almost to the bottom of the tube (cap end) and metering the product toward the open end.

Registration. Tubes are aligned either manually or, if the tube has a registration mark, by a photoelectric registration device. The registration mark should be about 0.3-in. (7-mm) long and 0.1-in. (2-mm) wide, normally on the back side of the

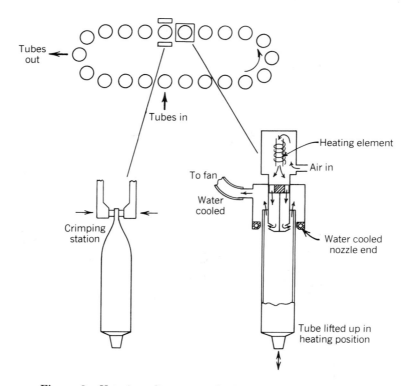

Figure 6. Hot-air sealing system for laminate and plastic tubes.

tube. Accuracy depends on the quality of the mark, which should provide good contrast with tube color. Special tube holders are sometimes required for more accurate photoelectric registration.

Sealing. Different tube materials require different sealing methods. Machines are normally set up for one type of tube, but the same machine can handle other types of tubes with spcial attachments.

Metal tubes. The open ends of metal tubes are normally folded with one of a number of types of crimp folds (see Fig. 1). The strongest fold is a #3 saddle fold, which is also ideal for double-side coding of the tube. The next best would be #2 or #4 double folds.

Plastic and laminated tubes. These tubes are sealed by heat (see Sealing, heat). The four established methods are radiant heat, heated jaws, high frequency, and ultrasonic. Radiant heat and heated jaws are commonly used for plastic tubes (see Figs. 2 and 3). Contamination of the sealing area can be a problem with these methods, though not with ultrasonic sealing.

High frequency sealing (see Fig. 4) is the preferred method for laminated tubes. This method provides excellent seal strength and highest speed output. Tube contamination can be a problem with stringy products and ultrasonic sealing can overcome that (see Fig. 5), but at a sacrifice in speed.

A fifth method of sealing plastic and laminated tubes has become available in recent years: hot-air sealing (see Fig. 6). Hot air (approximately 932°F (500°C)) is blown into the open end of the tube in an upwards motion that does not interfere with the product or the cap end of the tube. This is done with a special patented energy-saving water-cooled nozzle. Tubes sealed in this way are trimmed after sealing and can even be die cut with a hole for rack display.

Controls. Modern machines are equipped with microprocessor controls that check tube infeed, defective tubes, cap tightening, tube registration, filling accuracy, etc. Digital readouts show machine speeds, fault statistics, production output, etc. As many as 15 machines can be tied to a central data terminal.

D. J. White
Norden Packaging Corporation

TUBES, COLLAPSIBLE

The collapsible tube is a unique package designed specifically for viscous products. It is simple to use, allows metered expulsion of product by squeezing, and with a closure it can protect its contents for long periods of time. Two billion (10^9) tubes are used in the United States each year for a wide variety of products.

Until the 1950s, only metal tubes were available. With the invention of the plastic tube, cosmetics and similar products requiring nominal product protection found a package having adequate barrier properties and aesthetic appeal. Rather than compete with the metal tube, it expanded the total tube market.

The higher barrier laminated tube, which began extensive commercialization in the early 1970s, was a direct competitor. It has now replaced over 95% of the metal tubes in the U.S. dentifrice market.

Metal Tubes

The metal tube has been manufactured in the United States since 1885. The original tubes were made of tin or lead, which were easly formed by impact extrusion. Colgate adopted lead tubes for dentifrice in 1896 and in a few years became the world's largest toothpaste producer. Lead remained the primary metal for tubes until the early 1930s, when aluminum alloys were developed. The aluminum tube quickly became popular because of its superior strength and resistance to cracking. Aluminum now accounts for about 95% of the metal-tube market; tin represents the balance.

Metal tubes are made by the impact extrusion process. The process is called one hit backward extrusion because the slug is formed in one impact and the resulting sleeve is extruded backward over the male impacting tool.

Aluminum tubes are usually made from 1100 alloy, which is 99.5% pure aluminum and the easiest to form. The operation begins with round slugs produced by either the aluminum company or the tube producer. These are punched from hot-rolled sheet and annealed. The slug is slightly smaller than the diameter of the final tube and the volume is designed to produce a tube that is ca ¾ in. (19 mm) longer than the final tube length. This is because the leading edge of the impacted tube body is irregular and must be trimmed to the exact length. The profile of the slugs is generally dished in the United States and flat in Europe, although either form produces good tubes. Before use, the slugs are tumbled in zinc stearate which acts as a lubricant during the impacting operation. They are then automatically fed into the impact press.

No metal tube lines are made in the United States. Herlan & Co., Karlsruhe, FRG, one of the principal producers of metal tube lines builds horizontal toggle lever presses for this application. The toggle design allows the long stroke necessary to strip the finished tube from the tool and a more gradual application of force without a decrease in speed. A press designed for an aluminum tube 1 in. (25.4 mm) in diameter requires a rating of 80–90 short tons (72.6–81.6 metric tons). The ram motion is parallel to the floor. Speeds of 150 tubes per minute can be attained with one tool. It is relatively simple, consisting of a shallow female cavity and a male tool. A section of a typical tooling setup is shown in Figure 1. The male tool head matches the inside profile of the final tube. The largest diameter is only 1/32–1/16 in. (0.8–1.6 mm) long, after which it tapers back to a diameter which runs around 0.008 in. (0.2 mm) smaller than the tip. The clearance between the male tool and the female cavity establishes the wall thickness of the tube sleeve.

The slug is automatically fed into the cavity and is compressed by the male tool as it moves down. The first portion of the stroke seats the slug and takes up the bearing slack of the press. The balance of the stroke cold-works the slug, forming the head of the tube. As the plunger continues moving down, the excess metal is extruded out around the sides of the plunger, forming the body. Since the metal never reaches its melting point, the body is cold enough to maintain its shape and additional support from the tooling beyond the head area is not required. As the plunger moves back at the end of the stroke, the completed tube is stripped off, falling into a transfer conveyor which takes it to the trimming operation. Here, the body is trimmed to exact length, cap threads rolled onto the neck, and the mouth of the tube trimmed.

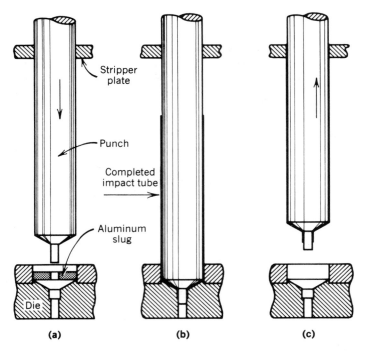

Figure 1. Impact extrusion of a metal tube: (**a**) beginning impact stroke; (**b**) bottom of stroke, (**c**) beginning to strip. Courtesy of Herlan & Co.

By now the tube has been severely work-hardened and has the feel of an aluminum can. To relieve this hardness, it is annealed in an 1100°F (593°C) oven and cooled, after which a spray application of a product-resistant lining is applied. A number of tough epoxy–phenolic or acrylic lacquers are available which not only resist the attack of the product but can be creased without chipping or peeling from the base metal. After the lining has been baked on, the tube is roller-coated with a white acrylic coating which gives an opaque printing base to the sleeve. For special applications such as pharmaceutical tubes (see Pharmaceutical packaging), this coating is designed not to adhere to the sleeve, permitting the pharmacist to strip off the coating with its subsequent printing before applying a prescription label. After application, the coating is dried and the tube moved to a turret for printing.

Metal tubes are printed by a dry offset process (see Decorating; Printing) using either thermally or UV-cured inks (see Curing; Inks). The tube is pushed on a mandrel and rolled past a curved printing blanket, which applies up to five colors simultaneously. The tube now goes through a final drying cycle to cure the ink, after which the cap is applied and the tube automatically packed in partitioned boxes for shipment to the filler. The tube remains in a horizontal position throughout the entire operation.

Tapered tubes were introduced in Europe in the 1970s. This design permits tubes to be stacked like paper cups. Tube tapering is done mechanically after the tube has been completed. The tubes are tapered about 2°, just enough to allow them to nest within each other, and coatings and inks must be tough enough to permit this stretch without cracking or delaminating. Both manufacturers and packers prefer this design since it requires less care in packing and shipping and increases shipping density. Equipment is available to nest tubes auto-

matically at the tube lines and denest them at the tube-filling location (see Tube filling). This process is being introduced into the United States by Herlan.

Plastic Tubes

The first practical plastic-tube patent was issued in 1954 (1). The patent covered the process of making a thin plastic sleeve by an extrusion method (see Extrusion) and then injection molding (see Injection molding) a head on one end to produce a tube. Modern equipment has improved but not changed the basic Strahm concept.

Many plastics can be used to make plastic tubes, but LDPE is the primary material used today (see Polyethylene, low density). It has high moisture barrier properties, low cost, and good appearance. Its lack of oxygen and flavor barrier has been improved with barrier coatings. It was the development of a barrier coating by American Can Co. that made the plastic tube a practical container for general packaging.

HDPE (see Polyethylene, high density) is used for packaging some hydrocarbon-based products such as grease, and PP (see Polypropylene) is used for applications requiring non-staining, better perfume barrier, or higher temperature resistance. Both HDPE and PP are much stiffer than LDPE for tube sidewalls, and are not as popular.

The selection of a suitable plastic for producing a tube is critical to its performance. Dupont Alathon 2020T, a 0.92 g/cm³-density resin with a melt index of 1.0 is the primary LDPE resin used by all tube manufacturers in the United States. It has good processibility, excellent stress-crack resistance to product attack, and an extremely low gel content, which reduces surface irregularities that would affect printing quality.

Plastic tubes are produced by two principal methods in the United States: the Strahm method and the Downs method. Both processes make excellent tubes.

Production of the tube begins with the extrusion of continuous thin-walled tubing (2). This has a wall thickness of .014–.018 in. (0.35–0.46 mm), depending on the diameter. A standard extruder is used with a thin-walled tubing die. As the hot plastic emerges, it is corona-treated (2) (see Surface modification) for later printing-ink adhesion. At the same time, it is drawn over a chilled internal forming mandrel and cooled on the outside with cold water. The tube cools and shrinks to an accurately controlled diameter as it is drawn off. After it has passed through the drive rolls of the haul-of unit, it is cut to exact length with a rotary knife cutter. This piece is called a sleeve to differentiate it from the completed container, which is called a tube.

Printing can be done before or after heading. The location is based on the layout and the relative scrap generated in printing and heading. All tubes are printed by dry offset printing. The same type of printer used for metal tubes is used for plastic sleeves. Thermally dried and UV-dried inks are available in a full range of colors. Good-quality process printing is possible, and for cosmetic applications, postdecoration with hot-stamp foils or silk-screening is popular (see Decorating).

After the ink has cured, the sleeves are roller-coated with a high gloss, oxygen- and flavor-barrier coating. High barrier, two-component amine-cured epoxy coatings are available which reduce the overall tube permeability by a factor of ten. Special coatings more resistant to product staining or having lower coefficients of friction are also used. Most coatings are thermally cured, but UV-curable coatings are beginning to be used. These provide equally high barriers to oxygen and essential oils and cure more quickly. After the coating is cured, the completed sleeves are headed.

The head of a plastic tube must be compatible with the sleeve in order to produce a good bond. Sleeves made of LDPE can be headed with LDPE or HDPE, but PP sleeves must have a PP head. Head thickness is 0.030–0.065 in. (0.76–1.65 mm), depending on tube diameter and application.

The Strahm heading process traps the top end of the sleeve in an injection mold and injects plastic into the cavity to form the head and bond it to the sleeve. The process is done with multiple tools at each station. Slower machines have a female tool which remains in the injection station, and the tube is held there until it has cooled sufficiently before moving to the next position. The higher speed machines have locked-die tooling, as shown in Figure 2. The male tool containing the sleeve is pushed up into the shoulder cavity. The end of the sleeve enters the shoulder cavity, rolling in slightly as it touches the radius of the shoulder. At the same time, the support sleeve, which moves with the male tool, contacts the shoulder cavity, lifting it up against the thread plates. This forces the plates together in the position shown in Figure 2. The injection nozzle now closes on the assembly, and molten plastic from a small reciprocating-screw extruder is injected to form the head and bond it to the sleeve. Injection-cavity pressures are low since excessive pressure causes "blowby," forcing plastic past

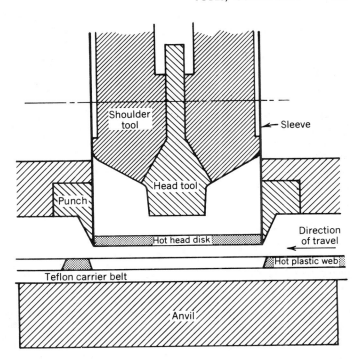

Figure 3. Downs process; head disk just after blanking out.

the head area and down the sleeve wall. Melt temperature of the plastic is generally over 500°F (260°C), high enough to ensure a good fusion seal to the sleeve wall. After the injection cycle, the nozzle retracts. The tooling remains locked together and continues to the next station. After the head cools, the male tool drops, the thread plates open, and the completed tube is released, remaining on the male tool. The die body and cavity support move off to one side to allow the completed tube to be extracted. A new sleeve is then placed on the tool, the female tooling moves over the top of the male tools, and the cycle is repeated.

After cooling, the completed tube is transferred to a snipper-capper. Here, the sprue is removed and a cap applied, after which the tube is packed in unpartitioned trays for shipment to the customer.

In the Downs process (3,4), the sleeve production is the same but the heading process considerably different. As shown in Figure 3, the sleeve on its male tool enters a punch which is placed just above a continuous strip of LDPE heated to well above its softening point. This strip is ca 2 in. (51 mm) wide and 1/8 in. (3.2 mm) thick. When the leading edge of the sleeve is about flush with the cutting edge of the punch, both parts move down into the semimolten mass below. The punch forms a disk of plastic the diameter of the sleeve, and the sleeve inside the punch immediately adheres to it. The tool moves up carrying the sleeve bonded to the hot plastic disk and indexes to the next station. Here, the female tool closes on the sleeve and head disk, compression-molding the head into its final shape. The compression-molding method eliminates the sprue and the snipping operation necessary with the Strahm process.

More recent commercial Downs machines have multiple heads at each station to make a number of tubes simultaneously.

A third process, developed by KMK (Karl Magerle AG Hinwil, Switzerland) (5), injects a "donut" of molten plastic in

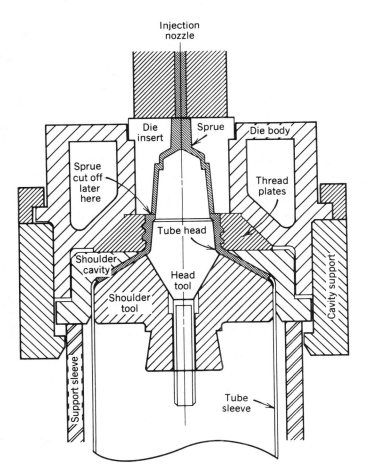

Figure 2. Strahm heading method.

the female cavity before it closes on the male cavity. The head is formed by compression molding. The Magerle machines are small, with only one tool per station, but they also incorporate capping on the same turret.

Although these are the most common methods, other processes have been developed. In Europe, the Valer Flax process (6) uses a premolded head which is spin-welded to the sleeve. Several companies have developed blow-molding methods (7) where the tube is blown as a bottle and the bottom trimmed off. This approach generates a high amount of scrap which must be put back into the process, and care must be taken to maintain uniform wall thickness for the later printing step.

Laminated Tubes

In 1971, Procter & Gamble Co. switched their Crest dentifrice to American Can's Glaminate tube. This tube was a laminated tube containing plastic, foil, and paper with barrier properties far superior to the plastic tube (see Laminating machinery).

The chief difference between this package and the plastic tube is the sleeve, made from preprinted laminate web (8,9). It contains up to ten layers, each contributing to the function of the structure. Figure 4 shows the material in each layer and describes its use. The complete laminate, 0.013 in. (0.33 mm) thick, is produced on an extrusion laminating line (10). This web is made in large rolls which are slit into the proper width for the tube being made. The slit rolls are shipped to the tube plant, where they are formed into sleeves. The sleeve has an impermeable aluminum layer, so the only permeation is along the seam overlap.

The seaming process takes the flat web, folds it into cylindrical form and seams it to form a continuous tube (11). The seam must protect the raw edge of the foil; this is achieved by overlapping the edges of the sleeve, then heating and compressing the overlapped portion to squeeze some of the plastic out around the raw edge of the foil and paper. After the sleeve has been seamed, it is cut to length using the print registration marks on the web.

The presence of aluminum foil in the sleeve permits use of r-f energy for induction heating (12). This is preferred in the seaming process because it permits better control of heat distribution.

To take full advantage of the vastly improved barrier properties of the laminated sleeve, an oxygen and flavor barrier is necessary in the head area. This is accomplished with a premolded insert (13) of polybutylene terephthalate (PBT), which has good oxygen- and flavor-barrier properties, can withstand the injection pressures and temperatures of the heading plastic, and does not crack in the head. Urea inserts are also used. The insert is placed on the heading tool before the sleeve is placed in position, and the head plastic locks the insert in position as it bonds to the sleeve. The resultant tube has some permeation windows along the sleeve seam and the area where the head bonds to the sleeve, but the result is still a very low gas and flavor permeability.

Alternative methods of producing laminated stock and tubes have also been developed, as well as alternative materials.

The American Can approach for seaming requires the axis of the tube to be parallel with the length of the laminated web (11). KMK has developed another system (14), where the axis of the tube is perpendicular to the web. In this process, the blank is cut from the web and wrapped around a fixed mandrel and sealed. KMK uses the heading process described under the plastic tube section above.

Another Swiss manufacturer, AISA (Automation Industrielle SA), has produced a machine using premolded heads which are bonded to the sleeve with r-f induction (15). To ensure a good bond, a foil laminate disk, called a rondelle (16), is placed over the shoulder of the tube. The sleeve is butted against the rondelle and the r-f bond made. This approach still requires a urea or PBT insert to act as a flavor barrier.

New Tube Concepts

The relatively recent appearance of ethylene–vinyl alcohol (EVOH) (see Ethylene–vinyl alcohol) as a clear barrier plastic has made possible the development of an all-plastic barrier tube (see Barrier polymers). Coextrusion methods are available for producing laminates, and the Japanese have produced coextruded sleeves and blow-molded barrier tubes (7). The latter tubes have a continuous barrier, ie, there are no higher permeation areas such as side seams or head bonds. At present, the all-plastic barrier tube is not economically competitive with a laminated structure. EVOH costs about the same as aluminum foil, so there are no material cost savings involved. The advantage of being able to print flat web is an important factor in the cost of the laminated tube which would appear to limit the appeal of this new concept. However, a significant drop in the price of EVOH relative to aluminum foil or the development of a new market where higher barrier properties are required could quickly change its importance.

BIBLIOGRAPHY

1. U.S. Pat 2,673,374 (Mar. 30, 1954), A. Strahm (to American Can Co.). Describes basic heading concept.
2. U.S. Pat. 3,849,286 (Nov. 19, 1974), R. Brandt and J. Piltzecker (to American Can Co.). Details the sleeve extrusion process.

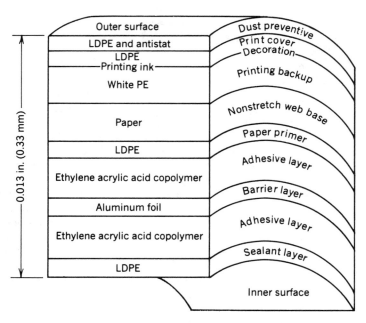

Outer surface	Dust preventive
LDPE and antistat	Print cover
LDPE	Decoration
Printing ink	
White PE	Printing backup
Paper	Nonstretch web base
	Paper primer
LDPE	Adhesive layer
Ethylene acrylic acid copolymer	Barrier layer
Aluminum foil	Adhesive layer
Ethylene acrylic acid copolymer	
	Sealant layer
LDPE	
	Inner surface

0.013 in. (0.33 mm)

Figure 4. Typical structure of a Glaminate toothpaste tube.

3. U.S. 3,047,910 (Aug 7, 1962), M. Downs (to Plastomer Development Corp.). Offers good description of basic Downs heading machine.

4. U.S. Pat. 3,591,896 (July 13, 1971), R. Tartaglia (to Peerless Tube Co.). Contains good drawings of the basic commercial machine using the Downs process.

5. U.S. Pat. 4,352,775 (Oct. 5, 1982), K. Magerle. Shows Magerle heading process.

6. U.S. Pat. 3,824,145 (July 16, 1984), V. Flax (to Continentalplastic AG). Contains good drawings of spin-welding method of heading.

7. U.S. Pat. 4,261,482 (Apr. 14, 1981), M. Yamada, T. Sugimoto, and J. Yazaki (to Toyo Seikan Kaisha, Ltd. and Lion Hanigaki Kabushiki Kaisha). Describes blow-molded tube with EVOH barrier.

8. U.S. Pat. 3,260,410 (July 12, 1966), R. Brandt and R. Kaercher (to American Can Co.). Shows basic laminate structure.

9. U.S. Pat. 3,347,411 (Oct. 17, 1967), R. Brandt and N. Mestanas (to American Can Co.). Basic article patent of laminated tube.

10. U.S. Pat. 3,505,143 (Apr. 7, 1970), D. Haas (to American Can Co.). Shows laminating process.

11. U.S. Pat. 3,388,017 (June 11, 1968), A Grimsley and C. Scheindel (to American Can Co.). Describes sideseaming of laminated sleeves.

12. U.S. Pat. 4,210,477 (July 1, 1980), W. Gillespie and H. Inglis (to American Can Co.). Illustrates r-f induction seaming coil design.

13. U.S. Pat. 3,565,293 (Feb. 23, 1971), A. Grimsley (to American Can Co.). Describes head inserts.

14. U.S. Pat. 4,200,482 (Apr. 29, 1980) (to KMK AG). Shows Magerle sideseamer.

15. U.S. Pat. 4,123,312 (Oct. 31, 1978), G. Schmid and R. Jeker (to Automation Industrielle SA). Shows AISA seaming and heading process.

16. U.S. Pat. 4,448,829 (May 15, 1984), A. Kohler (to Automation Industrielle SA). Shows AISA rondelle.

General References

N. L. Ward, "Cold (Impact) Extrusion of Aluminum Alloy Parts," in T. Lyman, ed., *Metals Handbook,* 8th ed., American Society for Metals, Metals Park, Ohio, 1969, vol. 4, pp. 490–494. Discusses impact extrusion of aluminum in general, with little specific information on tubes.

ASTM Committee on Aluminum and Aluminum Alloys, "Introduction to Aluminum," *Metals Handbook,* 9th ed., American Society for Metals, Metals Park, Ohio, 1979, vol. 2, pp. 3–23. Briefly refers to aluminum alloys.

Packaging in Plastic and Glaminate Tubes, Washington Technical Center, American Can Co., Washington, N.J. A nontechnical presentation of the manufacture of tubes and a good write-up on the problems of packaging in tubes and what to look for.

Patents describing the Strahm process of heading, all ultimately assigned to American Can Co.

U.S. Pat. 2,713,369 (July 19, 1955), A. Strahm. Illustrates sections of tube head and sleeve.

U.S. Pat. 2,812,548 (Nov. 12, 1957), A. Quinche and E. Lecluyse. Shows the concept of split-thread plates.

U.S. Pat. 2,994,107 (Aug. 1, 1961), A. Quinche. Describes first stationary-die heading machine.

Roger Brandt
American Can Company

U

UNIT-DOSE PACKAGING. See Pharmaceutical packaging.

UNIVERSAL PRODUCT CODE. See Bar code.

UNSCRAMBLING

In the packaging industry, an "unscrambler" is a machine that orients packaging components from random bulk storage to deliver them in an ordered fashion to the production line. The term *unscrambling* usually refers to the handling of relatively large components such as plastic containers, some metal cans, and glass bottles as opposed to sorting of relatively small components such as closures, caps, spray pumps, actuators, applicators, etc. With the higher speeds of today's production lines, limited production-line floor space, and the immense range of container shapes and sizes, it is not practical to unscramble containers from bulk manually. Unscrambling offers a means of delivering oriented containers to a packaging line with high speed and reliability, while taking advantage of the lower cost of purchasing the containers in random bulk lots. Containers can be packed in boxes or gaylords directly from the molding, extrusion, or forming machines and shipped directly to the packaging plant. The cost of preorienting into reshippers or onto pallets is eliminated. In some cases, shipping costs are also reduced because of the greater number of components that can be shipped in the same space, with no need for special handling.

The traditional method of shipping glass bottles and metal cans was on pallets or in boxes with the components preoriented and stacked in "tiers" or layers, usually separated by sheets of chipboard or cardboard. The feeding of these containers to the packaging line was accomplished by the use of depalletizers or feed tables for handling boxes of containers (see Palletizers). These feed tables worked basically on the same principle as the depalletizer in that a layer of containers would be pushed off an elevator table onto a feed chain or turntable. Both of these feeders required a full time operator and were limited to simple container shapes.

With the conversion to plastic containers within the packaging industry, more efficient feed systems were needed to accommodate the higher speeds and often exotic shapes and sizes available. The advent of plastic containers also opened the possibility for bulk shipments. Unscramblers, therefore, were created to sort out bulk shipments of the containers and deliver them to a packaging line. Other applications for unscramblers involve integration with printing or decorating equipment, labelers, sleeving or overwrap equipment, cleaning or sterilizing equipment, and pucking or base-cupping equipment.

There are several types of unscramblers on the market today and they employ different means of orientation. There are, however, some similarities:

1. The unscrambler is provided with, or can accommodate, a bulk storage/delivery system. This usually consists of a conventional hopper–elevator combination. Hopper size can vary depending on the type of container being handled and the desired time of unattended operation based on container size and line speed. Actual available floor space is, of course, another consideration when sizing and selecting the hopper-elevator package. The elevator or conveyor discharging the containers from the hopper must be selected to deliver the containers, still in random orientation, at a relatively constant rate to interface with the unscrambler-orienter unit (see Conveying). Most unscramblers provide level sensing through pneumatic sensors, electric photo-eyes, or electromechanical level sensors to meter the operation of the hopper–elevator.

2. Another similarity of unscramblers is that the process is carried out in three stages.

   A. Since all containers requiring orientation have an opening for filling or distribution, one end or side of the container is different from the opposite end or side. It is this difference that is used to orient the container. The first stage of unscrambling is to "preorient" the container into one of two conditions: neck leading or neck trailing. If the cross section of the containers is rectangular or oval-shaped, this first stage of orientation can also preorient width with respect to the minor or major container dimension (see Fig. 1).

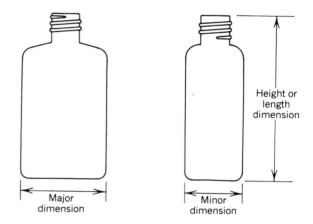

Figure 1. Typical use of terminology for describing container size.

   By presizing a channel, pocket, flight track, or lane to just more than the minor dimension, but less than the major dimension, containers can fit into this space only in the minor dimension. This first stage is accomplished by allowing the containers to tumble into the presized channel, pocket, flight track, or lane in which the containers will lie. All other containers which are skewed or standing are cleared away. Air jets, agitators, clearing wheels or other mechanisms are used to move the bulk of containers away from the preoriented neck-leading or neck-trailing containers.

   B. The second stage or orientation is accomplished by turning the containers so that they are all in the same orientation. Turning is accomplished by differentiating the neck from the base and dropping all containers base down, or by mechanically catching the neck of the container and rotating those containers which are misoriented. Another method of orienting is to simply reject all containers which are not in the correct orientation and return them back to bulk storage. This method, however, reduces the overall efficiency of the orienter system and presents the possibility of recirculating and handling the container several times, which may be detrimental to scuff-sensitive containers.

C. The third and final common stage of unscrambling is to deliver the container or component to the next successive packaging operation. In most cases, the container is delivered to and stabilized on a flat-top conveyor. This is a very important part of the unscrambling operation because of the relative instability of an empty container on a running conveyor. It is essential that the unscrambler discharge mechanism be synchronized with the discharge conveyor chain. Discharge is usually accomplished through synchronized pocket- or side-belt release onto the conveyor.

In many cases the unscrambler discharge is used for other secondary operations such as coding, marking, or container cleaning. It should be noted however, that the operation performed by most unscramblers is unique with respect to other packaging operations in that the container is empty and is interfacing at a random container-per-minute rate to a constant rate. When installing a new unscrambler or when laying out a new packaging line, allowances have to be made for both "surge" and "backlog" distance between the unscrambler and the next successive packaging machine. "Surge" distance is the amount of conveyor length required between the unscrambler and the next machine on line.

Consider, for example, that the next machine is the filler. When the filler starts, containers on the conveyor before the filler will begin to move downline. The line of containers will pass the unscrambler backlog detector (usually a photo eye) and start the unscrambler. As the unscrambler starts and accelerates to full operating speed, containers will be discharged onto the conveyor. The line of containers before the filler must not pass the filler low-level infeed sensor before the containers being discharged from the unscrambler "catch-up". If the filler low-level sensor is allowed to clear, the filler will stop and there will not be a synchronous automatic start-up of the packaging line, resulting in poor line efficiency and lost production. The solution to the "surge" distance problem is more than moving the unscrambler backlog sensor closer to the unscrambler to make it start sooner, because this would lead to a poor "backlog" condition. In this condition, when the filler stops, containers will backlog on the conveyor after the unscrambler and the unscrambler backlog detector will not detect the stoppage soon enough to decelerate and stop the unscrambler and avoid a jam.

All unscrambler manufacturers work with the designer of the packaging line to recommend proper spacing of the equipment for optimum performance. For high speed applications, for example, unscramblers are manufactured with variable speed settings to "ramp up or down" in speed rather than start and stop in automatic operation. Unscramblers are presently used in all of the major markets for plastic bottles, and their use is increasing along with the new capabilities of plastic bottles. Particular areas of growth are in the petroleum and beverage industries.

L. F. BYRON
New England Machinery, Inc.

V

VACUUM-BAG COFFEE PACKAGING

The concept of vacuum-bag packaging for ground coffee was introduced in the United States market just a few years ago. It had been developed in Europe years before, but except for nonhermetic paper bags still used in the southeastern states, ground coffee in the U.S. has remained in the three-piece sanitary can with plastic covercap that replaced the key-opening can about 40 years ago. The new vacuum-bag package has been described as a "flexible can," since it delivers the same vacuum-packed fresh coffee that the consumer buys in the can, but that is where the similarity stops. The "vac-bag" is rectangular rather than round, and it can be shelved standing up, or lying on its side like a "brick," which is one of its less glamorous names.

Coffee characteristics. Coffee has some unique characteristics. After coffee beans are roasted, they slowly evolve CO_2 amounting to over 1000 cm^3/lb (2200 cm^3/kg) of product depending on the blend and roasting color. During grinding, the volume of CO_2 is reduced to about half and, depending on the grind size, it requires varying holdup times before packing. Grind size varies greatly depending on the type of coffee-brewing equipment used and individual taste preference. Typical fine grind for most European countries is 500–600 μm; North American coarser grinds are 800–1000 μm. Vacuum-bag packaging of coarser grinds present more of a problem due to slower degassing. At 6–8 in. Hg (20–27 kPa) a vac-bag becomes soft and it can even balloon like a football (positive pressure) if it has insufficient degassing. Because rigid metal cans are capable of withstanding full vacuum as well as positive pressure, they require a relatively short holdup time even for coarse grinds. Because ground coffee also varies in density, the height of fixed cross-section bags varies as well. This means that the equipment and materials must be adaptable to a limited automatic height change during a production run. In addition, coffee must be protected from oxygen to maintain the freshness. Vacuum packing eliminates most of the oxygen from the package by the nature of the process.

The vacuum bag. The first commercial "vac-bag" packages were made in the early 1960s, probably in Sweden. At that time the package was a bag-in-box style (see Bag-in-box, dry product) that later evolved to a lower-cost bag-in-bag style, which is still the most widely used type in Europe and Canada. A still lower-cost style, the single-wall bag, is being used to some extent in Europe and by the major coffee producers in the United States. The bag-in-bag style and the single-wall style function equally well in protecting the product, but they differ in appearance. High quality graphics on the paper or foil/paper outer wrap of a bag-in-bag package gives a smooth surface appearance. Most single-wall bags have a rough or "orange peel" surface when under full vacuum. As with any package concept with more than one style, there are pros and cons for each; but the current United States market is predominately single-wall vac bag and is likely to stay that way.

Equipment. Since Europe spawned the vac-bag's popularity, it is no surprise that all the automatic high speed equipment is made there. There are three major equipment suppliers who compete in the coffee-packaging area: SIG, Hesser, and Goglio (1,2,3). All have automatic high-speed (40–120 bpm) 1–lb (0.45-kg) size machines capable of forming bags from roll stock material, filling, evacuating, sealing, and folding the top of the bag to deliver a finished vac-bag. Bags can be made in different ways from roll stock. Mandrel- and tube-forming machines form the bag, which is carried horizontally into the filler. After filling it is transferred into a vacuum chamber where it is evacuated and sealed. It is then transferred out of the vacuum chamber to a trimmer and final folding of the top. Tape or hot-melt glue are two methods used to hold the top folds in place. The sealing of the bag is what keeps it hard and under vacuum.

Materials. Each of the major equipment suppliers provides basic information on what packaging material properties are essential for efficient vac-bag production. There are some differences but there are many similarities for achieving a low level of defects from the machine. Each vac-bag must be durable enough to survive the rigors of transportation, warehousing, and retail distribution to remain vacuum-packed at point of purchase. Although many European packers use a three-ply structure made of 48 ga polyester film (OPET)/35 ga aluminum foil (AF)/300 ga sealant (see Film, polyester; Foil, aluminum; Sealing, heat), most United States coffee roasters employ a more durable four-ply structure. This can be accomplished by the addition of one of several films such as biaxially oriented polypropylene (BOPP) (see Film, oriented polypropylene) or biaxially oriented nylon (BON) (see Nylon). These structures can be constructed in various ways but most typically as follows:

48 ga OPET/75 ga BOPP/35 ga AF/300 ga sealant
48 ga OPET/35 ga AF/60 ga BON/300 ga sealant.

When these structures are designed for a single-wall bag, the OPET is reverse-printed. Gravure printing (see Printing) on OPET provides excellent graphics, high gloss, and a scuff-resistant package. Vacuum metallized films (see Metallizing) have been used successfully in Europe and Canada as lower-cost foil replacements. Since their oxygen barrier may be lower than foil, each coffee packer must determine if nonfoil structures provide adequate barrier.

Currently the best method to join these materials into a finished structure is by adhesive lamination (see Laminating; Multilayer flexible packaging). To produce a finished structure with uniform-gauge control, high internal bond strength, minimum curl, and a controlled coefficient of friction, adhesive lamination has proved to be an excellent converting system. Both solvent- and solventless-type adhesive laminators (see Adhesives; Coating equipment) are now converting materials in the United States and Canada for coffee packers. These new high-speed automatic packaging machines require consistently high quality materials to operate efficiently. All of the material converters acknowledge the sometimes difficult learning experience on vac-bag films and the need to set up a very thorough quality-control system.

Valves and sorbents. Gas-off of CO_2 from ground coffee is a problem for vac-bag packagers that can be handled in a number of ways. The simplest solution is to grind the coffee beans very fine, as in Europe. The fine grind evolves gas more quickly and requires only a short holdup time (1½–2 h). The North American markets require coarse grinds, however, that need up to 24 hours holdup time and considerable storage capacity. There are two methods to eliminate all holdup time. One is the one-way coffee valve, which allows CO_2 out of the bag without allowing O_2 into the bag (see Valves). The bag, however, is soft. This concept was tested in the United States by two major roasters but it was abandoned in favor of the hard vac bag that consumers seem to prefer. The other method involves a new technology from Canada, whereby a CO_2-sorbent pouch is placed in the coffee bag much like a dessicant pouch is used for moisture-absorbing applications. This new coffee vac-bag technology allows the roaster to package freshly ground coffee without any holdup time and still obtain a hard vacuum bag. This is not possible with any other system.

A growing coffee market has been found in specialty and gourmet stores where whole beans are purchased. The beans are generally not hermetically packaged, so CO_2 gas-off is not an issue. In this case, the use of a one-way valve is an excellent way to achieve an air-tight package and keep the beans fresh. The valve can be heat-sealed into the bag film as done by Goglio and SIG. It can also be attached to the outside surface using a pressure-sensitive valve concept developed by Hesser.

Advantages. Vac-bag is seen as a low-cost package compared with metal cans. For some coffee packers, this may be true if bag packing line speeds are equal to or greater than those for cans. For others, where the reverse is true, the material-cost advantage is not as large. Current 1-lb-size (0.45-kg-size) metal cans, including plastic overcap, cost about $0.17–0.19. A high quality four-ply vac bag costs about $0.08–0.10, which provides a material-cost advantage of about 9 cents per 1-lb (0.45-kg) package. Additionally, there are advantages in storage, handling, and shelving a vac-bag due to its compact shape. The reduction in cube is quite dramatic (35%). This cube reduction is beneficial to both manufacturer and retailer, since warehousing and shelving space is always a premium. As vac-bags become more widely distributed, consumers will determine the future for this new package.

BIBLIOGRAPHY

1. Product literature. SIG Swiss Industrial Company, Represented by Raymond Automation Company, Inc., 508 Westport Avenue, Norwalk, CT 06856

2. Product literature. Hesser Division of Bosch Package Machinery, 121 Corporate Blvd., S. Plainfield, NJ 07080

3. Product literature. Goglio, represented by Fres-co System USA, Inc., 10 State Road, Telford, PA 18969

F. J. NUGENT
General Foods Corp.

VACUUM PACKAGING.   See Controlled atmosphere packaging.

VALVES, PACKAGE

The technique of using heat-sealable multilayer flexible packaging materials (see Multilayer flexible packaging; Sealing, heat) has made it possible to manufacture airtight containers, or in other words, containers that completely insulate the contents from the external environment. It is sometimes necessary to separate the inside from the outside of the package by means of diaphragms which control the flow of gas in one direction or the other. In many cases, this requirement can

be met by applying a one-way discharge valve to the container (see Vacuum-bag coffee packaging).

By the use of a polyethylene valve, it is easy to obtain an airtight seal between the valve and the flexible bag. The operation of the valve is based on both mechanical and physical principles. It exploits the pressure difference that is generated between the interior and the exterior, so that when the internal pressure exceeds the external pressure by a given amount, the valve opens and the gas inside is let out until the internal pressure decreases to a set level. When this level is reached, the valve closes. The limit level is slightly higher than the pressure balance between the interior and the exterior to ensure that the flow of gas takes place in one direction only; hence, its characteristic feature as a "one-way" valve.

Valve structure. The valve consists of three basic parts: a circular base plate with a small hole; an elastic element; and a cap with a reaction strut. A hole is made on the surface of the bag for the insertion of the valve. The edge of the base plate is sealed to the heat-sealing part of the bag (normally the inner surface of the bag), and the valve becomes an integral part of the bag. On the inner raised ring of the base plate rests an elastic element (normally a rubber disk) which exerts pressure on the plate due to the force exerted on the elastic element itself by the reaction strut fixed to the cap. The cap also acts as an element of reaction to the forces exerted on the diaphragm and it protects the elastic diaphragm. The cap is fixed to the base plate by means of a snap-fit system. A filter, normally paper, is inserted over the base plate to keep out tiny corpuscles which impair the functioning of the valve when they lodge between the base and the diaphragm. Also, a layer of viscous liquid interposed between the base and the elastic element improves the seal of the valve itself.

Models and dimensions. The dimensions of the valve are normally as shown in Figures 1–3 (scale ~3:1), which illustrate the three possible models: (Fig. 1) a valve with a fully protruding cap, (Fig. 2) a valve with the cap half-protruding, and (Fig. 3) a valve with the cap completely sunk into the base plate.

For the application of the valves in Figures 1 and 2, the hole made on the material must be of the same diameter as the cap. With the valve in Figure 3, on the other hand, since it does not protrude, it can be hidden inside the bag and needs only a small hole, with a diameter less than the diameter of the cap.

Operating principle. When the pressure inside the bag balances the sum of the external pressure and the elastic reaction of the diaphragm, the diaphragm rises and lets out the gas through the aperture on the cap. When the internal pressure of the valve decreases and is lower than the sum of the two contrasting forces, the valve closes again. The opening and closing phases are critical phases for the valve. When the

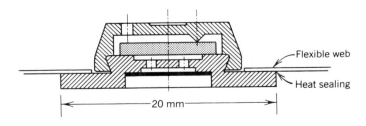

Figure 1. Valve with fully protruding cap.

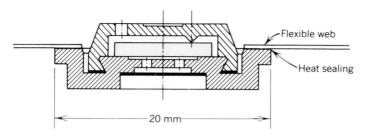

Figure 2.   Valve with half-protruding cap.

valve opens or closes, the forces are balanced, and the "one-way" feature of the valve is put at risk. To eliminate the risk, a layer of a viscous liquid is inserted to introduce viscous forces and thus avoid the danger of reversibility.

Valve insertion.   The valves are applied to the packaging material as the web unwinds from a roll, before the bag-making stage using simple- or multiple-valve-applicator units according to the speed at which the machine is operating. The valve-applicator unit can be situated "on-line" with both intermittent or continuous-movement machines. On intermittent machines, the applicator can be inserted on the bag-production line if the valve applicator time is shorter than the time during which the web intermittently stops. On continuous-motion machines, it is necessary to insert a device to stop the web for the length of time it takes for a valve to be applied. This is achieved by means of dancing rolls, which stop and then release the necessary amount of material. A photocell system makes it possible to register the web at adjustable, preset positions. The unit can house multiple applicators according to the required speed and the printing repeat, and thus up to 110 valves can be applied per minute. The valves are heat-sealed to the inner surface of the material. After the bag is formed, the valve becomes an integral part of the bag, and the only way that gas can get in or out is through the valve itself.

Applications.   There are many possible applications for these valves, but the major application thus far has been for the vacuum-packing of coffee. Coffee is extremely perishable when it comes into contact with air and when it is packaged just after roasting (ie, after its freshest point) it lets off a variety of gases (see Vacuum coffee packaging).

The valve can also be used for products that give rise to similar problems, such as cheese and chemical products. Another use of the one-way valve is as a means of irreversibly letting out the air contained in airtight packages that must be piled or stacked. If such bags contain air when they are

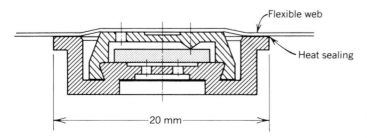

Figure 3.   Cap sunk into base plate.

stacked, the piles become unstable and the bags are liable to break at the seams. If there is an air discharge valve on the bag, not all of the air is let out through the valve. At the same time, since it is a *one-way* valve, no outside air is allowed to get into the bag.

LUIGI GOGLIO
Fres-Co System USA, Inc.

VIALS.   See Ampuls and vials.

VINYLIDENE CHLORIDE COPOLYMERS

Vinylidene chloride (VDC) is copolymerized with various other monomers to form thermoplastic resins useful in packaging because of their exceptionally low permeabilities to gases and liquids and their resistance to attack by chemicals and foodstuffs. All commercial VDC resins are copolymers with compositions that consist, by definition, of at least 50% vinylidene chloride. These materials are generally called poly(vinylidene chloride) or PVDC, but these terms are erroneous because they describe a homopolymer not used in commerce.

Worldwide use of VDC copolymers in 1984 totaled about 165 million (10^6) lb (74,800 t). Of that, extrudable resins accounted for 96×10^6 lb (43,500 t). Coating resins accounted for 69×10^6 lb (31,300 t): 48×10^6 lb (21,800 t) deposited from latices, and 21×10^6 lb (9500 t) deposited from solutions. More than 90% of the total was used in food and medical packaging. Other applications include fibers and specialty binders.

Types of Copolymers

Extrudable and moldable resins.   Because homopolymer PVDC has a melting point 9–18°F (5–10°C) below a rapid-decomposition temperature of 410°F (210°C), it is difficult to process at typical thermoplastic-fabrication temperatures. Comonomers such as vinyl chloride (VC), methyl acrylate (MA), and others depress the melting point to the range of 284–347°F (140–175°C), which makes melt-processing feasible. The most common form for melt processing is a powder of 8–12-mil (200–300-μm) dia. A pellet form of 0.1-in. (2.5-mm) dia has been introduced as well.

Coatings.   Homopolymer PVDC is not soluble in water or organic solvents. Comonomers impart solubility in organic solvents.

Latex coatings.   Latex coatings are aqueous dispersions of polymers produced by the copolymerization of VDC with other monomers. The comonomer(s) are usually MA or methyl methacrylate (MMA).

Solvent coatings.   Resins designed for solution in solvents, such as mixtures of tetrahydrofuran (THF) or methyl ethyl ketone (MEK) with toluene, and coating on cellulose substrates or plastic films are copolymers of VDC with acrylonitrile (AN), methacrylonitrile (MAN), MMA, or combinations thereof.

Coating selection.   VDC copolymer coatings are used on many different substrates for many purposes, and a variety of grades are required to fill these different needs. The choice between a coating applied from a latex and one applied from a solution depends primarily on the substrate, but equipment availability plays a role as well. On porous substrates, eg, uncoated paper and paperboard, water-based latices are preferred. Any water or solvents that have penetrated must be removed in a subsequent step. On nonporous substrates, eg, highly refined glassine, polyethylene-coated papers and boards, and plastics, either latices or solution coatings may be applied. On cellophane, the preferred coating is a solvent-borne VDC copolymer.

Once the choice between latex dispersion and solvent solution has been made, grade selection depends on the substrate and the performance required from the finished product, particularly with respect to gas and moisture barrier. Latices for coating porous webs differ from those best suited for coating plastics. Solvent coatings providing the best heat seals have chemical and physical properties differing considerably from those that provide maximum barriers.

General Principles

All VDC copolymers have certain common characteristics. Commercial copolymers have molecular weight ranges of ca 65,000–150,000. Molecules of lower molecular weights tend to be brittle and have limited commercial value. The physical and chemical properties of copolymers depend largely on the VDC content, which ranges from 72 to 92 wt % in most commercial products. The axial symmetry of the VDC molecule permits tight packing of molecular chains with concomitant formation of crystals constituting 25–45 vol % of the total structure. Specific gravities range from 1.65 to 1.75.

$$CH_2{=}C\overset{\textstyle Cl}{\underset{\textstyle Cl}{\big<}}$$

These characteristics of VDC copolymer resins are not significantly altered by the small amounts of processing and stabilizing aids used to facilitate extrusion and molding. A representative extrusion resin is a VDC–VC copolymer with ca 85% VDC and a molecular weight of 120,000, with plasticizer sufficient for ease of extrusion.

Note that the barrier and strength properties of VDC copolymers depend on the chemical compositions of the molecules, the degree of molecular orientation, and the direction of orientation in the finished form. Relatively high VDC content imparts relatively high barrier properties and gas resistance. A comparatively high degree of orientation imparts tensile strength. Higher crystallinity correlates with lower permeability.

Producers

The principal producers of VDC copolymers in their various forms are listed in Table 1. Dow Chemical produces two forms under the Saran trademark, as well as Saran Wrap household and commercial films and Saranex coextruded films.

Applications for Extrudable Resins

Flexible films. Monolayer films are used for household wrap, as unit-measure containers for pharmaceuticals and cosmetics, and as drum liners and food bags. Multilayer films, generally coextrusions with polyolefins, are used to package meat, cheese, and other moisture- or gas-sensitive foods. The structures, which contain 80–90% polyolefin with an inner

Table 1. Suppliers of VDC Copolymers[a]

Producers	Extrudable resins	Latices	Solvent coating resins
Dow Chemical	X	X	X
Asahi Kasei	X	X	
BASF		X	
Kureha	X		
ICI		X	X
Morton Chemical		X	
Solvay & Cie	X		X
Union Oil Chemicals		X	
W. R. Grace		X	

[a] Jan. 1986.

layer of VDC copolymer, are usually shrinkable films (see Films, shrink) that provide a tight barrier seal around the food product.

Semirigid containers. VDC copolymers are used as barrier layers in semirigid thermoformed containers. The sheet can be produced by coextrusion (see Coextrusion machinery, flat; Coextrusions for semirigid packaging) or by laminating monolayer or coextruded VDC copolymer films to semirigid styrenic or olefinic substrates. VDC copolymers can also be used as barrier layers in blow-molded bottles (see Blow molding).

Applications for Coatings

Paper and paperboard. Papers and paperboards (see Glassine, greaseproof paper, and parchment; Paper; Paperboard) coated with VDC latex copolymers are used where moisture resistance, grease resistance, oxygen barrier, and water-vapor barrier are required.

Cellophane. About 90–95% of all cellophane (see Cellophane) produced in North America is coated with VDC solution coatings to render the films moisture-resistant (thereby retaining the high gas barrier inherent to dry cellophane) and provide the needed moisture barrier.

Plastic films. Latex coatings on plastic films provide barrier to gases, moisture, flavors, and odors, and, in some packages, heat-seal capability (see Multilayer flexible packaging; Sealing, heat). Heat-seal latices are not especially good gas barriers, and, conversely, the best VDC barriers do not usually provide the best seals. When both heat sealability and barrier are required, it is best to apply two coatings, each designed for one purpose.

Semirigid containers. Latex coatings impart barrier properties to thermoformable plastic sheet used to produce high barrier food containers. Latex coatings on poly(ethylene terephthalate) (PET) bottles (see Polyesters, thermoplastic) impart barrier to oxygen, carbon dioxide, water, and flavors and odors.

Generally preferred forms of VDC copolymers for coating various substrates are shown in Table 2. Pretreatment with primers or electrotreatment (see Surface modification) may be required on some substrates. The form of the package, ie, film, thermoformed sheet, or blown bottles, influences the choice as well, but the key differentiating elements are the porosity and chemical composition of the substrate.

Table 2. Suggested Coatings for Various Substrates

| Substrates | Coatings | |
	Latices	Solutions
cellophane		X
nonporous papers[a]	X	X
porous papers	X	
polyolefins	X	X
polyesters	X	X
polyamides (nylons)	X	X
styrenics	X	
vinyls (PVC)[b]	X	

[a] Includes coated and dense forms, such as glassine.

[b] Plasticizers in highly plasticized PVC may migrate to the VDC copolymer coating, thereby damaging the coating's effectiveness as a barrier.

Properties

The properties of VDC copolymers most pertinent to food packaging include a unique combination of low permeability to atmospheric gases, moisture, and most flavor and aroma bodies and stress-crack resistance to a wide variety of agents. In addition, the ability to withstand the rigors of hot filling and retorting is important in the commercial sterilization of foods in multilayer barrier containers.

Extrudable resins. The range of barrier properties available in extrudable resins is shown in Table 3.

Higher barriers, ie, compositions with lower permeabilities, are being developed for commercial introduction. These materials, which use comonomers other than VC, are processed by extrusion and coextrusion using conditions similar to those used in processing VDC–VC copolymers.

The barrier of VDC copolymers can also prevent the undesirable gain or loss in a package wall of the packaged contents of many organic compounds. Table 4 shows how two VDC copolymer compositions compare with polyethylene (PE) in film forms as barriers to the permeation of some common materials

Table 3. Barrier Properties of VDC–VC Copolymer Extrudable Resins in Film or Sheet Form

Property	Value	
gas permeabilities at 73°F (23°C), cm³·mil/(100 in.²·d·atm) (cm³·μm/(m²·d·kPa))		
oxygen	0.13–0.16	(0.51–0.62)
carbon dioxide	0.25–0.30	(0.97–1.16)
water-vapor permeability at 73°F (23°C), g·mil/(100 in.²·d·mm Hg) (g·μm/(m²·d·kPa))	0.00077–0.0023	(2.27–6.79)
water-vapor transmission rate (WVTR) at 100°F (38°C), 90% rh, g·mil/(100 in.²·d) (g·μm/(m²·d))	0.034–0.10	(13.38–39.37)

Table 4. Transport of Certain Organic Molecules Through Selected Polymer Films

| | Permeability at 77°F (25°C), g·mil/(100 in.²·d·mm Hg) (g·μm/(m²·d·kPa)) | | |
Material	d-Limonene	n-Heptane	Styrene
VDC–VC film[a]	5.25×10^{-5} (0.155)	2.34×10^{-4} (0.691)	4.67×10^{-5} (0.138)
VDC–VC household film[b]	2.57×10^{-3} (7.6)	9.35×10^{-4} (2.76)	8.77×10^{-3} (25.9)
HDPE[c]	0.498 (1469)	0.278 (821)	0.064 (190)

[a] Composition equivalent to film or sheet described in Table 3.

[b] Saran Wrap household film.

[c] High density polyethylene.

encountered in packaging, as constituents of contents or the structural elements of packages themselves.

The contrast between general-purpose and high barrier films is shown in Table 5.

Coatings. The bulk mechanical properties of VDC-copolymer-coated papers and films and structures such as formed containers and blown bottles depend almost entirely on the properties of the substrate material. The substrate makes essentially no contribution to the specialized attributes provided by the coating, such as barrier to permeation, chemical resistance, or heat or dielectric sealability (see Table 6).

Regulatory Status

The regulatory status of VDC copolymers for food-contact applications has been reviewed (1). VDC–VC copolymers containing ca 10–27 wt % VC are considered to comply with the food-additive provisions of the Federal Food Drug and Cosmetic Act on the basis of "prior sanction" (see Food, drug, and cosmetic packaging regulations). A variety of specific regulations govern VDC copolymers other than those of vinyl chloride. These are included in Ref. 1 and summarized below.

1. As the food-contact surface, alone or as a coating on a substrate, or as a nonfood contact layer in a multilayer structure:

    VDC–MA copolymers with no more than 15% copolymerized MA may be used up to 250°F (121°C).

    VDC–MA–MAA copolymers with no more than 2% copolymerized MA and no more than 6% copolymerized MAA may be used up to 250°F (121°C) and at no more than 2 mil (0.005 cm) in thickness. Some extractive limitations must be met.

2. Coatings on films, wherein the coating meets certain extractive limitations:

    VDC copolymerized with a variety of other monomers as coatings on polyolefin films.

    VDC copolymerized with acrylic acid, acrylonitrile, MA, MMA, 2-sulfoethyl methacrylate, or other monomers as coatings on nylon films.

    VDC copolymerized with acrylonitrile, MA, and acrylic acid as a coating on polycarbonate (PC) films.

3. As adhesives wherein the adhesive is separated from the food by a functional barrier or as an adhesive in packages for dry food.

Table 5. Properties of Vinylidene Chloride Copolymer Films[a] 1 mil (25.4 μm) thick

Property	ASTM test method	Type of film	
		General-purpose	High barrier
specific gravity	D 1505	1.60–1.71	1.73
yield, in.^{2}/(lb·mil) (cm^2/(g·m))		17,300–16,200 (6.25–5.85)	16,000 (5.78)
haze, %	D 1003		2–3
light transmittance, %		85–88	80–88
tensile strength, psi (MPa)	D 882		
MD		7,000–14,500 (48.3–100)	12,000–12,500 (82.8–86.2)
XD		13,000–13,500 (89.7–93.1)	20,000–21,500 (138–148)
elongation, %	D 882		
MD		40–100	95–100
XD		40–100	50–60
tensile modulus, 2% secant, psi (MPa)			
MD		50,000–110,000 (345–759)	160,000–165,000 (1103–1138)
XD		45,000–105,000 (310–724)	135,000–150,000 (931–1034)
tear strength, gf/mil (N/mm) propagating	D 1922	10–<100 (3.9–<38.6)	10–<100 (3.9–<38.6)
folding endurance, cycles	D 2176	>500,000	>500,000
change in linear dimensions at 212°F (100°C) for 30 min, %	D 1204		
MD		12–22	6–7
XD		6–18	3–4
service temperature, °F (°C)		0–275 (−18 to 135)	0–275 (−18 to 135)
heat-seal temperature, °F (°C)		250–300 (121–149)	250–300 (121–149)
oxygen permeability at 73°F (23°C) and 50% rh[b], cm^3·mil/(100 in.^{2}·d·atm) (cm^3·μm/(m^2·d·kPa))	D 1434	0.8–1.1 (3.1–4.3)	0.08 (0.31)
water-vapor transmission rate (WVTR) at 100°F (38°C) and 90% rh, g·mil/(100 in.^{2}·d) (g·mm/(m^2·d))		0.2 (0.079)	0.05 (0.02)
COF face-to-face, back-to-back, at 73°F (23°C) and 50% rh	D 1894	0.3 to no slip	no slip

[a] Not to be confused with Saran Wrap household brand film.
[b] Humidity has no effect on the permeability of VDC copolymer films.

Table 6. Properties of VDC Copolymer Films[a] Made From Aqueous Latices and From Solutions

Properties	Latices[b] selected for		Solution resins[d]
	High barrier	High seal strength[c]	
oxygen permeability at 73°F (23°C), cm^3·0.1 mil/(100 in.^{2}·d·atm)) (cm^3·μm)/(m^2·d·kPa))	0.34–1.0 (0.132–0.388)	1.0 (0.388)	0.15–0.70 (0.058–0.27)
water-vapor transmission rate (WVTR), at 100°F (38°C) and 90% rh, g·0.1 mil/(100 in.^{2}·d) (g·μm/(m^2·d))	0.49–0.60 (19.3–23.6)	0.55–1.5 (21.7–59.1)	0.18–0.90 (7.1–35.4)
minimum heat-seal temperature, °F (°C)	270 (132)	180 (82)	220–265 (104–129)

[a] Oven-dried films.
[b] Cast on PET film from dispersions.
[c] Heat- and dielectric-sealing grades available.
[d] Cast from solutions of up to 20 wt % solids in 65% THF or methyl ethyl ketone/35% toluene (wt/wt) at 73–95°F (23–35°C).

4. Coatings on cellulosic substrates.

VDC copolymers with a variety of other monomers as coatings on cellophane.

VDC copolymers with a variety of other monomers as components of paper and paperboard in contact with aqueous and fatty foods providing the structure meets some extractive limitations.

VDC copolymers as components of paper and paperboard in contact with dry foods.

5. Coatings comprising copolymers of VDC with vinyl chloride applied to fresh citrus fruit.

6. Coatings of VDC with various other monomers applied as coatings on PET plastics must meet some extractive limitations and may be used for nonalcoholic foods at temperatures up to 250°F (121°C).

Processing

Extrusion. The temperature range over which VDC copolymers may be used without damage varies, as it does for other polymers, with the conditions and service expected. For example, continuous exposure of a typical VDC copolymer extrusion resin in air at 113°F (45°C) does generally not result in any significant loss of tensile strength for at least 10 yr. The time for an equivalent loss in strength is reduced to 5 yr if the temperature is raised to 120°F (49°C). The T_g of VDC–VC copolymer compositions is in the range of 14–32°F (−10 to 0°C) (see Polymer properties). Films and other fabricated forms can easily withstand freezer and cold-storage temperatures, providing they are not abusively handled. Compounding with appropriate plasticizers or alloying with elastomeric material yields film with toughness adequate to survive abusive handling at temperatures as low as −26°F (−32°C).

VDC copolymers can be processed by all of the conventional thermoplastic processing methods and some special variations. Because the melt strength of commercial VDC copolymer compositions is insufficient to support a molten stream, the copolymers must be quenched to temperatures below the melting point before drawing or forming of extruded tubes, sheets, or filaments. That provision is obviated if the resin is part of a multilayer structure in which a second material provides the melt strength required. The most common current method for melt processing incorporates the resin as a portion of extruded multilayer sheet to be formed or film to be cast or blown (see Coextrusion articles, Extrusion).

Melt processing should maintain VDC copolymers below 392°F (200°C) to prevent decomposition. Key considerations in melt-processing equipment are the use of corrosion-resistant alloys, particularly nickel alloys and high alloy steel, wherever hot polymer contact occurs, and the minimization of polymer residence times in the equipment. Low concentrations of heavy metals such as iron, copper, and zinc must be avoided because they catalyze the thermal decomposition of VDC copolymers. The corrosivity of molten VDC copolymers is in a class with other halogen-containing polymers such as PVC and chlorinated PE.

Solvent coating. Solution processes for the application of resins to cellophane and other substrates generally involve the solution of 20% by weight of noncrystalline solids of resins in methyl ethyl ketone at room temperature or crystalline solids of resins in THF at temperatures up to 140°F (60°C). A variety of diluents can replace some of these relatively expensive solvents to lower costs, reduce flammability, and improve sprayability. Solutions can be applied by dipping plus doctoring with a knife or wire-wound rod or by direct rotogravure, brushing, spraying, or flexographic techniques (see Coating equipment). The most common substrate film for lacquer coating is cellophane. Solutions can also be applied to plastic films (see Multilayer flexible packaging).

Latex coating. Water-based latices are applied to plastics, paper, and paperboard by methods similar to those used for lacquers in coating plastic films and sheets. Film formation takes place through coalescence of the spherical latex particles as water evaporates from the latex. The resulting coatings are clear, smooth, and glossy, with properties characteristic of VDC copolymers. Latices are commonly supplied in 50 wt % of water, though sometimes diluted to facilitate application or to control coating thickness, and usually only with those additives necessary to preclude coagulation for periods of 6–12 mo or longer. Thus, it is not unusual for films from latices to have lower permeabilities to gases than the equivalent melt-processible copolymers because the latter require additives such as plasticizers and heat stabilizers to facilitate processing. Latex particles typically range from 0.0003–0.001-in. dia (0.08–0.25 μm).

Uses in Packaging

VDC copolymers are used in applications requiring resistance to the flow of moisture, atmospheric gases, and constituents, eg, flavorants, along with associated properties such as grease resistance and stress-crack resistance. In most instances, the VDC resin is a part of the total package, having been combined with other materials. The principal exception to the compositing of VDC copolymers exists in household and commercial Saran Wrap films. Some medical packaging also utilizes VDC copolymers alone.

Vinylidene chloride copolymers can be sealed to themselves and to other materials with heat generated by sealing bars or by the application of r-f excitation (see Sealing, heat). Some compositions are capable of being sealed under pressure alone, ie, cold-sealed, or with very little additional heat.

BIBLIOGRAPHY

1. M. L. Rainey *Regulatory Status of Vinylidene Chloride Copolymers for Food Contact Applications, Bulletin,* The Dow Chemical Co., Midland, Mich., 1984, *Form No. 190-320-84.* For updates, see the current issues of *Code of Federal Regulations,* Title 21, published annually; *Federal Register,* published daily, both available from The Government Printing Office, Washington, D.C.

General References

D. S. Gibbs and R. A. Wessling in M. Grayson, ed., *Kirk-Othmer Encyclopedia of Chemical Technology,* 3rd ed., Vol. 23, Wiley-Interscience, New York, 1983, pp. 764–798.

C. J. Benning, *Plastic Films for Packaging: Technology, Applications and Process Economics,* Technomic Publishing Co., Inc., Lancaster, Pa., 1983.

P. T. DeLassus, *J. Vinyl Technol.* **1,** 14 (1979).

W. E. Brown and P. T. DeLassus, *Polym. Plast. Technol. Eng.* **14**(2), 171 (1980).

W. E. Brown in J. Agranoff, ed., *Modern Plastics Encyclopedia 1985–1986,* Vol. 62(10A), McGraw-Hill, Inc., 1985, pp. 99 and 100.

W. E. Brown
Dow Chemical USA

W

WAXES

Waxes in packaging are generally associated with paper and paperboard because they add properties and capabilities that paper and paperboard alone cannot provide: moisture and oxygen barrier, grease resistance, and heat sealability. The waxes used in packaging, derived from petroleum, fall into two categories: paraffin waxes and microcrystalline waxes.

Paraffin waxes. These are composed mainly of straight-chain hydrocarbon molecules with a few isoparaffins and few, if any, cyclic structures. The crystals are large and platelike, and the waxes are hard and brittle. Fully refined paraffin wax has an oil content of ca 0.5%, with a melting point range of 120–160°F (49–71°C). Used alone, paraffin wax provides an excellent moisture and oxygen barrier, but it also has some drawbacks. It is not durable and can be easily marred or scuffed; and in the low melting-point range, there can be problems with blocking in summer transport. Paraffin wax heat-seals easily, but with low seal strength (see Sealing, heat). Sometimes adhesives must be used as well. The higher melting-pointing grades have the best gloss and blocking resistance of the group, but for better performance, paraffin waxes are modified.

Modified paraffin waxes contain up to 10% of one or more modifiers: microcrystalline waxes, butyl rubber, polyisobutylene, polyethylene, ethylene copolymers, and/or other resins. If modifier content is below 10%, the barrier properties of the paraffin are not affected adversely, but heat-seal strength, gloss, and durability are all substantially improved. Most waxes for coating purposes are slightly modified or greatly modified paraffins. If the modifier content is over 10%, the product is called a hot melt. The modifiers in hot melts can be any of those mentioned above, and/or ethyl cellulose, butadiene rubber, and cyclized rubber. In some hot melts, there is no wax at all, but most contain over 50% wax. Hot melts can provide very strong heat seals, flexibility, and abrasion resistance, along with a very good moisture and gas barrier.

Microcrystalline waxes. These are complex mixtures of straight-chain hydrocarbons, isoparaffins, and monocyclic hydrocarbons. In contrast to paraffins, their crystals are extremely small and needle-shaped, and the waxes are soft and ductile. Melting points range from 140 to 195°F (60–90°C), with a broad melting-point range of 10–20°F (−12 to −7°C). The oil content can vary 0.5–7%. Barrier properties are good, but not as good as those of the paraffin waxes, and they are used differently. As noted above, microcrystalline waxes can be used to modify paraffin waxes for coating purposes. There are two groups of microwaxes: those for coating or blending and those for laminating (see Laminating machinery).

Microwaxes for coating or blending have a relatively high proportion of branched chains, and relatively little cyclic material. The composition gives them hardness and scuff resistance along with a moisture barrier. Coating grades generally have a melting point range of 160–195°F (71–91°C), with an oil-content range of 0.5–2%. Laminating microwaxes have some branched chains with a relatively high proportion of cyclic material. The composition gives them toughness and good low temperature flexibility and ductility, as well as good cohe-

sion and adhesion. Used to laminate paper or films, microwaxes serve not only as the adhesive, but as a moisture barrier as well. The melting points of the laminating microwaxes usually range between 140–175°F (60–79°C).

Waxed Paper and Paperboard

Waxed paper and board come in a number of varieties (see Paper; Paperboard). The chief categories are wet-waxed, dry-waxed, and wax-laminated (see Fig. 1). Paraffin waxes, plain or modified, or hot melts are used for wet and dry waxing. Microwaxes are generally used for wax lamination. Wet-wax applications include folding cartons (see Cartons, folding), carton overwraps, wraps for bread, soaps, and cakes, and household waxed paper. Dry-wax applications include deli and butcher papers, florist tissue, and razor-blade wraps. Products packaged in wax-laminated papers include some soaps, powdered drinks, and tobacco. Some waxed papers are coated on one side only, eg, gum wrappers and dessert-powder papers, for proper performance on the packaging machinery. Wax-laminated and wet-waxed glassine is used in bag-in-box packages for cereals and crackers (see Bag-in-box, dry product).

Wet waxing refers to waxed papers on which there is a continuous surface film on one or both sides of the paper. The key to successful wet waxing lies in "setting" the waxed surface immediately after waxing (see Fig. 2). This can be achieved by shock-chilling the waxed web either by a cold-water quench or by passing over a series of chilled idler rolls. Sudden chilling also imparts a high degree of gloss on the coated surface.

Dry waxing leaves no continuous film on the surfaces, for the wax is driven into the body of the substrate covering the fibers and filling the interstices. The basic difference between wet and dry waxing is that heated rather than chilled rolls are

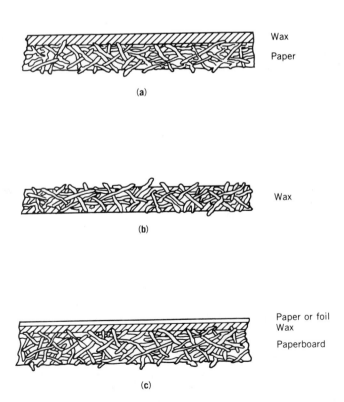

Figure 1. Cross sections of waxed paper and paperboard: **(a)** wet-waxed paper; **(b)** dry-waxed paper; **(c)** laminated paperboard.

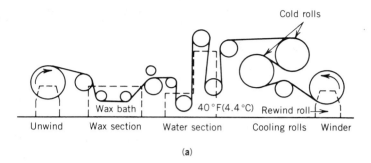

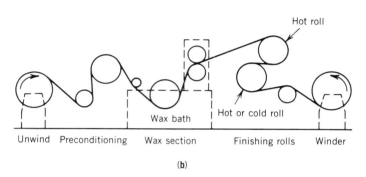

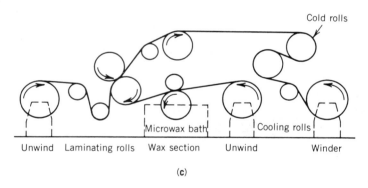

Figure 2. Waxing processes: **(a)** wet waxing; **(b)** dry waxing; **(c)** wax laminating.

used. Entirely moistureproof sheets cannot be produced by dry waxing; since the surfaces are not protected, the exposed and connecting fibers act as a moisture wick at the surfaces.

In *wax laminating,* composites of paper to paper, paper to board, or paper to foil are bonded with a continuous wax film acting as an adhesive. Laminating microwaxes are used as the base material for many laminates (see Laminating machinery; Multilayer flexible packaging). The primary purpose of the wax is to provide a moisture-vapor barrier and a heat-sealable laminant. Papers are also laminated with adhesives that do not contain wax. Special resins are added to improve adhesion and low temperature delamination; polymers impart special characteristics for specific laminations. In the laminating process, the webs of paper are simultaneously removed from two separate roll-unwind stands. One web passes through a wax bath and then contacts the unwaxed web. The joined webs pass through heated rolls to effect a seal that is often stronger than paper.

Tests. In paper coating, the main concern is moisture-vapor resistance. The standard test, TAPPI T-464-SM-45, for the moisture-vapor transmission rate (WVTR) requires that a coated sheet be sealed over a desiccant and subjected to an atmosphere of 90% rh at 100°F (38°C). Any moisture passing through the coated paper is absorbed by the desiccant. The coated paper may be either creased or uncreased, and the results are so recorded. A more recent test method eliminates the desiccant. In laminating, WVTR is important as well, but also strike-through (bleed), sealing strength, and adhesion.

In the *bleed test,* five droplets (240–310 mil or 6–8 mm dia) are placed on filter paper (Whatman #1). When the droplets have solidified, the paper is placed in an oven at 130°F (54°C) for 24 h. The average value (mil or mm) is obtained by subtracting the diameter of droplets from that of the oil rings. This is the strike-through (bleed) number.

The *sealing-strength test* (ASTM D 2005-65) measures the cohesive strength of a wax used to seal two sheets of porous paper together. The sealed sheets are pulled apart at a specified rate of speed in a manner such that the force required is recorded in gf/in. (N/m).

The *adhesion test* measures the force required to separate two sheets of nonporous paper, such as glassine, that have been laminated with wax. Since little or no wax impregnates the glassine, the test clearly measures the adhesive properties of the laminating wax.

Cups and Containers

Waxed cup stock is a hard-sized, bleached-sulfite fourdrinier paper supplied to converters in large rolls. In the cup plants, these rolls are slit, printed, die-cut, formed, and glued on mandrels in a continuous operation. The formed cups are then fed to the waxer conveyor, where the cups are either wet- or dry-waxed. In wet waxing, the cups pass directly from waxing to a cooling chamber; in dry-waxing, the cups are fed through a heat chamber before moving to a cooling cycle. In either case, all latent heat must be eliminated to avoid blocking when force nested. Cups and containers are waxed by the flood-and-spin method or a spray-coating system. Waxes formulated for coating cups and containers, ie, mainly paraffins and modified paraffins, are designed to give complete coverage to provide good stiffness, gas and odor barrier, and resistance to penetration of liquid food components. Converters look for a good satin finish, free of pinholes, surface blemishes, or orange-peel effects.

Folding Cartons

The term folding carton refers to the single-layer, die-cut, scored, and printed board stock used to package items such as butter, ice cream, bacon, and frozen foods (see Cartons, folding). In the early days, these packages were cold-water-waxed using a simple wax blend, and the flaps were dewaxed to seal the cartons. To solve this problem, "pattern" waxers were developed to apply the wax only where needed, leaving the flaps free for gluing. These waxers could not handle high viscosity blends or high operating speeds, for the blanks would either wrap around the applicator rolls or become warped at the finishing station. Today, third-generation coaters (see Fig. 3) are available that handle the new wax, ie, hot-melt, blends at viscosities of 15,000–20,000 cP (15–20 Pa·s) and operate at speeds of 300–350 ft/min (91–107 m/min). The waxes are applied to both surfaces in a single pass. With these new coaters,

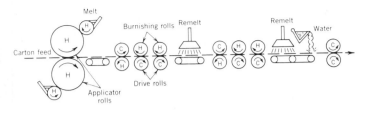

Figure 3. Third-generation folding-carton coater.

the coating thickness can be controlled to 5–6 lb/1000 ft² (24–29 g/m²) of board. At this rate, the full-color printing stands out without distortion, WVTR remains low, and the surface holds the gloss and has excellent scuff resistance. Because these coatings are generally heat-sealable, the coated cartons can be automatially sealed at the speed of the food processor's filling lines. Folding-carton converters use performance tests to measure WTVR, gloss, gloss retention, scuff resistance, sealability, and coating continuity.

Corrugated Containers

A principal use for wax in paper converting is in the corrugated board or container field (see Boxes, corrugated). The waxed corrugated container has replaced hydrocooled or top-iced wooden boxes for packaging and shipping produce, fish, and poultry by rail and truck, where stacking strength under high moisture conditions is imperative. The materials used are generally paraffin wax or modified paraffin wax blends, which can be tailor-made for different applications or box-performance requirements.

Wax application. Depending on the application, the box blanks can be surface-treated by roll coating or curtain coating, impregnated, saturated, or treated with a combination of methods.

Impregnation

This is usually done by on-machine techniques with a wax pickup of <25%, which is a partial saturation and does not offer the best water resistance. Because of this, the blanks are usually surface-coated by curtain coating after impregnation.

Saturation

Saturation requires 45–50% pickup, which doubles the crush strength and imparts excellent water resistance. Wax pickup, score-line flexibility, and surface finish can be controlled. Most corrugated boxes are saturated by dipping. This

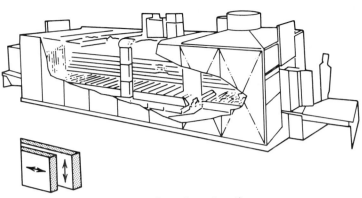

Figure 5. Cascader saturation.

is accomplished either by batch dipping, in which baskets are immersed, or by conveyors on which the blanks are suspended and carried through the wax tank and then oven-conditioned (see Fig. 4). Except for a few installations, the industry has used batch-type dippers built by the individual plants.

The latest development in saturation is the shower (or cascading-wax) method, which employs a cascader. This method of saturating boards provides better control of wax pickup than dipping. In a closed tunnel (see Fig. 5), the blanks are fed through vertical slots with guide bands and conveyed on edge under weir-head fountains. As the boards emerge into the conditioning zone, air knives at the top and bottom remove excess wax from the surface and flutes. As the blank passes into the final zone, cool air is applied to prevent stack blocking in shipping.

Curtain Coating

The wet-strength requirement of corrugated containers in hydrocooled shipments of produce, poultry, and seafood can be provided by curtain coating. There are two general types of equipment utilized in curtain coating (see Figs. 6 and 7) (see Coating equipment). One type, the pressure-head coater, forces the hot melt under pressure through a controlled orifice. The second type, the weir-head, merely circulates the melt to a holding tank where it overflows and falls by gravity onto the board passing underneath. The recent trend is toward the pressure-head equipment owing to its versatility and adaptability to high speed operation.

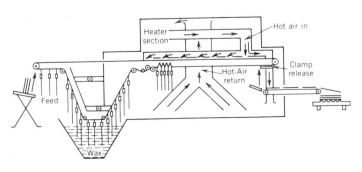

Figure 4. Dip saturation.

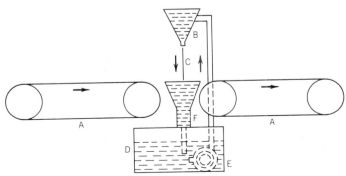

Figure 6. Pressure-head curtain coater. A, conveyor; B, V head; C, liquid curtain; D, circulating tank; E, pump; F, return trough.

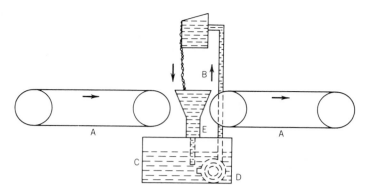

Figure 7. Weir-head curtain coater. A, Conveyor; B, liquid curtain; C, circulating tank; D, pump; E, return trough.

Tests. Converters of corrugated board test for wax pickup, crush strength, and glue-line adhesion.

Wax Pickup

Wax pickup can directly affect compression strength of the board. Wax pickup should run about 45% of the board weight. To test, uncoated and saturated samples of equal size are uniformly conditioned by holding for 2 h in a set-temperature room. The samples are then weighed to 0.01 g. A predetermined factor for this sample size (4.5 × 4.5 in. or 11.4 × 11.4 cm) is 15.67 and from this, board weight can be calculated to lb/1000 ft² (g/m²). Percent wax pickup can be calculated by the weight difference between the uncoated and saturated samples.

Crush Strength

Crush strength is determined by comparing uncoated and saturated board samples immersed in water. After shaking and drying, the samples are weighed and are individually subjected to the column crush test. The value determined after this conrolled test is reported as lbf/in. or N/m.

Glue-line Tests

These measure the effectiveness of the glue and gluing equipment. After the column crush test, each sample is pulled

Table 1. Quality-control Standard Wax Tests

Property	ASTM test
melting points	D 87 Cooling Curve
	D 127 Drop Point
	D 938 Congealing Point
oil content	D 721 Oil Content, %
color	D 156 Saybolt Chromometer
	D 1500 ASTM Scale
penetration	D 1321 Needle
	D 937 Cone
odor	D 1833 Rating
	D 1832 Peroxide Number
viscosity	D 88 Saybolt (SUs)
	D 445 Kinematic
blocking point	D 1465
flash point	D 92
gas chromatography	

apart by hand and the adhesion rated on percentage of fiber tear. This should be done after a 3-h immersion in water.

Just as important as these performance tests are the quality-control tests run as a check on the supplier's material-specification limits (see Table 1). These tests cover the pertinent typical physical properties of the waxes or wax blends and are as important to the converter as the performance tests are to the consumer.

Petroleum products developed for the food-packaging industry are specifically formulated to comply with the FDA and USDA regulations. Petroleum waxes, wax blends, and wax-based hot melts are biodegradable. They can be incinerated or deposited in landfills with no risk of noxious fumes.

BIBLIOGRAPHY

General References

H. Bennett, *Industrial Waxes,* Vols I and II, Chemical Publishing Co., Inc., New York, 1975.

J. J. Devlin
H. M. Farnham
Moore and Munger Marketing, Inc.

WEIGHING.   See Checkweighing; Filling machinery, dry.

WELDING, SPIN

Spinwelding is a process which joins plastic container parts together by heat from friction. This is accomplished by rotating either the top or the bottom container part, while holding the other part stationary. The heat produced from friction melts the surface of the two parts where they are in contact with each other, and then the melt solidifies to form the weld. For prototypes or specialty low volume items, manually operated equipment designed to weld one container at a time can be as simple as a drill press with a tooling drive and a clamp to hold the stationary part. For production runs, fully automated in-line equipment is available that can weld containers at rates exceeding 300 containers per minute.

Several requirements exist for spinwelding that warrant attention. One requirement is that the two container parts to be welded must have round cross sections at the weld band (see Fig. 1) existing in the same plane. Another requirement is that the container parts must be manufactured to extremely close tolerances, by thermoforming, injection molding, or other process. If a container part is out-of-round, too large, or too small, the integrity of the weld is severely jeopardized. A third requirement is that the parts to be welded must be made of compatible materials (ie, polymer type, melt flow, etc). Although dissimiliar materials can be used, they generally result in lower strength welds.

Spinwelding has many important applications in today's markets. The first and probably most obvious is as a product line extension for thermoformers (see Thermoforming). Spinwelding provides a cost-effective entry into plastic-bottle markets traditionally served by blow molding (see Blow molding) and it is of particular interest today as a cost-effective way to produce high barrier containers. Another advantage is the ability to combine container parts of different colors, or a

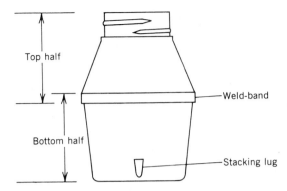

Figure 1. Container parts must have round cross sections at the weld band.

transparent part with an opaque part, providing unique appearance and marketing appeal. A third application involves the ability to apply tamper-evident closure systems (see Tamper-evident packaging). This represents just a few examples of the potential provided by spinwelding.

To automatically spinweld containers at high rates of speed, the following functions are performed sequentially: denest container parts; spinweld container parts together; and transfer finished containers to filling or other process system.

Denesting. Container parts are shipped to the spinwelding location (usually the filling location) in nested stacks. Container tops and bottoms are loaded into their respective magazines in tall vertical stacks. These magazines can be as simple as a single stationary stand, or as complex as a multiple-rotating stack fixture. Vacuum-equipped shafts pull the container bottoms down and the container tops are brought down with picker heads. These shafts then feed the container parts into infeed starwheels, which transfer them to the spinwelder. This process is aided by carefully locating the stacking lugs (see Fig. 1) on the container during design. These lugs act to separate the parts a predetermined amount when stacked in order to ease denesting.

Spinwelding. The spinwelding section of the machine accepts parts from the denester system and deposits each half of the container into closely machined tooling. These parts are held in the tooling in many ways, dependent on container design. Existing stacking lugs can be grasped, for example, and/or welding lugs (flutes) can be designed into the container solely for this purpose. The grasping is accomplished either by teeth or by a rubber facing and is usually vacuum assisted. Either part can be held in the tool to be rotated, but usually it is the bottom or the smaller half, again depending on container design.

The rotating tool is brought up to a predetermined speed by a spinner apparatus (see Fig. 2). In some machinery, the container parts are already in contact before the spinning begins. In these systems, the rotation is stopped abruptly by a brake after a very brief period of time, but the pressure is maintained to ensure that the melt solidifies. In other systems, the container parts are brought together only after the rotating tool has reached desired speed. This is accomplished rapidly utilizing a clutch and flywheel. Once the desired speed is attained, the clutch disengages, the parts are brought together, and pressure is maintained as the spinning is stopped by the solidifying melt.

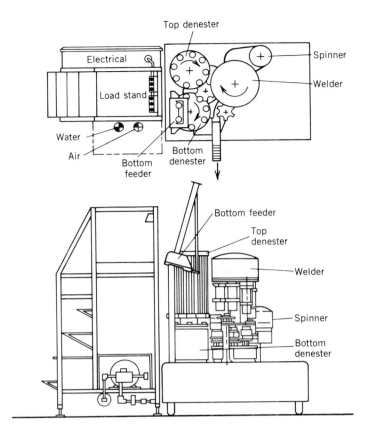

Figure 2. Spinwelding equipment.

Exit transfer. After welding, the finished containers are removed from the tooling by another starwheel, which transfers them either directly to a filler or onto a conveyor for further processing steps such as weld inspection and rinsing.

Weld integrity. Depending on the desired result, welds can be designed to be no more than a tack, or as much as a full penetration. If the latter is chosen, the weld-band has hoop strength, and is the strongest part of the container. Weld integrity is governed by several variables, including material choice, material thickness, joint pressure, driving force, rotation speed, tolerances of the parts, cycle time, and weld band design. The number of weld-band design choices, limited only by the imagination, includes butt, tongue and groove, and sleeve. Depending on design, a trimming tool may be required to remove flash created during the weld. One way around this is to design the weld-band incorporating a reservoir for the flash. This is most easily accomplished with a sleeve design as shown in Figure 1. It is sometimes necessary to provide a system to check the weld integrity, but modern commercial units can weld hundreds of millions of containers annually with less than 1% rejects.

BIBLIOGRAPHY

General References

Modern Plastics Encyclopedia, 1985–1986, McGraw-Hill, Inc., New York, 1985, pp. 354, 356.

Packaging Encyclopedia and Yearbook, 1985, Cahners Publishing Company, Boston, Mass., 1985, p. 260.

Technical Manual for Vercon Inc., Autowelder-10, Vercon Division of Fina Oil and Chemical Co., Dallas, Texas, 1983.

Scot R. Mitchell
Don MacLaughlin
Bert Peterson
Vercon, Inc.

WIREBOUND BOXES.   See Boxes, wirebound.

WOOD BOXES.   See Boxes, wood.

WOOD PALLETS.   See Pallets, wood.

WRAPPING MACHINERY

The history of packaging automation began in 1869 with the introduction of the first automatic machinery for making paper bags. It was not until the early 1900s that demand for other types of automatic packaging machinery resulted in the formation of a number of small companies devoted to the manufacture of packaging machinery. Early designs of machines for wrapping loaves of bread, cookies, food products, cereal boxes, candy bars, soap tablets, and tobacco products incorporated much of the established bag-making technology.

The 1936 introduction of cellophane (see Cellophane) as a new transparent wrapping material created new demands, and many new types of wrapping and wrapping-related machines were developed and put into production (see Films, plastic). Progress continued at a fast pace until World War II when war-related work took precedence over commercial activities. The growth of supermarkets during the war years created a need for new kinds of packaging and new kinds of automated equipment. Firms in the United States and abroad responded to this need with a wide range of new wrapping and wrapping-related machines. These new models featured higher operating speeds, increased efficiency, greater flexibility as to package size, and in many cases, the ability to convert both older and newer models for use with thermoplastic films.

Throughout the history of packaging automation there have been examples of equipment design changes to accommodate marketing trends. Three recent examples are the development and refinement of tamper-resistant packaging, the introduction and growth of generic packaging, and the expansion of promotional packaging programs.

Packaging Terms

Many of the trade terms that pertain to wrapping and wrapping-related operations are listed below.

Wrapper/wrap.   A cut-to-size sheet of monolayer or multilayer flexible packaging material used to wrap a product or package.

Loose wrap/removable wrap.   The forming of a wrap around a product or package in which the folds (overlaps/seams) are sealed to themselves and not to the object being wrapped.

Tight wrap/wet wrap.   The forming of a wrapper around a carton or container in which the wrapping material is adhered or sealed to all of the surfaces of the package through the application of wet adhesive or by activating a heat-sealing type of coating preprinted on the underside of the overwrap by the packaging material converter.

Intimate wrap/conforming wrap.   These terms describe the type of wrap that is in direct contact with the product, as required in the dairy, candy, and meat industries (see Films, shrink).

Bunched wrap/formed wrap.   Describes a style of wrap for irregular-shaped products in which the wrap is either draped over the product or the product is pushed through a die-box so that the wrapping material can be gathered (tucked/folded/pleated) and then sealed to itself on the bottom panel of the package.

Bundle wrap/parcel wrap.   A low-cost method of wrapping packages together by first combining (collating) a predetermined number of small packages into larger units (collations) preparatory to overwrapping with a heavy-duty packaging material.

Multipack wrap/supermarket multiple-pack wrap.   A variation of bundle-wrapping wherein identical products are arranged or collated into multiples of 2-, 6-, 10- or 12-pack units prior to overwrapping (see Bundle-Wrapping Machines).

Multipack band wrap/supermarket banded multiple-pack.   Multiple units of identical-size packages arranged or collated together in a pattern that will accommodate a band of stretch or shrink film to be positioned around four sides of the multiple units.

Bag wrap.   Prefabricated bags are individually fed into position, opened, and then loaded manually or automatically, prior to being closed in some conventional manner (stapling, crimping, or twist-tying).

Flow pack/tube wrap.   Type of style and seal produced on a horizontal form/fill/seal wrapping machine (see Form/fill/seal, horizontal) generally associated with wrapping irregular-shaped items such as candy bars, bakery items, etc.

Accumulator/collator.   Machine attachment designed to collect, assemble, and orient packages in specific patterns or collations, generally in preparation for bundling and multipack operations.

Wrapping, Banding, and Bundling Machines

Packaging machinery designed for wrapping, banding, and bundling operations can be divided into eight classifications:

Wrapping, general purpose.   Packagers faced with problems of short production runs, frequent changeovers, and the need for a high degree of flexibility in meeting changing packaging specifications can select from a wide range of new and rebuilt wrapping machines. Examples are supermarket packaging of meats and fruits, cosmetics and perfume packages, pharmaceuticals, and contract packaging (see Contract packaging).

Wrapping, shrink.   Standard-model wrapping machines are now available with conversion kits to handle new plastic packaging films, including shrink films (see Films, shrink). The actual shrinking of the film overwrap is accomplished by means of controlled heat supplied by a shrink tunnel or similar attachment (see Wrapping, shrink). Examples are candy multi-packs, video and audio cassettes, general range of packages needing protection against tampering, insect infestation, etc.

Wrapping/banding (multipacks).   In "multipack" packaging operations, there is some overlap between wrapping and banding. By combining or collating two or more similar-size packages into a multipack, the packager can choose between wrapping and banding techniques. In some cases the band is adhered directly to the surfaces of the individual packages within the multipack to prevent their sale as separate items. Examples are multipacks of soap tablets, candies, gum, tobacco products.

Wrapping/bundling. Bundling, parceling, and multi pack wrapping are terms applied to packaging machinery designed for collating or accumulating identical-size packages into a larger single unit (dozen count, for example) prior to being overwrapped in a heavy-duty flexible packaging material such as kraft paper or one of several plastic films. Examples are cookies and crackers, food products, drug packages, candy and confections, tobacco products, etc.

Wrapping/bunch or die-fold. Special machines designed for wrapping irregular-shaped products wherein the product is "pushed" through a sheet of packaging material into a die-box where the wrapper is gathered and sealed on the underside of the product. Fragile items, such as baked goods, generally require U-cards or bottom boards for protection and sealing of the film wrap. Examples are small baked items, soap, typewriter ribbons, friction tape, bearings, sardine cans, etc.

Wrapping/bagging. A simple approach to the packaging of irregular-shaped items is taken in equipment that dispenses prefabricated bags, one at a time, to a station where the bag is opened and loaded (manually or automatically), and then moved to a bag-closing station (twist-tying, stapling with a header-label, crimping, carding, etc.). Examples are bagging of bread, baked goods, textile products, multiples of plastic products such as clothespins, etc.

Wrapping/pouch-style: wrap/tube and wrap/pillow-pack wrap. Widely used for the packaging of unusual and irregular-shaped items, these horizontal form/fill/seal machines call for a continuous "tube" of packaging material to be formed around a mandrel or tube. As the "tubed" product moves forward through the machine, the overwrap material is tucked, sealed,

Table 1. Basic Fold Patterns

Figure 1, fold patterns		Description	Where used
(a)	pouch type (pillow style)	horizontal form/fill/seal style with extended end-folds	irregular shapes
(b)	pouch type	horizontal form/fill/seal style with end-folds tucked under	irregular shapes
(c)	standard elevator	double-point end-fold style with long overlap on bottom panel	cartons, boxes, trays
(d)	special elevator	standard elevator fold pattern with end-folds extended and plowed under and sealed on bottom panel	extension edge boxes, low profile items (hosiery boxes)
(e)	cigarette (old style)	special fold pattern (with cross-feed tucks, folds, seals made last)	small packages
(f)	cigarette (new style)	special fold pattern (with long-side overlap made first plus double-point end seals)	small packages

Table 1. (Continued)

Figure 1, fold patterns		Description	Where used
(g)	twist wrap	special fold pattern	individual candy pieces
(h)	cigar wrap	special fold pattern	cigars
(i)	snack-wrap	cigarette style wrap with end label	crackers, cookies
(j)	bread wrap	long overlap on bottom panel plus progressive end-lock folds secured by end labels	bread, baked goods
(k)	underfold wrap (die-fold wrap) (envelope wrap)	product placed on bottom board or U-board and then wrapped with all folds tucked and folded under product prior to sealing	candy bars, gum packs, crackers, tissues
(l)	underfold wrap (diagonal sheet feed)	irregular-shaped product loaded into a tray (plastic, foil, pulp or board material) and then wrapped using a diagonally positioned sheet wherein all four corners are folded and sealed against bottom of tray	trays of meat, poultry, produce, bakery items
(m)	underfold wrap (straight sheet feed)	tube is formed with overlap bottom seam prior to forming end tucks and plowing end folds down and under package before sealing (note: see Shrink packaging)	soft goods, books, bacon, trays of food products
(n)	underfold wrap (bunched wrap) (formed wrap)	irregular-shaped product is "pushed" up and through a sheet with extended edges brought together with a series of folds and tucks prior to sealing against the bottom of the package (note: an identifying label is generally applied to the bottom panel to help hold the folds in position)	paper plates, pot pies, pizzas, heads of lettuce, mints, bearings, tape rolls
(o)	roll wrap (turret wrap)	product is pushed through sheet into turret where long side overlap is made first followed by making end-lock folds progressively	mints, candies, cookies, biscuits
(p)	roll wrap (fin-seal wrap)	generally formed on a horizontal form/fill/seal type of wrapping machine	cookies, biscuits, tray packs

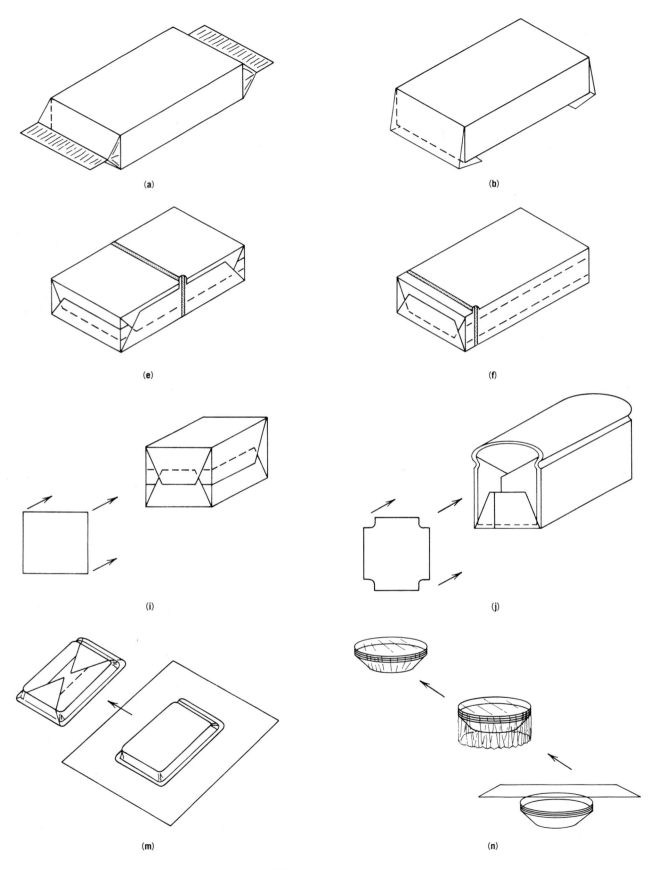

Figure 1. Fold patterns for wrapped packages.

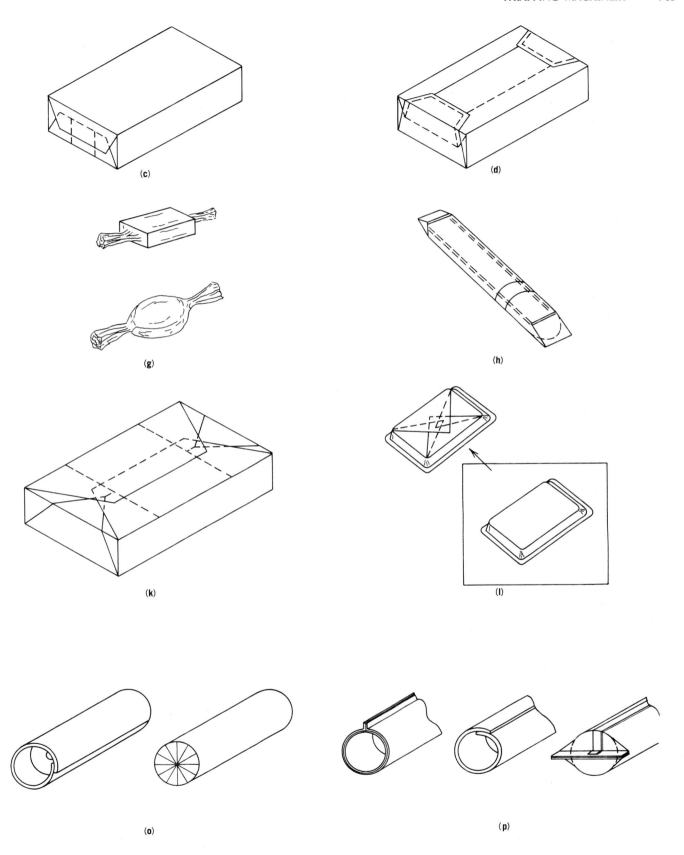

(c)

(d)

(g)

(h)

(k)

(l)

(o)

(p)

Figure 1. (*Continued*)

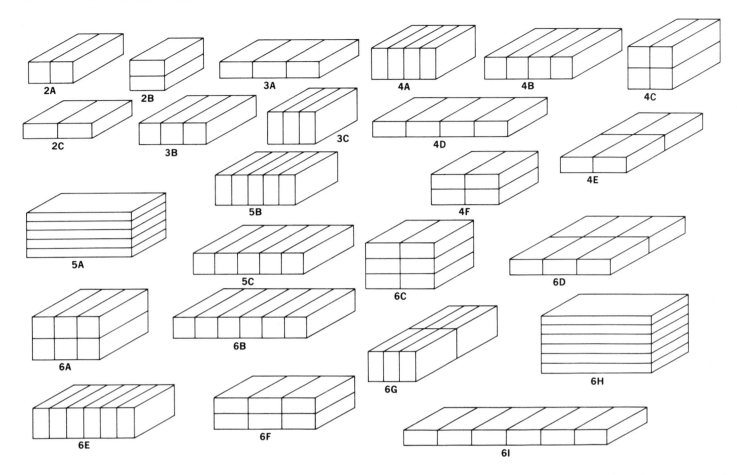

Figure 2.   Collation patterns used by Bundle-Wrapping Machines (Note: Collation specifications are based upon quantity and arrangement of packages to be bundle-wrapped, eg, 9 patterns for bundles of **6** packages, 6 for **10**, and 12 for **12**.)

and cut-off between packages. Variations include wire seals, fold-under of the ends, header crimps, etc. Examples are tray packages, baked goods, hardware items, paper products, textiles, foods, dairy products, etc.

Wrapping, special purpose.   There is a trend today towards the development of special-purpose wrapping machines. These machines are generally identified by the product being packaged (eg, cigar wrapper, soap wrapper, roll wrapper, gum wrapper, etc).

Wrapping Machine Fold Patterns

Since many wrapping machines are built specifically for a product (eg, bread, cigarettes, candy, etc) the packaging professional must be familiar with the various fold patterns and their use for certain shapes, products, and range of sizes. Some patterns are particularly suitable for higher speeds and improved package protection, as well as better economy in the selection of packaging materials and reduced changeover time where many sizes of product are to be wrapped.

Listed in Table 1 are many of the basic fold patterns presently in wide use. Attention is directed to the sketches (Fig. 1) illustrating each of the fold patterns. (It is important to note that although the fold patterns are similar, the tucking-folding-sealing operations can be produced using different mechanical means such as elevators, dies, turrets and the like.)

Wrapping Machine Attachments

Table 2 lists a number of standard and special attachments available for use on wrapping and wrapping-related packaging equipment.

"Multipack" Wrapping Machines

A very popular variation of overwrapping is the group wrapping of more than one unit into what is called a multipack. Retail multipacks provide an important merchandizing tool for a number of reasons:

1. Most multipacks can be produced on standard-model wrapping machines equipped with standard attachments (opening-tape, electric-eye registration of printed packaging materials, etc) or special attachments if necessary (eg, automatic collating attachments, coupon feeds, etc).

2. Many packaging materials are available to provide different combinations of appearance, tamper resistance, shelf-life, etc.

3. Multipacks offer the packager the opportunity to launch special promotions by simply changing the number of packages within the multipack. Examples are a "bonus" pack added to the regular six-pack of candy, a special promotion of "Buy one-get one free!", a one-cent sale for an extra tablet of bar soap, and cereal six- vs eight-packs. The promo-

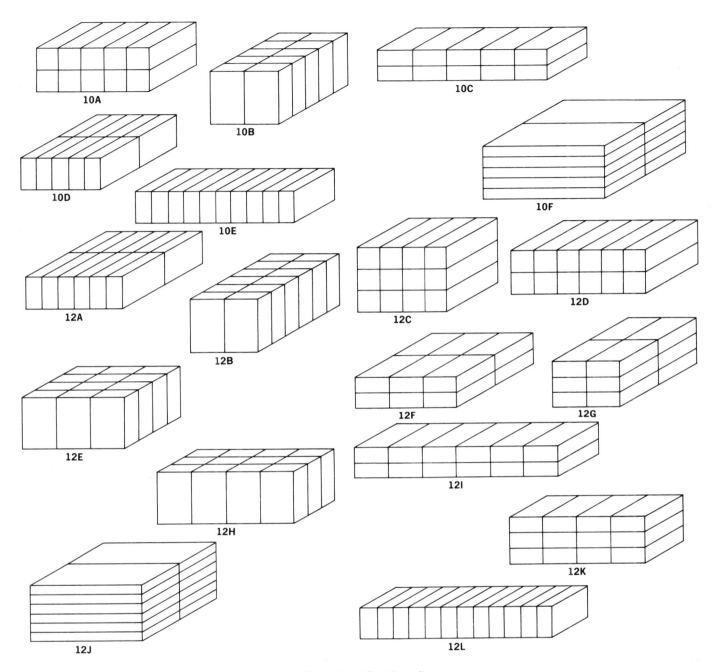

10A 10B 10C 10D 10E 10F 12A 12B 12C 12D 12E 12F 12G 12H 12I 12J 12K 12L

Figure 2. (*Continued*)

Bundle-Wrapping Machines (Bundlers and Parcelers)

Another variation of overwrapping is bundle-wrapping (called "parcel-wrapping" outside the United States) of a fixed multiple or group of like-size unit packages, using a tough, heavy duty, flexible kraft paper or plastic film. Most bundling (parceling) operations are replacements for more expensive operations using corrugated cases, chipboard shelf-containers, or display containers. Most food and drug products sold in the United States are bundled using the subdivisions of a gross (144 packages), generally in quarter-dozen, half-dozen, or one dozen counts. For products to be shipped outside of the United States, the most popular metric counts are 10, 20, or 25.

Many of the standard attachments used on standard-model wrapping machines are also available for use on bundling machines; such as electric-eye registration units for printed wrapping materials, easy opening tapes, panel printers, end labelers, perforators, and shrink tunnels for shrink films. There is also a large selection of special attachments, especially in the area of automatic infeed converyor and collating attachments designed to automatically receive and collate, and then feed the collation (see Fig. 2) into the infeed station of the bundle-wrapping machine.

tion can be set up on short notice and generally with a minimum investment in change-parts for the wrapping machine.

Table 2. Standard and Special Attachments

Attachment	Description	Designed for
package feeds	magazine single place/column	manual feeding of single packages
	magazine multiple-place/column	manual feeding of multipacks
	conveyor lug type infeed	manual loading of tray-type packages
	conveyor automatic infeed	links previous operation to wrapper
	conveyor automatic infeed with collator	receiving packages and then collating into groups of multipacks or collations
package materials	conversion kits (necessary in order to wrap with new-generation packaging materials such as polypropylene, polyethylene, polyester-PVC, and numerous laminations)	special kits designed for converting new or older model wrapping machines to handle various types of flexible packaging materials
	opening-tape (tear-tape, zip tape)	variety of attachments available for installation on new or older models
	printing attachments (electric eyes)	electric-eye registration units used to register the design on printed materials
	printing attachments (printers)	code-marker and price-markers or multipanel printing attachments available
special feeds	roll-feed attachments sheet-feed attachments index feeders (board forms)	feeding coupons, labels, revenue stamps, price tags, collars, bottom boards, hand-up tabs, header labels, etc
discharge units	various designs (rotary, reciprocating, elevator, shingle-style, etc)	wrapped packages that require collating preparatory for the next operation

Regarding collators, the packager should keep in mind that bundle-wrapping machines are subject to two speeds of operation. The first relates to the speed flow of individual packages (as high as 900 packages/min) to the infeed station of the collator attachment. The second speed pertains to the bundler and is determined by dividing the incoming line speed by the number of packages within the bundle. For example, a line speed of 600 packages/min translates into 60 bundles/min for a 10-count bundle, or 50 bundles/min for a 12-count bundle.

Two recent bundling developments warrant special mention, both involving the use of biaxially oriented polypropylene film (see Film, oriented polypropylene) as the bundling material. The first is a full six-sided wrap that is subjected to controlled shrinking in order to obtain a neat bundle that resists tampering and is strong enough to minimize damage during transit. The second development is a five-sided outgrowth of the first, with one side left unwrapped. The reason for the open side is to permit the price-marking of the individual packages within the bundle at retailer level. Shrinking the five-sided wrapped bundle completes the packaging operation with the contents of the bundle locked into position despite the open side of the bundle.

WILHELM BRONANDER, JR.
Scandia Packaging Machinery Company

WRAPPING MACHINERY, SHRINK FILM

The shrink-wrapping process was first introduced in 1948 as a protective-packaging technique for frozen poultry. It was 15 years later, however, before the process took the form in which it exists today. The concept of shrink wrapping is fairly simple. It involves the use of thermoplastic films that have been stretched (oriented) during the manufacturing process and shrink with the application of heat (see Films, plastic; Films, shrink). These films are sealed around a product or group of products and passed through a hot air tunnel, where the heated air shrinks the film tightly around the configuration of the product. A hot water bath is used for shrinking bagged poultry. The shrink-wrapping process involves two steps: (1) wrapping the product in either a full or sleeve wrap and (2) applying heat to tighten the film snugly around the product. Shrink-packaging lines generally require a wrapper and a shrink tunnel. Variations include bagging, use of a hot plate, heat gun, or hot-water bath.

L-Type Seal Machinery

The simplest wrappers used for shrink packaging are L-type sealers. The sealing bars of an L-type sealer are in the shape of a right angle, hence the name L-sealer. L-sealers always operate with centerfold shrink film. Because the film is folded on one side, it is possible for the L-shaped sealing bar to produce a fully sealed package with each stroke of the sealing head. In effect, the side seal of the next package is formed simultaneously with the preceding side seal. Each front seal is independent. In operation, the product is inserted between two layers of the folded film and moved into the sealing section. When the sealing head of the machine is depressed, a fully wrapped package results (see Fig. 1). The loosely wrapped package is then transferred to the shrink tunnel.

Most L-type sealers employ a bead (also called trim-heat) sealing method. A bead seal is made by applying heat in a fine

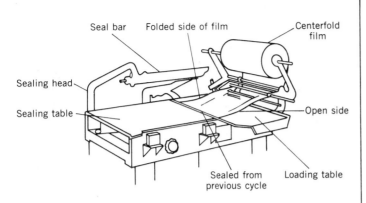

1. Ready for packaging. Open side of centerfold film, end sealed from previous cycle, passes over and under loading table

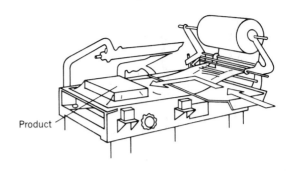

2. Operator slides product on loading table, into opening in film and into position on sealing table in one continuous motion.

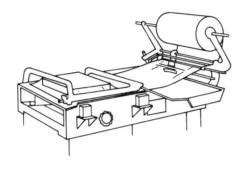

3. "L" shaped sealing bar seals two open sides, completing a bag around produce and leaving end of film sealed for next package.

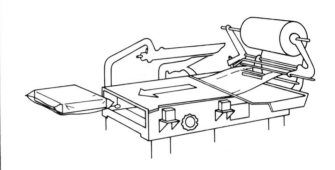

4. Completely bagged and sealed product moves out of machine into a shrink tunnel. "L" sealer is ready for next package.

Figure 1. How an "L" sealer works.

line to two layers of film such that the layers are completely melted along the line of heat application and at the same time severed. The heat causes the film to fuse, separate, and shrink back, forming a bead and trimming off the excess material above the seal line. Seal strength and quality are determined by three interrelated factors: the length of time the seal bars dwell together during the sealing cycle, the pressure exerted by the seal bars, and the temperature of the sealing medium. Most L-sealers are equipped with controls designed to regulate the amount and duration of heat flow as well as the seal cooling duration or dwell time. Control settings can be changed to achieve the best sealing results with different shrink films. Bead sealing systems include hot-blade, hot-wire, thermal-impulse, or radiant-impulse (see Fig. 2) (see Sealing, heat). Hermetic seals are also possible. These methods provide different degrees of flexibility in sealing the wide range of available shrink films. Generally, the impulse method provides the greatest seal strength with the broadest range of films.

A great number of L-sealers are available, ranging from manual to fully automatic. The most basic, hand-operated sealers require the user to pull down the sealing head, firmly compressing the seal bars to complete the seal. Some models have magnetic clamps that hold the sealing head down and free the operator to insert the next product into the film while the previous package is automatically sealing. Other models

feature electromagnetic clamping combined with a powered discharge conveyor that automatically transfers the wrapped product to the shrink tunnel. Models with push-button, pneumatically powered sealing heads are also available. The addition of a powered infeed conveyor and film unwind results in a fully automatic L-sealer. Each step up in automation increases the packaging speed as well as the cost. Manual L-sealers are capable of wrapping 6–10 packages/min. Semiautomatic sealers can attain speeds of 15/min and automatic models can produce 25–30 packages/min. Manual and semiautomatic L-sealers are used for low volume applications where the labor factor is not significant, or where product configuration and handling make automatic equipment impractical. Automatic L-sealers are often used for short runs of different sized products because changeover of these machines is fast and simple.

Form/Fill/Seal Machinery

Form/fill/seal (ffs) packaging machinery was first introduced in the early 1900s as a method of automatically wrapping products with hard machineable films such as cellophane (see Cellophane; Films, plastic). The term "form/fill/seal" describes the manner in which the wrap is made. The film is formed or shaped by a plow into a tube. The tube is filled with the product and then the tube is sealed around the product. The ffs process (see Form/fill/seal, horizontal; Form/fill/seal,

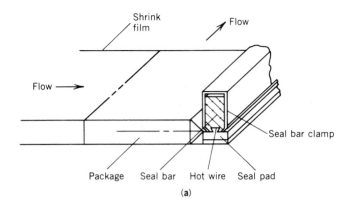

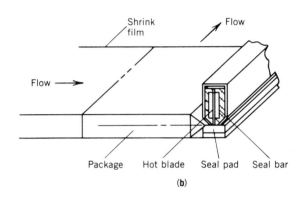

Figure 2. Bead-sealing systems. (**a**) Hot-wire type, (**b**) hot-blade type.

vertical) was not adapted to shrink packaging until the early 1960s when the technology evolved to the point where these machines could run soft shrink films. Form/fill/seal machinery is now commonly used for high-speed, automated shrink wrapping. These machines use flat single-wound film which is less costly than the centerfolded film used by L-sealers. In operation, the film roll, usually mounted on an overhead assembly, feeds over a shaping plow which forms a horizontal tube of film around the product (see Fig. 3).

The inch (2.5-cm) bottom lap of the film tube is longitudinally sealed, either thermally or electrostatically (see Fig. 4). The sealing mechanism is located at the base of the shaping plow. For thermal sealing, a hot-wired drag mechanism provides a single point of contact, producing a continuous seal between the two layers of shrink film. An ionic probe discharge mechanism is used for electrostatic sealing. The device places a negative charge on one layer of the film and a positive charge on the other. The polarity difference causes the layers to attract each other and lock together continuously across the full width of the overlap. This produces a seal that is almost invisible and is as strong in shear as the tensile strength of the film itself.

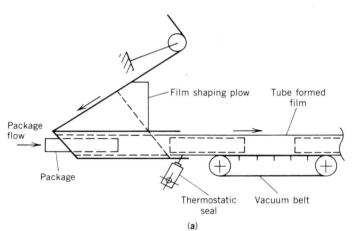

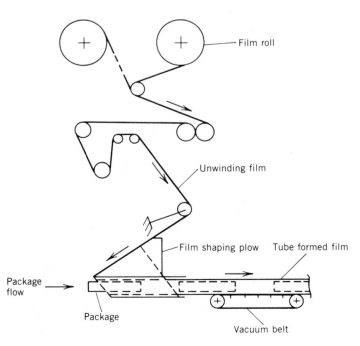

Figure 3. Film roll feeds over a shaping plow.

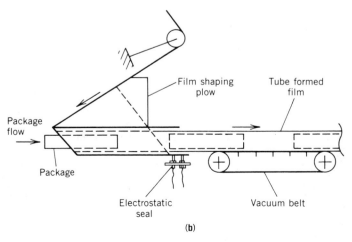

Figure 4. Bottom lap of film tube is longitudinally sealed. (**a**) Thermal-seal type, (**b**) electrostatic-seal type.

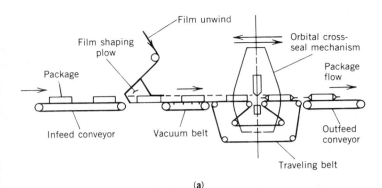

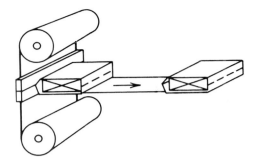

Figure 6. Sleeve-wrap-sealing system.

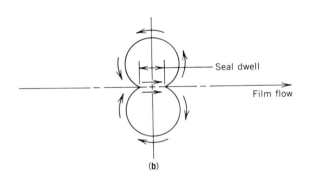

Figure 5. Orbital cross-sealing mechanism to ensure sufficient seal dwell for strong seals at high operating speeds. (**a**) Continuous-flow orbital cross-seal assembly, (**b**) orbital cross-sealing-motion diagram.

After the longitudinal seal is made, product and film advance by vacuum belt or crimp wheel conveyor to a sealing station where the transverse (cross) seal is made, either by hot wire or heated bar, and the packages are separated. The completed package is then automatically transferred to the shrink tunnel. Some ffs wrappers feature orbital cross-sealing mechanisms designed to ensure sufficient seal dwell for strong seals at high operating speeds (see Fig. 5). Effective cross sealing is the most critical aspect of high-speed, ffs shrink wrapping. The latest ffs shrink wrapping systems are microprocessor controlled and feature programmable electronic controls capable of data accumulation, self-diagnostics and preventive maintenance alerts. Programmable controllers also facilitate system integration into existing production lines.

Sleeve-Wrap Machinery

L-sealers and ffs shrink machines are used primarily to package consumer goods, because of the film's unique ability to enhance a product's merchandising appeal and provide protection from pilferage and tampering as well. Sleeve-wrap machines are used for industrial tray wrap and bundling applications, in which the primary function of the shrink overwrap is unitization. Tray wrap generally refers to the unitization of loose cans or bottles in a tray with a full overwrap of shrink film. There is also a partial film encapsulation system in which the film is applied only to the top and sides of the tray and affixed with adhesive. Sleeve-wrap bundling is used to unitize many kinds of products such as books, office supplies, newspapers, and ice cream cartons.

Sleeve wrappers use two rolls of single-wound film, horizontally mounted. The two webs are welded together (fused) by a seal bar. In operation, product is conveyed or manually placed onto the infeed section of the wrapper and moved into a sleeve of film created by the two webs of film, fed from top and bottom rolls. After the product moves past the seal-bar mechanism, the film webs cover the product and are cut off and sealed to complete the sleeve (see Fig. 6). The package is then transferred to the shrink tunnel. Exceptionally strong seals are a necessity for the heavy industrial applications of tray wrap and bundling.

Many sleeve wrapping machines employ a fin-sealing system which consists of a hot wire for cutting and two sealing ribbons. A double seal bar forms two ⅛-in. (3.2-mm) seals and separates them with the cutting wire (see Fig. 7). After sealing, the fin seal retains more than 60% of the film's original tensile strength. Recent advances in seal-bar technology have simplified the sealing process for sleeve wrapping. Now a single thermal seal bar can form a bead seal with the strength of a fin seal (see Fig. 8). The thermal bar is less costly and requires less maintenance.

Bundling sleeve wrapping machines are generally of the intermittent-motion type with practical speed capabilities from 25–35 wraps/min. Tray wrappers are generally more sophisticated continuous-motion machines capable of speeds to 75 cases/min.

Shrink Tunnels

Most shrink tunnels operate on the hot-air principle. The successful performance of a shrink tunnel is primarily a func-

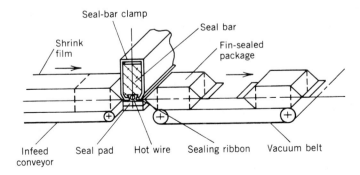

Figure 7. Fin-sealing system. A double seal bar mechanism forms two ⅛-in. (3.2-mm) seals and separates them with a cutting wire.

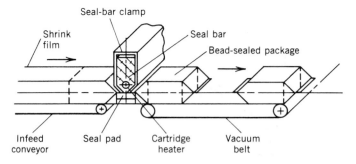

Figure 8. Bead-sealing system with a single thermal-seal bar. A solid-tip seal bar with cartridge heater for thermal bead-sealing system.

tion of the speed and effectiveness with which it can transmit sufficient heat to the film to initiate the shrink action. Because product exposure in the shrink tunnel is usually only two or three seconds, the heat has no effect on the product. The hot air acts only on the film, causing it to shrink snugly around the product. Different films have different shrink characteristics, however, so the tunnel should have a high degree of adjustability.

A typical shrink tunnel consists of a conveyor, a heat chamber, and a recirculating hot-air system. Conveyors are either flat belts or rollers and are generally equipped with variable speed controls. Belt conveyors are Teflon (DuPont Company), fiberglass, silicone, or metal mesh. Roller conveyors are generally steel with silicone rubber sleeves and often feature rotate or non-rotate action. Roller conveying systems can operate at high speeds, support substantial weight and provide complete exposure of the package to the hot air because the entire bottom surface is exposed to the air jets. Heat chambers vary as to the method of directing the hot air, the air pattern control system, the number and configuration of air ports, and the insulation and its resultant efficiency.

The significance of heat capacity depends on the film being used. Polyethylene, for example, requires a high temperature and a long duration in the tunnel. Therefore, a tunnel with large heater banks and a chamber long enough to permit the required duration must be selected for polyethylene shrink applications. Tunnels designed specifically for sleeve wrap applications usually have a heat capacity of 20–40 J/K. In addition, air flow should be directed in such a way as to produce adequate end-lock results. Because only a sleeve of film is used, it is essential that the ends of the film lock tightly around the product. A tunnel without the proper air-flow design will not produce the required results on a sleeve-wrapped package.

The heart of the shrink tunnel is its recirculating hot-air system. Accurate thermostatic control and high capacity, variable-velocity air systems are very important for effective use with different shrink films. For optimum results with polypropylene, for example, the velocity of air impinging on the film should be at least 3000 ft/min (915 m/min). In contrast, polyethylene film actually loosens and becomes molten prior to shrinking. Thus, high velocity air streams would form unwanted folds and wrinkles in the film. Proper shrink with polyethylene, therefore, requires a considerably lower air ve-

locity. There are many special types of tunnels available including those designed with cooling chambers or preheat sections.

Pallet-shrink tunnels. Shrink tunnels are also used for pallet-load unitization. This application requires a film bag, usually polyethylene, to be placed over the pallet load prior to shrink wrapping. Premade bags may be purchased or made in-plant with bag-forming devices. There are several types of equipment for shrinking the pallet cover. One system employs a movable rectangular shrink frame with infrared heating elements. In operation, the frame descends down around the load and shrinks the wrap on the way up. However, the most popular tunnels for pallet-shrink wrapping use recirculating heated air. Most of them are gas-fired, although some are electric. Heated air tunnels offer the greatest versatility in terms of pallet sizes and shapes as well as the types of films which can be used. The most efficient systems make the heated air impinge tangentially onto exposed pallet surfaces for controlled shrinking of the film overwrap.

BIBLIOGRAPHY

General References

R. J. Kelsey, *Packaging in Today's Society,* St. Regis Paper Co. (now a division of Champion International), New York, 1978.

M. Siegel, "Shrink and Stretch Equipment," in *The Packaging Encyclopedia,* **25**(4), Cahners Publishing, Boston, Mass., 1980.

C. A. Sherman, *Flexible Packaging,* Allied Automation Inc., Chicago, Ill., 1974.

W. C. Simms, "Wrapping, Shrink and Stretch," in *The Packaging Encyclopedia,* **29**(4), 306, Cahners Publishing, Boston, Mass., 1984.

W. M. CORBETT
Weldotron Corporation

WRAPPING MACHINERY, STRETCH FILM

In the United States today, about 250 million (10⁶) unitized loads per year are shipped with stretch-wrap protection. The concept of this unitizing technique is to wrap the load with a web of film that is stretched and under tension. The force of the film is transmitted over the entire surface of the load, holding it securely together as one unit. During the handling and shipping cycle, the film displays a "rubber-band" effect, ie, expanding and contracting in response to various forces.

Load Dynamics

Basically, there are two types of forces that the load experiences during handling and shipping: vibration and sway, and high impact forces.

Vibration and sway. These forces generate changing frequencies (resonance) causing internal movement of the load. A truck traveling on the highway, for example, constantly encounters roughness in the road surface. This constant vibration of product load can cause cartons to begin to shift or to slide laterally unless restricted. A relatively low holding force is required to restrict this type of movement.

High impact forces. These are generated from rapid starts and stops, extremely rough road conditions, and freight-car bumping or collision. In some cases, these forces cause total

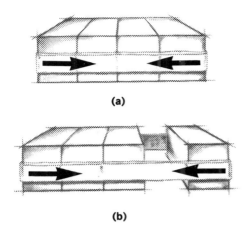

Figure 1. Vibration and sway are controlled by (**a**) stretch-force and (**b**) restretch-force properties. Low force is needed to restrict movement.

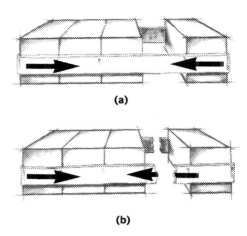

Figure 2. High impact forces are controlled by (**a**) restretch force and (**b**) breaking strength. High force is required to restrict movement.

load shift. Because of the high force generated in this situation, relatively high restrictive forces are required to counteract shifting and prevent damage to the product load. The sooner the restraining forces are applied, the less chance for momentum to build and destroy the load during transit.

Mechanical Load-holding Properties of Stretch Film

The resins used in the manufacture of stretch-wrapping films are linear low density polyethylene (LDPE), ethyl vinyl acetate (EVA), poly(vinyl chloride) (PVC), and conventional low density polyethylene (LDPE), or a combination (see Films, stretch). Different films have different load-holding characteristics. The relevant features are stretchability, stretch force, restretch forces, and breaking strength.

Stretchability (elongation). This is a measure of how much a given film type can be stretched over a specific load profile. The typical film properties contributing to stretchability are elasticity, tear resistance, puncture resistance, and film consistency. Irregularity of the load contours or profile taxes stretchability, and the overall percent of stretch must be decreased to prevent film breakage during the stretch-wrapping process. Stretchability is a measure of how far the film can be stretched during the wrapping process without distorting the load. Increasing stretch levels means less film is used to unitize the load, resulting in lower material costs.

Stretch force. This is a measure of the static force exerted on the four-sided surface of the load by the stretch film. It is expressed in terms of pounds per single wrap of 20 in. (51 cm) wide film. On a straight, square load, this force is typically highest at the corners of the load. It is also the force that can crush or distort the load during the wrapping if not controlled.

Restretch force (elasticity). This is a measure of the force available to restrict further movement of the load during transit once it has been wrapped. It is the force that restretches the film during vibration, sway, and high impact forces; the film expands and contracts using the "rubber-band" effect to hold the load together.

Breaking strength. This is a measure of the ultimate force that the film can withstand without failure. Expressed in pounds per single wrap of 20-in. (51-cm) film, it is the force

required to rupture a 20-in. (51-cm) clamped film sample separated at a rate of 14 ft/min (4.3 m/min), measured on an Instron Tester. The interaction between the stretch film and the various forces is shown in Figures 1 and 2.

The Stretch-wrapping Process

There are two basic techniques employed to stretch the film in the unitization of a product load.

Conventional method. The product load is placed on a turntable that rotates. The rotating action exerts the stretching force that pulls the film from the roll (see Fig. 3). The film stretching is accomplished by restricting the film unwind motion with some type of braking device at the film roll as the load is rotated. With this method, all of the force to stretch the film is exerted between the film roll and corner of the load.

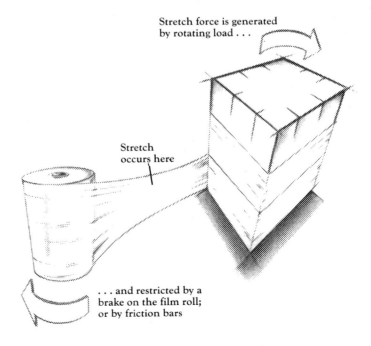

Stretch force is generated by rotating load . . .

Stretch occurs here

. . . and restricted by a brake on the film roll; or by friction bars

Figure 3. Conventional stretch.

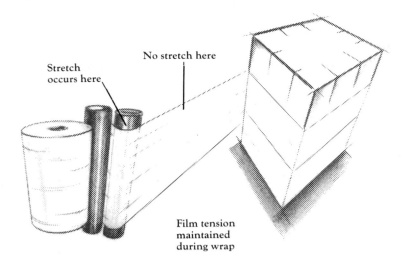

Figure 4.   Pre-stretch method.

Attempts to achieve high stretch levels can result in crushing the load or shifting of the load on the turntable, and protrusions can tear the film. With this system, the film can never reach the yield potential and high stretch levels that exist in the film. At best, under actual production conditions, 10–55% stretch is typical on a roll of film having a potential stretchability of 200–600%, depending on film-resin type. This technique is still proving to be an effective unitizing method, but the cost per load is high because of the relatively low yield of the film.

Pre-Stretch.   The pre-stretch concept, invented by Lantech, Inc. in 1979, reduces the effective cost to unitize a product load by 20–30%. The pre-stretch method isolates the stretching process away from the corner of the load. The film passes between two mechanical or electronically interconnected rollers before it is applied to the load. The exit roller rotates faster than the entry roller, thus stretching the film on smooth surfaces with consistent force (see Fig. 4). Pre-stretching typically increases stretch levels to 250%. The increased stretch reduces the amount of film consumed per pallet load and also makes the film stronger. At high stretch levels, the film is oriented to produce material stronger than the original film (stress–strain curve). The pre-stretch system also allows the operator to independently control how much force the film applies to the load during wrapping and it can assure consistent stretch levels in a three-shift operation. The stretching process is not dependent on the load corners, but takes place between the two rollers. With pre-stretching, all the mechanical properties of the film hold the load together, in contrast to the conventional method, which uses only stretch force.

Stretch-wrapping Methods

Pallets are wrapped with narrow or wide widths of stretch film.

Spiral wrapping method.   The spiral wrapping method wraps successive loads of widely varying length, width, or height with a single film size, typically 20 or 30 in. (51 or 76 cm, respectively) wide (see Fig. 5). The number of wraps required is influenced by load weight, load profile, and method of shipment. As the turntable rotates the load, the roll of film moves up and down the mast of the machine to wrap the load from bottom to top and returns to bottom. The operator can program the number of film layers to be applied to top and bottom of each load. The resulting selective distribution of film protects the areas of greatest load stress. Production speeds with this method are 20–80 loads/h, depending on the degree of equipment automation. Spiral wrapping is the dominant method of unitizing because it offers the flexibility to adapt to changing product sizes and shapes.

Full-web wrapping method.   The full-web method wraps loads of relatively constant height with a film width (typically 55–70 in. or 140–178 cm) which covers the full face of the load

Figure 5.   Spiral wrapping method.

Figure 6.   Full-web wrapping method.

Figure 7. A platform wrapper (*SVS Lan-wrapper System*) for operations unitizing 5,000 to 15,000 loads/yr. It features a pre-stretch film delivery system.

(see Fig. 6). Because the load is entirely wrapped with each turntable revolution, production speeds are faster than with the spiral wrapping method. Operation of the equipment is simplified and film costs can be lower on loads under 50 in. (127 cm) in height compared to the spiral wrapping method. Production speeds exceeding 100 loads/h can be achieved.

Equipment

Stretch wrappers are available for operations that wrap only a few pallets a day, as well as for high volume operations with throughput requirements of 2400/d.

Hand-held units. *Hand-held reels* for spiral wrapping or caster-mounted reels for full-web wrapping are available for low volume output. The operator tucks the film to the load and walks around the load to wrap the load as needed.

Platform systems. A *platform system* is loaded with a fork-lift truck. The operator tucks film into the load and activates the wrapping cycle as programmed. After wrapping, the operator cuts the film and removes the load from the turntable. Models are available to wrap in spiral or full-web mode at speeds of 20–25 loads/h (see Fig. 7). Variations include low profile systems that can be loaded by forktruck or pallet jack, as well as automatic platforms that can be controlled remotely by the forktruck operator.

A number of other systems are available as well. In a *conveyorized system,* loads enter the system by way of a conveyor, are wrapped according to programmed cycle, and exit the system via the conveyor. The entire operation can be controlled by a programmable logic controller to interface with upstream and downstream systems. In a *Straddle Lan-wrapper system,* the load remains stationary on a conveyor or automated guide vehicle (AGV) and the film roll rotates around the load, wrapping in the spiral mode. This is especially suited for light-weight, unstable loads or extremely heavy loads which would be difficult to rotate in the normal stretch-wrapping method. In *pass-thru systems,* the load is pushed into a web of film which is heat-sealed at the front and back of the load. The film is 1–3 mil (25–76 μm) thick, stretched 3–8%. These systems are designed for consistent-size loads and are available in semiautomatic and automatic models.

Stretch Bundling

Stretch bundling uses the same mechanical film properties to bundle relatively small batches of products or to overwrap products such as roll goods. The same films used for pallet-load unitization are used for stretch bundling. In the process, the film is rotated in the vertical plane around the product. Compared with shrink bundling, operating costs are low because no heat is required to finish the package.

P. R. Lancaster
Lantech, Inc.

X

X-Y RECORDERS

X-Y recorders are instruments designed to plot graphs in a Cartesian-coordinate system. A Cartesian-coordinate system is a two-dimensional system having an abscissa (X axis) and an ordinate (Y axis). The recorder plots the relationship between two variables which are represented by either d-c or low-frequency a-c voltages. Voltage corresponding to one of the variables is applied to one of two servomotors. The signal representing the other variable is applied to the other motor. One of the motors moves the pen along the horizontal axis and the other moves it in the vertical plane. Because the amount of pen movement is proportional to the amplitude of the applied voltage, and movement in one plane is independent of movement in the other, the system is capable of plotting the two variables in a two-dimensional system. X-Y recorders are frequently used in package testing.

Z

ZERO-CRUSH CONCEPT

The term zero-crush concept (ZCC) applies to a double-facer of a corrugator in which the conventional ballast rolls are replaced by an air cushion (1). Originally developed to reduce maintenance costs, the ZCC configuration has also improved the process of adhesive application and bonding. Because the air cushion minimizes the loading on the board in the double facer, it also minimizes crushing of the flutes. Caliper is improved by an average 1.5%.

BIBLIOGRAPHY

1. D. Whitman, "Double-factor bonding using the 'Zero-crush' concept," *Corrugated Containers Conference Proceedings,* TAPPI, Atlanta, Ga., 1983, pp. 21–26.

INDEX